THE OXFORD HANDBOOK OF

ANIMAL ETHICS

T0201996

THE OXFORD HANDBOOK OF

ANIMAL ETHICS

Edited by

TOM L. BEAUCHAMP

and

R. G. FREY

OXFORD

UNIVERSITY PRESS

OXFORD
UNIVERSITY PRESS

Oxford University Press is a department of the University of Oxford.
It furthers the University's objective of excellence in research, scholarship,
and education by publishing worldwide.

Oxford   New York

Auckland   Cape Town   Dar es Salaam   Hong Kong   Karachi
Kuala Lumpur   Madrid   Melbourne   Mexico City   Nairobi
New Delhi   Shanghai   Taipei   Toronto

With offices in

Argentina   Austria   Brazil   Chile   Czech Republic   France   Greece
Guatemala   Hungary   Italy   Japan   Poland   Portugal   Singapore
South Korea   Switzerland   Thailand   Turkey   Ukraine   Vietnam

Oxford is a registered trade mark of Oxford University Press
in the UK and certain other countries.

Published in the United States of America by
Oxford University Press
198 Madison Avenue, New York, NY 10016

Library of Congress Cataloging-in-Publication Data
The Oxford handbook of animal ethics.
p. cm. — (Oxford handbooks)
Edited by Tom L. Beauchamp and R. G. Frey.
Includes index.
ISBN 978–0–19–537196–3 (hardcover : alk. paper); 978–0–19–935197–8 (paperback)
1. Animal welfare—Moral and ethical aspects.   2. Animal rights.   I. Beauchamp, Tom L.
II. Frey, R. G. (Raymond Gillespie)   III. Title.   IV. Series.
HV4708.094 2011
179'.3—dc22   2010049021

CONTENTS

....

Preface, ix

Contributors, xi

Introduction, 3
Tom L. Beauchamp

PART I. HISTORY OF PHILOSOPHY

1. Animals in Classical and Late Antique Philosophy, 35
   Stephen R. L. Clark

2. Animals and Ethics in the History of Modern Philosophy, 61
   Aaron Garrett

PART II. TYPES OF ETHICAL THEORY

3. Interacting with Animals: A Kantian Account, 91
   Christine M. Korsgaard

4. Virtue Ethics and the Treatment of Animals, 119
   Rosalind Hursthouse

5. A Humean Account of the Status and Character of Animals, 144
   Julia Driver

6. Utilitarianism and Animals, 172
   R. G. Frey

7. Rights Theory and Animal Rights, 198
   Tom L. Beauchamp

8. The Capabilities Approach and Animal Entitlements, 228
   Martha Nussbaum

Part III. Moral Status And Person Theory

9. The Idea of Moral Standing, 255
   Christopher W. Morris

10. Animals, Fundamental Moral Standing, and Speciesism, 276
    David Copp

11. Human Animals and Nonhuman Persons, 304
    Sarah Chan and John Harris

12. Are Nonhuman Animals Persons? 332
    Michael Tooley

Part IV. Animal Minds And Their Moral Significance

13. Animal Mentality: Its Character, Extent, and Moral Significance, 373
    Peter Carruthers

14. Mindreading and Moral Significance in Nonhuman Animals, 407
    José Luis Bermúdez

15. Minimal Minds, 441
    Bryce Huebner

16. Beyond Anthropomorphism: Attributing Psychological
    Properties to Animals, 469
    Kristin Andrews

17. Animal Pain and Welfare: Can Pain Sometimes Be Worse
    for Them Than for Us? 495
    Sahar Akhtar

18. Animals That Act for Moral Reasons, 519
    Mark Rowlands

19. The Moral Life of Animals, 547
    Michael Bradie

PART V. SPECIES AND THE ENGINEERING OF SPECIES

20. On the Origin of Species Notions and Their Ethical Limitations, 577
    Mark Greene

21. On the Nature of Species and the Moral Significance of their Extinction, 603
    Russell Powell

22. Are All Species Equal? 628
    David Schmidtz

23. Genetically Modified Animals: Should There Be Limits to Engineering
    the Animal Kingdom? 641
    Julian Savulescu

24. Human/Nonhuman Chimeras: Assessing the Issues, 671
    Henry T. Greely

PART VI. PRACTICAL ETHICS

25. The Moral Relevance of the Distinction between Domesticated
    and Wild Animals, 701
    Clare Palmer

26. The Moral Significance of Animal Pain and Animal Death, 726
    Elizabeth Harman

27. The Ethics of Confining Animals: From Farms to Zoos to Human Homes, 738
    David DeGrazia

28. Keeping Pets, 769
    Hilary Bok

29. Animal Experimentation in Biomedical Research, 796
    Hugh LaFollette

30. Ethical Issues in the Application of Biotechnology to Animals
    in Agriculture, 826
    Robert Streiffer and John Basl

31. Environmental Ethics, Hunting, and the Place of Animals, 855
    Gary Varner

32. Vegetarianism, 877
    Stuart Rachels

33. The Use of Animals in Toxicological Research, 906
    Andrew N. Rowan

34. What's Ethics Got to Do with It? The Roles of Government Regulation
    in Research-Animal Protection, 919
    Jeffrey Kahn

35. Literary Works and Animal Ethics, 932
    Tzachi Zamir

Index, 957

PREFACE

When Oxford University Press issued an invitation to us to prepare a volume on animal minds and the ethics of human uses of animals for its series of Handbooks of Philosophy, we enthusiastically accepted. This multifaceted and rapidly growing area of scholarly activity deserves sustained and careful work from philosophers and professionals in several disciplines. We have been fortunate as editors to successfully plead this case with the outstanding group of contributors to this volume. All chapters are original essays in either already prominent or largely unexplored areas of animal ethics.

The *Handbook* is aimed at an audience that seeks accessible and high-level philosophical work. We have asked contributors not to scale down their contributions to make them more accessible to a wide audience, but one exception is made: The Introduction to this volume is intended for a multidisciplinary audience, including readers who may for the first time be encountering complex philosophical arguments. The Introduction presents the organizational structure of the volume and each author's main arguments and conclusions.

The preparation of this volume has taken several years. We deeply appreciate the willingness of several of our authors who have encountered personal difficulties during this period to plough ahead and complete the work. We are likewise grateful for their willingness to go through numerous drafts and to exhibit great patience as this book was brought to completion.

We gratefully acknowledge the help of Yashar Saghai in the final stages of the delivery of this manuscript to the publisher. Without his help, the book might have been delayed for several weeks. We also thank Peter Ohlin for his efficient arrangement of impartial reviews during the early stages of our planning and of course to the anonymous reviewers themselves.

Washington, D.C.   TLB
Bowling Green, Ohio   RGF

Contributors

....

Sahar Akhtar, Department of Philosophy, University of Virginia-Charlottesville

Kristin Andrews, Department of Philosophy, York University, Toronto, Canada

John Basl, Department of Philosophy, University of Wisconsin-Madison

Tom L. Beauchamp, Department of Philosophy and Kennedy Institute of Ethics, Georgetown University

José Luis Bermúdez, Dean of the College of Liberal Arts, and Department of Philosophy, Texas A&M University

Hilary Bok, Department of Philosophy and Berman Institute of Bioethics, The Johns Hopkins University

Michael Bradie, Department of Philosophy, Bowling Green State University, Ohio

Peter Carruthers, Department of Philosophy, University of Maryland, College Park

Sarah Chan, Institute for Science, Ethics and Innovation, The University of Manchester, UK

Stephen R. L. Clark, Department of Philosophy, University of Liverpool, UK

David Copp, Department of Philosophy, University of California-Davis

David DeGrazia, Department of Philosophy, The George Washington University, Washington

Julia Driver, Department of Philosophy, Washington University in St. Louis

R. G. Frey, (1941–2012) Department of Philosophy and The Social Philosophy and Policy Center, Bowling Green State University, Ohio

Aaron Garrett, Department of Philosophy, Boston University

Henry T. Greely, Center for Law and the Biosciences, Stanford University Law School

Mark Greene, Department of Philosophy, University of Delaware

Elizabeth Harman, Department of Philosophy and The University Center for Human Values, Princeton University

John Harris, The Institute for Science, Ethics and Innovation, The University of Manchester, UK

Bryce Huebner, Department of Philosophy, Georgetown University

Rosalind Hursthouse, Department of Philosophy, The University of Auckland, New Zealand

Jeffrey Kahn, Center for Bioethics and Department of Medicine, University of Minnesota

Christine M. Korsgaard, Department of Philosophy, Harvard University

Hugh LaFollette, Cole Chair in Ethics, University of South Florida St. Petersburg

Christopher W. Morris, Department of Philosophy, University of Maryland-College Park

Martha Nussbaum, Law School, Divinity School, and Department of Philosophy, University of Chicago

Clare Palmer, Department of Philosophy, Texas A&M University

Russell Powell, Uehiro Centre for Practical Ethics, Oxford University, UK

Stuart Rachels, Department of Philosophy, University of Alabama-Tuscaloosa

Andrew N. Rowan, The Humane Society of the United States, Washington, D.C.

Mark Rowlands, Department of Philosophy, University of Miami-Coral Gables

Julian Savulescu, Uehiro Centre for Practical Ethics, Oxford University, UK

David Schmidtz, Department of Philosophy, University of Arizona

Robert Streiffer, Department of Philosophy, University of Wisconsin-Madison

Michael Tooley, Department of Philosophy, University of Colorado-Boulder

Gary Varner, Department of Philosophy, Texas A&M University

Tzachi Zamir, English Department, The Hebrew University of Jerusalem, Israel

THE OXFORD HANDBOOK OF

ANIMAL ETHICS

INTRODUCTION

TOM L. BEAUCHAMP

HUMANS use animals in a stunning variety of ways. All reflective persons find it difficult to determine precisely which uses of animals are morally justified and which unjustified, and many are confused about how to make their moral views comprehensive and consistent. Since ancient times philosophers have been sporadically interested in questions of animal minds, but only beginning around the last quarter of the twentieth century has a significant philosophical literature developed on the ethics of our use of animals. Scholars then began to view the subject matter as needing sustained scholarly attention. This period saw a dramatic growth of the fields of biomedical ethics and the philosophy of biology, both of which for the first time became important and flourishing fields of philosophy. Philosophers also became attracted to investigations of the nature of animal minds and the ethics of many forms of human interaction with animals. The literature came to have a primary focus on animal psychology, the moral status of animals, the nature and significance of species, and a number of practical problems about our utilization of animals.

This book presents the issues as they stand today and looks to the next stages of this still young and developing field. It is remarkable how much has been achieved during the last thirty-five years in the study of moral and scientific questions about animals. A subject that could be described as virtually moribund four decades ago, and of virtually no interest to philosophers, has come to occupy a considerable place in philosophy. The field continues to expand every year. Several chapters in this volume explore matters that, to the editors' knowledge, have never previously been examined by philosophers.

The authors of the thirty-five chapters come from a diverse set of philosophical interests and explore an array of concerns about animal products, farm animals, hunting, circuses, zoos, the entertainment industry, safety-testing on animals, the status and moral significance of species, environmental ethics, the nature and significance of the minds of animals, and so on. They also investigate what the future may be expected to bring in the way of new scientific developments and new moral problems. The contributors draw on one or more of the following seven areas:

(1) History of Philosophy
(2) Philosophy of Mind
(3) Philosophy of Biology
(4) Philosophy of Cognitive Science
(5) Philosophy of Language
(6) Ethical Theory
(7) Practical Ethics

The cryptic term "animal ethics" is a convenient label for the diverse literature on animals that has emerged from these areas of philosophy as well as from disciplines other than philosophy. The term "animal rights" is the most common term used today to refer to views supportive of the protection of animals against human misuse, and it might be thought that this book should be entitled *Oxford Handbook of Animal Rights*. However, this title would be presumptive, insufficiently comprehensive, and in general a poor choice to capture the range of issues and positions that philosophers have been investigating. Given the history and power of rights language, many framers of declarations about protections for animals understandably chose rights language as the basic terminology. However, many others interested in animal welfare and the nature of animal species have chosen not to use the language of animal rights; and some philosophers are actively opposed to this terminology. Both the title of this volume and many of the authors within leave open questions about which terminology and types of theory are most suitable for discussion of the range of moral problems about animals that have arisen.

The remainder of this introduction is an orientation to each of the six parts and each of the thirty-five chapters. The section numbers in the introduction correspond exactly to the numbering of the parts in the volume, as displayed in the table of contents.

PART I. THE HISTORY OF PHILOSOPHY

Although most of the moral and scientific questions about animals that dominate current literature are of recent origin, some date back to the ancients, several of whom were aware of psychological and moral problems about animals. The most extensive and imposing ancient account was that of the Neoplatonist philosopher Porphyry (3rd–4th c. AD) in his treatise *On Abstinence from Animal Food* (also translated *On Abstinence from Killing Animals*). This work analyzes how we should and should not use animals, especially when we treat them as means to our own ends.

Porphyry sought an impartial philosophical approach, which he realized was rarely taken. He was against killing animals when nonanimal food sources would do as well or better, and he was concerned to improve the welfare of what he saw as highly intelligent creatures. Perhaps the most revolutionary ancient source was Plutarch's "Whether Land or Sea Animals are Cleverer" and "Beasts are Rational." He graphically depicted how clever many creatures are, how we should approach evaluation of the practice of eating another animal, and ways in which animals may be more advanced than humans. Quite different and also very influential in the ancient world were Aristotle's thesis that humans alone have reason and the Stoics' claim that significant language, reason, virtue, and even real emotion cannot correctly be attributed to animals.

This imposing ancient tradition of reflection on animals was carried into modern philosophy. The landmark, early modern *Dictionary Historical and Critical*, authored by Pierre Bayle, engages the thought of many figures throughout the history of the discussion. Bayle identified and analyzed authors, ancient and modern, who discuss whether animals have souls, whether they deserve some form of moral consideration, and whether there is some form of reason in animals. Less comprehensive, but ultimately more influential reflections on animals are found in the writings of several seventeenth- to nineteenth-century philosophers, most notably René Descartes, David Hume, Immanuel Kant, and Jeremy Bentham.

It is unsurprising that throughout this vast history the notion of animal rights went almost unnoticed. Until the seventeenth century, there were no clear doctrines even of universal natural rights. However, with the advent of theories of international rights and natural rights developed by philosophers such as Hugo Grotius, Thomas Hobbes, and John Locke—today often restyled as *human* rights—the way was paved for an account of animal rights. The first significant such theory seems to be that of Francis Hutcheson in the eighteenth century. However, truly pathbreaking developments in moral philosophy in the theory of animal rights would not occur until many decades after Hutcheson.

Only a handful of historians of philosophy have turned sustained attention to the centuries of literature that have discussed animals, but several scholars have in recent years begun to recover the details of this history. The essays in part I of this volume are devoted to this pursuit. (The reader interested in this history will also profit from the discussion of several leading figures, including Descartes, Hume, and Kant, in chapters after part I.)

In "Animals in Classical and Late Antique Philosophy," **Stephen R. L. Clark** explores a large array of conceptions and theories in the ancient world, with an emphasis on what the ancients thought of both themselves and the other "animals." Though Classical and Late Antiquity in today's framing in philosophy consists primarily of Greek Philosophy, Clark's survey is broader. The scope is immense in terms of leading schools of philosophy: the Pre-Socratics, the Golden Age of Athens, the Hellenistic period, and the Late Antique period (including Christian thought under the category of "Patristics"). But even these categories are not entirely satisfactory, in Clark's assessment, because cultural and philosophical diversity was so vast in the ancient world. Clark argues that while it is true that the so-called Western tradition inherited the view that plants are for animals and animals for dominion by humans, the conceptions of the ancients are more nuanced than such a generalization captures. One generalization that does seem to hold is that non-human animals commonly were viewed as foils—beastly in habits and without minds of moral significance. Clark assesses the ancient, classical, Greek, and Mediterranean attitudes as complicated and often contradictory. In general, animals were seen as entirely unlike us, but humans also were seen as capable of a descent into beastly behavior—to the point that humans were in effect seen as no more than animals. Animals were also seen in the ancient world as propelled by external stimulation and habit, as are those humans who fail to transcend the fundamentally animal side of their nature.

In "Animals and Ethics in the History of Modern Philosophy," **Aaron Garrett** examines the history of early modern philosophy, principally in the seventeenth and eighteenth centuries. He explains why early modern philosophers and jurists seldom reflected deeply about animal life and why the arrival of a decent theory of animal rights in early modern philosophy was a remarkable development. He begins with the general background of rights theory as it was developing in political philosophy. He discusses why these conceptions were not fruitful in acknowledging that animals might have rights. He uses as an instructive example eighteenth-century experimentalist Robert Boyle and his thesis that there is a *duty* to experiment on animals. This study of Boyle leads Garrett to an explanation of the philosophical conceptions that prevented development of a theory of animal rights. He describes the steady movement toward both a rejection of Boyle's view and toward the view that we have moral duties to animals. Eventually, Garrett argues, this historical trend leads to the "invention" of animal rights at the hands of Scottish moral philosopher Francis Hutcheson. However, just as Hutcheson came to his anticipation of later theories of animal rights, the theoretical framework of natural law and natural rights on which Hutcheson had relied itself came under attack. Garrett also discusses how animal welfare concerns in the late eighteenth century and early nineteenth century produced some remarkable ideas about the interests and needs of animals that hastened the arrival of animal welfare legislation in Britain and Germany.

PART II. TYPES OF ETHICAL THEORY

The terms "ethical theory" and "moral philosophy" are typically used to refer to reflection on the nature and justification of moral rights, moral obligations, and standards of moral character. Philosophers use their theories to introduce clarity, substance, and precision of argument into the discussion of moral rightness and wrongness, the virtues, social justice, moral obligations, and the like. They attempt to justify moral standards or some moral point of view by reasoned analysis and argument.

Each chapter in part II contains reflection on moral problems about animals and our treatment of animals that descend from one or more prominent and influential types of ethical theory. Knowledge of these general theories is indispensable for reflective study in the ethics of our use of animals, because the field's literature frequently draws on the terminology, arguments, methods, and conclusions of these theories. Each chapter in part II provides an overview of the characteristic features of the type of theory under examination, often with an assessment of the theory's value for today's discussions of the human uses of animals.

Several types of theory are discussed: Kantian theory, virtue theory, Humean theory, rights theory, and capabilities theory. With the possible exception of capabilities theory, each of these types of theory has been developed and refined for at least two centuries of philosophical thought. It is safe to say that no one theory can at present be considered the received theory or the most popular theory. We are living through a period in philosophy in which there is no dominant theory.

Kantian Theory. In "Interacting with Animals: A Kantian Approach," **Christine Korsgaard** discusses a theory that has often been called deontological (i.e., a theory that some features other than or in addition to consequences and good outcomes make actions obligatory), but now increasingly called Kantian because of its origins in the theory of Immanuel Kant. Korsgaard starts with a characterization of two ways in which differences between human beings and nonhuman animals might be drawn in moral theory: (1) thinking about what is good and (2) thinking about right and obligation. Two general types of argument have therefore been used by philosophers in their attempts either to justify or criticize our uses of animals: "First, there are arguments based on similarities or differences between the ways in which things can be good or bad for human beings and the ways in which they can be good or bad for the other animals. Second, there are arguments based on the grounds of right and obligation." Korsgaard's concern is to attend to an argument of the second kind, in particular Kant's argument that we have no obligations to nonhuman animals, because obligation derives from a reciprocal relation among rational beings. Korsgaard argues the somewhat surprising theses that Kant's theory (or at least contemporary Kantian theory) can accommodate duties to other animals and that his theory can also show why we have these duties. She defends the view that "Kant's principle requires that when we enter into an interaction with another, we must act in a way that makes it possible for him to consent." Since animals cannot give consent, we should adopt the norm that we should "interact with other animals as long as we do so in ways to which we think it is plausible to think they would consent if they could—that is, in ways that are mutually beneficial and fair, and allow them to live something reasonably like their own sort of life." Using this approach, we can justify our uses of animals as companions, aides to the handicapped and the police, search-and-rescue workers, guards, and the like. We can perhaps also justify using animals as providers of wool, dairy products, or eggs, but it is not plausible to maintain that a nonhuman animal would consent to being killed before the end of a natural course of life in order to be eaten or in order to create food and the like, and it is implausible to hypothesize that an animal would consent to painful scientific experimentation.

Virtue Ethics. In "Virtue Ethics and the Treatment of Animals," **Rosalind Hursthouse** considers a theory that most philosophers view as deriving historically from the work of the ancient Athenian-Macedonian philosopher Aristotle. Hursthouse shows how virtue ethics promotes the paradigm that we should think about moral rights and wrongs in our treatment of animals in terms of virtues and vices rather than in terms of consequences (as utilitarians do) or rights and duties (as Kantians do). She argues that two leaders in the field of ethics and animals, Peter Singer and Tom Regan, each implicitly picks out one virtue, but one virtue only—a too concentrated focus that renders their moral theories unsatisfactory. According to her virtue ethics theory, we ought to be thinking in terms of all of the virtues and vices pertinent to the moral problems that arise in human uses of nonhuman animals. She thinks that many theories have made this path difficult because of an undue focus on the concept of moral status. Hursthouse rejects this agenda and argues that virtue ethics has no need for the concept of moral status and that the quality of discussion in ethical theory would improve if we eliminated this notion from our discourse. Hursthouse is direct and blunt in her rejection of one of the mainstays of the literature on ethics and animals: "Moral status is a concept that moral philosophy is better off without." Much of her argument is directed at showing why this concept is superfluous and why virtue ethics offers a superior approach. Virtue ethics does not need recourse to moral status because its "virtue- and vice-rules," such as "do what is compassionate" and "do not do what is cruel," direct persons to appropriate conduct without any need to discover to which groups these rules apply or what the moral status of the members of the group is.

Humean Theory. In "A Humean Account of the Status and Character of Animals," **Julia Driver** presents a theory inspired by and rooted in the work of the Scottish philosopher and historian David Hume. Hume anticipated some features of Darwinian thinking about animal minds. In particular, Hume believed that when the term "understand" is used properly, animals can understand many features of the world. Hume attributed rationality, or at least the capacity to reason, to some animals, on grounds that these animals are significantly like humans in the principles of their nature, their patterns of learning, and their powers of inference. Driver interprets Hume to hold that animals resemble human beings both in a variety of behaviors and in critical aspects of their mental lives. She finds that these behavioral and psychological similarities form the basis of a Humean argument that animals have moral status, though she acknowledges that Hume is less interested in moral status questions and more interested in animal minds. Nonetheless, Driver argues that the Humean theories of animal character (a part of his theory of mental qualities) provides the basis for "an account of moral standing that acknowledges the moral considerability of animals as resting on a continuum with the considerability of human beings." Driver interprets Hume to hold that animal species are spread out across a continuum of reason and emotion—a continuum constructed on the basis of the similarities and differences between human beings and animals. Chief among the Humean categories is that animals both reason and feel, which grounds the claim that we owe them moral consideration. Driver judges this continuum of rational and affective capabilities between animals and human beings to be among the most original, distinctive, and defensible aspects of Hume's theory. She also argues that even if Hume overlooks certain deficiencies that would prevent animals from having full moral agency—for example, that they lack the right sort of meta-cognitive state—this lack is insufficient to undermine claims that Humeans make about moral status and duties of humanity.

Utilitarian Theory. In "Utilitarianism and Animals," **R. G. Frey** notes that, of all the traditional, mainstream ethical theories, none has been more disposed over the centuries to sympathetic consideration of the pains of animals than utilitarian-ism. By using a sentiency criterion of moral standing, Jeremy Bentham ensured that the pain and suffering of animals count in the moral calculus. Their pains confer on them moral considerability; and every utilitarian since Bentham has endorsed the sentiency criterion—to an extent that this criterion has become virtually identified with utilitarian approaches to the moral status of animals. Nonetheless, Frey does not maintain that a utilitarian, to be such, cannot provide a different account of the moral standing of animals, and Frey himself proposes such an atypical utilitarian account based on the comparative value of human life and animal life. Frey is con-cerned with why utilitarians up to the present day have accepted the sentiency crite-rion and yet almost routinely failed, with specificity, to include animals within the calculus of utility in their general moral philosophies. Bentham was the first to exhibit this shortcoming. Frey judges that in the last four decades, Peter Singer was the first utilitarian in the 1970s to take animals seriously, and Frey tries to show why a utilitarian can and should view Singer's theory as seriously deficient from the utili-tarian point of view. Frey argues that the appeal to sentiency has limitations that go to the very core of debates about our use and abuse of animals. At that core, he argues, lies an unpalatable choice that a consistent utilitarian must face. Frey's theory is intended to make the choice transparent, and to do so in a nontraditional way.

Rights Theory. In "Rights Theory and Animal Rights," **Tom L. Beauchamp** presents a theory of animal rights developed from basic categories in moral and political philosophy about the nature and sources of rights. Though he acknowl-edges the importance of historically influential rights doctrines, including John Locke's, Beauchamp concentrates on aspects of contemporary rights theory that are suited to the analysis and justification of animal rights. He argues that rights are justified claims that individuals, groups, and institutions can press upon others or upon society. If an individual or group possesses a right, others are validly con-strained from interfering with the exercise of that right. "Animal rights" in this way give an animal or group of animals valid claims against the harm-causing activities of humans. Beauchamp defends the claim that animals have rights, but he does not treat these rights in terms of the remarkably strong protections of animal interests found in some leading theories of animal rights, namely those that prohibit most of the ways in which humans use animals. Beauchamp nonetheless proposes a robust theory of animal rights that would significantly alter many current practices. A crit-ical part of his argument is that there is a firm correlativity between rights and obligations: all rights entail obligations and all obligations entail rights. Therefore, if we have any obligations at all to animals (e.g., an obligation to feed a farm animal, an obligation to provide exercise opportunities for zoo animals, and the like), they have correlative rights. Finally, Beauchamp uses his theory to generate what he calls a catalogue of the rights of animals.

Capabilities Theory. In "The Capabilities Approach and Animal Entitlements," **Martha Nussbaum** begins with a sketch of our current knowledge about animal thinking. She inquires into what this knowledge suggests for ethics and for public policy. She finds that it challenges what has been the most influential approach to the ethics of animal treatment, namely classical utilitarianism. After rejecting this theory, she proposes a theoretical approach that has two nonutilitarian elements as centerpieces. First, the approach has a Kantian element—"a fundamental ethical starting point [stressing] that we must respect each individual sentient being as an end in itself, not a mere means to the ends of others." Nussbaum regards this demand as an extension of Kant's approach to human beings, so that we can make it work for animals. Second, the approach has a neo-Aristotelian, capabilities-theory element: "the Aristotelian idea that each creature has a characteristic set of capabilities, or capacities for functioning, distinctive of that species, and that those more rudimentary capacities need support from the material and social environment if the animal is to flourish in its characteristic way." If we combine the Aristotelian view with the Kantian element, Nussbaum argues that we can justify the claim that we are obligated to respect sentient creatures as ends. Treating an animal as an end requires (negatively) *not obstructing* the animal's attempts at flourishing through acts of violence or cruelty and also requires (positively) support of efforts to flourish. Nussbaum believes that this theory can, over time, become the basis for an overlapping consensus in ethics among those who differ on many other moral matters. She thinks that "even an ethical Utilitarian can accept political principles based on neo-Kantian or (related) neo-Aristotelian insights," once they are properly articulated.

PART III. MORAL STATUS AND PERSON THEORY

Intelligence and adaptation in animals seem incomprehensible unless we attribute to them some form of understanding, intention, imaginativeness, or skill of communication. To attribute these capacities to animals is to credit them with capacities analogous to human capacities, an assessment that many philosophers find merited. They also think this assessment supports the view that many animals have some degree of moral status in the sense of a grade or rank of moral importance.

In Western history, animals have typically been treated as having, at best, only a low-level moral standing or status—and perhaps no moral status at all. If the latter view were correct, then humans owe nothing to animals and can do with animals as they wish. But this claim has been challenged in recent philosophical accounts that recognize animals as having a significant level of moral status. This commitment need not mean, however, that animals have rights or some similarly elevated form of moral protection. Here we enter disputed territory.

To navigate this territory, one needs a solid starting point that does not beg the central questions about moral status. That there is such a neutral starting point has been difficult to show. One problem is that there are several attractive starting points, and no decisive reason to start from one rather than another. For example, the following is one general and attractive place where we might start: moral status turns on whether a creature is an experiential subject with an unfolding series of experiences that can make that creature's life go well or badly. Such a creature has a welfare that can be positively or negatively affected by what we do or do not do to it. Rodents, dogs, and chimps are such subjects, each with a welfare and quality of life that our actions can cause to go up or down. If one accepts this view as a solid and acceptable starting point, then it is not difficult to identify some of the central concepts and norms in need of development and defense in a general theory of moral status. For example, a defense of this point of view presumably should start by carefully analyzing the notions of "welfare" and "quality of life."

Another attractive starting point is the view that the more an animal is like a human being in those qualities that capture the essence of humanity, the more the animal's moral status is upgraded. Its status would be still further enhanced if it could correctly be said that the animal has either personhood or autonomy, or both. A category such as "person" or "autonomous agent" seems to elevate the animal to a position approximating that occupied by individuals who have human rights. Not surprisingly, then, person theory plays a vital role in several contributions here in part III.

The mainstream approach to the question of which kinds of entity deserve a significant moral status has been to ask which *properties* an entity must possess in order to qualify for moral protection. Some say that there is one and only one property that confers moral status. For example, some say that this property is rationality, understood as a certain kind of cognitive capacity. Others say that another property or perhaps several other properties are required in order to reach a significant level of moral status—for example, properties of sentience or properties of moral agency. These different views about the relevant properties and capacities have left philosophers at odds about which theory of moral status is the best theory. Every selection in part III of this *Handbook* deals with some aspect of these moral and conceptual problems.

Finally, although most theories about the ethics of using animals have relied heavily on the concept of moral status, these theories have proved difficult to apply to many practical problems. It could be that the objective in theories of moral status of showing which creatures matter morally may not be as fruitful in advancing practical (by contrast to theoretical) issues of how animals should be treated. Perhaps we do not need any kind of account of moral status in order to address the moral problems. This issue is considered primarily in section VI below—and in the chapter by Hursthouse in part II.

In "The Idea of Moral Standing," **Christopher W. Morris** investigates the idea of moral status for nonhuman animals and how that status compares to the moral status of most humans. He also considers whether some artifacts and natural objects have a significant moral status. He finds much of the philosophical literature confusing, so he starts anew with an innovative distinction between the notion of status and that of standing. He regards standing as a special status that humans and perhaps a limited number of other beings have. Morris takes the idea of "duties to" other beings as critical for his account of moral status. He thinks that something has moral standing if it is owed any moral consideration or duty whatever. He notes that "very few people doubt that we have some duties *regarding* non-human animals, but there is a major controversy as to whether we have any direct duties *to* any animals." This issue is critical for Morris, who is tempted to use a conventionalist or constructivist account of justice to support his theory. However, he thinks that this understanding of moral standing does not provide an adequate account by itself. He argues that moral standing turns on obligations owed *to* a being, such that *it* is wronged when these obligations are disregarded. Certain duties of charity and benevolence, in particular, require us to aid animals for their own sake, not that of another.

In "Animals, Fundamental Moral Standing, and Speciesism," **David Copp** considers whether we have moral duties that are owed directly to the animals, or whether all duties regarding animals are derivative from duties we have to human beings. Copp maintains that we do have moral duties directly toward nonhuman animals, not merely duties regarding them, and that this claim can be adequately grounded in what he calls *the thesis of the fundamental standing of animals* (which basically means that we have morally fundamental duties to treat nonhuman animals decently). Nonetheless, Copp finds that the thesis of the fundamental standing of animals is in tension with a very different and intuitively plausible thesis that Copp calls *the thesis of the fundamental concern of morality*: morality is fundamentally concerned with advancing *human* welfare *by* enabling human beings to live together successfully in societies. The problem is that these two theses seem to be in conflict generally and in direct conflict in many circumstances. Copp argues that, despite the appearance of direct conflict, the two theses can be shown to be compatible; even the apparent "speciesism" of the second thesis is compatible with recognizing the fundamental standing of animals. The key to their compatibility is to understand the second thesis as about the nature of morality (a non-normative conceptual claim), whereas the thesis of the fundamental standing of animals is a normative moral claim. In short, the normative thesis that we have morally basic duties to animals is compatible with the metaethical claim that morality is fundamentally concerned to advance human welfare by enabling humans to live cooperatively together in societies. Copp adds that an appropriately structured society-centered conception of morality supports the thesis that there are morally fundamental requirements as to how animals should be treated. He offers the attention-getting thesis that "the content of our fundamental duties depends on facts about human psychology and the circumstances of human life. These facts plausibly imply that our obligations to treat animals decently are as fundamental as any of our obligations toward human beings."

In "Human Animals and Nonhuman Persons," **Sarah Chan** and **John Harris** start with the assumption that the term "person" in the everyday sense is generally taken to be synonymous with the term "human," whereas philosophers tend to use the word "person" in the more abstract sense of "those entities who possess a particular moral status and about whom particular moral claims may be made on the basis of that status." They note that the close association commonly made between the two concepts "human" and "person" produces the widely accepted view that humans are the most important type of creature to whom moral status is accorded, while downgrading nonhuman creatures from consideration as persons and as having moral status. But can such a view be justified? Chan and Harris consider whether a nonhuman animal or any other entity could, under certain specifiable conditions, be a person. They approach this subject through personhood theory and why certain attributes are thought so important to being a person. They then inquire what this account requires of nonhuman animals in order to be deemed persons. They next explore the implications of nonhuman animal personhood—politically, legally, and philosophically. They argue for a conception of personhood that deemphasizes the importance of "being human." They think that the question we should be asking, including about future generations of humans (perhaps also post-humans) is not whether they will still be human but whether they qualify as *persons*. In this way the concept of persons becomes the critical conceptual and moral issue surrounding problems of moral status, thereby lowering the importance of all species designations.

In "Are Nonhuman Animals Persons?" **Michael Tooley** likewise thinks of the concept of persons as the central issue in moral status debates. He finds questions about whether members of one or more nonhuman species of animals are persons among the most difficult philosophical questions we face today. He locates the difficulty in two sources: (1) how the concept of a person should be analyzed, especially the properties that give an entity a right to continued existence; and (2) how to determine which psychological capacities and which forms of mental life adult members of nonhuman species have. Tooley's concern is primarily with the first of these sources and issues. He begins by attempting to capture a purely descriptive, non-normative sense of "person" that is appropriate to the formulation of a fundamental moral principle concerning when the destruction of something is wrong. He argues that the fact that something is a continuing subject of experiences and has mental states that are psychologically connected over time is crucial to creatures having moral status and having a right to continued existence. Tooley then considers four leading arguments that have been offered by philosophers in support of the view that nonhuman animals do *not* have moral status: (1) contractarian approaches; (2) moral agency approaches; (3) approaches based on an absence of beliefs, desires, and thoughts; and (4) self-consciousness approaches. Tooley finds the first three theories generally unsuccessful, though variants of the third and fourth he finds difficult to evaluate. Tooley then examines three arguments in support of the view that nonhuman animals *do* have rights: (1) mere consciousness is sufficient to give something moral status (or at least sentience warrants moral status); (2) being a subject-of-a-life gives higher animals critical rights; and (3) imagistic thoughts of the right sort provide a sufficient basis for moral status. Tooley finds these arguments unsuccessful. In the end Tooley reaches the following firm and yet contingent conclusion: "On the one hand, if the idea of thinking that involves only images, and no use of language, is untenable, then very few non human animals are persons, or have a right to continued existence, or have moral status. On the other hand, if imagistic thinking is logically possible, these issues are very much open, and much further philosophical reflection is called for, though I am inclined to suspect that, in the end, the conclusion may not be all that different." This leaves the moral status of animals somewhat dependent upon what we learn through further study of the nature of the thought of animals.

Part IV. Animal Minds and Their Moral Significance

It has often been said in the literature on person theory that an individual is a person if and only if the individual possesses one or more of the cognitive properties mentioned previously. The term "cognition" here refers to processes of awareness and knowledge, such as perception, memory, thinking, and linguistic ability. In these theories the possession of these properties, above all other properties, elevates a being's moral standing. As a corollary, anything lacking these properties lacks moral standing. In these accounts there is a clear connection between having a type of mind and being morally significant. The hypothesis is that if animals lack some critical form of cognitive capacity, they lack significant moral status.

Critics of theories that make cognitive properties central take one of two alternative approaches. The first is to argue that some nonhuman animals in fact do have significant cognitive capacities, whereas the second is to argue that some animals have noncognitive mental capacities that are sufficient to confer some measure of moral status. The most frequently invoked properties in the second approach are those of sensation—especially pain and suffering—but also mentioned are properties of emotion, such as fear and anger. These two approaches are not mutually exclusive, and indeed can in principle be used together to present a very strong case for a high level of moral status for animals such as the great apes. Combining the two yields the view that animals have both morally relevant cognitive capacities such as intelligence and morally relevant noncognitive mental capacities such as experiencing pain and suffering—and that these two conditions are jointly sufficient to confer a high level of moral status.

A range of problems about how to understand animal minds is at the root of these moral problems. Most observers of animal behavior today agree that many animals have significant capacities to understand (which does not imply that they have a capacity to understand propositions in a language) and have developed complicated, sometimes elaborate forms of social interaction and communication. Little agreement exists, however, about the levels and types of mental activity or about their ethical significance. Humans understand relatively little about the inner lives of animals, or about how to connect many forms of observable behavior with other forms of behavior. Even the best scientists and the closest observers of animal species have difficulty understanding intention, communication, and emotion in animals. Attributions of emotion, intention, and the like have been criticized by some as an unscientific abandonment of critical standards and precise measurements, as well as an importing of an unsupported anthropomorphism.

It is not surprising that our ability to penetrate animal minds has proven profoundly difficult. Neither evolutionary descent nor the physical and functional organization of an animal system (the conditions responsible for its having a mental life) gives us the depth of insight we would like to have in understanding animal mentality. We also have only a weak idea of what constitutes a minimal mind in a creature. Some living creatures seem literally to have no mind. The more we are in doubt about an animal's mental life, including whether it has one, the more we are likely to have doubts about its moral status, and therefore about whether it has rights. If a theory requires high-level mental abilities in order to qualify for moral status, very few if any nonhuman animals will qualify, but if less demanding cognitive capacities are employed, animals might acquire a significant range of moral protections. For example, if a high-level qualifying condition such as speaking a human language is not required and conditions such as intention and intelligence are substituted, then many animals may have a significant moral standing.

These questions have played a role in the history of philosophy mentioned in part I. We can reach back to the classical philosophers, including Aristotle, who held that animals have intelligent minds, but lack reason, a defect regarded as sufficient to exclude them from the moral community. Other philosophers, among them David Hume, held that animals have both mind and reason and may have moral emotions, but still lack capacities of moral judgment and agency. Finally, some philosophers, including René Descartes, assess animals as lacking all mental capacity—all feeling and even consciousness. Descartes suggests that animals, being devoid of minds, are like plants. His views are still frequently discussed in literature on animals and ethics, and his theories find their way into some of the chapters here in part IV.

The term "morality" is often used to refer to learnable standards of right and wrong conduct that are widely shared so that there is a secure communal consensus. Many have thought that we can find morality in analogous ways in some parts of the animal world. Here we would have moral animal minds. Clearly some animals live in communities that require conformity to basic norms of communal life, much as human communities do, but whether there is good evidence of real morality in nonhuman animals is highly controversial. Some observers believe that it is little more than human fantasy to suppose that animals act morally, but other observers find close similarities between the behavior of humans and nonhuman animals—for example, in altruistic behavior. These theses are examined in the last two chapters in part IV.

In "Animal Mentality: Its Character, Extent, and Moral Significance," **Peter Carruthers** continues the discussion of whether animals possess moral standing, which he understands to be the question of whether they are deserving of our sympathy and concern and whether they possess moral rights. Carruthers thinks that the question of moral rights should receive a negative answer, even though he believes firmly in the evolutionary and cognitive continuities between humans and other animals. The first half of his chapter argues that pain and suffering of a great many animals do appropriately make them objects of sympathy, and it shows that they have minds with structures often similar to those of humans. He thinks that some moral theories are therefore forced to the conclusion that all such creatures thereby gain moral standing, a view Carruthers finds indefensible. He also judges that those philosophers who require extremely demanding conditions of consciousness, rationality, or spoken language as conditions of our moral concern for animals are making excessive and indefensible demands. His assessment is that a great many animals are genuinely *agents* with a perception-belief-desire psychology and with goals that can be frustrated. It follows that these animals are properly the objects of sympathy and concern. However, in the final half of his chapter Carruthers turns to a defense of a contractualist perspective, which is that all humans, and probably no other animals, possess moral standing. He acknowledges that this position seems counterintuitive, but he thinks the problem goes away once we see that we can still have indirect duties towards animals. From his contractualist perspective, morality is the outcome of an idealized contract among agents who can then constrain and guide their relations with others. The upshot is that, at Carruthers' hands, almost all humans, and no other animals, possess what really should be meant by moral standing.

In "Mindreading and Moral Significance in Nonhuman Animals," **José Luis Bermúdez** starts with the current state of the discussions about the moral significance of nonhuman animals and the basic capacities that have been stressed in the leading theories: sentience, basic forms of consciousness, animals' capacity for making long- or short-range plans, self-awareness, and the like. The goal of his chapter is "to extend the territory of this debate by exploring the moral significance of what researchers in cognitive science, developmental psychology, and comparative psychology term mindreading—that is, the ability to understand the mental states of others." His work is almost entirely centered on mindreading as a cognitive ability that intersects with the basic capacities put forward in theories of moral status. He thinks we have much to learn from empirical studies of both nonhuman animals and human children. Bermúdez is not primarily working as a *moral* philosopher. His main aim is to make it easier for moral philosophers to incorporate experimental and ethological work on mindreading into their discussions of moral significance. He does so by presenting and analyzing the relevant experimental results and observational data, and then by drawing taxonomical distinctions that are relevant to our thinking about moral significance. One central argument is that propositional attitude mindreading is language-dependent, because it involves metarepresentation.

After reviewing experimental findings on mindreading in primates, he tries to show that the data can be interpreted without assuming that the mindreaders are engaged in propositional attitude mindreading. However, his claim is not that primates have no capacity of mindreading. They can enter into basic mindreading and sophisticated forms of social cognition that do not involve attributing propositional attitudes. "Nonlinguistic mindreading," Bermúdez concludes, "is a powerful tool in the animal kingdom."

In "Minimal Minds," **Bryce Huebner** starts out by observing that we often speak in a loose manner as if beings all the way down to viruses have something like beliefs, desires, and intentions. Even some scientific discourse suggests desires, wants, and objectives in very low-level beings. Huebner sets aside these ways of speaking to see how far down the chain of being there might be true mentality—minimal mindedness. In the case of viruses and many kinds of beings, Huebner argues that it is unlikely that they have the computational and representational capacities necessary for cognitive states or cognitive processes. They are much more like "complex biological robots." Nonetheless, profound theoretical issues confront attempts to provide a general strategy for distinguishing mentality from witless and mechanical behavior. Huebner tries to establish the *minimal conditions*. He argues that the explanatory resources provided by contemporary cognitive science provide compelling reasons for thinking that even invertebrates such as cockroaches, termites, and honeybees possess genuine cognitive capacities. Huebner recognizes that many may be skeptical of this claim, but he supports the idea that invertebrates possess minimal minds in two ways. First, he argues that minimal mindedness can be realized without linguistic representations, so that minimal cognitive states and processes need not be beliefs or desires—and are unlikely to be so in invertebrates. Second, he argues that the evidence for invertebrate mindedness warrants the extraordinarily unusual and underexplored theses that some *collective behaviors* are best explained in terms of collective mental states and processes. Put another way, if invertebrates have minimal minds, then so do some *groups* of invertebrates have a mind. Even if this conclusion can be supported, Huebner notes, we would still face ethical questions about the kinds of minds that have moral status and warrant our compassion.

In "Beyond Anthropomorphism: Attributing Psychological Properties to Animals," **Kristin Andrews** discusses "anthropomorphism" in the sense of the attribution of uniquely human mental characteristics to nonhuman animals. One philosophical problem is to figure out how we can identify which properties are uniquely human. Humans and nonhuman animals share a vast number of biological, morphological, relational, and spatial properties—as well as psychological properties such as the ability to fear or desire. Andrews maintains that one goal of animal-cognition studies is to determine which cognitive abilities animals use and whether some identifiable cognitive properties are found only in the human species. If the properties are uniquely human, then asserting that some other animal has that property would be false and an example of anthropomorphism. In the empirical and the philosophical literatures, features that have been described as uniquely human include psychological states such as beliefs and desires, personality traits such as confidence or timidity, emotions such as happiness or anger, social-organizational properties such as culture or friendship, and moral behavior such as punishment or rape. But are these in fact uniquely human "psychological properties"? Several connected questions arise: first, is it scientifically respectable to claim to be able to examine questions about the mental, psychological, cultural, and other states of animals? Andrews argues that there is no insurmountable problem in asking and answering such questions. Second, questions surround how investigation into these questions is to be conducted. Andrews proposes that we use an approach to animals based on parallel assessments of prelinguistic children. A specific psychological attribution will be warranted if it takes into account the species and cultural-normal behavior, it has predictive power, and it mirrors the attribution of a similar property in prelinguistic infants. Andrews does not maintain that nonhumans are limited to psychological properties of human infants, but she does maintain that this general approach, modified to be species appropriate, can productively be used in the study of animal psychology. She also argues that there remain many dangers of false attributions of psychological properties to nonhuman animals.

In "Animal Pain and Welfare: Can Pain Sometimes Be Worse for Them Than for Us?" **Sahar Akhtar** probes the widely held view in philosophy and the biological sciences that the amount and ways in which a nonhuman animal can experience pain, by comparison to the human animal, is limited to the feeling of physical pain. The justification for this view is often said to be that animals are less cognitively sophisticated than humans because they lack awareness of self and a sense of the past and the future. If pain is inflicted on animals, it is thought that, while animals may be able to feel the pain itself, they are not capable of the higher order suffering that may accompany the feeling of pain in humans. This view suggests that pain for animals is not as bad as pain is for us. Akhtar presents a notably different approach to the understanding of animal pain. She uses welfare analysis and decision-making frameworks to argue that pain may be worse for animals than the comparable amount of pain is for humans. She hypothesizes that pain may not always play a significant role in human welfare both because the intensity of pain can often be mitigated through expectations, memories, and the consideration of and attention to other interests and because humans are able to engage in calculations about their interests, so that they can discount pain in order to achieve longer term interests.

In "Animals that Act for Moral Reasons," **Mark Rowlands** argues that some animals are moral *subjects* in the sense that they can be, and sometimes are, motivated by moral considerations. Although tradition has it that animals in many well-known scenarios are not acting compassionately, Rowlands challenges the claim. He notes that it does not follow from such claims that we have no reason to think that animals cannot be morally evaluated for what they do. Rowlands argues that there are no empirical or conceptual obstacles to regarding some animals as motivated by moral concerns. To suppose otherwise, he thinks, is to fall victim to certain views that invest quasi-magical properties in "meta-cognition"—properties that afford humans a status of a sort possessed by no other beings. Perhaps these animals cannot self-reflect and scrutinize their motivations and cannot ask themselves whether they are engaging in appropriate responses or whether what they are doing is something that they *should* or *ought* to be doing. Perhaps they are only pushed this way and that by their sentiments. But Rowlands strongly resists these perspectives. He thinks that the sentiments of animals can be genuinely moral ones and that there are no compelling reasons to suppose that these animals are not moral subjects that can be morally evaluated—even if they cannot be moral agents.

In "The Moral Life of Animals," **Michael Bradie** raises two main questions: "Do non human animals have minds?" and "What implications, if any, does the answer have for their moral status?" Bradie argues that a former tide against animal mentality and moral status has changed over the past 150 years, leading us to our present muddled set of opinions. He thinks that the prevailing opinion now is that the question "What determines mental status?" is likely to determine the answer to the question "What determines moral status?" This situation drives philosophers to be concerned about numerous questions about animal minds. Do animals have qualitative, internal experiences? What do we mean by attributing a "mental life" to anything? It is widely agreed that animals communicate with each other, but not widely agreed that animals possess a language relevantly comparable to human languages. Bradie argues that several lines of empirical evidence now support the claim that animals possess enough mindedness to count as moral patients or partial members of a moral community. On some views there is even sufficient evidence to count animals as moral agents and full members of a moral community. He thinks that evolutionary evidence supports the claim that there are no significant qualitative differences between humans and other animals. For example, neuro-physiological evidence supports this conclusion insofar as homologous brain structures and systems implicated in the cognitive and affective capacities of human beings are widespread in the animal kingdom. Also, evidence from cognitive ethology indicates that many animals possess the neural architecture necessary for sophisticated cognitive and affective behavior, and that they manifest moral sensibility. Despite what critics say, Bradie thinks that the default presumption ought to be that animals experience emotions and act on moral sentiments, properly understood. His assessment is that the burden of proof is on those who reject the attribution of cognitive and emotional states, as well as moral sensibilities, to animals. Bradie reaches this conclusion: "[h]uman beings are one among the animals [and] from this, human cognitive capacities are one among the cognitive capacities of animals. Finally, human moral systems are one among the moral systems of animals."

PART V. SPECIES AND THE ENGINEERING OF SPECIES

Prior to the work of Charles Darwin, many biologists and philosophers argued that some of the above-discussed problems about moral status and cognitive capacities are resolvable in terms of critical dissimilarities between the human species and other species—primarily differences of reason, speech, and moral sensibility. Darwin thought that animals often exhibit considerable powers of deliberation and decision-making, excellent memories, and imagination in their actions He wrote about the intelligence, sympathy, pride, and love found in many animals. Darwin criticized the hypothesis that only humans have significant cognitive powers. In general, Darwin was suspicious of the ways in which members of the human species distance themselves from the members of other species, without carefully examining the empirical evidence.

Darwin assessed the idea of species through the lens of his evolutionary biology rather than in terms of rigid and fixed differences between species. In *The Descent of Man*, he catalogued numerous similarities in mental ability between humans and apes. He then argued that despite "enormous differences" in degree of "mental power," no fundamental difference exists *in kind* between humans and many forms of animal life. The differences are *in degree*. The nonhuman apes are highly intelligent, similar to humans in emotional responses such as terror, rage, shame, and maternal affection, similar in character traits such as courage and timidity, and even similar in the use of systems of communication that approximate human language and human conceptual abstraction. Like some contemporary thinkers, Darwin proposed multiple levels of mentation that are shared across species, from basic pain receptors to intentionality.

This continuity-across-*many*-species model can coexist with what might be called a continuity-across-*individual*-species model. The latter holds that nature does not divide sharply into biological species in the sense that the members of each species all share some property or set of properties (so-called species essentialism). If "dog" is a species, it may not be the case that all members of the species share some property or properties of dogness. There may only be family-resemblance-like properties. This point can be generalized to the conclusion that there are no sharp or absolute species divisions in nature: A species is merely a group of individuals who interbreed. Across the population, they may share no defining property. Unlike dog shows where judges try to find the best representative of the animals on display, the species non-essentialist says that there is no paradigm dog and no perfect member of any species. Occasionally some group of the members of a single "species" becomes isolated from other members, and, when the path of evolution is subsequently cut, the two groups become two different species; and yet they may still have many properties in common.

In light of the evolution and disappearance of species, questions arise about whether we have general duties of species preservation, especially when humans cause the disappearance. Environmental ethics has brought a diverse array of concerns about ecosystems and with these concerns has come the idea that there is a general duty to preserve species. But do we have a duty of species preservation? If so, what is the basis of this duty? Sometimes, goals of species preservation come into conflict with goals of protecting the interests of other animals. It seems unreasonable to propose that species preservation always merits the highest priority. Even if an obligation of species preservation is stringent, it will not be strong enough to prevail in all circumstances of conflict with other obligations. Moreover, it is doubtful that there are any moral obligations of species preservation merely because a species would be preserved (by contrast to preserving species because they have some instrumental value). Perhaps, then, there are no general duties of this description. Several problems of this sort are considered in part V.

Another widely discussed moral problem about species and our preference(s) for certain species is labeled "speciesism." A speciesist is one who believes that the interests of members of one or more particular species, usually the species *Homo sapiens*, are to be favored over the interests of the members of other species. Species membership therefore is a factor in determining whether a creature has a high or a low level of moral status. The term "speciesism" is often used pejoratively and by analogy to racism and sexism. In this usage, speciesism represents an improper failure to respect the lives and rights of animals merely because they are other than human. Speciesism in this sense is the name of a form of bias or discrimination on the basis of species; that is, it takes the sheer biological fact that baboons and humans, for example, belong to different species as a reason to draw moral differences between them. But just as gender, race, IQ, accent, national origin, and social status are not relevant properties in morals, neither is species. Species membership, according to critics of speciesism, should not be a factor in making moral judgments. It is not the species, but certain qualities that make an animal morally considerable.

However, speciesism need not be understood in this pejorative manner. Some speciesists willingly, and even enthusiastically, accept the label speciesism. They declare themselves speciesist, meaning that they place a priority on moral relationships with members of the human species *because* they are human. They point out that humans have a natural feeling of kinship and closeness with members of their own species, just as some human family members feel closer to other family members and just as many animals strongly prefer to associate with members of their species and to have nothing to do with others. Such natural feelings create stronger obligations to members of the relevant group—the family, on the one hand, and the preferred species on the other. Another defense of speciesism is that certain properties associated with the human species—in particular, the cognitive properties discussed previously—give humans a special moral status; it is not species membership alone that justifies special treatment for humans. From this perspective, pro-speciesists maintain that it just so happens that those who possess these special properties are of one and only one species.

Finally, among the most compelling scientific developments in recent years is the human creation of genetically modified animals that have been engineered for certain human purposes. We can now alter animals and create new life forms—new species—in extraordinary ways. These may be transgenic animals, hybrids, or chimeras. A newly created animal could be in large part a human being—part-human and part-animal. But how adequate are our modes of ethical inquiry and oversight to deal with these scientific developments? Do we need more demanding strategies of ethical evaluation? The final two contributors in part V explore these frontier issues.

In "On the Origin of Species Notions and Their Ethical Limitations," **Mark Greene** probes the difference between the concerns we have about using individual animals, such as laboratory animals, and the concerns found in environmental ethics, which focus not on individual beings but on entities such as ecosystems, the wilderness, or even the whole biosphere. A concern that cuts across environmental ethics and ethical issues about animals is whether we have a duty to preserve species, most notably those threatened by human activities. Greene's objective is to critically examine the thesis that species themselves, not merely individuals of the species, have morally considerable value—that is, moral status. Greene thinks it is fairly easy to figure out why a species would have instrumental value because of the contributions it makes to other things of value, such as the contributions made to human well-being. Dogs and farm animals might be so viewed. However, this strategy can ground only limited and contingent duties of species preservation, and only for species that make such an instrumental contribution. Greene thinks that once defenders of species preservation have realized that their arguments are entirely contingent on whether animal species make contributions to humans, the defenders of a general duty of species preservation will want to dig in and argue that species must have non-derivative value in their own right. Greene finds this strategy implausible. He thinks that evolutionary biology calls into question the intelligibility of defending the non-derivative value of species. The main lines of Greene's argument are these: because "(1) a general duty of species preservation requires that species have non-derivative value and (2) species lack non-derivative value, it follows that (3) there is no general duty of species preservation."

In "On the Nature of Species and the Moral Significance of their Extinction," **Russell Powell** begins by noting that in the history of life, every species up to presently existing species has become extinct. Complex life itself has been on the brink of annihilation at various points in the evolutionary process. Over time species do not seem to get better at not going extinct. Except for certain accidents of history, even mammals might still be small creatures struggling to survive under challenge by dinosaurs. Powell's problem, given this history, is whether we should regard the causing or the permitting of the extinction of species as a bad outcome to be avoided. Like Greene, Powell queries the intuition that species qua species have moral value. He sees difficulties in justifying this view, and he asks whether it can be made robust enough to structure an ethical framework that might appropriately influence public policy. Powell thinks that so-called common-sense intuitions about these matters are not trustworthy and often do not hold up to theoretical scrutiny. Using the evolutionary and ecological sciences, Powell takes the currently received view—and the one he supports—to be that species should be analyzed in terms of individual lineages and not as atemporal natural kinds. Regarding species in this way will have critical implications for the value that we place on species. Powell then argues, like Greene, that a species does not have the properties that make for intrinsic moral value, but only for instrumental value. He then considers whether there is a morally important difference between human-caused extinctions and those that result from "natural" evolutionary processes. He argues that the degree of "badness" that occurs because a culpable moral agent was involved in the process is minimal, and thus should not affect our conservation priorities or obligations, which are themselves the most important matter.

In "Are All Species Equal?" **David Schmidtz** considers the defensibility of what he calls "species egalitarianism"—the position that all living things have equal moral standing and therefore all species command our respect. Schmidtz challenges the view that there are good reasons to believe that all living things have moral standing in even a minimal sense. Schmidtz explains why members of other species understandably and justifiably command our respect, but also why they cannot command *equal* respect. He thinks we can adequately respect species without embracing species egalitarianism. For example, we can hold with most vegetarians that it is worse to kill a cow than to kill a carrot, but we cannot hold this view consistent with true species egalitarianism. Schmidtz thinks that if we treat a chimpanzee no better than we would treat a carrot, we exhibit a failure of respect, not an instance of respect. He argues that we can agree that trees and chimpanzees share equally in the value we place on being a living and flourishing being, but this value cannot support the thesis that trees and chimpanzees have equal value. Schmidtz also argues that there is reason to doubt that species egalitarianism is compatible with true respect for nature. The theory improperly suggests that the moral standing of dolphins is no higher than that of tuna, and that the standing of chimpanzees is no higher than that of mice. Such a view does not give dolphins and chimpanzees the respect they deserve. From this perspective, "species egalitarianism not only takes humans down a notch. It takes down dolphins, chimpanzees, and redwoods, too. It takes down any species we regard as special."

In "Genetically Modified Animals: Should There Be Limits to Engineering the Animal Kingdom?" **Julian Savulescu** considers the extraordinary ways in which it is now possible to alter animals and to create new life forms by transgenesis and by the creation of hybrids and chimeras. Transgenic animals are created by transferring genes from one species to another. Hybrids are created by mixing the sperm of one species with the ovum of another. Chimeras are created by mixing cells from the embryo of one animal with those of a different species. In each case, one source of an animal so engineered could be a human being, that is, scientists could use the genes, gametes, or embryonic cells from a human to create an animal with human genes or a human-animal hybrid or chimera. Savulescu uses the term "genetically modified animals," or GMAs, to refer to these transgenic animals, hybrids, and chimeras. Our capacity to create these novel life forms suggests that as biotechnology progresses, the power to use these life forms will increase. He questions the adequacy of the ethics we currently have in place to exert proper control over and evaluate the creation of GMAs. Savulescu argues for three theses: (1) There should not be an overall general normative evaluation of the acceptability or unacceptability of the creation of GMAs; instead, we should evaluate proposals and new developments on a case-by-case basis. (2) We need richer strategies of ethical evaluation of technology than we now possess. (3) There are no good arguments against the creation of GMAs in general, though there may be such arguments in particular cases.

In "Human/Nonhuman Chimeras: Assessing the Issues," **Henry T. Greely** picks up on one area of the problems Savulescu mentions: human/non human chimeras. The term "chimera" has many meanings, but Greely's concern is restricted to living organisms that have, as part of their bodies, some living tissues, organs, or structures of human origin and some of nonhuman origin. Greely confines his analysis to creatures that are, or are viewed as, nonhuman creatures to which human tissues are added. His concern is with the ethical arguments made or implied about—generally *against*—human/nonhuman chimeras. He points out that most of the discussion took off from scientific study of the Human Neuron Mouse and focused on nonhuman creatures with "humanized" brains. He then describes the arguments and policies that have developed in regard to human/nonhuman chimeras, followed by a discussion of three particularly sensitive types of chimeras. Overall, Greely proposes that we take a pragmatic view about most of the scientific possibilities and the most sensitive types and uses of human/nonhuman chimeras. These developments can be used to create important human knowledge and medical treatments, but they need to be employed only for such good reasons, always being mindful of the evil purposes to which they might be put and the possible negative social reactions. In his conclusion, Greely points out that the Human Neuron Mouse that started much of this discussion still remains not born, but the idea of creating it is still not dead, which may be "for better or for worse."

Part VI. Practical Ethics

The term "applied ethics" and its near synonym "practical ethics" came into use in the 1970s when philosophers and other academics started to address pressing moral problems in society and in professional ethics. They almost from the start used ethical theory and political philosophy in their work. At the time philosophers seemed to say that they were putting theory to practical work by "applying" theories. Concentrated work of this sort began first in medical ethics, then quickly spread to business ethics, and then on to ethical issues regarding animals. Factory farms and animal subjects in biomedical research were widely discussed subjects in the early stages of interest in animals, and ethical theories were soon brought to bear on such problems.

Although moral philosophers have long discussed various questions about animals, it is arguably the case that no major philosopher throughout the history of moral philosophy developed a program of, or a method for, what is today called practical ethics. Moral philosophers traditionally formulated theories of the right, the good, and the virtuous set out in the most general terms, but a practical price is paid for theoretical generality: it is usually hazy whether and, if so, how a general theory is to be applied to quite concrete problems in order to generate public policy, settle moral dilemmas, and reduce social conflict and controversy. Many have thought that practical ethics is a "field" or "method" that can fill this gap, and ought to fill it, though many have also pointed out that some of the best pieces of work in practical ethics use arguments aimed directly at pressing and emerging moral problems, with relatively little actual reliance on a general ethical theory.

"Practical ethics" and "applied ethics" are today employed in philosophy to refer to the use of a mixture of philosophical methods and theories to deliberate about moral problems, practices, and policies in professions, institutions, and public policy. Reasoned argument is used to reach policy proposals and practical solutions. The chapters in part VI largely grow out of this conception. This practical commitment makes these chapters notably different in aim and in use of theory than the chapters in part II, which concentrate more on the theoretical side than on the practical side. Nonetheless, the actual differences between theoretical ethics and practical ethics may turn out to be more a matter of the degree of theory used than a real difference in the kind of moral philosophy being done.

Although there remain suspicions in many quarters of philosophy about the goals and methods of so-called practical philosophy, the importance of the issues treated in practical ethics is rarely disputed by anyone.

In "The Moral Relevance of the Distinction between Domesticated and Wild Animals," **Clare Palmer** considers whether a morally relevant distinction can be drawn between wild and domesticated animals. This distinction is seldom analyzed in the ethics literature, which leads Palmer to start with conceptual analyses of the terms "domesticated" and "wild." The term "wildness" she finds to be used in several different ways, only one of which (*constitutive* wildness, meaning an animal that has not been domesticated by being bred in particular ways) is generally paired and contrasted with "domesticated." Domesticated animals are normally deliberately bred and confined. One of Palmer's main arguments concerns human initiatives that establish *relations* with animals and thereby change what is owed to these animals. The main relations of interest in ethics are the *vulnerability* and *dependence* in animals that are created when humans establish certain relations with them on farms, in zoos, in laboratories, and the like. Domestication is a pervasive way in which humans make animals vulnerable, and thereby duties of animal care and protection arise in a persistent way. These duties are not similarly owed to wild animals that live independently of the human community. Palmer's argument is not that domestication is the sole way in which humans make animals vulnerable. Wild animals too can be made vulnerable through actions such as habitat destruction. The point is that wherever we have deliberately created animal vulnerability or dependence, we are obligated to care for or assist the animals affected.

In "The Moral Significance of Animal Pain and Animal Death," **Elizabeth Harman** addresses the question, "what follows from the claim that we have a certain kind of *strong* reason against animal cruelty?" Her particular concern is with what follows about the ethics of killing animals. She finds the following common assumption highly puzzling and problematic: despite our obligations not to commit animal cruelty, there is no comparably strong reason against painlessly killing animals in the prime of life. She argues that anyone who accepts this view is committed to the moral position that either we have no reasons against such killings or we have only weak reasons. The thesis that Harman takes to be at once puzzling and widely accepted has two parts: (1) "we have strong reasons not to cause intense pain to animals: the fact that an action would cause intense pain to an animal makes the action wrong unless it is justified by other considerations"; and (2) "we do not have strong reasons not to kill animals: it is not the case that killing an animal is wrong unless it is justified by other considerations." An example of holding this view is the common belief that while there is something deeply morally wrong with factory farming (because of the suffering caused), there is nothing morally wrong with "humane" farms on which the animals have a high level of welfare until they are killed (because the farms do not subject animals to suffering). Harman closely scrutinizes this "surprising claim" and argues that it is not true. She also considers four views that are defenses of the view that the surprising claim is true, and she argues that each of these views is false.

In "The Ethics of Confining Animals: From Farms to Zoos to Human Homes," **David DeGrazia** centers his work on basic interests that animals have in *liberty*— the absence of external constraints on movement. He takes liberty to be a benefit for sentient animals that permits them to pursue what they want and need. Obviously farms, zoos, pets in homes, animals for sale in stores, circuses, and laboratories all involve forms of confinement that restrict liberty. DeGrazia wants to know the conditions, if there are any, under which such liberty-limitation is morally justified. He first lays out the harms caused by confinement. He then examines and evaluates five possible standards for the justification of confinement: (1) a basic-needs requirement; (2) a comparable-life requirement; (3) a no-unnecessary-harm standard; (4) a worthwhile-life criterion; and (5) an appeal to respect. He reaches the following conclusions about these five standards: (1–3) are acceptable and often useful standards, though (3) is vague and in need of supplementation; (4–5), by contrast, either set an unjustifiably low standard or leave us without adequate guidance. After this theoretical examination of standards, DeGrazia provides a practical moral assessment of the ways in which animals are kept in factory farms, traditional farms, zoos and aquariums, and human homes. His conclusions are that factory farming cannot meet appropriate standards for the confinement of animals and is therefore indefensible, that traditional animal husbandry can meet appropriate standards, that there can be acceptable zoos and aquariums for many species, and that appropriate standards for confinement can be satisfied for some domesticated species of pets. DeGrazia cautions, however, that satisfying the relevant conditions can be far more demanding than many people realize, including even experienced pet owners.

In "Keeping Pets," **Hilary Bok** defines "pets" as nonhuman animals that people take into their homes and accept as members of the household. She makes a number of assumptions before she considers the major moral problems. For example, she assumes that cats and dogs are happy living in human households and have better lives than they would if they were in the wild. There is welcomed companionship and often a genuine friendship across species. There can of course be many failures and problems such as animal aggression and human cruelty and neglect, but Bok thinks these problems involve clearly wrongful behavior. They are not the real moral problems, which concern how we should accommodate the interests of the animals we control and the conditions of wronging that it is not easy to see. In some circumstances, the fact that humans can understand in ways their pets cannot creates a situation in which we should require more of ourselves in the way of care and patience than we would require in the case of a competent adult human. Bok thinks it is a mistake to regard duties to nonhuman animals as situated in a hierarchy of moral status in which animals with high-level capacities occupy a higher rank. Sometimes higher capacities translate into an increased array of rights, but the fact that an animal does not have some capacity may require that the animal receive more consideration, not less. Because sacrifice on our part can often promote a better relationship with a pet, we can expect to require such sacrifices of ourselves, and we cannot reasonably expect a comparable level of sacrifice by the pet. Meeting pets' needs conscientiously requires meeting their needs for attention, affection, and training, which rarely call for serious sacrifice, but require devotion and vigilance, as well as "opening our hearts to animals who are so willing to open theirs to us."

In "Animal Experimentation in Biomedical Research," **Hugh LaFollette** discusses the conditions under which it is permissible and advisable to use animals in biomedical experimentation. He is confident most people think that, in general, we should so use animals, which commits them to what LaFollette calls "the Common View" that (1) there are moral limits on what we can do to (some) nonhuman animals, but (2) humans can use them when doing so advances significant human interests. This view entails that animals have some moral status, but not a demandingly high status. He also thinks that most people believe that medical experiments using animals do wind up benefiting humans. In the end, the view is that, when these conditions hold, the practice of animal experimentation is morally justified. LaFollette distinguishes the Common View from both more lenient views about using animals in biomedical experimentation and stringent views about their use. "The Lenient View holds that even if animals have moral worth, their worth is so slight that humans can use them virtually any way we wish and for any reason we wish. The Demanding View holds that the moral worth of animals is so high that it bars virtually all uses of animals in biomedical research." Two major moral considerations seem at work in these disputes. First is the question of moral status, and second is the question of the extent to which animal research benefits humans. LaFollette finds that animal experimentation can withstand what the toughest critics say, but the benefits of the practice are also not as compelling as defenders claim. Moral arguments defending the practice have some merit, but there are significant moral costs of the practice and LaFollette thinks that defenders of the practice carry the moral burden of proof, given what we know about the moral status of animals and given that much research often does not yield human benefits. Defenders need to provide more evidence than we usually see that the value of biomedical research exceeds its moral costs.

In "Ethical Issues in the Application of Biotechnology to Animals in Agriculture," **Robert Streiffer** and **John Basl** discuss moral problems about the use of modern biotechnology in agriculture that emerged in the early 1990s over recombinant bovine growth hormone, a chemical produced using genetically engineered micro-organisms and then injected into dairy cows to increase milk yield. Then there came genetically engineered soybeans, corn, canola, and cotton, and recently genetically engineered animals and cloned animals intended as food or breeding stock in agriculture. Streiffer and Basl provide a moral framework for evaluating these new applications of modern biotechnology as they affect the food supply. They note that issues about genetically engineered livestock are focused on cattle, sheep, goats, pigs, chickens, and fish and that feed efficiency, growth rates, fat-to-muscle ratio, and resistance to pests and diseases are the major aims of the research. They assess the public's interest as focused much more on animal biotechnology than on plant biotechnology. They note that all of the livestock are sentient beings with determinable welfare levels, which assures them of some degree of moral status. They point out that the moral importance of animals takes on a massive significance in light of the number of animals in the livestock sector. The livestock sector also is one of the most significant contributors to global environmental problems. Streiffer and Basl address this complex situation by considering issues of animal welfare and whether animal biotechnology will improve or worsen the animal welfare issues that now arise in agriculture. To assess this problem requires them to go into the nature of animal welfare and what improves or worsens it, as well as how various applications of animal biotechnology can be expected to impact welfare and environmental problems such as water pollution, water use, and biodiversity loss. In order to consider immediate practical effects of their analyses and recommendations, they look at problems about Enviropigs and AquAdvantage salmon.

In "Environmental Ethics, Hunting, and the Place of Animals," **Gary Varner** notes that deep philosophical differences divide environmentalists and animal welfarists. Environmental ethicists often see the practical implications of animal welfare and animal rights views as anti-environmental; they are seen as sometimes harming ecosystems and as sometimes inexcusably banning interventions that environmentalists support. Varner outlines these differences as well as the status of the current debate. He argues that environmental ethicists have sometimes zealously caricatured animal welfare and animal rights views, and that a close examination of the interests of each group shows that there is more of a convergence of values than many have thought. While it is true that ecosystems and nonanimal species have only instrumental value from an animal welfare or animal rights perspective, an appropriately structured environmental ethic need not deny this claim and can still attribute an appropriate level of value to ecosystems and non-sentient organisms. Varner thinks it is a mistake to reason from the fact that sound environmental policy should focus on species and ecosystems to the conclusion that these holistic entities have more than instrumental value. Likewise, one who emphasizes the sentience of animals and believes that only the conscious experiences of sentient animals have intrinsic value can still appreciate that a sound environmental policy for humans and other animals requires that attention be focused on ecosystems and species. Accordingly, sentientists and holists should be able to agree that species and ecosystems are of paramount importance, and they need not commit to intrinsic value for holistic entities.

In "Vegetarianism," **Stuart Rachels** takes up a series of moral problems about industrial farming, which he assesses as a form of cruelty to animals, as well as environmentally destructive, harmful to rural regions, and with serious consequences for human health (because it spreads infectious disease, causes environmental pollution, and the like). He lays out and attempts to document the pertinent facts behind these claims, especially facts about the farming of pigs, cows, chickens, turkeys, and seafood. He attempts to explain why the industrial system has been remarkably successful as a dominant economic force. He argues both that we should boycott industrially produced meat and should not kill animals for food even if the means are humane. He presents what he calls the "argument for compassionate eating," which is the following: it is wrong to cause suffering unless there is a good reason to do so. Industrial farming causes billions of animals to suffer without good reason. Therefore, industrial farming is wrong. And therefore, you should not buy factory-farmed meat. This argument is not by itself strong enough to support vegetarianism, because one could obtain meat without buying factory-farmed meat. However, Rachels eventually concludes that in practice we should all be vegetarians. He concludes with the observation that "the philosophical arguments for vegetarianism are easy. What's hard is getting people to stop eating meat."

In "The Use of Animals in Toxicological Research," **Andrew Rowan** addresses practical problems about our use of animals to assess the toxicological risks posed by various drugs, cleaning agents, pesticides, cosmetics, and the like. Rowan presents scientific doubts about the usefulness of much of the animal test data for human risk assessment. He argues that there is now a consensus that animal tests are not particularly effective in predicting harms to either humans or the environment. This consensus is rapidly causing interest in toxicology to shift away from animal tests and toward quicker and cheaper alternative test systems, but Rowan thinks we need to move still more quickly away from using animals in testing. Rowan is also critical of the current system of institutional review committees (Institutional Animal Care and Use Committees in the United States), which in principle are comprised of people who are chosen as reasonable representatives of the various parties who have a stake in the decisions of these committees. Rowan argues that theoretically such a system could work well to protect the interests of test animals, but in practice these systems do not work. He offers several reasons how and why they fail. He also argues that the current lack of data on animal pain and distress entails that even if a review committee determined to make a serious attempt to assess the cost/benefit ratio for a toxicity study, little empirical data would be available to make the assessment in a suitable manner. Accordingly, most review committees focus their attention on experimental design questions rather than ethics issues. They tend to approve those studies that they assess as carefully designed by investigators, but then they rely uncritically on what investigators report to the review committee about how the study minimizes animal distress and suffering. Rowan finds this a morally unsatisfactory situation that suffers from inadequate information and investigator bias.

In "What's Ethics Got to Do with It? The Roles of Government Regulation in Research Animal Protection," **Jeffrey Kahn** begins with the question of whether ethics has much to do with normative questions of government policy for the oversight of animal research. He thinks that ethics is the normative backbone of such policies—both for animal subjects and human subjects. He finds that numerous parallels have evolved with respect to government-mandated oversight regimes in both the human and the nonhuman animal domains: rules regarding acceptable risk, inappropriate treatment of research subjects, prospective review and approval of proposed research, and institutional oversight committees. He thinks that the history of research ethics amply shows that moral argument is influential far beyond standing law and policy in establishing what may and may not be done in both human and animal research. From this perspective, ethics is the anchor of human and animal research oversight, providing the principles at their foundation. Nonetheless, the rules for research on animals depart from those for human research in fundamental ways, and ones that raise moral concerns about whether policies for animal research review are adequate. Policies for human research protections follow and are based on well-articulated moral principles. But the case of animal research has no such clear connection between policies and principles. "For example, assuming that animals do not have the capacity to consent to participate in research, what would be a reasonable substitute for consent in biomedical research that uses them as subjects?" This question is almost completely ignored. Though policies often require minimization of pain and suffering, overall the rules and restrictions are primarily about maximizing human benefit, even at the cost of harm to animals. Protecting the rights and interests of animal subjects drops into the background. It seems, then, that there are moral inconsistencies between human research and animal research policies. If ethics is important for how we think about the acceptability and proper limits of animal research, then the connection between ethical principles and policy must be made more explicit than it has thus far been made. Kahn finds that even though some policies regarding animal research seem to be mature and well developed, the moral rationales for our patchwork of regulations have received almost no serious attention and analysis. Kahn thinks we need to go back to the drawing board and rethink research animal protection policies and the ethical principles underlying them.

In "Literary Works and Animal Ethics," **Tzachi Zamir** notes that a number of Anglo-American moral philosophers have turned to literature for insights into moral reflection on animals. This "literary turn" in moral philosophy finds that some sensitivities or aspects of moral reflection are deepened by literary works. The main literary work focusing on animals that has attracted substantial interest from philosophers is J. M. Coetzee's *Elizabeth Costello*. Philosophers such as Stanley Cavell, Cora Diamond, Stephen Mulhall, and Peter Singer have all found it a repository of moral insight. Zamir's assessment is that moral philosophers are only beginning to mine rich literary descriptions of animals to gain moral insight and to explore the ways in which the invocation of animals awakens morally relevant dimensions in literary works. The point of Zamir's chapter is to show how literary-oriented animal ethics is capable of *unique* insights into our dealings with animals. He chooses and analyzes several literary examples that adumbrate the ways in which literature can unsettle a practice, such as the practices of slaughter houses. Zamir tries to show how and why literature can broaden moral perception. Some texts he finds to bring home our deep dependence on animals, and still others probe aspects of the ways in which we perceive animals—and indeed the way we are perceived by them. Zamir argues that a "central motif running through all of these works is how literature questions what one sees when one is looking at an animal." He is unsure how to gauge the impact of these works and the future of such work. He hopes that literary works can have a lasting and powerful effect, but he knows that, in his own case, the influence brought home to him "the enormity of the animal issue."

PART I

HISTORY OF PHILOSOPHY

CHAPTER 1

....

ANIMALS IN CLASSICAL AND LATE ANTIQUE PHILOSOPHY

....

STEPHEN R. L. CLARK

CLASSICAL AND LATE ANTIQUE PHILOSOPHY
....

To understand how the ancients thought of "animals" it is necessary first to understand who those ancients were, and what they thought of themselves and of the world. And to do even this much it may be necessary to put aside many present-day beliefs about those ancients, and about "animals" as we now conceive them. It is also necessary not to expect too much consistency in what they thought.

Classical and late antique philosophy, by present-day convention, is mostly Greek philosophy, though the Greeks themselves were conscious that they had borrowed much from older traditions and though, especially in the later centuries, many of the better known philosophers were not themselves of Greek descent or mother tongue. Again by present-day convention, commentators now divide the history of Greek philosophy into periods: the pre-Socratic, the Golden Age of Athens, the Hellenistic Period, and the late antique. That last period has in the past incorporated only "pagan" thought, with Christian thought studied under the title of "Patristics": the distinction is unhelpful. So also is the very attempt to isolate "Greek" thought from the rest of the Mediterranean (which includes both Egypt and Israel), and to propose that whatever went on in "Greek" philosophy was unaffected by the wider world.[1]

Some generalizations are still possible. All these eras, except the late antique, may be reckoned to have had a "naturalistic" outlook: things in general don't happen because they *should*, nor because someone supernatural wants them to. Even in the earlier, mythological accounts, gods, people, and animals "just happened," and had to live their lives in a world evolving from simpler to more complex substance, and obedient to an indifferent fate. Even in the later period, it was more usual for pagan philosophers—except the more literal-minded Platonists—to think that the world and all its creatures were moved by an immanent drive toward perfection than that it was, in any literal sense, created by a transcendent God, though it was also unusual for any to suppose that everything happened quite at random and without intelligence. In most ages, philosophical accounts preferred to explain what happened, in nature and in human society, by referring to "natural" and familiar properties. There were no sacred texts—not even the Homeric epics, significant though these were—to tell us what the world and our duties were; and all arguments, whether about the way things work or the way we ought to work, had to draw their premises from ordinary, non-mystical experience. Common moral tropes were simple prudence: nothing too much; never go bail; remember that you're mortal. Or at a slightly higher level: keep your word; remember the laws of hospitality; honor the gods and the laws of your city. Rather more uncommon advice—and a signal that there is another history of the world, another way of living, hidden within the common history— was to take care for our souls.[2] Later, in the late antique, it was easier than before even for pagans to believe and say that it was God or the gods who ruled the world and give us laws and divinely inspired texts in Greek and non-Greek traditions.

Popular accounts of "Greek" thought routinely suggest that "the Greeks" were dualists, dividing souls from bodies and Heaven from Earth, and that it is this heritage—assisted by the supposed Hebraic dictate that we have dominion over all the earth—that has blighted Western attitudes to animals and the wider biological environment. "The Greeks," we are to suppose, believed that we human beings are *rational* souls temporarily inhabiting bodies, while animals have only *animating* souls. Because we are superior beings, things are arranged for our benefit. Plants, so Aristotle once observed, are for animals, and animals for human beings.[3] Christian exegesis of Genesis 1:26 was taken to support the pagan teaching that the rational should rule the non-rational (masters should rule slaves, men should rule women, adults should rule children) and that animals are made "for us." There is some truth in the story, although, as I shall contend, the philosophical connections are more complex than this.

Oswald Spengler, despite his many errors of scholarship and reasoned argument, gave a more plausible account of conventional Greek theory when he suggested that the essence of "Classical Culture" was corporealist and presentist.[4] The real self was the present, corporeal self. Maybe something survived its death, but that was only a shadow, a breath, a partial memory. Anything beyond an easy journey or a living memory vanished into myth or fable. It was the dualists like Plato (428–348 B.C.) who resisted this, and also took more care of animals. Those who make the Platonic *Dialogues* their central reference will usually conclude that he is arguing for a clear distinction between Soul and Body. There is some truth in this as well—though exactly which truth isn't so easy to decide—but it should be obvious that the dialogues were contentious, that Socrates's interlocutors were astonished and disbelieving when these doctrines, whatever they meant, were even suggested to them. Dualisms of this sort were easily reckoned "foreign," whether they came from southern Italy, or Egypt, or the Scythian north, or India. They were as foreign and contentious as the claim that God or the gods were good, and not just very powerful. If there were a "typically Greek" concept of our situation, it is given in Sophocles's *Oedipus Rex*, not in Plato's *Phaedo*: we are all subject to fate, and must find such dignity as we can in bearing it. Complaint is useless, escape impossible, and virtue lies in knowing it. I shall return to Plato, and to what he meant: it is better to begin, though, with the more conventional outlook.

Conventional Greek opinion, with all the necessary caveats against supposing that all Greeks, of all periods and places, were agreed, was indeed *conventional* and conservative. We are who we think we are: rational, political animals, members of some particular tribe or city, with the power to conquer and domesticate other creatures, but always ourselves subject to superior powers, destiny, or the capricious gods. We should always bear in mind that we are mortal, and poets, philosophers, and historians all remind us that "good fortune" does not last forever. If we abuse others (other human beings, that is) we can expect to be abused. Nonhuman animals, if they feature at all in the histories, do so as foils: *animals* are beastly in their habits, mindless beings, dependent wholly on their senses. Stoics and Epicureans (both Hellenistic schools being firmly, though differently, naturalistic) were convinced that "justice" only applied to relations between *human* animals, as only they could make or keep a bargain. "We cannot act unjustly towards creatures which cannot act justly towards us,"[5] especially as this would be very inconvenient.

Like many other philosophers, they were not entirely consistent in this: the great Epicurean poet Lucretius (99–55 B.C.), for example, acknowledged duties of care to domesticated animals as a reward for their service.[6] Human and nonhuman animals alike were *bodies*, but this did not mean they mattered equally. Even late antique philosophers, who mostly believed in non-corporeal souls, still usually preferred to reckon that nonhumans weren't like us: they lacked the most important sort of soul, the intellectual. Quite what this meant, and the arguments they advanced for it, will concern me later. The *practical* reason for their rejection of an earlier Platonic or Pythagorean tradition was that they wanted to defend the practice of animal sacrifice as part of a compact with the divine which was by then entangled with issues of imperial power. The emperor Julian, for example (A.D. 331–363), in seeking to reverse the Constantinian adoption of Christianity as the imperial religion, combined a high philosophical Platonism with multiple sacrifices to appease the abandoned gods, even though he recognized that "the eating of meat involves the sacrifice and slaughter of animals who naturally suffer pain and torment."[7]

Explicit and detailed studies of how we should treat "animals" were rare in antiquity. The most significant text was Porphyry's *On Abstinence from Killing Animals*, in which the pagan Neoplatonist author (A.D. 234–305?) seeks to persuade his friend that he ought not give up the Pythagorean life. This text is a compendium of arguments for and against treating animals merely as things meant for our use, and our principal source for the Stoic arguments that, for better or worse, the Western Church at least adopted. Porphyry's chief concern may have been to preserve the proper "philosophical" spirit, which he does not expect *all* human beings to possess. "The contemplative holds to simplicity of lifestyle....He is content with simple things, so he will not seek to feed on animate creatures as if inanimate foods were not enough for him."[8] But he was also concerned for the lives and welfare of animals, emphasizing their intelligence as well as their other amiable qualities. When, centuries later, Thomas Taylor wrote his satirical response to Mary Wollstonecraft's argument for the rights of women, *A Vindication of the Rights of Brutes*,[9] he drew on Porphyry's arguments, and included a translation of *On Abstinence* in his later *Select Works of Porphyry*.[10]

An earlier Platonist, Plutarch of Chaeronea (A.D. 46–122), wrote on the relative cleverness of land and sea creatures, on the eating of flesh, and the dialogue *Gryllus*, in which a talking pig explains why animals are far superior to humans.[11] Other discussions of the way we deal with animals have to be pieced together from fragments and asides. Plutarch and Porphyry between them have been the original source of many later arguments about the intelligence of animals, and whether we are justly entitled to make use of them. Both authors deserve their place in the history of what was once called the Pythagorean way, but they did not speak for all the Classical world, and any introduction to how animals were perceived and treated in that world must range more widely.

Of recent work, Ingvild Gilhus's study, though she is principally concerned with the early years of the Roman Empire, is the most detailed and useful. Gary Steiner's longer account of the place of animals in Western philosophy gives several chapters over to a clear account of Classical ideas, from Homer to Porphyry. Patricia Cox Miller examines the metaphorical uses of animals in antiquity. Richard Sorabji discusses the philosophical arguments distilled from Porphyry and earlier sources. Stephen Newmyer gives special attention to Plutarch, with a view to enhancing modern ethical theory. Daniel Dombrowski and Catherine Osborne both make use of other ancient arguments, but don't pretend to give a complete account of them. Barbara Cassin and Jean-Louis Labarrière have edited a collection of papers on many aspects of ancient relationships with animals, as *L'animal dans l'antiquité*. There are also several excellent papers in *A Cultural History of Animals in Antiquity*, edited by Linda Kalof as the first volume of a longer *Cultural History of Animals*.

I have drawn on all these authors in what follows, but cannot expect them always to recognize their contributions, nor to agree with my interpretation of the data. The chief moral of this introduction, after all, is that our evidence is often as confused and confusing as present-day human attitudes to the nonhuman. Thinking ourselves back into the mind-frame of Classical Antiquity is an exercise in imaginative sympathy. Different aspects of that time and place attract different scholars' attention, especially as they must, inevitably, bring their own preoccupations with them, however detached they strive to be.

"ANIMALS" AND "ETHICS"

How we should treat "animals" depends on at least two factors: what animals actually are, and what governs what we should do. Neither question is easily answered. Neither Greeks nor Romans could point to a god-given list of commandments and prohibitions, though most would have agreed that it would at least be *prudent* to do what God or the gods require and prudent not to assume that we are invincible. There was no universal code, beyond the acknowledged duties of piety and hospitality, or prohibitions on incest, cannibalism, and patricide. Particular animals might be "sacred," off-limits for any ordinary use, and particular acts of cruelty might excite disgust and anger even in otherwise callous spectators. Even the Roman audience, for example, was upset when Pompey brought elephants to be killed in the arena in 79 B.C.[12] The domestication and mutilation of animals was as unquestionably a part of human life in ancient times as in modern, but the actual *killing* of the larger domestic animals had to be surrounded with ritual: they were sacrifices (on which more below).

Philosophical argument was often employed, as in modern times, to rationalize these practices: animals were made to serve us, or at any rate we do no wrong in requiring them to serve. Sometimes these efforts amounted to an ethical methodology: what we should do or refrain from doing is determined by an implicit contract, what we—but only rational beings—would all agree to, or have agreed to. This may either be a merely pragmatic rule, or else a covert appeal to what we *would* be required to do by an impartial lawgiver, though it is not clear why such a lawgiver would only be concerned with us.

So what did "the ancients" think that animals were, and what guidance did they hope to have in excusing how we treat them? Strictly, the question can have no single answer. The idea that "all the Greeks," or even all the philosophical Greeks— from Hesiod in the eighth century B.C. till the (supposed) closing of the School of Athens by Justinian in A.D. 529—in all the lands from Spain to Palestine were of one mind and manners is absurd. Even if we speak only of the Golden Age of Athens, from the defeat of Persia (479 B.C.) through its own brief empire and military defeat, to the death of Alexander (323 B.C.), we can only really conclude that the Athenians were a disputatious people, agreeing on almost nothing. Nor did answers to this or any other question change in easily described ways from one era to another. The Greeks themselves, like other nations, were sometimes inclined to seek their common identity by contrast with another: they felt themselves to be newcomers, governed as free people by a law they had some influence in constructing, by contrast with the ancient, slavish peoples of the East, the impulsive barbarians of the north, the weirdly impressive (and impressively weird) Egyptians, and, in later years, the philistines of Rome. They were often eager to learn how strange those Others were, how free and commonsensical, by contrast, their own lives could be, at least if they were freeborn males of a free city. And they were usually conscious that they had learned a lot from Egypt—or even from the barbaric north.

Nor was it only "philosophers" or the philosophically inclined who had a view. Indeed, it may be easier to identify the real assumptions of the classical world by considering what people in general did, and how (if at all) their behavior differed from humanity at large, or from Greeks of other ages. The way we think about ourselves and the world is the context within which our treatment of animals, and our thoughts about them, make what sense they do. How we think about ourselves is often revealed especially in how we think about those most obviously *not* ourselves, and how we treat them. In the earlier phases of classical history the most obvious Other was Egyptian. People did things differently there:[13] they worshipped cats, crocodiles, and ibises, permitted incestuous relationships, believed their souls immortal, and had a detailed history (or at least a list of pharaohs) stretching back into an otherwise legendary past. In a later period, it was India that filled a similar role: people did things differently there too and had an even longer imagined history.

"Greeks" (with all the preceding caveats) idealized the youthful human body, expected nothing in the afterlife, and let the past beyond immediate living memory dissolve into myth and heroic legend. At the same time, they mostly endorsed the precepts of civil religion. Law, decency, and civil religion were their defense against monsters, hybrids, and chaotic passion. Cosmos was at war with chaos, and its continuing victory cost the lives of heroes. Decency depended on creating barriers against chaos, but also on finding a place for Dionysus, Aphrodite, even the avenging Furies, even cleverness. Zeus, the overarching sky from which lightning strikes down the over-mighty, swallows up mere cleverness or cunning intelligence (the Titan Metis) and gives birth to Athena, who is intelligence at the service of the larger order.[14] Zeus also gives second birth to Dionysus, the god of masks and fluid boundaries, allowing even him a space, a regular festival, within the ordered commonwealth. The mortal sons and daughters of Zeus (and other gods) achieve great things, but rarely get much reward: their destiny is to suffer and endure. Their enemies, the enemies of order, are either half-human monsters (Typhon, Scylla, the Minotaur) or else such humans as have broken faith, abused the laws of hospitality or proper piety towards their kin.

What we now easily forget is that the Greeks were the survivors of a world-shattering catastrophe, recalled in many folklore variants, and visible to them in the ruins of more ancient cities.[15] What we now easily forget is that everyone needed both to remember and to put out of mind the simple truth that pain, disease, defeat, and madness could not be avoided by any amount of virtue: virtue lay in dealing with that truth, remembering we are mortal. What we forget—as the heirs to a long tradition of *supernatural* religion—is that it is possible to be both "naturalistic" and "religious": to feel a profound respect for the way things are arranged, without supposing that Anyone arranged them, and without any hope that they can be rearranged.

Where do nonhuman animals (hereafter "animals") fit into this account? At the basis of classical society, unsurprisingly, lies agriculture and animal husbandry, the latter being the domestication and exploitation of tamable animals, including human captives known as slaves. In creating agriculture and husbandry, the Neolithic revolution (before recorded history), created a master class, with claims of ownership, dependent on the muscle power of its animal and human servants. The original version, we can now imagine, amounted only to including the crops and the cattle within an existing family. We did not then exactly "own" them in the modern sense, any more than we own our children, since there could be no alienating sale of chattels. Even our slaves, the captives whom we did not kill, were part of our households then, and had a place not far removed from our dogs (perhaps a little lower). Oxen, as Aristotle (384–322 B.C.) remarked, are the poor man's slaves[16]—or slaves the rich man's oxen.[17] Animals not in our households were "wild animals," and by definition undisciplined—like barbarian tribes; it is easy for us to forget that human beings often had cause to fear them. Civil society was formed, originally, to defend its members against wild beasts, human and nonhuman.[18] Even the milder ones might "belong" to powerful presences—Artemis, mistress of wild things, or the Sun (whose cattle Odysseus's crew made the mistake of killing).[19] Even if they could not defend themselves, they might turn out to have powerful protectors.

Even in a more secular understanding, when what counts in hunting is the enjoyment of the chase (and practice for that other form of hunting, war), and there was no conscious reference to the gods who attended both to the prey, the hounds, and the hunter, a god, of a sort, was present, in the excitement, splendor, and pity of the hunt. "Gods," for the Greeks, are identical with those brighter moments that enliven the everyday: "Man is a shadow's dream (*skias onar anthropos*), but when god sheds a brightness then shining light is on earth and life is sweet as honey."[20] The brightness *is* the god, and there is a god of the hunt.[21] That unimaginatively decent man, Xenophon (ca. 430–ca. 350? B.C.), cared for his hounds and horses, and even for his prey, declaring in his work, *On Hunting*, that "so charming (*epichari*) is the sight that to see a hare tracked, found, pursued and caught is enough to make any man forget his heart's desire." And he acknowledged limits even in the chase: when hunting on cultivated land, avoid growing crops and let pools and streams alone. It is unseemly (*aiskhron*) and wrong (*kakon*) to interfere with them, and there is a risk of encouraging those who see to set themselves against the law. On days when there is no hunting, all hunting tackle should be removed.[22]

How he found it possible simultaneously to admire and pity the prey and also to set gin-traps and caltrops for deer or boar, is a question only sportsmen have much chance of answering. How he rationalized the suggestion, in chapter 13 of the same work, that the huntsman is showing and learning virtue in defending his community by attacking dangerous beasts like hares, deer, and pigs, is also obscure. Pigs may indeed be dangerous, especially when they defend their young—which is often cited as an example of natural bravery, even if not quite moral courage (hares less so).[23] But it can be admitted that Xenophon was, in his own way, honest in his appreciation of the creatures that he hunted, as well as in his excitement in the chase: he would have regretted a world in which there were no more hares, deer, or pigs as much as a world in which there were no more heady risks to take.

Wild animals might be hunted, with the usual precautions, but there were legends to remind even the hunter that the hunt might be reversed: Actaeon, for example, spotted the goddess Artemis bathing with her nymphs, and was transformed into a stag to be torn apart by his own hounds.[24] A much later legend speaks of the moment (notionally A.D. 118) when Placidus, later to be St. Eustace, had a vision of Christ crucified between the antlers of the white stag he had pursued, and turned Christian.[25] He is the patron saint, somehow, of hunters.

Domesticated cattle, kept for their milk, their wool, and their skins, were "sacrificed" rather than slaughtered: made over to the gods whose creatures they were. Even in the early years of the Roman Empire, domestic meat (cows, sheep, pigs) was from sacrificial victims:[26] how else should we dare? How should we dare not to, when sharing in that sacrificial meat was a way of dining with the gods of our city and the empire? Openly refusing the invitation was tantamount to treason (if anyone should bother to make the charge).

SLAVES AND OTHER ANIMALS

It is not only "animals" that we conquer and domesticate. If the slave is the poor man's ox, and if the earlier inhabitants of Laconia were turned into human cattle, "helots" (which is to say, captives), by the invading Spartans, what difference is there to be between the human serf and the animal? Both human and nonhuman animals may be alienated from their original attachments and affections, transformed into the masters' instruments, even bred for their convenient qualities. One obvious, social, difference is that we *eat* animals. How about eating people? It is easy to imagine that the answer is simply obvious, but this is not so. "If the Juju had meant us not to eat people, he wouldn't have made us of meat."[27] Why don't we?

One answer may simply be that the tribes that did eat people died of various horrible infections (the Egyptian, Jewish, and Muslim aversion to pork may have similar origins). But though this may explain why the aversion to "long pig" has been sustained, it does not explain how it happened that our ancestors gave up the practice and began to anathematize it. It may be that the social revolution that gave us slavery also spared us cannibalism. In our beginnings, enemies were there to be defeated, killed (perhaps also in sacrifice) and eaten—as they were in Aztec America. If they could instead be used to work for their captors, or be sold on to the current rulers, it would be worth not killing and eating them (or at least, not eating them). But this does not necessarily imply that there was a deep division between animal and human: human slaves were not to be eaten, maybe, but neither were dogs. Peoples who ate just anything that was physiologically edible were savages (especially if they ate it raw), or possibly Cynic philosophers. The cannibalistic Cyclops of Homer's *Odyssey* displayed his barbarism also in drinking "unmixed milk" (*akraton gala pinon*)—a deliberate parody of another, only a little less barbaric, habit (drinking unmixed wine: *akraton methu pinon*)!

In this the classical Greeks were like most other human tribes: we are very rarely omnivores (even those who eat prawns or snails will usually draw back from cockroaches or slugs), and the world we live in is structured by historical and personal meanings. But perhaps there are some early signs that they were readier to construct *humanity* than were many others. Just because *anyone* could become a slave (even the philosopher Plato, 428–347 B.C., it was said, was once sold by an irritated tyrant, Dionysius of Syracuse),[28] there was less reason to construct slave-castes, of supposedly different natures. Even in Sparta there was *some* chance that a helot might, by extraordinary efforts, be inducted into the élite (though the offer to let him do so might merely be a trick to identify helots who might someday be dangerous).[29]

Aristotle argued that only those who were *naturally* slavish should ever be enslaved,[30] but the actual practice was different. Just because we were all to be considered "human" the boundaries around humanity were "edgy"! On the one hand, the greatest (legendary) law-giver Minos was the son of the nymph Europa and Zeus (in the form of a bull). On the other, one of the most notorious of monsters was the Minotaur, son of a bull and the crazed princess Pasiphae (herself a daughter of Minos; perhaps she was hoping to repeat her grandmother's achievement). Animals might embody gods, but in general bestiality was a serious error! Again, Cheiron the Centaur, son of Cronos, taught all the heroes, from Asclepius to Achilles (or so Xenophon tells us).[31] But centaurs in general were prone to violence, and their overthrow—the Battle of Centaurs and the human Lapiths—decorated the Parthenon, with Apollo stretching out his arm above the conflict to contain or bless it. The gods—especially Zeus—might manifest themselves as animals, as they did in Egypt, but their *proper* form was human, and actual animals were not to be taken as gods, though the fact that Zeus might visit us in *human* form was a reason to be kind to strangers! Even the Pythagorean, Apollonius of Tyana (A.D. 3?–97?), or at any rate his biographer Philostratus (A.D. 170–247), thought the Egyptians were making the gods ridiculous by giving them an animal or hybrid form.[32] There were others, including Porphyry, who thought otherwise: that the Egyptians realized the virtues of those animals.

Others thought it not just ridiculous but dangerous. In the eighth book of Virgil' s *Aeneid* (70–19 B.C.), Aeneas is given a shield on which the Battle of Actium is prophetically depicted, that battle in which Caesar Augustus was to defeat his rival Antony and the dreadful gods of Egypt ("every kind of monstrous god and barking Anubis too"). The Egyptians, for the more doctrinaire Greek or Roman, were the deluded worshippers of *animals*, and predictably bestial in their habits (though at least they did not worship, or eat, *pigs*, who are the omnivores we aren't). Animals, animal-lovers, and barbarians alike must be beaten into submission. Human beings should know their place, which is to be different from "animals." Even Plutarch, who attempted a more sympathetic analysis of Egyptian thought, thought that portraying the gods as animals must lead "the weak and innocent into 'superstition' (*deisidaimonia*), and the cynical and bold into 'atheistic and bestial reasoning' (*atheos kai theriodes logismos*)."[33] Hybrids, by exposing the artificiality of these boundaries, were very dangerous. The advice was not merely political. Every one of us, as an individual, should be careful to suppress our animality in favor of human reason.

Stoic moralists agreed that all animals, including ourselves, sought to survive by making and maintaining emotional attachments to food, territory, and the company, even the well-being, of their peers. Our own emotional attachments to family, friends, and city grow from the same roots as a pig's defense of her young, or a wolf's loyalty to the pack. But our primary loyalty should be to the truth, as it is revealed to us through reason. The Stoic moralist, Epictetus (A.D. 55–135; a sometime slave of one of the emperor Nero's more notorious freedmen) gave the following warning:

> What is a man? The answer is: "A rational and mortal being." Then, by the rational
> faculty, from whom are we separated? From wild beasts. And from what others?
> From sheep and like animals. Take care then to do nothing like a wild beast; but if
> you do, you have lost the character of a man; you have not fulfilled your promise.
> See that you do nothing like a sheep; but if you do, in this case the man is lost.
> What then do we do as sheep? When we act gluttonously, when we act lewdly,
> when we act rashly, filthily, inconsiderately, to what have we declined? To sheep.
> What have we lost? The rational faculty. When we act contentiously and harm-
> fully and passionately, and violently, to what have we declined? To wild beasts.[34]

Human children, of course, have not yet attained "the age of reason," but are loved and cared for nonetheless, in the hope that they will soon grow out of their initial beastliness. Those who have lost "the good of intellect," by disease or injury or mortal sin, are only a little worse than the rest of us, by strict Stoic standards, since it is only "the wise" who are really free, really "friends of God," and the only real authorities on earth. This might have had very alarming moral consequences, were it not that Stoic moralists mostly preferred to emphasize that we share a human nature, a potential, which excites affection and deserves respect: so Cicero insisted that every human being would recognize and feel at least an initial affection for any other human being, however unfamiliar. "We are by nature suited to form unions, societies and states."[35] They chose to ignore the obvious fact that we also form such unions with other creatures.

We may not all, actually, be rational, in our beliefs or our behavior, but we could and should be. So ancient philosophers, poets, and statesmen all agreed that human beings must understand their own original motives, and act instead "according to reason," for the sake of some larger good. Sheep and other domesticated animals are led by desire or fear. Wild beasts act in pride and anger. Plato had a similar belief— though he was more optimistic about the possibility of restraining our irrational desires with the help of our semi-rational *thumos* (sometimes, confusingly, trans- lated "spirit"), and more inclined to accept that nonhumans also might be more or less "intelligent." Our real immortal selves are equipped with lesser selves, horses to pull the chariot of our lives, which needed a controlling hand to guide them, but those selves might find themselves in nonhuman bodies too. Aristotle also supposed that there are many motivators in the human soul: not everything that we do "will- ingly" (knowing what we are doing, and without being compelled) is done "deliber- ately" (as the right thing to do). We *should* act virtuously, to do the right (*to kalon*, the noble), but we don't always, even if in some sense we know we should, since we could be preempted by those other parts of the soul that we share with other animals: desire and angry pride. Quite what "right reason" is, whether a prudent calculation of the larger good or an inspiration, is too large an issue here (though I shall refer to the distinction later).

Of more importance is that it is crucial to a Classical humanity that we are to disown our merely "animal" motivations. Neither style—to act out of desire or fear, or to act in angry pride—is appropriate for us, and nothing thus "animal" is to be admired or worshipped. Only human beings, so Aristotle seems to suggest (though not quite consistently), have reason, and it is this doctrine that was dominant in much philosophizing of the Hellenistic era. Those who worship animals "are merely brutes in human form"[36]—a claim made without any actual attempt to see if they are really bestial. "Reason" must rule. The role of Rome, so Virgil tells us, is *parcere subiectis et debellare superbos* ("to spare the obedient and beat down the proud"): so also the role of humans, or at least of *proper* humans, neither barbarians nor slaves:

> Conscious of thought, of more capacious breast,
> For empire form'd, and fit to rule the rest.[37]

This was one excuse for the Roman games, in which wild beasts were "hunted" and killed to prove, dramatically, that human beings, and particularly *Roman* human beings, control the world.[38] Less imperially minded poets and philosophers had fewer global ambitions, confining their disciplinary endeavors to their own "animal impulses," the "sportive monsters" of the human soul.[39] But the moral, so far as dealing with actual animals, was not so far removed: just because we should not give way to those "lower" impulses, we should not share the goals of actual animals, nor be diverted from the human good by any thought about *their* good, any more than we should be moved by pity for merely material losses. Happiness, *eudaimonia*, was for human beings alone, and especially for adult, freeborn, male human beings, citizens of some real city: even if animals got what they wanted, they could not be *eudaimones*, precisely because we aren't *eudaimones* when *we* get what, as animals, we desire. Neither bodily pleasures nor high social standing are enough to make us "happy," or enough to make our lives worthwhile. But there is always a real risk for us, that we be turned into "animals," as Circe transformed Odysseus's crew, or as worshippers of Zeus Lycaeus in Arcadia are turned for a while into werewolves (in memory of the cannibal feast Lycaon offered Zeus, which prompted Zeus to wipe out most of humankind in the Flood).

It is wrong, on this account, even to give respect to animals, let alone worship, because this would be as much as to respect animal life, to say that what they want is really valuable and to turn aside from properly *human* goals. For much the same reason we have no duties toward them, even to the domesticated sort, since "animal life" is incompatible with action for the sake of doing the right thing. Animals may be affectionate, or even, in a way, faithful and courageous, but they can't act from any sense of justice, or else they wouldn't be animals, and so can't be supposed to make any bargain for us to honor too. Even Lucretius' duty of care does not strictly imply that the cattle have a right to it. Dogs might be dear to us (and we to them), but they could not—so Stoics and Epicureans alike insisted—really be our *friends*, a claim that other moralists, then and now, would certainly dispute. The same applies to slaves and barbarians: because we cannot think that they would keep their word, or even understand that words are to be kept, we need not keep *our* word to them. Only human beings—or only civilized, adult, male human beings—can ever lay claim to property, and only they can be considered, as a later philosopher called them, "ends in themselves." Only human beings can feel *shame*. Only human beings need to.[40]

Stoic philosophers in the Hellenistic period argued much more firmly than Aristotle that animals were entirely without reason, that they had no real plans or intentions, and perhaps not even any feelings or perceptions: they only looked as if they did. Pigs, according to Chrysippus, should be reckoned locomotive meals, with souls instead of salt to keep them fresh.[41] Because they couldn't talk, the argument went, they couldn't reason either, nor recognize their situation, nor even, strictly, have any sensations or perceptions. So also in the *Hermetic Corpus* (a late antique compilation of supposedly ancient Egyptian wisdom): "there is no evidence of understanding without reasoned speech....And without understanding it is impossible to have sensation."[42]

The argument rested on an analysis of what it is to perceive, to feel emotion or to suffer. To see a cat, it was supposed, was to judge, on the basis of visual data, that there was a cat to be seen. To be angry was to judge that one had been wronged, and was entitled to redress. To suffer is to suppose that one's state is evil. But none of these judgments could be made without access to any way of *asserting* them, or wondering whether or not they were really true. Creatures that cannot say what they mean (a class, of course, that includes human infants and older victims of aphasia) cannot be supposed to mean anything, and though they do have *sensations*, they do not "really" perceive or feel. Common sense, then as now, usually thought otherwise.

Porphyry, for example, insisted that nonhumans can and do communicate with each other, and with us:

> I myself reared [he writes] at Carthage, a tame partridge which flew to me, and as time went on and habit made it very tame, I observed it not only making up to me and being attentive and playing, but even speaking in response to my speech and, so far as was possible, replying, differently from the way that partridges call to each other. It did not speak when I was silent; it only responded when I spoke.[43]

And as Plutarch remarked, "it is extraordinary that [those who deny any intelligence in animals] obviously fail to notice many things that animals do and many of their movements that show anger or fear or, so help me, envy or jealousy. They themselves punish dogs and horses that make mistakes, not idly but to discipline them; they are creating in them through pain a feeling of sorrow, which we call repentance."[44]

The argument against "animal sentience" is still used by some modern thinkers,[45] but it had a particular resonance in the ancient Greek world: the ability to speak, publicly and plausibly, was vital to a citizen's self-esteem and public reputation. Those not allowed to speak, or denied the chance to learn, or assumed to be incapable, were of lesser worth.[46] Denying that they were sentient at all was a gambit familiar in human history.

The commoner outlook still was simply that human slaves and domestic animals alike were property, without proper goals of their own, though they had both desires and feelings. The gap was not between animal and human, but between slave and free. Only a few philosophers—and those inconsistently—urged that human beings, just as such, were owed a special respect (as being at least *potentially* rational). Barbarians, it was easy to suppose, were "natural slaves." Human enemies, like wild animals, were there to be killed or tamed (or sometimes ignored). And the particular character that proper humans had was not strictly *intelligence* (*metis* or *deinotes*), but willing service of a greater good even than that of families and friends. The ordinarily "natural affections" aren't enough to make us care for strangers, even our fellow citizens, let alone more distantly related creatures. Most of us were *shamed* into that service; the saints (or true philosophers) among us *saw* that it was right. Neither slaves nor animals had that insight, or that sense of shame. Neither could seek to do what was right or honorable or for the city's or the empire's good. "For the son of Cronos has ordained this law for men, that fishes and beasts and winged fowls should devour one another, for right [*dike*] is not in them; but to mankind he gave right which proves far the best."[47] The life of the sea, especially, was a war of each against all,[48] but fishermen and philosophers alike knew well that fishes were clever predators and evasive prey, just as they were well aware that slaves could be smarter than their masters were. Reason, or intellect, "is divine, and is the guiding principle of those who are willing to follow justice and [the gods]," not just the skillful adaptation of means to ends.[49]

Even, or especially, Stoics did not suppose that "reason" was merely the skillful adaptation of means to ends: it was the ability to see our acts in context, to see that more mattered than what we *wanted* or than what felt good. Nor is it merely the capacity to *speak* (and so to think and feel), as their arguments against nonhuman sentience suggest. It is a sense of justice. Strangely, our very capacity to understand and admire a world larger than our own sense and appetite becomes a reason to disregard the appetites of "lesser" creatures. Because they can't keep faith with us—though folk stories are full of faithful hounds and horses—we need not keep faith with them (which was also the pretext that Greek colonists used when dealing with the natives they displaced). Because they don't feel shame, we can be shameless in our use of them. Even more strangely (as intimated above), Stoics also argued that our moral sense was founded in the same sort of emotional attachments as influence other animals: "they postulate that love of one's offspring is the very foundation of our social life and administration of justice, and observe that animals possess such love in a very marked degree, and yet they assert and hold that animals have no part in justice."[50] Our difference from them, our superiority, was apparently only that we could extend those attachments further than they can.

> For just as in uncultivated plants, such as wild vines and figs and olives, Nature has implanted the principles, though crude and imperfect, of cultivated fruits, so on irrational animals she has bestowed a love of offspring, though imperfect and insufficient as regards the sense of justice and one which does not advance

beyond utility; but in the case of man, a rational and social animal, Nature, by introducing him to a conception of justice and law and to the worship of the gods and to the founding of cities and to human kindness, has furnished noble and beautiful and fruitful seeds of all these in the joy we have in our children and our love of them, emotions which accompany their first beginnings; and these qualities are found in the very constitution of their bodies.[51]

ANOTHER SIDE TO THE STORY

These attitudes and beliefs about "animals," emphasizing our power, our right, even our *duty* as rational beings to conquer, subdue, and domesticate the merely "animal," in ourselves and in the outer world, have had an enormous influence on later Western thought and practice. But there is another side to the story, or at least a different strand of thought and practice, interwoven with the common Classical one, and one that I have been gesturing toward. Oswald Spengler's other insight was that there was another culture alive in the Mediterranean milieu, mingled with "the Classical," though it never achieved a complete and undistorted political expression (so he supposed). The focus of Classical life and thought was the ideal human in the here-and-now, and our duties were mostly *civil* duties. By contrast, the "Magian," a title chosen without any reference to any historical Magi, had as its focus a transcendent, incomprehensible One, for which all things were possible, including transmutation. For the Classical mind, the tangible was the real; for the Magian, the imaginable. For the Classical mind, our hopes and fears are only for *this* life; for the Magian, we have eternal life to consider—a life we share with the nonhuman. For the Classical mind, merely irrational affection is—as Spinoza also supposed— "groundless superstition and womanish compassion."[52] For the Magian, God is love.[53] For the Classical mind, humanity, expressed in the free adult individual and his associates, was what mattered most; for the Magian, it was the community of faith.[54] Both might agree that we are the gods' toys,[55] but this for the Classical mind is a flat fact, something to endure; for the Magian, a boast. For the Classical mind, Law governed all—for the Magian, someone's Will, a will not confined by "Nature."

A probably accurate response to Spengler is that "Classical" and "Magian" are no more than two roughly delineated aspects of the real Mediterranean, reified into opposition. Both strands, and many others, went to make up historical reality, and neither has any claim to be the essence of a distinct, organic culture with a normal course of life from which we might learn to predict the dying years of our own "Western" culture (as Spengler suggested). Nor is it possible to distribute individual writers and actors to the different cultures. Even though, for example, most Stoics were mostly "Classical," some were sometimes "Magian." And Epicureans were maybe something else entirely.[56] But it is nonetheless of interest to reconstruct at least this one other way of thinking, and its attitude to "animals," though I shall not entirely follow Spengler's lead in characterizing it.[57] The Classical view is that we are mortal organisms, subject to the same law as cattle, but is contemptuous of the animal lives with whom we share the world. The "Magian" view, on the other hand, identifies us as immortal souls, and is generally more careful, or at least a lot more aware, of the lives of others.

I have already sketched a little of the attitude that Classically minded people had to Egypt, a land where "animals" were worshipped and the gods portrayed as hybrids. It was also a land of sects, where people identified with creeds and rituals, rather than civic obedience (to the horror or amused contempt of Roman officials). There—and also in that more distant Other, India—there were "gymnosophists," naked ascetics who were seeking a way away from everything that Classically minded persons thought most real. Even within the Greek tradition there were "Cynics," so-called because they chose to live like dogs—not comfortable domestic dogs, but the feral dogs who lived on charity or theft, without shame and without luxury (just possibly, the Cynics were conscious that dogs, along with snakes, were sometimes the agent of the cures offered in Asclepius's shrines: they too were offering, they thought, a cure for a bloated, ignorant society). Circe, according to Homer, had transformed Odysseus's unwary crew to pigs; one pig at least, so the Platonist Plutarch imagined in the dialogue *Gryllus*, refused to be turned back, insisting that pigs were honest, decent creatures, much to be preferred to human beings.[58] Not every "Magian" really abandoned all the social graces (Plutarch, after all, did not), but the idea that we should dissociate ourselves from the normal, civil ordering of society was an influence even on corporealist, Classically minded philosophers like the Stoics and (with some qualifications) the Epicureans. "Living according to nature" is an ambiguous slogan, and might lead one to accept the naturalness of cities, slavery, and economics as easily as to deny them. "Living like an animal" is no less ambiguous. But whereas the Classically minded would either reject such a life, or else find some small place within the system where an occasional release from civic duties could be possible, the Magian sort might reckon, with the earlier philosopher Heracleitos (540–480 B.C.), that real life went on out there, away from cities, and that it was possible to hope that we could live like that. It was not only the Christian Gospels that advised us to consider "the lilies of the field";[59] Cynics and Pyrrhonian Sceptics alike were inspired by the sight of mice scampering for their food, without any thought for tomorrow.[60]

In this other mind-set, ideal beauty is not to be found only among the biologically human, nor ideal justice only in cities or empires. Back in the Age of Cronos, the Golden Age before recorded history and the rule of law, it could be imagined, we lived as animals among animals, following our necessities, but without the delusions of grandeur and manifest destiny that now afflict us. According to Euripides's *Theseus*, it was a "mixed-up and bestial existence" that the gods saved us from, giving us understanding, language, agriculture, shelter, sea voyages, trade, and augury.[61] But not everyone approved of the change. Our ancestors' desire for luxury, and status, led them away from Plato's "city of pigs,"[62] not necessarily for the better. It was not *necessity*, but greed that led them (and us) to turn animals into meat.

> Nothing abashed us, not the flower-like tinting of the flesh, not the persuasiveness of the harmonious voice, not the cleanliness of their habits or the unusual intelligence that may be found among the poor wretches, No, for the sake of a little flesh

we deprive them of sun, of light, of the duration of life to which they are entitled by birth and being. Then we go on to assume that when they utter cries and squeaks their speech is inarticulate, that they do not, begging for mercy, entreating, seeking justice, each one of them say, "I do not ask to be spared in case of necessity; only to spare me your arrogance! Kill me to eat, but not to please your palate!"[63]

Much the same point is made in imagining ourselves as "souls," which could animate, which have animated, any number of different-seeming sorts of creatures: "I have been a boy and a girl, and a bush and a bird, and a dumb fish in the sea," declared Empedocles (490–430 B.C.).[64] All these things are equally expressions of a single identity, without much difference between them so far as the gods can see. And back in the beginning, only Aphrodite ruled: "their altar was not drenched by the unspeakable slaughters of bulls, but this was held among men the greatest defilement—to tear out the life from noble limbs and eat them."[65] There was a tradition even in Athens that Triptolemus, one of the ancient lawgivers, had laid three laws down for the Athenians: "respect parents, honor the gods with crops, do not harm animals," and Xenokrates (396–314 B.C.) worried whether the third law was founded only on the special horror of kin-killing, or on the realization that there were better ways of working with animals than sacrificing them for food.[66] Maybe it was based on both.

We are all animated by the same life. The distinctive feature of *humanity* is that we can recognize this, and so bridge the gap between corporeal and spiritual life. "Humankind is the only living thing that is twofold," according to the late antique treatise *Asclepius*,[67] with a remit to care for earthly beings in the light of eternity. We are to consider ourselves "spiritual amphibia," native both to the world of competing animals and atoms, and to the world of comprehensive beauty. "Reason" is not only the social intelligence we need to live together in kindness (for a while), but also a direct insight, call it a revelation, into the real world of which this is a defective copy. Here, so far as our senses go, we live among figments, in a dream that may be a delirium. There, in the original real world we are, as it were, theorems of a universal science, aspects of a single beauty, bound by universal sympathies. That beauty is visible, for those who care to look, even in the smallest and most vile of creatures.[68] Every visible creature is, in a way, an image of the divine, which can be "seen" more clearly in abstraction from immediate sensation.

Biological and mythological hybrids, in turn, reveal a truth, not a danger: that Soul stretches without a break from the seemingly inanimate all the way up to the star-gods and beyond. Those of us born human may have a special opportunity, a special responsibility (as also in Buddhist tradition): we have the chance to glance aside from immediate, parochial and sensual concerns, and can care for every animate creature without too much consideration of our own bodily welfare. Precisely because we are equipped to love and serve the Father (which is to say, the origin of all things), we must care for all His children: "for anyone who feels affection for anything at all shows kindness to all that is akin to the object of his affection, and to the children of the father that he loves. But every soul is a child of That Father."[69]

It is in this context that we can locate the anecdote about Pythagoras (ca. 580–ca. 490 B.C.), that he rebuked a man who was beating a dog, with the words "That's a friend of mine—I knew him by his voice."[70] The story is told to illustrate, maybe to mock (but why?), Pythagorean belief in metempsychosis, the thought that souls migrate into new bodies when their present body dies. It is a belief found all across the world, though not every believer—not even Porphyry—thinks that his or her soul might end up in a nonhuman body (that, as before, might raise too many political problems). But there is a simpler interpretation of the remark: Pythagoras acknowledged the dog's howling as a *complaint*, and so as a communication that placed an obligation (an experienced command) on those who heard it.[71] Animals can after all be our friends: consider Odysseus's old dog, who survives just long enough to recognize his master.[72] Consider the immortal horses of Achilles, briefly allowed to speak to him before the Furies who protect cosmic order close their mouths again.[73]

Even the Stoics, who were mostly adamant that animals could not talk or think or feel, also suggested that a dog, for example, could reason syllogistically: in pursuit of a scent, a dog, on approaching a parting in the road would take a scent down one path, conclude that his prey had not passed that way, and immediately go down the other.[74] (Whether this is a true story, we may reasonably doubt.) Aesop's stories (620–560 B.C.) as we now possess them, are no more than moralistic allegories, but perhaps they have their origin in earlier folk tales, where it was natural to suppose that "animals" had personalities and adventures, and could speak with a human tongue.[75] Even Epictetus, though he advised us *not* to be like "animals," also spoke of the pride they take in their natures, not in their reputations (and advised us to follow that example). He also spoke of caged birds as follows:

> Look at the birds yonder and see what lengths they go in striving to escape, when they are caught and reared in cages; why, some of them actually starve themselves rather than endure that sort of life; and even those that do not die, pine away and barely keep alive, and dash out if they find any chance of an opening. So strong is their desire for natural freedom, an independent and unhindered existence. Why, what ails you in your cage? "What a question! I am born to fly where I will, to live in the air, to sing when I will; you take all this away, from me, and say, 'What ails you?'"[76]

Things, including us and animals, are to be *themselves*. The ideal humanity is something that understands this, and lets things be. Perhaps it even allows the constant transformation of one thing into another, without complaint. Pythagoreans need not have been literally right about metempsychosis, nor mythmakers right to suppose that people are readily turned into animals or plants, but perhaps they were both correct in essence: the world of our experience is in flux, and even if there are eternal Forms to guide us it is of the essence of *our* world that no single Form holds sway forever. The eternal Beauty is made manifest in the diversity of creatures, and their swift changes one into another.

There was a common misunderstanding of Platonic Forms and Aristotelian species. Aristotle studied animals, as he also studied human constitutions, as "natural" organisms whose characters owed nothing to any value they might have as symbols. To avoid unnecessary repetition, he grouped them into distinct classes, which were broadly alike: sea-creatures or land quadrupeds or birds. But there were always "in-betweens" and ambiguous entities that could be classed as one thing or another. Living things here-and-now form a continuum, without absolute divisions, and there are significant homologies between, say, a seal's flippers and a horse's legs. A seal, indeed, by Aristotelian standards, is a malformed quadruped, having no external ears and no long legs.[77] Even without hybrids, lineages don't always breed entirely true—usually, in fact, they don't: the immanent form transmitted by the male parent doesn't always fully control the matter, and some new thing is born that might perhaps begin to manifest some different coherent form and way of life. Evolutionary change is no great surprise for either Platonists or Aristotelians. The human soul itself, like nature, is a continuum, that can be divided, in thought, in indefinitely many ways: it is the complete human being, the human animal, that reasons, reflects, and chooses in ways not so far removed from the way that nonhuman animals also reason, reflect, and choose. Species, in brief, are not natural kinds, and there is as good a reason to consider ourselves members of "lifekind" as of "humankind," whether we are merely natural organisms or embodiments of an eternal and incorporeal soul.

ANIMALS ALL TOGETHER

The point is that both in mythological and biological thought it was possible to think of people not as creatures of a radically different sort than animals, but *as animals*, as animated creatures, all together. As Plato remarks, to divide the world into the human and nonhuman is as silly as dividing it into cranes and non-cranes.[78] And Xenophanes (570–478 B.C.) before him spoke scathingly of those who suppose the gods have human form: "if cattle or horses or lions had hands and could draw, and could sculpt like men, then the horses would draw their gods like horses, and cattle like cattle."[79] The point is not a simple rejection of anthropomorphic imagery, but of an entire rationalist tradition, which supposes that reality is of the same form as human reason, that we as *rational* beings are especially close to God. "For my thoughts are not your thoughts, nor are your ways my ways, says the Lord."[80] This other, "Magian" tradition is clear that God is nothing like any creature (though all creatures, in a way, manifest something of God's power) and is encountered not through "reason" in the usual sense, but by revelation. We cannot tell much of God's nature and purposes merely by reflecting on first principles, nor yet by simple observation: we need, so "Magians" would suppose, to be told the truth by some authoritative voice, natural or visionary. Even "the voice of reason," from this point of view, might be an intrusive *daimon*, as it was (so it seems) for Socrates.[81]

But that, unfortunately, is not the end of the matter. Classical humanism placed people (especially adult, male, freeborn people) at the center and warned us away from sharing anything with animals. The "Magian" or Pythagorean strand identified us instead as souls, and at least in principle understood all living creatures as kin. *Nous*, in that strand, meant Spirit much more than Reason. But "animals" were still, all too often, images of depraved or damaged souls: this indeed is how Plato chooses to explain their existence in his cosmogonic myth, in the *Timaeus*. They might even be demons in the more familiar sense, enemies of the light. Classical humanism typically conceived the real as the tangible, whereas Magians imagined a world full of spirits and intangible influences—and these might be actively malevolent. In place of the typically Classical assumption that evil was only an error, a falling away from perfection, this other strain of thought was easier to convince that evil was something positive, an equal and opposite goal, the Enemy.

> The holy man's powers were shown in his relations with the animal kingdom, which had always symbolized the savagery and destructiveness of the demons: he drove out snakes and birds of prey, and he would settle down as the benign master of jackals and lions.[82]

Late Antique thought and practice, which might possibly have developed as Porphyry would have wished, instead insisted that animals were either there for us to use at will, or that they were the creatures of the Enemy.[83] To attempt the "Pythagorean" life (which later came to be called the "vegetarian")[84] was either to imply that human beings weren't worth more than animals, or to express disgust at being contaminated by animals. Christian thinkers, with some few exceptions, decided that the Pythagorean way was wrong: wild beasts at least were in the service of the Devil, and domestic beasts were there for us to use. There were exceptions: the Desert Fathers seem to have tried to return to paradise, where wild animals would be our friends again. Arnobius of Sicca (d. 330), an African convert, objected to animal sacrifice out of compassion for the animals, not simply in contempt for any gods or devils who would demand it.[85]

Christians did eventually put a stop to the games in which wild beasts were pitted against each other, or against convicted criminals (such as Christians), or simply slaughtered in a pretended "hunt" (which Xenophon would have despised). But they did so mostly for reasons familiar to pagan philosophers: that such sights excited atrocious passions in the human soul (and certainly did not inspire the audience to *courage*—as has been the favorite excuse ever since for bull-baiting, dog fights, cock fights, and the like). Animal sacrifices were also, after Julian, gradually abandoned—but this is only to say that animal slaughter was secularized, and even the ritual acknowledgement that the animals were not *ours*, but belonged entirely to God or the gods, was abandoned. Christians no longer needed to worry that the meat available had been sacrificed to idols: they could eat it with a good conscience, ignoring the pain and torment. They no longer saw even the image of a god looking out at them from the animal. Nor did they need to distinguish what was edible by anything but physiological features (though in actual practice very few people are literally omnivores): there were no longer to be clean and unclean animals, but all were available for use. Animals might still be used as symbols (consider the Lamb of God), but not as icons of the ancient sort: they might remind us via an arbitrary code of particular virtues, but they did not truly resemble them, since they were either (for those with a more Stoic outlook) things, or (for those of a more Magian sort) alien spirits whom we should disown.

Things might have worked out differently. Some early Christians, including James the brother of Jesus,[86] were vegetarian. In replacing animal sacrifice with the literally bloodless sacrifice of the Eucharist, the Church had fulfilled Porphyry's dream of a return to the bloodless sacrifices of the earlier age imagined by Empedocles and others. To join God's "table fellowship," we no longer needed animals to be our messengers or mediators. Even in taking a "secular" view of animals, the Church had opened the way to better treatment: animals used as symbols, even as incarnate gods, don't necessarily get treated as they themselves would wish. On the contrary, being used as a symbol of imperial power or a messenger to the gods may result in being imprisoned, mutilated, mauled, and killed for ritual purposes.

The Hebrew scriptures had included rules about treating animals well: not to muzzle the ox that treads out the corn,[87] nor take mother and young from any nest,[88] nor take a calf, lamb, or kid from its mother till seven days after its birth,[89] nor boil a kid in its own mother's milk,[90] nor leave a beast trapped in a well on the pretext that today is holy,[91] nor yoke ox and ass together,[92] nor plough up all of the fields, in every year, and so deprive the wild things of their livelihood.[93] "God hates nothing that He has made: why else would He have made it?"[94] These commands might have fitted well with the Pythagorean and Platonic tradition as that was developed by Plutarch or Porphyry. Unfortunately, it was early decided by the mainstream Churches, following a remark of Paul,[95] that the animal welfare laws were only allegorical, that we are linked by no "community of law,"[96] and even that a concern for animals was a sign, as it had been for the more doctrinaire of pagan thinkers, of a poor moral sense. If they are there for our use, as the Stoics had thought, we can make of them what we will, but there is, at least according to the Platonist tradition, a price to pay: "There is a price to be paid for fabricating around us a society which is as artificial and as mechanized as our own; and this is that we can exist in it only on condition that we adapt ourselves to it. This is our punishment."[97]

It need not have happened that way.

> These were the words of the Lord to me: Prophesy, man, against the shepherds of Israel; prophesy and say to them, You shepherds, these are the words of the Lord God: How I hate the shepherds of Israel who care only for themselves! Should not the shepherd care for the sheep? You consume the milk, wear the wool, and slaughter the fat beasts, but you don't feed the sheep. You have not encouraged the weary, tended the sick, bandaged the hurt, recovered the straggler, or searched for the lost; and even the strong you have driven with ruthless severity....I will dismiss those shepherds: they shall care only for themselves no longer; I will rescue my sheep from their jaws, and they shall feed on them no longer.[98]

Ezekiel, or the Lord, here takes it for granted that true shepherds care for sheep. "A righteous man cares for his beast, but a wicked man is cruel at heart."[99] So did Plato. Literally, of course, as Thrasymachus pointed out,[100] shepherds care for sheep so that they may profit from them in the end, but perhaps this was not so in the beginning, need not be wholly so even now, and is at least not the criterion of the *good* shepherd. The later Platonic tradition long preserved a memory of the better way.

> For any man who is just and good loves the brute creatures which serve him, and he takes care of them so that they have food and rest and the other things they

need. He does not do this only for his own good but out of a principle of true justice; and if he is so cruel toward them that he requires work from them and nevertheless does not provide the necessary food, then he has surely broken the law which God inscribed in his heart. And if he kills any of his beasts only to satisfy his own pleasure, then he acts unjustly, and the same measure will be measured out to him.[101]

The Ancient, Classical, Greek, and Mediterranean attitude to the nonhuman was as convoluted, as contradictory, as that of any age or region. Animals were utterly unlike us, but we might come to be like them. They were insentient and unfeeling, driven entirely by external stimuli, and yet—according to the very same thinkers—capable of loyalty and syllogistic reasoning. We should put aside all merely "animal" affections, but also follow their example in complaining about and escaping from our chains. We might be reborn as animals, or else their souls were of another, mortal, sort. Animals were both the actual, tangible creatures with whom we share a world, and on whose life and labor we often depend, and also symbols of human possibility, reminders that we don't *have* to live as usually we do, for better or for worse. They are both objects of our practical and symbolizing gaze, good to eat and also good to think with,[102] and also, sometimes, creatures who look back at us, remind us of a common heritage, and are vehicles of a commanding Spirit. Pythagoras heard his friend, and knew him by his voice.

NOTES

1. See Thomas McEvilley, *The Shape of Ancient Thought* (New York: Allworth Press, 2006) for a scholarly and persuasive account of that wider context.

2. Of which tradition the "Pythagorean" (later to be called the "vegetarian") way is an early manifestation. Oddly, to modern eyes, the original Pythagorean diet prohibited *beans* as well as meat (according to Diogenes Laertius, *Lives of the Philosophers* 8.34): the prohibition may have been a sectarian, ritual precaution, or—just possibly—a *medical* precaution, since favism, a form of haemolytic anemia brought on in susceptible persons by eating fava beans, may have been endemic in the local population. See Frederick J. Simoons, *Plants of Life, Plants of Death* (Madison: University of Wisconsin Press, 1998). Alternatively, as Plutarch suggested, it may be a warning not to get involved in state affairs: Plutarch "Education of Children," 12D: *Moralia*, trans. Frank Cole Babbitt, Loeb Classical Library (London: Heinemann, 1927), vol. 1, p. 59.

3. Aristotle, *Politics* 1.1256b15; cf. *Physics* 2.194a34.

4. O. Spengler, *The Decline of the West*, trans. C. F. Atkinson (NewYork: Knopf, 1926), vol. 1.

5. Porphyry, *On Abstinence from Killing Animals* 1.8, trans. Gillian Clark (London: Duckworth, 2000), p. 32. All subsequent quotations from Porphyry are taken from this translation.

6. Lucretius, *On the Nature of Things*, trans. W. H. D. Rouse and Martin F. Smith, Loeb Classical Library (London: Heinemann, 1924), 5.864–867. See also Gordon Campbell, "'And Bright was The Flame of Their Friendship' (Empedocles B130): Humans, Animals, Justice, and Friendship, in Lucretius and Empedocles," *Leeds International Classical Studies* 7:4

(2008): 1–23. Campbell suggests that Lucretius sought to reconcile the strict Epicurean doctrine with Empedocles's suggestion that we used to live in friendship with the other animals, and might do still.

7. "Hymn to the Mother of Gods" 174a–b, in *Works of the Emperor Julian*, trans. W. C.Wright, Loeb Classical Library (London: Heinemann, 1962), vol. 1, cited by Ingvild Saelid Gilhus, *Animals, Gods and Humans: Changing Attitudes to Animals in Greek, Roman and Early Christian Ideas* (London: Routledge, 2006), p. 147.

8. Porphyry, *On Abstinence* 1.56, p. 54.

9. Thomas Taylor, *A Vindication of the Rights of Brutes*, introduced by Louise Schutz Boas (Gainesville, Fla.: Scholars' Facsimiles & Reprints, 1966; originally published 1792).

10. Porphyry, *On Abstinence from Animal Food*, ed. Esmé Wynne-Tyson, trans. Thomas Taylor (New York: Barnes & Noble, 1965). The translation was first published in 1823.

11. Otherwise entitled "Beasts are Rational": Plutarch, *Moralia*, trans. Harold Cherniss and William C. Helmbold, Loeb Classical Library (London: Heinemann, 1957), vol. 12, pp. 492–531. The same volume includes "Whether Land or Sea Animals are Cleverer" and "On the Eating of Flesh."

12. Cicero, *Letters* 7.1.3; Gilhus, *Animals*, p. 25.

13. Herodotus, *Histories* 2.35: "the Egyptians themselves in their manners and customs seem to have reversed the ordinary practices of mankind"—by which Herodotus meant the ordinary practices of his fellow Greeks. In particular "they live with their animals—unlike the rest of the world, who live apart from them."

14. See Marcel Detienne and Jean-Pierre Vernant, *Cunning Intelligence in Greek Culture and Society*, 2nd ed., trans. J. Lloyd (Chicago: University of Chicago Press, 1991).

15. As the Egyptian priest is said to have told Solon, according to Plato's *Timaeus* 22B, trans. Benjamin Jowett: "There is no opinion or tradition of knowledge among you which is white with age.…Like the rest of mankind you have suffered from convulsions of nature, which are chiefly brought about by the two great agencies of fire and water.…The memorials which your own and other nations have once had of the famous actions of mankind perish in the waters at certain periods; and the rude survivors in the mountains begin again, knowing nothing of the world before the flood."

16. Aristotle, *Politics* 1.1252a12.

17. See Keith Bradley, "Animalizing the Slave: The Truth of Fiction," *Journal of Roman Studies* 90 (2000): 110–125, for an examination, partly through Apuleius's *Golden Ass*, of slavery as a form of animal domestication. Bradley argues that slave owners were conscious that human beings, by implication unlike more tameable animals, could not in the end be successfully domesticated: someday they would rise in fury. That consciousness was not enough to end the practice.

18. Porphyry, *On Abstinence* 1.10, after Hermarchus (fl. 270 B.C.; Epicurus's successor): see Thomas Cole, *Democritus and the Sources of Greek Anthropology*, American Philological Association Monograph 25 (Cleveland, Ohio: Case Western Reserve University, 1967), pp. 71–75.

19. *Odyssey*, Bk.12.

20. Pindar (553–443 B.C.), *Pythian* 8.95–97 (446 B.C.).

21. Aristaeus, son of Apollo and the nymph Cyrene (and the father of Actaeon). According to Plutarch ("On Love") 14: *Moralia*, trans. Edwin L. Minar Jr., F. H. Sandbach, and W. C. Helmbold, Loeb Classical Library (London: Heinemann, 1961), vol. 9, "even those who hunt gazelles and hares and deer have a sylvan deity who harks and halloos them on, for to Aristaeus they pay their vows when in pitfalls and snares they trap wolves and bears."

22. Xenophon, *On Hunting* 5.33–36: *Scripta Minora*, trans. E. C. Merchant, Loeb Classical Library (London: Heinemann, 1925), p. 401. Xenophon was one of Socrates's friends.

23. Xenophon, *On Hunting* 10.23: *Scripta Minora*, p. 441.

24. The story came in many versions, but the one chiefly remembered, is Callimachus's *Hymn 5* (see also Ovid, *Metamorphoses* 3.138–255), displayed in Titian's *Diana and Actaeon*, in the National Gallery in London.

25. The episode, from Jacobus de Voragine, *The Golden Legend* (c.1260), is mostly now recalled in the painting by Antonio Pisano (c.1440), also in the National Gallery. Eustace, according to the legend (which has no contemporary authority), was martyred by Hadrian with his family in a heated bronze bull (an instrument of torture invented long before by the tyrant Phalaris). The stag story is also told of St. Hubert.

26. Whether this was true of *all* such meat is uncertain: Paul at least had reason to agree that Christians couldn't be sure that the meat wasn't from such a sacrifice (and had therefore been offered to idols); see Paul, 1 Corinthians 8.

27. Michael Flanders and Donald Swann, "The Reluctant Cannibal," *At the Drop of a Hat* (London: Parlophone, 1960; reissued by EMI 1992).

28. Diogenes Laertius, *Lives of the Philosophers* 3.19.

29. Thucydides, *History of the Peloponnesian War* 4.80 records such an episode.

30. Aristotle, *Politics* 1.1255a1–1255b15.

31. Xenophon, *On Hunting*, 1.1–4: *Scripta Minora*, p. 367.

32. Philostratus, *Life of Apollonius* 6.19; Gilhus, *Animals*, p. 55.

33. Gilhus, *Animals*, p. 98, after Plutarch, *On Isis and Osiris* 379E: *Moralia*, vol. 5, trans. Frank Cole Babbitt, Loeb Classical Library (London: Heinemann, 1936), p. 165.

34. Epictetus, *Discourses* 2.9. See also 2.10, "Consider who you are. In the first place, you are a man; and this is one who has nothing superior to the faculty of the will, but all other things subjected to it; and the faculty itself he possesses unenslaved and free from subjection. Consider then from what things you have been separated by reason. You have been separated from wild beasts: you have been separated from domestic animals."

35. Cicero, *On Ends* 3.63. A. A. Long & D. Sedley, eds., *Hellenistic Philosophers* (Cambridge: Cambridge University Press, 1987), 57F, vol.1, p. 348.

36. Philo, *Decalogue* 80: cited by Gilhus, *Animals*, p. 101.

37. Ovid, *Metamorphoses* 1.68, trans. John Dryden. http://classics.mit.edu/Ovid/metam.1.first.html, describing the creation of humankind, though Ovid's narrative allows more transmutations back and forth than classical anthropocentrism wished. Ovid (43 B.C.–A.D. 18), judging from the stories he chose to tell, had Pythagorean sympathies.

38. See R. Auguet, *Cruelty and Civilization* (London: Routledge, 1994), pp. 112 113; see Gilhus, *Animals*, p. 34. See Yi-Fu Tuan, *Dominance and Affection: Making of Pets* (New Haven, Conn.: Yale University Press, 1984) for a further exploration of the impulse to *dominate* even those animals for whom we claim to feel affection.

39. Origen, cited by Patricia Cox Miller, *The Poetry of Thought in Late Antiquity* (Aldershot: Ashgate, 2001), p. 42; see also Gilhus, *Animals*, p. 207.

40. See Plato, *Protagoras* 322: shame (*aidos*) and a sense of justice are needed if people are to live together peaceably in cities. The later, Stoic suggestion was that cities should be united instead by a network of loving relationships, *eros*, on which idea see Malcolm Schofield, *The Stoic Idea of the City*, 2nd ed. (Chicago: University of Chicago Press, 1999). This may be impractical.

41. Chrysippus (280–207 B.C.), according to Porphyry, *On Abstinence* 3.20.1: Long and Sedley, *Hellenistic Philosophers* 54P: vol.1, p. 329.

42. *Hermetic Corpus* 9.1–2: *Hermetica*, trans. Brian P. Copenhaver (Cambridge: Cambridge University Press, 1992), p. 27.

43. Porphyry, *On Abstinence*, 3.192, p. 82.

44. Plutarch "The Cleverness of Animals," 961D: *Moralia*, vol. 12, p. 333.

45. For example, Peter Carruthers, *Language, Thought and Consciousness: An Essay in Philosophical Psychology* (Cambridge: Cambridge University Press, 1996).

46. See John Heath, *The Talking Greeks: Speech, Animals, and the Other in Homer, Aeschylus, and Plato* (Cambridge: Cambridge University Press, 2005) for an examination of the importance and significance of *speech* for the Greeks, and their denial of the capacity for reasoned speech not only to animals, but (in various ways) to slaves, foreigners, and women.

47. Hesiod, *Works and Days* 1.275–277, trans. Hugh G. Evelyn-White, Loeb Classical Library (London: Heinemann, 1914).

48. See Oppian of Corycus (fl. A.D. 170), *Halieutica* 2.43–45. "Among fishes neither justice is of any account, nor is there any mercy nor love; for all the fish that swim are bitter foes to one another"; they use "cunning wit and deceitful craft" against fishermen–and each other (Oppian, *Halieutica* 3.92): see Gilhus, *Animals*, p. 19.

49. Plato, *Timaeus* 41d.

50. Plutarch, "The Cleverness of Animals" 962A: *Moralia*, vol.12, p. 337.

51. Plutarch, "On Affection for Offspring" 3C: *Moralia*, vol.6, trans. Frank Cole Babbitt, Loeb Classical Library (London: Heinemann, 1939), pp. 343–345.

52. Benedict Spinoza, *Ethics* 4p37s1, blaming this for "the [supposed] requirement to refrain from slaughtering beasts": *Ethics and Selected Letters*, trans. S. Shirley (Indianapolis: Hackett, 1982), p. 175; see my *Animals and their Moral Standing* (London: Routledge, 1997), pp. 87–96.

53. 1 John 4:16; see also Plotinus, *Ennead* 6.8 [39].15.

54. Spengler, *Decline*, vol. 2, p. 235: "whereas the Faustian man [that is, the Western] is an 'I' that in the last resort draws its own conclusions about the Infinite; whereas the Apollonian man [the Classical], as one *soma* among many, represents only himself; the Magian man, with his spiritual kind of being, is only a *part of an pneumatic 'We'* that, descending from above, is one and the same in all believers." This characterization of differing styles of humanity rather obscures the fact that Classical, Apollonian man also conceived himself a member of a whole—but in his case, of the city rather than the sect.

55. See Plato, *Laws* 7.803c.

56. Epicureans, I suggest, were an offshoot of Western Buddhism, persuaded that the road to salvation, to the end of pain, lay in deconstructing our idea of self-hood, reducing our desires, and joining in a community of the like-minded. But this is another story.

57. For Spengler, the "Magian" or "Arabian" culture is typified in architecture by the dome (as against the temple or the spire) and in psychology by the division between Soul and Spirit: see especially Spengler, *Decline*, vol. 2, pp. 233–261.

58. In other contexts, Plutarch preferred the familiar trope that Circe matched their outward form to their inner greediness: Circe's cup causes "grossness and forgetfulness and fatuity" (Plutarch, "How to tell a Flatterer from a Friend" 52E: *Moralia*, vol.1, p. 283).

59. Matthew 6:28–30.

60. Early Christian missionaries probably looked and sounded very like Cynics: see Gerald F. Downing, *Christ and the Cynics* (Sheffield: Sheffield Academic Press, 1988).

61. Euripides, *Suppliants* 201–213.

62. See Plato, *Republic* 2.372d; see my "Herds of Free Bipeds," in *Reading the Statesman: Proceedings of the Third Symposium Platonicum*, ed. C. Rowe (Sankt Augustin: Academia Verlag, 1995), pp. 236–252, reprinted in *The Political Animal* (London: Routledge, 1999), pp. 134–154.

63. Plutarch, "On the Eating of Flesh" 994E: *Moralia*, vol. 12, pp. 549–551.

64. Diogenes Laertius, *Lives of the Philosophers* 8.77: *The Presocratic Philosophers*, 2nd ed., ed. G. S. Kirk, J. E. Raven, and M. Schofield (Cambridge: Cambridge University Press,

1983), p. 319. Empedocles was also the author of an evolutionary theory not unlike the Darwinian.

65. Empedocles, quoted by Porphyry, *On Abstinence* 4.21: Kirk, Raven, and Schofield, *Presocratic Philosophers*, p. 318.

66. Porphyry, *On Abstinence* 4.22: Clark, p. 118.

67. *Asclepius 7*: Brian P. Copenhaver, *Hermetica*, p. 70.

68. Aristotle, *De Partibus Animalium* 1.645a15–25.

69. Plotinus, *Ennead* 2.9 [33].15, 33–16.10.

70. According to Xenophanes (Diogenes Laertius 8.36): Kirk, Raven, and Schofield, *Presocratic Philosophers*, p. 219. I have discussed this issue at greater length in *Understanding Faith: Religious Belief and its Place in Society* (Exeter: Imprint Academic, 2009), pp. 110–24.

71. Obligations, I suggest, have the same form as the Celtic *geis*, which may be incurred simply by being who we are, or else by an external curse or dare: but that too is another and much longer story.

72. *Odyssey* 17.294–327. See Gary Steiner, *Anthropocentrism and Its Discontents: The Moral Status of Animals in the History of Western Philosophy* (Pittsburgh: University of Pittsburgh Press, 2005), p. 42.

73. *Iliad* 19.391–416.

74. Porphyry, *On Abstinence* 3.6.3; see also Sextus Empiricus, *Outlines of Pyrrhonism* 1.69.

75. Plato tells us that Socrates passed the time in prison by versifying the fables: Plato, *Phaedo* 61b.

76. Epictetus, *Discourses* 4.1, trans. P. E. Matheson (1916), http//www.sacred-texts.com/cla/dep/dep087.htm.

77. Aristotle, *De Partibus Animalium* 2.657a22–24: see my *Aristotle's Man: Speculations upon Aristotelian Anthropology*, 2nd. ed. (Oxford: Clarendon Press, 1983), pp. 29–30, for other examples of this trope.

78. Plato, *Statesman* 263d.

79. Xenophanes, cited by Clement, *Stromateis* 5.109.3: Kirk, Raven, and Schofield, *Presocratic Philosophers*, p. 168.

80. Isaiah (probably "Deutero-Isaiah," ca. 550?) 55:8.

81. Plutarch, "On the Sign of Socrates" 588D–589F: *Moralia*, Loeb Classical Library (London: Heinemann, 1949), vol. 7, pp. 372–509: see especially pp. 451–459. Plutarch suggests, in passing, in this passage that *daimones* don't need *words* to convey their thoughts. "Recognition of one another's thoughts through the medium of the spoken word is like groping in the dark" (*On the Sign* 589B: p. 455): he does not explicitly link this claim to thoughts about animal intelligence.

82. Peter Brown, *The World of Late Antiquity: AD 150–750* (London: Thames & Hudson, 1971), p. 101.

83. Even the Egyptians, so Plutarch says, reckoned most animals sacred to Typhon (which is to say, Set, the Enemy): "On Isis and Osiris" 380E: *Moralia*, vol. 5, p. 171. Terrance Callan discusses the various ways that Hellenistic Jews and Christians used the animal comparison in "Comparison of Humans to Animals in 2 Peter 2, 10b–22," *Biblica* 90 (2009): 101–113.

84. See my "Vegetarianism and the Ethics of Virtue," in *Food for Thought: The Debate over Eating Meat*, ed. Steven F. Sapontzis (New York: Prometheus Books, 2004), pp. 139–151.

85. Gilhus, *Animals*, pp. 151–154, citing *Against the Gentiles*, a work in seven volumes in defence of Christianity, owing something to Porphyry's writings.

86. Eusebius, *Ecclesiastical History* 2.23.5 (after Hegesippus), cited critically by Richard Young, *Is God a Vegetarian?* (Chicago: Open Court Press, 1998), p. 95.

87. Deuteronomy 25:4.

88. Deuteronomy 22:6–7; see Leviticus 22:28.

89. Leviticus 22.26–28.

90. Deuteronomy 14:21.

91. Deuteronomy 22:4; see Luke 14:5.

92. Deuteronomy 22:10.

93. Leviticus 19:9–10, 23:22, 25:6–7.

94. Wisdom of Solomon 11.14.

95. 1 Corinthians 9:8–9.

96. Augustine, *Morals of the Manichaeans* 2.17.59, cited by Gilhus, *Animals*, p. 268.

97. P. Sherrard, *The Eclipse of Man and Nature* (West Stockbridge: Lindisfarne Press, 1987), 71–72.

98. Ezekiel 34:1–4, 9–10.

99. Proverbs 12:10; cf. 27:23–27.

100. Plato, *Republic* 1.343b–344d.

101. Anne Conway, *Principles of the Most Ancient and Modern Philosophy*, ed. A. P. Courdert and T. Corse (Cambridge: Cambridge University Press, 1996), p. 35. The book was originally published, posthumously, in 1690.

102. Claude Lévi-Strauss, *Totemism*, trans. Rodney Needham (Boston: Beacon Press, 1963), p. 89. See also G. E. R. Lloyd, *Science, Religion and Ideology* (Cambridge: Cambridge University Press, 1983), p. 12. The point had also been made by Plutarch, "On Isis and Osiris" 380F–381F: *Moralia*, vol. 5, pp. 171–177.

SUGGESTED READING

CASSIN, BARBARA, and JEAN-LOUIS LABARRIÈRE, eds. *L'animal dans L'antiquité*. Paris: Vrin, 2000.

DOMBROWSKI, DANIEL. *The Philosophy of Vegetarianism*. Amherst: University of Massachusetts Press, 1984.

GILHUS, INGVILD SAELID. *Animals, Gods and Humans: Changing Attitudes to Animals in Greek, Roman and Early Christian Ideas*. London: Routledge, 2006.

HEATH, JOHN. *The Talking Greeks: Speech, Animals, and the Other in Homer, Aeschylus, and Plato*. Cambridge: Cambridge University Press, 2005.

JENNISON, GEORGE. *Animals for Show and Pleasure in Ancient Rome*. Philadelphia: University of Pennsylvania Press, 2005; 1st published, 1937.

KALOF, LINDA, ed. *A Cultural History of Animals in Antiquity*. Oxford and New York: Berg, 2007.

MILLER, PATRICIA COX. *The Poetry of Thought in Late Antiquity: Essays in Imagination and Religion*. Aldershot: Ashgate, 2001.

NEWMYER, STEPHEN. *Animals, Rights and Reason in Plutarch and Modern Ethics*. London: Routledge, 2005.

OSBORNE, CATHERINE. *Dumb Beasts and Dead Philosophers: Humanity and the Humane in Ancient Philosophy and Literature*. Oxford: Clarendon Press, 2007.

Porphyry of Tyre. *On Abstinence from Killing Animals*. Translated by Gillian Clark. London: Duckworth, 2000.

SORABJI, RICHARD. *Animal Minds and Human Morals: The Origins of the Western Debate*. London: Duckworth, 1994.

STEINER, GARY. *Anthropocentrism and Its Discontents: The Moral Status of Animals in the History of Western Philosophy*. Pittsburgh: University of Pittsburgh Press, 2005.

CHAPTER 2

....

ANIMALS AND ETHICS IN THE HISTORY OF MODERN PHILOSOPHY

....

AARON GARRETT

ARGUMENTS for animal rights are sometimes seen as unique to our century. Take the following quote from a recent book on French thought: "The twentieth century began with Vladimir Lenin's observation that making an omelet meant breaking eggs; it ended with the assertion of the rights of chickens."[1] It's a funny remark, but it's clear that the author wishes to suggest that what began as a tragedy ended as a farce. It also suggests the reader draw the false inference that discussions of animal rights and even animal welfare are a pathological product of the unserious late twentieth century. In this chapter, I'm going to try to explain why it was difficult for early modern philosophers and jurists to talk about animals as having rights, and why, therefore, a somewhat cogent theory of animal rights was a remarkable achievement.

"Rights talk" is ubiquitous in our culture, and we could all give many stories of questionable, confused, or incoherent invocations of rights. We could also give examples of the abuse of rights language. Philosophers can get quite precise about rights, but they also are guilty of a lot of confusion. It is not entirely philosophers' fault, or anyone's fault, that they talk about rights in a confused manner. "Rights talk" bridges law, politics, and morals, and what is taken as the definition of a right or a particular right in one area may not hold in another. Even internal to these different arenas there are conflicts. There are natural rights, acquired rights, negative rights, positive rights, Hohfeldian rights, rights that are right and rights that make a wrong, and on and on.

All said, it is unsurprising, given the confusion about rights, that ascribing a confusing concept to nonhumans would be difficult and even highly confusing. In addition to all the confusion, there were also specific conceptual difficulties for early modern philosophers, jurists, and political theories. I will begin by giving some general background on how early modern philosophers thought about rights and explaining why their conceptions made for specific problems in ascribing rights to animals. I will use the example of the great eighteenth-century experimentalist Robert Boyle and his grisly claim that we have a *duty* to experiment in order to motivate my discussion and focus it on three specific problems that confronted the development of a coherent theory of animal rights. I will then describe two ways in which early modern moral philosophers who wanted to argue that we have moral duties toward animals tried to sidestep these problems. I will next turn to the Scottish moral philosopher Francis Hutcheson's "invention" of animal rights and show how it responded to the problems I outlined previously.

Just as Hutcheson developed his account of animal rights, the assumptions of the general framework for natural law and natural rights on which he relied began to be challenged. In considering the challenges, I will take a side excursus into a few philosophers' considerations of the animal afterlife. I will conclude by discussing the direction modern animal welfare discussions took in the late eighteenth century and into the nineteenth century toward animal revolution in parallel with the French Revolution and toward utilitarianism and practical ethics leading to the rise of animal welfare legislation in Britain and Germany. Most of the story I will tell is British. With a few exceptions, the most interesting modern European philosophical literature about animal welfare (that I know of) was British.

"An Occasional Reflection on Dr. Charlton's Feeling a Dog's Pulse at Gresham-College"

In 1665, the statistician John Graunt gave the Royal Society what he claimed was a rare and particularly lethal poison from Macassar.[2] Walter Charleton, the medical experimentalist and advocate of modern Epicureanism, was sufficiently excited at the prospect of using the poison for comparative anatomical experiments that he pocketed the box when no one was looking (against the explicit orders of the Royal Society) and began to use it in experiments on dogs in his rooms.[3] Once they discovered what he had done, the Royal Society reprimanded Charlton, recovered the poison and then undertook its own experiments.

This peculiar incident was sufficiently noteworthy that the poet Samuel Butler wrote a parody—"An Occasional Reflection on Dr. Charlton's Feeling a Dog's Pulse at Gresham-College. By R. B. Esq."—making fun of Charleton, the great Boyle ("R. B. Esq."), and by extension the more pompous side of the Royal Society:

> Dr. Charleton with his judicious Finger examines the arterial Pulsation of its left Foreleg; a civil Office, wherein both Doctor and Dog, Physician and Patient with equal Industry contest, who shall contribute most to the experimental Improvement of this learned and illustrious Society—Little doth the innocent Creature know, and as little seems to care to know, whether the ingenious Dr. doth it out of a sedulous Regard of his Patient's Health, or his own proper Emolument; 'tis enough to him that he does his Duty; and in that may teach us to resign ourselves wholly to advance the Interests and Utility of this renowned and royal Assembly.

All things considered, the dog that Charleton experimented on may have been fortunate. To take a representative sample of dog experiments in the Royal Society from 1661–1667: dogs were bit by vipers, poisoned by arrows, had their spleens removed, were cut open alive with a bellows "thrust into the windpipe" to see how long they might survive (including removing the ribs and diaphragm), had holes drilled into them through which their nerves could be directly tickled, had nutritive broths injected into their jugular veins, had their blood transfused and mixed with the blood with other dogs and sheep as well as had quantities of blood combined with sugared milk, had their *vena cava* and jugulars tied off so their heads swelled and more.[4] Later the experiments became more involved.

Experimentation on live animals was not new, as it went back to antiquity. The Royal Society was by no means the only modern proponent. William Harvey, Marcello Malpighi, and others had made important discoveries about the circulation system and about physiology through experiments on animals. Furthermore, experimentation on living animals in the seventeenth century was not distinctively connected to the new science; its most influential early modern proponent, Harvey, was an early modern Aristotelian.[5] Nor were the Royal Society's experiments the most ghoulish. There were vivisectionists in Port Royale who believed animals to be senseless beast machines incapable of feeling pain and commensurate with this belief nailed animals to boards while mocking those who were sympathetic to the apparent suffering.[6] Finally, some members of the Royal Society and spectators at the public demonstrations were put off by the crueler experiments.[7]

So what was Samuel Butler parodying? By writing in Boyle's grandiose style, he seemed to be mocking the framework that Boyle and other members of the Royal Society used to justify their experiments. Boyle followed Francis Bacon's program laid out in the *Novum Organum* of providing careful inductive and experimental histories of the natural and human world in order then to organize and conquer nature and to extend human power.[8] And he conceived of natural scientific experimentation as a means to understand creatures that God had created so that the experimentalist could contribute to the mastery of nature and acquire "a Power that becomes Man as a Man."[9] In other words, God had made animals for experiments in that they are there to excite our intellectual faculties so that we can master the creation as God intended.[10] Charleton's clandestine experiments were a violation of the Royal Society's edicts, but they also showed an abundant enthusiasm for the duties of the naturalist.

To reject "dissecting *Dogs, Wolves, Fishes* and even *Rats* and *Mice*" due to "effeminate squeamishnesse"[11] was to pridefully look down on and diminish the purpose of a part of God's creation. Not to experiment was to manifest the worst vices from a Christian viewpoint. Conversely, the investigation of and experimentation on creatures was a moral duty of a believing inquirer.[12] If God created man in part to be the master of nature and nature to be mastered by man, then it was a duty to do whatever was necessary to properly master nature. That animals were created for man and that men were their masters was a commonplace reading of SCRIPTURE.

Virtue, rights, and duties were the default way of conceptualizing morals in the seventeenth and eighteenth centuries. In the picture tacitly accepted by many philosophers and jurists, rights, duties, and obligations pertained to relations between superiors and inferiors. God was the superior, God's creation the inferior, and this distinction anchored all other duties. Superior rulers had rights over inferior subjects, fathers governed the household and exercised rights over wives, servants, and children. Children, wives, servants, and the governed all gave up something in their relationships to superiors—the ability to act as they might choose without being governed—but they were also getting something (at least in principle) that motivated them to perform their duties and submit to superiors.

Duties had a close connection to rights—to have a duty that one ought to discharge implied the rights necessary to discharge that duty. A father has a duty to raise his children well and so he has a right over the children in order to discharge the duty. On some accounts, the rights could only be exercised insofar as they helped to satisfy the duty—sending a child to their room in order that they could become a responsible adult was as far as the right went. In this case, one might have a right against one's father if he went beyond what was permitted qua discharging his duties. On other accounts, the right was less circumscribed: it was a permission acquired by office or station. Conversely, the presence of a right tended to imply some sort of duty—the right to property might imply the duty to make it fruitful—but not always. This was the normal sense of right held by countless early modern moral philosophers, albeit with great disagreements on the details.

The early modern natural law works of Hugo Grotius, Samuel Pufendorf, Richard Cumberland, and many others were structured around fundamental duties and rights (some natural, some not) which anchored more and more fine-grained rights and duties. The web of rights, duties, and obligation that arose through human interaction made up the natural law and natural rights view that allowed humans to get along socially and to prosper. The best civil laws, that is, legal codes of actual states, expressed, backed, and promoted these natural laws. With the advent of John Locke's *Second Treatise on Government*, this way of talking about rights was supplemented with the language of natural rights—for example, that a child had a basic right to the conditions that would allow him or her to become a rational moral adult or a man had a right to property insofar as it extended his natural right to person. Lockean natural rights were closely connected to this picture. To have a natural right was to have a set of rights, duties, and obligations linked to the relation between God and man; or, to put it differently, natural rights were also given to fulfill duties and obligation (I will return to this below when I discuss the right to happiness).

A desire to unify moral virtues, rights, duties, religion, and natural philosophy was not unique to Boyle. It was and would continue to be the obsession of many if not most of the brightest lights in the European Republic of Letters. Cutting-edge natural philosophy had more than a whiff of impiety at best. For example, Spinoza and Hobbes were proponents of the new science and held notoriously unorthodox religious views; their mechanistic accounts of nature seemed to imply a world at odds with the God of scripture.

THE DIVIDING LINES BETWEEN HUMANS
AND ANIMALS

As just hinted, interest in the moral standing of animals was connected with the wider project of modern philosophers in trying to understand the mental capacity of animals—from cockles to orangutans and often tacitly presuming a great chain of being—order to capture what was distinctively human. Animals were the touchstone for discovering what human nature is.[13] Human nature was (mostly) considered what is not animal, or at least not merely what is animal. Certain humans—women and members of ethnic or racial groups—were further categorized according to whether they were more or less human. Unsurprisingly, when advocates for the equality of women and for abolition pleaded their cases, they often invoked animals, sometimes arguing for a common cause through suffering and sometimes stressing that it was abhorrent that animals were treated better than women and slaves.[14] Conversely, the standing of animals was discussed in relation to other "inferiors."

With the growth in understanding of natural history, in particular the publication and dissemination of Buffon's massive and massively influential *Histoire Naturelle*, animal ethology had become a more and more sophisticated diagnostic of and justification for complex theories of human nature. In the second half of the eighteenth century, Jean-Jacques Rousseau, and following him Lord Monboddo (James Burnett) and John Oswald (who I will discuss later), argued that the basic sentiments prior to the perfection of human mental faculties are either crucial to natural morality (Oswald) or lead to immorality when suppressed or effaced (Rousseau and Monboddo).[15] For Monboddo, following Rousseau, fanciful travel reports of orangutans (a catch-all term for higher primates) became the decisive test case for the human essence. Orangutans were speechless humans, so language was not essential, and orangutan women were modest, so humans were originally modest and only became depraved through society. Descriptions of "wild children"— children who spent their formative years in the wild or wholly unsocialized—provided confirmatory evidence.

Monboddo's interpretation of the evidence was idiosyncratic and wishful to say the least, but the pivotal diagnostic role of animals (or in the case of orangutans barely-not-animals) for understanding humans' mental and moral nature was (and still is) ubiquitous. The extreme Cartesian position maintained by Nicolas Malebranche and Antoine Arnauld was mostly viewed as an outlier example of excessive metaphysical enthusiasm. Most authors of abiding influence—Michel Montaigne, Thomas Hobbes, Pierre Bayle, Henry More, Ralph Cudworth, G.-W. Leibniz, Christian Wolff, John Locke, David Hume, and many others (including Scholastic authors and Aristotelians)—ascribed some sort of mental capacity to animals that was similar to or possibly even identical to lower human mental capacities. All used the mental and social capacities of animals to call into question human special standing. Almost all held that humans had superior capacity, although some, notably Montaigne, Pierre Charron, Hobbes, Benedict Spinoza, Bernard Mandeville, Hume, Baron D'Holbach, Denis Diderot, and Julien Offray de LaMettrie, criticized (in differing degrees) the ranking of different capacities as drawing on illicit anthropomorphic prejudice and by extension conventional morality.

For a good example of this last, consider Mandeville's parable of the Roman merchant and the lion in his *Fable of the Bees*. The merchant, having washed up on a foreign shore, encounters a lion, who agrees that he will not eat the merchant if a good reason is given why he shouldn't. After the merchant argued that due to his greater reason and his immortal soul he is superior, the lion responded:

> If the Gods have given you Superiority over all Creatures, then why beg you an Inferior? … Savage I am, but no Creature can be call'd cruel but what either by Malice or Insensibility extinguishes his natural Pity.…'Tis only Man, mischievous Man, that can make death a sport.…Ungrateful and perfidious Man feeds on the Sheep that clothes him, and spares not her innocent young ones, whom he has taken into his care and custody. If you tell me the Gods made man Master over all other Creatures, what Tyranny was it then to destroy them out of Wantonness.[16]

The lion went on to skillfully dispose of all of the merchant's arguments and concluded by underscoring his first criticism in the quote above by eating the merchant.

Mandeville's viewpoint was exceptional. The game for most, as it was for Rousseau (who cited Mandeville in his discussion of natural pity) and Monboddo, was in defining wherein the difference lay between human and nonhuman animals, them, it was likely to lie in a particular mental faculty.[17] Bayle's famous article "Rorarius" in his *Dictionary Historical and Critical* was one of the most celebrated examples of line dividing. Nominally, he treated the life of Hieronymous Rorarius[18] and his arguments that animals were more rational than humans. Where was the line? Memory? Language? Or, closer to Rousseau, a refined moral sense? Self-consciousness and reflection? If animals lacked the moral sense—as Hume held—but morals rested on a sense, then perhaps what was distinctive about us was not being elevated far beyond the animal world to a transcendent realm. If they lacked the capacity for reflection—as Shaftesbury, Leibniz, and many others held—then perhaps what was distinctively human transcended the animal realm.

Notably, our moral duties toward animals might look similar in practice whether one accepted Hume's or Kant's or Locke's or Leibniz's distinction (although perhaps not Rousseau's), but the justification of our duties to animals and other humans, and the account of human morals and reason on which it rested, would of course be drastically different. So although these discussions had consequences for what sort of standing animals had, the standing of animals and our moral duties merited barely a mention until the middle of the eighteenth century—and even until very recently these matters were still peripheral.

This was for a clear reason. As previously mentioned, most authors accepted that to have an obligation or a duty demanded that one understand the natural or positive law from which the obligation or duty derived. As Hobbes stated concerning the civil law:

> the law is a command, and a command consisteth in declaration or manifestation of the will of him that commandeth, by voice, writing, or some other sufficient argument of the same, we may understand that the command of the Commonwealth is law only to those that have means to take notice of it. Over natural fools, children, or madmen there is no law, no more than over brute beasts.[19]

Even if animals had some sort of mental capacity, they did not have sufficient intellect to understand rules and laws much less their justifications. Infants did not understand laws and rules either, but we had rights over them insofar as they would grow up to be capable of understanding. Animals were at best like eternal infants and at worst natural enemies, so they did not fit well into this scheme. Consequently, the discussions about the differences between men and animals did little to explain our moral duties to animals insofar as they did little to fit animals into the prevalent natural rights picture. Admittedly, this was not what the discussion of the differences between animals and humans was intended to do!

INDIRECT DUTIES, SKEPTICISM, AND THE SENSE OF HUMANITY

As I've noted, Boyle viewed experimentation on animals as a central duty of man and of a Christian, and he attempted to harmonize it within the framework I've just described. Men were superior to animals by virtue of reason and their rank in creation, and so they had rights over animals and permissions and duties to experiment. The permissions and rights were held in order to discharge our duties to God and to others.

Notice that all this holds of a general right over creation; it does not hold specifically of animals. Human relations to animals, in specific, were much harder to bring within the virtue, duty, and rights language, whether it was a right to experiment on animals or an animal right. The general problems for early modern philosophers with applying rights to human relations with animals can be seen by returning to Butler's parody of Charleton and Boyle and particularly Butler's claim that "Doctor and Dog, Physician and Patient with equal Industry contest, who shall contribute most to the experimental Improvement of this learned and illustrious Society." I will point to three problems, which I will call the cognition problem, the motivational problem, and the relational problem.

First, a dog would only "contest" to be experimented on if it could understand the aims of the experiments and was sympathetic with the Royal Society. But dogs do not understand the aims of experiments and are presumably not sympathetic with learned societies. Much of the line dividing described in the section above focuses on the cognitive differences between animals and humans—that animals lacked understanding or language. One of the tacit presumptions of the natural law picture I have just described is that one was only obligated if one was able to understand the obligation. And to have an obligation or a duty demanded that one understand the natural or positive law from which the obligation or duty derived. If one does not understand a law, one cannot be held responsible for breaking it or upholding it. Animals were presumed to be incapable of breaking or upholding laws, of having moral duties and obligations, and consequently of having rights.

The second problem, which I call the motivational problem, is also brought out by this passage from Butler. Butler cast the relation of dog to experimenter as between physician and patient. A patient allows a physician rights over his or her body in order to get well; patients are motivated by health to grant the physician the right to undertake a painful procedure. The natural law picture was almost always anchored by self-interested motivations. In this world you obey the laws of sociability because they educe to your happiness. You do not disobey civil laws for fear of punishment. And you obey the laws of morality because if you do you will be rewarded in the next life, and if you do not you will be punished. The dog would only "contest" to be experimented on if motivated, but clearly Macassar poison was not conducive to its health. More generally, if animals were to be brought into the natural law picture as bearers of rights and dischargers of duties then they would need to have a proper motivation.

Finally, Butler's doctor/patient analogy points to the difficulty of specifying the relation between man and animals. The basic natural law duties were God to man, men to other men in civil society, men to self, father to child, husband to wife, and master to slave or servant. These relations were all presumed either to be basic, rationally provable furniture of the world, or ubiquitous to any and all human society. There were also rights acquired over property, and animals were sometimes property.

But most moral philosophers of the period, including Boyle (like many scientists today), held that painful experiments on animals were only warranted insofar as they were useful to or necessary to human interests, and this theory pointed to the fact that there was a difference in kind between duties to animals and duties to shrubs or cornfields. The restriction of the experimenter's right over the animal had something to do with the welfare, feelings, or experiences of the animal which were unlike a restriction on the use of stone or a plant. But then how do we make sense of the relation between men and animals? Parents to children? Masters to servants?

Given these problems, it would seem reasonable to give up entirely on considering our moral relations to animals within a natural law theory. There were two common approaches that attempted to sidestep the three problems. The first was to argue that we only have indirect duties to animals as an extension of or in order to discharge our direct duties to humans and God. This would deny that animals have any rights (only their owners and the like do), while at the same time restricting our conduct toward them. And it would circumvent the problems I have mentioned above.

The position was usually justified like this: if we are cruel to animals without warrant, we will be more likely to be cruel to other humans. Therefore, we have an indirect obligation to not be cruel to animals in order discharge our duties to not be cruel (or become cruel) to humans to whom we have a direct obligation. On the indirect-duty account, there are no more problems of cognition, motivation, or relation than there are of any obligation that humans have to one another.

If the obligations consequently are indirect, and justified through obligations to other humans (or to fully rational beings), then animals' inability to understand the natural law is unproblematic.

Today this position is widely associated with Immanuel Kant, whose specific arguments rested on unique moral justifications in terms of his notion of rational agency and duty; but the claim that we have an indirect duty toward animals that can be justified via a duty to humanity could be found in a wide range of authors.[20] Allestree, Pufendorf,[21] and other writers criticized non-justified cruelty to animals to rule out *pointless* cruelty. Allestree, for example, argued for a close connection between cruelty to animals and the vice of pride, and so cruelty to animals undermined virtuous conduct insofar as it undermined the virtues connected with successful discharge of our duties to self, God, and neighbor.[22]

This line of argument was usually premised on a decisive mental and moral difference between men and animals. Kant suggested that animals lack reason, and so lack freedom, and are wholly in the world of necessity and sentiment. But if some animals, perhaps not cockles but dogs or primates, do not drastically differ from us mentally and morally, then this similarity might have consequences for their moral standing. Perhaps animals can be moral and even acquire virtues? Perhaps we have obligations due to shared concerns that have to take their standpoints into view? Or we might have direct obligations to animals by virtue of the fact that they suffer, and thus avoid what seemed implausible in the indirect-duty view—that the suffering might have no direct moral value in and of itself in guiding our duties. This latter thesis was central to utilitarian arguments concerning animal welfare standing.

The problem with this position for many of the authors who sought a different sort of justification is that there seemed to be a gap between the manner of justification and the moral motivation. I want to stop animals from suffering because I feel compassion for their suffering and wish to alleviate it, and not solely because I want to treat other humans well. We seem not to be recognizing what is morally important in responding to animal suffering. In the words of Christian-utilitarian Soame Jenyns, "the carman drives his horse, and the carpenter his nail, by repeated blows,"[23] but, as Jenyns recognized, there is a salient difference between the two acts of driving. One impacts the welfare of a sentient creature and the other does not, and this difference would seem to be a crucial part of what we are responding to. To paraphrase a famous recent expression from Bernard Williams, the indirect justification is "one thought too many."

A second, less common approach argued that animals were capable of exhibiting, or even of acquiring, moral virtues and also side-stepped the natural law picture. Up to the mid-eighteenth century, arguments for animal virtue were almost always associated with skepticism. The early modern skeptic's use of animal virtue had its origin in the "argument from differences in animals," the first of the ten modes for undermining dogmatism taught by Sextus Empiricus in the *Outlines of Pyrrhonism*.[24] Sextus' argumentative techniques mainly drew on the differences between humans and animals to undermine dogmatic naïve realist theories of perception, an argument technique used to great effect by Berkeley. However, Sextus also noted the similarities between animal and human practical reasoning and behavior to show that animals were capable not only of reason but virtues such as bravery as well. Sextus leveled the arguments from similarity, particularly against the Stoics, and used them to undermine dogmatisms about the divinity of reason and human special standing. Hume would use similar arguments in a few pivotal portions of *A Treatise on Human Nature* and *An Enquiry Concerning Human Understanding*.[25]

The primary purpose of Sextus' first mode was to undermine dogmatic philosophical beliefs and attain a state of *ataraxia* or carelessness. Hume, and others of a skeptical temper like Diderot and LaMettrie, on the other hand, used skeptical arguments and techniques at least partially in the service of positive claims about the nature of human inference and the psychology of the emotions and passions. Insofar as humans are similar to animals, an explanation of human behavior that does not outstrip the naturalistic explanations given of animal behavior is all that is warranted methodologically and can become the basis for a naturalistic account of man. In this way early modern skepticism became closely allied to naturalism. For Sextus to establish any facts about animal reason or morals would be dogmatism and counter to the goal of *ataraxia*. Montaigne's famous "Apology for Raymond Sebond" and Pierre Charron's *De La Sagesse* were the touchstones of and repository for the modern appropriation of Sextus' technique, drawing on our similarities with animals to destabilize human pride. The "Apology" in particular was deeply equivocal about securing knowledge, but elsewhere Montaigne made it clear that he had a positive position concerning the moral standing of animals that displayed a strong affinity for animals and a deep-seated distaste for cruelty. He went so far as to assert that:

> There is a certain respect and an obligation [Fr. *devoir*] of humanity in general which link us not only to beasts, which have life and feelings, but even to trees and plants. We owe [Fr. *devons*] justice to men: and grace and benignity to the other creatures who are able to receive them. If there is some interchange between them and us and some mutual obligation. I am not afraid to admit that my nature is so puerile that I cannot easily refuse an untimely gambol to my dog wherever it begs one.[26]

The condemnation of cruelty and the argument for compassion is also apparent in the passage quoted above from Mandeville, who appears to have been deeply influenced by Montaigne. Mandeville concluded his discussion of the lion and the merchant with a description of the violent slaughter of a "large and gentle Bullock":

> What Mortal can without Compassion hear the painful Bellowings intercepted by his Blood, the bitter Sighs that speak the Sharpness of his Anguish, and the deep sounding Grones with loud Anxiety fetch'd from the bottom of his strong and palpitating Heart....When a Creature has given such convincing and undeniable Proofs of the Terrors upon him, and the Pain and Agonies he feels, is there a Follower of Descartes so inur'd to Blood, as not to refute, by his Commiseration, the Philosophy of that vain Reasoner?[27]

As with the indirect-duties model, both Mandeville and Montaigne avoided the three problems as well. The cognition problem was made beside the point insofar as the emotion of compassion was felt by humans and animals alike, was sufficient for the obligation, and did not require any higher level cognition. The motivation problem was solved insofar as the motivation to fulfill one's duties was the natural motivation to compassion shared by humans and animals, which created a moral bond between animals and humans. There was no obvious problem with this relation. Although Mandeville followed Montaigne in giving compassion pride of place in morals, he like most others did not take up the notion of a natural duty and basic respect for trees and plants (Charron is a notable exception). Montaigne's distinction between justice as owed to men and gentleness and kindness as derived from the sense of humanity was taken up, influentially, by Hume. According to him:

> Were there a species of creatures, intermingled with men, which, though rational, were possessed of such inferior strength, both of body and mind, that they were incapable of all resistance, and could never, upon the highest provocation, make us feel the effects of their resentment; the necessary consequence, I think, is, that we should be bound, by the laws of humanity, to give gentle usage to these creatures, but should not, properly speaking, lie under any restraint of justice with regard to them, nor could they possess any right or property, exclusive of such arbitrary lords. Our intercourse with them could not be called society, which supposes a degree of equality; but absolute command on the one side, and servile obedience on the other. Whatever we covet, they must instantly resign: Our permission is the only tenure, by which they hold their possessions: Our compassion and kindness the only check, by which they curb our lawless will:

And as no inconvenience ever results from the exercise of a power, so firmly established in nature, the restraints of justice and property, being totally *useless*, would never have place in so unequal a confederacy.

This is plainly the situation of men, with regard to animals; and how far these may be said to possess reason, I leave it to others to determine.[28]

Hume goes on to compare this status to that of barbarous nations and women. Hume, like Montaigne, admits that animals are rational and, again like Montaigne, sees that fact as incidental to our gentle usage of them. Unlike Montaigne, the relation is not strictly moral in the sense that a relation of justice is, nor is it rooted in a general fellow feeling.

Hume incorporates another important and surprising theme in this passage. Hobbes and Spinoza held that we had absolute dominion over animals and they had absolutely no moral claims or standing as against us. According to Hobbes, although we are very similar in reason to animals they lack language and so we cannot make covenants with them. Since morality arises from covenant, we cannot have any moral obligations to them. Unlike Hobbes, and like Montaigne, Hume held that our compassion, kindness, and the "laws of humanity" should stop us from making grim use of animals.

Many philosophers and popular theologians would have agreed that cruelty was vicious and impermissible. The difference is that for Montaigne kindness to living creatures as backed by the sense of humanity was a fundamental, perhaps the fundamental, obligation and we owed benevolence to animals, not just for us but due to a standing they had qua living beings. Hume was more equivocal. A century later, Lawrence Sterne would personify Montaigne's attitude in *Tristram Shandy's* Uncle Toby. Finding himself in a room with a fly, Toby chooses not to swat it and opens the window exclaiming "This world surely is wide enough to hold both thee and me."

Montaigne's notion of an obligation of humanity in general anchored by the unity of all living beings had resonances with Hellenistic authors such as Porphyry but was independent of the early modern rights, duties, and virtue talk explicated above. Mandeville's centering of compassion was and is compelling, but it seems to provide little in terms of directives besides "be compassionate." So the problem remained. How could one reconcile the theologically tinged special standing of man and the framework of specific rights and duties with an argument that we have *direct* duties to animals or that animals even have rights? For this we must turn to the Scottish moral philosopher Francis Hutcheson.

RIGHTS

To make sense of "animal rights" within a natural rights theory, one must explain how animals could have standing in and of themselves—a precondition of their possessing natural or acquired rights as opposed to humans simply having rights over them—and how humans could have direct obligations to animals by virtue of this standing. Also the theory would, optimally, not undermine human standing and would fit in well with our ordinary moral intuitions about animals and about ourselves. As should now be clear, this is no means an easy set of conditions to satisfy.

The natural place to start was by drawing a distinction between domesticated and wild animals. I take "domesticated" in a broad sense as including all animals that are kept by humans for human benefit—whether pets kept for pleasure or farm animals kept for meat, wool, milk, and labor.[29] Most of the animals that the Royal Society experimented on were dogs, sheep, and cats, presumably either owned by the experimenters or caught strays. There was concern in the eighteenth century for wild animals, in particular there was outcry against bear baiting (as well as the baiting of domestic animals such as bulls) in England—tormenting tied up bears with dog packs and weapons—and the barbarity of the practice was seen to have a bad moral effect. To again quote Samuel Butler: "For if bear-baiting we allow, what good can reformation do?"[30] But domestic animals presented a more obvious basis for a rights argument. Insofar as animals were domesticated, they formed a domestic community with man. It was relatively common to argue that animals gained a benefit by virtue of being domesticated, and some argued further that their sacrifice at our tables was just recompense.[31] But to argue that their inclusion in our household changed their standing was novel, that they were part of the household just as wives, children, and servants were. Domestic status would allow them to be treated as discharging duties to the domestic unit and consequently as possessing some sort of right against their masters. This was an important, although we shall see not the sole, basis for Francis Hutcheson's arguments for ascribing rights to animals by virtue of their standing.[32]

Hutcheson viewed domestication as providing adventitious or acquired rights by virtue of the animals' contribution to the domestic economy allowing them to attain the rank of servant or natural slave. The idea of acquiring domestic standing and becoming a form of natural servant was beautifully expressed by Thomas Reid:

> Animals that are created by the bounty of heaven to partake of the Entertainment but on account of the Inferiority of their Nature may be considered as Servants to Man and to be served after Man....But even the brute Animals who serve us at this entertainment must not be neglected. They must have their Entertainment and Wages for the Service they do us.[33]

Reid was tendering a further suggestion, also made by Hutcheson, that all animals, not just domestic animals, were servants to man and thus should be allowed their entertainments. In other words, we have no right to interfere with the pleasures or needs of animals if they do not conflict in a significant way with our own needs, and we have a positive duty to reward animals for their contribution to the welfare of the household.

First, though, one might ask whether these analogies are anything more than analogy. An iron skillet may contribute in some sense to the welfare of our household but it does not merit moral standing. The moral standing of animals must be due to the fact that they have desires and a need of welfare that is similar to that of humans. In a note appended to a revised version of his *Inquiry*, in response to the criticism that if our moral evaluations depended on a sense then animals, insofar as they possessed senses, might be capable of virtue, Hutcheson admitted that "'tis plain there is something in certain Tempers of Brutes, which engages our Liking, and some lower Good-will and Esteem, tho' we do not usually call it Virtue, nor do we call the sweeter dispositions of Children Virtue; and yet they are so very like the lower Kinds of Virtue, that I see no harm in calling them Virtues."[34] In other words, a courageous dog possesses a virtue, albeit in a manner far lower than the way in which an adult human possesses a virtue. The dog is more akin to a courageous child.

That animals were like children capable of lower virtue, although toddlers *in perpetua*, and thus of being part of a moral community gave Hutcheson a way to combine the traditional image of man as steward of nature with the radical idea of animal virtue now put to a non-skeptical purpose. It also had strong intimations of one of the central Messianic passages in scripture, *Isaiah* 11:6, that "the wolf will lie down with the lamb…and a little child shall lead them." The importance this passage for radicals can be seen in William Blake among others. As in Reid's example, man governed the domestic community with his superior virtues and moral sense and guided it toward moral ends. Through man's stewardship and domestication, animals thus became incorporated in something larger with moral significance. Their talents and virtues are developed toward a superior goal. Of course, some of these talents might be for producing wool, some for making milk, and some for being delicious when roasted and served with mint jelly (here Blake might dissent).

Still, this approach gave Hutcheson a response to both the cognition and the motivation problems. Animals were motivated by the benefits of community, and they were stewarded due to their lack of cognitive capacity in ways that promoted their interests. To put it differently, animals could avail themselves through the community of human guidance and human cognitive capacity. Hutcheson was not unique in stressing the importance of community between humans and animals. The Danish philosopher Friederik Christian Eilschov also developed arguments "for a moral community of man and animals under natural law."[35] Like Hutcheson, he argued that it was by virtue of community that moral relations were governed by natural law.

In his *System of Moral Philosophy*, a massive posthumously published work, Hutcheson develops his argument for animal rights most extensively. Adventitious rights governed in a moral community are only one justification; Hutcheson also provides more straightforwardly Christian-utilitarian[36] arguments. God desires that all creation be happy and there be a minimum of pain—that there be the greatest happiness for the greatest number—and this holds of animals as well.

> These considerations would clearly shew that a great increase of happiness and abatement of misery in the whole must ensue upon animals using for their support the inanimate fruits of the earth; and that consequently it is right they should use them, and the intention of their Creator.[37]

The welfare of creation dictated that animals too possessed a right to happiness, or at least that it was right that they sought happiness. Hutcheson invoked right explicitly in the correlative discussion of the avoidance of pain:

> Brutes may very justly be said to have a right that no useless pain or misery should be inflicted on them. Men have intimations of this right, and of their own corresponding obligation, by their sense of pity. 'Tis plainly inhuman and immoral to create to brutes any useless torment, or to deprive them of any such natural enjoyments as do not interfere with the interests of men. 'Tis true brutes have no notion of right or of moral qualities: but infants are in the same case, and yet they have their rights, which the adults are obliged to maintain.[38]

It was commonly held that "no useless pain or misery" should be inflicted on animals, but Hutcheson expressed this as a right commensurate with the rights of infants and derived from a right to happiness, all things being equal, which all sentient beings share. Men recognize this obligation via their moral sense of pity and recognize the obligations it entails. In violating the right, they deny the clear testimony of their moral sense. Hutcheson thus argued that domestic animals have adventitious rights, but ultimately backed by (or rooted in) the natural right to happiness shared by all sentient beings.

Brute Theodicy

Hutcheson stressed that "no useless pain or misery should be inflicted on" animals. But what was the guide to usefulness? It was commonly and often tacitly assumed that there was a hierarchy of beings, "the great chain of being," and that although no suffering was good in and of itself, the suffering of creatures on the lower rungs of being could be permissible if it was "useful" to a creature of the higher ranks.[39] It also fit well with a common response to a variant of the theodicy problem: why should anyone suffer in a world guided by divine providence and governed by a benevolent and omnipotent God? The potential answer was that the degree of suffering corresponded to rank and that suffering of a creature of lower rank for the interests of a creature of higher rank was permissible under some or many circumstances.

This problem was highlighted by Samuel Johnson in a review of Soame Jenyns's *Free Inquiry into the Nature and Origin of Evil*. In the 1780s, Jenyns published one of the most powerful essays against animal cruelty: "On Cruelty to Inferior Animals."[40] In the essay, Jenyns attacked all manners of sport that derived pleasure from animal suffering, including hunting, bear and bull baiting, and cockfighting: "the majestic bull is tortured by every mode which malice can invent, for no offence, but that he is gentle, and unwilling to assail his diabolical tormentors."[41] On Jenyns's account, self-defense and humane slaughter for food purposes were permissible, but delight in torment was savagery.

Earlier in *A Free Inquiry into the Nature and Origin of Evil* (1757), Jenyns had argued that it is permissible, or even necessary, for subordinate creatures on the great chain of being to suffer for the pleasure of those higher up if they promoted utility—a Christian-utilitarian variant on utilitarian sacrifice. Jenyns saw that this was an unruly consequence and tried to explain it by an analogy with taxes. All must pay their taxes in order that the good of the whole might be served, and any one set of payments may be substituted for another.

In a review in the *Literary Magazine* now much more famous than Jenyns's book, Samuel Johnson savaged Jenyns on the grounds that if inferiors might suffer merely for the pleasure of superiors this would ramify up to angelic intelligences in a sort of grand guignol:

> As we drown whelps and kittens, they amuse themselves, now and then, with sinking a ship, and stand round the fields of Blenheim, or the walls of Prague, as we encircle a cockpit....As they are wiser and more powerful than we, they have more exquisite diversions...which, undoubtedly, must make high mirth, especially if the play be a little diversified with the blunders and puzzles of the blind and deaf.[42]

In Johnson's hands, Jenyns became the advocate of a sort of proto-Sadean theodicy where the wise gain pleasure from torturing the weak, and torture them well, due to their superior wisdom and skills. Given that Jenyns is now best known for his pamphlet attacking the British tax policy in the colonies, it is more than a bit ironic that here he pleaded for a kind of taxation without representation.[43]

Some philosophers sought a measure of divine rectificatory justice for animals, given the quantity and degree of their suffering, and argued for an animal afterlife. In his *Palingénésie philosophique, ou idées sur l'état passé et sur l'état futur des êtres vivans* (1769–70), the Swiss naturalist Charles Bonnet argued that in this best of all possible worlds all beings would eventually attain perfected states. Just as humans would become spiritually elevated, so also animals would acquire perfected intellects and perhaps be able to hold their tormenters accountable in the next world. This rested on the assumption that animals had eternal souls, which Bonnet argued for at length despite the controversy surrounding the claim.

Bonnet's argument pointed out the conflict between two strongly held beliefs about the natural world—the metaphysical principle of continuity governing its grading by degree from top to bottom and the religious belief in human singularity, that the difference between men and animals is a radical distinction of kind. Bonnet's Leibniz-inspired stress on continuity of natural degrees led him to abandon the denial of an animal afterlife that reinforced the differences in kind. These arguments of Bonnet's were championed in the English-speaking world by no less than John Wesley, demonstrating one point of the great breadth of interest in theological questions concerning animals.[44]

Lancashire clergyman Richard Dean provided powerful negative arguments against La Mettrie, Descartes, Malebranche, and other advocates of brute mechanism to establish that animals had souls and positive arguments that divine justice and benevolence demanded an animal afterlife. Dean had a particularly keen sense of the "many pinching hardships" of the poor and the weak and the cruelty of the Calvinist doctrine of original sin for men and animals born to suffer.[45] Shortly before he wrote *An Essay on the Future Life of Brutes*, he had taken up a farm and both his father and young wife had died,[46] so front and center in his own life were the afterlife, animals, and divine justice. In defending his position, Dean claimed, like Hutcheson, that animals were capable of morality: "there are Brutes which would sooner be hanged than pilfer and steal under the greatest Temptations…that take up Attachments, and profess Friendships."[47] More powerfully, he said that:

> Certain it is, that a future Life of Brutes cannot be absolutely denied, without impeaching the Attributes of God. It reflects upon his Goodness, to suppose that he subjects to Pains, and Sorrows, such a Number of Beings, whom he never designs to beatify.…It reflects on his Justice, to suppose that he destroys without a Recompense, Creatures that he has brought into such a State of infelicity, and in some Measure capacitated for everlasting Happiness.[48]

Utility and Practical Ethics

Dean's argument rested on a presumption of divine justice, that justice for innocent brutes must be forthcoming if God is good. Hutcheson's arguments also rested on a providentially governed world unified by a good God. Divine benevolence sought the happiness of all creatures, and divine justice allowed for moral communities such that all creatures might aspire to happiness and perfection. What Dean credited to the afterlife, Hutcheson optimistically found in our social world.

Consequently, Hutcheson maintained that animal pleasure and pain, though meriting less consideration than human pleasure and pain and not connected to reason or higher intellectual faculties, ought to be taken account of as part of the general happiness and that we naturally recognize the moral importance of animal pain and pleasure through our sense of pity. Furthermore, domesticated animals also acquired standing through incorporation in moral communities that gave them further special adventitious rights as well as the status of servant. The three problems were thus resolved. The cognition problem was again deemed irrelevant. Rights were either due to the capacity to feel pain and pleasure or to a being's contributions to the domestic community for which that being should garner wages and entertainments. Animals were motivated to be part of these communities insofar as they benefited and their virtues were perfected through them. And finally, the relation between man and animal was as widely recognized as any traditional natural law relation and had a similar social genesis through the rise of sociable communities furthering interests.

The dominance of the natural law picture was coming to an end at the time that Hutcheson was formulating these arguments in response to it. Hutcheson was one of the major figures in this shift, as was Hume. In fact, Hutcheson did not published *A System* in his lifetime, although he did publish a shorter version of some of the main arguments, and it is sometimes speculated that the non-publication was due to the young Hume's criticism of Hutcheson's use of providence to stitch together the natural law doctrines, his moral sense theory, and his version of utilitarianism.

One can also see these conflicts in a tension in Hutcheson's idea of animal rights. In the end isn't it the natural right to happiness and the sense of pity that does most of the work, as it had for Mandeville? The whole argument seems to rely on the "right to happiness," which is supported by a benevolent deity and ordered by providence. But if there is no surefire way to know that providence is benevolent, or is at all, then how can we be sure of this general right to happiness? Aren't the acquired rights just additional entitlements as dictated by communities bootstrapping on the basic right to and desire for happiness? Hutcheson admitted that in times of scarcity the rights could disappear. Is this in essence not a natural law picture but rather a confused precursor of utilitarianism with a lot of fancy footwork and a heavy dose of sentimentalism?

Today arguments for animal welfare are commonly associated with Jeremy Bentham and his famous footnote in the Introduction to the *Principles of Morals and Legislation*:

> The day has been, I grieve to say in many places it is not yet past, in which the greater part of the species, under the denomination of slaves, have been treated by the law exactly upon the same footing, as, in England for example, the inferior races of animals are still. The day may come, when the rest of the animal creation may acquire those rights which never could have been withholden from them but by the hand of tyranny. The French have already discovered that the blackness of the skin is no reason why a human being should be abandoned without redress to the caprice of a tormentor. It may come one day to be recognized, that the number of the legs, the villosity of the skin, or the termination of the os sacrum, are reasons equally insufficient for abandoning a sensitive being to the same fate. What else is it that should trace the insuperable line? Is it the faculty of reason, or, perhaps, the faculty of discourse? But a full-grown horse or dog, is beyond comparison a more rational, as well as a more conversible animal, than an infant of a day, or a week, or even a month, old. But suppose the case were otherwise, what would it avail? The question is not, Can they reason? nor, Can they talk? but, Can they suffer?[49]

Unlike Hutcheson's defense of animal rights, Bentham's arguments rely on nothing beyond the primacy of seeking happiness and avoiding pain, and that all pain avoiders and happiness seekers should be taken account of by rational legislators. Henry Salt, the important animal reformer and author of *Animals' Rights Considered in Relation to Social Progress* (1892), a book which Peter Singer referred to as "the best of the eighteenth- and nineteenth-century works on the rights of animals,"[50] claimed that "to Jeremy Bentham, in particular, belongs the high honour of first asserting the rights of animals with authority and persistence."[51]

Bentham used "right" in the sense of a law or rule backed by sanctions, not in the natural law sense. Salt saw in Bentham the germ of a notion of animal rights which he identified with Herbert Spencer's Millian liberty right—that each individual was at liberty to do what they wished provided they did not interfere with the equal liberty of others.[52] This is not what Bentham was arguing for, but as we have seen, passages in Hutcheson and Reid seemed to embrace something very close to this notion of right. Indeed, Salt maintained the distinction between wild and domestic animals in a manner very similar to Hutcheson. Although he seemed to be unaware of Hutcheson's work, he argued that all animals, wild and domestic, had a basic right to non-interference and liberty in seeking their happiness.[53]

Salt clearly wished to ally himself with the most secular elements of animal rights advocacy, but his position seems very close to the work of the Church of England clergyman Humphry Primatt. During the years in which Bentham wrote the *Principles*, and well before its publication, Primatt provided a parallel set of arguments in *The Duty of Mercy and the Sin of Cruelty to Brute Animals*. The work was a founding work of the animal advocacy movement of the early nineteenth century pioneered by Lewis Gompertz (founding member of the SPCA and the Animals' Friend Society for the Prevention of Cruelty to Animals),[54] Richard Martin (animal activist and namesake of one of the first pieces of British animal welfare legislation),[55] the great abolitionist William Wilberforce and others.

Primatt argued in no equivocal terms that animals' capacity to feel pain and their desire for happiness delimited our conduct toward them:

> we are all susceptible and sensible of the misery of *Pain*; an evil, which though necessary in itself, and wisely intended as the spur to incite us to self-preservation, and to the avoidance of destruction, we nevertheless are naturally averse to, and shrink back at the apprehension of it. Superiority of rank or station exempts no creature from the sensibility of pain, nor does inferiority render the feelings thereof the less exquisite. Pain is pain, whether it be inflicted on man or on beast; and the creature that suffers it, whether man or beast, being sensible of the misery of it whilst it lasts, suffers Evil.[56]

Many of Primatt's arguments were found in his predecessors. Like Dean, Primatt pointed out that the unjustifiable belief that animals were made for man, together with bad religion, were heavily to blame. Like Reid and Hutcheson, he argued that animals had a right to wages (that is, resources including food, rest, ease, and comfort) and we had a reciprocal duty not to deny them their due in the sense of not interfering with the satisfying of their natural enjoyments and needs.[57] All of these arguments were ably backed by scriptural quotations, and Primatt did assume a general providentialism with a benevolent deity. But Primatt's novelty was due to the fact that his core arguments, as above, needed no providentialist justification. Pain is pain whoever experiences it, and that alone is sufficient for moral obligation and legislation.

Primatt also, provided a novel twist on the logic of the great chain of being that pulled him away from utilitarian justifications. As men were superior to animals, to treat them cruelly was worse "than the cruelty of Men unto Men" and was the functional equivalent to "fury and barbarity on a helpless and innocent Babe."[58] In other words, a moral act was reprehensible in proportion to the power, capacity, and reason of the agent and the distance between the agent's power, capacity, and reason and the power, capacity, and reason of the sentient object of the agent's action. This justification drew on a basic moral intuition (as well as on scripture) that there is something deeply wrong with harming the weak, which the Christian duty of mercy disallowed in a novel and compelling way.

Primatt's work was disseminated by John Toogood in *The Country Clergyman's Shrovetide Gift to his Parishioners. The Duty of Mercy and the Sin of Cruelty to Brute Animals* went through multiple editions, and then was republished in association with the founding of the SPCA (1824). It was also connected to the political activism concerning the inhumane conditions of farm animals, test subjects, and pets which resulted in Martin's Act (1821–22), the Cruelty to Animals Act (1835), and eventually a vivisection act (1876). Salt put Primatt with Bentham, Mandeville, and a number of the other authors I have discussed as a foundational figure of animal rights and drew on him directly in his discussions of the rights of domesticated animals, although as I have suggested above he did not fully see the importance and novelty of Primatt (as he had not seen the importance of Hutcheson).[59]

As noted, Bentham was interested in concrete animal legislation, and Primatt was as well. Lawrence's *A Philosophical and Practical Treatise on Horses and on the Moral Duties of Man towards the Brute Creation* was perhaps the great extended masterpiece of arguments for the need for animal rights, not as philosophical fancies but as enshrined in laws. Unlike many of the works discussed above, it grew out of detailed first-hand knowledge of the lives of animals and offered practical means to alleviate their sufferings. Little is known about Lawrence, other than from his writings.[60] Lawrence discussed everything from shoeing, to veterinary medicine, to the equestrian arts, to hunting and hacking, to horse physiology, to purchase and sale in *A Philosophical and Practical Treatise*. In "On the Rights of Beasts," he argued that there could only be one justice for men and animals, and explicitly criticized Hume because he failed to argue that if men deserved justice then so did animals.[61]

Lawrence argued that even if philosophers might accept arguments for animal rights or animal standing, "the bare acknowledgment of the right, will be but small avail to the unfortunate objects of our solicitude, unless some mode of practical remedy can be devised."[62] Instead of philosophical arguments, a change in laws was needed, according to Lawrence:

> The grand source of the unmerited and superfluous misery of beasts, exists, in my opinion, in a defect in the condition of all communities. No human government, I believe, has ever recognized the *jus animalium*, which surely ought to form a part of the jurisprudence of every system, founded on the principles of justice and humanity....Experience plainly demonstrates the inefficacy of mere morality to prevent aggression, and the necessity of coercive laws for the security of rights. I thereby propose that the Rights of Beasts be formally acknowledged by the state, and that a law be framed upon that principle, to guard and protect them from acts of flagrant and wanton cruelty, whether committed by their owners or others.[63]

Like Bentham, Lawrence stressed the inefficacy of rights talk unless enacted through legislation. Lawrence, though, proposed specific laws and the creation of inspectors of animal markets to enforce them. Among the striking positions Lawrence argued for were that "experimental tortures" on live animals were absolutely impermissible[64] and that euthanizing animals was preferable to "the mistaken humanity of those tender-hearted persons, who turn adrift a poor dog or cat, which they choose not to keep."[65]

Thomas Young's *An Essay on Humanity to Animals* was also a practical work, a conduct manual intended to prescribe in detail how one ought to treat animals, including bees. Young concentrated his discussion on teaching children to respect animals. Like William Blake, Young was concerned by the mistreatment of pets: "What right have we to tame such animals as birds, squirrels, and hares, and to cage and confine some of them; thus debarring them from the unrestrained exertion of the energies of their natures, and depriving them of many enjoyments which a benevolent Creator had provided for them, and all this merely for the sake of amusement?"[66] Young supplemented arguments with careful descriptions of cruelty and rousing passages from literary works in order to educate his readers.

With Young and Lawrence, we see a transition toward a practical animal rights movement that would flourish in the nineteenth and twentieth century. This transition was connected to the practical turn in discussions of abolition, women's rights, and other issues—in Lawrence's case, spurred on by the French Revolution and the belief that through legislation the lot of animals can be bettered.

ANIMAL REVOLUTION

Yet although Lawrence and Young argued for humane slaughter, they did not argue for ethical vegetarianism. They were what we might call animal liberals, not animal radicals. The most radical animal welfare argument of the later eighteenth century, and perhaps until the late nineteenth or twentieth century, was offered by the Scots Jacobin John Oswald, who saw the change of the standing of animals as part of a wholesale transformation of society.

Oswald served as a lieutenant in India, returning to Britain by land and in the process picking up a few languages and becoming acquainted with the "Hindoo religion" and the vegetarian philosophy. As a journalist in London he published the radical journal the *British Mercury*, became friends with Tom Paine and James Mackintosh, moved to Paris in 1790, became a French citizen soon after and the commander of the Parisian pikemen in Vendée where he died in action in 1793.[67]

Unlike almost all of the authors I have described, Oswald was an atheist who was deeply influenced by Rousseau. His arguments in his work *The Cry of Nature; or, An Appeal to Justice, on behalf of the Persecuted Animals* (1791), which he signed

"John Oswald, Member of the Club des Jacobines," are accordingly not derived from Christian duties to mercy and also do not regularly invoke rights. Instead, Oswald provides a history of human self-alienation as arising from the denial of the universal sentiment of pity. The argument draws on a wide range of sources, from Plutarch and Porphyry to translations of Indian texts to Rousseau. According to Oswald, the connection between religious animal sacrifice and the establishment of political authorities led to the interconnection of undemocratic regimes, cruelty to animals, and bad religion. Through the religious use of animal sacrifice men came to deny their pity—the "rooted repugnance to the spilling of blood." Over time, this sacrifice became custom and the customary subordination of animal suffering—for example the practice of keeping abattoirs and butchers away from everyday commerce and life so that people could eat meat without having to feel the pity—paved the way for the animal experimentation practiced by the Royal Society.[68] By denying animal suffering, and pushing suffering animals out of sight, men deny their true nature and engage in unsatisfying cruelty toward men and animals. Through alienating animal suffering, men eventually alienated themselves in much the terms Rousseau described in the *Second Discourse*. Abattoirs and vivisection went hand in hoof with war, class stratification, and many other ills. The only solution was wholesale transformation of present society so that humans could live in peace with animals, one another, and themselves.

Oswald also argued at length for the nutritional superiority of a vegetarian diet and that meat eating resulted in physical debilitation and decay. The only solution would be social revolution, and any human social revolution must necessarily be animal revolution as well. *The Cry of Nature* directly inspired the printer George Nicholson's *On the Conduct of Man to Inferior Animals* (1797), which drew on a wide range of medical and anthropological evidence (as well as literary example) to argue, like Oswald, that humans' primeval and healthiest diet was vegetarian.

By the end of the eighteenth century, discussions of ethical vegetarianism and animal welfare merited parody. The great Plato translator Thomas Taylor—who was apparently quite fond of animals[69]—penned the satirical *A Vindication of the Rights of Brutes*.[70] It drew on some of Porphyry's strongest arguments for vegetarianism comingled with calls for humans to learn animal language and chapters entitled "That Magpies are naturally Musicians; Oxen Arithmeticians; and Dogs Actors."

Taylor concluded that it was only a short step from animal rights to the rights of "vegetables, minerals, and even the most apparently contemptible clod of earth" and Godwinian anarchy wherein "government may be entirely subverted, subordination abolished, and all things every where, and in every respect, be common to all."[71] Taylor's primary target was Godwin's wife Mary Wollstonecraft and her *A Vindication of the Rights of Woman*. Taylor's point is that if Wollstonecraft's questionable arguments for women's rights were accepted, then animal rights (and literal mineral rights) would soon follow as well, and this was a reductio ad absurdum.

As Peter Singer notes in his "Preface" to Salt's *Animal Rights*, many of the arguments that Taylor thought were so ridiculous as to result in a reductio are now widely accepted.[72] As was often the case throughout the history of the discussion of the moral standing of animals, they were connected with the standing of women, children, and non-European peoples. The early advocates of animal rights such as Martin and Wilberforce were also abolitionists, and in some cases feminists, notably Lewis Gompertz. Bentham's footnote makes explicit reference to racism, and Hutcheson's *System* also included arguments for the equality of women and against slavery. If animal welfare was the last to be a widespread social movement, although it was by the nineteenth century, the basic arguments for it were of a piece with the advocacy for the rights, welfare, and standing of subaltern groups that rose in the middle and late eighteenth century. It is notable that Henry Salt begins his bibliography of animal rights with Mandeville in the eighteenth century. Most every original argument for animal welfare and animal rights has its roots in this ferment.[73]

NOTES

1. Julian Bourg, *From Revolution to Ethics: May 1968 and Contemporary French Thought* (Montreal: McGill-Queens University Press, 2007), p. 38. Thanks to Josh Cherniss for this quotation.

2. An expert later denied that it was the right poison, which may explain why, when the Royal Society undertook the experiment, the "poison" did not affect the dog to whom it was administered. See Thomas Birch, *The History of the Royal Society of London*, vol. 2 (London, 1760), pp. 47, 318.

3. Emily Booth, *"A Subtle and Mysterious Machine": The Medical World of Walter Charleton (1619–1707)* (Dordrecht: Springer, 2005), p. 125.

4. Birch, *History of the Royal Society of London*, vol. 1, pp. 31, 288, 486, 509; vol. 2, pp. 84, 118, 164, 203.

5. See Anita Guerrini, *Experimenting with Humans and Animals: From Galen to Animal Rights* (Baltimore, Md.: Johns Hopkins University Press, 2003), chap. 2.

6. Nicolas Fontaine, *Memoires pour servir à l'histoire de Port Royal*, vol. 2 (Cologne, 1738), pp. 52–53. Fontaine's memoire is very sympathetic to the Port Royaliens so there's little reason to doubt his gruesome descriptions of vivisection. Cited in Leonora Rosenfield, *From Beast-Machine to Man-Machine: Animal Soul in French Letters from Descartes to La Mettrie* (Oxford: Oxford University Press, 1940), p. 54.

7. Hooke carried out many of Boyle's experiments. On Hooke's distaste for vivisection, see Anita Guerrini, "Animal Experimentation in Seventeenth-Century England," *Journal of the History of Ideas* 50 (1989): 401–2.

8. Francis Bacon, *The New Organon*, ed. Lisa Jardine and Michael Silverthorne (Cambridge: Cambridge University Press, 2000), pp. 33–34.

9. "Some Considerations Touching the Usefulness of Experimental Natural Philosophy," in Robert Boyle, *Works*, vol. 1 (London, 1744), p. 429.

10. Boyle, *Works*, vol. 1, p. 432.

11. Boyle, *Works*, vol. 1, p. 429.

12. There is a particularly strong statement to this effect in "An Essay Inquiring Whether, and How, a Naturalist Should Consider Final Causes," in Boyle, *Works*, vol. 4, p. 519.

13. See Aaron Garrett, "Human Nature," in *The Cambridge History of Eighteenth-Century Philosophy*, ed. Knud Haakonssen (Cambridge: Cambridge University Press, 2006), pp. 160–233.

14. See "Sophia," in *Woman not Inferior to Man* (London, 1739).

15. See Jean-Jacques Rousseau, "Discourse on Inequality," in *The Discourses and other Early Political Writing*, ed. and trans. V. Gourevitch (Cambridge: Cambridge University Press, 1997), p. 152; James Burnett (Lord Monboddo), *Of the Origin and Progress of Language*, 2nd ed., 6 vols. (Edinburgh, 1774–92); John Oswald, *The Cry of Nature; or, An Appeal to Mercy and Justice, on behalf of the Persecuted Animals* in Garrett, ed., *Animal Rights and Animal Souls*, vol. 6.

16. Bernard Mandeville, *Fable of the Bees*, ed. F. B. Kaye, vol. 1 (Indianapolis: Liberty Fund, 1988), pp. 178–80.

17. Indeed, Rousseau was criticizing the tendency of many of his predecessors to locate what was distinctively human in cognitive capacity and social achievements and arguing that what is best about us is closer to the wholly emotional world of animals.

18. Hieronymus Rorarius, *Quòd Animalia bruta ratione utantur meliùs Homine* (1648). Bayle, *The Dictionary Historical and Critical of Mr. Peter Bayle*, ed. and trans. Pierre Des Maizeaux, 2nd ed., 5 vols. (London, 1734–38; fac. New York: Garland, 1984), "Rorarius."

19. Hobbes, *Leviathan*, ed. Edwin Curley (Indianapolis: Hackett, 1994), ch. 26.

20. Immanuel Kant, *Lectures on Ethics*, trans. Peter Heath and ed. Peter Heath and J. B. Schneewind (New York: Cambridge University Press, 1997), 27: 413 (p. 177), 27: 458–60 (pp. 212–13), 27: 710 (pp. 434–35); Kant, *The Metaphysics of Morals* [Part II. *The Metaphysical Principles of Virtue*],trans. Mary J. Gregor, as in Kant, *Practical Philosophy*, ed. Paul Guyer and Allen W. Wood (Cambridge: Cambridge University Press, 1996), 6: 443 (p. 564); and see the chapters by Tom Beauchamp and Christine Korsgaard in the present *Handbook*.

21. Cf. Pufendorf, *De jure naturae et gentium* [*On the Law of Nature and Nations*], ed. and trans. C. H. Oldfather and W. A. Oldfather, 2 vols. (Oxford: Clarendon Press, 1934), vol. 1 fac. of 1688 ed.; vol. 2 translation, at 4.4.4.

22. Richard Allestree (attributed author), *The Whole Duty of Man* (London, 1713), p. 215.

23. Soame Jenyns, "On Cruelty to Inferior Animals," in *Disquisitions on Several Subjects* (London, 1782), p. 14.

24. Sextus Empiricus, *Outlines of Pyrrhonism*, trans. R. G. Bury (Cambridge, Mass.: Harvard University Press, 1939), 1.1.14.

25. *A Treatise of Human Nature*, ed. David Fate Norton and Mary J. Norton, vol. 1 (Oxford: Clarendon Press, 2007), 1.3.4, 2.1.12, 2.2.12; and *An Enquiry Concerning the Principles of Morals*, ed. Tom L. Beauchamp (Oxford: Clarendon Press, 1998), sect. 9.

26. "On Cruelty," *Complete Essays*, trans. M. A. Screech (London: Penguin Books, 1987), p. 488 (translation modified). Throughout the essay, Montaigne condemns humans' cruelty to one another and to animals, but does not argue that cruelty to animals is less of a vice than cruelty to humans (or that it is a vice insofar as it leads to cruelty to humans).

27. Mandeville, *Fable*, pp. 180–81. The force of Mandeville's condemnation may be in part autobiographical. Mandeville was a Leiden-trained physician, and his college dissertation was a Cartesian argument that animals do not feel; it was entitled "Disputatio Philosophica De Brutorum Operationibus" (1689). See Mandeville, *Fable*, p. 181 n. 1.

28. David Hume, *An Enquiry Concerning the Principles of Morals*, ed. Tom L. Beauchamp (Oxford: Oxford University Press, 2006), sect. 3, p. 18.

29. The distinction between wild and domesticated can get complicated very quickly. On the complexities involved, see Clare Palmer's chapter in this *Handbook*.

30. Samuel Butler, *Hudibras* II.516–17; first published, London, 1684. And further—"Now a sport more formidable/ Had rak'd together village rabble:/ 'Twas an old way of recreating, Which learned butchers call bear-baiting." I.675–78.

31. For example, the widely read dissenting theologian Phillip Doddridge: "in...beasts being designed for the use of men, we may as well grant, that species might be debased to a lower kind of life for his instruction and comfort, as that such multitudes of individuals should be daily sacrificed to his support," *A Course of Lectures on the Principal Subjects in Pneumatology, Ethics, and Divinity* (London, 1763), p. 348.

32. For a much more detailed development of these arguments, see Aaron Garrett, "Francis Hutcheson and the Origin of Animal Rights," *Journal of the History of Philosophy* 45 (2007): 243–65.

33. Knud Haakonssen, ed., *Thomas Reid: Practical Ethics* (Princeton, N.J.: Princeton University Press, 1990), pp. 204–5.

34. Francis Hutcheson, *An Inquiry into the Original of Our Ideas of Beauty and Virtue in Two Treatises*, rev. ed., ed. Wolfgang Leidhold (Indianapolis, Ind.: Liberty Fund, 2008), p. 254.

35. Haakonssen, ed., *Thomas Reid*, p. 379 n. 3. See also Carl Henrik Koch, "Man's Duties to Animals: A Danish Contribution to the Discussion of the Rights of Animals in the Eighteenth Century," *Danish Yearbook of Philosophy* 13 (1976): 11–28.

36. It is anachronistic to call Hutcheson a utilitarian, even a Christian utilitarian, but by this label I mean that he uses arguments consistent with a theologically interpreted greatest happiness principle. Early "utilitarians" such as John Gay were both influenced by and critical of Hutcheson. See John Gay, "Preliminary Dissertation. Concerning the Fundamental Principle of Virtue or Morality," *Utilitarians and Religion*, ed. James E. Crimmins (Bristol: Thoemmes Press, 1998), pp. 34–35.

37. Francis Hutcheson, *A System of Moral Philosophy* (Glasgow: A. Foulis, 1755), I/II.vi.ii.

38. Hutcheson, *System of Moral Philosophy*, I/II.vi.iii.

39. That there were settled ranks in the human world as well as the natural world was a pervasive belief in eighteenth-century thinking as well. See Emma Rothschild, *Economic Sentiments: Adam Smith, Condorcet and the Enlightenment* (Cambridge, Mass.: Harvard University Press, 2001) on two of the few philosophers who criticized this picture.

40. The essay was republished in extract in 1826 by Lewis Gompertz, the founding member of the Society for the Prevention of Cruelty to Animals and activist vegan, two years after the pioneering society's founding.

41. Jenyns, *Disquisitions on Several Subjects*, p. 17.

42. See Samuel Johnson, "Review of Soame Jenyns," in *Animal Rights and Animal Souls*, ed. Aaron Garrett, 6 vols. (Bristol: Thoemmes Press, 2000), vol. 4, pp. 64–65.

43. Soame Jenyns, *The Objections to the Taxation of our American Colonies, by the Legislature of Great Britain, Briefly Consider'd* (London, 1765).

44. See Charles Bonnet, *Conjectures concerning the Nature of Future Happiness* (London, 1792), A2.

45. Richard Dean, *An Essay on the Future Life of Brutes* (London, 1768), 2nd ed., I, p. 73, in Garrett, ed., *Animal Rights and Animal Souls*, vol. 2.

46. David Allan, "Dean, Richard (*bap.* 1726, *d.* 1778)," *Oxford Dictionary of National Biography* (Oxford: Oxford University Press, 2004).

47. Dean, *Essay on the Future Life of Brutes*, II, p. 69, in Garrett, ed., *Animal Rights and Animal Souls*, vol. 2.

48. Dean, *Essay on the Future Life of Brutes*, II, pp. 72–74, in Garrett, ed., *Animal Rights and Animal Souls*, vol. 2.

49. Jeremy Bentham, *An Introduction to the Principles of Morals and Legislation*, ed. J.H. Burns and H. L. A. Hart (Oxford: Oxford University Press, 1982), chap. 17, §1, 4, note.

50. Singer adds that Salt's book "anticipates almost every point discussed in the contemporary debate over animal rights." Singer, "Preface" to Henry S. Salt, *Animal Rights Considered in Relation to Social Progress* (1892; reprint, Clarks Summit, Pa.: Society for Animal Rights, Inc., 1980), pp. 136–53.

51. Salt, *Animal Rights Considered*, p. 5.

52. Salt, *Animal Rights Considered*, p. 2.

53. "Yet surely an unowned creature has the same right as another to live his life unmolested and uninjured except when this is in some ways inimical to human welfare." Salt, *Animal Rights Considered*, p. 47.

54. Lucien Wolf, "Gompertz, Lewis (1783/4–1861)," rev. Ben Marsden, *Oxford Dictionary of National Biography* (Oxford: Oxford University Press, 2004), online ed., http://www.oxforddnb.com/view/article/10934.

55. Richard D. Ryder, "Martin, Richard (1754–1834)," *Oxford Dictionary of National Biography* (Oxford: Oxford University Press, 2004), online ed., http://www.oxforddnb.com/view/article/18207 (accessed May 2008).

56. Humphry Primatt, *The Duty of Mercy and the Sin of Cruelty to Brute Animals* (London, 1776), p. 7. In Garrett, ed., *Animal Rights and Animal Souls*, vol. 3.

57. Primatt, *Duty of Mercy*, pp. 150–51.

58. Primatt, *Duty of Mercy*, pp. 35, 46.

59. See Salt, *Animal Rights Considered*, pp. 136–53.

60. He wrote a number of pamphlets supporting the French Revolution, and he was consulted by Richard Martin in connection with the introduction of Martin's Act. See Sebastian Mitchell, "Lawrence, John (1753–1839)," *Oxford Dictionary of National Biography* (Oxford: Oxford University Press, 2004), online ed., http://www.oxforddnb.com/view/article/16181.

61. John Lawrence, *A Philosophical and Practical Treatise on Horses and on the Moral Duties of Man towards the Brute Creation* (London, 1796/1798), 1: 127. This is not to imply that his criticism of Hume was on target.

62. Lawrence, *Philosophical and Practical Treatise*, 1: 122–23.

63. Lawrence, *Philosophical and Practical Treatise*, 1: 123.

64. "It has been said that the world could not have either gold, sugar or coals, but at the expense of human blood, and human liberty. The world, in that case, ought not to have either gold, sugar, or coals. The principle admits of no qualifications," Lawrence, *Philosophical and Practical Treatise*, 1:133.

65. Lawrence, *Philosophical and Practical Treatise*, 1:160.

66. Thomas Young, *An Essay on Humanity to Animals* (London, 1798), p. 178, in Garrett, ed., *Animal Rights and Animal Souls*, vol. 5.

67. T. F. Henderson, "Oswald, John (c.1760–1793)," rev. Ralph A. Manogue, *Oxford Dictionary of National Biography* (Oxford: Oxford University Press, 2004), online ed., http://www.oxforddnb.com/view/article/20922 (accessed May 2006).

68. Oswald condemned the experimentalists, whom he referred to as the "sons of science"—with very strong rhetoric—"ye that with ruffian violence interrogate trembling nature, who plunge into her maternal bosom the butcher knife, and, in quest of your nefarious science, the fibres of agonizing animals, delight to scrutinize." John Oswald, *The Cry of Nature*, p. 33 in Garrett, ed., *Animal Rights and Animal Souls, vol. 6.*

69. Andrew Louth, "Taylor, Thomas (1758–1835)," *Oxford Dictionary of National Biography* (Oxford: Oxford University Press, 2004), online ed., http://www.oxforddnb.com/view/article/27086.

70. Porphyry and Taylor are both discussed in the chapter by Stephen R. L. Clark in this volume.

71. Thomas Taylor, *A Vindication of the Rights of Brutes* (London, 1792), p. 103.

72. Singer, "Preface," in Salt, *Animal Rights Considered*, vii.

73. Thanks to Knud Haakonssen, Charles Griswold, Irina Meketa, Carol Gruber, the audience at the Three Arrows Schmooze, and the editors of this volume.

SUGGESTED READING

BODDICE, ROB. *A History of Attitudes and Behaviours toward Animals in Eighteenth- and Nineteenth-Century Britain: Anthropocentrism and the Emergence of Animals.* Lampeter: Edwin Mellen, 2008.

GARRETT, AARON. *Animal Rights and Animal Souls.* 6 vols. Bristol: Thoemmes Press, 2000. (Many of the texts discussed, including Oswald, Primatt, Dean, and Young, are reprinted here.)

———. "Francis Hutcheson and the Origin of Animal Rights." *Journal of the History of Philosophy* 45 (2007): 243–65.

GUERRINI, ANITA. "Animal Experimentation in Seventeenth-Century England." *Journal of the History of Ideas* 50 (1989): 391–407.

———. *Experimenting with Humans and Animals: From Galen to Animal Rights.* Baltimore, Md.: Johns Hopkins University Press, 2003.

HARRISON, BRIAN. "Animals and the State in Nineteenth-Century England." *English Historical Review* 88 (1973): 786–820.

HARRISON, PETER. "The Virtues of Animals in Seventeenth-Century Thought." *Journal of the History of Ideas* 59 (1998): 463–84.

HUTCHESON, FRANCIS. *A System of Moral Philosophy.* Glasgow: A. Foulis, 1755.

JENYNS, SOAME. "On Cruelty to Inferior Animals." In *Disquisitions on Several Subjects.* London, 1782.

KEAN, HILDA. *Animal Rights: Political and Social Change in Britain since 1800.* Chicago: University of Chicago Press, 1998.

LAWRENCE, JOHN. *A Philosophical and Practical Treatise on Horses and on the Moral Duties of Man towards the Brute Creation.* 2 vols. London, 1796/1798.

MAEHLE, ANDREAS-HOLGE. "The Ethical Discourse on Animal Experimentation, 1650–1900." In *Doctors and Ethics: The Earlier Historical Setting of Professional Ethics,* ed. Andrew Wear, Johanna Geyer-Kordesch, and Roger French, pp. 203–51. Amsterdam: Rodopi, 1993.

MANDEVILLE, BERNARD. *Fable of the Bees.* Edited by F. B. Kaye. 2 vols. Indianapolis, Ind.: Liberty Fund, 1988.

MONTAIGNE, MICHEL DE. *Complete Essays.* Translated by M. A. Screech. London: Penguin Books, 1987.

ROSENFIELD, LEONORA. *From Beast-Machine to Man-Machine: Animal Soul in French Letters from Descartes to La Mettrie.* Oxford: Oxford University Press, 1940.

SALT, HENRY S. *Animal Rights Considered in Relation to Social Progress.* Clarks Summit, Pa.: Society for Animal Rights, Inc., 1980.

THOMAS, KEITH. *Man and the Natural World: Changing Attitudes in England, 1500–1800.* New York: Pantheon, 1983.

WOLLOCH, NATHANIEL. "Christiaan Huygens's Attitude toward Animals." *Journal of the History of Ideas* 61 (2000): 415–32.

———. *Subjugated Animals: Animals and Anthropocentrism in Early Modern European Culture.* Amherst, Mass.: Humanity Books, 2006.

PART II

TYPES OF ETHICAL THEORY

CHAPTER 3

INTERACTING WITH ANIMALS: A KANTIAN ACCOUNT

CHRISTINE M. KORSGAARD

1 ANIMALS AND THE NATURAL GOOD

Human beings are animals: phylum: Chordata, class: Mammalia, order: Primates, family: Hominids, species: *Homo sapiens*, subspecies: *Homo sapiens sapiens*. According to current scientific opinion, we evolved approximately two hundred thousand years ago in Africa from ancestors whom we share with the other great apes.[1] What does it mean that we are animals? Scientifically speaking, an animal is essentially a complex, multicellular organism that feeds on other life forms.[2] But what we share with the other animals is not just a definition: it is a history—that is, it is a story—and a resulting set of attributes, and an ecosystem, and a planet.

What is the story? Living things are homeostatic systems—they maintain themselves through a process of nutrition that enables them to work constantly at replacing the fragile materials of which they are composed. Living things also work at reproducing, or contributing to the reproduction of, other living things that maintain themselves in essentially the same way.[3] To engage in those activities—to feed and reproduce—is essentially what it means to be alive. And in order to engage in those activities, a living thing must be, in some way, responsive to conditions in its environment. Plants, for instance, respond to dryness by growing deeper roots, or to sunshine by turning their leaves in its direction. Even a unicellular organism is drawn to some things, and recoils from others, in ways that promote its survival.

But once upon a time—about 600 million years ago—some of the living things on this planet became responsive in a particular way. They began to become *aware* of their surroundings, to form some sort of a representation of the environment in which they live. Presumably, this was because of the evolutionary advantages of such awareness, which enables a living thing to monitor the relationship between its own condition and the conditions in its environment. Perhaps there is no hard and fast line between that distinctive power we call *perception* and the kind of responsiveness exhibited by, say, a plant that turns its leaves toward the sun. But as responsiveness evolved into perception, something new began to appear in the world. A bare theoretical awareness of the environment, all by itself, could not do an organism any good: if perception is to help an organism to survive and reproduce, it must be informed or accompanied by something like motivational states. That is, the organism's awareness must be accompanied by experiences of attraction and aversion that direct its activities in ways that are beneficial to its survival and reproduction. And so the evolution of perception brought with it the capacity for negative and positive experiences—of hunger and thirst, and enjoyment in satisfying them; of pain and pleasure; and of fear and security. And as these organisms themselves became more complex, more complex feelings evolved out of these simpler ones: of interest and of boredom, of misery and delight, of family or group attachment and hostility to outsiders, of individual attachment, of curiosity, and eventually, even, of wonder.

What all of this means is that an organism who is aware of the world also characteristically experiences the world and his own condition in a positive or negative way, that is, as something that is, in various ways, good or bad for himself, or from his own point of view. And so there came to be living beings, homeostatic organic systems, *for whom* things can be good or bad. I will call goodness in this sense the "natural good." It is because there are beings for whom things are naturally good or bad, I believe, that there is such a thing as "good" and "bad" in what I will call the "objective" or "normative" sense—the sense that is morally significant, the sense that gives us reasons.[4] The beings who share this condition are the animals, and you and I are among them. And that gives rise to a moral question. How should we treat the others?

2 HUMAN ATTITUDES TOWARD THE OTHER ANIMALS

I have just suggested that what we share with the other animals—the condition of being beings *for whom* things can be naturally good or bad—is morally significant. Most people would seem to agree, for most people think it is morally wrong to hurt a nonhuman animal for a trivial reason. No one is more readily condemned than someone who kicks a dog out of anger or skins a cat for the sheer malicious fun of inflicting pain. On the other hand, we have traditionally felt free to make use of the other animals for our own purposes, and we have treated any use we may have for them, or any obstacle they present to our ends, as a sufficient reason to harm them. We kill nonhuman animals, and inflict pain on them, because we want to eat them, because we can make useful products out of them, because we can learn from experimenting on them, and because they interfere with agriculture or gardening or in other ways are pests. We also kill them, and inflict pain on them, for sport—in hunting, fishing, cockfighting, dogfighting, and bullfighting. We may even kill them because, having done some sort of useful work for us, they have outlived their usefulness and are now costing us money. Obviously, we think that we ought not to treat our fellow human beings in these ways.

What could make sense of the way we treat the other animals—or, alternatively, what could show that the kinds of actions I have just mentioned are wrong? Since human beings and the other animals share a morally significant attribute, what is the morally relevant difference between human beings and the other animals that is supposed to justify this difference in the way that we treat them? Obviously, not every attribute that people have claimed uniquely singles out human beings could be morally relevant. Many scientists and philosophers would single out language as the most important difference between human beings and the other animals. Faced with the fact that some nonhuman animals have been taught the rudiments of language, these thinkers have sometimes responded that true language requires a complex syntax. But it is not tempting to believe that it is all right to treat the animals as mere means and obstacles to our own ends simply because they lack a complex syntax.[5]

Essentially, there are two ways a difference between human beings and the other animals could be morally relevant: it could be relevant to our thinking about the good or relevant to our thinking about right and obligation. Accordingly, there are two general types of arguments that people have used, to try either to justify or criticize the way we treat the other animals. First, there are arguments based on similarities or differences between the ways in which things can be good or bad for human beings and the ways in which they can be good or bad for the other animals. Second, there are arguments based on the grounds of right and obligation. My main topic in this chapter is an argument of the latter kind: Kant's argument that we cannot have obligations to the other animals, because obligation is grounded in a reciprocal relation among rational beings. I am going to argue not only that Kant's theory can accommodate duties to the other animals, but also that it shows why we do indeed have them. But before I do that, I want to say something about the kind of argument that appeals to similarities and differences between what is good or bad for people and what is good or bad for the other animals.

3 Human and Nonhuman Good

The most effective critics of the way we treat animals to date have been the utilitarians, and their argument is essentially an appeal to the point I started out with: the other animals can experience pleasure and pain, therefore things can be good or bad for them in much the same way they can be good or bad for us.[6] Utilitarians believe, speaking a bit roughly, that the right action is the action that maximizes good results. Since, according to the utilitarians, the business of morality is the maximization of the good, the other animals plainly fall within its orbit.

But appeals to the way in which things can be good or bad for the other animals have also been used to justify some of the more questionable ways in which we treat them. According to the type of argument I have in mind, there are differences between the character of human experience and the character of the experiences of the other animals that justify us, at least sometimes, in putting our own interests first. The most extreme view along these lines is the Cartesian view that the other animals have no conscious experiences, so that nothing can really be good or bad for them in a morally relevant way. But some non-Cartesians hold a view that seems not far behind this. They believe that because the other animals (as they claim) lack any sense of their existence as extended in time, all that their consciousness *can be* is a series of discrete, disconnected experiences, which can be pleasant or painful, or perhaps frightening or comforting, but only in a local way. Such experiences could not be connected, in the way they are in us, by memory and anticipation, to long-term hopes or fears, or to any concern for one's own ongoing life. On this basis, some people have suggested that although we do have reason not to hurt the other animals, there is no special reason not to kill them when that suits our convenience and can be done without inducing fear or pain.

Somewhat surprisingly, in my view, some of the utilitarians who have been such powerful champions of animal rights hold a view of this kind. In his commentary on J. M. Coetzee's *The Lives of Animals*, Peter Singer, for example, voices the common view that the fact that human beings anticipate and plan for the future means that human beings have "more to lose" by death than the other animals do.[7] Singer imagines an interlocutor—his daughter Naomi—protesting that death for a non-human animal—her example is their dog Max—would mean the loss of everything *for that animal.* And Singer replies that although there would be no more good experiences for Max, they could arrange for the breeding of another dog, and then this other dog could be having good canine experiences in Max's place. In other words, what matters is not the goodness of Max's experiences *for* Max, but just that there be some good canine experiences going on in the world somewhere.

The trouble with this argument is that it depends on a more general utilitarian assumption, which has nothing special to do with the nature of nonhuman consciousness. Utilitarians regard the subjects of experience in general essentially as *locations* where pleasure and pain, that is, good and bad experiences, *happen*, rather than as beings *for whom* these experiences are good or bad.[8] To put it another way, they think that the goodness or badness of an experience rests wholly in the character of the experience, and not in the way the experience is related to the nature of the subject; so it is not essential to the goodness or badness of the experience that it is good or bad for the subject who has it.[9] This view of the relationship between subjects and the value of their experiences is essential to utilitarianism, because it is what makes it possible to think that you can accumulate value by adding pleasures and pains across the boundaries between different subjects of experience. If the badness of pain is, as I will put it, *tethered* to its badness *for* the subject who experiences it, the badness cannot coherently be added or subtracted across the boundaries between subjects in that way. For such "aggregation," as philosophers call it nowadays, requires cutting the tether.

As I said before, the view that a subject's relationship to her experiences is essentially one of *location* is a quite general feature of utilitarianism and doesn't have anything special to do with the nature of nonhuman consciousness. And this makes me wonder why Singer thinks that the fact that we have hopes and plans for the future makes death worse *for us*—or perhaps why he thinks it matters morally if it does. In an earlier paper, "Killing Humans and Killing Animals," Singer argues that because we are self-conscious, and aware of our lives, we are not replaceable in the way the other animals are. Each of us has a desire to live, which will not be fulfilled if we are killed.[10] But self-conscious experiences of memory and anticipation are in themselves just more experiences. If a person is just a place where these experiences happen, then we can always replace one human being who experiences, say, satisfaction at the thought that his plans have worked out, or worry about the fate of a loved one, with another human being whose experiences have a similar content. And a person whose desire to live is not fulfilled may be replaced with a person who will develop a desire to live that then will be fulfilled, at least for as long as he lives. In order to make the argument that there is a disanalogy between the death of a human being and the death of another animal, Singer would have to argue that because we human beings experience memory and anticipation, and have a desire for life, death can be good or bad *for us* in a way it cannot for a less self-conscious animal. And perhaps such an argument could be made. But in order for it to be made, Singer would have to grant that things can be good or bad for us in a way that goes beyond our being the mere location of good or bad experiences. And in that case, it also seems to me that Singer would have to give up utilitarianism, at least as applied to human beings.

Utilitarians, and consequentialists more generally, believe that the way to determine what is right is by adding up the goods and harms done by an action, and choosing the action that does the most good. So if death is worse for a human being, they think, human loss of life figures more largely in the calculus. And although the utilitarians themselves don't do this, we can imagine someone trying to generalize this argument to show that human goods and bads are always *so* much more significant than those of the other animals that human interests should always outweigh the interests of the other animals. Making things good for humans, someone might suppose, is then the way of doing the most good. But what I am suggesting here is that there is a conceptual problem with the idea of what "does the most good." If it seems plausible that everything that is good or bad is so in virtue of being good or bad for someone (some person or animal), then it is also plausible that the goodness or badness of experiences—or of anything else for that matter—is tethered to the subjects *for whom* they are good or bad. In that case, it may be that the goods of different subjects can't be added at all: what's good for me plus what's good for you isn't *better*, because there is no one *for whom* it is better.

The position I have just voiced is controversial, because it blocks all forms of aggregation.[11] And we do have intuitions that support aggregation. For instance, many people believe that if you can save either two lives or only one, you should save the two. And many people would agree that if we have only one dose of a pain-killer, and no one has a particular claim on it, we should give it to the person who is suffering the most. Of course, it is an open question whether the reason we should make these choices is because that way we "do more good," but that is a very natural thought.[12] So it is worth noting that it may still be intelligible, consistently with the idea that good and bad are always tethered to subjects, to claim that we "do more good" by choosing a course of action that benefits more different subjects, *so long as no one is harmed* by that course of action (in economists' jargon, by doing what is Pareto optimal). And it may even be intelligible to claim that we "do the most good" by giving a resource to the party who will benefit from it the most, *so long as we are not thereby harming the other parties among whom we are choosing.*[13] What most obviously becomes unintelligible on the view that good and bad are tethered to subjects is the idea that we can "do more good" by balancing the good of one subject against the good of another subject, say by taking pleasure away from Jack because that way we can give an even greater pleasure to Jill. That is good for Jill but bad for Jack, and if the goodness or badness must be tethered to a subject, that is all there is to say: there is no third party for whom the situation is better overall, and therefore no sense in which it is better.

These ideas, if they are right, may explain some of our intuitions about aggregation, in a way that is consistent with the idea that goods and bads are tethered to those for whom they are goods and bads. But if the intelligibility of the claim that an action does more good depends on the rider that it does no harm, then the intelligibility of such claims depends on where the parties concerned start from and what they have to lose. And this matters. Supposing it is true that human beings have "more to lose" by death than the other animals, we might "do more good" by saving a starving human being than we do by saving another starving animal, and so we might choose to do the former. But if we view things this way it is a quite different kind of question whether we should *kill* the animal to save the starving human being, for now there is someone to be harmed. I am not necessarily saying that we shouldn't, only that if we should, it is not simply because that is what does the most good. And it is a different question *altogether* whether we are justified in harming and killing animals in great numbers, either for food or in experiments, simply so that human beings can have a greater span of our supposedly "more valuable" lives.[14]

If goodness and badness are, as I have claimed, tethered to the subjects for whom things are good and bad, then we cannot be utilitarians, and we cannot generally weigh the interests of the other animals against the interests of human beings. We need another way of thinking about how we should treat them.

4 A KANTIAN APPROACH TO OUR RELATIONSHIPS WITH THE OTHER ANIMALS

Kant's work may seem an unpromising place to turn for help in thinking about our relations with the other animals, for in the philosophical literature on this topic, Kant is often cast as the villain of the piece. At the center of Kant's ethics is his "Formula of Humanity," the requirement that we should treat every human being as an "end in itself" who is never to be used as a mere means to another person's ends. The idea has found its way into our moral culture: "You are just using me!" is one of our most familiar forms of moral protest. But Kant did not only assert that all human beings should be treated as ends in themselves: he is also one of the few philosophers ever to have said bluntly that the other animals are mere means, and may therefore be used for human purposes. Before I examine his argument, I want to say a little about why, if Kant is wrong about this, a Kantian approach can throw important light on questions about how we should treat the other animals.

There are many ways in which people have characterized the essential difference between utilitarian or "consequentialist" and Kantian or "deontological" approaches to ethics. Often these characterizations proceed from the consequentialist point of view. For example, people sometimes say that consequentialists think it is *always* right to do what maximizes the good, while Kantians and other deontologists think (perversely, it is implied) that sometimes we should not do that, because there are "side-constraints" on the promotion of the good. Less polemically, people sometimes say that deontologists think that some actions are intrinsically right, apart from their consequences, and that this kind of value has priority over the good. But there is another way of thinking of the difference between these two kinds of theories that I think is deeper and goes more to the heart of the matter. This way of thinking about the difference is made available when we reflect on the practical implications of Kant's principle that we should treat human beings as ends in themselves.

Kant takes it to follow from this principle that you should never treat another human being as a mere means to your own ends, nor should you allow yourself to be treated that way. He thinks that the value of humanity requires us to avoid all use of force, manipulation, and coercion, because we must respect the rational choices of others and their free use of their own power of rational choice. He takes it to follow that all human interaction, as far as possible, should be on terms of voluntary cooperation, and aimed at ends that can be shared by all concerned. And finally, he takes it to follow that we should help each other when we are in need and promote each other's chosen ends when we can easily do so. For if you take it that your own chosen ends are worthy of pursuit, you must take the ends of others to be so as well.[15]

The fact that this formula expresses a *basic* requirement shows that consequentialists and Kantians have a different view about what the *subject matter* of ethics is. Consequentialists take the subject matter of ethics to be the results produced by our actions, and take the main questions of ethics to be things like: "What results should we aim to bring about? What should we make happen? How can we make the world the best possible place?" Kantians on the other hand take the subject matter of ethics to be the quality of our relationships and interactions, both with ourselves and with each other. So Kantians take the main questions of ethics to be things like: "How should I treat this person? What do I owe to him, and to myself, in this matter? How can I relate to him properly? What should our interactions be like?"[16]

Of course, I am not saying that either view ignores the other view's main questions, but the order of dependence is different. It is a notorious fact, much discussed in the critical literature, that consequentialists try to derive what we might call the values of relationship and interaction from considerations about what does the most good. If you should be just and honest and upright in your dealings with others, according to the consequentialist, that is because that is what does the most good. If you are allowed to be partial to your own friends and family, and not required always to measure their interests against the good of the whole, that is because it turns out, in some oblique way, that people maximize the good of the whole more efficiently by attending to the welfare of their own friends and family. It is less often noticed, but just as true, that in a Kantian theory, the value of producing the good is derived from the values of interaction and relationship. The reason that pursuing the good of others is a duty at all in Kant's theory is that it is a mark of respect for the humanity in another that you help him out when he is in need, and more generally that you help him to promote his own chosen ends when you are in a position to do that. This is why it is a serious mistake to characterize Kantian deontology as accepting a "side-constraint" on the promotion of the good. Kant does not believe there is some general duty to maximize or even promote the good that is then limited by certain deontological restrictions. Rather, he believes that promoting the good of another and treating her justly and honestly are two aspects of respecting her as an end in herself.

So when we turn our attention to our ethical relationships with the other animals, this means that the focus of our questions will be rather different than it is in a consequentialist account. If we are to treat animals as ends in themselves, then the relevant question is not whether human interests outweigh the other animals' interests, because of the special kinds of goods and evils to which we are subject, or conversely, whether the other animals' interests, because of their sheer number or intensity or gravity, sometimes do outweigh human interests after all. The relevant question is rather what, given their nature and ours, *each* of us can do in order to be related as well as possible to *each* of them.[17] These questions, as I will argue later, point us toward duties to the other animals that, although in some ways are more demanding, in other ways are more tractable, than the duties to which utilitarian accounts naturally point us.

But before we can discuss what those duties are, we must ask whether the other animals are indeed to be treated as ends in themselves. So let us turn now to Kant's own views.

5 Kant's Views on the Treatment of the Other Animals[18]

The views about the other animals that have made Kant notorious find their most famous expression in the *Groundwork of the Metaphysics of Morals* in his argument for the Formula of Humanity. In that argument, Kant first establishes that if there is a categorical imperative—that is, a principle of reason that prescribes duties with categorical force—there must also be, as he says, "something the existence of which in itself has an absolute worth, something which as an end in itself could be a ground of determinate laws."[19] He then proceeds to consider various candidates for the end in itself, and in the course of the discussion he says:

> Beings the existence of which rests not on our will but on nature, if they are beings without reason, still have only a relative worth, as means, and are therefore called *things*, whereas rational beings are called *persons* because their nature already marks them out as an end in itself, that is, as something that may not be used merely as a means.[20]

Following the tradition of Roman law, Kant divides his metaphysical world into two categories that are supposed to be exhaustive, persons and things. And if there were any doubt about whether Kant intends here to include nonhuman animals in the category of "things," those doubts are dispelled by things he says elsewhere. In his essay, "Conjectures on the Beginnings of Human History," a speculative account of the origin of reason in human beings, Kant explicitly links the moment when human beings first realized that we must treat one another as ends in ourselves with the moment when we realized that we do not have to treat the other animals in that way. Kant tells us that:

> When [the human being] first said to the sheep, "the pelt which you wear was given to you by nature not for your own use, but for mine" and took it from the sheep to wear it himself, he became aware of a prerogative which, by his nature, he enjoyed over all the animals; and he now no longer regarded them as fellow creatures, but as means and instruments to be used at will for the attainment of whatever ends he pleased.[21]

But actually, in spite of this remark, Kant himself did not think that it is morally permissible to treat the other animals in whatever way we please. Kant's own ethical views are reported in the records of his course lectures and in his book *The Metaphysics of Morals*. Kant does think we have the right to kill the other animals, but it must be quickly and without pain, and cannot be for the sake of mere sport. He does not say why we should kill them, and the subject of eating them does not come up in his discussion, but presumably that is one of the reasons he has in mind. He does not think we should perform painful experiments on nonhuman animals "for the sake of mere speculation, when the end could also be achieved without these."[22] He thinks we may make the other animals work, but not in a way that strains their capacities. The limitation he mentions sounds vaguely as if it were drawn from the Golden Rule: we should only force them to do such work as we must do ourselves.[23] And if they do work for us, he thinks that we should be grateful.

In his course lectures, Kant sometimes told a story about the philosopher Leibniz carefully returning a grub he had been studying to the tree from which he had taken it when he was done, "lest he should be guilty of doing any harm to it."[24] And both in his lectures and in *The Metaphysics of Morals*, Kant has hard words for people who shoot their horses or dogs when they are no longer useful.[25] Such animals should be treated, Kant insists, with "gratitude for…long service (just as if they were members of the household)."[26] He remarks with apparent approval that "[i]n Athens it was punishable to let an aged work-horse starve."[27] And he tells us that "[a]ny action whereby we may torment animals, or let them suffer distress, or otherwise treat them without love, is demeaning to ourselves."[28]

Kant says little about how he derives these duties from principles. But it is worth noting that the duties Kant mentions here are not just duties to be kind or avoid cruelty. Rather, they concern the ways in which we interact with the other animals, and the standards they set are as much standards of reciprocity as they are of kindness. We can make the other animals work, but no harder than we would expect of ourselves, and we should be grateful for their services when we do make them work. If they have worked for us all their lives, we should compensate them by a comfortable retirement. Leibniz, we may be sure, did not go through the world making general efforts on behalf of the welfare of grubs, but he did want to make sure that the grub with whom *he* interacted was not harmed by the transaction. Kant plainly thought it is necessary to kill or hurt animals for some reasons, and we might disagree with him about what exactly those reasons are. But he apparently thought that otherwise we should interact with the other animals, so far as it is possible, as we would interact with human beings, on terms of reciprocity and mutual benefit.

But as the last phrase I quoted suggests, Kant thinks that these moral duties are not owed *to* the other animals, but rather to ourselves: treating animals without love is "demeaning to ourselves." He says:

> violent and cruel treatment of animals is…intimately opposed to a human being's duty to himself…for it dulls his shared feeling of their suffering and so weakens and gradually uproots a natural disposition that is very serviceable to morality in one's relations with other people.[29]

In his course lectures, Kant made the same point by saying that nonhuman animals are "analogues" of humanity, and that we therefore "cultivate our duties to humanity" when we practice duties to animals as "analogues" of human beings.[30]

But if the other animals are indeed analogues of human beings, why don't we have obligations to them?

6 THE HUMAN DIFFERENCE

Earlier, I noted that anyone who thinks we do not have duties to the other animals must think there is a morally relevant difference between us and them. Kant thought that difference is that we are rational and therefore moral animals, and the other animals are not. These days, many philosophers and scientists argue that we can discern the roots of morality in the tendencies to altruism and cooperation found among the other social animals, just as we can discern the roots of language in their communication systems or the roots of technology in their manufacture of simple tools. Similarly, some philosophers and scientists have argued that the more intelligent animals exhibit a kind of rationality when they figure out how to solve problems. Are these then after all matters of degree and not of kind? To understand why Kant would reject that conclusion, we must understand what he means by "reason."

It is sometimes said that human beings are the only animals who are self-conscious. Animals are aware of the world but not of themselves. But actually the issue is more complicated than that, for self-consciousness, like other biological attributes, comes in degrees and takes many different forms. One form of self-consciousness is revealed by the famous mirror test used in animal studies. In the mirror test, a scientist paints, say, a red spot on an animal's body and then puts her in front of a mirror. Given certain experimental controls, if the animal eventually reaches for the spot and tries to rub it off, or looks away from the mirror toward that location on her body, we can take that as evidence that the animal recognizes herself in the mirror and is curious about what has happened to her. Apes, dolphins, and elephants have passed the mirror test, in some cases moving on to use the mirror to examine parts of their bodies that they can't normally see—apparently with great interest. Other animals never recognize themselves, and instead keep offering to fight with the image in the mirror, or to engage in some other form of social behavior with it. An animal that passes the mirror test seems to recognize the animal in the mirror as "me" and therefore, it is thought, must have a concept of "me." I will come back to the mirror test later on.

In any case, I think it can be argued that some animals who can't pass the mirror test have rudimentary forms of self-consciousness. You have self-consciousness if you know that one of the things in your world is *you*. A tiger who stands downwind of her intended prey is not merely aware of her prey—she is also locating *herself* with respect to her prey in physical space, and that suggests a rudimentary form of self-consciousness. A social animal who makes gestures of submission when a more dominant animal enters the scene is locating himself in social space, and that too suggests a kind of self-consciousness. So perhaps does a domestic animal's rivalry with other domestic animals for human attention. ("Don't play with her. Play with *me!*") Knowing how you are related to others involves something more than simply knowing about them.

Parallel to these abilities would be a capacity to locate yourself in mental space, to locate yourself with respect to your own experiences, thoughts, emotions, beliefs, and desires, and in particular, to know them as your own. This is what we more commonly think of as self-consciousness. Do the other animals have this ability to locate themselves in subjective, mental space? Scientists have sometimes taken the mirror test to establish this kind of self-consciousness, but it is a little difficult to articulate why. The animal grasps the relation between the image in the mirror and her own body, and in so doing, she seems to show that she grasps the relationship between herself and her own body. But what exactly does that mean? She grasps the relationship between two things, a certain physical body and—well, what? We can say "and herself," but what exactly is the "herself" that she identifies with that body?

Perhaps the idea is that what she identifies as *herself* is the self that is the subject of her own experiences, of which she must then have some awareness. That is, she must be aware not just of pain but that *she* feels pain, or not just of the smell of food but that *she* smells the food. And it is that "she," the subject of those experiences, which she correctly identifies with the body she sees in the mirror. Some such idea must be behind the thought that the mirror test reveals an inner self-consciousness.

However, even if this is right, it does not yet seem to show that the animal must be aware of herself as the subject of her *attitudes*—that is, of her beliefs, emotions, and desires. And this suggests a further division within this form of self-consciousness. An animal might be aware of her experiences and of herself as the subject of those experiences, and yet her attitudes might be invisible to her, because they are a lens *through* which she sees the world, rather than being parts of the world that she sees.[31] In fact, it seems likely that the way an animal's instincts function is by providing exactly that sort of lens.[32] As I said earlier, a bare theoretical awareness of the world would not do an animal—especially an intellectually primitive animal—any good unless it were accompanied by appropriate motivational states. So we may suppose that an animal instinctively perceives things *as* aversive or attractive in particular ways—*as* food, that is, as appetizing, or *as* threat, that is, as frightening— without being aware that it is a fact about herself that *she* is hungry or frightened. You don't need to know of yourself that you are hungry in order to respond to food correctly: you only need to perceive it as appetizing, as food.

Of course, more intelligent animals *might* also be aware of their own attitudes. Some of the language-trained animals seem able to express the idea "I want"—Koko the gorilla and Alex the African gray parrot, two famous language-trained animals, could both do this—so perhaps they have the ability to think about their own attitudes as well as about their experiences.[33] But however that may be, we human beings are certainly aware of our attitudes. We know of ourselves that we want certain things, fear certain things, love certain things, believe certain things, and so on. And we are also aware of something else—we are aware of the potential influence of our attitudes on what we decide to do. We are aware of the *grounds* of our actions. What I mean is this: a nonhuman animal may perceive something in his environment as, say, frightening. And that may induce him to run away. We can say that his fear, or his perception of the object as frightening, is the ground of his action—it is what causes him to run. We can even say, by analogy with our own case, that it is his *reason* for running, although he does not know that about himself. But once you are aware of the influence of a potential ground of action, as we human beings are, you are in a position to decide whether to allow yourself to be influenced in that way or not. As I have put it elsewhere, you now have a certain reflective distance from the impulse that is influencing you, and you are in a position to ask yourself, "but *should* I be influenced in that way?" You are now in a position to raise a *normative* question, a question about whether the action you find yourself inclined to perform is *justified*.[34] Kant, of course, held a particular view about how you answer such a question. You ask whether the principle of acting in the way you are considering could serve as a universal law, whether you yourself can will that everyone should act in that way.[35] You ask the question posed by the categorical imperative.

I believe that this form of self-consciousness—consciousness of the grounds of our beliefs and actions—is the source of reason, a capacity that I think is distinct from intelligence. Intelligence is the ability to learn about the world, to learn from experience, to make new connections of cause and effect, and to put that knowledge to work in pursuing your ends. Animals who solve problems do exhibit intelligence, but reason is not the same as intelligence. Intelligence looks outward, to the connection of cause and effect. Reason looks inward, and focuses on the connections between our own mental states and attitudes and the effects that they tend to have on us. It asks whether our actions are *justified* by our motives or our inferences are *justified* by our beliefs. I think we could say things about the beliefs of intelligent nonhuman animals that parallel what I have said about their actions. Nonhuman animals may have beliefs and may arrive at those beliefs under the influence of evidence; by analogy with our own case, we may say that they have reasons for their beliefs. But it is a further step to be the sort of animal who can ask yourself whether the evidence really justifies the belief, and can adjust your conclusions accordingly.

If this is correct, the difference between human beings and the other animals is not that we are self-conscious and they are not. It is, as it were, both smaller and bigger than that. Human beings have a particular form or type of self-consciousness: consciousness of the grounds of our beliefs and actions. But that little difference makes a very big difference. For it means that human beings are both capable of, and subject to, *normative self-government*, the ability to direct our beliefs and actions in accordance with rational norms. And normative self-government, according to Kant, is the essence of morality. Morality does not rest simply in being altruistic or cooperative, although it certainly does demand those things. It rests in being altruistic or cooperative or honest or fair or respectful *because you think you should be*: because, that is, you yourself would will that everyone should act in those ways. To be capable of normative self-government is to be in Kant's language "autonomous"—capable of governing yourself in accordance with the laws you make for yourself. And as far as we know, although it is an empirical question, no other animal does that.[36] If that is so, human beings are rational and moral animals, and the other animals are not.[37]

Of course, what most obviously follows from that is not that we have no duties to the other animals. Rather, what most obviously follows is that they have no duties to *us*. So now we must ask why Kant thought that their lack of moral agency disqualifies the other animals from being regarded as ends in themselves.

7 THE RECIPROCITY ARGUMENT

Kant is one of the main proponents of a kind of argument that purports to show that we cannot have obligations to the other animals at all. This kind of argument is not grounded in the nature of an animal's experiences or in some supposed difference between the human good and that of the other animals, but rather in the grounds of obligation. The argument comes in various forms, but the basic idea is that morality is a system of reciprocal relationships—a system in which human beings mutually impose obligations on each other, or at least one in which human beings have reciprocal rights and obligations. I have rights as against you insofar as you have obligations to me, and vice versa. But the other animals, because they are not moral beings, cannot have obligations to us, and therefore cannot participate in the system. They are out of the scope of morality. I will call this type of argument a "reciprocity argument."[38]

Kant, as we have seen, thinks we have duties to treat the other animals with compassion and even with a certain kind of reciprocity. But he does not think that we owe this *to* them. What exactly does it mean to be obligated *to* someone else? Ordinarily, we think you are obligated to someone when you would wrong *her* by not acting in the way you are obligated to, and therefore she can claim your acting that way as her right. But this can happen only under certain conditions. According to one view, versions of which have been put forward by Stephen Darwall and Michael Thompson in recent work, in order for me to owe a certain kind of treatment *to* you, it is not enough that I am under the authority of a law saying that I should treat you in a certain way.[39] Perhaps, for instance, I am under the authority of a law saying that I should not deface beautiful paintings, but I do not owe that *to* the paintings. Presumably, I owe it to those whose aesthetic heritage includes the paintings. Nor is it enough to add that you are the kind of creature to whom things *can* be owed. Rather, for me to be obligated to you, we must both be under the authority of the same laws, in the name of which we can make claims on each other.

To see why, suppose that I am a Christian, and the Bible says I should be kind to all people; you are a Muslim, and the Koran says the same, and each of us believes that our duties are somehow grounded in our faiths. Then I am obligated to treat you with kindness and you are obligated to treat me the same way. But do we owe this *to* each other? I cannot sincerely claim kindness from you as my right in the name of the laws of the Koran, since I do not concede any authority to those laws; and your position with respect to me is the same. So it may seem as if I owe it to myself and perhaps to my God that I should treat you with kindness, but I do not owe it *to* you; and you are in the same position with respect to me. In this way, we arrive at the idea that for one person to owe something to another, in the sense that makes it claimable by that other as a matter of right, they must conceive themselves as being under shared laws grounded in some authority acknowledged by both, be it political, moral, or religious.

There are at least two possible objections we could make to this argument as it stands. On the one hand, we might object that the argument does not establish that it is *necessary* for us to be under *exactly* the same laws in order to have duties to each other. For we still have something in common: we are moral beings, beings who recognize the authority of laws that concern the ways you are supposed to treat people, and who are capable of acting in accordance with that recognition. And as such we might reasonably expect treatment in accordance with those laws from each other. So for instance if I am visiting a country with a culture in which it is thought that the laws of hospitality require one to take in strangers for the night,

I might reasonably resent one of the inhabitants' refusal to treat me that way, even though I do not believe that the laws of hospitality require this. Perhaps it would be a little odd for me to demand it as my right, yet my resentment expresses the sense that I have been wronged. The thought here would be something like, "I treat you according to my standards of respect; you should treat me according to yours. We owe it to each other to treat each other in the ways we ourselves believe that people should be treated."

On the other hand, we might object that this argument does not show that it is *sufficient* for obligation to each other that we are under the authority of the same laws. If the authority of those laws rests in divine command, say, perhaps it is really only God to whom we owe our obedience. I am obligated to you only if you are the authority that stands behind the law, that is, only if I acknowledge the authority of your will over mine in this matter.

I mention these objections because reflecting on them can help us to see the force of Kant's own, somewhat more demanding, version of the argument. Although they seem to bear in opposite directions, both objections remind us of the connection that Kant makes between morality and autonomy. We have autonomy when the laws we are under are laws that we make for ourselves. Kant believed that the reason we are under *moral* laws is precisely because we are autonomous—or as I put it earlier, normatively self-governed—beings. Moral laws are exactly those laws that we ourselves would will for everyone to act on, and their authority for us springs from that fact—and so from our own will. The first objection suggests that the common law that all of us are under just insofar as we are moral beings is that of treating people in whatever way *we ourselves* think that people should be treated. In a sense, it suggests that the common law we are under is the categorical imperative itself, the law of acting according to the principles we ourselves will as laws. The second objection insists that the authority behind that common law must be the wills of those who make it, if they are to be obligated *to* each other, and it reminds us of the idea, familiar from political philosophy, that the authority of a law must ultimately come from the will of the people whom it governs. Putting these points together, we arrive at the idea that for one person to make a claim on another, they must be under common laws that spring from their own shared authority: laws that they make—that they autonomously will—together. And they suggest that insofar as we are moral beings, we are in a community governed by laws of that kind.[40]

These ideas give us a way of understanding the role, in Kant's moral philosophy, of what he called the "Kingdom of Ends"—a community of ends in themselves. In order to understand how this community arises and what it involves, we must back up and look more closely at the idea that people are ends in themselves.

When Kant introduces the Formula of Humanity, he argues that it is a "subjective principle of human actions" that each of us regards himself or herself as an end in itself.[41] A rational being, he says, "necessarily represents his own existence in this way."[42] What does Kant mean by that? Kant believes that insofar as we are rational, we will pursue an end only if we take it to be what I earlier called "objectively" or "normatively" good—something that there is *reason* to pursue. But in fact the things that we are motivated to pursue are the things that are, in the sense I set out at the beginning of this chapter, naturally good, *good for ourselves*. That doesn't mean that our pursuits are self-interested; rather, it means that we tend to pursue those things that we are naturally inclined to respond to and evaluate positively: our own lives, health, and happiness, and the lives, health, and happiness of those whom we love; freedom from pain and suffering; the exercise of our natural faculties; the satisfaction of our natural curiosity; and so on. Our standing as ends in ourselves is a "subjective principle" of our actions, because we choose to pursue the things that we judge to be, in this sense, naturally good for ourselves, and it is only rational to choose to pursue things that are objectively, normatively good. So when we act, we *take* our natural good to be an objective good. It is as if each of us said to herself, "I take the things that matter to me to be important, because I take myself to be important." When we take our own concerns to be important and worth doing something about, we take ourselves to be capable of conferring objective value on our ends through rational choice.[43]

There is an ambiguity in what I have just said, however. One might understand "taking" ourselves to be of value to mean that we *recognize* that, as a matter of metaphysical fact, we are valuable. But Kant did not believe that human beings have knowledge of metaphysical matters that are beyond the reach of empirical science. Rather, I think we should see "taking" ourselves to be important as a kind of original normative *act* that brings objective value into the world. Kant pictures valuing as an act of legislation: you make it a law for yourself and everyone else that what is naturally good for you should be taken to be objectively good. I don't mean you make it a law that every other person should find the same things to be good for him as you find to be good for you, of course, but rather that you make it a law that every other person must regard it as a good end—and so as a source of reasons—that you should achieve what is naturally good for you.[44] The idea of a Kingdom of Ends arises when we realize that, as Kant puts it, "every other rational being represents his existence in this same way on just the same rational ground that also holds for me."[45] Each of us asserts her standing as an end in itself for herself and all others, so together we form a Kingdom of Ends in which we legislate moral laws together. And in that way we become obligated to one another.

Kant thinks we have no obligations to the other animals because they are not members of the Kingdom of Ends. But Kant does not exclude the other animals from the Kingdom of Ends because he does not regard them as ends in themselves. Instead, it goes the other way: he thinks they are not ends in themselves because they are excluded from the Kingdom of Ends. He says: "Morality is the condition under which alone a rational being can be an end in itself, since only through this is it possible to be a lawgiving member in the Kingdom of Ends."[46] So the other animals are excluded from the Kingdom of Ends because they cannot make laws for themselves and cannot participate in reciprocal legislation—because they are not rational, normatively self-governing beings.

To summarize the argument: Kant supposes that as moral beings we are under a common law, the law of treating people as we ourselves think that people should be treated, or more technically, the categorical imperative: to act only on a maxim that we ourselves can will as universal law. And he thinks that the authority of that law, and the laws that follow from it, springs from our own wills. These facts place us in a community in which each of us has the right to claim morally good treatment from every other, and that is why we can be obligated to each other. But the other animals are not moral, therefore they cannot be "lawmaking members of the Kingdom of Ends," and therefore they cannot place us under obligations in the name of its laws. Only rational and moral beings, Kant thinks, can do this, and so obligations and duties can be owed only *to* rational and moral beings.[47]

8 Assessing the Reciprocity Argument

According to Kant's argument, we can make normative claims on each other, and so have obligations to each other, only if we conceive ourselves as bound by common laws that we make together. The question is whether that is really the *only* way in which one being can have an obligation to another. And that doesn't seem to be the case. A perfectly respectable sense of "obligated to" might be given by the reasons for making the law, by whom or what it is meant to protect: I am obligated to you if you are the source of the interests that the law I am under was made to protect.[48] Indeed, Joel Feinberg, in his work on ethics and animals, takes it for granted that this is what we *mean* when we say we are *obligated to* someone (some person or animal): some law protects his interests by giving him a right, and as the right-holder he makes a claim on us. If we make a civil law against homicide in order to protect people, assigning people the right not to be killed, then violating that law wrongs *people*, and we owe it to people not to kill them—not just to the citizens whose authority stands behind the law. So our obligation, say, not to murder foreigners on our soil is owed to the foreigners themselves.[49] So Feinberg argues that laws made for the protection of the interests of nonhuman animals quite straightforwardly makes it possible to be obligated to them.

But it also looks as if there is a difficulty in using this concept of "owed to" to define a sense in which we can be *morally* obligated to the other animals. The argument I just sketched depends on our having reasons for or against the laws we make—interests we have decided are worth protecting. In the case of positive or civil laws, those reasons are perhaps to be derived from morality itself. But moral laws themselves are not laws made for instrumental reasons, or to protect interests whose value has been independently determined. Moral laws are the *fundamental* laws in terms of which we identify which interests are worth promoting and protecting.

But I think that Kant's theory has the resources for addressing this difficulty. According to Kant's theory, as I have argued, there is a Kingdom of Ends because each of us claims a kind of standing, as an end in himself or herself, and that standing gives us duties to ourselves and each other. As we've seen, Kant's story about that original claim is that it is built into the very nature of rational action; there is a way in which it is almost impossible not to make it. As rational beings, we need to have reasons for what we do, and we find those reasons in the things that are naturally good for us. So we almost inevitably treat what is naturally good for us as normatively and objectively good. In a sense, just by the act of making a rational choice, you confer normative value not only on the end that you choose, but also on yourself.[50]

But as that formulation shows, the self that confers the value and the self on whom it is conferred are not precisely the same. The self that confers value is your autonomous rational self, the chooser as such, the lawmaking self. But the self on whom the value is conferred is not just the autonomous rational chooser; it is rather the self whose interests are in question, the self for whom, or from whose point of view, things can be naturally good or bad. For the way that you assert your standing as an end in itself *for yourself* is by legislating that what is naturally good for you is to be counted as an objective or normative good. And the self for whom things can be naturally good or bad is not merely your rational self. It is also, or rather it is, your animal self.

There are two ways to make this argument, one weaker and one stronger. The weaker way is to observe that *among* the interests we protect through our own moral legislation are interests that spring from our animal nature, not from our rational nature. Our objection to pain and suffering is an obvious example. You will a law, say, against being made to endure suffering as a mere means to the ends of someone else. This is not just because respect for your own autonomy demands that *you* should be allowed to decide for which ends you are prepared to suffer. It is also because you object to suffering. After all, when you decide for yourself that some end is not worth suffering for, your own decision cannot be based on respect for your own rational choice, for at that point you have not yet made any choice. Instead, it has to be based on the fact that you take your suffering, which is a bad thing for you, to be a bad thing objectively, and bad enough not to be endured for the sake of the end in question. So although being rational autonomous beings—moral lawmakers—is what enables us to *make* normative claims on ourselves and each other, the content of those claims cannot be given completely by respect for rational autonomy itself. That view necessarily leaves out the relation in which you stand to yourself, and the way that it governs your own choices. To respect yourself is to take your natural good to be objectively and normatively good. Although what is naturally good for a human being is different in some respects from what is naturally good for another animal, there are also common elements. Part of what we confer value on when we respect ourselves is certain interests that we have, not as rational beings, but simply as sentient ones, such as the interest in avoiding suffering. Those interests are ones we share with the other animals.[51]

The stronger way to make the argument is just to say that because the original act of self-respect involves a decision to treat what is naturally good or bad for you as something good or bad objectively and normatively, the self on whom value is conferred is the self for whom things can be naturally good or bad. And the self for whom things can be naturally good or bad is your animal self: that is the morally significant thing we have in common with the other animals. It is on ourselves *as possessors of a natural good*, that is, on our animal selves, that we confer value.[52] Since our legislation is universal, and confers value on animal nature, it follows that we will that all animals are to be treated as ends in themselves.

According to the argument I have just presented, the sense in which we owe duties to the other animals *is* slightly different from the sense in which we owe duties to other human beings. If obligation to another is understood as the acknowledgment of a claim under the authority of laws we make together, we are not obligated to the other animals. The other animals do not make claims on us in the name of common laws that we will together. Rather, we see them as falling under the protection of our laws, and we make claims on ourselves on their behalf. To that extent, Kant was right to think that our duties to the other animals arise by way of our duties to ourselves. But if obligation to another is understood as the acknowledgment that that other has a claim under laws whose authority we recognize, because they spring from our own will, then Kant was wrong to think that it follows that these duties are not owed to the other animals. For the act of taking ourselves to be valuable that brings the moral world, the Kingdom of Ends, into existence, and our acknowledgment of the claims of the other animals, are both responses to the same thing. They are responses to the predicament of being a being *for whom* things can be naturally good or bad.

As animals, we are beings for whom things can be good or bad: that is just a natural fact. When we demand to be treated as ends in ourselves, we confer normative significance on that fact. We legislate that the things that are good or bad for beings for whom things can be good or bad—that is, for animals—should be treated as *good or bad objectively and normatively*. In other words, we legislate that animals are to be treated as ends in themselves. And that is why we have duties to the other animals.

9 INTERACTING WITH ANIMALS

But we might think that there is still a problem. Is it even possible to treat a nonhuman animal as an end in herself? Kant's injunction forbids using another person as a "mere" means, not using another person as a means at all. Human beings use each other as means, in the sense that we avail ourselves of one another's services, all the time. According to Kant, speaking a bit roughly, what makes the difference between using someone as a "mere" means, and using him as a means in a way that is morally permissible, is whether you have his (informed and uncoerced) consent. We serve each other's interests, consenting to do so, from motives of profit, love, friendship, or a general spirit of cooperation. Speaking a little more strictly, as I have argued elsewhere, Kant's principle requires that when we enter into an interaction with another, we must act in a way that makes it possible for him to consent. For this reason, we must avoid force, coercion, and deception, since someone who is forced, coerced, or deceived has no opportunity to consent to what is happening to him.[53] But the other animals cannot give us their informed and uncoerced consent; they cannot choose, in the sense in which we choose, to engage in interaction with us. And many of even our most benign interactions with them involve force and even deception. The cat who is trapped in a baited cage, even if it is for her own benefit, is both tricked and forced. And when you bring her home from the shelter, she has no choice but to go and live with you.

I suppose someone might conclude that since we can't get their consent, we should try to avoid interacting with the other animals at all. But this is not an option as conditions stand at present, for the fate of most animals will inevitably be determined by what human beings do. And in any case, I see no reason to take such an extreme position. We may interact with the other animals as long as we do so in ways to which we think it is plausible to think they would consent if they could— that is, in ways that are mutually beneficial and fair, and allow them to live something reasonably like their own sort of life. If we provide them with proper living conditions, I believe, their use as companion animals, aides to the handicapped and to the police, search-and-rescue workers, guards, and perhaps even as providers of wool, dairy products, or eggs, might possibly be made consistent with this standard. But it is not plausible to suppose a nonhuman animal would consent to being killed before the term of her natural life is over in order to be eaten or because someone else wants the use of her pelt, and it is not plausible to think she would consent to be tortured for scientific information.

Earlier, I said that a Kantian story about our duties to the other animals is in some ways stricter but in some ways more tractable than a utilitarian account. We've just seen one way in which the Kantian account is stricter. Some utilitarians think it would be perfectly acceptable to kill animals in order to eat them, if only we kept them humanely while they were alive and then killed them painlessly. I do not think that is consistent with regarding animals as ends in themselves. But let me now say why I think that the Kantian account is nevertheless in some ways more tractable.

Part of the reason why people resist the idea that the other animals might also be ends in themselves is that it can look as if the duties that would result would be enormously burdensome, preposterous, or even impossible. Should we try to ensure the happiness or comfort of every rat and rabbit on the planet? Find a way to "disinfect" malarial mosquitoes so we don't need to kill them? What do we do when the interests of one species are at odds with the interests of another? I was once at a conference on ethics and animals where there was a lively discussion of the question whether, were it in our power, we should try to eliminate predation from the world.[54] So long as the interests of animals are at odds with each other, we can't make them all happy and comfortable. But of course the interests of animals *are* at odds with each other—that's the way that nature works. So should we try to fix nature?

Such questions arise naturally for utilitarians, who think that the main question of ethics is how we can make the world as good as possible. And I'm not saying we shouldn't ask such questions. It would indeed be better if every animal on the planet could be happy and comfortable, and live the term of his or her natural life.[55] But I suspect that focusing on such questions is what makes some people so anxious to believe that the other animals are in some way less valuable than we are. What, they wonder, would we be committed to if we thought otherwise? So it's worth remembering that on a Kantian account, the subject matter of morality is not how we should make the world; it is how we should interact with and relate to others. Even if we can't remake the world into a place without predation, we can avoid being predators; even if we can't ensure the comfort of every rat and rabbit on the planet, we can avoid experimenting on rats and rabbits for the sake of our own comfort. What makes it worth acting in these ways is not just that it has a good result. It is worth it for its own sake, as an expression of respect for, and solidarity with, the creatures on this planet who share our surprising fate—the other beings *for whom* things can be naturally good or bad.[56]

NOTES

1. Colloquially, people often contrast human beings with apes, saying for example that human beings "evolved from" them, but saying that humans "evolved from" apes makes no more sense than saying that humans "evolved from" animals—which is another odd thing that people sometimes say. Even in the scientific literature, the term "human" is often used contrastively with "ape" or "primate" or "animal." Of course one may argue that such locutions are mere conveniences—it is so tedious always to say "the *other* animals" or "the *other* primates." But when thinking about the ethical treatment of animals, it is useful to remind ourselves that we are among them, so I will say "the other animals" when that is what I mean in this chapter.

2. The remark in the text mainly singles out the way animals differ from plants. Animals are also distinguished scientifically from bacteria and other unicellular life forms, and from several other groups no longer counted as plants, such as fungi, by various structural features that are not especially germane to the discussion here.

3. I say "contributing to the reproduction" because in some animal species—bees, say—there are sterile members or non-reproducing members who nevertheless serve the reproductive process.

4. I believe that the "natural" sense of good does not give us reasons until we take up a certain attitude toward it, which I shall describe in this chapter. Obviously, this is a controversial view of the good, along several dimensions. Many philosophers, including many of the utilitarians whom I will discuss below, think that normative or objective goodness and badness are just intrinsic properties that certain objects or states of affairs or experiences have. G. E. Moore famously argued that we must understand good and bad in that way in his *Principia Ethica* (Cambridge: Cambridge University Press, 1903). Some philosophers who disagree with Moore favor an "attributive" account of the good. They argue that "good" is a functional notion: to say that something is good is to say that it is good of its kind (for the locus classicus of this view see Peter Geach, "Good and Evil," *Analysis* 17, no. 2 [December 1956]: 33–42) or that it has the properties it is rational to want

in an object of its kind (see John Rawls, *A Theory of Justice* [Cambridge, Mass.: Harvard University Press, 1st ed., 1971; 2nd ed., 1999], chapter 7). But something can be good or bad of its kind without being objectively or normatively good. Other philosophers (myself included) claim that nothing is good unless it is good *for* someone or something, or at least that this is the central notion (see for instance, Judith Jarvis Thomson, *Goodness and Advice* [Princeton, N.J.: Princeton University Press, 2001]). Aristotle combined elements from these two views: he provides for an explanation of why things can be good or bad for living organisms by giving a functional account of what a living organism is. A living organism has the function of self-maintenance and reproduction, and things are good or bad for it insofar as they enable it to stay alive and healthy and reproduce. (Aristotle, *Metaphysics*, books 7–10; *On the Soul*, books 2–3; and *Nicomachean Ethics*, book 1, section 7, all in *The Complete Works of Aristotle*, edited by Jonathan Barnes [Princeton, N.J.: Princeton University Press, 1984]). A view along Aristotelian lines has been defended by some contemporary philosophers (e.g., Philippa Foot, *Natural Goodness* [Oxford: Clarendon Press, 2001]; Richard Kraut, *What is Good and Why* [Cambridge, Mass.: Harvard University Press, 2009]). On that view, things can be naturally good or bad for plants as well as animals. I am drawing on the Aristotelian view in the text, but I think that the sense in which things can be naturally good for animals, or at least for conscious animals who function by responding positively or negatively to their condition, is slightly different than the sense in which things can be naturally good for plants, and it is that sense that I think becomes "morally significant" when we take up the attitude I am going to describe.

5. Of course, the lack of a complex syntax could result in or be attached to other psychological or intellectual differences that would matter in the ways I am about to describe. The point is just that that difference by itself has no clear moral relevance.

6. See Peter Singer, *Animal Liberation*, 3rd ed. (New York: HarperCollins, 2002; 1st ed., 1975; 2nd ed., 1990).

7. In his commentary in J. M. Coetzee, *The Lives of Animals* (Princeton, N.J.: Princeton University Press, 1999), Singer uses the expression "more to lose" on p. 87.

8. Richard Kraut also criticizes this conception of the nature of the value of pleasure in *What is Good and Why*, pp. 81–88.

9. The point I am trying to make here is obscured both by an obscurity in the notions of "pleasure" and "pain" and by an ambiguity in the notion of "experience." Some philosophers, the classical utilitarians among them, think that pleasure and pain are the name of certain sensations, differing significantly only in their intensity and duration. I myself think that we use the words "pleasure" and "pain" to refer to the welcomeness or unwelcomeness of experiences of quite heterogeneous kinds, including but not limited to sensations. If we use "pleasure" and "pain" that way, when we say an experience is pleasant we are *already* saying that the experience is at least prima facie "good-for" the subject: that is, we are already saying something relational. If you transfer that implication to the view that pleasure is a particular sensation, which is not itself relational, then it is easy to suppose that the goodness of "pleasure" is inherent in the sensation, because the "good-for" relation has now been, as it were, inserted into the sensation. But that just confuses the two notions of "pleasure." I want to sum that up by saying: experiences are not good or bad in themselves; they are only good or bad relative to the nature of an agent. But here we run into the parallel ambiguity in the notion of "experience," for we sometimes use the term in a way that refers to the object experienced ("the experience of personal loss") and some-times in a way that refers to the relation between the subject and the object experienced ("the experience of grief"). When I say, "experiences are not good or bad in themselves," I am using the term in the way that refers to the object experienced. Again, by confusing these two notions we can transfer the essentially relational character of the second usage

into the first, and make it seem as if the badness of grief is in the object experienced rather than in its relation to the subject. That's why it sounds peculiar when I say that experiences are not in themselves good or bad. For a defense of my views about pleasure and pain, see Korsgaard, *The Sources of Normativity* (Cambridge: Cambridge University Press, 1996), §§4.3.1–4.3.11, pp. 145–56 and *Self-Constitution: Agency, Identity, and Integrity* (Oxford: Oxford University Press, 2009), §6.2.5, p. 120.

10. Peter Singer, "Killing Humans and Killing Animals," *Inquiry* 22 (1979): 145–56. See especially pp. 151–52. In this paper, Singer notices the fact that utilitarianism (or at least "total" utilitarianism) treats the subjects of experience as locations: he says, "It's as if sentient being are the receptacles of something valuable and it does not matter if a receptacle gets broken, so long as there is another receptacle to which the contents can be transferred without getting any spilt" (p. 149). He wants to argue that this is the right way to view non-self-conscious beings but not the right way to view self-conscious ones. But he does not seem to see that if we do not regard people as receptacles of value, there is a problem about how we can add and subtract value across their boundaries.

11. The classical challenge to aggregation is found in John Taurek, "Should the Numbers Count?" *Philosophy and Public Affairs* 6, no. 4 (1977): 293–316.

12. There are other ways to explain these intuitions. Some of them can be explained by the idea that the right thing to do is what is dictated by principles we would have chosen under what Rawls calls "a veil of ignorance," in which we do not know our particular fates (Rawls, *A Theory of Justice*, §24), or by the general principles of what Kant called "the Kingdom of Ends," an idea I will discuss later in this chapter. The idea is that the intuitions in question are explained by principles to which we would all agree when we are thinking in general terms. For example, Thomas Hill, Jr., suggests that we might all agree to a principle that directs saving more lives rather than fewer in certain circumstances, on the grounds that such a general principle improves each person's chances of survival. See "Making Exceptions without Abandoning the Principle: or How a Kantian Might Think about Terrorism," in his *Dignity and Practical Reason in Kant's Moral Theory* (Ithaca, N.Y.: Cornell University Press, 1992).

13. This is complicated, since the question whether you are harming someone simply by giving a resource to someone else depends on circumstances, in particular on whether those among whom you are distributing the resource have any prior claim on it or not. If I am giving a certain amount of money to charity, no one has a particular claim on me, and I do not have enough for all the worthy causes, it seems plausible to say that I am not "harming" the parties to whom I simply fail to give. If I am a parent distributing resources among my children or a government official distributing them to the citizens, that may be another matter.

14. Singer uses the phrase "more valuable" in *Animal Liberation*, p. 20. Obviously, the idea of an untethered "valuable" has the same problems as an untethered "good." For whom are our lives more valuable? Perhaps responding to this worry, in *The Lives of Animals*, Coetzee imagines a professional philosopher who says: "It is licit to kill animals because their lives are not as important to them as ours are to us" (p. 64). But if "importance" is also tethered, it would not follow that our lives are more important than theirs.

15. For citations and an account of Kant's arguments for these conclusions, see my "Kant's Formula of Humanity" and "The Right to Lie: Kant on Dealing with Evil," both in Korsgaard, *Creating the Kingdom of Ends* (New York: Cambridge University Press, 1996), pp.106–32 and 133–58.

16. Another, perhaps more controversial way to describe the difference concerns the source of normativity. On a consequentialist view, what is normative for us—what makes a claim on us—is the goodness of the good, which calls out to us to promote it. On a

Kantian account, what is normative for us is people, and, if I am right, the other animals. See my *Sources of Normativity*, §4.5.5, p. 166.

17. The idea is not that our *purpose* in acting is to be well related to animals *rather* than to do them some good. Instead, the idea is that what makes it necessary to have the purpose of doing them some good is the relation in which we stand to the animals themselves, rather than a general relation in which, as agents, we stand to the good.

18. The passages in this section are drawn from my "Fellow Creatures: Kantian Ethics and Our Duties to Animals," in *The Tanner Lectures on Human Values*, ed. Grethe B. Peterson, vols. 25/26 (Salt Lake City: University of Utah Press, 2005); and on the Tanner Lecture website at http://www.tannerlectures.utah.edu/lectures/documents/volume25/korsgaard_2005.pdf.

19. Immanuel Kant, *Groundwork of the Metaphysics of Morals*, trans. and ed. Mary Gregor (Cambridge: Cambridge University Press, 1998), 4: 428, p. 36. Page references to Kant's works are given in the standard way by using the page numbers of the relevant volume of *Kants gesammelte Schriften* which appear in the margins of most translations, followed by the page number of the translation used.

20. Kant, *Groundwork* 4: 428, p. 36.

21. Kant, "Conjectures on the Beginnings of Human History," in *Kant's Political Writings*, 2nd ed., trans. H. B. Nisbet and ed. Hans Reiss (Cambridge: Cambridge University Press, 1991.), 8: 114, p. 225. I have changed Nisbet's rendering of the German *Pelz* from "fleece" to "pelt" although the German can go either way, because I think that the rendering "fleece" softens Kant's harsh point.

22. Kant, *The Metaphysics of Morals*, trans. and ed. Mary Gregor (Cambridge: Cambridge University Press, 1996), 6: 443, p. 193.

23. Kant, *Metaphysics of Morals*, 6: 443, p. 193.

24. Kant, *Lectures on Ethics*, trans. Peter Heath and ed. Peter Heath and J. B. Schneewind (New York: Cambridge University Press, 1997), 27: 459, pp. 212–13.

25. Kant, *Metaphysics of Morals*, 6: 433, p. 192; *Lectures on Ethics*, 27: 459, p. 212.

26. Kant, *Metaphysics of Morals*, 6: 443, p. 193

27. Kant, *Lectures on Ethics*, 27: 710, p. 435.

28. Kant, *Lectures on Ethics*, 27: 710, p. 434.

29. Kant, *Metaphysics of Morals*, 6: 443, pp. 192–93.

30. Kant, *Lectures on Ethics*, 27: 459, p. 212.

31. It's easier to understand what I mean here when you are thinking about practical, evaluative attitudes. It sounds odd to think of beliefs as a lens through which we see the world. But they are, in the sense that an animal could be moved by one belief to take up another without having any awareness of making an inference. Unlike a person, a non-human animal can think "X" without commitment to "I believe X" or "X is true," because he (probably) has no commitments of that sort.

32. "Instinctive" is sometimes used in contrast to "learned"; I am not using it that way here. As will become clearer in the discussion that follows, I am using it in contrast to "rational." As I am using the term, an animal can learn new responses that still count as "instinctive" rather than rational because the animal does not reflect on the grounds of the response. See my *Self-Constitution*, §6.1, pp. 109–16.

33. Of course, it is also possible that they have just learned that such utterances will produce the desired effect, by a sort of conditioning. For a defense of the claim that Alex the parrot understood "wants," see Irene Maxine Pepperberg's account of teaching Alex to use "wants" and her own conclusions about what exactly he learned when he learned it. *The Alex Studies: Cognitive and Communicative Abilities of Grey Parrots* (Cambridge, Mass.: Harvard University Press, 1999), pp. 197–208. Koko has a sign for "wants."

34. Korsgaard, *Sources of Normativity*, §§3.2.1–3.2.3, pp. 92–98.

35. Kant believed that the principles, "Act only on a maxim you can will to be a universal law" and "Always treat humanity, whether in your own person or in the person of any other, not merely as a means but as an end in itself" are equivalent. Speaking very roughly, the reason is that the former principle enjoins us to act in ways that are acceptable to all others, and therefore respects their choices. But it is a matter of debate among Kant scholars whether the two formulas actually come to the same thing.

36. Interestingly, Charles Darwin agreed with Kant about this. In *The Descent of Man* (Princeton, N.J.: Princeton University Press, 1981), he wrote: "I fully subscribe to the judgment of those writers who maintain that of all the differences between man and the lower animals, the moral sense or conscience is by far the most important. This sense, as Macintosh remarks, 'has a rightful supremacy over every other principle of human action;' it is summed up in that short but imperious word *ought*, so full of high significance" (p. 70). A moral being, Darwin remarks later, is one who is capable of approving or disapproving his own past and future actions, and, he adds "we have no reason to suppose that any of the lower animals has this capacity" (p. 88).

37. It's perhaps worth saying that I am not arguing that human beings are "superior" to the other animals (whatever that means). Rather, I am arguing that we are subject to normative standards, to the standards of normative self-government, and the other animals are not. Whether we are good or bad of our own kind depends on whether we meet those standards. But even if we do, that does not provide a dimension along which we are superior to the other animals, since they are not subject to those standards.

38. Kant is not the only philosopher to advance this sort of argument. It is a commonplace to suppose that only beings with duties have rights, although there is disagreement about how much of the moral territory "rights" takes in. Another version of the reciprocity argument is found in David Hume, in *An Enquiry Concerning the Principles of Morals*, ed. L. A. Selby-Bigge and P. H. Nidditch, 3rd ed. (Oxford: Clarendon Press, 1975), pp. 190–91. Hume argues that we have no duties of justice to the other animals, because we have such absolute power over them. Since we can force them to serve us, we do not require their voluntary cooperation, and therefore we need not "cooperate" with them. Hume still considers it a virtue to be kind to them, however.

39. Stephen Darwall, *The Second-Person Standpoint: Morality, Respect, and Accountability* (Cambridge, Mass.: Harvard University Press, 2006). Michael Thompson, "What Is It to Wrong Someone? A Puzzle about Justice," in *Reason and Value: Themes from the Moral Philosophy of Joseph Raz*, ed. Michael Smith, Philip Pettit, R. Jay Wallace, and Samuel Scheffler (Oxford: Oxford University Press, 2004).

40. The suggestion here is that the Kingdom of Ends involves a reciprocal constraint of each of us by all of us. Certainly Kant envisions the political community in that way. I should note that the argument as given in the text might seem to slur over a difficulty—it is not clear how we get from the fact that each of us wills the moral law autonomously to the fact that we will it together. This is one version of the puzzle referred to in Thompson's "What Is It to Wrong Someone: A Puzzle about Justice." I think that the answer rests in what I call the essential "publicity" of reason, but a discussion of this difficult issue would take us too far afield here. For my views on publicity, see *Sources of Normativity*, §§4.2.1– 4.2.12, pp. 132–45; and *Self-Constitution*, chapter 9, pp. 177–206. Interestingly, in *Religion within the Boundaries of Mere Reason*, trans. and ed. George di Giovanni and Allen W. Wood (Cambridge: Cambridge University Press, 1998), Kant himself seems aware of some difficulty about this, for he suggests that until the Kingdom of Ends is institutionalized in something like a church, "as long as each prescribes the law to himself," we are in an "ethical state of nature" (6: 95, p. 107). I ignore any complications this creates here.

41. Kant, *Groundwork* 4: 429, p. 37.

42. Kant, *Groundwork* 4: 429, p. 37.

43. I defend this interpretation of Kant's argument in "Kant's Formula of Humanity," in *Creating the Kingdom of Ends* (New York: Cambridge, 1996).

44. Part of what's at issue here is the order between value and valuing: do we value things because we *recognize* value and respond to it, or do things have value because we value them? I think it is the latter. However, I think the argument of this chapter still goes through on a "recognitional" view. See note 52.

45. Kant, *Groundwork* 4: 429, p. 37.

46. Kant, *Groundwork* 4: 435, p. 42.

47. At this point, many champions of animal rights would suggest that Kant's argument also implies that human infants, the insane, people in comas, and so on, have no rights since they are "not rational." Since we plainly don't think that, we should reject the argument. Although I am about to challenge Kant's argument myself, I don't think it has this implication. An entity, certainly a living entity, is not a mere collection of properties or capacities: it is a functional unity, and the idea of a species *is* important to the extent that it involves the idea of a certain way of functioning. An infant, or a severely insane person, is helpless or poorly functioning in part because his or her natural way of functioning, which essentially involves reason, is as yet undeveloped or defective. These are rational beings in whom reason is in an undeveloped or defective condition. The sense in which such people "lack reason" is entirely different from the sense in which a nonhuman animal "lacks reason," for the nonhuman animal functions perfectly well in his own way without it. And different moral responses are suitable to these different kinds of "lack." In the case of an infant, or those only temporarily in a nonfunctional state, it is also pertinent to note that a right is held by a person, and a person is not a mere time-slice of a person; nor is a human infant merely a "potential" rational being in the same sense that a lump of clay is a "potential" pot. The permanently insane or otherwise severely defective may present somewhat different problems than the infant or the temporarily nonfunctional person. But the issue about them is still not the same as the issue about animals, because there is a difference, morally as well as metaphysically, between being a defective being of a certain kind and being a different kind. (See my *Self-Constitution*, §2.1.8, pp. 33–34.) The reasons why we should accord moral respect to human beings at a stage of development when rationality is not fully expressed, human beings who are temporarily non-rational, human beings who are not rational by virtue of permanent defect, and to non-human animals are, in my view, heterogeneous. So I think it is a mistake to appeal to the so-called "marginal cases" argument. My own argument involves no such appeal.

48. See Joel Feinberg, "Human Duties and Animal Rights," in *On the Fifth Day: Animal Rights and Human Ethics*, ed. Richard Morris and Michael W. Fox (Washington, D.C.: Acropolis Books Ltd., 1978) and Joel Feinberg, "The Rights of Animals and Future Generations" in *Philosophy and Environmental Crisis*, ed. William T. Blackstone (Athens: University of Georgia Press, 1974).

49. This may seem as if it is just a denial of an argument I made earlier, to the effect that our obligation must be owed to the authority that stands behind the law. Kant, I think, believes that, since his most straightforward remark about why we can't be obligated to animals is "a human being has duties only to human beings (himself and others), since his duty to any subject is moral constraint by that subject's will" (*Metaphysics of Morals*, 6: 442, p. 192). So although I don't think Feinberg meant it this way, I am proposing Feinberg's account as giving a slightly different *sense* of "obligated to" than we find in Kant.

50. This is a summary of my argument in *Sources of Normativity*, §§3.1.1–3.4.10, pp. 90–125, and §§4.4.1–4.4.2, pp. 160–64.

51. People sometimes reply to this argument that perhaps it is only the suffering of rational beings that we object to. But that reply seems unmotivated, or rather, it seems motivated only by the desire to block the conclusion that we have duties to the other animals. And it also seems disingenuous. When you object to your own pain and suffering it is not only because the pain and suffering of a rational being seems like a bad thing to you. It is because the pain and suffering of a being who can suffer seems like a bad thing to you.

52. In note 44, I claimed that my argument would work just as well on the "recognitional" view that valuing is a response to value. What I had in mind is that at this point, one could substitute the thought that what we *recognize* to have value is the good of a being for whom things can be naturally good or bad. Of course, since this recognition must presumably be a matter of some sort of intuition, it is always open to the philosopher who favors that sort of view to insist that all that *his* intuition tells him is that rational beings are valuable. I hope at least to have convinced you that this is not the sort of argument Kant was making.

53. See my "The Right to Lie: Kant on Dealing with Evil," in *Creating the Kingdom of Ends*.

54. Just for the record, I don't think we should try to eliminate predation from the planet. We couldn't do it without eliminating the predators. Changing them into something so unrecognizably different from what they are would just be a way of killing them off. If as I suggest in the text we are only promoting the good when we improve someone's condition without harming someone else, then this wouldn't be a way of promoting the good.

55. I claim the good is tethered, so you will ask, "better for whom?" In making the remark in the text, I am availing myself of the idea behind one of the two principles of aggregation that I suggested are compatible with the tetheredness of good: that we could make things "better" by making them good for more creatures, so long as we could do so without harming any.

56. An earlier version of this chapter was delivered as the Dewey Lecture at the University of Chicago Law School in 2008. I would like to thank the audience on that occasion for helpful discussion. I also would like to thank Charlotte Brown for useful commentary on the manuscript.

SUGGESTED READING

Kant's moral philosophy is principally found in:

KANT, IMMANUEL. *Critique of Practical Reason*. Translated and edited by Mary Gregor. Cambridge: Cambridge University Press, 1997.

————. *Groundwork of the Metaphysics of Morals*. Translated and edited by Mary Gregor. Cambridge: Cambridge University Press, 1998.

————. *Lectures on Ethics*. Translated by Peter Heath and edited by Peter Heath and J. B. Schneewind. New York: Cambridge University Press, 1997.

————. *The Metaphysics of Morals*. Translated and edited by Mary Gregor. Cambridge: Cambridge University Press, 1996.

Some recent interpretations of Kantian ethics include:

HERMAN, BARBARA. *The Practice of Moral Judgment*. Cambridge, Mass.: Harvard University Press, 1993.

————. *Moral Literacy*. Cambridge, Mass.: Harvard University Press, 2008.

KORSGAARD, CHRISTINE M. *Creating the Kingdom of Ends.* New York: Cambridge
    University Press, 1996.
O'NEILL, ONORA. *Constructions of Reason: Explorations of Kant's Moral Philosophy.*
    Cambridge: Cambridge University Press, 1989.
RAWLS, JOHN. *Lectures on the History of Moral Philosophy.* Cambridge, Mass.: Harvard
    University Press, 2000. See esp. pp. 143–328.
SULLIVAN, ROGER. *Immanuel Kant's Moral Theory.* Cambridge: Cambridge University Press,
    1989.
WOOD, ALLEN W. *Kant's Ethical Thought.* New York: Cambridge University Press, 1999.

*More about my own views of human/animal differences, and the ethical
treatment of the other animals, may be found in:*

KORSGAARD, CHRISTINE M. "Fellow Creatures: Kantian Ethics and Our Duties to Animals."
    In *The Tanner Lectures on Human Values,* vol. 25/26, ed. Grethe B. Peterson (Salt Lake
    City: University of Utah Press, 2005); and available
    on the Tanner Lecture website at http://www.tannerlectures.utah.edu/lectures/
    documents/volume25/korsgaard_2005.pdf.
———. "Just Like All the Other Animals of the Earth." *Harvard Divinity Bulletin* 36, no. 3
    (Autumn 2008) and available on the web at http://www.hds.harvard.edu/news/
    bulletin_mag/articles/36-3/korsgaard.html.
———. *Self-Constitution: Agency, Identity, and Integrity.* Oxford: Oxford University Press,
    2009. See esp. chapters 5 and 6.
———. *The Sources of Normativity.* Cambridge: Cambridge University Press, 1996. See esp.
    chapters 3 and 4.

Some recent discussions of Kant's views on animals include:

DENIS, LARA. "Kant's Conception of Duties Regarding Animals: Reconstruction and
    Reconsideration." *History of Philosophy Quarterly* 17, no. 4 (2000): 405–23.
O'NEILL, ONORA. "Kant on Duties Regarding Nonrational Nature II." *Proceedings of the
    Aristotelian Society:* Supplementary 72 (1998): 211–28.
PYBUS, ELIZABETH, and BROADIE, ALEXANDER. "Kant's Treatment of Animals." *Philosophy*
    49 (1974): 375–83. This paper was followed by an exchange between Tom Regan and the
    authors, "Broadie and Pybus on Kant" in *Philosophy* 51 (1976): 471–72, and "Kant and
    the Maltreatment of Animals," *Philosophy* 53 (1978): 560–61.
TIMMERMAN, JENS. "When the Tail Wags the Dog: Animal Welfare and Indirect Duty in
    Kantian Ethics." *Kantian Review* 10 (2005): 128–49.
WOOD, ALLEN W. "Kant on Duties Regarding Nonrational Nature I." *Proceedings of the
    Aristotelian Society:* Supplementary 72 (1998): 189–210.

CHAPTER 4

....

VIRTUE ETHICS
AND THE TREATMENT
OF ANIMALS

....

ROSALIND HURSTHOUSE

VIRTUE ethics directs us to think about the rights and wrongs of our treatment of nonhuman animals in terms of virtues and vices rather than in terms of consequences, or rights and duties. From its perspective, the two main contenders in the field, Peter Singer and Tom Regan, each implicitly picks up on one of the virtues. However, each picks up on one virtue only, and this concentrated focus is in part why their approaches are unsatisfactory.[1]

It is clear from Amazon.com reviews of Singer's *Animal Liberation*[2] that its unique contribution to converting people to the animal liberation cause is largely due to the way it calls forth its readers' virtue of compassion. Many reviewers are aghast at how much animal suffering is caused by commercial farming and scientific experimentation and shocked to discover what is involved in the testing of something as trivial as cosmetics on animals. They write such things as "After reading *Animal Liberation* I was appalled. I really had no idea the situation was this bad," "I never knew how horrible the factory farm conditions were," and "This book made me a vegetarian because it made me aware of all the cruelty that is imposed on animals." However, it is equally clear that many of Singer's non-philosophical readers are not thinking of the moral significance of animal suffering in his utilitarian terms. It is common to find them roughly summarizing *his* view as, "denying that animals have rights is no different from racism," although this statement better describes Regan's explicitly anti-utilitarian rights-based theory.

It is not surprising that so many of them think Singer is defending animal rights, because the powerful analogy drawn between speciesism on the one hand and racism and sexism on the other immediately calls on the widespread—if unarticulated—idea of the virtue of respect or justice. The moral significance many of Singer's readers attach to the causing of the animal suffering they deplore is the disrespectful attitude they see manifested in it. They take it to be the same arrogantly assumed superiority and dominance manifested in racism and sexism. They also deplore the exploitation of animals, or the view of them as objects there for us to use as we choose.

From the perspective of virtue ethics, they are right to do so. From its perspective, we can identify in our current practices both forms of wrong, as so many of Singer's non-philosophical readers do, without insisting that one is more fundamental than the other. In many instances, the actions are both cruel and disrespectful. We can, as some of these non-philosophical readers do, go further and identify other forms of wrongdoing beyond those two. Quite a few, for example, speak of our selfishness, while others say it is irresponsible of us not to be aware of how our dinner gets on our plate. Others, responding to reviewers hostile to *Animal Liberation*'s message, note the dishonesty inherent in evading the issue of animal suffering by focusing exclusively on purely theoretical issues in philosophy.

According to the perspective of virtue ethics, this is just what we ought to be doing, namely, thinking about our treatment of nonhuman animals in terms of any of the virtues and vices that can appropriately be applied to it. However, for readers more philosophically sophisticated than most of those writing reviews on Amazon. com, thinking this way is surprisingly difficult to do. Both the utilitarian and deontological approaches to the issue share a structural reliance on the concept of moral status. When that is what one is familiar with, it is natural to assume that virtue ethics too must begin by saying something about the moral status of animals, only to find it impossible to work out what this could be. I shall argue, however, that virtue ethics has no need for the concept of moral status and that this is a strong point in its favor. "Moral status" is a concept that moral philosophy is better off without.

THE CONCEPT OF MORAL STATUS
AS SUPERFLUOUS

Let us consider what the concept of moral status is supposed to do in animal ethics literature. It is supposed to divide everything into two classes: things that have moral status and are within "the circle of our moral concern" and things that do not, which are outside the circle. Things within the circle matter morally; they are the entities our moral principles apply *to*. Things outside the circle do not matter morally at all in themselves, though sometimes they may matter incidentally, through some relation to something that does have moral status. Almost anything can come to matter morally in this incidental way. For example, some object that is worthless in itself may be of great sentimental value to you and therefore become an object of moral concern for me, as your friend.

So which things are within "the circle of our moral concern"? It is assumed that all decent people think that their fellow human beings have moral status. We all, at least at first glance, think it wrong to kill our fellow human beings, exploit them for our own ends, or not consider their interests when we are thinking about the consequences of our actions. So, human beings have moral status; our moral principles govern our treatment of them. But, it is argued, restricting the application of these principles to our fellow human beings is speciesism, which is akin to racism and sexism. There is no special feature that all and only human beings have which could justify such a restriction. Therefore, we should expand our circle of concern, recognizing that there are some other things with moral status because they have some feature that they share with us, such as the capacity to suffer or being an experiencing subject of a life. Having thus established that all sentient animals, not just humans, have moral status, the utilitarian and deontological "animal liberationists" are able to apply their moral principles to our treatment of the nonhuman sentient animals.

Mary Midgley, a contributor to animal ethics literature who is neither a utilitarian nor a deontologist and who eschews any use of the concept of moral status, pinpointed early on what is wrong with the attractively simple picture of "the expanding circle of concern"[3]: it drastically underestimates the range of features that may, in context, be morally relevant in decision making. When two beings within our circle of moral concern make competing claims on us, we are not always flummoxed. It may well be the case that one has a feature other than sentience that makes giving it priority justifiable or, indeed, compelling.

Even Singer and Regan found it necessary to draw a distinction within the class of things that had moral status, namely between those sentient animals that had the further feature of being persons and those that did not. This distinction allowed them to maintain the attractive view that in the lifeboat cases any of their readers, who are bound to be persons, could toss the dog overboard to preserve their own lives. It also allowed them to maintain their anti-speciesist position. They could say that any human in the lifeboat who was not a person, such as a baby or an adult with the same level of psychological capacities, would be equally up for sacrifice.

Many people flinch from this latter conclusion but cannot bring themselves to reject it outright, so powerful has the analogy between racism and speciesism proved to be. Nevertheless, Midgley also pinpoints the flaw in this analogy. "The term 'racism,'" she says, "combines unthinkingly three quite distinct ideas—the triviality of the distinction drawn, group selfishness, and the perpetuation of an existing power hierarchy....In the case of species the first element does not apply at all. The distinction drawn is not trivial; it is real and crucial."[4]

I have elsewhere[5] argued that speciesism is like familyism.

Speciesism involves:

(a) drawing a distinction between members of one's own species and others; and
(b) sometimes, in some ways, giving preference to the interests of members of one's own species.

Similarly, familyism involves:

(a) drawing a distinction between members of one's own family and others; and
(b) sometimes, in some ways, giving preference to the interests of members of one's own family.

Familyism is, obviously, sometimes wrong and sometimes right. In some circumstances, benefiting my own child rather than another's would be nepotism and therefore unjust; or callous, if the neighbor's child were alone and in distress; or dishonest, if I were to cover up my own child's wrongdoing instead of apologizing for it and making amends. But I have special responsibilities for my own children, and in some circumstances it would be irresponsible, unloving, or even cruel not to benefit my own child rather than another's. For example, I give my own child a birthday party and help my own child with her homework, not the neighbor's.

The distinction between members of my own family and others is not trivial. Nor, as Midgley notes, are the distinctions of age and sex within the class of human beings. "Serious injustice can be done to women or the old by insisting on giving them exactly the same treatment as men or the young,"[6] she argues. Neither is the distinction between humans and the other sentient animals trivial. In none of these cases does employing the distinction need to form part of "group selfishness" or "the perpetuation of an existing power hierarchy," though we know it can. In all of these cases, a real feature is highlighted that *in certain circumstances* will be relevant and crucial to making the correct moral decision about what to do. So, on the analogy with familyism, speciesism—giving preference to the interests of members of one's own species—is sometimes wrong but sometimes right. This is why I emphasize the qualification "in certain circumstances." Leaving a dog in a burning building to rescue a baby is one thing. Leaving it to avoid hurting someone's feelings by missing a birthday party is quite different.

Midgley's view is that moral claims can certainly be made on behalf of sentient nonhuman animals, and she contrasts this position with what she calls "absolute dismissal," or the view that animals don't matter at all and that "animal claims are just nonsensical, like claims on behalf of stones."[7] However, she sticks to expressing her position as the straightforward denial of the "absolute dismissal" position rather than in terms of moral status because she believes the set of moral claims is ineliminably complex. For example, even within the class of humans we will, in some circumstances, find it necessary to distinguish, as a relevant moral feature, those that are persons from those that are not. However, we need not agree with Singer and Regan that the former always have the strongest moral claims. Sometimes they do—when, for example, we have to respect their autonomy. Sometimes they do not—when, for example, the nonpersons need special care and protection: in the overburdened lifeboat scenario, we keep the baby on board and the persons have to draw straws, whether they want to draw them or not.

To Midgley, it seems obvious that there are many circumstances in which non-person nonhumans may also need special care and protection, even if this thwarts the exercise of some persons' autonomy. Moreover, it seems just as obvious that this need may well not be limited to the sentient.

It is probably no accident that Midgley both rejects the concept of moral status and has been a notable contributor to both the animal and the environmental ethics literature.[8] She is rare on both counts. Most of the prevailing utilitarian and deon-tological literature on animal ethics still proceeds as though the issues with which environmental ethics is concerned form no part of animal ethics, despite the fact that environmentalists pointed out the lack as far back as 1980.[9] This is probably connected to their adherence to the concept of moral status.

It is, I believe, generally unrecognized what a devastating effect the upsurge of environmental ethics has had on the concept of moral status. Once we move beyond focusing exclusively on the particular "animal liberation" issues of commercial farming and experimentation to the more environmental issues, we find it neces-sary to draw many more distinctions. The class of sentient animals contains not only humans, some laboratory animals, some livestock, and pets, but also the urban feral, the wild, the introduced and indigenous, and the rare and all-too-numerous.[10] Any of these may be a feature we want to cite as a morally relevant and perhaps decisive reason for why I may, or may not, or ought to, do something with respect to a member of the class of sentient animals which has it, but not the others.

From the perspective of anti-anthropocentric environmental ethics, not only the other sentient animals, but also the members of the millions of species of non-sentient animals, and of plants, ecosystems, and the like, all matter morally as well. The feature conferring moral status that all these groups have in common has to be something extremely broad—being alive, having a good of its own, being an inte-gral part of an ecosystem, or being part of nature—and thereby introduces count-less conflicting claims. Unsurprisingly, it is increasingly common for environmental philosophers to abandon the concept, following the example of two of the most famous environmentalists (Aldo Leopold and Arne Naess) who never used it.

One way to do without the concept of "moral status" is to use the concept of "intrinsic value" instead. Given the close connection between attributing moral sta-tus and attributing intrinsic value, it is occasionally possible to read environmental ethics literature couched in terms of the latter as though it were using the former. However, such an interpretation is difficult to maintain when one finds an author such as Holmes Rolston III maintaining the plausible thesis that there are different sorts of incommensurable intrinsic value to be found among natural things, that different things have intrinsic value for different reasons, and finally that some sorts of things may have different intrinsic value in different contexts.

Relying on different sorts of incommensurable intrinsic value is not the only way to go. We can also go Midgley's way. Looking at the various criteria for moral status that have been offered, we can see each as highlighting a "real and crucial" feature, one that can indeed be put forward as a moral reason, and, in certain circumstances, a decisive moral reason, for acting or refraining from acting in a particular way. "She (he, it) is a fellow human being, fully rational, a moral agent, sentient, alive …"—any of these might be supportive or decisive in relation to moral decision. But sometimes they will not be, and then, without undermining their significance, one can acknowledge with Midgley that a large number of further features can play the same role.

Either way, the approach is pluralistic, context sensitive, and open ended, rejecting the idea that there is just one monolithic set of principles concerning just one form of value that can be fruitfully used to address the great variety of moral issues we encounter when we extend our moral concern beyond ourselves. Its proponents also tend to acknowledge the impossibility of providing decisive arguments for one right solution rather than another in many cases, preferring to recognize that neither is one that it is easy to accept, but stressing the need—especially given the looming environmental crisis—for urgent action. This pluralism should come as no surprise, because a similar approach has come to seem obviously necessary in the comparatively limited area of medical ethics. There, too, reaching a decision about a real case is often a matter of balancing a range of considerations in the light of principles or norms that can be applied to cases in different ways, and recognizing that an equally good case could be made for deciding one way or another.

Closely allied to Midgley's approach is applied virtue ethics, which, as I noted at the outset, basically amounts to thinking about what to do in terms of the virtues and vices.[11] Long before any philosopher invented the notion of moral status, Hindus, Buddhists, and some of the ancient Greeks deplored cruelty to animals and espoused vegetarianism simply on the grounds that it was required by the virtue of compassion or love.[12] The anti-anthropocentric literature which calls for us to make a radical change in our attitudes to the living world characteristically deplores our greed, self-indulgence, materialism, insensitivity, short-sightedness, wantonness, cruelty, pride, vanity, self-deception, arrogance, and lack of wisdom.[13]

Virtue ethics has never needed the concept of moral status. Its principles are the "v-rules," that is, "virtue- and vice-rules" such as "Do what is compassionate, do not do what is cruel," and the answer to the question of what groups these apply to is given by the meaning of the terms. No group of beings has to be rubber-stamped as belonging "within the circle of our moral concern" before we know whether our v-rules apply to its members; we know that if we know how to use the terms. Doing what is just or unjust, if we choose to be really strict about the notion of rights, applies only to actions involving human persons, with forgiving being similarly restricted, but any of the vices listed above and the virtues opposed to them may be manifested in relation to our treatment of and attitudes to sentient animals—as Singer's non-philosophical readers readily recognize. So the concept of moral status is superfluous, and, as I shall eventually argue, when moral philosophers insist on using it, it can be dangerously divisive.

WHAT IS VIRTUE ETHICS?

If we are going to think about the rights and wrongs of our treatment of nonhuman animals in terms of the virtues and vices, we have to be clear about these moral notions. Although I have implied above that some of Singer's readers have an intuitive grasp of compassion and respect as virtues, and of cruelty, irresponsibility, selfishness, and dishonesty as vices, understanding of these and other concepts in modern virtue ethics merits careful discussion. So we must spend some time on the details.

Virtue and vice words can be used to assess both people and actions, and although the words "virtue" and "vice" are no longer common in ordinary conversation, we still employ a large and rich vocabulary of virtue and vice words. For example, we praise and admire people for being benevolent or altruistic, unselfish, fair, responsible, respectful, caring, courageous, honest, just, honorable, and the like. Similarly, we condemn, despise, or criticize people for being malevolent, selfish, callous, unfair, irresponsible, uncaring, cowardly, reckless, cruel, unjust, dishonorable, and the like. Neither list is even close to complete, and some people will not agree with every example, but it gives the general idea of the language of virtue and vice. Applying these words to actions is often a fairly straightforward business. Applying them, especially the virtue words, to people in exactly the way virtue ethics does—as ascribing a virtue to them—calls for a bit of philosophical expertise.

In moral philosophy, we think of a virtue as a morally good, admirable, or praiseworthy character trait, the sort of thing that is cited in a character reference. Conversely, a vice or defect is a morally bad, despicable, or regrettable character trait, the sort of thing we condemn, despise, or deplore people for having. Another way to think of the virtues is as the ways we aspire to be or hope we are if we want to be a morally good person, or as what we try to instill in our children when we are giving them a moral education.

One can see immediately that few, if any, of the virtue words always operate as virtue terms in ordinary language. That is, they do not always function as terms that refer to morally good character traits. This is one of the main reasons why the intuitive grasp of virtue terms needs to be sharpened. Possession of the virtues is what makes an agent a morally good person, or someone who reliably and consistently does what is right; they enable their possessor to act well. However, many of the virtue words just listed can be used in ordinary language to pick out a character trait that actually enables or prompts its possessor to act wrongly.

Suppose I am the boss of a protection racket, looking to replace my right-hand man. I certainly do not want someone who is compassionate, caring, or honest. Nevertheless, I might well say that I do need someone who is responsible rather than irresponsible, industrious rather than lazy, and, most especially, courageous rather than cowardly or timid or reckless. This example might lead one to think twice before admitting "responsible" and "industrious" as virtue words, but it seems beyond dispute that "courage" is one, and equally indisputable that a character trait that enables someone to face danger without giving in to fear will be what enables a member of a protection racket to act wrongly. So when, engaged in moral philosophy, we explicitly use "courage" as a virtue term, we need to restrict its application to facing danger *for a worthwhile end*. We may then say that those involved in protection rackets and other desperados may be daring, but do not possess the virtue of courage. Alternatively, we could say, as some philosophers do, that they do not have "virtuous courage."

However, this is not the only sort of case in which the ordinary use of virtue words does not pick out virtues. It is natural to say, for example, that some people who possess compassion could too easily be prevented by it from doing what they should. Might they not, for instance, find themselves unable to kill the bird their cat has mauled? Or might not their kindness prompt them to allow their dog too many treats, making him unhealthily fat? They are, we might say, "too compassionate, too kind, too virtuous."

We seem to have got into a muddle. Do we think that compassion can sometimes be a virtue and sometimes a fault, so that someone can be a compassionate person but thereby not be a morally good, admirable person? Or do we think that, if they are compassionate, of course they must be morally good, but that morally good people may be prompted by what makes them morally good to act wrongly?

We can get clear of this muddle by recalling that "virtue" is but one translation of the ancient Greek word *arête*. The other, more accurate, translation is "excellence," and the virtues are not just morally good character traits but *excellent* character traits. Here is how the "excellence" translation makes a difference: something that is excellent is as good of its kind as it could be reasonably expected to be, and we do not say that anything is "too excellent." However, someone whom we initially describe as "too compassionate or kind, too virtuous" is not as good as anyone could be. He would be better if he were the sort of person who could quickly wring the bird's neck or put his dog on a diet when he ought to. That is, he would be better if he had the *excellences* of compassion and kindness. Though he is, it seems, well on the way to being excellent in these ways, with his heart in the right place, we can readily imagine people who are compassionate without being squeamish or kind without being over-indulgent, and he is not as good as they are. So we should think of a virtue as an excellent moral character trait and, rather than describing someone as too compassionate, honest, or the like, find longer but more accurate descriptions. We now turn to considering what is involved in a virtue being a character trait.

An excellent character trait is a strongly entrenched dispositional state of a person or a certain sort of way they are all the way down, and the disposition is of a complex sort. It is, for a start, a disposition to act in certain ways, that is, in accordance with the virtue in question. As we noted above, the virtue terms are used to assess not only people but also actions, and people who can rightly be described as benevolent, honest, just, and so on consistently and reliably perform actions that can also rightly be described by those words. The benevolent consistently do what is benevolent, the honest consistently do what is honest, the just consistently do what is just, and so on. In general terms, someone with a particular virtue, V, consistently and reliably does what is V.

In looking to virtue ethics for action guidance, we can often simply employ the v-rules. What it is right to do is what a virtuous agent would do in the circumstances, and the virtuous agent typically does what falls under a virtue-rule and abstains from doing what falls under a vice-rule. Nevertheless, the connection between being virtuous and doing an action that can be correctly described by a virtue word is not entirely straightforward, because life presents us with moral dilemmas, that is, forced choices between actions such that no virtuous agent would willingly do either.

This "conflict problem" is one that all normative ethical theories, barring the simplest forms of act utilitarianism, face. All the theories employ basically the same strategy to deal with moral dilemmas, namely give an account of how they can be resolved correctly. A discriminating understanding of the virtues or moral principles in question, possessed only by those with practical wisdom or provided by moral philosophers, will perceive that there is at least one feature of the situation that justifies doing one thing rather than the other. (If there are any irresolvable dilemmas, proponents of any normative approach may point out reasonably that it could only be a mistake to offer a resolution of what is, ex hypothesi, irresolvable.[14]) So, having resolved a dilemma correctly, a virtuous agent may well do something that in different circumstances would fall under the prohibition of a vice-rule, or that, in the very circumstances in which she does it, may appear to the naïve to do so, as a child may be shocked by my apparent callousness when I wring the wounded bird's neck.

Returning to the straightforward cases, people with the virtues or excellences of character not only typically do the V things, or act in virtuous ways, but also do so for the right reasons. They do not do what is compassionate merely on emotional impulse, nor do they do what is benevolent or honest for ulterior reasons. Thereby, they contrast with Kant's famous tradesman who does what is honest because it's to his advantage. Virtuous people do what is honest or benevolent because, in some sense, they think that the action is right, or worth doing for its own sake.

I say "in some sense" because of two points. One is that, notoriously, we do not think much of the agent who visits a friend in hospital for the avowed reason that she thinks it is right to do so. We think she should have gone because she thought her friend might like some company, might need cheering up, or just because he *is* a friend; these are what count as "right reasons" in this sort of case, and they show why the agent thinks the action is worth doing.

The second, less familiar point, is that in some *other* sense it is obvious that nice, well-brought-up children may typically do what is kind or honest and sincerely give as their reason for doing so, "Because it was right." Yet, when nice children do this they do not have, and act for, that reason in exactly the same way as virtuous adult rational moral agents have and act on it. If we rely on Aristotle, we can use his technical term *prohairesis*, usually translated as "rational choice," note that children do not "rationally choose" their actions because they have not yet reached the age of reason, and stipulate that the virtuous agent rationally chooses her virtuous actions and rationally chooses them for their own sake. However, we do not want to go into Aristotelian technicalities here. What we can say instead is that adults' reasons manifest *their* values or *their* views on what is worth having and pursuing, protecting and preserving, avoiding and doing, and that these are views for which they can be held morally responsible, whereas the values children manifest in the reasons they have and give, though they may become their own, are not yet so.

The fact that an agent who is V consistently and reliably does what is V for the right reasons brings out two things. The first is that the values manifested in the right reasons will be manifested in other ways too, not exclusively in the doing of V actions. So, for example, someone who has the virtue of compassion is someone who thinks that alleviating suffering is something worth doing. Her circumstances may be such that few occasions arise in which she can actually do this herself, but she contributes to effective organizations dedicated to alleviating suffering. So a virtue, V, is not only a disposition to act in V ways for right reasons, but also to act in various *other* ways, for related reasons. The second is that virtuous activity is rational activity, specifically the activity of practical rather than theoretical reason. The virtuous agent is ipso facto a rational agent, an agent in whom reason rules and rules well.

To those trained in modern Anglo-American philosophy of mind, this second claim will sound Kantian and anti-Humean. Does it not simply amount to the claim that the morally good person acts from reason, not from passion or desire? The answer is no, it does not. Modern virtue ethics has largely taken over Aristotle's moral psychology, and this is significantly different from either Kant's or Hume's, particularly with respect to the distinction between emotion (or "passion") and reason.

The Aristotelian idea is that in humans, because of our rationality, our desires in general and our emotions in particular, become something different from what they are in nonhuman animals. In us, they are shaped or informed by the reason of adults when we are young and, as we grow older, they become informed by our own reason, for good or ill. If we have, for example, been brought up to act and feel emotions well in relation to the family animals, we can make the reasons for so acting and feeling our own and, by reflecting on them or relating them to things we have read, come to act and feel in new ways. For instance, we may become vegetarian even if not brought up to be so, like so many of Singer's readers.

In the preface to the first edition of *Animal Liberation*, Singer describes someone he met at a tea party to which he had been invited because his hostess had heard that he was planning to write a book on animals. The other guest had already written a book on animals and while, as Singer notes, chomping on a ham sandwich, told him how much she loved them. In thus pillorying her, Singer may have done her an injustice, for who knows but that all she needed to become vegetarian was to have her eyes opened to the realities of commercial farming. Be that as it may, his point is well taken. To be a *virtuous* emotion, the makings of the character trait of compassion that enables its possessor to act well, the emotion of compassion or love must be informed and shaped by reason.

The above sketch of modern virtue ethics may seem a little abstract. For some people, it is only when they see virtue ethics illustrated that they see the point of some of the abstract account. So I turn now to applying it to some areas of our treatment of nonhuman animals, beginning with the familiar example of vegetarianism.

VEGETARIANISM

People sometimes suppose that virtue ethicists would say that vegetarianism is a virtue. Of course it isn't, because virtues are character traits and vegetarianism is not a character trait. It is a practice. So would they, perhaps, say that it is a virtuous practice? If we take that as meaning that those who practice it manifest virtue, then obviously not. One can be a vegetarian for a variety of reasons, not necessarily moral ones. However, if we take it as meaning that it is a practice that the virtuous, as such, tend to go in for, then the answer is yes, for the sorts of reasons that I noted Singer's non-philosophical readers tended to give.

Anyone who is reading this book knows, as I do, that in regularly eating commercially farmed meat we are being party to a huge amount of animal suffering, and it would be dishonest and hypocritical to pretend otherwise. Some people may not know this fact, but unless they are very young or mentally incapacitated, this is culpable ignorance; it is irresponsible not to think about how your dinner got onto your plate. Knowing that, we know that shrugging off such suffering as something that doesn't matter is callous. Knowing that our use of animals for food is not only unnecessary but positively wasteful, can we deny that our commercial farming practices are cruel? No, we cannot. So we should not be party to them.

Note that this apparently simple move, from "the practices are cruel and cause unnecessary suffering" to "we should not be party to them"—is not one that the act, or "direct," utilitarianism Singer espoused in the first edition of *Practical Ethics* allows him to make, though he made it. It needs the rule, or "indirect," utilitarianism he added in the second edition.[15] However, within virtue ethics, given that a virtue is a disposition of a very complex sort that is expressed in a variety of sorts of actions, it is a simple move. The compassionate are not willingly party to cruelty any more than the just are willingly party to injustice or the honest to chicanery.

I want now to introduce a virtue the ancient Greeks made much of, *sophrosyne*, for which the usual translation is, unfortunately, "temperance." It is unfortunate because someone who possesses this virtue and thereby characteristically does what is temperate is unlikely to abstain totally from alcohol. Nor, though "moderation" is another possible translation, is it quite right to say that the temperate pursue the pleasures of food, drink, and sex "in moderation." Some people might do so solely because they thought it was healthy and would enable them to live longer, but still behave selfishly. Rather, the temperate characteristically pursue these physical pleasures in accordance with reason, which in this context entails not only in such a way as to maintain their health, but also in ways that are consistent with all the other virtues.

If the virtue is fully developed, the temperate characteristically abstain from such pleasure when they have reason to with no inner conflict, having brought their desires for food, drink, and sex into harmony with their reason. In this respect, temperance is the paradigm illustration of the Aristotelian point that our desires, even our most "animal" desires, are shaped and informed by our reason, for good or ill. Many people who convert to vegetarianism as adults are familiar with this phenomenon. Some find it remarkably easy, but for others, at first, it is really hard; one's mouth waters at the thought of bacon or a juicy hamburger, and one eats one's tofu without enthusiasm. Gradually, however, the desires come into line.

Part of what goes into bringing such desires into line is coming to see them and their satisfaction in a different light. Midgley says, strikingly, "To himself, the meat-eater seems to be eating life. To the vegetarian, he seems to be eating death."[16] I have not yet managed to get myself that far, but I can at least now see my desires for meat as simply greedy and self-indulgent. "Self-indulgence" is a fortunate translation of the ancient Greek word for the vice opposed to temperance. I suspect it may be what those of Singer's readers who spoke of our selfishness in eating meat actually had in mind, for indulgence of one's own desires to the exclusion of regard for others' suffering is certainly a form of selfishness. Nonetheless, regard for others' suffering, though often the most relevant reason for restraining one's desires for certain foods, is not the only reason. We know that bottom trawling for fish is hugely destructive; we know that trawling in general is hugely wasteful because so much of the catch is thrown away. Should we not regard satisfying our desire for affordable fresh fish as self-indulgent when we have such good reason to boycott those practices, regardless of whether fish can feel pain?

Do the preceding paragraphs imply that, from the perspective of virtue ethics, one should never eat meat or fish? Consult your own grasp of the virtue and vice terms and you will surely find that the answer is no. After all, what would you think of someone who, accidentally stranded in the Australian outback without food, killed a rabbit and ate it rather than sitting down and waiting to die of starvation? Would you take it that they must be deficient in compassion? Callous? Self-indulgent? No. Of course, they might have all those vices, but their action, in *those* circumstances, gives us no reason to suppose that they have. When we try to imagine what a virtuous person, even an ideally virtuous person, would do in those circumstances, we do not usually imagine that she just resigns herself to death.

Although it has become common to say that one has a right to kill "innocent threats," the willingness to lay down one's life for other human beings in certain circumstances has always been a paradigmatic virtuous act. The deontological category of the supererogatory is, in virtue ethics, the category of actions in which, because of the particular circumstances, virtue is, in Foot's words, "severely tested" and comes through.[17] If we fail the test, we may nevertheless be fairly virtuous, but we should be humbly aware of the fact that better people than ourselves would have passed it, rather than self-righteously thinking of ourselves as "within our rights."

Nothing in our concepts of the individual virtues rules out our taking on board the idea that there may well be circumstances in which an ideally virtuous agent *would* lay down her life for a nonhuman animal. The concept of courage is just waiting for us to recognize the saving of a nonhuman animal as, sometimes, a worthwhile end, and that will enable us to say that that is what the ideally virtuous agent would do. As it is, there are some circumstances in which people at least *risk* their lives to save animal lives or rescue them from suffering, and we think they are compassionate and courageous to do so, not just cranks. As I write, those who are running interference on the Japanese whalers in the southern ocean are doing just that.

Comparing oneself unfavorably with the ideally virtuous agent is one thing; comparing others is quite different. In this connection, I should mention what Lisa Tessman has identified as one form of "moral trouble spawned by oppressive conditions," namely the form in which agents are "morally damaged [and] prevented from … exercising some of the virtues."[18] For millions of people, the business of getting hold of enough food to feed themselves and their children is a continual struggle. Whether they are getting it by killing for "bushmeat," scavenging in the rubbish of the wealthy, buying the cheapest of the limited range on offer, or surviving on the handouts of NGOs, they are in no position to pick and choose what they eat or give to their children and in no position to do much in the way of exercising compassion.

The above collection of examples is intended to illustrate the point that an action such as eating meat, which is exactly what a virtuous agent characteristically refrains from doing in many circumstances, may nevertheless be something that, in other circumstances, a virtuous agent does do. Hence, virtue ethics shares much of act utilitarianism's flexibility when applied to particular practical issues. However, "circumstances" are not just the same as "consequences." Whether my eating meat is compassionate or cruel or neither, temperate or self-indulgent or neither, is bound to depend on the circumstances in which I do it, and often (though not always) on my reasons, but only sometimes on its consequences.

SOME OTHER EXAMPLES

I now turn to considering some less familiar examples of actions that involve nonhuman animals to illustrate how virtue ethics engages with them. I begin by introducing a further virtue, namely love.

The virtue of compassion is a particular form or aspect of the virtue of love (which also gets called "benevolence" or "charity") because that virtue concerns itself with the good or well-being of others, and freedom from suffering, the concern of compassion, is part of the well-being of any sentient creature. As such it is, as noted above, the virtue called forth by Singer's utilitarian approach and supports the usual objections to commercial farming. But, remembering Regan, we should note Kant's sapient observation that the virtue of love needs to be tempered with the virtue of respect.[19]

In the Kantian context, limited to our relations with our fellow autonomous agents, love thus tempered—respectful love[20]—is mindful of others' right to make their own choices, to live or die the way they want to, even if, in our view, they thereby jeopardize their own well-being. It is thus a corrective to the arrogance of paternalism, an undue even if well-intentioned assumption of authority or knowledge. Contrary to Kant, we may say that love, or benevolence, as a virtue is not limited in its concern to our fellow human beings, autonomous or otherwise. We may think of it, for example, as also governing our treatment of pets and other animals, such as horses, that we personally love. There is no doubt that many people love their cats, dogs, and horses, delighting in their company, mourning their deaths, and willingly, on occasion, going to considerable trouble on their behalf. However, following Kant, we may also say that, as a virtue, love has to be respectful love; our love for our pets should be shaped and informed by our recognition of the ways in which their needs and their lives are their *own*, peculiar to the sorts of animals they are.

Conscientious veterinarians frequently lament the fact that many pet-lovers have a view of their animal's good which, far from being informed and shaped by reason in this way, is largely informed by anthropomorphism or, worse, their own desires. It leads them, for example, to expend their resources on expensive food or treatment that their pet would be better off without, and even if their ignorance is not the result of arrogance or selfishness, it is culpable ignorance nonetheless. It is a form of folly, stupidity, or thoughtlessness, which leads them, unwittingly, to perform actions that are cruel, inconsiderate, and disrespectful.

Veterinarians also lament, rightly, the inbreeding of cats and dogs that produces animals which win "best in breed" at shows but have been bred to have traits that give them congenital health problems. No one with the virtue of respectful love or benevolence would be party to such a practice; buying such animals is vain and self-glorifying.

From the perspective of the major non-anthropocentric environmentalists, such as Aldo Leopold and Arne Naess, respectful love is something that should extend well beyond the limits of our fellow autonomous agents and sentient others to any living thing, because they all have a good of their own. In this context, respectful love loses any connection with rights, though it is still the corrective to our arrogant assumption not only of authority but also of superiority. Arrogance is not only displayed by riding roughshod over others' rights. We are not *the* beauty of the world, nor the paragon of animals. Our sublime rationality should enable us, especially now that we know as much as we do, to recognize that the living world contains a myriad of wonders that should make us humble as well as be a source of delight.

This anti-anthropocentric, "biocentric" claim, that *any* living thing is an appropriate object of respectful love, is more familiar in Eastern than in Western philosophy and one that we may find very hard to take on board. I do, but I can certainly find much that is true in it. One obvious point the claim is making is that "being alive" is a standing reason for not killing the thing in question that has to be overridden by another reason, and that the killing should never be done triumphantly or with glee or pride. This rules out all killing for sport as well as idly swatting flies or mindlessly slashing at plants as you walk through the countryside. Like the appeal to compassion, this claim supports the usual objections to commercial farming. Less obviously, with the emphasis on the "respectful," it correctly identifies what is wrong with the idea that we might, ingeniously, produce genetically engineered versions of the familiar farm animals that were non-sentient, thereby allowing ourselves to have our compassionate cake and eat it too. Or, to take, alas, a real-life example, it correctly identifies what is wrong with the creation, as a work of art, of a transgenic rabbit that glows green in the dark.[21]

Creating non-sentient animals for no better reason than to give ourselves guilt-free meat is, at the very least, arrogant and self-indulgent, as is the creation of the transgenic rabbit, though in neither of these cases could it be said that any animal's right has been violated. (No normal rabbit has had its putative right to enjoy a good rabbit life violated by the production of a non-sentient or glowing green rabbit.)

The virtue of respectful love may also help us on at least one issue concerning wild animals, namely what, if anything, we should do about the suffering that carnivores inflict in the wild. This is rarely discussed in the animal ethics literature. When it is, some theory-driven philosophers argue that we should interfere with the lives of wild animals to prevent the pain they inflict on other animals. This is sometimes given as a justification for keeping wild animals in zoos, sometimes proposed as part of a very-long-term project: "[T]he gradual supplanting of the natural by the just."[22] However, from the perspective of environmental ethics, the latter is yet another manifestation of anthropocentric arrogance. Respectful love of wild animals wishes them the good that really is their *own*, just as respectful love of our pets does. The difference is that the lives of most wild animals are red in tooth and claw, unlike those of our pets, but respect still entails leaving them to live their own form of life, not one that we, playing God, create for them.[23]

Wild animals cannot live their forms of life if we destroy their habitats, but discussing that issue would take us squarely out of animal ethics into environmental ethics. I will say a little more about the connection between the two fields below, but for the moment let us consider some of the animals that are neither what is called "wild" nor "domestic," for example, urban and suburban feral cats and dogs, rats, foxes, rabbits, and raccoons. One salient feature these animals have in common is that they live alongside human beings, dependent on us for food and sometimes for shelter, though all are capable of surviving in the wild. Another salient feature, not one always shared for instance by commensal animals such as some birds, is that they are often, to some degree, a nuisance to us. The virtue of respectful love rules out taking *any* degree of nuisance to be a good enough reason to kill them, and similarly forbids killing them in such a way that they suffer unnecessarily, or mindlessly condemning them all as "dirty," or killing them with triumphant glee. But does it give us any more guidance?

Our concept of a virtuous agent is one of a human being with many virtues, perhaps even all the virtues, leading a human life into which many concerns must be fitted. Consult your own concept again, and ask yourself how you think a virtuous agent would treat such animals if she were aware of the fact that they were living alongside her. The answer I expect you to come up with is that there is no telling, because it would depend so much on her circumstances, and that is the right answer from a virtue ethics perspective.

Living in a rat-infested apartment block where your neighbors fear for their children's safety is one sort of circumstance; living on the edge of a suburb where you can often catch the rats and return them to the wild is quite different. Again, you might live in such a place and be busy with other concerns, and that is different from living a life in which concern for animals is fairly central. Although a virtuous agent must have many virtues, that is not to deny that two equally virtuous agents might lead very different lives wherein the exercise of one virtue might leave less room for the exercise of another. In otherwise similar situations they may, consistently with their virtue, decide to do different things. A civil rights lawyer, fighting vigorously for justice, may rarely exercise kindness or compassion on a personal level, and someone who runs a cat sanctuary may seldom have occasion to exercise justice. If both are virtuous, each must surely be glad to know that someone is doing what the other is doing, and, if an occasion offers, be willing to do something similar. The lawyer can find a moment to ring a suitable local organization about the stray cat raiding her rubbish bins, and the cat carer to sign a petition about civil rights, but human life is short. Even the most idealized, unrealistic concept of "a virtuous life" would still have to allow many different versions of it. So, the answer to what one ought to do with respect to this class of animals is "It all depends."

EXPERIMENTATION

Much experimentation on nonhuman animals can be straightforwardly condemned as cruel. Some authors[24] in animal ethics take the strange view that the act of inflicting unnecessary suffering on the sentient is cruel only if the agent feels about it in a certain way. That is, if the agent enjoys or is indifferent to the animal's suffering, then it is cruel, but if he feels neither, it is not. Perhaps these authors have been misled by the ambiguity of the phrase "an act of cruelty." It can indeed mean an act from the character trait of cruelty, or a characteristic act of a cruel person, and a cruel person is indeed someone who takes pleasure in or is indifferent to inflicting suffering on others. However, it can also mean "a cruel act," and a cruel act is nothing but the infliction of unnecessary suffering.

"Unnecessary" means "not necessary to secure some good which is worth the cost of, or suitably proportionate to, the suffering or evil involved," and hence much experimentation fails on at least one of four different counts. The first three of these are familiar in the animal ethics literature. The first is that what the suffering secures is not a good at all. The painful testing of new cosmetics on animals is the obvious example; new cosmetics are not a good, something worth pursuing or having. The old ones with new packaging will do just as well, as Singer noted decades ago. The second is that it has little or no chance of securing the good which is supposed to justify it. Closely related to this is the third—that we can, or are more likely to, secure the good in other, better ways, such as avoiding duplication of the same experiments in different institutions and pooling any knowledge gained, or by using computer modeling of human organs. The fourth count is that the good in question, even when all other criteria are satisfied, is not suitably proportionate to the suffering or evil involved.

Much experimentation on animals is cruel and thereby, in virtue ethics terms, wrong, to be condemned, and not to be done. Thinking in virtue ethics terms makes us notice something about how the issues of animal experimentation and vegetarianism differ. Concluding that eating meat is, in many circumstances, not what a virtuous agent would do because she is not willingly party to cruelty, does not do what is self-indulgent, and so on, gives one clear guidance. However, the same is not true when we conclude that most animal experimentation is cruel. The question of abstaining from experimentation does not arise for most of us, and what counts as not being a party to this often cruel practice is somewhat obscure. It rules out buying cosmetics that have been tested on animals, and it could indicate that one should refuse to save or prolong one's life by accepting an organ transplant from a nonhuman animal as one should refuse to accept a kidney commercially obtained from someone in a developing country. But this is not an issue that commonly arises. Short of refraining from any of the benefits modern medicine offers, there does not seem to be anything that most of us could do that would count as "refraining from being party to the practice."

The lack of action guidance in this case is not peculiar to virtue ethics; proponents of the other approaches, having concluded that at least much of the experimentation on animals is wrong, leave it at that. But this is unsatisfactory, because normative ethics is supposed to provide action guidance; its point is to be practical.

So we should think about the question, what would a virtuous person do when he has recognized that most animal experimentation is cruel and therefore wrong? Given that he is conscientious and responsible, he thinks about what he could do to prevent it, but he immediately encounters a problem. Animal experimentation is a practice entrenched in powerful, well-established institutions, including medical schools, medical research centers, corporate research centers and laboratories aiming for profit, and universities training students to work in such institutions. The existence of this set of interlocked institutions is something that no one person has the power to change, not even heads of state.

Does the virtuous agent therefore give up in despair? Not if she has fortitude, or even just the commonsense that is part of practical wisdom. We know that people have been similarly situated, living in a society engaged in a wrong practice that they are quite powerless, individually, to change, and we know that the good ones haven't just given up in despair. They have looked around to see what individually insignificant but collectively influential contribution they could make to bring about change.

One may be so situated that there is very little one can do as an individual, as was the case for many white people in the 1960s who deplored racism. Still, good people opposed to institutionalized racism did what they could, unremarkable as it was; they signed petitions, joined pressure groups, voted for politicians who spoke against racism, and any of us can do at least the first two to combat cruel animal experimentation.

Some people are in a position to do more. A growing number of science students refuse to experiment on or dissect animals, and in some universities, the system has been changed so that they are no longer required to do so. Many universities and corporations now have ethics committees that regulate the use of animals in research, and a significant number of experiments that are done in institutions that do not have such committees are no longer done in those that do. So some of us can join an ethics committee. However, there is no point in doing so determined to argue against every proposed experiment that uses animals; we will just be thrown off it and achieve nothing. Might it be said that that is what a truly virtuous agent would do? If you let some of the experiments through without protest, wouldn't this be hypocritical and lacking in integrity? No. Consult, again, your own understanding of the virtue and vice terms and think of Schindler, who consorted with the Nazis and "allowed" many Jews to go to the death camps without protest while he schemed to get some into the protection of his factory.[25] Far from condemning him for being hypocritical or lacking integrity, we regard him as admirable and his actions as exceptionally virtuous. Situated as he was, what could possibly count as acting better than he did? How could he have shown more virtue? By not saving anyone and getting himself killed?

More to the point, we can reflect on the career of Henry Spira,[26] one of the most effective animal rights activists of the twentieth century. His effectiveness is an excellent example of the practical wisdom that, according to Aristotle, is inseparable from the possession of any virtue in its perfected form. People with practical wisdom get things right in action, not only because they are virtuous and hence try to do what is right, but also because they are excellent at practical reasoning, at finding really good means to their virtuous ends. "Really good" in this context does not connote "especially virtuous," though given that practical wisdom is inseparable from virtue, it does connote "not vicious." "Really good" means are really *effective* means, the ones that secure one's end in the most efficient (but non-vicious) way.

Full possession of practical wisdom, and hence perfected virtue and always getting things right in action, is no doubt an unrealizable ideal. Nevertheless, as an ideal it is the standard we rely on when, having failed to get things right and do what we intended to do, we say "I wish I had realized or known that or thought of so and so; then I wouldn't have done what I did." We say this—or should—not only when our error was the result of culpable ignorance, but even when it was not and we are not blameworthy, for our concern should be not the mere avoidance of wrongdoing, or being blameless, but being effective in good action.

We admire practical wisdom in people such as Spira in particular realms of action, without having to suppose they are perfect in every way. The general end he devoted himself to from his forties until his death was not the abolition of commercial farming and experimentation on animals as totally wrong, but reducing the amount of animal suffering caused. He always looked for really effective means to that end. Among the means he rejected as useless was the "self-righteous ... hollering 'Abolition! All or nothing.'"[27] Among those he rejected as comparatively ineffective were the adversarial vilifications of animal-user industries in contrast to working with them in a friendly and cooperative way. This cooperative approach has been astonishingly effective.

It is not quite so astonishing when one looks at his unique campaigning methods and how he reasoned about securing his general end, the details of which are described in Singer's book on him.[28] One striking detail is that for each campaign, he set a specific and in some ways modest end that he thought was achievable. For example, in his perhaps most famous campaign, he did not set out to stop all testing of cosmetics on animals, but just the Draize test; not to stop all companies using it, but largely Revlon; and, indeed, not to stop Revlon from using it entirely, but to get them to take the step of putting money into seeking alternatives. This final restriction on his target exemplifies another striking detail—his non-adversarial approach. Having dealt Revlon a hefty blow with a full page *New York Times* advertisement showing a rabbit with black patches over its eyes, he gave them the let out. He met with the vice president of Revlon who had been assigned the task of coping with the fallout from the advertisement and its follow-ups, and convinced him that he was someone who, in the vice president's own words, "would listen to us and be prepared to work something out that we could live with too."[29] Taking the step that was Spira's target was, it turned out, something Revlon could live with—unsurprisingly, because he had carefully selected it to be just that.

Spira's targets were, I think, modest in two ways: in being so limited in comparison with the enormous amount of animal suffering to be combated and also in being the targets of a modest man. When I read the details of what he did, I am struck not only by how dynamic and energetic he was—a real force to be reckoned with—but also by how modest and how lacking in self-centeredness he was. He was not interested in telling people about his moral views or in proving that they were right, or in what other people thought of his integrity, except insofar as that would have a bearing on the effectiveness of his campaigns. He was utterly lacking in the vanity that makes people think *they* can achieve great things that obviously no one can achieve: For over twenty years, he was focused on getting just *something* done about animal suffering, content to, in his own words, "move things on a little."[30]

At this point, I return to the topic of moral status, because I think Spira is an object lesson to moral philosophers who write on animal ethics and insist on making moral status the central issue. As I noted above, the fourth count on which experimentation on animals may be deemed unnecessary and hence cruel is that the good secured by the experiments is disproportionate to the suffering or evil involved. With respect to clinical research specifically directed toward the good of saving and improving the lives of human beings, the question of moral status is assumed to be crucial. If the animals used in medical experiments have the same moral status as human beings, then either we should be using brain-damaged orphans instead or we should stop it entirely, as it is all wrong.

However, insisting on making this claim, or insisting that rational argument shows that it must be accepted to avoid the speciesism that is just like racism, alienates many people who might otherwise be willing to contribute to campaigns against particular sorts of experiments. I have found, when teaching animal ethics, that many students who are reluctantly beginning to feel uncomfortable about our treatment of animals and drawn toward vegetarianism, seize on this claim when they come to it, finding in it an excuse to toss out everything that has gone before. They had feared Singer had a point about animal suffering, they had feared Regan had a point about exploitation, but now they can see that it is all just silly and they need not worry about it. So, having lost all the ground I had gained, I then have to try to show them that the initial points stand independently of any claims about moral status, that much medical experimentation *is* wrong, and that they should be thinking about what they can do about it. It is uphill work. I claimed previously that the concept of moral status drastically underestimated the range of features relevant to good decision making about what to do and that we did not need to use it. In the context of medical experimentation, I would claim further that the use of moral status as an evaluative concept is actually pernicious. It is the equivalent of hollering "Abolition! All or nothing," which is guaranteed to impede progress rather than to further it.

ANIMAL ETHICS AND ENVIRONMENTAL ETHICS

I have claimed that environmental ethics has a powerful effect on the notion of moral status and that this is underestimated by authors of animal ethics literature who isolate themselves from the issues environmentalism raises. I now conclude with a discussion of the impact of environmental ethics on contemporary moral philosophy as a whole. This connection is extremely challenging, for at least two reasons.

The first reason is the difficulty about moral status in a different guise. It is no accident that some of the early environmentalists, writing at a time when utilitarianism and deontology were the only viable candidates in normative ethics, said that what was needed was an entirely new ethic.[31] They rightly saw that we cannot get rid of anthropocentrism by simply declaring that our moral principles apply to beings other than ourselves, because the fact is that neither the general utilitarian principle nor the familiar deontological principles were formulated to do so. They were formulated to govern our treatment of, primarily, each other as rational adults. The apparent ease with which the general utilitarian principle and some of the deontological principles can be applied to some of the sentient animals makes it look as though we can just keep going from there. However, this overlooks the point that, insofar as the extension is easy, the animals in question really are *very* like us.

It is not only that most of them look like us insofar as they have faces and looking into their eyes is like looking into their souls, but also that, with most of them, we can quite often read their body language and recognize what they are feeling and doing. Able to do that, we can also see, especially if we become familiar with individual animals, that like us they have lives of their own, personal preferences, and are individuals. So we can make some sense of the wrongness of causing these animals suffering and of thwarting their desires by exploiting them. When we limit our attention largely to commercial farming, hunting for sport, and cruel experimentation, the extension of some utilitarian and deontological principles seems to work.

However, as soon as we turn to other cases where the ways in which the other animals are not like us become salient, it is not so easy. When one group of human beings is disrupting the lives of another group, we find this difficult enough to sort out, but at least we have the possibility of negotiation and a compromise solution. We can't negotiate with any nonhuman animals. What can Singer and Regan say about the chimpanzee in the lifeboat? That we are to assume it would consent to our all drawing straws and agree to be the one that dies if the straw we draw on its behalf is the short one? What compromise can we reach with the elephants, who need an enormous amount of space if they are not to have a fatally destructive effect on their own habitat and thereby the other species in it?[32] We cannot do anything *with* them but cull them; we have to do something *for* them by changing ourselves and the world.

Moreover, the fact that sentient animals are not like us in some salient ways is only the beginning of the problem highlighted by environmental ethics. If we are going to look to the familiar principles of utilitarianism and deontology for the terms in which we should think about the moral significance of what we are doing, we will have to apply them to our treatment of things that are like us only in being alive, millions of which we can see only under a microscope. Indeed, we may need to go even further and apply these principles to things that are only alive in a broad sense of the word, such as ecosystems, lakes, and species.

There is a second, and related, reason why environmental ethics is so challenging. As we have recently discovered, we have, quite unintentionally, been changing the natural world on a devastating scale. We thought we were just changing the social human world to bring it about that human beings treated each other and some of the other sentient animals better and had better lives. Similarly, we thought we were only changing the natural world in very minor ways, for example by preventing various diseases and managing the way it yields food more efficiently, all of which seemed mandated or at least permitted by our moral principles. However, with industrialization the changes we have wrought in the natural world, many under the guidance of those very moral principles, have been major and disastrous. As Leopold says, we have been "remodeling the Alhambra with a steam shovel."[33]

The Enlightenment frame of thought that shaped utilitarianism and modern deontology prepares us ill for this discovery of our destructiveness, for it was shaped by Christianity. However avowedly secular the modern ethical theories may be, they presupposed that if we were all to adhere to their principles, the world was bound to become a much better place. The problem of the evil of natural disasters would remain, but everything else could be improved if we abided by the principles. How could we have known, ignorant of ecology, blind to "the tragedy of the commons,"[34] that our individually innocent actions, permitted or sometimes mandated by the principles, would collectively amount to our digging our own graves and killing off thousands of species of animals and plants?

One quick philosophical fix to this startling new discovery is anthropocentric environmental ethics, which requires no more than extending the application of our moral principles to future generations of human beings, since our obligations to them then place heavy restrictions on what we can permissibly do. However, it seems that in the areas with which animal ethicists are usually concerned, the restrictions would not be heavy. Although farming animals in order to eat them can be ruled out, because it is such a waste of resources, other activities, such as hunting, keeping animals in zoos, and medical experimentation might be positively encouraged, and cockfighting and bullfighting could continue unchecked.

The more commonly pursued approach in environmental ethics is to abandon the idea that one set of moral principles can guide us in our thinking about what to do and to use the concept of incommensurable, context-dependent, intrinsic value instead. This approach is closely allied to that of virtue ethics, as I noted above, especially if we make it clear that being as good as a human being can be reasonably expected to be involves being able to recognize things that have intrinsic value as such.[35] It may be that rather than our needing a "new ethic," virtue ethics will stand us in good stead in environmental as well as animal ethics.

Of course, to make this claim is not to deny that our understanding of many of the familiar virtues and vices will have to change, for our concepts of them have also been mostly used in relation to our treatment of, and more general responses to, each other. However, the very facts that the virtues are character traits and that someone with a virtuous character trait acts for certain reasons make the concept of a virtue open ended. Having been taught to restrain and moderate our own desires so as not to be cruel to others, including animals, in such a way that we do not even want to, some of us later find that we have reason to become vegetarian and come to recognize our meat eating as greedy and self-indulgent. In the same way, having learned more about the world we inhabit, we cease to want fur coats, modern furniture made of woods such as mahogany, new cosmetics, new cars, and the like. We acquire a new understanding of what is involved in being compassionate and temperate, rather than greedy, or in having respectful love for the natural world, rather than being arrogant.

It may be that we need to resurrect some forgotten virtues too, or take some over from Eastern thought. Roger Scruton claims that the feeding of meat products to cattle in the United Kingdom, which led to the outbreak of mad cow disease, was contrary to piety.[36] This comment initially struck me as odd, but I came to see it as insightful. In its secular conception, piety indicates proper humility, the virtue recommended by Hill as an appropriate one for environmental ethics,[37] and as such is the virtue opposed to the vice of arrogance. It also connects with the virtue of reverence, which along with "awe, ... respect, wonder (or) acceptance," is one of the terms Midgley offers as suitable to speak of things in nature which, unlike the sentient animals, are very different from us.[38] In relation to such things, reverence, like piety, can readily be part of a secular virtue ethics.

Midgley offers this list in relation to her claim that moral philosophers who employ utilitarianism and deontology have "the tendency to insist on conceptual monoculture—on bringing all moral questions together under simple headings, reducible ultimately to a few key terms"[39] and says we need more words, not fewer. In my view, she is right; given the looming environmental crisis, we need all the help we can get, and the rich vocabulary of virtue ethics is waiting for us to use it.

NOTES

1. I reiterate, in this paper, much that I have said in "Applying Virtue Ethics to Our Treatment of the Other Animals," in *The Practice of Virtue*, ed. Jennifer Welchman, (Indianapolis, Ind.: Hackett Publishing Company Inc., 2006), pp.136–55.

2. Peter Singer, *Animal Liberation* (New York: New York Review/Random House, 1975); rev. ed., New York Review/Random House, 1990; reissued with a new preface, New York: HarperCollins, 2001.

3. Mary Midgley, *Animals and Why They Matter* (Athens: University of Georgia Press, 1983), pp. 28–30.

4. Midgley, *Animals and Why They Matter*, p. 100.

5. Rosalind Hursthouse, *Ethics, Humans and Other Animals* (London: Routledge, 2000), pp. 127–32.

6. Midgley, *Animals and Why They Matter*, p. 100.

7. Midgley, *Animals and Why They Matter*, p. 10.

8. An instructive example is her strikingly titled "Duties Concerning Islands," in *Environmental Philosophy*, ed. R. Elliot and A. Gare (St. Lucia: University of Queensland Press, 1983), pp. 166–81.

9. J. Baird Callicott, "Animal Liberation: A Triangular Affair," *Environmental Ethics* 2 (1980): 311–38.

10. I owe this illuminating list to Clare Palmer, "Rethinking Animal Ethics in Appropriate Context: How Rolston's Work Can Help," in *Nature, Value, Duty*, ed. C. J. Preston and W. Ouderkirk (Dordrecht: Springer, 2007), pp. 183–201.

11. Without the label "virtue ethics," which did not exist at the time, this was the approach used by Stephen R. L. Clark in his book *The Moral Status of Animals* (Oxford: Oxford University Press, 1984), first published by the Clarendon Press in 1977. Notwithstanding his title, Clark's Aristotelian view is that "what it is right to do is found by considering what someone of sound moral character … would do" (p. viii); it is the book that converted me to vegetarianism.

12. Richard Sorabji, *Animal Minds and Human Morals* (Ithaca, N.Y.: Cornell University Press, 1993).

13. Louke van Wensveen, *Dirty Virtues* (Amherst, N.Y.: Humanity Books, 2000).

14. For a much more thorough discussion of this issue, see my *On Virtue Ethics* (Oxford: Oxford University Press, 1999), chapters 2 and 3.

15. Compare Singer, *Practical Ethics* (Cambridge: Cambridge University Press, 1979), p. 12, and *Practical Ethics* (Cambridge: Cambridge University Press, 1993), pp. 13–14.

16. Midgley, *Animals and Why They Matter*, p. 27.

17. Philippa Foot, "Virtues and Vices," in her *Virtues and Vices and Other Essays in Moral Philosophy* (Oxford: Oxford University Press, 2002).

18. Lisa Tessman, *Burdened Virtues* (New York: Oxford University Press, 2005), p. 4.

19. Kant, *The Metaphysics of Morals*, trans. and ed. Mary Gregor (Cambridge: Cambridge University Press, 1996), AK nos. 6: 449–6: 450, pp.198–99.

20. I owe this illuminating term to Christine Swanton, who introduces the idea in her *Virtue Ethics: A Pluralistic View* (Oxford: Oxford University Press, 2003), chapter 5.

21. http://www.genomenewsnetwork.org/articles/03_02/bunny_art.shtml.

22. Martha Nussbaum, *Frontiers of Justice* (Cambridge, Mass.: Harvard University Press, 2006), pp. 399–400.

23. See Rebecca L. Walker, "The Good Life for Non-Human Animals: What Virtue Requires of Humans," in *Working Virtue*, ed. Rebecca L. Walker and Philip J. Ivanhoe (Oxford: Oxford University Press, 2007).

24. Mary Anne Warren (agreeing with Regan, in her otherwise-critical discussion of his view), "Difficulties with the Strong Animals Rights Position," *Between the Species* 2 (1987): 163–73.

25. See Thomas Keneally, *Schindler's Ark* (London: Hodder and Stoughton, 1982).

26. Peter Singer, *Ethics in Action: Henry Spira and the Animal Rights Movement* (Lanham, Md.: Rowman and Littlefield, 1998).

27. Singer, *Ethics in Action: Henry Spira*, p. 53.

28. Reading Singer's whole book is best, but how Spira reasoned is encapsulated in pp. 184–192, under the heading "Ten Ways to Make a Difference," available at http://www. utilitarian.net/singer/by/1998----02.htm.

29. Singer, *Ethics in Action: Henry Spira*, p. 102.

30. Singer, *Ethics in Action: Henry Spira*, p. 198.

31. Richard Routley (later publishing as "Richard Sylvan"), "Is there a Need for a New, an Environmental Ethic?" in *Proceedings of the 15th World Congress of Philosophy* 1 (Sophia, Bulgaria: Sophia-Press, 1973), pp. 205–10.

32. Iain and Oria Douglas Hamilton, *Battle for the Elephants* (London: Doubleday, 1992).

33. Aldo Leopold, "The Land Ethic," in *A Sand County Almanac* (Oxford: Oxford University Press, 1981).

34. Garrett Hardin, "The Tragedy of the Commons," *Science*, n.s. 162 (1968): 1243–48.

35. See Harley Cahan, "Against the Moral Considerability of Ecosystems," *Environmental Ethics* 10 (1988): 195–206, n. 6.

36. Roger Scruton, *Animal Rights and Wrongs* (London: Demos, 1996).

37. Thomas Hill Jr., "Ideals of Human Excellence and Preserving Natural Environments," *Environmental Ethics* 5 (1983): 211–24.

38. Mary Midgley, "Sustainability and Moral Pluralism," in *The Philosophy of the Environment*, ed. T. D. J. Chappell (Edinburgh: Edinburgh University Press, 1997), p. 149.

39. Midgley, "Sustainability and Moral Pluralism."

SUGGESTED READING

ARISTOTLE. *Nicomachean Ethics*. Translated by Roger Crisp. Cambridge: Cambridge University Press, 2000.

CLARK, STEPHEN R. L. *The Moral Status of Animals*. Oxford: Oxford University Press, 1984.

DIAMOND, CORA. "Eating Meat and Eating People." *Philosophy* 53 (1978): 465–79.

GRUEN, LORI. "The Moral Status of Animals." In *The Stanford Encyclopedia of Philosophy*, ed. Edward N. Zalta. Spring 2009. Available at http://plato.stanford.edu/archives/spr2009/entries/moral-animal/ (accessed March 10, 2011)

HURSTHOUSE, ROSALIND. *Ethics, Humans and Other Animals*. London: Routledge, 2000.

————. "Applying Virtue Ethics to Our Treatment of the Other Animals." In *The Practice of Virtue*, ed. Jennifer Welchman, 136–55. Indianapolis, Ind.: Hackett Publishing Company Inc., 2006.

————. "Virtue Ethics." In *The Stanford Encyclopedia of Philosophy*, ed. Edward N. Zalta. Spring 2009. Available at http://plato.stanford.edu/archives/spr2009/entries/ethics-virtue/ (accessed March 10, 2011).

JAMIESON, DALE. "Zoos Revisited." In *The Philosophy of the Environment*, ed. T. D. J. Chappell, 181–92. Edinburgh: Edinburgh University Press, 1997.

MIDGLEY, MARY. *Animals and Why They Matter*. Athens: University of Georgia Press, 1983.

PALMER, CLARE. "Rethinking Animal Ethics in Appropriate Context: How Rolston's Work Can Help." In *Nature, Value, Duty*, ed. C. J. Preston and W. Ouderkirk, 183–201. Dordrecht: Springer, 2007.

SAGOFF, MARK. "Animal Liberation and Environmental Ethics: Bad Marriage, Quick Divorce." *Osgoode Hall Law Journal* 22 (1984): 297–307.

SCHMIDTZ, DAVID. "Are All Species Equal?" As revised and published in this *Handbook*. New York: Oxford University Press, pp. 628–40.

SINGER, PETER. *Ethics in Action: Henry Spira and the Animal Rights Movement*. Lanham, Md.: Rowman and Littlefield, 1998.

SWANTON, CHRISTINE. *Virtue Ethics: A Pluralistic View*. Oxford: Oxford University Press, 2003.

WALKER, REBECCA L. "The Good Life for Non-Human Animals: What Virtue Requires of Humans." In *Working Virtue*, ed. Rebecca L. Walker and Philip J. Ivanhoe, 173–90. Oxford: Oxford University Press, 2007.

CHAPTER 5

....

A HUMEAN ACCOUNT
OF THE STATUS
AND CHARACTER
OF ANIMALS

....

JULIA DRIVER

DAVID Hume believed that human beings resembled animals in a variety of important behaviors.[1] This resemblance led him to infer that animals possessed a mental
life analogous to that of human beings. These behavioral and psychological premises form the basis for an argument that animals have some form of moral standing.
Even though Hume was not as clear on this point as he should have been, my view
is that the Humean account of animal character provides the basis for an account of
moral standing that acknowledges the moral considerability of animals as resting
on a continuum with the considerability of human beings. Nonhuman animals
developed along the same continuum as human beings in terms of reason and emotion. Even if they had not, however, and even if nonhuman animals simply sprang
into existence as they now are, they would still possess moral standing on the basis
of their psychology.[2] Animals reason and feel. Because they reason and feel, we owe
them consideration on the basis of duties of humanity, though not, in Hume's very
special sense, on the basis of justice. All beings that have moral standing deserve
moral consideration.[3]

What is distinctive in Hume's view of animal mentation is not that animals
have mental lives. That view had a long history prior to Hume, but Hume also had
distinctive views about what this similarity commits us to in terms of various forms
of normativity. His views are the result of arguments about the resemblance of

humans and other animals together with a naturalistic worldview. This account of animal status is in keeping with contemporary views of mind and normativity that eschew the mysterious or supernatural as the basis for the higher moral standing of human beings. My view is that the Humean is committed to regarding nonhuman animals as similar in important respects to humans. Further, the similarities are sufficient to ground moral standing for animals, which, in turn, grounds human obligations to treat animals with compassion.

Importantly, animals resemble human beings with respect to their mental qualities. Animals possess primitive capacities for reason, and they also possess emotions. Notably, they possess sympathy. However, what prevents them from being full moral agents is their lack of the right sort of metacognitive state regarding their sympathy and how it is actually realized. They cannot approve or disapprove of these states. This is the crucial difference between humans and animals, but it is insufficient to undermine moral standing.

BACKGROUND: AGAINST HUMAN EXCEPTIONALISM ON MORAL STANDING

That human beings differ psychologically from the other animals is obvious. What I will refer to as "human exceptionalism" goes beyond this claim, however. Human exceptionalism is the view that the differences we observe between human beings and the other animals account for a very special status that human beings have, a moral status that is of a fundamentally different kind than that of the other animals, not merely a difference of degree.[4] This difference in kind accounts for the inherent superiority of human beings and sets human beings apart from the natural world.

Historically, the human exceptionalism thesis has frequently explained the difference between humans and animals by postulating something about human beings that is a mark of divine favor. Human beings alone are created in God's image; they alone have souls.[5] Augustine, toward the beginning of *The City of God*, laid out the orthodox view: "when we read 'You shall not kill' we assume that this does not refer to bushes, which have no feelings, nor to irrational creatures, flying, swimming, walking, or crawling, since they have no rational association with us, not having been endowed with reason as we are, and hence it is by a just arrangement of the Creator that their life and death is subordinated to our needs."[6] This view, I believe, Hume found unsupported by facts. There is no spark of the divine in human beings that would warrant an exceptionalist claim.[7] Of course, even if there were, another argument would have to be given as to why this would be relevant to the issue of moral standing.

The more contemporary version of human exceptionalism holds that human beings are radically different from animals in a way that renders the lives of animals fairly insignificant in themselves. What marks the difference seems to shift in writings on the subject with the growing body of evidence we have about animal capabilities. There are clearly cognitive dissimilarities between human and non-human animals, but the question that is raised with the exceptionalism issue has to do with the *significance* of those dissimilarities, assuming we abandon the outdated view that they are evidence of some sort of divine aspect that humans possess and animals do not.

Hume believed in psychological continuity between animals and human beings. Given that our psychological qualities are the qualities that confer moral standing, then the Humean will be comfortable with the view that animals, as well as human beings, have moral standing. My view is that Hume held this view. However, regardless of Hume's own views, the theoretical framework he provides allows us to argue that this is the sort of view he could have and should have articulated.

The exceptionalist view is based on a flawed methodology in which a biased sample of evidence is used as providing the basis for the view. In the case at hand, the method focuses on differences rather than similarities, and does not adequately consider the similarities in accounting for status. This general methodological bias is one that Hume took explicit exception to in other areas as well. For example, he critiques proponents of the design argument for a one-sided diet of evidence. In the case of animals, Hume believes that if we consider a balanced diet of evidence, we will see the similarities as well as differences, and the similarities are remarkable indeed.

In this chapter, I will agree with the view that human beings have higher moral status than that of the other animals, but this agreement will be based upon considerations having to do with the more sophisticated cognitive abilities human beings possess. The aspect of exceptionalism that I will disagree with is that there is some dramatic difference in kind between human beings and animals that marks us as apart from the natural world and renders animals devoid of moral standing. There are many animals who possess reason, and David Hume was one of the earlier philosophers to recognize the significance of this fact. In arguing for this position, I will defend the Humean approach against the frequently made criticism that it fails to account for animal welfare. This criticism was leveled against Hume based on his remarks on animals and the norm of justice. However, the limited scope of Hume's comments has led to misinterpretations of his claims. His account can easily be expounded, or at least expanded, to include the view that animals have moral standing, and that human beings owe animals consideration in virtue of that standing. Hume's view that animals possess many of the same kinds of mental states as human beings forms an important part of the argument.

That animals were capable of reason is a view that various influential writers held long before Hume. Two hundred years before Hume wrote on the nature of mind, a Catholic priest, Jerome Rorarius (1485–1566), wrote an essay, "That Animals Use Reason Better Than Man."[8] Rorarius was struck by the fact that animals seemed to learn from experience more adeptly than human beings (a view notably similar to that of his near contemporary, Michel de Montaigne [1533–92]).

Rorarius's essay was read and discussed by Pierre Bayle in his *Dictionary Historical and Critical*, in the well-known entry, "Rorarius." Bayle, who was carefully read by Hume, was interested in Rorarius's work because he thought that Descartes' view of animal thought was mistaken, and Hume may well be following Bayle's lead. Descartes was, and is still today, well known for the view that animals lack souls, consciousness, and mentation beyond mere mechanical reflex. Bayle noted that the Cartesian system—while it was able to preserve a sharp distinction between animals and human beings—could not account for the cases of animal intelligence that had motivated Rorarius to write his essay. For example, Rorarius reported that he had seen, in Germany, wolves who were hung as an example to other wolves, as a deterrent against killing sheep. Bayle writes of Rorarius's observation: "Rorarius...saw two wolves hung on the gallows in the duchy of Juliers; and he observes that this makes a stronger impression on other wolves than branding with a hot iron, the loss of ears, [etc.]" make on human criminals.[9] Animals seemed to be sensitive to such warnings, if one were to judge merely by their behavior.

Rorarius wasn't alone in observing that animals seemed to act with intelligence and consideration. Montaigne, in his "Apology for Raymond Sebond," recounts many ordinary examples of animals exhibiting intelligent behavior. One behavior that particularly impressed him was that of seeing-eye dogs: "I have noticed how they stop at certain doors where they have been accustomed to receive alms, how they avoid being hit by coaches and carts....I have seen one, along a town ditch, leave a smooth, flat, path and take a worse one, to keep his master away from the ditch." These sorts of observations on intelligent behavior on the part of animals were much discussed in the seventeenth century, particularly in light of the controversy surrounding Descartes' view of the mind. Descartes explicitly disagreed with Montaigne, challenging Montaigne's interpretation of the observations and arguments presented in his *Essays*.[10] Bayle wanted to introduce into the debate early observations of animals at odds with the view that they lacked any mental states whatsoever. There was a very lively debate on animal mentation, because a good many people disagreed with Descartes' views that animals were mechanistic creatures utterly incapable of thought—that is, that they were automatons, which exhibited behaviors associated with thought in human beings, yet with no thoughts of their own.[11]

The other Sentimentalists also viewed animals as possessing mental states. Sentimentalists held, roughly, that moral judgment requires emotion, and that reason alone is insufficient ground for moral judgment. This commitment had to do with a view about motivation: unless the emotions were somehow engaged, an agent would not be moved to perform an action. Reason, the province of belief, was insufficient in itself to provide this motivational force. However, the Sentimentalists did not deny reason *a* role in moral judgment and action.

In the theory of Lord Shaftesbury (1671–1713), animals lack moral qualities such as virtues because they cannot make the reflective judgments necessary to ascribe virtue to others.[12] Shaftesbury notes a distinction between goodness and virtue.

Animals can be good as animals, and as members of a given species, as long as they contribute to the well-being of the species, or even broader categories such as the "system of all animals." They cannot have virtue, however. Note that Shaftesbury ties the possession of virtue itself to the capacity to make judgments of virtue. These would be separated by Hume. In Shaftesbury's view, virtue requires a second-order affection—that is, an affection regarding another affection. Suppose that one witnesses an act of benevolence as someone pulls a small child away from a speeding car. This makes one feel good. Further, human beings have the capacity to reflect on this feeling, and form yet another feeling about the feeling itself. Not only does one feel good, but after reflection, one feels good about feeling good. When we act on this second-order feeling, we act virtuously. The person performing the benevolent action experiences a kind of self-approval and acts accordingly. This capacity to judge about and ascribe virtue is what Shaftesbury considers the "moral sense."

For Hume, being judged morally and being a moral judge should be pulled apart. One can be the object of approbation without being able, oneself, to judge and approve of the mental qualities of others. Because animals do have a mental life, it seems reasonable to hold that they possess "qualities of the mind" that constitute traits of character. These traits of character can be pleasing or not, from the general point of view; thus, it seems that animals, like persons, can possess their own form of virtue. Hume explicitly notes that we often approve of certain traits in animals. This is an unusual feature of Hume's account, and falls out of his third-person account of moral evaluation. A third-person account of moral evaluation treats a judger's act of evaluating as primary in explaining moral evaluation. For example, a view that holds that virtue is determined by what an ideal observer approves of is a third-person account. By contrast, a view holding that the virtuous action for me to perform is the one that I approve of is not a third-person account; it would be a first-person account. On this first-person account, the criterion for virtue is my approval, not the approval of an idealized third party. The relevance to the animals issue is that, for Hume, it won't matter whether animals themselves approve of their qualities of the mind. What matters to the status of those qualities as virtues is whether they are approved of from the "general" or "common" point of view. Animals may not be able to make their own judgments of virtue, because they lack the crucial sort of reflectivity, but they may be the *objects* of evaluation in terms of whether they possess virtue.

That animals possess rich mental lives was a significant part of the background for Hume's own writings on mind and animals, though only a piece of it. He was taking sides in a debate that already had a long history that stretched much farther back than Montaigne and in scope reached far beyond the writings of philosophers.[13] However, Hume was cognizant of relevant differences between human beings and animals, differences that would be reflected in his account of moral agency. The Humean, then, has a nice way of noting intuitive differences. I shall argue that animals have moral standing and that they possess pleasing "qualities of the mind," but that we have no evidence, yet, that there are animals who are themselves moral agents.

HUME ON ANIMAL AND HUMAN PSYCHOLOGY

In Hume's view, we can learn a good deal about our own nature by looking at animals and learn about animals by looking at ourselves. Any general account of mental processes must include that of animals to be complete, and Hume was convinced that nature is not radically bifurcated in terms of human-nonhuman. Simple observation of human and animal behavior leads us to infer similar psychological mechanisms.

> Tis from the resemblance of the external actions of animals to those we ourselves perform, that we judge their internal likewise to resemble ours; and the same principle of reasoning, carry'd one step farther, will make us conclude that because our internal actions resemble each other, the causes, from which they are deriv'd, must also be resembling. When any hypothesis, therefore, is advanc'd to explain a mental operation, which is common to men and beasts, we must apply the same hypothesis to both; and as every true hypothesis will abide this trial, so I may venture to affirm, that no false one will ever be able to endure it.[14]

The mental similarities had chiefly to do with the way we perceive and reason through experience. Animals have mental faculties and are subject to forming habits, just as people are. This is not to say that members of nonhuman species deliberate or engage in arguments.[15] They remember past events and develop habits of thought on the basis of these memories. Remembering and developing habits of thought is what underlies reasoning from cause to effect, and animals are capable in various degrees of this reasoning—and may be very skilled at it.

There are differences—and there is some debate in the literature as to whether or not the differences are matters of degree or kind. This issue can only be resolved when we determine what we mean by "kind." Hume clearly thought that human beings were superior to animals in terms of the quality of reasoning. He was not a follower of Rorarius on that score. He believed that animals were impressive, and at least when it comes to non-demonstrative reasoning, their capacities are on a continuum with ours. Annette Baier has argued that Hume seems to defend a continuum view of animal and human nature.[16] In this view, rather than human beings marking a dramatic shift in animal development, we mark simply another point along the same continuum of development. There could be many continua on the way to analyzing human and nonhuman nature, but the capacities that are relevant here are those of reason and sympathy.

For Hume, the function of reason is to discover matters of fact, to enable the being to arrive at true beliefs about the world around her. Human beings have highly developed capacities for investigation. We try to discover truths that are not obvious, for example, by experimenting—by engaging in complex manipulations. We try to determine why something happened, and not merely note that it happened. This capacity is useful. Noting that hurricanes tend to come in late summer and fall is useful, but it is also useful to know why that is the case. There is a depth and scope to our use of this capacity that exceeds that of the nonhuman animals.

A similar point can be made about sympathy. Human beings and animals both possess this capacity. However, human beings are capable of reflecting on the capacity and forming metalevel emotional states of the sort that Shaftesbury discusses. Hume placed a great deal of importance on this capacity with respect to moral agency. However, most important is that our sympathetic capacities are shared with animals, albeit in a more primitive form, where animals are spread out along the lower end of the continuum.

Thus, we can make some feature of our use of reason and sympathy that distinguishes the mental lives of humans from those of animals. My claim will be that there is a special kind of metacognition involving an awareness that we have no evidence animals possess, though a great deal of evidence that humans possess such metacognitive abilities. This is important because it marks an ability to think about one's own cognitive states in a critical way. And this is crucial to enhanced moral agency in the Humean view. Nonetheless, developmentally, it is still along the same continuum of the development of reason that we share with animals.

What evidence do we have that Hume himself regards human mental states as on the higher end of the same continuum with animals? Hume attributes all sorts of mental states to animals that are shared with human beings. For example, he attributes pride and love to some animals.[17] Baier notes that he makes much of similarities and continuities between human beings and animals, so that human nature is not so much different in kind from animal nature: "Indeed, one might say of Hume's version of human nature, in all its aspects, that it presents us as not radically different from other animals....Both in our cognitive habits and our emotional range, human nature as Hume sees it, is a special case of animal nature."[18]

Others take a different view. For example, Tom Beauchamp argues that Hume *did* believe that there were "kind" differences between animals and humans—that humans are capable of demonstrative reason, whereas animals are not.[19] In demonstrative reason, the conclusion of a line of inquiry follows with certainty from the preceding premises in that line of inquiry. When it comes to the sort of reason that underlies our knowledge of matters of fact, non-demonstrative reason is what is used. Non-demonstrative reason involves the sort of inference that yields probability rather than certainty, and is based on features of past experience. Again, clearly, Hume believed, animals engaged in this sort of reasoning, because they engage in cause and effect reasoning. Reason regarding matters of fact is causal inference. While there are some difficulties with this classification, the crucial point in the Humean account is that the way animals learn about the world is the same as the way human beings do. Animals may not be capable of a priori knowledge because they do not have the right cognitive structures for reasoning about ideas themselves. Beauchamp is correct that Hume believed human beings possess a kind of reason animals lack. However, this capacity will not be determinative for the issue of whether or not an animal possesses *some* moral status.

To note that animals engage in cause-and-effect reasoning, and that they have many of the same capacities to form mental habits as human beings, can deflate human pretensions to possess amazing intellectual capacities that transcend instinct. Hume notes in the same passage that,

> Reason is nothing but a wonderful and unintelligible instinct in our souls, which carries us along a certain train of ideas, and endows them with particular qualities, according to their particular situations and relations. This instinct, 'tis true, arises from past observation and experience; but can any one give the ultimate reason, why past experience and observation produces such an effect, any more than why nature alone shou'd produce it? Nature may certainly produce whatever can arise from habit: Nay, habit is nothing but one of the principles of nature, and derives all its force from that origin.[20]

However, Hume did hold that the reasoning capacity of humans was superior to that of animals.[21] This is because we developed different traits for the purposes of survival.

> There is this obvious and material difference in the conduct of nature, with regard to man and other animals, that, having endowed the former with a sublime celestial spirit, and having given him an affinity with superior beings, she allows not such noble faculties to lie lethargic or idle; but urges him, by necessity, to employ, on every emergence, his utmost *art* and *industry*. Brute-creatures have many of their necessities supplied by nature, being cloathed and armed by this beneficent parent of all things: And where their own *industry* is requisite on any occasion, nature, by implanting instincts, still supplies them with the *art*, and guides them to their good, by her unerring precepts. But man, exposed naked and indigent to the rude elements, rises slowly from that helpless state, by the care and vigilance of his parents; and having attained his utmost growth and perfection, reaches only a capacity of subsisting, by his own care and vigilance. Every thing is sold to skill and labour; and where nature furnishes the materials, they are still rude and unfinished, till industry, ever active and intelligent, refines them from their brute state, and fits them for human use and convenience.[22]

As this passage reveals, Hume believed that human beings differ from animals in that we need to employ artifice in order to survive. Our superior intellect is utterly necessary to our survival, given how little nature has provided us. Men make tools, build their shelters, weave cloth, all to make up for nature's "frugality" in allocating natural resources.[23] I believe that Hume would have been pleased to discover what current scientists find in animal behavior: some animals do employ tools. Capuchin monkeys use rocks to open nuts.[24] New Caledonian crows use leaves that they have modified to harvest insects from holes.[25] There are other examples, but this tool use is more primitive than that in humans. Our capacity for instrumental reason outstrips animals' and it's a good thing for us that it does, otherwise we couldn't survive. Hume would have been very much at home with Darwin. The Darwinian continuum view is that the fact that we are surviving, in spite of our feeble physical capacities and gifts, is evidence of the benefits of instrumental reason.[26]

For the Humean, these benefits include *artificial* virtue. Hume distinguishes artificial virtue from natural virtue. Artificial virtues, unlike natural virtues, are essentially conventional. Just as we weave cloth to make up for our lack of fur, we construct social systems to regulate behavior such as the behavior surrounding acquisition and transfer of goods, or property. We create such virtues in the course of creating social systems. The Humean regards the development of artificial virtues as representing the emergence of norms as a way of solving cooperation and coordination problems. We benefit from living in groups where certain traits, and corresponding rules, are approved of. For example, any society needs a system for the governance of property—such a system establishes norms of property acquisition and transfer. Without such a system, people would be in constant conflict over property.

The standard for evaluating the artificial norms will be pure social utility. To take one of Hume's cases, he argues that chastity in women is not something that is immediately agreeable, and, indeed, can be disagreeable in many ways. However, without a norm that censured lack of chastity, fathers would fail to support off-spring out of a concern that they were not the actual fathers of the children. This situation would be socially disruptive. In the case of property norms, animals are not included in the class of beings that require consideration. They are not included simply because solving the specific problem does not require a consideration of what animals will do. This insight is one that underlies Hume's own controversial view of justice, which will be explored in the next section.

The mental life of animals differs from that of humans in other respects that are significant to the capacity for moral judgment. Hume was sensitive to both relevant differences and relevant similarities between human beings and animals. Animals lack the kind of reflective capacity necessary for making their own moral evaluations. Animals have mental states, but they do not reflect *on* those mental states, nor the mental states of others. They cannot survey the character of other beings. They cannot have thoughts about the thoughts of other beings. They are incapable, on this view, of *metacognition*.

There is current debate about whether metacognition is possible for those animals that seem to display attitudes such as uncertainty, which suggests that they believe that they do not know something to be the case. Uncertainty is by no means the only metacognitive state animals are capable of, but it is one that has been extensively researched and shows that animals possess a general capacity to have mental states about other mental states. Uncertainty would indicate a kind of intrapersonal metacognition rather than interpersonal metacognition, but it would still be highly significant because it is committed to attributing thoughts about thoughts to other animals. This data was not available to Hume, but it is to the modern Humean, and it is worth considering how other kinds of metacognition are relevant to moral agency. The Humean would regard metacognition as important to agency and moral standing, but not solely determinative of moral standing. It is in virtue of metacognitive capacities that human beings have greater moral standing than animals, but these capacities are not solely sufficient to determine whether or not a being has some form of moral standing. In the case of moral agency, the requisite metacognition might more properly be termed "meta-affect." It involves *approving* one's mental states and the mental states of others.

Much of the contemporary work on animal metacognition has focused on the attitude of uncertainty. In human beings, we understand uncertainty to consist in a failure to know that one knows, or, weaker, a failure to believe that one knows. To experience uncertainty one needs metacognitive capabilities—one needs the capacity to form mental states about whether or not one has other mental states. One needs to be able to think about one's own thoughts or the thinking process itself. There is evidence that some animals can become uncertain about what they believe to be the case. The most famous experiment involves macaque monkeys. In these experiments, the monkeys were required to choose among several options that were represented on a computer screen. They were required to make discriminations on the basis of how many pixels a box on the computer screen contained—over 2,950 pixels, and choosing the box was correct; less than that, then choosing the box was mistaken. They could also decline the choice by making another selection, the "UR," or "Uncertainty Response." The upshot of the trials was that the monkeys "assessed correctly when they risked a discrimination error and they make URs to selectively decline these trials."[27] Further, these results matched the results for humans, leading the experimenters to conclude that "there is strong cross-species isomorphism in the use of the UR...that produces some of the closest existing human-animal performance correspondences."[28] When humans engage in the same test, they report conscious uncertainty. Similar uncertainty results are available in a dolphin trial that required the dolphin to discriminate between sounds.[29] After reporting the data however, the researcher goes on to caution against overinterpretation. Still, the evidence is highly suggestive that some higher animals, such as monkeys and dolphins, do have *beliefs* about their own mental states. Being able to have beliefs about one's mental states is important for self-regulation. Generalizing to beliefs about the mental states of others is important for evaluation, particularly on the Humean view where the main focus of moral evaluation is motive.

In order to engage in reflective endorsement, a kind of self-approval that involves approving one's own commitments, we need to have beliefs about them. As far as we know, human beings are the only beings capable of reflective moral endorsement.[30] However, metacognitive capacities are a precondition to being able to engage in reflective endorsement of one's moral sense. Animals possess this precondition. What we see in some animals in terms of their mentality are various early stages of development in the continuum leading up to the fully rational and sympathetically engaged human. We need metacognitive faculties to engage in "reflective endorsement" of our own mental states and abilities. Some animals apparently have sympathy and the ability to think about their own mental states, though there is no evidence yet that they can approve or disapprove of the sympathy they feel, or any other mental state. It is this that is required for moral agency of the sort that the Humean focuses on: judgments of virtue require being able to evaluate, and not just think about, mental qualities.

Psychologists have speculated that metacognitive capabilities evolved so that animals would be able to regulate their internal operations.[31] This ability indicates an executive level of cognition that monitors lower level cognition. The identification of metacognitive capabilities in animals has led some to further speculate on the cognitive differences between humans and animals. Domain-generality of knowledge is one identified difference: Sara Shettleworth notes that candidates for human cognitive uniqueness include the findings that animals lack "domain-general abilities to form abstract concepts, make transitive inferences, plan, and teach," though humans possess all of these cognitive capacities.[32] These domain-general capacities are important because they demonstrate an ability to infer that something that works in one specific context can be used differently in another context—presumably, though, on the basis of the same features. This requires a good deal of imaginative abstract reasoning. The stick that helps me probe a hole also helps me lever a rock. This particular difference may be of interest to Humeans because it underlies something that Hume stressed in our psychology—our penchant for generality in various forms.

While there certainly is this sort of difference in the reasoning mechanisms, it does not seem to follow that the Humean model of continuity between animal and human reason is disrupted. The continuity model presupposes that there will be differences, but the differences will be understood in terms of degrees. There is a difference between being tall and being short, but the difference is one of degree measureable along the parameter of height. An ant and an elephant differ from each other dramatically along a wide variety of measures. One is height, and along this measure the difference is one of degree rather than kind, even though it is a dramatic degree difference. Humans and plovers both engage in use of deceptive strategies, though humans are far more sophisticated about it.[33] Both creatures use deception, but the human being is capable of using it more often and more effectively. Consider a person who uses a hammer to hammer nails, but does not see that it can also be used to hammer other things, or to prop open a door, or to dig up an ant hill, and so forth. Generalizing the use of a strategy is itself a helpful strategy that animals may well lack (or may possess by degrees), but it can be represented along the same continuum of strategy utilization. However we decide to model the difference, there is no doubt there is a difference and one that a Humean will recognize. Hume himself repeatedly points to the tendency human beings have to generalize and to prefer general rules. His reason typically has to do with pragmatic concerns: it would be too inefficient and unworkable in complex social settings to do anything else.

In his essay "Of the Immortality of the Soul," Hume explicitly holds that animals have wills: "Animals undoubtedly feel, think, love, hate, will, and even reason, tho' in a more imperfect manner than man."[34] This passage provides evidence that he viewed animals as possessing *qualities of the mind*, and it is only qualities of the mind that are either useful or agreeable that qualify as virtues. Hume himself didn't seem to hold that animals make *judgments* of virtue. Though they do possess mental states, their metacognitive abilities seem to be limited in the way discussed previously. They aren't able to approve from the general point of view, and therefore cannot grasp whether another being possesses a virtue. However, the moral standing of these animals does not depend on the ability to approve from the general point of view. To hold otherwise without further argument is to conflate the capacity for moral judgment with the capacity to have morally significant qualities. The Humean approach to animals was a precursor to the utilitarian approach in that it distinguished being a moral patient—that is, being the sort of creature that can be harmed, that can be acted on morally—from being a moral agent—that is, being the sort of creature capable of moral agency. The Humean goes further than just making this particular distinction by ascribing to animals virtue traits. Hume, for example, notes animals are capable of kindness, to each other as well as to human beings.[35] It would then be a separate issue as to whether some animals perform virtuous *actions*. Anthony Pitson believes they do not in Hume's view, though he admits to some "diffidence" on this issue. He holds that, for Hume, virtue requires that the possessor exercise it with some discrimination:

> To possess a trait as a virtue, the agent must be able to assess its likely effects so as to exercise it in favor of those who will truly benefit from it. I imagine that this is a capacity which Hume would regard as belonging exclusively to persons on account of their superior understanding; and, if so, this would provide some rationale for his reluctance to treat animals as moral agents, capable of virtue and vice.[36]

However, Hume frequently noted how animals seem to exhibit various sorts of discrimination in their actions. They care properly for their young, for example. This behavior might be from instinct or habit, but then so is cause-and-effect reasoning, in Hume's account, and social animals should be capable of the sorts of traits that suit them to social living as well. This is quite in keeping with the spirit of Hume's theory. Thus, Pitson's conclusion seems unduly speculative and unwarranted. Of course animals do not possess all the virtues that humans possess, and it does not follow that all animals possess some virtue. It simply follows that animals who possess "qualities of the mind" that are agreeable or have utility from the right perspective, which is the general and impartial point of view, have the virtue in question.

Some recent empirical work on animal personality traits is supportive of the view that animals do, indeed, have characters of the relevant sort. Researchers such as Samuel Gosling and Oliver John, in studying animal psychology, examined research that tested certain animals along the same parameters humans are tested to determine personality traits. Their finding was that "Extraversion, Neuroticism, and Agreeableness showed the strongest cross-species generality, followed by Openness; a separate Conscientiousness dimension appeared only in chimpanzees, humans' closest relatives."[37] To consider one example, agreeableness, the authors looked at studies that displayed behavior in animals that demonstrated trust, tender-mindedness, cooperation, and lack of aggression. Agreeableness was displayed over a wide range of animals tested, including nonprimates, with the exception of guppies and octopi.[38] They write: "These remarkable commonalities across such a wide range of taxa suggest that general biological mechanisms are likely responsible."[39] Of course, different species display personality differently. Their example involved the extroversion/introversion set: human introverts (those who score low on the extroversion measure) tended to stay home on Saturday night; the octopus keeps to his den and tries to blend in with the surroundings. Interestingly, and relevantly, they found no markers for what they term conscientiousness in their study of other animals, except for chimpanzees. Conscientious behaviors are ones that are taken to display "deliberation, self-discipline, dutifulness, order."[40] These are the behaviors that we normally associate, in humans, with moral awareness and agency. The study indicates that traits associated with conscientiousness evolved fairly recently, and evolved in the subgroup of primates that includes chimps and humans.

There is, in principle, a difference between a personality trait and a character trait, though the two overlap. Traditionally, character traits, which include the virtues, are those qualities of the mind that are thought to involve the exercise of a will. As noted earlier, Hume explicitly holds that animals possess wills. Given that animals possess character traits that correspond to what we think of as virtues (e.g., pleasing in a certain way), then they can possess virtues as well, though not to the degree or in as sophisticated a form as human beings.

We should also be careful to distinguish the possession of a virtue trait and the performance of virtuous action. We should think in terms of three theses rather than two:

(1) Animals are moral patients, that is, they have some moral status.
(2) Animals possess mental qualities that are pleasing from the general point of view (i.e., virtues).
(3) Animals are moral agents.

The Humean holds that (1) and (2) are true. At this time, my view is that (3) is unsupported by the evidence, even in the case of higher animals. Animals engage in behaviors that we can call "good" behaviors for those animals, for example, a mother bird who risks her life to defend her babies. However, there is no evidence that the mother bird reflects on and endorses her own motives in so acting. However, the vital point is that (2) and (3) are, in principle, separable on the Humean account of virtue and moral agency. One can have virtues without being a moral agent because possessing pleasing mental qualities does not entail *moral* agency. There is the entailment of some agency or other, perhaps, but moral agency is distinctive because it calls for guidance by norms of approval and disapproval of a particular sort, of the sort that can be generalized. So far, we do not have evidence of this in animals.

Hume is clear that one can possess a virtue as a natural ability. That is, the possession of the virtue itself, the disposition to act a certain way, may not be voluntarily acquired at all. In this way, Hume's account of virtue departed dramatically from the classical conception of virtue in Aristotle, where virtue is portrayed as something people need to work at and to decide to acquire. It does not follow from the fact that the virtues may not be voluntarily acquired that the actions that are performed in accordance with the virtue are not, themselves, voluntary. A person who is naturally generous still performs *voluntary* generous actions. For example, a person who never *tried* to develop the habit or disposition of generosity because she was by nature always generous is still giving voluntarily to the poor when she makes a charitable donation. She is acting on her own desires, even if she did not train herself to have those desires. This form of action can apply to animal characters as well. Simply because an animal does not train himself to have a certain disposition does not mean that the disposition is not a virtue and does not mean that the actions the animal performs from that disposition are not, in some sense, voluntary.

Hume's view of the significance of animals' resemblance to humans contrasts sharply with Kant's.[41] Kant also thought that animal behavior resembled human behavior, but that animals *lack* reason, at least of the sort relevant to moral agency and moral standing. For Kant, the sort of reason relevant to moral decision making and moral standing requires the ability to represent rules and to conform one's behavior to a rule. Animals, in his view, lack this capacity. Kantians and Humeans share the view that animals lack certain sorts of metacognitive abilities associated with moral approval of one's own mental states and the states of others. They disagree, however, on the issue of the moral significance of the difference. The Kantian holds that the difference indicates a lowered moral status (if any) and that animals may be used as *mere* means, at least in principle. The resemblance of human and animal behavior had *some* significance for Kant, though, in terms of actual moral practice. But the significance is indirect. That is, the resemblance of animal behavior to human behavior explains why we have *indirect* duties to animals.[42] That is, hurting animals is, morally, to be avoided, but if and only if it would lead the agent to undermine his character in such a way that he would become more likely to hurt human beings. The direct duty is still only to other rational beings.

However, when we look at what Hume actually wrote on the issue of what we owe to animals, his view is murkier than one might hope. He never directly engages with the issue of moral status. The Humean must work to develop an account of moral standing using Hume's views about the nature of animal cognition. This Humean view will hold, as I have argued in this section, that moral standing is an issue that should be considered separately from the issue of the possession of morally significant properties such as virtue or agency. One can be an appropriate object of moral consideration even if one is not a moral agent.

MORAL SENSE AND MORAL STANDING

Hume does not attribute moral sense, or the ability to engage in moral judgment, to animals.[43] They cannot take the general point of view required for warranted moral approval, which has sometimes been called the moral point of view. Beauchamp believes that the premise that animals have no such capacity was Hume's considered view.[44] Setting aside matters of textual interpretation, this claim is compatible with the claim that there is a kind of proto-moral sense in animals. Hume noted that, though animals do not possess the moral sense, they do possess sympathy, which is one requirement for the moral sense. He seems to regard the sympathy as a kind of contagion—the feelings of one animal are communicated to another through the medium of sympathy. For example, fear, anger, and courage are communicated via sympathy in such a way that the animal will feel it without perceiving the original cause—thus, there must be sympathetic engagement of a sort, on Hume's view. Another bit of evidence he uses is the phenomenon of dogs hunting in packs. He notes that, though they have "little or no sense of virtue and vice," they do experience pride and humility, love and hatred; it is just that they lack the mental abilities to reflect on these passions. The causes of the passions, for them, he judges, are in the body.

The modern Humean has more empirical evidence available on animal sympathy. Scientist Frans de Waal has written extensively, and controversially, on animal empathy and sympathy.[45] His use of "empathy" is similar to Hume's "sympathy" in that it refers to the capacity to respond to the emotions of others—and de Waal focuses on the emotion of distress. Sympathy, on his view, is more focused than empathy, and, specifically, is other directed—motivating individuals to try to alleviate the other individual's distress. Other forms of empathic engagement may be self-directed, in the sense that perception of the sadness of others might make the individual try to alleviate his distress that the empathic engagement generates. Researchers visited homes in which family members were supposed to pretend to sadness—in the form of crying—or some other distress. Not only did the children in the households, as young as one year of age, display comforting behavior, but some pets in the household did as well: they "appeared to be as worried as the children by the 'distress' of a family member, hovering over them or putting their heads in their laps."[46]

The issue of empathy and sympathy *deficits* in animals is an engaging issue. In human beings, we know that empathy deficits are associated with psychopathology, for example. There are also empathy deficits that are associated with autism spectrum disorder, though the Humean would note that in the case of autism there is a *higher-order concern* to be a good person and to act appropriately that is lacking in psychopaths. Some writers, such as Jeannette Kennett, have used empathy deficit disorders as evidence against a Humean account of moral agency. However, the existence of these disorders and the ways in which, for example, autistic moral agents are able to act on norms regardless of the empathy deficit, does not settle the question as to whether Hume or Kant is right on whether or not moral action requires antecedent desire, that is, *basic* antecedent desire that is not itself the result of the independent recognition of a moral reason. Humeans are warranted in arguing that autistic moral agency can be accounted for on their account via more general desires (such as a desire to be a good person) or higher-order desires (such as a desire not to have malicious desires). The speculative issue raised for this chapter has to do with whether there are equivalents in the animal kingdom. Are there autistic chimps, for example, and how does the empathic deficit display in terms of their behavior?[47] That research might shed some light on cognitive difference between humans and animals in terms of empathic behavior.

Hume recognized that animals have likes and dislikes that they can communicate to others. There is behavior that they do not like; a dog does not like being struck, and a mother dog will not like her puppies being struck. "The affection of parents to their young proceeds from a peculiar instinct in animals, as well as in our species."[48] There is no telling what goes on in the dog's mind when he avoids the person who tends to strike him. He may just regard that person as something to be avoided, like a hole in the ground or a raging current. It would seem that in Hume's view, the animal cannot recognize the person as having a morally bad *character*, as opposed to simply possessing a bad quality. Thus, simply judging by the animal's behavior, we cannot tell if the animal is making even a primitive evaluation of character.

Given his enlightened view of the continuity between animal nature and human nature, one might have thought that Hume would have been clear, and enlightened, on the moral status of animals as well. I noted earlier that the issue of moral status is distinct from the issue of moral agency. I argued in the previous section that on Hume's account it should be possible for animals to possess virtue and that it is reasonable to suppose that some in fact do, particularly the social animals. But one *could* argue that even though they can be virtuous, they lack moral standing in the sense that they are not owed any particular form of treatment, or consideration, from rational beings such as ourselves who *are* capable of moral judgment. What little Hume actually said on this subject seems to indicate that, in terms of *justice*, we owe nothing to animals:

> Were there a species intermingled with men, which, though rational, were possessed of such inferior strength, both of body and mind, that they were incapable of all resistance, and could never, upon the highest provocation, make us feel the effects of their resentment; the necessary consequence, I think, is that we should be bound by the laws of humanity to give gentle usage to these creatures, but should not, properly speaking, lie under any restraint of justice with regard to them, nor could they possess any right or property exclusive of such arbitrary lords.[49]

At the end of the passage he then invites comparison with animals. The disparity in power is such that they would never be able to press any claims of justice against us. I promised earlier in the essay to show how the Humean could develop an account of animal moral standing while remaining largely faithful to what Hume said. The key is that it does not follow from what Hume was saying specifically about *justice*, that we have no duties to animals or that animals are deserving of no moral consideration.[50] We simply have no duties of justice in the sense required in his theory of justice.

In Hume's view, virtues such as justice displayed a triumph of human ingenuity. It would be surprising to see this replicated in animals who lack the social organization necessary to give rise to the social concerns that made justice helpful, perhaps even necessary, for human well-being. Hume famously distinguishes the natural from the artificial virtues. The virtues animals may possess at least at the primitive, or proto-virtue, level in his view would be natural, but not likely the artificial. Justice is an artificial virtue; its usefulness rests on convention: "there are some virtues, that produce pleasure and approbation by means of an artifice or contrivance, which arises from the circumstances and necessities of mankind."[51] Examples of artificial virtues are promise keeping, justice, and chastity. There is nothing naturally pleasing in these qualities. Rather, to be pleased regarding these traits one needs to reflect on the systematic usefulness of the traits. Their usefulness renders them virtues. In the case of natural virtues, however, such as benevolence, we are pleased by the individual instances in which the virtuous agent displays the disposition. Again, the underlying mechanism is sympathetic engagement, though in the case of the natural virtues, it is unmediated by consideration of systematic usefulness. In the case of justice, the idea is that norms of justice develop among social creatures such as ourselves because they are needed to regulate our behavior in nondestructive ways that are beneficial overall. Given other features of our natures—the facts that we tend toward selfishness and favoritism and that we are roughly equally matched in terms of strength and abilities—the rules of justice are necessary to maintain any sort of social order with respect to property. Human beings and animals, however, are not equally matched in Hume's view. Thus, we have no need, or necessity, to respect property rights for animals in order to maintain a stable society.[52] Hume's view that justice is *limited* to contexts in which interacting beings can make their demands felt on each other has struck many writers as implausible, not simply as it regards animals.[53] Whether we are talking about animals, or some other type of being with equal, even superior, cognitive capacities, it seems too harsh to claim that justice is a norm that applies only to our dealings unless those beings can make their demands felt. This claim seems to deny standing to sentient creatures simply on the basis of *weakness*, which is surely not a morally relevant consideration.

However, a careful reading shows that Hume is not committed to the view that animals have no moral standing when it comes to what we today think of as requirements of justice. He is simply committed to the view that they have no property rights in justice. Any misreading of Hume is due to the fact that when we currently use the word "justice," we have in mind something much broader than he had in mind. His view of justice is focused on the acquisition and transfer of property within society. Beings that would sit by and not be able to defend their "property" are not, as a matter of justice, entitled to it. This does not however mean that such beings lack moral standing, that it is impossible to treat such beings unfairly or in a morally reprehensible way. A dog that has worked long and hard for his human owner deserves benefits, not a kick.

This raises an interesting issue in interpreting Hume. What is it to "make one's demands felt"? Is the bear that kills a man over a loaf of bread making his demands felt? Hume's example in the above passage seems to indicate that it is force of physical strength that matters, and that in order to maintain human society against people exerting this strength to make their demands felt, we need a system of justice and rights; otherwise, there would be chaos. We have nothing like this to fear from animals as a group. There may be an occasional conflict between a person and an animal over an object in which the animal wins, but that is rare. As a group, animals are no threat to humans. Of course, as a group, humans are no threat to some animals. We've tried very hard to eliminate rats, roaches, and some other pests as a group, at least in certain locations, so if that were what establishes the relevant criteria, we have failed. This does not seem to make the account consistent with what we know about other animals.

In comparing Hume and Kant on the treatment of animals, Christine Korsgaard has taken the following line on what she regards as Hume's considered view:

> I noted that most people seem to hold that we should not kill or hurt the animals unless we have a good reason, but also that any reason except malicious fun is probably good enough. In the same way, Hume's "laws of humanity" do not clearly forbid us to use the other animals in any way that we might find convenient. We are, after all, as Hume says, stronger than they are, and able to outsmart them. And since we do not need their willing cooperation, and can extort their services, Hume thinks we owe them nothing in return. Although he urges kindness when it is not inconvenient, his view is apparently that when one kind of animal is able to control the others, as human beings are, that is the end of the matter.[54]

I do not believe the texts support this interpretation of Hume. Arthur Kuflik argues that Hume really could not have meant that it is by dint of physical force that we owe justice to those who can exert such force against us.[55] For Hume, justice, as an artificial virtue, is itself justified on the basis of social usefulness. Under some circumstances, such as extreme material deprivation or a utopian plenty, justice does not come into play because it serves no purpose. This raises an issue. As Kuflik notes, if there are groups within a society that the society could "do without," then would the remainder owe them any considerations of justice?

Kuflik argues that much attention has been paid to the literal reading of the passage on justice and animals, without adequate attention to what Hume says elsewhere. The best view, he argues, is that Hume is overstating his case in this passage; to hold otherwise is to uncharitably attribute to him a very implausible view, the view that justice would cease to be a virtue under circumstances in which injustice can be "perpetrated with impunity."[56] It is not simply that this view is implausible: it conflicts with other claims Hume made, notably in the *Treatise*, to the effect that there must be a distinctive motive to honesty and justice, and not simple *private interest*—because simple private interest itself is the source, frequently, of violence and injustice.[57] It cannot be the case that no injustice exists where one's private interests will be enhanced by a vicious action. Thus, we violate norms of justice at least some of the time when we act to promote private interest. Simply being able to "get away with it" does not nullify the norms of justice.

An important distinction exists between public interest and private interest. In discussing the artificial virtues, Hume is careful to note that the best strategy for promoting public interest, in general, which itself is crucial for preservation of private interest, is to hold to conventions, or rules, that one might determine do not yield the best consequences in some situations.[58] This is due to two factors—the first is that private interest varies from person to person, even though public interest does not. However, individuals have differing opinions about what, in particular, is in the public interest. "We must, therefore, proceed by general rules, and regulate ourselves by general interests, in modifying the law of nature concerning the stability of possession." Further, individual acts of artificial virtue may not promote either public or private interest. Rather, it is adherence to the norms in general that systematically promotes both public and private interest. As animals do not fit into either public or private spheres of interest, artificial virtues have no scope with respect to the animals' interests.

Kuflik notes that Hume does not make the same argument with groups such as Native Americans and women with respect to European males, even though these groups at the time seemed quite vulnerable to exploitation by European males. In the case of women, Hume indicated that their "charm" and "insinuations" could come to their rescue, so to speak.[59] Kuflik notes, however, that many women lack charm. For every Mme. de Pompadour there are numerous women with no rights or privileges granted them by stronger men. And Hume describes their situation as a "severe tyranny." This leads Kuflik to the reasonable view that for Hume it wasn't physical equality—equality in physical power—that mattered, but, rather, mental or intellectual equality. Animals are clearly mentally inferior to autonomous humans, as they cannot argue with us.[60] Korsgaard notes as well that it is significant (for Hume's view) that we can outsmart animals. If Kuflik's interpretation is correct, then what Hume means by making one's "demands felt" has to do with the ability to argue for one's interests, not the ability to physically force others to respect them.

This interpretation would make Hume similar to Kant: the reason we make a sharp distinction between animals and humans—in Hume's case when it comes to justice—has to do with their inability to reasonably press claims. Mental inferiority of certain sorts is responsible for diminished moral standing. Though, for Hume, while mental inferiority may explain why animals will not be subject to the norms of *justice*, it does not lead to complete lack of moral standing. Hume's emphasis on the laws of humanity can provide a plausible response to Korsgaard. I grant that Hume was not as clear on this point as he could have been. His remarks, however, indicate a strategy for today's Humean. There is private interest and public interest, and these are often discussed by Hume in explicating what is at stake in artificial virtue. Another distinction would be that between private interest and common interest. Common interest consists of the combination of individual interests (i.e., private interests). Common interest is either honored or promoted via the natural virtues. We have a duty of humanity to promote the common interest, and animals' interests count as part of the common interest.

In virtue of their sentience, they fall into the class of beings with interests. All animals have welfare interests that are tied to their good. These are interests associated with living a good life for them. They have, for example, interests in avoiding hunger and pain. Some animals may have more sophisticated interests that go beyond living minimally good lives in terms of survival and bare comfort—for example, they may have conscious desires and aims, such as a desire to advance through a social system (as with apes).[61] The duty of humanity obtains independent of any social conventions whatsoever. A person who causes misery to an animal is bereft of the natural sentiment of sympathy. Not to the extent that an abusive parent is, not to the extent that a murderer is, and so forth, but nevertheless, if a person who is thoughtful and rational, and able to reflect on the suffering of others, fails to see a reason not to cause suffering to an animal, this fact does denote a failure of character.

The Humean can soften the criticism by holding that the suffering of an animal only provides us with an easily defeasible reason to avoid causing it. This position raises another set of familiar problems. Why *easily* defeasible? I take it that this is at the heart of Korsgaard's problem with the Humean picture. If it is "very easily" defeasible, then the animal's suffering can be outweighed by a human whim, a desire for mild amusement. This form of defeasibility would make the reason to refrain from harming animals much too weak. The contemporary Humean could argue, however, that the duty of humanity with respect to animals is not so easily defeated. It is defeated only in conflict with *duties* to other human beings—duties to refrain from causing pain to other human beings, and duties to alleviate their misery. In this way, a person's mere whim would not outweigh the animal's suffering. The framework is there in Hume's writings.

Further, there are animals that have a kind of society with us—pets and other domesticated animals that human beings interact with. Might it be the case that this relationship sets up a separate class of artificial norms governing interactions between humans and animals in the same society? As a group, again, they lack the strength and coordination to make demands against human beings, but they possess other capacities. These capacities are what make them likely candidates for domestication in the first place. It is likely, for example, that dogs became domesticated by human beings in part because their own pack social structure made them amenable to cooperative social structures.[62] In the case of pets, they served a useful function of some sort, which was either manifested in their behavior or their appearance. In the case of dogs, they helped to herd other animals, they kept other predators away, and they displayed a childlike affection that appealed to us. When they were hungry, we were motivated to feed them—we wanted to. What of, then, the individual dog who is too old to work, mangy, and decrepit? The Humean could argue that, in the case of animals that are more than simply intermingled with us, but also have actual society with us, our obligations extend further. It may not be justice, and it may be more specific than beneficence. There is a kind of debt of gratitude or friendship that does not exist in the case of animals bearing no connection to humans.

Hume was committed to the force of generality in our thoughts. So, the reference to the "*laws* of humanity"[63] is not off the cuff for Hume. Unlike Adam Smith, who seemed to believe that the duty of humanity was completely unnecessary for social cohesion, Hume believed in a very substantive duty of humanity, or, more generally, benevolence.[64] In interpreting Hume, one does need to tread lightly here, because Hume seldom writes of duty, and when he does use the term, it is not clear that he meant something corresponding to a strict obligation. Indeed, in the *Treatise* he clearly separates acting from a motive of benevolence from strict obligation, but then speaks of a "duty" of humanity. In the passage in question Hume writes: "Tho' there was no obligation to relieve the miserable, our humanity would lead us to it; and when we omit that duty, the immorality of the omission arises from its being a proof, that we want the natural sentiments of humanity."[65]

Hume is talking about rendering positive aid to those who are suffering. It is generally thought that positive duties to aid are harder to justify than negative duties to avoid causing harm. In the case of animals, the Humean could argue that there is a duty to avoid causing harm, even if there is no positive duty to relieve misery. In both cases, failure to abide by norms of benevolence is evidence of a flawed character, though the failure of the duty to avoid causing harm may be much more serious. The significance of this should not be underestimated. For Hume, moral evaluation centers on character evaluation. Actions are evidence of character. A failure of benevolence is a serious moral failure. Humanity provides a kind of correction for our behavior and is of special interest in understanding normative relations between rational beings of unequal power.

To continue the above-quoted passage: "A father knows it to be a duty to take care of his children: But he has also a natural inclination to it. And if no human creature had that inclination, no one cou'd lie under any such obligation."[66] Hume seems to be saying that in order for there to be a duty to take care of children, there has to be a natural "inclination" to do so. This inclination marks a crucial difference between the natural and artificial. If we fail to perform actions that reflect this duty, we give evidence of immoral, vicious motivation, the lack of "natural sentiments." And note the example he uses—that of a father's duty to his child. Children are another class of vulnerable beings. Humanity requires of us that parents take care of their children, and that they treat them well. We have this duty independent of human convention. Correspondingly, one could argue for a duty to treat animals well, too, that exists independently of human convention. And this is the force of his claim regarding the duty of humanity as it applies in his thought-experiment regarding our treatment of rational, though nonhuman, beings. Far from a cold, chilling view of duty, it is actually very expansive. We do have duties to the vulnerable that are independent of convention. Of course, the basis for the duty is inclination, sympathy, the propensity that we care for those who suffer. It is clear that Hume believes animals are covered by this duty or law, otherwise he would not appeal to it in the thought-experiment.

This seems right to me. The Humean would note, plausibly, that the moral responsibility for taking care of children is not *merely* conventional, even though there may be conventional ways in which it is satisfied—for example, in some places by sending one's child to school, in other places by teaching hunting techniques, and so forth. Taking care is here a positive duty. In the case of animals that bear no connection to us, notably wild animals, there may not be positive duties to aid them, though benevolence and good moral character may dictate to us that we do so. In any event, there will be negative duties to not harm them, or at least not harm them for trivial reasons.

Joyce Jenkins and Robert Shaver argue that Hume's intention is to put humanity before justice, and in the above passage, to hold that justice can get in the way of humanity.[67] For example, one might, as a matter of justice, force a hard-working, poverty-stricken individual to pay back a small debt to a fabulously wealthy individual. This rule of justice runs counter to benevolence, and may even run counter to a more expansive sense of justice as fairness, if we think, for example, the wealthy person has less personal merit or virtue than the poverty-stricken person. Rigid rules may be necessary when it comes to institutionalizing property rights. However, when beings of unequal power interact, benevolence is better realized on a case-by-case basis. They use the example of taking a bone away from a dog when continuing to play with the bone might harm the dog. "Justice" does nothing to help the dog in these circumstances. However, acting to promote the dog's interests, generally, that is, acting on the dictates of humanity, will help the dog.

Humanity might also play an important role in shoring up convention. Shaver elsewhere argues that humanity *is* needed for society—indeed, that justice is not independent of humanity, but presupposes it:

> Justice presupposes humanity, where humanity is understood as the concern for others that operates between unequals and especially between parents and children....If so, we have as strict an obligation to humanity as to justice, since both obligations derive their strictness from their instrumental role in keeping society running.[68]

Does this claim support Korsgaard's dim view of Hume on the moral status of animals? Do we need animals to "keep society running?" Yet the details of *how* humanity supports justice are very important. Hume held that trust is crucial to sustaining justice in a society. Instilling trust, and trustworthiness, in one's children is a very important duty of parents. They need to do this in order to sustain an orderly, productive society. Social order is beneficial, thus humanity in treating one's children well coincides with interest; the humanity underlies conditions necessary for artificial virtues to actually work. There is a distinction to be made as well between "interest" and "private interest," which elsewhere Hume notes in *contrast* to justice. Because of his views on human nature, and our universal capacity for sympathy, "interest" should be understood broadly.

If this is the case, and we take "interest" broadly, then it is in the "interest" of members of society to treat each other well, according to norms of benevolence, in order that conditions amenable to artificial justice take root. This situation spills over into our relations with animals, which constitute another vulnerable category of beings with whom adult humans interact. Thus, for animals that have a connection with us, there are positive duties to aid.

Animals that exist "intermingled" with humans are owed humanity in spite of their mental inferiority because the *general* concern for humanity—the general concern for the vulnerable—is important to keep society running smoothly. Annette Baier noted that in Hume we see that natural virtues such as kindness, and kindness to the vulnerable, are important to sustaining society.[69] True, only some animals figure into human households. However, Hume often noted that our minds worked according to *general* principles, and in my view he would be, or at least should be, friendly to the thought that we respond sympathetically to animals as well as people. We owe animals humanity, and virtuous individuals treat animals well, even when those animals do not pose a threat to them and even when there is no point to negotiating with them because they lack the necessary cognitive capacities. What is not owed to animals is respect for property rights, which are the sorts of artificial consideration that are of no concern to them.

Conclusion

Hume is often credited with a radically different view of animals that departed significantly from previous writers. I have argued that this claim is not entirely true, because many writers before Hume clearly attributed mentation to animals and vociferously disagreed with Descartes on this issue, often anticipating central claims made by Hume. However, the details of Hume's account are distinctive in his theory of a continuum of rational and affective capabilities between animals and human beings. This theory led him to a view of animals that depicted them as having moral status as sensitive but intellectually vulnerable creatures. If one mistreats an animal, one has not violated a norm of justice, in his view, but the norms of justice are understood so narrowly that this fact does not pose a problem for the claim that animals deserve moral consideration through a duty of humanity.

Acknowledgments

I would like to thank Tom Beauchamp and Yashar Saghai for their comments on drafts of this chapter.

Notes

1. Most references to David Hume's work today cite David Fate Norton and Mary J. Norton's critical edition of *A Treatise of Human Nature*, vol. 1 [THN] (Oxford: Clarendon Press, 2007) and Tom L. Beauchamp's critical edition of *An Enquiry Concerning the Principles of Morals* [EPM] (Oxford: Clarendon Press, 1998). They use a universal system of reference, which is adopted here.

2. Note that this standard only commits one to a sufficient condition for moral standing, not a necessary condition. There may be other qualities that are also sufficient for moral standing.

3. Moral standing and being worthy of moral consideration are not conceptually identical, even if it turns out that the only beings we ought to treat well are those possessing moral standing. I would like to leave open the possibility that we owe good

treatment to beings that lack moral standing on their own (perhaps non-sentient living organisms, for example).

4. "Human exceptionalism" can be understood relative to different claims. For example, one could be a human exceptionalist regarding tool use, and hold that only human beings use tools. The type I consider here, however, is that relative to the issue of moral standing. José Bermúdez discusses the contrast between human exceptionalism and the continuity thesis with respect to language and thought. I use the term with respect to moral standing. See his discussion in "Animal Language and Thought," *Routledge Encyclopedia of Philosophy* (London: Routledge, 1998), vol. 1, pp. 269–272. In addition to believing in a continuity between animals and humans regarding thought, and the development of crucial features of our psychology, Hume believed this continuity underlay moral standing. As I hope to argue, the text seems to support this view since the fact that they can feel makes them appropriate objects of benevolence.

5. *The Bible: Authorized King James Version*, ed. Robert Carroll and Stephen Prickett (New York: Oxford University Press, 2008), p. 2. Genesis 1:26: "And God said, Let us make man in our image, after our likeness: and let them have dominion over the fish of the sea, and over the fowl of the air, and over the cattle, and over all the earth, and over every creeping thing that creepeth upon the earth."

6. Augustine, *The City of God*, trans. Henry Bettenson (New York: Penguin Books, 1986), pp. 31–32.

7. Consider what Hume writes, for example, on animal souls in "On the Immortality of the Soul," in *Dialogues Concerning Natural Religion, with Of the Immortality of the Soul, Of Suicide, and Of Miracles*, ed. Richard Henry Popkin (Indianapolis, Ind.: Hackett Publishing Company, 1980), p. 95: "The souls of animals are allowed to be mortal; and these bear so near a resemblance to the souls of men, that the analogy from one to the other forms a very strong argument."

8. For an extremely interesting discussion of Rorarius and the Rorarius entry in Bayle, see Dennis Des Chene, "*Animal* as Category: Bayle's 'Rorarius,'" in *The Problem of Animal Generation in Early Modern Philosophy*, ed. Justin E. H. Smith (New York: Cambridge University Press, 2006), pp. 215–234. According to Des Chene, Rorarius's editor likely omitted a qualifier, *seape*, which means "often," from the title, p. 215. This omission makes Rorarius's claim look more radical than he likely intended.

9. Pierre Bayle, *Historical and Critical Dictionary: Selections*, ed. and trans. Richard Henry Popkin and Craig Brush (Indianapolis, Ind.: Hackett Publishing Company, 1991), pp. 230–31.

10. Descartes discusses Montaigne's views in a letter to the Marquess of Newcastle, 23 November 1646, translated and reproduced in *The Philosophical Writings of Descartes*, ed. and trans. John Cottingham, Dugald Murdoch, Robert Stoothoff, and Anthony Kenny (New York: Cambridge University Press, 1997), vol. 3, p. 302. The Complete Essays of Montaigne, ed. Donald M. Frame (Stanford, CA: Stanford University Press, 1965), p. 340.

11. It should be noted that Descartes himself never advocated wantonly cruel practices towards animals, as is sometimes thought. Indeed, he had a pet dog, "Monsieur Grat," of whom he was quite fond. For discussion of this issue, see Justin Leiber's "'Cartesian' Linguistics?" *Philosophia* 18 (1988): 309–46.

12. Anthony Ashley Cooper, 3rd Earl of Shaftesbury, *An Inquiry Concerning Virtue, or Merit*, ed. David Walford (Manchester: Manchester University Press, 1977).

13. See Stephen Clark and Aaron Garrett's contributions in this volume for more information on historical context.

14. THN 1.3.16.3.

15. See section 9 of Hume's *An Enquiry Concerning Human Understanding*, ed. Eric Steinberg (Indianapolis, Ind.: Hackett Publishing Company, 1977).

16. Annette Baier, *Postures of the Mind* (Minneapolis: University of Minnesota Press, 1985), p. 147.

17. THN 2.1.12 and THN 2.2.12.

18. Baier, *Postures*, p. 147.

19. Tom L. Beauchamp, "Hume on the Nonhuman Animal," *Journal of Medicine and Philosophy* 24 (1998): 322–35.

20. THN 1.3.16.9.

21. Hume, "Of the Immortality of the Soul," p. 92.

22. David Hume, "The Stoic," *Essays Moral, Political, and Literary*, ed. Eugene Miller (Indianapolis: Liberty Classics, 1987), p. 23.

23. I imagine Hume is being sarcastic here. A contrast with Hume's *Dialogues Concerning Natural Religion* comes to mind. The view of nature we get there is not pretty. Nature accounts for all the good in our lives, but all the bad as well, and it can get very bad indeed. It isn't that nature deserves credit for what's good; nature deserves credit for *everything*. Even the way we human beings think and live. It conforms completely to his view of nature as a blind force, not guided by the hand of an intelligent designer— and certainly not an intelligent, benevolent designer.

24. Elisabetta Visalbergh and Dorothy Fragaszy, "What is Challenging About Tool Use? The Capuchin's Perspective," in *Comparative Cognition: Experimental Explorations of Animal Intelligence*, ed. Edward A. Wasserman and Thomas R. Zentall (New York: Oxford University Press, 2006), pp. 529–554.

25. Sara J. Shettleworth discusses this and similar examples in *Cognition, Evolution, and Behavior* (New York: Oxford University Press, 2010), pp. 399ff.

26. See Robert J. Richards, *Darwin and the Emergence of Evolutionary Theories of Mind and Behavior* (Chicago: University of Chicago Press, 1987), pp. 106ff.

27. J. David Smith, "The Study of Animal Metacognition," *Trends in Cognitive Science* 13 (2009): 389.

28. Smith, "Study of Animal Metacognition," p. 389.

29. Smith, "Study of Animal Metacognition," p. 390.

30. Christine Korsgaard has developed a line of interpretation of Hume in which reflective endorsement figures prominently. I agree with the general tenor of the framework that Korsgaard develops, though I disagree on certain features of the line she develops for Hume. Basically, there are two strands that run through Hume on reflective endorsement, and my view is that the endorsement is based on our sympathetic responses *when those responses have properly been corrected by reason*. I agree that this is not consistent with everything that Hume said on the subject, but I believe this line is the best way to pursue a Humean form of constructivism that is substantive rather than merely formal. For more on Korsgaard's views, see her excellent discussion in *The Sources of Normativity* (New York: Cambridge University Press, 1996), pp. 51ff.

31. For example, see Diego Fernandez-Duque et al. "Executive Attention and Metacognitive Regulation," *Consciousness and Cognition* 9 (2000): 288–307.

32. Shettleworth, *Cognition, Evolution, and Behavior*, p. 556.

33. See Shettleworth, *Cognition, Evolution, and Behavior*, for more on animal deception.

34. Hume, "Of the Immortality of the Soul," p. 92.

35. EPM, Appendix 2.

36. Anthony Pitson, "The Nature of Humean Animals," *Hume Studies* 19 (November 1993): 311.

37. Samuel Gosling and Oliver P. John, "Personality Dimensions in Nonhuman Animals: A Cross-Species Review," *Current Directions in Psychological Science* 8 (1999): 69–75.

38. The animals studied also included dogs, cats, donkeys, pigs, rats, hyenas, as well as other monkeys. Amazingly, even guppies and octopi displayed personality commonalities with humans along the Extroversion and Neuroticism ranges. The authors also note that the evidence needs to be interpreted cautiously. It may be that some animals score low in certain traits simply because the appropriate environmental prompts have not been identified.

39. Gosling and John, "Personality Dimensions," p. 70.

40. Gosling and John, "Personality Dimensions," p. 70.

41. Immanuel Kant, "Duties Towards Animals and Spirits," in *Lectures on Ethics*, trans. Louis Infield (Indianapolis: Hackett Publishing Company, 1930), pp. 239–41.

42. Kant is explicit about regarding any duties we have to animals as indirect. That is, we ought to treat animals well but only because treating them well makes it more likely that duties to human beings will be respected. In his *Lectures on Ethics* in the section entitled "Duties Towards Animals and Spirits" (pp. 239–240), he writes that, "so far as animals are concerned we have no direct duties. Animals are not self-conscious and are there merely as a means to an end. That end is man....Our duties towards animals are merely indirect duties towards humanity....If a man shoots his dog because the animal is no longer capable of service, he does not fail in his duty to the dog, for the dog cannot judge, but his act is inhuman and damages in himself that humanity which is his duty to show towards mankind."

43. THN 2.1.12.5, where he writes that "animals have little or no sense of virtue and vice."

44. Beauchamp, "Hume on the Nonhuman Animal," p. 328.

45. Frans B. M. de Waal, "On the Possibility of Animal Empathy," in *Feelings and Emotions: The Amsterdam Symposium*, ed. A. S. R. Manstead, Nico H. Frijda, and Agneta Fischer (New York: Cambridge University Press, 2004), pp. 381–398.

46. De Waal, "On the Possibility of Animal Empathy," p. 381.

47. There are studies of empathy deficits in mice that seem to indicate a genetic basis for empathy—mice lacking the relevant genes did not engage in empathic displays. See Qi Liang Chen et al., "Empathy is Moderated by Genetic Background in Mice," *PLoS One* 4 (2009): e4387; available at http://www.plosone.org/article/info%3Adoi%2F10.1371%2 Fjournal.pone.0004387 (accessed April 1, 2011).

48. THN 2.2.12.5.

49. EPM 3.18–19.

50. In her *Memoirs*, Lady Anne Lindsay recounted a lovely story about Hume's moral sensitivity. The story is set at a party at her grandmother's house in Edinburgh when Hume was a young man, and quoted by Ernest Mossner in his biography of Hume: "As a boy he was a fat, stupid, lumbering Clown, but full of sensibility and Justice—one day at my house, when he was about 16 a most unpleasant odour offended the Company before dinner....'O the Dog...the Dog' cried out everyone 'put out the Dog; 'tis that vile Beast, Pod, kick him downstairs, pray.'—Hume stood abashed, his heart smote him....'Oh do not hurt the Beast' he said '...it is not Pod, it is Me!' How very few people would take the evil odour of a stinking Conduct from a guiltless Pod and wear it on their own rightful Sleeve." Lady Anne noted something of the young Hume. It wouldn't have been fair to Pod to have him blamed for Hume's "odour." See E. C. Mossner, *The Life of David Hume*, 2nd ed. (Oxford: Clarendon Press, 1980), p. 65.

51. THN, 3. 2. 1. 1.

52. Hume notes, in his discussion of justice as an artificial virtue, that there is no intrinsically negative feature of actions we call "unjust" that accounts for the injustice. Instead, it is relational, having to do with an "external relation...which we call *occupation or first possession*" which "has only an influence on the mind, by giving us a sense of duty

in abstaining from that object, and in restoring it to the first possessor. These actions are properly what we call justice" (THN 3.2.6.3). Here he does seem to equate the rules of justice with the rules of possession. Further, at THN 3.2.2.28 he notes that, as regards the state of nature, "I only maintain, that there was no such thing as property; and consequently cou'd be no such thing as justice or injustice."

53. See, for example, Joyce Jenkins and Robert Shaver, "Mr. Hobbes Could Have Said No More," in *Feminist Interpretations of David Hume*, ed. Anne Jaap Jacobson (University Park: Penn State University Press, 2000), pp. 137–155.

54. Christine Korsgaard, "Just Like All the Animals of the Earth," *Harvard Divinity Bulletin* 36 (2008), available at www.hds.harvard.edu/news/bulletin_mag/articles/36-3/korsgaard.html.

55. Arthur Kuflik, "Hume on Justice to Animals, Indians and Women," *Hume Studies* 24 (1998): 53–70.

56. Kuflik, "Hume on Justice," p. 56.

57. THN 3.2.1.10.

58. THN 3.2.1.11.

59. EPM 3.19.

60. It might reasonably be argued that Kuflik is being very charitable to Hume. Hume in numerous places laments both the physical and mental inferiority of women to men.

61. Joel Feinberg discusses the desire to advance, in higher animals, as an example. See his discussion of animal interests in *Harm to Others* (New York: Oxford University Press, 1984), pp. 58–59.

62. For a fascinating discussion of the evolutionary history of the domesticated dog, see Juliet Clutton-Brock, "Origins of the Dog: Domestication and Early History," in *The Domestic Dog: Its Evolution, Behavior, and Interactions with People*, ed. James Serpell (New York: Cambridge University Press, 1995), pp. 7–20.

63. Emphasis added by author.

64. See Rudolph V. Vanterpool, "Hume on the 'Duty' of Benevolence," *Hume Studies* 14 (1988): 93–110. There is some issue over the extent of a duty of benevolence in Hume. Vanterpool argues that the extent is limited, that is, the duty to benevolence "consists of affirmative undertakings in response to human *needs* of well-being."

65. THN 3.2.5.6.

66. THN 3.2.5.6.

67. Jenkins and Shaver, "Mr. Hobbes Could Have Said No More."

68. Robert Shaver, "Hume on the Duties of Humanity," *Journal of the History of Philosophy* 30 (1992): 545–56.

69. See Annette Baier, "Knowing Our Place in the Animal World," in *Postures of the Mind* (Minneapolis: University of Minnesota Press, 1985).

SUGGESTED READING

BAIER, ANNETTE. "Knowing our Place in the Animal World." In *Postures of the Mind*, pp. 139–56. Minneapolis: University of Minnesota Press, 1985.

BEAUCHAMP, TOM L. "Hume on the Nonhuman Animal." *Journal of Medicine and Philosophy* 24 (1998): 322–35.

BRADIE, MICHAEL. "The Moral Status of Animals in Eighteenth-Century British Philosophy." In *Biology and the Foundation of Ethics*, edited by Jane Maienschein and Michael Ruse, pp. 32–51. New York: Cambridge University Press.

DE WAAL, FRANS B. M. "On the Possibility of Animal Empathy." In *Feelings and Emotions: the Amsterdam Symposium*, edited by A. S. R. Manstead, Nico H. Frijda, and Agneta Fischer, pp. 381–98. New York: Cambridge University Press, 2004.

FLACK, JESSICA, and FRANS B. M. DE WAAL. "'Any Animal Whatever': Darwinian Building Blocks of Morality in Monkeys and Apes." In *Evolutionary Origins of Morality: Cross-disciplinary Perspectives*, edited by Leonard Katz, pp. 1–30. Bowling Green, Ohio: Imprint Academic, 2000.

GOSLING, SAMUEL, and OLIVER JOHN. "Personality Dimensions in Nonhuman Animals: A Cross-Species Review." *Current Directions in Psychological Science* 8 (1999): 69–75.

JENKINS, JOYCE, and ROBERT SHAVER. "Mr. Hobbes Could Have Said No More." In *Feminist Interpretations of David Hume*, edited by Anne Jaap Jacobson, pp. 137–155. University Park: Penn State University Press, 1992.

KORSGAARD, CHRISTINE. "Just Like All the Animals of the Earth." *Harvard Divinity Bulletin* 36 (2008), available online at http://www.hds.harvard.edu/news/bulletin_mag/articles/36-3/korsgaard.html.

KUFLIK, ARTHUR. "Hume on Justice to Animals, Indians and Women." *Hume Studies* 24 (1998): 53–70.

PITSON, ANTONY. "The Nature of Humean Animals." *Hume Studies* 19 (1993): 301–16.

———. "Hume on Morals and Animals." *British Journal for the History of Philosophy* 11 (2003): 639–55.

SHAVER, ROBERT. "Hume on the Duties of Humanity." *Journal of the History of Philosophy* 30 (1992): 545–56.

SMITH, J. DAVID. "The Study of Animal Metacognition." *Trends in Cognitive Science* 13 (2009): 389–96.

TRANOY, KNUT ERIK. "Hume on Morals, Animals, and Men." *Journal of Philosophy* 56 (1959): 94–103.

CHAPTER 6

....

UTILITARIANISM
AND ANIMALS

....

R. G. FREY

THIS essay is about utilitarianism as a theory: its types, its history, and its moral importance for the discussion of animals. I shall not engage in disputes that this theory has had with other moral theories. Most of the readers of this volume will either be familiar with these disputes or can read about them in several chapters in this *Handbook* that treat these disputes and their continuing significance in philosophy.

The earliest significant utilitarian philosophical writings were those of Jeremy Bentham (1748–1832) and John Stuart Mill (1806–1873). I shall make much of Bentham's contribution to the discussion of animal welfare and moral standing— both his influence and where he may have misled us. For 150 years, his powerful views about animals dominated theoretical discussions in animal ethics—at least his views dominated thought among those inclined to a utilitarian defense of animal welfare. By the mid- to late-twentieth century, Mill's views were close to the canonical expression of utilitarianism in many discussions of utilitarian moral theory. At this time, utilitarianism had not changed all that much from what we inherited from Mill one hundred years earlier.[1] The model he gave us, at least on common interpretations, treats utilitarian theory as *consequentialist, welfarist, aggregative, maximizing,* and *impersonal.* The view is consequentialist in that it holds that acts are right or wrong in virtue of the goodness or badness of their actual consequences. The view is welfarist in that rightness is made a function of goodness, and goodness is understood as including both human and animal welfare. The view is impersonal and aggregative in that rightness is determined by impartial assessment of the increases and diminutions of the welfare or well-being of all affected by the act, and summing those increases across all affected. The view is a maximizing one in that a concrete formulation of the principle of utility, framed in light of welfarist considerations, is "Always maximize net happiness."[2]

In the late twentieth century, the received account of utilitarianism came under fierce attack. It found numerous opponents, though many of them could not agree among themselves exactly what was wrong with the theory. What drew, and continues to draw, much attention is the *consequentialist* condition. The main criticisms, including ones today having to do with integrity, separateness of persons, demandingness claims, and impersonal accounts of rightness, still turn on the claim that classical utilitarianism produces results strongly at odds with ordinary moral thinking. I find this central criticism to be an unacceptable way to criticize the theory, a matter I will turn to later in this chapter.[3]

More recently, the value component of the theory has spurred debate. Bentham had maintained that consequences are to be assessed by a standard of intrinsic goodness that requires maximization of the good. Today, the good in act-utilitarian theory is commonly construed to include both human and animal welfare, though precisely how terms such as "welfare" and "well-being" are to be understood remains contentious. Bentham was a hedonist, maintaining that all and only pleasure is intrinsically good; Mill spoke of both happiness and pleasure and tried (in order to reduce conflicts with ordinary morality) to introduce a distinction between higher and lower pleasures; and the later utilitarian G. E. Moore maintained that other things, such as beauty and friendship, as well as pleasure and/or happiness, were also good in themselves.[4]

All kinds of questions arise about utilitarianism, but I shall start by focusing on one having to do principally with the value component, namely, whether it encompasses animals. Bentham argued that animals, like humans, have the capacity to feel pain and therefore deserve moral protections. He reasoned that despite important differences between humans and animals, there are also important and relevant similarities, the chief being the capacity of sentience—that is, the capacity to experience pleasure, pain, and suffering. This claim was part of the received view that came under fire. Since roughly 1975, mainly as a result of the writings of Peter Singer, this thesis about sentience has once again come to be a central feature in controversies about utilitarianism and in current theoretical discussions of animal ethics,[5] and it has become no easier to formulate the controversies and to resolve them. I will emphasize that if it is difficult to say in what human welfare consists, it is still harder to say in what animal welfare consists.

When I first arrived as a student at Oxford University, in the same philosophical environment that Peter Singer would enter four or five years later, utilitarianism was much in vogue. Utilitarianism seemed to dominate discussion, even by those opposed to it. The feeling was that if only we worked long and hard enough we should eventually work out a theory that sufficed as an adequate normative ethical theory. The utilitarianism most discussed was act-utilitarianism,[6] but its defenders attached various apparently deontological restrictions to the theory, presumably restrictions that are grounded in a consequentialist justification.[7] The removal of clashes with ordinary morality by utilitarians and non-utilitarians alike became a major focus of debate. Rules, rights, duties, and so on were soon built into utilitarian theory—all presumably on consequentialist grounds.[8]

This great hubbub over utilitarianism took my fancy as a student, and I focused on working out an adequate form of act-utilitarianism, with Richard Hare as my supervisor.[9] I became convinced after a time to take on rule-utilitarianism, and then journeyed back to act-utilitarianism.[10] Remarkably, when twenty years later I departed Oxford, there was scarcely an act-utilitarian to be found, at least out in the open. I do not mean that the theory had ceased to be discussed, only that it was no longer thought to form the basis of an adequate normative ethical theory.[11] Clearly, the landscape of utilitarian theory and all of moral philosophy was changing, but utilitarianism also would continue to be the most widely *debated* of all general moral theories.

About the time these changes were occurring, applied, or practical, ethics was helping stimulate a new kind of interest in ethical theory, especially in the United States. To some extent, act-utilitarianism was revived with it.[12] A massive wave of literature in applied ethics was unleashed, and a new branch of ethics, so to speak, was born. On issue after issue, act-utilitarianism anchored one side of the debate, and when critics objected to utilitarian solutions to practical problems, it was almost always act-utilitarianism that was the object of their ire.

In *Moral Thinking*, Hare tried to quiet things and bring unity to utilitarian theory: he introduced a split-level theory that enables him to be something of a rule-utilitarian on the practical level and an act-utilitarian on the theoretical level, the level at which the rules to be followed are established and justified. I discussed this split-level theory many times with Hare, and he seemed not to grasp that his opponents feared that, under his theory, utilitarian thinking at the theoretical level would seep over into the thinking at the practical level and, as a result, purely consequentialist thinking would wind up determining what we ought to do.[13]

As it would happen, when applied ethics came alive, act-utilitarianism once again came alive with it. Biomedical ethics led the way, followed some distance behind by business ethics and environmental ethics. Both biomedical and environmental ethics brought with them new attention to human uses of and policies in regard to animals. Other articles in this volume deal with biomedical and environmental ethics. I will focus on only some underlying theoretical issues, sticking close to issues that pertain to animal ethics.

Bentham, Singer, and the Criterion of Sentiency

One reason that utilitarianism has been the most prominent ethical theory in the discussion of animals for 150 years is found in Bentham's claim that pain and suffering are sufficient conditions of moral standing and that we should not rely heavily on other conditions such as the ability to think or to reason or to deploy reasons or to engage in abstract thought or to be possessed of agency, autonomy, or to be possessed of rights.[14] This thesis has proved attractive far beyond utilitarian thinkers. Many who aggressively shun the company of utilitarians have looked favorably upon this thesis. I shall refer to Bentham's view as *the sentiency criterion* for the possession of moral standing, meaning that sentience is a sufficient condition of some significant level of moral standing.

The utilitarians who followed Bentham almost all adopted the sentiency criterion (as did many non-utilitarians) and, while there were disputes about which creatures were included—for example, worms, honeybees, and lobsters—all of the main food animals, such as pigs, cows, hogs, and chickens were included, as were many of the laboratory animals. This makes all the more remarkable the fact that none of these animals were enveloped into utilitarian theory. Consider Mill, Bentham, John Austin,[15] Hastings Rashdall,[16] W. T. Stace,[17] G. E. Moore,[18] J. J. C. Smart, and Hare: None so far as I am aware envelopes "the higher animals," including dogs, within their developed theories. Some of these philosophers, while paying lip-service to Bentham, had nothing deep to say about animals at all. The first solid attack within utilitarianism on the problem of including animals among those with moral standing was by Peter Singer in *Practical Ethics* in 1986.[19]

No specific argument was given even by Bentham in support of the criterion of sentiency. In the discussion of animals and their mistreatment, the fact of animal suffering was presumed sufficient for moral standing, without further argument. There is no reason to argue for what all can see to be the case. One might argue either that animals did not feel pain or that they felt pain in some lesser way, but little credibility was given to these options and little argument was directed against them. These are not attractive options, Bentham realized, and—though I am not a Cartesian scholar—I doubt that Descartes really adopted the first. Some more recent and sophisticated views by Peter Carruthers on animal pain have not garnered much support either.[20]

It is easy to see what the emphasis upon pain does in part to the argument: it simply seizes upon the pain involved, weighs it against the pain on the other side (though the method of doing so is not obvious and hardly ever discussed), and decides accordingly what ought to be done. This proposed methodology is part of the attraction of act-utilitarianism. If we can agree on the weighing principles and agree as well on the weights we decide to give those principles, we can decide what is right and what is wrong. There is an empirical element that is itself attractive, because we are presumably not left to make the decision in terms of abstract reason, appeal to rationality and autonomy, or to some theory of reasons or other.

Sentiency is thought to do so much work in animal ethics that the literature often reads as if Singer's view is just a restatement of Bentham. Singer describes in great detail the brutalities and horrors to which food animals are subjected and comes down strongly on the side of the critical importance of the criterion of sentiency. Sentiency is thought by some attracted to the criterion to ground a theory of animal rights, which in turn can be used to resist the appeals of defenders of general social welfare and to provide a barrier against meat-eaters. Singer, rightly, resists this ploy, and animal rights do not play a role in *Practical Ethics*.

I agree with this part of Singer, though he occasionally favorably reviews books that turn upon appeals to rights. These are theories that I resist. But his position in *Practical Ethics* is that it is not because they have rights that animals count, and true animal *rights* philosophers are not on the whole followers of Singer. Here is why: without rights for animals that can serve to block appeals to the human collective good from overriding the interests of animals, we shall continue unabated with our eating of and experimenting upon animals. Given the absence of rights, Singer can be seen as standing in that long line of thinkers, following Bentham, who used sentiency in order to gain moral access to the condemnation of the treatment of animals. The difference on this score is that Singer is better at showing what is available under a sentiency criterion. However, for all his philosophical skills and his opposition to a theoretical basis in rights, I will be challenging the sentiency criterion and the work it is supposed to do.

In important respects it is the absence of rights in Singer's theory that has helped spur growth of the rights piece of the animal rights movement. Those who prefer the language of rights see utilitarian contingency accounts as follows: if the pleasures and pains of humans and animals go one way, animals may not be eaten; but if conditions are different in the calculus of pains and pleasures, they may be eaten. Because one cannot see how commercial farming on a gigantic scale is possible without considerable pain, Singer thinks matters tilt his way. But in all smaller cases—for example, appropriately adjusted traditional farming where animals are free-ranging—they do not always seem to tilt his way, and the meat-eater has no barrier to his delicious supper. The clinical researcher has no barrier to harmful interventions using animals for the cure of human disease. What is needed, some have thought, is a theory of rights that bars appeals to the human collective well-being from outweighing the right of a sentient animal to life, the right not to be harmed, the right to liberty, and other rights of comparable importance to these.

One might try to build theories of rights into utilitarianism on utilitarian grounds, and more than one philosopher has tried to do so. But a pure rights-theorist will likely give them no credibility and will worry that these utilitarians offer only protections contingent on the circumstances. Of course, one could state the rights in such a strong form that nothing justifies overriding or setting them aside, but this kind of theory runs the risk of not being truly utilitarian. So, here we see the first obvious problem that arises in utilitarian theory about animals: how strongly are we utilitarians to treat sentiency and what place should it have in our theories?

Do Humans Deserve Higher Moral Standing?

A second problem for utilitarians is today less carefully inspected and frequently overlooked. Of the utilitarians I have mentioned, and many others of lesser note, some were adherents to Judaic/Christian beliefs, and some, perhaps many more, were not. Into this interplay between how we shall live (eat, experiment, have pets, and the like) and religious belief, moral standing falls into even further obscurity than the value of animal life. Some of the Christian utilitarians, including William Whewell, held straightforwardly that human life is more valuable than animal life, and they took the Bible as the inspiration for their views.[21] The extraordinary thing is that non-believing utilitarians also took human life to be more valuable than animal life—though they, of course, cited different grounds for their views. Herein lies the problem: utilitarians give the higher animals moral standing and they give human life a still higher moral standing, and when high standing collides with even higher standing, as we shall see, nonhuman animals do not win. But utilitarianism is a balancing theory that allows little room to stratify and rank levels of moral standing. Levels of sentiency are the critical matter. But in that case, nonhuman animals will not always lose in circumstances of contingent conflict. My view is that a properly constructed value-of-life view will allow value to win when moral standing collides with value. So just what is the right utilitarian theory, and how can it be defended?

Here is an explanation of why a hoary old chestnut has generally posed no problem for utilitarians (leaving aside some complications, some of which I shall mention later): a ship goes down, and a dinghy is afloat that can hold only one more creature, either the man on the left or the dog on the right. In the early discussions among utilitarians of this kind of problem, it was obvious to all that we are morally obligated to save the man. If we assign the dog no moral status, we can save the man; if we assign the dog lower moral status, we can also save the man. Either way the man has higher moral status. But a major difficulty lurks here: if utilitarians were to follow key elements of general utilitarian moral theory and assign all sentient creatures equal moral status, the equality could be broken only as we learn about the nature and condition of the man and the dog, and therefore that saving one became dubious and saving the other obligatory. We could in the theory assign a lower moral status to the dog whatever its condition and whatever the condition of the man, but utilitarianism as a general theory did not seem to do so or to provide premises that allow one to do so.[22]

Here is one of the main reasons that we do not find utilitarians prior to Singer enveloping animals within their theories or providing explicit hierarchies or comparative accounts of moral standing. In the dog case, the man does not seem positioned to win through a general utilitarian theory premised on sentiency, pain, pleasure, and the like. The man seems to be winning on hidden grounds that are no part of the earlier claims about sentiency: the dog is sentient and has standing; the man is sentient and has standing. In utilitarianism, it is not by appeals to the moral standing of humans and nonhumans that we will solve this problem. The key will be found in the comparative value of the lives involved. It is this absolutely critical problem of the comparative value of lives that I think Singer's discussion overlooks and that the earlier utilitarians simply assumed they could handle.

So we reach a curious impasse. The earlier utilitarians typically followed Bentham's line on sentiency, and Singer follows it too; but what difference does it make to the case, if, even granted sentiency, we always save the man? And on what moral grounds would we make a decision to do so? In a way, assigning no moral status or lower level moral status in our example did not matter at all: if the dog has moral status, we save the man; if the dog does not have moral status, we save the man. Of course it matters that we take animals to be members of the moral community and that their pains count, morally, in deciding what to do. To leave them out of consideration is to overlook the effects on all those creatures affected by the act, as Bentham demanded we do; but even when we do this it seems at the hands of the bulk of utilitarians that we still save the man. It became a problem for the utilitarians to put these thoughts together coherently.

It might be that we can decide the issue by adding features to it that color the example one way rather than another.[23] Suppose the man was Hitler or a vicious child-molester or a man with terminal AIDS: do we still save the man? Or suppose, as Baruch Brody once suggested to me, we simply assign the animal a lower moral status than the man and we do this in a systematic fashion: the man gets saved each time. But I do not believe moral status does or should work like this, nor did the earlier utilitarians in their general pronouncements: it was presumably to be an all-or-nothing affair; someone or something either has standing or not. Nonhuman creatures do not have little bits of it, while I have a great deal of it because I am human. There are various ways to try to avoid the kind of problem that the utilitarians confronted here, but all of them have, so far as I am aware, come to nothing. Still today we lack a general utilitarian theory of degrees of moral status that sorts creatures into higher and lower moral status, especially when it comes to comparisons made between nonhuman and human animals.

The Real Problem and the Claim
that Supports It

The real problem for a utilitarian theory of animal ethics has nothing essentially to do with pain and suffering, even if they are intrinsically evil. The right starting place is the fact that we are using animal lives for our own purposes and often using them up. Whether pain is or is not inflicted in the process, we are still using and often using up these lives. The real problem is to determine what, if anything, entitles us to do so. I argue in this section that the real problem is the comparative value of lives; this is the problem with which utilitarians should be dealing.

Some questionable moral premises are often at work in moral arguments about our uses of animals. In particular, the following premise (a carryover from the discussion of moral standing in the prior section) is questionable: we are justified in using animal lives because ours are of *higher value*. If we must take a life in a circumstance of contingent conflict between the parties, we should resolve the problem by taking the life that is of lower quality; and if we have to save a life in a conflict situation in which only one party can be saved, we should save the life that is of higher quality. These premises, it may be maintained against my formulation, are a set-up for us to take advantage of the weakest among us. To some extent this point may be correct, but I have in mind principally practices that are very common in our institutions and widely approved. For example, in a conflict situation where a choice must be made (e.g., as to who will be admitted to an intensive care unit), hospitals do not save the person with the worst prognosis of future life in preference to the person with a far better prognosis.

Comparing the *quality* of lives, and so comparing the *value* of lives, is a critical element in these decisions. We often make the assumption that we can compare lives and decide which of two lives has the prospect of going better if supported by medical treatment. Are these decisions about comparative prospects decisions about who will be made better off? Yes, they are, and this applies as much to the comparison of human lives as it does to the comparison of animal lives. We do not save the anencephalic infant in a pediatric intensive care unit at the expense of a newborn baby we can restore to perfect health. We compare lives in our hospitals on a daily basis despite any distaste we might have for doing so.

The mixture of questions of moral standing and those of the value of life can make some cases a bit misleading. Even in cases in which Singer found animals to have moral standing and to be harmed by a practice, we still have no guarantee that, in a conflict situation, the human's life would not be found to have a greater value. I see no reason for not treating value in human-animal conflicts as we do in the hospital cases of human-human conflict I just mentioned. The question is, "Which creature has the higher quality of life in a given case, the dog or the man?" If the dog, then we should save it and not the man.[24]

Many people will argue that we have no idea how to assess the quality of life of a dog, and often little access to quality of life for humans. However, doctors use talk of quality of life routinely and not in trivial ways. Quality-of-life considerations have played a major role in many of the most famous cases of death and dying in the bioethics literature. All other things being equal (e.g., you are not the daughter of a famous person)—again, in a conflict situation in which the prolongation of life is equal in both cases—the doctor opts for the person with the higher quality of life. There are numerous cases in which we know the outcome: a person is dying in the final stages of full-blown AIDS, or a person is in the final stages of pancreatic cancer. These cases easily come to mind. It is not merely physical ailments that fall into this category. Cases include people dying in the final stages of Alzheimer's disease or pancreatic cancer or amyotrophic lateral sclerosis (ALS) or children born with hypoplastic left-heart syndrome. If we have to make a choice between these lives and other lives with a far better prognosis, we do not save these lives. Money might make a difference in where we will prolong life in some of these tragic cases, but this consideration is beyond the present investigation.

I do not mean to minimize the well-known fact that quality-of-life judgments are notoriously difficult to make and involve at times highly subjective elements. Quality often depends not upon external ailments but upon internal factors that are difficult to assess. This problem also besets appeals to sentiency. We would like to get inside the animal's "mind" in order to discover what its internal states are, what the animal makes of them, and how they affect how the animal "sees" its life. Does a squirrel have toothache? This would not obviously be answered by showing that the squirrel gave indications of feeling pain, nor need we both agree whether the squirrel can grasp the concept of toothache. Our question is about how well or badly the squirrel experiences its life going, given that it is having miserable pain in one of its teeth. The problem is trying to see the squirrel's life from the inside to tell how well or badly its life is going. This inside point of view affects the judgment of quality of life in humans and animals, though often we only have access to an external and not an internal view. We often find it difficult to say how well our own lives are going, even in the midst of illness, because this judgment of quality of life has an internal factor that we find difficult to pinpoint. In the case of animals, we find it far more difficult to do so. The quality of human and nonhuman life is not determined solely by how well or badly the physical body is doing; there is also the attempt to see the life being lived from the inside, of how it seems and feels and affects the life in question, physical injuries or not.[25] However difficult it is to make these judgments and however difficult it is to be confident that scientific knowledge in this area is accurate and not merely a form of sophisticated anecdotes, I reject the view that we cannot discover anything about the interior lives of the higher animals.

Clearly animals do feel pain, but I want to reach beyond pain: What I am asking is whether we can determine how well or badly their lives are going. In our case as humans, pain makes a life miserable and can lower its quality immensely. The same is true of the higher nonhuman animals. Yet we often cannot determine the quality of life of an animal by observing what is happening to it physically. If a cat were dipped in petrol and set alight, I have no doubts about the quality of its life. But does the squirrel have a toothache, and does the gorilla caged alone in a zoo have depression? Higher animals have biological lives that consist of a series of experiences, of different kinds or varieties, of different intensities, with different effects on their lives, and so on. Philosophers, utilitarian or otherwise, who are concerned about sentiency in animals cannot afford to miss the interior life beyond pain, including the emotions of animals. These experiences profoundly affect how well or badly life is going. This is as true of human cases as it is of nonhuman cases.

Thus, both pain and other moral-bearing characteristics come into the picture of the quality of life. The problem with Singer's picture is not that pain is not a moral-bearing characteristic worthy of consideration in commercial farming and in other of our uses of animals. Of course it is. The problem is that pain and suffering are made to be the only important characteristics, thus ignoring other dimensions of quality of life.[26]

MORAL STANDING AND COMPARATIVE VALUE

Here is the bind utilitarians found themselves in: even if animals are ceded moral standing, we still, in the choice situation that I have imagined, are morally obligated to save the man. Only if we introduce various other very serious considerations do we move to save the dog. To be sure, these factors may obtain, but the determinative consideration that many people, including utilitarians, invoke in approaching these problems is that we take the man to have a higher value and to be more valuable than the dog. The reason is found in the comparative value of human and animal lives in the state we find those lives. The problem with focusing upon the pain inflicted by factory farms and slaughterhouses is that we can miss the central issues of comparative value. We use up animal lives in order to extend and enhance human lives. Whether the balancing of human interests and animal interests that is ubiquitous in our culture is merited is a major moral problem, well known to all utilitarians. However, this issue is entirely independent of the relative moral standing of the creatures involved. It has to do with the comparative value of lives, an issue to which the earlier utilitarians did not devote much space, if indeed any at all.

May we, then, use mice, in order to enhance and to extend our lives? We do so as a matter of course, and their numbers, especially given developments in genetic engineering, xenotransplantation, cloning, and the like have increased enormously in recent decades. Millions of colonies of mice are used every year. We use some, we discard others, perhaps in the search for "designer" mice that exhibit just those features that we are breeding them to exhibit. Vast numbers of mice as by-products are produced along the way, many to be sold to other labs, many to be destroyed. These "designer" mice are produced for human ends. The justification given—at least, most commonly among people who discuss this issue—is that we may use mice to extend and enhance our lives because our lives are more valuable than the lives of the millions of mice in question. We use up the lives of these mice, and often we use up those lives painfully.

Bare appeals to sentiency do not resolve these issues: the mouse is sentient; yet the question remains whether we are justified in using up its life. The Benthamite appeal to sentiency does not get at this more fundamental issue, but morality demands that we answer it. I do not have space here to give a detailed account of how I think this case for using the mouse goes and how we are to choose between animal and human lives, but I can give an indication of one of the important issues that bears upon the issue. It is controversial, but then much of the literature on the subject is highly controversial.[27]

One path for the resolution of these problems is to exclude all uses of animals and endorse anti-vivisectionism. This abolitionist proposal fails because the vehicle by which the case for it is to be made cannot bear the weight that is put upon it. The vehicle often attempted recently has been a robust account of moral rights. I emphasize how robust this particular account has been in the animal ethics literature (so vigorous that it cannot even be made a defensible account of moral rights for humans). Under any defensible conception of the rights of animals it will not follow that a case for human use of animals in experimentation will be barred. There will be merely a prima facie right on the animal's part,[28] and such rights can have countervailing concerns arrayed against them and so be outweighed in many circumstances. So, we would need a conception of a right that bars precisely this effect.

Tom Regan is an abolitionist who wants to confer upon animals rights along the rigid lines that he interprets Ronald Dworkin as defending—namely, the idea that rights are trumps.[29] Regan maintains that rights provide a trump playable against considerations of the general welfare. By giving animals rights in this robust sense, appeals to human benefit cannot serve as a justification for the use of animals in scientific research, because that would amount to using appeals to the general human welfare to justify infringements of animal rights. One cannot eat them, since that obviously violates their right to life; one cannot cage them, because their right to liberty is violated, and so on.

In Regan's philosophy, the only way moral perplexity can be handled is if some countervailing right comes into the matter, such that it then poses a conflict with the animal's right to life. One then is on the difficult terrain of a rights theorist having to deal with a conflict of (prima facie) rights. Conflicts, of course, pose problems, and their resolution is not always easy, but in the case of the mice previously discussed, absolutism entails the following claim: there is no countervailing right for humans to do research or to exterminate pests. These human conveniences and protections conflict with the rights of mice, and we lose in the absolutist's world. There is no way, then, to register the moral perplexity people feel between weighing and balancing human and animal lives and seeing whether there can be a case for using animals. The balancing of utilities so vital in the history of utilitarian systems of moral philosophy are not permitted. They are considered immoral.

For years Regan and I traveled around the United States holding debates on animal welfare/animal rights issues. These were rousing occasions, and Regan was a good debater. Enormous numbers of people appeared. Regan used the debates to set out his rights position, but this only occurred after he had spent some time informing the audience that I was an act-utilitarian, so that I could justify anything as right depending on the consequences, including, for example, rape or feeding Christians to the lions. As I looked out upon the audience, I often could see that I was being taken as little better than a child-molester.

Once Regan's position was clearly in the public view, I felt it was my turn, and I went after what I took to be the weakness of his positions that rights are powerful instruments and that, since animals have them, they have these rights in the *same respects* in which human persons do. Nothing gets fully to contend against such a right except another right, since anything else is not sufficiently weighty to be in contention. Thus, there is no way to portray the effects of polio vaccine in eliminating one of the scourges of human life in order to justify using monkeys in the research. All invasive (or, for that matter, non-invasive) research that is for human benefit must be stopped, preferably at once because high levels of human social benefit cannot trump rights.

Regan gives animals rights of such power that nothing is able to contend against them even in a circumstance in which there will be enormous human benefit. From this perspective, to cite benefit as a ground for vivisection is to fail to appreciate the force of the rights that animals possess. It is not possible even to state a pro-research position that involves overriding animal rights, because these animal rights trump utilitarian justifications based on the benefits of animal research. This argument will not wash: to convey to the animal a right so strong that one thereby ensures that no case from benefit can even register against it, and then to turn around and point to the fact that no case from benefit can overcome an animal's right (to life) seems like putting into the pot what one is now going to claim can be discovered in the pot. The consequences are devastating for human progress.

Moral Standing and Speciesism

What is it about animals that warrants our moral concern? The usual answers are their general welfare—in particular, avoidance of pain and suffering. Almost everywhere today the scientific research community has presented guidelines governing animal pain and suffering that insist that they be controlled, limited, mitigated where feasible, and justified in the research protocol and actual experiment. The care that many researchers bestow upon their animals shows that they take animal suffering seriously, as does the insistence by many that animals be painlessly euthanized under certain conditions. If this level of care should be absent, government and funding oversight committees can challenge—indeed, close down—research projects.

Both the suffering of animals and their death at our hands are moral concerns. Much of the worry over suffering, whether in the human case or the nonhuman case, is precisely owing to the way it can blight, impair, and destroy a life. If animal lives had no value, why should we care about ruining those lives? Why, in research protocols, do we go to such great lengths to justify their sacrifice? If, however, animal lives have some value, then we need to justify their confinement in cages, the intentional diminution of their quality of life, and their destruction.

I once saw a man in the audience for an animal ethics symposium who held that there is a genuine moral difference between the human and animal cases. He pointed to the difference between burning a child and burning a chimp and between infecting a man with a certain disorder and genetically engineering a chimp to be subject to that disorder. The man was maintaining that doing these things to the chimp is of no *moral* concern despite the fact that suffering occurs, quality of life is drastically lowered, and killing takes place. If done to the man, he said, these things are immoral, but not so with animals because their lives are not matters of moral concern.

The question that arises here is, "How can species membership make this difference?" It is not easy to see how mere species membership can constitute a moral difference between two relevantly similar acts of killing or lowering of quality of life. In the case of pain and suffering, I cannot see how they constitute a moral difference at all. Nor is my view any different if we substitute a mouse for the chimp—if, that is, we use a creature of a so-called lower species who experiences pain and suffering. In my view of moral standing, we can sensibly assign value to the lives of mice just as we can for chimps and humans.[30] This is not to say, of course, that they all have the same value.

I once encountered Bernard Williams in Regent's Park in London, and he proceeded to object rather fiercely to my views about speciesism. He suggested that I gave too much ground to Singer, of whose philosophical framework, he rightly observed, I was an opponent.[31] Fair enough, but on speciesism I agree with Singer, and I tried to show how this acceptance was compatible with my objections to Singer's conclusions. Honesty compels me to say that Williams accepted not a word I said; in his view, I failed to illuminate how morality is an essentially human endeavor.[32] His comments reminded me of the man in the audience at the animal ethics symposium. They are both speciesists who think I have a misguided view of the limits of moral concern.

But their view is too narrow because it tosses aside all considerability for animals. In my view, moral standing (moral considerability) turns upon whether a creature has an unfolding series of experiences that, depending upon their quality, can make that creature's life go well or badly. Such a creature has a welfare that can be positively and negatively affected, depending upon what we do to it, and with a welfare that can be enhanced and diminished, a creature has a quality of life. Mice and chimps are experiential subjects with a welfare and a quality of life that our actions can affect, and this is so whether or not they are agents (which, Williams insisted, they are not) and whether or not they are the bearers of rights (which he thought highly implausible). Agency and rights, to my mind, are not the right conditions of moral standing, though the ability to perform actions certainly makes life go better for all creatures. They have lives that consist of the unfolding of experiences they largely control and so have a welfare and a quality of life, and while there may be some creatures about which we all remain uncertain of their inner lives, mice in laboratories are not among these. Accordingly, laboratory creatures have moral standing and so are part of the moral community on the same basis that we are.

There is no reason to deny that mice and chimps feel pain, and I can see no a priori moral difference between burning a man and burning a mouse or chimp. Pain is pain, and species is irrelevant. What matters is that a creature is an experiential one and pain is evil when experienced. But if pain and suffering count morally, then so do animal lives. Just as what concerns us so much about pain and suffering in our lives' case is how these things can impair and significantly diminish the quality of life, so they can impair all creatures who can experience them.[33]

In conclusion, I am not a speciesist; I do not think that we can justify a practice such as vivisection by citing species as a morally relevant reason for using animals in what we are doing. Nor do I deny that animals are members of the moral community; they are. But not all members of the moral community have lives of equal value, and the threshold for taking lives of lesser value is lower than it is for taking lives of higher value.

LIVES OF DIFFERENT VALUE

It is deeply unpalatable to many to think that some lives are less valuable than others; they want it to be true that at least all human lives are equally valuable. When I speak of not all lives being equally valuable, I am not referring merely to the difference between animal and normal adult human lives; I refer also to the difference in value among human lives. A quality-of-life view of the value of a life makes the value of a life a function of its quality, and it is a commonplace today that not all human lives are of equal quality. Indeed, some people lead lives of such a low quality that even they themselves seek release from them, as some cases involving a right to die and physician-assisted suicide make clear. It would seem bizarre to these individuals to tell them that, according to some abstract theory of equality, they have lives as valuable as normal adult human lives. There are some lives we would not wish upon anyone, and it seems mere pretense to claim that these are as valuable as normal adult human lives. Some may find comfort in such abstractions that substitute for, or, indeed, even reflect the religious view that all human lives are equal in the eyes of God; but many people no longer find comfort in this venerable adage, and the claim also introduces a notably different sense of equality.[34] One can be equal in the eyes of God and still have a low quality of life.

The comparative value of lives is again critical. If we think that not all human lives have the same value, and if we think about the depths to which human life can tragically plummet, then it may well turn out that some animal lives have a higher quality than some human lives. And if we have to use lives in experiments (*if*, I emphasize, we must do so), then we also ought to use the life of lower quality in preference to the life of higher quality, irrespective of whether the life is human or nonhuman.[35]

I do not believe one can reject this position just because community standards here and there reject such use of humans. I am defending a utilitarian moral position that provides reasons that we can use to correct for just such indefensible community standards. The fact that X is a community standard does not by itself have moral weight: I grew up in the South, and community standards pretty well excluded blacks from most of our community facilities, a fact that clearly illustrates the way in which community standards can depart significantly from morality. If a life is unavoidably at stake and we must choose one life over another, we are justified in taking the life of lower quality; if that is the man's life instead of the dog's, we leave the man to his fate.

It may be objected that many people have strongly held intuitions that run against this view. People have intuitions that run against all kinds of claims; one rarely hears the rich demanding that they be taxed more in order to benefit the poor. So what? I do not find it compelling that those who devise moral codes or community standards devise them so that certain persons or groups benefit from them. What else would one expect? Nor do I find it compelling to maintain that my position takes advantage of the weakest among us. It simply applies a quality-of-life view to situations of the taking of life. One does not hear protests from the public that not saving the dog fails to save the weakest among us.[36]

In my view, what matters is not life but quality of life. The value of a life is a function of its quality, its quality of its richness, and its richness of its capacities and scope for enrichment; it matters, then, what a creature's capacities for a rich life are. Not all human lives have the same richness and potentialities for enrichment, and not all human lives are equally valuable. In fact, some human lives can be so blighted, with no or so little prospect for enrichment, that the quality of such lives can fall well below that of ordinary, healthy animals.

The question about the value of a life in this chapter is not whether a mouse's life or a pelican's life has value. They do have value. The mouse has an unfolding series of experiences and can suffer, and it is perfectly capable of living out a life appropriate to its species. My standard of richness asks whether the mouse's life is rich in ways that approach a normally rich adult human life in quality, given its capacities and the life that is appropriate to its species. Although animal life is typically not as rich and therefore not as valuable as human life, some animals have lives that are more valuable than other animals, and some animals have lives that are more valuable than some human lives. Value is contingent on richness, and a considerable gulf exists in the richness of animal life than can be invoked in cross-species comparative judgments. This comparative-value and quality-of-life analysis supports the view that normal adult human life is considerably more valuable than animal life, based on greater richness and greater potentialities for enrichment. However, my view makes no claim about species per se.

Autonomy and the Richness in a Life

It will be said, however, that my arguments neglect one of the main values in the discussion of morality, namely, autonomy. I think autonomy matters in this discussion, but in a way different from what I take to be Kant's view and the view of many contemporary writers on the subject of autonomy. Clearly the capacity and exercise of autonomy enhance the value of a life. However, it is instrumentally, not intrinsically, valuable (hence my difference with Kant over autonomy). Its value depends upon the uses made of it, and, in the case, at least, of normal adult humans, those possible usages significantly enrich a life. Autonomy allows one to direct one's life to secure what one wants; to make one's choices in the significant affairs of life; to assume responsibility over a domain of one's life and so to acquire a certain sense of freedom to act; to decide how one will live and to mold and shape one's life accordingly. These are the sorts of experiences that open up areas of enrichment in a life, with consequent effect upon that life's overall richness, quality, and value.

Arguably, these ways of augmenting the value of a life are not available to non-human animals. Even so, it does not follow that animal life has no value or that an animal life cannot have greater value than some human life (again, it is my view that it can). Animals may not act autonomously, but they do act purposively. In any event, what should be centrally at issue is the comparative richness and value of normal adult human life and some other form of life and how we go about deciding the matter.

It might be claimed that we can know nothing of the richness of animal lives. But ethologists and animal behaviorists, including some sympathetic to the animal rights cause, generally think otherwise; how else, for example, could the claim that certain rearing practices blight animal lives be sustained? That we cannot know everything about the inner lives of animals in no way implies that we cannot know a good deal. If we are to answer the question of the comparative value of human and animal life we must inquire after the richness of their respective lives. Intraspecies comparisons are difficult, as we know in our own cases in, say, medical ethics; but such comparisons are not completely beyond us. Interspecies comparisons of richness and quality of life are likely to be even more difficult, though again possible. To be sure, as we descend from the "higher" animals, we lose behavioral correlates that we use to gain access to animals' interior states. Yet, more and more scientific work is appearing that gives us a glimpse into animal lives, even if it is hard as yet to make a case for extensive knowledge of the richness of these lives.

Again, we should not think that criteria for assessing the richness of human lives apply straightforwardly to animals. We must use all that we know about animals to try to gauge the quality of their lives in terms appropriate to their species. Then, we must try to gauge what a rich, full life is for an animal of that species and, subsequently, try to gauge the extent to which this approaches what we should mean in the human case when we say of someone that he or she had led a rich, full life. A rich, full life for a mouse does not approach the richness of a rich, full life for a human; the difference in cognitive capacities is just too great. However, we also need evidence to figure out what it is in the mouse's case that compensates for its apparent lack of certain capacities.

If we are going to work a quality-of-life view of the value of a life of a mouse, then we must try imaginatively to place ourselves in the mouse's position, with the capacities and life of the mouse. This will be difficult and the outcome of the investigation limited, but it is not impossible. In the case of primates and animals closer to ourselves, we may well be able to overcome many difficulties that impede our understanding of mice and what we think of as lower creatures. Indeed, much work with chimps and other primates suggests that we are only beginning to understand the interior lives of these creatures.

Of course, one might just want to claim that humans and animals have different capacities and lives and that, so judged, each leads a rich and full, though different, life. This makes it appear that we are barred from comparative judgments, when, in fact, the central ingredients of the respective lives, namely, experiences, appear remarkably alike in critical respects, as ethologists and animal behaviorists are now helping us appreciate. Still, I need far more evidence than I have seen to make me believe that, for example, one of the mouse's capacities so enriches its life that it approaches the level of a normal adult human life in richness.

My argument takes us to an unpalatable outcome: we cannot be sure that human life will always and in every case be of higher richness and quality than animal life. Today, in medical ethics, we appeal routinely to concerns of quality of life, and we treat quality as if it determined the value of a life. So, what can be cited that in each and every case guarantees that human life will be more valuable than animal life? We would need to know what separates the cases by species properties. I have maintained that nothing does.

ONE CONCEPTION OF THE MORAL COMMUNITY

There are many cultural differences in our societies, and they incorporate many different views of our relations to animals and even of the comparative value of human and animal lives. But from the bare fact that different, possible accounts of this comparative value may arise in communities, nothing follows per se about the adequacy of the conceptions. Justificatory argument must establish the soundness of such accounts, if they are sound.

Talk of community in animal ethics, especially "the moral community," runs a risk of confusing two different notions. No one will deny that the patient in the final throes of Alzheimer's disease or the severely mentally enfeebled are members of the moral community *in the sense of having moral standing*. They remain experientially aware subjects, with a welfare and quality of life that can be augmented or diminished by what we do to them. This is true of all kinds of human beings who presently, as the result of disease or illness, have had the quality of their lives radically diminished, from those seriously in the grip of amyotrophic lateral sclerosis to those with Huntington's disease. All kinds of human beings presently live lives of massively reduced quality, reduced from the quality of life we find in presently healthy, normal, adult humans. Yet, they all remain members of the moral community. Bentham on sentiency got this right, and utilitarians are right to insist upon it.

By contrast, patients in a permanently vegetative state or anencephalic infants are more problematic candidates for membership in the moral community in this first sense of "moral community." Although what happens to them may well affect the welfare and quality of life of other people, such as their parents or their children, it is not obvious that they themselves have experiential states that would deem them members of the moral community in their own right.

This first sense of moral community—the one under discussion in this section—is that in which the creatures that figure within it are all those who are morally considerable in their own right. I have indicated how I think the "higher" animals get into the moral community in this sense. Almost all accounts of moral community in this first sense are accompanied by disclaimers that animals are moral agents or are capable, in the sense of agency that matters for the assessment of moral responsibility, of weighing and acting upon reasons. The idea is that they are not morally responsible for what they do, not because they fall outside the moral community in the first sense, but because they do not act for and weigh reasons for action. This may be so, but it should be remembered that some humans too are not morally responsible for what they do and cannot act on reasons and never will be able to do so.

A Second Conception of the Moral Community

One move in the arguments today is to insist that one can find the rudiments of both agency and morality in animals.[37] In *Good Natured: The Origins of Right and Wrong in Humans and Other Animals*, Frans de Waal writes:

> The question of whether animals have morality is a bit like the question of whether they have culture, politics, or language. If we take the full-blown human phenomenon as a yardstick, they most definitely do not. On the other hand, if we break the relevant human abilities into their component parts, some are recognizable in other animals.[38]

De Waal does not discuss agency as such, but he does discuss some of what might be taken to be involved in it. It is, I claim, precisely the way we grasp what he calls "component parts" that we learn to understand sources of value and their relative importance to the value in the "full-blown human phenomenon." On how much or little we share with animals, de Waal continues:

> To focus attention on those aspects in which we differ—a favorite tactic of the detractors of the evolutionary perspective—overlooks the critical importance of what we have in common. Inasmuch as shared characteristics most likely derive from the common ancestor, they probably laid the groundwork for much that followed, including whatever we claim as uniquely ours. To disparage this common ground is a bit like arriving at the top of a tower only to declare that the rest of the building is irrelevant, that the precious concept of 'tower' ought to be reserved for this summit.[39]

I have no desire to disparage our common holdings with animals. But is there not a problem here, namely, that to focus exclusively upon what "component parts" the breakdown of human phenomena can show to be in common between humans and animals runs the risk, so far as the value of a life is concerned, of overlooking the critical importance for philosophy of mind, moral psychology, and moral philosophy of the "full-blown human phenomenon"? In de Waal's terms, this would be the danger of taking one part of the tower as the tower. No account of the value of our lives can overlook the role agency plays in them. To be sure, we may think certain animals, such as the great apes, will be favored in this hunt for "component parts" of agency, and it is no part of my brief to suggest that we may not find some critical and relevantly similar "parts" of human abilities in the great apes. The question is whether this or that part forms the whole, of whether animals can fashion lives for themselves within communities of agents and so augment the value of their lives through their agency in ways relevantly similar to the ways in which intact human adults do.

Whatever the answer, I have argued that animals are morally considerable; what befalls them through our actions and affects their welfare counts morally. Not all the creatures that fall within the class of morally considerable beings, however, are alike: some are included as agents, some as patients, and there is a sense of moral community in the case of the former that there is not in the case of the latter. In this second sense of moral community, members have duties to each other, and reciprocity of action standards exist for the assessment of conduct. We understand the conditions under which deviation from standards occurs, and we understand when reasons are appropriately offered and received. The absence of the proffering and receiving of standards and reasons matters because those who cannot do these things are not appropriately regarded as moral beings, in the sense of being held accountable for their actions. To be accountable for what one does, in a community of others who are accountable for what they do, is not the same as being considerable in one's own right.

Utilitarians have generally overlooked the importance of this difference. Plainly, some humans are not members of the moral community in this second sense: they are incapable of putting forth standards for the evaluation of conduct, of conforming their conduct to those standards, and of receiving and weighing reasons for action. Disease and illness, for example, can cause a loss of agency. Also, perfectly normal children and many of the very severely mentally enfeebled are not members of the moral community in this second sense. In this sense, many more humans fall outside the moral community as a community of agents than fall outside the moral community as a community of morally considerable beings. Put another way, persons who are not in the moral community in the second sense are in the moral community in the first sense and therefore have moral standing. The Benthamite position does not obviously capture this fact, but it is critical that utilitarians capture this second sense of moral community and incorporate its implications in their theories.

Some humans, such as those in permanently vegetative states and anencephalic infants, are going to fall outside the moral community in both of these conceptions of moral community. Hence, much of the controversy about, say, whether the former may permissibly be removed from respirators or whether the latter may permissibly be used as organ donors. On the other hand, while a great many, if not all, animals arguably fall outside the moral community in the second sense, a great many fall inside the moral community in the first sense. Thus, some humans will be outside the moral community altogether, even while some animals are within the moral community in the first sense and conceivably some in the second sense as well. If one were going to select a creature upon which to perform clinical research, these distinctions will matter to all morally serious people.

The second conception of moral community actually has more in common with Kant's non-utilitarian theory than with Bentham, but I think a utilitarian must embrace it no matter its historical source or its neglect in utilitarian theory. A critical feature in the view I have been defending is that agency, construed as acting and weighing reasons for action in the light of standards of assessment, is not required in order to be morally considerable in one's own right, and this insight is relevant to augmenting and helping us to determine the value of a life.

In a quality-of-life view of the value of a life, being a member of the moral community in the second sense stands to enrich one's life and, therefore, enhance its quality. It does this by informing the relations in which we stand to others and so affecting how we live and judge our lives. The moral relations in which we stand to each other are part of the defining characteristics we give of who we are. We are wives, mothers, daughters, friends, and so on. These are important roles we play in life, and they are informed by a view of the moral burdens and duties they impose on us, as well as the opportunities for action they allow us. Seeing ourselves in these relations is often integral to whom we take ourselves to be. In these relations, we come to count on others, to entwine ourselves with the fate of at least some others, to be moved by what befalls these others, and to be motivated to do something about the fate of these others to the extent that we can. Our lives and how we live them are affected in corresponding ways. Being a functioning member of a unit of this kind is one of the great goods of life.

In being a functioning member of a community characterized by these moral relations, we learn to take the rules and duties that in part comprise these relations to be, at least prima facie, norms for judging our own and others' actions. That these standards take a normative form by which we can evaluate reasons and actions, whatever their substance, is the crucial point. It is a normative understanding of these roles that seems crucial (i) to how we see ourselves within them, (ii) to how we live our lives and judge many of our actions within those lives, and so (iii) to how we judge how well or badly those lives are going.

The relations in which we stand to each other add enormously to how well we take our lives to be going. Since our welfare is to a significant extent bound up in these kinds of pursuits, to ignore this fact is to give a radically impoverished account of a "characteristically" human life. We often cannot explain who we are and what we take some of our prized ends in life to be except in terms of these relationships in a moral community, and we often find it difficult to explain why we did something that obviously was at great cost to ourselves except through citing how we see ourselves linked to certain others.

Much that we do in our relationships is similar to the activities of other animals, but agency enables us to do much more to fashion a rich life; we live a life molded and shaped by choices that are of our own making and so reflect, presumably, how we want to live. These great goods of human life enrich all who enjoy them. Unfortunately, not all humans have these opportunities or have such enriched lives, and we cannot adequately understand moral standing without resolving issues about marginal cases.[40] In contemporary animal ethics, marginal human lives have rightly been at the center of the discussion. Some of these lives cannot be lived in the moral community in either of the two senses I have discussed. Nonetheless, it has been argued that we may not use humans of this description as we use animals, even if it turns out that the animal in question has a higher quality of life and is a member of the moral community. This strategy makes morality, construed in a certain light, look like a convenient tool for protecting humans and leaving animals unprotected, even at the cost of consistency of argument.

I have been suggesting that utilitarianism should not be identified solely with the Benthamite sentiency criterion, though this has been the most influential moral norm in the history of utilitarian animal ethics. Many have construed utilitarianism as implying that humans get to use up animal lives, provided pain and/or cruelty are adequately controlled, even if a human life is of far lower quality than the animal's and no matter which animal is in question. My suggestions in this chapter carry with them a steep and unpalatable price, one that is directly connected to the argument from marginal cases. We are all agreed that we need to conduct AIDS research. In this work we can use, let us suppose, either a perfectly healthy chimp or a fifty-five-year-old man with the mental age of three months. A quality-of-life theory of the sort I have developed is going to select the man. Many have claimed that this implication of my account is reason to cleanse ourselves of all such quality-of-life arguments.

So which moral arguments are to be put in the place of these arguments? Some appeal to the intrinsic or inherent value of human life? But what justifies these claims, and how far are they to be carried?[41] The severely mentally disabled man's life is as valuable as that of a normal man, it might be claimed; but if I offered you one hundred additional years of life, only you will remain forever with the mental age of a three-month old, who would choose such a life? My claim is very simple: whoever claims that we cannot determine the quality of a life must answer why they would not select the human life I offered them, no matter its quality?

I do not have space here to go into objective views of quality of life versus subjective views, but it is a subjective view for which I ultimately would opt. This makes even more difficult the task of trying to see a life from the inside, whether the life be that of a dog, chimp, or man. Difficulty, however, is not the same thing as impossibility.

Conclusion

Utilitarianism is the name of a group of theories that judges the rightness of acts, choices, decisions, and policies by their consequences for both human and animal welfare. It is far from clear how we are to understand welfare and well-being. If it is difficult to determine human well-being, how much more difficult will it be to determine animal well-being? Some utilitarians today are writing on this very subject in order to develop appropriate theories and to address some of the difficulties I have raised in this chapter. Though this task is difficult and elusive, I find it no more elusive than the claim that human life, unlike animal life, is intrinsically or inherently valuable. The latter claim is far from well understood, and I doubt that it can be understood in any way other than through the quality-of-life position that I have defended in this chapter.

NOTES

1. I leave aside here any disputed claims about exactly how Mill is to be interpreted. His legacy in part turns upon this matter of interpretation. I refer not merely to the dispute over whether he was an act- or rule-utilitarian, but disputes over quite specific parts of his views. The final chapter of *Utilitarianism* (the best edition being J. M. Robson, ed., *Essays on Ethics, Religion and Society*, vol. 10 [Toronto: University of Toronto Press, 1969]) contains a discussion of rights and justice that many utilitarians did not follow.

2. All aspects of this depiction of utilitarianism have been the subject of fierce dispute, and different views have been adopted by different utilitarians. Thus, some have thought it is not actual but expected consequences that matter; that minimizing unhappiness rather than maximizing it should be the focus; that we should opt for satisficing rather than maximizing views of utility; and so on.

3. The reason is obvious: such a claim privileges some of our moral intuitions over others. These, one says, I just know to be correct, and theories that produce results that clash with these intuitions are just mistaken. In fact, we lack any firm view everywhere in ethics, I think, of how to determine the adequacy of any normative ethical theory. A theory that today produced the result that slavery in certain cases might be morally permissible would be rejected out of hand by most people because they just know that their belief about slavery is correct.

4. G. E. Moore, *Principia Ethica*, rev. ed., ed. Thomas Baldwin (Cambridge: Cambridge University Press, 1993).

5. I refer here principally to *Animal Liberation* (New York: Avon Books, 1975) and *Practical Ethics* (Cambridge: Cambridge University Press, 1986). Any number of articles and discussion pieces on animals have followed and continue to appear today.

6. To be sure, all kinds of utilitarianism were on offer, from numerous forms of rule-utilitarianism to cooperative and motive utilitarianism, but the thought that dominated many people's thinking was that a form of act-utilitarianism could be worked out. All kinds of additions were added to the theory, sometimes on non-utilitarian grounds, in order to play down the number and severity of clashes of act-utilitarianism and ordinary moral convictions. This clash with ordinary morality was, and in a somewhat more sophisticated form remains, the central objection to the theory. I think this of the integrity objection, the separateness-of-persons objection, the demandingness objection, and the partiality objection. These are the names given today to perhaps the four most dominant criticisms of utilitarianism, and these criticisms would be assumed to hold even if act-utilitarianism did give a successful account of the moral standing of animals and the value of animal life. It would still be claimed, for example, that by summing utilities over persons one ignores the separateness of persons. The fact that there are numerous powerful

utilitarian replies to these objections on hand today seems in no way to have lessened their use as criticisms of the theory.

7. These included Kantian-like rules, Dworkinian-like principles, and all manner of restrictive duties that would enable the act-utilitarian to avoid many of the clashes with ordinary morality that seemed so devastating to the theory to some. Even Hare, in appealing to levels of moral thinking, left us following ordinary moral rules at one level, while at the other suggesting that it was precisely these rules of ordinary morality that act-utilitarian thinking would select for us to act upon. *Moral Thinking: Its Levels, Method, and Point* (Oxford: Oxford University Press, 1981).

8. This is a well-known fact of the recent history of utilitarianism. It was fostered by works such as L. W. Sumner, *The Moral Foundation of Rights* (Oxford: Oxford University Press, 1987) and James Griffin, *Well-Being* (Oxford: Oxford University Press, 1987).

9. Hare, *Moral Thinking*. Hare worked hard with me on utilitarianism, and a number of relevant figures were then visiting Oxford, including Jan Narveson, David Lyons, Henry West, and J. J. C. Smart. Smart greatly influenced my thinking.

10. Smart talked me out of this decision and back into some form of act-utilitarianism. To all but Smart, figuring out some way of avoiding clashes with ordinary morality seemed to be the nature of work on act-utilitarianism; with him, this most certainly was not the driving force behind the theory. The most commonly cited text on Smart is his reprinted piece with Bernard Williams in *Utilitarianism: For and Against* (Cambridge: Cambridge University Press, 1973). Smart had much more to say about utilitarianism in conversation and had replies to several of the criticisms that Williams makes. I tried to convince Williams to reprint this volume with expanded essays by both men, but he would not agree.

11. The theory was not dead, but it was out of fashion as the theory likely to be worked into the adequate normative ethical theory everyone was seeking. I worried at times that my dissertation had gone with the wind.

12. Different works by Jonathan Glover and Peter Singer were partially responsible for this revival.

13. I remained attracted to Smart's single-level theory, which could be improved without turning it into either a split-level act-utilitarianism and without embracing rule-utilitarianism. This is a common objection to Hare's split-level view today. Yet, Hare led the way with split-level theories of this kind, which today is largely reflected in the distinction between direct and indirect theories (in their application of consequentialism to practical affairs).

14. Jeremy Bentham, *An Introduction to the Principles of Morals and Legislation*, in *The Collected Works of Jeremy Bentham* (Oxford: Clarendon Press, 1996 [1789]), chapter 4. It is interesting to read William Godwin's *An Enquiry concerning Political Justice, and its Influence on General Virtue and Happiness* (London: Robinson, 1793) in connection with the centrality of sentiency to a utilitarian viewpoint.

15. John Austin, *The Province of Jurisprudence Determined*, ed. W. Rumble (Cambridge: Cambridge University Press, 1832).

16. Hastings Rashdall, *The Theory of Good and Evil*, 2 vols. (Oxford: Oxford University Press, 1907).

17. W. T. Stace, *The Theory of Knowledge and Existence* (Oxford: Oxford Unniversity Press, 1932).

18. G. E. Moore, *Principia Ethica*.

19. In *Animal Liberation* (1975), there was no philosophy, utilitarian or otherwise. Indeed, I doubt that any book could be filled with sophisticated philosophy and sell like that one. I even receive requests from high-school students in the United States to send

some of my pieces on Singer to them, something that when it first happened I found hard to credit.

20. See, for example, his essay in this volume.

21. William Whewell, *The Elements of Morality* (Cambridge: Cambridge University Press, 1845); and *Lectures on the History of Moral Philosophy in England* (Cambridge: Cambridge University Press, 1852).

22. Of course, it is not obvious that or why utilitarians must assign all sentient creatures equal moral status. Far from seeming obvious today, this generous assumption could come under rather fierce attack. At a conference, the late David Lewis told me that he thought this was a mere prejudice on the part of utilitarians, a hold-over from their past views.

23. Today, this is done standardly to introduce at once into the debate the argument from marginal cases. Examples of this argument can be found throughout this chapter.

24. Saying exactly in what the quality of life consists remains a difficult task. But it is not as if we have no knowledge of how to go about gaining answers. By our usage of the notion today, we give every indication of thinking the opposite.

25. To say that we cannot know everything about an animal's interior life is not to say that we can know nothing. One must be careful here: the view that we can just read off the animal's interior states from its external behavior needs to be carefully assessed, especially since so much of an animal's external behavior remains the same from situation to situation.

26. I do not suggest that Singer thinks otherwise.

27. I know that what follows is controversial, but I do not think that ordinary moral thinking or commonsense morality or the convictions of the average person constitute a test of a view's moral adequacy. This is a very important issue that I have no space to address here, but it is crucial, I think, that anyone putting forward an ethical theory put forward as well the justification by which we can tell that the theory is adequate (or, I suppose, some might say, correct).

28. See Tom Beauchamp's arguments to this effect in his article on rights theory in this *Handbook*.

29. See Tom Regan, *The Case for Animal Rights* (Berkeley and Los Angeles: University of California Press, 1983); and his article "Utilitarianism, Vegetarianism, and Animal Rights," *Philosophy and Public Affairs* 9 (1980): 309–24. See also his book *Defending Animal Rights* (Urbana: University of Illinois Press, 1993). This interpretation of Dworkin is common, but commentators often fail to note that his views were specifically targeted to political and legal philosophy. He did not express such a view about rights outside of this limited context.

30. I do not mean to suggest here that I would not in another context want to draw distinctions between different kinds of animals, for I would want to allow that a primate gets moral protection for the value of its life in a way that the mouse does not. Thus, I have a problem with AIDS research: while I would have no difficulty in using mice to conduct the research, using baboons or other primates in the research ups the argumentative price of the argument. A quality-of-life argument may well suggest that we may not use baboons in this research. I should need more space to make clear the kind of difference I am trying to draw here, but the fact that the United Kingdom has barred AIDS research in the great apes does not seem to me all that great a leap on my view. However, because the information derived from the research using the great apes can still be purchased by labs that themselves do not use primates may simply undo the actual ban.

31. The fact that I object to Singer's arguments on animals (and famine) does not show that I am not in agreement with him on much else. My writings on euthanasia, for

example, though the arguments are different from his, reach the same basic conclusions that Singer does.

32. I did not know that Williams was writing on speciesism until a recent, posthumous piece "The Human Prejudice" appeared. See Jeffrey Schaler, ed., *Singer Under Fire* (Chicago: Open Court, 2009). What Williams says in this essay reflects what he said to me.

33. Williams objected to this way of introducing quality-of-life talk into the discussion, which he thought imported an unfair advantage to me in the argument. The advantage he thought it gave me was that I could point to differences between lives so radical that it might (I say *might*) force one to rethink what one thought a human life was.

34. This explains why Kantians, even ones as distinguished as Christine Korsgaard in this volume, leave the argument from marginal cases still seeming to point to a view of moral standing and worth that turns upon the content of the life being lived, and this can vary mightily. Do we rethink our view of who or what is human? Do we exclude some of these marginal humans from the category of those with moral standing and value? How does the abstract discussion of rationality and autonomy as the basis of standing even hook up with the individuals to whom I have been referring.

35. Here, I allude only to the logic of the position, not to any side effects that might easily bar one from acting on that logic. I mean to include under side effects the fact that much of the population might be so revolted by the conclusions reached that they threaten riot and mayhem. Yet, is this not the same state in which we now find the abortion, active euthanasia, and physician-assisted suicide debates? Yet, in many places in different parts of the world these things have come to pass.

36. It should be obvious here why I think the debate over the adequacy of a normative ethical theory is so important. To what extent must that debate reflect what many people think to be the case? Once in a debate, Brad Hooker suggested that he was more confident that his intuition about it being wrong to gratuitously torture young babies was correct than he was of the correctness of any moral theory. Is this how we are to proceed, by finding out what each of us with our varying moral intuitions just cannot live with?

37. This view, incidentally, runs completely counter to Regan's philosophy, which argued for animals through their being moral patients, not agents.

38. Frans de Waal, *Good Natured: The Origins of Right and Wrong in Humans and Other Animals* (Cambridge, Mass.: Harvard University Press, 1993), p. 6.

39. De Waal, *Good Natured*, p. 7.

40. I disagree with the claim of Christine Korsgaard in this volume about the matter.

41. I have no idea what these notions mean. What are the criteria for determining whether something has inherent value?

SUGGESTED READING

BEAUCHAMP, TOM L., F. BARBARA ORLANS, REBECCA DRESSER, DAVID B. MORTON, and JOHN P. GLUCK. *The Human Use of Animals*, 2nd ed. New York: Oxford University Press, 2008.

CARRUTHERS, PETER. *The Animals Issue*. Cambridge: Cambridge University Press, 1992.

DAWKINS, M. S. *Through Our Eyes Only? The Search for Animal Consciousness*. New York: W. H. Freeman, 1993.

FREY, R. G. "Animals." In *Oxford Handbook of Practical Ethics*, edited by Hugh LaFollette, pp. 151–186. New York: Oxford University Press, 2003.

————. "Medicine, Animal Experimentation, and the Moral Problem of Unfortunate Humans." *Social Philosophy and Policy* 13 (1996): 181–211.

————, ed. *Rights, Killing, and Suffering*. Oxford: Blackwell, 1983.

GRIFFIN, DONALD. *Animal Minds*. Chicago: University of Chicago Press, 1992.

HARE, R. M. *Moral Thinking: Its Levels, Method and Point*. Oxford: Oxford University Press, 1981.

HOOKER, BRAD. *Ideal Code, Real World: A Rule-Consequentialist Theory of Morality*. Oxford: Oxford University Press, 2000.

KAGAN, SHELLY. *The Limits of Morality*. Oxford: Oxford University Press, 1989.

KORSGAARD, CHRISTINE. *Self-Constitution: Agency, Identity and Integrity*. Oxford: Oxford University Press, 2009.

MULGAN, TIM. *The Demands of Consequentialism*. Oxford: Oxford University Press, 2001.

NORCROSS, ALASDAIR. "Scalar Act-Utilitarianism." In *Blackwell Guide to Mill's Utilitarianism*, edited by Henry R. West, pp. 217–232. Boston: Blackwell, 2006.

PALMER, CLARE, ed. *Animal Rights*. Aldershot, England: Ashgate Publishing, 2008.

PORTMORE, D. "Can an Act-Consequentialist Theory Be Agent-Relative?" *American Philosophical Quarterly* 38 (2001): 363–377.

RACHELS, JAMES. *Created from Animals: The Moral Implications of Darwinism*. Oxford: Oxford University Press, 1990.

RAILTON, PETER. "Alienation, Consequentialism, and the Demands of Morality." *Philosophy and Public Affairs* 13 (1984): 134–171.

REGAN, TOM. *Defending Animal Rights*. Urbana: University of Illinois Press, 2001.

————. *The Case for Animal Rights*. Berkeley and Los Angeles: University of California Press, 1983; rev. ed., 2004.

SCHALER, JEFFREY A. *Peter Singer under Fire*. Chicago: Open Court, 2009.

SCHEFFLER, SAMUEL. *The Rejection of Consequentialism*. Oxford: Oxford University Press, 1982.

SINGER, PETER. *Animal Liberation: A New Ethics for Our Treatment of Animals*. New York: Avon Books, 1975.

SINGER, PETER. *Practical Ethics*. Cambridge: Cambridge University Press, 1986.

SLOTE, MICHAEL. *Common-sense Morality and Consequentialism*. London: Routledge, 1985.

SMART, J. J. C. "An Outline of a System of Utilitarian Ethics." In *Utilitarianism: For and Against*, by J. J. C. Smart and Bernard Williams. Cambridge: Cambridge University Press, 1973.

SMITH, J. A., and K. M. BOYD, eds. *Lives in the Balance: The Ethics of Using Animals in Biomedical Research*. Oxford: Oxford University Press, 1991.

VALLENTYNE, PETER. "Against Maximizing Act-Utilitarianism." In *Contemporary Debates in Moral Theory*, edited by James Dreier, pp. 21–37. Boston: Blackwell, 2006.

WILLIAMS, BERNARD. "A Critique of Utilitarianism." In *Utilitarianism: For and Against*, by J. J. C. Smart and Bernard Williams. Cambridge: Cambridge University Press, 1973.

CHAPTER 7

RIGHTS THEORY AND ANIMAL RIGHTS

TOM L. BEAUCHAMP

PHILOSOPHERS often defend a comprehensive and abstract ethical theory and then fill out the theory with weighty arguments against competing theories. I will not defend rights theory in this fashion. I will concentrate on aspects of rights theory suited to the analysis and justification of animal rights. My objective is to defend the claim that animals have rights. In some philosophical writings, these rights are treated as remarkably strong protections of animal interests that cannot be compromised and can never be overridden by human interests. The rights of animals in these theories are sufficiently strong that they prohibit most, and likely all, of the practices involving animals that are familiar features of modern society, including biomedical research, toxicological testing, factory farms, zoos, circuses, children's petting farms, hunting, cosmetic surgery of pets, aquariums, fishing, horse racing, clothes from animal products, pest control, and the like. The account of animal rights that I propose is not prohibitionist as these theories are, but it is still a robust defense of animal rights that would significantly alter many current practices.

I begin with landmark historical sources of rights discourse in philosophy (section 1). I then discuss the meaning of "animal rights," including why a distinction made between animal rights and animal welfare is unreliable (section 2). There follows an analysis of the meaning of the term "rights" that cuts across discussion of both animal rights and human rights (section 3). Philosophers disagree about the theoretical foundations of human rights, and yet they manage to converge to agreement about a list of rights that I will call *basic human rights*. I propose basic rights and basic obligations as the proper starting point for the analysis of animal rights (section 4), and I argue that there is a firm correlativity between rights and obligations (section 5). I then turn to the *specification* of obligations and rights, a method that renders rights less abstract and more practical (section 6). I also recommend a list of rights that I call a catalogue of animal rights (section 7). Finally, I consider ways in which rights are legitimately overridden by competing moral considerations (section 8).

1. HISTORICAL SOURCES OF RIGHTS DISCOURSE IN PHILOSOPHY

Only recently have philosophers, legislators, lawyers, and the general public been much interested in serious pursuit of the notion of animal rights. Indeed, the term continues to be used as a form of ridicule in many influential quarters, including cable-television news services and talk radio.

Historical neglect of the notion of animal rights is unsurprising. Until the seventeenth century, problems of political morality and political philosophy were rarely expressed in terms of rights even for humans, though historians have discovered a few comparable ideas about rights between roughly the twelfth and sixteenth centuries. This history was altered by the introduction of pioneering theories of universal rights, international rights, and natural rights—now restyled as *human* rights. These ideas first prospered in philosophy through the social and political theories of Hugo Grotius, Thomas Hobbes, John Locke, and their successors. They sought to delimit legitimate government authority and to defend the rights of individuals in the face of political abuse and undue restriction of liberty.

It would be historically inaccurate to suggest that theories of rights came into philosophy, political theory, and law only through political theories of natural laws and political states of the sort for which Grotius, Hobbes, and Locke are rightly celebrated. Rights were not, in the moral philosophy of the period, considered exclusively or even principally as legal or political. Moral-rights theories were developed alongside theories of moral obligations, and both types of theory were almost seamlessly connected to political philosophy. For example, property rights were discussed as moral rights, not merely as political or legal rights. In Locke's political writings, rights were referred to as "just and natural rights" and "common rights,"[1] language that I would happily embrace in the framework I develop in this chapter.

2. THE MEANING OF "ANIMAL RIGHTS"

In light of this history and recent moral theory that makes use of both "human rights" and "moral rights," does it make sense to assert that animals have rights? If so, what does the term "animal rights" mean, and what might it come to mean?

"Animal rights" is a generic term that in popular and casual use refers indiscriminately to a wide range of views about how animals should be protected against improper human practices. The term "animal rights movement" co-travels with it as an all-purpose label for social movements that work to protect the *interests* of animals. The term "interest" is common, and critically important, in rights theory. I here use "interest" to refer to that which is in an animal's interest—that is, what is to the animal's welfare advantage in a given circumstance. "In an animal's interest" might be interpreted as "in its *best* interest" (to its maximal advantage), but there are problems about using "*best* interest," and I shall not use "interest" with this demanding provision. Finally, the term "interest" does *not* mean what a particular being is interested in, seeks, or desires.[2]

Many theories that are concerned with the abuse and protection of animals cannot be comfortably situated under the label "animal rights," because the language of *rights* does not fit comfortably in these theories. To see why, we need to attend to the meaning of "animal rights."

Persons interested in the protection of welfare interests of animals have commonly been classified into two major types: (1) those who believe that animals have an extensive array of robust rights (animal rightists) and (2) those who believe that humans have obligations to protect basic welfare interests of animals (animal welfarists). Animal rightists are typically portrayed as endorsing rights for animals such as the right to a life, the right to an uncontaminated habitat, the right to not be constrained in cages or pens, the right to not be hunted, the right to not be used in biomedical research, and many other rights enjoyed by humans. These animal rights theories are said, by their proponents, to prohibit human utilization of animals: animal rightists seek to abolish the use of animals in biomedical research, the production of food, the entertainment industry, zoos and aquariums, and the like. Animal welfarists, by contrast, are depicted as holding utilitarian or pragmatic perspectives that acknowledge human obligations to not cause avoidable harm or undue loss of liberty to animals, while allowing that many uses of animals can be justified if there is a net benefit once the interests of all parties are taken into consideration. The welfarist presumably promotes the humane treatment of animals and stands opposed to unsanitary facilities, abuse, neglect, exploitation, cruelty, and the like.[3]

The most widely discussed and the most demanding and unyielding animal rights theory in philosophy presents animals as having significant value because they are, like humans, "subjects of a life." Tom Regan's "rights view" is the best known among the various theories that are entirely opposed to almost all human uses of animals. His "postulate of inherent value" of animals places humans under strict obligations of respectful treatment.[4] Animals have rights to respectful treatment, and those rights cannot be infringed by human utilitarian interests.[5] His theory is a categorically abolitionist philosophy that gives animals rights to not be hunted, exploited in sports activities, involved in scientific research, consumed as food, confined in zoos, and the like.[6]

The distinction between animal rightists and animal welfarists is understandable, but it is a crude tool for dividing up the world of protective support for animals and for sorting out types of theory. The many theories that afford protection to animals are better analyzed as a spectrum of accounts spread across a continuum that ranges from, on one end, a minimal set of human obligations to animals (e.g., "do not treat animals cruelly" and "do not slaughter animals inhumanely") to, on the other end, a maximal and prohibitionist set of human obligations to animals (e.g., "do not kill animals" and "do not utilize animals in laboratories"). An ample and diverse range of theories stretches across this continuum.[7]

"Animal rights" has often functioned as a polarizing term, suggesting both that there is inherent conflict or an inseparable gulf between "rightists" and "welfarists" and that a conflict inherently exists between animal rightists and certain professional groups such as research scientists, veterinarians, and directors of zoos. In this chapter, I make no such presumption. I reject the polarizing distinction between rightists and welfarists in favor of the continuum account that I have proposed in this section.

3. THE CONCEPT OF RIGHTS

I turn now to analysis of the concept of rights. The term "rights" does not have different meanings when we distinguish between human rights and animal rights. Different individuals and species have different sets of rights, and there will be different criteria for selecting this or that set of rights; but the meaning of the term "rights" should be invariant across the contexts of its usage.

Rights as Justified Claims

H. J. McCloskey has defended a well-known, and instructively flawed, analysis of rights in terms of what he calls *entitlements to*: "A right is an entitlement to do, to demand, to enjoy, to be, to have done for us....We speak of rights as being *possessed*, *exercised*, and *enjoyed*....We speak of our rights as being *rights to*—as in the rights to life, liberty and happiness—not as *rights against*, as has so often mistakenly been claimed."[8] But it is McCloskey who is putting forward the mistaken claim. Both of these theses about rights are correct. The language of "rights against" captures the fact that rights are held *against* others, a language that has long prevailed in rights discourse in the law. For example, the right to privacy, which has principally been invoked to prevent governmental interference with personal decision-making, is a right against the state. The best analysis of the concept of rights is that a right gives its holder a justified claim *to* something (an entitlement) and a justified claim *against* another party. Claiming is a mode of action that appeals to moral norms that permit us to demand, affirm, or insist upon that which is due to us. "Rights," then, may be defined as "justified claims to something that individuals and groups can legitimately assert against other individuals or groups." A right thus positions one to determine by one's choices what others must or must not do.[9]

Rights claiming is a rule-governed activity in each domain in which there are rights. The rules may be moral rules, legal rules, institutional rules, or rules of games. All rights exist or fail to exist because the relevant rules allow or disallow the appropriate claiming. The rules distinguish justified from unjustified claims. *Legal* rights are justified by normative structures in law and *moral* rights are justified by normative structures in morality.

How Robust Are Rights?

Some writers use the language of rights in a more robust sense than I am proposing. They maintain—following the suggestive language of Ronald Dworkin—that particularly critical interests are firmly protected by rights that have the force of trump cards.[10] However, this trump metaphor does not work well for situations in which one moral right conflicts with another moral right—a critical problem in the treatment of animal rights. It is questionable in many circumstances whether individual rights that conflict with the public interest, the national interest, and the like are truly trumps against other parties. It is still more doubtful that there are absolute rights.[11] The major value of the trump metaphor is to remind us both that rights powerfully protect individuals from having their interests balanced or traded off and that attempts to override them in the public interest need the most careful inspection and justification. Dworkin gives a notably concise and limited account of rights as trumps: "Rights are best understood as *trumps over some background justification for political decisions* that states a goal for the community as a whole."[12] In effect, Dworkin thinks that rights are stronger—much stronger—than the moral claims created by community goals and preferences. Rights trump such background justifications. Dworkin makes no stronger claim.

Interpreting rights as trumps is an appealing view in contexts in which individuals are vulnerable to serious harms or in which minority populations might be oppressed by majority preferences. However, models from trumps, absolute shields, and uninfringeable deontological protections are often more misleading than insightful in moral philosophy. This trumping thesis is not the best way to understand either the force of rights or what makes rights special. Rights are special, and especially cherished, because individuals hold justified claims that they can exercise at their discretion. The having of rights provides a reason for personal self-respect and security that is not provided by moral categories such as the obligations and virtues of others. This particular thesis holds, of course, only for those who are aware that they have rights. Because animals lack the awareness required for self-respect, rights do not promote self-respect for them, but rights still may be the animals' only secure moral basis of protection against serious wrongdoing.

Rights Exercised by Surrogates

Possession of a right is independent of being in a position to *assert* the right or to *exercise* the right. A right-holder need not be the claimant in a particular case in order to have a justified claim. That one does not know that one has a right is no basis for asserting that one does not have it. For example, infants and the severely mentally handicapped do not know, assert, or claim their rights, but they still possess them, and claims can be made for them by authorized representatives. Many dependent humans and dependent animals (for example, animals on factory farms) have rights whether or not they have a surrogate or authorized representative in a position to exercise the rights.

4. Basic Rights and Obligations

Philosophers rarely agree about the philosophical foundations of substantive principles and human rights, but they do often agree about the particular human rights that we possess. Diverse philosophical theories acknowledge human rights such as rights of privacy, liberty rights, rights to life, equal political rights, rights to fair trials in courts, and rights to equal consideration of interests. Philosophical theories, as well as international legal agreements, generally converge on the bulk of human rights, though there is a more concentrated convergence on negative rights (justified claims to liberty—that is, rights to others' abstentions, noninterference, and forbearances) than on positive rights (justified claims to goods or services).

Henceforth, I will assume that there are some identifiable basic human rights and that we can locate the interests they protect. However, the connection between rights and interests, as well as the set of interests that deserve inclusion, are understandably disputed. Many philosophers also have a different starting point from mine, and they may recognize either a smaller or a larger body of interests and rights than I do. For example, in some moral theories the interests protected by rights are determined by an attribute or capacity the individual has, such as whether the individual is capable of autonomous agency (the capacity to voluntarily act on one's own chosen ends). Philosophers differ profoundly about the precise range of interests that deserve protection. I will focus on welfare (or well-being) interests, such as avoidance of pain and suffering and satisfaction of basic needs, but I acknowledge, of course, that there is no consensus at present in rights theory on which range of interests is protected or on precisely why it is protected, for humans or for animals.

The primary assumptions I carry into the discussion are these: (1) an animal has a right only if the individual has an interest that merits protection for the individual's *own sake*, not because the individual's welfare creates a benefit *for others*[13]; and (2) a compelling moral case can be made for protection of the interests of at least some animals who are vulnerable to unnecessary suffering and bodily assault.[14] In my view, the "compelling moral case" mentioned in this second assumption cannot be adequately made by appeal to theories that engage rights entirely through models such as autonomous agency, cognitive activity, or contractarian agreements among human individuals.

I will often proceed as follows: Once I have identified a relevant class of *human* interests and its connection to rights (e.g., interests in liberty and liberty rights), I will then investigate whether there are morally relevant differences between human and nonhuman animals in the interests protected by these rights. If there are no relevant differences, I will assume that it is unjustified to assert that humans hold this class of rights whereas all nonhumans do not (unless additional argument yields a justification). When we lack justification for distinctions we make between humans and others, it is morally indefensible to have social policies that protect the interests of humans by according them rights while denying those rights to relevantly similar species of animals.

This style of argument must proceed by identifying and examining classes of rights, not by lumping all rights together. Humans have many rights that animals do not hold, but many rights are also shared. I will eventually argue the perhaps surprising thesis that many dependent, domesticated animals have various basic rights that are not in the set of human rights. That is, some animals have rights that humans do not have.

Basic Moral Rules and Rights

The moral norms on which all persons committed to morality and all philosophical theories of morality tend to converge I will refer to as *basic* moral standards, whether these standards are rights, principles, rules, virtues, paradigm cases, or some other type of general moral norm. Philosophical theories tend to converge to agreement on truly universal norms because they are essential elements of morality. The fact that we often disagree about how to specify these basic moral norms to deal with particular problems does not undercut their status as basic.

It would be ridiculous to assert that all persons accept and act on these basic moral norms. Many amoral, immoral, and selectively moral persons do not care about or identify with morality. Nonetheless, almost all persons in all cultures who are genuinely committed to a moral way of life do accept the basic norms. This morality is not merely *a* morality in contrast to *other* moralities.[15] It is normative for everyone who can rightly be judged by its standards. The following are examples of rules of obligation in this shared morality: "do not cause pain or suffering to others," "tell the truth," and "keep your promises." Corresponding to each such rule is a basic moral right, including "the right to not have pain or suffering caused by others," "the right to be told the truth," and "the right to have promises kept."

Human rights are today the most frequently mentioned category of universal morality. Rights language is used to insist on justified claims that are not contingent on whether a culture or an international court confers or protects the rights. I am going to assume without argument that there are some *basic rights*—so-called human rights in today's international discourse—and that they do not have their foundations in courts, political states, religions, or the heads of philosophers. Basic rights transcend all such norms.[16]

Basic Rights for Animals and Human Obligations to those Animals

The natural reading of "human rights" is *rights for humans only* (and also *rights that all humans have by virtue of being human*). However, to assume that all basic rights are for humans only is presumptive and prejudicial. Rights that are basic protect fundamental interests. Some interests—for example, in not being in pain, not suffering, having freedom of movement, having basic needs met, and the like—are not interests of humans only. Pigs, cattle, and chimpanzees have similar, and in many cases identical, interests.[17]

In using the term "basic rights" instead of "human rights" I am shifting the focus away from our traditional preoccupation with human rights. The goal is to see what can legitimately be transported to animal rights from our considered judgments about human interests and rights. It is of course a mistake to assert that *all* rights that are basic in a defensible framework of human rights are transferable from the human to the nonhuman. Basic rights such as "the right to consent to medical treatment," "the right to have promises kept," and "the right to freedom of speech" would make little, if any, sense for nonhuman animals. Still, humans and many nonhumans share various interests that merit protection by rights.

Some basic rights of humans and members of other species derive from conditions of vulnerability and potential harm. Parents owe to children, and farmers owe to their animals, protection against starvation and disease. To say that an individual animal has a right to such protections is to claim that it (1) has moral status, that is, the individual's interests count morally, and (2) deserves basic moral protections that follow from the status held and the rights that co-travel with the status. Consider as a paradigm instance our uses of both humans and animals, in similar and dissimilar ways, as experimental subjects of scientific research. Government oversight of biomedical investigations requires that assessments of the ethical justifiability of research on humans and research on animals should be assessed differently under different systems of rules and forms of review. However, it is close to universally agreed that we are morally obligated to assess and justify risks of harm that we impose on both human and animal research subjects, because both classes of subjects are often in relevantly similar situations of vulnerability to pain and suffering and both have moral status.

The prime nonhuman candidates that qualify for basic rights to be protected against such harms are animals with a cognitive, sensory, or emotional life relevantly similar to human cognition, sensation, or emotion. The members of many species of animals perceive, experience pain, make assessments of what to do or not do, have memories from past experience, and the like, whereas the members of many other species lack one or more of these properties. Where we draw the line between creatures that qualify under a basic rights conception and those that do not is as complicated and difficult as deciding which creatures have which levels of moral status. Continued advances in evolutionary biology and other scientific fields are sure to alter further how these lines should be drawn.

To treat these problems and to locate which rights animals have, I need an additional theory, which I will call, following precedent, the correlativity of rights and obligations.

5. The Correlativity of Rights and Obligations

That there is a correlativity between obligations and rights is generally accepted in philosophy and law, but its precise nature and its complexity are disputed. The correlativity thesis has traditionally been used to connect only the obligations and correlative rights of humans. My claim is that it has considerable significance for the theory of animal rights.

Correlativity Theory and Animal Rights

Correlativity is best analyzed in terms of what it means to assert that "X has a right to do or have Y." At least part of what is meant is that X's right entails that some party has an obligation either to not interfere if X does Y or to provide X with Y. In all contexts of rights, a system of norms imposes an obligation to act or to refrain from acting so that X is enabled to do or to have Y. The language of rights is thus translatable into the language of obligations: a right entails an obligation, and an obligation entails a right. If, for example, a political state has a legal obligation to provide goods such as food or health care to needy citizens, then any citizen who meets the relevant criteria of need has a legal entitlement to food or health care.

Here is a brief schema of the idea that basic moral obligations and basic moral rights are correlative:

Basic Obligations	Basic Rights
1. Do not kill;	1. The right to not be killed;
2. Do not cause pain or suffering to others;	2. The right to not be caused pain or suffering by others;
3. Prevent harm from occurring;	3. The right to have harms prevented from occurring;
4. Rescue persons in danger;	4. The right to be rescued when in danger;
5. Tell the truth;	5. The right to be told the truth;
6. Nurture the young and dependent;	6. The right to be nurtured when young and dependent;
7. Keep your promises;	7. The right to have promises kept;
8. Do not steal;	8. The right to not have one's property stolen;
9. Do not punish the innocent;	9. The right to not be punished when one is innocent;
10. Obey the law.	10. The right to have others obey the law.

The rights shown are notably abstract, by contrast to specific and concrete. Most of these rights are for humans only. In this table, I am not considering whether these obligations and rights have broader application to nonhuman animals, but I turn now to this question.

Anti-cruelty and anti-torture statutes in law at first glance are paradigms of correlativity in the domain of animal protection. Because we are legally obligated to not be cruel to an animal in ways specified in applicable statutes, an individual animal would seem to have a correlative legal right to not be treated cruelly. However, the law in most nations does not recognize moral standing for animals, and therefore animals do not have legal rights. This traditional legal hurdle to correlativity is not a barrier, however, in the moral domain: Because we are morally obligated to not be cruel to animals, they have a moral right to not be treated cruelly. The obligation strictly entails the right. To inflict unnecessary suffering by capturing a wild monkey and placing it in a tiny enclosure in a zoo or laboratory is to make the animal suffer in ways comparable to human suffering under conditions of small and isolated enclosures in prisons. Just as humans have correlative rights against such abuse, so do monkeys.

Correlativity is morally protective of nonhuman animals in many domains. If a zoo has an obligation to provide food and roaming space to needy members of certain species of animals in order to meet the basic needs of the animals, then any member of a relevant species has a right to the food or to the roaming space. Likewise, because of obligations we owe to race horses, seriously injured horses have a right to either rehabilitation or euthanasia, depending on the severity of the injury. Similarly, farm animals have a right to veterinary care or to euthanasia under specified conditions of serious disease or injury.

The logic of the correlativity thesis is that whenever humans have obligations to animals, then, whatever the obligations are, a creature has correlative rights. All who accept obligations of humans to animals logically must recognize correlative rights for animals. In some circumstances, there are obligations even if no specific individuals can be identified as having rights. For example, professionals in veterinary public health have obligations to protect both animals and people against communicable diseases, though no specific animal or human is identifiable in many circumstances.[18] These right-holders are unidentified members of populations.

This correlativity should be understood in terms of *direct* rather than *indirect* obligations to animals—that is, in terms of obligations *to* an animal by contrast to obligations merely *regarding* the animal.[19] If zoo A borrows an animal from zoo B and then abuses the animal by neglect, an account of animal rights cannot be grounded in a system in which zoo A's obligations to take care of the animal are only to zoo B. The owners of zoo B may not care what happens to its animals or zoo B may be bankrupt by the time of the scheduled return of the animal. If the obligations were entirely to owners of animals, then only the owners of the animals have rights, and then there are no rights of animals, from which it follows that there are no bona fide obligations to animals of any sort. But clearly there are such obligations, as is now almost universally agreed.

Is the Correlativity Thesis Flawed?

The correlativity thesis has been challenged on the grounds that the correlativity between obligations and rights is untidy[20] in that (1) only *some* obligations entail rights and (2) only *some* rights entail obligations.[21] I find these two challenges to correlativity unconvincing, but I will discuss only the first challenge because it is now generally conceded by critics of the correlativity thesis that all *genuine* rights (by contrast to merely *proclaimed* rights and *aspirational* rights) carry correlative obligations. Also, the first challenge is the sole crucial matter for the theory that animal rights follow from human obligations to animals.

The objection is that some appropriate uses of the term "obligation," as well as the terms "requirement" and "duty," clearly show that some obligations do not imply correlative rights. Alleged examples come from the fact that we refer to obligations of charity, and yet no person can claim another person's charity as a matter of a right. Such obligations have no correlative rights. Obligations of love and obligations of conscience are also put forward as examples of obligations without correlative rights. The problem with these claims is that, although it is correct to say that such alleged norms of "obligation" express what we "ought to do" or are "required to do" in some sense, they do not obligate us from genuine moral obligation, but only from admirable moral ideals that exceed obligation. They are self-imposed rules of "obligation" that are widely admired and endorsed in the moral life but that are not obligations literally *required* by morality.

This difference is sometimes marked in moral philosophy by saying that perfect obligations have correlative rights, whereas imperfect obligations do not. I prefer the tidier approach that only perfect obligations are moral obligations. So-called imperfect obligations are moral ideals that allow for discretion. Also, I will not distinguish between special obligations and universal obligations—a distinction that introduces the categories (1) perfect and imperfect special obligations and (2) perfect and imperfect universal obligations. I am only interested in what is morally owed by individual or group X to individual or group Y. My proposal is that all genuine obligations are perfect and have correlative rights, and all rights likewise have correlative obligations.[22]

It is not always clear, however, whether an obligation is a genuine moral obligation. Consider a circumstance is which a fire has broken out in a large barn that is spatially contiguous to a small horse corral. A person who has been admiring the beauty of the horses can now save their lives by opening a conveniently situated corral gate that is forty feet from the fire. The gate happens to be located exactly where the onlooker is standing. The person can, without any risk to himself, release the latch on the gate or he can simply stand by and watch the horses burn when the barn falls on them. My view is that the observer has a moral obligation to rescue the horses, which are in distress and now their lives are in danger. The rescuer is at no risk and it simply takes five seconds to reach down and unlatch the gate, and thereby to rescue the horses.

This moral obligation is in a different class from self-imposed "obligations" of beneficence such as the rescue of horses from raging rivers that endanger the lives of rescuers as much as the lives of the horses. Such river rescues can be called a moral "requirement" only in an extended and misleading sense. Here there is no right of rescue, by contrast to the right of rescue of the corral-trapped horses. We could of course change the example to a different situation in which the person standing by the horses would be seriously endangered by the fire if he or she had to run through a blazing fire to a distant gate. As risks in circumstances of fires and raging rivers increase, it becomes increasingly less likely that there is an *obligation*, and a rescuer at some point in the risk index becomes a hero and not at all a person who discharges an obligation. There is no obligation for the hero except a self-imposed ideal adopted from a sense of being obliged.

Some writers on human uses of animals reject the thesis that the obligations are to the animals; obligations to animals, they say, are only indirect because all direct obligations are to humans only.[23] Carl Cohen defends a blunt version of this position and one that has been influential in the literature critical of animal rights. He argues that justified claims occur only within a community of moral agents who have the capacity to make claims against one another and are authorized to do so.[24] His view is that there are neither animal rights nor correlative human obligations to animals, and yet Cohen acknowledges that some medical research causes pain to animals and that experimenters have an obligation to cause only necessary pain: "Animals certainly can suffer and surely ought not to be made to suffer needlessly."[25] Cohen correctly asserts that requirements of humane treatment take precedence over any mere preferences that a medical researcher might have to experiment on an animal. On the account of correlativity I have presented, once one acknowledges that a research investigator has an obligation to animal subjects to feed them and abstain from extremely painful procedures during the conduct of research, animal subjects have a right to be fed and to not have the pain inflicted. Logically, one cannot assert an obligation and deny a correlative right, as Cohen appears to do.

I conclude that the correlativity thesis holds firmly for all cases of genuine moral obligations and rights. There are no exceptions, though I will argue in section 8 that both rights and obligations are sometimes legitimately overridden by other moral considerations.

6. The Specification of Obligations and Rights

I examined basic rights and the correlativity of rights and obligations in sections 4 and 5, but I did not discuss *derivative* rights and *specified* rights. This section treats these subjects.

Basic and Derivative Rights

A *derivative* right can be traced back to a *more basic* right (or perhaps to more than one more basic right), but the derivation might involve several steps of argument.

For example, the right of a farm animal to be given food might be derived from the more abstract right to be protected from starvation (by those who render the animal dependent), which in turn might be derived from the more basic right to not be caused harm by others. To "derive" means to show by reasoning that a right is supported by a source-right. In the ideal, a basic right is a non-derivative right; that is, it has status as basic because it does not derive from any other right. However, neither in theory nor in practice is it obvious that the most basic rights that we care about (e.g., rights of confidentiality) have no dependence on any other rights. Accordingly, whereas the dependence of derivative rights on other norms seems clear, the concept of a basic right needs more careful analysis than I can here provide. I will only assume that some rights are *more basic* than others and that a distinction can be made between them that provides the line of descent to the less derivative right.

More important than the distinction between basic rights and derivative rights is the closely related subject of the *specification* of rights. Both more basic rights and derivative rights can be progressively specified (i.e., made increasingly specific for contexts), and this method is critical for progress in the study of animal rights.

The Specification of Rights

James Griffin rightly points out that we are sometimes satisfied that a basic right exists and that there are correlative obligations, yet we may be uncertain about what precisely the basic right gives us a right to.[26] Basic rights are abstract moral notions that do not fix how to formulate detailed policies or resolve practical moral problems. I agree with Ronald Dworkin's assessment that "abstract rights…provide arguments for concrete rights, but the claim of a concrete right is more definitive [in political contexts] than any claim of abstract right that supports it."[27] General norms, whether moral or legal, must be made specific in content and scope to become practically effective. Specification is the process of reducing the indeterminate character of abstract norms and giving them more specific action-guiding content. Specifying the norms with which one starts, whether they are norms in the morality we hold in common or norms from a source such as a professional code for veterinarians, is accomplished by narrowing the scope of the norms for a context and then adding contextually suitable content.[28]

An example of specification is found in a provision in the *Principles of Veterinary Medical Ethics of the American Veterinary Medical Association* (hereafter, AVMA) that centers on emergency circumstances: "II. F. In emergencies, veterinarians have an ethical responsibility to provide essential services for animals when necessary to save life or relieve suffering, subsequent to client agreement. Such emergency care may be limited to euthanasia to relieve suffering, or to stabilization of the patient for transport to another source of animal care."[29]

Principle II.F is a specification developed for emergency situations. It specifies more general obligations that veterinarians have of saving life and relieving suffering. The final sentence in this specification is concerned with limitations of the care (the "service") to be provided when euthanasia or stabilization prior to transport is the most appropriate action. Immediately after the specification quoted above, the AVMA provides a second specification of a veterinarian's moral responsibility in a circumstance in which he or she is either inexperienced or does not have the equipment to handle the encountered emergency situation; this second specification states that the veterinarian should advise the "client" truthfully and offer to expedite a referral. Both of these specifications will need further specification for circumstances in which other problems arise. The AVMA has a separate document on euthanasia that spells out further the conditions under which various forms of euthanasia are appropriate, the kinds of methods that should be used, acceptable (as well as unacceptable) pharmaceutical agents and nonpharmaceutical means, and special measures to be used for certain species.[30] This document can itself be interpreted as a series of specifications.

All rights and obligations are subject to specification, and many already-specified rules will need further specification to handle new problems of conflict.[31] Progressive specification can continue indefinitely, as new situations arise that were not previously known or anticipated. At each step, the listed obligations are correlative to rights. In some cases they are *animal* rights (e.g., the rights of the *patient* mentioned in II.F), while in other cases the rights are those of humans (e.g., the rights of the *client* mentioned in the second specification).

Disagreements in the Process of Specification

Specification is a method that we should use in practical ethics in the formulation of institutional rules, public policies, charters, conventions, codes, legal statutes, and the like. Major statements and documents of international rights have almost uniformly been constructed in this manner.

When rights must be made specific and practical, we often encounter controversy about the proper formulation of the specified right. Intractable disagreements over specifications occasionally persist because different persons offer competing specifications. The method of specification should not be interpreted as suggesting that we can avoid such moral disagreement. There is no magical line to be drawn in the progressive specification of rights so that we can say that specification X1 is rightly specified and truly a right but that specification X2 is not truly a right. Such matters can be worked out only by a process of moral argument and evaluation. The lack of an evident line also holds of *human* rights as they become progressively specified; it is often difficult to say when, in a line of specification, the more concrete claim stops being a human right.

However, the following generalization seems safe: the more specific we become in the process of specifying animal rights, the more likely it is that a rights claim will fall into disputed territory as to whether it is genuinely a right. This is a general truth about specification in the moral life and has nothing to do with animal rights in particular.

7. A Catalogue of the Rights
of Nonhuman Animals

The members of many species of animals qualify as rights holders, but it is murky as to which species qualify and which rights the members of any given species have. Whatever the species, nonhuman animals do not possess the same array of rights that adult humans possess. They do not have the right to purchase automobiles, the right to vote, and the like. Animals likely do not have the right to be told the truth or the right to not have their property stolen, though they may have analogous rights in some contexts, such as the right to not have a habitat destroyed or the right to not have life-sustaining water diverted.

The primary question is whether we can confidently say that certain rights belong on the list of animal rights. In this section, I construct a catalogue of rights, or perhaps what is more accurately characterized as a sketch of the types of rights that deserve a place in the catalogue. I will not attempt to specify the rights—an enormously ambitious enterprise. My inspiration for this catalogue is drawn from the long history in moral philosophy of cataloguing the virtues and vices.[32] Complete agreement does not exist among moral philosophers who write on the virtues as to precisely which virtues and vices comprise the catalogue, but there has been agreement on most items that belong on the list (e.g., benevolence, truthfulness, justice, charity, prudence, discernment, patience, courage, conscientiousness, trustworthiness, politeness, caring, compassion, and integrity). My objective is to compile a meaningful schema of the rights of animals on which we might find an overlapping consensus, and, in this respect, general agreement—though it is unreasonable to expect complete agreement.

I propose four main categories, or parts, in the catalogue:

(1) Rights to Nonmaleficent Treatment;
(2) Rights to Have Basic Needs Met;
(3) Rights of Nonconstraint;
(4) Rights from Human Agreements.

Categories 2–4 may be derivative from 1, that is, the final three categories may be derivative rights drawn from rights of nonmaleficence. Such matters are theoretically interesting in moral philosophy, but I will not address them. Even if there is a relationship of dependence, derivation, or specification, it should not affect the central problem of determining which rights deserve a place in the catalogue.

Vulnerability, Harm, and Justifying the Causation of Harm

I start with preliminary observations about the concepts of vulnerability and harm—two key notions for construction of the catalogue. Vulnerability is among the most commonly mentioned notions in biomedical research ethics, where it refers to situations in which harms may befall human research subjects. Animals are often in such situations. Physical and mental harms are paradigm instances of harm, and I will restrict discussion to these types of harm.

When I use the term "harm," I do not mean wrongful injuring or maleficence. A "harm" is a thwarting, defeating, or setting back of the interests of one party by the actions of another party or by natural events, so that the affected party is rendered worse off than before the condition occurred.[33] A harmful invasion by one party of another's interests is not always a wrong, maleficent, or unjustified intervention or interference. It is harmful to arrest and place someone in jail, for example, but it may be a right, justified, and nonmaleficent action.

It might be questioned whether animals have interests, whether their interests can be set back, and so whether an animal can be harmed. However, I will here assume that Joel Feinberg is correct when he bluntly asserts that animals "most assuredly have welfare interests."[34] I take it to be beyond reasonable challenge that animals have basic interests in food, minimal pain, not being diseased, and the like. These welfare interests cannot, I think, be limited to *survival* interests, but I shall not address this question with arguments.

Rights to Nonmaleficent Treatment

The principle of nonmaleficence is a general moral norm requiring that we abstain from causing harm to others. Correlative to obligations of nonmaleficence are rights to not be caused serious harms. The principle of nonmaleficence is the primary principle that many have used to reflect on obligations that humans have to animals, principally obligations to avoid causing pain, suffering, and emotional distress. Rules of nonmaleficence require only intentionally refraining from actions or inactions that cause harm. Rights correlative to obligations of nonmaleficence therefore can be described as negative rights only.

Rules of nonmaleficence take the form "Do not do X." To adapt part of an earlier table, examples of rules of nonmaleficence and correlative rights in discussions of *human* rights are:

Basic Obligations	Basic Rights
1. Do not kill;	1. The right to not be killed;
2. Do not cause pain or suffering to others;	2. The right to not be caused pain or suffering by others;
3. Do not incapacitate;	3. The right to not be incapacitated;
4. Do not cause offense;	4. The right to not be caused offense;
5. Do not deprive others of the pursuit of the basic goods of life.	5. The right to not be deprived of the pursuit of the basic goods of life.

Each of these rules and correlative rights might be stated so that it is defensible as an animal right, not merely a human right. Writers who defend a comprehensive and powerful set of animal rights, such as Tom Regan, probably would endorse this approach. However, rules and rights 1, 4, and 5 involve complicated considerations when brought to a focus on animals. I cannot here range into these problems, and I note that some of these rules and rights—for example, rule 4 and right 4—make no clear sense when applied to animals. I will concentrate, then, on rules and correlative rights that fall in the key area 2. I will discuss these rights largely as they pertain to domesticated animals, not wild animals.[35] The latter present their own complicated challenges.

Nothing about the general principle of nonmaleficence or its correlative rights declares that these rights are confined to a particular species of being such as the human being. The principle of nonmaleficence places no restriction on the range of individuals affected. I approach this problem as follows: when animals are in fundamental ways like humans—for example, in the capacity to experience pain and suffering—we have obligations to them from the principle of nonmaleficence, and these obligations are correlative to rights. Many human activities such as research involving human and animal subjects place humans under requirements to avoid causing significant pain or suffering and to justify activities that do cause pain and suffering. Pain is pain and suffering is suffering wherever they occur—in animal laboratories no less than human cancer treatment centers. That the individual is not human or is not a mammal is morally irrelevant. The principle of nonmaleficence requires avoidance of misery that might be caused to the individual affected. It is irrelevant whether the individual is a Bengal tiger in a zoo, a stumptail macaque in psychological research, or a sow on a factory farm.[36]

To move beyond the general principle of nonmaleficence and into more practical territory, consider the following chart of laws, policies, and resolutions governing the use of great apes in biomedical research:[37]

Countries Banning or Limiting Chimpanzee [and Great Ape] Research

Belgium	Law banning great ape research, 2008
Spain	Resolution granting great apes legal rights, 2008
Balearic Islands	Resolution granting great apes legal rights, 2007
Austria	Law banning great ape research, 2006
Japan	Agreement ending invasive chimpanzee research, 2006
Australia	Policy limiting great ape research, 2003
Sweden	Regulation banning great ape research, 2003
Netherlands	Law banning great ape research, 2002
New Zealand	Law banning great ape research, 2000
United Kingdom	Policy banning licenses for great ape research, 1997

New Zealand was the first nation to pass an outright ban, in 2000. In the subsequent decade a domino effect occurred in other countries. A common body of moral reasons underlies this legislation and policy: the conviction is that for the great apes certain forms of biomedical and behavioral research cause unjustified harm of such depth and significance to sensitive creatures that it must be banned altogether. The United States is the only remaining large-scale user of chimpanzees in research. It appears that legislators in the United States have not sympathetically examined available evidence about the effects on these animals, and it also appears that there is little public recognition of a problem.

To conclude this section on rights of nonmaleficence, I am proposing that entries such as the following belong in the catalogue of rights, starting with one very general and basic right (1) and then moving to four more specific rights (2–5):

(1) The right to not be caused pain or suffering;
(2) The right to not be traumatized by controllable human activities (such as slaughterhouses);
(3) The right to be protected against the harmful side effects of controllable human activities (such as the spraying of pesticides and herbicides, farming that produces dangerous waste, and pollution-causing construction projects);
(4) The right to not be placed at risk of serious pain, injury, or death in "sport" activities (such as cockfighting, rodeos, bullfighting, and dogfighting);
(5) The right to not be placed at risk of serious pain, injury, or death in the testing of human products (including cosmetics, pharmaceuticals, and chemicals).

These rights are not trumps. Competing moral considerations, such as protecting against disease and injury to human and animal populations, may legitimately override these rights—a problem addressed in section 8 below.

Rights to Have Basic Needs Met

We should not think exclusively in terms of harm to animals, but also in terms of their basic needs, on which I concentrate in this second category in the catalogue. I again confine discussion to domesticated animals whose welfare is dependent on humans. It is not difficult to envision extensions to wild animals experiencing habitat destruction or water depletion as a result of human practices, but the rights of wild animals present tangled problems about how to specify the bearers of correlative obligations and how to deal with incremental, aggregative, and cumulative harms—problems that I cannot consider here.

One might ask why and to what extent the welfare of an animal should be an object of concern. It is not apparent, in the abstract, that we have general obligations even to promote the well-being of other persons. However, it is a moral certainty that we have obligations to another individual in many roles of parenting and guardianship where there is jurisdiction over a person. Domesticated animals stand in a directly analogous situation. When we deliberately create both dependence and vulnerability in these animals, and take caretaking and supervisory charge over them, we acquire moral obligations of care.

These situations have sometimes been described using the language of "special relationships," but this wording is not ideal. The better category is the *special* or *circumstantial obligations* humans can have owing to, for example, (1) humans having placed animals in the circumstances they are in (for example, farm owners or heads of laboratories); (2) humans having assumed a job that requires caretaking for animals (for example, caretakers in zoos, rodeos, or circuses); or (3) humans' acceptance of responsibility for animals formerly raised as companion animals (for example, bringing animals into pounds or laboratories).

Neglect of the special obligations so created is a moral, and often legal, transgression. Football player Michael Vick learned this lesson when he neglected and otherwise abused the dogs he had confined on his property. He was sentenced to twenty-three months in federal prison and three years' probation for his role in sponsoring dogfighting and for killing dogs (those who fought poorly) by use of electrocution, hanging, and drowning. In addition to the wrongs of sponsoring dogfighting, there was neglect in meeting the basic needs of dogs in the facilities, both in the way they were housed, prepared for fighting, and bred (using so-called "rape stands").[38]

As we have learned more about animal environments and behaviors, we have increased our understanding of animals' basic needs and levels of poor welfare for many species. Obvious needs of domesticated animals are food and drink, avoidance of distressing discomfort, sufficient rest as required for health, and freedom of movement that approximates the needs of the species, including opportunity for exercise that maintains well-being. Many animals will have basic psychological and social needs.[39] In the case of companion, farm, circus, rodeo, and zoo animals, those responsible for their care also have obligations of the alleviation or elimination of pain, suffering, injury, or disease, including obligations of veterinary care.[40] Here, then, is a brief list of some basic-needs rights of domesticated animals:

(1) the right to appropriate food and hydration;
(2) the right to appropriately protective housing;
(3) the right to both adequate exercise and sufficient rest;
(4) the right to have basic psychological and health needs met;
(5) the right to a decent level of basic species-typical social relationships.

Rights of Nonconstraint

I turn now to the third category. Although clearly connected to both causing harm and failing to meet basic needs, this category merits independent treatment. It might be called "liberty rights," but the narrower designation "rights of nonconstraint" is more precise. This category might be derivative from the above two categories of rights, but it merits individual attention in a catalogue of rights even if there is such a connection.

In philosophical discussions of human freedom, there has long been a concern to distinguish freedom of action from constraints that necessitate conduct. For example, coercive institutionalization constrains the freedom of prisoners; the tighter the constraining condition, the greater the loss of freedom. To not be subjected to controlling constraints by others is a basic right. There have been no more important rights than liberty rights in the long history of struggles over human rights. Although there are significantly different human interests in liberty, by contrast to the interests of nonhuman animals, the latter still have their own commanding interests in being free of constraints. Placing an animal in a small cage in a zoo, a factory farm, a laboratory, a rodeo, a dog-racing facility, or a circus can have an effect on the animal much like that of placing a human in a tiny jail cell. Cages, pens, stalls, fences, stakes, whips, electric shocks, and the like can significantly cause loss of freedom to roam, to exercise, and to engage in social activities.

Rights enter the picture when the imposed constraints are unreasonably confining. These rights should not be understood as rights to new options and opportunities. These rights are limited to undue constraint. It is difficult to determine precisely which enclosures or constraints are undue, unreasonable, or unjustifiable, but it is not difficult to locate examples of what is justifiable and what is not. For example, large wildlife sanctuaries that exceed the needs for roaming territory for the "enclosed animals" commonly have fences that keep the animals out of certain lands, but these constraints may be entirely reasonable. The National Elk Refuge in the United States is a good case in point. It was established in 1912 to provide a winter habitat and preserve for the Jackson elk herd, northeast of the town of Jackson, Wyoming. It originally provided a wintering ground of only 1,760 acres for a large elk herd, but it has grown today to about 25,000 acres. The refuge is sealed off in various areas by fences, but it would be absurd to say that its 7,500 elk are unduly constrained. Moreover, the welfare index of these animals is, overall, increased rather than decreased by the refuge.[41]

By contrast, the conditions under which millions of farm animals are contained are thoroughly unjustified. In the United States and many other countries, gestation crates used on hog farms often force sows to hunch down just to fit in them. Industry-standard crates measure 2 to 2.5 feet wide by 7 feet long by 3 feet high, yet breeding sows weigh 500 to 600 pounds. The sows cannot walk, turn around, or socialize. They eventually exhibit many atypical (stereotypic) behaviors including compulsive, repetitive actions such as bar biting, purposeless chewing, and digging and pawing at the floor and walls of the crates for up to 70 percent of the time. A second example is chicken farming, which commonly takes place in windowless sheds. In Europe (though not in all European countries) the space allowance for a bird is .5 square feet when stocked, but toward the end of the production time, when the birds have significantly increased in size, the recommended density is around seven pounds of bird mass for every square foot, which is approximately equivalent to one bird on a single sheet of legal-size paper.[42]

To eliminate such conditions, a number of countries have legally recognized the right to an adequate standard of space for movement, grazing, and exercise for farm animals. For example, in 1988 Sweden initiated a law that addresses unjust confinement and recognizes rights of the members of species such as cattle, pigs, and chickens. The normal crowded confinement found on factory farms is not permitted in Sweden, where cattle hold grazing rights, pigs hold rights to specific forms of bedding and space, and chickens have rights to free range.[43]

To conclude this subsection, here are some rights of nonconstraint that animals with the relevant interests have:

(1) the right to healthy space allowances when in confined areas;
(2) the right to graze and free-range when a farm animal;
(3) the right to not be prevented from adequate exercise;
(4) the right to not be put in forced isolation;
(5) the right to a decent level of opportunity to engage in species-typical patterns of social behavior.

These rights are subject to qualification and specification, perhaps most notably the fifth right listed. A pet owner or a farm owner would need to know a great deal more about this right even to begin to know what the right protects.

Rights from Human Agreements

Joel Feinberg presents an unambiguous example of a right for animals, though, as he says, it is a comparatively minor right:

> Claims can be made on behalf of specific animals to goods that belong to them as a result of agreements made between human beings. Because of the intellectual incompetence of animals, it is impossible for a human being to make a promise to an animal or for the animal to reply with promised quid pro quo required of a legal contract. But human beings can and do make promises to one another of which animals are the intended beneficiaries. I see no difference in principle between these arrangements and contractual agreements that confer rights on third-party human beneficiaries—for example, an agreement between a policy-holder and an insurance company to pay a given sum to the policyholder's children in the event of his death. Upon the death of the insured, the children have a valid claim that can be pressed in their behalf against the insurance company in a court of law....Similarly, human beings commonly make wills leaving money to trustees for the care of animals. Is it not natural to speak of the animal's right to his inheritance in such cases?[44]

Feinberg argues that if there is a violation of such an agreement, it is a violation of the animal's moral right. An example of such a violation, I suggest, is the following: if an employee of a bank embezzles money from an account established for an animal named as a beneficiary, the animal's right is violated. To press a claim on behalf of the animal is to invoke the animal's right, not merely a right of a guardian.

Although Feinberg's analysis descends from a legal context, such moral rights also spring from human agreements. Suppose there is an agreement among my several family members, each of whom owns a dog, about what to do if any one of us dies and leaves a companionless dog. We agree that it would be cruel to our dogs to allow any one of them to go to a pound, be used in medical experiments, or be euthanized. The survivors in the family, whoever they turn out to be, therefore agree to take in and take care of any such family dog for the remainder of the dog's life. The obligations of care, all parties agree, are to the dog, not to remaining family members or to the deceased. These contractually assumed obligations of care are stated in terms of type of food, exercise, veterinary care, and the like. Upon the death of a human companion in the family, a dog left behind has a moral right that can be pressed in the dog's behalf by a family member against another family member who agreed to house the dog and has failed to do so. The matter would be no different if someone left money to upgrade the substandard care of elephants in an elephant sanctuary. If a human agreement creates an obligation to protect members of the species, there is a correlative right of the elephants to be protected.

Contracts and agreements between humans therefore can establish animal rights. Conditional on such agreements, animal rights (whether or not stated as "rights" in the agreements) include:

(1) rights to receive benefits specified in wills;
(2) rights correlative to obligations set forth in government regulations;
(3) rights to have existing institutional rules protective of animals recognized and enforced;
(4) rights to veterinary care as authorized by human agreements;
(5) rights to have authorized third parties bring legal suits for cruel or abusive treatment, as authorized by human agreements.

8. Overriding Rights and Justifying Infringements of Animal Rights

We saw in section 3 that rights are not absolute standards or trumps. Justified claims not infrequently come into conflict with other justified claims. Not even basic rights are categorical trumps. An instructive example of contingent conflict occurs when proponents of a religion engage in the sacrifice of live animals. Human rights of freedom of religious exercise are in conflict with animal rights to not be brutalized and killed unnecessarily. In a precedent legal case, the City Council in Hialeah, Florida argued that the adherents of the Santeria religion do not have sufficiently strong rights of religious freedom to engage in the practice of the live sacrifice of animals and that anticruelty prosecution is the least restrictive means for the city to advance its compelling interest in animal protection. However, this balancing of interests was re-balanced and the decision overturned by the U.S. Supreme Court.[45]

In such circumstances of contingent conflict, *violations* of rights should be distinguished from *infringements* of rights.[46] By "violation" I refer to an unjustified and wrong action against an interest that is protected by a right, whereas by "infringement" I refer to a justified action that legitimately overrides a right. A reason many consider sport hunting unjustified is that an animal's life is taken merely for the sake of a person's personal enjoyment, which they take to be an insufficient justification. Lacking an adequate moral justification for the killing, this action is a violation of animals' rights. However, if one has a good reason for "hunting," it may justifiably override an animal's right. A good case can be made, for example, for culling an overpopulated herd to benefit the environment and to promote the subsistence of the herd (sometimes called a management hunting program). Here rights are infringed, not violated.

Prima Facie or Pro Tanto Obligations and Rights

For humans and animals alike, an action that harms, causes basic needs to go unmet, or produces extreme forms of confinement is often said to be wrong *prima facie* (i.e., wrongness is upheld unless the act is shown justifiable) or wrong *pro tanto* (i.e., wrong to a certain extent or wrong unless there is a compelling justification)—which is to say that the action is wrong in the absence of other moral considerations that supply a compelling justification. Very compelling justifications sometimes arise. For example, in circumstances of a severe swine flu pandemic, the forced confinement of humans through isolation and quarantine orders might be justified. Here a valid infringement of liberty rights occurs. Likewise, in a circumstance of disease in a community of animals—in a preserve or sanctuary, say—forced isolation and confinement, even death, might be a justified infringement of what I have called rights of nonconstraint.

W. D. Ross defended a distinction that I accept in principle between *prima facie* and *actual* obligations. A *prima facie* obligation must be fulfilled unless it conflicts with an equal or stronger obligation. Likewise, a *prima facie* right, I maintain (here extending Ross), must prevail unless it conflicts with an equal or stronger right (or conflicts with some other morally compelling alternative). Obligations and rights always constrain us unless a competing moral obligation or right can be shown to be overriding in a particular circumstance. Ross says that "the greatest balance" of right over wrong must be found.[47] However, the matter is more complicated than Ross suggests because we will in many circumstances want to see a structured system (1) in which some rights in a certain class of rights have a fixed priority over others in another class and (2) in which it is nearly impossible for some rights or morally compelling circumstances to "outweigh" basic rights. In Ronald Dworkin's memorable, albeit vague language, rights have "a certain threshold weight against collective goals."[48]

Justification

A process of moral deliberation and justification is required to establish an agent's actual obligation in the face of a contingent conflict. There are countless such problems about the justification of our uses of animals. The most commonly accepted justification in most circumstances of conflict is the substantial benefit to be gained for humans—for example, benefits from biomedical research, benefits of protecting against loss of farm animals, food benefits, and entertainment benefits. Productive medical research has been a particularly attractive justification of harm caused to animals because of many reported success stories,[49] though there are some acute problems about how to justify claims regarding what is necessary to medical progress.[50]

Human subjects could be substituted for animal subjects in biomedical research, and research results would thereby be improved. But the painful, invasive, and even lethal character of animal research has universally been considered to pose insuperable moral problems for use of human subjects.[51] The problem is that the absence of a justification for using human subjects does not justify using animal subjects. If the goals of research cannot be carried out using humans because of inflicted suffering, the justification for causing similar suffering to animals is made more difficult, not facilitated. If "harm" is understood, as I previously defined it (a thwarting, defeating, or setting back of a nontrivial interest), it is indisputable that many harms are suffered by animals in biomedical research, but it is understandably controversial whether these harms are justified.[52]

It is a prudent principle that the higher the degree of harm and the more questionable the benefits, the more difficult it is to justify causing the harm to animals. A stubborn problem of justification has been how to draw a line at the point where medical research exceeds a threshold or upper limit of pain, suffering, anxiety, fear, and distress that can be justified. In research with human subjects, such thresholds must be set, and we have rights against researchers for exceeding the threshold. There is no reason why animals should be handled differently, but they are.

9. An Objection from the Right to Life

A possible objection to my arguments throughout this chapter is that they contain a morally dangerous incompleteness. I will frame this objection in terms of both the right to life and the question "do humans have rights that others do not?"

If species per se is not a morally relevant consideration and if pain-managed killing of various sorts is morally permissible, as my claims either imply or assume, then humans (as well as all great apes) seem to have no special protection from being used or killed in the same ways that pigs, rats, and deer are used. Unless there is a constraint in the theory in which a human right to life functions as a trump against social utility, it would seem that humans could be farmed for their organs, be raised for medical experimentation, be objects of religious sacrifice, and the like. It is currently the practice that humans, and various other animals, are not killed or exploited in these ways, but on my general account there is no moral principle or moral right that would prevent a change in these practices if social conditions change. Because I have offered no special right to life, my account seems incomplete and morally dangerous.

Two connected replies are in order. First, nothing I have said prevents the members of some species from having rights that members of some other species lack. I have insisted that there will be different sets of rights for different groups contingent on the presence or absence of certain properties, such as cognitive properties and the ability to act as moral agents. Also, a strong right to life could be held by the members of more than one species, but there still could be various levels of protection built into the rights. The highest level might be occupied by only one species because of its unique properties. Finally, no absolute right to life exists even for humans, as we know from killing on the battlefields of war, killing in self-defense, and second-party authorizations of the withholding or withdrawal of life-sustaining medical technology. These problems derive from the problems of justification discussed in section 8.

Second, the objection under consideration in this section presents an opportunity to consider how general ethical theories might supply us with good reasons to aggressively protect humans by an uncompromising right to life that is more stringent than a right to life for the members of other species. For example, one might use utilitarianism to defend a stronger right based on the greater richness of experience found in humans by contrast to other species. Or one might employ a contractarian theory to build a stringent account of the right to life erected on consent and contractual agreements, which only humans can make. Or one might embrace a Kantian theory erected on the importance of moral law and moral autonomy, of which only humans are capable. Or one might adopt person theory to argue that persons and only persons (whatever their species) have certain classes of rights. I will not here attempt to select among these theories. The point is only that such moral accounts may appropriately be employed to justify special rights or especially stringent rights that would take the sting out of the objection raised in this section.

10. Conclusion

The account of "animal rights" defended in this chapter has led to a notably underspecified catalogue of animal rights. I have provided only the philosophical rudiments of this catalogue. I hope that philosophers will give increased attention to its development and its defense in upcoming years.

NOTES

1. Locke, *Two Treatises of Government*, ed. Peter Laslett (Cambridge: Cambridge University Press, 1960), preface and bk. 1.6.67. Locke's close association of moral, political, and legal rights is evident at bk. 2.7.87.

2. Criteria of animals' best interests should be determined by the highest net benefit to the animal among the available options that would advantage the animal. Death, among other outcomes, can be in an animal's best interest.

3. Tom Regan presents a pointed distinction between rightists and welfarists in "Animal Rights," *Encyclopedia of Animal Rights and Animal Welfare*, 2nd ed., ed. Marc Bekoff (Santa Barbara, Calif.: Greenwood Press, 2010), vol. 1, p. 36. For the general background of the distinction, see Lawrence Finsen and Susan Finsen, *The Animal Rights Movement in America: From Compassion to Respect* (New York: Twayne, 1994).

4. Tom Regan, *The Case for Animal Rights* (Berkeley and Los Angeles: University of California Press, 1983; rev. ed., 2004), pp. 178, 182–84.

5. Tom Regan, *Case for Animal Rights*, pp. 240, 243, 258–61; Regan, *Empty Cages: Facing the Challenge of Animal Rights* (Lanham, Md.: Rowman & Littlefield, 2004), chaps. 3–4.

6. See Regan's strongly worded conclusion in the essay "The Fate of Animals is in our Hands; God Grant We are Equal to the Task," available at http://www.animalsvoice.com/TomRegan//regan_rites.html (accessed July 24, 2010); and see the even stronger view in Gary L. Francione, *Animals as Persons: Essays on the Abolition of Animal Exploitation* (New York: Columbia University Press, 2008), chap. 1.

7. For an approach to animal rights situated between the polar ends of this continuum, and a position I find congenial, see Elizabeth Anderson, "Animal Rights and the Values of Nonhuman Life," in Cass R. Sunstein and Martha C. Nussbaum, eds., *Animal Rights: Current Debates and New Directions* (New York: Oxford University Press, 2004), pp. 277–98.

8. H. J. McCloskey, "Rights," *Philosophical Quarterly* 15 (1965): 118. For his application to animals, see McCloskey, "Moral Rights and Animals," *Inquiry* 22 (1979): 23–54, as reprinted in Clare Palmer, ed., *Animal Rights* (Aldershot, England: Ashgate, 2008), pp. 79–110.

9. My representations in this and the next paragraph are deeply indebted to the theory of rights in Joel Feinberg's *Rights, Justice, and the Bounds of Liberty* (Princeton, N.J.: Princeton University Press, 1980), esp. pp. 139–41, 149–55, 159–60, 187; and *Social Philosophy* (Englewood Cliffs, N.J.: Prentice-Hall, 1973), chaps. 4–6. See also Alan Gewirth, *The Community of Rights* (Chicago: University of Chicago Press, 1996), pp. 8–9.

10. Ronald Dworkin, *Taking Rights Seriously* (Cambridge, Mass.: Harvard University Press, 1977), pp. xi, xv, 92 (and, as reissued with "Appendix: A Reply to Critics," in 2002, pp. 364–66); and *Law's Empire* (Cambridge, Mass.: Harvard University Press, 1986), p. 160.

11. A clever and atypical attempt to find an absolute right is Alan Gewirth, "Are There Any Absolute Rights?" *The Philosophical Quarterly* 31 (1981): 1–16; reprinted as chapter 9 in Gewirth's *Human Rights* (Chicago: University of Chicago Press, 1982).

12. Ronald Dworkin, "Rights as Trumps," in Jeremy Waldron, ed., *Theories of Rights* (Oxford: Oxford University Press, 1984), p. 153 (italics added).

13. Cf. Frances Kamm, *Intricate Ethics: Rights, Responsibilities, and Permissible Harm* (New York: Oxford University Press, 2007), pp. 227–31, 234; Claire Andre and Manuel Velasquez, "Who Counts?" Markula Center for Applied Ethics, *Issues in Ethics* 4 (1991), available at http://www.scu.edu/ethics/publications/iie/v4n1/counts.html (accessed August 8, 2010).

14. James Griffin, *On Human Rights* (Oxford: Oxford University Press, 2008), pp. 54–55, 149ff. (using the language of a "more pluralist account"). See also the main line of argument in Anderson, "Animal Rights and the Values of Nonhuman Life."

15. Despite the convergence on norms, there is more than one *theory* of the morality we share in common: for a variety of theories, see Alan Donagan, *The Theory of Morality* (Chicago: University of Chicago Press, 1977); Bernard Gert, *Common Morality: Deciding*

What to Do (New York: Oxford University Press, 2007); and Tom L. Beauchamp and James F. Childress, *Principles of Biomedical Ethics*, 6th ed. (New York: Oxford University Press, 2009), chaps. 1, 10.

16. For arguments showing that basic rights should not be understood as rights formulated in a political state (as in Rawls' political conception), see Allen Buchanan, "Taking the Human out of Human Rights," in his *Human Rights, Legitimacy, and the Use of Force* (New York: Oxford University Press, 2010), pp. 31–49.

17. My argument is indebted to James Rachels, "Why Animals Have a Right to Liberty," in *Animal Rights and Human Obligations*, ed. Tom Regan and Peter Singer, 2nd ed. (Englewood Cliffs, N.J.: Prentice Hall, 1989), pp. 122–31 (in Palmer, *Animal Rights*, pp. 57–66), at 122–23 (Palmer, *Animal Rights*, pp. 57–58).

18. World Health Organization, "Zoonoses and Veterinary Public Health," available at http://www.who.int/zoonoses/vph/en/ (accessed August 1, 2010).

19. A substantially similar point is made by Feinberg, *Rights, Justice, and the Bounds of Liberty*, pp. 161–62.

20. See David Braybrooke, "The Firm but Untidy Correlativity of Rights and Obligations," *Canadian Journal of Philosophy* 1 (1972): 351–63; Feinberg, *Rights, Justice, and the Bounds of Liberty*, pp. 135–39, 143–44; Feinberg, *Harm to Others*, vol. 1 of *The Moral Limits of the Criminal Law* (New York: Oxford University Press, 1984), pp. 148–49; Griffin, *On Human Rights*, pp. 51, 96, 107–9; Joseph Raz, *The Morality of Freedom* (New York: Oxford University Press, 1986), pp. 170–72. Probing discussions of correlativity are found in Gewirth's *The Community of Rights*. See also Feinberg's wonderful explanation of the confusions that enter moral discourse owing to the ambiguity of the words "duty," "obligation," and "requirement," in his *Doing and Deserving: Essays in the Theory of Responsibility* (Princeton, N.J.: Princeton University Press, 1970), pp. 3–8.

21. See the standard objections by David Lyons, "The Correlativity of Rights and Duties," *Nous* 4 (1970): 45–55; Theodore M. Benditt, *Rights* (Totowa, N.J.: Rowman & Littlefield, 1982), pp. 6–7, 23–25, 77; Alan R. White, *Rights* (Oxford: Clarendon Press, 1984), pp. 60–66; Richard Brandt, *Ethical Theory* (Englewood Cliffs, N.J.: Prentice Hall, 1959), pp. 439–40.

22. Cf. the similar, and illuminating, conclusions in Joel Feinberg, *Rights, Justice, and the Bounds of Liberty*, pp. 138–39, 143–44, 148–49.

23. The paradigm is Immanuel Kant, *Lectures on Ethics*, trans. Peter Heath and ed. Peter Heath and J. B. Schneewind (New York: Cambridge University Press, 1997), 27: 413 (p. 177), 27: 458–60 (pp. 212–13), 27: 710 (pp. 434–35); Kant, *The Metaphysics of Morals* [Part II. *The Metaphysical Principles of Virtue*], trans. Mary J. Gregor, as in Kant, *Practical Philosophy*, ed. Paul Guyer and Allen W. Wood (Cambridge: Cambridge University Press, 1996), 6: 443 (p. 564). For an influential social contract theory that renders duties to animals indirect, see Peter Carruthers' chapter in this *Handbook*.

24. Carl Cohen, "The Case for the Use of Animals in Research," *New England Journal of Medicine* 315 (1986): 865–70, esp. pp. 865–66; and, similarly, McCloskey, "Moral Rights and Animals." See also Carl Cohen and Tom Regan, *The Animal Rights Debate* (New York: Rowman & Littlefield, 2001) for Cohen's later views and Regan's criticisms.

25. Cohen, "Case for the Use of Animals in Research," p. 867.

26. James Griffin, *On Human Rights* (Oxford: Oxford University Press, 2008), pp. 97, 110.

27. Dworkin, *Taking Rights Seriously*, pp. 93–94.

28. Henry S. Richardson, "Specifying Norms as a Way to Resolve Concrete Ethical Problems," *Philosophy and Public Affairs* 19 (1990): 279–310; and Richardson, "Specifying, Balancing, and Interpreting Bioethical Principles," as in *Belmont Revisited: Ethical*

Principles for Research with Human Subjects, ed. James F. Childress, Eric M. Meslin, and Harold T. Shapiro (Washington, D.C.: Georgetown University Press, 2005), pp. 205–27. For analysis of specifying *rights* in particular, see Russ Shafer-Landau, "Specifying Absolute Rights," *Arizona Law Review* 37 (1995): 209–25.

29. AVMA, *Principles*, as approved by the Executive Board, Nov. 2006; revised April 2008, available at http://www.avma.org/issues/policy/ethics.asp (accessed July 26, 2010).

30. AVMA, *AVMA Guidelines on Euthanasia* (formerly Report of the AVMA Panel on Euthanasia), June 2007, available at http://www.avma.org/issues/animal_welfare/euthanasia.pdf (accessed July 29, 2010).

31. Richardson, "Specifying Norms as a Way to Resolve Concrete Ethical Problems," p. 294.

32. Possibly to be dated from Aristotle's *Nicomachean Ethics*. See David Hume's comments on the catalogue in his *An Enquiry concerning the Principles of Morals*, ed. Tom L. Beauchamp (Oxford: Clarendon Press, 1998), beginning at 1.10 (sect. 1, par. 10); see also 6.21; 9.3; 9.12. In a letter to Francis Hutcheson of September 17, 1739 (*The Letters of David Hume*, ed. J. Y. T. Greig, 2 vols. [Oxford: Clarendon Press, 1932], 1:34), Hume said, "Upon the whole, I desire to take my Catalogue of Virtues from *Cicero's Offices*" (*De officiis*). My use of "catalogue of rights" is far from original. See L. W. Sumner, *The Moral Foundation of Rights* (New York: Oxford University Press, 1987), chap. 1 for a related use.

33. Feinberg, *Harm to Others*, pp. 32–36, 51–55, 77–78.

34. Feinberg, *Harm to Others*, p. 58.

35. On this distinction and its importance for animal rights, see the chapter by Clare Palmer in this *Handbook*.

36. See Judith Jarvis Thomson, *The Realm of Rights* (Cambridge, Mass.: Harvard University Press, 1990), pp. 292–93; and the chapter by Sahar Akhtar in this *Handbook*.

37. This chart is from the website of Release and Restitution for Chimpanzees in U. S. Laboratories, "R & R Project," available at http://www.releasechimps.org/mission/end-chimpanzee-research/country-bans (accessed July 27, 2010).

38. ESPN Sports, "Apologetic Vick gets 23-month sentence on dogfighting charges," ESPN.com news services, December 11, 2007, available at http://sports.espn.go.com/nfl/news/story?id=3148549 (accessed July 28, 2010).

39. A brief list of dogs' basic needs is found in Keith Burgess-Jackson, "Doing Right by Our Animal Companions," *Journal of Ethics* 2 (1998): 159–85, esp. 179–80.

40. See the chapter by Clare Palmer in this *Handbook* and Leslie Pickering Francis and Richard Norman, "Some Animals Are More Equal than Others," *Philosophy* 53 (1978): 507–27.

41. U.S. Fish and Wildlife Service, "National Elk Refuge," information available at http://www.fws.gov/refuges/profiles/index.cfm?id=61550 (accessed July 28, 2010). There are some moral issues about practices of culling in this refuge.

42. For these and further details, see Tom L. Beauchamp, F. Barbara Orlans, Rebecca Dresser, David B. Morton, and John P. Gluck, *The Human Use of Animals*, 2nd ed. (New York: Oxford University Press, 2008). I owe the facts mentioned in this paragraph to the research of David Morton.

43. Steven Lohr, *New York Times*, October 25, 1988, available at http://www.nytimes.com/1988/10/25/world/swedish-farm-animals-get-a-bill-of-rights.html (accessed July 26, 2010).

44. Joel Feinberg, "Human Duties and Animal Rights," in Richard Morris and Michael Fox, eds., *On the Fifth Day: Animal Rights and Human Ethics* (Washington, D.C.: Acropolis Books, 1978), pp. 45–69, as reprinted in Clare Palmer, ed., *Animal Rights* (Aldershot, England: Ashgate, 2008), pp. 399–423, at pp. 410–11 (in Palmer, *Animal Rights*, pp. 56–57). In a 2009 case, billionaire Leona Helmsley left $12 million (later reduced by a judge to

$2 million) to her treasured Maltese and millions to her brother and grandchildren for their assigned roles in caretaking. Associated Press, "Helmsley's Dog Gets $12 Million in Will," as published in *The Washington Post*, Wednesday, August 29, 2007, available at http:// www.washingtonpost.com/wp-dyn/content/article/2007/08/29/AR2007082900491.html (accessed August 8, 2010). It is now common for companion animals to be left resources in trusts in the United States, a recent development unavailable to Feinberg when writing his pioneering article.

45. *Church of the Lukumi Babalu Aye, Inc. v. City of Hialeah*, 723 F. Supp. 1467 (S.D. Fla. 1989), *rev'd*, 113 S. Ct. 2217 (1993); David M. O'Brien, *Animal Sacrifice and Religious Freedom: Church of the Lukumi Babalu Aye v. City of Hialeah* (Lawrence: University Press of Kansas, 2004). I owe these references and the facts of the case to Rebecca Dresser; see Beauchamp, Orlans, Dresser, Morton, and Gluck, "Animal Sacrifice as Religious Ritual: The Santeria Case," *Human Use of Animals*, chap. 9.

46. Thomson, *Realm of Rights*, pp. 122–24 and also 106–17, 149–53, 164–75; and Joel Feinberg, *Rights, Justice, and the Bounds of Liberty*, pp. 229–32.

47. W. D. Ross, *The Right and the Good* (Oxford: Clarendon Press, 1930), esp. pp. 19–36, 88. On important cautions about both the meaning and use of "*prima facie* rights," see Feinberg, *Rights, Justice, and the Bounds of Liberty*, pp. 226–29, 232; Thomson, *Realm of Rights*, pp. 118–29; and Carl P. Wellman, *Real Rights* (New York: Oxford University Press, 1995), pp. 250–51.

48. Dworkin, *Taking Rights Seriously*, p. 92.

49. However, see Hugh LaFollette's chapter in this *Handbook* for a skeptical challenge.

50. The problem of how to determine which harms count as necessary and as justificatory is considered in this *Handbook* by David DeGrazia. For worries about how such questions can be stated and addressed fairly, see Jeff McMahan, "Animals," in *A Companion to Applied Ethics*, ed. R. G. Frey and Christopher Wellman (Malden, Mass.: Blackwell Publishing, 2003), esp. pp. 532ff.

51. For aspects of this problem, see J. A. Smith and K. M. Boyd, eds., *Lives in the Balance: The Ethics of Using Animals in Biomedical Research*, Report of a Working Party of the Institute of Medical Ethics (Oxford: Oxford University Press, 1991).

52. See a forum and debate on this issue: Andrew Rowan, Neal D. Barnard, Stephen R. Kaufman, Jack H. Botting, Adrian R. Morrison, and Madhusree Mukerjee, "The Benefits and Ethics of Animal Research," *Scientific American* (February 1997): 79–93 (each article with a separate title).

SUGGESTED READING

BEAUCHAMP, TOM L., F. BARBARA ORLANS, REBECCA DRESSER, DAVID B. MORTON, and JOHN P. GLUCK. *The Human Use of Animals*. 2nd ed. New York: Oxford University Press, 2008.

BEITZ, CHARLES R. *The Idea of Human Rights* New York: Oxford University Press, 2009.

COHEN, CARL, and TOM REGAN. *The Animal Rights Debate*. Lanham, Md.: Rowman & Littlefield, 2001.

DEGRAZIA, DAVID. *Animal Rights: A Very Short Introduction*. New York: Oxford University Press, 2002.

———. *Taking Animals Seriously: Mental Life and Moral Status*. Cambridge: Cambridge University Press, 1996.

FEINBERG, JOEL. *Rights, Justice, and the Bounds of Liberty*. Princeton, N.J.: Princeton University Press, 1980.

FRANCIONE, GARY L. *Animals as Persons: Essays on the Abolition of Animal Exploitation*. New York: Columbia University Press, 2008.

FRANKLIN, JULIAN H. *Animal Rights and Moral Philosophy*. New York: Columbia University Press, 2005.

FREY, R. G., ed. *Rights, Killing, and Suffering*. Oxford: Blackwell, 1983.

GEWIRTH, ALAN. *The Community of Rights*. Chicago: University of Chicago Press, 1996.

GRIFFIN, JAMES. *On Human Rights*. Oxford: Oxford University Press, 2008.

McMAHAN, JEFF. "Animals." In *A Companion to Applied Ethics*, edited by R. G. Frey and Christopher Wellman, pp. 525–36. Malden, Mass.: Blackwell Publishing, 2003.

PALMER, CLARE, ed. *Animal Rights*. Aldershot, England: Ashgate Publishing, 2008.

REGAN, TOM. *The Case for Animal Rights*. Berkeley and Los Angeles: University of California Press, 1983; rev. ed., 2004.

ROWLANDS, MARK. *Animal Rights: Moral Theory and Practice*. New York: Palgrave Macmillan, 2009.

SHUE, HENRY. *Basic Rights: Subsistence, Affluence, and U.S. Foreign Policy*. Princeton, N.J.: Princeton University Press, 1980.

SUMNER, L. W. *The Moral Foundation of Rights*. New York: Oxford University Press, 1987.

SUNSTEIN, CASS R., and MARTHA C. NUSSBAUM, eds. *Animal Rights: Current Debates and New Directions*. New York: Oxford University Press, 2004.

THOMSON, JUDITH J. *The Realm of Rights*. Cambridge, Mass.: Harvard University Press, 1990.

WALDRON, JEREMY, ed. *Theories of Rights*. Oxford: Oxford University Press, 1984.

CHAPTER 8

THE CAPABILITIES APPROACH AND ANIMAL ENTITLEMENTS

MARTHA NUSSBAUM

1. ANIMAL THINKING AND ETHICS

HAPPY is an adult female Asian elephant who lives in the Bronx Zoo, one of the most humane of American zoos in providing large animals with a rich diverse natural habitat and a wide range of social interactions. Like adult females in the wild (though, as I shall later say, there are very few Asian elephants in the wild today, and whether there is such a thing as "the wild" is subject to doubt), she lives in a group with other adult females who enjoy one another's company and share rich social relationships. During several days this past fall, researchers Joshua Plotnik, Frans de Waal, and Diana Reiss set up a large mirror in the enclosed area shared by these three females as the nighttime "home" from which they roam outward during the day. All three females immediately took quite an interest in the mirror. They walked back and forth in front of it, and then walked up to it. All showed a marked interest in facing the mirror with open mouths, apparently studying their own oral cavities and poking their teeth. (Elephants lose at least four sets of teeth during a lifetime, and they are thus often in a state of dental discomfort.) In one case, Maxine put her trunk into her mouth, in front of the mirror, using it to touch parts of her teeth and mouth; she later used her trunk to pull her ear forward so that the inside of her ear cavity could be seen in the mirror. Maxine, Patty, and Happy also explored the back of the mirror with their trunks to see whether there was something over on the other side of it, quickly ascertaining that there wasn't.

On the second day, a visible large X mark was applied to the right side of each elephant's head, and an invisible sham mark was applied to the other side of the head, to forestall the possibility that the tactile sensation of applying the mark would account for the experimental result. The mark was visible only in the mirror. Maxine and Patty did nothing unusual; by that time they had become somewhat bored with the mirror. But Happy, still engaged, went up to the mirror and studied the reflection of her own head. Repeatedly she took her own trunk and scrubbed the mark with it, as if she were perfectly aware that what she saw in the mirror was a part of her own head, and she wanted to wipe away the unusual mark.[1] On the basis of this experiment, de Waal and his fellow researchers conclude that the Asian elephant is capable of forming a conception of the self; until now, this level of complexity has been found only in apes and humans, though there is one ambiguous experiment with dolphins.

We have long been learning that the elephant society is highly complex. Elephants exhibit complex forms of social organization, in which child care is shared among a group of cooperative females. Elephants also have rituals of mourning when a child or an adult member of the group dies, and appear to feel grief. Even when they come upon the corpse of a fellow species member that has been dead for a long time, they explore the body or bones for signs of the individual who has inhabited them.[2] All of this we have increasingly understood through the work of researchers working both in the field and in the better research zoos.

We might think, well, so now we know that there are a few species that have complex forms of social behavior, based on complex forms of cognition. These cases, however, are exceptional, and should not affect our assessment of the standard case. We have also, however, been learning recently about unexpected complexity in animal thinking in quite another region of the animal kingdom, a part of it that we're accustomed to think of as "lower." In June 2006, *Science* published an article entitled "Social Modulation of Pain as Evidence for Empathy in Mice" by a research team at McGill University in Montreal led by Jeffrey Mogil.[3] This experiment involves the deliberate infliction of mild pain, and is thus ethically problematic. Nonetheless, I ask you to forgive me for describing what we learn from it. The scientists gave a painful injection to some mice, which induced squealing and writhing. (It was a weak solution of acetic acid, so it had no long-term harmful effects.) Also in the cage at the time were other mice who were not injected. The experiment had many variants and complexities, but to cut to the chase, if the non-pained mice were paired with mice with whom they had previously lived, they showed signs of being upset. If the non-pained mice had not previously lived with the pained mice, they did not show the same signs of emotional distress. On this basis, the experimenters conclude that the lives of mice involve social complexity: familiarity with particular other mice prepares the way for a type of emotional contagion that is at least the precursor to empathy.

Human beings have gone through many phases in understanding the complexity of animal lives and animal thinking. The ancient Greeks and Romans believed that there were large areas of commonality between humans and animals. Most of the ancient philosophical schools (with the exception of Stoicism) attributed to animals complex forms of cognition and a wide range of emotions; some of them used their observation of these complexities to argue against cruel practices and also against meat-eating.[4]

Nor was this simply a specialized movement of elite intellectuals. There was widespread public awareness of the complexity of animal lives and of the implications this complexity had for human treatment of animals. When Pompey the Great introduced elephants into the gladiatorial games, there was a public outcry, described by Cicero, who notes that the people who saw elephants in the ring had no doubt that there were commonalities between them and the human species.[5] Large sections of the ancient Greco-Roman world were vegetarian. Meanwhile, in India, the Buddhist emperor Ashoka, in the third century B.C., made a long list of animals that should not be killed and wrote that he himself was attempting, with increasing success, to live a vegetarian life.[6] This tradition continues: India is one of the world leaders in legal protection for animals, and close to 50 percent of Indians are vegetarians.[7] Europe and North America have lagged behind, partly because we have lost the vivid awareness of the complexity of animal lives and animal thinking that people in other times and places have had, and partly because we have an ethical sensibility that is weakly and inconsistently developed in this area.

Why is the complexity of animal cognition important, and what does it mean for ethical thought and for action? That is the question I shall attempt to answer in this paper. First, I shall offer a very brief sketch of some of the high points in our current knowledge of animal thinking—including social and emotional aspects of cognition. Then I shall ask what this means for ethics and for public policy. I shall argue that the facts about animal lives, as we now know them, cause serious trouble for the most influential approach to the ethics of animal treatment in the modern Euro-American debate, classical Utilitarianism. Important though the ethical work of Utilitarians on this problem has been, the approach is oversimple and offers inadequate guidance for ethical thought and practice. I shall then argue that, although Kant's own approach to the ethics of animal treatment is not terribly promising, a type of Kantian approach with Aristotelian elements, recently developed by Christine Korsgaard, does much better and can offer an adequate basis for the ethical treatment of animals. Although I shall give some reasons why I myself prefer a rather closely related type of neo-Aristotelian approach containing Kantian elements, I do not think that they knock out Korsgaard's Kantian approach, and the two slightly different hybrid approaches can be allies in the ethical pursuit of good treatment for animals.

All this concerns ethics. When we turn to political principles, the argumentative terrain changes, since we will need to find arguments that citizens from a wide range of different comprehensive ethical approaches can share, and we will need to be certain that the arguments respect them and their conscientious commitments. We will only do this, as John Rawls argued, if we find arguments that avoid making commitments on some of the most divisive ethical and epistemological questions. For this task, I shall argue, Korsgaard's ethical approach as currently stated will not do, since its very strong insistence on the fact that humans are the only sources of value in the world is not a position that all reasonable citizens can share. (Korsgaard does not suggest that they could, and she never proposes her view as a political doctrine.) Nor would a comprehensive neo-Aristotelianism be adequate for political purposes, if it asserted, as I think it should, that value exists in the world independently of human legislation and that the lives of animals are valuable in this independent way. (That is a part of the view that I have never developed in writing on the subject, since I, unlike Korsgaard, have focused on defending a political rather than a comprehensive ethical doctrine.) Once again, realism about value is a metaphysical and epistemological view about which reasonable citizens can disagree.

What we need for political purposes, then, I shall argue, agreeing with John Rawls, is a stripped-down view that does not make claims on this and other divisive issues. For our political purposes in thinking about animal entitlements, either a stripped-down version of Korsgaard's Kantian view or a stripped-down neo-Aristotelianism of the sort that I have tried to develop can provide a good basis for reasonable principles, and, indeed, the two will converge to a large degree, since the metaphysical commitments that divide them will have been removed.[8] I shall argue that the view I describe can become, over time, the object of an overlapping consensus among citizens who differ on many other aspects of life. And, finally, I shall argue that even an ethical Utilitarian can accept political principles based on neo-Kantian or (related) neo-Aristotelian insights. (This is important, because Utilitarianism is among the reasonable comprehensive doctrines that will need to sign on to the political "overlapping consensus," even if there seem to me to be very good reasons to think that it's not the best ethical doctrine.)

2. ANIMAL THINKING: SOME MAJOR FINDINGS

The study of animal cognition is a vast network of research programs, each focused on particular cognitive capacities. As biologist Marc Hauser, one of the most careful and enterprising such researchers, comments, the abstract category "thinking" is not really all that useful in approaching animals: we need to ask "about mental phenomena that are more precisely specified, phenomena such as an animal's capacity to use tools, to solve problems using symbols, to find its way home, to understand its own beliefs and those that others hold, and to learn by imitation."[9] Each of these more precise questions, in turn, must be posed in the context of a detailed understanding of the animal's form of life as it has evolved within a particular set of environmental conditions and challenges. Only by such a detailed study of what the animal is "up against" can we avoid making errors of anthropomorphization.[10] The best work, therefore, is done by researchers who spend a long time with a particular species, studying it in its natural habitat as much as is possible. (Obviously some tests, such as the mirror test, need a controlled situation and cannot be done in "the wild."[11])

There is so much excellent work here, on birds and mammals of all sorts, that it would be foolish to pretend to summarize. Instead, a few select high points can be mentioned.

1. *Tool Use.* It took some time for researchers to become convinced that even chimpanzees were tool-users; Jane Goodall's painstaking observations both established beyond doubt that this was so and set research going on a whole set of questions about species of varying sorts. With Marc Hauser, we may define a tool as "an inanimate object that one uses or modifies in some way to cause a change in the environment, thereby facilitating one's achievement of the target goal."[12] (Thus using a part of one's own body to achieve a goal does not count as tool use.) Using this definition, we can now say with confidence that a wide range of creatures use tools, including crows, vultures, monkeys and apes of many types. Much of this use takes place in food contexts, but some occurs, as well, in social contexts, for example grooming. (Apes groom one another's teeth using a range of tools.) Monkeys prove quite flexible in recognizing the potential usefulness of a tool-like object placed in their environment, in their ability to distinguish between function-relevant and irrelevant characteristics of tools of a variety of kinds, and also in their ability to modify the object to make it more useful.

2. *Causal Thinking.* There are many ways in which nonhuman animals manifest an understanding of causal connections, but one of the most interesting has recently been demonstrated by a Harvard research team working with rhesus monkeys. The team showed that these monkeys, like humans, generate causal hypotheses "from single, novel events" together with their general knowledge of the physical world, rather than only through repeated experiences of a particular type of connection.[13]

3. *Spatial Thinking.* Animals, of course, manage to find their way from one point to another, navigating through tremendous environmental complexities. Their spatial mastery in many respects exceeds that of humans. What researchers have tried to ask is to what extent each species' behavior gives evidence that they are working with something like a representation of the spatial world as they find their way around. Not only primates but also birds, rats, and some types of insects appear to have rather complex abilities to form some type of "cognitive map." Chimpanzees may even be able to read and use spatial models made by humans.[14]

4. *Perspectival Thinking.* Some animals have the ability to distinguish between the way they see a situation and the way another person on the scene will see it. This ability, crucial in deceiving others, is found in chimpanzees,[15] and the new experimental evidence about elephants and the mirror suggests that we will find this ability in elephants as well. Dolphins are another species that

may have this ability. Obviously, this cognitive ability is a crucial precursor of empathy—the ability to inhabit the experiential perspective of another—and also, of sympathy or compassion, emotional pain at the hardship or pain of another.[16]

5. *Emotion.* Emotions involve cognition, so they are appropriately included in this discussion.[17] More or less all animals experience fear, and many experience anger of some type. Other emotions require a more specialized repertory of thoughts and are thus less widely distributed. Animals who are capable of perspectival thinking are capable of at least some type of compassion or sympathy; some animals appear to experience grief; shame can be observed in animals who have a conception of their proper role in a system of social rules. At least a rudimentary conception of guilt can be seen in animals who try to conceal an inappropriate act and who exhibit pain on its discovery.[18]

6. *Social Cognition.* Social cognition is a huge category, since much of the life of animals of many species is lived in interaction with other species members. There are hundreds of research programs in this area, focusing on dozens of more specific abilities.[19] Each such project aims at first describing in detail the social interactions characteristic of the species and then, within these patterns, isolating more specific phenomena: for example, rule-following and rule-violation; the awareness of hierarchy and group order; punishment for deviations; the acculturation of the young; reciprocity and some type of altruism.[20] We can certainly conclude that animals of many types exhibit at least some of these forms of complex cognitive ability, and, in general, are aware of an orderly group life with a division of labor, characteristic ways of doing things, and ways that deviate and are occasions for stigma. Particularly significant is widespread evidence of reciprocal altruism: animals follow complex rules about giving and giving back, favoring those who give by giving to them in return.

Since I have focused on elephants and will focus on them again at the end of this article let me give some examples of what we have learned about this remarkable species, which has been extensively studied. In general, creatures with larger bodies tend to have what are called slower life spans, meaning that life stages unfold more gradually and that the whole life span lasts longer. Slow life span is highly correlated with the ability to develop and exhibit complex forms of intelligence, and elephants are among the most long-lived of the nonhuman mammals, often living to sixty or so.[21] All three elephant species exhibit very similar patterns, so I shall not distinguish them in what follows.

Elephants form female-dominated groups consisting of adult females and both male and female young; these groups typically have one leading matriarch, usually the oldest female. All adult members of the group share child-rearing tasks, and the older females help younger females learn how to raise a young child. The matriarch takes the lead in moving the group from place to place, and also in initiating complicated communications about movement and food. (Meanwhile, bull elephants form their own hierarchical society, which lives apart from the females, plays no role in rearing the young, and has in general been less well studied.) A wide range of elephant calls has been analyzed, and we now know that these calls are a highly complex long-distance communication system, based on low-frequency sound, that not only enables each group, female and male, to stay together though widely dispersed for foraging, but also enables males and females to locate one another for mating.[22] "That such a system has evolved," writes Katy Payne of the mating call system, "is particularly striking in light of the fact that a female elephant typically spends only one period of two to five days every four or five years in estrus." Meanwhile, within the female group, young elephants are initiated into a wide range of appropriate behaviors, disciplined gently for being too rambunctious, and, in general, shown how to function as a member of the group.[23]

These are generalizations; but it is also important to note that elephant society is highly individualized; elephants have different personalities and tastes, and the presence of these individual differences is acknowledged by the group.

Particularly striking is the fact that elephants appear to have some understanding of death and to respond to death with something that seems to be at least akin to grief.[24] Here is a description by Cynthia Moss of the reaction of other elephants in Amboseli National Park to the death of a young female by a poacher's bullet (this sort of behavior has by now been widely observed in all three species):

> Teresia and Trista became frantic and knelt down and tried to lift her up. They worked their tusks under her back and under her head. At one point they succeeded in lifting her into a sitting position but her body flopped back down. Her family tried everything to rouse her, kicking and tusking her, and Tallulah even went off and collected a trunkful of grass and tried to stuff it into her mouth.[25]

The elephants then sprinkled earth over the corpse, eventually covering it completely before moving off. When elephants come upon the bones of elephants, even bones old and dry, they examine them carefully, something they don't do with the bones of other species, as if they are trying to recognize the individual who has inhabited them.

One thing that this new research surely ought to do is to awaken our ethical concern for animals, if we have previously been inclined to treat them as mere objects for our use or as automata devoid of experience. Once concern is awakened, however, we still need to ask what general theoretical approach to the ethics of animal treatment is likely to prove the best guide.

3. UTILITARIANISM: STRENGTHS AND PROBLEMS

The philosophical school that has, until now, made the largest contribution to thinking about the ethical treatment of animals is classical Utilitarianism. Both Jeremy Bentham and John Stuart Mill were passionately interested in the lives of animals, and both thought that human treatment of animals was ethically unacceptable. Bentham—noting that Hinduism and Islam are ahead of Christianity in their recognition of ethical duties to animals—famously predicted that a day would come when species difference would seem to all as ethically irrelevant, in the context of bad treatment, as race was by then beginning to be agreed to be:

> The day may come when the rest of the animal creation may acquire those rights which never could have been withholden from them but by the hand of tyranny. The French have already discovered that the blackness of skin is no reason why a human being should be abandoned without redress to the caprice of a tormentor. It may come one day to be recognized that the number of the legs, the villosity of the skin, or the termination of the os sacrum, are reasons equally insufficient for abandoning a sensitive being to the same fate. What else is it that should trace the insuperable line?…The Question is not, Can they reason? Nor, Can they talk? But, Can they suffer?[26]

Mill, noting that this passage, written in 1780, anticipates many valuable legal developments that make at least a beginning of protecting animals from cruelty, responds in similar terms to Whewell's dismissive statements concerning duties to animals. Whewell argues that it is "not a tolerable doctrine" that we would sacrifice human pleasure to produce pleasure for "cats, dogs, and hogs." Mill responds: "It is 'to most persons' in the Slave States of America not a tolerable doctrine that we may sacrifice any portion of the happiness of white men for the sake of a greater amount of happiness to black men." He adds a comparison to feudalism.[27] At his death, Mill left a considerable portion of his estate to the Society for the Prevention of Cruelty to Animals.

Both Bentham and Mill felt not only that large conclusions for our treatment of animals followed from their Utilitarian principles, but also that the ability of those principles to generate acceptable conclusions in this area was a point in favor of those principles—by contrast, for example, with the principles of vulgar Christianity (represented in Whewell's hostile reaction to Bentham), which made species difference all-important. For both Bentham and Mill, Utilitarianism, with its commitment to treat all sufferings and pleasures of all sentient beings on a par, had made decisive progress beyond popular ethics in just the way that abolitionist views were then making progress beyond popular racist views. Seeing how the view enabled one to cut through unargued prejudice, and to treat subordinated beings with due concern, one saw a strong reason, they thought, in the view's favor.

There is no doubt that Utilitarian thought has made valuable and courageous contributions in this area, and that it still does so today, in the work of preference-Utilitarian Peter Singer, one of the leading voices against cruelty to animals. I now want to argue, however, that Utilitarianism cannot meet the challenge of animal complexity, as we currently understand it. (It has related problems with human complexity).[28]

Utilitarianism can be usefully analyzed, as Bernard Williams and Amartya Sen have analyzed it, as having three parts.[29] The first is consequentialism: the best choice is defined as the one that promotes the best overall consequences. The second is "sum-ranking," a principle of aggregation: we get the account of consequences by adding up all the utilities of all the creatures involved. Third, the theory invokes some specific theory of the good: pleasure in the case of Bentham and Mill, the satisfaction of preferences in the case of Peter Singer. Looking at animals, Utilitarians begin from the understanding that they, like human beings, feel pleasure and pain, and they argue that the calculus of overall pleasure cannot consistently exclude them. The right choices will be those that produce the largest aggregate balance of pleasure over pain—or, in Singer's case, the largest net balance of satisfaction over dissatisfaction.

The Utilitarian approach has the merit of focusing attention on something of great ethical importance: suffering. Humans cause animals tremendous suffering, and much of it is not necessary for any urgent human purpose. Animals would suffer a great deal without human intervention, but there is no doubt that much of animal suffering in today's world is caused, directly or indirectly, by human activity. So the focus on animal suffering is valuable, and these philosophers deserve respect for the courage with which they put this issue on the agenda of their nations.

Five problems, however, can be seen, if we hold this theory up against the complex cognitive and social lives of animals. The first point is that pleasure and pain, the touchstones of Utilitarianism, are actually disputed concepts. Bentham simply assumes that pleasure is a single homogeneous type of sensation, varying only in intensity and duration. But is he correct? Is the pleasure of drinking orange juice, for example, the same sort of sensation as the pleasure of listening to a Mahler symphony? Philosophers working on this question, from Greek antiquity to the present day, have, on the whole, denied this, insisting that pleasures vary in quality, not just quantity. Moreover, Mill himself insists on this point in *Utilitarianism*. A second point on which Mill insists—along with many other philosophers, past and present—is that pleasure is a type of awareness very closely linked to activity, so that it may be impossible to separate it conceptually from the activities that are involved in it.[30] We do not need to resolve all these issues in order to realize that they arise in animal lives as well as human lives, once these lives are seen with sufficient complexity. Happy's pleasure seeing herself in the mirror seems unlikely to be the very same sensation as her pleasure when she eats some nice bananas or hugs her small baby elephant with her trunk.

Pleasures, second, are actually not the only things relevant to animal lives. These lives consist of complex forms of activity, and many of the valuable things in those lives are not forms of pleasure. Happy's self-recognition in the mirror, and the mourning of elephants for their dead, are not pleasures; the latter may even be deeply painful. Nonetheless, such meaningful elements in animal lives should, we intuitively feel, be fostered and not eclipsed—for example, eclipsed by raising animals in isolation so that they don't have contact with fellow group members and so are unable to mourn. Animals want much more than pleasure and the absence of pain: free movement, social interactions of many types, the ability to grieve or love. By leaving out all this, Utilitarianism gives us a weak, dangerously incomplete way of assessing our ethical choices.

Third, animals, like human beings, can adjust to what they know: they can exhibit what economists call "adaptive preferences."[31] Women who are brought up to think that a good woman does not get very much education may not feel deprived if they don't get an education, so Utilitarian theory would conclude that education is not valuable for them. This means that the theory is often the ally of an unjust set of background conditions. Much the same sort of thing can be said about animal preferences. If animals are given a very confined life, without any access to social networks characteristic of their species, they may not actually feel pain at the absence of that which they haven't experienced, but this does not mean that there is not an absence or that it should not be taken seriously. By refusing to recognize value where there is not pleasure or pain, Utilitarianism has a hard time criticizing bad ways of treating animals that have so skewed their possibilities that they don't even hope for the alternative.

Fourth, a familiar point in criticism of Utilitarian theories of human life, Utilitarianism's way of aggregating consequences doesn't treat each individual life as an end; it allows some lives to be used as mere means for the ends of others. If it should turn out that the pleasures of humans who exploit animals for their use are great and numerous, this might possibly justify giving at least some animals very miserable lives.

Finally, all Utilitarian views are highly vulnerable in respect of the numbers. If the goal is to produce the largest total pleasure or satisfaction, then it will be justified, in the terms of the theory, to bring into existence large numbers of animals whose lives are extremely miserable, and way below what would be a rich life for an animal of that sort, just so long as the life is barely above the level of being not worth living at all.

4. Two Strong Theoretical Alternatives

Seeing these problems helps us think: it informs us, I believe, that we need a theoretical approach in ethics that can do two things. First, the approach must have what I would call a Kantian element: that is, it must have as a fundamental ethical starting point a view that we must respect each individual sentient being as an end in itself, not a mere means to the ends of others. (I'm simply helping myself to an extension of Kant's approach to human beings at this point, not offering any story about how one might use Kant's own actual views to generate obligations to animals.) Second, the approach must have what I would call a neo-Aristotelian element, the ability to recognize and accommodate a wide range of different forms of life with their complicated activities and strivings after flourishing. I've suggested in writing about this that for this part of the view we can turn to a version of the Aristotelian idea that each creature has a characteristic set of capabilities, or capacities for functioning, distinctive of that species, and that those more rudimentary capacities need support from the material and social environment if the animal is to flourish in its characteristic way. But of course that observation only goes somewhere in ethics if we combine it with the Kantian part, the idea that we owe respect to each sentient creature considered as an end. Putting these two parts together, we should find a way to argue that what we owe to each animal, what treating an animal as an end would require, is, first, not to obstruct the creature's attempt to flourish by violence or cruelty, and, second, to support animal efforts to flourish in positive ways (an analogue of Kantian duties of beneficence.)

In the case of humans, as Kant and Aristotle would agree, our beneficence is rightly constrained by concerns about autonomy and paternalism: rather than pushing people into what we take to be a flourishing life, we ought to support, instead, ample space for choice and self-determination. In the case of animals, by contrast, although we should always be sensitive to considerations of choice, to the extent that we believe an animal capable of choice among alternatives, for the most part we must and should exercise informed paternalistic judgments concerning the good of the creature, and our duties of beneficence will be correspondingly more comprehensive, as they are in the case of human children.[32]

There are two recent ethical approaches that both contain these two elements. One is Christine Korsgaard's Kantian view, developed in her recent Tanner Lectures, "Fellow Creatures: Kantian Ethics and Our Duties to Animals."[33] Another is the extended version of the neo-Aristotelian capabilities approach that I have described in *Frontiers of Justice*.

Kant's own views on animals are not very promising. He holds that only humans are capable of moral rationality and autonomous choice, and that only beings who are capable of autonomy can be ends in themselves. Animals, then, are available to be used as means to human ends. Kant thinks that we do have some duties with regard to animals, but these, on closer inspection, turn out to be indirect duties to human beings. In particular, Kant holds that treating animals cruelly forms habits of cruelty that humans will then very likely exercise toward other human beings. This, rather than any reason having to do with respect for animals themselves, is his reason for imposing a range of restrictions on the human use of animals.

Korsgaard's view is subtle and difficult to summarize, but let me try to state its essential insight. For Korsgaard as for Kant, we humans are the only creatures who can be obligated and *have* duties, on account of our possession of the capacity for ethical reflection and choice. Korsgaard, however, sees that this fact does not imply that we are the only creatures who can be the objects of duties, creatures to whom duties are owed. She also puts this point another way. There are, she argues, two different senses in which a being can be an "end in itself": (a) by being a source of legitimate normative claims, or (b) by being a creature who can give the force of law to its claims. Kant assumes that these two ways in which something can be an end in itself pick out the same class of beings, namely all and only human beings. Korsgaard points out that a being may be an end in itself in the first sense while lacking the capacity for ethical legislation crucial for the second sense.

Korsgaard's conception of animal nature is Aristotelian: she sees animals, including the animal nature of human beings, as self-maintaining systems who pursue a good and who matter to themselves. She gives a fine account of the way in which we may see animals as in that sense intelligent—as having a sense of self and a picture of their own good, and thus as having interests whose fulfillment matters to them. We human beings are like that too, she argues, and if we are honest we will see that our lives are in that sense not different from other animal lives.

Now when a human being legislates, she does so, according to Kant and Korsgaard, in virtue of a moral capacity that no other animal has. This does not mean, however, that all human legislation is *for* and *about* the autonomous will. Much of ethics has to do with the interests and pursuits characteristic of our animal nature. When we do make laws for ourselves with regard to the (legitimate) fulfillment of our needs, desires, and other projects issuing from our animal nature, it is simply inconsistent, and bad faith, Korsgaard argues, to fail to include within the domain of these laws the other beings who are similar to us in these respects. Just as a maxim cannot pass Kant's test if it singles out a group of humans, or a single human, for special treatment and omits other humans similarly situated, so too it cannot truly pass Kant's test if it cuts the animal part of human life from the animal lives of our fellow creatures.

I have saved until last a part of Korsgaard's conception that lies at its very heart. We humans are the creators of value. Value does not exist in the world to be discovered or seen; it comes into being through the work of our autonomous wills. Our ends are not good in themselves; they are good only relative to our own interests. We take our interest in something "to confer a kind of value upon it," making it worthy of choice.[34] That, in turn, means that we are according a kind of value to ourselves, including not only our rational nature but also our animal nature. Animals matter because of their kinship to (the animal nature of) a creature who matters, and that creature matters because it has conferred value on itself.

Korsgaard's conception of duties to animals has what I demanded: a Kantian part and an Aristotelian part. It says that we should treat animals as ends in themselves, beings whose ends matter in themselves, not just as instruments of human ends. And it also conceives of animal lives as rich self-maintaining systems involving complex varieties of intelligence. So far so good. Now I shall describe the way in which my own conception articulates the relationship between the Kantian and the Aristotelian. Then, more tentatively, we can ask what reasons there may be in favor of choosing one rather than the other.

Because my view has been advanced as a political doctrine rather than a comprehensive ethical doctrine, I have not developed the view's metaphysical/epistemological side. In keeping with my espousal of a Rawlsian "political liberalism," I have expressed the relevant idea of intrinsic value in a nonmetaphysical and intuitive way. However, were I to flesh out the view as a comprehensive ethical view, I would insist that the lives of animals have intrinsic value. This value is independent of human choice and legislation, and it is there to be seen. If humans had never come into being, the lives of other animals would still be valuable. We humans are, fortunately, attuned to value, so we are capable of seeing what Aristotle's students saw, that there is something wonderful and awe-inspiring in the orderly systems characteristic of natural end–pursuing creatures. To this sense of awe, I suggested in *Frontiers of Justice* that we must add an ethical sense of attunement to dignity. What is wonderful about an animal life is its active pursuit of ends, so our wonder and awe before such a life is quite different from our response to the Grand Canyon or the Pacific Ocean: it is a response to the worth or dignity of an active being who is striving to attain its good. Wonder and awe before the dignity of such a life would be inappropriately aestheticizing, would fail to recognize what, precisely, is wonderful about the creature, if it simply said "Ooh!" "Aah!" and saw no implications for the ethics of animal treatment. If we have appropriate wonder before an animal life, wonder that homes in on what the creature actually is, a self-maintaining active being, pursuing a set of goals, then that appropriate wonder, I argue, entails an ethical concern that the functions of life not be impeded, that the life as a whole not be squashed and impoverished.

Let me put this point another way. When I have wonder at the Grand Canyon, it would seem that I have appropriate wonder, wonder that is appropriately trained on the relevant characteristics of the object, only if I form some concern for the maintenance of the beauty and majesty of that ecosystem; even here, then, wonder has practical consequences. If I say, "How wonderful the Grand Canyon is," and then throw trash around, I am involved in a contradiction: my actions show that at some level I really do not think that the Grand Canyon is very wonderful. With animals, all this is true, but also much more. Animals, because they are active sentient beings pursuing a system of goals, can be impeded in their pursuit by human interference. In *Frontiers of Justice*, I argued that it is this quality of active, striving agency that makes animals not only objects of wonder but also subjects of justice.[35] The way we wonder at the complexity of animals, if it is appropriate, really trained on what they are, includes a recognition that they are active, striving beings, and thus subjects of justice. The right sort of wonder (not pretending that an animal is like a fine chair or carpet, but seeing it for what it is) leads in that sense directly to an ethically attuned awareness of its striving.

With that intuitive picture as my starting point, I then go on in *Frontiers of Justice* to argue that our ethically attuned awareness of the value of animal striving suggests that we ought to promote for all animals a life rich in opportunities for functioning and lacking many of the impediments that we humans typically put in the way of animals' flourishing. Since my views on the content of our duties lie very close to Korsgaard's, I need not enumerate them here.

What might lead one to choose one of these views of our duties to animals rather than another? It is obvious that some people find realism about value implausible, and others find the idea that all value is a human creation implausible. The choice between the two views on this score must await the much fuller development of arguments for and against realism. Here Korsgaard has gone a lot further than I have, since I have deliberately avoided defending realism, given that I am trying to advance the capabilities view as a non-metaphysical political view. There would be a great deal of work for me to do were I to try to work out the view as an ethical doctrine comparable to Korsgaard's in its detail and completeness. The Aristotelian approach to value does involve a large measure of reliance on intuitions, as Korsgaard justly argues in *The Sources of Normativity*.[36] It will not satisfy all people. By contrast, the Kantian account of normativity is intricate and philosophically rich; it does not seem to rest on such a fragile empirical foundation.

On the other hand, I think that the Aristotelian view has at least some advantages, albeit subtle and not decisive. Korsgaard does not exactly make the value of animals derivative from the value of human beings. Instead, her picture is that when we ascribe value to ourselves, we ascribe value to a species of a genus, and then it is bad faith, having once done that, to deny that the other species of that genus, insofar as they are similar, possess that same action-guiding value. However, there still seems to be a strange indirectness about the route to animal value. It is only because we have similar animal natures ourselves, and confer value on that nature, that we are also bound in consistency to confer value on animal lives. Had we had a very different nature, let's say that of an android, we would have no reason to value animal lives. And, so far as I can see, the rational beings recognized by Kant who are not animal (angels, God) have no reason to value the lives of animals. For me this is just too indirect: animals matter because of what *they* are, not because of kinship to ourselves. Even if there were no such kinship, they would still matter for what they are, and their striving would be worthy of support. For Korsgaard, it's in effect an accident that animals matter: we just happen to be pretty much like them. But I think that the value of animal lives ought to come from within those lives; even if one doesn't think of value as eternal and immutable, one still might grant that it comes in many varieties in the world, and each distinctive sort is valuable because of the sort it is, not because of its likeness to ourselves.

So, while I agree with Korsgaard that we are the only creatures who have duties, and while I think that she has argued in a way that puts the Kantian view in its best form, and, indeed, in a very attractive form, I still feel that there's something backhanded about the route to animal value, and that it would be good to acknowledge that this value is there whether or not these creatures resemble us. (Whether one could acknowledge that without relying on intuitions as the source of normativity is a further question that I shall not try to answer here; clearly, I am less worried about reliance on intuition than Korsgaard is, or else I would not be willing to venture ahead at this point.)

There is another point of interesting difference, pertinent to our concern with animal thinking. Korsgaard, as I said, makes a very compelling case for recognizing in animals a range of types of awareness; even those who can't pass the mirror test are held to have a point of view on the world, and ends that matter to them. All this seems to me just right. So, while one might have expected that a Kantian view would draw a too-sharp line between the human and the animal, that seems not to be true of Korsgaard's view. In another way, though, I wonder whether there is not after all a bit too much line-drawing.

Korsgaard rightly says that we are the only truly moral animals, the only ones that have a full-fledged capacity to stand back from our ends, test them, and consider whether to adopt them. She does, however, say of children and people with mental disabilities that they too are rational beings in the ethical sense, it's just that they reason badly.[37] If she once makes that move, I do not see how she can avoid extending at least a part of ethical rationality to animals. Animals, as we saw, are aware of their place in a social group. Many of them have the capacity for a type of reciprocity, and some, at least, seem to be capable of positional thinking, thus understanding the impact of their actions on others. At least some varieties of shame and even guilt figure in some of these animal lives in ways related to their awareness of the rules that govern social interactions. So it seems that what animals most conspicuously lack is the capacity for *universal* ethical legislation, but that there's a part of ethical capacity that at least some of them have already. By splitting humans from all the others, as the only rational legislators, Korsgaard seems to have drawn a line that is not that sharp in reality.

Korsgaard will surely say at this point, as she does early in her lectures, that if the definition of rational being turns out to fit some nonhuman creatures, all very well and good, she is only focusing on our obligations to those whom it doesn't fit.[38] But I'm not altogether happy with that reply. It seems that we need to understand our moral capacities as well as we can, making use of all the scientific information that is available. Information about animals is very helpful to us, in showing us how the capacities we have, which we might have thought transcendent and quasi-divine, are a further development of some natural capacities that we share with the animals. In understanding ourselves that way, we also attain a fuller understanding of how what Korsgaard calls our animal nature is related to our moral nature: our moral nature is actually one part of our animal nature, not something apart from it. Our moral nature is born, develops, ages, and so forth, just like the rest of our capacities, and, like them, it has an evolutionary origin in "lower" animal capacities. I feel that Korsgaard has pushed Kant to the limit in giving her extremely sensitive and appealing picture of how Kant and Aristotle may cooperate, but there is really no way, without departing from Kant rather radically, to acknowledge that our moral capacities are themselves animal capacities, part and parcel of an animal nature. I think that any conception that doesn't acknowledge this is in ethical peril, courting a danger of self-splitting and self-contempt (so often linked with contempt for women, for people with disabilities, for anything that reminds us too keenly of the animal side of ourselves).[39] Although Korsgaard heads off this peril sagely wherever it manifests itself, she still doesn't altogether get rid of it; it is still lurking, in the very idea that we are somehow, in being moral, above the world of nature.

Those, then, are my reasons for tentatively preferring my own conception to Korsgaard's as a basis for the ethics of animal treatment. Nobody could doubt, however, that hers is considerably more finished than mine with respect to its metaphysical/epistemological side, which I've deliberately left uncultivated; nor should anyone doubt that her view provides a very good basis for thinking about our duties toward animals.

5. POLITICAL PRINCIPLES: AN OVERLAPPING CONSENSUS?

Now let us turn to political principles. Here, as I've said, agreeing with Rawls,[40] we want to be abstemious, not making controversial metaphysical or epistemological claims. We are seeking an overlapping consensus among citizens who hold a wide range of reasonable comprehensive doctrines—including comprehensive Korsgaardianism and neo-Aristotelianism, but also including Buddhism, Hinduism, Christianity, and much else. So, we do not say that the human being was created on the seventh day of creation, or that humans will be reincarnated into animal bodies. By the same token, we do not say, with Kant, that human beings are the sole creators of value or, with Aristotle, that human beings discover a value that exists independently.

At this point, then, the major difference between Korsgaard's view and my imagined comprehensive neo-Aristotelianism has been bracketed. There are subtle differences that may remain, concerning the relationship between ethical rationality and other aspects of animals' good. It is difficult to say whether these differences really do remain: the idea that we are the creators of value goes very deep in Korsgaard's view, and colors every aspect of it, so it is very difficult to know exactly what her view would look like when recast in the form of a political doctrine appropriate to grounding a form of political liberalism in Rawls' sense. Certainly there would remain the idea that every sentient being[41] has a good, consisting of a range of (non-commensurable) activities that are the activation of its major natural capacities, and that each animal is entitled to pursue that good. There would also remain (or so I believe) the sense that this good exacts something from human beings who are capable of choice: we have duties to protect and promote the good of animals. In these two respects, the imagined Korsgaardian view overlaps pretty completely with my neo-Aristotelian view, which borrowed the Kantian notion of dignity to ground ethical duties to the forms of life that wonder already singled out as salient. The emphasis on capacity and activity, the emphasis on a plurality of interrelated activities, the emphasis on ethical duty—all of this seems shared terrain between the two approaches.

In this case, then, we may not even need to talk, as Rawls did, of the overlapping consensus as consisting of a family of liberal political doctrines.[42] We may be able to agree on a single political doctrine.[43] If Korsgaard judges it important on balance to endorse a Kantian over a neo-Aristotelian political doctrine (Rawls' view, for example, over mine), the reasons for this difference would not come, I believe, from this particular area of the political doctrine, where Korsgaard has rightly seen the importance of invoking Aristotelian ideas.

The core idea of the political conception is the one I have already mentioned in talking about the ethical conception: animals have characteristic forms of dignity that deserve respect and give rise to a variety of duties to preserve and protect animal opportunities for functioning. With this starting point, I then go on to envisage the general shape of a constitution for a minimally just multispecies world.

The political conception I have articulated seems like one that will be able to achieve an overlapping consensus among neo-Aristotelians and Korsgaardian Kantians. I conjecture that many other reasonable comprehensive doctrines will also support it: Buddhism, Hinduism, and, with time and persuasion, many varieties of Judaism, Christianity, and Islam.

What, however, of Utilitarianism? I have argued strenuously against accepting Utilitarianism as a comprehensive ethical doctrine concerning animal treatment, but I do not think that we should, without extremely strong reasons (such as have not yet been presented) conclude that Utilitarianism is not among the reasonable comprehensive doctrines that should be part of any political consensus. Since, however, the political principles I advocate are grounded in Kantian and Aristotelian ideas that are not as such part of Utilitarianism, it might seem that Utilitarians will have difficulty accepting that political conception. John Rawls argued that Utilitarians could form part of an overlapping consensus supporting his own political doctrine,[44] and yet not all readers of that argument have been convinced by it. So we must ask ourselves: what reasons do we have to think that Utilitarians concerned with the ethical treatment of animals will accept the principles I have proposed?

The first point to be made here is that most of the points to which I've objected in Utilitarianism are already noted by John Stuart Mill, who proposed a variety of Utilitarianism in which qualitative distinctions among diverse life activities plays a central role, and in which activity is understood to be valuable in its own right, not simply as a means to pleasant sensations. Mill's Utilitarian view, notoriously, is rather Aristotelian; his arguments against simple Benthamism are so cogent that anyone who ponders them is likely to be strongly swayed in that direction. A Mill-style Utilitarianism can easily sign on to the overlapping consensus I have proposed.

Even were a Utilitarian to refuse to accept Mill's reformulations, another route of accommodation awaits us. Henry Sidgwick, while insisting that the correct ethical principle was the unmodified Utilitarian principle, also thought that this principle would not be a good one for most people to apply: better results, from the point of view of that principle itself, would be obtained by encouraging most people to follow a more conventional ethical code based on non-commensurable principles of virtue and vice. Now Sidgwick also thought that for this reason some top government officials should operate, meanwhile, with the correct principle, but his conception of government has been widely criticized for its undemocratic character and its insistence that we ought to conceal from most people the grounds of the political choices that govern their lives. If a modern Utilitarian believes, with Sidgwick, that most people should not try to use the Utilitarian principle, but also believes, unlike Sidgwick, that political principles should be based on ideas that can be publicly stated and that all citizens can understand and accept, then such a Utilitarian, while continuing to prefer the Utilitarian principle to others as the source of a comprehensive ethical view, will gladly accept my Aristotelian view for political purposes.

Much more would need to be done to show that, and how, each of the major reasonable comprehensive doctrines could support the political consensus proposed here. At this point, however, I believe we may conclude that there are no evidently overwhelming obstacles to that agreement. The transition from our current immoral situation to an ethical/political modus vivendi, and from that modus vivendi to a constitutional consensus, *and*, one may hope, from a constitutional consensus to a genuine overlapping consensus, is a development we may seek and foster without feeling that in so doing we are working in vain.

6. PRACTICAL CONSEQUENCES

In *Frontiers of Justice*, I proposed a rather detailed account of animal entitlements, giving an idea of how the capabilities view would be extended to deal with the lives of animals. Here, in keeping with my starting point, I want to narrow the focus and indicate how this view (or a related view) would take account of the new research on animal thinking. I shall focus on the two species with which I began, elephants and mice. So: what does our emphasis on diverse life-forms, and our new awareness of the complexity of these life-forms, suggest in these two cases?

I have said that elephant minds are highly complex. Elephants have capacities for self-recognition, for elaborate forms of social cooperation, for some sort of awareness of the death of an individual. They are, then, not simply sites of pleasant or painful sensation: they are complexly thinking and functioning organisms with an elaborate form of life.

Now to their situation. Elephants are highly endangered. Because an adult elephant needs to eat about 200–250 pounds of vegetation per day to stay healthy, elephants have to cover a lot of territory, and there can't be too many elephants in one territory. South Asia and Africa, where most elephants live, have rapidly growing human populations, and this growth has diminished the space where elephants can roam free. When they get too close to human habitations, moreover, things do not go well: groups of young males, particularly, mix badly with human villages. Added to these problems is the terrible problem of poaching: hundreds of elephants are killed every year for the ivory market, despite domestic laws and international agreements against this practice. In 1930, there were between five and ten million African elephants and somewhere around a million Asian elephants. Today, there are probably only about 35,000 to 40,000 Asian elephants left in the wild, and only about 600,000 African elephants.

Needless to say, all the major ethical approaches agree that gratuitous killing of elephants for sport or for luxury items like ivory is utterly wrong, and that laws against this should be vigorously enforced. In the light of our new understanding of elephant society, however, we have broader and deeper reasons for opposing this killing, seeing the way in which it tears apart the complex network of the group, threatening the upbringing of young elephant calves. Utilitarians will have grave difficulty making these facts matter as they should.

Beyond this point, if we were Utilitarians, we might think that all we need to do is not to inflict pain on elephants. Many zoos manage something like this. But if we adopt my more complex approach, we will think that what we should support is something much more complicated, a whole form of life that includes love, grief, self-recognition, and much more. This makes our practical task very complicated. It means that we must think much harder than we have so far about the habitat of elephants in the wild, trying to protect large tracts of land indefinitely for this purpose.

Should we permit elephants to be confined in zoos at all? This is a very difficult question. No zoo can supply elephants with the grazing space their typical form of life requires. Especially in Africa, elephants have lived in large open tracts of land for millennia; and there is some genuine hope that good policies will make that way of life possible into the future. In Africa, then, policy should focus on protecting a habitat within which elephants can live lives characteristic of their kind. This would include aggressive efforts to stop poaching, and it would also very likely include efforts to limit elephant population size through contraception, something that is increasingly being understood to be part of the solution to human-animal conflicts.

The case of Asian elephants, however, seems different. Asian elephants are a distinct species. They have become far more endangered than are African elephants, in part because the governments involved (India, Sri Lanka, Thailand) are overwhelmingly focused on questions of human survival and morbidity, and consider that the very creation of animal sanctuaries—which usually deprives some very poor rural people of their livelihood gathering firewood and leaves in the forest—is not a top priority. Thus the solution that seems best for African elephants is just not possible for Asian elephants: that train left the station long ago.

There is another difference between Asian and African elephants that may be significant. Asian elephants have been working in symbiosis with humans to a greater extent than have African elephants; a large proportion of those who survive are working animals, or sacred animals at temples, and so on. We don't have evidence of real domestication—that is of the evolution of a life-form in response to a symbiotic relationship—as we do in the case of dogs and horses. Elephants, being highly intelligent, can learn to work alongside humans, but it would be an exaggeration to say that they really live *with* humans in the way that dogs do. And yet, the most familiar form of life for elephants in Asia is a symbiotic form. Thus what we are contemplating when we think of Asian elephants in zoos is the substitution of one symbiotic form of life for another.

A good zoo, one that provides large open tracts of land for Asian elephants, can give them a better life, in terms of the characteristic life-form of that species, than they are likely to get anywhere in Asia today or at any time in the foreseeable future. Moreover, some excellent zoos, such as those in the Bronx, St. Louis, and San Diego, do a very good job with breeding programs, which are absolutely crucial if this species of elephant is not to become extinct. My approach, however, insists that the usual way elephants are kept in zoos is horrendous: one or two females in a tiny enclosure, in which they have more or less no room for movement or foraging, and no opportunity for the group life characteristic of their kind. We need to think very carefully about the needs of elephants in confinement for wide space, motion, and, above all, for complex social networks characteristic of elephant life. I would say that at minimum an elephant herd in a zoo ought to include four females with their young, and that such elephants should have a hundred acres of land around them. If people want to see them, clever devices can be found, such as the bridge in San Diego that permits visitors to walk high above the elephant habitat and see down into their land, or the complicated curvilinear path in St. Louis that brings spectators up close to various different parts of a large amoeba-shaped elephant habitat.

The same sort of thing holds across the board, so let me end by talking about mice. The level of complexity of a sentient creature does not, I believe, make one species "higher" and one species "lower," meaning that it's more permissible to inflict damage on one than on another. *Each* form of life demands respect, nor should we respect lives simply because they look somehow like our own. Level of complexity does, however, affect what can be a damage for a creature.[45] For a mouse not to have the freedom of religion is not as damaging for a mouse as it is for a human. However, new research on mice shows that even here the Utilitarian approach, regarding them simply as sites of pleasure and pain, would be incomplete. Social bonds and the ability to recognize individuals play a role in their lives too. So when we think how we should treat them, we have to think of all that.

Research mice have usually been treated as mere objects, too "low" to be respected, too simple for their lives to be considered worth carefully supporting. At best there has been an awareness of the importance of sparing mice unnecessary pain. However, the new research suggests that for mice as for larger research animals, quality of life means much more than the mere absence of pain: it means access to social relationships, the ability to live with familiar others over time, and the ability to form communities based on these recognitions.

Once again, then, in this case, the capabilities approach proves an improvement, both over a Benthamite Utilitarianism and over common-sense views, setting our political debate into a framework that focuses on each animal's entitlement to the conditions of a flourishing life characteristic of its kind. In *Frontiers of Justice,* I argue that this suggests an absolute ban on all killings of animals for sport, vanity, and unnecessary research. I also argue that most current research has a tragic aspect, in that wrongs are being inflicted on animals even when the conditions of research have been designed to be as humane as possible.[46] Even when the research in question is important for human and/or animal health, noticing this tragic aspect, this clash between right and right, should lead us to work as hard as possible toward research models that do not inflict harm on animals—models based on computer simulation, for example.

Each type of animal has its own cognitive complexity; each type has a story including at least some emotions or preparations for emotion, some forms of social cognition, often very complex, and complex forms of interactivity. We should learn a great deal more about these complexities, and we should test our ethical views to see whether they are adequate to them. We should then try to imagine ways of human life that respect these many complex forms of animal activity, and that support those lives—all of which are currently being damaged, almost beyond rescue, by our interference and our greed.

NOTES

1. Joshua M. Plotnik, Frans B. M. de Waal, and Diana Reiss, "Self-recognition in an Asian Elephant," *Proceedings of the National Academy of Sciences of the United States* (October 30, 2006), available at http://www.emory.edu/LIVING_LINKS/pdf_attachments/PlotniketalPNAS.pdf (accessed August 10, 2009). The reference number given on the title page is doi:10.1073/pnas.0608062103. By clicking on the link for "supplementary material," one can see the film version.

2. Good discussions of these matters can be found in Cynthia Moss, *Elephant Memories: Thirteen Years in the Life of an Elephant Family*, 2nd ed. (Chicago: University of Chicago Press, 2000) and Katy Payne, "Sources of Social Complexity in the Three Elephant Species," in *Animal Social Complexity: Intelligence, Culture, and Individualized Societies*, ed. Frans B. M. de Waal and Peter L. Tyack (Cambridge, Mass.: Harvard University Press, 2000), pp. 57–86. The latter book also contains valuable articles on many other species.

3. Dale J. Langford, Sara E. Crager, Zarrar Shehzad, Shad B. Smith, Susana G. Sotocinal, Jeremy S. Levenstadt, Mona Lisa Chanda, Daniel J. Levitin, and Jeffrey S. Mogil, "Social Modulation of Pain as Evidence for Empathy in Mice," *Science* 312 (2006): 1967–70.

4. See Richard Sorabji, *Animal Minds and Human Morals: The Origins of the Western Debate* (Ithaca, N.Y.: Cornell University Press, 1993).

5. The incident is discussed in Pliny, *Naturalis Historia* 8.7.20–21; Cicero, *Ad familiares* 7.1.3; see also Dio Cassius, *Roman History* 39, 38, 2–4.

6. See D. N. Jha, *The Myth of the Holy Cow* (London: Verso, 2002).

7. For an impressive court judgment, holding that animals are entitled to a life in accordance with dignity as protected by Article 21 of the Constitution of India, see *Nair v. Union of India*, No. 155/1999, at para. 13 (Kerala High Court, June 6, 2000). The case involved circus animals who were being ill treated and made to perform undignified tricks. The conclusion of the judgment is as follows:

In conclusion, we hold that circus animals…are housed in cramped cages, subjected to fear, hunger, pain, not to mention the undignified way of life they have to live, with no respite and the impugned notification has been issued in conformity with the…values of human life, philosophy of the Constitution….Though not homo sapiens, they are also beings entitled to dignified existence and humane treatment sans cruelty and torture….Therefore, it is not only our fundamental duty to show compassion to our animal friends, but also to recognise and protect their rights….If humans are entitled to fundamental rights, why not animals?

8. In my *Frontiers of Justice: Disability, Nationality, Species Membership* (Cambridge, Mass.: Harvard University Press, 2006), I argue that the neo-Aristotelian "capabilities approach" answers some political questions better than John Rawls' type of political Kantian view, "justice as fairness," but I do not claim that it does better overall, and, of course, I do not argue that my view does better than Korsgaard's, since she has not yet proposed a political doctrine.

9. Marc D. Hauser, *Wild Minds: What Animals Really Think* (New York: Henry Holt, 2001), p. xviii.

10. Hauser, *Wild Minds*, pp. xiii–xx. There are complex debates in this area, in particular between Hauser and de Waal; Hauser holds that many of de Waal's conclusions about similarities between animals and humans are infected by illicit anthropomorphism, whereas de Waal holds that the similarities are real and supported by masses of evidence. On some of the contested points about elephants, my sense is that de Waal has the better of the argument, and his project has over the years involved experts in many different species, whereas Hauser's close empirical work concerns monkeys only. In the light of Hauser's conviction for serious ethical violations in the conduct of his research, his findings must be regarded with skepticism until the extent of these violations has became clear.

11. These scare quotes reflect my belief that there is no part of the Earth that is not profoundly shaped by human activity, including the large tracts of land that appear to be wild.

12. Hauser, *Wild Minds*, pp. 33–34.

13. Marc Hauser and Bailey Spaulding, "Wild Rhesus Monkeys Generate Causal Inference about Possible and Impossible Physical Transformations in the Absence of Experience," *Proceedings of the National Academy of Sciences of the United States of America* 103 (2006): 7181–85.

14. Hauser, *Wild Minds*, chap. 4.

15. See Frans B. M. de Waal, *Good Natured: The Origins of Right and Wrong in Humans and Other Animals* (Cambridge, Mass.: Harvard University Press, 1996), pp. 71–78.

16. On empathy and compassion, see also Frans B. M. de Waal, *Primates and Philosophers: How Morality Evolved* (Princeton, N.J.: Princeton University Press, 2006), pp. 21–33.

17. See my *Upheavals of Thought: The Intelligence of Emotions* (New York: Cambridge University Press, 2001), especially chapter 2 on animal emotions; nothing I say here depends on the more controversial aspects of my view. All the major views acknowledge that emotions contain at least some cognitive components; usually they acknowledge, as well, that the cognitive components in emotion are important in defining the emotion type in question, and in distinguishing one emotion from other emotions.

18. See de Waal, *Good Natured*.

19. For an excellent recent summary, see de Waal and Tyack, *Animal Social Complexity* (above, n. 2).

20. See de Waal, *Good Natured*; de Waal and Tyack, *Animal Social Complexity*; Hauser, *Wild Minds*.

21. See Payne, "Sources of Social Complexity in the Three Elephant Species," pp. 57–86.

22. See Payne, "Sources," p. 76.

23. Some of this can be seen in the nice video *Kandula: An Elephant Story* (produced by the National Zoo in Washington D.C. and Rocket Pictures for the Discovery Channel, 2003), concerning the first five years of a young male Asian elephant by that name.

24. I put things this way so as to avoid having to adjudicate the debate between Hauser and de Waal, which lack of expertise makes it difficult for me to resolve, and also in order to avoid arguing for a precise definition of grief, which I do have ideas about (see my *Upheavals of Thought*, chap. 1), but don't want to get into in the context of the present argument.

25. Moss, *Elephant Memories*, p. 73.

26. Jeremy Bentham, *Introduction to the Principles of Morals and Legislation* (Oxford: Clarendon Press, 1996), p. 283. For a fascinating recent set of findings, see Marc D. Hauser, M. Keith Chen, Frances Chen, and Emmeline Chuang, "Give Unto Others: Genetically

Unrelated Cotton-Top Tamarin Monkeys Preferentially Give Food to Those Who Altruistically Give Food Back," *Proceedings of the Royal Society, London* 270 (2003): 2363–70.

27. John Stuart Mill, "Whewell on Moral Philosophy," in *Utilitarianism and Other Essays* by John Stuart Mill and Jeremy Bentham (London: Penguin Classics, 1987), p. 252.

28. Here I am summarizing some of the arguments of my *Frontiers of Justice*.

29. See Amartya Sen and Bernard Williams, Introduction to *Utilitarianism and Beyond*, ed. Amartya Sen and Bernard Williams (Cambridge: Cambridge University Press, 1982), pp. 1–22.

30. See my discussion in "Mill between Bentham and Aristotle," *Daedalus* 133 (2004): 60–68.

31. I discuss the question of adaptive preferences in *Women and Human Development: The Capabilities Approach* (New York: Cambridge University Press, 2000).

32. On paternalism, see my *Frontiers of Justice*, pp. 372–80.

33. Christine Korsgaard, "Fellow Creatures: Kantian Ethics and Our Duties to Animals," in *The Tanner Lectures on Human Values*, ed. Grethe B. Petereson, vol. 25/26 (Salt Lake City: University of Utah Press, 2004), pp. 79–110.

34. Korsgaard, "Fellow Creatures," p. 93.

35. I am grateful to a fine paper by Jeremy Bendik-Keymer, "Problems with the Ecological Extension of Dignity" (presented at the Human Development and Capability Association Annual Meeting, New York, September 18–20, 2007), for helpful reflections on the relationship between wonder and justice in my conception.

36. Christine Korsgaard, *The Sources of Normativity* (Cambridge: Cambridge University Press, 1996).

37. Korsgaard, "Fellow Creatures," p. 82.

38. Korsgaard, "Fellow Creatures," p. 82.

39. See the reflections on this question in my *Hiding from Humanity: Disgust, Shame, and the Law* (Princeton, N.J.: Princeton University Press, 2004), and, more recently, in my "Compassion, Human and Animal," forthcoming in a festschrift for Jonathan Glover, ed. Jeffrey McMahan (New York: Oxford University Press).

40. John Rawls, *Political Liberalism*, exp. paper ed. (New York: Columbia University Press, 1996).

41. Korsgaard actually doesn't draw a sharp line between the animal kingdom and the rest of nature, and she considers it an open question whether plants have entitlements connected with their good. My own sensibilities are more Utilitarian than Korsgaard's: I think that sentience is a minimum necessary condition of ethical considerability, but I admit that I do not have a good argument for this position. I do not pursue that difference here.

42. See John Rawls, "The Idea of Public Reason Revisited," in his *The Law of Peoples* (Cambridge, Mass.: Harvard University Press, 1999), pp. 129–80.

43. Here I speak only of the relationship between my view and Korsgaard's; whether, when we add the contributions of Hinduism, Christianity, and all the other reasonable comprehensive doctrines, we would have a single political doctrine in this area is far less clear; on the whole question of whether we can achieve an overlapping consensus in this area, see my *Frontiers of Justice*, pp. 388–92.

44. Rawls, *Political Liberalism*, pp. 169–71.

45. See my *Frontiers of Justice*, pp. 358–62.

46. See my *Frontiers of Justice*, pp. 401–5.

SUGGESTED READING

De Waal, Frans B. M. *Good Natured: The Origins of Right and Wrong in Humans and Other Animals*. Cambridge, Mass.: Harvard University Press, 1996.

————, and Peter L. Tyack, eds. *Animal Social Complexity: Intelligence, Culture, and Individualized Societies*. Cambridge, Mass.: Harvard University Press, 2000.

DeGrazia, David. *Taking Animals Seriously: Mental Life and Moral Status*. Cambridge: Cambridge University Press, 1996.

Hauser, Marc D. *Wild Minds: What Animals Really Think*. New York: Henry Holt, 2001.

Korsgaard, Christine. "Fellow Creatures: Kantian Ethics and Our Duties to Animals." In *The Tanner Lectures on Human Values*, edited by Grethe B. Petereson, vol. 25/26, pp. 79–110. Salt Lake City: University of Utah Press, 2004.

Nussbaum, Martha. *Frontiers of Justice: Disability, Nationality, Species Membership*. Cambridge, Mass.: Harvard University Press, 2006.

————. *Hiding from Humanity: Disgust, Shame, and the Law*. Princeton, N.J.: Princeton University Press, 2004.

————. *Upheavals of Thought: The Intelligence of Emotions*. New York: Cambridge University Press, 2001.

Plotnik, Joshua M., Frans B. M. de Waal, and Diana Reiss. "Self-recognition in an Asian Elephant." *Proceedings of the National Academy of Sciences of the United States* (October 30, 2006), available at http://www.emory.edu/LIVING_LINKS/pdf_attach-ments/PlotniketalPNAS.pdf (accessed August 10, 2009).

Sen, Amartya, and Bernard Williams, eds. *Utilitarianism and Beyond*. Cambridge: Cambridge University Press, 1982.

Sunstein, Cass R., and Martha C. Nussbaum, eds. *Animal Rights: Current Debates and New Directions*. New York: Oxford University Press, 2004.

Sorabji, Richard. *Animal Minds and Human Morals: The Origins of the Western Debate*. Ithaca, N.Y.: Cornell University Press, 1993.

PART III

MORAL STATUS AND PERSON THEORY

CHAPTER 9

THE IDEA OF MORAL STANDING

CHRISTOPHER W. MORRIS

WHAT is the moral status of nonhuman animals?[1] How does that status compare to that of most humans, to human fetuses, or to valuable artifacts or natural objects of different kinds? These questions are familiar ones in ethics and related fields. It is often thought that humans or human persons have a special status, that of *moral standing*, and the question is then whether some nonhumans also share it.

In much of the literature, the concept of moral standing is not as clear or as well characterized as it might be, and there seems to be considerable disagreement about how to understand it. Some even counsel that we dispense with it. In this chapter, I shall examine the idea of moral standing and look at what many contemporary thinkers have said about it. I shall offer an interpretation that I have developed in other writings and that captures important elements common to many accounts. I shall also discuss a way in which moral standing, as I have interpreted it, could be extended to animals in a conventionalist or constructivist account of justice, which is usually thought to be unlikely to extend moral standing to animals. But I have come to think that this understanding of moral standing does not tell us everything that we need to know about the status of animals, and that we need to broaden our understanding of the kinds of status beings can have. I argue that there may be two statuses that merit to be called moral standing. In this respect, I have come to agree with some thinkers who do not share the analysis of moral standing that I have until now adopted.

1. Moral Standing

There are differences of terminology in the literature. Many writers talk only of "moral status," using it narrowly, to mean the special status that some beings such as ourselves might have. Others distinguish between broad and narrow kinds of moral status. And others use "moral status" and "moral standing" interchangeably. One author even claims that "'Standing' is an anachronistic way of saying 'status.'"[2] These terminological differences may be confusing, but they are not as troublesome as the different characterizations of the notions in question.

In this chapter, I distinguish between "moral status" and "moral standing." I regard the latter as being the special status that we and perhaps other beings have, a status that remains to be explicated. I shall think of "moral status" as whatever status something might have morally. In this broad sense, we could sensibly ask about the moral status of a speck of dust and presume that the answer would be that it is insignificant.[3] This particular use of the terms should not distort the discussion, but it is different from that of many contemporary writers, and it will help to remember that in my usage, everything will have moral status but only some things will have moral standing.

Moral standing is a kind of moral status. There are other kinds. In our lives, we make many distinctions between the kinds of things or creatures protected by our moral norms and practices, as well as the ways they are protected. Our concerns range widely over different kinds of object. Consider a story that illustrates the variety of things that may concern us, as well as introducing many of the concepts that we'll need in this chapter.

> The protagonist of this tale is a villain who goes on a rampage. His motives are not important; our attention is on the status of the objects he trashes. He is evidently in a foul mood. As he walks down the street, he kicks a garbage can, denting it and spreading its contents all over the sidewalk. He picks up a discarded metal rod and strikes a small tree, breaking its trunk. He spots an American and a Canadian flag nearby and shreds them. He comes across a group of ladybugs, presumably just hatched, and stomps on them, killing large numbers. Nearby are some pigeons, and he throws some stones at them, injuring one of them quite badly so that it cannot fly. A dog wanders by, and our villain gives it a big kick, also injuring it severely. Next he runs into a young man and punches him, breaking some ribs. He does the same to some small children nearby.
>
> Walking by a hospital, our villain goes in and stumbles on the room in the basement where the bodies of deceased patients are kept; he mutilates some of them. He finds a permanently comatose patient and detaches her from life-support. Our villain then walks into the art museum where he hopes to have a good lunch at the restaurant. His strength restored by a quick but satisfying meal, he goes to the room where the museum's sole Vermeer is displayed. After the guard steps out of the room, he mutilates the masterpiece irreparably with a sharp knife.
>
> Our villain now turns his thoughts to more grandiose projects. He remembers Hitler's wish to destroy Paris in the summer of 1944 ("Is Paris burning?") and starts plotting to destroy the City of Lights. It takes him many

months of planning. His aim is to destroy the city, but not to injure the inhabitants. He manages to contrive a means of evacuating them as the city is burning. Succeeding in his mission of the destruction of Paris, he wonders what projects to take on next—perhaps the destruction of the Grand Canyon.

The details of our tale do not matter much; I will not be interested very much in the villain's motives; it is the objects of his rampage that concern me. They are: a trash can, a small tree, the flags, some ladybugs, a pigeon, a dog, a young man, some small children, some corpses and a permanently comatose woman, a Vermeer, and Paris. I have arranged the story and the objects so that the man's greatest crimes appear to involve valuable objects that presumably lack moral standing, in particular the Vermeer painting and the city of Paris. He did not kill the children and the young man, but the painting and the city he destroyed. I expect that the destruction of the painting and the city will seem to many of us a more serious crime than the injury to the young man or children.

No one should deny that the young man and children in this story have moral standing; many think that the dog does too and perhaps the pigeons. Virtually no one thinks that the garbage can, the tree, the ladybugs, the painting, and Paris have moral standing, however great their value may be. This is compatible with thinking that the villain's denting of the garbage can and destruction of the Vermeer and Paris are wrongs, indeed, in the last two cases, very serious wrongs. Importantly, these objects are not themselves wronged; no wrong is done *to* them. Our duties not to harm these objects are not duties *to them*. We would be concerned if we saw a small child kicking over the garbage can or killing the ladybugs; we would not want our children to do either. Our concern about the second act would be about the malevolence revealed by the destruction of the insects, even if we thought they lacked moral standing. I trust these judgments are familiar and not controversial.

What is to be learned about the status of nonhuman animals? The examples collected in this section are meant to illustrate the variety of things that may be wrongfully harmed and to introduce some of the concepts and distinctions we shall use. It is also to lay out some of the judgments or evaluations we make of acts of this kind, and in particular to point to the ways it can be wrong to harm something lacking moral standing. There is, needless to say, controversy about the status of nonhuman animals. A good number of people think the higher animals have moral standing, much like the small children and the adults in our story. Others deny this assessment. They think the status of animals is more like that of the ladybugs, the garbage can, the flag, the body on life-support, or a valuable artifact. We need to clarify the different types of moral status that entities can have.

2. WHO OR WHAT COUNTS?

Most people think that humans or human persons count in a way that ladybugs and lobsters don't, that inanimate objects are to be protected only insofar as they are valued by us or have certain relations to us. Dirt and bugs usually matter not at all (except extrinsically). Extraordinary works of art, cities like Paris, and natural wonders like the Grand Canyon are to be appreciated and protected. But they don't matter in the way that people do, even if we think them more valuable in some ways. These reactions are commonplace, and I intentionally express them imprecisely. The question is how exactly to express and to understand them. Having moral standing makes a difference. What is that difference?

One finds many different characterizations of moral standing in the literature, some more helpful than others. The most general interpretations sometimes have the virtue of being employable by a number of different moral traditions; the more specific and detailed ones often seem to be unavailable to at least some moral theories. I shall review a number of common characterizations, explaining my own conception later in the section. In many places, where an author eschews the term "moral standing" and talks exclusively of "moral status" I usually signal, as I do in the quotation that immediately follows, that I understand him or her to be talking about what others, such as myself, think of as moral standing.

"To have moral status [i.e., moral standing] is to be the sort of being whose interests must be considered from the moral point of view."[4] Characterizations like this one by Bonnie Steinbock are not uncommon. It has two central elements. The first is the familiar metaphor of "the moral point of view," the second the notion of interests. Insofar as the metaphor has significant content—there are competing "points of view" and morality has one—this characterization will be unavailable to moralists who eschew notions of points of view. If it does not have much content, there may be better ways of expressing the main thought. The second element is that of interests. Many explicate moral standing in terms of interests, and this seems a natural way to go for many contemporary moral thinkers. It is worth remembering, however, that some moral traditions such as Kantianism do not focus on interests, at least at this level of theorizing.

Often the metaphor of counting is employed: if something has moral standing then it "counts morally." This formulation is too broad, so it is usually supplemented: something must count *in its own right*. Wayne Sumner says that "Let us say that a creature has *moral standing* if, for the purpose of moral decisionmaking, it must be counted for something in its own right."[5] Allen Buchanan says "that a being has *moral standing* if it counts morally, in its own right."[6] And Francis Kamm introduces the idea of "Counting morally in their own right" early in her interesting discussion of moral status.[7] It seems clear that some things "count" morally in their own right whereas others do not. But the way in which the former are to "count" or to matter must be made clear.

The language of intrinsic value is often used here as a way of clarifying or specifying the meaning of "counting."[8] This familiar if not altogether clear notion is usually explained by contrast with its pair, instrumental value. Something has instrumental or extrinsic value if it is valuable as a means to something else; something has intrinsic value if it is valuable in itself. Many things have both intrinsic and instrumental value (e.g., a good meal, the company of a friend). It may be that humans or persons have moral standing by virtue of their intrinsic value. But it's not clear that intrinsic value *simpliciter* will explain the notion of moral standing. Vermeer's paintings and Paris certainly are valuable intrinsically, and much that has intrinsic value is not very valuable (e.g., a small pleasure). Francis Kamm notes Christine Korsgaard's claim that the contrast to mere instrumental value is not intrinsic value but having value as an end. Kamm says that being an end in this sense means that the thing's "condition can provide a reason (even if an overridable one) for attitudes or actions independently of other considerations."[9] Perhaps the notion that is needed here is that of an end in this sense.

Many thinkers, including Kamm, use the notion of something mattering "for its own sake." Claire Andre and Manuel Velasquez express the idea thus:

> What is moral standing? An individual has moral standing for us if we believe that it makes a difference, morally, how that individual is treated, apart from the effects it has on others. That is, an individual has moral standing for us if, when making moral decisions, we feel we ought to take that individual's welfare into account *for the individual's own sake* and not merely for our benefit or someone else's benefit.[10]

There is something right about this idea of *for something's own sake*. It is similar to "in its own right," introduced above. Kamm argues that counting in its own right may not be the notion needed to explicate the kind of moral status we are concerned with in these contexts. I claimed above that great works of art or cities have intrinsic value, but most people do not think they have moral standing, even if destroying them is a great or serious wrong. Kamm also thinks that "counting in its own right" may apply to works of art or natural objects. The distinctions she makes are important:

> A work of art or a tree may count in its own right in the sense that it gives us reason to constrain our behavior toward it (for example, not destroy it) just because that would preserve the entity. That is, independently of valuing and seeking the pleasure or enlightenment it can cause in people, a thing of aesthetic value gives us (I think) reason not to destroy it. In that sense, it counts morally. But this is still to be distinguished from constraining ourselves *for the sake of* the work or art or the tree. I do not act for its sake when I save a work of art, because I do not think of its good and how continuing existence would be good for it when I save it. (Nor do I think of its exercising its capacities or performing its duties. Acting for the sake of these might also involve acting for an entity's sake, though it need not involve seeking what is good for it.) Rather, I think of the good *of* the work of art, its worth as an art object, when I save it for no other reason than that it will continue to exist.
>
> By contrast, when I save a bird, I can do it for its sake, because it will get something out of continuing to exist, and it could be a harm not to continue.[11]

Kamm makes an interesting distinction between something counting in its own right (e.g., a work of art, a tree) and acting for the sake of something (e.g., a bird, a person). In both cases, we have reasons, she thinks, to act in certain ways, for instance, not to destroy the thing. But only the second notion brings out what is distinctive about what she calls the narrower sense of moral status (i.e., moral standing). She says:

> So, we see that within the class of entities that count in their own right, there are those entities that *in their own right and for their own sake* could give us reason so act. I think that it is this that people have in mind when they ordinarily attribute moral status [i.e., moral standing] to an entity.…I shall say that *an entity has moral status when, in its own right and for its own sake, it can give us reason to do things such as not destroy it or help it.*[12]

This is a helpful characterization of moral standing. It is not mine, but it may be better in some important respects than the one I have previously developed, as I shall argue in the last section of this chapter.

Kamm goes on to say that "we can have duties to behave in certain ways toward entities that count in their own right and, as a subset, to entities that have moral status [i.e., moral standing]. But this still does not imply that all of these are entities *to which we owe* it to behave in certain ways."[13] A duty *owed to* a being is a "directed duty"—what I call a "direct duty"—and it typically is correlative to a claim-right held by the being. Importantly, "there is a difference between doing the wrong thing…and *wronging* some entity in failing to perform the duty owed to her." There is a distinction thus to be made within the category of beings that have moral status in Kamm's narrow sense (which I have been interpreting as moral standing) between those who are benefited or protected by duties and those to whom duties are owed. "So just as only some entities that count in their own right are entities which have moral status (as I defined it) [i.e., moral standing], so it may be that only entities that have moral status are owed things and have rights against us." This too seems right, and I shall return to the point in the last section.

The distinction between duties and duties *to* will figure centrally in my interpretation of moral standing. Many explications do not privilege the concept of a duty, as we saw, and some that do still don't distinguish between duties and duties *to*. A good example of the latter is David DeGrazia's characterization of moral status (i.e., moral standing): "To say that X has moral status is to say that (1) moral agents have obligations regarding X, (2) X has interests, and (3) the obligations are based on X's interests."[14] This formulation may express how many moral thinkers wish to understand moral standing. But I think it is important to distinguish between duties and duties to, as this distinction may be central to understanding the status that many animals have.

Duties To and Moral Standing

The idea of considerations or duties *owed to* someone has figured centrally in many accounts of moral standing, including my own. And it is to this kind of account that I turn now. In a famous early essay on abortion, Mary Ann Warren says that "the question which we must answer to produce a satisfactory solution to the problem of the moral status of abortion is: How are we define the moral community…?" She immediately adds "the set of beings with full and equal moral rights."[15] The idea or metaphor that interests me is that of the moral community, one she seems to abandon in her later work.[16] We can take it as an interpretation of moral standing:

something has moral standing if it is a member of the moral community. This is the interpretation many have given of this notion,[17] and Warren's explication of the idea—"the set of beings with full and equal moral rights"—begs too many questions to be the right one. It equates moral standing with the status that persons have, and more controversially, it assumes that this status gives all its bearers a set of several important rights, including "the inalienable rights to life, liberty, and the pursuit of happiness." A weaker and less-question-begging interpretation is that a creature with moral standing is owed some types of moral duties or obligations. In her later work on moral status, Warren says that "to have moral status is to be morally considerable, or to have moral standing. It is to be an entity towards which moral agents have, or can have, moral obligations."[18] The thought here is that something is *owed to* another. This has long seemed to me a useful way of expressing what it is to count in the contexts where the question of moral standing is central. When raising certain questions about animals or about human fetuses, the brain dead, the irreversibly comatose, and so on, we often wish to know what we owe them, if anything.

Warren's metaphor of the moral *community* is an interesting one. It makes possession of moral standing analogous to the political status of citizenship. Like membership in the political community, membership in "the moral community" gives one a particular status that non-members lack, in particular a set of rights. This understanding of moral standing connects it with the notion of legal standing, which may well have been the source of the notion of moral standing in philosophy.[19] Legal standing is normally understood as the status needed to be entitled to have a matter settled by courts of law. It is the idea of a status which entitles the holder to something. Thus someone who was not a "member of the legal community," that is, was not a citizen, would not normally have legal standing. In his book on moral considerability, Mark Bernstein introduces the concept of "moral enfranchisement," which nicely extends this political analogy.[20] These political notions are helpful for making central the duties that are owed to others. They are also helpful for conceptualizing the complex relations of moral standing in theories that relativize moral norms to social communities smaller than the set of persons or humans.[21]

I have for some time thought of moral standing in this way, as involving considerations owed to other beings in the form of obligations, duties, and the like. It is not an uncommon understanding, I think. I shall now briefly describe this understanding of moral standing.

The basic idea is simple: something has moral standing if it is owed duties.[22] This understanding of moral standing connects it with the notion of legal standing; both are conceptions of a status that entitles the holder to something. The distinction between duties and duties to is helpful, and I shall now bring in the well-known distinction between duties *to* a being and duties *regarding* a being. In the case of duties regarding a being, the latter being is the beneficiary of a direct duty owed not to it, but rather owed to another being. Kant uses one form of this distinction to express his view of the ethics of animals: we can have only duties to humans, and protections for animals are always duties regarding them, owed to humans (for instance, oneself).[23]

I can now construct an important distinction between direct and indirect moral objects. If something itself has moral standing and consequently is owed duties, it is a *direct moral object*. By contrast, if something is protected solely by considerations owed to others, then it is a mere *indirect moral object*. We are both direct and indirect moral objects most of the time. But some things—presumably artifacts and possibly some or all animals—may only be indirect moral objects. Mere indirect moral objects might have great value, even moral value. Vermeer's paintings, Paris, and the Grand Canyon have great value, and arguably some of it is moral. I do not wish to make much of this last claim, as I am not entirely clear what distinguishes moral value from other kinds of value. But it is clear that someone who destroyed these structures would be committing wrongs. Our villain's destruction of Paris would be recorded among the great crimes of history.

An important feature of the debate about the moral status of animals, on the interpretation of moral standing I have offered, seems to focus on whether some of them are direct or mere indirect moral objects. If their status is the latter, that of a mere indirect moral object, then we have duties *regarding* them and no duties *to* them. This was Kant's view, as well as that of St. Thomas.[24] An indirect moral object is not wronged by someone who fails to observe a duty regarding it. For St. Thomas and Kant, we cannot wrong animals. They lack moral standing.

Very few people doubt that we have some duties *regarding* nonhuman animals, but there is a major controversy as to whether we have any direct duties *to* any animals. Many, like St. Thomas and Kant, conclude that these animals are merely indirect moral objects. In the last section I shall argue that the distinction between direct and indirect moral objects is misleading and not exhaustive.

3. WHAT HAS MORAL STANDING?

Many controversies about the ethical treatment of animals concern the matter of moral standing, in particular, the status of different species of animals. As we have just seen, some writers think that no nonhuman animal has moral standing; others think that many do, especially those that are sentient and minimally rational. Different moral traditions or thinkers point to different attributes or properties as giving their possessors moral standing. Adult humans typically are sentient, rational, self-conscious agents, and these properties figure in many accounts of moral standing. Many animals are sentient and exhibit some elements of rationality, consciousness, and agency, though not all of those possessed by most adult humans. While many animals have sophisticated means of communication, they seem to lack what we think of as the ability to use a language, which includes the capacity to combine words to form an infinite number of grammatical sentences. So it is natural for many to think that the chief property that gives them moral standing is their *sentience.*

This is what Jeremy Bentham, perhaps the founder of the modern animal liberation movement, thought:

> The day may come when the rest of the animal creation may acquire those rights which never could have been witholden from them but by the hand of tyranny. The French have already discovered that the blackness of the skin is no reason a human being should be abandoned without redress to the caprice of a tormentor. It may one day come to be recognised that the number of the legs, the villosity of the skin, or the termination of the os sacrum are reasons equally insufficient for abandoning a sensitive being to the same fate. What else is it that should trace the insuperable line? Is it the faculty of reason or perhaps the faculty of discourse? But a full-grown horse or dog, is beyond comparison a more rational, as well as a more conversable animal, than an infant of a day or a week or even a month, old. But suppose the case were otherwise, what would it avail? the question is not, Can they reason? nor, Can they talk? but, Can they suffer?[25]

To this day, sentience is the natural and the central criterion for those who think that (some) animals are direct moral objects. Utilitarians in particular favor this criterion,[26] but the claim is one that some non-consequentialists also make. Many thinkers, however, reject sentience as a criterion of moral standing, and they have formulated different criteria. I shall review a number of criteria now in order to display the range of positions commonly held on the question of who or what has moral standing. A basic criterion of moral standing looks something like this:

> Something has moral standing if (and only if?) it _____.

Two things distinguish different theories and their criteria: what is placed in the blank space (e.g., is sentient; is rational; knows a language; is an agent), and whether possession of the attribute is sufficient for moral standing or *both* necessary *and* sufficient. A weaker formulation would assert that the property put in the blank space is merely sufficient for moral standing: something has moral standing *if* it _____. The stronger formulation would say: something has moral standing *if and only* if it _____.

Here are three widely invoked criteria, familiar to many readers of this volume:

(1) Something has moral standing if it is sentient.
(2) Something has moral standing if it is rational.
(3) Something has moral standing if it is an agent.

The stronger versions of these criteria would be:

(1′) Something has moral standing if and only if it is sentient.
(2′) Something has moral standing if and only if it is rational.
(3′) Something has moral standing if and only if it is an agent.

Stronger still is use of these criteria to create the following account:

> Something has moral standing if and only if it is (1) sentient, (2) rational, and (3) an agent. All seven versions constitute substantially different accounts

of moral standing, and no one of these accounts can be reduced to any one of the other six.

Degrees of Moral Standing

These criteria could each be formulated so that moral standing is a matter of degree, in which case we might want to say that something has moral standing *insofar as* or *to the extent that* it is _____ (e.g., sentient, rational, or an agent). My interpretation of moral standing considers something to have moral standing if it is owed any moral consideration or duty whatsoever; so I have not thought it necessary to interpret criteria as allowing of degrees. However, there are two ways in which we might want to take into account degrees of moral standing. The first would be to construct the very notion of moral standing to allow for degrees. Two beings, for instance, might have different degrees of moral standing insofar as one is owed many more duties (and as having many more correlative rights) than the other. This is a common way of talking about moral standing; humans might be said to have "full moral standing" and some animals something less. My formulation of moral standing does not allow of degrees in this way; if something is owed at least one duty, it has moral standing. This has the virtue of simplicity, as well as that of distinguishing moral standing from the matter of how many duties something is owed and the difference this makes.

Another matter of degree is harder to know how to factor into an account of moral standing. The concepts of sentience, rationality, consciousness, language, and agency are complex and allow of degrees. Consider that the rationality or agency of a normal adult human is more developed than that of a child; the rational capacities of some animals are more advanced than others; the rational capacities of some exceed those of some humans.[27] How to take into account the differences between beings who possess the requisite property or properties (e.g., sentience, rationality) but to different degrees? This is a difficult question, and I do not wish to take it up here. I think it best, at least initially, to keep the notion of moral standing simple. It is not clear that the notion needs to do more than one job at a time, the job here being to designate the set of beings to whom some duties are owed. I suggest we assume there is a threshold level of sentience, rationality, and so on that is sufficient to give its bearer moral standing and not bring in this second kind of matter of degrees, though I recognize that it is not clear how to do this.[28]

Some thinkers prefer to deploy a cluster concept such as *person*, instead of using criterion 2 or 3. Persons are usually thought to be complex rational, self-conscious, language-speaking agents like us. The notion is hard to characterize except in terms of a cluster of attributes.[29] So we might have a personhood criterion:

(4)   Something has moral standing if it is a person.
(4′) Something has moral standing if and only if it is a person.

Kant may be interpreted as holding (4′), as did Mary Ann Warren in her early essay on abortion.

All of these criteria make possession of moral standing depend on possession of certain properties or attributes. Creatures who merely *potentially* possess these attributes—that is, they will come to possess them fully in the normal course of biological development—will lack moral standing according to criteria 1′–4′. Consequently, some favor a criterion of moral standing that will allow potential as well as actual persons to have moral standing:

(5) Something has moral standing if it is an actual or potential person.

(5′) Something has moral standing if and only if it is an actual or potential person.

Potentiality is usually connected to personhood (or humanity in a sense different from species membership discussed below). But conceptually, it is open to someone to favor a criterion of potential sentience or something else. Potentiality criteria are important mainly for the abortion debate, so I will not dwell on them.[30]

Others point instead to the importance of being a member of a kind of being such as the human species. Our moral debates and quarrels frequently make reference to the importance of humanity—for example, the abolitionist pointing to the humanity of the slave. So some favor a humanity criterion of moral standing, where species membership is central:

(6) Something has moral standing if it is human (i.e., a member of the species *homo sapiens*).

(6′) Something has moral standing if and only if it is human (i.e., a member of the species *homo sapiens*).

The humanity criterion is often criticized by pointing to humans who are not rational, self-conscious agents of the familiar sort, not even potentially. So some are tempted to modify it or to adopt another, which we may call the natural kind criterion:

(7) Something has moral standing if it is a member of a natural kind, whose mature members are normally persons.

(7′) Something has moral standing if and only if it is a member of a natural kind, whose mature members are normally persons.[31]

The thought here is that members of our species are special because they characteristically become persons and that persons have special value or significance. Even if particular humans are not and will never be persons, they have moral standing as members of this kind of being. The extension of this criterion, as far as we know, is the set of humans. One day other creatures may be added.

Agents and Cooperators

Many of these criteria will generally be familiar to readers of this volume. I now return to the criterion of personhood, and specifically to the matter of agency, which thus far has received scant attention. Why might agency be important for moral standing? In many familiar accounts, agents are owed a particular kind of respect, and in some accounts, agency is important because only agents can reciprocate. Why might reciprocation be important? There are a number of views, and I wish to mention one that gives rise to an interesting conception of moral standing.

On some views of morality, moral constraints do not make sense unless they are likely to be reciprocated in some way. This is not an unusual view of many legal constraints, and some theorists think it true of moral constraints as well. Suppose we think of large parts of morality as "a cooperative venture for mutual advantage," to adapt Rawls' famous phrase.[32] Cooperative ventures here are to be understood as mutually beneficial arrangements or practices requiring that participants share in the burdens of the endeavor. Hume's conventionalist account of justice is a good model of this position. In this view, agents are not constrained by justice toward others unwilling to reciprocate or cooperate. As Hume put it, suppose "that it should be a virtuous man's fate to fall into the society of ruffians, remote from the protection of laws and government.... His particular regard to justice being no longer of use to his own safety or that of others, he must consult the dictates of self-preservation alone, without concern for those who no longer merit his care and attention."[33] Assuming my conception of moral standing—a being has moral standing if it is owed duties—then an account of justice like this one would endorse another criterion of moral standing:

(8′) Something has moral standing if and only if it is an agent and willing to cooperate.

This criterion is not going to grant moral standing to animals, not even to young children. These implications are often cited by critics of the conventionalist or contractarian ethical tradition, and they seem to them sufficient to condemn this type of account.[34] But more needs to be said here.

Of the criteria previously mentioned, only the first, the sentience criterion, promises to give a clear and unrestricted moral standing to sentient, nonhuman animals. Defenders of other criteria usually conclude that animals don't have moral standing and are protected only by being indirect moral objects. Sometimes they try to show that some animals have enough of the relevant properties to have moral standing, and this may require the introduction of degrees of moral standing. Interestingly, there is a way for defenders of conventionalist accounts of justice such as Hume's to extend moral standing to some animals. They will need first to relax 8′, as follows:

(8) Something has moral standing if it is an agent and willing to cooperate.

The second step will then be to show how other beings can acquire moral standing.

Suppose that norms of justice are mutually beneficial conventions. This is, broadly speaking, a Hobbist and Humean idea, and it has been developed in contemporary moral theory in different ways.[35] Versions of the argument to follow can, I believe, be given for Rawls' or Scanlon's contractualist theories.[36] Now, given characteristic human attachments, it will be the case that few adult agents will cooperate with others unless their children or those to whom they are personally close are accorded some protections. In particular, agents may insist that others accord their children and some others moral standing and not merely the status of an indirect moral object.[37] Elsewhere I call this "secondary moral standing," an unfortunately misleading label insofar as it suggests that the children are indirect moral objects. What is meant by the expression is a second way of acquiring moral standing in this conventionalist framework. The idea is that something (e.g., children) can acquire (genuine) moral standing by being favored by other agents with moral standing.[38]

In principle many nonhuman animals could acquire moral standing in this way.[39] Domestic animals like dogs might so acquire moral standing. With the development of human attachments, other animals might be included. The argument sketched here and developed elsewhere challenges an element of Peter Carruthers' contractualist account of the reasons why animals lack moral standing.[40]

4. Two Kinds of Moral Standing?

My account of moral standing focuses on obligations owed *to* a being, such that *it* is wronged when these obligations are disregarded. We might call this and accounts like it the *juridical* interpretation of moral standing, not because it makes reference to law, but because it privileges the virtue of justice. This is justice in the broad sense, having to do with what is owed to others, that to which they have a right. In this broad sense, justice "covers all those things owed to other people: it is under injustice that murder, theft and lying come, as well as the withholding of what is owed for instance by parents to children and by children to parents, as well as the dealings which would be called unjust in everyday speech."[41]

Justice in this sense, concerning that which is owed to others, is a central moral virtue, if not the principal one. Creatures that are protected in this direct way by justice do have a status different from other things. One might still worry that something is missing from the juridical interpretation of moral standing. One need only to remember the virtue of charity or benevolence, the other central interpersonal virtue. This virtue attaches itself not to what others have a right to, but to their good. We act charitably or beneficently when we help someone in need when this is not required by justice. The same act may be contrary to both virtues, to justice and to benevolence, as when someone is deprived of what is his due and what he desperately needs.[42] Normally the question of charity or benevolence arises only when there is a question of helping someone in ways not required by justice. It might be that charity or benevolence is shown in our kind or considerate treatment of animals when this is not required by justice. If charity or benevolence is an important virtue, as important as many moral thinkers hold it to be (e.g., Aquinas, Hutcheson, Hume), then my juridical account of moral standing seems to leave something out.

My thought is that there are two important kinds of moral standing and that the juridical account must be supplemented by a second one. I shall motivate this second kind of moral standing by first examining Francis Kamm's account, then by looking at a feature of Hume's moral theory, and lastly with some remarks about duties of charity or benevolence.

I turn first to Kamm's discussion of "moral status in the narrow sense," which I have interpreted as moral standing. She argues that "within the class of entities that count in their own right, there are those entities that *in their own right and for their own sake* could give us reasons to act. I think that it is this that people have in mind when they ordinarily attribute moral status [i.e., moral standing] to an entity."[43] She notes that moral status in this sense does not entail that something is owed to a being or that depriving it of that treatment wrongs it. "We can have duties to behave in certain ways toward entities that count in their own right and, as a subset, to entities that have moral status [i.e., moral standing]. But this still does not imply that all of these are entities *to which we owe* it to behave in certain ways."[44] Kamm suggests that entities to whom we owe certain things may "have a *higher* moral status than other entities which also have moral status (as I defined it)."[45]

Kamm's view is that there are several kinds of moral status. The broadest sense of "moral status" is that "defined as what it is morally permissible or impermissible to do to some entity."[46] A narrower sense of moral status comes in when something "counts morally in its own right." Kamm interprets this sense as involving something having value as an end, but this is not yet what "people have in mind when they ordinarily attribute moral status to an entity." This moral status—what I have interpreted as moral standing—is to be attributed only to a proper subset of the set of entities that count in their own right, namely those which also for their own sake give us reasons to act in certain ways. In addition, a proper subset of this last class will be entities *to whom* we owe certain treatment.

We can illustrate Kamm's complex view with examples taken from her essay:

(1) something to which it is permissible to do anything, for instance, a rock;
(2) something which counts in its own right, for instance, a work of art or a tree;
(3) something which in its own right and for its own sake gives us reasons to act, for instance, a bird;
(4) something owed certain treatment, for instance, a person.

The first is said to have moral status in the broad sense. The second introduces a narrower sense. The third has moral status of the sort "people have in mind when they ordinarily attribute moral status to an entity," what I have called moral standing. And the fourth, she thinks, has a higher moral status than the third.

Comparing Kamm's account with my juridical interpretation of moral standing, they differ in their attributions of moral standing to 3 and 4 above:

I now think that Kamm, as I have interpreted her, is right to think that there are two kinds of moral standing. When one saves a bird or a dog from harm, one does

property	example	moral standing (Kamm)	moral standing (juridical acct)
counting in its own right	work of art, tree	no	no
+ for its own sake	bird	yes	no
+ is owed duties	person	yes	yes

so "for its own sake," and this is a consideration which is what "people have in mind when they ordinarily attribute moral status [i.e., moral standing] to an entity."

Let me motivate this view, that there are two kinds of moral standing, by drawing attention to an important feature of Hume's moral theory. He thought that we have no obligations of justice to animals. His account of justice is conventionalist, much like Hobbes's and some contemporary theories. Nevertheless, Hume notes, "we should be bound, by the laws of humanity, to give gentle usage to these creatures, but should not, properly speaking, lie under any restraint of justice with regard to them....Our compassion and kindness [is] the only check, by which they curb our lawless will."[47] Justice for Hume is not the only source of moral considerations. There is "humanity," as well as benevolence, which are important for Hume, unlike thinkers like Hobbes.

Hume does not talk about moral status or moral standing, but we can see in his kind of moral theory a motivation for eschewing a purely juridical interpretation of moral standing and introducing an additional kind of moral standing. If benevolence or "humanity" is a sufficiently central and even demanding virtue, why do not animals thereby have a status which is a kind of moral standing? They won't if we think of moral standing juridically, as privileging justice, but we can also think of moral standing in relation to benevolence. There is nothing in the meaning of "moral standing," a constructed notion, expressed by a technical term, to determine which way to go. Much depends on the importance of benevolence or other virtues or values, and these receive different interpretations in different frameworks. Many traditions will understand the virtues of charity and benevolence to be much more important than they are often thought to be by contemporary moral theorists, and these accounts may have room for a second kind of moral standing.

I might note here that a count against my juridical interpretation is that it is not neutral; it begs the question against some moral traditions. Moral consequentialists, for instance, have little use for this interpretation of moral standing. They understand the relation of *being owed* something, much like that of a right or any other deontic relation, as derivative from the object of morality (maximum good). On this view, someone is owed moral considerations when instituting duties (or rights) will increase the realization of greater good. The utilitarian understanding of moral standing might be expressed thus: something has moral standing if its interests are to be counted by the fundamental principle of morality. I do not wish to make this particular formulation (something has moral standing if its interests are to be counted by _____) canonical; some moral traditions don't have fundamental principles, much less privilege interests. It may not be an important objection to an interpretation of moral standing that it is not theory-neutral. Much depends on what we wish the concept to do, and it may be that it cannot do much without being embedded in a particular understanding of morality.

Some have recently counseled that we abandon the notion. James Rachels understands the concept the way I have: "Theories of moral standing try to answer the question: To whom do we have direct duties?"[48] And he thinks that these different theories "all assume that the answer to the question of how an individual may be treated depends on whether the individual qualifies for a general sort of status, which in turn depends on whether the individual possesses a few general characteristics. But no answer of this form can be correct."[49] Rachels thinks this is the wrong way to think about the question of how to treat different sorts of beings. "There is no characteristic, or set of characteristics, that sets some creatures apart from others as meriting respectful treatment....Instead we have an array of characteristics and an array of treatments, with each characteristic relevant to justifying some types of treatment but not others."[50] We should abandon talk of moral standing, he thinks, and just say, for instance, that the fact that a certain act would cause pain to a creature is a reason not to do it.[51]

I think that the notion of moral standing is still useful in moral philosophy. One use is to clarify the different kinds of status a being can have. But I am sympathetic to Rachels' dissatisfaction with the concept and to his thought that we should turn our attentions instead to the characteristics of beings that are reasons for our acting in certain ways. And here I think that the fact that the nature of some animals is a reason to act charitably or beneficently towards them is of some significance insofar as we regard the virtue of charity or benevolence as important. This fact about animals may make us think of them as having a moral status different from other beings such as trees or objects such as great paintings, as having a kind of moral standing, even if they are owed no duties.

I said earlier that the distinction I drew between direct and indirect moral objects may mislead. It does so insofar as it suggests that the two classes exhaust the possibilities. Some may assume that if an animal is not a direct moral object it must be an indirect one. The thought would be that if we do not owe duties to animals then the only way they could be protected by duties would be if they were third-party (or indirect) beneficiaries as a result of direct duties owed to humans or persons. Now it seems that animals could fail to be direct moral objects and still be the proper objects of charitable or beneficent acts. If there are, as I have been suggesting, two kinds of moral standing, animals who are the proper objects of other-regarding virtues different from justice could have a kind of moral standing. The distinction that I drew between direct and indirect moral objects may suggest a false dilemma.

I shall now briefly sketch a way in which animals would be thought to have this second kind of moral standing. Suppose that we have duties of charity or benevolence (or a similar other-directed virtue). This assumption is not uncontroversial, but it will simplify the case for understanding the beneficent treatment of animals as reflecting their possession of a kind of moral standing. Some consider charity or benevolence supererogatory. Much good done for others certainly is over and above the call of duty. But some isn't over and above the call of duty. It is not uncommon to think that we have duties of charity or benevolence; many give to the poor thinking they have a duty to do so. If one thinks that the recipient of such aid has a claim-right to it, then one has in effect understood these as duties of justice in the broad sense introduced earlier. Many positive duties—duties of service, as opposed to non-interference—are in fact duties of justice, for instance, duties to rescue such as the one we have on the high seas. But some duties to help others are not connected to correlative rights. As I said, many who give away part of their income or who volunteer certain services for the poor do so thinking they have a duty to help others, believing that it would be wrong for them not to do so.

Charity and benevolence seem to require us to help others in a number of ways. Duties to give to the poor may be imperfect duties, the choice of when and where to act being up to the agent.[52] So we might fulfill our duty to help by giving to some but not all needy people near us or elsewhere. But sometimes charity or benevolence seems to require us to refrain from doing something whenever we can or whenever we encounter severe suffering. It would be wrong to do something even though desisting from doing it is not owed to another.

As we saw, Hume thinks us constrained "by the laws of humanity to give gentle usage to these creatures." We think ill of or condemn someone who fails to act as he or she is "bound, by the laws of humanity," but we do not praise them for acting as they should if their acts were supererogatory. It is wrong not to refrain from inflicting certain harms on animals, we often think, even if they have no right against us that we desist.

The framework I deploy here may be controversial, and my appeal to a common opinion will not be shared by all. But suppose what I have said is right, that we have duties of charity or benevolence to help animals, that it is wrong to do certain things to them even if it does not wrong them. Benevolence thus requires us to act in certain ways, even if some beneficent actions are over and above the call of duty. If this is right, then it seems that the status of those animals we have duties to help, even if these duties are not owed to them, is a kind of moral standing. We act wrongly in failing to keep these duties, and this wrong is not to be explicated in terms of duties owed to others. Charity and benevolence require us to help some animals for their own sake and not for that of another.

This seems to be, to use Kamm's words, what "people have in mind when they ordinarily attribute moral status [i.e., moral standing] to an entity." If some animals are objects, as I have suggested, of our charity or benevolence, "in their own right and for their own sake," then they are objects of a possible wrong even if they cannot be wronged. Their status seems to merit being called moral standing. If the use of the term here worries some, then I can resort to another term, for instance Kamm's language of the different kinds of moral status. But I think that would be disingenuous. One of two kinds of moral status on her view seems to be what "people have in mind when they ordinarily attribute moral status to an entity." This is what many have called moral standing.

I do not know if supplementing the juridical account with a second kind of moral standing, as I have recommended, will resolve many substantial questions in practical ethics. My first order of business has been to clarify what we have been talking about the past several decades in our controversies about moral status and standing. As I said in regard to Rachels' challenge to the notion of moral standing, clarifying the different kinds of status beings can have makes something clearer than it was before, and I think it is helpful to distinguish the two kinds of moral standing identified in this chapter.[53]

NOTES

1. I am very grateful to Tom Beauchamp, Andrew I. Cohen, and Benjamin Sachs for helpful written comments on drafts of this chapter.

2. Jeremy Bendik-Keymer, review of James Rachels, *The Legacy of Socrates: Essays in Moral Philosophy*, *Ethics* 117 (2007): 781.

3. My distinction is the one Wayne Sumner made some years ago: "Every physical object has some status or other; it makes no more sense to say that a thing lacks moral status than to say it lacks shape or color.... To count for nothing is to have *no moral standing.*" L. W. Sumner, *Abortion and Moral Theory* (Princeton, N.J.: Princeton University Press, 1981), p. 26.

4. Bonnie Steinbock, *Life Before Birth: the Moral and Legal Status of Embryos and Fetuses* (New York: Oxford University Press, 1982), p. 9.

5. L. W. Sumner, "A Third Way," in *The Problem of Abortion*, 3rd ed., ed. Susan Dwyer and Joel Feinberg (Belmont, Calif.: Wadsworth Press, 1997), p. 99. See also his *Abortion and Moral Theory*, pp. 26–29, 195–200.

6. Allen Buchanan, "Moral Status and Human Enhancement," *Philosophy & Public Affairs* 37 (2009): 346. Buchanan notes that "The terms 'moral status' and 'moral standing' are sometimes used interchangeably," and he wishes to distinguish them.

7. Francis Kamm, "Moral Status," in *Intricate Ethics: Rights, Responsibilities, and Permissible Harm* (New York: Oxford University Press, 2006), pp. 227–28.

8. "A being has moral standing if it or its interests matter intrinsically, to at least some degree, in the moral assessment of actions and events." Agnieszka Jaworska, "Caring and Full Moral Standing," *Ethics* 117 (2007): 460.

9. Kamm, "Moral Status," 228. The reference to Christine Korsgaard is to her "Two Distinctions in Goodness," *Philosophical Review* 40 (2004): pp. 1367–86.

10. Claire Andre and Manuel Velasquez, "Who Counts?" available at http://www.scu.edu/ethics/publications/iie/v4n1/counts.html (emphasis added).

11. Kamm, "Moral Status," pp. 228–29.

12. Kamm, "Moral Status," p. 229.

13. Kamm, "Moral Status," p. 230.

14. David DeGrazia, "Moral Status As a Matter of Degree?" *Southern Journal of Philosophy* 46 (2008): 183, italics omitted.

15. Mary Ann Warren, "On the Moral and Legal Status of Abortion," *Monist* 57 (1973), reprinted in Dwyer and Feinberg, *Problem of Abortion*, 165.

16. Mary Ann Warren, *Moral Status: Obligations to Persons and Other Living Things* (Oxford: Clarendon Press, 1997).

17. "The question of who or what has moral standing, of who or what is a member of the moral community, has received wide exposure in recent years." R. G. Frey, "Moral Standing, the Value of Lives, and Speciesism," *Between the Species* 4 (1988): 191.

18. Warren, *Moral Status*, p. 3. She emphasizes that an "important feature of the concept of moral status [i.e., moral standing] is that the moral obligations that are implied by the ascription of moral status to an entity are obligations *to that entity*. To violate an obligation arising from A's moral status [i.e., moral standing] is to wrong A, and not merely a third party" (p. 10).

19. Christopher D. Stone, "Should Trees Have Standing? Toward Legal Rights for Natural Objects," reprinted in *Should Trees Have Standing? Law, Morality, and the Environment*, 3rd ed. (New York: Oxford University Press, 2010).

20. Mark H. Bernstein, *On Moral Considerability: An Essay on Who Morally Matters* (New York: Oxford University Press, 1998), p. 20.

21. Warren's "multi-criterial" conception of moral standing may have some relativistic implications insofar as the criteria include "certain relational properties, which sometimes include being part of a particular social or biological community," *Moral Status*, p. 21. Gilbert Harman's conventionalist moral theory has very clear relativist implications. See the essays reprinted in Part I of his *Explaining Value and Other Essays in Moral Philosophy* (Oxford: Clarendon Press, 2000). For a treatment of some questions about the justification of punishment and moral standing, see my "Punishment and Loss of Moral Standing," *Canadian Journal of Philosophy* 21 (1991): 53–79.

22. My ideas about moral standing and their implications in a contractarian moral framework have been developed in a number of essays. See especially my "Moral Standing and Rational-Choice Contractarianism," in *Contractarianism and Rational Choice: Essays on David Gauthier's* MORALS BY AGREEMENT, ed. Peter Vallentyne (Cambridge: Cambridge University Press, 1991), pp. 76–95, and "A Contractarian Account of Moral Justification," in *Moral Knowledge? New Readings in Moral Epistemology*, ed. Walter Sinnott-Armstrong and Mark Timmons (New York: Oxford University Press, 1996), pp. 215–42.

23. Immanuel Kant, *The Metaphysics of Morals*, trans. and ed. Mary Gregor (Cambridge: Cambridge University Press, 1996 [1797]), pp. 192–93.

24. Saint Thomas, *Summa contra Gentiles*, ed. Joseph Kenny (New York: Hanover House, 1955–57), Bk. III, ch. 112.

25. Jeremy Bentham, *An Introduction to the Principles of Morals and Legislation* (Oxford: Clarendon Press, 1907 [1823]), ch. 17, note 122.

26. See Sumner, *Abortion and Moral Theory*.

27. See Peter Singer, "Speciesism and Moral Status," *Metaphilosophy* 40 (2009): 567–81.

28. See the chapters in this *Handbook* by Bryce Huebner and Mark Rowlands that directly or indirectly address such threshold problems. For a good discussion of the question of matter of degrees, see David DeGrazia, "Moral Status as a Matter of Degree?"

29. For characterizations of a person, see Warren, "On the Moral and Legal Status of Abortion" and Daniel Dennett, "Conditions of Personhood," reprinted in *Brainstorms* (Cambridge, Mass.: MIT Press, 1978), pp. 269–97. See also Sarah Chan and John Harris's contribution to this *Handbook*.

30. For an interesting discussion of potentiality, as well as the matter of degrees of moral standing, see Elizabeth Harman, "The Potentiality Problem," *Philosophical Studies* 114 (2003): 173–98.

31. Alan Donagan, *The Theory of Morality* (Chicago: University of Chicago Press, 1977), p. 171; Sumner, *Abortion and Moral Theory*, pp. 96–99.

32. John Rawls, *A Theory of Justice* (Cambridge, Mass.: Harvard University Press, 1971), p. 4. Rawls, of course, did not intend this characterization of a society, much less a morality, to be interpreted as above. The interpretation in question is Hobbist and Humean (on justice). This type of account is developed by a number of contemporary thinkers, in particular David Gauthier and Gilbert Harman. For the first, see *Morals by Agreement* (Oxford: Clarendon, 1986); for Harman, see the references in note 21.

33. David Hume, *An Enquiry concerning the Principles of Morals*, ed. Tom L. Beauchamp (Oxford: Clarendon Press, 1998), 3.10 (sect. 3, part 1, para. 10), pp. 15–16.

34. See, for instance, the recent criticisms of moral contractarianism in Martha C. Nussbaum, *Frontiers of Justice: Disability, Nationality, and Species Membership* (Cambridge, Mass.: Harvard University Press, 2006). See also my "Justice, Reasons, and Moral Standing," in *Rational Commitment and Social Justice: Essays for Gregory Kavka*, ed. Jules L. Coleman and Christopher W. Morris (Cambridge: Cambridge University Press, 1998), pp. 186–207.

35. See the references in note 32. The argument I sketch here is developed in my "Moral Standing and Rational-Choice Contractarianism."

36. T. M. Scanlon, *What We Owe to Each Other* (Cambridge, Mass.: Harvard University Press, 1998).

37. Compare two caregivers (e.g., baby-sitters), one who believes your children have moral standing and a second who does not. The latter in other respects is just like the first and will respect all obligations to the parents, including those which benefit the children. Given the choice between these caregivers, few parents would hire the second.

38. For a passionate and moving expression of the view that we must recognize the moral personhood (i.e., moral standing) of severely cognitively disabled humans, see Eva Feder Kittay, "The Personal is Philosophical is Political: A Philosopher and Mother of a Cognitively Disabled Person Sends Notes from the Battlefield," *Metaphilosophy* 40 (2009): 606–27. Kittay's view can, I think, find expression in a contractarian theory.

39. This contractarian account of the moral standing of some animals has been developed by Andrew I. Cohen. See his "Contractarianism, Other-regarding Attitudes, and the Moral Standing of Nonhuman Animals," *Journal of Applied Philosophy* 24 (2007): 188–201; "Dependent Relationships and the Moral Standing of Nonhuman Animals," *Ethics & the Environment* 13 (2008): 1–21; and "Contractarianism and Interspecies Welfare Conflicts," *Social Philosophy & Policy* (2008): 227–57.

40. See his contribution to this volume. Wayne Sumner has argued that the beings accorded moral standing by contractarian moral theory are determined largely by the concerns of the agents. See L. W. Sumner, "A Response to Morris," *Values and Moral Standing*, Bowling Green Studies in Applied Philosophy 8, ed. Wayne Sumner et al. (Bowling Green, Ohio: Bowling Green State University, 1986), 22–23.

41. Philippa Foot, "Moral Beliefs," in *Virtues and Vices* (Oxford: Clarendon Press, 2002 [1978]), p. 125.

42. Philippa Foot, "Euthanasia," in *Virtues and Vices* (Oxford: Clarendon Press, 2002 [1977]), pp. 44–45.

43. Kamm, "Moral Status," p. 229.

44. Kamm, "Moral Status," p. 230.

45. Kamm, "Moral Status," p. 230.

46. Kamm, "Moral Status," p. 227.

47. David Hume, *An Enquiry concerning Human Understanding* 3.18 (sect. 3, part 1, para. p. 18).

48. James Rachels, "Drawing the Line," in *Animal Rights: Current Debates and New Directions*, ed. Cass Sunstein and Martha Nussbaum (New York: Oxford University Press, 2004), p. 164.

49. Rachels, "Drawing the Line," pp. 166–67.

50. Rachels, "Drawing the Line," p. 182.

51. A similar view is expressed by Rosalind Hursthouse in her chapter of this *Handbook*.

52. The second element in many traditional characterizations of an imperfect duty is that it is not owed to anyone. I use the distinction between perfect and imperfect duties to distinguish benevolence from justice in "The Trouble with Justice," in *Morality and Self-Interest*, ed. Paul Bloomfield (New York: Oxford University Press, 2008): pp. 15–30.

53. In an unpublished essay, "The Status of Moral Status," Benjamin Sachs defends a claim stronger than that of Rachels', that the notion of moral status (or moral standing) just brings confusion to discussions about the treatment of marginal cases, such as of individuals who do not have the full set of capacities possessed by typical adult humans. His argument is complex and points to the considerable lack of clarity about these concepts in the literature. This chapter could help to alleviate some of that. I also think, as I have in effect argued in a number of essays, that the notion of moral standing has uses broader than treatments in contemporary ethics of marginal cases. But I think he is right to worry about introducing a technical notion that is ill-understood and about which there is considerable disagreement.

SUGGESTED READING

BERNSTEIN, MARK H. *On Moral Considerability: An Essay on Who Morally Matters.* New York: Oxford University Press, 1998.

COHEN, ANDREW I. "Contractarianism, Other-regarding Attitudes, and the Moral Standing of Nonhuman Animals." *Journal of Applied Philosophy* 24 (2007): 188–200.

DEGRAZIA, DAVID. "Moral Status as a Matter of Degree?" *Southern Journal of Philosophy* 46 (2008): 181–98.

GOODPASTER, KENNETH. "On Being Morally Considerable." *Journal of Philosophy* 75 (1978): 308–25.

KAMM, FRANCIS. "Moral Status." In *Intricate Ethics: Rights, Responsibilities, and Permissible Harm.* New York: Oxford University Press, 2006.

RACHELS, JAMES. "Drawing the Line." In *Animal Rights: Current Debates and New Directions*, ed. Cass Sunstein and Martha Nussbaum, 162–74. New York: Oxford University Press, 2004.

SACHS, BENJAMIN. "The Status of Moral Status." Unpublished paper.

SINGER, PETER. "Speciesism and Moral Status." *Metaphilosophy* 40 (2009): 567–81.

WARREN, MARY ANN. *Moral Status.* New York: Oxford University Press, 2000.

CHAPTER 10

ANIMALS, FUNDAMENTAL MORAL STANDING, AND SPECIESISM

DAVID COPP

SOME people would claim that any moral duties we have regarding the treatment of nonhuman animals are not owed directly to the animals. Any duties regarding animals, they would say, are derivative from duties we have to human beings, such as our duty to promote human welfare. I believe, however, that we do have moral duties to nonhuman animals, not merely duties regarding them, and I believe that our moral duties to animals and to human beings have the same basis and are equally fundamental. I call this claim *the thesis of the fundamental standing of animals*. This thesis strikes me as intuitively attractive, and, for my purposes in this chapter, I shall assume it is true. It holds that we have morally fundamental duties to nonhuman animals, duties to treat them decently.[1] It implies that our duty not to cause an animal gratuitous pain is not derived from our duty to advance human welfare or from any other duty to human beings. Our duties to animals are due to the nature of the animals themselves and to what would undermine *their* welfare, not to considerations about human welfare.[2]

Unfortunately, the thesis of the fundamental standing of animals is in tension with another thesis, which I call *the thesis of the fundamental concern of morality*. According to this thesis, which also strikes me as intuitively attractive, morality is fundamentally concerned with advancing human welfare.[3] More specifically, morality is concerned to advance human welfare *by* enabling human beings to live together successfully in societies, despite competing interests that come into conflict.[4] For my purposes in this paper, I shall also assume that this thesis is true. The trouble is that these two theses seem to be in tension with each other and in direct conflict in many circumstances. How could they both be true?

It may seem that they cannot both be true. For if morality is fundamentally concerned with advancing human welfare and enabling humans to live peacefully and cooperatively together, then it may seem that any duty we have to treat animals decently must be defended ultimately by reference to human welfare, not to the needs or welfare of the animals. Conversely, if we have morally fundamental duties to treat animals decently, because of the nature of the animals and because of what they need, then, it may seem that the fundamental moral concern cannot be the advancement of human welfare. It must be something wider or different from that, such as perhaps the advancement of welfare, period, without special regard for human welfare. It may seem, moreover, that if one of these theses is to be rejected, it ought to be the thesis of the fundamental concern of morality. The idea that morality is fundamentally concerned with the good of humans may seem to be "speciesist."

My goal in this chapter is to argue that the two theses are in fact compatible so that the "speciesism" of the second thesis is compatible with recognizing the fundamental standing of animals. The key to seeing that they are compatible is to understand that this second thesis, the thesis of the fundamental concern of morality, is a thesis about the nature of morality rather than a normative moral claim. The other thesis, the thesis of the fundamental standing of animals, is a normative moral claim. The normative thesis that we have morally basic duties to animals is compatible with the metaethical claim that morality is fundamentally concerned to advance human welfare by enabling humans to live cooperatively together in societies. This is the view I shall be defending.

To be more specific, I will be arguing that the society-centered conception of morality that I have proposed in other places supports the thesis of the fundamental standing of animals. That is, it supports the idea that there are morally fundamental requirements as to how animals should be treated. This conception of morality takes the needs of human societies to determine the content of morality. It views morality as a "device" that makes society possible. It might seem that such a view could give animals at best a kind of derivative moral standing. I want to show that this is not so. I will argue that the society-centered theory supports a fundamental duty or virtue of compassion to animals as well as a fundamental duty to protect animal welfare. I thereby hope to show that the above two theses are compatible. The conviction many people have that an obligation to treat animals decently is fundamental to morality is compatible with the attractive idea that morality is basically centered on the human need to be able to live cooperatively together.

There are many questions that my discussion will prompt, but that I will not be able to address in any detail. Most important are a series of normative questions about our moral duties to animals. What exactly is morally required of us regarding the treatment of animals? I will argue that we ought to treat animals with compassion and protect their welfare, and that we have a defeasible duty not to kill animals.[5] Another question is which kinds of animal must we treat in which kinds of ways? Presumably we have different duties to different kinds of animals. And another question is whether we owe it to the animals that we treat them in certain ways, or whether instead it is merely that we ought to treat them in these ways. I will not be able to address these questions in much depth.

I begin by discussing the thesis of the fundamental standing of animals. I then consider some theories that might seem to account for it, and point out difficulties they face. Next, I briefly explicate the society-centered theory in order to show that it underwrites a version of the thesis of the fundamental concern of morality. Finally, I show how the society-centered theory can underwrite the thesis of the fundamental standing of animals. In the course of the argument I shall attempt to address the questions about the treatment of animals that I raised in the preceding paragraph. I will also point out complications in our thinking about animals. My attention throughout will be focused on the foundational issue raised by the apparent tension between the thesis of the fundamental concern of morality and the thesis of the fundamental moral standing of animals.

1. The Moral Standing of Animals: A Commonsense View

As I said, I assume that the thesis of the fundamental standing of animals is true. In this section, I contend that a widely shared moral view is committed to this thesis. I call this view "commonsense morality." As I will explain, the content and structure of commonsense morality commit those who hold it to the thesis of the fundamental moral standing of animals, or at least to central elements of the thesis. Commonsense morality holds, for example, that it would be wrong to torture a dog merely for our amusement.[6] Moreover, in commonsense morality, this proposition is not derived from any more fundamental moral proposition. Hence, commonsense morality is committed to the thesis that at least this duty regarding the treatment of animals is not derived from any duty we owe to human beings. It is therefore committed to a central element of the thesis of the fundamental standing of animals.

Consider whether it would be wrong to torture a dog merely for our amusement. To many people, it will seem obvious that this would be wrong. It might seem that there could be no good reason to doubt that this would be wrong. It might seem that no morally decent person would think otherwise. Even to raise the question whether it would be wrong might seem morally outrageous. If we were to ask *why* it would be wrong to torture a dog just to amuse ourselves, I think most people would be outraged and dumbfounded.[7] The question might seem unanswerable. Or it might seem that the answer is so obvious that anyone who raised the question would implicate that she doubts the obvious answer. And it is unclear how one could explain the force of the obvious answer to someone who doubted it.

The obvious answer is that the pain of an animal being tortured has more weight morally than our mere amusement. This is why we ought not to torture a dog just to amuse ourselves. If we were to ask why the pain of an animal being tortured matters more than the amusement of a human being, or why the pain of an animal matters at all, those who accept commonsense morality would be indignant and astonished.

These observations suggest that it is fundamental to commonsense morality that it would be wrong to torture an animal just for our own amusement. That is, I think, commonsense morality takes the claim to be true, yet the claim is not derived in commonsense moral thinking from any other moral claim. It is taken as "given," we might say. It can be inferred from other propositions that are accepted in common sense, such as the propositions that it would be wrong to inflict pain without good reason, that torture causes pain, and that amusement is not a good reason to inflict pain. But I think that commonsense morality does not accept it on the basis of any such inference.

The idea that some proposition is fundamental in commonsense morality is, then, the idea that this proposition is widely accepted, but it is not widely accepted on the basis of a derivation from any other propositions in commonsense morality. Consider the proposition that the pain of an animal has moral weight. I want to say that this proposition is fundamental to commonsense morality even though it can be inferred from the widely accepted proposition that pain has moral weight. For if pain has moral weight, then since the pain of animals is a kind of pain, the pain of animals has moral weight. But I do not think that people who accept that the pain of animals has moral weight typically accept it on the basis of this or any other inference. This inference is so trivial, immediate, and obvious that it would be question begging to offer it as a reason to believe that the pain of animals has moral weight. So I take it to be fundamental to commonsense morality that the pain of animals has moral weight.[8]

Because it is widely accepted that it would be wrong to torture a dog just for our amusement, it might seem pointless to consider the example. It might seem much more interesting to consider questions such as whether it is morally permissible to use animals for food, whether it is permissible to use animals to test cosmetics or in medical experiments, and so on. But the example is important because of what it suggests about the moral standing of animals in commonsense morality. It suggests that commonsense morality takes there to be certain duties toward animals that are fundamental or underived. It tells us that animals have a fundamental standing in commonsense morality.

It does not tell us, of course, whether commonsense morality takes our duties to animals to be more or less important than, or of equal importance to, our duties to human beings. Consider now whether it would be wrong to torture a human infant just for our amusement. I doubt that commonsense thinking about this matter is any more articulated than commonsense thinking about the torturing of animals just for fun. Most people would say that there is really no question, but that it would be wrong to torture an infant just for fun. If we asked why it would be wrong, I think most people would be outraged and dumbfounded. The obvious answer to the question is surely that the pain caused by the torture is morally much more weighty than our mere amusement. But again, if we asked why the pain of an infant being tortured matters at all, I think that most people would be indignant and astonished. It seems to be fundamental in commonsense morality both that it would be wrong to torture a child just for our amusement and that the pain of a child has moral weight.

Commonsense morality holds, I believe, that the duty not to torture a dog is not as important or weighty as the duty not to torture a human infant. If there were a situation in which someone were forced to violate one of these duties, commonsense would come down on the side of the human infant and prescribe torturing the dog rather than the infant, other things being equal. Suppose an evil genius took one hundred children hostage and threatened to torture every one of these children unless you torture either the dog Fido or the infant Ray. In this case, other things being equal, if the genius speaks truly and his threat is credible, you clearly ought to torture Fido. I believe that this is what commonsense would say. And I think that this would also be fundamental in commonsense morality.

The contrast is perhaps more clear if we consider duties not to kill. Common sense says it is wrong to kill an animal just for fun, at least if the animal is "sentient" and among the "higher" animals. It is wrong to torture a dog just for fun, and it is also wrong to kill a dog just for fun. And common sense says the same of human beings. But according to common sense, it may be permissible to cause an animal great pain or to kill an animal in the course of experiments that are hoped or intended to bring significant benefits to human beings. Animals of many kinds are used in medical experiments, including primates, and although there are many who protest this, it seems to be widely believed that it is permissible. Even those who protest might concede that the use of animals in medical experiments would be justified if the promised benefit to humans were large enough and reasonably expectable. But commonsense morality comes down very strongly against the view that it may be permissible to cause a human being great pain or to kill a human being in the course of experiments that are hoped or intended to bring significant benefits to human beings. It may allow exceptions in cases in which the experimental subjects have consented or volunteered; but since animals are not in a position to consent or volunteer, there remains a significant contrast in commonsense thinking between morally appropriate ways of treating human beings and morally appropriate ways of treating other animals.

Commonsense morality might leave unanswered most of the interesting questions about our duties to animals. That is, there might not be very widely shared answers to the most interesting questions. Commonsense might say that our duties to human beings are more weighty than our duties to animals, other things being equal. It might say that it would be worse to use humans in painful or risky medical experiments than to use animals in such experiments. But this does not settle whether it is permissible to use either humans or animals in this kind of experimentation.

There is the further point that commonsense morality is not always a reliable guide to the truth. It was once common sense in many nations that slavery was permissible. There are communities in which it seems to be widely accepted that there is nothing wrong with dog-fighting, and there are communities in which bull-fighting is widely viewed as permissible. Yet I would say that all of these pursuits are morally impermissible and ought to be outlawed.

Moreover, commonsense morality is perhaps not to be trusted when it seeks to *explain* the moral judgments that it accepts. In certain circles, it may be widely accepted that species membership explains why the human good has greater moral weight than the good of other animals.[9] The infant, unlike the dog, is a member of the species *Homo sapiens*. But on reflection, it may seem clear that species membership cannot matter morally by itself. It may seem clear that species membership is morally arbitrary.[10] Species membership obviously cannot help to explain why we think pain is morally significant since we think pain is significant whether it is experienced by a dog or by a human. If species membership is irrelevant to the question whether pain matters, why would it be relevant to explaining why the pain of an infant matters more than the pain of a dog? If species membership is irrelevant to this issue of weighting, then what is it about the difference between the dog and the human infant that can explain the greater moral weight of the child's pain such that we morally must give preference to the infant?[11]

We cannot take at face value everything that commonsense morality tells us to be true. Nevertheless, I do think that the thesis of the fundamental standing of animals is plausible, and I think that commonsense morality is committed to central elements of this thesis. This is merely a starting point.

2. THEORETICAL ACCOUNTS OF THE MORAL STANDING OF ANIMALS

Moral theorists will want to evaluate the pronouncements of commonsense morality since, as we have seen, common sense is not always reliable. Why is it that the pain of animals counts morally? Why would it be wrong to torture a dog just for fun? Why is it that, except perhaps in special circumstances, the pain of an animal is less morally significant than the pain of a human infant? In a forced choice between torturing Ray and torturing Fido, why is it that we ought to torture Fido, other things being equal? Why is it that if we were forced to kill one or the other, it would be wrong, other things being equal, to kill the infant rather than the dog? If these judgments are correct, can we explain why they are correct?

Ideally, we would like to answer these questions on the basis of an account of the deep structure or core truths of morality. I will argue that the society-centered theory can provide such an account. Given my purposes in this chapter, moreover, it is of central importance whether we can adequately support or ground the thesis of the fundamental moral standing of animals—the thesis that there are fundamental duties to treat animals decently. In order to underwrite this thesis, we would need to ground duties to treat animals decently *without deriving* them from any duties we have to human beings. I will argue that the society-centered theory can provide a suitable grounding for the thesis of the fundamental standing of animals, but it will be useful to consider some alternatives. I begin with welfarism. I then turn to theories that ascribe value to human lives, including theories that ascribe greater value to human lives than to the lives of other animals.

3. WELFARISM

Brad Hooker argues that we cannot adequately explain why it would be wrong to torture a dog just for our amusement unless we accept that the pain of animals is a matter of fundamental moral concern. In a famous passage where he discusses animal welfare, Jeremy Bentham said, "The question is not, Can they reason? nor Can they talk? but, Can they suffer?"[12] Following Bentham, Hooker contends that the central reason that it is wrong to torture a dog just for our amusement is that welfare is fundamentally important in morality and that animal welfare is a part of the overall welfare.[13] He says, "the overall good matters morally; animal welfare is at least some part of the overall good."[14]

The welfarist approach, therefore can treat the pain of animals as being of fundamental moral importance and it can explain *why* it is of fundamental importance. For if welfare is at least a part of what matters morally at the fundamental level, animal welfare matters fundamentally, since animal welfare is just a kind of welfare. And since pain detracts from the welfare of any sufferer, pain matters at a fundamental level. Moreover, since welfarism grounds our duties in considerations of welfare, it is a short step to the wrongness of torture, at least to its being defeasibly wrong,[15] whether or not the victim is an animal.

Welfarism has been ably explicated and defended by Wayne Sumner.[16] According to the welfarist, Sumner explains, "the ultimate point of ethics is to bring about intrinsically valuable states of affairs," and, because of this, "everyone has a reason to promote" such states of affairs, or to want them to come about. That is, there is "agent-neutral reason" to promote such states of affairs or to want them to come about. And according to the welfarist, welfare or well-being is the single foundational value.[17] Sumner contends, moreover, that a theory of welfare must account for the welfare of many nonhuman animals as well as the welfare of human beings: "Besides the paradigm case of adult human beings, our welfare vocabulary applies just as readily to children and infants and to many nonhuman beings." And he suggests that "exactly same concept of welfare applies" to his cat as to his friends. For a welfarist then, if we are given an adequate account of welfare, the "ground floor" of morality is the *value* of welfare, including the welfare of any being with a welfare, such as many nonhuman animals.

The welfarist view does not give moral standing to *animals* at a fundamental level. It is *welfare* that counts fundamentally. For a welfarist, it is not of fundamental concern *which* being enjoys a bit of welfare. That is, it is not of fundamental moral concern whether Fido has a warm and dry place to take shelter or whether, instead, Ray has a warm and dry place to take shelter. For a welfarist, it is not of fundamental concern whether a good goes to a human or to an animal of another kind. What is fundamentally important is simply that welfare be promoted and defended, not that animal welfare be promoted, not that human welfare be promoted, and certainly not that Fido's or Ray's welfare be promoted or defended in particular. The important point for my purposes, however, is that animals and humans have the same status in a welfarist theory. Their welfare counts fundamentally, and nothing else is of fundamental importance. Humans and animals alike have moral standing because we cannot promote welfare except by promoting the welfare of some sentient creature. All such creatures are on a par except insofar as promoting the welfare of some creatures might be a more efficient way of promoting overall welfare than promoting the welfare of some other creatures.

Welfarism is therefore a view that gives a fundamental moral standing to animals since it takes welfare to be the basic moral value and since animals are subjects of welfare. Animals and humans alike each possess a certain level of welfare, and it is of fundamental moral importance to promote welfare, which cannot be done except by doing things that promote the welfare of individual possessors of welfare.

There are many familiar objections to welfarism. For my purposes here, the two most interesting objections both suggest, as I will explain, that welfarism fails because it does not accord value to *living beings*. The objections both lead in the direction of a view I will call "valuationism." As I will argue, valuationism can underwrite the thesis of the fundamental moral standing of animals, and it does so in a more plausible way than welfarism.

4. VALUATIONISM

Valuationism gives a fundamental moral status to animals, for it holds that the value of living beings is fundamental to morality. In the valuationist view, it is fundamental to morality that humans have value, and also that many nonhuman animals have value.[18] Moreover, our duties are grounded in the value of living beings.

We can be led to valuationism if we take note of certain objections to welfarism. The first objection is that it does not give a foundational moral status or standing to *human beings*. It instead merely gives such a status to their *welfare*. We need a reason to suppose that the welfare of humans is of value. The objection maintains that the reason the welfare of humans is of value is that humans are themselves of value. The value of human beings is basic, not the value of welfare. The objection suggests, then, that an adequate theory would take it to be foundational that human beings have intrinsic value, not merely that their welfare has intrinsic value. We can think of this as a Kantian view, since Kant held that humanity has value as "an end-in-itself."[19] A valuationist would add that many nonhuman animals also have value, at least, presumably, the sentient animals.

The second objection to welfarism begins with the idea, suggested by Don Regan[20] and developed by Connie Rosati,[21] that a welfarist needs to explain *why* well-being might be a source of moral duties or obligations; that is, it needs to be explained why everyone morally ought to promote the well-being of anyone, or at least to want it to be promoted. It is commonly thought that each person has reason to promote her *own* welfare, since this will be good for her. But it does not follow from this that a person ought to promote the welfare of any *other* person, or even to want it promoted. The idea that welfare might be a source of moral duties or obligations should seem highly problematic unless we suppose that, in promoting the welfare of anyone, we would thereby be promoting or sustaining something of intrinsic value *simpliciter*, and not just of value to the person whose welfare is in question. Developing this idea, Rosati proposes that in order to explain why a person's welfare is a source of moral duties we must suppose that any person is "a being with value."[22] That is, we must suppose that persons are "valuable in the sense that their nature renders them sources of a legitimate demand for our considerate and respectful treatment."[23] We must suppose that persons have "ethical value."[24]

If Rosati is correct, then the idea that there is a moral duty to promote the welfare of animals presupposes that these animals have intrinsic value. This idea would be most plausible in the case of sentient animals, it seems to me. I find it hard to believe that a non-sentient living thing, such as the West Nile virus, has intrinsic value, such that everyone ought to promote or at least to want what would be good for this virus. It seems implausible that *all* nonhuman animals have intrinsic value of the sort that, in Rosati's words, would render them sources of legitimate moral demands.

Of course, there are different ideas about where the morally significant line comes between animals with intrinsic value and those with no intrinsic value. It would be arbitrary to suppose that only members of our own species have intrinsic value. Kantians would suppose that all and only rational agents have the relevant kind of value.[25] But the commonsense view, I believe, is that the morally important line falls between the sentient and the non-sentient. As Peter Singer has said, "the limit of sentience [or the capacity to suffer or experience enjoyment or happiness] is the only defensible boundary of concern for the interests of others."[26] Perhaps, to be sure, there is no sharp demarcation between the sentient and non-sentient; perhaps the difference is one of degree. But this is a complication I shall ignore. In any event, in what follows I will restrict attention to the idea that all sentient creatures have the kind of intrinsic value at issue.

A welfarist might think that all and only sentient animals have a welfare. This seems to be Sumner's view,[27] and the valuationist might agree. On the valuationist view that I am considering, moreover, all and only sentient animals have intrinsic value of the kind that renders them sources of legitimate moral demands. A valuationist and a welfarist might then agree that all and only sentient animals are such that there are moral duties to promote or protect their welfare. The important point, however, is that the valuationist seeks to *explain* why welfare is a source of moral duties by citing the intrinsic value of the kinds of beings that have a welfare.

This is not an interesting explanation, however, unless it is accompanied by a substantive and independently plausible account of the relevant kind of intrinsic value. Rosati suggests that having the relevant kind of value is sufficient for being a source of legitimate moral claims.[28] But the explanation of why there is a duty to promote or protect the welfare of all sentient creatures cannot simply be that all such animals are intrinsically valuable in that they are sources of legitimate moral claims to have their welfare promoted or protected. Why is there a moral duty to promote or protect the welfare of these animals? Because their value grounds a claim to have their welfare promoted or protected? This is not an interesting explanation as it stands. We need a substantive account of the value of lives.

Raymond Frey has proposed an account that rests on an idea of the "quality" of a life.[29] As he says, "the [intrinsic] value of a life is a function of its quality."[30] The quality of an animal's life depends among other things on the quality of its subjective experiences. But normal adult human beings can live lives of much greater value than can most sentient animals since they can have lives of much greater "richness." In addition to being sentient, normal adult human beings have a "scope and potentiality for enrichment" that greatly exceeds what we know of chimps, dogs, rabbits, and the like because they are capable of exercising their autonomy in pursuit of a conception of the good life.[31]

Frey uses this theory of the value of lives to explain much of what we have wanted to explain. First, he uses the theory to account for the moral standing of animals. He holds that all sentient animals have moral standing due to the fact that they have valuable lives, yet he holds that normal adult human beings have a more significant standing than other sentient creatures due to the fact that they have significantly more valuable lives.[32] Second, Frey uses his theory to account for the reasons and duties we have to promote and protect animal welfare. He proposes that we have reason to promote the well-being of sentient creatures just in virtue of the fact that the lives of sentient creatures have intrinsic value. Third, Frey proposes a valuationist explanation of why our duty not to kill a normal adult human being is more stringent than our duty not to kill animals of other kinds. He thinks the stringency of our duties to promote and protect the welfare of a creature, and to avoid taking its life, is proportional to the value of its life. In Frey's view, the relative stringency of our duty to avoid killing an animal depends on just how valuable its life is by comparison with the lives of other animals, and he thinks the life of a normal adult human being is more valuable than the life of a nonhuman animal.[33]

Frey seems clearly to be assuming, not implausibly, that we have reason and a duty to promote and protect whatever is of value and that the stringency of this reason or duty in any given case depends on the degree to which the thing is of value. Given Frey's theory of the value of lives, this assumption can explain why we have a duty to promote and protect the welfare of all sentient animals, including human beings. It can also explain why our duty to promote and protect the welfare of a normal adult human is more stringent, other things being equal, than our duty to promote and protect the welfare of a nonhuman animal.

Frey's position implies that the lives of some nonhumans might be of greater value than the lives of some humans. For the life of a human who is acutely developmentally disabled or who has permanent severe brain damage may be less rich than the life of a chimp. The chimp's subjective experiences might be more desirable, for the human might suffer from great anxiety or be in great pain. Hence, Frey concludes, "the lives of some perfectly healthy animals have a higher quality and greater value than the lives of some humans."[34] In Frey's view, this means that if, for example, we are morally permitted to use some animals in medical experiments in an attempt to benefit human beings, we may also be morally permitted to use some humans in such experiments.[35] There might be biological reasons to think that an experiment would be more useful in some cases if human beings were used rather than nonhuman animals. And Frey contends that if we are going to use living creatures at all, we must use a creature with a lower quality of life in preference to one with a higher quality of life.[36] We might argue, of course, that it would be wrong to use human beings in such experiments, regardless of the quality of their lives. But if so, then, on Frey's position, we should conclude that it would also be impermissible so to use an animal.

Unfortunately, Frey's position has some highly troubling implications. On his view, the lives of human beings are not all of the same degree of value, for the lives of some people have higher quality than the lives of other people. This is obvious if we compare the quality of life of a person who is in the grip of a highly degenerative brain disorder with the quality of life of a healthy and happy person in the prime of life.[37] Unfortunately, however, given the assumption that the stringency of our duties to promote the welfare and to protect the life of a being depends on the value of its life, Frey's view leads directly to the unpalatable conclusion that we have stronger duties to promote the welfare and to protect the lives of those human beings whose lives are more valuable. We must do more for those with a higher quality of life than for those with a lower quality of life. Call this the "unpalatable conclusion." We must reject it. A person with a serious case of arthritis has a lower quality of life than a person without arthritis, other things being equal. We must not be in the business of saying that we have stronger duties to promote the welfare and to protect the lives of those human beings who do not have arthritis than of those who do have arthritis, not even if we add the qualification, "other things being equal."

We cannot avoid the unpalatable conclusion by tying our duties, not to the *actual* quality of life, but to a being's *potential* for a life of high quality. This would help to explain why we cannot rightly treat a newborn child in the way that we might treat an animal with no potential for a life of greater richness.[38] But the move to potential does not address the fundamental problem. For people who are developmentally disabled or who have a incurable chronic painful condition not only have lives of lower quality than the life of a normal healthy and hearty adult human being, they also lack the potential for a life of higher quality.

We might be able to avoid the unpalatable conclusion if we abandoned Frey's quality-of-life account of the value of a life. We might insist that all human lives have the same value. But common sense says that our duties to animals are less stringent than our duties to human beings. In order to capture this idea, a valuationist position would need to say that the lives of animals are less valuable than the lives of human beings. It is unclear how we can make sense of this without adopting a view like Frey's, according to which the value of some humans' lives is lower than the value of some other humans' lives.

Of course, we might simply say that the degree to which a life has value is a matter of, first, the stringency of the duties we have to promote and protect the welfare of the being whose life it is, and second, the stringency of our duty not to kill that being. But on this approach, valuationism would not provide any substantive explanation of our duties to animals nor, for that matter, of our duties to human beings. If we want such an explanation, it is no help to be told that the explanation lies in the value of the lives of these beings if this is simply a matter of our having duties to protect and promote the welfare of these beings.[39]

To avoid Frey's unpalatable conclusion, I conclude, we ought to reject the underlying assumption of valuationism. I have not considered all possible valuationist positions, but I think the fundamental problem lies with the valuationist assumption that we have a moral reason and a moral duty to promote and protect whatever is of intrinsic value, and that the stringency of this reason or duty in any given case depends on the degree to which the thing is of value. This assumption is doubtful, especially in the absence of a clear account of what it is for a thing to be of intrinsic value. Some people have lives that are barely worth living. But our duty not to destroy these lives is no less stringent than our duty not to destroy the lives of people who are thriving and flourishing in the midst of great successes and joys. Indeed, we may have a more stringent duty to help the unfortunate people whose lives are barely worth living, especially if they are unable to meet even their basic needs, than to help those who are wonderfully happy and flourishing. To explain this, and to explain our duties to animals, we need to look elsewhere.

5. The Fundamental Concern of Morality: The Society-Centered Theory[40]

I begin with the thought that morality is fundamentally concerned to advance human welfare by enabling human beings to live together successfully in societies. Human beings are not self-sufficient. We have biological and psychological needs of various kinds, and we cannot generally meet these needs without interacting cooperatively with other people. We have goals and things that we value, and we also generally cannot achieve, protect, or realize these values without interacting cooperatively with other people. We are vulnerable to interference from other people. At various points in our lives, we need to be cared for by other people. For all of these reasons, we need the cooperation of others to achieve what we value. This is true no matter what we value, within at least a wide range of things we might value. We also need the existence of a minimum level of peace and stability. We need to live peacefully and cooperatively together. Unfortunately, as J. L. Mackie points out, people have conflicting interests, limited resources, and limited sympathy as well as a tendency to pursue their own advantage, so there is always a risk that peaceful cooperation will break down or that conflict will develop or that "what would be mutually beneficial cooperation" will not emerge.[41] Call this the problem of "sociality." It seems obvious that this problem can be ameliorated if the members of society subscribe to a suitable system of norms calling for a willingness to cooperate with others, for noninterference with others, and so on, as appropriate. According to my view, morality is just such a normative system. Mackie describes morality as a "device." I describe it as a normative system, the point of which is to enable us to deal with the problem of sociality. This is the basic idea that underlies the society-centered theory.

We can think of a moral code as a "system" of standards or norms, where a standard is a content expressible by an imperative. An example is the standard that calls on us not to torture anyone. Different systems of standards obviously would differ in how well their currency in society would deal with the problem of sociality. There are purely arbitrary standards, such as the standard that calls on us to do cartwheels in the street at midnight every day. There are also standards that seem to correspond to moral truths, such as the standard that prohibits torture. I will say that standards of the latter kind have "moral authority." A standard is not the kind of thing that is truth-apt, but there is presumably some important truth-related property that standards with moral authority possess, one that distinguishes them from the enormous variety of arbitrary standards that have no moral authority, such as the standard that requires us to do cartwheels at midnight. The crucial point is that the standard that prohibits torture corresponds to the fact that torture is wrong and other standards with moral authority correspond similarly to moral facts. At any rate, I will be assuming there are moral truths and moral facts such as the fact that torture is wrong. The property we are after is one that distinguishes the standards that correspond to moral truths from the merely arbitrary or otherwise putative moral standards. We can call the key property the "truth-grounding property." We are looking for a moral code where the standards included in it have the truth-grounding property.

Obviously, there is room for disagreement about what this property might be. But if we assume there is such a property, we can provide a "standard-based" schema for explicating the truth conditions of moral judgments. This schema says that a pure and basic moral proposition, such as the proposition that torture is morally wrong, is true just in case a corresponding moral standard has the relevant truth-grounding status.[42]

Let us assume that the point of morality is to enable us to deal with the problem of sociality. This suggests a way to think about the truth-grounding status of moral standards. For suppose that some moral code M is such that its currency would do most to ameliorate the problem of sociality. Call M the "ideal" moral code. It is tempting to take this code to have the relevant truth-grounding status. That is, it is tempting to think that its standards correspond to the moral truths—that if it includes a standard that prohibits killing animals, then it is true that killing animals is wrong. The idea, then, is that a standard has the relevant truth-grounding status if and only if it is included in, or derivable from, the ideal moral code M, the code whose currency in society would do most to enable the society to deal with the problem of sociality. To think otherwise one would have to think that even though the currency of M best addresses the problem of sociality, some other moral code determines what we are morally required to do or how we ought morally to live. Even if the ideal moral code requires us to do something, it might not be true that we are morally required to do it. But in that case, there would be little sense to the idea that the point of morality is to enable us to deal with the problem of sociality. The point of morality and the truth about how we must live would pull in different directions. This seems absurd. I conclude, then, that if we agree with society-centered theory about the point of morality, we should agree as well that the ideal moral code has the relevant truth-grounding status.

To make vivid the basic idea of society-centered theory, it will help to think as social engineers who are looking for a moral code whose currency would best address the problem of sociality. A system of moral standards has "currency" in a society just when the members of the society by and large "subscribe" to the standards. That is, they by and large are disposed to comply with the standards, to have negative attitudes toward themselves if they fail to do so, to have negative attitudes toward others who fail to do so, and so on. The currency of a moral code would significantly affect the motivations of the members of a society. This is why it might have an impact on the problem of sociality.

Given this, it seems clear that the currency of a moral code can help a society to deal with the problem of sociality provided that it calls on people to behave in appropriate ways or to have appropriate traits of character. What are these appropriate ways? Given what I have said, it seems that the ideal code will call on people to cooperate peacefully with one another. But we can say more than this. Thinking again as social engineers, we can see that to minimize the impact of the problem of sociality on their ability to achieve what they value, people need to live in societies that are well-functioning in certain obviously relevant ways, and any society must have certain characteristics in order to be well-functioning in these ways. Say that a society must meet certain *needs*, in order to be well-functioning in these ways. Most important, societies have a need for physical integrity, a need for cooperative integrity, a need for internal social harmony, and a need for peaceful and cooperative relationships with neighboring societies. It seems clear that to minimize the impact of the problem of sociality, we must live in a society that is able to meet these needs. So the kind of moral code we are looking for, as social engineers, would be such that its currency in society would contribute to the society's ability to meet these needs. Mackie said that "limited resources and limited sympathies together generate both competition leading to conflict and an absence of what would be mutually beneficial cooperation."[43] That is, limited resources and limited sympathies tend to undermine the ability of societies to meet the needs listed above. It is plausible, I therefore think, that a solution to the problem of sociality would tend to enable societies to meet these needs.[44]

If we continue to think as social engineers, then, our goal is to design a moral code whose currency would best enable a society to meet its needs. All societies have the same needs at the most fundamental level, but their circumstances differ, and because of this, the moral code that is best designed for one society might differ somewhat from the code that is best designed for another society. We should at least leave open this possibility.

A further point is that societies can have values—that is, a society can be such that virtually all of its mature members agree in valuing something where this state of affairs is stable and endorsed by people, so that they do not wish it to change—and the values of one society can differ from those of another society. Thinking again as social engineers with the goal of designing a moral code that would best address the problem of sociality for a given society, it becomes obvious that it would be counterproductive to ignore the society's values. To be sure, we need to discount societal values that work against society's ability to meet its needs. The theory is meant to be used in *assessing* the values of societies. But a good solution to the problem of sociality would be compatible with a society's values, insofar as the society's values are compatible with a moral code whose currency would do as well as any other in enabling the society to meet its needs. Suppose then that several moral codes are tied as best for a given society—suppose, that is, that there is a set of moral codes, M1, M2, and so on, such that, for each of these codes, Mi, (a) the currency of Mi in a society would better contribute to the society's meeting its needs than the currency of any code that is not in this set and (b) no other code in the set is such that its currency would better contribute to the society's meeting its needs than would the currency of Mi. In this case, the society's values can break the tie. If one of these codes would be more compatible with the society's values than any of the others, then it qualifies as the ideal code. Of course, respecting societal values in this way can introduce variation among the moral codes justified with respect to different societies. For example, differences between societies' values regarding the treatment of animals can make for different moral requirements regarding the treatment of animals.

In recent work where I have discussed the society-centered theory, I have tended to ignore this point about societal values. I wanted to simplify my exposition of the theory, and, for reasons I will explain below, in section 9, it seemed to me that in most cases societal values would not play a significant role in the theory. But paying attention to the role of societal values in the theory can help us to understand the theory's ability to address certain vexing moral issues, including, I think, the distinction between pets or companion animals and other animals. I will turn to questions about animals in the next section of the paper.

To summarize, society-centered theory says that a moral code has the relevant truth-grounding status if and only if it is the code whose currency would *best* serve the basic needs of society,[45] or, if several codes would serve equally well, it is the code among these that would be most compatible with the society's values.[46] This is the "ideal" moral code for the society. It is the code whose currency, I suggest, would best ameliorate the problem of sociality. If a standard that rules out eating meat is included in or implied by the code, then, the theory says, eating meat is morally wrong. That is, according to the theory, a pure and basic moral proposition is true if and only if a corresponding moral standard is included in, or implied by, the ideal moral code for the society.[47]

I do not believe that the society-centered theory is true a priori. It does not seem to me to be a conceptual truth. I do not claim to be providing an analysis of the concept of a moral judgment, nor do I claim to be giving truth conditions for moral judgments that would be seen to be true by anyone who was thinking clearly and had the relevant concepts. I am proposing here some basic elements of a theory that can only be assessed, I believe, by comparing it with alternatives and by assessing its implications. Compare beliefs about economic matters, such as the belief that the U.S. dollar is the reserve currency in many countries. We do not expect ordinary thinkers to be able to lay out the truth conditions of these beliefs, not even if they are sufficiently competent with the relevant concepts for ordinary purposes.[48] It is not in general the case that competent thinkers are capable of laying out the truth-conditions of their beliefs.

The society-centered theory is a metaethical theory. It leaves open what morality requires of us, although, to be sure, it has implications about this when it is combined with empirical premises. It has a similar structure to a rule-consequentialist theory, such as has been defended by Brad Hooker.[49] It faces similar objections, and it can be defended in similar ways.[50] But it is not a rule-consequentialist theory. First, rule-consequentialism is a normative theory that aims to set out the content of morality. The claims that it makes are moral claims, even though highly abstract ones. Society-centered theory, however, is a metaethical theory, a theory about morality. It tells us what moral reasons *are* rather than what moral reasons we have. It says that moral reasons are considerations entailed by the relevant ideal moral code. The claims made by society-centered theory are claims about morality rather than moral claims. Second, rule-consequentialism evaluates moral codes on the basis of whether their currency would maximize the good. Society-centered theory evaluates moral codes on the basis of whether their currency would best enable society to meet its needs, something that I would not identify with *the good*.

For present purposes, my claim is that the society-centered theory captures the intuitive idea that morality is fundamentally concerned with advancing human welfare by enabling human beings to live together successfully. To capture this idea, the theory asks us to think of the indefinite array of distinguishable abstract systems of standards to which we could in principle subscribe. We can see right away that the members of society must subscribe by and large to the same system in order to cooperate successfully and peacefully and to coordinate with each other. We can also see that it is not enough that people subscribe to the same system, since many of the systems we can imagine and describe would lead to disaster for society if they were widely accepted. If there is a moral code that is to enable human beings to live together successfully, it must be such that subscription to it by and large by the members of society would contribute to their ability to live together in society. But then subscription to it by the members of society must contribute to society's ability to meet its needs as I explained before, since societies need their members to work together cooperatively and productively. It is now plausible that the moral code whose currency would best contribute to society's ability to meet its needs is the moral code that is authoritative, or that has the truth-grounding status. This is the central thesis of society-centered theory.

This picture allows us to distinguish between fundamental moral considerations and derivative ones. Fundamental considerations are given by the principles included in the ideal code. Derivative considerations are derivable from these principles. On the society-centered theory, then, the question whether the moral standing of animals is morally fundamental is the question whether the ideal moral code would include principles regulating our treatment of animals or whether instead such principles would be derivative. It might seem obvious that they would have to be derivative. But let us look further.

6. The Moral Standing of Animals in Society-centered Theory

Even if we accept the society-centered theory, we are left with many questions about the content of morality. The important point for our purposes here is that, as we will see, the ideal moral code might include a standard that demands the "decent" treatment of animals. It might include a standard that implies the wrongness of factory farming, or eating sentient animals, or using them in testing cosmetics, or using them in medical experiments. Or it might include standards addressed primarily to our states of character, such as standards that call for people to be kind, charitable, and honest, and a standard calling for us to be considerate of animal welfare. The issues are left open. The theory does not entail that our only duties are owed to other members of our own society, nor does it follow that we owe nothing to animals.

The central issue for our purposes is whether the ideal moral code would include principles regulating our treatment of animals. And of course there is the question of exactly what principles regulating the treatment of animals it would include. To know how to answer these questions, one would need to know a great deal about how societies work and about human psychology. The question is basically: what kind of moral code is such that its currency would best contribute to society's ability to meet its needs? This is an empirical question. I doubt that anyone knows enough about such matters to have any certainty. But we can speculate, and our speculations can be more or less informed.

The kind of issue we face is analogous to certain issues in applied economics. For example, we might wonder what kind of regulatory regime would minimize the risk of recession. The answer to the question depends on what the economic facts are as well as on psychological and sociological facts about how people would respond to various regulatory regimes. People certainly speculate about such matters, even though I doubt that anyone knows enough to have any real certainty about what kinds of regulations would work best.

In thinking about the content of the ideal code, we need to bear in mind the cognitive and other psychological limitations of typical human beings since the ideal code is one that is to be subscribed to, by and large, by the members of society. We are not cognitive and psychological superheroes, and the code must be "designed" with this in mind. It cannot be too complex, nor can the derivations needed to determine which specific actions are required be too complex. This means I think that the ideal code will contain a plurality of principles, but not too great a plurality. It presumably will prohibit torture, for example, and its prohibition of torture might be fundamental, since it might be best to build in such a prohibition at a fundamental level. That is, the moral code whose currency would best contribute to society's ability to meet its needs might build in such a prohibition at a fundamental level.

In the next two sections of the paper, I will discuss two families of arguments that point us to views about the status of animals in society-centered theory. First are arguments from psychological boundaries. Second are arguments from societal values.[51]

7. Psychological Boundaries

In this section, I begin by arguing that the society-centered theory supports a fundamental moral duty or virtue of compassion toward animals. I then argue that the theory supports the idea that there is a fundamental, defeasible duty not to kill animals, although the duty not to kill a human is more stringent, other things being equal, than the duty not to kill an animal. The idea of psychological boundaries is central to both arguments.

As social engineers with the goal of designing a moral code whose currency will best enable society to meet its needs, we should bear in mind that the content of the moral code a person subscribes to will affect her motivational states. When we see a kind of behavior that, if widespread, would contribute to society's ability to meet its needs, we should aim to design our moral code to encourage or perhaps to require such behavior. A society needs to promote conditions in which social harmony and stability can flourish. This need will be better served if the members of society tend to show compassion and kindness in interacting with one another than if they do not. Because of this, and because we see that the content of the moral code we choose for society can affect people's sense of compassion, we will choose a moral code that calls for kindness and compassion.

Kindness and compassion are emotions or attitudes that are elicited by certain needs in others. Those who are kind and compassionate tend to respond with kindness and compassion to all beings that have the relevant needs, including animals, once the needs are noticed and made salient. Our psychology cannot easily and reliably be tuned so that we respond with compassion and kindness only to humans. True, our tendency to respond to others with kindness and compassion is acquired and shaped in childhood. It is shaped by the culture and by the expectations of our society's moral code. If the local culture is racist, for instance, or sexist or xenophobic, or if it ridicules those who respond with concern to the suffering of animals, then our kindness and compassion might be narrowed to beings who are members of our own race or our own society or our own species. Despite this, however, a person who is disposed to respond with compassion and kindness toward *some* of those whom she sees to be in need also has a tendency to respond with kindness and compassion to any being that she sees to be similarly in need. This tendency might be suppressed to some degree, by a racist culture, for example, but it is there. It is not easy for us to see the boundaries of our sex or our race or our society or our species as boundaries beyond which beings are not relevantly similar to us in their capacity to be in need. If we have *no* tendency to respond to those from outside with compassion and kindness, even if they are in need and this is salient to us, then our tendency to respond to those from inside our society with kindness and compassion must be weaker than it otherwise might be, other things being equal.

Kindness and compassion tend to be elicited by certain needs. Humans from outside our society can have these needs, as can (nonhuman) animals that are sentient and experience pain and suffering. And so, assuming we have any tendency to exhibit kindness and compassion in the face of need, animals tend to elicit compassion and kindness if we see them as sentient and capable of pain and suffering.

Kant claimed that "he who is cruel to animals becomes hard also in his dealings with men."[52] I would add that he who is cruel to people from outside his society becomes hard in his dealings with members of his society. Schopenhauer claimed that "boundless compassion for all living beings is the firmest and surest guarantee of pure moral conduct."[53] I would say it supports our compassion for members of our own society. These remarks, and the argument of the preceding paragraphs, imply a number of generalizations about "psychological boundaries."

These are empirical generalizations and they might be false. There are exceptions, as I have conceded, because there are cases in which people who, although compassionate toward members of one group, lack compassion toward members of other groups that are different. Some racists illustrate this. But it is enough if the generalizations are true or are sufficiently likely to be true. As social engineers, we are looking for a moral code that ideally would have currency among all the members of society. The question is whether to choose a code that permits people to treat those from outside the society as well as nonhuman animals differently from members. Such a code would permit us to tailor the boundaries of our sentiments of compassion and kindness so they coincide with the boundaries of our society or of the species. There would ordinarily be no reason to prefer a code of this kind, and since the currency of such a code might weaken the tendency to feel sentiments of kindness and compassion toward fellow members of society, there is ordinarily a reason to avoid such a code.

To summarize, a society needs its members to show compassion and kindness in their dealings with one another, but it is likely to do better at meeting this need if the societal code calls for compassion and kindness toward those in need *period* than if the code explicitly limits its call for compassion to human members of the society. If this is correct, there is reason to think that the ideal moral code would call on people to show compassion and kindness to all those in need, including animals. A society would have to be in especially difficult circumstances for its ability to meet its needs to be hurt by its members' compassion toward animals. Non-sentient animals are automatically excluded, because it is not possible to show compassion and kindness toward animals that one sees to be non-sentient any more than toward rocks. There is, then, an argument for a moral duty or virtue of compassion toward animals that relies only on considerations about societal needs. If we imagine ourselves as social engineers designing a moral code whose currency is to best enable a society to meet its needs, and remembering that societies can be expected to do better at this when their members show kindness and compassion to other members, we ought to choose a code that calls for kindness and compassion to those in need *period*. A restriction to members of the society, or to members of the species, would needlessly risk weakening or undermining the strength of people's disposition to show kindness and compassion to members of the society.

I turn now to the duty not to kill. The ideal moral code would include a prohibition on killing human beings *period*, including of course nonmembers as well as members of the society. In general, we should expect the ideal moral code to give nonmembers the same moral status as members to the extent that this does not undermine a society's ability to meet its needs. A society may need to restrict possession of some of the rights that it recognizes, so that only its members have rights against its resources.[54] But there is no similar need for a moral code according to which the members of society have *no* duties to nonmembers, and I have been arguing that considerations about psychological boundaries provide an argument against such a code.

What about animals? Again, imagine ourselves to be social engineers. As social engineers, we should see that we need members of society to have a strong inhibition against killing other members of the society. Ordinarily, however, people come into close contact mainly with other members of their own society and there is no need to risk weakening their inhibition against killing by choosing a code that permits killing members of other societies. And permitting killing members of other societies would risk undermining society's ability to achieve needed peaceful and harmonious relations with other societies. So as social engineers, we ought to choose a moral code that prohibits killing other people, with very few exceptions, such as killing in self-defense. Similarly, there is no need to risk weakening people's inhibition against killing by permitting killing members of other species, even though members of other species are physically different from members of our own species in ways that are salient to us. There is plausibly an exception, however, when a choice must be made between an animal and a human or when there is a significant prospect that the welfare of members of the society can be enhanced to some significant degree by putting an animal's well-being or life at risk. The theory therefore supports the idea that there is a defeasible duty not to kill animals, although the duty not to kill a human being is more stringent, other things being equal, than the duty not to kill an animal.

The arguments of this section have shown that a fundamental duty or virtue of compassion toward animals, and a fundamental, defeasibile, duty not to kill animals, can be supported by the society-centered theory. These duties plausibly are included in the ideal code, and so they are morally fundamental. Of course, as should be clear, the arguments turn on the truth of certain empirical generalizations. The important point, however, is that there is nothing in the society-centered theory that rules out the thesis that nonhuman animals have moral standing at a fundamental level.

8. Societal Values

In the preceding section, I argued that there is reason to think that the ideal code for a society would call for people to show kindness and compassion to animals, and this argument did not depend on invoking any premises about societal values. The same argument tends to support the idea that the ideal code would call on people to protect animal welfare, because kindness and compassion toward animals that are suffering or in need tend to motivate actions aimed at promoting or protecting animal welfare. As I will show in this section, a stronger argument can be made in cases in which a society values animal welfare. The point of this section is to explain the difference that cultural tradition can make to the implications of society-centered theory regarding our duties to animals.

Suppose that a society values animal welfare, in that, although its members give priority to promoting and protecting the welfare of human beings, they nearly unanimously want to promote and protect the welfare of animals as well. Suppose that this desire is stable and endorsed by people, so that they do not wish they could rid themselves of it. In this case, on the account I have given, the ideal code for the society would call on people to promote and protect animal welfare. For there is no reason to think that the currency of such a code would in any way undermine the society's ability to meet its needs by comparison with what would be the case if a different moral code were to have currency.

That is, this seems to be a case in which there is no conflict between societal needs and values. Perhaps codes M1 and M2 would serve the society's needs equally well, if either were the social moral code, but suppose code M1 would better serve the society's values because it calls for people to promote and protect animal welfare. In such a case, code M1 would qualify as the ideal code for the society. We can think of M1 and M2 as having a core in common, but M1 has an additional normative characteristic to reflect the society's values.

Different societies do of course have very different attitudes toward animals. To illustrate what this can imply regarding our duties toward animals, I will consider three examples. Two of the examples pull in the direction of requiring greater concern for the well-being of certain animals than do the arguments we have considered so far, and one pulls in the opposite direction.

Begin with the latter example. Some cultures appear to tolerate astonishing cruelty to animals,[55] and some, of course, tolerate killing animals for food. This does not gainsay the arguments of the preceding section, unless the existence of such cultures is taken to undermine my thesis that, other things being equal, as social engineers we ought to choose a moral code that rules out cruelty to animals or killing animals in order to avoid the risk of undermining people's reluctance to be cruel to or to kill other members of their own society. I think the existence of such cultures might instead, in some cases, illustrate situations in which "other things" are not "equal." Suppose a society has a strong cultural tradition that permits killing animals for food. There is a risk that, if the ideal code permitted killing animals for food, this circumstance would tend to undermine or weaken people's inhibition against killing human beings. If the cultural tradition is strong enough, however, the risk might be very low. So it is arguable that there are societies in which, despite the arguments I have been giving, society-centered theory implies that it is permissible to eat animals. These are societies in which there is a strong cultural tradition that tolerates killing animals for food.

For a second example, consider cultures that accord a special status to animals that fall into the category of "pets" or "companion animals." Even in a culture that does not frown on killing some animals for food, there can be a prohibition on killing dogs, cats, and horses for food. In such societies, the society-centered theory may say that it is not permissible to eat animals of these kinds. In such societies, it is arguable that the ideal code would include special protections for such animals.

The third example is a culture that values the welfare of certain kinds of animals *more* than human welfare. Consider a society in which cows are taken to have such a special status. Some parts of India might provide actual examples. In any event, imagine a hypothetical case in which people would prefer that, in cases of conflict, societal resources be used to further the welfare of a cow rather than the welfare of a human. Suppose that this preference is nearly unanimous and that it is stable and endorsed by nearly everyone. We can still say that, except for cases where there is conflict between the welfare of cows and the basic *needs* of humans, the story is just as I have been telling it so far. The ideal code for the society would be one whose currency would lead to the best overall realization of its needs, and the society's values play no role in setting moral requirements except insofar as these values do not lead to behavior that undermines the society's ability to meet its needs. However, if the ideal code for this society included a requirement to favor the welfare of cows over the welfare of humans, the society's ability to meet its needs *would* be weakened unless the requirement were restricted to cases in which no humans would be deprived of their ability to meet their basic needs. As I have argued elsewhere, a society's needs can best be met only if the basic needs of its members are met to a decent minimal level and with rough equality.[56] If I am correct, the society-centered theory would say that, in such a society, the culture is morally unacceptable if it calls for depriving a person of her ability to meet her basic needs for the sake of a cow. Despite this, however, the theory would say that, in the society of our hypothetical example, there is a moral requirement to favor the welfare of cows over the welfare of other animals.

9. A Simplifying Complication

One might object at this point that when we think about the moral status of animals we do not normally have any specific society in mind. We want to know the status of animals *period*. Moreover, if someone were to raise the general question whether the existing values of a society can make a difference to the status of animals in that society, we would ordinarily take the question to be a sociological question rather than a normative moral question. But there is no objection here to the society-centered view. To explain this, I need to introduce a complication.

Notice that societies can overlap and can be nested in larger societies.[57] Basque society overlaps both the Spanish and French societies, for example, and all three of these societies are contained in the larger European society. When we think about or discuss a moral issue, the society-centered theory says that the ideal code of *some* society is relevant to determining the truth value of the moral claims we make, but, according to the theory, which society is the relevant one depends on the context. It depends on our intentions and understandings and on the people we are addressing or whom we have in mind. According to the theory, the relevant society is the smallest one that includes all the parties to the conversation as well as everyone in the intended audience and everyone who is otherwise referred to or thought about and everyone who is "quantified over" in the conversation or in the thoughts of the parties.[58] So, for example, if a Frenchman and a Spaniard are discussing whether bullfighting is a morally acceptable practice, the relevant society might be the European society as a whole. It is more likely, however, to be the global society, the society of all human beings,[59] for the Frenchman and Spaniard likely have in mind everyone and every social context, without restriction. It is unlikely that they would dismiss as irrelevant someone's remark about an African practice that is similar to bullfighting. When philosophers discuss moral issues, they too typically have in mind an unrestricted quantification that has every human being in its scope. So if someone in such a discussion claims that bullfighting is morally impermissible, she likely intends to claim that it is impermissible for any person, without restriction. In cases of this kind, the global society is the relevant one.

Bearing in mind this complication about which society is the one whose ideal moral code is relevant in a given context, we can simplify what was said about societal values in the preceding section. Larger societies are less likely than smaller ones to have any societal values. Recall that societal values are stable preferences held and endorsed with near unanimity by the members of the society. Larger societies are likely to be more diverse, so their members are less likely to share values. It is likely, then, that societal values play no role in discussions of the moral status of animals that are of philosophical interest. For it is likely that the ideal code that determines the truth value of philosophically interesting claims about the moral status of animals is the one whose currency would best serve the global society's needs.

It nevertheless seems plausible that this ideal code would include a standard calling on people not to violate the cultural standards of specific local societies provided that this is compatible with compliance with the rest of the standards in the code. The ideal code plausibly would require people to respect local prohibitions on the eating of animals and local requirements that companion animals be shown special concern by comparison with other animals. For in these cases there is not likely to be any conflict between complying with the ideal code that best serves the society's needs and complying with local cultural standards.

10. Conclusion

Whether we are philosophers or lay-people, when we think about the moral status of animals, we do not take up the kind of social engineering perspective I have been recommending. We do not try to decide which moral code for our society would best meet the criteria proposed in the society-centered theory. To do so would be to take a metatheoretical point of view on moral reasoning rather than to engage in moral reasoning, which is what we normally do when faced with such questions. We reason from the moral point of view. We offer moral reasons or moral arguments for our view about the moral status of animals. This is to be expected in the society-centered view. Indeed, this point figures in my argument that society-centered theory supports the thesis of the fundamental moral status of animals. Let me explain.

I argued that the society-centered theory supports a duty or virtue of compassion toward animals. I argued as well that it supports the idea that there is a defeasible duty not to kill animals and a duty to protect animal welfare. That is, I argued that the ideal moral code that determines the truth value of philosophically interesting claims about the moral status of animals includes, at a fundamental level, moral standards that underwrite the virtue of compassion or the duty to show compassion to animals and the duties not to kill animals and to protect animal welfare. This means that moral reasoning about the treatment of animals that responds to the truth about these matters would take these duties to be underived from anything morally more basic. It follows that the theory supports the thesis of the fundamental moral status of animals. More important, it supports duties to treat animals decently, to show them compassion and to protect their welfare.

My central goal in this chapter has been to argue that the thesis of the fundamental moral status of animals is compatible with the thesis of the fundamental concern of morality. The thesis of the fundamental concern of morality can seem to be "speciesist," because it views morality as a device with the function of advancing human interests. It holds that morality is concerned to advance human welfare by enabling human beings to live together successfully in societies. This thesis might seem to imply that we have no basic or fundamental duties to animals. I have tried to show, however, that the society-centered theory offers a plausible interpretation of the thesis and, moreover, that the theory plausibly implies that there are fundamental moral requirements to protect animal welfare and to show animals compassion. The content of our fundamental duties depends on facts about human psychology and the circumstances of human life. These facts plausibly imply that our obligations to treat animals decently are as fundamental as any of our obligations toward human beings.[60]

NOTES

1. In what follows, unless I indicate otherwise, I will suppress the adjective "nonhuman" and use the unmodified term "animal" to speak of nonhuman animals. I use the vague term "decently" here as a placeholder. We will want to explain what "decent treatment" amounts to.

2. See Brad Hooker, *Ideal Code, Real World* (Oxford: Oxford University Press, 2000), pp. 66–70.

3. Joseph Raz, *The Morality of Freedom* (Oxford: Oxford University Press, 1986), p. 267.

4. Kurt Baier, *The Rational and the Moral Order: the Social Roots of Reason and Morality* (Chicago: Open Court, 1995). I also defended a view of this kind in David Copp, *Morality, Normativity, and Society* (New York: Oxford University Press, 1995).

5. A "defeasible duty" is one that can be overridden, so it might not be a duty all-things-considered. Consider the duty not to lie. This is a defeasible duty since there are circumstances in which it would be permissible to tell a lie, given everything that is at stake. For example, it might be permissible, all things considered, to tell a lie in order to save a person's life.

6. That is, there is a widely shared moral view that accepts this proposition. This is a sociological claim although, of course, the proposition that it is wrong to torture animals just for our amusement is a normative moral claim. I take the example from Hooker, *Ideal Code*, p. 66.

7. For a discussion of dumbfounding, see Jesse Prinz, *The Emotional Construction of Morals* (New York: Oxford University Press, 2007), pp. 29–32.

8. This account of what it is for a belief to be fundamental to commonsense moral thinking will suffice for my purposes. I do not claim that beliefs that are fundamental in this sense are "self-evident."

9. Kwame Anthony Appiah cites a story told by the historian Martin Gilbert in which a woman stops a group of villagers from throwing a girl down a well by saying, "She's not a dog after all." See Appiah, *Experiments in Ethics* (Cambridge: Harvard University Press, 2008), p. 160, citing Martin Gilbert, *The Righteous* (London: Black Swan, 2003), p. 11.

10. Is the difference between *Homo sapiens* and *Homo erectus* morally significant in itself? Given a Darwinian perspective, this is difficult to accept. For useful discussion, see R. G. Frey, "Moral Standing, the Value of Lives, and Speciesism," in *Ethics in Practice*, ed. Hugh LaFollette, 3rd ed. (Oxford: Blackwell, 2007), pp. 192–204. Originally published in *Between the Species* 4 (1988): 191–201.

11. For this useful framing of the issues, I am grateful to Frey, "Moral Standing, the Value of Lives, and Speciesism."

12. Jeremy Bentham, *Introduction to the Principles of Morals and Legislation*, 1st published, 1789, ch. 17. In J. H. Burns and H. L. A. Hart, eds., *The Collected Works of Jeremy Bentham* (Oxford: Oxford University Press, 1996), p. 283.

13. Hooker, *Ideal Code*, pp. 33, 37–43.

14. Hooker, *Ideal Code*, p. 68.

15. A kind of action is "defeasibly" wrong if there is a defeasible duty not to do actions of that kind. For example, it is wrong to tell a lie, but this is only defeasibly wrong since there are circumstances in which it would be permissible to tell a lie, given everything that is at stake.

16. L. W. Sumner, *Welfare, Happiness and Ethics* (Oxford: Oxford University Press, 1996), pp. 184–223.

17. Sumner, *Welfare, Happiness and Ethics*, p. 185. Following Sumner and many others who write about these issues, I will use "welfare" and "well-being" interchangeably.

18. This position is taken by Tom Regan, "The Case for Animal Rights," in LaFollette, *Ethics in Practice*, pp. 208–9.

19. Immanuel Kant, *Grounding for the Metaphysics of Morals*, trans. James. W. Ellington (Indianapolis, Ind.: Hackett, 1981), p. 35 (Ak., 428).

20. Donald H. Regan, "Why Am I My Brother's Keeper?" In *Reason and Value*, ed. R. Jay Wallace et al. (Oxford: Oxford University Press, 2004), pp. 202–230, at p. 203.

21. Connie Rosati, "Objectivism and Relational Good," *Social Philosophy and Policy* 25 (2008): 314–49, at pp. 318–20.

22. Rosati, "Objectivism and Relational Good," p. 343.

23. Rosati, "Objectivism and Relational Good," p. 344.

24. Rosati, "Objectivism and Relational Good," p. 344. Rosati is not thinking that we need to suppose that all persons are *morally* good. What we must suppose is that people have value in a sense that entails that they are sources of legitimate moral claims.

25. "Man, and in general every rational being, exists as an end-in-himself," and so "has an absolute worth." Kant, *Grounding*, trans. Ellington, p. 35 (Ak., 428).

26. Peter Singer, "All Animals are Equal," in LaFollette, *Ethics in Practice*, p. 175.

27. Sumner, *Welfare, Happiness and Ethics*, pp. 178–79.

28. Rosati, "Objectivism and Relational Good," p. 344.

29. Frey, "Moral Standing, the Value of Lives, and Speciesism."

30. Frey, "Moral Standing, the Value of Lives, and Speciesism," p. 197.

31. Frey, "Moral Standing, the Value of Lives, and Speciesism," p. 196.

32. Frey, "Moral Standing, the Value of Lives, and Speciesism," pp. 192–93.

33. Frey, "Moral Standing, the Value of Lives, and Speciesism," p. 194.

34. Frey, "Moral Standing, the Value of Lives, and Speciesism," p. 198.

35. Frey, "Moral Standing, the Value of Lives, and Speciesism," p. 198.

36. Frey, "Moral Standing, the Value of Lives, and Speciesism," p. 198.

37. Frey, "Moral Standing, the Value of Lives, and Speciesism," p. 197.

38. Frey does not himself go in this direction. Frey, "Moral Standing, the Value of Lives, and Speciesism," p. 198.

39. Tom Regan claims, for example, that all sentient creatures have "inherent value." The only explanation he offers of what he means by this is that all sentient creatures have the right not to be treated in certain ways. This is not an interesting explanation of why it is wrong to treat these creatures in these ways. Regan, "Case for Animal Rights," p. 209.

40. In this section, I mainly follow the exposition of my view in David Copp, "Toward a Pluralist and Teleological Theory of Normativity," *Philosophical Issues* 19 (2009): 21–37. I sometimes use wording from that paper. I originally presented the view in Copp, *Morality, Normativity, and Society*. I presented a better version, which I call the "basic society-centered theory," in Copp, *Morality in a Natural World* (Cambridge: Cambridge University Press, 2007), pp. 18–21. This chapter presents the so-called basic theory.

41. J. L. Mackie, *Morality: Inventing Right and Wrong* (Harmondsworth, England: Penguin, 1977), p. 111.

42. Here I follow Copp, *Morality in a Natural World*, pp. 14–18. A "pure" moral proposition has no non-moral entailments or presuppositions (other than those given by the standard-based theory itself). A "basic" moral proposition ascribes a moral property to something. The proposition that Smith was wrong to steal Jones's car is impure but basic. The proposition that theft is wrong is both pure and basic.

43. Mackie, *Morality*, p. 111. See also G. J. Warnock, *The Object of Morality* (London: Methuen, 1971).

44. This account raises a number of questions. What is a society? What are the needs of a society? Which is the "relevant" society in a given context? I will have to set aside most of these issues since they are not relevant to my concerns here. I have addressed them elsewhere. See Copp, *Morality, Normativity, and Society*, pp. 124–28, 192–200, 218–23; *Morality in a Natural World*, pp. 16–26. See also section 9, below.

45. It is the code whose currency in the relevant society would enable the society better to serve its basic needs than would the currency of other sets of rules and better than would be the case if no set of rules had currency in the society.

46. I follow the formulation of the "basic" society-centered theory in Copp, *Morality in a Natural World*, p. 17, but I add the qualification about values. For this, see Copp, *Morality, Normativity, and Society*, p. 206.

47. If several codes are tied as best, then society-centered theory would say that the truth conditions of a moral claim depend on the content of all the moral codes tied as best. We would say, for instance, that torture is morally wrong if and only if it is permitted by none of the moral codes tied as best. On ties, see Copp, *Morality, Normativity, and Society*, pp. 198–99.

48. I use the example of economic beliefs in Copp, *Morality in a Natural World*, pp. 73–75. See also p. 23.

49. Hooker, *Ideal Code*.

50. See Hooker, *Ideal Code*, chap 4. I address important and common objections in Copp, *Morality, Normativity, and Society*, pp. 213–45 and in *Morality in a Natural World*, pp. 25–26, 55–150, 203–83.

51. I follow the arguments in Copp, *Morality, Normativity, and Society*, pp. 204–7. I do not now agree with everything I said in those pages, but here I simply ignore my earlier mistakes.

52. Immanuel Kant, *Lectures on Ethics*, trans. by F. Max Muller (London: Methuen, 1930), p. 230.

53. Arthur Schopenhauer, *On the Basis of Morality*, trans. by E. F. J. Payne (New York: Bobbs Merrill, 1965), sect. 19. I owe this reference, and the reference in the preceding note, to Mary Midgley, *Animals and Why They Matter* (Athens: University of Georgia Press, 1983), pp. 51–52.

54. I have in mind the right that the resources of society be used to enable each person to meet her basic needs. Unless a society were very rich, it could not sustain a standard that called for its resources to be used to enable *every human without restriction* to meet her basic needs.

55. See Fuchsia Dunlop, *Shark's Fin and Sichuan Pepper: A Sweet-Sour Memoir of Eating in China* (New York: W.W. Norton, 2008), pp. 48–53.

56. For this argument, see Copp, *Morality, Normativity, and Society*, pp. 201–3.

57. For the concept of society that I am assuming here, see Copp, *Morality, Normativity, and Society*, pp. 124–43.

58. This is the default case, but circumstances can change things. I will not discuss this complication. See Copp, *Morality, Normativity, and Society*, p. 221.

59. For reason to think there is such a society, see Copp, *Morality, Normativity, and Society*, pp. 139–40.

60. I am grateful to Tom Beauchamp and Ray Frey for very helpful comments and suggestions.

SUGGESTED READING

Copp, David. *Morality, Normativity, and Society*. New York: Oxford University Press, 1995.
——— . *Morality in a Natural World*. Cambridge: Cambridge University Press, 2007.
——— . "Toward a Pluralist and Teleological Theory of Normativity." *Philosophical Issues* 19 (2009): 21–37.
Dunlop, Fuchsia. *Shark's Fin and Sichuan Pepper: A Sweet-Sour Memoir of Eating in China*. New York: W.W. Norton, 2008.
Frey, R. G. "Moral Standing, the Value of Lives, and Speciesism." In *Ethics in Practice*, edited by Hugh LaFollette, 3rd ed. Oxford: Blackwell, 2007.
Hooker, Brad. *Ideal Code, Real World*. Oxford: Oxford University Press, 2000.
Kant, Immanuel. *Lectures on Ethics*. Translated by F. Max Muller. London: Methuen, 1930.
Mackie, J. L. *Morality: Inventing Right and Wrong*. Harmondsworth, England: Penguin, 1977.
Midgley, Mary. *Animals and Why They Matter*. Athens: University of Georgia Press, 1983.
Regan, Donald H. "Why Am I My Brother's Keeper?" In *Reason and Value*, edited by R. Jay Wallace et al. Oxford: Oxford University Press, 2004.
Regan, Tom. "The Case for Animal Rights." In *Ethics in Practice*, edited by Hugh LaFollette, 3rd ed. Oxford: Blackwell, 2007.
Rosati, Connie. "Objectivism and Relational Good." *Social Philosophy and Policy* 25 (2008): pp. 314–19.
Singer, Peter. "All Animals are Equal." In *Ethics in Practice*, edited by Hugh LaFollette, 3rd ed. Oxford: Blackwell, 2007.
Sumner, L. W. *Welfare, Happiness and Ethics*. Oxford: Oxford University Press, 1996.
Warnock, G. J. *The Object of Morality*. London: Methuen, 1971.

....

CHAPTER 11

HUMAN ANIMALS AND NONHUMAN PERSONS

....

SARAH CHAN AND JOHN HARRIS

WHAT defines a person and the necessary attributes for personhood remain the subject of extended philosophical discussion. It is also not established that being "human" is either a necessary or a sufficient condition for personhood, universal rights, dignity, moral status, or basic moral and political protections. The term "person" in the everyday sense is often used interchangeably and is taken to be synonymous with the term "human." Philosophers, however, tend to use the word as a term of art, in a more abstract sense: persons are those entities who possess a particular moral status and about whom particular moral claims may be made on the basis of that status.

Nevertheless, "human" continues to be used widely as an indicator of privileged moral status and is regarded as a quality of morally significant beings—that is, of ourselves. Accordingly, we talk of "human dignity" as an essential property that should be respected or protected, while "human rights" are similarly assumed as the natural patrimony of human beings simply because they are human.

This usage perhaps reveals a deep-seated confusion resulting from a conflation of two concepts, "human" and "person." The assumed semantic equivalence of these terms implies that humans are the most important, even the only types of creatures to whom moral status should be accorded, and that all humans are entitled to claim such status by virtue of their membership in this—if not exclusive, at least excluding—club. Conversely, this construction automatically excludes nonhuman creatures from consideration as persons. The focus on humanness as an indicator of moral status draws an implicit line: entities which are human are different, morally speaking, from entities which are not. But why should this be the case? Are there any good moral reasons to limit the status of "person" to humans alone, or to include all humans in the privileges and protections usually embraced by concepts of human rights and dignity?[1]

In this chapter, we address the question of whether a nonhuman animal, or by implication, whether any other nonhuman organic creature or animate machine, could ever be considered a person, and explore the implications of the answer. In section 1, we revisit personhood theory and consider what attributes make a person, why those attributes are important, and hence why persons and personhood are important, morally speaking. In section 2, we consider what this account of personhood would require of nonhuman animals for them to be deemed persons, and whether and how we might determine if those requirements are satisfied, in order to answer the central question. Section 3 briefly explores some of the implications of animal personhood, politically and legally as well as philosophically. We also consider the implications of personhood theory and the possibility of nonhuman personhood in the context of so-called enhancement technologies, an especially pertinent issue because the potential of these technologies to create nonhuman persons makes it all the more imperative to address these issues as a matter of practical ethics, not just theory.

1. Personhood and Moral Status[2]

Consider the question: "Are there persons on other planets?" Although we do not know the answer to this question, we do know some of the things that would convince us that we had found such extraterrestrial *persons* by contrast to *nonpersons*. We have a good idea of what we are looking for when we look for people or evidence of people on other worlds. We will start with what we are *not* looking for.

First, we are not looking for other humans, however amazed we might be to find them. We do not expect persons on other planets, if there are any, necessarily to be members of our own species. Second, we are not confining our search to organic life forms; it may be that we will become convinced that, for example, self-constructing machines of sufficient intelligence would count as persons. Third, we are not looking for nonpersonal life forms, although we may also find these and be pleased, or at least interested, if we do. These observations show us that we do not, in fact, regard species membership as hugely significant in trying to understand what a person might be and in responding to the important similarities that might exist between them and us.

What, then, are we looking for? What should convince us that we had discovered persons, that is, morally significant individuals whose lives are of value and should be valued, on other planets?

Suppose that instead of us discovering them, they discovered us. Having demonstrated their vastly superior technology by arriving on Earth after traversing unimaginable interstellar distances, the extraterrestrials are hungry and tired after their long journey. What could we point to about ourselves that ought to convince the extraterrestrials that they had discovered persons, morally significant beings of special importance, on another planet? What could we say of ourselves that should convince them of the appropriateness of "having us for dinner" in one sense rather than another; what should convince them to treat us as dinner guests rather than the dinner itself? What makes for a moral distinction between ourselves and other edible life forms, lettuces or turnips, cats, canaries, or chickens?

Toward the end of the seventeenth century, in his *Essay Concerning Human Understanding*, philosopher John Locke attempted to answer this question in a way that has scarcely been surpassed. He wrote:

> We must consider what person stands for; which I think is a thinking intelligent being, that has reason and reflection, and can consider itself the same thinking thing, in different times and places; which it does only by that consciousness which is inseparable from thinking and seems to me essential to it; it being impossible for anyone to perceive without perceiving that he does perceive.[3]

It is beings possessed of these capacities, or something closely akin to them, that we are looking for when we ask the question, "Are there persons on other planets?" This account is neutral as to species, but it identifies those features, the presence of which for example, in space creatures should surely convince us that we had at last encountered persons elsewhere in the universe. We must hope that if it is others of vastly superior technology that are asking the question, that they also recognize us as being fellow creatures of moral standing, fellow persons.

Can someone be more or less of a person? All of the elements in Locke's definition—intelligence, the ability to think and reason, the capacity for reflection, self-consciousness, memory, and foresight—are capacities that admit of degrees. Does this lead us into a hierarchy of persons and hence of moral importance or value? Suppose you were asked to write down in rank order of importance, the hundred things that made life valuable, worth living, for you. Of course there would be no clear rank order in importance for many of the items, and many people would have died laughing long before the list had reached a hundred items. Some lists would tend toward the prurient, others toward the exalted. The philosophical interest of the exercise, contrasted with its "human" interest, lies not in the contents of the list, but rather in the fact that for anyone thinking about this exercise, or about meeting Locke's criteria, there are likely things that make life valuable and worth living, or valueless and hence worth not living.

The importance of the exercise lies not in what is on the list, nor some moral or objective evaluation of what is or might be on the list. The significance of the thought experiment lies in the fact that it identifies a particular sort of being, *a being that can value existence and hence can have a valuable life*. Our suggestion, then, is that if we ask which lives are valuable in the ultimate sense, and which lives are capable of being valued by the individuals whose lives they are, we are in effect asking, Which lives are the lives of persons? The answer will be: the lives of any and every creature, whether organic or not, who is capable of valuing his or her own existence.

The reasons why existence is valued, and the extent to which it is valued, are irrelevant to this question, although they may be relevant to other questions. Thus, the question as to which individuals have lives that are valuable in this sense is a threshold one: anyone capable of valuing existence, whether they do or not, and regardless of the extent or passion with which they do, is a person in this elementary sense.

2. WHAT DOES PERSONHOOD DO?

Personhood theory attempts a systematic account of the qualities that enable individuals who possess them to lead lives of value. That a life has intrinsic value, by which we mean value to the person whose life it is, is demonstrated by the fact that the individual in question is capable of forming a view about whether they wish the life to continue or not. If they wish the life to continue, it has positive value; if not, it has negative value (for them at least). For a creature to be capable of valuing existence, she needs to meet something like Locke's criteria. For to value existence is to have a view about the desirability of existence continuing, and to do that one needs to be aware of oneself as existing over time, to have a rudimentary awareness of what future existence might be like, and to know whether one wants to experience that future existence or not.

This account is the basis of a moral theory of personhood. The elements of that moral theory are in place when individuals understand that other people are relevantly like themselves in terms of moral status, and accept that not only do their own lives have value but that there are compelling reasons to respect the value that other lives have to the individuals whose lives they are.

This capacity for reciprocity and understanding is a function of the combination of language and of social life, and it is probable that no linguistically competent community could exist and survive without it. Such understanding may have its origins in primitive feelings of sympathy or empathy, or in merely prudential reciprocity. It may also be part of more complex conceptions of altruism, such as that found in the golden rule ("do unto others as you would have them do unto you"), and it is, moreover, consistent with an evolutionary account of the function and benefits of morality.

Thus, the principle of respect for persons derives from, rather than being a necessary adjunct to, the theory of personhood. It is a special case of application of a general moral theory to a particular class of beings. Personhood theory thus plausibly (although probably not definitively or exclusively) explains and justifies the special value accorded to certain types of beings. There is a danger here of anthropocentrism, but we are here guarding against it. Any theory of moral status or value must try to master this trick.

3. WHY PERSONHOOD?

The concept of "personhood" has been criticized as being too vague to be useful in normative moral analyses, and some writers have suggested that it ought to be bypassed in favor of an approach that addresses directly the various criteria on which different accounts of personhood are based.[4] So why *is* personhood useful to us—what are we concerned with in this analysis? Why are we asking if an animal could be a person, and what would be the implications of the answer?

Personhood theory argues that whatever moral status humans have cannot coherently derive from the fact of membership of a particular species but rather, if it exists at all in a form which allows us to make moral distinctions between living creatures, must flow from some properties that some members of that species have, which are lacked by those creatures judged to be either of different or of lesser status. For instance, Tom Beauchamp's view is that "it is fortunate for animals and humans who lack moral personhood that moral standing does not require personhood of any type. Some creatures have moral standing even though they do not possess even a single cognitive or moral capacity. The reason is that certain *noncognitive* and *nonmoral* properties are sufficient to confer a measure of moral standing."[5] Animals with this sort of moral standing may have a "right" not to have pain inflicted in virtue of their sentience, but unfortunately for them, this sort of moral standing can confer no right not to be killed painlessly. True, moral standing does not require personhood, but personhood confers a particular *level* of moral standing, which carries with it a much more "fortunate" set of rights than those available to nonpersons—including the right not to be killed at all, painlessly or otherwise. Thus, the alleged "failure of theories of personhood" asserted in the title of the article by Beauchamp turns out not to be a failure at all but a success with important consequences. It is useful because it makes plausible distinctions that are otherwise arbitrary and indefensible.

In *The Value of Life*, John Harris attempts to make plausible and defensible the moral content of the cognitive elements of personhood.[6] Personhood, although its most common application is in telling us which sorts of lives are valuable and hence which sorts of creatures might have rights to life, is useful as a threshold concept, because the ability to value existence in a hundred or more ways confers meaning not only on existence itself but on all the "ways of living" in which we might celebrate or experiment with our existence.

The use of "personhood" as a moral concept does not preclude us from considering moral problems according to their specific dimensions rather than solely in terms of persons; nor does it commit us to the position that personhood is the only form of moral standing and that nonpersons can have no moral claims on us. Although some might argue that this conception of personhood renders the concept itself redundant other than as a philosophical exercise, we nevertheless think that personhood remains a valuable moral concept that can contribute usefully to normative ethical analyses. This is because the use of "person" as a heuristic moral category, that is as philosophical shorthand for creatures who matter morally *in a certain way*, influences our attitudes toward the sorts of creatures who might be persons, in all frames of reference from the colloquial to the political, legal and philosophical.

Beauchamp maintains that the problem with most theories of species-neutral personhood is that either the criteria are sufficiently broad that we cannot but accept that some animals are persons, or sufficiently narrow that they exclude many humans.[7] The fact that this is perceived as problematic indicates both a persistent, underlying agenda that seeks to find reasons why humans should retain unique moral status and a reluctance to admit that some humans may not have the sort of lives that confer that unique status. This simply increases the imperative to investigate the question of animal personhood independent of bias in favor of creatures who manifestly benefit from the bias and have a powerful interest in its continuance.

It is true that we could consider all issues of animal ethics purely in terms of the specific properties of the animals involved and whether our actions toward animals are acceptable in light of those properties. The kinds of questions asked, however, tend to proceed from the assumption that animals do not require particular moral consideration and that special reason needs to be shown to accord this to them. The introduction of the personhood concept to animal-ethics discourse and the resulting idea that animals could be persons might encourage us to reframe these questions in a less "speciesist" way, to ask why certain acts should be acceptable rather than whether there is any reason why they are not.

With this in mind, we turn now to consideration of the central question in this chapter: Could a nonhuman animal in fact be a person?

4. ANIMALS AS PERSONS

Thus far, we have outlined a theory of personhood that sets out certain criteria by which we may determine whether a given entity is a person. Central to this definition of personhood is the possession of a certain cognitive capacity: the capacity to value one's own life.

When we ask whether a nonhuman animal could be a person, in a strict logical sense the answer is straightforward: since the criteria laid out for personhood do not include "human," there is no reason why it could not. The material question in which we are really interested, however, is whether it is possible or likely that a nonhuman animal might possess the qualities required for personhood, in particular the capacity to value its own life, or whether such qualities are unique to human animals. Are nonhuman animals capable of mental states sufficient to ground personhood?

Addressing this issue requires us to consider the concept of animal minds. Do animals have minds? What are animal minds like? These questions have long been the subject of popular, philosophical, and scientific speculation. Descartes, for example, maintained that animals lacked minds altogether, being mere automata.[8] Darwin, from the perspective of evolutionary biology, thought that animals must necessarily have minds, differing in degree but not fundamentally in kind from humans, in order to explain the gradual development of human minds.[9] And all of us must have wondered, at least idly, what it would be like to be a dog taking delight in the simplest of walks, or a bird soaring in flight or, if we are philosophers, perhaps we are more likely to wonder what it is like to be a bat.[10]

The combination of philosophical enquiry with empirical research as an inter-disciplinary endeavor directed toward sustained investigation of this subject, is relatively recent.[11] The convergence of animal behavioral studies with the development of theories of animal minds has resulted, over the past few decades, in the emergence of the discipline of cognitive ethology,[12] serving as a useful lens to focus enquiries in this area. Such research involves considerations in the philosophy of mind as well as studies of animal behavior, neuroscience, and physiology in order to form and test hypotheses and to draw conclusions about the nature of animal minds.

Functional evolutionary approaches to analyzing cognitive properties provide another valuable perspective by drawing attention to the similarities between humans and other animals. The difference between humans and animals has sometimes been regarded as a sticking place in considerations of animal minds. In asking what it is like to be an animal, we are apparently handicapped by the difficulties both that we can never find out through the experience of *being* an animal (of the nonhuman variety, that is),[13] and that they can never (or rather, thus far they have not been able to) tell us. However, our inability to gather firsthand experience about animals' minds and what the minds of animals are like by no means implies that they have none.

We can never experience what it is to be another human, but we do know what it is like; we can and do draw inferences from the fact that other humans are (to a greater or lesser degree) like ourselves, or, if we are of a Wittgensteinian turn of mind, we simply, and perhaps necessarily, act as if they are.[14] Because animals, then, are to some degree similar to humans, we should be able to understand their behavior in ways analogous to the way we do with humans, and draw reasonable inferences about their inner lives so far as those similarities hold—for example, similar neurological structures and physiological responses to pain give us prima facie reason to conclude that animals can feel pain as we do, even if they cannot reason or think about the experience in the same way. The evolutionary approach is therefore important in that its starting point for enquiry is not why animals should be assumed to be the same as us but why we should assume them to be different. It allows us to apply elements of the same reasoning process to the question of animal minds as to other humans.

To answer the problem of animal minds and moral theories about animals, philosophers and cognitive ethologists have used a range of approaches, treating various factors as indicative of mental states and cognitive capacities, which in turn might form a ground for moral status. The factors often considered in this respect include sentience, consciousness, the capacity to have beliefs and desires, intentionality, language, autonomy, and moral agency. Which (if any) of these might be either necessary or sufficient to establish the existence of a self-aware mind remains a subject of investigation in the philosophy of mind, as does the nature of consciousness and self-consciousness itself. The practical issue, which studies of animal behavior help to address, is how these attributes can be tested in a scientifically rigorous manner.

An in-depth treatment of these questions could form and has formed the basis of several volumes, so it is far beyond the scope of this chapter to recapitulate this work in detail. Borderline cases in particular pose thorny philosophical and ethological problems: for example, to what extent do the states of "belief," "desire," and "intention" either require or indicate the presence of an "I" to engage in those states, and in the absence of being explicitly told, how can we infer such states in animal minds? Rather than start from the bottom of the scale of morally relevant properties that animals may possess (mere life, sentience, and so forth) and work our way up through the hierarchy until we find the minimal threshold of personhood, a more useful approach may be to ask whether any animals clearly surpass whatever that threshold may be.

What features would convince us unambiguously that an animal was a person? As in the case of possible persons from other planets, we would be looking for evidence of Locke's "thinking, intelligent being," aware of itself and capable of valuing its own life. Most plausibly able to make the case for personhood would be any creature that could communicate this by telling us directly about its sense of self and its values and desires. Such animals are, of course, most commonly to be found around us in the form of other humans, but there is also evidence to show that some nonhuman animals may meet this test.

5. Language and Logic

The role of language has previously been granted significance in discussions of animals' moral status, often to assert that since animals lack language they cannot be capable of reasoning or self-awareness. Descartes, for example, wrote: "The reason why animals do not speak as we do is not that they lack the organs but that they have no thoughts."[15]

Although there is a lack of conclusive evidence thus far as to whether nonhuman animals might have complex verbalized languages of their own,[16] there are an increasing number of reports, *contra* Descartes' assertion, of animals being able to comprehend and use human language in a way that indicates beyond reasonable doubt that they not only have thoughts but can express them. Perhaps the best-known examples are those of chimpanzees who have been taught to communicate using sign language and who have proven able to use language in a manner equivalent to basic human abilities.[17] Even stronger evidence is presented in the case of gorillas, who have demonstrated abilities to use sign language to communicate, express feelings, emotions, and desires, generate novel linguistic constructs, and, most importantly, indicate their understanding of "self."[18] They are also able to understand spoken English and respond appropriately using sign. Other examples of animals who can understand and use language as we understand it, at least to some extent, include dolphins and even parrots.

These examples suggest that some animals have demonstrated the ability to use language both meaningfully and to indicate self-awareness, and the case for recognizing such animals as persons is correspondingly strong. Reading Patterson and Gordon's account of Koko the gorilla, and assuming their observations are accurately recorded and interpreted, it is difficult to find any justification to disagree with their conclusion that she is a person, save the discreditable reason that she is not human. But what of other animals of these species, who presumably have similar innate capacities but lack the training necessary to communicate in linguistic form?

It has previously been suggested that thoughts as complex as, for example, the idea of a future that one wishes to live to experience, require a vehicle of thought in the form of a language in which they can occur.[19] Whether or not the neurological structures underpinning self-awareness also underlie the ability to learn to use language is an intriguing neuropsychological question. Evidence suggests that language, both the mental ability to use language and the physical capacity for speech, coevolved with increased cognitive skills and emerging self-awareness; but whether this evolution was interdependent or language was an optional consequence of self-awareness is unclear.[20] Recent investigations have begun to uncover the genetic origins of speech and language, and in the future such research may reveal more about the link between the evolutionary emergence of language and the development of sociocultural practices that indicate self-awareness in our proto-human ancestors.

Philosophically, language is a good indicator but may be a poor prerequisite for the existence of a self.[21] In general, we can gain an understanding of the mental states of other humans, their mental lives, because they are able to use language to communicate such information directly. Most animals are unable to do so, but that does not necessarily mean that they lack a mental life. It is true that language provides the *easiest means* of ascertaining an individual's mental state and the existence of an "inner life."[22] Of those nonhuman animals who have acquired the tool of human language and can thus communicate directly with us, at least some have demonstrated a concept of self, an understanding of death and an ability to hold preferences about their future selves, indicating to us that these animals at least can be persons. The absence of language, however, need not by itself signify a corresponding lack of self-awareness. "Language," at least as we define the term here, is not a prerequisite for the exercise of logic and reasoning, of which animals have been shown to be capable.[23]

If we were to encounter a human who was unable, because of some impairment or even language barrier, to directly communicate her thoughts and feelings to us, we would still assume (all other things being equal) that she was capable of *having* thoughts and feelings. The simplest explanation for others displaying outward behavior that is consistent with an inner life comparable to our own is that they too have such a life. We have, of course, to be careful here, recognizing that it is tempting to assume excessive similarities with ourselves both in the case of other humans and in the case of animals; but just as we can often assume too close a similarity between ourselves and other humans, we may be equally reckless to assume the contrary in the case of animals.

Although a language-capable animal who could say "Cogito ergo sum"—and mean it—would provide incontrovertible evidence that animals are capable of self-awareness, the fact that animals so far have not done so is not powerful evidence of the contrary. A real possibility exists that some animals may be self-aware even in the absence of language, and that their awareness of self may extend to a capacity to value their continued lives. For an animal, any sort of animal, to be a person according to our account of personhood, certain mental and cognitive elements are required. In the following sections, we try to establish what these are and consider whether it can be claimed that nonhuman animals possess them.

6. SELF-AWARENESS

Self-awareness is a prerequisite for personhood in the account we have outlined, because having the capacity to value one's own life necessitates an awareness of the fact that one has a life to be valued. To put it another way, to be able to understand the possibility of discontinued existence and prefer continuation to discontinuation requires a conception of a self that exists and could cease to exist.

The mirror self-recognition test has been used as an indicator for the emergence of self-awareness in young human children and has also been applied to test whether nonhuman animals are self-aware. The test involves covert application of a visual marker, such as a spot of paint, to the subject's head in such a way that it is visible to the subject only on inspection in a mirror. Subjects are then exposed to a mirror in which they can view their own image, and their awareness of the marker as being on their own body is indicated by increased touching of the area. The ability to recognize and respond to this stimulus as perceived through a mirror, it is supposed, indicates an awareness of the physical self as well as an abstract awareness of a self that can be identified with the creature perceived in the mirror.

Mirror self-recognition trials on animals indicate so far that chimpanzees,[24] gorillas,[25] and other great apes, as well as dolphins[26] and elephants,[27] are capable of the degree of self-recognition necessary to pass the test. Human children, by comparison, acquire the ability to recognize their own reflections at the age of about two. But does this great-ape behavior indicate true self-awareness of the sort necessary to ground personhood as we have defined it? To answer this question, we need to break down the idea of self-awareness further into its component parts and examine these separately.

7. Consciousness and Self-consciousness

One challenge in determining whether animals are conscious is to address what consciousness is. The root of this problem spreads beyond considerations of animal consciousness and personhood to how we recognize and define conscious-ness in other humans and even in ourselves. Despite the fundamental philosophical difficulties involved in determinations of consciousness, however, moral philoso-phers are usually prepared to grant that other humans can be and generally are conscious. The inherently "applied" nature of our present inquiry suggests that a similar approach, as adopted by David DeGrazia, may prove useful:

> [T]here is excellent reason to think that many animals are conscious
> creatures.... [W]e know that humans are conscious. Given evolutionary
> continuity, behavioral and neurological analogues between humans and many
> animals support the common-sense claim that animals too have conscious
> mental states.[28]

This argument is plausible even in the absence of an agreed definition of "consciousness." As one of the presumptions inherent in the question of whether a nonhuman animal could be a person is that most human animals are persons—in other words, we conclude on the basis of analogy as well as evidence that other humans are conscious—we ought to accord the same "benefit of the doubt" to ani-mals who present similar evidence.

Mere consciousness may not be sufficient to ground personhood, however, and consciousness need not necessarily imply self-consciousness. We need some further means of distinguishing those animals who are simply aware and those who are self-aware. Here we can ask, "Does the apparent capacity of some ani-mals to form and act on desires and intentions, as indicated by a multitude of behavioral studies, mean that they are self-conscious?" Tom Regan has argued that it does:

> To recognize the status of mammalian animals as intentional agents paves the
> way for recognizing that they should also be viewed as self-conscious. For an
> individual, A, to act now in order to bring about the satisfaction of his desires at
> some future time is possible only if we assume that A is self-aware at least to the

extent that A believes that it will be *his* desires that will be satisfied in the future as a result of what he does now. In other words, intentional action is possible only for those who are self-conscious.[29]

This account seems inaccurate: recognizing animals as intentional agents implies that they are conscious, but not necessarily *self*-conscious. An animal that behaves in this manner must have some understanding of causality but not, however, necessarily of *herself* as the intentional agent. It is possible to know or perhaps even to reason that "if X then Y" without ever formulating the thought "if *I* do X, then Y." Similarly, it is possible to "perceive without perceiving that one does perceive." In the simple sense, at least, "perceive" just means "see," and flies can see. Simple action does not require a concept of the self as actor.

By contrast, Bernard Rollins's definition of "consciousness" seems too wide: "To say that a living thing has interests is to suggest that it has some sort of conscious awareness, however rudimentary."[30] Rollins imputes a mental life to any creature with a nervous system, physiological pain/pleasure responses and sensory capability.[31] However, taking into account the more demanding constructions of desire, intention, and self-awareness that we have argued for above, it seems highly implausible that earthworms have a mental life: they may have the capacity for experiences, but they are not themselves "subjects of a life."

We conclude that an animal can be conscious without being self-conscious.[32] Attributing desires to animals does not require that the animals be self-conscious, just "merely" conscious, and deliberate action does not require the formulation "I act," but only the desire leading to the action itself. Consciousness must come before self-consciousness, but something more than bare consciousness is required to establish personhood.

8. Psychological Unity

The capacity to value one's own life requires the ability to conceive of oneself not just in the present moment, but as the subject of a life—that is, Locke's "thinking intelligent being." A precondition for being able to conceive of oneself as the subject of a life is, self-evidently, to *be* the subject of a life. This property has been referred to as psychological unity, an attribute Jeffrey McMahan describes as "a complex notion, encompassing both psychological connectedness and continuity."[33] He builds on Derek Parfit's definitions to state that:

> *Psychological connectedness* is the holding of particular direct psychological connections.
> *Psychological continuity* is the holding of overlapping chains of strong connectedness.[34]

While the identification of psychological unity with personal identity is the subject of debate,[35] both Parfit and McMahan agree that it is a significant constituent of "what matters" to us about ourselves, or as McMahan puts it, of prudential unity relations: "the relations that ground rational egoistic concern."[36] He defines psychological unity further:

> The degree of psychological unity within a life between times t1 and t2 is a function of the proportion of the mental life that is sustained over that period, the richness or density of that mental life, and the degree of internal reference among the various earlier and later mental states.[37]

He goes on to note that

> substantial psychological unity within a life presupposes parallel conditions for psychological capacities.... [T]he degree of psychological unity within a life is a function of the richness, complexity and coherence of the psychological architecture that is carried forward through time.[38]

In this account, then, some animals can be said to have a degree of psychological unity through time, and the more complex their psychological capacities are, the stronger those relations will be. Thus, for example, humans with normal mental capacities will generally have strong psychological unity throughout most of their adult lives. It is reasonable to suppose that many nonhuman animals, especially higher primates and other mammals, may have some psychological unity, the degree of which will vary with their mental capacities, whereas animals with minimal psychological capacities such as earthworms and fruit flies might be said to have little or no psychological unity in their lives.

9. THE TEMPORAL SELF: MEMORY AND LEARNING

Another property that persons are likely to possess is the ability to relate past, present and future. To conceive of oneself as existing through time and therefore to see future life as something that can be valued requires a sense not only of time but of one's existence throughout time.

It is unarguable that many animals have the capacity to learn and by inference, to form some sort of memories or associations in relation to past experiences and events. One might suppose that the ability to remember presupposes the animal having some concept of itself existing through time. However, learning that stimuli of type X are generally associated with outcome Y is different than being able to recall the last time an event of type X happened, to travel mentally back to that occurrence and re-experience the associated outcomes. The latter is known as episodic memory[39] and is associated with the phenomenon of "autonoetic" (that is, self-knowing) consciousness, which requires a conception of the self as well as of subjective time.[40] Semantic memory, on the other hand, is the mere association of present ideas in the light of past experience, rather than the deliberate conceptualization of the self in the past and now present.

It has been argued previously that only humans possess episodic memory.[41] There is evidence, however, that some animals display behavior consistent with episodic memory,[42] though the existence of episodic memory itself is difficult to verify in the absence of language. The relatively few studies of animals with sophisticated language abilities do seem, however, to indicate that some animals are able to comprehend their past in terms of episodic memory (see below). Both episodic memory in relation to past events and episodic future thinking,[43] the ability to project one's thinking into the future to consider likely events and outcomes, rely on "the notion of a self which spans past, present and future…[and] has an enduring existence through time."[44]

Note that the reverse is not necessarily true: the ability to formulate a notion of the *temporally extended* self as opposed to the present self may not be necessary for self-consciousness. Some brain-damaged human patients who are unable to engage in episodic memory or recall are still clearly able to formulate the concept of a self; by their own description, they are not profoundly brain-damaged.[45] If animals demonstrate episodic temporal awareness, however, it is good evidence that they must also possess self-awareness.

This thesis raises an intriguing problem about whether humans incapable of episodic recall are able to conceive of their past and future existence, and if not, whether they meet the definition of personhood we have proposed. Remembering, though, that we are looking to find animals who count as persons, rather than defining precisely the minimal threshold of personhood, it is enough for this purpose to note that some animals possess this capacity to conceive of themselves as the same creature in different times and places, though it may not be strictly necessary to fulfill the definition of personhood.

10. Do Dolphins Wonder Why They Wonder? Metacognition and Psychological Unity

In addition to possessing psychological unity, some animals are capable of *conceiving of themselves* as a distinct entity with a continuous existence through time. This capability, it may be argued, is different from mere psychological connectedness or continuity in that it requires a kind of second-order awareness of psychological unity. It entails not just being a distinct entity but realizing that one is such a being. This criterion is not the tautology it may seem: if an animal can be conscious without being self-conscious, then it can have psychological unity without realizing that it does so or ever reflecting on its own psychologically connected state or progression through life.

The indicia of such a state of self-consciousness might include the abilities to reason abstractly—not only to have desires but to be aware of desire as a mental state and to attempt to modulate this and other mental states, or to have the capacity to reflect on actions and to be aware of doing so. This form of thinking, termed "metacognition," also implies and requires some degree of self-awareness. Richard Feynman penned the following humorous philosophical musings on self-awareness:

> I wonder why, I wonder why,
> I wonder why I wonder?
> I wonder *why* I wonder why I wonder why I wonder?[46]

Although his words were intended facetiously, they nevertheless illustrate that there may be a pertinent distinction to be drawn between animals who are merely capable of wondering why and animals who can wonder *why* they wonder. The lives of such animals contain what McMahan describes as a "complex narrative unity"[47] that makes them more than just the sum of their experiences and requires additional moral consideration. These animals include the majority of humans, and may also include other intelligent creatures, in particular higher mammals such as great apes, who evince behavior substantially consistent with a near-human level of self-awareness,[48] and possibly also some cetaceans.[49]

It is this level of abstract self-awareness that is required to satisfy the criteria for personhood in the account we are defending.

11. DESIRE, INTENTION, AND VALUING EXISTENCE

Any animal that meets the criteria set out so far—as self-aware, temporally contiguous, psychologically unified consciousness—has a strong case for being considered a Lockean person, a "thinking intelligent being, that ... can consider itself, the same thinking thing, in different times and places.[50] The account of personhood set out in section 1 above adds a crucial element, the dimension of "value" and the capacity for valuing life.

What does it mean to value something? Those things we value are typically characterized by our attitudes toward them. If we value something, then it is something we want or want more of, something we seek after and seek to protect. Value means more, however, than mere wanting: wants can be simply "of the moment," whereas valuing is indefinite and requires a level of abstraction to the concept of what one values. I may at this particular moment want a chocolate biscuit, but what I value is not the biscuit itself (in any sense, that is, other than the trivial) but the fact that when I want one I can have one. More generally, I value the capacity to have my desires satisfied, the circumstances that allow this satisfaction, and of course the life that is a prerequisite for having, let alone satisfying, any desires at all.

Valuing something requires that one both know that one has it or at least that it exists, and desire to keep, attain, or preserve it. The ability of animals to hold desires is therefore relevant insofar as it may indicate the capacity for mental states and preferences that may contribute toward being able to value anything, life included. Are nonhuman animals capable of having the relevant forms of desires and intentions? By this, we mean more than mere physiological response to circumstances: a bacterium will move away from negative chemical stimuli, but that does not mean it can be said to "want" or "intend" to escape them. How much of animal behavior can we attribute to true intention, as opposed to instinctive behavior (equivalent to nothing more than a complex form of bacterial chemotaxis) or conditioning?

This question and how to answer it are again constant problems within cognitive ethology, and the issues have both empirical and philosophical components. First, there is the difficulty of observation: a state of wanting can be recognized by trying to get,[51] but not all acts that might be interpreted as trying to get are necessarily caused by a state of wanting (depending, once more, on what we mean by wanting). How would we recognize intentional, rather than merely instinctive, behavior? How do we distinguish the two philosophically and what is the significance of doing so? What does it mean to attribute desire or intention to an animal?

The analysis of desire, belief, and intention has often played a part in considerations of animal minds and has been itself linked to the notion of self-awareness. These concepts are difficult to disentangle, in philosophy of human minds as well as animal ones. The complex form of desire and intention of which humans are capable, and that they can express through words, involves the capacity to form the concept of what it is that is desired or intended and to reflect critically on it. It would be a leap of inference to conclude merely on the basis of observing an animal "trying to get" that the animal is similarly capable of this level of conscious intent: does the cat chasing the bird desire the bird or intend to catch it? Perhaps predatory behavior is mere instinct and does not imply real intention.

Although it may be hard to determine whether animals share the capacity for intention in the abstract, they can plausibly be shown to be capable of behavior that is more than mere instinct. It is arguable that an animal who acts in a novel way to achieve an objectively desirable outcome cannot be acting on instinct or conditioning. Examples of this type of behavior have been reported in crows who demonstrate tool use and manufacture in problem-solving situations. A remarkable demonstration of behavior that cannot be explained by instinct alone is the example of a crow who, without trial and error or other conditioning-type learning, bent a straight tool to produce a hook when such a tool was required.[52] Crows natively use sticks and other straight tools in the wild and can learn to use hooked tools. The use of naturally occurring tools in wild crows is thought to be partly inherited and partly learnt.[53] However, the ability to fashion entirely new tools in response to new circumstances[54] can be explained by neither.

It is plausible—indeed, more likely than not, given the biological and evolutionary continuity of humans and other animals, and of other congruent properties we share—that complex intention of the sort displayed by humans has developed via a continuum that begins with reflex action such as a bacterium moving down a chemical gradient or a plant turning its leaves to the sun, and passes through simple instinct such as salmon swimming upstream, more complex instincts that may be involved in social or rearing behavior, and various levels of simple intention that overlap and go beyond this, from predatory behavior by cats to problem-solving by crows. This being the case, it would still be difficult if not impossible for most animals to prove that they share the capacity for complex intention with humans. Crucially, however, the question becomes one not of whether there are any reasons to suppose that they do, but whether there are sufficient reasons to conclude definitively that they do not.

We noted above that disentangling desire and intention from self-awareness is difficult philosophically, and that it cannot conclusively be said that desire (in the loose sense of attributing wants on the basis of actions) requires self-awareness or that behavior proves animals capable of desire in the rigorous sense. Once we have good evidence that animals are self-aware, however, further evidence of desire and intention on their part can reasonably be interpreted as a manifestation of preferences about representational states on the part of a creature capable of such. To prefer one state over another is to value states differentially, and the ability to have preferences about one's future existence, in particular that it continue, translates to the capacity to value one's own life.

12. What Is Not Required for Personhood?

So far, we have laid out some features that, if satisfied by nonhuman animals, would qualify those animals clearly as persons by our definition. We have tried to establish that there are at least some animals who possess all of these. These characteristics are *sufficient*, on our account, to ground personhood, but they may not all be *necessary*. There are, however, additional properties on which some authors have held personhood to be based. Although we would argue that they are irrelevant to determinations of personhood, it is worth mentioning some of them, if only to explain why they are irrelevant to our account. Additional attributes on which moral status might be based include autonomy,[55] rationality[56] and moral agency.[57] Although none of these permit of simple definition, it is probably reasonable to say that the exercise of any of them assumes self-awareness as a prerequisite.

It is hard to determine what would definitively constitute autonomy or autonomous action on the part of animals, not least because there are multiple competing philosophical accounts of what autonomy itself is. It is clear that many animals can act independently and with simple intention—that is, more than just reacting instinctively to their environment—but this would be an unrealistically low threshold by which to characterize autonomy, being nothing more than free action or mere liberty. Almost all accounts of autonomy pertain to the reasons underlying action, not the simple fact that it is unconstrained. Some authors have identified the ability to engage in higher order reasoning with the property of autonomy, along the lines of certain philosophical accounts.[58] In this account, though, it is difficult ever to say that an action is completely autonomous: what degree of reflecting on reflecting on one's preferences about one's preferences would be required? Rationality is similarly difficult to pin down: at one end of the spectrum, the basic capacity for reasoning that many animals have been observed to display might satisfy the requirements of simple rationality, whereas at the other extreme, it is doubtful whether any living creatures meet the definition of a perfect rational being.

The exact degree of autonomy, rationality, and moral agency that humans and animals can be said to possess remains a matter for debate. What is clear, however, is that some animals, notably nonhuman primates and cetaceans as mentioned above, may well have capacities in this area that overlap partially with the range of capacities possessed by humans. It would be intriguing to attempt to discriminate morally among humans on the basis of differing capacities for autonomy, rationality, and most particularly moral agency. In general, however, we refrain from doing so for human persons, because we believe either that self-awareness is a fairly low threshold or because we believe it would be invidious so to do, and ought therefore not to do so for animal persons either.

It may well be true that nonhuman animals make fewer autonomous decisions, or act with a lesser degree of autonomy, or are less capable of complex moral reasoning, than humans in general. However, personhood, if it is to serve as a measure of the inherent moral status of beings—moral status that they possess indefinitely and in virtue of the kind of beings they are rather than what they do—cannot be act-relative. It seems illogical that we should be persons when we are acting autonomously (according to whichever definition) but nonpersons when we make choices that are less than autonomous, or persons when we act on the basis of considered moral belief and can justify our actions by reasoned moral defense, but nonpersons when we fail to engage in such extended contemplation before acting. If we were to base a test for moral personhood in both animals and humans on either autonomy or moral agency, the test should be one of quality, not of degree: whether beings possess the qualities of autonomy and moral agency, not the degree to which they exercise them. Although acts are the outward manifestations by which we principally judge the inner capacities of others, it is not the extent to which they, and we, exercise the capacity for moral agency or autonomy that should matter but the fact of the capacity itself.

Beauchamp's analysis of personhood draws a distinction between metaphysical personhood, based on psychological and cognitive properties akin to those we have described in this account, and moral personhood, which he argues "indicates individuals who possess properties or capacities such as moral agency and moral motivation.... In principle an entity could satisfy all the properties requisite for metaphysical personhood and lack all the properties requisite for moral personhood."[59]

On the subject of moral agency, there is an emerging body of work in the field of cognitive ethology that examines the behavior of animals as moral or proto-moral agents,[60] and it seems at least possible that some animals might actually meet the definition of moral personhood required by Beauchamp. Moral awareness, however, is not required for either our definition of personhood as a characteristic that denotes those individuals whose lives are inherently valuable, or the normative application of this concept to determine how persons should be treated, and in particular, with regard to not being deprived unwillingly of their lives.

It is true that what we require of persons includes the ability to hold the concept of self and other; the concept of better or worse, not necessarily in the sense of moral judgment but as a matter of preference; and the concepts of action and cause and effect. All of these would also seem to be required for moral agency, but this does not imply that all persons are necessarily moral agents. To be capable of moral reasoning, it is not enough to have sympathetic engagement with others or some idea of reciprocity, and to be able to act intentionally to bring about desired outcomes. We need also to have a theory of, or at least reasons *why*, the "good" is good. In order to exercise moral agency, one must be a moral agent and in order to be a moral agent, one must have moral awareness, that is, at least a rudimentary understanding of right and wrong. Of course, quite a few humans might well fail this test!

We thus return to the "problem" with personhood that is not really a problem at all, namely that any sensible and non-species-based criteria will either include some nonhumans or exclude some humans. If most moral theories are ultimately self-interested, at least in part, in that they strive to show why creatures like us deserve moral consideration, then we need to be open to the prospect that our idea of "creatures like us" need not be limited to our own biological species.

13. Animal Personhood
and Its Implications

To summarize the argument thus far, a person is an entity capable of valuing its own existence, with all of the requisite capacities of temporal self-awareness and reflexivity that entails. In moral terms, if animals are persons then they are entitled to the same moral consideration as human persons, the most salient aspect of which is the prima facie right not to be killed, not to be deprived of the lives that they value.

Many animals are clearly not persons on this account: paramecia and nematodes, for example. At the other end of the scale, compelling evidence is emerging that some nonhuman animals, in particular great apes and cetaceans, may well be persons and it is therefore incumbent upon us to treat them as such. Between these extremes is a wide swathe of uncertainty within which it is difficult to discern with conviction whether or not animals possess the necessary attributes for personhood. How we treat these animals will depend partly on how we make that determination in each case, but not entirely.

Allowing that some nonhuman animals may be persons and that some human animals may not be persons does not commit us to a position that ignores the moral obligations we may have toward the latter as well as toward other nonperson animals. We are not arguing that personhood is the sole ground for moral standing. A much broader account of moral standing giving rise to reasons for moral action is possible and desirable. The possibility, and in some cases probability, of animal personhood, however, gives us particular reasons for particular moral actions—that is, to treat animal persons, where recognized as such, *as* persons—the moral imperative for which has previously received less recognition than deserved

Politically and legally, what would be the consequences of granting that animals could be persons? Some commentators have argued that the moral rights to which some nonhuman animals ought to be entitled, in virtue of the kinds of creatures they are, should also form the basis of certain legal rights: for example, the right to bodily integrity and bodily liberty for chimpanzees[61] and the right to freedom from torture or slavery.[62] Others have gone so far as to consider the possibility of political autonomy and sovereignty for great apes.[63]

Several jurisdictions now have legislation that protects great apes and in some cases nonhuman primates from being used as experimental subjects in medical research.[64] This is of course still a far cry from recognizing these animals as full persons under the law, but there have also been various attempts, through both the courts and legislatures, to secure recognition of legal rights or legal status for great apes. While these have met with mixed success so far, the issue of legal rights for apes and potentially other animals continues to receive increasing consideration, as it should, given the possibility of animal personhood as discussed.

It is worth noting that the status of "person" is not absolute in terms of the rights it creates under law. All living human beings are regarded as natural persons, but only a subset of these persons has the legal right to, for example, enter into binding contracts or make decisions about medical treatment. The recognition of animals as moral persons does not immediately commit us to allowing dolphins to vote, or permitting interspecies marriage between humans and chimpanzees. It does, however, mean that we might have to rethink the role of moral personhood and the rights to which legal persons are entitled.

Another eventuality with immense ramifications would be the possible extension of "human rights" to animals. Some "human rights," at least as they are expressed in political or legal instruments, are clearly irrelevant or inappropriate when applied to nonhuman animals. For example, the right to privacy, enshrined in Article 8 of the European Convention on Human Rights, has been used to illustrate why it is inappropriate to apply "human rights," at least as they are currently used and understood, to dolphins, whose courtship and mating behavior seems to display a flagrant disregard for what we consider to be privacy.[65] It would, however, be fallacious reverse logic to conclude from this that dolphins are not persons because some human rights to which we assume persons are entitled seem not to apply to them. Instead this should lead us to consider the degree to which "human rights" are genuinely appropriate for all humans![66] Privacy, to use the same example, is heavily culturally relative, and several human societies display behavior that is also discordant with the Western notion of privacy.

One "human" right that it should be morally unproblematic to extend to all persons, by definition, is the right to life. Yet again, this might create social and political dilemmas. If dolphins have the same right to life as humans, is it morally acceptable to kill dolphins as by-catch while fishing for tuna to sustain human lives? These sorts of seemingly insurmountable problems have been used as a practical reason to avoid the issue of animal personhood, because its consequences would be too difficult and too dire. How could we justify destroying chimpanzees' environment to build human habitations, or saving human lives over ape lives? How could we not?

The existence of these problems does not invalidate the truth of their cause— that we ought to recognize as persons those animals that qualify as such, no matter what difficulties result or how we resolve them—any more than the runaway trolley problem[67] and other utilitarian conundrums that pit the life of one human against many preclude us from recognizing that humans are persons. It would be invidious in the extreme to refuse animal persons this recognition merely because of the inconvenience it would create to do so!

Acknowledging that dolphins, or any sort of animals (human or nonhuman), are persons does not confer absolute protection on their lives, nor does it prevent us from asking the question of how to weigh one life against others. There will continue to be circumstances in which we are forced to weigh animal needs against human desires, and to ask how we should weigh animal lives against human needs. But it becomes a different sort of question when we ask whether we can countenance destroying the lives of other *persons* to suit our needs.

14. ANIMAL ENHANCEMENT AND PERSONHOOD

Another question of immense contemporary importance concerns the ethics of mixing human and animal parts. The most likely scenarios consist of starting with a human individual (embryo or more mature instantiation of a human genome) and adding animal bits (genes, cells, tissue, organs, etc.) or starting with animal and adding human bits. In either case, the ethics of doing such a thing turn on two main issues. Again, assuming that we are not planning on arresting development at the embryonic stage but allowing the resulting creature to develop into a mature individual, the main two ethical issues seem to be the following:

(1) Will the admixture of elements from another species be likely to prove beneficial or harmful to the individual? Would it, on the balance or probability, be a cruelty or a kindness to allow a creature like this (or as we expect a creature modified in this way) to grow to maturity?

(2) Are we prepared to accept the consequences? That is, are we as a society prepared to accept that if we "enhance" animals to the point at which they might count as persons, we should be prepared to accord them the rights and protections of interests, dignity, and status that go with personhood?

Enhancement of animals to a condition comparable to that of human persons (and hence to a moral and political status) is something that happens to (almost all[68]) human animals. Human individuals start, and remain at all stages up to and including that of neonate and probably beyond, as human nonpersons comparable in all essentials, save only that of their potentiality and genetic constitution, to animal nonpersons. Thereafter, through a combination of socialization, language acquisition, and education, they become enhanced to the point where they become "persons" properly so called. We do not usually balk at this form of animal enhancement; indeed, we would regard it as criminally negligent to fail to enhance such human animals to the point of personhood and beyond. If we reach a point at which we can be confident that animal enhancement, whether by human admixture or by other means, is as safe and reliable as the normal development of human children, the potentiality of both humans and animals for personhood will be comparable in the sense that just as all educable children are potential persons, all enhanceable animals will also be potential persons.[69]

If we could meet the objections encapsulated in ethical issue (1) above, and satisfy ourselves that we could avoid cruelty to the resulting creatures and that to enhance them in these ways would be beneficial, then, provided we could satisfy (2) above, it is not only arguable that we should do so, it would be difficult to find moral objections to doing so in any particular case. Doing so comprehensively, however, would have economic, population, and many other further ramifications that would have to be carefully thought through and would be unlikely to prove attractive. But the present, almost unbridled, expansion of the human population is almost equally unattractive from many perspectives.

Finally, the rare instances of apparently authentically reported so-called "wolf children"—children brought up in isolation from other humans by animals and who have as a result neither been socialized nor acquired language—give us further reasons to reflect on the moral reasons we have to provide the ingredients that might turn animals into persons when we find them lacking. Ludwig Wittgenstein provides an analogous case. In his *Philosophical Investigations*,[70] he takes up an example given by William James and speculates about whether it would be possible for a human deaf mute to have abstract thoughts before he had acquired a language which might provide the medium for such thoughts. Wittgenstein seems skeptical and asks (at section 344), "Would it be imaginable that people should never speak an audible language, but should still say things to themselves in the imagination?"[71] In both these sorts of cases, language acquisition functions as an enhancement tool that apparently makes possible a step change in development.

What we are concerned about in the case of enhancing animals or creating humanimals is the ethics of increasing the likelihood for some creatures that they may qualify as persons. Provided that once they do qualify we are prepared to recognize them as such, there should be no moral objections to doing so.

15. Nonhuman, Human, Transhuman?[72]

There is another context in which the idea that some nonhuman animals may be persons (and that some human animals may be nonpersons) has potentially significant implications. Concerns and considerations relating to human enhancement have included the possibility that the use of radical enhancement technologies such as genetic manipulation or cybernetic modification on humans might lead to the creation of individuals we would no longer regard as "human" in the present sense. It is probable that in the future there will be no more humans as we know them now, since the further evolution of our species, either Darwinian, or more likely determined by human choices,[73] will result in the emergence of new sorts of beings better able to cope with the intellectual and physical challenges of the future.

An example of the ways in which this is already happening is the sorts of cognitive enhancement that are already with us;[74] another is signaled by attempts at extended human lifespan. The potential evolution of future humans, or, as some like to call them, "transhumans,"[75] raises some interesting questions about how we would regard such beings and how they would regard us. Added to this is the possibility of the emergence of other forms of existence such as artificial or non-biological life. Our definition of "personhood" requires biographical, not necessarily biological, life. So far, all who have the former also have the latter, but this state of affairs may not remain the case. Just as we humans have not in the past needed or attempted to separate the concepts of "humans" and "persons" except in quite narrow contexts, so far we have not paid much attention (other than in science fiction) to the separation between biographical and biological life. But as animal capacities and human enhancement technologies have prompted us to reconsider the question of the relationship between personhood and humanness, developments in synthetic life and artificial intelligence may necessitate a reconsideration of that distinction.

As we have noted, the term "person" is applied as a sort of philosophical shorthand to indicate "creatures like us, who matter morally to the same extent we do," and "human" is often misapplied to mean the same thing. Moral concerns expressed as being about becoming something other than human or losing our essential humanity, then, cannot logically or consistently be about "humanness" per se, since "human" is an arbitrary biological property rather than a moral one, and "personhood" as we have defined it is not dependent on possession of that property.

We have argued in this chapter for a conception of personhood that does at least some useful work in normative analyses of how we should behave toward others. Acknowledging the possibility of animal personhood, by deemphasizing the importance of "being human" in discussions and recognizing humans as a species of animal, a product of evolution and a point on the evolutionary continuum, also paves the way for us to consider future points on this continuum in the same way.

The question we should be asking about our future heirs and descendants, be they genetic and biological descendants or descendants of the mind, is not whether they will still be human but whether they will be *persons*. Given that our nonhuman animal ancestors, at least some of our animal contemporaries and our possible future animal (including both human and nonhuman animal, and even humanimal) companions may well be persons, there is little reason to suppose that the answer will be anything other than "yes."

ACKNOWLEDGMENTS

Work on this chapter was supported by the Wellcome Trust Strategic Programme on *The Human Body: Its Scope, Limits and Future*. We thank the Editors of this *Handbook* for many helpful comments and suggestions.

NOTES

1. In this chapter, we leave aside questions of the forfeiture of such privileges and protections by due process of law or in armed conflict.

2. In this section, the argument follows closely lines developed in John Harris, *The Value of Life: An Introduction to Medical Ethics* (London: Routledge and Kegan Paul, 1985); John Harris, "The Concept of the Person and the Value of Life," *Kennedy Institute of Ethics Journal* 9 (1999): 293–308; and John Harris, *Enhancing Evolution: The Ethical Case for Making Better People* (Princeton, N.J.: Princeton University Press, 2007).

3. John Locke, *An Essay Concerning Human Understanding* (London: Oxford University Press, 1964), book 2, chapter 27, p. 188.

4. For example, see David DeGrazia, *Taking Animals Seriously: Mental Life and Moral Status* (Cambridge: Cambridge University Press, 1996); and Tom L. Beauchamp, "The Failure of Theories of Personhood," *Kennedy Institute of Ethics Journal* 9 (1999): 309–24.

5. Beauchamp, "Failure of Theories of Personhood."

6. See also Harris, "Concept of the Person and the Value of Life."

7. Beauchamp, "Failure of Theories of Personhood."

8. Rene Descartes, "Animals are Machines," in *Animal Rights and Human Obligations*, ed. Tom Regan and Peter Singer (Englewood Cliffs, N.J.: Prentice-Hall, 1976), 60–66.

9. Charles Darwin, *The Descent of Man* (Princeton, N.J.: Princeton University Press, 1981).

10. Thomas Nagel, "What is it Like To Be a Bat?" *The Philosophical Review* 83 (1974): 435–50.

11. Colin Allen and Marc Bekoff, *Species of Mind: The Philosophy and Biology of Cognitive Ethology* (Cambridge, Mass.: MIT Press, 1999).

12. Donald R. Griffin, *Animal Thinking* (Cambridge, Mass.: Harvard University Press, 1985); and Donald R. Griffin, *Animal Minds* (Chicago: University of Chicago Press, 1994).

13. An illusory difficulty. See Ludwig Wittgenstein, *Philosophical Investigations*, translated by G. E. M. Anscombe (Oxford: Basil Blackwell, 1953), 293.

14. "My attitude towards him is an attitude to a soul, I am not of the *opinion* that he has a soul." Wittgenstein, *Philosophical Investigations*, 293.

15. Descartes, "Animals are Machines," p. 64.

16. Vocal communication amongst animals is not an uncommon phenomenon; extensive studies have been carried out in some primate and cetacean species showing that these animals do communicate directly through vocalizations. However, the degree of complexity of these communications and whether they qualify as language as such are still open to debate. See R. Allen Gardner, Beatrix T. Gardner, and Thomas E. Van Cantfort, *Teaching Sign Language to Chimpanzees* (Albany: State University of New York Press, 1989).

17. Gardner et al., *Teaching Sign Language to Chimpanzees*; R. Allen Gardner, Thomas E. Van Cantfort, and Beatrix T. Gardner, "Categorical Replies to Categorical Questions by Cross-fostered Chimpanzees," *American Journal of Psychology* 105, no. 1 (1992): 27–57.

18. Francine Patterson and Wendy Gordon, "The Case for the Personhood of Gorillas," in *The Great Ape Project*, ed. Paola Cavalieri and Peter Singer (New York: St Martin's, 1993), pp. 58–77.

19. Harris, *Value of Life*.

20. See Constance Holden, "The Origin of Speech," *Science* 303 (2004): 1316–19.

21. As discussed in Tom Regan, *The Case for Animal Rights* (London: Routledge. 1984).

22. See Harris, *Value of Life*, pp. 19–21.

23. Reviewed in Shigeru Watanabe and Ludwig Huber, "Animal Logics: Decisions in the Absence of Human Language," *Animal Cognition*, 9 (2006): 235–45.

24. Gordon G. Gallup, "Chimpanzees: Self-Recognition," *Science* 167 (1970): 86–87.

25. Patterson and Gordon, "Case for the Personhood of Gorillas."

26. Diana Reiss and Lori Marino, "Mirror Self-Recognition in the Bottlenose Dolphin: A Case of Cognitive Convergence," *Proceedings of the National Academy of Sciences of the United States of America* 98 (2001): 5937.

27. Joshua M. Plotnik, Frans B. M. de Waal, and Diana Reiss, "Self-Recognition in the Asian Elephant," *Proceedings of the National Academy of Sciences of the United States of America* 103 (2006): 17053–57.

28. DeGrazia, *Taking Animals Seriously*, p. 114.

29. Regan, *Case for Animal Rights*, p. 75.

30. Bernard E. Rollin, *Animal Rights and Human Morality* (New York: Prometheus Books, 1981), p. 41.

31. Rollin, *Animal Rights and Human Morality*.

32. DeGrazia, *Taking Animals Seriously*, pp. 114–15.

33. Jeff McMahan, *The Ethics of Killing* (Oxford: Oxford University Press, 2002).

34. Derek Parfit, *Reasons and Persons* (Oxford: Oxford University Press 1986), p. 206.

35. See Parfit, *Reasons and Persons*, chapter 10, pp. 199–217; and Derek Parfit, "Reductionism and Personal Identity," in *Philosophy of Mind: Classical and Contemporary Readings*, ed. David J. Chalmers (New York: Oxford University Press 2002), pp. 655–61.

36. McMahan, *Ethics of Killing*, p. 42.

37. McMahan, *Ethics of Killing*, pp. 74–75.

38. McMahan, *Ethics of Killing*, p. 75.

39. As opposed to semantic memory, which is used to describe the first type of memory.

40. Thomas Suddendorf and Janie Busby, "Mental Time Travel in Animals?" *Trends in Cognitive Sciences* 7 (2003): 391–96.

41. Endel Tulving, "Memory and Consciousness," *Canadian Psychology* 26 (1985): 1–12.

42. N. S. Clayton, T. J. Bussey, and A. Dickinson, "Can Animals Recall the Past and Plan for the Future?" *Nature Reviews Neuroscience* 4 (2003): 685–91.

43. Cristina M. Atance and Daniela K. O'Neill, "Episodic Future Thinking," *Trends in Cognitive Sciences* 5 (2001): 533–39.

44. Atance and O'Neill, "Episodic Future Thinking," 129.

45. See Clayton et al., "Can Animals Recall the Past," 685–91.

46. Richard P. Feynman, *Surely You're Joking, Mr. Feynman! Adventures of a Curious Character* (London: Unwin Paperbacks, 1986), p. 49.

47. McMahan, *Ethics of Killing*, p. 197.

48. For descriptive evidence and analysis, see Frans de Waal, *Good Natured: The Origins of Right and Wrong in Humans and Other Animals* (Cambridge, Mass.: Harvard University Press, 1997); de Waal, *The Ape and the Sushi Master: Cultural Reflections by a Primatologist* (New York: Basic Books, 2001); and Peter Singer, ed., *In Defense of Animals* (Oxford: Blackwell, 1985).

49. See Richard C. Connor and Kenneth S. Norris, "Are Dolphins Reciprocal Altruists?" *American Naturalist* 119 (1982): 358–74; Wade Doak, *Dolphin, Dolphin* (New York: Sheridan House, 1982); and Bernd Würsig, "The Question of Dolphin Awareness Approached through Studies in Nature," *Cetus* 5 (1982): 4–7.

50. Locke, *Essay Concerning Human Understanding*, p. 188.

51. G. E. M. Anscombe, *Intention* (Oxford: Blackwell, 1957).

52. Jackie Chappell, "Avian Cognition: Understanding Tool Use," *Current Biology* 16 (2006): R244–R245; Jackie Chappell and Alex Kacelnik, "Tool Selectivity in a Non-Primate, the New Caledonian Crow (*Corvus moneduloides*)," *Animal Cognition* 5 (2002): 71–78; and Nathan J. Emery and Nicola S. Clayton, "The Mentality of Crows: Convergent Evolution of Intelligence in Corvids and Apes," *Science* 306 (2004): 1903–7.

53. Ben Kenward, Alex A. S. Weir, Christian Rutz, and Alex Kacelnik, "Behavioural Ecology: Tool Manufacture by Naive Juvenile Crows," *Nature* 433 (2005): 121.

54. See Alex A. S. Weir, Jackie Chappell, and Alex Kacelnik, "Shaping of Hooks in New Caledonian Crows," *Science* 297 (2002): 981; Alex A. S. Weir and Alex Kacelnik, "A New Caledonian Crow (*Corvus moneduloides*) Creatively Re-Designs Tools by Bending or Unbending Aluminium Strips," *Animal Cognition* 9 (2006): 317–34.

55. Regan, *Case for Animal Rights*, pp. 84–86.

56. Rollin, *Animal Rights and Human Morality*, pp. 22–24.

57. Regan, *Case for Animal Rights*, pp. 151–56.

58. For example DeGrazia, *Taking Animals Seriously*, pp. 204–6.

59. Beauchamp, "Failure of Theories of Personhood," 309.

60. See for example de Waal, *Good Natured*.

61. Stephen M. Wise, "Legal Rights for Nonhuman Animals: The Case for Chimpanzees and Bonobos," *Animal Law* 2 (1996): 179–86.

62. Deborah Rook, "Should Great Apes Have 'Human Rights'?" *Web Journal of Current Legal Issues* 1 (2009).

63. Robert E. Goodin, Carol Pateman, and Roy Pateman, "Simian Sovereignty," *Political Theory* 25 (1997): 821–49.

64. Andrew Knight, "The Beginning of the End for Chimpanzee Experiments?" *Philosophy, Ethics, and Humanities in Medicine* 3 (2008): 16.

65. Thomas I. White, *In Defense of Dolphins: The New Moral Frontier* (Oxford: Wiley-Blackwell, 2007).

66. John Harris, "Taking the 'Human' Out of Human Rights," *Cambridge Quarterly of Healthcare Ethics* 19 (2010): 61–74, published online Feb. 19, 2010, available at http://journals.cambridge.org/action/displayAbstract?fromPage=online&aid=7279820 (accessed Oct. 25, 2010).

67. Philippa Foot, "The Problem of Abortion and the Doctrine of the Double Effect," in *Moral Problems in Medicine*, ed. Samuel Gorovitz et al. (Englewood Cliffs, N.J.: Prentice-Hall, 1976), 267–76; and Judith Jarvis Thomson, "Killing, Letting Die, and the Trolley Problem," *The Monist* 59 (1976): pp. 204–17.

68. Save only those whose development is arrested or who die at a stage short of personhood.

69. Sarah Chan, "Should We Enhance Animals?" *Journal of Medical Ethics* 35 (2009): 678–83.

70. Wittgenstein, *Philosophical Investigations*, section 342.

71. Wittgenstein, *Philosophical Investigations*, section 344.

72. In this section the discussion follows lines developed in Harris, "Taking the 'Human' Out of Human Rights."

73. Harris, *Enhancing Evolution*.

74. Henry Greely, Barbara Sahakian, John Harris, Ronald C. Kessler, Michael Gazzaniga, Philip Campbell, and Martha J. Farah, "Towards Responsible Use of Cognitive-enhancing Drugs by the Healthy," *Nature* 456 (2008): 702–5.

75. See Max More, "The Overhuman in the Transhuman," *Journal of Evolution and Technology* 21 (2010): 1–4; Allen Buchanan, "Moral Status and Human Enhancement," *Bioethics* 23 (2009): 141–50; Andrew Edgar, "The Hermeneutic Challenge of Genetic Engineering: Habermas and the Transhumanists," *Medicine, Health Care and Philosophy*, 12 (2009): 157–67; James Wilson, "Transhumanism and Moral Equality," *Bioethics* 21 (2007): 419–25; Nick Bostrom, "A History of Transhumanist Thought," *Journal of Evolution and Technology* 14 (2005): 1–25.

SUGGESTED READING

ALLEN, COLIN, and MARC BEKOFF. *Species of Mind: The Philosophy and Biology of Cognitive Ethology*. Cambridge, Mass.: MIT Press, 1999.

CAVALIERI, PAOLA, and PETER SINGER. *The Great Ape Project*. New York: St Martin's, Griffin, 1993.

CHAN, SARAH. "Should We Enhance Animals?" *Journal of Medical Ethics* 35 (2009): 678–83.

DE WAAL, FRANS. *Primates and Philosophers: How Morality Evolved*. Princeton, N.J.: Princeton University Press, 2006.

————. *Good Natured: The Origins of Right and Wrong in Humans and Other Animals.* Cambridge, Mass.: Harvard University Press, 1997.

DeGrazia, David. "Great Apes, Dolphins and the Concept of Personhood." *Southern Journal of Philosophy* 35 (1997): 301–20.

————. *Taking Animals Seriously: Mental Life and Moral Status.* Cambridge: Cambridge University Press, 1996.

Gardner, R. Allen, Beatrix T. Gardner, and Thomas E. Van Cantfort. *Teaching Sign Language to Chimpanzees.* Albany: State University of New York Press, 1989.

Griffin, Donald R. *Animal Minds.* Chicago: University of Chicago Press, 1994.

————. *Animal Thinking.* Cambridge, Mass.: Harvard University Press, 1985.

Harris, John. *Enhancing Evolution: The Ethical Case for Making Better People.* Princeton, N.J.: Princeton University Press, 2007.

————. "The Concept of the Person and the Value of Life." *Kennedy Institute of Ethics Journal* 9 (1999): 293–308.

————. *The Value of Life: An Introduction to Medical Ethics.* London: Routledge and Kegan Paul, 1985.

Regan, Tom, and Peter Singer, eds. *Animal Rights and Human Obligations.* Englewood Cliffs, N.J.: Prentice-Hall, 1976.

Rollin, Bernard E. *Animal Rights and Human Morality.* New York: Prometheus Books, 1981.

Singer, Peter. *In Defense of Animals.* Oxford: Blackwell, 1985.

————. *Animal Liberation.* New York: Avon Books, 1977.

White, Thomas I. *In Defense of Dolphins: The New Moral Frontier.* Oxford: Wiley-Blackwell, 2007.

CHAPTER 12

ARE NONHUMAN ANIMALS PERSONS?

MICHAEL TOOLEY

THE questions of whether members of some nonhuman species of animals are persons, and—if so—which ones, are among the most difficult questions in ethics. The difficulty arises from two sources. First, there is the problem of how the concept of a person should be analyzed, a problem that is connected with the fundamental and challenging ethical question of the properties that give something a right to continued existence. Second, there is the problem of determining what psychological capacities, and what type of mental life, adult members of a given nonhuman species have.

My focus in this essay is upon the first of these issues, and my discussion is organized as follows. In section 1, I shall discuss the concept of a person and the concept of a right to continued existence. Then, in section 2, I shall examine some important arguments aimed at establishing that nonhuman animals do not have a right to continued existence—in some cases on the grounds that they have no moral rights at all. Finally, in section 3, I shall turn to an examination of arguments in support of the view that nonhuman animals do have rights, including a right to continued existence.

1. THE CONCEPT OF A PERSON

How is the concept of a person best understood? My suggestion is that in defining the concept of a person, one should be guided by two considerations. First, one should seek to minimize divergence from the ordinary understanding of the term "person." Second, one should attempt to define the concept in such a way that it can be used to formulate fundamental ethical principles—including a basic principle concerning when the destruction of something is not only prima facie wrong, and seriously so, but also (prima facie) intrinsically wrong, because it wrongs the thing destroyed. So defined, the concept of person is a purely descriptive concept, but the choice of the content is guided by moral considerations.

The basic moral principle in question would take the following form: the destruction of a person is prima facie (seriously) wrong. It might be, however, that this is not the only basic principle in this area, since some philosophers hold that the destruction of potential persons is also prima facie (seriously) wrong. Consequently, being a person will be a sufficient condition of having moral status, but it may not be a necessary condition.

1.1 Basic Moral Principles and the Importance of the Concept of a Person

Why should the concept of a person enter into basic moral principles concerned with identifying when and why the destruction of something is seriously wrong, not because of its impact upon other things, but because it wrongs the thing destroyed? For traditionally, it has often been thought that the basic principle in this area is that the killing (or, perhaps, the direct killing) of innocent members of the species *Homo sapiens* is prima facie intrinsically wrong, and seriously so. Thus formulated, however, the principle is not a satisfactory candidate for a basic moral principle, for three reasons. First, there are excellent grounds for holding that being an innocent member of the biologically defined species *Homo sapiens* does not itself make it prima facie wrong to kill a thing, because, if such an entity has suffered either whole-brain death, or upper-brain death, killing is not prima facie seriously wrong. So the claim that the killing (or the direct killing) of innocent members of the species *Homo sapiens* is prima facie intrinsically wrong, and seriously so, is not true.

Second, there are excellent reasons for holding that, even if one focuses upon the case of normally functioning members of the species *Homo sapiens*, the claim that the killing (or the direct killing) of innocent members of the species *Homo sapiens* is prima facie intrinsically wrong, and seriously so, is not a *basic* moral principle. Here the case of possible extraterrestrials is relevant. Suppose, in particular, that E.T.—from the movie *E.T.—The Extraterrestrial*—rather than being a fictional entity, were an actual living individual. If an enterprising owner of a restaurant decided to add E.T. steaks to the menu, what would one think? The answer, surely, is that most people would think that that was seriously wrong. Moreover, if asked why it was wrong, their answer would likely be that beings such as E.T., unlike other nonhuman animals, appear to have the same psychological abilities, and appear to have the same sort of mental life, as normal adult members of our own species. But then must it not also be the case that there is some general principle that deals with the morality of killing, and that, rather than referring to any particular species, refers instead only to properties that individuals of the E.T. sort, if they existed, would share with normal adult human beings? The conclusion, accordingly, is that, rather than being a *basic* moral principle, the principle that the killing (or the direct killing) of innocent members of the species *Homo sapiens* is prima facie intrinsically and seriously wrong, to the extent that it is true, is derived from a principle that does not refer to any particular species. So even if all human beings turned out to have moral status according to the principle, this would not be so because they are members of a species.

Third, many people believe in the existence of non-embodied subjects of experiences and other mental states—beings such as ghosts, angels, and gods. Such immaterial beings are normally thought to be such as cannot be harmed by us. But the idea that we could harm or destroy a non-embodied subject of mental states, or that one non-embodied individual might be able to destroy another, is not incoherent. Moreover, such an action would be morally comparable to killing an innocent human being. If that is right, then there is an additional reason why one needs, not only a general principle dealing with the destruction of individuals that does not refer to any particular biological species, but also a general principle that is not restricted to biological individuals: the principle should also cover non-embodied subjects of mental states.

The case of non-embodied persons also shows that the fundamental principle in this area should not involve the concept of *killing*, because that concept is tied to the idea of a living organism. But there are other important reasons why the fundamental principle should not involve the concept of killing. One is the case of upper-brain death. If someone does something that causes the death of a human's upper brain, one still has a living human organism, but, nonetheless, the action is surely seriously wrong. Moreover, if one thinks in terms of the harm that is done, does it seem that it is better to have one's upper brain destroyed than to be killed? Most people do not feel that the former outcome is less bad for a person than the latter. Is that not, then, a good reason for concluding that the one action is as wrong as the other?

The other case involves an action that cannot be performed now, but that seems in principle possible. It involves the idea of reprogramming an individual's brain, so that existing mental states are removed and replaced by new ones. Suppose, for example, that there were technological developments that allowed the brain of an adult human to be completely reprogrammed, with the result that the organism wound up with memories—or rather, apparent memories—beliefs, attitudes, and personality traits completely different from those associated with it before it was subjected to reprogramming. (The pope is reprogrammed, say, on the model of David Hume or Richard Dawkins.) In such a case, however beneficial the change might be for the world as a whole, most people would say that *someone* had been destroyed, that an adult human being's "right to life" had been violated, even though no biological organism had been killed.

Is there anything that all of these actions have in common, and that is absent in cases where one does not think that an action is seriously wrong, even when it involves killing something? It seems to me that there is. The correct description depends, however, on certain other matters. In particular, it depends on whether one thinks that death is the end of a human person, or whether one thinks that people survive death, either because they have immaterial minds or souls, or because a deity exists who will resurrect one after death. I think that both of these happy scenarios are, unfortunately, immensely improbable. If that is the case, then what all of the above actions have in common can be described as follows. When one kills a normal adult human being, one is destroying a person. But one is also destroying a person if one causes such a human being to undergo upper-brain death, or complete reprogramming. The same would be true if E.T. were real, and one were to kill him, or destroy his upper brain, or reprogram it. Similarly, if there were non-embodied minds whose psychological abilities and mental lives were in no way inferior to that of normal adult humans beings, and if one could destroy such beings, one would be destroying persons. In all of these cases, one has actions that are prima facie seriously wrong, and actions where a certain sort of continuing subject of experiences and other mental states is destroyed.

By contrast, if one considers the case where a brain-dead human being, or one who has suffered upper-brain death, is killed by directly stopping all life processes in the human being in question, one does not have an action that appears to be prima facie seriously wrong, and the action is one that does not destroy a continuing subject of experiences and other mental states: such an individual was destroyed by the earlier brain damage.

The upshot is that one fundamental moral principle in this area should be free of all biological concepts. It should not involve, therefore, the concept of any particular biological species, or the concept of biological species in general. Nor should it involve the concept of killing a biological organism. It should be formulated, instead, in terms of the idea of the destruction of a certain sort of continuing subject of experiences and other mental states.

The situation will be slightly more complicated if, contrary to what seems to me likely, humans do survive death, either because they have immaterial minds, or because they are going to be resurrected at some future point. For while it would still be true that the destruction of a certain sort of continuing subject of experiences and other mental states is prima facie seriously wrong, that principle would not explain the wrongness of killing a normal, innocent, adult human being, or of causing upper-brain death. Such actions would instead be wrong either because the action expelled a person from this world, or, alternatively, because the action prevented a person from living a normal life in this world, either by rendering the person totally unconscious, or by depriving the person of the ability to perform any bodily actions. This complication does not affect, however, the fundamental point that the correct fundamental moral principle (or principles) in this area should be formulated in terms of the concept of a person, and should be free of all biological concepts.

1.2 Defining the Concept of a Person: Overview

How, then, is the concept of a person to be defined? I shall argue that the concept must, on the one hand, itself be a purely descriptive concept that is free of all moral and evaluative elements, and, on the other hand, must be guided by moral considerations. In particular, what I want to suggest is that the definition of the concept of a person should be based upon the answer to the following question: what non-potential, intrinsic properties make it intrinsically wrong to destroy an entity, and do so independently of its intrinsic, axiological value? A person is to be defined, then, as an entity that possesses at least one of those non-potential, intrinsic properties.

Three questions might naturally be raised about this proposed approach to the concept of a person. First, why must the relevant properties be *intrinsic?* The answer is that anything can be such that it has extrinsic properties that make it wrong to destroy it. It may be wrong, for example, for Mary to destroy a certain stone, not because of the stone's intrinsic properties, but simply because it belongs to John. The concept of a person that I am defining here is intended, among other things, to capture the idea of a type of entity whose destruction harms that very entity, and to do that, whether something is or is not a person must depend upon its intrinsic properties, rather than on its relations to other things.

Second, what is the reason for not allowing *potentialities* to be among the intrinsic properties? Some people would argue that this restriction is highly contentious, on the grounds that human zygotes are persons, not because they have, say, the capacities for thought and self-consciousness, but because they have an active potentiality for acquiring such capacities. The answer to this question is that, in the first place, defining the concept of a person as an entity that possesses certain nonpotential properties does not beg any moral question, because one can go on to argue that while it is prima facie intrinsically wrong to destroy persons, and seriously so, the same is true with regard to the destruction of potential persons. Personhood can be a sufficient condition of having moral status, without being a necessary condition.

In addition, I think that most people use the term "person" to refer to a certain sort of continuing subject of experiences and other mental states. But unless a human zygote has an immaterial mind—and it can be shown, I believe, both that there is no sound philosophical argument for that view, and that there is extremely strong scientific evidence against it—a human zygote, not having a brain, is not only not associated with any *continuing* subject of experiences and other mental states, it is also not even the locus of momentary experiences. If human zygotes possess any moral status, then, it cannot be because they are persons in the ordinary sense of that term.

Finally, if the concept of a person is defined exclusively in terms of actual rather than merely potential properties, one can state, succinctly, what the central disagreement is between many of the philosophers who hold that abortion is morally unproblematic and many who hold that it is not: many of the former maintain that persons, and only persons, have a right to continued existence, whereas many of the latter hold, on the contrary, that potential persons also have such a right. By contrast, if one insists, as advocates of antiabortion views frequently do, in using the term "person" so that it also covers what would otherwise be referred to as "potential persons," no such crisp formulation of the disagreement is available.

The third and final question concerns the clause "independently of its intrinsic, axiological value." What is the reason for that clause? Do not many philosophers—such as Tom Regan,[1] for example—talk about "inherent value," or about "intrinsic value," in discussing moral status or in talking about a right to life? The answer is that it is crucial to distinguish between axiological questions and deontological ones. The former are concerned with good-making and bad-making properties, understood as properties whose possession by states of affairs makes the world, respectively, a better place or a worse place, other things being equal; the latter are concerned, instead, with right-making and wrong-making properties, where these are properties that make actions, respectively, either morally right or morally wrong, other things being equal.

Given this distinction, how is the term "value" best understood—as an axiological notion, or as a deontological one? I think that many philosophers working in ethics interpret it as an axiological notion: to say that something has value is to say that it is such as, other things being equal, makes the world a better place. In any case, I have spoken explicitly of "axiological value," and the crucial point is that the moral status of something does not depend upon that thing's having axiological value. For first of all, if one can destroy something that has axiological value, and immediately replace it with a qualitatively identical object, such an action is not morally wrong, other things being equal. By contrast, the destruction of something with a right to continued existence, followed by the creation of a qualitatively identical thing, is prima facie wrong, and seriously so. Moral status, therefore, is very different from value in the axiological sense.

Second, consider a human being who, throughout his life, has been a thoroughly unpleasant and unhappy individual and who is not going to change in either respect, but who has not seriously harmed anyone and who wants to go on living. If such an individual were to die, the world as a whole might be an intrinsically better place, but it would still be seriously wrong to kill such a person. So an individual's right to continued existence cannot be based on the individual's existence making the world a better place.

In defining the concept of a person, then, one's goal is to capture, not those properties whose possession endows something with axiological value, and whose existence therefore makes the world a better place, but, rather, those nonpotential properties whose possession makes it prima facie intrinsically wrong to destroy something, independently of whether its existence makes the world a better place.

1.3 Persons as Continuing Subjects of Consciousness

Which properties, then, should enter into the definition of the concept of a person? Many answers have been proposed. One natural starting point, very widely accepted by philosophers working in this area, is that for something to be a person it must, at some time during its existence, enjoy states of consciousness—although some philosophers have held that it is not actual consciousness, but merely an immediate capacity for consciousness, that is needed. But this apparent agreement on the importance of consciousness is deceptive, because the term "consciousness" may be construed differently, depending upon one's general views in philosophy of mind. On the one hand, philosophers who are logical behaviorists, or central-state materialists, will be inclined to analyze consciousness in terms either of a general capacity on the part of the individual to acquire information about events, both external and internal, or, more narrowly, in terms of a capacity to acquire information about one's own present mental states. Many dualists, on the other hand, will be inclined to offer an analysis of consciousness according to which states of consciousness are nonphysical states involving qualitative properties to which the individual in question has some sort of direct and epistemically privileged access.

These differing accounts of consciousness do not affect our ordinary ethical decisions, because behaviorists, mind-brain identity theorists, and dualists generally agree about what things are conscious and what things are not. However, this convenient agreement with respect to everyday moral decisions should not blind one to the underlying disagreement concerning fundamental moral principles—disagreements that in the not-too-distant future may well lead to serious moral disagreements about everyday matters. Consider, for example, the possibility of a robot whose "psychological" behavior and "intellectual" capacities were indistinguishable from those of human beings. What would be the moral status of such a being? Would it be just a complex, inanimate object that one could destroy at will, and otherwise treat as one pleased? Or would it possess consciousness? Would it have sensations, thoughts, and feelings? If its behavior were indistinguishable from human behavior, a behaviorist would have to say that such a robot enjoyed all the mental states that humans enjoy. A central-state materialist would, I think, be forced to take the same view. A dualist, on the other hand, would probably say that the fact that the robot's behavior was exactly like human behavior does not provide sufficient reason for holding that it enjoys states of consciousness, conceived of as states that are nonphysical and private. The upshot is that the logical behaviorist and the central-state materialist would classify the robot as a person, while the dualist probably would not.

As a result, it appears to be impossible to settle certain absolutely fundamental issues in ethics without taking a stand on certain equally fundamental questions in philosophy of mind. This important fact deserves to be stressed. While there has been some awareness of it among those working in philosophy of mind—for example, in discussions of what it is about pain that makes it so unappealing—moral philosophers have generally given little attention to it. Until it is attended to, there is little hope of arriving at a satisfactory set of basic moral principles.

If, for the moment, we ignore this disagreement about how "consciousness" is to be construed in the statement of moral principles, we can say that there is general agreement that something is not a person unless it possesses consciousness at some point. One can then ask what, if anything, must be added to consciousness in order for something to be a person, or at least to have moral status. Some people hold that nothing more is needed, that consciousness itself is sufficient to make it intrinsically wrong to destroy something. Many people, however, feel that mere consciousness is not itself sufficient to give something moral status, and several proposals have been advanced as to which additional properties are required. Among the more important suggestions are the following: (1) the capacity to experience pleasure and/or pain; (2) the capacity for having desires; (3) the capacity for remembering past events; (4) the capacity for having expectations with respect to future events; (5) an awareness of the passage of time; (6) the property of being a continuing, conscious self, or subject of mental states, construed, in a minimal way, as nothing more than a construct out of appropriately related mental states; (7) the property of being a continuing, conscious self, construed as a pure ego, that is, as an entity that is distinct from the experiences and other mental states that it has; (8) the capacity for self-consciousness, that is, for awareness of the fact that one is a continuing, conscious subject of mental states; (9) the property of having mental states that involve propositional attitudes, such as beliefs and desires; (10) the capacity for having thought episodes, that is, states of consciousness involving intentionality; (11) the capacity for reasoning; (12) problem-solving ability; (13) the property of being autonomous, that is, of having the capacity for making decisions based upon an evaluation of relevant considerations; (14) the capacity for using language; (15) the ability to interact socially with others. In some accounts, one or more of these properties are necessary conditions either of being a person or of having moral status, or both, and in some accounts one or more constitutes a set of sufficient conditions.

These alternatives, and various combinations of them, provide one with a bewildering selection of candidates for the properties that should enter into the definition of the concept of a person—that is, the delineation of the conditions of personhood. Many philosophers have been content simply to indicate their own preferred choices—usually rationality, autonomy, or self-consciousness. Only a small set of philosophers has seriously attempted to develop any sort of persuasive argument either in support of their own view, or against the more important alternatives. From a certain perspective, this is understandable. Most of our interaction is with human beings, who by and large either possess *all* of those properties, or will come to do so with the passage of time. As a result, the question of which of these properties are the morally significant ones may not have seemed pressing.

But there are a variety of reasons why it is a pressing question. In the first place, normal adult human beings do not, in their development, acquire all of the above properties at the same time. If the fact that one would later acquire a given property had moral significance comparable to the actual possession of that property, this would make no difference. But it seems clear that many people do not believe that potentialities have that sort of significance, because otherwise one could make no sense of their lack of compunction about the destruction of fertilized human egg cells. Consequently, such people are confronted with the problem of distinguishing morally between different stages in the development of a human being. Where the line is drawn will depend upon the stage at which the developing human acquires the morally crucial property or properties, and this in turn will depend upon which of the above properties, or combinations of them, are the properties that should enter into the definition of a person.

Second, and crucially in the present context, when we consider nonhuman animals, mature members of a range of species may share some of the above properties, but not others. Thus, many nonhuman animals are capable of experiencing pleasure and pain, so if that property is sufficient to make something a person, then many nonhuman animals belonging to many species will be persons, and will have a right to continued existence. By contrast, if something is not a person unless it possesses, say, the capacity for self-consciousness, that is, for awareness of the fact that it is a continuing, conscious subject of mental states, or unless it is able to reason, or unless it has the property of being autonomous, that is, of having the capacity for making decisions based upon an evaluation of relevant considerations, or unless it is able to use language, then it seems very likely, in each case, that very few if any nonhuman animals will be persons.

How, then, should one proceed? The ideal would be to establish precisely which non-potential, intrinsic properties make it intrinsically wrong to destroy an entity, independently of its intrinsic, axiological value. But that is a demanding task. Consequently, I shall have to confine myself here to some preliminary reflections concerning properties that I hope the reader will find plausible candidates for being either necessary or sufficient for something's being a person—and, in many theories, for an entity's having moral status.[2] The place to begin is with the idea of extremely rudimentary states of consciousness. Consider the visual experiences that one is currently having. One may be attending to some of those experiences. One may have thoughts about some of those experiences or beliefs about them. But can one have visual experiences to which one is not attending, that one has no thoughts or beliefs about? It seems to me that one can. Suppose that one is playing a game where the first person to spot something purple receives a reward. During that game, one gets immersed in a philosophical discussion. A purple object moves into a position in which light from that object will strike the outermost color cones on one's retina. Will there not be a purplish color patch in one's visual field? Yet is it not possible that, absorbed as one is in something else, one will not notice the presence of the purple object, nor have any relevant beliefs or thoughts?

It seems to me that this is possible. If so, one can have an experience of a certain sort, one can be in a certain state of consciousness, but not be conscious *of* that experience, nor have any thoughts or beliefs about it.

This, then, is what I mean by a rudimentary experience, and a number of careful scientific studies have provided strong support for the conclusion that it is likely that many animals have such sensory experiences.[3] Based on a survey of that evidence, Mary Anne Warren concluded that it was very likely that mammals and birds have such experiences, quite likely that fish, reptiles, and amphibians do, and, although more open to doubt, still more likely than not that many "complex invertebrate animals" also have rudimentary experiences.[4]

Such experiences are first-order states that it is natural to describe as states of consciousness, but they are not states that one is conscious *of*, that one is aware of, or that are the object of any belief or thought. In the case of humans, though it is possible to have such experiences, those experiences will typically be accompanied by other experiences of which one is conscious. But given that such experiences are possible, surely there could be a brain with neural circuitry that gave rise only to rudimentary experiences, and not to any consciousness of those experiences, nor to any beliefs or thoughts or desires or feelings, either about those rudimentary experiences, or about anything else.

Are there any animals whose experience is like this? It seems unlikely, for when it is reasonable to attribute rudimentary experiences to an organism, it also seems reasonable to attribute an ability to experience pleasure and pain, and the latter surely involve, or are connected with, desires.

Even if there are no such entities, however, it is still philosophically useful to ask what the intrinsic moral status of such an entity would be, and, as was in effect noted earlier, a helpful way of approaching questions concerning something's moral status is to consider the possibility of substitution. In the case of axiological value, if object A does not belong to anyone, and no one has any desires concerning A, then if one can replace A with an exact replica, B, there is nothing morally wrong with destroying A and replacing it with B. By contrast, if A has moral status, then destroying A and replacing it with an exact replica B is morally wrong.

Consider, then, a brain that can only ever give rise to rudimentary experiences. Would it be wrong to destroy such a brain, replacing it with an exact replica, in the same location? If one did this, the rudimentary experiences that would later arise when the new brain went into various physical states would be qualitatively the same as those that would have arisen if the original brain had not been destroyed. There will, of course, be no psychological connections between the rudimentary experiences associated with the new brain and those associated with the old brain, but, equally, there are no psychological connections between the rudimentary experiences associated with either brain on its own. Nor is it the case that there are any desires that would have been satisfied if the original brain had not been destroyed. Had the original brain not been destroyed, later rudimentary experiences would have been causally dependent upon the same persisting brain as earlier rudimentary experiences, but it is hard to see why that should be morally significant. It would seem that replacement is not wrong in such a case.

Let us now consider brains that can support other psychological states. First, consider a brain that has experiences, some of which are either pleasurable or painful. What exactly one is adding here to rudimentary experiences is controversial. Some philosophers and psychologists would say that pleasure and pain are qualitative properties of experiences; others would say that an experience is pleasurable by virtue of being intrinsically desired, and painful by virtue of its absence being intrinsically desired. Which view is correct does not seem crucial, because it does not appear to bear upon the issue of moral status. The question is whether replacement is morally problematic, once one has a brain that can give rise to pleasurable and painful experiences.

If replacement were not objectionable in the previous case, it is hard to see why it should now be problematic. For it is still true that there are no psychological connections between experiences at different times that would have existed were it not for replacement. Nor does replacement make any difference with respect to the amount of pleasure and pain in the world, or with respect to whether specific desires get satisfied. The only difference is that experiences of various sorts, rather than being causally dependent upon a single brain, are dependent upon two successive brains. If that difference is morally irrelevant when there are only rudimentary experiences, it is hard to see why it should be relevant when there are experiences that are either pleasurable or painful.

Second, consider a brain that supports, in addition to rudimentary experiences, some of which are either pleasurable or painful, what I shall refer to as "functionalist-style beliefs" and "functionalist-style desires." But what are these states, and why might the idea of such states be important in thinking about the moral status of animals?

To answer the first of these questions will require a brief detour into philosophy of mind, where one of the major divides concerns whether propositions about mental states can be analyzed in terms of other sorts of propositions, or whether, on the contrary, such propositions are analytically (or semantically) basic. A position that was popular in the 1940s and 1950s, known as logical behaviorism, claimed that propositions about mental states can be analyzed in terms of propositions about *behavior*—either propositions about the individual's actual behavior at a time, or else propositions about the individual's behavioral dispositions, where an individual's behavioral dispositions are a matter of the behavior that the individual *would* exhibit if circumstances *were* different in various ways. Consider, for example, a situation where a person, Mary, wants a cold beer. How can that proposition about Mary be analyzed? Logical behaviorists claimed that one could arrive at an answer by considering how Mary will behave, in various circumstances, on the one hand, if she wants a cold beer, and how that differs from how she will behave if she does not want a cold beer. It was then claimed that Mary's wanting a cold beer was nothing more than her having relevant dispositions.

Logical behaviorism was exposed, however, to a variety of objections, and another view gradually emerged, due to the work of U. T. Place,[5] J. J. C. Smart,[6] David Armstrong,[7] and others, which more or less completely replaced behaviorism as an alternative to dualist accounts of propositions about the mental. This view, known as functionalism, agrees with logical behaviorism that propositions about the mental are analyzable, but the functionalist offers a different account of the concept of the mental, and of concepts of specific types of mental states, than the logical behaviorist. The key to the functionalist approach is that, as a first approximation, mental states, rather than being constituted by actual behavior together with behavioral dispositions, are instead *inner* states that stand in certain *causal relations* to *behavior* on the one hand, and to *stimulation* of the individual on the other. Thus, as David Armstrong puts it, "the concept of a mental state is primarily the concept of *a state of the person apt for bringing about a certain sort of behaviour*"—though in the case of some mental states "they are also *states of the person apt for being brought about by a certain sort of stimulus.*"[8]

Consider, for example, perceptual beliefs. A perceptual belief, according to Armstrong, is the type of mental state that it is because, on the one hand, it is directly caused by the organism's immediate environment in certain ways, and, on the other hand, it plays a certain role in causing the organism's behavior. More generally, a belief, according to Armstrong, is a *map* of the organism's environment, and it is a map by which the organism *steers*.

Beliefs alone, however, do not give rise to action. States such as desires or preferences are needed. What account does a functionalist approach offer of mental states of the latter sorts? The answer is that desires are essentially states that, together with beliefs, causally give rise to behavior. So beliefs and desires, and, indeed, all mental states, are characterized by their causal connections to behavior, to sensory stimulation, and to other mental states.

Why are these functionalist concepts of beliefs and desires relevant to questions concerning the moral status of nonhuman animals? The answer is, first, that many nonhuman animals may have beliefs and desires only in the functionalist sense, and, second, it may be that such beliefs and desires are not relevant to moral status.

Why might one think that the second is the case? Consider a robotic device that scans its environment, thereby generating what might be called a "perceptual map" of its immediate surroundings. What is a perceptual map? The idea is that when the robot scans a spatial arrangement of properties in its vicinity, that gives rise to a corresponding spatial arrangement of other properties inside of the robot, such that, just as in the case of an ordinary map, there is a one-to-one correspondence between the properties that are being mapped and the properties of the map that represent the former properties. The robot could then employ that perceptual map in determining how to move around its environment.

The robot could, however, generate a map of its immediate environment in a slightly more complex way, by saving perceptual maps, and then using the information to generate a more informative map of its immediate environment. Suppose, for example, that the robot is programmed to move toward the hottest object in its immediate vicinity. Perhaps the hottest object is behind a screen. A robot using only a perceptual map would not move toward that object. But a robot that had "memory" maps—that is, earlier perceptual maps—could calculate that there was an object that was hotter than any on its current perceptual map, and that was located behind the screen, and it could use that information to move toward the hottest object in its immediate vicinity.

A robot that had such maps, and that was programmed to react in certain ways to things with certain properties—such as the property of being hotter than anything else nearby—would behave in ways that paralleled something that had genuine beliefs and desires. However, the robot's maps and dispositions would bear no relation to states of consciousness. I shall therefore speak of them as "functionalist-style beliefs" and "functionalist-style desires," because, while they are related to action in the same way as genuine beliefs and desires, they have *no relation to consciousness*.

Such a device would have no moral status. There would, for example, be nothing morally wrong about preventing the "satisfaction" of its functionalist-style desires. But now imagine something that, in addition to having functionalist-style beliefs and functionalist-style desires, also has rudimentary experiences, including pleasurable and painful ones. Suppose, further, that some of those rudimentary experiences arise via stimulation of the device by features of its immediate environment, and that those sensory experiences in turn give rise to its perceptual maps, so that its functionalist-style beliefs are causally dependent upon its sensory experiences. Would such a being, which had rudimentary experiences, including pleasant and unpleasant ones, along with functionalist-style desires and functionalist-style beliefs, including ones about past states of affairs, but *absolutely no thoughts at all*, have moral status? Would it be wrong to destroy such a being if one knew that, by doing so, an exact replica would immediately be produced?

My own inclination is strongly to think that such a being would not have moral status. First, to have moral status is to have a right to continued existence, and it seems to me plausible that the specific rights that something has depend upon what sorts of things can be in its interest, that is, what sorts of things can make its existence better. The latter depends upon the types of desires that something is capable of having. What one might call "functionalist-style interests"—interests based upon functionalist-style desires—do not count morally, and the satisfaction of such purely functionalist-style interests does not make something's existence better: it is only what might be called "genuine desires" that make it the case that something is in one's interest in a morally significant way—where a genuine desire is a mental state that involves, in addition to a functionalist-style desire, conscious awareness of that functionalist-style state.

Second, in the case of a being that had only rudimentary experiences, including pleasant and unpleasant ones, along with functionalist-style desires and functionalist-style beliefs, one would not have a *persisting* self that was the subject of the rudimentary experiences associated with the being in question at different times. What one would have, instead, would be a sequence of *momentary* subjects of experiences. Those momentary subjects of experience could enjoy better or worse momentary states, depending on whether the rudimentary experiences in question were pleasurable or painful, but painlessly destroying such a being would not make any momentary subject of experience worse off: it would simply prevent the existence of some momentary subjects of experiences that otherwise would have existed.

By contrast, if one has a continuing or persisting subject of experiences, then the nature of its experiences at different times can contribute to the overall goodness or badness of its existence. Painlessly destroying such a being can then make its existence worse by depriving it of future valuable experiences that it would have had if it had not been destroyed.

Third, consider how your existence would be if you no longer had genuine memories of past experiences. Imagine, for example, that you were having lunch. You would be having various sensory experiences—visual, gustatory, and so on—but you would not be able to have any thoughts about what you had had for breakfast.

You might have functionalist-style beliefs about breakfast, however, and that might influence your present behavior. Perhaps you have chosen a light lunch, because you had a large breakfast, but you would not have any thought that it was the case. Suppose that, while eating lunch, you use up all the milk. You might have a functionalist-style desire to have milk in the house, and that desire—together with the functionalist-style perceptual belief that there is no milk left—might give rise to a functionalist-style intention to pick up milk later on, but none of this would involve any thoughts or any conscious forming of a plan about when to do this. But there would be a functionalist-style plan, and later on it would result in your going out to get milk. That later action would not be accompanied, however, by any thought about what your goal was, or about why you were moving about in the way that you were.

How valuable would such an existence be, devoid of conscious beliefs, memories, desires, and intentions? Suppose that there were a disease that would, if not treated quickly, always result in permanent damage to one's brain that would make it impossible to have any genuine beliefs, memories, desires, or intentions, but leave one with their functionalist-style counterparts. Suppose, further, that a type of treatment were available that, when successful, was completely successful, but that, if not successful, resulted in death. Suppose, finally, that you had contracted the disease. How great would the probability of success have to be for you to choose to undergo treatment?

The smallest probability, p, such that you would choose to undergo treatment at some point before brain damage occurs is a measure of how much you value a continued existence devoid of all genuine beliefs, memories, desires, and intentions. In my own case, there is no smallest probability: as long as the probability, p, is greater than zero, I would choose to undergo treatment. So for me, a life devoid of absolutely all genuine beliefs, memories, desires, and intentions has no value at all.

I do not know whether my preferences on this matter are idiosyncratic, or if, on the contrary, they would be widely shared. If the latter turned out to be the case, and if, in addition, one could argue that there is a connection between the extent to which a certain type of existence is generally valued by an individual, and the wrongness of destroying something of that type, then one could conclude that the painless destruction of something that had rudimentary experiences, some of which were either pleasurable or painful, but that had only functionalist-style beliefs, memories, desires, and intentions, is not intrinsically wrong.

The answer to this question may well be crucial in determining which nonhuman animals are persons, which are continuing subjects of experiences and other mental states, and which have moral status, because it may be that many animals have only functionalist-style memories of past experiences. Their behavior may causally depend upon their earlier experiences without their present mental states being linked in any conscious psychological way to those earlier experiences.

If this line of thought is right, then only beings that are capable of conscious thought have moral status, and the crucial question is what is it to be capable of thought? Some philosophers, such as Donald Davidson[9] and R. G. Frey,[10] have argued that there is a necessary connection between the capacity for thought and the ability to use language. If that view is right, and if only beings that are capable of thought have moral status, then the conclusion will be that few nonhuman animals are such that their painless destruction is morally wrong. Some primates presumably will turn out to have moral status, given the strong evidence that they are capable of linguistic behavior, and the same may be true of whales and dolphins, depending on the outcome of further investigations, but this probably will be true of very few, if any, other animals.

One natural objection to the view that thought presupposes language involves the plausible claim that humans appear to have thoughts before they can use language. But this is open to the rejoinder that humans can *understand* language before they are able to speak, and thus that the explanation of how they are able to have thoughts at that point is that they are using language internally by having the relevant auditory images.

I think that this rejoinder is plausible. However, there is a deeper objection to the thesis that thought presupposes language. Think about some past event in your life. To what extent are your thoughts about that event *verbal?* The answer, I suggest, is that they are so only to a limited extent. In the main, what takes place when one thinks about a past event is that one enjoys a number of memory images related to one's original perceptual experiences of the event. But one could have those memory images even if one had no ability at all to use language, and if one did so, would it not be the case that, in having such images, one was thinking about a past event?

Similarly, consider thinking about how one can achieve some goal—such as getting an object that is out of reach. In order to have a genuine desire to achieve some goal, does the desire have to be such as one can express via some verbal thought? To engage in formulating a plan for achieving that goal, does one have to have a series of verbal thoughts?

Neither of these contentions seems right. Suppose that one is able to form an image of the state of affairs in which one has achieved the desired outcome. Suppose further that the formation of such an image causes one to form various series of other images that correspond to series of possible actions that one might perform, and that one arrives, finally, at a series of images that correspond to actions that will enable one to achieve one's goal. At that point, one proceeds to carry out the series of actions in question. In such a case, is it not reasonable to view the original image of the state of affairs in which one has achieved the outcome in question as expressing a desire for that state of affairs? Is it not also reasonable to view the formation of images corresponding to different sequences of possible actions as nonverbal thinking involved in an attempt to find a way of achieving one's goal?

If this is right, then experiments performed by Wolfgang Köhler[11] and other researchers on chimpanzees and other animals that provide grounds for thinking that such animals are arriving at solutions to problems via some sort of mental manipulation of images also give one reason for concluding that those animals are capable of nonverbal thinking.

These matters deserve extended discussion, and I shall return to them, albeit briefly, in the next section. Prima facie, however, it seems to me that reflection upon the roles that images play in the mental life of humans, and on corresponding roles that they could play in the mental life of nonhuman, intelligent animals that lack the ability to use language, provides a good reason for taking seriously the thesis that thought can take a nonverbal form.

1.4 Summing Up

The basic idea of this first section has been to define the concept of a person in a way that, without diverging significantly from our ordinary use of the term "person," captures a purely descriptive, non-normative concept that is needed for the formulation of a fundamental moral principle concerning when the destruction of something is intrinsically wrong. Or, more precisely, the idea was to define the concept of a person in terms of those non-potential, intrinsic properties that make it intrinsically wrong to destroy an entity, and do so independently of its intrinsic, axiological value.

I have attempted to offer support for an analysis of the concept of a person that involves the following five crucial claims. First, a person is an entity that has had, or that is now having, actual experiences. It is not merely a potential subject of experiences; things of the latter sort, with the appropriate properties, are potential persons, and their moral status needs to be settled by further argument. Second, a person is not a momentary subject of experiences that can only exist for a moment: a person is a continuing, or persisting subject of experiences that can exist at different times. Third, the experiences and other mental states of a person at different times must be psychologically connected. Otherwise, what would exist would not be a person, but simply a sequence of isolated, momentary subjects of consciousness. Fourth, the psychological connections cannot be based only on unconscious mental states, such as what I have referred to as functionalist-style beliefs, functionalist-style memories, functionalist-style desires, and functionalist-style intentions; at least some of the connections must involve conscious thoughts at one time about experiences and other mental states at other times, past or future. Finally, it should not be assumed that the conscious thoughts that serve to unify momentary subjects into continuing or persisting subjects of experiences and other mental states are necessarily verbal in nature. For while the matter is far from uncontroversial, there are reasons for thinking that there can be thoughts that are based upon images, rather than upon any internal use of language.

Is there anything more that should be part of the concept of a person? Though the matter seems difficult and uncertain, I am still somewhat inclined to think that there is. In particular, I am inclined to think that the thoughts that serve to tie together mental states at different times in a conscious way need to involve the idea that an experience or mental state that existed at some other time was *one's own*. If this is right, then something is not a person unless it possesses self-consciousness, or self-awareness.

This view has been rejected by others. James Rachels,[12] for example, does so on the grounds that the idea of awareness of a self is metaphysically suspect. Given some concepts of the self, this is certainly so. But no such concept is needed for the thought that experiences and other mental states existing at other times are one's own. The idea of persistence does not require occult entities; it can be explained in terms of the right sort of causal relations. To think of a past experience as one's own, then, is simply a matter of thinking of that experience as standing in certain causal relations to one's present mental states.

Although this shows that the idea of self-awareness is not metaphysically problematic, and thus that Rachels' objection is misguided, one is left with a serious *epistemological* hurdle, since when self-awareness is explained in terms of a capacity for thinking of a past experience as standing in certain causal relations to one's present mental states, it is going to be a challenging matter to provide evidence for self-awareness in the case of animals that lack the ability to use language.

In any case, I do not think that one should build the idea of self-awareness into the concept of a person at this point, simply because the claim that self-consciousness is a necessary condition for something's having a right to continued existence is far too controversial. But I shall return to this issue later, in sections 2.4 and 3.3.

2. Arguments against the Possession by Nonhuman Animals of a Right to Continued Existence

Let us now turn to arguments that philosophers have offered in support of the view that nonhuman animals do not have moral status. These arguments are typically formulated in terms of the idea of a right. What they attempt to establish is that nonhuman animals do not have a right to continued existence.

I shall consider four arguments, some of which, as we shall see, attempt to establish that nonhuman animals do not have a right to life by arguing for the much stronger thesis that nonhuman animals do not have any rights at all. The four types of arguments are as follows, First, Jan Narveson and others have appealed to contractarian approaches to morality to support the conclusion that nonhuman animals cannot have any rights. Second, there are a variety of views, advanced by Kant and others, involving the claim that there is a crucial relationship between being an actual and/or possible moral agent and possessing rights. Third, there are arguments in support of the view that nonhuman animals cannot have beliefs or desires, and thus cannot have interests, and so cannot have any rights. Finally, some philosophers have argued that even if nonhuman animals can have some desires, they cannot form the concept of a continuing subject of experiences, and thus cannot have desires about future states of themselves, and that, because of this, nonhuman animals cannot have an interest in their own continued existence, and so do not have the corresponding right.

2.1 Animal Rights and Contractarian Approaches to Morality

Contractarian approaches to morality come in different forms, but the basic idea is that the only sound approach to morality is one according to which an action is morally wrong if and only if it is contrary to a moral principle governing behavior such that rational, self-interested persons would freely agree to enter into a contract with one another under which they were bound to act in accordance with the principle in question.

In the present context, an important distinction within contractarian approaches is between ones in which those entering into the contract have full information about the abilities and other resources that they will have in the resulting society, and approaches where the decision concerning the form that the contract is to take is made behind a Rawlsian "veil of ignorance," where one does not know what position one will occupy in the resulting society, or what resources and abilities one will have.[13]

Jan Narveson has offered an argument for the conclusion that nonhuman animals have no rights that rests upon the former type of contractarian approach. Thus he says,

> On the contractarian view of morality, morality is a sort of agreement among rational, independent, self-interested persons, persons who have something to gain from entering into such an agreement. It is of the very essence, on such a theory, that the parties to the agreement know who they are and what they want—what they in particular want, and not just what a certain general class of beings of which they are members generally tend to want.[14]

Given this approach, it is clear, Narveson says, that the parties to such a contract will have the following two characteristics: "(1) they stand to gain by subscribing to it, at least in the long run, compared with not doing so, and (2) they are *capable* of entering into and keeping the agreement."[15]

Nonhuman animals, however, are not capable either of entering into such an agreement or of keeping it, while, on the other hand, humans have nothing to gain from entering into a contract that places restrictions on how they treat animals. Consequently, if the type of contractarian approach that Narveson favors is correct, it follows that no principle restricting how one can behave with regard to nonhuman animals can be correct. So nonhuman animals cannot possess any rights.

Certain features of Narveson's contractarianism are worth noting. First, if some people in society have much more power than others, the former may have nothing to gain by entering into a contract with the weaker, and may have a considerable amount to lose—such as the possibility of having the weaker as slaves. Second, rights are generally thought of as placing restrictions upon the behavior of all agents. But if some humans enter into a contract with other humans, that contract will place restrictions only upon those who enter into the contract, so that if powerful extraterrestrials appear on the scene, or if the universe contains powerful supernatural beings, humans who have entered into the contract will have no rights against such beings, and those beings can treat humans however they want, without doing anything morally wrong. Third, not all humans can enter into the contract. To begin with, there are humans that lack the capacity for thought: zygotes, embryos, anencephalic babies, humans suffering from extreme dementia, and others. It might be argued, however, that that is not troubling, because such humans are not persons. But there are humans who clearly are persons—for example, three-year-olds—who also will not be able to enter into any contract. On Narveson's approach, then, raising humans until three or four years old for food, if properly done, would not be morally problematic.

The three points just mentioned would not arise if one abandoned Narveson's preferred form of contractarianism in favor of a Rawlsian approach, because when one is deciding on moral principles behind a veil of ignorance, one does not know, for example, whether one is going to turn out to be among the strong or among the weak. So why not formulate the argument in terms of Rawls' approach?

Narveson's reason for not doing so is this:

> Now Rawls' theory has his parties constrained by agreements that they would have made if they *did not* know who they were. But if we can have that constraint, why should we not go just a little further and specify that one is not only not to know *which* person he or she is, but also whether he or she will be a person *at all*: reason on the assumption that you might turn out to be an owl, say, or a vermin, or a cow.[16]

The problem with this is that what Narveson describes as someone's turning out to be an owl is an impossibility: a person's body might be transformed into the body of an owl, but if owls are not persons—and Narveson, in his example here, is certainly assuming that that is so—then such a transformation is not one where a person turns out to be an owl. Persons cannot survive as nonpersons, and, from a self-interested point of view, persons have no reason to care about what happens to their bodies when they no longer exist.

A contractarian argument against animal rights should be recast, then, in terms of a Rawlsian form of contractarianism. When that is done, what is one to say about the resulting argument? One of the conclusions will be that when one is deliberating behind the veil of ignorance, it is not in one's self-interest to accept a moral principle that prohibits people from inflicting as much pain as they want upon sentient beings that are not persons. This conclusion, however, is on a collision course with both the axiological intuitions that pleasure is prima facie good, and pain prima facie bad, and the normative intuition that inflicting pain upon anything is prima facie wrong. These intuitions, moreover, are widely accepted indeed, and are generally firmly held. It would seem, therefore, that there is a strong prima facie case against the view that the correct moral principles are those that would be chosen by rational, self-interested persons deciding behind a veil of ignorance.

2.2 Moral Agency and Rights

A second important argument for the thesis that nonhuman animals are not possessors of rights appeals to the idea that there is a connection between moral agency
and the possession of rights. Different accounts have been offered of what precisely
the connection is. Of these, the simplest is this:

> **Version 1:** Being a moral agent is both a sufficient condition and a necessary
> condition for being a possessor of any basic rights.

Some philosophers who are attracted to the general idea, however, are troubled by
the fact that this thesis leads to the conclusion that some human beings do not have
rights. Thus Mary Anne Warren, for example, discussing the views of Kant, says,

> Kant ascribes full moral status only to rational moral agents; thus his community
> of moral agents would appear to exclude not only animals, but also human
> infants, young children, and human beings who are severely mentally disabled.
> Infants and young children are not yet capable of acting on general moral
> principles. Some human beings suffer from genetic or developmental
> abnormalities that preclude their ever becoming moral agents. And some persons
> suffer injury or illness that permanently robs them of the capacity for rational
> moral agency.[17]

As a result of such considerations, Warren views moral agency as a *sufficient* condition for the possession of all moral rights, and embraces the following
principle:

> **The Agent's Rights Principle:** *Moral agents have full and equal basic moral rights,
> including the rights to life and liberty* [18]

On the other hand, she rejects the view that moral agency is a *necessary* condition
for the possession of rights, arguing that there are six other principles concerning
when something has moral status, one of which is

> **The Human Rights Principle:** *Within the limits of their own capacities and of [the
> Agent's Rights Principle, above], human beings who are capable of sentience but not
> of moral agency have the same moral rights as do moral agents.*[19]

A number of other scholars who are attracted to the basic idea that there is a connection between moral agency and the possession of rights, such as Tibor Machan,[20]
also want to avoid the view that cognitively impaired members of *Homo sapiens* do
not have rights. So they embrace something along the lines of one of the following
versions of the general moral agency view:

> **Version 2a:** Being a moral agent is a sufficient condition for being a possessor of
> all basic rights, but it is not necessary, because all members of *Homo sapiens* that
> are capable of sentience possess the same rights.
> **Version 2b:** Being a moral agent is a sufficient condition for being a possessor
> of all basic rights, but it is not necessary, because all members of *Homo sapiens*
> possess the same rights.

> **Version 2c:** Being a moral agent is a sufficient condition for being a possessor of all basic rights, but it is not necessary, because all members of any species whose normal adult members are moral agents possess the same rights.

These three versions of the general view combine the claim that moral agency is relevant to the possession of rights with a principle that claims that species membership is also relevant in itself.

There are, however, a number of strong objections to the view that species membership is morally significant in itself,[21] including objections that focus on cases of brain death or upper-brain death—which tell against versions 2b and 2c. Version 2a escapes the brain-death and upper-brain-death objections because of the fact that it extends the right to continued existence only to members of our species that are capable of sentience. But consider a member of our species that, because of some genetic defect, is born with an abnormal brain that will support only rudimentary experiences, as delineated above. The mental life of such a human would be *significantly* inferior to that of a normal cat or dog. Yet according to version 2a, it has precisely the same rights—including a right to life—as a normal adult human being. This implies that if one is in a situation where, if one does nothing, millions will die, but where one can save those millions, either by killing a normal adult human being or by killing a human being whose mental life can never involve anything beyond the most rudimentary experiences, then, other things being equal, those two acts of killing are morally on a par—a conclusion that is surely implausible.

In addition to the strong moral agency view formulated by version 1, and the speciesist variants of versions 2a, 2b, and 2c, there is a third possibility in which the pure-agency view is revised by appealing to the idea that an active potentiality for moral agency is also relevant to the possession of moral rights. This third possibility can be expressed as follows:

> **Version 3:** Being a moral agent is a sufficient condition for being a possessor of all basic rights, but it is not necessary because anything that has an active potentiality for becoming a moral agent also possesses the same rights. What is necessary if something is to be a possessor of rights is that the being is either a moral agent or a potential moral agent.

The part of this principle that formulates a sufficient condition for having rights combines a rather uncontroversial claim—that being a moral agent is sufficient to make one a possessor of all basic rights—with a highly controversial claim—that an active potentiality for moral agency is also sufficient. That the latter is highly controversial is evident from the fact that it supports the conclusion that human zygotes have a right to continued existence that is fully on a par with that of normal adult human beings. Elsewhere, I have argued at length that an active potentiality for acquiring properties that give normal adult human beings a right to continued existence does not itself give an organism such a right.[22] This issue can, however, be ignored in the present context, because what is relevant with regard to the question of whether nonhuman animals have rights is the part of the principle that formulates a necessary condition for having moral rights.

The crucial question, then, is whether something can have moral rights only if it is a moral agent or at least a potential moral agent. One objection to this claim is that rights do not stand or fall together, and that, in particular, something does not have to have a right to continued existence in order to have a right not to be tortured. This objection, moreover, seems to me right, as I would argue that the function of rights is to protect interests, that desires can give rise to interests, and that there can be subjects of experiences and other mental states that have a desire not to be in a painful state, and therefore a corresponding interest, even though they do not have a type of mental life and psychological capacities that make it the case that continued existence is in their interest. But I shall not defend this objection here, because even if successful, it could be escaped by shifting to a claim that moral agency, rather than being connected to rights in general, is connected specifically with the right to continued existence. In particular, one might move on to the following claim:

> **Version 4**: Being a moral agent is a sufficient condition for having a right to continued existence, but it is not necessary, because anything that has an active potentiality for becoming a moral agent also has a right to continued existence. What is necessary if something is to have a right to continued existence is that that thing is either a moral agent or a potential moral agent.

This proposed necessary condition for having a right to continued existence seems unsound. Suppose that one encounters an adult human being who seems perfectly normal as regards psychological capacities, who can think and reason and who possesses self-consciousness, and who is also friendly, gentle, and sociable. It turns out, however, that the individual has no moral intuitions at all, and no beliefs about the rightness and wrongness of actions. If this were the case, the individual in question would not be a moral agent, for a moral agent is not simply something that deliberates about possible actions and makes choices based on such deliberations: a moral agent, in his or her deliberations, must assign weight to considerations of the moral rightness and wrongness of actions. So the individual I have described is not a moral agent. Nor would such an individual have any active potentiality for becoming a moral agent. Moreover, we can assume that many attempts have been made to draw the individual's attention to the existence of moral facts, all of which have been unsuccessful, so that even if one counted something as a potential moral agent if it had a passive potentiality for moral agency, the individual I am imagining would be neither a moral agent nor a potential moral agent.

If version 4 is right, then, such an individual, who would have all of the capacities of normal adult human beings except the capacity to acquire beliefs about the rightness and wrongness of actions, would not have a right to continued existence, and could be killed for trivial reasons. Is this at all plausible?

One possible move at this point is to revise version 4 with a speciesist addendum, holding that the individual in question has a right to continued existence because he or she belongs to a species whose normal adult members are moral agents. But even if one waives objections to such a speciesist revision, such a move fails to come to grips with the basic point, because one can shift to considering nonhuman extraterrestrials who have all the psychological capacities of normal adult humans, and to a higher degree, except that they have no moral intuitions at all, and no beliefs about the rightness and wrongness of actions. It might be suggested that, lacking moral beliefs, such beings must be dangerous. But we can suppose that, in contrast to what Mark Twain referred to as the "damned human race," all civilizations involving the extraterrestrials have been free of such things as wars, inquisitions, religious persecutions, and so on. Finally, when we attempt to convince them of the existence of moral facts, they tell us that they think that J. L. Mackie's Error Theory[23] is correct, that there are no objective values, and that what humans refer to as "moral intuitions" are nothing more than feelings they probably have because they were rewarded or punished for various actions when they were being raised. Conscience is, our extraterrestrials believe, either simply a conditioned reflex, or perhaps one that has been hardwired through evolution.

The thesis that a necessary condition of having a right to continued existence is being either a moral agent or a potential moral agent entails that such extraterrestrials would not have a right to continued existence, and there is no speciesist escape from this conclusion. But is it at all plausible that such beings would not have a right to continued existence? I submit that it is not, and that the desires such beings would have about future states of affairs, including their own continued existence, and the interest in continued existence that would supervene upon those desires, would make it intrinsically wrong to kill such beings, and thereby give them a right to continued existence. If so, then neither moral agency nor potential moral agency is a necessary condition for having a right to continued existence. The thesis that nonhuman agents lack moral status cannot be defended, therefore, by appealing to the claim that they are not moral agents.

2.3 Nonhuman Animals and the Absence of Beliefs and Desires

A third argument comes in two different versions. The first, advanced by R. G. Frey, attempts to show that nonhuman animals cannot have any beliefs and desires, and thus cannot have any rights. The second claims that nonhuman animals can have beliefs and desires, but not the kind necessary for a right to continued existence.

2.3.1 *R. G. Frey's Argument*

In a number of publications, but especially in his book *Interests and Rights: The Case Against Animals*,[24] R. G. Frey has set out and defended an important argument for the view that nonhuman animals do not have moral status—an argument that can, I think, be summarized as follows:

(1) There can be ways of treating something that are intrinsically wrong only if there can be states of affairs whose obtaining is either in the interest or contrary to the interest of the entity in question.

(2) There can be states of affairs whose obtaining is either in the interest or contrary to the interest of something only if the entity in question is capable of having desires.

Therefore,

(3) There can be ways of treating something that are intrinsically wrong only if the entity in question is capable of having desires.

(4) Something cannot have either beliefs or desires unless it has the ability to use language.

Therefore,

(5) There can be ways of treating something that are intrinsically wrong only if the entity in question has the ability to use language.

(6) Nonhuman animals do not have the ability to use language.

Therefore,

(7) There is no way of treating a nonhuman animal that is intrinsically wrong.

(8) Something cannot have any rights at all unless there can be some way of treating that entity that is intrinsically wrong.

Therefore,

(9) Nonhuman animals do not have any rights at all.

What is one to say about this argument? The premises here are (1), (2), (4), (6), and (8). Frey himself is skeptical concerning the usefulness of the idea of rights,[25] so he would view (8) and (9) as not really adding anything to the argument. But if one favors talk about rights, (8) seems unproblematic.

All of the other four premises can be questioned, but I shall ignore (1), and concentrate on (2), (4), and (6), all of which may strike one as problematic. In the case of (2), for example, one might object that animals can experience pleasure and pain—as Frey grants[26]—and that this is sufficient, even in the absence of desires, to make it the case that there are things that are in the interest of such animals.

Frey's discussion of this type of objection in chapter 11, "Pain, Interests, and Vegetarianism,"[27] does not seem to me satisfactory. But a possible reply, which may well be defensible, involves arguing that the idea of an experience as being painful can be analyzed in terms of its absence being intrinsically desired. But one problem is that that will then lead, if the rest of Frey's argument is sound, to a conclusion that Frey does not want to embrace, namely, that nonhuman animals do not experience pain.

In the case of (6), the ability of some primates to learn a sign language is a natural basis for an objection. But, in the first place, the significance of this ability has been questioned, and Frey himself appeals to Chomskian ideas concerning language to support the conclusion that in this scenario one does not have a case of a genuine use of language.[28] Second, and more important, even if one can sustain the claim that one does have genuinely linguistic behavior in such cases, if the rest of Frey's argument is sound, one still has the conclusion that higher animals that cannot use language have no rights or moral status at all, and this is a strong conclusion, because it entails that such animals do not even have a right not to be tortured.

Finally, what about (4)—which will, I think, strike most people as the most problematic premise in the argument? How does Frey argue in support of (4)? There are two main lines of argument. First, Frey appeals to Quine's account of belief, according to which belief involves a relation to *sentences*: "In expressions of the form 'I believe that...', what follows the 'that' is a sentence, and what I believe is that the sentence is true."[29] The upshot is then that only beings that can use and understand sentences can have beliefs. Second, Frey argues that the only ground that one can have for attributing beliefs to an animal that is incapable of using a language is the animal's behavior. That ground, however, is inadequate, because any piece of behavior is compatible with quite different beliefs. There is thus no criterion for attributing one belief, rather than a number of others, to an animal that cannot use language.

Both of these arguments reflect a behaviorist approach to belief, according to which the beliefs an organism has must be analyzable in terms of the organism's behavior. But that approach is unsatisfactory, and once one shifts to a functionalist account of belief and other mental states, and adopts something like Armstrong's view that beliefs are an inner map by which the organism steers, then, first of all, there is no need at all to bring in language, and, second, the beliefs a dog is acting upon at a given moment, rather than being determined by its behavior, are determined by the inner states—brain states, according to Armstrong and other central-state materialists—that are causing the dog's behavior, and those inner states, in turn, are identified with different beliefs, based on the types of sensory stimulation that causally give rise to them. The conclusion, in short, is that while Frey's arguments may be plausible given a behaviorist approach to belief, such an approach is untenable, and once a functionalist account of beliefs and other mental states is on the table, the arguments can be seen to be unsound.

Finally, the idea involved in functionalist analyses of mental states—namely, that at least some mental states are to be identified on the basis of their causal relations to other things, such as sensory stimulation and behavior—not only undercuts Frey's arguments for a crucial premise, it can also be used to show that the argument as a whole cannot succeed. The reason is that one can use the basic functionalist idea to analyze what it is for a sensation to be pleasant or painful: a sensation's being painful, for example, can be analyzed in terms of an animal's having a functionalist-style desire for the absence of the sensation. The upshot is that there appear to be good grounds for holding that there are things that are in a nonhuman animal's interest, because there are good reasons for holding that nonhuman animals do have sensations, and for attributing to such animals functionalist-style desires for the presence or absence of certain sensations.

2.3.2 *Thought and Conscious Beliefs and Desires*

If the preceding objection to premise (4) in Frey's argument is right, the attempt to show that nonhuman animals have no rights at all, because they have no interests, is doomed. But the door is still open for a more modest argument, the goal of which is to show that animals do not have a right to continued existence. Recall that earlier, in section 1.3, I argued that the psychological unification of momentary experiences and other mental states into a single, continuing subject of experiences requires beliefs, memories, desires, and intentions that are not merely functional states. If so, what more is required? The claim was that one needs *conscious* beliefs and desires. But if that is right, then one may be able to argue, for example, that conscious beliefs are *thoughts*, and that thoughts, in contrast to functionalist-style beliefs, involve the internal use of language.

In response, however, I suggested that there is reason for holding, for example, that images are crucial to our memory-thoughts about past experiences, and this suggests the possibility that some thoughts might consist simply of images, rather than of internal linguistic episodes. Of course, one can have images both when one is remembering some event, and when one is merely imagining something. Here too, however, the functionalist idea can enable one to make the relevant distinction. Images that are memory images play a causal role that images that are mere imaginings do not: the former, unlike the latter, play a role in generating the map by which one steers.

Let me mention two grounds for doubt that might be raised about my appeal to the possibility of imagistic thinking. The first concerns the very idea of imagistic thinking. Earlier, I tried to render it plausible by appealing to the use of images in memory. But it might be objected that though we use images in memory, those images are accompanied by a verbal thought, to the effect that the image represents the way things really were. But if so, is it really the case that images, *on their own*, can be thoughts?

An answer to this question needs a more extended defense than I can offer here. The general line along which I would argue, however, is as follows. What is *important* about verbal thoughts is that they are *conscious* mental states that *express beliefs*. They express beliefs because they involve an internal use of language, and the particular belief that a given thought expresses is the belief that would be expressed by an utterance of the corresponding sentence. In turn, however, that sentence has the meaning it does because of the particular functionalist-style belief that disposes one to utter the sentence.

The upshot is that a verbal thought is a conscious state that is causally related to a belief, where the causal connection involves the intermediary of language. Nonverbal images could stand, however, in the same causal relations to beliefs, the only difference being that the relation would be a direct one, rather than an indirect one. If that were so, then such images would express beliefs, just as much as verbal thoughts do. Consequently, if what is important about thoughts is that they are conscious states that express beliefs, then, given that nonverbal imagistic states can have those two properties, imagistic thinking is possible.

The other point concerns whether, if images on their own can be thoughts, it is reasonable to attribute such thoughts to nonhuman animals. On the one hand, when a dog wags its tail when a door is about to open, it seems natural to think that the dog has some sort of image of its owner entering the house. In addition, certain problem-solving abilities, mentioned earlier, seem to point to the presence of imagistic thinking. But on the other hand, if higher animals are capable of having such nonverbal thoughts, why are they not also capable, not of learning to speak a human language, but of learning to understand it? For even if they cannot produce the relevant sounds, they have auditory experiences, and if they are capable of thinking via visual images, why should they be unable to do so by means of auditory images, and so be capable of verbal thought?

Moreover, it is striking that even most primates that have the capacity to learn a sign language lack this ability. There is, however, an exception, namely, bonobos (or pygmy chimpanzees). This species of primate is, of extant species, the one most closely related to humans, and two bonobos, Kanzi and Panbanisha, have learned to understand spoken language. The case of Kanzi, described in detail in the book *Kanzi*,[30] is especially impressive, and more recently it has been claimed that Kanzi understands about three thousand spoken English words.[31]

The upshot is that it does not appear possible at present to come to a confident judgment concerning the present argument. On the one hand, I think that it is plausible, on philosophical grounds, both that conscious thoughts are necessary if something is to have a right to continued existence and that nonverbal, imagistic thinking is logically possible. On the other hand, reasons can be offered both for and against the view that higher animals are capable of imagistic thinking and, in the end, careful scientific investigation will be needed to determine what the fact of the matter is. At the very least, the apparently defensible idea of nonverbal thinking would seem to provide a promising avenue in thinking about certain arguments bearing upon the moral status of nonhuman animals.

2.4 Self-consciousness and the Right to Life

The final argument that I want to consider for the claim that nonhuman animals— perhaps with a very few exceptions, such as Kanzi—do not have a right to continued existence, can be put as follows:

(1) One cannot have a right to continued existence unless one's continued existence is in one's interest.

(2) One's continued existence cannot be in one's interest unless one is capable of having conscious desires about one's existence as a continuing subject of experiences and other mental states.

(3) It is impossible to have a conscious desire that p be the case unless one can have a thought that p is the case.

(4) The continued existence of a subject of experiences is not something that can be represented via an image.

(5) The only types of thinking are verbal thinking and imagistic thinking.

Therefore, in view of (3), (4), and (5),

(6) Verbal thought is necessary if a thing is to be capable of having a desire concerning its own existence as a continuing subject of experiences and other mental states.

Hence, given (1), (2), and (6), we have

(7) Something cannot have a right to continued existence unless it is capable of verbal thought.

(8) Virtually no nonhuman animals are capable of verbal thought.

Therefore,

(9) Virtually no nonhuman animals have a right to continued existence.

The crucial premise here is, I believe, (2), and I do not know what sorts of arguments one can offer either for or against it. So let me simply describe what I take to be the main alternative to (2) and leave further reflection about this issue to the reader.

If imagistic thinking is possible, then, to begin with backward-directed psychological connectivity, one can have imagistic memories of past experiences, and also imagistic memories of past experiences occurring in a body reasonably similar to one's present body. But one cannot have imagistic memories of past experiences as belonging to a certain continuing subject of experiences—namely, oneself—or as associated with that persisting physical object that is one's body. Similarly, if one considers forward-looking psychological connectivity, one can have functionalist-style beliefs and intentions and desires about future experiences, along with imagistic thoughts and conscious desires that correspond, in a certain way, to those functionalist states, but there cannot be anything in the images themselves, that represent certain possible experiences as being experiences that will belong to a certain continuing subject of experiences—namely, oneself—nor even as experiences that lie in the future, for there can be no imagistic representation of the concept of the future. Moreover, while one can have functionalist-style plans of action, and sequences of images that correspond to successive steps in the functionalist plan, there cannot be any image that represents even the temporal order of the steps, let alone that represents each step as a stage in the mental life of a continuing subject of experiences.

The question, in short, is whether the types of backward and forward psychological connections that consist of those sorts of imagistic thoughts plus related functionalist states constitute something that has an interest in its own continued existence, or whether, on the contrary, and as (2) claims, it is essential to be able to have conscious desires about one's own existence as a continuing subject of experiences and other mental states. I shall return to this question in section 3.3.

To sum up: philosophers have offered a number of arguments for the view that all or almost all nonhuman animals do not have moral status. I have attempted to show that arguments that appeal to contractarian approaches to rights, or to relations between rights and moral agency, or to the thesis that nonhuman animals do not have beliefs and desires, are unsuccessful. A variant on the third of these arguments, which appeals instead to the thesis that nonhuman animals do not have thoughts, is more difficult to evaluate, and I argued that whether it is sound depends upon whether nonhuman animals are capable of imagistic thinking. Finally, and even more difficult to evaluate, is the argument that self-consciousness is a necessary condition for having moral status.

3. Arguments for the Possession of a Right to Continued Existence by Nonhuman Animals

In this third and final section, I shall consider two arguments that philosophers have offered in support of the view that nonhuman animals do have moral status, along with a third argument that is suggested by the last negative argument just considered. First, there are arguments that claim either that mere consciousness is sufficient to give something moral status, or that sentience, understood as the capacity to experience pleasure or pain, is sufficient to do so. Second, there is the view, defended especially by Tom Regan, that being a subject-of-a-life gives higher animals a right to continued existence as well as many other rights. Third, there is an argument that imagistic thoughts of the right sort can provide a psychological connection between experiences at different times that is a sufficient basis for a continuing subject of experiences, and therefore sufficient for moral status.

3.1 The Arguments from Consciousness and from Sentience

Some philosophers have defended the view that nonhuman animals have moral status by appealing to one of the following claims:

(1) It is intrinsically wrong to destroy something that has states of consciousness.
(2) It is intrinsically wrong to destroy something that is sentient, where sentience is understood as the capacity to experience pleasure and pain.

Dale Jamieson,[32] for example, offers a defense of the view that nonhuman animals have moral status that appears to involve (1).

A question that immediately arises with regard to any appeal to (1) or (2) is whether everything that has moral status based on (1) or (2) has *the same* moral status. This would entail the implausible consequence that it is intrinsically just as wrong to kill an organism that has no mental life beyond the most rudimentary experiences, possibly accompanied by pleasure and pain, as it is to destroy a normal adult human being. So it would seem that if an appeal to (1) or (2) is to have any plausibility, one will have to hold that the right to continued existence is a matter of degree. Here the idea would be that the property that gives something a right to continued existence is one that admits of degrees and that this is relevant to the strength of the right that is based upon it. So, for example, one might hold that it is wrong to destroy anything that has states of consciousness, but that the wrongness is greater as the states of consciousness are more complex.

Earlier, I offered an argument for the view that consciousness plus a capacity for pleasure and pain, even when combined with functionalist-style beliefs, memories, desires, and intentions, is not sufficient to give something a right to continued existence. If so, (1) and (2) cannot provide a basis for a right to continued existence.

What argument can be offered for (1) or (2)? In the case of (1), Jamieson claims that "consciousness is itself a good and that we implicitly recognize it as such."[33] Now, if this is right, then that provides support for (1), but not the type of support that enables one to use (1) as grounds for a claim about moral status, or about a right to continued existence. The reason is that the claim that consciousness is itself a good is an axiological claim, and, as such, it does not follow that there is anything wrong with destroying one organism that has states of consciousness if, at the same time, one creates a comparable organism that has similar states of consciousness. So Jamieson's claim, even if one grants it for the sake of argument, cannot provide any ground for a right to continued existence, because if something has such a right, it follows that replacement is morally wrong.

3.2 Regan's Argument: The Concept of the Subject-of-a-Life

One of the best-known and most-discussed defenses of the view that nonhuman animals have a right to continued existence is an argument advanced in a number of essays by Tom Regan, and defended most fully in his book *The Case for Animal Rights.*[34] There Regan claims that anything that is a "subject-of-a-life" has what he refers to as "inherent value," and he formulates his subject-of-a-life criterion as follows:

> To be a subject-of-a-life is to be an individual whose life is characterized by those features explored in the opening chapters of the present work: that is, individuals are subjects-of-a-life if they have beliefs and desires; perceptions, memory, and a sense of the future, including their own future; an emotional life together with feelings of pleasure and pain; preference and welfare interests; the ability to initiate action in pursuit of their desires and goals; a psychophysical identity over time; and an individual welfare in the sense that their experiential life fares well or ill for them independently of their utility for others and logically independent of their being the object of anyone else's interest.[35]

As mentioned earlier, talk about inherent value in this context seems to me best avoided, and the term "value" reserved for axiological discussions. But in any case, it certainly seems to me plausible that something that is a subject-of-a-life, thus defined, has a right to continued existence. Indeed, given that such individuals have "a sense of the future, including their own future," it seems that they must possess self-consciousness, and so such individuals satisfy the strong requirement that was appealed to in the final argument—set out in subsection 2.4—for the claim that nonhuman animals do not have a right to continued existence.

How, then, does Regan attempt to establish that some nonhuman animals—specifically "mammalian animals"[36]—meet the requirements for being a subject-of-a-life? It is in chapter 2—"The Complexity of Animal Awareness"—that Regan tackles this task, and his approach, in outline, is as follows. First, he claims that the combination of evolutionary theory with information about the behavior of mammals provides initial grounds for explaining the behavior of such animals in the same way that one explains human behavior—namely, in terms of beliefs and desires.[37] Next, Regan considers whether that prima facie case can be overthrown, and he devotes pages 35–73 to a detailed discussion of objections directed against the attribution of beliefs to animals, such as dogs, that have been advanced by Stephen Stich in his article "Do Animals Have Beliefs?"[38] and by R. G. Frey,[39] discussed above. Regan argues that those objections are unsuccessful, and thus that it is reasonable to hold that mammals do have beliefs and desires.

Finally, Regan focuses on a specific type of belief that he thinks it is reasonable to attribute to a dog—what he refers to as a "preference-belief"—where, for example, this might be a belief that "a bone is to be chosen if a given desire is to be satisfied,"[40] and he asks what consequences follow if one is justified in attributing such a belief to an animal—say, Fido—and his answer is that all of the following conclusions are justified:

(1) Fido is able to form general beliefs connecting, for example, gnawing on a bone with experiencing a certain taste.[41]

(2) Fido is able to remember past experiences, because otherwise he would be unable to form the general belief in question.[42]

(3) Because Fido could not form the general belief unless he had concepts, Fido "must have the ability to abstract from individual cases, from general concepts (e.g., the concept of a bone), and apply this concept to particular cases."[43]

(4) Given that we are justified in attributing beliefs and desires to Fido, we are justified in viewing Fido as being capable of intentional action, such as getting a bone "*in order that* he may satisfy his desire for the flavor bones provide."[44]

(5) Next, Regan argues that once we have shown that animals such as Fido are capable of acting intentionally,

we have additional reasons for crediting Fido and similar animals with beliefs about the future, if we do so in the human case, since to act in the present with the intention of satisfying one's desire in the future (as Fido does when he acts in a way that leads us to let him out in order that he may satisfy his desire by getting his mouth on the bone he believes we have buried) requires that Fido and these other animals have such beliefs.[45]

(6) Lastly, Regan argues that it follows that we are justified in viewing mammalian animals such as Fido as possessing self-consciousness:

> To recognize the status of mammalian animals as intentional agents paves the way for recognizing that they should also be viewed as self-conscious. For an individual, A, to act now in order to bring about the satisfaction of his desires at some future time is possible only if we assume that A is self-aware at least to the extent that A believes that it will be *his* desires that will be satisfied in the future as a result of what he does now. In other words, intentional action is possible only for those who are self-conscious. [46]

Let us start with Regan's last conclusion—that Fido and other mammals possess self-consciousness. Consider what is involved in Fido's believing that "it will be *his* desires that will be satisfied in the future as a result of what he does now" in a case where Fido is described as acting to get a bone such that, when he gnaws on it, *he* will have certain taste experiences that satisfy the desires that *he* has at the later time in question. While some controversial metaphysical issues are involved, there are strong arguments in support of the general claim that one has a persisting entity only if some of the property instances present at later times are causally dependent, and in an appropriate way, upon property instances present at earlier times. Moreover, accounts of personal identity typically bring in states, such as memories, that involve causal relations to earlier states, such as experiences.[47] Consequently, for Fido to believe that what he is doing now will bring it about that he has experiences that satisfy desires that he has at that later time, Fido must believe that his present actions will lead to experiences that will stand in certain causal relations to his present mental states, and which will satisfy future desires that are also causally related to his present mental states. Conceptually, this seems like a fairly complex belief, and it would be surprising if Fido has any such belief.

One can, however, say much more than that this is surprising. In my opinion, Regan's argument goes off the rails at the start. The basic source of the problem is that, on the one hand, philosophy of mind and empirical psychology bear crucially upon this argument, and, on the other, Regan fails to bring in the relevant ideas. As with Frey's arguments, one needs to consider seriously the possibility of a functionalist account of mental concepts, and especially of the concepts of belief and desire. Once that possibility is on the table, one can ask whether the initial belief that Fido has about bones is merely a functionalist-style belief—something like an Armstrongian map by which Fido steers—or whether it is the sort of state that humans have when they believe things, and which involves a disposition to have a corresponding *thought*.

Regan, surprisingly, seems never to address the question of whether mammalian animals are capable of having thoughts, even though that is a crucial issue. As a consequence, it is unclear what concept of belief he is employing. When he attributes beliefs to Fido, do those beliefs involve dispositions to have corresponding thoughts? Or are the beliefs merely functionalist-style maps that need have no relation to states of consciousness?

Whatever concept of belief Regan is employing, his argument will run aground. First, suppose that the concept of belief that Regan is using involves a disposition to have corresponding verbal thoughts. Then the claim that evolutionary history together with the behavior of mammals provides good grounds for attributing such beliefs to mammals is not at all plausible. Given that, aside from Kanzi and company, and perhaps a few other species, such as whales and dolphins, mammals lack the capacity for understanding speech, we have no grounds for attributing a capacity for verbal thought to animals such as Fido.

Second, the simplest hypothesis, given the behavior of mammals, is that they have functionalist-style beliefs and desires. But if that is all that they have, then unless Regan has formulated his subject-of-a-life criterion in a misleading fashion, mammals will not have beliefs or desires, or engage in intentional action, in the relevant sense, all of which involve conscious thought, let alone have a sense of the future, or self-consciousness.

As noted earlier, however, there is an intermediate possibility, namely, that according to which to have a belief is to be in a functionalist-style belief state while having a disposition to have images of a relevant sort. So perhaps what Regan would claim is that the hypothesis that is justified by the combination of evolutionary history and mammalian behavior is that mammals have beliefs and desires that involve imagistic thoughts.

If that is Regan's view, the first point to note is that one needs to offer a reason for accepting the hypothesis that mammals have beliefs and desires that involve imagistic thoughts, rather than the ontologically more economical hypothesis that mammals have only functionalist-style beliefs and desires. Regan has not offered any such argument.

Second, suppose that a good argument could be advanced in favor of the more complex, intermediate hypothesis. How would Regan's attempt to justify animal rights then stand? The answer is that it would still fail. Consider one crucial belief that Regan attributes to Fido, the belief, namely, that "it will be *his* desires that will be satisfied in the future as a result of what he does now." This belief involves the temporal concepts of the present and the future; it also involves the concept of causation; and, finally, it involves the idea that one and the same thing can both perform actions and have desires. There are, however, no images that correspond to any of these concepts. Try, for example, forming images of three possible events, one past, one present, and one future. Consider those images. Is there anything in them that marks out one as being of a possible past event, another as being of a possible present event, and the third as being of a possible future event? It is hard to see what such a marker could be. It is only if an image is accompanied by a verbal thought to the effect that, say, this is an image of a past event, that there are any differences. This means that, unless an individual has the capacity for verbal thought, the only type of belief that it can have about a future event is a functionalist-style belief, because no imagistic thought can represent an event as lying in the future.

It can be shown, in similar fashion, that there can be no image that represents one event as causing another. But to believe that certain future experiences will be one's own and will satisfy desires that one will have at that time, one needs to have both a belief about the future, and a belief that certain future experiences and desires stand in causal relations to one's present mental states. The upshot is that an animal that is unable to understand language cannot have any conscious thoughts about the future, or about future states of itself. This in turn means that such an animal cannot have any sense of the future, or any self-consciousness, for both a sense of the future, and self-consciousness, involve the having of thoughts.

The conclusion, accordingly, is that unless Regan is using expressions such as "sense of the future" and "self-consciousness" in a misleading sense in which mere functionalist-style beliefs and desires can suffice for one to have a sense of the future, and to be self-conscious, there are good reasons for holding that mammalian animals, lacking, as almost all of them do, an ability to understand spoken language, must also lack both a sense of the future and self-consciousness. They fail to satisfy, then, Regan's subject-of-a-life criterion.

3.3 Continuing Selves, Psychological Connectivity, and Imagistic Thinking

The final argument to be considered for the view that many nonhuman animals have moral status arises out of a rejection of the claim, advanced in the final negative argument, that self-consciousness is necessary if one is to have a continuing subject of experiences. The argument can be put as follows:

(1) Whenever experiences at different times can be linked via thoughts, one has a continuing self, or subject of experiences.

(2) Imagistic thinking is logically possible, and it can serve to link experiences existing at different times: an experience at a later time can be part of the same overall state of consciousness as an image of an earlier experience, and that image, moreover, may be part of an imagistic thought about the earlier experience.

(3) It is reasonable to believe that many nonhuman animals have imagistic thoughts about earlier experiences.

Therefore,

(4) It is reasonable to believe that many nonhuman animals are continuing subjects of experiences.

(5) Continuing subjects of experiences are persons, and have moral status.

Therefore,

(6) It is reasonable to believe that many nonhuman animals are persons, and have moral status.

An advocate of this argument can accept the claims, first, that potentialities aside, only continuing selves have moral status; second, that one has a continuing self only when experiences at different times are psychologically linked; third, that the psychological connections must, at least sometimes, be conscious; and fourth, that those conscious psychological connections must involve thoughts. But a defender of this argument rejects the claim that those thoughts must involve concepts such as causation and identity over time—concepts that, I have argued, cannot be represented via images.

Consider what one's mental life would be like if one were incapable of thought that involved an internal use of language. One would have memory images of past experiences, and those memory images would be accompanied by functionalist-style beliefs that would play a role in constructing one's map of how the world around one presently is. One would have the ability to move through sequences of images that represented possible courses of action, and terminal points in those sequences, together with functionalist-style preferences and desires, would then determine what if anything one would do at a given point. But there would be no thoughts to the effect that the experiences of which one had memory images were experiences that lay in the past, and which one had had. Nor would there be thoughts to the effect that other images represented experiences that one could have in the future.

What moral status would such an existence have? One method of attempting to answer this question involves, once again, the hope that some connection exists between the wrongness of destroying a certain type of subject of experiences and the value that a person would assign to existing in that way. As for myself, on the one hand, I have some inclination to say that such an existence would be at least somewhat better than none at all. On the other hand, however, if I imagine that I am suffering from a disease that will lead to that outcome, and that there is only one treatment available, which leads to a complete cure with probability p, and otherwise to death, I think that I would opt for the treatment even if the value of p were very low indeed.

Perhaps the link in question cannot be forged. Perhaps my own preferences concerning the type of existence in question are idiosyncratic. I shall have to leave it to the reader to decide about those matters. If there is such a connection, however, and if my own attitudes are not too idiosyncratic, the conclusion will be that the psychological unification of experiences over time that can be generated by imagistic thinking results in a being with no significant moral status. This final argument, then, does not provide grounds for assigning significant moral status to nonhuman animals.

4. Summing Up

We have been considering three closely related questions. To what extent are nonhuman animals persons? To what extent do they have a right to continued existence? To what extent do they have moral status? I have attempted to show that most arguments that have been advanced in this area are unsound. Thus, as regards arguments for the views that animals are not persons, that they do not have a right to continued existence, and that they do not have moral status, I have argued that neither contractarian approaches to rights, nor Kantian views of morality that link personhood or rights with moral agency, nor attempts to establish that animals are not persons and do not have moral status because they do not have beliefs and desires, provide successful defenses of the negative conclusions in question.

On the other hand, it seems to me that familiar attempts to show that animals do have moral status, or are persons, or have a right to continued existence, are also unsuccessful. Thus both the claim that mere consciousness gives one a right to continued existence or moral status, and the claim that the ability to experience pleasure and pain does so, seem to me open to very strong objections. Among other things, it seems to me very plausible that it is the fact that something is a continuing subject of experiences and other mental states that are psychologically connected over time that is crucial to things having moral status and having a right to continued existence.

A much more challenging and promising argument for the view that many nonhuman animals have moral status is that set out and defended in great detail by Tom Regan. As I argued above, however, it is crucial to distinguish between interpretations of the concepts of belief, desire, and intention in which there is an essential reference to thought, and interpretations where there is no such reference. When this is done, it can be seen that Regan's argument is also unsuccessful.

In addition to criticizing the above arguments, I also attempted to provide support for the view that, potentialities aside, it is persons and only persons who have a right to continued existence, where a person is a continuing subject of experiences whose mental states at different times are psychologically connected. That connectivity over time, moreover, must, at least in part, be through conscious mental states: purely functionalist-style memories, desires, and intentions are insufficient. Finally, the conscious connections in question must involve thoughts connecting experiences at different times.

If the capacity for thought presupposed the ability to use language, it would follow very quickly that very few nonhuman animals were persons, and had a right to continued existence. Bonobos, who can understand spoken language, other primates who are capable of learning a sign language, together with whales and dolphins, if it turns out that they also communicate linguistically, would qualify, but probably very few other species. In the above discussion, however, I attempted to make plausible the idea that, in addition to thought that involves an internal use of language, thought that involves images is also possible. If that idea is sound, then the class of nonhuman animals that are persons, and that have moral status, may be much larger.

The crucial question that then emerged was whether the content that such thoughts would have to have to provide a type of psychological connectivity needed to give one a continuing subject of experiences whose destruction would be seriously wrong could be expressed by imagistic thoughts. If, as I am inclined to think, the thoughts needed must involve such concepts as those of causation, and of a persisting self, it seems to me unlikely that such thoughts can be expressed imagistically. In addition, however, I suggested that one should consider what life would be like if one were capable of thinking only through images, and the extent to which one would judge such an existence to be one that is valuable.

The situation seems to me as follows. On the one hand, if the idea of thinking that involves only images, and no use of language, is untenable, then very few non-human animals are persons, or have a right to continued existence, or have moral status. On the other hand, if imagistic thinking is logically possible, these issues are very much open, and much further philosophical reflection is called for, though I am inclined to suspect that, in the end, the conclusion may not be all that different. Even if a continuing self whose mental life is unified over time by thinking that involves only images does have moral status, that moral status is not likely to be very significant.

ACKNOWLEDGEMENTS

"Tom Beauchamp provided extremely helpful critical comments and suggestions that led me to improve my discussion at certain points. I very much appreciate his close editorial work."

NOTES

1. Tom Regan, *The Case for Animal Rights* (Berkeley and Los Angeles: University of California Press, 1983), p. 235.

2. Cf. Bryce Huebner's contribution to this *Handbook* in his chapter, "Minimal Minds."

3. See, for example, Donald R. Griffin, *The Question of Animal Awareness: Evolutionary Continuity of Mental Experience*, 2nd ed. (New York: Rockefeller University Press, 1981); Denise Radner and Michael Radner, *Animal Consciousness* (Amherst, N.Y.: Prometheus Books, 1986); Rosemary Rodd, *Biology, Ethics and Animals* (Oxford: Clarendon Press, 1990); Bernard E. Rollin, *The Unheeded Cry: Animal Consciousness, Animal Pain and Science* (New York: Oxford University Press, 1989).

4. Mary Anne Warren, *Moral Status* (Oxford: Clarendon Press, 1997), pp. 59–60.

5. U. T. Place, "Is Consciousness a Brain Process?" *British Journal of Psychology* 47 (1956): 44–50.

6. J. J. C. Smart, "Sensations and Brain Processes," *Philosophical Review* 68 (1959): 141–56.

7. David M. Armstrong, *A Materialist Theory of the Mind* (London: Routledge & Kegan Paul, 1968).

8. Armstrong, *Materialist Theory of the Mind*, p. 82.

9. Donald Davidson, "Thought and Talk," in *Mind and Language*, ed. S. Guttenplan (Oxford: Oxford University Press, 1974).

10. R. G. Frey, "Why Animals Lack Beliefs and Desires," in *Animal Rights and Human Obligations*, 1st ed., ed. Tom Regan and Peter Singer (Englewood Cliffs, N.J.: Prentice Hall, 1976), pp. 115–18, and R. G. Frey, *Interests and Rights: The Case Against Animals* (Oxford: Clarendon Press, 1980).

11. Wolfgang Köhler, *The Mentality of Apes* (New York: Harcourt Brace & Company, 1925).

12. James Rachels, "Do Animals Have a Right to Life?" in *Ethics and Animals*, ed. Harlan B. Miller and William H. Williams (Clifton, N.J.: Humana Press, 1983), pp. 277–88.

13. John Rawls, *A Theory of Justice* (Cambridge, Mass.: Harvard University Press, 1971).

14. Jan Narveson, "A Defense of Meat Eating," in *Animal Rights and Human Obligations*, ed. Tom Regan and Peter Singer (Englewood Cliffs, N.J.: Prentice Hall, 1976), p. 192.

15. Narveson, "Defense of Meat Eating," p. 193.

16. Narveson, "Defense of Meat Eating," p. 192.

17. Mary Anne Warren, *Moral Status* (New York: Oxford University Press, 1997), pp. 101–2.

18. Warren, *Moral Status*, p. 164.

19. Warren, *Moral Status*, p. 156.

20. Tibor R. Machan, "Rights, Liberation and Interests: Is There a Sound Case for Animal Rights or Liberation?" in *New Essays in Applied Ethics*, ed. Hon-Lam Li and Anthony Yeung (New York: Palgrave Macmillan, 2007), pp. 52–53.

21. Michael Tooley, "Abortion: Why a Liberal View is Correct," in *Abortion: Three Perspectives*, ed. James Sterba (New York: Oxford University Press, 2009), pp. 21–35.

22. Tooley, "Abortion," pp. 35–51.

23. J. L. Mackie, *Ethics: Inventing Right and Wrong* (Harmondsworth: Penguin, 1977).

24. R. G. Frey, *Interests and Rights*.

25. Frey, *Interests and Rights*, pp. 4–17, and Frey, "On Why We Would Do Better to Jettison Moral Rights," in *Ethics and Animals*, ed. Harlan B. Miller and William H. Williams (Clifton, N.J.: Humana Press, 1983), pp. 285–301.

26. Frey, *Interests and Rights*, pp. 39 and 139.

27. Frey, *Interests and Rights*, pp. 139–67.

28. Frey, *Interests and Rights*, pp. 93–96.

29. Frey, *Interests and Rights*, p. 87.

30. Sue Savage-Rumbaugh and Roger Lewin, *Kanzi: The Ape at the Brink of the Human Mind* (New York: John Wiley & Sons, 1994).

31. Paul Raffaele, "Speaking Bonobo," *Smithsonian Magazine*, November 2006, www.smithsonianmag.com/science-nature/10022981.html#ixzzoblJlMWCL (accessed January 4, 2010).

32. Dale Jamieson, "Killing Persons and Other Beings," in *Ethics and Animals*, ed. Harlan B. Miller and William H. Williams (Clifton, N.J.: Humana Press, 1983), pp. 135–46.

33. Jamieson, "Killing Persons and Other Beings," p. 145.

34. Regan, *Case for Animal Rights*.

35. Regan, *Case for Animal Rights*, p. 243.

36. Regan, *Case for Animal Rights*, p. 34.

37. Regan, *Case for Animal Rights*, pp. 34–35.

38. Stephen Stich, "Do Animals Have Beliefs?" *Australasian Journal of Philosophy* 57 (1979): 15–28.

39. Frey, *Interests and Rights*.
40. Regan, *Case for Animal Rights*, p. 58.
41. Regan, *Case for Animal Rights*, p. 73.
42. Regan, *Case for Animal Rights*, p. 73.
43. Regan, *Case for Animal Rights*, p. 74.
44. Regan, *Case for Animal Rights*, p. 74, Regan's emphasis.
45. Regan, *Case for Animal Rights*, p. 75.
46. Regan, *Case for Animal Rights*, p. 75.
47. Compare Derek Parfit's essay, "Personal Identity," *Philosophical Review* 80 (1971): 3–17.

SUGGESTED READING

ALLEN, COLIN. "Animal Consciousness." In *The Stanford Encyclopedia of Philosophy*, ed. Edward N. Zalta, Spring 2009, http://plato.stanford.edu/entries/consciousness-animal/.

ANDREWS, KRISTIN. "Animal Cognition." In *The Stanford Encyclopedia of Philosophy*, ed. Edward N. Zalta, Spring 2009, http://plato.stanford.edu/entries/cognition-animal/.

ARMSTRONG, DAVID M. *A Materialist Theory of the Mind*. London: Routledge & Kegan Paul, 1968.

ARMSTRONG, SUSAN J., and RICHARD G. BOTZLER, eds. *The Animal Ethics Reader*, 2nd ed. London and New York: Routledge, 2008.

BEKOFF, MARC, COLIN ALLEN, and GORDON M. BURGHARDT, eds. *The Cognitive Animal: Empirical and Theoretical Perspectives on Animal Cognition*. Cambridge, Mass.: MIT Press, 2002.

CAVALIERI, PAOLA, ed. "The Great Ape Project." *Etica & Animali*, Special Issue (1996).

————, and PETER SINGER, eds. *The Great Ape Project*. New York: St. Martin's Press, 1993.

CLARKE, STEPHEN R. L. *The Moral Status of Animals*. Oxford: Oxford University Press, 1977.

COHEN, CARL, and TOM REGAN. *The Animal Rights Debate*. Lanham, Md.: Rowman & Littlefield, 2001.

FREY, R. G. *Interests and Rights: The Case against Animals*. Oxford: Clarendon Press, 1980.

McMAHAN, JEFF. *The Ethics of Killing: Problems at the Margins of Life*. New York: Oxford University Press, 2002.

MILLER, HARLAN B., and WILLIAM H. WILLIAMS, eds. *Ethics and Animals*. Clifton, N.J.: Humana Press, 1983.

PERRY, JOHN, ed. *Personal Identity*, 2nd ed. Berkeley and Los Angeles: University of California Press, 2008.

REGAN, TOM. *The Case for Animal Rights*. Berkeley and Los Angeles: University of California Press, 1983.

————, and PETER SINGER, eds. *Animals Rights and Human Obligations*. Englewood Cliffs, N.J.: Prentice Hall, 1976.

ROLLIN, BERNARD E. *Animal Right and Human Morality*. Buffalo, N.Y.: Prometheus Books, 1981.

SAVAGE-RUMBAUGH, SUE, and ROGER LEWIN. *Kanzi: The Ape at the Brink of the Human Mind*. New York: John Wiley & Sons, 1994.

SAPONTZIS, S. F. *Morals, Reason, and Animals*. Philadelphia: Temple University Press, 1987.

STICH, STEPHEN. "Do Animals Have Beliefs?" *Australasian Journal of Philosophy* 57 (1979): 15–28.

WARREN, MARY ANNE. *Moral Status*. New York: Oxford University Press, 1997.

PART IV

ANIMAL MINDS AND THEIR MORAL SIGNIFICANCE

ANIMAL MENTALITY: ITS CHARACTER, EXTENT, AND MORAL SIGNIFICANCE

PETER CARRUTHERS

Do animals possess *moral standing*? That is to say, are they deserving of our sympathy and concern, and do they possess moral rights, in their own right? This chapter will claim that such questions should receive negative answers. The first four sections will argue that the pains and sufferings of almost all animals (including many insects and crustaceans) are of the right sort to make them possible objects of sympathy, figuring within a mind whose basic structure is similar to our own. In this case, utilitarian ethical theories will have a hard time denying that all such creatures have standing, thereby adding to the problems that such theories face. The remaining five sections will argue from a contractualist perspective that all humans, but probably no animals, possess moral standing. This conclusion, too, is admittedly counterintuitive, but some of the sting will be drawn from it by acknowledging that we can nevertheless have indirect duties toward animals.

Although humans are, of course, a kind of animal, when I speak of animals in this chapter, I should be understood as referring to nonhuman animals only. This is merely for ease of expression, and should not be taken as a commitment to any sort of "Cartesian divide" between ourselves and members of other species. On the contrary, I believe firmly in the evolutionary and cognitive continuities between humans and other animals.

1. Do Animals Have Conscious Experiences?

There are a variety of different uses of the term "conscious," many of which are surely applicable to at least some animals. All animals spend time awake, for example, and are therefore conscious as opposed to unconscious during these periods. However, one usage has seemed especially pertinent to the question of the moral standing of animals. This is so-called *phenomenal* consciousness, which is the property that perceptions and bodily sensations possess when there is something that it is *like* for a creature to undergo those events, or when the events in question possess a subjective *feel*. For on the one hand, it can seem very plausible that only creatures whose pains are phenomenally conscious—and which thereby *feel like* something—truly suffer, and are worthy of sympathy and moral concern. But on the other hand, there are theories of phenomenal consciousness that probably imply that few if any animals besides humans are phenomenally conscious.[1] I shall discuss the latter first, before turning in section 2 to consider the implications of such theories for the moral standing of animals.

Most philosophical theories of phenomenal consciousness belong to one of two broad classes, either first-order[2] or higher-order.[3] First-order theories maintain that phenomenally conscious states possess a distinctive kind of representational content (analog, perhaps, or nonconceptual) as well as occupying a distinctive sort of functional role (such as being *poised* to have an impact on the belief-forming and decision-making processes of the creature in question). They are described as "first-order" because all of the mental states appealed to in the account (perceptions, beliefs, and so forth) are representations of properties of the world or body. Such theories probably entail that phenomenal consciousness is widespread in the animal kingdom. For as we will see in section 3, the evidence suggests that even navigating insects like bees and wasps have perceptual states located within the right sort of first-order cognitive architecture. Higher-order theories likewise agree that phenomenally conscious states possess a distinctive sort of content, but they claim that these states must be ones *of which the subject is aware*. It is by being aware of our perceptual states that they acquire their subjective dimension or "feel," on this sort of account. Higher-order theories are so called because they crucially appeal to representations of other mental states (namely the mental states that are thereby rendered phenomenally conscious).

What higher-order theories, as such, should claim about the distribution of phenomenal consciousness in the animal kingdom isn't entirely straightforward, since they come in significantly different varieties. Some say that the higher-order states in question are themselves nonconceptual in character, resulting from perception-like *monitoring* of our own mental states.[4] Others claim that they are fully conceptual *thoughts* about our current perceptual states.[5] One might naturally think that higher-order *perceptions* could be found in creatures that are as yet incapable of higher-order *thought*, in which case the largely negative evidence of any capacity for the latter in most other animal species might grossly underestimate the extent of higher-order representations in the animal kingdom. However, I have argued elsewhere that this natural thought is likely false, because there would be no need for a creature to evolve a capacity to monitor its own mental states unless it were also capable of entertaining thoughts about those states.[6]

If we suppose that this latter claim is correct, then the prospect of widespread phenomenal consciousness among animals appears quite bleak, from a higher-order perspective. Admittedly, the evidence suggests that humans are by no means unique in possessing some mental state concepts and higher-order thoughts. For monkeys and apes appear to possess simple forms of mindreading ability, and can ascribe perceptual and knowledge states (but not beliefs and appearances) to other agents.[7] But there is no convincing evidence that these animals are capable of applying mental-state concepts to themselves in the sort of way that would be required for phenomenal consciousness, according to higher-order theories of the latter. While some comparative psychologists have claimed to find evidence of capacities for self-directed higher-order thoughts in monkeys and chimpanzees,[8] my own view is that the data admit of simpler explanations.[9] And although it is easy to be tempted by the Cartesian idea that introspective (higher-order) access to our own mental states is a necessary precursor for attributing such states to others,[10] such a view is by no means mandatory and faces significant difficulties.[11] Indeed, from an evolutionary perspective, the main pressure toward developing a capacity for higher-order thought is likely to derive from the third-person uses of such thoughts in forms of social competition and cooperation.[12]

Although these issues are still highly controversial, it is worth exploring the implications for ethics if it should turn out that phenomenally conscious states are restricted to human beings (and perhaps also to members of some closely related species). Would it follow that sympathy for the pains and apparent sufferings of other creatures is inappropriate? It is natural to think so, because if animals aren't phenomenally conscious, then their mental lives are all "dark on the inside." If their perceptual states (including their pains) lack phenomenal properties, then their pains won't be *like* anything for them to undergo and will lack any subjective "feel." Indeed, their pains would have the same sort of status as the perceptual states of so-called "blindsight" patients.[13] In that case, it would seem that animals would be beyond the pale of possible sympathy (or so I once claimed).[14] Section 2 will argue that these temptations should be resisted, however.

2. Does Consciousness Matter?

I shall argue that the question whether or not a creature's pain is phenomenally conscious should be irrelevant to the question whether that pain is a possible—or *appropriate*—object of sympathy and concern.[15] Whether we are *required* to be concerned about animal pain is another matter, and is a topic for ethical theory to pronounce on. (See the discussion that follows in later sections of this chapter.)

To fix ideas, let us suppose that some version of the higher-order monitoring account of phenomenal consciousness is correct. (Nothing substantive turns on this assumption. Similar points can be made with respect to other higher-order approaches.) If this is so, then phenomenally conscious experiences of pain will have a dual aspect. One is objective, or body-representing, while the other gives the experience its subjective feel. On the one hand, there will be a first-order representation of a state of the body (normally involving cellular damage of some sort). This can be thought of as a perceptual representation of a secondary quality of the body, much as first-order visual perceptions represent secondary qualities of external objects such as colors. However, in addition, this representation will be monitored to produce a higher-order awareness that one is experiencing pain. It is this that is responsible for the subjective, *feely*, qualities of the state, according to the higher-order theorist.

We can now ask which of these two aspects of our pain experiences makes those pains *awful* (in the sense of being bad or unwelcome from the perspective of the subject). The answer is that it is the first-order representation of a state of the body rather than our higher-order awareness of the representing event. Imagine a case where you have just been stung between the toes by a bee while walking barefoot through the grass, and you are experiencing intense pain. The focus of your concern—and what it is that you want to cease—is surely the event that is represented as occurring in your foot. A naïve subject, such as a child, might gesture toward his foot saying, "Mommy, make *that* go away" (meaning the pain represented). What the child wants to stop is *the pain*, not his experience of the pain. The content of his desire is first-order, not higher-order. It is only when one knows about analgesics and their effects that one comes to care about the *experience* of pain. For now one might say, "I don't care whether you get rid of *that* [meaning the pain represented], I want you to give me something to stop my experience of the pain."

Many animals experience pain, of course, including invertebrates such as hermit-crabs.[16] They are also strongly motivated to avoid the relevant represented properties of their bodies, or to make those properties stop. (However, if a higher-order theory is correct then they aren't *aware* that they are experiencing pain, and their pains aren't phenomenally conscious.) Therefore, they have what *we* have when we find our own pains to be awful. In each case, it is the same represented first-order property of the body that is the object of the motivation of avoidance. So if sympathy is appropriate in the one case, it is also at least possible in the other.

Admittedly, we don't know how to imagine a pain that isn't phenomenally conscious. This is because any pain that we imagine will carry with it a higher-order awareness that a representing of pain is occurring, and the imagining will therefore be a phenomenally conscious one. However, this shouldn't obscure the theoretical point, which is that the animal has the same sort of property occurring as we do when we find our own pains to be awful. If the position sketched above is sound, then it won't actually be misleading to imagine the animal's pain as conscious. Although strictly false, if the subjective, phenomenally conscious aspect isn't what makes pain bad in our own case, then the introduction of such an aspect into our imagining of the experience of the animal shouldn't be leading us astray.

Suppose, however, that someone is unconvinced and thinks that the phenomenally conscious properties of one's pain are intrinsic to what one finds awful, in such a way that the pain would not be awful (to us) without them. Still, there remains another argument that phenomenal consciousness is irrelevant to the possibility of sympathy. For what really makes pain bad, in any case, isn't its phenomenal properties as such, but rather the fact that the state of being in pain is *unwanted*. I shall argue that even if the phenomenal properties of pain are generally intrinsic to what one finds awful and wishes to see cease, this isn't always the case. Indeed, I will suggest that it is goal frustration that is the proper object of sympathy, not pain sensations as such.

Pain perception in mammals (at least) is underlain by two distinct nervous pathways. The so-called "new path" is fast, projects to a number of different sites in the cortex, and is responsible for pain discrimination, location, and feel. It is this pathway that gives rise (in humans) to the felt qualities of pain. The "old path," in contrast, is comparatively slow, projects to more ancient subcortical structures in the limbic system of the brain, and is responsible for pain motivation. It is this that makes one want the pain to stop. Some analgesics like morphine suppress the old path while leaving the new path fully functional. Subjects will say that their pain feels just the same to them, but that they no longer care. What they are aware of is now just a sensation. It is no longer an *awful* sensation. Such people are no longer appropriate objects of sympathy, surely. Of course one might be sympathetic for any physical damage that has occurred, because of its likely future effects on the life of the agent; but that is another matter. Not only is there no obligation on us to try to make the remaining pain sensation stop, but it seems that doing so would be morally completely neutral. Making the pain sensation stop wouldn't be doing the subject any sort of favor.

What makes pain (or anything else) awful, then, is that it is the object of a negative desire. Phenomenal consciousness is irrelevant to its status. (Note that these points would motivate some kind of *preference* utilitarianism over any form of *hedonistic* utilitarianism in moral theory.) It would be absurd to insist that the person in our example above is undergoing something bad (at least assuming that there is no physical damage in addition to the pain sensation), despite the fact that he doesn't care. And the claim made here probably generalizes. For there are powerful arguments for thinking that what things and events count as valuable depend ultimately on our desires, values, and preferences.[17] The question that we need to ask, therefore, isn't whether animals are capable of phenomenally conscious experience, but whether they are subjects of propositional attitudes, especially desires and goals.[18]

3. How Many Animals Have Attitudes?

What does it take for a creature to be capable of attitude states? Some philosophers have placed conditions on genuine attitude possession that are extremely demanding, such as a capacity for consciousness,[19] rationality,[20] and/or spoken language.[21] Such demands seem to me quite excessive. Section 2 has already shown that the consciousness condition is irrelevant to the question whether or not sympathy for animals is appropriate. The rationality condition probably doesn't even apply to human beings, much of the time.[22] Moreover, the argument for claiming that language is necessary for attitude possession conflates an epistemic condition with a metaphysical one. It runs together the question of how one might *know* of the existence of a fine-grained attitude in the absence of language (such as the difference between believing that the cat is up this tree and believing that the furry animal is up the biggest tree in the yard) with the question of what it takes to *possess* such an attitude. In addition, three decades of careful work by comparative psychologists has shown us how we can make significant progress even on the epistemic question.

Other philosophers, in contrast, have placed extremely weak conditions on attitude possession. They have claimed, for example, that it is enough that a creature's behavior should allow it to be *interpreted as* possessing beliefs and desires.[23] I shall assume that such weak claims are likewise incorrect, for two reasons. The first is that people are intuitive realists about mental states. We are therefore open to the possibility of being mistaken in our interpretations, even under ideal conditions.[24] The second is that our concern should be with the real mental properties that animals possess (in the sense that those properties are acceptable to science), not whether it is pragmatically useful to think of them in such terms. We therefore need to know whether there is a real, scientifically valid, distinction between the belief states and the desire states of animals (and between these and their perceptual states). We also need to be assured that these states are compositionally structured out of concepts or concept-like elements, interacting with one another in inference-like processes in virtue of their compositional structures. These are demanding conditions. Nonetheless, I shall argue that even insects can meet them.[25]

The dominant position in both philosophy and psychology throughout much of the twentieth century was that animals aren't capable of genuine thought (although they can be interpreted as such, anthropomorphically). Animal behavior was believed to be the product of conditioning, resulting from learned associations among stimuli, and between stimuli and behavioral responses. Anyone espousing such a view has a ready-made reason for denying moral standing to animals, if attitude-possession is a necessary condition for such standing. But the adequacy of the account has been crumbling rapidly since at least the 1980s. Animals engage in many forms of learning that cannot be accounted for in associationist terms, and even conditioning itself is better explained by the operations of a computational rate-estimation system, as I shall now briefly explain.

Gallistel and colleagues have shown that animals in conditioning experiments who are required to respond to randomly changing rates of reward are able to track changes in the rate of reward about as closely as it is theoretically possible to do.[26] Thus, both pigeons and rats on a variable reward schedule from two different alcoves will match their behavior to the changing rates of reward. There is a lever in each alcove, each set on a random reward schedule of a given probability. However, these probabilities themselves change at random intervals. It turns out that the animals respond to these changes *very* rapidly, closely tracking the random variations in the immediately preceding rates. They aren't averaging over previous reinforcements, as associationist models would predict. On the contrary, the animals' performance comes very close to that of an ideal Bayesian reasoner, and the only model that can predict the animals' behavior is one that assumes that they are capable of calculating the ratio of the two most recent intervals between rewards from the two alcoves. It is therefore hard to resist the conclusion that the animals are genuinely reasoning, rather than learning by association.

Another dramatic example of nonassociationist learning has been provided by Balci and colleagues.[27] They tested swift and intuitive assessments of risk, using similar experiments in both humans and mice with very similar results. All subjects were set the task of capturing an object in one of two positions for a reward. (For the humans, this occurred on a computer screen. The mice had to press a bar to obtain a reward in one of two alcoves.) There were two types of trial, short latency and long latency, whose probability of occurring varied from one series of trials to the next. If the trial was a short one, the target could be captured in the left-hand position within two seconds of the stimulus onset. If the trial was a long one, the target could be captured in the right-hand position during the third second. Subjects were therefore required to estimate the optimum time to switch from the short-latency strategy to the long-latency strategy, an estimate that depends on two factors. The first is the objective chance (set by the experimenters in each series of trials) that the interval would be either short or long. The second is the accuracy of each subject's estimate of elapsed time (which varies from individual to individual, but is normally in the region of ±15%). Balci and colleagues were able to compute the optimum switch time for each subject, combining both sets of probabilities. This was then compared with actual performance. Human subjects came within 98% of optimum performance, whereas the mice were at 99%. Moreover, very little learning was involved. In most series of trials, subjects were just as successful during the first tenth or the first quarter of the series as they were during the final tenth or the final quarter.

In addition, Gallistel has demonstrated that conditioning behavior itself is best explained in rule-governed computational terms, rather than in terms of associative strengths.[28] He points out that there are many well-known conditioning phenomena that are extremely puzzling from an associationist perspective, but that fall out quite naturally from a computational account. To give just a single example: the number of reinforcements that are necessary for an animal to acquire an intended behavior is unaffected by mixing *un*reinforced trials into the learning process. One set of animals might be trained on a 1:1 schedule: these animals receive a reward every time that they respond when the stimulus is present. But another set of animals might be trained on a 10:1 schedule: here the animals only receive a reward once in every ten trials in which they respond when the stimulus is present. It still will, on average, take both sets of animals the same number of rewarded trials to acquire the behavior. It will take the second set of animals *longer* to acquire the behavior, of course. If it takes both sets of animals forty rewarded trials to acquire the behavior, then the first set might learn it in eighty trials, whereas the second set will take eight hundred. However, the number of *reinforcements* to acquisition is the same. This is extremely puzzling from the standpoint of an associationist. One would expect that all those times when the stimulus *isn't* paired with a reward ought to weaken the association between stimulus and reward, and hence make learning the intended behavior harder. However, it doesn't, just as Gallistel's computational model predicts.

Moreover, many forms of animal learning give rise to stored informational states that can interact with a variety of different goals to guide the animal's behavior, just as can our own beliefs. To offer a single illustrative example: chimpanzees can acquire detailed information about the spatial (and other) properties of their forest environment. They use this information in the service of a variety of foraging goals (such as seeking out one sort of fruit that they can predict to be ripening rather than another), but also when patrolling their territory or launching an attack on a neighborhood troupe. Depending on their goals, they travel in a straight line to their desired location, which they can approach in this way from many different directions.[29] The animals are therefore engaging in a form of practical reasoning, accessing their beliefs to achieve the satisfaction of a current goal.

Generalizing from data of the sort considered above, together with a range of other forms of evidence, we are warranted in concluding that mammals and birds, at least, share a perception/belief/desire cognitive architecture much like our own. Moreover, it should be stressed that to attribute beliefs and goals to animals is not just to give a redescription of their behavior. On the contrary, it is to ascribe real underlying states to them as the *causes* of their behavior, doing so on the basis of an inference to the best explanation. The question remains, however, how widespread minds of this sort are within the animal kingdom. I shall now briefly argue that navigating invertebrates like bees, wasps, and spiders share a similar sort of mental architecture, and are likewise capable of propositional attitudes.[30]

For brevity, I shall focus on honeybees. Like many other insects, bees use a variety of navigation systems. One is dead reckoning, which involves integrating a sequence of directions of motion with the distance traveled in each direction, to produce a representation of one's current location in relation to the point of origin.[31] This requires that bees can learn the expected position of the sun in the sky at any given time of day, as measured by an internal clock of some sort. Another mechanism permits bees to recognize and navigate from landmarks, either distant or local.[32] Moreover, some researchers have shown that bees will also construct crude mental maps of their environment from which they can navigate.[33] (The maps have to be crude because of the poor resolution of bee eyesight. But they can still contain the relative locations of salient landmarks, such as a large freestanding tree, a forest edge, or a lake shore.) Furthermore, in addition to learning from their own exploratory behavior, bees famously also acquire information from the dances of other bees about the spatial relationships between the hive and various desired substances and objects (including nectar, pollen, water, and potential new nest sites).

Although basic bee motivations are, no doubt, innately fixed, the goals that are activated on particular occasions (such as whether or not to move from one foraging patch to another, whether to finish foraging and return to the hive, and whether or not to dance on reaching it) would appear to be influenced by a number of factors.[34] (Note that similar claims can be made about humans.) Bees are less likely to dance for dilute sources of food and they are less likely to dance for the more distant of two sites of fixed value. They are less likely to dance in the evening or when there is an approaching storm, when there is a significant chance that other bees might not be capable of completing a return trip. Moreover, careful experimentation has shown that bees scouting for a new nest site will weigh up a number of factors, including cavity volume, shape, size and direction of entrance, height above ground, dampness, draftiness, and distance away from the existing nest.[35]

Most important for our purposes, bees' goal states and information states interact with one another in flexible ways, and in a manner strongly suggestive of an underlying constituent structure. Thus, the very same information about the direction and distance between the hive and a newly discovered source of nectar can be used to guide a direct flight to the hive when the bee's goal is to return there (often flying a route that has never previously been traversed by that individual); or it can be used to guide the orientation and number of waggles in the bee's dance to inform others of the location; or it can be used to guide a straight flight back to the nectar from the hive once the bee has been unloaded. Moreover, the flexibility of bee learning and navigation suggests that their informational states are compositionally structured, with some having the following form: "[object or substance] is [measure of distance] in [solar direction] from [object or substance]." The distance and direction information will be utilized differently depending on whether the bee's goal is represented in the first position or the last one (or on whether the goal is to dance). Moreover, we also know that bees are capable of computing a novel bearing (from a known landmark to a feeder, for example) from two others (from the landmark to the hive and from the feeder to the hive).[36]

From this and much other data, we can conclude that not only do bees have distinct information states and goal states, but that such states interact with one another in the determination of behavior in ways that are sensitive to their contents and compositional structures. In this case bees really do exemplify a perception/belief/desire cognitive architecture, construed realistically. There are also many things that bees can't do, of course, and there are many respects in which their behavior is inflexible. However, this inflexibility doesn't extend to their navigation and navigation-related behavior. On the contrary, the latter displays just the right kind of integration of goals with acquired information to constitute a simple form of practical reasoning. Similar points can be made with respect to other species of navigating invertebrates.

None of this is to deny that there are significant differences between the minds of humans and other animals, of course. Indeed, many psychologists have converged on the idea that humans employ two distinct types of systems for reasoning and decision making.[37] The first consists of a set of quick and intuitive systems that are largely shared with other animals. The second is a more reflective system that is thought to be unique to humans, and which employs a stream of inner speech and visual imagery to direct and control our mental lives and (indirectly) our behavior. It should be stressed, however, that on what I take to be the best account of the operations of the reflective system, the latter is *realized in* the workings of the intuitive systems (hence being parasitic upon them).[38] They are also dependent upon motivations provided by the latter to achieve their effects. Moreover, it is very doubtful whether the reflective system really contains, itself, any propositional attitude states.[39] Furthermore, the case of pain, discussed in section 2, suggests that it isn't the presence of some sort of reflective mind that determines the appropriateness of sympathy. For it isn't by conscious reflection that we determine that our pains are awful, of course. Rather, a powerful desire to get rid of them is generally forced on us as part of the painful episode itself.

4. The Extent of Warranted Sympathy for Animals

Our conclusion in section 2 was that the frustration of an agent's goals constitutes the most basic object of sympathy, irrespective of anything phenomenological. However, our conclusion in section 3 was that navigating invertebrates (including bees and wasps, together with many kinds of ants and spiders) are genuinely *agents* with a perception/belief/desire psychology and with goals that can be frustrated. Putting these two conclusions together, it follows that many invertebrates are at least possible objects of sympathy and altruistic concern.

There is a famous story about the medieval Scottish rebel leader, Robert the Bruce. Hiding in a cave while on the run from the English following a defeat in battle, he is said to have watched a spider repeatedly try to spin its web across a section of the cave, eventually succeeding after many failed attempts. Robert the Bruce is said to have been inspired by the spider's persistence to resume his war against the English. It is not reported whether or not he felt sympathy for the spider, but had he done so, it now appears that he would not have been making any sort of metaphysical mistake. For it is quite likely that the spider was genuinely an agent with its own beliefs and goals, making the frustration of one of its goals an appropriate target of sympathy.

However, it is one thing to say that sympathy for the frustrated goals of an animal is possible or appropriate, and another thing to say that it is required, or that the animal's situation makes any sort of moral claim on us. (These correspond to two different senses in which sympathy for an agent can be *warranted*.) Compare the following. Most people feel disgust at the thought of incest between siblings, even when consensual and guaranteed to be reproductively barren.[40] This is probably a tendency that is innate to human psychology.[41] Moreover, this sort of incest has many properties in common with cases where we almost certainly *should* feel disgust, such as incest between a father and his 16-year-old daughter. Nevertheless, many of us have come to believe that sibling incest doesn't in itself deserve our moral disapproval. Similarly, then: although sympathy for the frustrations of any creature correctly categorized as an agent can occur, and may also result from an innate psychological tendency,[42] it is another matter to claim that such sympathy *should* be felt, or that it gives rise to any kind of moral obligation when it is. This can only be resolved by considerations of moral theory.

Bentham is famous for having claimed that the only factor relevant to determining whether we have moral obligations toward a creature is whether it can suffer and feel pain.[43] Updated in light of our discussion in section 2, the claim should now be that the only relevant factor is a capacity for goal frustration. If this is what utilitarianism requires, then it appears that we owe equal moral consideration to the goals of many invertebrates. No doubt some will rush to embrace such a conclusion. Many, however, will not. Most of us don't feel that we do anything wrong when we kill the ants that enter our kitchens, although the costs of not doing so may be quite minimal. Still less do we feel that we would be wrong to prevent them from entering in the first place. Furthermore, I think most people would feel that it would be a serious moral error to allow the goals of even millions of ants or bees to outweigh the goals of a single human child, or even to be put into the same equation with the latter. (I should stress that I am referring to individuals here, not entire species. For of course there are powerful reasons for maintaining biodiversity and preserving endangered species, having to do with the long-term interests of humans themselves.) Admittedly, this might be because people don't really consider bees to be genuine agents with desires of their own. It may be that anyone convinced by the arguments sketched in section 3 would immediately regard the animals in question as possessing moral standing, although I doubt it. The real question before us is what the best moral theory would support.

In what follows, I will focus on forms of utilitarian and contractualist moral theory, discussing the former briefly here before exploring the implications of the latter throughout the remainder of the chapter. (Forms of virtue theory are best pursued and accounted for within the framework of one or the other of these two approaches, in my view. See section 8 for a sketch of a contractualist treatment.) There are many different varieties of utilitarianism, of course, which have been defended in a number of different ways. So it is hard to know what utilitarianism as such entails on any particular issue such as the moral standing of animals, or to find ways of evaluating all forms of utilitarianism at once. One of the most acceptable utilitarian theories has been provided by Singer.[44] This is distinctive in providing a highly plausible—and naturalistically acceptable—account of the origins of morals and moral motivation. In Singer's telling, morality takes its start from natural (innate) human sympathy. Initially this is focused on family and tribe members, all of whom will be individually known to the agent. However, the impact of rational considerations thereafter forces the moral circle to be gradually widened to include members of other tribes and nations. For people can be brought to see that there is no rational difference between the sufferings and frustrations of someone whom they know and can see, and the similar suffering of someone who is a member of another group living elsewhere in the world. This issues in a principle of equal consideration of interests that is applicable to all people and that should, on similar grounds, be extended to include members of some other species of animals, Singer believes.

It is difficult to see how any such sympathy-based ethical theory can have the resources to deny moral standing to insects. For it seems that bees and humans are relevantly alike in the one respect that matters: both are agents with goals that can be frustrated; and as we saw in section 2, the most fundamental target of sympathy is goal frustration. This isn't to say that utilitarians of this sort can't find *any* relevant differences between humans and other animals, of course. For they can allow numbers of desires to count (humans will generally have many more of them), as well as normal life expectancy, in addition to indirect effects on humans who are friends or relatives of the person in question, and so on. However, just as Singer has claimed that there can be no defensible grounds for according standing to all humans while denying it to animals, so it seems he can't grant the standing of *some* animals without also granting it to almost all (including individual members of many species of insect).

Other forms of utilitarianism may be able to avoid this consequence. For example, one might make a commitment to the intrinsic value of *endorsed* desires (of the sort that a human might have) in contrast with the *mere* desires of animals. Such an account would face problems of its own, requiring us to give up on naturalism and accept the mind-independence of value.[45] In what follows, however, I propose to set utilitarian moral theories to one side. They confront two pervasive sets of problems. These are sufficiently deep to render utilitarian theories unacceptable, in my view, provided that there are alternatives that are both viable and sufficiently plausible.

The first of these problems is to provide adequate protection for individuals against the tyranny of the utility of the majority. Differently put, utilitarian theories face notorious difficulties in accounting adequately for principles of *justice*. The second problem is more theoretical: it is to provide a non-moral space, a domain of action in which individuals can be free to do what they wish. For common sense divides actions into three basic kinds: those that are duties (and are required), those that are against duty (and are forbidden), and those that are neither (which are discretionary). The third category then further subdivides into those that are morally admirable (but supererogatory) and those that are morally indifferent (where agents can please themselves). Utilitarianism, in contrast, is apt to consider pleasing oneself to be either a duty (in the right sorts of utility-increasing circumstances) or against duty.

I do not pretend to have refuted utilitarian moral theories in these few comments, of course. That has not been my goal. Taken together with the brief discussion of contractualist moral theories that follows in section 5, my aim is simply to explain why the remainder of the chapter will approach the question of the moral standing of animals from the perspective of contractualism.

5. CONTRACTUALIST MORAL THEORY

In the discussion that follows, I shall assume that some or other version of contractualist moral theory is correct. All contractualists of the sort that I am concerned with agree that moral truths are, in a certain sense, human constructions, emerging out of some or other variety of hypothetical rational agreement concerning the basic rules to govern our behavior.

In Rawls' version of contractualism, moral rules are those that would be agreed upon by rational agents choosing, on broadly self-interested grounds, from behind a "veil of ignorance."[46] (For these purposes, rational agents are agents who have the necessary mental capacities to consider, reason about, and implement systems of universal rules.) On this account, we are to picture rational agents as attempting to agree on a set of rules to govern their conduct for mutual benefit in full knowledge of all facts of human psychology, sociology, economics, and so forth, but in ignorance of any particulars about themselves, such as their own strengths, weaknesses, tastes, life plans, or position in society. All they are allowed to assume as goals when making their choice are the things that they will want *whatever* particular desires and plans they happen to have—namely, wealth, happiness, power, and self-respect. Moral rules are then the rules that would be agreed upon in this situation, provided that the agreement is made on rational grounds. The governing intuition behind this approach is that justice is fairness: since the situation behind the veil of ignorance is fair (all rational agents are equivalently placed), the resulting agreement must also be fair.

In Scanlon's version of contractualism, in contrast, moral rules are those that no rational agent could reasonably reject who shared (as his or her highest priority) the aim of reaching free and unforced general agreement on the rules that are to govern behavior.[47] On this account, we start from agents who are allowed full knowledge of their particular qualities and circumstances (as well as of general truths of psychology and so forth). However, we imagine that they are guided, above all, by the goal of reaching free and unforced agreement on the set of rules that are to govern everyone's behavior. Here each individual agent can be thought of as having a veto over the proposed rules, but it is a veto that they will only exercise if it doesn't derail the agreement process, making it impossible to find any set of rules that no one can reasonably reject.

In what follows, I shall often consider arguments from the perspective of *both* of these forms of contractualism. In this way, we can increase our confidence that the conclusions are entailed by more than just the specifics of a particular account. In addition, it should be stressed that for contractualists of these sorts, rational agents aren't allowed to appeal to any moral beliefs as part of the idealized contract process. This is because moral truths are to be the output of the contract process, and hence cannot be appealed to at the start. In other words, since morality is to be constructed through the agreement of rational agents, it cannot be supposed to exist in advance of that agreement. It should be acknowledged, however, that not all varieties of contractualism satisfy this constraint. For some allow the contracting agents to appeal to antecedent moral *values*, and in such cases the implications for the question of the moral standing of animals are much more difficult to discern.[48] Nonetheless, the constraint is justified, in my view, by the goal of providing a comprehensive moral theory that would be naturalistically acceptable, requiring us to postulate no properties and processes that would be unacceptable to science. In consequence, in the discussion that follows I shall, for simplicity, restrict my use of the term "contractualism" to *only* the two kinds described at the outset of this section.

The two main theoretical advantages of contractualism are the converse of the two major difficulties for utilitarianism. For one of the main goals of the contracting agents is to agree on a set of principles of justice, and individuals will veto any proposed rules that allow the vital interests of one to be compromised for the benefit of others (without adequate compensation). Individuals should therefore receive adequate protection. Moreover, contracting self-interested agents will be concerned to preserve as much freedom for themselves to pursue their own projects and goals as possible under the contract. Hence a significant space for non-moral action is nearly guaranteed.

It should be noted, however, that according to each of the forms of contractualist account that we will be considering, some moral rules will be mere local conventions (whereas others will be universally valid). This will happen whenever the contract process entails that there should be *some* moral rule governing a behavior or set of circumstances, but where there are no compelling grounds for selecting one candidate rule over the others. (By way of analogy, the rule requiring people in the United States to drive on the right suggests that there should be a rule requiring people to drive on one side of the road or the other, or chaos will ensue. But it doesn't much matter which side is chosen, and in the United Kingdom, in contrast, one should drive on the left.) It may be that rules governing the treatment of animals are of this general sort, as we shall see later.

One important theoretical challenge remains to be addressed before we can turn to the question of the implications of contractualism for the moral standing of animals. This is the question of the sources of moral motivation. Why should anyone care what rational agents *would* agree to? Why should agents take the results of any sort of hypothetical agreement to be binding on their actual behavior? In contrast with the difficulties that contractualism apparently faces on this question, some utilitarian approaches to moral theory have a plausible story to tell in this regard, for they can postulate an innate tendency to sympathize with the sufferings and struggles of other agents, as we have seen. Such a claim is actually very plausible,[49] and it is therefore quite easy to see why people should care about increasing utility, for this connects directly with one of their basic motivations.

In reply, contractualists should postulate an innate desire to be able to justify oneself to others in terms that the latter can freely accept. Given that no one will freely accept justifications that require them to believe falsehoods or to reason irrationally (by their own lights), this will amount to a desire to be able to justify one's conduct to other agents in terms that they can freely and *rationally* accept. It then seems very plausible that any set of rules that would enable one to satisfy this desire would be one that no one could reasonably reject who shared the aim of reaching free agreement.[50]

How plausible is it that humans have an innate desire of this sort? It makes a good deal of sense that the desire to justify oneself to others should be quite basic. Even hardened criminals will characteristically offer attempted justifications of their conduct. Moreover, we know that young children will begin making complaints and offering justifications to one another at quite an early age, without any training or encouragement from adults.[51] In addition, the existence of such a desire is just what one would predict evolution should have produced, given that punishment for unjustified breaches of societal norms (often resulting in death or exclusion from the group) has been a fundamental part of human society from time immemorial.[52]

I no more claim to have established the truth of contractualist moral theory here than I claimed to have refuted utilitarianism in section 4. But I hope to have shown that it can be reasonable to assert (as I do) that contractualism forms the best framework for moral theorizing (at least, modulo its commitments concerning the rights of animals and infants, which we consider in the sections to come—like many others, I endorse the method of reflective equilibrium in moral theorizing).

6. ALL HUMANS HAVE STANDING

In the present section, I will argue that all, or almost all, human beings have moral standing, irrespective of their status as rational agents. I will argue first that all rational agents have standing, and will then show that the same basic sort of standing should be accorded to human infants and senile (or otherwise mentally handicapped) adult humans. Since these arguments don't extend to animals (as we will see in section 7), they constitute a reply to Singer's main challenge.[53] He claims that contractualism can't consistently deny moral standing to animals without *also* withholding it from infants and mentally defective humans. This section and the next will demonstrate that Singer is mistaken.

The topic of the present section is important, since it is often taken as a reductio of contractualism that it can't adequately accommodate the moral standing of humans who aren't rational agents. If contractualism can't give a convincing account of the moral standing of infants and senile elders, then that will conflict with powerful and deeply held intuitions. This would mean deep trouble for the entire theoretical approach. I should note that I don't, however, see the question of the moral standing of human fetuses as belonging in the same "make or break" category, and so won't attempt to tackle that topic here. For this is an issue that is already deeply controversial, on which people have wildly differing intuitions. It is consequently difficult for everyone. (I actually believe, however, that contractualists should grant moral standing to fetuses in the later stages of development while denying such standing in the early stages, such as during the first trimester.)

The contractualist framework plainly entails that all rational agents should have the same moral standing. This is because moral rules are conceived to be constructed *by* rational agents *for* rational agents. Rational agents behind a veil of ignorance would opt to accord the same basic rights, duties, and protections to themselves (that is to say: to all rational agents, since they are choosing in ignorance of their particular identities). And likewise within Scanlon's framework, any proposed rule that would withhold moral standing from some subset of rational agents could reasonably be rejected by the members of that subset.

Contractualism accords the same basic moral standing to all rational agents as such, and not merely to the members of some actual group or society. On Rawls' approach, contracting agents don't even know which group or society they will turn out to be members of once the veil is drawn aside. On Scanlon's account, although we are to picture rational agents seeking to agree on a framework of rules in full knowledge of who they are and the groups to which they belong, those rules can be vetoed by *any* rational agent, irrespective of group membership. It follows that if Mars should turn out to be populated by a species of agent possessing the right sort of psychology, then contractualism would accord the members of that species full moral standing.

However, it seems that rational contractors wouldn't automatically cede moral standing to those human beings who are *not* rational agents (such as infants and senile elders), in the way that they must grant standing to each other. Nevertheless, there are considerations that should induce them to do so. The main one is as follows.[54] Notice that the fundamental goal of the contract process is to achieve a set of moral rules that will provide social stability and preserve peace. This means that moral rules will have to be *psychologically supportable*, in the following sense: they have to be such that rational agents can, in general, bring themselves to abide by them without brainwashing. (Arguably, no rational agent would consent to the loss of autonomy involved in any form of the latter practice.) But now the contractors just have to reflect that, if anything counts as part of "human nature" (and certainly much does)[55] then people's deep attachment to their infants and aged relatives surely belongs within it. In general, people care as deeply about their immediate relatives as they care about anything (morality included), irrespective of their relatives' status as rational agents—in which case contracting agents should accord moral standing to all human beings, and not just to those human beings who happen to be rational agents.

Consider what a society would be like that denied moral standing to infants and/or senile old people. The members of these groups would, at most, be given the same type of protection that gets accorded to items of private property, deriving from the legitimate concerns of the rational agents who care about them. That would leave the state or its agents free to destroy or cause suffering to the members of these groups whenever it might be in the public interest to do so, provided that their relatives receive financial or other forms of compensation. For example, senile elders might be killed so that their organs could be harvested, or it might be particularly beneficial to use human infants in certain painful medical experiments. We can see in advance that these arrangements would be highly unstable. Those whose loved ones were at risk would surely resist with violence, and would band together with others to so resist. Foreseeing this, contracting rational agents should agree that all human beings are to be accorded moral standing. Note that this doesn't mean, of course, that all humans are given the same *rights*. While normal human adults might be given a right to autonomy, for example, it will make little sense to accord such a right to a person who isn't an autonomous agent.

It might be replied against the argument from social stability that there have been many communities in the world where infanticide and the killing of the old have been sanctioned, without any of the predicted dire consequences for the stability and peacefulness of those societies. Thus, in many traditional societies the smaller of a pair of twins, or any infant born deformed, might be abandoned by its mother to die.[56] Moreover, certain Inuit tribes are said to have had the practice of forsaking their elders to die in the snow when the latter became too infirm to travel.

One point to be made in response to this objection is that all of the communities in question were sustained and stabilized by systems of traditional belief (often religious belief: "the gods require it" might be the justification given). This is no longer possible for us in conditions of modernity, where it is acceptable for any belief, no matter how revered and long-standing, to be subjected to critical scrutiny. In addition, the contract process envisaged by contractualism can't make any appeal to such traditional beliefs.

Another point to be made in response to the objection is that all of the communities in question were teetering on the edge of survival for their members; or at least the costs to individuals for acting differently would have been *very* high. In such cases, it is not obvious that the practices we are considering involve the denial of moral standing to infants and/or the old. This is because in these communities death occurs from failure to support, or from the withdrawal of aid, rather than by active killing. We, too, accept that it can be permissible to withdraw support, allowing someone to die, when the costs to oneself become too great. Think, for example, of someone in the process of rescuing another person from drowning who has to give up their effort when they realize that the current is too strong and that they themselves are in danger of drowning.

Infants and senile old people aren't by any means accorded "second-class moral citizenship" within contractualism, it should be stressed. Although it is only rational agents who get to grant moral standing through the contract process, and although the considerations that should lead them to grant moral standing to humans who aren't rational agents are indirect ones (not emerging directly out of the structure of the contract process, as does the moral standing of rational agents themselves), this has no impact on the product. Although the considerations that demonstrate the moral standing of rational agents and of nonrational humans may differ from one another, the result is the same: both groups have moral standing, and both should have similar basic rights and protections.

It probably isn't true that contractualism should accord moral standing to *all* human beings, however. Consider anencephalic infants. These are undoubtedly human beings. Yet they are born without a cortex, and although they sometimes possess a rudimentary brainstem, this lacks any covering of skull or skin. They are blind and deaf, and incapable of feeling pain, although reflex actions such as breathing and responses to touch and sound may occur. If not stillborn, most die within a few hours or days of birth. There is no cure or treatment.[57] The argument from social stability appears to have no application in such cases. Most parents will grieve at the *birth* of an anencephalic infant rather than its death, and will make no requests that the infant's life should be prolonged. What would be the point? Moreover, if the state were to legislate to permit harvesting organs from such infants, vigorous debate would ensue, no doubt, but it seems very unlikely that serious social instability would result. It is true that *some* people will care deeply about the lives of their anencephalic infants, as (famously) did the mother of Baby K.[58] But, in contrast with normal or handicapped infants, the vast majority of parents will not. Moreover, it is likely that the attachments of those who do care don't result from the normal operations of an innate human nature, but are produced, rather, by prior moral or religious beliefs.

We can conclude the following. If, as I claim, contractualism is the correct framework for moral theorizing, then it follows that almost all human beings—whether infant, child, adult, old, or senile—should be accorded moral standing. They should also be provided with a similar basic structure of protections (depending on their powers and capacities). In section 7, I will show, in contrast, that contractualism leaves all animals beyond the moral pale, withholding moral standing from them altogether.

7. No Animals Have Standing

In the present section, I will maintain that the argument just given for according moral standing to all humans doesn't extend to animals. I shall then consider two further attempts to secure moral standing for animals within contractualism, showing that they fail. The upshot can be captured in the slogan: "Humans in, animals out." But first I propose to argue that no animals count as rational agents in the sense that is relevant to contractualism—in which case they don't *automatically* acquire moral standing through the contract process.

What does it take to qualify as a rational agent from the perspective of contractualist moral theory? A rational agent is a potential contractor, which means that such a person should be capable of proposing and examining normative rules, as well as reasoning about the consequences of their adoption. It also means having the sort of motivational and emotional systems necessary to comply with and enforce such rules (at least some of the time) and to constrain one's behavior in accordance with previous agreements. Therefore, emotions like guilt and indignation are plausibly part of what it takes to be a rational agent.

The evidence suggests that rational agency is a distinctively human adaptation. Animals are certainly agents, and possess many remarkable cognitive capacities. For example, apes seem to possess at least some of the ingredients of human moral psychology, such as sympathy for others and engagement in reciprocal social interactions.[59] But there is no reason to believe that apes are capable of thinking in terms of normative rules, or that they would be motivated to comply with such rules if they could. On the contrary, evidence is beginning to accumulate that humans are unique in possessing an innate moral faculty that was selected for in evolution because of its role in sustaining complex cooperative societies.[60] This means that we are (at least for the present) warranted in assuming that only human beings are rational agents in the sense relevant to contractualism.

The argument of section 6 was that human beings who aren't rational agents should nevertheless be accorded moral standing to preserve social stability, since people's attachments to their infants and aged relatives are generally about as deep as it is possible to go. Someone might try presenting a similar argument to show that animals, too, should be accorded moral standing, citing the violence that has *actually* occurred in western societies when groups of people (like members of the Animal Liberation Front) have acted in defense of the interests of animals. Such an argument fails, however, because members of these groups are acting, not out of attachments that are a normal product of human emotional mechanisms, but out of their moral beliefs (which they take to be justified, of course, but which aren't the product of the contract situation).

Rational agents engaging in the contract process are forbidden from appealing to any antecedent moral beliefs, whether their own or other people's. This is because moral truth is to be the outcome of the contract, and shouldn't be presupposed at the outset. Therefore, contracting rational agents should *not* reason that animals ought to be accorded moral standing on the grounds that some people have a moral belief in such standing and may be prepared to kill or engage in other forms of violence in pursuit of their principles. The proper response is that such people aren't entitled to their belief in the moral standing of animals unless they can show that rational agents in the appropriate sort of contract situation should agree to it.

Many people care quite a bit about their pets, of course, which rational contractors might be expected to know. Could this give rise to a social-stability argument for moral standing? The answer is "no," for at least two distinct reasons. One is that it is far from clear that the phenomenon of pet keeping and attachment to pets is a human universal (in contrast with attachment to infants and aged relatives). It may rather be a product of local cultural forces operating in some societies but not others. If the latter is the case, then such attachments aren't a "fixed point" of human nature, which should constrain rational contractors in their deliberations. They might appropriately decide, instead, that society should be arranged in such a way that people don't develop attachments that are apt to interfere with correct moral decision making.

A second problem with the suggestion is that attachment to pets is rarely so deep as attachments to relatives, in any case. Because of this, people should have little difficulty in coming to accept that pets can only be accorded the sorts of protections granted to other items of private property. Most of us would think that it would be foolish (indeed, reprehensible) to continue to keep a pet that threatens the life of a child (e.g., through severe allergic reactions). And when the state declares that the public interest requires that someone's dog be put down (e.g., because it is vicious), it would surely be unreasonable to take up arms to defend the life of the animal, just as it would be unreasonable to kill to preserve a house that has been condemned for demolition.

It is true that *some* people care more for their pets than for their relatives, and might well go to great lengths to preserve the lives of the former. Here too, however (as in the example of anencephalic infants discussed earlier), numbers matter. That such strengths of attachment are relatively rare means that the argument from social stability fails to apply. Moreover, to the extent that deep attachments to pets are increasing in our society, this is likely to be the product of more widespread beliefs in the moral standing of animals, combined with increases in individual social alienation. These are not the kinds of factors that can be appealed to legitimately in the construction of the moral contract.

While the argument from social stability fails to show that animals should be accorded moral standing, other arguments could still be successful. One suggestion would be that some rational agents behind the veil of ignorance should be assigned to represent the interests of animals, much as a lawyer might be assigned to represent the interests of a pet in a court of law in a case involving a disputed will.[61] If it was the job of those representatives to look out for the interests of animals in the formulation of the basic moral contract, then they might be expected to insist upon animals being granted at least enough moral standing to protect their interests from invasive human harms.

This suggestion, however, is plainly at odds with the guiding idea of contractualism. For what possible motive could there be for assigning some agents to represent the interests of animals in the contract process, unless it were believed that animals *deserve* to have their interests protected? But that would be to assume a moral truth at the outset: the belief, namely, that animals deserve to be protected. We noted above, in contrast, that contractualism requires that the contracting parties come to the contract situation either without any moral beliefs at all, or setting aside (taking care not to rely upon) such moral beliefs as they do have.

The point is even easier to see in Scanlon's version of contractualism. Real individual agents with knowledge of their own particulars, but who either lack moral beliefs or have set aside their moral beliefs while trying to agree to rules that no one could reasonably reject, could have no reason to assign some of their number to represent the interests of animals. For to do so would be tantamount to insisting at the outset that animals should be accorded moral standing, preempting and usurping the constructive role of the contract process.

Another suggestion is that people behind the veil of ignorance should be selecting moral rules in ignorance of their species, just as they are ignorant of their life-plans, age, strength, intelligence, gender, race, position in society, and so on.[62] Then just as rational agents might be expected to agree on rules to protect the weak, since for all they know they might end up *being* weak, so rational agents might be expected to agree on a system of fundamental rights for animals, since for all they know they might end up *being* an animal.

One problem with this suggestion is that Rawls' veil of ignorance is designed to rule out reliance upon factors that are widely agreed to be morally irrelevant. Among the intuitions that a good moral theory should preserve is the belief that someone's moral standing shouldn't depend upon such factors as their age, or gender, or race. In contrast, we don't (or don't all) think that species is morally irrelevant. On the contrary, this is highly disputed, with (I would guess) a clear majority believing that differences of species (e.g., between human and horse) *can* be used to ground radically different moral treatment.

The veil of ignorance is a theoretical device designed to ensure that deeply held moral beliefs about what is, or isn't, morally relevant should be preserved in the resulting theory. So although the contracting agents aren't allowed to appeal to any moral beliefs in the contract process, the moral theorist has relied upon his prior moral beliefs in designing the surrounding constraints. Scanlon's version of contractualism, in contrast, digs deeper. It has the capacity to *explain why* the properties mentioned in the veil of ignorance are morally irrelevant. This is because one should be able to see in advance when one comes to the contract situation that if one proposes a rule favoring men, then this will be vetoed by those rational agents who are women, and vice versa; and so on for differences of age, intelligence, strength, race, and so on. Therefore, if we are motivated by the goal of reaching free and unforced general agreement among rational agents, we should abjure proposals that might favor one group over another. For we can foresee that these will be vetoed, and that others could equally well suggest proposals favoring other groups in any case, which *we* would need to veto. In contrast, there is no reason for us to abjure rules that favor humans over animals.

The idea of choosing rules in ignorance of one's species isn't even coherent within the framework of Scanlon's form of contractualism, in which agents are supposed to have full knowledge of their own particular qualities and circumstances, as well as of general truths of psychology, economics, and so forth. So there is no way to argue for the moral significance of animals from such a standpoint. One should be able to see in advance that a proposed rule that would accord moral standing to animals would be vetoed by some, because of the costs and burdens that it would place on us.

I conclude that while contractualism entails the moral standing of almost all humans (including infants, the handicapped, and senile old people), by the same token such standing should be denied to animals. However, even if this position is theoretically impeccable it faces a serious challenge. This is that most people believe strongly that it is possible to act wrongly in one's dealings with animals (especially by displaying cruelty). Most people also believe that it is something about what is happening *to the animal* that warrants the moral criticism. These are powerful intuitions that need to be explained, or explained away. This will form the topic of section 8.

8. INDIRECT MORAL SIGNIFICANCE FOR ANIMALS

Imagine that while walking in a city one evening you turn a corner to confront a group of teenagers who have caught a cat, doused it in kerosene, and are about to set it alight. Of course you would be horrified. You would think that the teenagers were doing something very wrong, and the vast majority of people would agree with you. It would be a serious black mark against contractualist moral theories in general, and against the line that I am pursuing in this chapter, if this intuition could not be accommodated.

To meet this challenge, we should claim that while we do have duties toward animals, they are *indirect*, in the sense that the duties are owed to someone other than the animal, and that they fail to have any corresponding rights in the animal. According to one suggestion, they derive from a direct duty not to cause unnecessary offense to the feelings of animal lovers or animal owners, and it is to them that we have the duty. Compare the above scenario with this one: while walking though a city you come across a pair of young people, stark naked, making love on a park bench in broad daylight. In this case, too, you would be horrified, and you would think that they were doing something wrong. But the wrongness isn't, as it were, intrinsic to the activity. It is rather that the love-making is being conducted in a way that might be disturbing or distressing to other people: namely, in public. Likewise, it might be said, in the case of the teenagers setting light to the cat: what they are doing is wrong because it is likely to be disturbing or distressing to other people.

This particular proposal isn't at all promising. While it can explain why the teenagers are wrong to set light to a cat in the street (since there is a danger that they might be observed), it can't easily explain our intuition that it would be wrong of them to set light to the cat in the privacy of their own garage. Admittedly, there is some wiggle room here if one wanted to defend the account. For animals, having minds of their own, are apt to render public a suffering that was intended to remain private. The burning cat might escape from the garage, for example, or might emit such ear-piercing screams that the neighbors feel called upon to investigate.

We can demonstrate the inadequacy of this whole approach through an example in which such factors are decisively controlled for, such as the example of Astrid the astronaut. You are to imagine that Astrid is an extremely rich woman who has become tired of life on Earth, and who purchases a space rocket for herself so that she can escape that life permanently. She blasts off on a trajectory that will eventually take her out of the solar system, and she doesn't carry with her a radio or any other means of communication. We can therefore know that she will never again have any contact with anyone who remains on Earth. Suppose now that Astrid has taken with her a cat for company, but that at a certain point in the journey, out of boredom, she starts to use the cat for a dart-board, or does something else that would cause the cat unspeakable pain. Astrid does something very wrong, but the grounds of its wrongness can't be the danger that animal lovers will discover and be upset, because we know from the description of the case that there is no such danger.

Quite a different approach, which I shall spend most of the remainder of this section developing and defending, would be to claim that the action of torturing a cat is wrong because of what it shows about the moral character of the actor, not because it infringes any rights or is likely to cause distress to other people. Specifically, what the teenagers do in the street and what Astrid does in her space rocket show them to be *cruel*, which would be our ground for saying that the actions themselves are wrong. In order for this account to work, however, it needs to be shown more generally that we sometimes judge actions by the qualities of moral character that they evince (without necessarily being aware that we are doing so), irrespective of any morally significant harm that they cause or of any rights that they infringe.

Return to the example of Astrid the astronaut, but now suppose that, in addition to a cat, she has taken with her another person. In one version of the story, this might be her beloved grandfather. In another version of the story (to avoid contaminating our intuitions with beliefs about family duties) it might be an employee whom she hires to work for her as a lifetime servant. Now at a certain point in the journey, this other person dies. Astrid's response is to cut up the corpse into small pieces, thereafter storing them in the refrigerator and feeding them one by one to the cat.

What Astrid does is wrong. But why? It causes no direct harm of any sort because her companion is dead and can't know or be upset, and nor can any harm be caused indirectly to others. In the nature of the case, no one else can ever know and be offended, nor are any rights infringed. Even if one thinks that the dead have rights (which is doubtful), Astrid might know that her companion was a non-believer who took not the slightest interest in ceremonies for the dead. He might once have said to her, "Once I am dead I don't care what happens to my corpse; you can do what you like with it," thus waiving any rights that he might have in the matter. But still one has the intuition that Astrid does something very wrong.

Why is what Astrid does wrong? I suggest it is because of what it shows about *her*. Just as her treatment of her cat shows her to be cruel, so her treatment of her dead companion displays a kind of disrespectful, inhuman, attitude toward humanity in general and her companion in particular. (Note that practices for honoring the dead, and for treating corpses with respect, are a human universal. They are common to all cultures across all times.)[63] In each case, we judge the action to be wrong because of the flaw that it evinces (both manifesting and further encouraging and developing) in her moral character, I suggest.

Consider a different sort of example. Suppose that lazy Jane is a doctor who is attending a conference of other medical professionals at a large hotel. She is relaxing in the bar during the evening, sitting alone in a cubicle with her drink. The bar is so arranged that there are many separate cubicles surrounding it, from each of which the bar itself is plainly visible, but the insides of which are invisible to each other. Jane is idly watching someone walk alone toward the bar when he collapses to the floor with all the signs of having undergone a serious heart attack. Jane feels no impulse to assist him, and continues calmly sipping her martini.

Plainly what Jane does (or in this case, doesn't do) is wrong. But why? We can suppose that no harm is caused. Because the man collapses in plain view of dozens of medical personnel, expert help is swift in arriving, and she had every reason to believe that this would be so in the circumstances. Nor are any rights infringed. Even if there is such a thing as a general right to medical assistance when sick (which is doubtful), the man had no claim on her help in particular. If he had still been able to speak, he could have said, and (perhaps) said truly, "Someone should help me." But he surely wouldn't have been correct if he had said, "Jane, in particular, should help me." Since our belief in the wrongness of Jane's inactivity survives these points, the explanation must be the one that we offered in connection with Astrid the astronaut: it is wrong because of what it reveals about *her*. Specifically, it shows her to be callous and indifferent to the suffering of other people; or at least it shows that she lacks the sort of spontaneous, emotional, non-calculative, concern for others that we think a good person should have.

My suggestion is that our duties toward animals are indirect in just this sort of way. They derive from the good or bad qualities of moral character that the actions in question would display and encourage, where those qualities *are* good or bad in virtue of the role that they play in the agent's interactions with other human beings. On this account, the most basic kind of wrongdoing toward animals is *cruelty*. A cruel action is wrong because it evinces a cruel character, but what makes a cruel character bad is that it is likely to express itself in cruelty toward *humans*, which would involve direct violations of the rights of those who are caused to suffer. Our intuition that the teenagers and Astrid all act wrongly is thereby explained, but explained in a way that is consistent with the claim that animals lack moral standing.

I shall return to elaborate on this idea shortly. But first we need to ask how, in general, qualities of character, or virtues, acquire their significance within a contractualist moral framework. This question needs to be answered before the position sketched above can be considered theoretically acceptable.

Contracting rational agents should know in advance that human beings aren't calculating machines. We have limited time, limited memory, limited attention, and limited intellectual powers. In consequence, in everyday life we frequently have to rely on a suite of "quick and dirty" heuristics for decision making, rather than reasoning our way slowly and laboriously to the optimal solution.[64] Contracting rational agents should also realize the vital role that motivational states and emotional reactions play in human decision making.[65] Hence, they should do far more than agree on a framework of rules to govern their behavior. They should also agree to foster certain long-term dispositions of motivation and emotion that will make right action much more likely (especially when action is spontaneous, or undertaken under severe time constraints). That is to say: contracting agents should agree on a duty to foster certain qualities of character, namely, the *virtues*.

For example, contracting agents should agree on a duty to develop the virtue of beneficence because they should foresee that more than merely rules of justice are necessary for human beings to flourish. (Such rules are for the most part negative in form: "Don't steal, don't kidnap, don't kill, etc.") Humans also need to develop positive attachments to the welfare of others, fostering a disposition and willingness to help other people when they can do so at no important cost to themselves. For there are many ways in which people will inevitably, at some point in their lives, need the assistance of others if they are to succeed with their plans and projects, ranging from needing the kindness of a neighbor to jumpstart one's car on a frosty morning, to needing someone on the river bank to throw one a life-buoy or rope when one is drowning. It is important to notice, moreover, that this does *not* mean that actions undertaken out of generosity are really self-interested ones. On the contrary, generous people are people who feel an impulse to help others simply because they can see that the other person needs it. It only means that self-interest enters into the explanation of why generosity is a virtue. This is because self-interested rational agents attempting to agree on a framework of rules that no one could reasonably reject would agree on a duty to become a generous sort of person.

Rational contractors should also agree that people's actions can be judged (that is, praised or blamed) for the qualities of character they evince, independently of the harm caused, and independently of violations of a right. This is because people *should possess*, or should develop, the required good qualities. Although these good qualities *are* good, in general, because of their effects on the welfare and rights of other people, their display on a given occasion can be independent of such effects. Hence we can and should evaluate the action in light of the qualities of character that it displays, independently of other considerations. It is for this reason that we can blame Astrid for her actions, even though she will never again have the opportunity to interact with other human beings.

If the account given above of the reasons why it is wrong for the teenagers to set light to the cat is to be successful, then cruelty to animals needs to be psychologically and behaviorally linked to cruelty to humans. To a first approximation, it must be the case that there is a single virtue of kindness, and a single vice of cruelty, that can be displayed toward either group. How plausible is this? The American Society for the Prevention of Cruelty to Animals claims on its website to have amassed voluminous evidence that people who are cruel to animals are also likely to engage in cruelty that involves human beings.[66] The United Kingdom's RSPCA makes a similar claim on its "information for professionals" website, citing a number of empirical studies,[67] and prior to the Animal Welfare Act (which came into force in April 2007), the Society's prosecutions for cruelty to animals were almost always built upon this premise.

It certainly appears that attitudes toward the sufferings of humans and animals are quite deeply linked, at least in western culture. This is because many of us have pets whom we treat as honorary family members, toward whom we feel filial obligations. Our practices of child-rearing also make central use of animal subjects in moral education. A child's first introduction to moral principles will often involve ones that are focused upon animals. A parent says, "Don't be cruel—you mustn't pull the whiskers out of the cat," "You must make sure that your pet gerbil has plenty of water," and so on and so forth. It would not be surprising, then, if attitudes toward the sufferings and welfare of animals and humans should thereafter be pretty tightly linked. This will warrant us in saying that the teenagers who are setting light to a cat are doing something wrong, not because the cat has moral standing, but because they are evincing attitudes that are likely to manifest themselves in their dealings with human beings.

It seems possible, however, that the linkages that exist between attitudes to human and animal suffering depend upon local cultural factors. Hence it might be questioned whether these links reflect properties of a universal human nature. In cultures where pets aren't kept, where people's interactions with animals are entirely pragmatic (e.g., through hunting or farming), and where animals aren't used as exemplars in moral education, it is possible that these attitudes are pretty cleanly separable. At the very least, since cruelty involves causing *unjustified* suffering (just as murder is unjustified killing), we would expect cultures to differ a great deal in the circumstances in which cruelty is displayed toward an animal, because the virtue in question will have been molded by cultural assumptions and expectations. Thus consider someone in another culture who hangs a dog in a noose, strangling it slowly to death (perhaps because this is believed to make the meat taste better). This might not display cruelty under local conditions (at least in the sense of evincing a quality of character that is likely to generalize to that person's treatment of human beings), although in someone from our culture who behaved likewise it would do so.

If these speculations are correct, then our western moral attitudes toward animals should be thought of as forming part of the *conventional* content of our morality. If there is nothing in our human nature that links causing suffering to animals with cruelty to humans, then contracting rational agents would have no reason to insist upon a rule forbidding harsh treatment of animals, or a rule mandating a virtue of kindness that extends to animals. But contracting agents have to settle upon some or other way of bringing up their children, and cultural practices (such as pet-keeping) may be adopted for reasons having nothing to do with the moral contract itself, but which nevertheless have an impact upon morals. Given such facts, we can become obliged not to be cruel to animals. However, the question whether the wrongness of (what we take to be) cruelty-evincing behavior toward animals is either a conventional component of morality, on the one hand, or depends on universal facts about human nature, on the other, isn't the main issue. This is because, on either account, such wrongness will be consistent with the denial of moral standing to animals.

In either case, moreover, it is important to see that someone with the right sort of kindly character who acts to prevent suffering to an animal will do so *for the sake of the animal.* This is required for having the right sort of sympathetic attitude. The latter involves a spontaneous upwelling of sympathy at the sight or sound of suffering (at least in certain circumstances). Likewise, it is something about the animal itself (its pain) that forms the immediate object of the emotion, and of the subsequent response. Certainly, someone acting to ease the suffering of an animal won't be doing it to try to make himself into a better person! Nevertheless, the reason why this attitude is a virtue at all will be because of the way in which the behavior is likely to manifest itself in the person's dealings with other human beings.

We can therefore explain away the commonsense intuition that when we are morally obliged to act to prevent suffering to an animal, we are required to do so *for the sake of the animal* (where this would be understood to entail that the animal itself has standing). As a theoretical claim about what grounds our duties toward animals, this is false, since animals lack standing. But as a psychological claim about the state of mind and motivations of the actor, who has acquired the right kind of kindly attitude, it is true. While agents should act as they do for the animal's sake (with the animal's interests in mind), the reason why they are required to do so doesn't advert to facts about the animal (which would then require animals to have standing), but rather to the wider effects on human beings.

9. THE EXPANDING CIRCLE

We have seen how contractualism can explain why cruelty to animals is wrong while denying that animals have moral standing. However, a final challenge remains: How are the changing attitudes toward animals (at least in western cultures) to be explained? Why do so many more people today think that animals have moral standing? Singer has a plausible story to tell, which makes the change in question appear progressive.[68] According to Singer, as we have seen, morality is sympathy based. Initially, feelings of sympathy were confined to members of one's own family or tribe. However, rational considerations have forced the moral circle to expand because one can see that there is no relevant moral difference between the suffering of someone in one's own social group and the suffering of someone from another tribe or nation state. On Singer's telling, the same rational movement of thought has now (for many people) caused the moral circle to expand still further to embrace animals.

One aspect of this challenge has already been addressed above. For we have shown in section 7 that the divide between humans and animals is by no means arbitrary from the perspective of contractualism. Moreover, there are reasons (briefly reviewed in sections 4 and 5) for preferring contractualist moral theories to utilitarian ones. But why, then, have so many people come to feel that animals have moral standing, if really they don't? I am forced to deploy a form of error-theory. I claim that people have been seduced by faulty arguments and false theoretical assumptions, as well as by psychological tendencies that are apt to get reinforced in our culture. In fact, I can offer two distinct lines of explanation. Since these are consistent with one another, both may actually be at work.

We have already noted in section 8 how we use people's treatment of animals as an indicator of moral character. It may be that such a tendency is innate, and is partly a product of sexual selection. We know that what people want most in a marriage partner all around the world is someone who is kind.[69] Hence, kindly behavior toward animals (as well as other people's children) may be an honest indicator of fitness. (A thought experiment: you see a stranger in the street stop to lift up a cicada from a place where it is likely to get crushed, putting it safely on a nearby tree. Wouldn't you be inclined to feel warmly toward that person, even if, like most people, you don't think that insects have moral standing?) We also know that infants as young as six months of age show a preference for helpful over neutral agents, as well as a preference for neutral agents over unhelpful ones.[70] (This, too, may be adaptive, given how vulnerable infants still are at ages when they would normally start interacting with strangers.) Moreover, these preferences are displayed in respect to anything that gives off cues of animacy, including cartoon squares and triangles (on which a pair of eyes may have been drawn) that appear to be capable of self-motion. It would seem, therefore, that humans possess an innate tendency to prefer people who behave in a kindly fashion toward other agents, even when those agents are quite minimally characterized as such.

In previous eras, such an innate tendency would presumably have been prevented by social learning from issuing in a belief in the moral standing of all agents. Children would have observed adults interacting with animals in the context of hunting, fishing, and farming (as well as listening to adults talk). However, in our own culture there are few opportunities for such correction to take place, except in respect of adults' treatment of household and garden pests (which are insects, for the most part). Most children today have no experience of hunting, and little experience of farming beyond visits to a petting zoo and whatever they learn from television and books. Most children's only contact with vertebrate animals is with pets, who are generally treated in our culture as honorary members of a family. With nothing to prevent them from doing so, children's natural inclinations to feel warmly toward people who are kind, and not unkind, to other agents, leaves them wide open to a tendency to *moralize* such feelings, resulting in a belief in the moral standing of vertebrate animals. But in my view this is an error, comparable to the manner in which people in many cultures have tended to moralize their initial feelings of revulsion toward consensual incest between siblings or toward homosexuality.[71]

A second explanation of the "expanding circle" in our culture is suggested by the literature on *dehumanization*.[72] In what ways do humans tend to conceive of other groups of humans when they deny them moral standing and think that they may kill or harm them with impunity? Interestingly, and counting against

Singer's sympathy-based ethic as an account of our moral psychology, people don't usually deny that dehumanized groups feel pain, or fear, or other emotions that humans share with animals. Rather, what is denied is that those groups are subject to *distinctively human* emotions of love, guilt, indignation, shame, and so forth. Indeed, it appears that the upshot of dehumanization is a denial that members of the other group possess some of the main ingredients of rational agency, in the sense discussed in section 7. By parity of reasoning, then, one might expect that widening the moral circle to include some animals would be associated with a tendency to attribute human-like emotional states to them. In this connection, it is surely no accident that representations of animals as undergoing such states are now *rife* in children's storybooks, movies, and in popular culture more generally. But again, the result is a moral mistake. By overhumanizing the psychological states of animals in the fantasy lives of young people we create a tendency (which mostly remains unconscious, no doubt) to think of them as rational agents and potential collaborators, and hence as possessing moral standing in their own right.

If the position defended in this chapter is correct, in contrast, then the increasing moral importance accorded to animals in our culture can be seen as a form of creeping moral corruption and should be resisted. Particular attention would need to be paid to the moral education of our young, correcting each of the corruptive tendencies identified above.

10. Conclusion

This chapter has defended a number of important claims. One is that the kind of mindedness that makes sympathy appropriate is *extremely* widespread in the animal kingdom, extending to individuals belonging to many species of invertebrate. This presents utilitarian moral theories with a challenge: either to somehow persuade us of the moral standing of bees, spiders, and ants, or to find some morally relevant difference between the sufferings of invertebrates and those of mammals. (Moreover, the latter would need to be done in a naturalistically acceptable way, in my view, without making a commitment to the mind-independence of value.) Another claim defended in this chapter is that we possess at least one cognitive adaptation that sets us apart from other animals. This is a psychology that enables and supports cooperation and norm-governed behavior. From a contractualist perspective, morality is the outcome of an idealized contract among agents who share such a psychology, undertaken to constrain and guide their relations with one another. If contractualism provides the best framework for moral theorizing, as I have suggested, then the upshot is that almost all humans, but no other animals, possess moral standing.

NOTES

1. Both premises are endorsed by Peter Carruthers, "Brute Experience," *Journal of Philosophy* 86 (1989): 258–69.

2. See among others Fred Dretske, *Naturalizing the Mind* (Cambridge, Mass.: MIT Press, 1995); Michael Tye, *Ten Problems of Consciousness* (Cambridge, Mass.: MIT Press, 1995), and Tye, *Consciousness, Color, and Content* (Cambridge, Mass.: MIT Press, 2000).

3. See among others William Lycan, *Consciousness and Experience* (Cambridge, Mass.: MIT Press, 1996); Peter Carruthers, *Phenomenal Consciousness* (Cambridge: Cambridge University Press, 2000), and *Consciousness: Essays from a Higher-Order Perspective* (Oxford: Oxford University Press, 2005); and David Rosenthal, *Consciousness and the Mind* (Oxford: Oxford University Press, 2005).

4. Lycan, *Consciousness and Experience*.

5. Rosenthal, *Consciousness and the Mind*.

6. Carruthers, *Phenomenal Consciousness*.

7. See: Brian Hare, Josep Call, and Michael Tomasello, "Do Chimpanzees Know What Conspecifics Know?" *Animal Behaviour* 61 (2001): 139–51; Brian Hare, E. Addessi, Josep Call, Michael Tomasello, and E. Visalberghi, "Do Capuchin Monkeys, *Cebus paella*, Know What Conspecifics Do and Do Not See?" *Animal Behaviour* 65 (2003): 131–42; Michael Tomasello, Josep Call, and Brian Hare, "Chimpanzees Understand Psychological States—The Question is Which Ones and To What Extent," *Trends in Cognitive Sciences* 7 (2003): 153–56; and Laurie Santos, Aaron Nissen, and Jonathan Ferrugia, "Rhesus Monkeys (*Macaca mulatta*) Know What Others Can and Cannot Hear," *Animal Behaviour* 71 (2006): 1175–81.

8. John Smith, Wendy Shields, and David Washburn, "The Comparative Psychology of Uncertainty Monitoring and Metacognition," *Behavioral and Brain Sciences* 26 (2003): 317–73.

9. Peter Carruthers, "Metacognition in Animals: A Skeptical Look," *Mind and Language* 23 (2008): 58–89.

10. Alvin Goldman, *Simulating Minds: The Philosophy, Psychology, and Neuroscience of Mindreading* (Oxford: Oxford University Press, 2006).

11. Shaun Nichols and Stephen Stich, *Mindreading: An Integrated Account of Pretence, Self-Awareness, and Understanding of Other Minds* (Oxford: Oxford University Press, 2003).

12. See Richard Byrne and Andrew Whiten, eds., *Machiavellian Intelligence* (Oxford: Oxford University Press, 1988); and Andrew Whiten and Richard Byrne, eds., *Machiavellian Intelligence II: Evaluations and Extensions* (Cambridge: Cambridge University Press, 1997).

13. Laurence Weiskrantz, *Consciousness Lost and Found* (Oxford: Oxford University Press, 1997).

14. Carruthers, "Brute Experience"; and see also Peter Carruthers, *The Animals Issue: Moral Theory in Practice* (Cambridge: Cambridge University Press, 1992), chapter 8.

15. For an extended discussion of the arguments sketched here see Carruthers, *Consciousness*, chapters 9 and 10.

16. Robert Elwood and Mirjam Appel, "Pain Experience in Hermit Crabs?" *Animal Behaviour* 77 (2009): 1243–46.

17. Sharon Street, "A Darwinian Dilemma for Realist Theories of Value," *Philosophical Studies* 127 (2006): 109–66.

18. Most theorists think that propositional attitudes aren't phenomenally conscious per se, although some attitudes may give rise to phenomenally conscious effects. There will be phenomenally conscious bodily sensations (e.g., a dry throat) causally associated with

one's desire to drink, for example. But that desire itself isn't phenomenally conscious, because it lacks the right kind of fine-grained nonconceptual content. Admittedly, some philosophers claim that some attitudes, too, are phenomenally conscious. See, for example, Galen Strawson, *Mental Reality* (Cambridge, Mass.: MIT Press, 1994); and Charles Siewert, *The Significance of Consciousness* (Princeton, N.J.: Princeton University Press, 1998). For a critique, see Peter Carruthers and Bénédicte Veillet, "The Case Against Cognitive Phenomenology," in *Cognitive Phenomenology*, ed. Tim Bayne and Michele Montague (Oxford: Oxford University Press, 2011). But in any case, it is doubtful whether the phenomenally conscious properties of an attitude would have any bearing on the value or disvalue of the *objects* of that attitude. What makes pain a negative value for me is surely that I don't want it, not anything about the phenomenal properties of my desire for it to cease (supposing that there are any).

19. John Searle, *The Rediscovery of the Mind* (Cambridge, Mass.: MIT Press, 1992).

20. See Donald Davidson, "Rational Animals," *Dialectica* 36 (1982): 317–27; and Jennifer Hornsby, *Simple Mindedness* (Cambridge, Mass.: Harvard University Press, 1997).

21. Donald Davidson, "Thought and Talk," in *Mind and Language*, ed. Samuel Guttenplan (Oxford: Oxford University Press, 1975), 7–24.

22. See Daniel Kahneman, Paul Slovic, and Amos Tversky, eds., *Judgment under Uncertainty: Heuristics and Biases* (Cambridge: Cambridge University Press, 1982); and Dan Ariely, *Predictably Irrational: The Hidden Forces that Shape Our Decisions* (New York: HarperCollins, 2009).

23. Daniel Dennett, *The Intentional Stance* (Cambridge, Mass.: MIT Press, 1987).

24. William Ramsey, Stephen Stich, and J. Garon, "Connectionism, Eliminativism, and the Future of Folk Psychology," in *Philosophical Perspectives*, vol. 4, ed. J. Tomberlin (Atascadero, Calif.: Ridgeview Publishing, 1990), 499–534.

25. For a more extended discussion of many of the points made in this section, together with others, see Peter Carruthers, *The Architecture of the Mind: Massive Modularity and the Flexibility of Thought* (Oxford: Oxford University Press, 2006), chapter 2; and Peter Carruthers, "Invertebrate Concepts Confront the Generality Constraint (and Win)," in *The Philosophy of Animal Minds*, ed. Robert Lurz (Cambridge: Cambridge University Press, 2009), 89–107.

26. Charles Gallistel, Terrence Mark, Adam King, and P. Latham, "The Rat Approximates an Ideal Detector of Rates of Reward," *Journal of Experimental Psychology: Animal Behaviour Processes* 27 (2001): 354–72.

27. Fuat Balci, David Freestone, and Charles Gallistel, "Risk Assessment in Man and Mouse," *Proceedings of the National Academy of Sciences* 106 (2009): 2459–63.

28. See Charles Gallistel and J. Gibbon, "Time, Rate and Conditioning," *Psychological Review* 108 (2001): 289–344; and Charles Gallistel and Adam King, *Memory and the Computational Brain* (Chichester: Wiley-Blackwell, 2009).

29. Emmanuelle Normand and Christophe Boesch, "Sophisticated Euclidean Maps in Forest Chimpanzees," *Animal Behaviour* 77 (2009): 1195–1201.

30. For elaboration of this and related material, see Carruthers, *Architecture of the Mind*, chapter 2, and Carruthers, "Invertebrate Concepts Confront the Generality Constraint."

31. C. R. Gallistel, *The Organization of Learning* (Cambridge, Mass.: MIT Press, 1990).

32. T. Collett and M. Collett, "Memory Use in Insect Visual Navigation," *Nature Reviews: Neuroscience* 3 (2002): 542–52.

33. R. Menzel, U. Greggers, A. Smith, S. Berger, R. Brandt, S. Brunke, G. Bundrock, S. Hülse, T. Plümpe, S. Schaupp, E. Schüttler, S. Stach, J. Stindt, N. Stollhoff, and S. Watzl,

"Honey Bees Navigate According to a Map-Like Spatial Memory," *Proceedings of the National Academy of Sciences* 102 (2005): 3040–45.

34. Thomas Seeley, *The Wisdom of the Hive* (Cambridge, Mass.: Harvard University Press, 1995).

35. James Gould and Carol Gould, *The Honey Bee* (New York: Scientific American Library, 1988).

36. Menzel et al., "Honey Bees Navigate According to a Map-Like Spatial Memory."

37. See the papers contained in Jonathan Evans and Keith Frankish, eds., *In Two Minds: Dual Systems and Beyond* (Oxford: Oxford University Press, 2009).

38. See Keith Frankish, *Mind and Supermind* (Cambridge: Cambridge University Press, 2004); and Peter Carruthers, "An Architecture for Dual Reasoning", in *In Two Minds*, ed. Evans and Frankish, pp. 109–27.

39. See Peter Carruthers, "The Illusion of Conscious Will," *Synthese* 159 (2007): 197–213; and "How We Know Our Own Minds: The Relationship Between Mindreading and Metacognition," *Behavioral and Brain Sciences* 32 (2009): 121–38.

40. Jon Haidt, "The Emotional Dog and Its Rational Tail," *Psychological Review* 108 (2001): 814–34.

41. Daniel Fessler and Carlos Navarrete, "Third-Party Attitudes Toward Sibling Incest," *Evolution and Human Behavior* 25 (2004): 277–94.

42. Kiley Hamlin, Karen Wynn, and Paul Bloom, "Social Evaluation by Preverbal Infants," *Nature* 450 (2007): 557–59.

43. Jeremy Bentham, *An Introduction to the Principles of Morals and Legislation* (1789).

44. Peter Singer, *Practical Ethics* (Cambridge: Cambridge University Press, [1979] 1993); and *The Expanding Circle* (Oxford: Oxford University Press, 1981).

45. Street, "Darwinian Dilemma for Realist Theories of Value."

46. John Rawls, *A Theory of Justice* (Cambridge, Mass.: Harvard University Press, 1971).

47. Thomas Scanlon, "Contractualism and Utilitarianism," in *Utilitarianism and Beyond*, ed. A. Sen and B. Williams (Cambridge: Cambridge University Press, 1982), pp. 103–28.

48. This is true of the later work of both Rawls and Scanlon, for example. See John Rawls, *Political Liberalism* (New York: Columbia University Press, 1993); and Thomas Scanlon, *What We Owe to Each Other* (Cambridge, Mass.: Harvard University Press, 1998).

49. See M. Hoffman, *Empathy and Moral Development* (Cambridge: Cambridge University Press, 2000); and Paul Bloom, *Descartes' Baby* (New York: Basic Books, 2004).

50. Scanlon, "Contractualism and Utilitarianism."

51. E. Turiel, *The Development of Social Knowledge* (Cambridge: Cambridge University Press, 1983); and *The Culture of Morality* (Cambridge: Cambridge University Press, 2002).

52. See Robert Boyd and Peter Richerson, "Punishment Allows the Evolution of Cooperation (or Anything Else) in Sizable Groups," *Ethology and Sociobiology* 13 (1992): 171–95; and Peter Richerson and Robert Boyd, *Not by Genes Alone* (Chicago: University of Chicago Press, 2005).

53. Singer, *Practical Ethics*.

54. For other arguments for the same conclusion, see Carruthers, *Animals Issue*, chapters 5 and 7.

55. See Donald Brown, *Human Universals* (New York: McGraw-Hill, 1991); and Steven Pinker, *The Blank Slate* (New York: Viking Press, 2002).

56. Sarah Hrdy, *Mother Nature* (New York: Pantheon Press, 1999).

57. Information retrieved from the National Institutes for Health website at http://www.ninds.nih.gov/disorders/anencephaly/ (accessed August 27, 2009).

58. For a brief account of the case, see the entry in Wikipedia, available at http://en.wikipedia.org/wiki/Baby_K (accessed August 27, 2009).

59. Frans de Waal, *Good Natured* (Cambridge, Mass.: Harvard University Press, 1996).

60. See Richerson and Boyd, *Not by Genes Alone*; Richard Joyce, *The Evolution of Morality* (Cambridge, Mass.: MIT Press, 2006); Chandra Sripada and Stephen Stich, "A Framework for the Psychology of Norms," in *The Innate Mind*, vol. 2, ed. P. Carruthers, S. Laurence, and S. Stich (Oxford: Oxford University Press, 2006), 280–301; Chandra Sripada, "Adaptationism, Culture, and the Malleability of Human Nature," in *The Innate Mind*, vol. 3, ed. P. Carruthers, S. Laurence, and S. Stich (Oxford: Oxford University Press, 2007), 311–29; and John Mikhail, *Elements of Moral Cognition* (Cambridge: Cambridge University Press, 2011).

61. Tom Regan, *The Case for Animal Rights* (London: Routledge, 1984).

62. Regan, *Case for Animal Rights*.

63. Brown, *Human Universals*.

64. Gerd Gigerenzer, Peter Todd, and the ABC Research Group, *Simple Heuristics that Make Us Smart* (Oxford: Oxford University Press, 1999).

65. Anton Damasio, *Descartes' Error* (London: Papermac, 1994).

66. See the documents, "Domestic Violence and Animal Abuse" and "Animal Cruelty Prosecution," linked under "Resources" from the ASPCA's Fighting Animal Cruelty page, available at http://www.aspcapro.org/fighting-animal-cruelty/ (accessed on August 27, 2009).

67. See the report, "Animal Abuse amongst Young People Aged 13–17: Trends, Trajectories and Links with Other Offending," linked from the RSPCA's Science Group page, http://www.rspca.org.uk/servlet/Satellite?pagename=RSPCA/RSPCARedirect&pg=sciencegroup (accessed on August 27, 2009).

68. Singer, *Expanding Circle*.

69. David Buss, "Sex Differences in Human Mate Preferences: Evolutionary Hypotheses Tested in 37 Cultures," *Behavioral and Brain Sciences* 12 (1989): 1–49.

70. Hamlin, Wynn, and Bloom, "Social Evaluation by Preverbal Infants."

71. See Jon Haidt, Paul Rozin, C. McCauley, and S. Imada, "Body, Psyche, and Culture: The Relationship of Disgust to Morality," *Psychology and Developing Societies* 9 (1997): 107–31; and Haidt, "Emotional Dog and its Rational Tail."

72. Nick Haslam, "Dehumanization: An Integrative Review," *Personality and Social Psychology Review* 10 (2006): 252–64.

SUGGESTED READING

CARRUTHERS, PETER. *The Animals Issue: Moral Theory in Practice.* Cambridge: Cambridge University Press, 1992. See especially chapters 1–7.

———. *The Architecture of the Mind: Massive Modularity and the Flexibility of Thought.* Oxford: Oxford University Press, 2006. See especially chapters 2 and 3.

CHENEY, DOROTHY, and SEYFARTH, ROBERT. *Baboon Metaphysics: The Evolution of a Social Mind.* Chicago: University of Chicago Press, 2007.

FRANKISH, KEITH. *Consciousness.* Milton Keynes: The Open University, 2005.

LOCKWOOD, RANDALL. "Animal Cruelty Prosecution," (2006), accessible at the ASPCA website at http://www.aspcapro.org/fighting-animal-cruelty/.

LURZ, ROBERT, ed. *The Philosophy of Animal Minds*. Cambridge: Cambridge University Press, 2009.

MIKHAIL, JOHN. *Elements of Moral Cognition: Rawls' Linguistic Analogy and the Cognitive Science of Moral and Legal Judgment*. Cambridge: Cambridge University Press, 2011.

RAWLS, JOHN. *A Theory of Justice*. Cambridge, Mass.: Harvard University Press, 1971.

REGAN, TOM. *The Case for Animal Rights*. London: Routledge, 1984.

SCANLON, THOMAS. "Contractualism and Utilitarianism." In *Utilitarianism and Beyond*, edited by Amartya Sen and Bernard Williams, 103–28. Cambridge: Cambridge University Press, 1982.

SINGER, PETER. *The Expanding Circle*. Oxford: Oxford University Press, 1981.

———. *Practical Ethics*. 2nd ed. Cambridge: Cambridge University Press, 1993.

CHAPTER 14

MINDREADING AND MORAL SIGNIFICANCE IN NONHUMAN ANIMALS

JOSÉ LUIS BERMÚDEZ

DISCUSSIONS about the moral significance of nonhuman animals have tended to focus on a small number of basic capacities. There has been extensive discussion of the moral significance of sentience, for example, and basic forms of consciousness of the environment.[1] Other authors have focused on the significance of animals' sensitivity to the past and the future—their capacities for different types of memory, for example, or for making long- or short-range plans.[2] The idea that moral significance might rest upon some form of self-awareness has also been much discussed.[3] My aim in this paper is to extend the territory of this debate by exploring the moral significance of what researchers in cognitive science, developmental psychology, and comparative psychology term mindreading—that is, the ability to understand the mental states of others.

Mindreading is a complex cognitive ability that intersects in interesting ways with some of the capacities that have already been highlighted in discussions of moral significance. As we will see in the next section, it is also a cognitive ability that has been extensively studied both experimentally and ethologically in nonhuman animals and in human children. These investigations provide a rich vein of data to structure and frame debates about moral significance.

At the same time, the data need to be carefully interpreted. Animal ethics is an emotive arena in which advocacy often trumps accuracy. There are clear distinctions to be made between different levels and types of mindreading. These distinctions can be extremely important for conclusions that are drawn about moral significance. I am not a moral philosopher, and this paper will not draw normative conclusions about moral significance. My aim is to make it easier for moral philosophers to incorporate experimental and ethological work on mindreading into discussions of moral significance. I plan to achieve this goal, first, by presenting some of the most significant experimental results and observational data, and, second, by drawing some taxonomical distinctions that are particularly relevant to thinking about moral significance.

Section 1 explores some background issues, explaining why understanding mindreading in nonhuman animals is potentially important for animal ethics. In section 2, I embark upon the taxonomical task of giving a basic framework for thinking about experimental studies of animal mindreading. This section introduces the following three fundamental distinctions:

(1) Minimal mindreading versus forms of social behavior that do not involve mindreading;

(2) Minimal mindreading versus substantive mindreading; and

(3) Perceptual mindreading versus propositional attitude mindreading.

Section 3 argues that propositional attitude mindreading is language-dependent, because it involves metarepresentation. This conclusion places significant constraints on how we can conceptualize mindreading in the animal kingdom. In section 4, I review some key experimental findings on mindreading in primates, and show how they can be interpreted in terms of the conceptual framework developed in sections 2 and 3—and, in particular, how they can be interpreted without assuming that the mindreaders are engaged in propositional attitude mindreading. Finally, in section 5, I draw all of these arguments into a unified body of conclusions about mindreading in animals.

1. Why Mindreading Is Important for Thinking about Animal Minds and Animal Ethics

Discussions of the moral significance of nonhuman animals tend to focus on claims about the continuity or discontinuity between the cognitive and affective lives of human and nonhuman animals. Mindreading is a promising area in which to pursue this debate, for three reasons:

(1) There has been extensive research on the continuities and discontinuities between human and nonhuman mindreading. Mindreading in general has been a growth topic in the cognitive sciences in the last two decades or so, and comparative questions have come to the fore.

(2) Mindreading and social cognition have seemed to many either to involve, or to be necessary preconditions of, cognitive capacities that have been thought morally significant. These include metarepresentation and higher-order cognition, self-awareness, and collaborative action.

(3) The close studies that have been made of the developmental progression in human infants and young children, together with the development of experimental paradigms that can be applied to both humans and non-humans, allows a close calibration to be made between the mindreading achievements of particular species of nonhuman animals and the mindreading abilities of humans at different stages of development. A more focused and precise discussion of comparative considerability is thereby facilitated.

Each of these reasons will be explored in turn in the discussion that follows.

1.1 Background: Debates about Nonhuman Mindreading

Premack and Woodruff's much-discussed 1978 paper "Does the Chimpanzee Have a Theory of Mind?" was the first contribution to the ongoing debate about mindreading in the great apes.[4] They answered their own question affirmatively. Other researchers were more skeptical.[5] I will look at some of these experiments in more detail in later sections.

Experimental explorations of mindreading in great apes have proceeded more or less in parallel with, and independently of, ethological observations of animal behavior in the wild. Ethologists have described many forms of animal behavior in ways that have sometimes been taken to implicate mindreading abilities. The most frequently discussed example is tactical deception, which has been observed in many different species. One famous example is the injury-feigning behavior of plovers, which has been well studied by Carolyn Ristau and others.[6] There are few, if any, commentators who would see this activity as involving any kind of mindreading, but mindreading claims have been made about deception behaviors in primates, particularly in the great apes.

Richard Byrne and Andrew Whiten have developed a database of ethological observations of tactical deception in primates, together with a sophisticated taxonomy for interpreting those observations.[7] Their operational definition of tactical deception is: "An act from the normal repertoire of the agent, deployed such that another individual is likely to misinterpret what the acts signify, to the advantage of the agent."[8] Many behaviors falling within this general category do not plausibly implicate mindreading in any shape or form. But Byrne and Whiten identify a significant number of behaviors within their database that fall in one of the following two categories, which they term level 1.5 and level 2 deception, respectively:

(1) Behaviors in which the deceiver seems to be able to appreciate how the world appears from a conspecific's perspective.

(2) Behaviors in which the deceiver seems to be able to represent a conspecific's mental states.

Both types of behavior seem prima facie to count as mindreading

Important background for debates about mindreading, particularly in the primate case, comes from the so-called Machiavellian intelligence hypothesis, according to which the evolution of intelligence was a function of the demands of an increasingly social existence. This hypothesis, originally proposed by Nicholas Humphreys and developed in two influential collections edited by Byrne and Whiten, provides a unifying framework for an impressive array of ethological and anthropological data showing correlations across different primate species between, for example, increasing size of social groups, increasing brain size, and increasing degrees of cognitive competence, as measured both experimentally and observationally.[9]

Discussions and studies of mindreading in nonhuman animals have focused primarily, but not exclusively, on the great apes. There have also been suggestive and thought-provoking studies of a range of other nonhuman animals, including rhesus monkeys,[10] domestic pigs,[11] western scrub-jays,[12] domestic dogs,[13] ravens,[14] and dolphins.[15]

1.2 The Cognitive Requirements and Possibilities of Mindreading

One of the main reasons that mindreading has inspired so much research is that it brings together many different cognitive and social capacities. Another is that it seems to be an enabling condition for other, even more complex, forms of social understanding and social coordination. In both senses, mindreading is a rich mine for thinking about moral significance, because many of the capacities and skills that it either involves or makes possible have been taken to be sources of intrinsic moral significance.

Mindreading is often taken to be an example of second-order representation or metarepresentation, which occurs when a thinker takes as an object of thought not some state of affairs in the world, but rather another thinker's representation of that state of affairs. We will look in more detail at the structure of metarepresentation in later sections. For the moment, we can simply note that metarepresentation provides a possible connection between mindreading and certain types of self-awareness, such as thinkers' ability to reflect upon and monitor their own mental states. These forms of self-monitoring clearly exploit metarepresentational abilities. Moreover, this type of self-monitoring has often been thought to be an important source of moral significance.

The link between mindreading and self-awareness is potentially even closer. A number of philosophers have argued for some version of the symmetry thesis, according to which the capacities for self-awareness and other-awareness are mutually dependent.[16] If this is broadly correct, then mindreading can be used as an index of self-awareness. Just as there are different levels and gradations of mindreading, so too are there different levels and gradations of self-awareness.

Finally, mindreading skills seem to be necessary for many normal kinds of social interaction. Social coordination does not always rest upon social understanding. The animal kingdom is full of examples of social coordination that are completely "mindblind"—flocking behaviors, for example. But the types of social interaction that might plausibly be thought morally significant (such as taking part in caring relationships) do seem to require forms of mindreading—at the very least, for example, they require sensitivity to another subject's emotional states, desires, and needs. Again, this area has many levels and gradations—not just in the basic phenomena, but also in moral significance. Thinking about the different types of mindreading is a useful way of making sense of the complexities.

1.3 Calibrating Mindreading across Species

There are two possibilities for thinking about the moral significance of mindreading. It might be the case, first, that certain types of mindreading themselves bestow intrinsic moral significance on the mindreader. Alternatively, it might be that forms of mindreading are indices of other cognitive or affective capacities that are themselves intrinsically significant. No matter which of these two possibilities holds, there is a pressing need for ways of calibrating mindreading capabilities across species. It is not useful to have criteria of moral significance that cannot be practically measured and applied. It is relevant, then, that there are experimental and conceptual frameworks for measuring the mindreading skills of human infants and young children, for measuring the mindreading skills of nonhuman animals, and for thinking about the relation between human and nonhuman mindreading.

The general developmental trajectory of mindreading in human children has been extensively studied through experimental techniques such as the "Sally-Anne" or "false belief" test (which is designed to test children's understanding that other subjects can have and act on false beliefs, and which we will examine in more detail in section 4) and different versions of the "dishabituation paradigm" (which tests nonlinguistic infants' expectations about how objects and people will behave by measuring their looking times for anomalous as opposed to non-anomalous types of events). These experimental paradigms have been much discussed and refined by developmental psychologists. As we will see further below, comparative psychologists have spent much time and ingenuity working out ways of extending and modifying these experimental paradigms so that they can be applied to nonhuman animals, and hence used to calibrate the mindreading abilities of different species.

We will look in detail at studies of mindreading in nonhuman animals in section 4. For the moment it is enough to note that mindreading is in better shape from an experimental point of view than other cognitive capacities that might be taken either to bestow or to index moral significance. The experimental study of self-awareness in nonhuman animals remains dominated by the mirror self-recognition test and other forms of mirror-related behavior. Despite some of the claims that have been made about the significance of mirror self-recognition, I find it hard to disagree with Michael Tomasello and Josep Call when they write:

> Primates know a good bit about their bodies. Some parts they know visually, and probably all parts they know tactually and/or proprioceptively. In the most conservative accounts of mirror self-recognition, the organism is simply using the mirror as a tool to gain visual access to a part of its body (i.e., the face or rump) that it previously could perceive only tactually or proprioceptively. Mirror self-recognition is thus about perception of the body, which all primates are likely skillful at with no special training or experiences. Why vision should be privileged over other forms of self-perception and why the face should be privileged over other parts of the body, so that some researchers have begun talking about mirror self-recognition as "self-awareness" in a humanlike sense, is a question that researchers should address more directly.[17]

The study of mindreading is on a much stronger experimental (and theoretical) footing than the study of self-awareness in nonhuman animals. Again, we have a reason for viewing questions of moral significance through the prism of mindreading.

2. A Basic Taxonomy: Three Distinctions

This section introduces a basic taxonomy for thinking about mindreading. This taxonomy will be used in later sections to systematize experimental and ethological studies of mindreading in nonhuman animals. Subsection 2.1 identifies the subset of social behaviors that involve what I call minimal mindreading. The distinction between minimal mindreading and other social behaviors is drawn with explicit reference to nonhuman animals. Subsection 2.2 distinguishes minimal mindreading from substantive mindreading, while in subsection 2.3, I distinguish two types of substantive mindreading—perceptual mindreading and propositional attitude mindreading. These last two distinctions are drawn in the abstract, leaving it an open question whether and how they can be applied to nonhuman animals.

2.1 Social Behaviors and Minimal Mindreading

Minimal mindreading, as I define it, is implicated in behaviors that display two components. The first component is a degree of social interaction. The second component is that an organism's behavior depends systematically upon the psychological states of other participants in the interaction. In order to appreciate the significance of these two components occurring together, we can look at some examples of how they can occur independently of each other.

Many species of animals, even those generally thought to be cognitively impoverished, are capable of sophisticated social behaviors. Some of these behaviors have no psychological dimension at all. Schooling behaviors in fish and flocking behaviors in birds are good examples. These behaviors can display a high degree of coordination, but this coordination is achieved even though each participant in the social behavior need only be aware of basic properties of the other participants—their distance, trajectory, and velocity, for example.

Conversely, there are behaviors with a psychological dimension that do not involve any social coordination. Emotional contagion is a case in point. Instances of emotional contagion have been identified in nonhuman primates. In the 1960s, Robert Miller and his colleagues tested the sensitivity of rhesus monkeys to the facial expressions of conspecifics. They found that observing signs of fear in another monkey led monkeys to pull a lever that would spare both monkeys an electric shock.[18] Approaching the same phenomenon rather differently, another group of experimenters found that although naïve captive-born rhesus monkeys do not show fear behavior when confronted with snakes, they can acquire a fear response by watching films of a wild monkey interacting with a snake.[19] So, there can be sensitivity to the psychological states of others without social coordination and social coordination without psychological sensitivity. Neither of these can plausibly be taken as instances of mindreading. Even the simplest kind of mindreading requires both elements. In order to make this condition more precise, we can start with an example.

A well-known set of experiments by Brian Hare and his collaborators has revealed that domestic dogs are strikingly successful in tasks that involve selecting objects as a function of social cues.[20] In a standard object choice task, an experimenter hides a food reward in one of two opaque containers. The subject, who did not see the food being hidden, has to choose between the two containers. Before the animal is presented with the choice the experimenters "signal" which container the food is in by using one of a range of communicative cues (such as pointing to, marking, or looking at the correct container). Hare et al. found that domestic dogs master object choice tasks quickly, often without any learning.

These results are interesting for a number of reasons. From a comparative psychological perspective, they are striking because, despite the sophisticated social-cognitive skills typically attributed to them, most primates seem unable to perform above chance on object choice tasks (for exceptions, see the experiments described in section 4.4 below). The success of domestic dogs in picking up and exploiting social cues to solve object choice tasks is a paradigm example of what I am describing as social coordination with a psychological dimension.

That the experiments illustrate social coordination is obvious. What makes it social coordination involving sensitivity to psychology is that the dogs behave in ways that depend systematically upon changes in the psychological states of the other participant in the interaction. This behavior contrasts with coordinated group behaviors, such as schooling or flocking, where (to simplify somewhat) an individual's behavior depends simply upon changes in the behavior of other participants in the coordinated group behavior.

A natural question to ask is: what benefits are there to thinking about this type of behavior in terms of mindreading—as opposed, for example, to thinking of it as a sophisticated form of behavior-reading? The principal advantage is the significant explanatory and predictive leverage that it provides. The dogs in the object choice tasks are able to respond to different visual cues. These cues all have something in common. They all have a common cause in the psychological profile of the experimenter—namely, the experimenter's intention to signal to the animal the location of the reward. Identifying this common cause and extrapolating allows us to make predictions about how the dogs will behave in future tests—namely, that they will respond to visual cues that have the same cause and origin. In essence, we assume that the dogs are responding to cues in the abstract, rather than to the physical gestures by which those cues are made—a "multi-track" sensitivity to a psychological state that can be physically manifested in different ways, as opposed to a set of contingencies between particular responses and particular stimuli.

The phenomenon of psychological sensitivity is here being understood in a *quasi-operational* sense. It is quasi-operational because, while it goes beyond observed behavior in making reference to the psychological states of the experimenter, it does not go beyond the observed behavior of the experimental subject. The experimental subject is characterized in purely behavioral ways. We can say that a nonlinguistic creature displays psychological sensitivity in this quasi-operational sense without attributing to it any psychological states. Displaying sensitivity to psychology requires only behaving in ways that are suitably dependent upon the psychological states of another participant in the interaction. The notion of dependence is being understood in a thin sense, as requiring only some form of systematic mapping between the animal's behavior and the relevant psychological states. The question of what secures that dependence is left entirely open. It is neither implied nor required that an animal displaying sensitivity to psychology should represent (or even be capable of representing) the psychological states of the other participant(s) in the exchange.

2.2 Substantive Mindreading

Here is an official definition of minimal mindreading, as discussed in section 2.1:

> A creature engages in *minimal mindreading* when its behavior is systematically dependent upon changes in the psychological states of other participants in the interaction.

Let me stress again that characterizations of minimal mindreading in this sense are *descriptive* rather than *explanatory*. To say that an animal is engaged in minimal mindreading is simply to assert that certain contingencies hold between its behavior and the psychological states of the creatures with which it is interacting. It is not in any sense to say *why* those contingencies hold. The mechanism by which they are achieved, representational or otherwise, is left open.

Minimal mindreading needs to be distinguished from substantive mindreading. This *is* an explanatory notion. Attributions of substantive mindreading are made in order to explain how and why an animal's behavior depends systematically upon the psychological states of other participants in the interaction. It is typically claimed by those who identify substantive mindreading in the animal kingdom that what explains the dependence is the fact that the animal engaged in a social interaction is mentally representing the psychological states of other participants in the interaction.

Here is the matching official definition of substantive mindreading:

> a creature engages in *substantive mindreading* when its behavior is systematically dependent on its representations of the psychological states of other participants in the interaction.

Although the notion of systematic dependence features in the definition both of minimal and of substantive mindreading, it is doing different work in each.

In the definition of minimal mindreading, the systematic dependence is not intended to be causal. Minimal mindreading is purely and simply a matter of covariation or correlation. In contrast, claims of substantive mindreading make claims about causation: the animal's behavior is caused by (and hence can be explained by appeal to) how it represents the mental states of others.

A final point: minimal mindreading and substantive mindreading can dissociate in both directions. It is not hard to see how we can have minimal mindreading without substantive mindreading. This would be a psychological version of the "Clever Hans" effect, as in the famous case of the horse supposedly doing arithmetic that turned out to be responding to unconscious cues from its trainer. It may well be that this is the best way to understand the object choice experiments discussed in the previous section.

But we can also have substantive mindreading without minimal mindreading. This will occur when a creature's mindreading representations are not accurate. If a creature is incorrectly representing the psychological states of other participants (and behaves accordingly), then it may be that its behavior does not depend systematically upon changes in the psychological states of other participants. In fact, this may be an important clue that substantive mindreading is going on. The false belief test (to be discussed further in section 4) proposes a criterion for possession of the concept *belief* that exploits the possibility of just such a divergence between minimal mindreading and substantive mindreading. The guiding idea behind the false belief test is that a subject's inaccurate representation of another creature's mental state leads it to behave in ways that do not display the type of sensitivity to psychology required for minimal mindreading. Put another way, the best explanation for the breakdown of coordination is that the subject is inaccurately representing another creature's mental state.

2.3 Two Types of Substantive Mindreading: Perceptual Mindreading and Propositional Attitudes Mindreading

Substantive mindreading occurs when a creature behaves in ways that depend systematically upon how it represents the psychological states of other participants in the interaction. Substantive mindreading is not a unitary phenomenon. Because there are different types of psychological state, there are different types of substantive mindreading. In this section, I articulate a distinction between the type of mindreading that involves representing perceptions and the type that involves representing what philosophers standardly term propositional attitudes (psychological states such as belief, desire, hope, and fear). These two types of mindreading involve importantly different types of representation, ultimately deriving from the different logical and psychological structures of perceptions and propositional attitudes.

I begin with the differences between perceptions and propositional attitudes. There is a standard way of thinking about these differences in the philosophy of mind that seems to me to be correct in its essentials: perception is a relation that holds between individuals, on the one hand, and objects or states of affairs, on the other. Specifying the exact nature of this relation is no easy matter. There has been much discussion, for example, of whether the relation needs to be causal. But for present purposes what is important is not the relation itself, but rather its relata (the things between which it holds)—and, in particular, the fact that the objects of perception are things in the world. This is where the contrast lies with propositional attitudes.

As their name suggests, propositional attitudes are standardly interpreted in terms of two components—a propositional content and a thinker's attitude toward that content. If I am correctly described as believing that p, for example, this is typically taken to mean that there is a propositional content (the content that p) to which I take the attitude of belief. I might have taken a different attitude to that same content— I might hope that p while believing that p does not currently hold. Or I might take the same attitude to a different content—I might believe that q for example. This general framework for thinking about propositional attitudes has proved its utility many times, although there is considerable controversy about how to work out the details. There are different accounts, for example, of what propositions are. My argument in the following does not depend upon any particular one of the candidate accounts. All it rests upon are some general ideas about what it is to represent a proposition.

We can get traction on the difference between perceptual mindreading and propositional attitude mindreading. Perceptual mindreading has a tripartite structure. Representing another creature as perceiving, say, that the food is under the tree involves representing

    (i) a particular individual as
    (ii) perceiving
    (iii) a particular state of affairs (the state of affairs of the food being under the tree).

Propositional attitudes mindreading also has a tripartite structure. Representing another creature as believing that the food is under the tree involves representing

    (i) a particular individual as
    (ii) bearing the propositional attitude of belief to
    (iii) a particular proposition (a representation of the food being under the tree).

Despite this similarity in structure, however, the representations required for propositional attitudes are far more complex than those required for perceptual mindreading.

Some authors have argued that understanding propositional attitudes is intrinsically more complex than understanding perceptual states—and so that perceptual and propositional attitude mindreading come apart at (ii). Davidson proposed a well-known argument to the effect that only language-using creatures can properly possess the attitude of belief.[21] He argued that understanding belief requires understanding the distinction between correct and incorrect belief, something that can in turn only be understood through participation in a public language. Davidson's argument has failed to find general acceptance, however.[22] What I want to stress, in contrast, is the difference at (iii). There is a fundamental distinction between representing a particular proposition and representing a particular object or state of affairs.

Consider a perceptual mindreader M representing another agent α as perceiving a state of affairs S. The perceptual mindreader is already perceiving S. In order to represent α as perceiving S, M needs simply to add to its representation of S a representation of a relation between α and S. In many cases of perceptual mindreading, this additional representation can be straightforward. It can be simply a matter of representing S as lying in α's line of sight. The representational skills required are basic geometric skills, on a par with those involved in working out possible trajectories through the environment.

Now consider a propositional attitude mindreader M* representing another agent β as believing a proposition P. Here it is not typically the case that P corresponds to a state of affairs in the distal environment that M* is already perceiving. In many of the cases where propositional attitude mindreading is identified in the animal kingdom, P is false. This is certainly the case in instances of tactical deception where the aim (as these behaviors are standardly interpreted) is to generate false beliefs in another agent—and hence where the deceiver must intend to bring it about that an agent believe that p where p is false. So, the question arises: what is it to represent a proposition?

One answer here is that propositions just are states of affairs, and so representing a proposition is no more and no less complicated than representing states of affairs. On this interpretation, propositional attitude mindreading does not come out as fundamentally different in kind from perceptual mindreading. It is true that propositional attitude mindreading can involve representing states of affairs that do not exist (as in tactical deception cases), but it is widely accepted that many types of nonhuman animals can represent nonexistent states of affairs. After all, we can only explain the behavior of nonlinguistic creatures in psychological terms if we attribute to them desires, and having a desire often involves representing a nonexistent state of affairs.

The principal difficulty with this view is that states of affairs lack some of the fundamental characteristics of propositions. In particular, propositions are true or false, while states of affairs are not the sorts of thing that can be either true or false. In many standard ways of thinking about propositions and states of affairs, states of affairs are the things that make propositions true or false. (There are some technical issues here that need to be mentioned, although not resolved. These arise because propositions can be logically complex and it is not clear that there are logically complex states of affairs. Even if one thinks that the proposition that *the table is red* is made true by the state of affairs of the table being red, it is far from obvious that the proposition *the table is not red* is made true by the state of affairs of the table not being red. Many philosophers would deny that negative states of affairs exist. Nonetheless, the fact remains that when *the table is not red* is true, its truth plausibly consists in the holding of some state of affairs—the table being black, for example.)

The "truth-aptness" of propositions is absolutely fundamental to the whole enterprise of propositional attitude mindreading. Propositional attitude mindreading allows us to explain and predict the behavior of other subjects in terms of the representational states that generated it. What makes it such a powerful tool is that it works both when other subjects represent the world correctly and when they misrepresent it. In order to use the tool, however, a propositional attitude mindreader has to be aware that propositional attitudes such as belief can be either true or false. This is the genuine insight in Davidson's argument mentioned earlier, and it is the key idea behind the false belief task.

To summarize the point as neutrally as possible: my claim is that propositional attitude mindreading involves representing another agent's representation of a state of affairs. I will use the term "proposition" to abbreviate "representation of a state of affairs."

The expression "metarepresentation" is often used to describe what goes on in representing a representation. So, the difference between perceptual mindreading and propositional attitude mindreading can also be put in terms of metarepresentation. Propositional attitude mindreading involves metarepresentation, whereas perceptual mindreading typically does not. The different structure of propositional attitude and perceptual mindreading can be represented diagrammatically (see figures 14.1 and 14.2).

This distinction between two types of substantive mindreading is very important for thinking about mindreading in the animal kingdom. As we will see in the

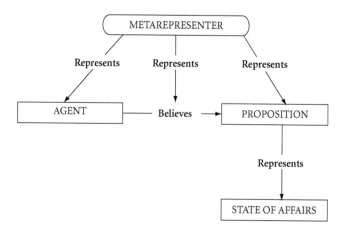

Fig. 14.1

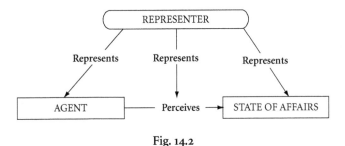

Fig. 14.2

next section, there is a powerful argument that propositional attitude mindreading is not available to nonhuman animals. This argument does not extend to perceptual mindreading, however.

3. The Limits of Mindreading in the Animal Kingdom: The Argument from Metarepresentation

This section presents an argument I have developed elsewhere.[23] The basic claim of the argument is that metarepresentation, in precisely the sense identified in the previous section, is a distinctively language-dependent cognitive achievement—and so unavailable to nonhuman animals. I will give the argument in schematic form, and then develop three crucial points.

(A) Unlike perceptual mindreading, propositional attitude mindreading involves representing another creature's attitude to a proposition.

(B) Propositional attitude mindreaders must represent propositions in a way that allows them to work out how the relevant propositional attitudes will feed into action.

(C) Working out how a particular set of propositional attitudes will feed into action depends upon the inferential relations between those propositional attitudes and the agent's background psychological profile (including other propositional attitudes).

(D) The representations exploited in propositional attitude mindreading are consciously accessible constituents of a creature's psychological life.

(E) The representational format for propositional attitude mindreading must be either language-like or image-like.

(F) The representational format for propositional attitude mindreading must exemplify the canonical structure of the represented propositions.

(G) Representing propositions in a pictorial or image-like format does not reveal their canonical structure.

(H) The canonical structure of a proposition is only revealed when propositions are represented in a linguistic format.

(I) The representations required for propositional mindreading must involve natural language sentences.

The first claim that needs further discussion is (C). The starting point here is that the most basic application of propositional attitude mindreading is in predicting behavior. But the move from propositional attitude attributions to predictions of behavior is rarely straightforward. It is true that organisms generally act in virtue of their beliefs and desires, but any given belief-desire pair can feed into action in many different ways. It all depends on what other beliefs and desires the organism has. A hungry organism that believes there to be food under a tree can typically be relied upon to seek out the food—unless it believes there to be a predator nearby, or unless it believes there to be a better food source elsewhere, or unless its desire for water outweighs its desire for food, and so on.

A third party with good reason to think that the organism believes there to be food under the tree needs to consider whether the organism has any such potentially countervailing beliefs or desires in order to generate an accurate prediction. The predictor needs to represent salient aspects of the organism's background psychological profile. Moreover, the predictor needs to work out how the propositional attitudes it is attributing relate to those salient aspects of the background psychological profile. These relations are inferential, entailing that a propositional attitude mindreader must be able to represent propositions in a way that makes it possible to evaluate the inferential relations between propositions. This requirement is the basis for the rest of my argument.

Turning now to (D), the background idea is that mindreading is not a self-contained activity. When a propositional attitude mindreader forms beliefs about the mental states of another agent, those beliefs are integrated with the rest of the mindreader's propositional attitudes—with the mindreader's beliefs about the distal environment, and with their short-term and long-term goals, for example. This type of integration is required for the results of propositional attitude mindreading to feature in a creature's practical decision making.

As far as (D) is concerned, I am taking practical decision making to involve some form of conscious deliberation—or, more accurately, I am taking the capacity for practical decision making to be closely bound up with the capacity for conscious deliberation. No doubt there are instances of practical decision making that do not involve conscious deliberation, but I claim that no creature that is not, at least on occasion, a conscious deliberator can properly be described as a practical decision maker.

The upshot of (D) is that propositions must be represented by propositional attitude mindreaders in a way that allows them to be objects of conscious deliberation and reflection. In particular, they must be represented in a way that allows them to be objects of the particular type of conscious deliberation and reflection required by (C)—that is to say, deliberation that exploits and explores the inferential connections between propositions. The remainder of my argument explores the constraints that this imposes on how propositions are represented by propositional attitude mindreaders.

Claim (F) is particularly important. If state of affairs must be represented in a way that allows the mindreader to reason about the logical and inferential connections between them, then they must be represented in a way that makes clear their internal structure. This is because many of the most basic logical and inferential relations between propositions hold in virtue of their structure. It is hard, for example, to see how a mindreader could reason about the propositional attitudes and background psychological profile of another agent without attributing to him or her conditional beliefs. These conditional beliefs might, for example, track dependencies between particular behaviors and particular features of the environment. An ethologist interpreting the behavior of a wolf might attribute to it the following belief: if the prey goes into the woods then I will follow it. Alternatively, a conditional belief might record regularities and contingencies in the environment. A chimpanzee might be interpreted as believing that if it rains then there will be more insects on the leaves. Such conditional beliefs have a basic composite structure—they are attitudes to a complex proposition that relates two other propositions. A conditional belief has both an antecedent (if it rains...) and a consequent (...then there will be more insects on the leaves). The inferential role of a conditional belief depends upon this structure. One way of exploiting a conditional belief is to come to believe the antecedent. If the chimpanzee comes to believe that it is raining, it will draw the conclusion that there will be more insects on the leaves. This very basic pattern of inference (generally called *modus ponens*) crucially depends upon conditional beliefs having an antecedent-consequent structure.

The structure requirement carries over to the propositional attitude mindreader, who must be able to recreate the reasoning that might be gone through by the organism it is predicting. This means that the conditional belief must be represented in a way that reflects its structure—in a way that reflects the fact that it is an attitude to a complex proposition that relates two other propositions. Unless the conditional belief is represented in this way it will be impossible to reason, for example, that because the agent has seen the prey going into the woods it will follow it.

To clarify, there is a fundamental difference between the reasoning carried out by a decision-making organism and the type of reasoning engaged in by a propositional attitude mindreader. There is no requirement that the first-order reasoner (as opposed to the metarepresenting mindreader) represent propositions. According to the standard picture, having a particular belief involves standing in the belief relation to a proposition. But standing in the belief relation to a proposition does not require representing that proposition. It requires representing some actual or possible state of affairs. The theoretical notion of a proposition is at bottom a tool for characterizing what it is to represent an actual or possible state of affairs.

To put it another way, a thinker stands in the belief relation to a proposition in virtue of representing an actual or possible state of affairs in a particular way. So, simply having propositional attitudes does not require metarepresentation. In contrast, engaging in propositional attitude mindreading does require metarepresentation because representing another organism's propositional attitudes is a matter of representing their representations of actual or possible states of affairs. As figures 1 and 2 show, the different objects of representation involved is the crucial distinguishing feature between propositional attitude mindreading and perceptual mindreading.

To summarize the argument so far: I have argued that propositional attitude mindreading presupposes a very distinctive type of representation. The propositional attitude mindreader must be able to represent another creature's propositional attitudes in a way that reveals their canonical inferential structure. This in turn means that propositional attitude mindreaders must represent propositional attitudes in a linguistic form.

Let me add two more highly plausible assumptions. The first is the assumption that the type of linguistic abilities implicated in metarepresenting the structure of propositions and the inferential relations between them require a medium with the recursive and combinatorial power of human language. The second is that there is nothing approaching, or even remotely approximating, human language in the animal kingdom.

When these two assumptions are combined with the main argument of this section, it follows that nonhuman animals cannot be propositional attitude mindreaders. There are clear limits to the scope of mindreading in the animal kingdom. These limits are derived from the fact that propositional attitude mindreading is fundamentally metarepresentational. But, let me emphasize, this conclusion does not extend either to perceptual mindreading or to the simple ability to have propositional attitudes and to reason about how things are and how they might turn out. Neither of these basic cognitive abilities either depends upon or involves metarepresentation.

4. Mindreading in Nonhuman Primates: Interpreting the Evidence

In this section, I show how the distinctions and arguments thus far mentioned bear on experimental work that has been done on mindreading in nonhuman animals. I will focus on mindreading in primates, because this area has seen the most sustained and systematic investigation.

4.1 Applying the Distinction between Minimal and Substantive Mindreading

Gaze following in primates has been extensively studied, both ethologically and experimentally. There are many reports of primates observed in the wild behaving in ways that seem to involve manipulating and exploiting the direction of a conspecific's gaze. This work sometimes presents paradigm examples of tactical deception.

Here are two sample reports from Richard Byrne and Andrew Whiten's field studies of baboons:

> (23 May 1983) Subadult male ME attacks one of the young juveniles who screams repeatedly pursuing ME while screaming (this is common when aid has just been successfully solicited). Adult male HL and several other adults run over the hill into view, giving aggressive pantgrunt calls. ME, seeing them coming, stands on hindlegs and stares into the distance across the valley. HL and the other newcomers stop and look in this direction; they do not threaten or attack ME. No predator or baboon troop can be seen through 10 x 40 binoculars.[24]

> (15 July 1983) Adult male JG displaces female PK from her feeding patch. She responds by immediately enlisting JG, by characteristic rapid flicks of gaze from potential ally to target, to attack juvenile PA. PA (who is not PK's offspring) was merely feeding 2m away, but is threatened and chased by JG. PK meanwhile continues to feed on her patch.[25]

Here is another more elaborate example from Hans Kummer, also observing baboons:

> An adult female spent 20 min gradually shifting in a seated position over a distance of about 2m to a place behind a rock about 50 cm high where she began to groom the subadult male follower of the group—an interaction not tolerated by the adult male. As I was observing from a cliff slightly above [the animals] I could judge that the adult male leader could, from his resting position, see the tail, back and crown of the female's head, but not her front, arms and face: the subadult male sat in a bent position while being groomed, and was also invisible to the leader. The leader could thus see that she was present, but probably not that she groomed.[26]

These reports are suggestive. If we take them at face value, the baboons seem to be exploiting the fact that other members of their troupe can be relied upon to follow the gaze of conspecifics. But, as has often been observed, the reports are consistent with two rather different ways of thinking about gaze following—and, consequently, two different ways of thinking about the kind of deception going on here.

On one reading, the baboons have complex behavior-reading abilities. They are disposed to look where others' eyes are directed. This disposition might be learned, or it might be hard-wired. Either way, it has become established because looking where others are looking generally pays off—it is a good way of finding food, locating predators, and so on. But it is not obvious that this type of complex behavior-reading involves any type of psychological sensitivity. Nor need any psychological sensitivity be required in order to exploit this behavior-reading disposition in tactical deception. If a baboon knows that other baboons are disposed to follow conspecifics' gazes, then it can manipulate this disposition just as it can manipulate any other behavioral disposition.

According to a second and less parsimonious interpretation of what is occurring, gaze-following is not simply behavior-reading. Rather, gaze-following is really perception-tracking. Baboons look where others are looking because they understand that the act of looking is a way in which their conspecifics can find out about the physical and social world. By the same token, tactical deception involves more than simply manipulating behavioral dispositions. It involves considered attempts to control and limit what others can find out—either by hiding their own activities (as in Kummer's female baboon) or by misdirecting the attention of others elsewhere (as in ME and JG described by Byrne and Whiten).

On the first interpretation, thinking about these behaviors in terms of the conceptual framework developed earlier raises two questions. The first is whether gaze following on the first interpretation counts as mindreading. We can make this question more precise by using the definition of minimal mindreading. Do these examples of gaze following involve a creature's behavior being systematically dependent upon changes in the psychological states of other participants in the interaction? The contrast being drawn is with systematic dependence on non-psychological states of the other participants. One candidate would be head direction. Looking in the direction in which another's head is pointed would not count as psychological sensitivity in even the most minimal sense.

Exploring this question further would be hard to do in the wild, but it has been pursued experimentally with chimpanzees. There is some evidence that chimpanzees can track eye direction independently of head position and head movement, as first reported, for example, by Daniel Povinelli and Timothy Eddy in 1996.[27] Similar findings were reported more recently by Sanae Okamoto and colleagues.[28] They trained a thirteen-month-old chimpanzee infant to look at one of two objects indicated by a range of cues. The cues included the experimenter glancing at one of the objects without moving his head. The study showed a reliable following response to glancing, even though successful responses to the glancing cue were not rewarded. These experiments suggest that chimpanzees can reliably track eye direction, as opposed to head orientation. Because eye direction is reliably correlated with seeing, this result would be enough to establish that gaze following counts as minimal mindreading, but it does not indicate any richer sense of mindreading.

In order to demonstrate substantive mindreading, as opposed to minimal mindreading, an experiment would need to show that experimental subjects are representing the mental states of other subjects—as opposed to simply behaving in ways that covary with another subject's mental states. How might this be achieved? What is needed is some sort of experimental index that shows a basic understanding of perception as a mental state. Perhaps the most basic characteristic of perception is that it places the perceiver in epistemic contact with a particular object or a particular state of affairs—an epistemic contact that has certain predictable and exploitable implications for how that perceiver behaves. At a minimum, one might plausibly think, a perceptual mindreader should understand that a perceiver's attention will be focused on things that are potentially significant for it, such as food or a predator. One way of testing for this understanding would be to see what efforts an organism makes to locate and identify the object of another's gaze.

Michael Tomasello, Brian Hare, and Bryan Agnetta proposed some interesting ways of operationalizing this understanding of attention in a set of experiments reported in 1999.[29] One experiment looked to see how chimpanzees reacted when confronted with an experimenter looking around a barrier. Would the animals move around in order to follow the experimenter's gaze around the barrier? If their gaze following were simply a matter of looking for interesting objects in the general direction in which the experimenter is looking, then the prediction was that the chimpanzees would not investigate behind the barrier. But they would look behind the barrier if they thought that the experimenter's gaze was focused on an interesting object.

Another experiment explored how chimpanzees would react when a distractor object was interposed in the experimenter's general line of sight. The experimenter captured the chimpanzee's attention and then looked conspicuously at the back of the cage. In the test condition, another experimenter shook a doll when the animal started to follow the experimenter's gaze. Again, if the chimpanzees were simply looking for interesting objects in the general direction of the experimenter's line of sight then they would stop at the doll and not follow the experimenter's gaze to the back of the cage.

The results of both experiments suggest perceptual mindreading. In the first experiments, the chimpanzees reliably looked around the barrier, while in the second experiment they reliably looked past the distractor to follow the experimenter's gaze to the back of the cage. These results were extended to the other three great ape species by the same group of experimenters in a paper published in 2005. Juliane Brauer, Josep Call, and Michael Tomasello found that bonobos, gorillas, and orangutans also consistently followed an experimenter's gaze around barriers and past distractors.[30] Experimenters have shown that macaque monkeys do not follow the gaze of experimenters, but there is some evidence that rhesus macaques do behave like the great apes when it comes to following the gaze of conspecifics (as opposed to an experimenter). The experiments here involve presenting monkeys with videotapes of another monkey with its attention directed to one of two identical objects and then analyzing the first monkey's eye direction.[31]

In sum, there is strong evidence that the mindreading abilities of great apes and perhaps other primates go beyond what we termed minimal mindreading. They behave in ways that do not simply covary with the perceptual states of conspecifics and experimenters, but that seem to involve their representing those perceptual states.

4.2 Testing for Understanding of False Belief in Chimpanzees

I turn now to propositional attitude mindreading. This is an area where there seems to be potential tension between the argument of section 3 and some of the interpretations of experimental and ethological data. Some ethologists and comparative psychologists have claimed to find evidence for propositional attitude mindreading in primates. But this conflicts with the argument for the language-dependence of metarepresentational mindreading. In this section, I reconcile the tension in two ways. First, I describe experiments that seem to show that primates fail a nonlinguistic version of the false belief test. Second, I argue that the evidence for claims of propositional attitude mindreading in primates is best interpreted as evidence for perceptual mindreading.

Let us start with the negative evidence. As mentioned earlier, the standard test of propositional attitude mindreading in developmental psychology is the false belief test, originally developed by Joseph Perner and Franz Wimmer.[32] There are many versions of the test. Here is one made famous by Simon Baron-Cohen and known as the Sally-Anne task.[33] A child sits in front of an experimenter, who has two puppets, Sally and Anne. There is a table with a basket and box between the child and the experimenter. In full view of the child, Sally places a marble in the basket and then leaves the room. While she is away, Anne transfers the marble from the basket to the box. Sally then returns. The experimenter asks the child: "Where will Sally look for her marble?" (or, in some versions of the test, "Where does Sally think the marble is?"). The point of the experiment is that, although the child saw the marble being moved, Sally did not. If the child has a clear grip on the concept of belief and understands that it is possible to have false beliefs, then she will answer that Sally will look in the basket—because nothing has happened that will change Sally's belief that the marble is in the basket. If, on the other hand, the child fails to understand the possibility of false belief, then she will answer that Sally will look for the marble where it in fact is—namely, in the box.

Josep Call and Michael Tomasello explored whether nonhuman animals can understand false belief by developing a nonverbal false belief task, which they applied to human children between four and five years of age as well as to chimpanzees and orangutans.[34] The task was directly modeled on the Sally-Anne task. The children and apes were familiarized with a scenario in which one experimenter (the hider) hid an object in one of two identical containers while a second experimenter (the communicator) helped the participant keep track of where the object was by marking the relevant container. The first stages of the experiment established that the participants understood the procedure, the role of the markers, and so on—and were able to use the relevant cues to find the hidden object. Only then was the crucial false belief condition applied. In this condition, the communicator left the room and while she was gone the hider switched the object to the other container. On returning the communicator marked the container in which the object had originally been hidden and then the participant was allowed to search for the hidden object.

The reasoning was exactly the same as in the Sally-Anne test. Participants who understood that the communicator had a mistaken belief about the location of the reward would ignore the marked container and search in the other one, while participants who failed to understand that the communicator's belief was false would search in the marked container. The experimenters further explored the parallels between the tests by giving the children versions of the Sally-Anne test. The results of the experiment were unambiguous. None of the apes performed above chance in the false belief condition, although they were all able to track the location of the hidden object successfully in the control conditions. At the same time, there was a strong correlation between children's performance on the nonverbal false belief task and on the Sally-Anne task.

One potential problem with applying Call and Tomasello's paradigm to great apes, however, is that it is essentially cooperative. Orangutans are solitary animals, and subsequent experiments have shown that in certain contexts, chimpanzees perform better on competitive tasks than cooperative ones. This reflects the basically competitive nature of chimpanzee social interactions in the wild. Perhaps chimpanzees fail Tomasello and Call's nonverbal false belief tasks for reasons that have nothing to do with false belief.

Carla Krachun and Melinda Carpenter, working together with Tomasello and Call, have addressed this issue in recent work.[35] They developed a competitive version of the nonverbal false belief task, which they applied both to children and to apes (both chimpanzees and bonobos). The experiments (for the apes) took place in a three-windowed testing booth. The ape was in one enclosure, in front of an experimenter, while a second experimenter (the competitor) was at a side window. In the basic set up, the experimenter placed food rewards in one of two identical containers, which were then slid across the table toward the ape. While the containers were moving across the table the competitor reached out for the container with the reward inside. In some of the familiarization trials the competitor succeeded in getting the food. The aim was to teach the ape that the competitor's movements were a good guide to the location of the food. The context was competitive because the competitor's getting the reward left none for the ape.

In the experimental trials, an opaque screen blocked the ape's view of the experimenter hiding the reward, so that the competitor's behavior was its only guide to the location of the reward. Once the reward was hidden the screen was removed, so that the ape could see both containers. In both True Belief and False Belief conditions, the competitor either left the room or turned around after the reward was hidden. In the True Belief condition, the experimenter sat still while the competitor was out of the room and then, when the competitor returned, started sliding the containers over toward the participant, with the competitor reaching for the correct container as before. In the False Belief condition, in contrast, the experimenter switched the containers around while the competitor was out of the room, so that the competitor ended up reaching for the incorrect container. Success in the False Belief condition would be selecting the container for which the competitor did not reach.

The results in this competitive version of the nonverbal false belief test were similar to those in the earlier collaborative version. Both chimpanzees and bonobos consistently selected the same container that the competitor reached for, even when the containers had been switched while the competitor was out of the room. This means that they failed to adjust for the competitor's false belief about the containers. The children, in contrast, performed much better, showing that in the crucial cases they could distinguish between where the reward really is, and where the competitor believes the reward to be. There was significant correlation between the group's performance on the nonverbal false belief test and on the Sally-Anne test.

In addition to the behavioral tests just outlined, the experimenters also measured where participants looked during the test(s). The background here is the extensive use of looking-time measures in developmental psychology (particularly in the dishabituation paradigm used to explore violation of expectations in nonlinguistic infants). So, for example, a much-discussed set of experiments published in

Science in 2005 by Kristine Onishi and Renée Baillargeon claimed to have found an understanding of false belief in fifteen-month-old human infants.[36]

Krachuk, Carpenter, et al. did not measure looking time. Instead, they tested whether infants and apes in the False Belief condition looked at least once at the correct container in the brief period during which the two containers were being slid across the table. The idea here was that looking at the correct container would show at least an implicit understanding of the true location of the food reward (and hence, they argued, of the fact that the competitor had a false belief about where it was). The results here were suggestive, but ultimately inconclusive. There was a higher proportion of looking responses to the container to which the competitor was not reaching in the False Belief condition than in the True Belief condition, but not sufficiently high to be statistically significant. As the authors note in their discussion, this could be a reflection of the ape's level of uncertainty, rather than of any appreciation of the competitor's having a false belief.

So, even if we take the nonverbal false belief test to be a nonlinguistic equivalent of the Sally-Anne test, the evidence seems clear that the chimpanzees and bonobos tested did not have the level of understanding of false belief manifested in the young children tested at the same time. This conclusion is entirely consistent with section 3's conclusion that propositional attitude mindreading is restricted to language-using creatures.

4.3 Interpreting Claims about Knowledge of Knowledge

Although the false belief test is the most discussed index of propositional attitude mindreading, belief is far from being the only propositional attitude. A number of researchers who have argued that there is no evidence of false belief understanding in primates have identified other types of propositional attitude mindreading in primates. So, for example, in a recent survey article in *Trends in Cognitive Science*, Call and Tomasello write, "We believe that there is only one reasonable conclusion to be drawn from the totality of the studies reviewed here. Chimpanzees, like humans, understand that others see, hear, and know things. We have many different methodologies involving several different experimental paradigms and response measures all leading to the same conclusion."[37]

As standardly discussed by philosophers, knowledge is a propositional attitude. To know something is to stand in an attitude to a proposition. One illustration of this is that knowledge has the following *opacity* property: someone can know that *a* is F without knowing that *b* is F, even when *a* is *b*. Standard illustrations of this phenomenon use the Morning Star and the Evening Star; Robert Zimmerman and Bob Dylan; or Superman and Clark Kent. Exactly the same phenomenon holds for belief. One can believe that the Morning Star rises in the morning while believing that the Evening Star does not rise in the morning. In fact, this type of substitution failure is often taken to be a defining characteristic of propositional attitudes. In this respect, propositional attitudes such as belief or knowledge contrast with perceptual states. If you hear Bob Dylan then you hear Robert Zimmerman (even if you are unaware that

Bob Dylan is Robert Zimmerman, or even if you believe them to be two completely different people). Likewise, if you see the Morning Star then you see the Evening Star, even if you believe them to be two completely different heavenly bodies.

The so-called opacity of knowledge is one reason for thinking that knowledge is a propositional attitude. Another reason comes from reflection on the structure of knowledge. Many philosophers, dating to Socrates, have analyzed knowledge as true belief that has certain extra qualifying features. The precise nature of these extra qualifying features has not yet been fully clarified, but the important point is that most extant accounts of knowledge start out from the assumption that knowledge is a species of belief—and hence that states of knowledge are states of belief.

But if knowledge is a propositional attitude, then understanding the knowledge states of another individual have to count as an instance of propositional attitude mindreading. The opacity of knowledge is matched by the opacity of attributions of knowledge. My knowing that the Morning Star rises in the morning does not entail my knowing that the Evening Star rises in the morning. By the same token, my judgment that Carla knows that the Morning Star rises in the morning does not commit me to judging that Carla knows that the Evening Star rises in the morning—I know that the Morning Star is the Evening Star, but Carla may not.

This means that thinking about another person's knowledge involves thinking about their relation to a proposition. If I judge that Carla knows that the Morning Star rises in the morning then (to put it rather ponderously) what I am judging is that Carla stands in the knowledge relation to a proposition—the proposition that would typically be expressed by the sentence "The Morning Star rises in the morning." That proposition is very different from the proposition that would typically be expressed by the sentence "The Evening Star rises in the morning."

So, something has to give. If knowledge is a propositional attitude, then understanding someone else's state of knowledge must be an instance of propositional attitude mindreading. This creates a tension between, on the one hand, the claim that nonlinguistic creatures such as chimpanzees can understand that others know things and, on the other, the central contention of this chapter, which is that propositional attitude mindreading is language-dependent. Clearly we need to look more closely at the experimental evidence.

Much discussion has focused on a well-known study by Brian Hare, Michael Tomasello, and Josep Call. The title of the paper poses the question "Do chimpanzees know what conspecifics know?"[38] The authors answer this question in the affirmative, based on how chimpanzees perform in different versions of a food-competition paradigm. In each of the versions, a dominant and a subordinate chimpanzee competed for food placed on the subordinate's side of two opaque barriers. The subordinate was always able to see the food being placed, whereas the experimenters controlled the dominant's visual access to the location of the food. The aim of the experiments was to track the subordinate's sensitivity to the dominant's visual access to the food being hidden.

As the experiments are often reported, the subordinates did indeed show sensitivity to the dominant's visual access, and this is interpreted as an index that they understood what the dominant did and did not know about the location of the food. However, an alternative interpretation also suggests itself. Hare and collaborators actually performed three different experiments. The "knowledge" interpretation is derived from the chimpanzees' performance in the first two experiments. I shall argue, though, that the mindreading displayed in these two experiments is best viewed as perceptual, rather than propositional attitude, mindreading. Subordinate chimpanzees in the third experiment did not act in ways that showed the right sort of sensitivity to the dominant's visual access. Success on the third experiment would require something much closer to propositional attitude mindreading.

The basic set-up was the same for all three experiments. It is illustrated in figure 14.3. The dominant and subordinate chimpanzees are in cages separated by a room that they can both access. The room contains two occluders. Food was placed behind one of the occluders in full view of the subordinate. The experiment began when the door to the subordinate's cage was opened—at which time the dominant's door was closed. The door to the dominant's cage was only opened once the subordinate had entered the central room, so that the subordinate was not reacting to the dominant's behavior. Its decision whether or not to leave its cage had to be based only on its observations of what had gone on during the hiding process.

The first experiment directly tested the subordinate's sensitivity to the dominant's visual access to the food being hidden. It turned out that the subordinate went for food significantly more often in conditions when the dominant had not seen the food being hidden. And, in the cases where the dominant did see the food being hidden, the subordinate left the cage when the food was subsequently moved behind the other occluder without the dominant seeing. The ability of the subordinate chimpanzees to exploit their understanding of the dominant's visual access was further displayed in the second experiment, which tested how the subordinates would react when a dominant chimpanzee who had seen the food being hidden was replaced by another dominant who had not witnessed the baiting. Again, the subordinates went for food more often when they were competing against the dominant who had not seen the food being hidden.

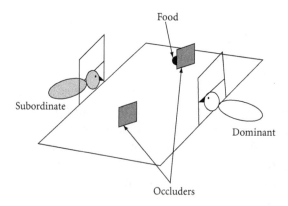

Fig. 14.3

As described, these two experiments can be interpreted in two ways. On one interpretation, the subordinates are engaged in propositional attitude mindreading. They go for the food when they think that the competitor does not know where the food is hidden, and they stay in their cage when they think that the competitor does know where the food is. In sum, their behavior is explained by how they represent the dominant's knowledge of the food's location. This interpretation is perfectly consistent with the data, but so too is an interpretation in which the subordinates are engaged in perceptual mindreading. In the perceptual interpretation, the subordinates go for the food when they think that the competitor did not see where the food is hidden. They stay in the cage when they think that the competitor saw the food being hidden.

Aside from the obvious difference that one invokes propositional attitude mindreading while the other invokes perceptual mindreading, the two interpretations differ in that the first involves the subordinates behaving on the basis of what they take to be the dominant's mental state at the moment that they (the subordinates) are choosing whether or not to go for the food. The propositional attitude mindreading interpretation has a subordinate representing the dominant's occurrent mental state. As a consequence it requires the subordinate to engage in a form of rudimentary practical reasoning about how the dominant will behave. A subordinate will need to reason that the dominant's knowledge of the food's location will lead it to go after the food itself. Why is this? Because the dominant's knowledge state is unobservable and so there are no observable connections between a subject's being in that mental state and the subject going after the food.

If what is going on is perceptual mindreading, then the subordinate can simply trade on its lengthy experience of the observable connection between perception and action—between a conspecific seeing where food is and going after it. But working out the connections between knowledge and action is a different order of cognitive achievement. As I tried to bring out in the argument for the language-dependence of propositional attitude mindreading, predicting how an organism will behave on the basis of its propositional attitudes is much less direct than using its perceptions to predict how it will behave. This type of prediction typically involves recreating the other organism's own practical reasoning. The two experiments so far described are not complex enough to distinguish between the two interpretations. One reason for this is that the task demands are too straightforward. On what basis could one decide whether the subordinate is reasoning about unobservable knowledge states or trading on observable connections between perception and action?

One way of working out whether the subordinate chimpanzee is engaging in practical reasoning is to increase the task demands in a way that can only be solved by practical reasoning. This is where the third experiment comes into the picture. In the third experiment two pieces of food were hidden. The subordinate saw both being hidden, but the dominant's visual access was manipulated. In the Uninformed condition, the subordinate and the dominant both saw one piece of food being hidden behind one occluder and then only the subordinate saw the second piece of food being hidden behind the other occluder. In the Misinformed condition, the subordinate and the dominant both saw two pieces of food being hidden behind one of the occluders and then, while the door to the dominant's cage was closed, the subordinate observed one of the pieces being moved to behind the other occluder.

The results of the third experiment were different from the results of the first two. The subordinate chimpanzees showed no statistically significant tendency to go after the pieces of food that the dominant was less likely to go after—that is, the ones that had either been hidden or moved without the dominant seeing. The experimenters offer some possible explanations for this, including a change in the incentive structure of the experiment (rendering it less competitive by having food available behind each of the two occluders) and information overload (the difficulty of keeping track of a conspecific's visual access to two different objects).

An alternative explanation, however, is that the cognitive demands of the third experiment are too complicated to be solved through perceptual mindreading. Successfully "outwitting" the dominant chimpanzee in either the Informed or the Misinformed condition requires more than simply trading on an observed connection between perception and action. In the first two experiments, the subordinate chimpanzee needs to come up with a Yes-No prediction—will the dominant go for the food or not? In the third experiment, it needs to predict something more complicated—which of the two locations will the dominant select? This prediction would require recreating the dominant chimpanzee's practical reasoning, which in turn would seem to involve much more complex thinking about the dominant's overall psychological profile than in either of the first two experiments.

If this alternative explanation is correct, it would give us a strong reason for thinking that the results of the first two experiments should be interpreted in terms of perceptual mindreading. In any event, though, those results are certainly consistent with the perceptual-mindreading interpretation. There is nothing here that would compel us to attribute propositional attitude mindreading to the subordinate chimps, and so no empirical threat to the earlier claim that propositional attitude mindreading does not exist in the animal kingdom (because propositional attitude mindreading requires linguistic abilities of a kind that do not exist in the animal kingdom).

4.4 How to Think about Understanding Intentions

Up to now, I have focused primarily on nonhuman animals' understanding of what might be termed informational states, such as perceptions, beliefs, and knowledge. But what about their understanding of goal-directed states, such as desires and intentions?

There is significant experimental evidence that different species of nonhuman animal can understand goal-directed states and the intentional actions to which they give rise. For a way of operationalizing this type of understanding, we can look to some much-discussed experiments reported by Gyorgy Gergely and his collaborators in *Nature* in 2002.[39] Gergely's group analyzed infants' understanding of goal-directed intentional action in terms of two distinguishable types of sensitivity—sensitivity to the goal and sensitivity to the means used to achieve the goal. Their insight was that understanding an action as an attempt to achieve a goal requires understanding the "fit" between means and end. In any successful goal-directed action, the means must be appropriate for achieving the end, given the constraints imposed by the environment and by the agent's own situation. For that reason, a good test of whether a mindreader understands goal-directed action is whether he or she is sensitive to that appropriateness.

Gergely tested this in fourteen-month-old infants by exploring whether infants' imitation of actions varied as the appropriateness of the behavior for achieving the required end varied. The infants were presented with what must surely have been a completely unfamiliar action—an experimenter leaning forward and using her head to illuminate a light box. The appropriateness of the action for the end was varied in two conditions. In the first condition, the experimenter's hands were occupied, so that using her head was the only way of achieving the goal of switching the light on. In the second condition, the experimenter's hands were empty and clearly in view at the side of the box. What Gergeley found was that infants imitated the head action much more often in the hands-occupied than in the hands-free condition—69% as opposed to 21%. In some sense, therefore, the infants seem to be sensitive to the appropriateness of the head action relative to certain environmental constraints, but not others.

It is no doubt complex to spell out quite what an understanding of appropriateness might consist in. We do not need to explore these complexities. What matters is simply that we have a workable operational criterion for assessing the capacity for understanding goal-directed actions—namely, imitation behaviors that are suitably sensitive to how means are determined by environmental constraints. In effect, Gergely's experiment gives an operational criterion for mindreading in this sphere. Subjects who display the appropriate sensitivity in their imitation behavior can plausibly be said to understand that the behavior they are imitating is goal-directed. This means that they are sensitive to the desire/intention driving the behavior.

Versions of the operational criterion developed by Gergely and his collaborators have recently been applied to chimpanzees, rhesus macaques, and cottontop tamarin monkeys by a group of researchers from Harvard and Boston Universities in an article published in *Science* in 2007.[40] In place of imitation, the experimenters used a version of the object-choice paradigm discussed in section 2.1 above. The apes were presented with two food containers. The experimenter signaled one of the containers and then the animal was allowed to inspect the containers. One set of experiments tested for the apes' ability to distinguish between different signals according to their appropriateness relative to environmental constraints. The experimenter used the unfamiliar action of touching a container with his elbow.

As with Gergely's original experiments, there were two conditions. In one condition, the experimenter's hand was occupied (the hand on the same arm as the elbow). In the second condition, the relevant hand was free. The results were comparable to the infant experiments. In the hand-empty condition the macaques and tamarins were only at chance in selectively inspecting the signaled container, while the chimpanzees were below chance. In the hand-occupied condition, in contrast, all species inspected the signaled container at levels significantly above chance (88% of trials for the macaques, 76% for the chimpanzees, and 69% for the tamarins). The authors interpret this experiment as follows:

> If these species take into account the environmental constraints facing the experimenter, then only the hand-occupied condition should be perceived as a rational, goal-directed action; given that the experimenter's acting hand was occupied at the moment of gesturing, his elbow provides an alternative means to both indicate and contact the target goal. Accordingly, the hand-empty condition would not be perceived as a rational, goal-directed action because, at the time, the experimenter could have used his unoccupied acting hand to grasp and indicate the target container, leaving the subject uncertain as to the target goal. Therefore, subjects should not infer that the experimenter's goal was to contact the box with the potentially concealed food.[41]

This interpretation seems plausible. If correct, it illustrates how these primates are capable of a distinctive type of mindreading. They are able to identify goal-directed actions and the desires and intentions that they manifest and reveal.

That nonhuman primates are capable of this type of mindreading is widely accepted among researchers into animal cognition. Tomasello and Call's 2008 survey of research into chimpanzee mindreading reviews ten different studies on chimpanzees' understanding of actions (not including the study just described, which was published after their article went to press). They conclude:

> We believe that there is only one reasonable conclusion to be drawn from the ten studies reviewed here: Chimpanzees, like humans, understand the actions of others not just in terms of surface behaviors but also in terms of the underlying goals, and possibly intentions, involved. Behavioral or contextual rules might be concocted to explain the results of any of the seven studies in which the chimpanzees react to or predict the behavior of others, but this requires many different ad hoc behavioral and contextual rules for which there is absolutely no positive evidence. Indeed, consistent use of this explanatory strategy would also deny human children an understanding of goals and intentions because most of the chimpanzee studies are modeled on child studies. Moreover, the three imitation studies would not seem favorable to behavioral-rules explanations at all.[42]

This type of mindreading may be very important for thinking about the moral considerability of nonhuman primates. The ability to recognize goal-directed behaviors opens up all sorts of possibilities for social coordination and interpersonal interactions that may prove to be morally significant.

However, for the purpose of this essay, I wish to focus on the question of how we should conceptualize this type of mindreading. In particular, what is it to understand desires and intentions? This question is particularly important given the earlier discussion of propositional attitude mindreading and the argument that this type of mindreading is language-dependent. Philosophers standardly take intentions and desires to be propositional attitudes, but the earlier argument clearly is incompatible with thinking of primate understanding of intentional action in these terms. So, is there an alternative way of thinking about intentions and desires?

In *Thinking Without Words*, I distinguished two ways of thinking about desire.[43] One can desire a particular thing, or one can desire that a particular state of affairs be the case. I termed this the distinction between *goal-desires* and *situation-desires*. When philosophers talk about desires as propositional attitudes, they are talking about situation-desires. We can appreciate this by looking at how the distinction is marked in how one might go about attributing desires. There are two different ways of completing the phrase "X desires ___" in order to obtain a sentence characterizing X's psychological state. In order to get a sentence ascribing a goal-desire, we would add the name of an object or the name of a kind of stuff (e.g., 'food'). In order to get a sentence ascribes a situation-desire, in contrast, we would need to add a "that ___" clause in which the blank is filled by a complete sentence specifying the state of affairs that X desires to obtain. This might be the state of affairs of there being food on the table, for example.

Situation-desires are propositional attitudes. To have a situation-desire is to desire that a particular proposition hold. The sentence following the "that___" clause in an attribution of a situation-desire picks out the proposition that is the object of the desire. Here understanding another agent's situation-desires involves metarepresentation. It is a matter of representing the proposition that is the object of their desire. But whereas situation-desires are directed at propositions, goal-desires are directed at objects, features, or properties. For this reason, understanding another agent's goal-desires is not a metarepresentational ability. Goal-desires are relations between a subject and objects/features/properties, rather than between a subject and a proposition. Understanding goal-desires is on a par with perceptual mindreading.

My claim is that we can conceptualize the type of mindreading involved in nonhuman animals' understanding of goal-directed actions in terms of the ability to represent another agent's goal-desires. Several of the studies we have already examined can be profitably interpreted in terms of an agent's understanding of goal-desires. Consider, for example, the competitive version of the false belief task. We saw that nonhuman primates consistently perform badly on this task. They fail to adjust their behavior to compensate for the difference between where the food reward is and where the competitor believes it to be. But, on the other hand, they do interpret the competitor's behavior as a cue to the location of the food reward (even if they do not exploit this cue very well). This interpretation depends upon their understanding that the competitor is motivated by the food reward, and hence upon an understanding of the competitor's goal-desire for the reward.

A creature capable of attributing goal-desires will be able to make the basic distinction between purposeful behaviors, on the one hand, and random movements and instinctive reactions on the other. A purposive action is an action for which a motivating goal-desire can be identified. And moreover, representations of goal-desires are rich enough to permit the type of reasoning about the appropriateness of means to end displayed in the experiments discussed above.

5. Conclusion: Nonlinguistic Folk Psychology

A complex picture of nonlinguistic mindreading is beginning to emerge. On the one hand, there are definite limits to the mindreading capabilities of nonhuman animals. Those types of mindreading that involve metarepresentation are, I have argued, language-dependent. Representing beliefs and other propositional attitudes (such as knowledge states) can only be done through a linguistic medium, for reasons explored in section 3. But, on the other hand, even taking into account the language-dependence of propositional attitude mindreading, it remains the case that nonhuman animals are capable of a rich psychological understanding of other subjects. This was apparent in the experiments on primate mindreading reviewed in section 4. There are sophisticated forms of social cognition that do not involve attributing propositional attitudes. As we have seen, primates can go a long way with perceptual mindreading and a basic understanding of purposive and goal-directed actions. In effect, primates have a nonlinguistic folk psychology.

There is an important lesson here for how we think about mindreading in the case we are most familiar with—the case of language-using humans. Most philosophical discussions of human mindreading assume that it is primarily based on propositional attitude mindreading. So, for example, many philosophers and cognitive scientists have claimed that human mindreading should be understood in essentially theoretical terms. On this view social cognition and social coordination rest upon tacit mastery of a folk psychological theory with generalizations connecting propositional attitudes and behavior. There are different ways of thinking about what this theory might look like when made explicit. On one approach, the fundamental folk psychological principle is the belief-desire law, according to which people can generally be relied upon to do what they think will best satisfy their desires. Construing folk psychology in this narrow way has the advantage of potentially being modeled by some form of expected utility theory. On a different approach, folk psychology is composed of a tacitly known network of principles that connect specific propositional attitudes with specific behaviors (to the effect that people with a given set of beliefs and desires will typically behave in certain ways).

Looking at the nonhuman case suggests that it may be unwise to focus exclusively on propositional attitude mindreading in the human case. Nonlinguistic mindreading is a powerful tool in the animal kingdom. It may well be that it explains far more of human social cognition and social coordination than philosophers and cognitive scientists standardly assume.[44] Any form of folk psychological understanding of another subject's behavior must have two central components. The first is an understanding of the information that the subject has about its environment, while the second is an understanding of what motivates the subject. On the propositional attitude model we should think about the first component in terms of identifying the subject's beliefs and about the second in terms of identifying situation-desires. Looking at nonhuman mindreading opens up the possibility of thinking about the first in terms of perceptual mindreading, and about the second in terms of identifying goal-desires.

The basic question this hypothesis poses is: How far can we get in making sense of the mindreading skills that underpin human social understanding and social coordination in terms of perceptual mindreading and goal-desires? This is a question of fundamental importance for how we think about the moral significance of nonhuman animals. For the reasons set out in section 1, there are close connections between mindreading and moral significance. And yet, if the argument of section 3 is sound, there is a basic type of mindreading that is unavailable to nonlinguistic creatures.

How significant is this conclusion for thinking about the moral significance of nonhuman animals? Plainly, it all depends on how central propositional attitude mindreading is to human social life. This issue raises a host of questions that cannot be addressed here. So I will close by simply recording my personal suspicion that we will not have a full understanding of the moral significance of nonhuman animals until we look at human mindreading through the lens of nonlinguistic folk psychology.

NOTES

1. On sentience, basic forms of consciousness, and the environment, see the essays in this *Handbook* by Huebner, Carruthers, Andrews, Greene, and Powell.

2. On capacities for different types of memory and making long- or short-range plans, see the essays in this *Handbook* by Huebner, Carruthers, Chan-Harris, Tooley, Rowlands, and Savulescu.

3. On the idea that moral significance might rest upon some form of self-awareness, see the essays in this *Handbook* by Andrews, Chan-Harris, Tooley, Carruthers, Morris, and DeGrazia.

4. D. Premack and G. Woodruff, "Does the Chimpanzee Have a Theory of Mind?" *Behavioral and Brain Sciences* 1 (1978): 515–26.

5. A number of the experiments initially provoked by Premack and Woodruff's study are discussed in section 10.3 of M. Tomasello and J. Call, *Primate Cognition* (Oxford: Oxford University Press, 1997). See also C. M. Heyes, "Theory of Mind in Nonhuman Primates," *Behavioral and Brain Sciences* 21 (1998): 101–34. For a more recent survey, see J. Call and M. Tomasello, "Does the Chimpanzee Have a Theory of Mind? 30 Years Later," *Trends in Cognitive Science* 12 (2008): 187–92.

6. C. A. Ristau. "Aspects of the Cognitive Ethology of an Injury-feigning Bird, the Piping Plover," in *Cognitive Ethology: The Minds of Other Animals: Essays in Honor of Donald R. Griffin*, ed. C. A. Ristau (Hillsdale, N.J.: Laurence Erlbaum, 1991).

7. R. W. Byrne and A. Whiten, "Tactical Deception in Primates: The 1990 Database," *Primate Report*, Whole Volume 27 (1990): 1–101.

8. R. W. Byrne and A. Whiten, "Cognitive Evolution in Primates: Evidence from Tactical Deception," *Man* 27 (1992): 609–27, here 612.

9. R. W. Byrne and A. Whiten, eds. *Machiavellian Intelligence: Social Expertise and the Evolution of Intellect in Monkeys, Apes, and Humans* (Oxford: Oxford University Press,

1988); R. W. Byrne and A. Whiten, *Machiavellian Intelligence II: Extensions and Evaluations* (Cambridge: Cambridge University Press, 1997).

10. J. I. Flombaum and L. R. Santos, "Rhesus Monkeys Attribute Perceptions to Others," *Current Biology* 15 (2005): 447–52.

11. S. Held, M. Mendl, C. Devereux, and R. W. Byrne, "Behaviour of Domestic Pigs in a Visual Perspective Taking Task," *Behavior* 138 (2001): 1337–54.

12. N. J. Emery and N. S. Clayton, "Effects of Experience and Social Context on Prospective Caching Strategies in Scrub Jays," *Nature* 414 (2001): 443–46.

13. J. Call, J. Brauer, J. Kaminski, and M. Tomasello, "Domestic Dogs (*Canis familiaris*) are Sensitive to the Attentional States of Humans," *Journal of Comparative Psychology* 117 (2003): 257–63.

14. T. Bugnyar and B. Heinrich, "Food-storing Ravens Differentiate between Knowledgeable and Ignorant Competitors," *Proceedings Royal Society London Series B* 272 (2005): 1641–46.

15. A. A. Pack and L. M. Herman, "Dolphin Social Cognition and Joint Attention: Our Current Understanding," *Aquatic Mammals* 32 (2006): 443–60.

16. P. F. Strawson, *Individuals* (London: Methuen, 1959); G. Evans, *Varieties of Reference* (Oxford: Oxford University Press, 1982), J. L. Bermúdez, *Thinking Without Words* (New York: Oxford University Press, 2003).

17. Tomasello and Call, *Primate Cognition*, p. 337

18. R. E. Miller, J. Banks, and N. Ogawa, "Role of Facial Expression in 'Cooperative Avoidance Conditioning' in Monkeys," *Journal of Abnormal and Social Psychology* 67 (1963): 24–30.

19. S. Mineka and M. Cook, "Observational Conditioning of Fear to Fear-Relevant versus Fear-Irrelevant Stimuli in Rhesus Monkeys," *Journal of Abnormal Psychology* 9 (1988): 448–59.

20. B. Hare, M. Brown, C. Williamson, and M. Tomasello, "The Domestication of Social Cognition in Dogs," *Science* 298 (2002): 1634–36.

21. D. Davidson, "Thought and Talk," in *Mind and Language*, ed. S. Guttenplan (Oxford: Oxford University Press, 1975).

22. J. Heil, *The Nature of True Minds* (Cambridge: Cambridge University Press, 1992).

23. Bermúdez, *Thinking Without Words*, chap. 9, and most recently in J. Bermúdez, "Mindreading in The Animal Kingdom," in *The Philosophy of Animal Minds*, ed. R. Lurz (Cambridge: Cambridge University Press, 2009).

24. R. W. Byrne and A. Whiten, "Tactical Deception of Familiar Individuals in Baboons (*Papius ursinus*)," *Animal Behaviour* 33 (1985): 669–72, here 671.

25. Byrne and Whiten, "Tactical Deception of Familiar Individuals in Baboons," p. 671.

26. Report by Hans Kummer quoted in R. W. Byrne, *The Thinking Ape: Evolutionary Origins of Intelligence* (Oxford: Oxford University Press, 1995), p. 106.

27. D. J. Povinelli and T. J. Eddy, "What Young Chimpanzees Know about Seeing," *Monographs of the Society for Research into Child Development* 61 (1996).

28. S. Okamoto, M. Tomonaga, L. Shii, N. Kawai, M. Tanaka, and T. Matsuzawa, "An Infant Chimpanzee (*Pan troglodytes*) Follows Human Gaze," *Animal Cognition* 5 (2002): 107–14.

29. M. Tomasello, B. Hare, and B. Agnetta, "Chimpanzees Follow Gaze Direction Geometrically," *Animal Behaviour* 58 (1999): 769–77.

30. J. Brauer, J. Call, and M. Tomasello, "All Great Ape Species Follow Gaze to Distant Locations and Around Barriers," *Journal of Comparative Psychology* 119 (2005): 145–54.

31. N. J. Emery, E. N. Lorincz, D. I. Perrett, M. W. Oram, and C. I. Baker, "Gaze Following and Joint Attention in Rhesus Monkeys (*Macaca mulatta*)," *Journal of Comparative Psychology* 111 (1997): 286–93.

32. H. Wimmer and J. Perner, "Beliefs about Beliefs: Representation and Constraining Function of Wrong Beliefs in Young Children's Understanding of Deception," *Cognition* 13 (1983): 103–28.

33. S. Baron-Cohen, A. M. Leslie, and U. Frith, "Does The Autistic Child Have a 'Theory Of Mind'?" *Cognition* 21 (1985): 37–46.

34. J. Call and M. Tomasello, "A Nonverbal Theory of Mind Test: The Performance of Children and Apes," *Child Development* 70 (1999): 381–95.

35. C. Krachun, M. Carpenter, J. Call, and M. Tomasello, "A Competitive Nonverbal False Belief Task for Children and Apes," *Developmental Science* 12 (2009): 521–35.

36. K. H. Onishi and R. Baillargeon, "Do 15-Month-Old Infants Understand False Beliefs?" *Science* 308 (2005): 255–58.

37. Call and Tomasello, "Does the Chimpanzee," p. 190.

38. B. Hare, J. Call, and M. Tomasello, "Do Chimpanzees *Know* What Conspecifics *Know*?" *Animal Behaviour* 61 (2001): 139–51.

39. G. Gergely, H. Bekkering, and I. Király, "Rational Imitation in Preverbal Infants," *Nature* 415 (2002): 755.

40. J. N. Wood, D. D. Glynn, B. C. Phillips, and M. D. Hauser, "The Perception of Rational, Goal-Directed Action in Nonhuman Primates," *Science* 317 (2007): 1402–5. After serious question were raised about the integrity of the experiments reported in this paper, the journal *Science* published online a replication with full documentation by Justin Woods and Marc Hauser. See http://www. sciencemag.org/content/suppl/2011/04/25/317.5843.1402.DC2

41. Wood, Glynn, Philipps and Hauser, "The Perception of Rational," p. 1404.

42. Call and Tomasello, "Does the Chimpanzee," p. 189.

43. See Bermúdez, *Thinking Without Words*, pp. 48–49.

44. For further discussion of alternatives to propositional-attitude-based models of folk psychology, see chapter 7 of J. L. Bermúdez, *Philosophy of Psychology: A Contemporary Introduction* (London: Routledge, 2005).

SUGGESTED READING

ALLEN, C., and M. BEKOFF. *Species of Mind: The Philosophy and Biology of Cognitive Ethology*. Cambridge, Mass.: MIT Press, 1997.

BERMÚDEZ, J. L. *Thinking Without Words*. New York: Oxford University Press, 2003.

BERMÚDEZ, J. L. "Thinking Without Words: An Overview for Animal Ethics." *Journal of Ethics* 11 (2007): 319–35.

BRAUER, J., CALL, J., and M. TOMASELLO. "All Great Ape Species Follow Gaze to Distant Locations and around Barriers." *Journal of Comparative Psychology* 119 (2005): 145–54.

BYRNE, R. W. *The Thinking Ape*. Oxford: Oxford University Press, 1995.

CALL, J., and M. TOMASELLO. "A Nonverbal False Belief Task: The Performance of Children and Great Apes." *Child Development* 70 (1999): 381–95.

EMERY, N. J., LORINCZ, E. N., PERRETT, D. I., ORAM, M. W., and C. I. BAKER. "Gaze Following and Joint Attention in Rhesus Monkeys (*Macaca mulata*)." *Journal of Comparative Psychology* 111 (1997): 286–93.

HARE, B., M. BROWN, C. WILLIAMSON, and M. TOMASELLO. "The Domestication of Social Cognition in Dogs." *Science* 298 (2002): 1634–36.

HARE, B., and M. TOMASELLO. "Do Chimpanzees Know What Conspecifics Know?" *Animal Behavior* 61 (2001): 139–51.

HAUSER, M. D. *Wild Minds: What Animals Really Think.* London: Penguin Books, 2000.

HEYES, C. M. "Theory of Mind in Nonhuman Primates." *Behavioral and Brain Sciences* 21 (1998): 101–34.

HURLEY, S., and M. NUDDS, eds. *Rational Animals?* Oxford: Oxford University Press, 2006.

OKAMOTO, S., M. TOMONAGA, K. ISHII, N. KAWAI, M. TANAKA, and T. MATSUZAWA. "An Infant Chimpanzee (*Pan troglodytes*) Follows Human Gaze." *Animal Cognition* 5 (2002): 107–14.

POVINELLI, D. "Chimpanzee Theory of Mind." In *Theories of Theories of Mind*, edited by P. Carruthers and P. K. Smith. Cambridge: Cambridge University Press, 1996.

POVINELLI, D., and J. VONK. "We Don't Need a Microscope to Explore the Chimpanzee's Mind." In *Rational Animals?* edited by S. Hurley and M. Nudds. Oxford: Oxford University Press, 2006.

PREMACK, D., and G. WOODRUFF. "Does the Chimpanzee Have a Theory of Mind?" *Behavioral and Brain Sciences* 1 (1978): 515–26.

TOMASELLO, M., and J. CALL. *Primate Cognition.* Oxford: Oxford University Press, 1997.

———. "Do Chimpanzees Know What Others are Seeing—Or Only What They Are Looking At?" In *Rational Animals?* edited by S. Hurley and M. Nudds. Oxford: Oxford University Press, 2006.

———. "Does the Chimpanzee Have a Theory of Mind? 30 Years Later." *Trends in Cognitive Science* 12 (2008): 187–92.

WHITEN, A., and R. W. BYRNE, ed. *Machiavellian Intelligence II.* Cambridge: Cambridge University Press, 1997.

CHAPTER 15

MINIMAL MINDS

BRYCE HUEBNER

We sometimes speak, in a "loose and popular" way, as though viruses have human-like beliefs and desires. For example, a recent article in *Time Magazine* that detailed our understanding of the HIV virus suggested that,

> All the virus wants to do is break inside a healthy cell, steal its genetic machinery, and start profiting from the intrusion. To stop a thief, you need to throw a monkey wrench—or several—into his plans. That's exactly what anti-HIV drugs, known as antiretrovirals (ARVs) are designed to do.[1]

It comes as no surprise when such intentional idioms occur in the hyperbolic prose of the mainstream media. However, it is more surprising when scientific discourse relies on anthropomorphizing claims when discussing the nature of viruses. In recent papers, it has been suggested that if the HIV "virus wants to resist AZT, it needs to make a specific mutation at codon 215 and at another position, such as position 41"[2] and that the Hepatitis C virus must try to coexist with its host by reducing its visibility.[3] Yet, philosophical reflection suggests that although we *can* use intentional idioms whenever we find an entity that exhibits robust and systematic patterns of behavior, a plausible theory of mentality should not treat viruses as even minimally minded. After all, viruses are nothing more than packets of RNA (or sometimes DNA) encased in protein. So, although they are likely to be mechanically complex, the type of complexity that we find in viruses is unlikely to implement the computational and representational capacities that are necessary for being in cognitive states or carrying out cognitive processes. Viruses are not the kind of entities that can *want to* threaten our wellbeing; they do not *make plans* or *adopt deceptive strategies* for navigating our immune systems; and they have not *learned* to outsmart our best antivirals. Although many viruses exhibit systematic and predictable patterns of behavior, we can and should treat them as complex biological robots.[4] However, difficult theoretical issues arise in providing a more general strategy for distinguishing genuine mentality from witless mechanical behavior.

In this chapter, my goal is to establish the *minimal conditions* that must be satisfied for an entity to be in cognitive states or carry out cognitive processes. I begin with an examination of the philosophical assumptions that underlie the Cartesian assertion that linguistic capacities are necessary for genuine mentality. I argue that the explanatory resources provided by contemporary cognitive science undercut the force of these assumptions and provide compelling reasons for thinking that even invertebrates (e.g., crickets, cockroaches, termites, and honeybees) possess genuine cognitive capacities. While this claim may seem familiar, I extend the arguments for treating invertebrates as possessing *minimal minds* in two ways. First, I argue that minimal mindedness is likely to be realized by computations that operate over nonlinguistic representations; thus, minimal cognitive states and processes are unlikely to be beliefs or desires. Second, I argue that the theoretical and empirical evidence that supports treating invertebrates as minimally minded also warrants the more intriguing conclusion that some sorts of *collective behavior* are best explained in terms of collective mental states and processes; hence, I argue that if invertebrates have minimal minds, then so do some *groups* of invertebrates.

1. Minds Like the Human Mind

Our commonsense strategies for making claims about the mental lives of other entities are a heterogeneous mixture of genuine ascriptions of mental states, ungrounded anthropomorphic assertions, and idiosyncratic expressions of unjustified theories. This makes it difficult to know how to proceed in addressing ontological and epistemological questions about other kinds of minds. As Bertrand Russell rightly notes:

> We observe in ourselves such occurrences as remembering, reasoning, feeling pleasure, and feeling pain. We think that sticks and stones do not have these experiences, but that other people do. Most of us have no doubt that the higher animals feel pleasure and pain, though I was once assured by a fisherman that "Fish have no sense nor feeling." I failed to find out how he had acquired this knowledge. Most people would disagree with him, but would be doubtful about oysters and starfish. However this may be, common sense admits an increasing doubtfulness as we descend in the animal kingdom, but as regards human beings it admits no doubt.[5]

However, the problem is not merely that we become less sure about the existence of other kinds of minds "as we descend in the animal kingdom." More importantly, the diverse range of strategies that we deploy in ascribing mental states and processes also yields deep theoretical disagreements across philosophical traditions and scientific methodologies. Indeed, there is no consensus about how to draw the distinction between genuine mentality and witless mechanical behavior.

Even if we must begin from commonsense presumptions about what minds really are, a philosophically respectable theory of the nature of the mind must also triangulate these commonsense presumptions against a wide variety of scientific data and philosophical theories about the nature of mental states and processes.[6] Where commonsense intuitions about some cognitive states or processes conflict with a philosophical or scientific theory, these intuitions provide defeasible evidence against the truth of that theory. However, commonsense ascriptions of mental states and processes are also anthropomorphic in a way that licenses treating all entities that behave in an apparently intentional manner "as if they were just like us—which of course they are not."[7] So, developing an adequate understanding of what it means to be minded, and taking the results of scientific inquiry seriously, may sometimes require deep revisions to our commonsense understanding of mentality.

To make this point clear, consider a case from cognitive ethology.[8] Marc Hauser once observed a dominant male macaque who attacked a subordinate male for mating with a high-ranking female. Hauser reports that the subordinate male remained quiet as the dominant monkey ripped off one of his testicles, never even exhibiting a facial expression indicative of pain. More surprisingly, the subordinate was again attempting to mate with a high-ranking female after little more than an hour. Although we readily assume that monkeys experience pain in some way that is relevantly similar to the pain felt by humans, such examples suggest that we should not assume that this is the case without further evidence to support this intuition. In short, a plausible account of macaque pain must draw upon commonsense intuitions about what pain is, but it must then utilize careful ethological observations, neurological data, and evolutionarily informed theories about the relationship of these neural structures to those responsible for implementing the human experience of pain.

Still, Russell is right to note that our minds are the only ones about which we have knowledge first-hand. Whenever we ask whether an invertebrate feels the sort of pain that warrants our compassion, or remembers where it has stored food, we are asking—even if only implicitly—whether it has memories or feels pains that are like our own in some significant respect; and this fact opens up a philosophical puzzle about the nature of minimal minds: in what respect can an invertebrate (e.g., a fly, a honeybee, or a spider) be in mental states or carry out mental processes that are enough like ours to justify the claim that it has a mind?

One answer to this question adopts the conservative suggestion that *genuine* mentality requires the capacity for linguistic representation. From this perspective, invertebrates cannot be in mental states or carry out mental processes, and neither can cats or dogs. This may initially seem implausible. However, the presence of robust linguistic capacities in *Homo sapiens* yields an enormous increase in behavioral flexibility and cognitive sophistication well beyond what can be exhibited by any nonhuman animal. Although there is little consensus about the precise role that is played by language in human cognition, maximal minds like our own exhibit numerous cognitive capacities that are deeply embedded in linguistic and social practices. At bare minimum, the capacity to use language augments and transforms the cognitive strategies that are employed to remember the past, coordinate future actions (both individually and collectively), deliberate about counterfactual possibilities, and form fine-grained beliefs and desires about the world in which we live.[9] Moreover, language underwrites our capacities for reasoning together and engaging in collaborative activities in ways that are not likely to be present among the members of any other species.

For these reasons, our ascriptions of mental states and processes to other humans also rely on a practice of cooperative mind-shaping that allows us to establish and maintain strategies for cooperative engagements, a practice that we cannot adopt in making sense of nonlinguistic minds.[10] Consider the mental states and processes that are ascribed during a collaborative writing project. Such writing often takes place without considering what a collaborator believes or desires; but, where a branchpoint suggests no clear agreement about how to proceed, it is often necessary to project the sorts of beliefs and desires that a collaborator is likely to have formed in light of the way that the world has been disclosed to her. For example, in the face of a slight conflict over theoretical commitments, it is often necessary to draw inferences about the judgments that a coauthor is likely to express regarding a certain lab's results, or about the theoretical tools that should be applied in addressing an issue. However, adopting this interpretive strategy is rarely a matter of attempting to predict or explain a collaborator's future behavior.[11] Instead, this strategy is typically adopted to recalibrate beliefs and desires by giving and asking for reasons, in hopes of retrieving a shared set of background assumptions that can underwrite further collaboration.

We can and do stretch these interpretive strategies to include human beings who live in radically different situations from our own. For example, when I have a conversation with a homeless person in my neighborhood, I ascribe a variety of mental states and processes over the course of the conversation in an attempt to negotiate a shared understanding of his world—for the world has been disclosed to him in a radically different way as a result of his living situation. However, as we move even further away from our own situation, it becomes harder to adopt this strategy. Even highly trained dogs who spend every day with a human being acquire only a narrowly constrained ability to communicate and never develop the capacity to flexibly respond to the *meaning* of what is said to them. (I return to this worry in section 2.1.)

Ascribing mental states and processes to even the most familiar nonhuman animals, thus, requires a different kind of mentalizing strategy. In attempting to understand the behavior of a nonhuman, nonlinguistic animal, we adopt an interpretive and explanatory strategy that treats their mental states and processes as functionally specified "black boxes." The existence of these black boxes is inferred from robust and predictable behavioral dispositions. While the supposition that an entity utilizes such states and processes is often useful for explaining its dispositions to behave in predictable ways, this thin strategy for ascribing mental states and processes also ushers in a host of theoretical difficulties. Everything from self-replicating macromolecules, to pine trees, thermostats, amoebas, plants, rats, bats, people, and chess-playing computers *can* be treated as an intentional system that has mental states and processes of this sort.[12] However, little philosophical ground can be gained by demonstrating that flies, honeybees, and spiders are minded in the same way as a pine tree that can be tricked into "believing that it is summer" by lighting a fire beneath its boughs. So we must look deeper to find a theoretically and empirically plausible foundation for distinguishing between genuine cognition and witless mechanical behavior.

2. MAXIMAL CARTESIAN MINDS

In the opening pages of the *Principles of Psychology*, William James argues that although the boundary between genuine cognition and witless mechanical behavior seems vague on first blush, there is a plausible intuitive test for determining whether a nonhuman entity has genuinely cognitive capacities. James extracts the details of this test by contrasting the behavior of soap bubbles blown through a straw into a pail of water and the behavior of a frog who is placed at the bottom of the same pail. Although the bubbles rise to the surface in a way that can be "poetically interpreted as a longing to recombine with the mother-atmosphere above," they continue to display this behavior even when a glass dome stands in the way of their "goal."[13] However, although the frog will also swim upward out of a desire to reach the mother-atmosphere above, "if a jar full of water be inverted over him, he will not, like the bubbles, perpetually press his nose against its unyielding roof, but will restlessly explore the neighborhood until he has discovered a path around its brim" to reach the desired goal.[14]

James contends that for the frog the goal remains fixed and the means of achieving this goal are varied in light of the salient environmental contingencies of a particular situation. This fact, he claims, justifies the claim that the frog mentally represents its goal in some way. Building from this case, James argues that the "pursuance of future ends and the choice of means for their attainment, are thus the mark and criterion of the presence of mentality."[15] However, in spite of James's protestations to the contrary, it is not easy to know whether frogs, flies, honeybees, and spiders possess the representational capacities that would allow them to satisfy these criteria, or whether they merely behave as if they do. If mental states and processes must pick out "real, intervening internal states or events, in causal interaction, subsumed under covering laws of causal stripe," then something could hop like a frog and croak like a frog, without really thinking like a frog.[16] In short, adopting methodological principles that are too permissive risks leading us to posit mental states and processes where nobody is home. So, how can we draw a sharp line between minded and nonminded entities?

2.1 Cartesian Arguments

René Descartes famously argues for a fundamental and unbridgeable gap between genuine mentality and the merely mechanical activity of nonhuman entities. He claims that the capacity to use language provides the only evidence that an organism's *goal-directed* behavior is not fully determined by its design, and he argues that the ability to use a public language cannot be explained merely by appeal to the operation of complex physical mechanisms. In this section, I briefly examine each of these claims.

Descartes' argument relies on a version of the Principle of Sufficient Reason—the claim that for everything that is the case, it must be possible to provide a causal explanation or reason *why* it is the case.[17] Descartes' version of this principle suggests that an explanation of why a sandbar occurs at a particular point in a river is provided in terms of the structural properties of the grains of sand and the mechanistic principles governing their composition and aggregation.[18] Because Descartes also views biological bodies as complex machines, he advances a similar strategy for describing the behavior of biological organisms.[19] Briefly, to give an adequate explanation of a biological organism's behavior, it is sufficient to appeal to the organization of complex biological mechanisms, which are composed of bones, muscles, nerves, arteries, veins, and "animal spirits," and to explain how they are designed to carry out particular tasks in particular environments. With this model of explanation in hand, Descartes claims that the "high degree of perfection displayed in some of their actions makes us suspect that animals do not have free will."[20]

According to this Cartesian view, we can exhaustively explain why it is difficult to swat a fly in mid-air by treating the fly as a complex machine that relies on a fixed set of action patterns to rapidly change directions when it is faced with a looming object. Ascribing free will or attempting to work out what the fly believes about the looming object would seem to provide no additional explanatory advantage. Similarly, we can explain why cockroaches are efficient at scuttling away when the lights come on by appeal to the fixed action patterns that allow them to monitor the structure of their immediate environment and rapidly change directions without re-presenting anything in a cognitively significant way. These invertebrates are "designed" to navigate particular environmental contingencies. Where such design-based explanations prove successful, they seem to preempt appeals to intentional explanations. So, even if they often seem to outwit us in their native environments, we should not take this to be a matter of thought.

Descartes argues that while such mechanistic explanations always suffice for claims about the behavior and structure of complex physical bodies, they leave something to be desired in the case of human behavior. Here, we must adduce an additional reason or cause to explain the capacities that derive from will and judgment. Descartes contends that, as rational agents, we possess irreducible capacities for deliberating and withholding judgment, and although we do not have control over every behavior we exhibit, we experience ourselves as possessing the capacity to freely will and consciously control at least some of our actions. These facts call for an explanation, and since mechanistic principles could only explain human behavior by explaining away our capacity to freely will and consciously control our own actions, Descartes contends that a sufficient explanation of these capacities must advert to the existence of a thinking substance that obeys rational, as opposed to physical, laws. In light of this argument, Descartes raises well-known epistemic worries about the existence of other human minds by suggesting that the bodies crossing a square *could be* mere automata.[21] Since we cannot tell from the outside whether these bodies possess minds, we must find some behavior that can only be explained by adverting to the presence of a thinking substance.

To resolve this worry, Descartes proposes a proto-Turing test, suggesting that the capacity to flexibly respond, in language, to the meaning of any unexpected question provides incontrovertible evidence that a mind is associated with a body.[22] Of course, Descartes acknowledges that an incredibly complex biological machine could be constructed so as to exhibit a finite and situationally constrained capacity to use particular words and expressions in response to a specified range of stimuli. However, he claims, it is inconceivable "that such a machine should produce different arrangements of words so as to give an appropriately meaningful answer to whatever is said in its presence, as even the dullest of men can do."[23] After all, over the course of an ordinary conversation, we encounter everything from jokes and sarcasm, to unexpected communicative breakdowns, absurd suggestions that require further elaboration, and even semantically ambiguous sentences that make sense only in light of shared background assumptions. Our flexible and open-ended capacity for thought allows us to seamlessly deploy a variety of cognitive resources that help us to reason through such situations, and the fact that we can reestablish a shared ground for communicating in light of such breakdowns does seem to provide us with incontrovertible evidence that other people have context-sensitive capacities for thought just like our own. Descartes contends that the capacity to produce context-sensitive representations in a public language suggests an unbridgeable gap between genuine cognition and the fixed action patterns displayed by non-human animals, and on this basis concludes that language provides the only clear way of distinguishing intentional action from witless and mechanical behavior that is exhaustively explained in terms of the design of some machine.[24]

Of course, this does not establish that these capacities for genuine mentality cannot be constructed from simpler, less sophisticated parts. However, Descartes suggests that any attempt to articulate similarities between the capacities of human beings and those of nonhuman animals will find that "the actions of the brutes resemble only those which occur in us without any assistance from the mind." He continues:

> when the swallows come in spring, they operate like clocks. The actions of honeybees are of the same nature; so also is the discipline of cranes in flight, and of apes in fighting, if it is true that they keep discipline. Their instinct to bury their dead is no stranger than that of dogs and cats which scratch the earth for the purpose of burying their excrement; they hardly ever actually bury it, which shows that they act only by instinct and without thinking. The most that one can say is that though the animals do not perform any action which shows us that they think, still, since the organs of their bodies are not very different from ours, it may be conjectured that there is attached to these organs some thought such as we experience in ourselves, but of a very much less perfect kind. To this I have nothing to reply except that if they thought as we do, they would have an

> immortal soul like us. This is unlikely, because there is no reason to believe it of some animals without believing it of all, and many of them such as oysters and sponges are too imperfect for this to be credible.[25]

In other words, since simple invertebrates like sponges and oysters possess mechanistic capacities to successfully navigate their environments, we must either suppose (1) that they have minds of a sort that cannot be explained in purely mechanistic terms; (2) that some high degree of mechanical complexity is sufficient for genuine mentality; or (3) that all nonhuman behavior can be exhaustively explained in mechanistic terms. Descartes rules out (1) by claiming that it is implausible to suppose that sponges and oysters have the capacities for will and judgment that are indicative of minds—as organisms, they are too simple to exercise conscious control over their behavior. He rules out (2), if only implicitly, by suggesting that any moderate criterion for ascribing mentality to nonhuman entities will be a matter of arbitrary line drawing. He sees no principled reason why we must draw the boundary between genuinely cognitive and merely mechanical systems in one place rather than another; however, his commitment to the Principle of Sufficient Reason entails that there must be a principled reason why some behavior can be explained mechanistically while other behavior cannot. The only distinction that Descartes sees as tenable is the capacity for fluidly and flexibly responding, in language, to the meaning of what is said. Put simply, we must either claim that the behavior of oysters, sponges, and human beings can all be exhaustively explained in mechanistic terms, or we must adopt the restrictive claim that treats our linguistic capacities as providing the only incontrovertible evidence that an entity is minded.

2.2 Where Does the Cartesian Argument Go Wrong?

Descartes' avowedly materialist contemporaries saw that his arguments rested on problematic metaphysical presuppositions. Thomas Hobbes argued that *deliberation* and *will* were nothing more than *capacities* to manipulate internal representations, and because the capacity to represent an "alternate Succession of Appetites, Aversions, Hopes and Fears, is no lesse in other living Creatures then [*sic*] in Man," he thought it was obvious that nonhuman animals had these capacities as well.[26] Similarly, Baruch Spinoza argued that people fall prey to Cartesian assumptions about the mind because they have not "learned from experience what a body can and cannot do, without being determined by mind."[27] He argued that although commonsense psychology searches for purpose in the world, "final causes are but figments of the human imagination."[28] Causal and natural forces govern the behavior of all bodies; so, human minds are governed by the same laws that govern the mind of the snail.[29]

The naturalistic worldview of the cognitive sciences takes up this anti-Cartesian perspective, suggesting that the complexity and sophistication of an entity's behavior must be explicable in terms of interaction and aggregation of simpler, mechanistic systems. These presumptions derive from two "inversions of reason" that emerged in the nineteenth and twentieth centuries. Charles Darwin's opponents aptly captured the first inversion of reason, noting that "in order to make a perfect and beautiful machine it is not requisite to know how to make it."[30] Indeed, the mechanisms of natural selection explain how evolution can witlessly produce complex organisms out of uncomprehending parts. But mental states and processes should not be seen as a "lone exception" to the paradigm of mechanistic explanation.[31] Behavioral and cognitive capacities must also be explained by demonstrating that entities that possessed this capacity were more likely to survive and reproduce in their natural environments.

This brings us to the second inversion of reason, Alan Turing's argument that a mechanistic explanation of information processing can be given that does not advert to mechanisms of top-down or intentional control. Just as evolution by natural selection replaces the trickle-down theory of design by divine will, the theory of computation displaces the assumption that a computational device must understand the meaning of the symbols it manipulates to compute a function. In short, Turing provided a theoretical apparatus that can explain how computations can be carried out by a system that consists exclusively of uncomprehending and mechanistic components.

Together, these inversions of reason suggest plausible resources for resisting Descartes' arguments. Rather than assuming discontinuity between the physiological capacities exhibited by human and nonhuman animals, we should begin from the methodological assumption that every behavioral or cognitive *capacity* can be explained mechanistically. Complex physical structures emerge from simpler structures though a process of modification by natural selection; and as Turing argues, even the *capacity* for flexible linguistic engagement can be given a mechanistic explanation. Thus, psychological explanation, like all mechanistic explanation, must begin by decomposing complex cognitive systems into simpler parts, and repeating this process until a level of explanation is reached that requires nothing more than the equivalent of on-off switches. Such explanations look "backward" to an organism's evolutionary history, and "downward" to the underlying computational architecture that implements its capacities to behave in various ways.

Thus, genuine mentality is always *implemented* by some mechanically complex computational system, and simpler minds are merely simpler mechanistic systems. So, extending mentality to invertebrates is not as dire a response as Descartes supposed, at least so long as a plausible boundary can be drawn between genuine cognition and witless mechanical behavior. Turing's suggestion is that the capacity to utilize representations in the production of goal-directed behavior serves to mark this boundary. However, to undercut the force of Descartes' arguments, it is necessary to offer a fuller theory of minimal mentality and to demonstrate that representational capacities provide a plausible strategy for distinguishing genuine cognition and mere mechanistic activity. I turn to these tasks in section 3.

3. A Theory of Minimal Minds

As I noted above, the interpretive strategy that we adopt in describing intentional behavior can be readily deployed even where a simple entity displays nothing more than a fixed-action pattern (e.g., in describing the behavior of a virus). With this in mind, both cognitive ethologists and philosophers of cognitive science have tended to adopt a conservative perspective on nonhuman minds. Specifically, they have tended to adopt "Morgan's Canon," a special application of Occam's razor according to which the behavior of a nonhuman animal is not "to be interpreted in terms of higher psychological processes, if it can be fairly interpreted in terms of processes which stand lower in the scale of psychological evolution and development."[32] I too agree that we should attempt to avoid the empty anthropomorphism that arises in a careless deployment of the intentional stance; treating viruses and thermometers as believers is at best crude behaviorism, and at worst flatly misleading. Thus, we need a more careful philosophical analysis of the problem of minimal minds.

Starting near the bottom of the phylogenetic tree, we find that all eukaryotic cells (i.e., cells that contain a distinct membrane-bound nucleus) possess the capacity to flexibly arrange and rearrange their internal structure in response to the current state of their environment.[33] Tecumseh Fitch has recently argued that eukaryotes, including single-celled bacteria, have the capacity to flexibly respond to novel circumstances in ways that are not explicitly encoded in their DNA, using trial-and-error methods to discover new responses to novel difficulties, and recording these discoveries for future use. A simple example of this biological capacity is the healing response: "a damaged organism can often stem the loss of precious bodily fluids, stitch itself up, and (with some scar perhaps) continue living."[34] However, the possession of this capacity is insufficient to license a claim of genuine mentality. It only shows that simple organisms are capable of engaging in some degree of semi-autonomous functional reorganization. Because there is no "dedicated information-processing machinery even vaguely equivalent to a vertebrate nervous system," it doesn't make sense to say of bacteria that they have mental circuits of the sort that would allow them to represent the world in any way whatsoever.[35] In short, bacterial cells possess the capacity to record the strategies that they have used in responding to their environment; however, these recordings are fixed by particular interactions with particular bodies. Every interaction requires a new attempt at functional reorganization, because bacteria are not internally complex enough to re-deploy previously honed skills in coping with novel situations.

From a Darwinian perspective, such capacities are likely to be necessary preconditions for the emergence of the flexible "strategies" that are required to achieve even the simplest of goals (e.g., feeding and mating). However, simple bacteria do nothing more than take in sensory inputs, and witlessly and mechanically translate them into the sorts of internal changes that allow them to survive (another day, another hour, or another minute). Given the Darwinian history that structures their pattern of responses, we can truly say that these bacteria are not fully designed from birth. There are elements of their design that can be adjusted by the events that occur over the course of their interactions with their environment.[36] However, these adjustments do not *mean* anything to the bacteria themselves. Genuine cognition requires enough mechanical complexity to represent the world in a way that can facilitate being in cognitive states and carrying out cognitive processes.

Crude behaviorism and empty instrumentalism do little more than operationalize "belief" and "desire," radically redefining these terms to the point that they share little in common with the familiar states and processes of other human minds. Empty anthropomorphism, by contrast, threatens to deploy our familiar mental terminology in cases where the use of such terms is completely unwarranted. To avoid these worries, discussions of other kinds of minds often begin by providing relatively conservative criteria for discriminating genuine cognition from witless mechanistic behavior. For example, even the most strongly naturalist positions on the existence of invertebrate minds, such as that of Peter Carruthers, depend on the claim that a minimally minded entity must have "distinct belief states and desire states that are discrete, structured, and causally efficacious in virtue of their structural properties"; these beliefs and desires must be "construed realistically," and because these states must "be both discrete, and structured in a way that reflects their semantic contents," it has been claimed that genuine mental states and processes must be implemented in a system of symbols that obey systematic transformational rules (e.g., in a language of thought).[37]

I believe that there is a great deal of value to Carruthers' proposal. However, there is something strange about the claim that the states and processes that must be present for an entity to count as minded must be beliefs and desires. Carruthers attempts to defuse this worry by adverting to *spatial* beliefs and desires as opposed to *causal* beliefs and desires. He holds that many kinds of invertebrate beliefs are likely to represent by encoding the spatial relationships between salient features of their environment rather than the causal relations that obtain between various objects and entities in their environment. However, this seems to miss the crucial worry about positing invertebrate beliefs and desires: Why should the high-level computations that underlie the capacities for believing and desiring play such a central role in philosophical and scientific theorizing about the capacity for genuinely intelligent action? Our cognitive capacities depend to a great extent on our linguistic capacities, and I agree with Carruthers that any entity that can be in cognitive states or carry out cognitive processes must be able to "act with flexibility and forethought, choosing between different courses of action and anticipating future consequences. These abilities seem to demand representations that stand in for external objects."[38] However, it is not obvious what it takes to be able to represent in this way.

As my discussion of bacteria suggests, genuine cognition requires internal states and processes that *represent* rather than *merely record* the way that the world is. Put differently, genuine cognition requires internal states or processes that have the *function* of *conveying information* about the way that the world is, in a way that can provide the necessary resources for guiding the behavior of that organism.

As John Haugeland puts the point, an internal state need only be the result of particular processes in order to count as a recording, but "representing is a functional status or role of a certain sort, and to be a representation is to have that status or role."[39] Here we must ask, what sort of functional role has to be filled if something is to count as a representation? I propose to follow Haugeland in accepting the following rough-and-ready desiderata on the sort of functional organization required for representation.[40] These are only intended as rough-and-ready criteria, but they should seem highly intuitive and it should be clear that they play a dominant role in structuring philosophical and empirical approaches to questions about representations.[41] I suggest that the possession of a minimal mind requires that an entity:

(1) Possess internal states and processes that have the function of adjusting the entity's behavior in ways that allow it to cope with features of its environment in ways that are not fully determined by the design of the system;

(2) These states and processes must be capable of standing in for various features of the environment that are significant for the system, and they must be capable of doing so even in the absence of immediate environmental stimuli;

(3) These states and processes must be part of a larger representational scheme that allows for a variety of possible contents to be represented (in a systematic way) by a corresponding variety of possible representations; and,

(4) There must be norms governing the proper production, maintenance, and use of these representations under various environmental conditions.[42]

I suggest that any entity that possesses such capacities will have sufficient representational complexity to be in genuinely cognitive states and to carry out genuinely cognitive processes. However, it is important to acknowledge that there are a number of ways in which the second clause of (2) can be construed.

On the most robust construal, such states and processes would have to be able to represent things that do not exist. For example, my representation of a DEMOCRATIC SOCIETY regulates much of my behavior, in ways that allow me to cope with my environment. It stands in for a nonexistent structure of political organization, plays an important role in my overall understanding of the world, and is embedded in a system of social norms that regulate my claims about whether some society can reasonably be said to be a genuine democracy.

On a much weaker construal, however, these criteria might license the claim that there are some sorts of entities that possess these sorts of representations without having the kinds of discrete beliefs and desires that Carruthers requires. Indeed, it may be that for the most minimal minds, all of the relevant representational capacities are implemented in perception-action circuits that preclude the complete decoupling of the indicative and imperative components of their representations. That is, minimal minds might *only have* pushmi-pullyu representations.[43] Such states and processes both represent the world as being a particular way and function to direct immediate action. However, they are not mere couplings of belief-like states and desire states that are filtered through a general-purpose practical reasoning mechanism. These are a more primitive type of representation, which immediately yield changes in behavior as a direct function of changes in the environment.

According to Morgan's Canon, "we should not trust psychological explanations of behavior unless we are convinced that those explanations are indispensable— that is to say, unless we are convinced that the behavior in question cannot be explained in nonpsychological terms."[44] Indeed, we *should be* dubious about theories that accord minimal minds the capacity to entertain complex thoughts that are "linguistically vehicled," or capacities for reasoning that require the manipulation of abstract propositions.[45] Much more than this, I hold that without the capacity to use linguistic representations, it is overwhelmingly unlikely that minimal minds will possess capacities to form episodic memories, deliberate about counterfactual possibilities, or form anything resembling a fine-grained belief or desire. This is why José Luis Bermúdez is right to argue that nonhuman animals are likely to possess only capacities for thought and reasoning that are tightly coupled to the perceptible features of their environment.[46] Yet, explanations in terms of the cognitive capacities of fairly simple invertebrates might still be indispensable from the perspective of cognitive science, even if such entities are not complex enough to have states or processes that are like beliefs and desires in any interesting sense. Examining this point, however, requires a turn to the empirical data that support the claim that some invertebrates can be in cognitive states and carry out cognitive processes.

3.1 Darwinian Minimal Minds

In his final book, *The Formation of Vegetable Mould, through the Actions of Worms with Observations on Their Habits,* Darwin turns his ethological eye to the behavior of earthworms. While the majority of this work is dedicated to examining the impact of worms on their physical environment, Darwin also examines the possibility that worms are intelligent. His observations begin with the fact that worms tend to plug the openings of their burrows with leaves, sticks, and petioles of varying sizes. Darwin found that these objects were pulled into the openings of burrows in a way that was too uniform to be attributed to mere chance, but at the same time they seemed to be carried out in a way that is not unvarying or inevitable enough to be called a mere instinct.[47] By carefully examining the diverse strategies for manipulating these objects, Darwin finds that there are significant differences in the techniques used for pulling various leaves (e.g., rhododendron leaves, pine needles, and artificial leaves made from triangles of folded paper) into the opening of the burrow; and with unfamiliar artificial leaves, he found that worms often attempted several distinct strategies for pulling the leaf, using a methodology of trial and error before settling on a preferred strategy for handling these new objects.

With these observations in hand, Darwin argues that worms should be seen as possessing a capacity to learn from experience to discriminate between various shapes of leaves, twigs, and petioles. So, adopting an argument from parity, he argues that "if worms have the power of acquiring some notion, however rude, of the shape of an object and of their burrows, as seems to be the case, they deserve to be called intelligent; for they can act in a manner as would a man under similar conditions." Darwin thus attempts to show that earthworms adjust their behavior in ways that allow them to cope with novel features of their environment, and he argues that these states are not fixed by, though they are clearly structured by, the evolutionary history of these organisms. Darwin is keenly aware of the initial oddity of this claim, and he carefully articulates a series of reasons for why his observations of worm intelligence cannot be generalized to explain the behavior of other invertebrates without further empirical investigation. However, through the careful examination of worm behavior, and by ruling out all of the available alternative explanations, Darwin argues that his views "will strike everyone as improbable; but it may be doubted whether we know enough about the nervous system of the lower animals to justify our natural distrust of such a conclusion."[48]

Darwin's groundbreaking examination of the mental life of worms is important not for its substantive conclusions—which fail to demonstrate a capacity for mental representation, even if worms might have such a capacity—but for its methodological conclusion. Instead of arguing that there is no reason to deny the possibility of insect minds, Darwin set out to examine the evidence in favor of this claim. As Ellen Crist notes, he demonstrates that claims about invertebrate intelligence are

> not ontologically problematic because it is not rational to presume, prior to inquiry, that the existence of conscious action is unlikely, even among invertebrates; not epistemologically problematic because once the question of conscious action is allowed to be posed, the scientific imagination finds fascinating ways to address it; and finally, not semantically problematic because writing off "conscious action" as anthropomorphism commits the deeper fallacy of anthropocentrism.[49]

Darwin demonstrates that careful observation and experimentation can be turned on the behavior of invertebrates, and he shows that it is possible to investigate the question of invertebrate minds scientifically. While he recognizes that this claim will seem questionable from the standpoint of commonsense psychology, he hopes to displace our unfounded prejudices about the nature of cognition with sound scientific inquiry so that the commonsense understanding of mentality can be revised to construct a plausible scientific understanding of what it takes to have a minimal mind.

3.2 The Mind of the Honeybee

To further develop these Darwinian assumptions about minimal mentality, it will be useful to consider the research on the intelligence of honeybees. It has long been known that honeybees possess the capacity to communicate the distance and direction of food sources by way of a set of ritualized dancing movements.[50] When a high-quality food source is found near the hive, honeybee scouts use a "round" dance that informs the other bees that they should search for the odor associated with the scout upon leaving the hive. However, for more distant food sources a more elaborate "waggle dance" is used to convey information about the direction, distance, and quality of the food source. Direction is indicated by the angle of the movements of the abdomen across the center of a figure eight (which represents the angle from the position of the sun); distance is indicated by the duration of the dance (and sounds); and quality is indicated by the vivacity of the dance.

Honeybees have also been shown to possess capacities for navigating and encoding new memories acquired during explorations. Bees that travel long distances from their nest site seem to construct cognitive maps that rely on simple geometric representations of space, and vector information that can be recovered using "dead reckoning" strategies.[51] They also appear to integrate landmark information that is "experienced en route" to recalibrate distances, reduce the potential navigational errors, and to encode procedural information about what to do next.[52] Moreover, Randolf Menzel argues that honeybees discriminate between different colors, shapes, patterns, odors, and textures while foraging to construct accurate mappings of landmarks.[53] Finally, Shaowu Zhang, James Bartsch, and Mandyam Srinivasan have shown that bees can navigate complex and unfamiliar mazes by learning to see particular colored disks as meaning either "turn left" or "turn right."[54]

Honeybees also develop long-term expectations that cannot be derived from the strength of an association with a predicting signal, using these expectations to guide behavior even after a relatively long delay following an initial reinforcement.[55] This capacity plays a critical role in determining the quality of a food source; but it is also deployed to evaluate the threat of predation by crab spiders that change color to match their environment. Foraging bees that face a simulated risk of predation by robotic crab spiders that are camouflaged (yellow spiders on yellow flowers) or easily detectable (white spiders on yellow flowers) initially land on flowers that are inhabited by predators randomly. However, after a simulated attack, the likelihood that they will land on an inhabited flower falls off rapidly, at the same rate for camouflaged and conspicuous spiders. Using tracking software, however, Thomas Ings and Lars Chittka have shown that bees that are exposed to camouflaged threats trade off speed for accuracy in threat detection, increasing the duration of inspection flights prior to landing on potentially occupied flowers. This levels out predation risk by increasing the amount of time that they spend foraging. However, bees do not alter their foraging speed when they expect to be able to detect predators easily.[56]

Finally, honeybees also extract within category similarities and between category differences for various kinds of stimuli. For example, bees that are presented with a variety of gratings that differ in both the distances between lines and in edge orientation can extract edge orientation as a category that indicates the presence of a reward. When 45-degree gratings indicate a high-reward sucrose solution in a training condition, bees generalize to predict that 45-degree gratings will indicate the presence of a reward in an experimental condition that contains novel stimuli that differ in all structural properties other than edge orientation.[57] Additionally, honeybees who are trained to fly down the corridor in a Y-maze marked with the same (or a different) color than an entry room reflexively transfer this skill to a novel domain in which corridors and entry rooms are marked with smells instead of colors.[58] Indeed, numerous experiments confirm that honeybees can successfully learn match-to-sample and don't-match-to-sample rules—suggesting a capacity to encode differences and not just similarities.[59]

The communicative activity displayed in the waggle dance, as well as the individual representations of location, landmarks, predators, and the like, require a set of internal states and processes that have the function of carrying information about salient features of their environment. Some of these internal states and processes are then expressed as communicative signals that are encoded in the precise structure of the "waggles" in the waggle dance. These states and processes carry information about the world that is not completely determined by the design of the system, but encoded in the interaction of the bee with its environments. The internal states and processes that are expressed in the "waggles" stand in for various features of the environment that are significant for the bees, and they do so even for the bees who have not yet foraged in the relevant environment. The "waggles" also express internal states and processes that are part of a larger representational scheme that allows a variety of possible contents (e.g., different distances, directions, and food sources) to be systematically represented. Finally, the experiment on predation demonstrates that the representational system can misrepresent a robotic spider as a genuine threat. This much is sufficient to establish the existence of honeybee minds that include "a suite of information generating systems that construct representations of the relative directions and distances between a variety of substances and properties and the hive, as well as a number of goal-generating systems taking as inputs body states and a variety of kinds of contextual information, and generating a current goal as output."[60]

It is important to note, however, that these capacities do not involve arbitrary signals that can be *robustly decoupled* from the salient features of the environment. For example, because the waggle dance is an iconic representational system, the content of a particular waggle is necessarily tied to the particular qualities of the environment that are at issue in a particular dance. For each dimension of variation in the environment (e.g., distance or direction), there is a single transformation rule that maps the variation in that parameter onto a parallel variation in the signal. This being the case, the consumer of the signal need only apply an inverse mapping to decode the signal.[61]

This iconic symbol system still allows for a variety of foraging behaviors that take honeybees incredible distances away from their hives.[62] Yet, many of the inferential capacities that are present in language-using animals like us are wholly lacking in honeybees. Specifically, the representational states and processes that we find in honeybees do not appear to satisfy the *generality constraint*. In explicating what the generality constraint comes to, Gareth Evans notes that there is an important sense in which thoughts like ours are complex and structured representations; this being the case, anyone who can have the thoughts SASHA IS HAPPY and RAMON IS SAD will, thereby, be able to think RAMON IS HAPPY and SASHA IS SAD.[63] Put more formally, "If a subject can be credited with the thought that A is F, then he must have the conceptual resources for entertaining the thought that A is G, for every property of being G of which he has a conception."[64] The problem, as Carruthers aptly notes, is that it appears as though bees can be in mental states with the following content: THERE IS SOME HIGH QUALITY NECTAR 115 METERS DUE EAST OF THE HIVE and THERE IS SOME POLLEN 45 METERS DUE WEST OF THE HIVE. However, it seems highly unlikely that possessing these capacities can allow the bee to think THERE IS SOME NECTAR 190 METERS DUE EAST OF THE POLLEN.[65] This being the case, we might suppose that the honeybee representations are not belief-like in a robust, conceptual sense. Carruthers dismisses this worry by noting that because ecologically valid situations in which such a complex representation as this is likely to be required are highly unlikely, the fact that bees never do produce such a representation does not imply that they cannot.[66] However, there are further empirical data suggesting that bees cannot draw the sorts of inferences that we would expect from an entity that had robust conceptual abilities like our own.

Bees do not possess the inferential capacities that allow for transitive inferences.[67] Although they can be taught that the reward at point A is greater than the reward at point B, and that reward at B is greater than the reward at C, they cannot infer that the reward at A is greater than the reward at C; "bees do not establish transitive inferences between stimuli but rather guide their choices by the joint action of a recency effect…and by an evaluation of the associative strength of the stimuli" (i.e., they rely on the most recent item encountered and the strongest conditioned associations).[68] This fact cuts against the claim that bees have representational states and processes that resemble the conceptual representations possessed by language users, and it seems plausible that because bees do not possess the capacities using words in a natural language, their mental states and processes will be both more restricted and less inferentially promiscuous than human beliefs and desires.

I contend that it is a mistake to treat the states and processes of honeybees as beliefs and desires, as Carruthers suggests they must be. Yet, Carruthers is right to note that an argument grounded in the generality constraint fails to undercut the claim that invertebrates have simple mental states and processes. Carruthers' concerns derive from a desire to resist empty instrumentalism. However, this argument neglects the fact that there are many ways of being in mental states and carrying out mental processes that do not depend on having beliefs and desires like our own. Honeybees do possess cognitive capacities that cannot be explained unless we advert to representational capacities.[69] However, the mistake is to assume that mental representations must have a proto-linguistic structure; in fact, many mental representations are embodied and skill-based representations that underwrite the fast-and-frugal coping behavior in rapidly changing environments.[70]

The quasi-Cartesian assumption that genuine mentality requires conceptual representations in a language of thought ignores the representational capacities that are most likely to be implemented in biological systems that must rapidly cope with changes in their environments. For example, sensory systems have been selected for the way in which they provide information to motoric systems to facilitate rapid coping behavior in the face of danger, food, or potential mates. Important relationships may well obtain between such representations and the properties, relations, and things in the world; however, because sensory systems have been selected to yield fast-and-frugal action, it is unlikely that they will depend on symbolic relations whose content is isomorphic with the content of the semantic representations that we find in language. To put this point another way, the most "primitive" representations in sensory systems are likely to be constructed in a contextually sensitive and narcissistic fashion (*sensu* Akins). More complex representations in language and thought are then likely to be the result of triangulating these low-level representations on the basis of competitive or quasi-competitive algorithms (see the concluding section of this chapter).

With this fact in mind, I suggest that there are likely to be many kinds of mental states and processes that are not completely decouplable from their immediate causes, and that may not satisfy Evans's generality constraint. However, such states and processes can be genuinely representational and can play an important role in the mental lives of human and nonhuman animals alike. To clarify, consider the population of neurons in the rat's parietal cortex that *represent* (in a rich mental sense) the direction of the rat's head. As Andy Clark notes, we gain a great deal of explanatory power by acknowledging that these neurons represent the position of the rat's head; unless we do so, we cannot understand how information flows through the rat's cognitive system as a whole.[71] Yet, it would be a mistake to assume that these neurons implement beliefs and desires; doing so would be nothing short of revisionary semantics given that they are nothing more than information-bearing structures that represent the direction of the rat's head. However, provided that they are integrated into a larger cognitive system that allows for the representation of a variety of facts about the world (in a variety of ways), they ought to be treated as mental states that are empirically tractable from the perspective of the cognitive sciences.

The internal states and processes that we find in the neural architecture of a honeybee are likely to be carried out independently of anything like central cognition. Many of these states and processes will be simple sensory-motor loops that reflexively yield particular behaviors depending on the current state of the environment. Others may be more decouplable from immediate presentations, allowing for more-generalized inferences on the basis of category membership and cross-modal integration. However, the minimal minds of these invertebrates are unlikely to have anything like robustly conceptual, amodal, and quasi-linguistic representations that can be completely decoupled from their environmental triggers. The fact that these representations cannot be decoupled from environmental triggers is the crucial thing that distinguishes relatively maximal from relatively minimal mental representations. In short, different kinds of mental states and processes are likely to lie along a continuum, with pushmi-pullyu representations lying at one pole, and the richly amodal representations of linguistically structured thought at the other. While it is unlikely that we share our capacity to represent amodally with any non-human animals, many of our capacities to represent the world by way of pushmi-pullyu representations are likely to be shared far down the phylogenetic tree.

4. COLLECTIVE MINIMAL MINDS

In this final section, I wish to redeem the promissory note that I offered at the end of the introduction. There, I suggested that an adequate understanding of minimal mindedness would also provide evidence for the existence of collective mental states and processes in eusocial insects such as bees. Thomas Seeley has argued that honeybee colonies should be seen as unified systems, and that these systems rely on the iconic representations of the waggle dance to propagate information in a way that allows the colony to respond to stimuli that are salient to the colony *as such*.[72] There is a growing consensus that colonies of eusocial insects should be treated as single units of selection for the purposes of biological research. Even Richard Dawkins, a rabid "smallist" about explanation, agrees.[73] However, more is required to establish that honeybee colonies have a cognitive life beyond that of the individual bees.

Seeley contends that observations of foraging bees suggest that monitoring the location and richness of food sources and evaluating the relative quality of various food sites can only take place in the distributed representations of the colony.[74] Seeley argues that foragers act as a diffuse sensory extension of the colony. Each bee begins with a random search for foraging sites. These initial foragers map the surrounding environment, finding patches of food as far as 10 km away.[75] However, for foraging sites within 2 km, it seems that bees can engage in comparisons of richness in just the way that should not be possible given the results reported by Menzel and Guirfa.[76] This process takes place by way of an aggregation of information at the level of the colony that cannot be carried out by any of the individual bees.[77] When employed foragers return to the hive, they advertise the distance, direction, and quality of a foraging site with their waggle dance. However, while individual bees follow only one other bee's dance, the likelihood of being recruited to a foraging site is determined by the duration and vivacity of a forager's waggle dance.[78] Those foragers who have visited desirable worksites dance longer and more vivaciously than bees who have visited less desirable foraging sites. The information about quality is distributed across the employed foragers in a way that does not require a centralized decision-making structure to allocate unemployed foragers to new foraging sites.

By modulating the quality and quantity of a pair of artificial food sources, Seeley demonstrated that honeybee colonies become more selective when food sources are abundant.[79] However, under conditions of scarcity, foragers are allocated to even low-profit nectar sources. The mechanism of this selectivity can be specified in terms of the modulation of the threshold at which waggle dances occur.

When a forager returns to the hive, its first task is to find a receiver bee who will accept nectar for storage. When resources are scarce, returning foragers rapidly find receiver bees, and so even a short dance will find an audience. This being the case, bees are recruited to less profitable foraging sites. When food is abundant, by contrast, the search for a receiver bee takes a longer amount of time, so only a longer and more vivacious dance will find an audience. Only high-profit food sources are then exploited. Although there is no central cognition dedicated to monitoring the abundance or scarcity of food, the colony is capable of evaluating the relative abundance of food sources at various foraging sites even though none of the individual foragers or receivers is capable of representing this.

Still, there are complications. Nectar collection and processing sometimes fall out of sync. When this happens, foragers who have found incredibly rich food supplies need to help boost nectar collection rates to collect as much nectar as possible, but they also have to increase the rate at which nectar is processed to allow the bees who are returning from a high-quality foraging site to find receivers for their pollen. When a forager returns from an incredibly high-quality foraging site and finds that it has an extensive search time for finding a receiver, it executes a "tremble dance" that carries the information that unemployed bees should immediately begin processing nectar. For bees who have been foraging, this dance also carries the information that they should refrain from recruiting additional foragers, thereby acting as a suppression signal to inhibit the waggling of other bees. The execution of a tremble dance thus updates the rate at which nectar is processed so that the quantity and quality of nectar can be recalibrated.

Finally, Seeley and his colleagues turn to the process by which new nest sites are selected.[80] When a colony outgrows its hive, it splits in two. One colony swarms around a tree branch and sends out scouts (approximately 5 percent of the swarm) to find a new nest site. During the initial search, as many as a dozen potential nest sites are selected, and each is evaluated by a scout according to six desiderata: cavity volume; entrance size, height, direction, and proximity to the cavity floor; and presence of combs in the cavity.[81] As the scouts return, they waggle dance to indicate the presence and quality of these features of the potential nest site, and although each scout only dances for one site (rarely, if ever, dancing for another site after having made their initial selection), a collective decision emerges and there is a consensus on one site (I address this claim more fully below). Strikingly, the swarm reliably chooses the site that best satisfies the six desiderata listed above (rather than settling on the first adequate site, for example), and the swarm only moves where there is complete consensus on that site. Seeley and his colleagues demonstrate that this consensus emerges because after the initial scouts dance for their chosen site, those bees that have found a mediocre or passable nest site dance less vigorously than those bees that have found a high-quality site.[82] Heavier recruitment of additional scouts occurs for higher quality nest sites, and eventually this leads to the cessation of dancing for lower quality nest sites. Lower quality sites lose support until only the highest quality site is being danced for, leading eventually to the reliable selection of the highest quality nest site without requiring any of the individual bees to have a broad knowledge of all of the alternative possible nest sites that are under consideration by the swarm.

Seeley's data provide good reason for thinking that the specialization of function in a honeybee colony can facilitate the propagation of representational states (e.g., states that represent the location of nectar, the quality of a foraging site, and the location of a nest site) between bees with very different functionally specified tasks. As these representations are propagated between the members of a colony, a complex comparative evaluation emerges that cannot be made by the individual bees on their own. Thus, he suggests that the comparative judgments are carried out by distributed computational architecture that is realized by the colony as a whole, rather than by any of the computational nodes (i.e., the individual bees). By positing cognitive states and processes that are properly attributable to the honeybee colonies as such, Seeley can explain such diverse phenomena as the decision to build a nest in one site rather than another and the decision to allocate more resources to collecting or storing nectar. Such predictions are only possible on the assumption that there are cognitive states and processes that are properly attributable to the collectivity. The choice of a nest site is a striking demonstration of this fact. The colony chooses the best nest site possible even though none of the individuals has the capacity to choose or even represent any of the nest sites as better or worse than any other. It is only through the coordinated activity of a number of bees, and only through the representation of particular facts about particular nest sites across various bees, that this capacity can emerge. This coordination gives us good reason to think that there is a sufficient amount of emergent phenomena here to give the collectivity a rich life of its own. This research suggests two questions. First, do these states and processes satisfy the four rough-and-ready desiderata on representation? And if they do, should the relevant states and processes really be called "decisions" or "judgments"?

I suggest that the colony seems to have internal states and processes with the function of adjusting the hive's behavior to facilitate skillful coping with changes in the environment. Consider the mechanisms that implement a decrease in foraging when too much food is coming into a hive too quickly. Here, none of the individual bees represents a need for a decrease in foraging, but the colony is designed to be sensitive to the relation between incoming nectar and nectar storage. When the rate at which nectar is being returned to the hive exceeds the rate at which it is stored, the system is designed to decrease the amount of nectar that is coming into the system. Moreover, this evaluation is not a matter of absolute quantity of input or output; rather, the evaluation of the amount of nectar coming into the hive is carried out by examining the relationship between the current state of a honeybee colony and the current state of the foraging sites in the area. Seeley clearly demonstrates that it is only by way of such internal states and processes that this sort of behavior, which is sensitive to changes in the environment, can be produced. Of course, there is still an important sense in which the behavior of the honeybee colony is fully a function of its evolutionary design. Unlike a human mind, a honeybee colony does not possess the sorts of representations that can be used to preselect behaviors on the basis of internal models. Indeed, the states and processes that we find in a honeybee colony share far more in common with the context-bound, and action-oriented, pushmi-pullyu representations that we find in the case of individual bees. So, along the spectrum of mental states and processes that I mentioned at the end of the last section, honeybee colonies are likely to have minimal minds, even though they are quite large in size.

Finally, let me close on a more technical note by suggesting that these colony-level representations share much in common with the computational structures that we find in a human brain. Each neuron in a human brain constantly updates its structure in light of the behavior of the other neurons to which it is connected (e.g., by modulating the production of neurotransmitters, extending and pruning dendritic branches to increase connectivity with preferred neighbors, adjusting firing patterns to compensate for the flow of neurotrophins, and dying when unable to integrate into the local environment). As Daniel Dennett (in personal correspondence) has recently argued, neurons are likely to have taken on much of the Darwinian "research-and-development" carried out by the eukaryotic cells from which they are descended. The individual mind is a pandemonium that consists of numerous layers of demons, sub-demons, and sub-sub-demons, all competing for control of neural resources (e.g., by modulating the amount of neurotransmitter that is available, extending and retracting dendritic branches, and by adjusting firing patterns). However, these competitions also modulate the structure of the brain's overall neural architecture—and unwittingly drive the computations that are carried out at higher levels. Because neurons inhabit a highly integrated, hierarchically organized, and massively parallel computational system, their competitive interactions yield computational outputs that can be consumed by the computational structure that they compose.

Research in the neurobiological sciences has suggested the presence of attentional mechanisms in the parietal cortex that depend on extreme winner-take-all computations. For example, Christof Koch and his colleagues have argued that the allocation of "bottom-up" attention is the result of visual features (e.g., color, orientation, and movement) being encoded in separate feature maps that, when subjected to a competitive algorithm, can be integrated into a unified representation that encodes the strength of each feature in a topographically oriented saliency map that represents multiple values in a single multidimensional space.[83] Similarly, Robert Desimone and John Duncan offer a winner-take-all model of visual attention to explain how multiple competitions can eventually result in consciousness and subsequent reportability.[84] As Dennett has often argued, higher level information is often processed in a way that "contributes to the creation of a relatively long-lasting Executive, not a place in the brain but a sort of political coalition that can be seen to be in control over the subsequent competitions for some period of time."[85] In each of these cases, we find competitive algorithms that integrate various sources of information to create unified representational structures that can readily be deployed in the production of action. The relevant representations might eventually come to be categorized in map-like or language-like structures in a semantically transparent language of thought; however, what is important for understanding their role in cognition has very little to do with representational genera, and a whole lot to do with the way that they can be rapidly deployed in skillful coping behavior.

I suggest that the competitive algorithms that are operative in honeybee colonies should not be seen as decisions, since decisions are the things that have to be aggregated out of beliefs and other cognitive sophisticated representations. However, as representational structures they do share much in common with the perceptual-motor and attentional structures in a human brain. While waggle and tremble dances, as well as search times, can stand in for features of the environment (specifically the location of a food source and the rate or consumption by the system), they do so only when the system is immediately presented with raw data about the natural environment. The dance times as well as the vigorousness of an individual bee's dance are fully determined by features of the world, and the behavior of unemployed bees and collectors are fully determined by the dances of the returning forager bees. However, honeybee colonies are incapable of engaging in behavior that is anywhere close to being as rich as our own cognitive behavior. This being the case, even though I hold that we should recognize that there are collective mental states and processes in a honeybee colony, the mental life of a honeybee colony is far more impoverished than the mental life of a human being. This collective mind must itself be seen as a minimal mind, along some dimensions even more impoverished than the minimal mind of a single honeybee. The key point is that these emergent phenomena suggest an interesting range of cognitive phenomena that can only be studied by examining the behavior of honeybee colonies *as such*. These states and processes are cognitive to the same extent that the states of the neurons in my parietal cortex are cognitive states—they are not beliefs or desires, but they play a crucial role in structuring the way that the world is disclosed to me.

The upshot is that the range of explanatory projects that ought to be studied within the cognitive sciences is likely to outstrip the commonsense understanding of the mind, and this is why psychology is best served by dissociating cognitive states at the sub-personal level from the core cases of cognition that might be present only in language users like us. There are many sorts of states that are important for explaining behavior, but we must be sure not to get carried away in ascribing states such as beliefs and desires to honeybee colonies—positing structures of collective mentality does not require assuming that honeybee colonies have beliefs, desires, hopes, wishes, and dreams.

5. CONCLUSION

If the arguments in this chapter are successful, they demonstrate that there is at least one respect in which commonsense assumptions about mentality should be revised: some groups of organisms—as such—can be in genuinely cognitive states and carry out genuinely cognitive processes. This conclusion stands in stark contrast to the commonsense assumption that cognitive systems are always organism bound. However, I have attempted to show that this conclusion is not nearly as counterintuitive as it might seem.

I first argued that adopting the computational approach to cognition that is advanced within the cognitive sciences provides us with strong reason to reject the assertion that the capacity for using language yields an unbridgeable gap between genuine cognition and witless mechanical activity. Contemporary approaches to the study of the mind begin from the assumption that even the most cognitively sophisticated capacities should receive a mechanistic explanation in terms of the representational and computational mechanisms by which they are implemented. On the basis of this argument, I suggested that different kinds of minds are likely to lie along a continuum running from minds that consist exclusively of pushmi-pullyu representations to human minds that deploy the richly amodal representations that are indicative of beliefs and desires of linguistically structured thought at the other end of the continuum. On the basis of these arguments, I suggest that the decision to treat some collectivities as minimally minded is no more and no less reasonable than the decision to treat an ordinary invertebrate as minimally minded. In both cases, these minimal minds are likely to represent the world by way of pushmi-pullyu representations. They are not likely to be populated by beliefs, desires, hopes, dreams, and wishes.

Unfortunately, this argument leaves open difficult philosophical questions. Given that minimal minds like these are likely to be quite different from our own, my arguments do not have straightforward ethical implications regarding the use of poisons to kill cockroaches, the consumption of honey, or the use of silk. To answer such questions, one would have to move beyond these questions about minimal mentality and ask difficult empirical questions about the range of mental states and processes that populate the minds of invertebrates. With these data in hand, we would still face hard philosophical questions about the kinds of minds that warrant our compassion. Even offering cursory answers to such questions, however, would take me far beyond the scope of this chapter.

NOTES

1. Alice Park, "Beefing Up the Arsenal Against AIDS," *Time Magazine* (2007), http://www.time.com/time/health/article/0,8599,1595377,00.html (accessed August 9, 2010).

2. David Ho, "Therapy of HIV Infections: Problems and Prospects," *Bulletin of the New York Academy of Medicine* 73 (1996): 37.

3. Vito Racanelli and Barbara Reherman, "Hepatitis C Virus: Infection When Silence is Deception," *Trends in Immunology* 24 (2003): 456–64.

4. Cf. Daniel C. Dennett, *Kinds of Minds* (New York: Basic Books, 1996), pp. 21–26.

5. Bertrand Russell, *Human Knowledge: Its Scope and Limits* (London: Allen & Unwin, 1948), p. 486.

6. Cf. José Luis Bermúdez, *Philosophy of Psychology: A Contemporary Introduction* (London: Routledge, 2005), chapter 1.

7. Dennett, *Kinds of Minds*, p. 33

8. Marc D. Hauser, *Wild Minds* (New York: Henry Holt, 2000), p. 9.

9. For arguments about the role of language in human cognition, see José Luis Bermúdez, *Thinking Without Words* (New York: Oxford University Press, 2003); Peter Carruthers, *The Architecture of The Mind* (Oxford: Oxford University Press, 2006) and "How We Know Our Own Minds," *Behavioral and Brain Sciences* 32 (2009): 121–38; Donald Davidson, "Thought and Talk," in *Inquiries into Truth and Interpretation* (Oxford: Clarendon Press, 2001); Daniel C. Dennett, *Brainstorms* (Cambridge, Mass.: Bradford Books, 1978), *Consciousness Explained* (New York: Penguin, 1992), and *Kinds of Minds*; and Ray Jackendoff, *Language, Consciousness, Culture: Essays on Mental Structure* (Cambridge, Mass.: MIT Press, 2007).

10. Tadeusz Zawidzki, "The Function of Folk Psychology: Mind Reading or Mind Shaping?" *Philosophical Explorations* 11 (2008): 193–210.

11. Cf. Adam Morton, "Folk Psychology Is Not a Predictive Device," *Mind* 105 (1996): 119–37; and Zawidzki, "Function of Folk Psychology," pp. 193–210.

12. Dennett, *Kinds of Minds*, p. 34.

13. William James, *The Principles of Psychology* (Cambridge, Mass.: Harvard University Press, 1890), p. 7.

14. James, *Principles of Psychology*, p. 7.

15. James, *Principles of Psychology*, p. 8.

16. Jerry Fodor, as quoted by Dennett, *The Intentional Stance* (Cambridge, Mass.: MIT Press, 1987), p. 53.

17. As Descartes suggests in the *Second Set of Replies* (p. 135) "there is nothing in the effect which was not previously present in the cause, either in a similar or in a higher form is a primary notion which is as clear as any that we have"; he says that this is just the claim that "Nothing comes from nothing." All references to Descartes' work refer to *The Philosophical Writings of Descartes*, 3 vols., translated by John Cottingham, Robert Stoothoff, and Dugald Murdoch (Cambridge: Cambridge University Press, 1985).

18. Janet Broughton, *Descartes' Method of Doubt* (Princeton, N.J.: Princeton University Press, 2003), p. 155.

19. Descartes, *Discourse on the Method*, vol. 1, p. 56.

20. Descartes, *Early Writings*, vol. 1, p. 219.

21. Descartes, *Meditations*, Meditation II, vol. 2, p. 32.

22. Alan Turing suggests an imitation game where a judge asks any question that she wishes to a sophisticated computer and an ordinary person; both the computer and the human attempt to convince the judge that they are the person. Turing argues that any machine that could regularly convince a discerning judge that they were a person by responding to the sense of her questions would display any plausible mark of genuine intelligence. Turing, "Computing Machinery and Intelligence," *Mind* 59 (1950): 433–60.

23. Descartes, *Discourse on the Method*, Part 5, vol. 1, pp. 57ff.

24. In the *Discourse on the Method* (Part 5, vol. 1, p. 58), Descartes claims that it is "a very remarkable fact that although many animals show more skill than we do in some of their actions, yet the same animals show none at all in many others; so what they do better doesn't prove that they have intelligence, for if it did then they would have more intelligence than any of us and would excel us in everything. It proves rather that they have no intelligence at all, and that it is nature which acts in them according to the disposition of their organs. In the same way a clock, consisting only of wheels and springs, can count the hours and measure time more accurately than we can with all our wisdom."

25. Descartes, *Fourth Set of Replies*, vol. 2, p. 230.

26. Thomas Hobbes, *Leviathan* (Indianapolis, Ind.: Hackett Publishing, 1994), p. 33; similarly, David Hume, *A Treatise of Human Nature* (Oxford: Oxford University Press, 1978), p. 176: "no truth appears to be more evident, than that beast are endow'd with thought and reason as well as men."

27. Spinoza, *Ethics*, trans. Samuel Shirley (Indianapolis, Ind.: Hackett, 1991), p. 105.

28. Spinoza, *Ethics*, p. 59.

29. Yitzhak Melamed, "Spinoza's Anti-Humanism," in *The Rationalists*, ed. Carlos Fraenkel, Dario Perinetti, and Justin E. H. Smith (Amsterdam: Kluwer, in press), p. 17.

30. Robert MacKenzie, as quoted in Dennett, "Darwin's 'Strange Inversion of Reasoning'" *Proceedings of the National Academy of Science* 106 (2009), p. 10062.

31. "Darwin's 'Strange Inversion of Reasoning,'" p. 10062.

32. C. Lloyd Morgan, *An Introduction to Comparative Psychology*, 2nd ed. (London: W. Scott, 1903), p. 59.

33. W. Tecumseh Fitch, "Nanointentionality," *Biology and Philosophy* 23 (2008): 158.

34. Fitch, "Nanointentionality," p. 162.

35. Fitch, "Nanointentionality," p. 163.

36. Dennett, *Kinds of Minds*, p. 83.

37. Carruthers, *Architecture of Mind*, pp. 67–68; and this volume.

38. Jesse J. Prinz, *Furnishing the Mind* (Cambridge, Mass.: MIT Press, 2004), p. 4.

39. John Haugeland, "Representational Genera," in *Having Thought* (Cambridge, Mass.: Harvard University Press), p. 180.

40. Haugeland, "Representational Genera," p. 172.

41. Andy Clark, *Being There: Putting Brain, Body and World Together Again* (Cambridge, Mass.: MIT Press, 1997), pp. 143–77.

42. This does not require misrepresentation in the fullest sense of the term. Familiarly, you can fill a frog with buckshot by shooting BBs past it in the lab. This is a representational failure, but it is not a genuine misrepresentation. The frog's visual system is just too impoverished to reliably discriminate between food and near-non-food.

43. Ruth Millikan, *Language, Thought, and Other Biological Categories* (Cambridge, Mass.: MIT Press, 1988).

44. Bermúdez, *Thinking without Words*, p. 7.

45. Bermúdez, *Thinking without Words*, p. ix.

46. Bermúdez, *Thinking without Words*.

47. Charles Darwin, *The Formation of Vegetable Mould through the Action of Worms* (Chicago: University of Chicago Press, 1985), p. 93.

48. Darwin, *Formation of Vegetable Mould*, p. 97.

49. Ellen Crist, "The Inner Life of Earthworms," in *The Cognitive Animal*, ed. Marc Bekoff, Collin Allen, and Gordon Burghardt (Cambridge, Mass.: MIT Press, 2002), pp. 7–8.

50. Karl Von Frisch, *The Dance Language and Orientation of Bees* (Cambridge, Mass.: Harvard University Press, 1967); James Gould and Carol Gould, *The Honey Bee* (New York: Scientific American Library, 1988).

51. "Dead reckoning" is the process of integrating direction, velocity, and time traveled to produce a representation of current location relative to initial location. See Randy Gallistel, *The Organization of Learning* (Cambridge, Mass.: MIT Press, 1990); and James Gould, "The Locale Map of Honey Bees," *Science* 232 (1986): 861–63. A skeptical response to these data is provided by Margaret Wray, Barrett Klein, Heather Mattila, and Thomas Seeley, "Honeybees Do Not Reject Dances for 'Implausible' Locations: Reconsidering the Evidence for Cognitive Maps in Insects," *Animal Behaviour* 76 (2008): 261–69.

52. Randolf Menzel and Martin Giurfa, "Dimensions of Cognition in an Insect, the Honeybee," *Behavioral and Cognitive Neuroscience Reviews* 5 (2006), p. 26.

53. Menzel, "Searching for the Memory Trace in a Mini-Brain," *Learning and Memory* 8 (2001): 53–62.

54. Shaowu Zhang, James Bartsch, and Mandyam Srinivasan, "Maze Learning by Honeybees," *Neurobiology of Learning and Memory* 66 (1996): 267–82.

55. Mariana Gil de Marco and Randolf Menzel, "Learning Reward Expectations in Honeybees," *Learning and Memory* 14 (2007): 491–96.

56. Thomas Ings and Lars Chittka, "Speed-Accuracy Tradeoffs and False Alarms in Bee Responses to Cryptic Predators," *Current Biology* 18 (2008): 1520–24.

57. Uwe Homberg and John Hildebrand, "Serotonin-Immunoreactive Neurons in the Median Protocerebrum and Suboesophageal Ganglion of the Sphinx Moth *Manduca sexta*," *Cell and Tissue Research* 258 (1989): 1–24.

58. Randolf Menzel and Martin Giurfa, "The Concepts of 'Sameness' and 'Difference' in an Insect," *Nature* 410 (2001): 930–33.

59. Reviewed in Menzel and Giurfa, "Dimensions of Cognition in an Insect, the Honeybee," pp. 24–40.

60. Carruthers, *Architecture of the Mind*, p. 74.

61. Bermúdez, *Thinking Without Words*, p. 288.

62. Fred Dyer and Thomas Seeley, "Distance Dialects and Foraging Range in Three Asian Honey Bee Species," *Behavioral Ecology and Sociobiology* 28 (1991): 227–34.

63. Gareth Evans, *Varieties of Reference* (Oxford: Oxford University Press, 1982), p. 101.

64. Evans, *Varieties of Reference*, p. 104.

65. Carruthers, *Architecture of Mind*, pp. 78–81.

66. Carruthers, *Architecture of Mind*, p. 79.

67. Menzel and Giurfa, "Dimensions of Cognition in an Insect, the Honeybee."

68. Menzel and Giurfa, "Dimensions of Cognition in an Insect, the Honeybee," p. 36.

69. Carruthers, *Architecture of Mind*, p. 73.

70. Kathleen Akins, "Of Sensory Systems and the 'Aboutness' of Mental States," *Journal of Philosophy* 93 (1996): 337–72; Clark, *Being There*.

71. Clark, *Being There*, p. 145.

72. Thomas Seeley, *The Wisdom of the Hive* (Cambridge, Mass.: Harvard University Press, 1995).

73. Richard Dawkins, *The Selfish Gene* (Oxford: Oxford University Press, 1989).

74. Seeley, "Social Foraging by Honeybees: How Colonies Allocate Foragers among Patches of Flowers," *Behavioral Ecology and Sociobiology* 19 (1986): 343–54; "The Tremble Dance of the Honey Bee: Message and Meanings," *Behavioral Ecology and Sociobiology* 31 (1992): 375–83; and "Honey Bee Colonies Are Group-Level Adaptive Units," *American Naturalist* 150 (1997): 22–41.

75. Seeley, "Honey Bee Colonies Are Group-Level Adaptive Units."

76. Menzel and Giurfa, "Dimensions of Cognition in an Insect, the Honeybee," pp. 24–40.

77. Seeley, "Division of Labor Between Scouts and Recruits in Honeybee Foraging," *Behavioral Ecology and Sociobiology* 12 (1983): 253–59; "Social Foraging by Honeybees"; "Honey Bee Colonies are Group-Level Adaptive Units"; and Corinna Thom, Thomas Seeley, and Jürgen Tautz, "Dynamics of Labor Devoted to Nectar Foraging in a Honey Bee Colony: Number of Foragers Versus Individual Foraging Activity," *Apidologie* 31 (2000): 737–38.

78. Thomas Seeley, Scott Camazine, and James Sneyd, "Collective Decision-Making in Honey Bees: How Colonies Choose among Nectar Sources," *Behavioral Ecology and*

Sociobiology 28 (1991): 277–90; and Seeley and William Towne, "Tactics of Dance Choice in Honey Bees: Do Foragers Compare Dances?" *Behavioral Ecology and Sociobiology* 30 (1992): 59–69.

79. Seeley, "Honey Bee Colonies are Group-Level Adaptive Units," pp. 28–31.

80. Madeline Beekman, Robert Fathke, and Thomas Seeley, "How Does an Informed Minority of Scouts Guide a Honeybee Swarm as It Flies to its New Home?" *Animal Behaviour* 71 (2006): 161–71; Kevin Passino and Thomas Seeley, "Modeling and Analysis of Nest-Site Selection by Honey Bee Swarms: The Speed and Accuracy Trade-Off," *Behavioral Ecology and Sociobiology* 59 (2006): 427–42; Thomas Seeley and Susannah Buhrman, "Nest-Site Selection in Honey Bees: How Well do Swarms Implement the 'Best-of-N' Decision Rule?" *Behavioral Ecology and Sociobiology* 49 (2001): 416–27; and Thomas Seeley and P. Kirk Visscher, "Choosing a Home: How the Scouts in a Honey Bee Swarm Perceive the Completion of Their Group Decision Making," *Behavioral Ecology and Sociobiology* 54 (2003): 511–20.

81. Seeley and Buhrman, "Nest-Site Selection in Honey Bees."

82. Beekman et al., "How Does an Informed Minority of Scouts Guide a Honeybee Swarm as it Flies to its New Home?"; Passino and Seeley, "Modeling and Analysis of Nest-Site Selection by Honey Bee Swarms"; Thomas Seeley, "Consensus Building During Nest-Site Selection in Honey Bee Swarms: The Expiration of Dissent," *Behavioral Ecology and Sociobiology* 53 (2003): 417–24; Seeley and Buhrman, "Nest-Site Selection in Honey Bees"; and Seeley and Visscher, "Choosing a Home."

83. Christof Koch and Shimon Ullman, "Shifts in Selective Visual Attention: Towards the Underlying Neural Circuitry," *Human Neurobiology* 4 (1985): 219–27; and Ernst Niebur and Koch, "Control of Selective Visual Attention: Modeling the 'Where' Pathway," *Neural Information Processing Systems* 8 (1996): 802–8.

84. Robert Desimone and John Duncan, "Neural Mechanisms of Selective Visual Attention," *Annual Review of Neuroscience* 18 (1995): 193–222.

85. Daniel Dennett, *Sweet Dreams: Philosophical Obstacles to a Science of Consciousness* (Cambridge, Mass.: MIT Press, 2005), p. 141.

SUGGESTED READING

BERMÚDEZ, JOSÉ LUIS. *Thinking Without Words*. New York: Oxford University Press, 2003.

CARRUTHERS, PETER. *The Architecture of the Mind*. Oxford: Oxford University Press, 2006.

DARWIN, CHARLES. *The Formation of Vegetable Mould through the Action of Worms*. Chicago: University of Chicago Press, 1985.

DENNETT, DANIEL C. *Kinds of Minds*. New York: Basic Books, 1996.

FITCH, W. TECUMSEH. "Nanointentionality." *Biology and Philosophy* 23 (2008): 157–77.

GALLISTEL, RANDY. *The Organization of Learning*. Cambridge, Mass.: MIT Press, 1990.

HAUSER, MARC D. *Wild Minds*. New York: Henry Holt, 2000.

JAMES, WILLIAM. *The Principles of Psychology*. Cambridge, Mass.: Harvard University Press, 1890.

MILLIKAN, RUTH. *Language, Thought, and Other Biological Categories*. Cambridge, Mass.: MIT Press, 1988.

SEELEY, THOMAS. *The Wisdom of the Hive*. Cambridge, Mass.: Harvard University Press, 1995.

CHAPTER 16

....

BEYOND ANTHROPOMORPHISM: ATTRIBUTING PSYCHOLOGICAL PROPERTIES TO ANIMALS

....

KRISTIN ANDREWS

In the context of animal cognitive research, "anthropomorphism" is defined as the attribution of uniquely human mental characteristics to nonhuman animals. Those who worry about anthropomorphism in research are confronted with the question of which properties are uniquely human. As animals, humans and nonhuman animals[1] share a number of biological, morphological, relational, and spatial properties. In addition, it is widely accepted that humans and animals share some psychological properties such as the ability to fear or desire. These claims about the properties animals share with humans are often the products of empirical work.

Prima facie, one might think that in order to justify the claim that a property is uniquely human, it would be necessary to find empirical evidence supporting the claim that the property is not found in other species. After all, the goal of animal cognition is to determine what sort of cognitive abilities animals use. If scientists were to discover that a cognitive property wasn't found in any species except the human species, then the claim that some other animal had that property would be a false charge, and would be an example of anthropomorphism.

However, in practice anthropomorphic worries play a pre-empirical role. Research programs are charged with being anthropomorphic because they are examining whether some species has some feature that the critic believes only animals can have, based on some pre-empirical consideration. This charge is sometimes defended by theoretical arguments about the nature of the ability or property being examined.

A number of features have been described as uniquely human on theoretical grounds, including psychological states such as beliefs and desires, personality traits such as confidence or timidity, emotions such as happiness or anger, social organizational properties such as culture or friendship, moral behavior such as punishment or rape. For convenience, I will refer to the members of the class as "psychological properties." J. S. Kennedy, a visible critic, includes feeling, purpose, intentionality, consciousness, and even cognition in his list of psychological properties that are incorrectly attributed to animals.[2] Among the critics, there is considerable disagreement about what counts as an anthropomorphic attribution, and this alone should raise questions about the charge.

We can identify two different questions about the attribution of psychological properties to animals in scientific contexts. First we can ask whether it is scientifically respectable to examine questions about the mental, psychological, cultural, and other such states of animals. Those who bemoan anthropomorphism think that we have no warrant for asking such questions. I will look at these worries and will argue that there is no special problem inherent in asking and answering such questions.

The second question arises with an affirmative answer to the first. After establishing that it is scientifically respectable to investigate whether an animal has a psychological property, we must then ask how such an investigation is to be carried out. In answer to the question of how we can study the psychological properties of animals, I will propose that we use an approach to the attribution of psychological features to animals that is based on the approach we use for prelinguistic children. A specific psychological attribution will be warranted if it takes into account the species and cultural normal behavior, it has predictive power, and it mirrors the attribution of a similar property in prelinguistic infants. This is not to say that nonhumans can have only the psychological properties that infants have. It *is* to say that the general approach, modified so as to be species appropriate, and the degree of evidence we use when studying infant psychology should be used when we study animal psychology. I will show how this method can be used to examine different kinds of psychological properties.

In some current research programs, researchers are following methods that fall on the side of the methods I will propose, but in other programs violations of these methods lead to what I will argue are false attributions of psychological properties to nonhuman animals.

Can We Study Animal Psychology?

One worry about allowing scientists to ask about the psychological states of animals is that the scientists' own subjective biases may affect the work. The worry has been stated in different ways. For one, psychological attributions in general might be thought to be subjective interpretations of behavior. If psychological properties are in the eye of the beholder, then they are not appropriate objects of scientific study—except, of course, at a metalevel; an anthropologist might examine the behavior of attributing psychological states as a cultural practice. But if this worry is well-grounded, it holds just as well for research on humans as it does for animal research. If the critics have this in mind, they would be forced to reject most of contemporary cognitive approaches to human psychology since human psychology research works to find real human psychological properties. Giving up human psychology in order to avoid giving in to animal psychology is a price few would want to pay.

Another way of understanding the concern that scientists ought not examine the psychological properties of animals because it will lead to biased results is that humans are unable to control their tendency to see psychological properties wherever they look, so if they look for psychological properties in animals they will certainly "find" them. Humans begin to attribute intentionality at a young age, and overattribution is ubiquitous among small children. Those scientists who are willing to see animal behavior as intentional and explained by reference to psychological properties might be stuck in such a youthful developmental stage. This bias seems to be what G. H. Lewes had in mind when he criticized his contemporaries Charles Darwin and George Romanes for talking about animal psychology. He wrote that "we are incessantly at fault in our tendency to anthropomorphize, a tendency which causes us to interpret the actions of animals according to the analogies of human nature."[3] Kennedy writes that "anthropomorphic thinking about animal behavior is built into us. We could not abandon it even if we wished to,"[4] though he also believes that it needs to be corrected.

The critics seem to suggest that the scientist must avoid this bias by moving far in the other direction: the bias toward seeing all animal behavior as intentional can only be confronted by denying that any animal behavior is intentional. While I do not deny the existence of the bias, I do deny some features of the proposed response to the bias, which is overreactive. Humans are replete with biases that affect our ability to make accurate judgments, such as the gambler's fallacy (e.g., thinking that repeated losses in roulette indicate that there will be future repeated wins), observer-expectancy effect (e.g., reinterpreting your past expectation so that it matches with reality), or the primacy effect (e.g., accepting as most plausible the first explanation you hear). The critic who says that the existence of a bias makes it impossible for us to do science related to that bias would be forced to deny the possibility of science at all! Thus, while scientists need to acknowledge the bias, its existence does not entail the impossibility of scientifically investigating animal psychological properties. Rather, it speaks to the need for a scientific methodology designed to counter the bias.

There are two other theoretical concerns that motivate anthropomorphic worries. One is that having language is necessary for having many if not all psychological properties. The other is that all behavior can be explained by Thorndike's laws, associative learning, or classical conditioning. Both of these concerns, I think, are unjustified, or at least are limited in scope and potentially misleading. Let's take the second concern first.

The behavioristic principles of learning are used to explain behavior—for example, we can explain why Pavlov's dog salivated at the sound of a bell by indicating that Pavlov presented the dogs with the bell before he presented the dogs with the food, that it is a natural reflex that food produces salivation in dogs, and that such training is an example of classical conditioning. The critic thinks that all animal behavior can be likewise explained by reference to one of the behavioristic laws. However, to defend that claim, behavior types must be examined one by one, and that requires that we first have a catalog of every behavior type for each species. Biologists, psychologists, and anthropologists regularly uncover new behaviors, and so any claim that behavioristic principles can explain all animal behavior are premature. In addition, there are many behaviors, such as chimpanzee insight learning[5] and capuchin monkey finger-in-eye games[6] that do not appear to admit of non-psychological explanations.

The worry that language is necessary for (many) psychological properties is similarly flawed. There is little concern about avoiding psychological research with prelinguistic human infants due to adultomorphic concerns (i.e., attributing adult psychological properties to children), and if we can ask about psychological properties in some individuals who lack language, having language cannot be a necessary condition for having any psychological properties. One might object, however, that since the child is a potential language-user, a scientist is more justified in ascribing psychological traits to an individual who will eventually use language. But there are at least two reasons to reject this response: not all infants gain language and using language is just one kind of behavior.

In addition, the critic who says that language is necessary for thought may be relying on an argument from ignorance by claiming that language is the only possible vehicle to support the cognitive processes required to explain how thinkers are able to make logical inferences between propositions. For example, a familiar argument against the view that animals have beliefs is that to have a belief one must be able to represent a propositional attitude, and the only way to represent a propositional attitude is through language. But this claim is based on a number of controversial assumptions. For one, it assumes that an external spoken, written, or gestural language is necessary to have an internal language of thought. It also assumes that belief requires representation, which is a view that has been challenged by recent work in philosophy and cognitive science.[7] Finally, the argument assumes that there are no alternative representational vehicles other than language, a view that has similarly been challenged.[8]

Rather than starting with these theoretical commitments, a scientist can remain agnostic and examine whether there are target behaviors that seem to be explicable only in terms of an animal having a belief. Such empirical work can help to promote the theoretical research by providing a larger class of relevant data. The critics who see animal cognition research as anthropomorphic want to end such research; they do not promote it. And while it is true that if scientists stopped investigating whether animals have psychological properties, they would be less likely to make false claims, they would also be doing less in the way of science. The general principle that we should avoid false claims should not cause us to stop making claims altogether, since that would also result in making fewer true claims. The best scientific methods are those that will maximize the number of true claims over the number of false ones, not the methods that will avoid false claims altogether.

Finally, related to the above discussion there is a methodological worry about anthropomorphism in animal cognition research. The field of psychology has long embraced a methodological rule of thumb that might be seen as a conservative principle: one should always avoid the risk of making a type-1 error in favor of the risk of making a type-2 error. The errors are defined as follows:

> Type-1 error – Rejecting a null hypothesis when it is in fact true.
> Type-2 error – Failing to reject a null hypothesis when it is in fact not true.

The null hypothesis is what is assumed unless and until investigation shows it to be false. In the case of animal cognition, the null hypothesis is that animals lack the particular psychological property under investigation. For example, in what is known in psychology as the theory of mind research program, the null hypothesis is that animals do not have the ability to consider others' mental states, or to attribute beliefs and desires to themselves or others. In the literature, this ability to attribute mental states is called a "theory of mind." So, a type-1 error in this context can be seen as a false positive, whereas a type-2 error would be a false negative. If in fact chimpanzees do not have a theory of mind, and some researcher concludes that the chimpanzee does have a theory of mind, then the researcher is committing a type-1 error. Some critics of animal cognition studies take this methodological principle as reason not to accept animal psychological properties; because we fail to have the required evidence that, for instance, the chimpanzee has a theory of mind, we conclude instead that the chimpanzee does not have a theory of mind.

There are several problems with this line of reasoning. First, the methodological principle does not permit the inferences to the nonexistence of chimpanzee theory of mind; rather, it requires that we remain agnostic about chimpanzee theory of mind. From this it would follow that we don't know whether or not having a theory of mind is uniquely human, and hence, we don't know whether it is anthropomorphic to attribute a theory of mind to an animal.

Second, it has been argued that the acceptance of the methodological rule of thumb has resulted in a behavioristic bias for animal cognition research. One piece of evidence for the supposed behaviorist bias is that while false positives in animal cognition research have a widely recognized name ("anthropomorphism"), false negatives do not. Some have argued that not having a well-established name for false negatives in animal cognition research leads researchers to have a behavioristic bias, and terms have been introduced as an attempt to combat this worry. Frans de Waal calls the false negative error "anthropodenial,"[9] while Maxine Sheets-Johnstone calls it "reverse anthropomorphism."[10]

In his discussion of the role that type-1 and type-2 errors play in animal cognition research, Sober says that both errors are:

> maxims of "default reasoning." They say that some hypotheses should be presumed innocent until proven guilty, while others should be regarded as having precisely the opposite status. Perhaps these default principles deserve to be swept from the field and replaced by a much simpler idea—that we should not indulge in anthropomorphism *or* in anthropodenial until we can point to observations that discriminate between these two hypotheses. It is desirable to avoid the type-1 error of mistaken anthropomorphism, but it is also desirable that we avoid the type-2 error of mistaken anthropodenial.[11]

While I agree with Sober's analysis, I think that the worst error here is anthropodenial because it hinders the progress of science. As a part of the scientific process, one must be willing to make a claim that turns out to be false, whether that claim is one that is antecedently accepted or not. For progress in science to be possible, one must be open to being wrong, one must ask questions even when the answer turns out to be no, and one must challenge the null hypothesis in order to examine its accuracy. The willingness to be wrong is a willingness to make type-1 errors in the course of the acquisition of new knowledge. Scientific progress does not take a linear path; there are bumps and errors along the way. He who wants to avoid error at all cost ought not be a scientist.

This concludes my discussion of the common criticisms brought against the animal cognition research program and its investigation into the psychological properties of animals. In responding to those criticisms, I intend to have defended the scientific respectability of empirically studying the psychological properties of animals. Now that it is established that we *can* study animal psychological properties, the question that we must answer is *how* we can engage in such study.

How Can We Study Animal Psychology?

Given that we can and should investigate issues in animal cognition, are there general methodological principles that we can use to do so? To help answer this question, we can look at a respected field of psychology that shares many of the challenges of research in animal cognition discussed previously. Developmental psychology research on prelinguistic infants also deals with subjects who cannot tell them what they think or how they feel. When devising research programs, both investigators who study human infants and investigators who study nonhuman animals propose to examine the minds of their subjects without relying on linguistic behavior.

The fact that we can't talk to these subjects might be seen as a limitation of the research programs, but it should not be so regarded. In studying older children and adults, psychologists rely heavily on linguistic behavior. The measures are more subtle than the introspective methods of the nineteenth century, but current research still assumes that language permits us a more direct window into the mind than does nonverbal behavior. Of course, linguistic behavior is still behavior, and the relationship between language and thought is still hazy at best. The idea that language unproblematically gives us a window to the mind ignores both the worries of the philosophers and the psychologists. W. V. O. Quine's discussion of radical translation points to the difficulty of coming to understand, from the perspective of one language-user, what someone from another language group is thinking.[12] Donald Davidson's notion of radical interpretation is based on similar worries about our ability to understand what others mean.[13] For both Quine and Davidson, to understand others we must begin by accepting the principle of charity and take the behavior we observe to be rational, noncontradictory, and derived from the same sorts of causes as our own behavior is. That is, to get the interpretative task off the ground we must see others as like us in an important sense; we must observe from a particular interpretive stance.

Psychologists too worry about using linguistic utterances as a window into the mind. In contrast to the old philosophical principle of privileged access, according to which individuals have private and privileged access to the contents of their minds, and in contrast to the old psychological method of introspection, the New Unconscious research program in social psychology is finding that in many cases, our reports of our own mental states are confabulations. In one of the first research papers in this field, Richard Nisbett and Timothy Wilson presented evidence that people do not know why they do the things they do and will make up stories to account for how they solved physical problems.[14]

If we want to learn about the mechanisms of mind, then both the philosophers and the psychologists warn us away from giving linguistic behavior too much of a privileged position. Rather than taking research on infant and animal minds to pose a special challenge, we could equally well treat it as more straightforward than research on humans, since when studying animals and pre-linguistic infants, we will not be misled (either intentionally or unintentionally) by the participant's linguistic utterances.

Despite several critics (e.g., John Newson's worry about "adultomorphism,"[15] and Robert Russell's worry that our attribution of psychological properties to children is a non-universal cultural practice not warranted by science[16]), research on infant cognition is flourishing, and scientists express little concern about ascribing psychological properties to human children. As an example of suitable ascription to infants, it is generally accepted in development psychology that children have emotions, beliefs, and desires and can communicate some of these mental states by one year of age.[17] This is so despite the fact that one-year-old children typically do not have the ability to string words together to form propositions, and often have not yet begun to produce any words, which they typically do between twelve and eighteen months.

Why is it that the adultomorphism concern has less effect on child development research than the anthropomorphism concern has on animal cognition research? One possible justification for the difference may be that most human infants do eventually develop language, and this potentiality could be exploited by the researcher as justification for making an attribution to a preverbal infant. This attempt at justification fails, however, because the mere potential for language doesn't help either with the concerns about the vehicle of the mental state or with concerns about specifying the content of the mental state. If one does not yet have external language, one cannot be thinking in language—unless external language is not necessary for thinking in language, and in that case there would be no reason to exclude animals from the class of thinkers.

I think there is an explanation for the different standards for studying human infants and nonhuman animals, but that this explanation doesn't justify it. The difference can be explained in term of the kind of relationships investigators have with their subjects. The traditional approach to science involving animal subjects has been to keep a distance between the scientist and her research. Researchers who violate this principle are often thought not to be objective. For example, Jane Goodall's insistence on naming the chimpanzees she was observing was unconventional and caused some worry about her ability to remain objective, as did her use of gender pronouns to refer to the chimpanzees.[18] While naming ape subjects has now become standard, the rationale behind the criticism remains. Even today, the quest for scientific rigor and objectivity still strongly encourages researchers to take a position of detachment and neutrality toward their research subjects. The degree of success one manages depends at least partially on how the scientist interacts with her subject. When the subject is in a cage, there is metaphorical distance between the caged and the free individual that can have affective consequences in the researcher. When the subject is across a field being observed using binoculars, the physical distance can also cause a certain emotional distance. But when a researcher is working with a human child, it is almost impossible to avoid all forms of emotional response to the subject. Humans are wired to have emotional responses to infants (and, as Konrad Lorenz pointed out, to animals that resemble human infants by having big eyes, big heads, and little noses).[19]

The relationship between human researchers and human subjects is strengthened due to their shared physical and social world. Psychologists see human infants in their normal physical and social environment, and often have spent much time interacting with infants socially or as caregivers, teachers, or other similar roles during practical aspects of their training. Psychologists who plan to work on infants typically have a lot of lay expertise with children and develop commonsense views about infants that inform their research. I will argue that this lay expertise forms an undeniable and beneficial starting position for the researcher's future work.

For many working in animal cognition, there does not exist the same sort of shared social and physical environment between researcher and subject. Researchers who focus on experimental laboratory research may never see the subject in its typical ecological and social environment. They may not spend time with their subjects outside of the research context. They may, indeed, work hard not to develop an emotional or sympathetic relationship with their research subjects. In addition, those who are working with species who exist in very different contexts from us, such as water-living mammals or avian species, are limited in their ability to develop the same sort of folk expertise given the difficulty with spending large chunks of time with individuals of those species outside of the research context. In support of this view is the finding that fieldworkers are more likely to attribute psychological properties to animals than are those working in a controlled environment such as a zoo or a lab.[20] While some might see this as evidence that field researchers are biased, it may also be evidence that field researchers have better access to the cognitive and affective capacities of their subjects than do researchers on captive animals.

What difference in methodology between field research and experimental research of a captive animal accounts for this difference? It may be that those who choose to do fieldwork are more prone to attribute psychological states to begin with. But there is another possibility: experience in the field may involve the development of a skill that makes such researchers more likely to understand their subjects, just as investigators who study infants develop skills associated with handling and regulating infants. Fieldworkers who engage primarily in observational studies typically spend much more time with their subjects than do experimentalists, and they have to learn how to observe before they begin to see what is going on. It isn't until after an observer learns how to see, and learns the typical behaviors of the group being studied that she can develop an ethogram—a catalog of species-normal behaviors, and the functional roles associated with them—and only then can she conduct the formal observational study. The pre-study period of observation allows the scientist to get to know her subjects and understand the individual differences in a group, so it also gives her a baseline of normal behavior. I suspect that this sort of experience results in the development of a skill that allows the fieldworker to notice intentional behavior, much as experts across fields come to notice saliencies that otherwise would have been perceived as noise. Researchers who are working with students know that when a student first enters the field she has to learn how to see, much as X-ray technicians have to learn how to read X-rays. Graduate students who are collecting data in the field for the first time will discard their first weeks or months of data, or not take data during that time, because they are still developing the skill of observing. In classical ethology, this preliminary stage of observation is called "reconnaissance observation," and new students are given exercises to develop skills in the art of seeing.[21]

Others who are not explicitly trained how to see can also come to develop an understanding of what behaviors mean by implicitly recognizing the context of the behavior. This is true both of species-typical behaviors, and individual differences. For example, when I was at Samboja Lestari Orangutan Rehabilitation Centre, there was a young male orangutan named Jovan who had a special trick: he would suck on his thumb to disarm the human working with him, and then grab something from the caregiver (a pen, a backpack, etc.) and run away with it. Someone who didn't know Jovan would fail to interpret his thumb sucking correctly and fall for the trick, whereas a caregiver who was familiar with Jovan knew to protect her gear, because she knew what he *wanted*. We might think of caregivers and nonacademic observers of animal behavior as being "trained" by the situation insofar as they receive feedback from the animals that may or may not match their expectations.

The fieldworkers' and caregivers' experiences are notably similar to the sort of experiences humans who work with children have. We live in the field, as far as the study of infant cognition goes. There are few difficulties conducting naturalistic observations of children; there is no need to travel, live in makeshift camps, or deal with unfamiliar environments. Children are everywhere, and child development researchers may have thousands of hours of watching before they even begin their formal studies. It is this experience with children that explains in part why there is little scientific worry about investigating the psychological properties of human infants, and I propose that what fieldworkers and infant researchers have greater access to than laboratory experimentalists is what I have called *folk expert opinion*.[22]

Folk expertise develops with experience with a taxon, a developmental stage, or an individual. It is what one has when one knows one's subject well. Most humans are folk experts on human behavior given their experience with others. Parents are folk experts of their children, and career nannies have folk expertise about children generally. Nurses and caregivers become experts about their charges with dementia or other geriatric mental disabilities. Caregivers have acquaintance knowledge of their charges, a kind of knowledge that scientists or other formal experts may lack. Folk experts on animal behavior include human caregivers, technicians, and others who work with captive animals, as well as individuals who have spent a significant amount of time observing the behavior of individual animals in the field.[23] A folk expert can also be an academic expert, who has studied the species formally, but for an academic expert to become a folk expert, she needs to gain additional knowledge through direct experience observing or interacting with members of the species. A researcher's stock of anecdotes can be seen as part of her folk expertise. For example, while I was interning as a dolphin trainer at The Kewalo Basin Marine Mammal Laboratory in the 1990s, I observed a male dolphin force copulation on a protesting female, and this observation became part of my knowledge about dolphins. Having seen that forced copulation happens in dolphins, and seeing that the female dolphin was struggling to obstruct the sex act, I saw that sex under these conditions was aversive to the female. With this kind of knowledge, caregivers can take precautions to minimize the risk that the female is subjected to the experience again.

Though most humans are not experts in the behavior of exotic animals, most humans with children are experts in child behavior, and most humans are experts in some areas of adult human behavior. We gain this status as experts through our experiences interacting with people rather than through explicit instruction or formal training; expertise is something we have to some degree even before taking our first psychology class. We know that people have psychological properties, and we know something about how these psychological properties are related to one's environment and behavior. The academic expertise that is gained through formal education builds upon the folk understanding of human psychology, and while students learn about mechanisms and breakdowns of normal mental events, and while they may learn that some parts of commonsense psychology are false, the science that led to the discovery of mechanisms, deficits, and failures of folk psychology is itself based on the lay expertise humans have about human minds. Starting at a relatively young age, we come to learn that classes of behaviors can be described using a particular term, and that application of the term can help us to formulate predictions about future behavior, as well as to make sense of the behavior by embedding it into a larger explanatory network. A child can soon come to think that a mean child is one who will not share his toys, who pulls hair, and who doesn't wait his turn for the slide. Calling this individual "mean" helps the child to understand how to deal with him and to predict what his future actions are likely to be.

In research on humans, folk expertise is sometimes explicitly recognized and used, for example, in some psychological assessment instruments. Parents, teachers, and caregivers answer questions about the target individual's behavior, emotional state, and so forth, and this information can be used by researchers to make judgments about, for example, personality or social adjustment. Since not all the folk expert's knowledge is directly available to her, psychologists interested in this knowledge design instruments to extract the knowledge. Psychological instruments are calibrated in part on the basis of their functionality. The results of these instruments are functional if they produce novel accurate predictions, and if the prediction bears out, the attribution is deemed accurate.

For example, the "Caregiver-Teacher Report Form for Ages 1 1/2–5" from the *Child Behavior Checklist*, which is designed to measure children's emotional and social development, is a checklist that is presented to parents or teachers. The caregivers are taken to be folk experts with knowledge that can be extracted using these measures, and they are asked to rate children's behavior and traits.

Having folk expertise is only the first stage of doing good science with animals. Folk experts can be wrong, just as parents can be wrong about their children. My suggestion is not to forgo science in favor of the folk experts' common sense. The starting point for controlled study of infant behavior is much more robust than is the starting point for controlled study of animal behavior, and I propose that the science of animal cognition research will progress only if we are able to improve its foundations.

Nonetheless, some may feel quite uncomfortable with the role I am giving to folk expertise as the foundation for doing good science. The worry is that relying on folk expertise is a bias, and will lead to false conclusions about animals' cognitive abilities. For example, one might worry about people's folk expertise regarding their pets as a prime example of unwarranted psychological attribution. The dog owner who sees that her dog destroyed the furniture may interpret her dog's head-hanging behavior as expressing guilt, despite the possibility that there exists a more parsimonious explanation from associative learning: that the dog had been conditioned to expect a scolding after similar acts in the past. Parents are notorious for suffering from similar delusions as pet owners, and making over-attributions when simpler explanations suffice. Given these obvious problems with folk expertise, shouldn't we rather try to eliminate it from our research on human children, rather than bring it into our research on animals?

I answer this question with an emphatic "No," given the fecundity this method has had for our study of infant behavior. From our science, we have gained a greater understanding of when attributions are false and when they are accurate, and we have done this by using methods that capitalize on our folk understanding of infants. Consider for example the habituation-dishabituation method of studying infants.[24] The method involves showing a human infant a stimulus until she is habituated to it, as indicated by either eye-gaze or by reduced sucking on a pacifier. The infant is then shown a new stimulus that differs from the original in some subtle way. If the infant's eyes move back toward the stimulus, or if the sucking rate increases, researchers conclude that the infant notices the difference. This method gives us interesting results only because our folk expertise of infants allows us to conclude that children are interested in things they look at, and that a high rate of sucking indicates interest in the stimulus. This appeal to the child's interest is an appeal to a basic mental process that has not been determined by additional scientific investigation, but from our infant folk psychology.[25]

In order to make similar progress on animal cognition research, we must work toward first achieving folk expertise in the species to be studied. Fieldworkers, who spend years observing individuals, come close to gaining the kind of folk expertise that infant researchers gain so easily. Folk expertise about a species will include knowledge and understanding about stages of development, culture, and species-normal behavior. Folk expertise about an individual will include knowledge about the individual's typical behaviors, and the extent to which those behaviors reflect individual differences in the species or developmental stage. My claim is that animal cognition researchers should develop folk expertise of their subjects, and base their research programs on knowledge gained from experience with the species in its natural physical and social environment, just as human-infant researchers do.

I turn now to examples of animal cognition research to examine research programs that both followed and did not follow the advice I am giving.

Two Problematic Cases: Gorilla Mirror Self-Recognition and Chimpanzee Economic Games

Not having folk expertise can lead to studies that don't take into account species-normal behavior. I will discuss two cases of studies that suffer from this problem.

One example of this problem can be seen in the early research on mirror self-recognition in gorillas. The research program on mirror self-recognition began with the work of Gordon Gallup. While there was anecdotal evidence that chimpanzees recognized themselves in mirrors, there was no formal test until Gallup introduced the mirror test for chimpanzees. He exposed four juvenile chimpanzees to a mirror for eighty hours, during which time the chimpanzees first responded socially to the mirror image before they began using the mirror to explore their own bodies. Chimpanzees would examine their teeth and other parts of their body that are not accessible without the aid of a mirror. After this initial exposure, Gallup gave the chimpanzees the mark test by placing red marks on their faces while the animals were anesthetized. After they woke from the anesthesia, the chimpanzees were observed for some time and then exposed to a mirror. Gallup found that chimpanzees began to touch the marks on their face after being given the mirror, and that they touched the mark significantly more often in the presence of a mirror than when no mirror was present. Given that chimpanzees passed the mirror task in this way, Gallup concluded that chimpanzees understand that the image in the mirror is a reflection of the self and hence they have an understanding of self.[26]

This study was devised for use with chimpanzees, but the methods of Gallup's mark test were also used on other primates, including human children. While children of eighteen months and orangutans passed the task, early research on gorilla mirror self-recognition concluded that adult gorillas, unlike the other great apes, do not respond to their reflection. Given this negative finding it was suggested that gorillas might "be the only great ape which lacks the conceptual ability necessary for self-recognition."[27] And while subsequent studies confirmed the negative result, there is clear evidence that one gorilla does recognize herself in the mirror.[28] Koko, a gorilla who started learning sign language at one year of age, started to spontaneously show mirror-guided self-directed behaviors when she was about three-and-a-half years old, picking her teeth, combing her hair, and dressing up in wigs, hats, and makeup in front of the mirror. Using a variant of Gallup's task, Koko was tested when she was nineteen years old, and she, like the chimpanzees, touched the mark when exposed to the mirror significantly more often than when there was no mirror present. These findings were cross-validated by asking her questions about her reflection. In response to the "Who is that?" question, Koko signed in response, "ME THERE KOKO GOOD TEETH GOOD."

This evidence strongly supports the claim that Koko recognizes herself in the mirror and thus raises the question of why other gorillas fail the mark task. One might initially think that Koko's language training provided her with the kinds of concepts necessary for passing the mark task, but since human children can pass the test before they have language, it isn't clear why having a symbolic communicative system should aid in this task.

Drawing on their folk expertise with gorillas, Patterson and Cohen suggest another explanation for the failure of other gorillas to pass the mirror test: The gorillas found the presence of unfamiliar experimenters aversive. They write:

> It has been our experience that the presence of strangers profoundly affects gorilla behavior. We have found that it can take from several months to a full year for Koko and Michael to habituate to the presence of a new caretaker....In each of the previous formal self-recognition studies with gorillas, experimenters who were not the gorillas' caretakers were in the room with them in very close proximity to the mirror....Averting their gaze from strangers is a common

behavior in gorillas. Observed social responses to the mirror may have been elicited by the experimenters, whereas mirror gazing and self-directed behaviors may have been inhibited by their presence.[29]

This explanation points to a difference between chimpanzees and gorillas: only gorillas tend to avoid the gaze of strangers. When first exposed to a mirror, the chimpanzees treated it socially, as if the reflection was a stranger. If gorillas find the gaze of strangers aversive, they have a strong motivation to avoid interacting with the mirror long enough to realize its function. That is, there is a difference in species-normal behavior between gorillas and other apes that can account for the early suggestions that gorillas don't have an understanding of the self. Modifications of the mark task, to account for the problem of motivation, found that gorillas do recognize themselves in mirrors.[30]

The lesson from the gorilla mirror self-recognition studies is that understanding the species is essential for devising studies to examine the psychological properties of an animal. Negative results are as important as positive results, and both need to be disseminated as part of the project of determining what psychological properties are attributable to a species or individual. But negative results are only valuable if they are based on a foundation of folk expertise. If the experimenter doesn't know that gorillas suffer from xenophobia and find the gaze of strangers aversive, then she might not take this variable into account when designing her study. But in such a case, the negative findings that result do not tell us anything about the gorilla.

A second example of a study that fails to follow the proposed method is the work on economic games in chimpanzees, research modeled on the groundbreaking work of Werner Güth and his colleagues.[31] In the original studies with humans, Güth and colleagues found that the traditional view of economic decision making, according to which people act according to the goal of maximizing resources for themselves and act rationally in pursuit of that goal, is false. Rather, by using an ultimatum game paradigm, the authors found that humans value the norm of fairness in the distribution of resources. In the standard version of the ultimatum game, two individuals are randomly assigned the roles of proposer and responder. The proposer is offered a sum of money and can decide to offer some portion of it to the responder. If the responder accepts the offer, both parties keep the money. However, if the responder does not accept the offer, then neither player gets anything. In humans, divisions that are perceived as unfair are often rejected, and the explanation for this is that humans have goals other than maximizing resources. Importantly, humans are sensitive to the interests and goals of others, and will make personal sacrifices in order to follow norms of fairness and cooperation, and punish transgressors.

It was found that chimpanzees and humans respond differently in ultimatum games. In order to test for fairness in apes, Keith Jenson and colleagues gave a version of the ultimatum game to a group of eleven chimpanzees in a controlled laboratory setting.[32] They found that the chimpanzees are more like the idealized rational man of traditional economic theory than are humans because chimpanzees, but not humans, fail to reject unequal divisions of resources. The conclusion is that chimpanzees, unlike humans, are not concerned with fairness, and are much closer to Adam Smith's *Homo economicus* than humans turn out to be.

However, the normative conclusion of this research doesn't take into account species-normal behavior. For one, the conclusion that chimpanzees are not concerned with fairness is inconsistent with evidence from ethology and other research programs.[33] Just to give one example, Frans de Waal writes, "I once saw an adolescent female interrupt a quarrel between two youngsters over a leafy branch. She took the branch away from them, broke it in two, then handed each one a part."[34] Interventions such as this are common among chimpanzee societies, as is punishment of negative actions.[35]

In addition, in the human studies the experiments are based on species- (or at least cultural-) normal behavior. Plausibly, in our society there is a norm that when you fall into unexpected wealth, you share that wealth with others. For example, we seem to expect lottery winners to share their winnings, and in fact one survey of U.K. lottery winners found that 83% of those who won over £50,000 in the lottery gave money to family members.[36] The existence of this norm is also demonstrated by the controversy surrounding Bill Gates's unexpected mega-wealth. Before he started the Bill and Melinda Gates Foundation, he was called a miser, and there was discussion about how the current class of the superwealthy differed from those of another time.

Given the existence of the norm "Unexpected wealth should be shared," when humans play the ultimatum game and their partner violates that norm, we should expect the actual result—that human players punish the partner who violates the norm. The poor offer to the partner is seen as unfair because it violates the norm of unexpected wealth, but if there were no such norm among humans, then we shouldn't expect the behavior to be seen as unfair or to be punished. The sentiment of fairness is based on a background expectation about normal behavior.

To claim that the chimpanzees do not have a sense of fairness simply because they fail a test based on the human norm of unexpected bounty is to assume that this human norm can be translated into chimpanzee societies. That is, we need to know whether species-normal behavior at the stage of development of the chimpanzee participants involves a norm about sharing unexpected bounties. And it seems that, for chimpanzees, there is no norm about sharing food resources; it is not part of their natural interactions.[37] While chimpanzees do share food in some circumstances, such as the meat that is acquired through cooperative hunting,[38] the ultimatum game does not reflect a norm about sharing jointly earned resources, and so the chimpanzee meat-sharing behavior cannot be seen as evidence for the existence of an unexpected wealth norm. That is, if this research had been based in folk-expertise on chimpanzee behaviors, researchers would never have asked whether chimpanzees have a concept of fairness by examining whether they accept inequitable distributions of goods.

When a human wins the lottery, she is expected to share some of her winnings with others. Nevertheless, when a human wins the sexual lottery and finds a good mate, she is not expected to share those winnings. To test the chimpanzee concept of fairness by examining whether they share food is like testing the human concept of fairness by examining whether they share sexual access to mates. If an extraterrestrial researcher were to study the human sense of fairness by examining whether humans share their mates, the researcher might hastily conclude that humans don't have a sense of fairness. But the more sensitive researchers might try the study with different goods. Humans don't generally share things like sexual access to partners, toothbrushes, and so forth. Chimpanzees don't generally share food. If we want to know whether the chimpanzee has a sense of fairness, we first need to see whether there appears to be any relevant chimpanzee norm that could be tested or otherwise examined, and to do that relies on having some folk expert understanding of the species.

The failures of the early gorilla mirror-recognition work and the chimpanzee fairness studies both stem from a lack of knowledge about the species. To avoid such problems, I suggest that research on animal psychological properties must begin with folk expert opinion, just as our infant cognition studies are. What's good for the infant studies is good for the animal studies, so far as it goes. That is, if starting with folk expert opinion is a warranted starting point for human studies, it should be a warranted starting point for animal studies. Opinion based on folk expertise is a largely unacknowledged starting point for the human studies, and should be seen as comprising an important aspect of the methodology of infant studies. If the method counts as good science for infant cognition research, then it should count as good science for animal cognition research as well. Correspondingly, if the method doesn't count as good science for animal cognition research, then we must be very skeptical of its use with nonverbal humans. I'm suggesting that acceptance of the methodology should be based on the same considerations, whether the subject is an infant human or a member of another species.

Successful use of this approach can be seen in some research programs. Let me present one area of research that begins with knowledge of species-normal behavior and relies on folk expertise in the attribution of psychological properties to animals: the research on personality traits.

A SUCCESSFUL CASE: ANIMAL PERSONALITY TRAITS

In human psychology, it is taken for granted that there are individual differences, and that these differences can be seen in terms of differences in personality traits. The Five-Factor Model (FFM) of human personality was developed to describe the way attributions of trait terms group together into statistically significant clusters, and it organizes personality into five domains: Neuroticism, Extraversion, Openness to Experience, Agreeableness, and Conscientiousness.[39] While there are some theoretical and methodological worries about the FFM, supporters of this approach have argued that most individual differences can be described using this model[40] and that there are underlying genetic factors related to these domains.[41]

An individual's personality traits can be assessed using an instrument such as the Five-Factor Personality Inventory (FFPI). While the FFPI can be administered to the subject who is asked to make self-ratings, it is thought to be more accurate when it is given to a number of individuals who know the subject well.[42] Administering the instrument to people who know the individual well is seen as more accurate than administering the instrument to the target subject, because the responses to questions about oneself invoke social goals such as image control.

The use of third parties to assess a target subject is a common approach of psychological instruments. Children's emotional and social development is assessed using *The Child Behavior Checklist*.[43] *The Vineland Adaptive Behavior Scales*[44] is used to assess the personal and social skills of children and low-functioning adults. Geriatric patients' social and functional impairments are assessed using the *Social-Adaptive Functioning Evaluation*.[45] All these instruments rely on third-party judgments in order to evaluate a target subject, and since they require that the third parties know the subject well, they explicitly rely on the knowledge of a folk expert. In some cases, there is no option but to use a third party's response to make judgments about the target subject; from children with delayed language skills to elderly people suffering from dementia, such tests are relied upon in order to assess the personality, social development, intelligence, emotional adjustment, communication skills, and other psychological factors of children who cannot speak for themselves and adults who are low-functioning or suffering from dementia.

These instruments rely on the judgments of caregivers who do have language and the relevant concepts, and who are folk experts on the individual being examined. The assumptions behind these instruments are that caregivers have knowledge of their charges, and that this knowledge can be extracted. However, the instruments do not take the views of the caregivers at face value. Instead, the folk expert opinion is used as raw data. The use of such instruments follows the proposed methodology for animal cognition research: Begin with folk expert opinion, and then use established scientific methods to determine whether a psychological property is attributable to an individual.

Given the widespread use of these kinds of instruments to assess psychological properties of individuals based on the folk expertise of those who know the target subject well, it is a natural extension to use this method to investigate personality in nonhuman species. Several species have been studied, but I will focus on the research on personality in great apes. To assess the existence of ape personality traits, researchers spoke with folk experts such as zookeepers and others involved in daily husbandry or training activities in order to develop an instrument for assessing personality using the same methods used in developing the human FFM,[46] and used this method to assess personality in chimpanzees[47] and orangutans, respectively. As with the development and implementation of the FFM, raters are given lists of adjectives and asked to rate an animal on a 7-point Likert scale (according to which

1 indicates total absence of the trait and 7 indicates extremely large degrees of the trait). Adjectives and descriptions on the orangutan scale include:

Defiant: Subject is assertive or contentious in a way inconsistent with the usual dominance order. Subject maintains these actions despite unfavorable consequences or threats from others.
Protective: Subject shows concern for other orangutans and often intervenes to prevent harm or annoyance from coming to them.[48]

After administering the instrument to a number of raters, their responses were assessed for statistical reliability both within and between raters. It was found that the individual differences in chimpanzees and orangutans are grouped together by factor analysis just as they are in the case of humans. However, differences were found between species in the content of the factors. For one, six personality factors were found in chimpanzees; they found correlates for all the human factors, plus an additional factor for dominance.[49] Orangutans, on the other hand, showed only the orangutan correlates for Neuroticism, Extraversion, Agreeableness, and Dominance, but also showed a factor that is a combination of Conscientiousness and Openness that was called Intellect.[50] Here we have one example of a research program that follows the general methodological suggestion outlined above: start with folk expertise, and then develop research paradigms along the lines of other paradigms used with pre-linguistic children.

Despite the elegant simplicity of the animal personality research, there are a number of concerns about this research program. First, one might have general problems with the factor analysis of personality. However, this problem is not for animal research per se, but rather a criticism that is equally applicable to human research, so I here set that concern aside.

Another worry is that since instruments such as the Orangutan Personality Trait Assessment rely on the judgments of individuals who are familiar with the target subjects, one might worry that the folk judgments of a number of people could all be wrong; there may be concerns about justifying ascriptions based solely on consensus. In fact, the coherence of the folk experts' opinions might indicate the existence of an implicit collusion. The experts might, in their discussions of their charges, begin to speak about individuals in a certain way, and thus jointly construct narratives of individuals that consist of developed personality when no such personality actually exists. Given the existence of this socially constructed narrative, when the instrument is administered the caregiver might think more about the narrative than the animal.

However, this scenario is unlikely, because the narratives that are constructed about the individual animals are constructed because they are useful. With continued research that examines traits across sites, such worries can be eliminated. For example, the caregivers might talk of one orangutan as the policeman because he always intervenes in unbalanced fights and protects the more vulnerable members of the community. But this picture of the individual is given because it allows the caregivers to predict what the individual will do, and, for example, might be relied upon when deciding whether or not human intervention is required. The impressions caregivers have of their charges is based on a familiarity through association with their charges. In the case of the intersubjective expert opinion of caregivers, consensus is not reached because the experts trained together in how to correctly rate individuals on personality surveys. The instruments are used to capture the kind of knowledge that is developed through hands-on experience with the subject. As a further protection against this worry, the researchers collect data from different groups of humans in order to minimize the danger of collecting shared interpretive frameworks.

Caregivers gain knowledge of their charges by looking for correlations, by implicitly making inductive generalizations, and testing predictions—the same thing humans do in the development of their everyday folk psychology. Our adult human folk psychology, just like a parent's folk child psychology or a zookeeper's folk orangutan psychology, is not something learned at a teacher's knee. It is a strategy for understanding behavior that is adopted because it is pragmatically useful; it allows us to make predictions we couldn't make before, and it allows us to understand our charges, to find creative solutions to an individual's emotional or social problems. If it is useful to apply a personality trait to an animal, then it is meaningful to do so, just as it is in the case of human beings. But this emphasis on pragmatism shouldn't be interpreted as anti-realism about the traits being identified. Rather, such pragmatically useful methods of classification are a means of uncovering objective features in the world; for example, we know that there are underlying genetic factors related to the identified traits in humans.[51] Further research may uncover the same in other species.

IMPLICATIONS FOR OTHER STUDIES: THEORY OF MIND

Understanding the species or the individual is the starting point for studies of human cognition, and if we hope to make progress in studies of animal cognition, we must attempt to gain the same degree of expertise with regard to the species under examination. Research on chimpanzee theory of mind is an area that could benefit from greater attention to the folk expertise of those who work closely with chimpanzees.

First, I will present a brief history of the theory of mind research program. This program began with David Premack and Guy Woodruff's investigation into whether a chimpanzee can attribute states of mind to others.[52] While research on chimpanzee theory of mind languished for two decades thereafter, it became an active research program with human children, where the emphasis was placed on the ability to attribute belief. The false-belief task, which was designed to test for belief attribution in children, was undoubtedly inspired by folk expertise. In his commentary on Premack and Woodruff's article, Daniel Dennett pointed out that young children watching Punch and Judy puppet shows squeal for joy when Punch is about to push a box over a cliff; though Punch thinks that Judy is still in the box, the children know that Judy snuck out when Punch wasn't looking.[53] This folk knowledge of children was used to devise the false-belief task, in which participants watch a puppet show of Maxi hiding a piece of chocolate before leaving the room. While Maxi is out, his mother finds the chocolate and moves it to another location. Maxi returns to the scene, the show is stopped, and the participants are asked to predict where Maxi will go to look for his chocolate.[54]

While research on children's theory of mind thrived, the chimpanzee research largely failed to find experimental evidence that chimpanzees understand belief.[55] However, this negative result may be a result of not relying on folk expertise as a starting point. That is, the criteria of evidence need to be reconceived so that it is salient for chimpanzee subjects. Dennett, in his layman's interaction with children, noted that children find the false belief in the Punch and Judy show very entertaining. Tests of chimpanzee theory of mind that are modeled on the false-belief task and that fail to find evidence that chimpanzees understand belief are based on folk expertise about *children*, not chimpanzees. And chimpanzees are quite different from human children.

To construct a good chimpanzee theory of mind task, one that is founded on folk expertise about chimpanzees, researchers can look for natural behaviors in chimpanzees that are correlative to the children's behavior in the Punch and Judy show. Some research on theory of mind did just that, and recognized the importance of beginning with species-normal behavior. After countless studies suggesting that chimpanzees have nothing resembling a theory of mind, Hare and colleagues designed a competitive task to test whether chimpanzees understand what others can and cannot see, and he found that chimpanzees do understand such seeing.[56] Earlier research on the chimpanzees' understanding of seeing and theory of mind relies on cooperation with a human researcher.[57] To explain the difference in findings, Hare and colleagues write,

> Perhaps the communicative situations of these latter [cooperative] studies may be unnatural for chimpanzees, who have not evolved for this kind of cooperative communication over monopolizable food resources and who do not normally experience in their individual ontogenies others helping them to find food....Chimpanzees' most sophisticated social-cognitive abilities may emerge only in the more natural situations of food competition with conspecifics.[58]

Here, the authors are acknowledging the importance of starting with species-normal behavior and using our knowledge of that behavior in devising experiments to test for psychological properties.

While the food-competition studies do not offer a definitive answer to the question of whether chimpanzees understand others' perceptual states, they are examples of how folk expertise can inform experimental design. While there are not yet experiments concerning chimpanzees' understanding of belief that are based on the expertise of caregivers, researchers who have worked closely with apes are often willing to attribute something like a theory of mind to their charges. Savage-Rumbaugh writes, "there can be no doubt that Kanzi attributes intentions and feelings to others and that he recognizes the need to communicate things about his own mental state to others."[59] This judgment isn't based on the results of formal studies, but rather is the result of her thirty-five years of close work with bonobos. The judgment comes from what one might deride as anecdotes: that Kanzi plays pretend games, that he takes advantage of new caregivers by getting them to allow him to do things he is not normally allowed to do, and that he doesn't believe everything he is told. For example, Savage-Rumbaugh reports one incident in trying to educate Kanzi about the danger of electrical outlets. She writes:

> I have just tried to tell Kanzi that "shocks" come out of the wall—that the small
> hole in the wall is dangerous and can hurt him badly. It is clear that he
> understands something of what I have said to him, because he approaches the
> outlet with extreme caution, his hair on end. He smells it, he looks at it, he even
> throws something at it gingerly. The outlet just sits there. Kanzi stares at me with
> a rather incredulous look on his face—why, he wonders, do I think this thing is
> dangerous, and why did I lecture him so when he started to stick a screwdriver
> into it?…Waiting until I was not looking, he carefully hid the screwdriver under a
> blanket. Then, when I was thoroughly occupied…he removed the screwdriver
> from its hiding place and placed it directly in the outlet.…He stood ramrod
> straight, and his hair rose two inches. He yanked the screwdriver out of the socket
> and immediately burst into a series of emphatic "Waa" sounds.[60]

As a symbol-trained bonobo who is able to communicate with his human caregivers, Kanzi is certainly a special case, and the point is not to insist that Kanzi has a theory of mind or that a study of theory of mind in chimpanzees should start with an understanding of a single bonobo. Rather, the point is that Savage-Rumbaugh thinks that Kanzi has something like a theory of mind because she has a relationship with him that she describes as one of mutual empathy, and she knows who he is as an individual. Savage-Rumbaugh's special relationship with Kanzi may be used as a basis for hypothesis generation, and it can be used to design a formal test of theory of mind that would be appropriate for Kanzi. For tests of non-enculturated chimpanzees, the same sort of folk expertise is required.

CONCLUSION

I have argued that we must investigate hypotheses about the psychological properties of animals without prejudice. Researchers should report both positive and negative findings in order to determine whether we ought to reject the null hypothesis that animals don't have psychological properties. But negative findings are only valuable if they are based on a foundation of folk expertise. Experiments for determining whether or not an animal has some psychological property must be informed by the same folk expertise that informs our creation of experiments on human children. We ought not assume that an experimental paradigm that works on one species or at one developmental stage will work on other species or at other stages.

Using folk expertise and knowledge of species-normal behavior as a starting point acknowledges the fact that psychology is a product of evolution, and that it evolved to cope with the natural social and physical environment of the species. If an animal has a psychological property, we should expect to see evidence of it in the animal's naturalistic interactions. And if the methods of ethology are not sufficient for determining the existence of a psychological mechanism, then we can use an experiment based on the naturalistic event in order to formally test for it. Naturalistic observations and folk expertise go hand in hand. Psychologists working with infants are utterly dependent on folk expertise, and animal cognition researchers should be as well.

NOTES

1. Henceforth, all mention of "nonhuman animal" will be shortened to the less accurate but stylistically preferred term "animal."

2. John S. Kennedy, *The New Anthropomorphism* (New York: Cambridge University Press, 1992).

3. George Henry Lewes, *Seaside Studies at Ilfracombe, Temby, the Scilly Isles, and Jersey* (Edinburgh: William Blackwood & Sons, 1860), p. 385.

4. Kennedy, *New Anthropomorphism*, p. 5.

5. Wolfgang Kohler, *The Mentality of Apes* (New York: Harcourt, Brace, 1925).

6. Susan Perry, Mary Baker, Linda Fedigan, Julie Gros-Louis, Katherine C. MacKinnon, Joseph H. Manson, Melissa Panger, Kendra Pyle, and Lisa Rose, "Social Conventions in Wild White-Faced Capuchin Monkeys: Evidence for Traditions in a Neotropical Primate," *Current Anthropology* 44 (2003): 241–68.

7. For arguments against representationalism from cognitive science, see, e.g., Rodney A. Brooks, "Intelligence without Representation," *Artificial Intelligence* 47 (1991): 139–59; Esther Thelen and Linda B. Smith, *A Dynamic Systems Approach to the Development of Cognition and Action* (Cambridge, Mass.: MIT Press, 1994). Philosophical arguments against the representational nature of cognition include Andy Clark and Josefa Toribio, "Doing without Representing," *Synthese* 101 (1994): 401–31; Tim Van Gelder, "What Might Cognition Be, If Not Computation?" *Journal of Philosophy* 92 (1995): 345–81.

8. For arguments in favor of a cartographical representational system, see, e.g., Elisabeth Camp, "Thinking with Maps," *Philosophical Perspectives* 21 (2007): 145–82; David Braddon-Mitchell and Frank Jackson, *The Philosophy of Mind and Cognition* (Oxford: Oxford University Press, 1996); Michael Rescorla, "Cognitive Maps and the Language of Thought," *British Journal for the Philosophy of Science* 60 (2009): 377–407.

9. Frans B. M. de Waal, "Anthropomorphism and Anthropodenial: Consistency in Our Thinking about Humans and Other Animals," *Philosophical Topics* 27 (1999): 225–80.

10. Maxine Sheets-Johnstone, "Taking Evolution Seriously," *American Philosophical Quarterly* 29 (1992): 343–52.

11. Elliott Sober, "Comparative Psychology Meets Evolutionary Biology: Morgan's Canon and Cladistic Parsimony," in *Thinking with Animals: New Perspectives on*

Anthropomorphism, ed. Lorraine Daston and Gregg Mitman (New York: Columbia University Press, 2005), pp. 85–99.

12. Willard Van Orman Quine, *Word and Object* (Cambridge, Mass.: MIT Press, 1960).

13. Donald Davidson, "Radical Interpretation," *Dialectica* 27 (1973): 313–28.

14. Richard E. Nisbett and Timothy D. Wilson, "Telling More Than We Can Know: Verbal Reports on Mental Processes," *Psychological Review* 84 (1977): 231–59.

15. John Newson, "The Growth of Shared Understandings between Infant and Caregiver," in *Before Speech: The Beginning of Interpersonal Communication*, ed. Margaret Bullowa (London: Cambridge University Press, 1979), pp. 207–22.

16. Robert L. Russell, "Anthropomorphism in Mother-Infant Interaction: Cultural Imperative or Scientific Acumen?" in *Anthropomorphism, Anecdotes, and Animals*, ed. Robert W. Mitchell, Nicholas S. Thompson, and H. Lynn Miles (Albany: State University of New York Press, 1997), pp. 116–22.

17. John H. Flavell, Patricia H. Miller, and Scott A. Miller, *Cognitive Development*, 4th ed. (Upper Saddle River, N.J.: Prentice Hall, 2002).

18. Mary Midgley, "Being Objective," *Nature* 410 (2001): 753.

19. Konrad Lorenz, "Part and Parcel in Animal and Human Societies," in *Studies in Animal and Human Behaviour*, vol. 2 (Cambridge, Mass.: Harvard University Press, 1971), pp. 115–95.

20. Alexandra Horowitz and Marc Bekoff, "Naturalizing Anthropomorphism: Behavioral Prompts to Our Humanizing of Animals," *Anthrozoos* 20 (2007): 23–35.

21. Philip N. Lehner, *Handbook of Ethological Methods* (Cambridge: Cambridge University Press, 1996).

22. Kristin Andrews, "Politics or Metaphysics? On Attributing Psychological Properties to Animals," *Biology and Philosophy* 24 (2009): 51–63.

23. Others also point out that there is great knowledge in those who work closely with animals. Bernard Rollin writes, "usually the best source of information about animal pain are farmers, ranchers, animal caretakers, trainers—in short those whose lives are spent in the company of animals and who make their living through animals" ("Anecdote, Anthropomorphism, and Animal Behavior," in *Anthropomorphism, Anecdotes and Animals*, ed. Robert W. Mitchell, Nicholas S. Thompson, and H. Lynn Miles [Albany: State University of New York Press, 1997], p. 128). However, we shouldn't conclude that all those who work with animals are sensitive to their behavior. Being a folk expert requires having a certain quality of relationship, and it is likely that those caregivers who have a nurturing relationship with their charges are the ones who have developed the greater understanding of their personality, capability, and intentions.

24. Renée Baillargeon and Julie DeVos, "Object Permanence in Young Infants: Further Evidence," *Child Development* 62 (1991): 1227–46; Elizabeth S. Spelke, "Physical Knowledge in Infancy: Reflections of Piaget's Theory," in *Epigenesis of Mind: Studies in Biology and Cognition*, ed. Susan Carey and Rochel Gelman (Hillsdale, N.J.: Erlbaum, 1991).

25. The Habituation-Dishabituation Method has also been used to study primate concepts; for example, the number concept was studied in rhesus monkeys (Marc D. Hauser, Pogen MacNeilage, and Molly Ware, "Numerical Representation in Primates," *Proceeding of the National Academy of the Sciences* 93 [1996]: 1514–17) and cotton-top tamarins (Claudia Uller, *Origins of Numerical Concepts: A Comparative Study of Human Infants and Nonhuman Primates* [Cambridge, Mass.: MIT Press, 1997]).

26. Gordon G. Gallup, "Chimpanzees: Self-Recognition," *Science* 167 (1970): 341–43.

27. David H., Ledbetter and Jeffry A. Basen, "Failure to Demonstrate Self-Recognition in Gorillas," *American Journal of Primatology* 2 (1982): 307–10.

28. Francine G. P. Patterson and Ronald H. Cohn, "Self-Recognition and Self-Awareness in Lowland Gorillas," in *Self-Awareness in Animals and Humans: Developmental Perspectives*, ed. Sue Taylor Parker, Robert W. Mitchell, and Maria L. Boccia (Cambridge: Cambridge University Press, 1994), pp. 273–90.

29. Patterson and Cohn, *Self-Recognition and Self-Awareness in Lowland Gorillas*, p. 286.

30. Robert W. Shumaker and Karyl B. Swartz, "When Traditional Methodologies Fail: Cognitive Studies of Great Apes," in *The Cognitive Animal: Empirical and Theoretical Perspectives on Animal Cognition*, ed. Marc Bekoff, Colin Allen, and Gordon M. Burghardt (Cambridge, Mass.: MIT Press, 2002), pp. 335–43; M. Allen and Bennett L. Schwartz, "Mirror Self-Recognition in a Gorilla (Gorilla Gorilla Gorilla)," *Electronic Journal of Integrative Biosciences* 5 (2008): 19–24.

31. Werner Güth, Rolf Schmittberger, and Bernd Schwarze, "An Experimental Analysis of Ultimatum Bargaining," *Journal of Economic Behavior and Organization* 3 (1982): 367–88.

32. Keith Jensen, Josep Call, and Michael Tomasello, "Chimpanzees Are Rational Maximizers in an Ultimatum Game," *Science* 318 (2007): 107–9.

33. Marc Bekoff and Jessica Pierce, *Wild Justice: The Moral Lives of Animals* (Chicago: University of Chicago Press, 2009).

34. Frans B. M. de Waal, *The Age of Empathy: Nature's Lessons for a Kinder Society* (Toronto: McClelland & Stewart, 2009).

35. Frans B. M. de Waal and Lesleigh M. Luttrell, "Mechanisms of Social Reciprocity in Three Primate Species: Symmetrical Relationship Characteristics or Cognition?" *Ethology and Sociobiology* 9 (1988): 101–18.

36. Gary Finn, "Lottery Winners Reveal How Much They Gave Away," *The Independent*, November 15, 1999 at http://www.independent.co.uk/news/lottery-winners-reveal-how-much-they-give-away-1126242.html.

37. Brian Hare, Josep Call, Bryan Agnetta, and Michael Tomasello, "Chimpanzees Know What Conspecifics Do and Do Not See," *Animal Behaviour* 59 (2000): 771–85.

38. Christophe Boesch, "Cooperative Hunting in Wild Chimpanzees," *Animal Behaviour* 48 (1994): 653–67; Christophe Boesch, "Cooperative Hunting Roles among Taie Chimpanzees," *Human Nature* 13 (2002): 27–46.

39. Samuel D. Gosling and Oliver P. John, "Personality Dimensions in Nonhuman Animals: A Cross-Species Review," *Current Directions in Psychological Science* 8 (1999): 69–75.

40. John M. Digman, "Personality Structure: Emergence of the Five-Factor Model," *Annual Review of Psychology* 41 (1990): 417–40.

41. T. J. Bouchard, Jr. and J. C. Loehlin, "Genes, Evolution, and Personality," *Behavior Genetics* 31 (2001): 243–73.

42. A. A. Jolijn Hendriks, Willem K. B. Hofstee, and Boele De Raad, "The Five-Factor Personality Inventory," *Personality and Individual Differences* 27 (1999): 307–25.

43. Thomas M. Achenbach and Craig Edelbrock, *Manual for the Child Behavior Checklist and Revised Child Behavior Profile* (Burlington, Vt.: Queen City Printers, 1983).

44. Sara S. Sparrow, David A. Balla, and Domenic V. Cicchetti, *Vineland Adaptive Behavior Scales—Survey Form* (Circle Pines, Minn.: American Guidance Service, U.S. Census Bureau, 1984).

45. Philip D. Harvey, Michael Davidson, Kim T. Mueser, Michael Parrella, Leonard White, and Peter Powchik, "The Social Adaptive Functioning Evaluation: An Assessment Measure for Geriatric Psychiatric Patients," *Schizophrenia Bulletin* 23 (1997): 131–45.

46. James E. King and Aurelio José Figueredo, "The Five-Factor Model Plus Dominance in Chimpanzee Personality," *Journal of Research in Personality* 31 (1997): 257–71.

47. Alexander Weiss, James E. King, and Lori Perkins, "Personality and Subjective Well-Being in Orangutans (*Pongo pygmaeus* and *Pongo abelii*)," *Journal of Personality and Social Psychology* 90 (2006): 501–11.

48. Alexander Weiss, in a personal communication.

49. King and Figueredo, "Five-Factor Model Plus Dominance in Chimpanzee Personality," pp. 257–71.

50. Weiss et al., "Personality and Subjective Well-Being in Orangutans," pp. 501–11.

51. Thomas J. Bouchard and John C. Loehlin, "Genes, Evolution, and Personality," *Behavior Genetics* 31 (2001): 243–73.

52. David Premack and Guy Woodruff, "Does the Chimpanzee Have a Theory of Mind?" *Behavioral and Brain Sciences* 1 (1978): 515–26.

53. Daniel C. Dennett, "Beliefs about Beliefs," *Behavioral and Brain Sciences* 4 (1978): 568–70.

54. Heinz J. Wimmer and Josef Perner, "Beliefs about Beliefs: Representation and Constraining Function of Wrong Beliefs in Young Children's Understanding of Deception," *Cognition* 13 (1983): 103–28.

55. Josep Call and Michael Tomasello, "Does the Chimpanzee Have A Theory Of Mind? 30 Years Later," *Trends in Cognitive Science* 12 (2008): 187–92.

56. Hare et al., "Chimpanzees Know What Conspecifics Do and Do Not See," p. 771–85.

57. Daniel J. Povinelli and Timothy J. Eddy, "What Young Chimpanzees Know about Seeing," *Monographs of the Society for Research on Child Development* 61 (1996): 1–152; Josep Call and Michael Tomasello, "A Nonverbal False Belief Task: The Performance of Children and Great Apes," *Child Development* 70 (1999): 381–95.

58. Hare et al., "Chimpanzees Know What Conspecifics Do and Do Not See." p. 783.

59. Sue Savage-Rumbaugh, Stuart G. Shanker, and Talbot J. Taylor, *Apes, Language, and the Human Mind* (New York: Oxford University Press, 1998), p. 56.

60. Savage-Rumbaugh et al., *Apes, Language, and the Human Mind*, p. 8.

SUGGESTED READING

ANDREWS, KRISTIN. "Politics or Metaphysics? On Attributing Psychological Properties to Animals." *Biology and Philosophy* 24 (2001): 51–63.

BOESCH, CHRISTOPHE. "Cooperative Hunting Roles Among Taïe Chimpanzees." *Human Nature* 13 (2002): 27–46.

———. "Cooperative Hunting in Wild Chimpanzees." *Animal Behaviour* 48 (1994): 653–67.

CALL, JOSEP, and MICHAEL TOMASELLO. "Does the Chimpanzee Have a Theory of Mind? 30 Years Later." *Trends in Cognitive Science* 12 (2008): 187–92.

DASTON, LORRAINE, and GREGG MITMAN, eds. *Thinking With Animals: New Perspectives on Anthropomorphism*. New York: Columbia University Press, 2005.

DE WAAL, FRANS B. M. "Anthropomorphism and Anthropodenial: Consistency in Our Thinking about Humans and Other Animals." *Philosophical Topics* 27 (1999): 225–80.

FISHER, JOHN ANDREW. "The Myth of Anthropomorphism." In *Interpretation and Explanation in the Study of Animal Behavior*, vol. 1, edited by Marc Bekoff and Dale Jamieson, pp. 96–116. Boulder, Colo.: Westview Press, 1990.

GALLUP JR., GORDON G. "Chimpanzees: Self-Recognition." *Science* 167 (1970): 341–43.

HARE, BRIAN, JOSEP CALL, BRYAN AGNETTA, and MICHAEL TOMASELLO. "Chimpanzees Know What Conspecifics Do and Do Not See," *Animal Behaviour* 59 (2000): 771–85.

HOROWITZ, ALEXANDRA, and MARC BEKOFF. "Naturalizing Anthropomorhism: Behavioral Prompts to Our Humanizing of Animals," *Anthrozoos* 20 (2007): 23–35.

KEELEY, BRIAN L. "Anthropomorphism, Primatomorphism, Mammalomorphism: Understanding Cross-Species Comparisons." *Philosophy and Biology* 19 (2004): 521–40.

KENNEDY, JOHN S. *The New Anthropomorphism*. New York: Cambridge University Press, 1992.

KING, JAMES E., and AURELIO JOSÉ FIGUEREDO. "The Five-Factor Model Plus Dominance in Chimpanzee Personality." *Journal of Research in Personality* 31 (1997): 257–71.

MITCHELL, ROBERT W., NICHOLAS S. THOMPSON, and H. LYNN MILES, eds. *Anthropomorphism, Anecdotes, and Animals*. Albany: State University of New York Press, 1997.

POVINELLI, DANIEL J., and JENNIFER VONK. "We Don't Need a Microscope to Explore the Chimpanzee's Mind." *Mind and Language* 19 (2004): 1–28.

PREMACK, DAVID, and GUY WOODRUFF. "Does the Chimpanzee Have a Theory of Mind?" *Behavioral and Brain Sciences* 1 (1978): 515–26.

SAVAGE-RUMBAUGH, SUE, STUART G. SHANKER, and TALBOT J. TAYLOR. *Apes, Language, and the Human Mind*. New York: Oxford University Press, 1998.

SHUMAKER, ROBERT W., and KARYL B. SWARTZ. "When Traditional Methodologies Fail: Cognitive Studies of Great Apes." In *The Cognitive Animal: Empirical and Theoretical Perspectives on Animal Cognition*, edited by Marc Bekoff, Colin Allen, and Gordon M. Burghardt, pp. 335–43. Cambridge, Mass.: MIT Press, 2002.

SOBER, ELLIOTT. "Comparative Psychology Meets Evolutionary Biology: Morgan's Canon and Cladistic Parsimony." In *Thinking With Animals: New Perspectives on Anthropomorphism*, edited by Lorraine Daston and Gregg Mitman, pp. 85–99. New York: Columbia University Press, 2005.

WEISS, ALEXANDER, JAMES E. KING, and LORI PERKINS. "Personality and Subjective Well-Being in Orangutans (*Pongo pygmaeus* and *Pongo abelii*)." *Journal of Personality and Social Psychology* 90 (2006): 501–11.

WIMMER, HEINZ, and JOSEF PERNER. "Beliefs About Beliefs: Representation and Constraining Function of Wrong Beliefs in Young Children's Understanding of Deception." *Cognition* 13 (1983): 103–28.

CHAPTER 17

ANIMAL PAIN AND WELFARE: CAN PAIN SOMETIMES BE WORSE FOR THEM THAN FOR US?

SAHAR AKHTAR

THERE is a widely held view in philosophy and the biological sciences that the amount and ways in which an animal can experience pain as compared to us is limited to the feeling of sensory pain.[1] The primary reason for this view is the belief that animals are less cognitively sophisticated than we are, and in particular that animals lack awareness of self and a sense of the past and the future. If pain is inflicted on animals, it is thought that while animals may be able to feel the pain itself, they are not capable of the higher order suffering that may accompany the feeling of pain in us. This view suggests that pain for animals is not as bad as pain is for us.

A considerable literature in philosophy of mind, ethics, cognitive ethology, and biology examines the ability of animals to feel pain and examines the nature of their pain. In this chapter, I will present a different approach to the understanding of animal pain, largely relying on the framework of welfare analysis and practical rationality, although the arguments can be broadened beyond this framework. I will argue that due to the same reasons for which animals are thought to be incapable of sophisticated forms of suffering—namely, a lack of self- and time-awareness—a given amount of pain may actually be worse for animals in certain respects than many have thought and may even be worse for them than the comparable amount of pain is for us. The argument that pain may be worse for them follows mainly as a consequence of reflecting on the reasons philosophers, psychologists, and economists typically give for thinking that pain and discomfort are sometimes discounted.

Specifically, I will argue that at least within a moderate range of severity, pain may not always play a significant role in our welfare or well-being for two reasons. The first is that the intensity of pain for us can often be mitigated through expectations, memories, and the consideration of and attention to other interests. The second and main reason is that we are able to engage in inter-temporal calculations with respect to our interests and frequently discount pain in order to achieve other, higher order or longer term interests. The arguments in this chapter suggest that we are moving too fast in thinking that because we are capable of suffering, and various animals are not, a given amount of pain is worse for us than it is for animals. I will argue that at least for some animals, a given measure of pain can be worse for them than a comparable measure of pain is for us.

VIEWS OF PAIN AND SUFFERING

Some philosophers have challenged the idea that animals are capable of experiencing pain. René Descartes famously denied consciousness to animals on the grounds that they lack language and he believed animals to be automata.[2] Much more recently, Peter Harrison has suggested that animals do not feel pain.[3] He argues that in order for a being to feel pain, there must be reasons for him to bear pain; that is, he argues that pain operates as a kind of consideration in the rational decision-making processes of free agents. Without such rationality, pain is superfluous. Harrison writes, "We are free, in painful situations, to damage our bodies if we believe there is a higher priority....Pain frees us from the compulsion of acting instinctively; it issues harsh warnings, but they are warnings which may be ignored."[4] He thinks that we do not conceive of animals as possessing free will and as acting from reasons, and therefore that there is a reason to be skeptical of thinking that animals experience pain. Harrison's view does not seem to be compelling even if one were to adopt a broadly functionalist account of pain. Pain would serve other functions besides figuring in rational decision-making, such as to indicate to the being that its body has been damaged or that it is in danger. Aside from these worries with Harrison's view, however, we will see that the point about rational decision-making can be turned around to show that pain can actually be worse for animals than for us.

Despite the presence of a few accounts that deny pain experience to animals, most philosophers do attribute such experience to animals. This is even true of Peter Carruthers, although some commentators seem to interpret him incorrectly.[5] The misinterpretation is perhaps partly due to the fact that Carruthers' position seems to have been somewhat modified. In his earlier papers, he maintained that animals do not feel anything, because all of their experiences are unconscious.[6] His view, then and now, can be characterized as falling into a broad category of consciousness accounts that entail what is called "higher order" consciousness. These theories propose that a perceptual state, such as a bodily sensation, is conscious if and only if it is represented in higher order consciousness. For Carruthers, representation in higher order consciousness requires that the perceptual state is available to be thought about. He maintained that animals cannot think about their experiences with acts of thinking that can be scrutinized, and so "while animals *have* pains, they do not *feel* pains."[7]

However, Carruthers has recently modified and clarified his earlier position. In a 2004 paper, he claims that animals have first-order experiential states.[8] He denies to most animals the higher order sort of "phenomenal consciousness" that humans have. The latter sort of consciousness, according to Carruthers, is distinct because of its subjective quality and availability to introspection. Although some commentators continue to read Carruthers as maintaining that most animals do not experience pain, this does not seem to be supported. Carruthers believes that most animals do experience pain in the first-order sense, that of perceiving pain as located in some specific region of their body, even if their pains are not conscious in the subjective and introspective sense.[9]

Furthermore, it is the first-order perception of pain, according to Carruthers, that is aversive or awful. He writes that "reflecting on the phenomenology of our own pain perceptions is yet another way of appreciating that it is the first-order content of pain-states that carry their intrinsic awfulness. Suppose that you have just been stung by a bee while walking barefoot across the grass, and that you are feeling an intense sharp pain…the focus of your attention, is the property that is *represented* as being present in your foot, not the mental state of *representing* that property. And this would have been the same even if your pain-state had lacked a higher-order analog content, and hence hadn't been a phenomenally conscious one."[10] Thus, despite not having phenomenal consciousness of pain, most animals can find their pains to be awful or aversive. Carruthers further concludes that it is this first-order pain experience that is the appropriate object of sympathy in both animals and us.[11]

Many prominent authors in the philosophy of mind literature maintain that there is a distinction between pain and suffering. While granting that many animals may experience pain, they either deny that animals suffer or suggest that animals are at best capable of an attenuated form of suffering, at least as compared to us. Daniel Dennett has prominently made such a distinction. He argues that in order for some being to suffer, there must be a subject or self to whom the pain is occurring. To satisfy this condition, he believes we need the kind of consciousness that is to be found in adult humans. Dennett writes that "in order to be conscious—in order to be the sort of thing that it is like something to be—it is necessary to have a certain sort of informational organization that endows that thing with a wide set of cognitive powers (such as the powers of reflection, and re-representation). This sort of internal organization does not come automatically with so-called sentience."[12] He says of creatures who do not have a complex internal organization that they are "constitutionally incapable of undergoing the *sort* or *amount* of suffering that a normal human can undergo."[13]

What is the suffering that a normal human can experience? Although he never expressly defines it, Dennett gestures toward it in a few passages. For instance, in one passage he asserts that if someone were to step on your toe, despite the intense pain that this may cause you, if there is no long-term damage to your foot then the person's action has not harmed you, because it is too brief to matter. However, he adds that "if in stepping on your toe I have interrupted your singing of the aria, thereby ruining your operatic career, that is quite another matter."[14] A few paragraphs later, he discusses suffering in the following way: "The anticipation and aftermath, and the recognition of the implications for one's life plans and prospects, cannot be set aside as the 'merely cognitive' accompaniments of the suffering."[15] This suggests that the former are what constitute suffering. This is supported a few sentences later in the same text when Dennett asserts that in order to understand suffering, we need to study a creature's life, not its brain.[16] Putting all of these remarks together, it seems that suffering consists in the thwarting or frustrating of one's desires and goals, and the recognition of such frustration, as well as the memory and anticipation of it.

Michael Tye has recently argued that very simple creatures are phenomenally conscious and are capable of experiencing pain, but may not be capable of suffering. Suffering, he argues, requires cognitive awareness of one's experience of pain. More specifically, simple creatures lack higher order consciousness, which means that "they have no *cognitive* awareness of their sensory states. They do not bring their own *experiences* under concepts."[17] For Tye, suffering is characterized by awareness of and attention to one's pain. This view implies, however, that when awareness and attention diminish, so does suffering. Indeed, he writes that "the person who has a bad headache and who is distracted for a moment or two does not suffer at that time. The headache continues to exist—briefly not noticing it does not eliminate it—but there is no cognitive awareness of pain and hence no suffering."[18]

Robert Hanna similarly makes a distinction between pain and suffering. He argues that pain amounts to something like the conscious perception of pain located in the body—what he calls "bodily nociperception"—whereas suffering amounts to "self-conscious or self-reflective emotional pain."[19] Finally, the distinction between pain and suffering also seems to figure in Harrison's work when he states that "animals do not experience pain as we do" and that animals are mere bundles of sensations that cannot remember pain because they have no concept of self.[20]

To now collect the views of all of these authors: in order to suffer, beings must be capable of being aware that they are the subject of pain; they must be able to anticipate and remember pain; and they must recognize that their desires or goals will be frustrated by pain. We can reduce these requirements to two general capacities that must be present for suffering to occur, namely *self-* and *time-awareness*. We have seen that some philosophers maintain that animals lack self-awareness. Self-awareness, for many of them, requires the possession of language. But it also seems plausible that they would tie time-awareness to language. Indeed, some philosophers have specifically argued that animals do not have a sense of past or future because they lack language.

Immanuel Kant famously argued that to move mentally beyond the sensations of the present, one must be capable of having thought. Jonathan Bennett argues that it is only with a sophisticated language that one can have thoughts that are not about the present time or circumstances. He writes that "to express beliefs about the past, or to express general beliefs, one needs a language of the right kind: roughly speaking, language with a past-tense operator (in the one case) and with something like quantifiers (in the other)."[21] Bennett adds that the behavior of animals lacking this kind of language can of course be *affected* by the past, but this does not imply that these animals can consider other times—it does not remove them from the "prison of what is present and particular."[22]

Against the few authors who deny any pain experience to animals, I will proceed on the plausible view that there is pain that animals are capable of experiencing and that, following Carruthers, this pain experience, as opposed to the higher order experience of pain, is what is aversive and awful. I will call this the *feeling* of pain or, simply, *pain*.[23] For the sake of the present analysis, though, I will also assume that many of the views described above are correct in that there is a distinction between pain and suffering, and that many animals are incapable of experiencing suffering. The experience of suffering requires awareness of self and time, or cognitive sophistication, and, at least initially, I will grant that many animals lack such awareness.[24] Additionally, I will assume that there are two distinct categories of beings, normal adult humans and animals—that is, cognitively sophisticated (hereafter, sophisticated) and cognitively unsophisticated (hereafter, unsophisticated) beings—and that there are no degrees in awareness of self and time.[25] However in a later part of the chapter, I will discuss research that shows that many primates and higher mammals possess at least a rudimentary form of cognitive sophistication. I will examine the implications of this research for pain experience in these animals.

In the remainder of this chapter, I will argue that the absence of self- and time-awareness may mean that the feeling of pain for animals is much worse in certain respects than has previously been thought. The absence of awareness may mean that in some cases pain is worse for animals than a comparable measure of pain for us.[26]

ANIMAL PAIN AND WELFARE

In order to illuminate how pain in certain contexts may be worse for unsophisticated beings than it is for us, I will frame the arguments first in the terms of welfare analysis.[27] Even though some of these terms are not commonly employed in philosophical discussions, there is sufficient similarity in the concepts to enable us to draw philosophical implications. For instance, even though welfare analysis relies on the ideas of welfare and interests or preferences and many philosophers speak instead of desires and well-being, there is considerable overlap between these concepts, at least in some accounts. For example, welfare depends on the extent to which one's interests or preferences are satisfied, and this is also one common way to define "well-being."[28] Moreover, it may be more appropriate to speak in terms of interests for discussions involving animals because the possession of desires may be more cognitively demanding, and therefore it may be thought that animals lack desires. Some authors deny that animals are capable of having desires on the grounds that desires require language, however it is not generally denied that animals have interests; in order for it to be said that a being has interests it simply requires that certain things can be good or bad for that being.[29] On the reasonable view that animals can have the feeling of pain and that pain is aversive or bad, animals have an interest in avoiding pain.

Saying something meaningful about the strength of an interest is not straightforward, and it is often viewed as difficult if not impossible to compare strength of an interest across individuals.[30] However, we can discuss how much value or weight an interest has in some being's welfare—or more generally, how significant it is to the being's welfare—relative to the welfare of others.[31] Perhaps the most trivial way to think about the weight of any interest has to do with the proportion of welfare affected by that interest. In particular, all other things being equal, the greater the *amount* of similarly important interests the satisfaction of which comprises welfare, the smaller the change in welfare when any one of these interests is satisfied or frustrated. That is, the less is the interest's proportion of total welfare. Considering the interest in avoiding pain, in an extreme case in which a being has only this one interest, to violate it by inflicting pain would be to reduce that being's total welfare to zero, at least for a period of time. To take less extreme cases, animals such as mice or rats will have few interests other than the avoidance of pain and discomfort, such as subsistence, shelter, and rearing offspring. For these animals, the infliction of pain would present a large change in welfare.

When speaking of sophisticated beings such as adult humans, it may be appropriate to shift the terminology from interests to desires, for the reasons stated above and for further reasons that will be illuminated below. However, for the sake of comparison between different kinds of beings, I will continue to use the terminology of *interests*, but "concerns," "desires," or "goals" can be substituted for "interests" for most of what follows. Given, as we will see, that we typically have numerous significant interests beyond that of avoiding pain, when the interest in avoiding pain is violated because of pain infliction, there will be a relatively smaller change in our welfare than for many animals, all things being equal. For instance, compare the animal to someone who has interests in pursuing hobbies, in cultivating his talents, in public esteem and his reputation, and in living in greater harmony with his loved ones. Because of his other significant interests, pain would seem to constitute a smaller share of his total welfare compared to the animal, all things being equal.

Of course, all things are not equal; how much weight pain has in the being's total welfare will not only be determined by the amount of interests the being has, but also by at least two other important and related factors. First, the weight that pain has in a being's welfare will depend on the extent to which the interest in avoiding pain is satisfied or violated, or the extent to which the being experiences pain.

Second, it seems that the weight that pain has will also depend on the value or weight that the being himself assigns to pain—something, as we will see, that may only be possible for sophisticated beings. Having more sophisticated and a larger amount of interests, I will argue, may reduce both the weight that we assign to pain and the intensity of pain, thereby decreasing the change in welfare that results when pain is experienced. More specifically, I will show that cognitive sophistication gives rise to certain psychological abilities, such as the ability to engage in inter-temporal calculation, and increases the variety and kinds of interests that may serve as substitutes for focusing or concentrating on pain. Both mechanisms may operate to diminish emphasis on oneself and on one's short-term and present circumstances, including on the feeling of pain.

Weighting Pain: Discounting and Trading-off

In this section, I will argue that as sophisticated beings we possess psychological abilities that may decrease the role that a given amount of pain plays in our welfare. This argument follows as a consequence of reflecting on the reasons philosophers, psychologists, and economists typically give for thinking that we make inter-temporal calculations about our interests and, especially, that pain and discomfort are sometimes discounted. Before making this argument, I will discuss how awareness of self and time expands the kinds of interests that we have and enables us to engage in inter-temporal interest calculation.

In addition to being able to make temporal distinctions, we know that *we* exist over time. Understanding that one exists over time is distinct from merely temporally differentiating events as the former requires both temporal- and self-awareness— that is, a concept of the passage of time that includes an idea of the self along that time horizon. As suggested above by the work of authors like Dennett, Harrison, and Bennett, time-awareness enables reflection on and evaluation of one's goals and one's life, but in order to do this, one must see the different goals and stages of life as belonging to the same continuing subject. Awareness of ourselves as existing over time allows us to engage in a familiar kind of perspective-shifting so that we abstract from a moment in time and consider our interests from the perspective of different periods of our lives or from some meta-perspective.

Perspective-shifting gives rise to complex and long-term interests. By "complex," I am referring to a category of interests that includes what are commonly called higher order and global interests. Harry Frankfurt and Richard Jeffrey separately make a distinction between simple preferences, what they call first-order preferences—which are for ordinary objects, such as for pizza, to play sports, or to read a book—and the more sophisticated preferences that we have regarding our first-order preferences, or higher order preferences.[32] Whereas a preference to smoke would be a first-order preference, having a preference not to have the preference to smoke would be a higher-order preference.[33] Our higher-order preferences are thought to be our more considered concerns and to reflect what we care about or what is important to us.[34] A similar kind of complex interest may be called a global interest. Parfit has illustrated the differences between global and local interests with the example about drug addiction.[35] Even if one's drug addiction could be satiated all the time, which would satisfy a local interest, one would prefer not to be a drug addict, at all, which is a global interest. The sum total of each local interest fulfillment—here every time one is given the drug—could be greater than the non-fulfillment of the one global interest—here not being a drug addict—but Parfit argues that this is not how most of us think about our lives.[36]

In addition to the distinction between higher order or global interests—complex interests—and first-order or simple interests, there is an important distinction between long-term and short-term interests.[37] Complex interests may require more reflection and metaevaluation than simple long-term interests, but even simple long-term interests are still only possible with a sense of oneself existing over time. In order to formulate an interest for the long-term, one needs to recognize that there will be a future self that the interest will serve. This recognition of possible benefit in the future requires that one has a sense of oneself as a continuing single subject.[38] Additionally, there is a conceptual overlap between complex and longer term interests—complex interests are often formulated with regard to more than one period of time and are often long-term interests. For instance, an interest in breaking a drug addiction might arise from an interest in being healthy or productive in the future.

Authors such as Frankfurt, Jeffrey, Parfit, and Nozick suggest that many of us place great weight on long-term and complex interests, and in the process our short-term and first-order interests are discounted. The latter are discounted because we engage in inter-temporal interest calculation, which can be described as assigning weight or value to one's different interests and concerns across time, including discounting some of them or valuing them less, and formulating trade-offs among them.[39] Making trade-offs among one's interests involves comparing them and favoring the satisfaction of some over others.[40] Inter-temporal trade-offs and weighting, and discounting especially, can play an especially important role in reducing emphasis on one's short-term and first-order interests. In particular, these interests are often minimized in an effort to live a more enriched life, to insure against future financial or health concerns, to be more productive or successful, or to live morally—for the things we care about.[41]

Our ability to form reflective, long-term, and complex interests places pressure on us to avoid living moment to moment. Most of us do not have lives of unconnected moments of interest-satisfaction without any regard to the long term, even if it is a rather myopic long term. We bring interests across different time periods into relation with one another when we compare them, weight them, and favor the satisfaction of some over the satisfaction of others. There is ample research in psychology and behavioral economics that demonstrates that we do in fact shift between time frames and levels of interests when making decisions in a variety of contexts.[42] According to some of this research, we are even capable of anticipating the emotions that will accompany bad choices, and this anticipation consequently influences our decision making and subsequent behavior.[43] Indeed we do not even require empirical research to confirm that we engage in these cognitive activities; perspective-shifting and anticipating the results of our decisions and how we will feel about them is a familiar feature of our lives.

We have seen that as sophisticated beings we are capable of forming long-term and complex interests and of engaging in inter-temporal interest calculation. We are now in a position to place this discussion together with the discussion of welfare in order to see what kind of impact these considerations may have on the role of pain in the welfare of animals. Recall that there are at least three ways in which the weight or value that pain has in a being's total welfare would be determined. First, and most trivially, the weight that pain has depends on the amount of other interests which, if satisfied, comprise welfare. In general, the greater the variety and amount of similarly or equally important interests, the smaller the change in welfare when any one interest is frustrated or violated. Second, and the topic of the next section, the weight of pain depends on the extent to which it is experienced. Third, its weight depends on the weight the being himself assigns to the pain. It is through this last mechanism in particular that inter-temporal interest calculation becomes important.

To put it simply, we are often willing to violate short-term and first-order interests, including the interest in avoiding pain, in order to satisfy our long-term and complex ones. That is, we often discount pain when doing so favors long-term and complex interests. Numerous examples support this claim. For instance, even something as commonplace as rigorous exercise can be painful, but when we have a concern for our health or physical appearance, we are often willing to discount the pain that we experience during and after exercise. We trade the interest in avoiding pain in this context for the sake of satisfying a complex or long-term interest to be physically fit. If we consider professional athletes, such as football players and ballerinas, it is even more obvious that many discount pain in order to achieve complex goals.

Consider also the cultural and religious practices that involve pain. These include the more severe forms of mortification of the flesh in Christianity; walking on hot coals in Hinduism; and a range of extremely painful rites of passage.[44] Then there are a variety of painful practices that countless people in various cultures engage in for the sake of aesthetic benefits. Many people often voluntarily endure blisters and bleeding so that they can wear particular shoes. Even some demanding diets can be painful. Several cosmetic procedures and techniques can similarly be very painful for people, such as getting tattoos or piercings and having certain facials. Even deep tissue massages can be painful. Other everyday ways in which we discount pain include working through an illness or headache in order to be productive. Despite the fact that we know that these things can worsen if we do not rest, we often choose to work anyway. Perhaps the most pervasive examples of choosing to endure pain for the sake of more complex interests are found in medical exams and procedures. Getting blood drawn, receiving injections, undergoing mammograms and prostate exams, and the aftereffects of most surgeries can be quite painful.

In all such cases, we discount the weight of pain in order to achieve other goals. This gives us reason to think that at least in the face of more long-term and complex interests, the role that pain plays in determining our welfare may not always be significant. Of course, much of the analysis concerning how we discount pain may only apply to pain within some moderate range. In other words, perhaps we only discount the pain of exercise, of headaches, or of medical procedures if the pain is moderate. Although this is not always the case—consider rigorous exercise, extreme cultural and religious rituals, and severe pain after surgery—this leaves open the possibility that pain outside of some moderate range of intensity would still be an important contributor to our welfare. However, we will see in the next section that there is also reason to think that pain's intensity might sometimes be less for us than for animals.

Someone might object to the foregoing analysis on the grounds that it relies too much on an interest-satisfaction view of welfare. On that account, welfare consists in the satisfaction of subjective interests. However, in some of the major alternatives to the interest-satisfaction view, welfare (or well-being) consists in the realization of certain objective interests or the attainment of certain objective goods.[45] According to this worry, the reason that pain is frequently discounted is because agents themselves do so, but if we were to adopt a more objective account of welfare, it would not be obvious that the value or weight of pain should be discounted so often. Nevertheless, there are many hybrid views of welfare that include a component of subjective interest-satisfaction, or something similar, even if they also value other goods.[46] More importantly, we need not adopt the present account of welfare or even a hybrid view to make sense of the fact that pain can be and ought to be discounted frequently. For instance, good health would be on any plausible list of objective goods and it often requires us to endure the pain of exercise and of medical procedures and surgeries.[47]

Another potential worry with the above analysis is that it incorrectly paints a picture of us as beings that rationally maximize our interests. It is true that we do not always discount present costs, broadly speaking, including pain.[48] However, as Michael Bratman has stressed, we are planning creatures who have complex goals that require us to form and then stick to plans.[49] A consequence of our planning nature seems to be that even if we are not always rational in the sense of favoring long-term and complex interests over short-term and first-order ones, there is still pressure to discount the latter, or conversely to place greater weight on future time periods. It seems plausible that the greater our ability to reflectively think about long-term and complex interests, the more we experience pressure to discount present costs, including pain. Furthermore, it is not obvious that we must assume that we choose rationally in order for the point that we frequently discount pain to be effective. It is plausible to view many of the practices and actions used as examples in the preceding paragraphs as irrational in some way.

I have argued that as sophisticated beings we are capable of discounting pain and frequently do so for the sake of achieving longer term and complex interests.

However, many animals are not capable of discounting their pain. Both long-term and complex interests require a developed sense of time and self, and in particular self-reflection or introspection, and we have seen that the authors above who discuss the nature of animal pain—including Carruthers, Dennett, Harrison, and Bennett—claim that most animals do not possess these qualities.

Inter-temporal interest calculation is something that many animals cannot do because they lack long-term and complex interests. That is, inter-temporal calculation entails self-reflection and evaluation of one's different interests over time or from a perspective that is abstracted from any period of time, and this kind of reflection and evaluation requires the recognition that the different interests belong to the same continuing self. In addition to research in biology and ethology that denies self- and time-awareness to many animals, other research suggests that many animals cannot engage in inter-temporal decision-making, or at least not nearly to the extent that we can.[50]

Animals without a sense of self and time are not able to choose to endure pain for the sake of satisfying long-term interests, and there is no global or higher order perspective to consider.[51] Unlike us, an animal that cannot see itself as existing over time cannot reflect on the value or meaning of its life taken entirely, cannot form interests for life taken as a whole, and cannot formulate lifelong objectives. Thus for animals, it makes far less sense to think of pain as something that can be discounted for the sake of other long-term or complex interests. We thus have reason to think that a given measure of pain can be a larger detriment to their welfare than a comparable measure of pain is to ours.

Intensity of Pain

I will now discuss ways in which the pain experience that results from a given measure of pain *can be more intense and worse for animals* than it is for us. There are different bases for why this might happen. In some cases, the discounting of pain might directly bear on its intensity and, specifically, lower it. In other cases, the same cognitive conditions required for inter-temporal interest calculation might, for different reasons, also lower the intensity of pain. Many of the points I will make in this section are speculative, and I offer them only to demonstrate that it is not obvious that pain experience for us is worse in all ways than it is for animals.

The discounting of pain should not be assumed to affect the intensity of pain. How much value or weight one places on the interest in avoiding pain does not always affect the intensity of the experience of pain; I might discount the pain I will experience tomorrow after a strenuous run today, for example, but this discounting does not necessarily mean that the pain I will feel tomorrow will be less intense.

However, in circumstances when we discount *present* pain for the sake of long-term or complex interests, discounting might in fact bear on the intensity of pain.

Although I have been unable to find empirical research on this specific topic, a small thought experiment illuminates the point: imagine that you are sitting at your desk reading a book and your friend comes up beside you, inserts the tip of a three-inch needle in a vein in your arm, holds it there for few seconds, and then removes it. Now, instead imagine that you are at a doctor's office and the doctor inserts the tip of a needle of the same length, in the same spot, withdraws blood, and then removes it. It is not hard to imagine that the pain in the first scenario would be more intense than in the second scenario. Knowing that there would be a good reason for having the needle inserted in the second scenario—that there is a long-term or higher order interest in health, for instance—would likely lessen the intensity of the pain felt from the stick of the needle.[52]

Without the belief that there are good reasons to endure their pain—even if there are good reasons, such as a painful but necessary veterinary procedure—animal pain would likely be more intense. There are other ways in which pain's intensity might sometimes be more for animals than for us. Namely, intensity seems to require a certain focus or attention, and cognitive sophistication may decrease the extent to which one focuses on one's pain. Consider the times you fail to notice the pain of a headache, backache, bruise, thorn prick, paper cut, or hangnail until you stop engaging in some activity, such as working, biking, gardening, or watching a movie. It seems that the more ways in which we have substitutes for focusing on or attending to pain, the more ways in which the intensity of a given painful experience might be diminished. Although we have many substitutes for focusing on pain because we have many other interests besides avoiding pain, this easing of pain may not be available for most animals.

Awareness of self and time both seem to be important to having a variety of interests. For instance, a sense of self is necessary for thinking abstractly about the self as well as about things that exist independently of the self. Interest in other persons, activities, projects, relationships, careers, general world or societal conditions, and higher abstract principles are some of the interests that may only follow from cognitive sophistication. A highly developed sense of self seems to paradoxically give rise to a host of interests that reduce the focus on oneself by fostering interests that are not directed toward the self. For example, people care very much about the long-term welfare of their children and sometimes even view their children as partial extensions of themselves. People in pain may take comfort in thinking about their children or other interests, including hobbies and passions. This is related to the point about discounting pain. When we are thinking about or reflecting on our other, more complex interests, we might not be focusing on pain as much.

Additionally, we can often will ourselves to refrain from focusing on or attending to our pain, such as through meditative practices or by simply concentrating on other things. This willing, according to authors such as Harrison, is also something that depends on cognitive sophistication and is not a willing of the sort animals are likely capable.[53] While cognitive sophistication may be necessary for having the freedom to choose what we focus on, and for having different kinds and a large variety of interests, it would not seem to be necessary for the ability to focus on or attend to one's circumstances, as this is needed to perform a variety of simple tasks, such as finding food. Thus, it may not be plausible to think that animals are less able to focus on their pain than we are.

Even aside from engaging in inter-temporal interest calculation, the mere ability to make inter-temporal *distinctions* might reduce the intensity of pain. Animals that have a limited temporal sense would seem to be unable to escape, or get outside of, their present mental state, including a present painful state. Greater time horizons in either direction might lessen the intensity of one's pain: being able to think about the past can allow us to recollect and take comfort in painless past experiences. Similarly, by thinking about future periods we may be able to take relief in the thought of future periods without pain and we are capable of recognizing that the pain will end. An animal may not be able to remember times without pain or imagine future periods without the pain.

Rollin proposes that pain might be more intense for animals that cannot anticipate its cessation or remember its absence. He writes, "If they are in pain, their whole universe is pain; there is no horizon; they *are* their pain."[54] If the pain is sufficiently extreme, even knowing that death is a possible future ending may help alleviate it for us, as compared to animals. Furthermore, we can often anticipate when a pain will likely end because it, or pain similar to it, has occurred in the past and we remember the duration of the experience. When we have a headache, we can often anticipate when the headache will end based on past experiences, and this anticipation might reduce the intensity of the headache. Animals without an awareness of self and time would not be able to anticipate the cessation of the pain or take relief in the thought that it may or will end.

The ability to conceive of temporal distinctions is also necessary for having hope, which can be characterized as a positive attitude directed toward some possible future outcome. If one can hope that one's pain will end, this might help alleviate its intensity. On the other hand, fear might not always be future-directed. It seems one can be afraid of a present adverse situation without also having the additional fear of it getting worse or some other thing happening. It is possible for an animal to experience fear of a present situation without being capable of hoping the situation will improve. On a more general note, hope in the face of extreme pain might mean the difference between a life worth living and one not worth living. Severe pain without the expectation or hope of it ending, even if it will end, can drastically reduce someone's welfare, and animals lacking self- and time-awareness would not be able to hope that their pain will end.

The point about how expectations can affect the intensity of pain can perhaps best be illustrated by considering the so-called "placebo effect." Many placebos seem to work by giving one the expectation that one's pain or condition will lessen. In one prominent example, an orthopedic surgeon who was skeptical about the benefits of using a particular surgery for alleviating the pain of arthritis in the knee conducted an empirical study to test for the placebo effect. In a sample of 180 patients, he performed the standard surgery on some and a placebo surgery on others. The latter group of patients received anesthesia along with incisions in order to simulate the appearance of the standard surgery. Following the real and simulated surgeries, both groups of subjects reported relief from the pain to the same extent.[55] While there is conflicting evidence on the role of placebos, many studies demonstrate a placebo effect.[56]

I have discussed how a given amount of pain might sometimes be more intense for animals than it is for us. Through many of the same mechanisms I have suggested—namely, memory, expectations, and sophisticated emotions like hope—it also seems plausible that pain might be more intense for us: while in pain, we might recall other previous painful episodes in our lives and we might not expect or anticipate that the pain we are experiencing will end, or have hope that it will end. The point of my discussion has not been to argue definitively that animal pain is more intense than our pain, but to cast some doubt on the idea that our pain, in virtue of our cognitive sophistication, is always or even generally worse. Even though some of the reasons offered for thinking that our pain might sometimes be less intense can also be used to argue that our pain might be more intense, other reasons cannot be so easily employed to support the argument that our pain might be more intense. In particular, enduring pain for the sake of complex goals and our ability to focus on other things besides our pain, including by willing ourselves to do so, seem only to be ways in which our pain might be *less* intense than animals'.

PAIN AND SUFFERING

It might be maintained that many of the considerations discussed in the previous section do not impact pain but rather affect how we should conceive of suffering. In order to assess this sort of argument, we need to return to the distinction between pain and suffering. The ability to suffer, according to the views of authors such as Dennett, Hanna, and Tye, would only apply to sophisticated beings. The ability to feel pain, however, would pertain to both sophisticated and unsophisticated beings. This means that for each type of being, sophisticated and unsophisticated, there is some overall pain experience caused by a given amount of pain.[57] We can describe the constituents of the respective experiences in the following way, where P stands for the feeling of pain, S for suffering, and Ep is the overall experience of pain:

Cognitively Sophisticated (adult humans): $P + S = Ep$
Cognitively Unsophisticated (most animals): $P = Ep$

While Ep for animals consists only of P, Ep for humans consists in both P and S. S was said to consist of a variety of cognitive accompaniments to pain including the anticipation and memory of pain, the emotions caused by pain, and the recognition that one's interests and goals will be thwarted by pain. Since many of the mechanisms suggested above by which the intensity of pain might be altered, such as attention, memory, expectations, and sophisticated emotions like hope, arguably involve cognitive dimensions, it could be argued that it is not P that they affect but rather S. Consider again Tye's comments about the person who is distracted from her bad headache. He argues that while the person is distracted there is no cognitive awareness of the pain and hence no suffering at that time. For Tye, it seems more accurate to say that avoiding focusing on one's pain diminishes S and not P. Similarly, someone might argue that when we expect the cessation of pain or when we consider moments without pain, this decreases S and not P.

However, even if we do not see the above mechanisms as decreasing the intensity of P, insofar as they diminish S they would still decrease Ep, or the overall experience of pain, for us. Furthermore, whether these mechanisms impact P or S, it would still be the case that they are unavailable to animals that lack an awareness of self and time. In other words, even if these mechanisms affect S and not P for us, this point by itself does not show that their absence does not affect P for animals. It does not show that P for animals would not be more intense than P for us, possibly leading to an overall worse experience for animals from a given amount of pain.

Consider the contrast between someone who has a deep gash on his leg and a dog, or some other animal with a lack of cognitive sophistication, who has a deep gash on his leg. The injuries are equally serious in terms of tissue damage and the like, and neither is given any medication to relieve the pain. Even if it is a familiar fact that in such situations we feel frustrated with the way the injury is interrupting certain activities, it is also a familiar fact that we recognize that the interruption is temporary. Furthermore, it is not hard to imagine that we would force ourselves to concentrate on other circumstances, ideas, or activities and that we would recognize that the wound will eventually heal and the pain will begin to diminish. These are all familiar things we do and believe when we are pain. Conversely, the dog, if lacking cognitive sophistication, is not able to take relief in these sorts of ways. What can he do but sit there and feel pain? This dog may be unable to escape what Bennett referred to as the "prison of what is present and particular," and this would apply to his experience of pain as well.[58] The contrast between the respective pain experiences of a person and the dog would be stronger if we assume that the wound of each has been caused for the sake of preventing some long-term tissue damage. Again, while the person would know that there is a good reason for his pain, the dog would not, even if there is a good reason. Knowing there is a good reason would further reduce the overall pain experience for the person.

The facts, if they are facts, that many animals do not have many substitutes for focusing on their pain and cannot will themselves to focus on other things, cannot form expectations about the ending of pain, think about other times without pain, or consider more complex interests for which pain may be a necessary means, provide us with reasons for thinking that the overall pain experience caused by a given measure of pain might sometimes be worse for animals than it is for us. Placing this discussion back into the context of welfare, these considerations suggest that in virtue of the fact that we can suffer and animals cannot, it is not obvious that the same measure of pain is always worse for us than for animals—where worse is defined in terms of a change in welfare—at least over the period of time during which the pain is felt.

This last qualification is important. It seems plausible that, through memory and anticipation, the effects of pain may last longer for us than for animals. For instance, we might anticipate a painful experience before it occurs or we might endure lingering emotional consequences as a result of remembering pain, and these are of course both instances of suffering according to most of the philosophers discussed above. Through suffering, the consequences of pain may have a longer duration for us than for animals. This may mean that while pain to an animal can present a relatively greater change in their welfare at the time pain is experienced, pain to us can present a relatively greater change in our welfare over a period of time.[59]

I can now discuss some research that has so far been absent from the discussion. Throughout this chapter, I have assumed that there are two distinct categories of beings, sophisticated and unsophisticated, because I have assumed for the sake of analysis that there are no degrees in awareness of self and time. Furthermore, I have not challenged the assumption that most animals are cognitively unsophisticated because I have wanted to explore what the absence of sophistication in animals might imply about their pain. However, it is now useful to note that a variety of empirical studies suggest that many primates as well as other higher mammals, such as dolphins and elephants, have some capacity for self- and time-awareness.[60]

Returning to the different perspectives by which to measure welfare changes caused by pain—during the period of time in which pain is experienced versus over time—even if we are interested in the latter perspective, there may be an upper limit to the welfare change that can occur over a period of time from a given measure of pain. All that may be needed to experience suffering over time from pain is anticipation and memory. On the other hand, a highly developed sense of time is necessary to think of the extended past or the extended future, to know that the pain will at some point end, to will oneself to refrain from focusing on pain, and especially to engage in inter-temporal trade-offs and weighting. Also, a highly developed sense of time and self greatly increases the number of other interests that may not be affected by pain. After a certain level of self- and time-awareness that is necessary for memory, anticipation, and hence the psychological suffering that might occur from pain, self- and time-awareness might once again serve to reduce the role of pain in determining welfare by increasing the number of other, more complex interests and by enabling the ability to discount and to choose to avoid focusing on pain.

Over both dimensions, beings with only a basic or rudimentary sense of time and of self may be at the greatest disadvantage from pain. They may possess enough self- and time-awareness to suffer from the anticipation and memory of pain, but not enough to be able to discount pain, choose to refrain from focusing on pain, form expectations about the cessation of pain, or to consider other interests or times without pain. Some empirical research suggests that many primates and higher mammals possess just this sort of rudimentary capacity to see themselves as entities existing over time.[61]

CONCLUSION

I have argued that through psychological abilities such as weighting and through the role of mechanisms such as attention and expectations, a given amount of pain might not always be as bad for us as it is for many animals. I have argued that cognitive sophistication may sometimes reduce the role that pain plays in one's welfare. While I have not shown conclusively that pain is worse for animals than for us, the considerations I have given suggest that it is not obvious that, in virtue of the fact that we can suffer and many animals perhaps cannot (itself a contested matter), pain is always worse for us than for animals—a view many have held. Additionally, the analysis I have offered suggests that out of all beings, those with a rudimentary cognitive sophistication, such as many primates, may undergo the most significant welfare change from the experience of pain.

ACKNOWLEDGMENTS

I am grateful to the editors of this *Handbook* for offering careful and insightful feedback on this chapter, and to Guven Guzeldere, Geoff Sayre-McCord, and William Wojtach for giving me very helpful advice on an earlier version of the chapter.

NOTES

1. Those who have argued expressly for this view include Bob Bermond, "A Neuropsychological and Evolutionary Approach to Animal Consciousness and Animal Suffering," in *The Animal Ethics Reader*, 2nd ed., ed. Susan J. Armstrong and Richard G. Botzler (New York: Routledge, 2008); Peter Carruthers, "Brute Experience," *Journal of Philosophy* 86 (1989); Daniel C. Dennett, *Brainchildren: Essays on Designing Minds* (Cambridge, Mass.: MIT Press, 1998) and *Kinds of Minds: Towards an Understanding of Consciousness* (New York: Basic Books, 1997); and Robert Hanna, "What Is It Like To Be a Bat in Pain? The Morality of Our Treatment of Non-Human Animals" (draft chapter, quoted with permission from author). Michael Tye argues that simple creatures, like honey bees and fish, do not suffer in the way that we do. See his *Consciousness, Color, and Content* (Cambridge, Mass.: MIT Press, 2000).

2. Questions concerning the interpretation of Descartes as denying consciousness to all animals have been raised, for example, in Tom Regan, *The Case for Animal Rights* (Berkeley and Los Angeles: University of California Press, 1983), pp. 3–12.

3. Peter Harrison, "Theodicy and Animal Pain," *Philosophy* 64 (1989): 79–92. "Do Animals Feel Pain?" *Philosophy* 66 (1991): 25–40.

4. Harrison, "Theodicy and Animal Pain," p. 85.

5. See for instance Colin Allen, "Animal Pain," *Nous* 38 (2004): 617–43. Despite the differences in our reading of Carruthers, Allen's paper provides a rich overview and compelling arguments for the experience of pain in animals.

6. Carruthers, "Brute Experience."

7. Peter Carruthers, "Can Animals Feel Pain in the Morally Relevant Sense? No," *The Ag Bioethics Forum* 4 (1992): 2.

8. Peter Carruthers, "Suffering Without Subjectivity," *Philosophical Studies* 121 (2004): 99–125.

9. Carruthers, "Suffering Without Subjectivity," p. 10.

10. Carruthers, "Suffering Without Subjectivity," p. 12 (original emphasis).

11. Carruthers, "Suffering Without Subjectivity," p. 14.

12. Dennett, *Brainchildren*, p. 347.

13. Dennett, *Kinds of Minds*, p. 164. He attempts to demonstrate how a being that is capable of experiencing pain may nonetheless be incapable of suffering by highlighting cases of children who are dissociated in the presence of great pain. He writes that "whatever this psychological stunt of disassociation consists in, it is genuinely analgesic—or, more precisely, whether or not it diminished the pain, it definitely obtunds suffering" (p. 164).

14. Dennett, *Kinds of Minds*, p. 166. Specifically, he claims that stepping on your toe is "of vanishing moral significance" (p. 166). Earlier in the book, he attributes moral significance to suffering and not pain: he writes of pain that "for such states to matter—whether or not we call them pains, or conscious states, or experiences—there must be an enduring subject to whom they matter because they are a source of suffering" (p. 163; emphasis added).

15. Dennett, *Kinds of Minds*, p. 167.

16. Dennett, *Kinds of Minds*, p. 167.

17. Tye, *Consciousness, Color, and Content*, p. 182 (original emphasis).

18. Tye, *Consciousness, Color, and Content*, p. 182.

19. Hanna, "What Is It Like To Be a Bat in Pain?" p. 15.

20. Harrison, "Theodicy and Animal Pain," p. 91.

21. Jonathan Bennett, "Thoughtful Brutes," *Proceedings and Addresses of the American Philosophical Association* 62 (1988): 199.

22. Bennett, "Thoughtful Brutes."

23. I will also frequently refer to an objective understanding of pain, such as when making statements about the comparative intensity of the *same measure* of pain that may be involved with an equally bad injury and the like.

24. Many researchers have attempted to demonstrate that animals are more cognitively sophisticated in these ways. See note 60. While this is a fruitful and important vein of research, my concern in this chapter is to examine whether cognitive sophistication always makes pain worse for the being experiencing it.

25. The analysis in this chapter regarding how to conceive of the pain of sophisticated beings may not apply to human children. Up to a certain age, children may be more similar to animals in the way that they experience pain. The same may be true of some of the cognitively disabled. See note 61.

26. The arguments I will advance will not bear on positions that maintain that a sophisticated self- and time-awareness are necessary for *any* experience of pain. For those positions can grant that it might be true that self- and time-awareness can serve to reduce the experience of pain for humans, but maintain that this says nothing about the comparative point for animal pain since animal pain would be denied. Again, however, most authors do attribute pain experience to animals.

27. In economics, welfare analysis often involves making interpersonal comparisons and judgments about the satisfaction or frustration of interests and preferences in order to make assessments about the total welfare of either an individual or a group.

28. This is commonly called a "desire-satisfaction" view of well-being. Two other common views of well-being are "hedonic" and "objective lists" views. For an excellent discussion of these different accounts of well-being, see Martha Nussbaum and Amartya Sen, eds., *The Quality of Life* (Oxford: Clarendon Press, 1993).

29. For a discussion of these differences, see R. G. Frey, *Interests and Rights: The Case against Animals* (Oxford: Clarendon Press, 1980). Also, see Stephen P. Stich, "Do Animals Have Beliefs?" *Australasian Journal of Philosophy* 57 (1979): 15–28. While it is unclear whether Stich believes that most nonhuman animals lack beliefs and desires, he explores the reasoning behind such a view.

30. For a good discussion and criticism on some attempts at cardinality and comparability, see Amartya Sen, *Collective Choice and Social Welfare* (San Francisco: Holden-Day, 1970), especially chapters 7 and 8. Cardinal comparisons of this sort would require, at a minimum, knowing how much a certain individual values or disvalues an object compared to another individual on an objective or absolute scale. There may not be any additional conceptual difficulty when comparing across species with considerably different cognitive abilities, although there may be greater practical difficulties, yet cardinal comparability remains a theoretical difficulty even across remarkably similar individuals of the same species.

31. For instance, marginal or relative welfare comparison would entail comparing the change in welfare across individuals if they are given one more unit of some good object or if they are relieved of one more unit of some bad object, given their relative amounts of the good or bad object. So if individual A has one unit of X, and individual B has ten units of good X, there will be a smaller relative change in the welfare of B if he were given one more unit of X (we do not assume diminishing marginal utility for this point. All we need to assume is constant marginal utility). This approach is similar to the extended sympathy approach of Kenneth Arrow where interpersonal comparability can consist in judgments of the following kind: it is better (or worse) to be person I in state X than to be person J in state Y (Kenneth Arrow, *Social Choice and Individual Values*, 2nd ed. [New York: John Wiley & Sons, 1963]). This approach is inclusive of the particular characteristics and traits of each individual without committing to measuring precise comparability in common units between different individuals. Relative welfare analysis seems to be a particular case of the more general extended sympathy. Extended sympathy analysis may include both statements such as "It is better to be Niels Bohr as a physicist than to be Mr. Jones as a car insurance salesman" and "It is worse for Niels Bohr to be unable to perform analyses in atomic physics due to some illness than it is for Mr. Jones to lose his ability to sell car insurance due to some illness." Marginal welfare comparison only permits the second type of statement. It only makes judgments in regard to proportionate welfare changes.

32. Harry Frankfurt, "Freedom of the Will and the Concept of a Person," *Journal of Philosophy* 68 (1971): 5–20; Richard C. Jeffrey, "Preference among Preferences," *Journal of Philosophy* 71 (1974): 377–91.

33. Jeffrey in "Preferences among Preferences" argued that the preference to continue to live is a second-order preference since living entails wanting to have preferences. Since it is claimed that animals are not capable of being reflective and are not self-aware, then it would seem to follow on this view that animals cannot care whether they live or not.

34. In addition, Frankfurt has also distinguished between wanting or having an interest in something, on the one hand, and caring about something, on the other. He famously described caring about something as consisting of guiding oneself along a certain course in life and as related to the formation of a person's will (Harry Frankfurt, "The Importance of What We Care About," *Synthese* 53 [1982]). He writes of something that is cared about that "insofar as the person's life is in whole or in part devoted to anything, rather than being merely a sequence of events whose themes and structures he makes no effort to fashion, it is devoted to this" (p. 260, original emphasis).

35. Derek Parfit, *Reasons and Persons* (New York: Oxford University Press, 1984).

36. Similarly, Robert Nozick proposed that few people would prefer to be hooked up for life to an experience machine that would make them think and feel that they were experiencing a series of events, even if all the events were happy. *Anarchy, State, and Utopia* (New York: Basic Books, 1974). Nozick's main point was that we would prefer to have actually achieved the things we virtually accomplish in the machine, and thus we value something more than experience, no matter how good that experience is. We could say a person's global or higher order interests in actually achieving or *living* a life would be violated or compromised in the experience machine.

37. Long-term and complex interests are not necessarily the same thing. Someone could have a first-order interest for the long-term, such as an interest in satisfying a drug addiction over a long period of time. The distinction between short- and long-term refers to interests over different time periods rather than to different *levels* of interest.

38. Jan Narveson argues that because animals are not as capable of formulating long-range plans and have less intelligence, or sophistication more generally, their disutility of pain will be less. From his perspective, the pain of a sophisticated being is more valuable than that of an unsophisticated being. Narveson also believes that animals are less able to form long-term plans. "Animal Rights," *Canadian Journal of Philosophy* 7 (1979): 161–78.

39. In welfare economic theory, discounting generally refers to the idea of assigning a lower value to benefits received in future periods, but the same idea can be applied more broadly to benefits received in present time periods and to local benefits. For a thorough discussion of different approaches to time, especially on the plausibility of being temporally neutral when we are concerned with our own welfare, see Parfit, *Reasons and Persons*, pp. 149–86.

40. In addition, we not only make trade-offs across time periods, but also within time periods, such as when one favors eating ice cream rather than cake for dessert, but these sorts of trade-offs are less relevant to the discussion in this chapter.

41. See Frankfurt, "Importance of What We Care About."

42. Terry Connolly and Jochen Mattias Reb, "Omission Bias in Vaccination Decisions: Where's the 'Omission'? Where's the 'Bias'?" *Organizational Behavior and Human Decision Processes* 91 (2003): 186–202; Richard P. Larrick and Terry L. Boles, "Avoiding Regret in Decisions with Feedback: A Negotiation Example," *Organizational Behavior and Human Decision Processes* 63 (1995): 87–97.

43. René Richard, Joop Van Der Pligt, and Nanne De Vries. "Anticipated Regret and Time Perspective: Changing Sexual Risk-Taking Behavior," *Journal of Behavioral Decision Making* 9 (1996): 185–99; Jochen Matthias Reb, *The Role of Regret Aversion in Decision Making* (PhD diss., University of Arizona, 2005).

44. Including one among an Amazon tribe, the Satere-Mawe, where for ten minutes young men place their hands in gloves filled with bullet ants, which have a powerful sting. Steve Backshall, "Bitten by the Amazon," *The Sunday Times*, January 6, 2008. Available at

http://www.timesonline.co.uk/tol/travel/holiday_type/wildlife/article3131030.ece (accessed July 30, 2010).

45. As an example of an objective interests-satisfaction view, see James Griffin, *Well-Being: Its Meaning, Measurement, and Moral Importance* (Oxford: Oxford University Press, 1986). Thomas Scanlon argues that the former type of view reduces to an objective-list account of well-being. "Value, Desire and Quality of Life," in *The Quality of Life*, ed. Nussbaum and Sen, pp. 185–200.

46. For views that might be conceived of as hybrid accounts of well-being see Joseph Raz, "Personal Well-Being" in his *The Morality of Freedom* (Oxford: Oxford University Press, 1986), pp. 288–320; and Amartya Sen, "Capability and Well-Being," in *Quality of Life*, pp. 30–54.

47. Note that self-destructive behavior is often characterized by an unwillingness or inability to discount present costs or place greater weight on future time periods. It is often described as favoring short-term or first-order interests over long-term or complex ones. An example might be someone who continues to eat well after the point of being full despite the long-term health costs. For some, disordered eating can be associated with trying to seek a current pleasure or avoiding a current pain. This suggests that even if we were to adopt, at least for humans, a more objective conception of welfare than interest-satisfaction, pain might still not play a significant role in many circumstances. The value of good health would give us reason to discount the current pain of the overeater for the sake of his long-term health and welfare. There are numerous other examples to support the idea that these other conceptions of welfare would discount pain. To take just two, we should endure exercise for the sake of health and we should work through a mild headache for the sake of productivity.

48. There is also a tendency to be biased toward the satisfaction of short-term and first-order interests; all things being equal, many of us prefer getting benefits sooner rather than later, and this is sometimes true even when the benefits to be had sooner are smaller than the later benefits. See George Ainslie, *Picoeconomics: The Strategic Interaction of Successive Motivational States within the Person* (Cambridge: Cambridge University Press, 1991); and "Précis of Breakdown of Will," *Behavioral and Brain Sciences* 28 (2005): 635–73. This seeming fact has given rise to a number of so-called dynamic-choice problems. However, empirical studies also demonstrate that the rate at which we discount future benefits is inversely related to the amount of the benefit. See Gretchen B. Chapman, and Arthur S. Elstein. "Valuing the Future: Temporal Discounting of Health and Money," *Medical Decision Making* 15 (1995): 373–86; Leonard Green, Joel Myerson, and E. McFadden, "Rate of Temporal Discounting Decreases with Amount of Award," *Memory and Cognition* 25 (1997): 715–23; Kris N. Kirby, "Bidding on the Future: Evidence against Normative Discounting of Delayed Rewards," *Journal of Experimental Psychology: General* 126 (1997): 54–70. Additionally, when the receipt of neither benefit is imminent, we rationally prefer the later, larger benefit (see Kirby, "Bidding on the Future"). This research suggests that there is significant regard for long-term and complex interests, even if only at certain thresholds.

49. Michael Bratman, *Intentions, Plans, and Practical Reason* (Cambridge, Mass.: Harvard University Press, 1987).

50. George Ainslie, "Impulse Control in Pigeons," *Journal of the Experimental Analysis of Behavior* 21 (1974): 485–89; Melissa Bateson and Alex Kacelnik, "Rate Currencies and the Foraging Starling: The Fallacy of the Averages Revisited," *Behavioral Ecology* 7 (1996): 341–52; A. Ramseyer, M. Pelé, V. Dufour, C. Chauvin, and B. Thierry, "Accepting Loss: The Temporal Limits of Reciprocity in Brown Capuchin Monkeys," *Proceedings of the Royal Society,*

Biological Sciences 273 (2006): 179–84; William Roberts, "Are Animals Stuck in Time?" *Psychological Bulletin* 128 (2002): 473–89; H. Tobin, A. W. Logue, J. J. Chelonis, and K. T. Ackerman, "Self-Control in the Monkey *Macaca fascicularis*," *Animal Learning and Behavior* 24 (1996): 168–74.

51. Indeed, we saw before that this point was explicitly made by Harrison when he wrote that unlike animals "We are free, in painful situations, to damage our bodies if we believe there is a higher priority." Harrison, "Theodicy and Animal Pain," p. 83.

52. It might be thought that expectations are doing a lot of work to reduce intensity in the second scenario. That is, the pain here is less because we expect the doctor to stick the needle, whereas we do not expect the friend to do so. However, having such an expectation still requires cognitive sophistication that animals may lack. I return to the role of expectations below.

53. Harrison, "Theodicy and Animal Pain."

54. Bernard E. Rollin, *The Unheeded Cry: Animal Consciousness, Animal Pain, and Science* (Ames: Iowa State University Press, 1998), p. 144.

55. They also reported improvements in walking to the same extent. J. Bruce Moseley, Kimberly O'Malley, Nancy J. Petersen, Terri J. Menke, Baruch A. Brody, David H. Kuykendall, John C. Hollingsworth, Carol M. Ashton, and Nelda P. Wray, "A Controlled Trial of Arthroscopic Surgery for Osteoarthritis of the Knee," *New England Journal of Medicine* 347 (2002): 81–88.

56. For a good overview of many of these studies, see Dan Ariely, *Predictably Irrational* (New York: HarperCollins Publishers, 2009), especially chapter 10.

57. We do not need to assume any discrete phylogenetic shift toward cognitive sophistication in order to divide up pain experience in this way. Below, I examine the plausible view that there are degrees of cognitive sophistication across different species.

58. Bennett, "Thoughtful Brutes," p. 199.

59. The unsophisticated being's Ep may be more intensely bundled in a moment, but the sophisticated being's Ep might be spread over a period of time. If we believe that the intensity of pain matters, the first being may be worse off; conversely, if we believe that duration of pain matters, the second being may be worse off.

60. M. D. Hauser, J. Kralik, C. Botto, M. Garrett, and J. Oser, "Self-Recognition in Primates: Phylogeny and the Salience of Species-Typical Traits," *Proceeding of the National Academy of the Sciences* 92 (1995): 10811–14; Lori Marino, Diana Reiss, and Gordon G. Gallup, "Mirror Self-Recognition in Bottlenose Dolphins: Implications for Comparative Investigations of Highly Dissimilar Species," in *Self-Awareness in Animals and Humans*, ed. Sue Taylor Parker, Robert W. Mitchell and Maria L. Boccia (Cambridge: Cambridge University Press, 1994), pp. 380–91; Joshua Plotnik, Frans de Waal, and Diana Reiss, "Self-Recognition in an Asian Elephant," *Proceedings of the National Academy of Sciences* 103 (2006): 17053–57; Diana Reiss and Lori Marino, "Mirror Self-Recognition in the Bottlenose Dolphin: A Case of Cognitive Convergence," *Proceedings of the National Academy of Sciences* 98 (2001): 5937–42; Alexandra Rosati, G. Jeffrey, R. Stevens, Brian Hare, and Marc D. Hauser, "The Evolutionary Origins of Human Patience: Temporal Preferences in Chimpanzees, Bonobos, and Human Adults," *Current Biology* 9 (2007) 1663–68; Rosemary Rodd, *Biology, Ethics, and Animals* (New York: Oxford University Press, 1990), especially pp. 64–73; Robert W. Shumaker, and Karyl B. Swartz, "When Traditional Methodologies Fail: Cognitive Studies of Great Apes," in *The Cognitive Animal: Empirical and Theoretical Perspectives on Animal Cognition*, ed. Marc Bekoff, Colin Allen, and Gordon Burghardt (Cambridge, Mass.: MIT Press, 2002), pp. 335–45; Michael Tomasello, Josep Call,

and Brian Hare, "Chimpanzees Understand Psychological States—The Question is Which Ones and To What Extent," *Trends in Cognitive Science* 7 (2003): 153–56.

61. Likewise, the arguments about limited sense of time and self can also be applied to infants (or more generally human children) and cognitively impaired adult humans. Many psycho-physiological conditions associated with aging drastically change sense of time and sense of the self. For instance, patients with Korsakov's syndrome quickly forget recent events and impressions. Oliver Sacks describes how a patient became extremely agitated and upset when he could not remember the name of something in a satellite photo. "I had already, unthinkingly, pushed him into panic, and felt it was time to end our session. We wandered over to the window again, and looked down at the sunlit baseball diamond; as he looked his face relaxed, he forgot the *Nimitz*, the satellite photo, the horrors and hints, and became absorbed in the game below. Then as a savory smell drifted up from the dining room, he smacked his lips, said 'Lunch!', smiled, and took his leave. And I myself was wrung with emotion" (Oliver Sacks, *The Man Who Mistook His Wife for a Hat* [New York: Summit Books, 1970], p. 27). Almost immediately afterward, the patient lost all memory of having been upset. At the same time, the patient could be intense and steady in his attention and concentration of a single moment. Sacks describes the world of this kind of patient as well as the mentally disabled in general as concrete, vivid, intense, yet simple (p. 167). Allen Edwards describes Alzheimer's and dementia patients as people who lose awareness of the self and of others. Allen Jack Edwards, *When Memory Fails: Helping the Alzheimer's and Dementia Patient* (New York: Plenum Publishing, 1994). The patient will care about "creature comforts" and will become highly self-centered in a child-like manner (pp. 146–47).

SUGGESTED READINGS

ALLEN, COLIN. "Animal Pain." *Nous* 38 (2004): 617–43.

ARMSTRONG, SUSAN J., and RICHARD G. BOTZLER, eds. *The Animal Ethics Reader*. 2nd ed. New York: Routledge, 2008.

BEKOFF, MARC, COLIN ALLEN, and GORDON BURGHARDT, eds. *The Cognitive Animal: Empirical and Theoretical Perspectives on Animal Cognition*. Cambridge, Mass.: MIT Press 2002.

BEKOFF, MARC, and DALE JAMIESON, eds. *Readings in Animal Cognition*. Cambridge, Mass.: MIT Press, 1996.

BENNETT, JONATHAN. "Thoughtful Brutes." *Proceedings and Addresses of the American Philosophical Association* 62 (1988): 197–210.

CARRUTHERS, PETER. "Brute Experience." *The Journal of Philosophy* 86 (1989): 258–69.

———. "Suffering Without Subjectivity." *Philosophical Studies* 121 (2004): 99–125.

DENNETT, DANIEL C. *Kinds of Minds: Towards an Understanding of Consciousness*. New York: Basic Books, 1997.

———. *Brainchildren: Essays on Designing Minds*. Cambridge, Mass.: MIT Press, 1998.

GOODALL, JANE. *The Chimpanzees of Gombe: Patterns of Behavior*. Cambridge, Mass.: Harvard University Press, 1986.

GRIFFIN, DONALD. *Animal Minds*. Chicago: University of Chicago Press, 1992.

HARRISON, PETER. "Theodicy and Animal Pain." *Philosophy* 64 (1989): 79–92.

———. "Do Animals Feel Pain?" *Philosophy* 66 (1991): 25–40.

POVINELLI, DANIEL. *Folk Physics for Apes.* Oxford: Oxford University Press, 2000.
POVINELLI, DANIEL and TIMOTHY EDDY. "What Young Chimpanzees Know About Seeing."
    Monographs of the Society for Research on Child Development 61 (1996): 1–189.
RODD, ROSEMARY. *Biology, Ethics, and Animals.* New York: Oxford University Press, 1990.
ROLLIN, BERNARD E. *The Unheeded Cry: Animal Consciousness, Animal Pain, and Science.*
    Ames: Iowa State University Press, 1998.

CHAPTER 18

ANIMALS THAT ACT FOR MORAL REASONS

MARK ROWLANDS

1. PRELIMINARIES

Nonhuman animals (henceforth, "animals") are typically regarded as moral *patients* rather than moral *agents*. Let us define these terms as follows:

(1) X is a moral *patient* if and only if X is a legitimate object of moral concern: that is, roughly, X is an entity whose interests *should* be taken into account when decisions are made concerning it or which otherwise impact on it.

(2) X is a moral *agent* if and only if X can be morally evaluated—praised or blamed (broadly understood)—for its motives and actions.

Nothing in (1) and (2), of course, rules out one and the same individual being both a moral agent and a patient. Most humans are both. In addition, the notion of a moral agent is typically run together with that of a moral *subject*:

(3) X is a moral subject if and only if X is, at least sometimes, motivated to act by moral considerations.

However, (2) and (3) are not equivalent: the motivation for an action is one thing, the evaluation of the action or the motivation quite another. Though motivation and evaluation are conceptually distinct, as a matter of practice these are, for obvious reasons, run together. Many will regard the claim that motivation and evaluation are conceptually distinct with surprise. The bulk of this paper can be regarded as an attempt to defend this claim as it applies to the case of animals. In these opening paragraphs, I merely offer some preliminary remarks in its defense.

Suppose that my wife, worn down by the domestic squalor that goes with two young boys, a very large dog, and a slovenly husband, has me (unknowingly) hypnotized. Now whenever she utters the word "Rosebud," I experience an uncontrollable desire to mop the floor. This desire is, it seems, a motivational state, one that when combined with relevant cognitive states (the belief that this is a mop, the belief that this is a floor, and so on) will, ceteris paribus, result in a certain sort of behavior on my part. This floor mopping, however, is not something for which I can be praised or blamed. Ex hypothesi, it is the result of a motivational state that is outside my control. In other words, my behavior is motivated by a certain state, but this state is not the sort of thing for which I can be evaluated. The motivation for my behavior is one thing; the evaluation of that behavior is quite another. If hard determinism—in effect, a generalization of this case—were true, then no one could be morally evaluated for what they do, but it would not follow that they were not the subjects of motivational states. The distinction between moral subjects and moral agents follows from this distinction between the motivation of an individual's action and the evaluation of that action.

One objection is likely to proceed as follows: if someone has no control over the motivational state that provides the (whole or partial) cause of their behavior, then that state cannot be thought of as providing a *reason* for their behavior. This objection, however, is difficult to sustain.

First, my uncontrollable desire to mop the floor does seem to be (part of) the reason why I am mopping the floor. If we want to deny this, we must be building much into the idea of a reason beyond that contained in common sense. If so, the burden of proof, it seems, is going to be on those who want to bulk up the notion in this way. This is the subject of this paper. Suppose someone were to ask: "Why is Rowlands mopping the floor?" An appropriate answer would be: "He has an uncontrollable desire to do so." In response, one might comment: "That's most unlike him," and then the explanation might be continued: "His wife had him hypnotized." The latter explanation cites a cause of my uncontrollable desire. But the first explanation cites what certainly seems to be a reason for my action. It is the content of my desire that I mop the floor which, when combined with the beliefs that what I have in my hand is a mop and that what lies beneath my feet is a floor, that explains why I am acting in the way that I do. This deployment of content seems to be the hallmark of reason-giving explanations.

Nevertheless, one might object, while the uncontrollable desire to mop the floor is a reason (or, rather, a component of a reason—I will henceforth ignore this complication), it is not *my* reason. It is not *my* reason because I have no control over it. However, if we accept that the desire is (a component of) a reason, then the idea that it is not my reason seems difficult to sustain. Whose reason would it be if not mine? There are two possibilities:

(1) It is someone else's reason.
(2) It is nobody's reason.

The first option need not detain us. It is, after all, I and not anyone else who is mopping the floor. So, the relevant alternative is (2). The problem with (2) is that it commits us to the possibility of *ownerless* reasons: that there can exist a reason that is a reason for no one. However, there are no ownerless reasons. All reasons are reasons for someone.[1] Therefore, if the uncontrollable desire to mop the floor is *a* reason then it is *my* reason. But it is not the sort of thing for which I can be evaluated. To have a reason does not entail the possibility of moral evaluation. And this means that moral subjects are not the same as moral agents. This paper is concerned only with the former.

I am going to argue that some animals are moral *subjects* in the sense expressed by (3): they can be, and sometimes are, motivated by moral considerations. It does not follow from this that they can be morally evaluated for what they do. To be a moral subject is a necessary condition of being a moral agent, but it is not sufficient.

2. ANIMALS AS MORAL SUBJECTS

The claim that animals can be objects of moral concern is, by now, a reasonably orthodox one. Disagreements persist about the nature and extent of our moral obligations to animals, but the idea that we have *no* obligations whatsoever to them—neither direct nor indirect—is endorsed by few.[2] In contrast, the claim that animals can be moral subjects—in the sense expressed in (3)—as well as objects of moral concern is distinctly unorthodox. However, it is certainly not unheard of. For example, David DeGrazia claims: "These examples support the attribution of moral agency—specifically, actions manifesting virtues—in cases in which the actions are not plausibly interpreted as instinctive or conditioned. On any reasonable understanding of moral agency, some animals are moral agents."[3] This claim is echoed by Steven Sapontzis[4] and by Evelyn Pluhar, who writes:

> Is it really so clear, however, that the capacity for moral agency has no precedent
> in any other species? Certain other capacities are required for moral agency,
> including capacities for emotion, memory, and goal-directed behavior. As we have
> seen, there is ample evidence for the presence of these capacities, if to a limited
> degree, in some nonhumans. Not surprisingly, then, evidence has been gathered
> that indicates that nonhumans are capable of what we would call "moral" or
> "virtuous" behavior.[5]

Both DeGrazia and Pluhar express their claim in the language of agency, and both employ the concept of virtue. To act virtuously is to act from a virtuous motive. To be thus motivated, one need only be a moral subject, as characterized in (3), rather than a moral agent, as described in (2). Therefore, one can plausibly maintain that DeGrazia's and Pluhar's target is the claim that animals can be moral subjects, rather than moral agents, in the sense described in this paper. Similar claims, although in varying forms, can be found in the work of Vicki Hearne, Jeffrey Moussiaeff Masson, Susan McCarthy, Stephen Wise, Frans de Waal, and Marc Bekoff.[6] Indeed, Darwin claimed that animals can be motivated by the "moral sentiments."[7]

An important and systematic defense of the claim that some animals qualify as moral subjects is to be found in Marc Bekoff and Jessica Pierce's recent book, *Wild Justice*.[8] Bekoff and Pierce argue that at least some animals exhibit a broad repertoire of behaviors that can correctly be regarded as moral. These include being fair, showing empathy, exhibiting trust, and acting reciprocally. These behaviors, Bekoff and Pierce argue, are the causal result of a complex range of emotions. Such animals, they conclude, are moral subjects, and the property of being a moral subject is, therefore, one that humans share with at least some other animals.[9]

In the Bekoff and Pierce thesis, we find both the strengths and weaknesses of the thesis that animals can be moral subjects. The strengths lie in the breadth and depth of the empirical work they deploy. The weaknesses lie in certain logical gaps in their case—gaps that would almost certainly allow one skeptical of animal agency to resist their conclusion. The aim of this paper is to exploit the strengths of their position while eliminating the weaknesses. What Bekoff and Pierce have done, in effect, is to issue an invitation: why do you not think of moral subjecthood—of what it is to be a moral subject—in this way? The problem with invitations is that they can always be declined. The goal of this paper is to convert this invitation into something more like an offer that can't be refused.

3. Happy Dog and Sad Dog

I shall organize the discussion around an example cited by Bekoff and Pierce. The example is a nonstandard one because Bekoff and Pierce do not endorse it as an unequivocal example of moral behavior. I am going to use it largely for expository purposes: it effectively highlights the weaknesses of the Bekoff/Pierce position, and throws into sharp relief the issues to be addressed in this paper. The authors write:

> Yvette Watt, an artist and animal advocate in Hobart, Tasmania, told Marc the following story about two dogs. One is a well-fed and happy canine, the other a sad dog always tied to a rope. The happy dog's daily walk takes him by his unfortunate neighbor. One night, the happy dog eats his usual dinner, but saves his meaty bone. The next morning he carries the meaty bone on his walk, and delivers it to his tethered friend.[10]

If you doubt the veracity of the story, or are skeptical about the deployment of anecdotal evidence more generally, then you are at liberty to think of this case simply as a thought experiment. My primary interest is in identifying precisely why a skeptical moral philosopher might deny that this is a case of an animal exhibiting moral agency—even if events did unfold exactly as described in the story.

The example is typical in the type of moral motivation it attributes to Happy Dog. He is motivated by a *sentiment* or *feeling*. Let us accept, for the purposes of developing the skeptic's response, the following claims: Happy Dog's behavior is (a) compassionate, and (b) is the causal result of compassionate feelings toward Sad Dog.

There are two distinct stances toward these claims available to the skeptic. First, the skeptic can simply accept both (a) and (b), yet still deny that Happy Dog's motivation is a moral one. That is, the skeptic can argue that the truth of (a) and (b) do not entail that Happy Dog is a *moral* subject. Second, the skeptic can choose to view compassion as an essentially moral phenomenon, and therefore deny that Happy Dog is exhibiting compassion as opposed to some nonmoral facsimile of this emotion. In the discussion that follows, I shall highlight this distinction when it proves important and ignore it when it does not. In either of its forms, we can identify two sources for this skepticism. Our first task is to identify these.

4. Kant and Normative Self-control

Bekoff and Pierce gesture toward one possible source of the skeptical denial of Happy Dog's moral subjecthood. Here, the skeptic in question is Kant, ably represented in this instance by Christine Korsgaard. Bekoff and Pierce quote Korsgaard:

> [T]he capacity for normative self-government and the deeper level of intentional control that goes with it is probably unique to human beings. And it is in the proper use of this capacity—the ability to form and act on judgments of what we ought to do—that the essence of morality lies, not in altruism or the pursuit of the greater good.[11]

Bekoff and Pierce counter this claim as follows:

> Even if there are bona fide differences in kind, this does not mean that many aspects of morality aren't also shared, or that there aren't significant areas of continuity or overlap. We view each of these possibly unique capacities (language, judgment) as outer layers of the Russian Doll, relatively late evolutionary additions to the suite of moral behaviors. And although each of these capacities may make human morality unique, they are all grounded in a much deeper, broader, and evolutionarily more ancient layer of moral behaviors that we share with other animals.[12]

This reply, however, misses Korsgaard's point, and the point of the Kantian moral tradition she represents. Korsgaard claims that the ability to reflect on or form judgments about what we ought to do is the *essence* of morality. Any behavior that is not subject to this sort of normative self-reflection is not moral behavior. If a creature is unable to reflect on what it does, and ask itself whether this is, in the circumstances, a morally good thing to do or ask itself whether what it feels in these circumstances is the morally right thing to feel, it is not a moral creature. Normative self-control is the essence of morality. Anything—for example, a compassionate feeling—that is not subject to such control is not a moral phenomenon, appearances notwithstanding. Therefore, if we assume that Happy Dog is not able to subject his feelings and actions to this sort of normative self-scrutiny, neither his action nor the feeling that motivates this action qualifies as a moral phenomenon.

That Happy Dog's feeling even qualifies as compassion is, therefore, also doubtful. If, as the second form of skepticism identified earlier asserts, compassion is an essentially moral sentiment, and if Happy Dog's sentiment is not a moral sentiment, then it follows that Happy Dog's feeling is not, in fact, the feeling of compassion. Presumably, this feeling would then have to be regarded as a sort of nonmoral facsimile of compassion: a feeling that resembled compassion in its phenomenological profile—in terms of what it is like to have that feeling—but lacks the essential moral core that would convert this phenomenology into the properly moral sentiment that compassion is taken to be. Is Happy Dog able to ask himself whether bringing the bone to Sad Dog is the right thing to do in the circumstances? Is he able to ask himself whether the sympathy (we have assumed) he feels for Sad Dog's plight is the right thing to feel in the circumstances? If the answer is to these questions is "no"—and Korsgaard and Kant presumably think that it is—then, for them, Happy Dog's action is not an example of a moral behavior, and the sentiments we assume he bears toward Sad Dog are not examples of moral sentiments.

With the Russian Doll analogy, Bekoff and Pierce have, in effect, issued an invitation: why do you not think of morality in this way? Korsgaard and Kant would likely respond with a firm "No thanks." I am going to argue against the Korsgaard/Kant picture of moral subjecthood. I shall argue that normative self-control, in the Kant/Korsgaard sense, is not a necessary condition of being a moral subject. At this point, however, I outline their position (i) to illustrate an example of the sort of view of morality that needs to be undermined if the claim that animals are moral subjects is to be accepted, and (ii) to note that as yet Bekoff and Pierce have not succeeded in undermining it.

5. ARISTOTLE AND THE VIRTUES

Skepticism concerning the moral status of Happy Dog's motivation is not restricted to those who adhere to the Kantian vision of morality as constituted by normative self-control over one's actions and sentiments. What is, for our purposes at least, a cognate view can be found in what is generally thought of as a quite different tradition: Aristotle's account of the virtues.

In this passage from the *Nicomachean Ethics*, Aristotle emphasizes the psychological complexity of the virtues:

> But for actions in accord with the virtues to be done transparently or justly it does not suffice that they themselves have the right qualities. Rather, the agent must also be in the right state when he does them. First he must know that he is doing virtuous actions; second, he must decide on them, and decide on them for themselves; and, third, he must also do them from a firm and unchanging state.[13]

For an action to be an expression of a virtue, it must not simply be an example of what would commonly be regarded as a virtuous action. In addition, the agent must (a) know that he is performing a virtuous action, and (b) perform the action because it is a virtuous action and (c) make the decision from a stable disposition. This does not exhaust the psychological complexity of the virtues. A more complete characterization would at least sketch the cognitive and emotional setting in which the action is embedded. The agent must deplore the corresponding lack of virtue, in herself and others, and so on. For the purposes of this paper, these complications can safely be ignored.

It is the conjunction of conditions (a) and (b) that are important for our purposes, because it is these that underlie hostility to the idea that, in bringing the meaty bone to Sad Dog, Happy Dog is acting virtuously. Together, in a manner not dissimilar to that identified in Kant, they impose a minimal condition of *reflection* on the virtuous agent. To satisfy these conditions, the agent must understand what a virtue is, and be motivated by this understanding to perform a certain action because it would be expressive of this virtue. I shall call this—the conjunction of (a) and (b)—the *reflection condition* on virtue:

> For action φ, performed by agent A, to be an expression of virtue, V, it is necessary that A (i) be able to understand that φ is an instance of V, and (ii) A must perform φ because he understands that φ is an instance of V and wishes to be virtuous.

Condition (i) is not unambiguous. The notion of understanding that something is a virtue can be interpreted in two ways, one more cognitively demanding than the other. The cognitively demanding way interprets understanding φ to be a virtue as "understanding φ to fall under the concept of virtue." This requirement seems unreasonably strong. If correct, someone could not know whether something was a virtue unless she were in possession of the correct concept of a virtue, and so must, among other things, be familiar with the debates surrounding the definition of this concept, and have (correctly) adjudicated between the competing positions. Few—animal or human—could understand that φ is a virtue in this sense.[14] According to a weaker—and more reasonable—interpretation, understanding that φ is an instance of a virtue only requires that agent A be able to distinguish with a reasonable though not necessarily infallible level of success, those things that are virtuous from those things that are not.[15] I shall understand (i) in this weaker sense.

Why think that virtues must conform to the reflection condition? More precisely, why think that the reflection condition constitutes a necessary condition of the possession of a virtue? Aristotle closely ties the reflection condition to what we might think of as the *normative grip* of a virtue.

Happy Dog, we have supposed, is caused to bring the bone to Sad Dog by his sentiment of compassion. However, a sentiment, in itself, will simply cause whatever it causes. Bringing the bone to Sad Dog is something that Happy Dog does, but can we give any sense to the claim that Happy Dog was morally right to do so? This question actually divides into two: (i) Can we make sense of the idea that Happy

Dog was *right*—morally right—to have this sentiment in these circumstances? (ii) Can we make sense of the idea that Happy Dog's behavior was a morally appropriate response to this sentiment? The sentiment, we are supposing, exerts *causal* pressure on the action—in the sense of being the triggering cause of it. But can we make sense of the idea that it also exerts *normative* pressure? Is the sentiment one that Happy Dog *should* have in these circumstances and is his resulting behavior something that he *should* perform? An affirmative answer to (i) and (ii) requires that there be standards of correctness governing both the having of the sentiment and the relation between the sentiment and the action.

For Aristotle, the second element of normativity is explained by the fact that, when they occur as constituents of a virtue, sentiment and action are internally, not externally, related. In other words, the normative character of the relation between sentiment and action is explained by way of the normative structure of virtue itself. To explain the first element of normativity, Aristotle needs to show that the virtue is something that is appropriate to the circumstances. That is, he needs to show that the given context requires the exercise of the virtue: if the virtue is not exercised in this context, then something has gone wrong. If he can explain this first element of normativity, then the second comes for free—as a consequence of the normative structure of virtue.

Aristotle has two, mutually reinforcing explanations of the first element of normativity. First, he emphasizes that the virtues are things that must be *acquired*:

> The virtues arise in us neither by nature nor against nature. Rather, we are by nature able to acquire them, and we are completed through habit....Virtues, by contrast [to the senses], we acquire, just as we acquire crafts, by having first activated them. For we learn a craft by producing the same product that we must produce when we have learned it; we become builders, for example, by building, and we become harpists by playing the harp. Similarly, then, we become just by doing just actions, temperate by doing temperate actions, brave by doing brave actions.[16]

A virtue is something that must be acquired, and so can be acquired well or poorly, properly or improperly—and this possibility introduces the required element of normativity into the possession of a virtue. This is why we can be praised or blamed for their possession or lack thereof. If virtues were like the senses, whose possession is a matter of nature and not our efforts, we could not be morally evaluated for their possession or lack—whether we have them or not is nature's "choice," not ours.

Second, and for our purposes more importantly, a virtue is something that can be exercised well or poorly. In a well known passage from the *Nicomachean Ethics*, Aristotle remarks:

> So also getting angry, or giving and spending money, is easy and everyone can do it; but doing it to the right person, in the right amount, at the right time, for the right end, and in the right way is no longer easy, nor can everyone do it. Hence, doing these things well is rare, praiseworthy, and fine.[17]

Whether or not an action counts as the expression of a virtue is a context-sensitive matter: only if the action is done to the right person, in the right amount, at the right time, and so on, does it count as the expression of a virtue. But whether the person is the right one, whether the amount is correct, whether the time is right, and so forth, are difficult matters that call for good judgment. This judgment is something that can be executed well or poorly—mistakes are always possible. And it is the possibility of *mistake* in the application of a virtue that underlies the normative grip of a virtue on the world.

So, with regard to Happy Dog, Aristotle would recommend to us the following sorts of question: (1) Does Happy Dog Know that what he is doing is virtuous? (2) Does Happy Dog take the bone to Sad Dog because he recognizes that it would be virtuous to do so and wants to be virtuous? (3) Does Happy Dog think that Sad Dog is the right dog to be taking the bone to? (4) Does Happy Dog think that it is the right time to do this? (5) Does Happy Dog think that taking the bone to Sad Dog is an act that is in proportion to Sad Dog's suffering? (And so on, and so forth.)

If Happy Dog is unable to entertain these (and related) thoughts, then we can make no sense of the idea that this sentiment is something Happy Dog *should* have in these circumstances (the first element of normativity). And, so, such a sentiment would exert only causal, and not normative, pressure on his behavior (the second element of normativity). Therefore, for Aristotle, neither Happy Dog's sentiment nor the behavior that it occasions would qualify as moral phenomena.

In Aristotle, therefore, we see the close connection between two phenomena that are central in the attempt to deny animals the status of moral subjects or agents. On the one hand, there is the idea of *reflection*. In order to qualify as a moral agent, a creature must be able to critically scrutinize both its behavior and its motivation—to subject these to moral evaluation. On the other hand, there is the idea of praise and blame: a moral being can be praised or blamed for what it does. In Aristotle, the connection between these is intimate. Critical reflection on one's behavior and motivation is required because without it there is no possibility of praise and blame. Without it, Happy Dog is simply the subject of a feeling on which he cannot reflect, and so over which he has no control. Happy Dog has not worked to acquire the feeling—this feeling is simply there or not as the case may be. This feeling causally compels him to take the bone to Sad Dog, but he cannot reflect on whether this is the right thing to do. Thus, no sense can be given to the idea that Happy Dog is acting in a morally praiseworthy manner, that he is doing what he *should*. Happy Dog simply does what he does.

I have argued earlier that the ideas of praise and blame are, for the purposes of this paper at least, red herrings. Our concern is with whether animals can be moral subjects—in the sense identified earlier—and not whether they can be moral agents in the sense of the legitimate subjects of praise and blame. In deciding whether animals can qualify as moral subjects, therefore, a key idea for some major moral philosophers is the idea of reflection. Animals do not qualify as moral subjects, according to both Kantian and Aristotelian views, because they are unable to morally scrutinize their actions and motivations.[18]

6. THE IDIOT

To claim that animals can be moral subjects is to claim that they can be motivated by moral considerations. The denial that animals can be moral subjects is underwritten by a certain picture of the moral subject as a *reflective scrutinizer* of her moral motivations and actions—as we see in Aristotelian and Kantian theories. Animals can reflectively scrutinize neither their motivations nor their actions, the argument goes. Therefore, the motivations for their ostensibly moral actions—typically sentiments of some sort—do not qualify as moral and neither do their actions.

In the rest of this paper, I am going to challenge the condition of the reflective moral subject, and therefore I will be challenging the Kantian and Aristotelian accounts laid out previously.[19] I will argue that it is implausible to suppose that the existence of a reflective moral subject is a *necessary* condition of the possibility of moral motivation. To suppose that it is betrays a set of implausible assumptions concerning the nature of moral motivation. If this argument is correct, there is no reason to suppose that we *must* be reflective moral subjects in order to be motivated by moral considerations.

Consider a person who I am going to call Mishkin after the prince in Dostoevsky's *The Idiot*. Prima facie, Mishkin has the soul of a prince: throughout his life, he performs many acts that seem to be kind or compassionate. He performs these acts because he is the subject of sentiments that—again, at least prima facie—seem to be kind or compassionate ones. When he sees another suffering, he feels compelled to act to end or ameliorate that suffering. When he sees another happy, he feels happy because of what he sees. If he can help someone get what she wants without hurting anyone else, he will help because he finds that he enjoys doing it. In short, Mishkin deplores the suffering of others and rejoices in their happiness. His actions reflect, and are caused by, these sentiments. What Mishkin does not do, however, is subject his sentiments and actions to critical moral scrutiny. Thus, he does not ever think to himself things like: "Is what I am feeling the right feeling in the current situation— that is, is what I am feeling what I *should* be feeling?" Nor does he think to himself things like: "Is what I propose to do in this circumstance the (morally) correct thing to do (all things considered)?" Let us suppose he does not do this because he is incapable of doing it. He operates on a much more visceral level. What we might think of as Mishkin's moral profile looks something like this:

> (M1) (i) Mishkin performs actions that *seem* to be good, and (ii) Mishkin's
> motivation for performing these actions consists in feelings or sentiments that
> *seem* to be good, but (iii) Mishkin is unable to subject these actions and
> sentiments to critical moral scrutiny.

The question is: would we want to claim that the combination of (i)–(iii) disqualifies Mishkin as a moral subject—that is, does it disqualify him as someone whose motives for action are moral ones? If we are convinced by Aristotle or Kant, we would answer in the affirmative because in the absence of critical moral scrutiny, Mishkin's sentiments exert no normative pressure on his actions. But normative pressure is essential to something counting as a reason. If Mishkin has reason to φ then, ceteris paribus (e.g., Mishkin has no countervailing reasons, Mishkin is able to φ, and so on) Mishkin should φ. If he does not φ then, ceteris paribus, something has gone wrong.[20] Accordingly, these sentiments do not even provide reasons for Mishkin's actions; a fortiori, they are not moral reasons. But how reasonable is this denial? Mishkin spends his life helping others, and does so because he delights in the happiness of others and deplores their suffering. We need to distinguish the question of (a) whether someone qualifies as a moral subject, in the sense of being motivated by moral concerns, and (b) whether something qualifies as a moral subject in precisely the same way we do. Perhaps, having reasons in the sense adduced by Kant and Aristotle is only one way of being a moral subject? Could Mishkin not be a moral subject, merely in a somewhat different way than other persons are?

This last question, however, is simply a variation on the Bekoff/Pierce invitation to expand our conception of morality: why not think about morality—specifically moral subjecthood—in such a way that Mishkin turns out to be a moral subject, merely one somewhat different from the way we are (or take ourselves to be)? I will attempt to strengthen this invitation into something more compelling.

There is a clear sense in which (M1) does a disservice to Mishkin. The disservice stems from the characterization of his actions and feelings as ones that *seem* to be good. The motivation for this characterization is fairly clear. Someone who rejected the idea that Mishkin was a moral subject would also reject the idea that Mishkin's feelings and actions counted as good (e.g., kind or compassionate)—such a description would be true only of a moral subject. His feelings exert no normative grip on his actions and, therefore, do not qualify as reasons, let alone moral reasons, for his actions. It is this idea that Mishkin's sentiments do not amount to reasons that does most of the damage to the claim that Mishkin could be a moral subject. So let us work on providing Mishkin with reasons.

Suppose someone—Marlow, after the skilled scrutinizer of motivations who narrates some of Joseph Conrad's novels—is capable of the sort of critical moral reflection of which Mishkin is incapable. And suppose, on the basis of this reflection, Marlow were to endorse the same sorts of sentiments and actions that Mishkin has and performs. That is, suppose for any given circumstance C, Mishkin feels F and, as a result, performs action A. Marlow, a critical and skilled moral scrutinizer of his sentiments and actions, independently comes to the conclusion that in circumstance C, it is morally correct to feel F and perform action A. Suppose, further, that Marlow—who we might think of as a rational and ideal spectator—invariably reaches the correct moral conclusion (or, if you prefer, does so at least most of the time). In such circumstances, we can strengthen (M1) as follows:

> (M2) (i) Mishkin performs actions that *are*, in fact, good, and (ii) Mishkin's motivation for performing these actions consists in feelings or sentiments that *are*, in fact, the morally correct ones to have in the circumstances, but (iii) Mishkin is unable to subject these actions and feelings to critical moral scrutiny.

The transition from (M1) to (M2) is a significant one, and we can use a familiar distinction (introduced earlier) to explain why. The strengthening of (M1) to (M2), in effect, gives Mishkin *external* reasons for his actions. This is in contrast to an internal reason, which is, roughly, a reason that an agent *has* for his action. *Having a reason* is open to a variety of interpretations but in the moral domain it is typically understood in terms of reflection upon and endorsement of a reason: an internal reason is one of which the agent is aware, or upon which he has reflected, and which he consequently endorses. An external reason is a reason that there is for an agent to perform an action—whether or not he is aware of it. If doing X promotes A's important long-term interests, then A has a reason to do X even if he is unaware of this fact.

In the case of (M1), it is open to someone to argue that Mishkin possesses no moral reasons for his actions: his feelings simply exert causal pressure on his actions but do not exert normative pressure. Therefore, they do not qualify as reasons of any sort. The switch to (M2), however, means that Mishkin now has external moral reasons for his actions. That is, given (M2), there *are* now reasons for Mishkin's actions.[21] Mishkin cannot entertain those reasons—that is, they are not internal reasons—but nonetheless they are reasons that exist for Mishkin to do what he does.

So, now, we can repeat our question: would we want to say that Mishkin—as characterized by (M2)—is not a moral subject? He does things that are, in fact, morally good, and he does them from a motivation that is, in fact, a morally commendable one. There are reasons—moral reasons—that connect his sentiments and actions: it is just that these are external ones. Should we really be so confident in the claim that Mishkin does not qualify as a moral being? This is still an invitation to think of morality in a certain way; but is it still an invitation that can be reasonably declined?

Maybe, but a skeptical moral philosopher who still wishes to decline the invitation is committed to the idea that a necessary condition of being a moral subject—of being something that can act for moral reasons—is that one possess *internal* reasons for one's actions. These internal reasons are available to rational scrutiny; they are reasons that one not only can endorse but that one has, at some point, in fact endorsed. The question we must now consider, then, is: why would someone want to claim this?

The motivation for this claim is going to come from a perceived deficit that accompanies Mishkin's lack of internal reasons. There is a clear sense in which, although Mishkin gets things "right"—that is, in given circumstance C, he feels the morally correct feelings and performs the morally correct actions—he gets it right by *accident*. Mishkin acts on the basis of feelings he possesses, and he has scrutinized neither the feelings, nor the actions, nor the connection between the feelings and the actions. So, while there is a sense in which Mishkin gets things right, morally speaking, he does not get things right *for the right reasons*. And that is what is needed for Mishkin to be a moral agent. There are indeed reasons for what Mishkin does, but since he is unaware of those reasons, he can't be said to do what he does *for* those reasons.

The picture of a reflective moral subject as someone who can get things right—that is, identify the correct moral conclusion—for the right reasons is one that many will find compelling. So, let us try to remove the offending element of contingency in Mishkin's moral decision making. We can do this by, to some extent, *internalizing* Mishkin's reasons—although, crucially, not in a way that converts him into a full-blown reflective moral subject.

Suppose that Mishkin's getting things right—by which I mean feeling the right feelings and performing the right actions in given circumstances—is not as accidental as it seems. Suppose there are good (i.e., rational) reasons why he has these feelings in these circumstances, and suppose there are equally good reasons connecting feelings and actions. It is just that these reasons are not available to his conscious, rational scrutiny. They are reasons that have been embodied in Mishkin's subpersonal, nonconscious, inferential processing operations. For example, if our tastes ran to the modular (although, I emphasize, it is strictly optional to think in this way), we might suppose that Mishkin has a "moral module," the operations of which are *cognitively impenetrable*—that is, cannot be penetrated by subsequent belief- and concept-forming operations. Nonetheless, the operations of the module are reliable ones and lead to correct moral conclusions with the same level of success as our ideal, rational spectator, Marlow. In such circumstances, we could amend (M2) to the following:

> (M3) (i) Mishkin performs actions that are good, and (ii) Mishkin's motivation for performing these actions consists in feelings or sentiments that are the morally correct ones to have in the circumstances, and (iii) Mishkin *has* his own good reasons for having these feelings and performing these actions in these circumstances, and (iv) Mishkin is not aware of these reasons.

I argued that (M2) gave Mishkin external reasons for his actions. In one clear sense, (M3) internalizes those reasons. They are now not simply reasons that exist for Mishkin. They are reasons that Mishkin *has*. He has them not in the sense that they are available for his critical moral scrutiny, but in the sense that they are embodied in the nonconscious, subpersonal processing operations that occur in his moral module. These operations culminate in his conscious, personal-level moral sentiments and actions.

So, once again, we have our question. Would we want to deny that Mishkin is a moral subject? Or would it not be more reasonable to suppose that Mishkin is a moral subject, merely a somewhat different one than the critical moral appraisers we take ourselves to be? My intuition is that, at this point, the denial of moral subjecthood to Mishkin has become unreasonable. Mishkin is indeed a moral subject, just a somewhat different subject than the one we—or at least Kant and Aristotle—take ourselves to be. Let us see if we can drive this intuition home.

7. Moral Super Blindsight

What is the precise nature of the difference between Mishkin and a reflective moral subject? Mishkin's moral profile, we have seen, is characterized by (M3). But a reflective moral subject—represented by Marlow—would be characterized by (M4):

> (M4) (i) Marlow performs actions that are good, and (ii) Marlow's motivation for performing these actions consists in feelings or sentiments that are the morally correct ones to have in the circumstances, and (iii) Marlow has good reasons for having these feelings and performing these actions in these circumstances, and (iv) Marlow is aware of these reasons.

The difference between Mishkin and Marlow thus pertains only to condition (iv). In characterizing this difference, we could stay with the idiom of modularity: the operations of Marlow's moral module are, whereas those of Mishkin's module are not, *available* to subsequent operations of (conscious) belief-formation. There are two ways in which we might explain this idea of lack of availability. First: (1) the operations occurring within Mishkin's moral module are not *phenomenally conscious*. That is, there is nothing that it is like to be the subject of these operations. For Marlow, in contrast, there is something that it is like to engage in moral deliberation. This, according to the present suggestion, constitutes the difference that condition (iv) tries to capture. The second way of understanding lack of availability is in terms of Mishkin's inability to form *higher order thoughts* about his moral deliberations: (2) when Mishkin is engaged in moral deliberations he has no *higher order thoughts* to the effect that this is what he is doing. This section will be concerned with (1). The rest of the paper will then focus on (2).

Mishkin has a certain sort of *access* to the operations of his moral module. As characterized by (M3), these operations are responsible for Mishkin's sentiments, and they connect those sentiments to his subsequent actions by way of providing reasons—inaccessible to Mishkin—for those actions. In other words, the operations of his moral module determine his occurrent moral decision making. This, in the terminology associated with Block, is sufficient to show that these operations are *access-conscious*: where, very roughly, a state or process is access-conscious if it is playing an active role in shaping occurrent psychological processes and resulting behavior.[22] However, what these operations are not is *phenomenally conscious*.[23] There is *nothing that it is like* for Mishkin to undergo the operations occurring in his moral module. Marlow, in contrast, also has access-consciousness to the operations of his module. But there is also something that it is like for Marlow to undergo these operations: his operations are phenomenally conscious ones.

If this is the difference between Mishkin and Marlow, then our question is a straightforward one: does the fact that Marlow's moral decision making is phenomenally conscious whereas Mishkin's moral decision making is not entail that Mishkin is not, whereas Marlow is, a moral subject? Does the presence or absence of phenomenal consciousness provide a legitimate way of demarcating moral subjects from nonsubjects? The idea that it does is extremely implausible. The rest of this section will try to show why.

Here is a useful way of thinking about Mishkin, as described by (M3). Mishkin is characterized by ("suffering from" would be tendentious) a form of *moral super blindsight*. In its ordinary sense, blindsight refers to a deficit of phenomenal consciousness.[24] Due to lesions of the visual cortex, blindsighted subjects will claim to have no awareness of an object in a certain portion of their visual field. Yet, when asked to guess what the object is, or where the object is, they succeed in guessing accurately at rates far above chance. It would be wrong to think of blindsighted subjects as functionally equivalent to those with normal vision. They are not, and will often have great difficulty navigating their way around the environment, at least under normal conditions. Nevertheless, the phenomenon of blindsight seems to be indicative of the possibility that phenomenal consciousness and the sort of functional abilities that constitute access-consciousness can come apart. Subjects can identify an object in front of them, even though they profess to have no awareness of that object.

Based on this, we can imagine a case of what we can call *super blindsight*. People with super blindsight have no visual phenomenal consciousness. If you were to ask them what they see, they will tell you: nothing. Nevertheless, they easily navigate their way around, avoiding objects in their path, and so on. If you ask the subject: how did you know that object—the one he just avoided—was there? He would respond: I just had a feeling there was something there. You toss a baseball in the direction of these subjects, and they easily catch it. You ask: How did you do that? They respond: I just had a feeling there was something there. That is how they manage to get around their environment: on the basis of feelings. But these feelings are utterly reliable—or at least as reliable as the visual impressions had by sighted people. As a result, the super blindsighted subjects are able to navigate their way around their environment with a level of success that is roughly the same as sighted people.

Can the super blindsighted subject see? This question is ambiguous. On the one hand, it could mean: does the super blindsighted subject have visual consciousness? A negative answer to this question is plausible—although it would be rejected by proponents of *enactive* or *sensorimotor* accounts of visual consciousness.[25] However, the question could also mean: is someone with super blindsight a visual subject? A negative answer to this question is far less plausible. Thus, when we ask the subject what is the object in front of him, he would point his eyes in the appropriate direction. Light would strike his eyes, and a message would be sent via the optic nerve to the visual cortex. Various information-processing operations would occur there, and the result would be a feeling with, for example, the content: "It's a cabbage!" Note also that the processes in question are sensitive only to the visually salient properties of objects. It is not as if, for example, the processes give him access to the innards of the cabbage, but only to the properties that could be seen by someone of normal sight.

It would be implausible to say that, in this case, the super blindsighted subject is not seeing. It is not simply *that* he detects objects, but the *way* in which he detects them that is decisive. Thus, for example, he is not locating objects by echolocation. He is using his eyes, and the deliverances of his eyes are processed in the visual cortex of the brain. It is just that these processes that usually result in a perception possessing visual phenomenology now result in a feeling of, let us suppose, a more visceral character. Rather than denying that he is seeing, I think it is far more plausible to claim that this is what seeing consists in for a super blindsighted subject.

In effect, (M3) provides a characterization of someone suffering from what we might call moral super blindsight. Mishkin performs the right (i.e., morally correct) actions, feels the right feelings, and even has the right reasons for these actions and feelings, but he is unaware of these reasons, and so has no idea why he feels the way he does, or why he should do the things he does. Morally speaking, though, he just keeps getting things right. The question is: would we want to say that Mishkin is not a moral subject, that he is not and cannot be motivated by moral considerations?

Mishkin, as a morally super blind subject, is a moral subject in the way that a visually super blind person is a visual subject. The capacity to see is compatible with differing visual phenomenologies. And so too, I think we should accept that the capacity to be moral—to be motivated by moral considerations—is compatible with differing moral phenomenologies. Rather than denying that the morally super blind person is a moral subject, it is far more plausible to claim that, for him, this is precisely his way of being a moral subject. This is what, for him, being a moral subject consists in. Being a moral subject is not so closely tied to phenomenology that we can deny someone the status of moral subject simply because his associated moral phenomenology differs from ours. Moral phenomenology—the phenomenology associated with moral decision making—is not a decisive determinant of moral subjecthood.

8. MORAL SUBJECTS AND METACOGNITION

We have looked at one of the ways in which an individual, Marlow, characterized by (M4), differs from Mishkin, an individual characterized by (M3). The moral decision-making processes exhibited by Marlow are phenomenally conscious: there is *something that it is like* for Marlow to engage in moral deliberation. I have argued that this difference cannot plausibly be regarded as decisive in qualifying Marlow as a moral subject while disqualifying Mishkin from this status. To suppose that it is decisive ties moral subjecthood closely to moral phenomenology in a way that is, ultimately, unconvincing.

There is, however, another difference between Marlow and Mishkin, one that might seem prima facie more significant. Marlow has, while Mishkin lacks, *higher-order thoughts* about his motivations and decision-making processes. That is, Marlow is able to entertain thoughts of the following sort: "I am being motivated to perform act A by reason R." To have this thought, Marlow must be able to entertain other thoughts—for example, "I am motivated to perform act A" and "I am the subject of reason R" (or, equivalently: "Reason R is a reason that I endorse"). These are higher-order thoughts—thoughts about other mental states (in this case motivations and reasons). Mishkin is incapable of entertaining higher-order thoughts in this sense.

In one view, the issue of whether Marlow possesses higher-order thoughts about his motivations and decisions and the issue of whether his motivations and decisions are phenomenally conscious are equivalent. According to higher-order thought (HOT) models of consciousness, mental state M is phenomenally conscious if and only if creature C, the possessor of M, has a thought to the effect that it has M. I am going to look at HOT models of consciousness in more detail shortly because, I shall argue, they fall victim to a certain sort of error that is especially pertinent to our concerns in the remainder of this paper. At present, however, I shall treat the issue of higher-order thoughts about motivations as distinct from the issue of phenomenal consciousness. So, our question is this: is a higher-order thought about a motivation, M, the sort of thing that can convert M into a properly moral motivation? Can Marlow be correctly regarded as the subject of moral motivations—and, therefore, as a moral subject—because, and only because, he has higher-order thoughts which take these motivations as their content? Is it metacognition about one's motivations that makes these motivations moral ones?

The idea that thoughts about one's motivations and decisions is a necessary, and perhaps sufficient, condition of being a moral subject is one strongly associated with Kant. Here, again, is that lucid representative of the Kantian position, Christine Korsgaard:

> Kant believed that human beings have developed a specific form of self-consciousness, namely, the ability to perceive, and therefore to think about, the grounds of our beliefs and actions *as grounds*. Here's what I mean: an animal who acts from instinct is conscious of the objects of its fear or desire, and conscious of it as *fearful* or *desirable*, and so as *to-be-avoided* or *to-be-sought*. That is the ground of its action. But a rational animal is, in addition, conscious *that* she fears or desires the object, and *that* she is inclined to act in a certain way as a result. That's what I mean by being conscious of the ground *as a ground*. So as rational beings, we are conscious of the principles on which we are inclined to act. Because of this, we have the ability to ask ourselves whether we *should* act in the way we are instinctively inclined to. We can say to ourselves: "I am inclined to do act-A for the sake of end-E. But should I?"[26]

"Rational" beings, in Kant's restricted sense identified in this passage, are ones who are aware of the principles on which they act. That is, roughly, they are aware of their motivations for action and so can ask themselves, among other things, whether these motivations are good ones and whether they are the sort of things that *should* be acted upon. A subject that can do this is, for Kant, a rational, and therefore moral, subject.

In this paper, I have deployed a fairly simple sense of moral subject with which Kant might, or might not, agree. To be a moral subject is to be motivated by properly moral considerations. So, without being drawn into interpretative issues concerning whether Kant could or could not agree with this account of what it is to be a moral subject, we must simply be very clear on the question we have to ask ourselves. The awareness of motivations with which we are concerned is metacognitive: it takes the form of thoughts about motivations (broadly construed to cover various ways in which this motivation might be implemented—desire, character, commitment, etc). And our question is: is this form of metacognition about motivations *necessary* for those motivations to be properly moral ones? That is, for a creature C to be motivated by properly moral motivation, M, is it necessary that C have thoughts about M?

It is fairly clear why someone might think that metacognition about motivations is necessary for those motivations to be moral ones. Without such metacognition, one might argue, motivations lack an appropriate element of *normativity*. We have already identified this line of thought in Aristotle. This also seems to be what Korsgaard has in mind. It is precisely because we are aware of the principles on which we are inclined to act, Korsgaard claims, that we can ask ourselves whether we *should* act according to these principles. Conversely, if we can't ask ourselves whether we should be motivated to act on these principles, we can give no content to the idea that we should, or should not so act.

Mishkin is clearly the subject of motivations of various sorts. However, because he is unaware of these motivations, he is, in one sense, at their "mercy." These motivations push him this way and that—causing him to act in one way or another. But these motivations are always causes: they are states that belong, as Sellars once put it, to the *space of causes*. They have no normative dimension. Marlow, on the other hand, has metacognitive abilities that allow him to survey and critically evaluate his motivations. Because of these abilities, he can think "I am motivated to perform act A"; "is act A something that I should, morally speaking, perform in these circumstances?"; "is the intended beneficiary of act A a morally worthy beneficiary?" and so on. Mishkin's motivations are merely causes because he cannot think such thoughts. Marlow, however, is able to distinguish between what a given motivation actually inclines him to do and whether he should do what this motivation inclines him to do. Because of this, it might be thought, Marlow's motivations have a normative dimension that Mishkin's must lack. Marlow's motivations belong to the *space of reasons*, not the space of causes.

In short, the argument that Mishkin is incapable of being motivated by properly moral considerations runs as follows: Moral motivations are essentially normative items. Given his lack of metacognitive abilities, Mishkin's motivations have no normative dimension. Therefore, Mishkin's motivations are not moral motivations.

In the argument to follow, I will not question whether moral motivations are essentially normative. I will assume they are, since the view I want to undermine is predicated on this assumption. I will argue that it is a mistake to suppose that metacognition about motivations is the sort of thing that can underwrite the normativity of those motivations. To suppose that metacognition can confer normative status on motivations is to fall victim to an illicit picture. I shall call this picture the *miracle of the meta*.

9. Consciousness and Higher Order Thoughts

The remainder of the paper is aimed at showing just why metacognition cannot confer normative status on motivations. I am going to approach this issue indirectly. In this section, I am going to examine the *higher-order thought* (HOT) model of consciousness.[27] I shall argue that this model falls victim to a certain sort of mistake—a version of the miracle of the meta. The idea underlying the HOT model of consciousness is that metacognition, in the form of higher-order thoughts, confers consciousness on mental states. The idea under examination in the remainder of this paper is that metacognition, again in the form of higher-order thoughts, can confer normativity on motivations. I want to exploit the parallels here, because I think the problems with the HOT account of consciousness are clear.[28] I shall use this to throw some light on the problems with the idea that metacognition can confer normativity on motivations.

In order to properly understand HOT models of consciousness, we need two distinctions: (i) the distinction between *creature* and *state* consciousness, and (ii) the distinction between *transitive* and *intransitive* consciousness.

Creature versus state consciousness. We can ascribe consciousness both to *creatures* (e.g., John is conscious as opposed to asleep) and to mental *states* (my belief that Ouagadougou is the capital of Burkina Faso is, at this point in time, a conscious belief). HOT accounts are attempts to explain state consciousness, not creature consciousness.

Transitive versus intransitive consciousness. We sometimes speak of our being conscious *of* something (e.g., of how the sun dances on the bright, blue water). This is *transitive* consciousness. Transitive consciousness is a form of creature consciousness. Mental states are not conscious of anything. Rather, creatures are conscious of something in virtue of the mental states they have. *Intransitive* consciousness, on the other hand, can be ascribed both to creatures and states. A creature is intransitively conscious when it is conscious, as opposed to asleep, knocked out, or otherwise unconscious. A state is intransitively conscious when we are consciously entertaining it—as when I think to myself that Ouagadougou is the capital of Burkina Faso.

The core idea of HOT models is that intransitive state consciousness can be explained in terms of transitive creature consciousness. This idea can be divided into two claims. First, and roughly, a mental state M, possessed by creature C, is intransitively conscious if and only if C is transitively conscious of M. Second, a creature, C, is transitively conscious of mental state M if and only if C has a thought to the effect that it has M. Thus, intransitive state consciousness is to be explained in terms of transitive creature consciousness, and transitive creature consciousness (at least in this context) is to be explained in terms of a higher order thought—a thought about a mental state.

HOT accounts face a reasonably obvious dilemma, best introduced by an example. Suppose I am in pain. According to the HOT model, this pain is intransitively conscious if and only if I have a higher order thought about this pain—a thought to the effect that I am in pain. However, either the higher order thought to the effect that I am in pain is itself intransitively conscious or it is not. If it is, and if its being intransitively conscious is what grounds the intransitive consciousness of the pain, then the HOT model clearly faces a regress. The nature of intransitive state consciousness has not been explained; this explanation has merely been deferred. This is the first horn of the dilemma. To avoid this, the HOT account is committed to the claim that the higher order thought that confers intransitive consciousness on a mental state is not, itself, intransitively conscious.

This, however, leads directly to the second horn of the dilemma. Suppose that the higher order thought that (allegedly) confers intransitive consciousness on my pain is not itself intransitively conscious. Then, among other things, I will not be aware of thinking that I am in pain; I will, in effect, have no idea that I am thinking this. But how can this thinking that I am in pain make me aware of my pain if I have no idea that I am thinking that I am in pain?

Take my belief that Ouagadougou is the capital of Burkina Faso. When I entertain this belief it makes me aware of a fact—the fact that Ouagadougou is the capital of Burkina Faso. However, it is only rarely that I entertain this belief; most of the time this belief is one of my unconscious mental states. What makes it unconscious? Well, precisely that it does not make me aware of the fact that Ouagadougou is the capital of Burkina Faso. I still possess the belief because I am disposed to become aware of the fact under certain eliciting conditions (e.g., someone asks me what the capital of Burkina Faso is). However, when it is in unconscious form, as it is most of the time, it is unconscious precisely to the extent that it does not make me conscious of anything. Intransitively unconscious states do not make the creature that has them transitively conscious of anything at all—that is precisely what makes them intransitively unconscious. Conversely, if they do make the creature that has them transitively conscious of something, that is because they are not, in fact, intransitively unconscious.

The HOT theorist can object that the higher order thought that confers intransitive consciousness on my pain is an *occurrent* thought, and this makes it crucially different from my belief that Ouagadougou is the capital of Burkina Faso, which does not make me aware of the relevant fact because it exists in *dispositional* form.

What we must try to understand, then, is (i) how we can entertain a thought in a nonconscious manner, and (ii) where doing so makes us aware of what the thought is about. How can a nonconscious thought that we are currently entertaining make us aware of what that thought is about?

We can, I think, make sense of the idea of occurrently entertaining an intransitively nonconscious thought. Suppose I think unconsciously—perhaps due to various mechanisms of repression—the thought that someone very close to me is seriously ill. What would this mean? We might explain it in terms of various puzzling feelings of melancholy that assail me when I am talking to them, or a vague sense of foreboding to which I can't quite give shape. The thought is occurrent because it is playing an active role, or is poised to play such a role, in shaping my psychological life. The truth of this account can be contested, but it does at least make sense. However, what we can make no sense of is the idea that this unconscious but occurrent thought makes me aware of what it is about. If it were to do this, then I would have to be aware of the fact that my friend is seriously ill. But as soon as I am aware of this fact, then I am consciously thinking that my friend is seriously ill. That is, the thought has become intransitively conscious. To undergo the dawning realization that my friend is seriously ill is precisely to think—to consciously think—that my friend is seriously ill.

Therefore, while we might be able to make sense of the idea of occurrently entertaining a nonconscious thought, what we cannot make sense of is the idea that this thought should make us aware of what it is about. That it does not make us aware of what it is about is precisely what it is for the thought to be nonconscious. Conversely, as soon as it does make us aware of what it is about, it is a conscious thought—because making us aware of what it is about is precisely what it is for the thought to be conscious.

The moral is clear. Intransitively nonconscious thoughts do not make us transitively conscious of what they are about. That is precisely what it means for them to be intransitively nonconscious. The HOT account tries to explain the intransitive consciousness of a mental state—say my pain—in terms of our being transitively conscious of that state. However, making me transitively aware of my pain is precisely what an intransitively nonconscious higher order thought cannot do. If it does, it is thereby an intransitively conscious higher order thought. If it is an intransitively conscious higher order thought, then we are back to the problem of regress.

10. The Miracle of the Meta

The idea that a higher order thought could confer intransitive consciousness on a mental state is one example of what I am going to call the *miracle of the meta*—the attribution of miraculous powers to metacognition or metacognitive abilities. This attribution faces a problem whose general form looks like this: we are to suppose that metacognitive item, MC, confers property P on mental state M. Does MC have P? If yes—then MC provides no explanation of P. If no—then MC cannot confer P on M.

With regard to the HOT model of consciousness, "P" denotes intransitive state consciousness. In the case that interests us, "P" denotes normativity. The claim is that metacognition about one's motivations can confer normative status on those motivations. In this section, I want to cast doubt on this idea.

Implicated in the idea that metacognition can confer normative status on motivations is a certain pre-theoretical picture of the difference in the psychological profile of an agent brought about by metacognitive abilities. In the absence of these abilities, Mishkin is at the "mercy" of his motivations. They push him in this way and that, and he has no awareness of the role they play and so no control over that role. Metacognitive abilities, however, transform Marlow. He can sit above the motivational fray, observing, judging and evaluating his motivations and deciding the extent to which he is to allow them to determine his decisions and actions. Marlow can ask himself whether he *should* follow the dictates of a given motivation. And this normative element is what transforms the motivation into a moral phenomenon. The concepts of normativity and control are closely intertwined in this view of the moral subject.

What if the metacognitive abilities were not at all like their portrayal in this pre-theoretical picture of the moral subject? What if they were, in certain crucial respects, like the motivations that they take as their objects? Marlow can undoubtedly think to himself thoughts such as, "what is it that is motivating me to do this?" and "is this motivation something that I *should* act upon, or something that I *should* decline?" However, suppose the answers he gives to these questions are ones over which he has little control: being, for example, crucially sensitive to features of the situation in which he finds himself—features of which he has only a dim awareness and whose influence on him largely escapes his conscious grasp. Marlow could still observe and evaluate his motivations, but this observation and evaluation is both clouded and shaped by contextual factors of which he has little awareness and over which he has little control.

In the initial description of the pre-theoretical picture of the moral subject, the lack of metacognitive abilities puts Mishkin at the "mercy" of his motivations—they, in fact, pull him in this way or that. However, in the revised version of the picture it seems that Marlow is similarly at the mercy of his metacognitive assessments of his motivations. These assessments pull him in this way or that—make him evaluate his motivations in one way or another—but they work in subterranean ways. He has little grasp, and just as little control, over their workings.

In the second picture, I am alluding to recent influential work in psychology of a broadly *situationist* orientation.[29] Situationist accounts of the moral subject can be contrasted with dispositionalist accounts. According to the latter, very roughly, the moral subject is constituted, in part, by a set of dispositions that are internal to the subject and the subject is responsible for their formation and preservation. This set of dispositions, again in part, makes up what we might call the *character* of the person. Included in these dispositions will be ones pertaining to the evaluation of motivations. Thus, the character of the moral subject is, in part, made up of dispositions to classify given motivations as ones that should be acted upon or ones that should be resisted. Situationist accounts, however, see things very differently. According to a strong version of situationism, a person's character is as malleable as the situations in which she finds herself. Change the situation, and the subject's dispositions to morally evaluate motivations in one way rather than another will also change. Thus, we might imagine Marlow as a participant in a Stanford Prison Experiment, or as a guard at Abu Ghraib. If the situationist account is correct, we should expect Marlow's evaluation of his motivations to vary with these variations in circumstances.

Therefore, the idea that Marlow's metacognitive abilities confer normativity on his motivations—a normativity that Mishkin's motivations therefore lack—seems a hostage to empirical fortune. If Marlow is, as the situationist argues, at the mercy of his metacognitive assessments of his motivations in the way that Mishkin has been portrayed as being at the mercy of his motivations, then it is unclear how possession of metacognitive abilities can confer normative status on Marlow's motivations in the way imagined. If so, then the corresponding attempt to deny that animals can be moral subjects—subjects of properly moral motivations—is, similarly, a hostage to empirical fortune.

My primary interest is not with these empirical concerns but with a deeper conceptual worry that underlies them. We appealed to higher order thoughts because we thought it might ground the normativity of our motivations and actions. Without metacognition, so the idea goes, we are at the mercy of our motivations: they, in fact, pull and push us this way and that, but since we are unable to scrutinize them, it would not be possible for us to assess whether we *should* be thus pulled this way and that. With metacognitive abilities we can scrutinize our motivations and actions, and thus, for the first time, assess whether we should act on them or reject them. It is only when metacognition enters the picture that we can meaningfully attribute normative properties to our motivations and actions.

This account is an example of the pull exerted on us by the miracle of the meta, and conveniently overlooks that precisely the same reasons why normativity was thought to be problematic at the first-order level—the level of motivations and actions—will be reiterated at the higher order level. Higher order cognition was supposed to allow Marlow to sit above the motivational fray and calmly pass judgment on his motivations, thus transforming them into normative items. However, there is reason to think that higher order cognition is *not* above this motivational fray. If first-order motivations can pull Mishkin this way and that, then second-order evaluations of those motivations can do exactly the same to Marlow. If Mishkin is indeed at the "mercy" of his first-order motivations, as the Kant/Korsgaard picture would have us believe, then, Marlow is similarly at the "mercy" of his second-order evaluations of these motivations. In short, second-order evaluation of our first-order motivations cannot lift us above the motivational fray that we think endemic to the first-order motivations for the simple reason that we can be motivated to evaluate our motivations in one way rather than another.

The *miracle of the meta* would have us believe that something miraculous happens in the jump from first- to second-order cognition. But nothing miraculous happens here. Whatever features, patterns and, most importantly for our purposes, shortcomings we find at the first-order level, these are merely reiterated at the second-order level. If higher order cognition were indeed unproblematically normative, then it might indeed explain the normativity of first-order motivations and actions. However, the very considerations that have led us to believe in a lack of normativity at the first-order level will simply be reiterated at the second-order level. The appeal to higher order cognition, therefore, is incapable of achieving what it was supposed to achieve: to confer normativity, and hence moral status, on motivations.

11. CONCLUSION

Here is an account of apparent elephant empathy, extracted from Ian Douglas-Hamilton and colleagues (2006), but cited in Bekoff and Pierce.[30] Eleanor, matriarch of the First Ladies family, is ailing and unable to stand. Grace, of the Virtues family, touches her with her trunk and foot, and then lifts her back to her feet. Douglas-Hamilton writes: "Grace tried to get Eleanor to walk by pushing her, but Eleanor fell again....Grace appeared very stressed, vocalizing, and continuing to push and nudge Eleanor with her tusks....Grace stayed with her for at least another hour as night fell."

Was Grace acting compassionately? Tradition would tell us "no"—at least if we understand compassion as a moral emotion. Grace can't scrutinize her motivations and actions. She can't ask herself questions like: is what I am doing an appropriate response to the situation? Is it a proportionate response to the suffering of Eleanor? Fundamentally, she cannot ask herself if what she is doing in these circumstances is what she *should* or *ought* to be doing in the circumstances. Therefore, no sense can be made of the idea that she should or ought to be doing this. Grace is simply pushed this way and that by her sentiments. She is not the author or master of those sentiments. Therefore, she is not a moral subject.

I have argued, on the contrary, that there are no empirical or conceptual obstacles to regarding Grace, in this instance, as being motivated by moral concerns. Her sentiments can be genuinely moral ones. Grace is a moral subject. Or, at least, we have no compelling reasons to suppose that she is not a moral subject. To suppose otherwise is to fall victim to the *miracle of the meta*. It is the investing of quasi-magical properties in metacognition—properties that afford it unproblematically normative status of a sort possessed by no other sort of cognition.

My suggestion is this: there are no miracles. We should resist attempts to deny animals the status of moral subjects when those attempts rest, ultimately, on an appeal to magic. Grace is a moral subject. Mishkin is a moral subject, as well as Marlow. In taking his bone to Sad Dog, Happy Dog was not only a happy dog, he was also a good dog.

NOTES

1. Even so-called justifying or normative reasons—as opposed to motivating reasons—are reasons for someone. If I know nothing of tooth decay, there is still a reason for me to brush my teeth—even if my ignorance means that it is not a motivating reason. But this is a reason *for me* to brush my teeth. If my teeth were metallic, like Jaws in the old James Bond movies, then tooth decay might not be a justifying reason for me to brush my teeth (gum disease would presumably be a different matter). Anything that is a reason—whether motivating or justifying—is a reason for someone. There are justifying but non-motivating reasons, of course. But there are no ownerless reasons.

2. Disagreements persist concerning whether these obligations are direct or indirect. For defense of the idea that our obligations to animals are both direct and more extensive than is commonly thought, see, among many others, Tom Regan, *The Case for Animal Rights* (Berkeley and Los Angeles: University of California Press, 1983); Mark Rowlands, "Contractarianism and Animal Rights," *Journal of Applied Philosophy* 14 (1997): 235–47; Mark Rowlands, *Animal Rights: A Philosophical Defence* (Basingstoke: Macmillan, 1998); Mark Rowlands, *Animals Like Us* (London: Verso, 2002); Steven Sapontzis, *Morals, Reason, and Animals* (Philadelphia: Temple University Press, 1987); Peter Singer, *Animal Liberation* (New York: Avon Books, 1975).

3. David DeGrazia, *Taking Animals Seriously* (New York: Cambridge University Press, 1996), p. 203.

4. Sapontzis, *Morals, Reason, and Animals.*

5. Evelyn Pluhar, *Beyond Prejudice: The Moral Significance of Human and Nonhuman Animals* (Durham, N.C.: Duke University Press, 1995), p. 2.

6. Vicki Hearne, *Adam's Task: Calling Animals by Name* (New York: Vintage Books, 1987); Jeffrey Moussiaeff Masson, *Dogs Never Lie About Love: Reflections on the Emotional World of Dogs* (New York: Three Rivers Press, 1997); Jeffrey Moussiaeff Masson and Susan McCarthy, *When Elephants Weep: The Emotional Lives of Animals* (New York: Delacorte, 1995); Stephen Wise, *Rattling the Cage: Toward Legal Rights for Animals* (Cambridge, Mass.: Perseus Books, 2000); Frans de Waal, *Good Natured: The Origins of Right and Wrong in Humans and Other Animals* (Cambridge, Mass.: Harvard University Press, 1996); Marc Bekoff, ed., *The Smile of a Dolphin: Remarkable Accounts of Animal Emotions* (New York: Discovery Books, 2000); Marc Bekoff, *Minding Animals: Awareness, Emotion, and Heart* (Oxford: Oxford University Press, 2002).

7. Charles Darwin, *The Expression of Emotions in Man and Animals* (London: John Murray, 1872).

8. Marc Bekoff and Jessica Pierce, *Wild Justice: The Moral Lives of Animals* (Chicago: University of Chicago Press, 2009).

9. Bekoff and Pierce do not distinguish between moral subjects and moral agents, instead preferring the rather vague suggestion that agency is species-specific (*Wild Justice*, p. 144). The failure to draw this distinction, I think, distracts significantly from the plausibility of their case.

10. Bekoff and Pierce, *Wild Justice*, p. 60.

11. Christine Korsgaard, *Fellow Creatures: Kantian Ethics and Our Duties to Animals*, Tanner Lectures on Human Values, ed. G. Peterson (Salt Lake City: University of Utah Press, 2004), p. 140.

12. Bekoff and Pierce, *Wild Justice*, pp. 140–41.

13. Aristotle, *Nichomachean Ethics*, trans. T. Irwin, 2nd ed. (Indianapolis: Hackett, 1999), 1105a27–35.

14. In such circumstances, virtue, it seems, would be the preserve of a smallish number of professional moral philosophers. In my view, this, in itself, constitutes a reductio ad absurdum.

15. Consider an analogous case: the notion of understanding what water is. On the one hand, this might mean understanding something as falling under the concept of water—which (at least on one way of thinking about concepts) someone can do only if they are in possession of the concepts of hydrogen, oxygen, etc. On the other hand, it can mean being able to differentiate, with a reasonable though not necessarily infallible level of precision, which things count as water and which things do not. Children and the uneducated can understand what water is even if they don't possess the concept of water (in this sense).

16. Aristotle, *Nichomachean Ethics*, 1103a19–03b2.

17. Aristotle, *Nichomachean Ethics*, 1109a27–30.

18. Essentially the same conclusion has been endorsed by B. A. Dixon, *Animals, Emotion and Morality* (New York: Prometheus Books, 2008).

19. Kant and Aristotle are not the only famous philosophers who have been concerned with this issue. I am grateful to Tom Beauchamp for pointing out that the position I am going defend is Humean in spirit (if not in details).

20. I do not, of course, even pretend to know how to complete the specification of the ceteris paribus clause. As far as I know, neither does anyone else.

21. Marlow need not have these feelings himself, of course, nor need he perform the actions. The crucial point is that Marlow—our imagined ideal spectator—judges, correctly, that in these circumstances, feeling F and action A are morally appropriate.

22. Ned Block, "On a Confusion about the Function of Consciousness," *Behavioral and Brain Sciences* 18 (1995): 227–47.

23. This is extraordinarily rough. Block characterizes access consciousness as follows: a state is access conscious if, in virtue of a subject's having the state, a representation of its content is (i) inferentially promiscuous (i.e., poised for deployment in reasoning), (ii) poised for rational control of action, (iii) poised for rational control of speech. The last claim is tendentious and we can ignore it for our purposes. Block emphasizes that (i)–(iii) provide a sufficient rather than necessary condition for access consciousness. He is, moreover, quite comfortable with the claim that animals can possess states characterized by access consciousness.

24. Lawrence Weiskrantz, *Blindsight: A Case Study and Implications* (Oxford: Oxford University Press, 1986).

25. J. Kevin O'Regan & Alva Noë, "A Sensorimotor Account of Vision and Visual Consciousness," *Behavioral and Brain Sciences* 23 (2001): 939–73; Mark Rowlands, *Body Language* (Cambridge, Mass., MIT Press, 2006).

26. Korsgaard, *Fellow Creatures*, pp. 148–49.

27. David Rosenthal, "Two Concepts of Consciousness," *Philosophical Studies* 49 (1986): 329–59.

28. See Rowlands, "Consciousness and Higher-Order Thoughts," *Mind and Language* 16 (2001): 290–310; and *The Nature of Consciousness* (Cambridge: Cambridge University Press, 2001), chapter 5.

29. See, for example, Philip Zimbardo, *The Lucifer Effect: How Good People Turn Evil* (New York: Random House, 2007).

30. Ian Douglas Hamilton, S. Bhalla, G. Wittemyer, and F. Vollrath, "Behavioural Reactions of Elephants towards a Dying and Deceased Matriarch," *Applied Animal Behaviour Science* 100 (2006): 67–102. This case was cited by Bekoff and Pierce, *Wild Justice*, pp. 103–4.

SUGGESTED READING

ARISTOTLE. *Nicomachean Ethics*. Translated by Terence Irwin. 2nd ed. Indianapolis: Hackett, 1999.

BEKOFF, MARC, ed. *The Smile of a Dolphin: Remarkable Accounts of Animal Emotions*. New York: Discovery Books, 2000.

———. *Minding Animals: Awareness, Emotion, and Heart*. Oxford: Oxford University Press, 2002.

———, and JESSICA PIERCE. *Wild Justice: The Moral Lives of Animals*. Chicago: University of Chicago Press, 2009.

DIXON, B. A. *Animals, Emotion, and Morality*. New York: Prometheus Books, 2008.

DARWIN, CHARLES. *The Expression of the Emotions in Man and Animals*. London: John Murray, 1872.

DEGRAZIA, DAVID. *Taking Animals Seriously*. New York: Cambridge University Press, 1996.

DE WAAL, FRANS. *Good Natured: The Origins of Right and Wrong in Humans and Other Animals*. Cambridge, Mass.: Harvard University Press, 1996.

HEARNE, VICKI. *Adam's Task: Calling Animals by Name*. New York: Vintage Books, 1987.

KORSGAARD, CHRISTINE. *Fellow Creatures: Kantian Ethics and Our Duties to Animals*. Tanner Lectures on Human Values, edited by G. Peterson. Salt Lake City: University of Utah Press, 2004.

MASSON, JEFFREY MOUSSIAEFF. *Dogs Never Lie About Love: Reflections on the Emotional World of Dogs*. New York: Three Rivers Press, 1997.

———, and SUSAN MCCARTHY. *When Elephants Weep: The Emotional Lives of Animals*. New York: Delacorte, 1995.

PLUHAR, EVELYN. *Beyond Prejudice: The Moral Significance of Human and Nonhuman Animals*. Durham, N.C.: Duke University Press, 1995.

REGAN, TOM. *The Case for Animal Rights*. Berkeley and Los Angeles: University of California Press, 1983.

ROWLANDS, MARK. "Contractarianism and Animal Rights." *Journal of Applied Philosophy* 14 (1997): 235–47.

———. *Animal Rights: A Philosophical Defence*. Basingstoke: Macmillan, 1998. Second edition published as *Animal Rights: Moral Theory and Practice* (Basingstoke: Macmillan, 2009).

————. "Consciousness and Higher-Order Thoughts," *Mind and Language* 16 (2001): 290–310.

————. *The Nature of Consciousness*. Cambridge: Cambridge University Press, 2001.

————. *Animals Like Us*. London: Verso, 2002.

SAPONTZIS, STEVEN. *Morals, Reason, and Animals*. Philadelphia: Temple University Press, 1987.

SINGER, PETER. *Animal Liberation*. New York: Avon Books, 1975.

WISE, STEPHEN. *Rattling the Cage: Toward Legal Rights for Animals*. Cambridge, Mass.: Perseus Books, 2000.

ZIMBARDO, PHILIP. *The Lucifer Effect: How Good People Turn Evil*. New York: Random House, 2007.

CHAPTER 19

THE MORAL LIFE
OF ANIMALS

MICHAEL BRADIE

Do nonhuman animals have minds? What implications, if any, does the answer have for their moral status? At one point it was popular, at least in some circles, to think of animals as having no experiences, no consciousness, no feelings, and no particular moral status worth considering. The earliest history of changing Western attitudes toward animals and their moral status is well documented in Richard Sorabji's *Animal Minds and Human Morals*.[1] However, the tide has changed over the past 150 years, and the mental and moral status of nonhuman animals has become muddled and controversial. Since Charles Darwin's *Origin of Species*, the under-standing that human beings are just one species among the animals has become commonplace. The moral status of human animals is, in some degree or form, no longer problematic. Therefore, I will use the term "animals" to mean "nonhuman animals" unless the context makes it clear that humans are part of the referent.

The problem with the connection between the moral status of animals and their mental status divides neatly into two sub-problems: What determines mental status? and What determines moral status? The answer to the first question may heavily determine the answer to the second.

There are a number of questions that need to be addressed vis-à-vis the question of animal minds. Do animals have qualitative, internal experiences? Are animals conscious? Do animals have a mental life? What do we mean by attributing a "mental life" to anything? Finally, what is the basis for drawing any conclusions about any of these questions? Animals clearly communicate with each other. Nevertheless, with some controversial exceptions it has been argued that, given the fact that animals do not possess language in a manner comparable to humans, we are not in any position to reliably and directly determine what if anything animals are experiencing or feeling. We cannot ask a mournful-looking dog, how are you feeling? One of the questions that must be addressed is whether there is a way around this profoundly important linguistic barrier.

There are also a number of questions that need to be addressed concerning the nature of moral status. It is convenient to divide this question into two sub-questions and consider separately whether animals qualify as *moral agents* or only as (possible) *moral patients*. Moral agency, it is often argued, requires a high level of cognitive capacity that most animals appear to lack. In the past, it was common to argue that the lack of language was sufficient to reject the idea that animals could qualify as moral agents. Whether they qualify as moral patients is sometimes presumed to rest on satisfying a less stringent requirement, namely, whether animals can feel.

In addition, and perhaps more basically, there is the question of what should count as the benchmark of moral status. Traditionally, it has been presumed that human morality is the gold standard and that the moral status of other animals is to be judged in terms of how close they come to being "like humans" in whatever ways are deemed important for human moral status. Aristotelians might judge other animals in terms of their capacity for virtue, Kantians in terms of their capacity for rationality, Benthamites in terms of their capacity for feeling, and so on. Recently, however, this presumption has come under attack for being myopic. Evaluations of moral status that arbitrarily place human morality as the standard have come to be labeled anthropocentric. Non-anthropocentric alternatives have been proposed, and it remains to be seen to what extent these alternative viewpoints are supported by scientific data while at the same time preserving their claim to be true *moral* alternatives to traditional views.

EMPIRICAL DATA AND MORAL STANDING

There are three major issues concerning the impact of empirical data on the moral status of animals. The first is the question of the relevance of empirical data for ascertaining the moral status of animals. I take this to be relatively uncontroversial. Even those who reject the view that animals have any significant moral status do so on the basis of alleged facts that show that animals are incapable of being members of any moral community. On the other hand, there are three lines of empirical investigation that seem to yield results that are relevant to determining the moral status of animals. First, there are arguments based on Darwinian or evolutionary considerations. Second, there are the neuroscientific data. Finally, there is the work of cognitive ethologists, both in the laboratory and in the field, that bears on the moral status of animals.

The second major issue concerns the interpretation of the scientific data. This is more problematic and controversial, especially with respect to the neuroscientific findings and the data collected by cognitive ethologists. The difficulties in interpreting the data arise primarily from the language barrier that prevents us from sitting down with animals and asking them what they are thinking and feeling. Are they, in fact, even thinking or feeling anything? We are faced with a radical form of the problem of other-minds. Finally, there are many who argue that any attempt to describe the behavior of animals in terms of how they feel, what they believe, or what their intentions might be is to adopt a dangerous, misleading, and anthropomorphic point of view that cannot be *justified* by the observed data even if the view were correct.

The third major issue concerns the nature of morality itself. Given the relevance of the empirical findings, in what ways do they affect our understanding of what it means to be a moral being? How one answers this question depends on whether one adopts an anthropocentric perspective, taking human moral systems as the basic and perhaps only standard against which to judge the moral capacities of animals, or whether one adopts a non-anthropocentric perspective. It remains to be seen what the implications of adopting a non-anthropocentric view might be. Needless to say, these are large questions that I cannot hope to do full justice to in the present format.

My principal aim in this chapter is to argue that locating human beings as one animal species among others presents a serious challenge to traditional anthropomorphic understandings of what it means to be an animal capable of moral behavior. In particular, our understanding of the shared evolved neural architecture of human beings and related lineages suggests that we should evaluate the moral status of non-human animals in terms that are intrinsic to their capacities, rather than in terms of the extent to which they do or do not satisfy conditions that may be deemed relevant from the point of view of some traditional human understanding of morality.

The plan of the chapter is as follows. I first review the relevant scientific data that suggest that animals have sufficient cognitive and affective capacities to be worthy of moral consideration—that is, that to a greater or lesser extent they share with human beings the fundamental capacities and behaviors that deserve to be called "moral." That done, it remains to be established what kind of consideration they are due. Are animals merely at most moral patients, as some have suggested, or can some of them be considered moral agents of a sort? If so, what are the implications of the answer? Finally, I will argue that if we take the scientific evidence seriously and explore the moral implications that the evidence suggests, we are led to a radical reconceptualization of the nature of the province of morality. Once we recognize the affinity of human beings with other animal species in light of their shared evolutionary heritage, we can question the traditional presumption that morality is the special province of being human and that which marks us off as different in kind from other animals.

THE SCIENTIFIC EVIDENCE

The scientific evidence that establishes the affinity between human beings and other animals comes from three sources. First, there are the arguments offered by Charles Darwin and his colleagues in the nineteenth century supporting the thesis that human beings and other animals have a shared evolutionary history. Second, neuroscientific findings in the latter part of the twentieth century reveal that the brain structures and neural networks that underlie the human capacity for moral behavior exist in homologous forms, that is, have a common ancestral source, across a wide range of nonhuman organisms. Finally, there is a growing body of evidence from cognitive ethologists that supports the view that animal social behavior is permeated by norms and values.

The beginning of our modern evolutionary understanding of the affinity between human beings and animals was the publication of Darwin's *Origin of Species* in 1859. This was followed by his *Descent of Man* in 1871, and his *The Expression of Emotions in Man and Animals* in 1872.[2] Another important figure, George Romanes, published a number of books in the 1870s and 1880s on the status of animal minds. A continuity argument can be constructed on the basis of such investigations—an argument that suggests that most differences between humans and other animals are a matter of degree and not of kind.

In *Origin of Species*, Darwin was concerned with establishing two major theses. First, he argued that the diversity of plants and animals, and the similarities and differences between them, could be explained on the basis of the theory of descent with modification. In effect, extant organisms were modified versions of ancestral, extinct species. Second, he argued that the major, but not exclusive, modifying influence was a process of natural selection whereby favored variations enjoyed a reproductive advantage over their less favored conspecifics. In time, through gradual changes, this process would lead to new varieties and ultimately to new species. For any group of species, one could in principle find a common ancestor whose descendants would share some ancestral traits in modified form.

Darwin made only one passing allusion in the *Origin* to the potential implications of his theory for the evolution of human beings, but others were quick to draw the obvious conclusion: Human beings were just one among the animals. The oft-observed similarities between monkeys, apes, and human beings had a natural explanation: At some time in the remote past, they shared a common ancestor. Modern anthropoid species were modifications of some ancestral form, where the modifications were the result of slow gradual changes over enormous times. The source of physical similarities between the species was clear. However, did the hypothesis of shared similarities extend to those features which traditionally were held to mark a qualitative difference between man and beast? I refer here to the mental and moral features that seem to present an unbridgeable gap between human beings and the lower animals.

Darwin addressed these questions first in the *Descent of Man* and then in *Expression of Emotions*. In *Descent*, Darwin argued that differences in the mental and moral capacities between human beings and animals were largely differences of degree and not differences of kind. This conclusion was a straightforward extension of his argument about the evolution of physical characteristics. In chapter 2 of *Descent*, Darwin looks at the similarities and differences between the mental powers of human beings and the lower animals. On the phylogenetic origin of mental powers, Darwin is silent. We cannot conclude that mental powers evolved because they confer selective advantage on the organisms which possess them. It is clear that Darwin thinks that the possession of mental powers is an advantage that, once obtained, will be maintained and sharpened by selection or other evolutionary forces. Strictly speaking, without some hard empirical or experimental data, all we are entitled to believe is that the possession of mental powers is not selectively disadvantageous.

The evidence for the continuum of mental capacity across the animal kingdom includes the following considerations. Humans and animals share senses of sight, touch, smell, hearing, and taste. These shared senses imply shared fundamental intuitions.[3] In addition, Darwin argues that animals experience pleasure and pain, happiness and misery, and emotions much as human beings do.[4] These conclusions are based on behavioral observations and anecdotal evidence. Darwin occasionally throws caution to the wind and makes some questionable anthropomorphic assumptions, for example, attributions of jealousy, shame, and magnanimity to animals.[5] The blatantly anthropomorphic interpretations by Darwin of animal behavior led some observers to regard the *Descent* as a secondary work. The tide has turned, however, and some modern commentators are not as loathe to draw qualified anthropomorphic conclusions from animal data as the behaviorist psychologists who dominated the field of animal studies in the mid-twentieth century were. We return to this point below. Finally, Darwin points out that animals possess the ability to imitate, attend, remember, and use imagination; he also argues that they possess the rudiments of reason.[6] Darwin notes, however, that these displays are not always clearly distinguishable from instinctual behavior.

In *Expression of Emotions*, Darwin continues this line of reasoning by pointing out that expressions of emotion in human beings have a cross-cultural validity and are manifested by the same or similar muscular structures and nervous pathways that can be found in animals. The implications seemed clear: Not only are the roots of the expression of emotions in humans to be found in common ancestors shared with other animals, but these other animals were experiencing emotions as well. Of course, this conclusion was equally anthropomorphic and rejected by those who argued that in light of the fact that there was no independent way to determine what an animal might or might not be experiencing or feeling, the attribution of emotions and feelings to animals was mere gratuitous speculation. Of the three Darwinian classics, the reputation of *Expression of Emotions* ran a distant third to *Origin* and *Descent* until recently. New empirical data have prompted some to reexamine the book and look more favorably on Darwin's interpretations. Paul Ekman's introduction, afterword, and commentary on the third edition attest to the renewed interest in Darwin's work on the emotions.[7]

Finally, although Darwin thought that the roots of mind and morality were to be found in animals, he also thought that a true moral sense required a capacity for self-reflection that, for all that was known at the time, was possessed by human beings but not by animals. However, as we shall see, this reservation is more a reflection of Darwin's adherence to a traditional anthropocentric conception of morality than an implication of his empirical findings.

In the 1880s, Romanes drew on his own research and papers on animals, as well as Darwin's notes on animal intelligence, and wrote two works, *Animal Intelligence* and *Mental Evolution in Animals*.[8] In *Animal Intelligence*, Romanes appealed to Darwin's theory of common descent to lay the groundwork for a comparative psychology that he saw as complementary to and just as important as comparative anatomy. Just as the latter explored the comparative physical structures of different animals, so the former would study the comparative mental structures of different animals.

Romanes took himself to be providing a systematic organization of the evidence of animal mentality and intelligence with the ultimate aim of generating principles of animal and comparative psychology. As such, his work, as he himself noted, might appear to be not much more than a collection of anecdotal stories lacking in scientific rigor. Romanes sought to deflect this criticism by adopting a strict method for evaluating the facts that he found in reports and through correspondence. Romanes used three principles to sift through the stories and observations that were sent to him. First, he relied most heavily on observers with known reputations for accuracy. Second, he accepted observations offered by unknown individuals if the alleged fact was significant and there was sufficient collateral evidence to indicate that the observations were accurate. Finally, he looked for corroborating data from different observers who were unknown to him on the presumption that "where statements of fact which present nothing intrinsically improbable are found to be unconsciously confirmed by different observers, they have as good a right to be deemed trustworthy as statements which stand on the single authority of a known observer."[9]

In *Mental Evolution*, Romanes moved to provide a theory of mental evolution that the collected facts in the first volume supported. The starting point of his speculations was the "truth of the general theory of Evolution" which he understood to entail that

> as a matter of fact, we do not find anyone so unreasonable as to maintain, or even to suggest, that if the evidence of Organic Evolution is accepted, the evidence of Mental Evolution, within the limits of which I have named, can consistently be rejected. The one body of evidence therefore serves as a pedestal to the other, such that in the absence of the former the latter would have no *locus standi* (for no one could well dream of Mental Evolution were it not for the evidence of Organic Evolution, or the transmutation of species); while the presence of the former irresistibly suggests the necessity of the latter, as the logical structure for the support of which the pedestal is what it is.[10]

Not all researchers were as sanguine as Romanes in seeing the connection between organic evolution and mental evolution. While the anatomical evidence supporting the theory of organic evolution was manifest to the plainest of observers once placed in an evolutionary context, the case for mental evolution was not so straightforward, given the relative inaccessibility of the mental states of animals. One such critic was Conway Lloyd Morgan, who sought to put the study of animal behavior on an experimental basis. Originally opposed to Romanes view that animals possessed a psychology, he formulated his famous canon: "In no case may we interpret an action as the outcome of the exercise of a higher mental faculty, if it can be interpreted as the exercise of one which stands lower in the psychological scale."[11]

According to Alan Costall, Morgan intended his canon to be a constraint on research but not as a bar to the imputation of consciousness or mental processes to animals tout court.[12] Twentieth-century behaviorists, however, appealed to the canon as a justification for their refusal to treat attributions of mental processes to animals as serious scientific conjectures. Morgan himself warned against this interpretation, but his cautions were ignored. Part of the reason that the behaviorist misinterpretation of the canon prevailed for much of the twentieth century was that the term "animal behavior" was ambiguous. As Morgan understood it, it apparently did not preclude the possibility that the behavior in question was conscious.[13] However, as the behaviorist paradigm came to dominate much of twentieth-century psychology, the term came to be understood to denote *mere* behavior as opposed to consciousness, a reflection in a sense of the neo-Cartesian basis of behaviorism. The dichotomy between animal consciousness and animal behavior was only reinforced by the interspecies language barrier that prevented "behaving" animals from reporting on their inner feelings and emotions.

For example, a dog chasing a rabbit comes to a fork in the road. He sniffs at one branch and presumably finding no rabbit scent immediately bounds down the second branch in pursuit of his prey. He is behaving as if his actions were guided by the following line of reasoning. "Either the rabbit went this way or that way. He didn't go this way because there is no telltale scent there. So, he must have gone that way." How are we to understand the dog's behavior? Did he just perform the actions or did he perform some canine calculation as well? Not being conversant in canine thought, we cannot ask him.

Thus, while there was growing empirical evidence at the end of the nineteenth century that animals were sentient, the language barrier loomed large as an effective obstacle to drawing conclusions about the "inner life" of animals. The rise of behaviorism in the early twentieth century put an effective end to speculations about the mental life of animals on the basis of the fact that they were merely speculations.

Although recent challenges to behaviorism by cognitive scientists and neuroscientists have somewhat reestablished the scientific credibility of questions about the inner mental lives of animals, the mainstream consensus still seems to be that such inferences or speculations are somehow illicit or at least highly untrustworthy.[14] Neil Berridge alludes to what he calls the "fig leaf of objectivity," or the use of terms, such as "limbic system," that serve to sanitize the findings of relevant brain activity in animals that might otherwise be used as evidence that such animals might be exhibiting emotions or feelings.[15]

The Evidence from Neuroscience

In the second half of the twentieth century, advances in neuroscience enabled researchers to begin to identify brain structures and neural systems that were involved in affective experience and behavior. In 1990, Paul MacLean put forward the "triune brain" hypothesis. The basic idea is that the evolved mammalian brain can be conveniently represented as the product of three developmental stages: A primitive reptilian brain located in the basal ganglia, an old mammalian brain located in the limbic system, and a new mammalian brain located in the neocortex.[16] The triune brain thesis argues for deep homologies, that is, structures that emerge from a shared evolutionary lineage, between the brains of animals and the brains of human beings. Neurological evidence points to deep structural similarities between ancient brain systems that we share with other animals. In particular, the ancient structures are the neural source of basic qualitative *feels* or *affects*. Jaak Panksepp has identified seven primary limbic emotional action systems which, he argues, are the basis of animal responsiveness and lie at the foundation of both emotional and cognitive states. In addition to this shared affective neurostructure, he has recently argued that mammals share brain structures that constitute what he calls "proto-selves" and "core selves." Further study, he suggests, may reveal the basis for attributing a sense of self to a wide range of animals. It stands to reason, he argues, that animals with brain structures similar to those in humans not only react in ways that make them appear to have qualitative experiences similar to those of humans when the homologous brain structures are stimulated, but also that they do in fact have those experiences.[17]

Is this theory an unjustified anthropomorphism? Frans de Waal, among others, argues that it is not. De Waal argues that there is a double standard at work.[18] On the one hand, researchers take cognitive differences to justify the non-attribution of emotional and mental capacities to animals while, on the other hand, they ignore evolutionary evidence that suggests that animals and human beings have shared inherited brain structures associated with emotional and mental capacities. De Waal labels this blind spot "Anthropodenial," which he characterizes as the a priori rejection of the importance of the fact that although nonhuman animals are not human, humans are animals.

De Waal is critical of much of the research designed to determine whether apes, for instance, have a "theory of mind," that is, the capacity to anticipate and understand what a conspecific (or any other animal) is thinking or experiencing. The evolutionary explanation for the development of a "theory of mind" in social organisms is the need for cooperation among conspecifics. The problem with much of the animal research in this area, according to de Waal, is that interactions between humans and chimps, for example, face a species barrier. To the apes, he suggests, we appear as "all-knowing gods." Experiments designed to discover whether apes have theories of mind, that is, whether apes are capable of attributing intentions and expectations to fellow apes, he argues, often only yield results about an ape's ability to fathom the intentions and expectations of human experimenters. What we need, he suggests, is to refocus our attention on trying to discover the extent to which apes are capable of attributing intentions and expectations to conspecifics. When we remove human beings from the experiments, he argues, we get positive results that suggest that apes do, in fact, have a suitably relativized theory of mind.[19]

Still, one might object to the projection of mental attributes onto animals that lack the linguistic ability to make their feelings, intentions, and beliefs, if any, known to us. The other-minds problem for animals can be viewed as an extension of the other-minds problem for humans. Individual human beings have first-person cognitive and emotional experiences that lead them to behave in characteristic ways. Hitting one's thumb with a hammer may bring tears to one's eyes, or an uttered oath, along with a swollen thumb and pain. Human beings recognize or identify other humans as members of the same species. On this basis they assume, infer, presuppose, or just take for granted that others in similar circumstances have the same or similar experiences or feelings. We have no direct access to the feelings or emotions of other human beings. Of course, we can ask whether they are in pain or feeling sad and take their positive verbal responses as affirmation of our judgment. For nonhuman animals, however, this is not possible. We can only indirectly determine that their behavioral responses are or are not indicative of some inner experience. Nevertheless, the neurological evidence of appropriately similar brain structures provides a second, albeit also indirect, line of evidence that animals are having inner experiences similar to those that humans experience under similar circumstances.

The neuroscientific evidence thus supports and bolsters the conclusions derived from evolutionary considerations.

The Evidence from Cognitive Ethology

The third line of relevant scientific findings comes from investigations by cognitive ethologists. In their book *Wild Justice*, Marc Bekoff and Jessica Pierce argue from the perspective of cognitive ethology that animals exhibit behaviors that are best interpreted as manifestations of empathy, cooperation, and a sense of fairness. In essence, "animals have morality."[20] They understand morality to be "a suite of interrelated other-regarding behaviors that cultivate and regulate complex interactions within social groups."[21] However, these behaviors do not constitute morality in themselves; a certain level of cognitive and emotional sophistication is necessary. Bekoff and Pierce's approach is data-driven, and they emphasize the need and importance of expanding research beyond nonhuman primates to other social mammals: hunting predators such as wolves, coyotes, and lions, as well as elephants, mice, rats, meerkats, and whales, among others. In addition, they emphasize the importance of studying animals in their natural habitats and not merely in the confines of laboratories where they are often asked to perform in accordance with the interests of animal behaviorists, which may or may not reflect the interests of the animals themselves.

Where is the line to be drawn between animals that evince morality in this limited sense and those that do not? Bekoff and Pierce suggest that the line is shifting as more empirical evidence becomes available and as our philosophical understanding of what it means to be moral is modulated by reflection on the scientific data. Although their focus is on social mammals, there is a widening body of evidence that suggests that some birds have the wherewithal to constitute a moral community, in the sense of relevant emotions, cooperation, and the like.

Although they argue that the data strongly support the attribution of morality to animals, Bekoff and Pierce also suggest that what constitutes morality has to be understood as species specific. Thus, what counts as morality for human beings may not apply well to wolves, for instance. Nevertheless, they claim, the fact that human standards of morality are not appropriate for wolves does not mean that wolves do not possess some sense of moral relationships that is exhibited in their own manifestations of empathy, cooperation, and a sense of fairness. The net effect is that there is not one sense of moral community and that we humans, as proto-typical moral agents, may expand our understanding of morality to include some organisms and exclude others. The proper way to understand animal morality, they suggest, is to see that there are a number of distinct species-specific moral communities. Within these diverse communities, what counts as moral needs to be attuned to the characteristic features of the species themselves. Indeed, even within species, different communities may develop different social practices, so that what is acceptable in one wolf pack, for example, may not be acceptable in another.

Insofar as animal communities share a common grounding in the need to develop social mechanisms that enable and promote the social welfare necessary for their well-being, one can attribute some sense of morality to them however much their moral systems may differ from the human paradigm. That granted, the appeal to the peculiar circumstances of human moral systems as the standard for judging who or what deserves moral consideration becomes problematic. The idea that we should not judge the adequacy or inadequacy of the moral—or perhaps moral-like—systems of animal societies by the standards that we use to judge human communities is a liberating insight. It sets in sharp perspective how anthropocentric the efforts of expanding-circle defenders such as Peter Singer and James Rachels are. Singer's utilitarian argument is that animals deserve our moral consideration in virtue of the fact that they can suffer. Similarly, Rachels argues for a view he calls "moral individualism" on the grounds that what constitutes a morally relevant criterion is situation specific. However, both of these views seem to be committed to the idea that there is a single moral community, that is, a community either delimited by the application of a single criterion—for example, for utilitarians, the capacity to suffer, or, for Rachels the application of contextually determined criteria that are derived from anthropomorphic standards of what is relevant for moral consideration.

I suggest that the position endorsed by Bekoff and Pierce, by contrast, supports a *speciocentric* view. Bekoff and Pierce are aware that many of the conclusions they ask their readers to draw are subject to the "surely, you're joking" rejoinder. They deliberately adopt as conservative an interpretative stance as they feel the data allows, and they are very cognizant of the fact that inferences from an animal's behavior to its mental state are somewhat problematic. So, while Bekoff and Pierce argue that cooperative behavior is pervasive in nature, they do not conclude from this that moral behavior is equally pervasive. Rather, they limit the behaviors that they consider to be moral to the animals "in which there is a certain level of cognitive complexity and emotional nuance."[22] Exactly what that level is and how nuanced the behaviors have to be remains a work in progress as far as I can see. I do not take this to be a powerful criticism, but only a reminder and recognition that a sophisticated and nuanced view of the extent to which animals can be construed as members of local and species-specific moral communities will, as the authors themselves realize, not be forthcoming without difficult, extensive, labor-intensive, and expensive field research combined with a philosophical reassessment of the plausibility of trans-specific moralities.

The focus of *Wild Justice* is "moral behavior in social mammals," although the suggestion is that further research is needed to reveal the extent to which such behavior is prevalent throughout the animal kingdom. One of the book's major messages is that we need to look at the survival behavior of a wider range of animals beyond nonhuman primates. We might call this a plea against primatocentrism.

For example, spindle cells in humans and apes are associated with the production of empathy, emotional reactions, and a sensitivity to the feelings of others. Bekoff and Pierce report on recent research showing that whales have spindle cells as well. Indeed, they have three times as many as human beings. One implication Bekoff and Pierce point out is that the more we learn about "animal cognition and emotions, the more ethically problematic neuroscience research becomes."[23] The basic idea is that the more we come to see animals as organisms deserving of intrinsic respect the less we will be able to justify using them as mere objects or tools to advance our research agendas.

The more general question raised by treating morality as species-specific is this: How are we to understand and deal with cross-species interactions? Given that wolves have a moral code of their own, attacking prey animals is not out of bounds for them. Assuming that at least some of their prey animals have a modicum of emotional sensitivity, it seems ludicrous to suggest that wolves have obligations to, at the very least, minimize the suffering of their victims. That they might consider becoming vegetarians is clearly out of the question.

This line of thought raises the following questions: Are human beings the only animals who have cross-species moral obligations, and does this mark a distinctive separation between human morality and animal morality? Whether other animals have a sense of obligation to non-conspecifics may be problematic. However, Bekoff and Pierce cite examples of what they claim to be manifestations of cooperation and empathy between animals of different species. If these are genuine expressions of moral concern, then the moral circles generated by the species-specific social behaviors of different animals will sometimes intersect.[24]

The need to study animals in their natural habitats makes animal research difficult, time consuming, and expensive. The function of moral behavior seems to be connected to the need for social animals to regulate behaviors in order to maintain the stability and viability of the groups in which they live. This is consistent with a suggestion by Phillip Kitcher that morality is an evolved response to the social tensions and conflicts that arise in the context of group living.[25] On this account, norms evolve to promote social cohesion and reduce conflicts that would otherwise rip social groups apart. The particular forms these norms take will vary from society to society, but there is an underlying evolutionary rationale for them all. Kitcher's suggestion was with respect to the evolution of *human* moral norms. The argument by Bekoff and Pierce makes it clear that such an idea, to the extent it is viable, applies to animal societies as well. Animal moral norms, such as they are, may and predictably will vary from species to species, and from community to community within species, but they are all manifestations of designs for living that will enable the members of the community to flourish.

Bekoff and Pierce, who look on a species-by-species basis, nevertheless point to the fact that the fundamental features that underlie morality, namely feelings of empathy and a sense of cooperation, are often exhibited across both community and species lines. The authors argue that a sense of justice and fairness is an evolved characteristic manifested in a wide range of social animals, although what constitutes fairness for a wolf differs from what constitutes fairness in a chimpanzee or human being.[26] In this regard, they cite the role of play in helping to establish and inculcate a sense of what is and is not acceptable behavior. When animals play, they often mimic aggressive, mating, and predatory behaviors that will stand them in good stead when they are adults. In play situations, however, these actions are coupled with behavioral cues that signal "I'm playing, don't take this seriously." Animals that signal they are playing and then bite too hard or attempt to mount their playmates engender surprise in their playmates. Habitual violation of the norms of play can lead to avoidance by other members of the community.

In any case, Bekoff and Pierce hold, along with Frans de Waal, that it is not the behaviors as such that are relevant to whether animals are behaving morally. Rather it is the underlying "cognitive and emotional capacities" that enable animals to behave in the ways that they do that are the relevant factors. To the extent that evolutionary considerations, observations, and experiments by cognitive ethologists and results from neuroscience point to shared capacities between humans and other animals, it seems provincially anthropocentric to reject the obvious conclusion that the rudiments of morality are at the very least not unique to human beings.

Moral Implications

What are we to conclude morally from this data? In particular, how can we sort out the various implications and underlying suppositions that follow from these empirical studies and results to determine their moral implications? In the first place, it is clear that the interpretation of the empirical evidence with respect to the mental and moral capacities of nonhuman animals constitutes a problematic set that is tentative and subject to reevaluation. There are as yet no definitive "aha!" results that definitely establish that animals are or are not like-minded creatures capable or incapable of moral behavior. Accordingly, one must ask what if anything is the relevance of the empirical data for the question of the moral status of animals? While I have argued that the empirical evidence suggests that animals have both mental capacities and at least a kind of proto-morality, these conclusions are controversial and in need of further evidence and philosophical evaluation.

Two key points need to be addressed. Do animals have the appropriate neural architecture to qualify as potential moral agents or moral patients? In effect, are they sentient and is it plausible to attribute affective states to them? If these questions are answered in the affirmative, what criteria for being a moral agent or moral patient should we deploy? Are these criteria to be drawn from traditional human conceptions of what is morally relevant, or should we treat the human model as merely one manifestation of a shared evolved heritage? If the latter, then we need to explore the extent to which animals may be subject to and act in accordance with species-specific norms and values.

First, and most obviously, it cannot be denied that animals, for the most part, are sentient. That in itself should be sufficient to give them some moral standing. That is, as sentient beings they are deserving of some moral consideration, the scope of which remains to be determined. Precisely what moral standing this capacity affords them will be a function of what one takes to be the fundamental basis of moral consideration. Secondly, there is overwhelming neurophysiological evidence that a wide range of animals, including mammals and birds, have neurological structures that are homologous to the structures in the human brain responsible for affect, emotional responses, and reactions.

Do such organisms feel and experience affects and emotions? Do they have an inner life? This is not as easily determined as the existence of relevant brain structures. It could be the case that animals with appropriate brain structures are merely responding to external cues and internal processing, and somehow are not aware of what they are doing or what is happening to them. The prevalent behavioristic view of twentieth-century animal psychology was that no inner awareness needed to be or could be attributed to animals that were incapable of expressing the contents of their thoughts and feelings in terms that human investigators could understand. Nevertheless, in light of our current knowledge of the shared neural structures, the workings of the brain, and the Darwinian arguments for common descent, it seems implausible to think that animals do not have some semblance of an inner life or mindfulness.

From the position of traditional anthropocentric conceptions of morality, animals are candidates for moral standing to the extent that animals are sentient creatures and capable of experiencing emotions. They are entities toward which at least some moral consideration is due. They are at least moral patients. Fully fledged members of the moral community are agents as well as patients. In order to qualify as moral agents, animals not only need to be aware, they also need to be self-aware. They need to have the capacity to deliberate and reflect on their actions and desires. Without self-awareness, the argument goes, even if one attributes desires and motives to animals, they are not in a position to reflect on these desires and motives and hence not in a position to judge whether their actions are right or wrong.

The moral implications of the sentience of animals and their capacity for suffering will depend on how one comes down on the following question. What are the fundamental criteria for moral status? To the extent that animals are sentient, experience affect, have a mental life, and are capable of solving problems posed by their environments and their circumstances they have interests. Having interests means that the animals have states, for example, feelings of pleasure and absence of pain, and activities, for example, successfully finding food, shelter, mates, and companions, that are good for them and are such that a failure to realize them results in a diminished quality of life. However, without the capacity for reflective understanding, animals cannot enter into agreements or assume contractual obligations. Thus, those who, like Rosemary Rodd, argue that pet owners have contractual obligations to their pets have taken the continuity and kinship between animals and humans too far by presuming that pet owners have implicit contracts with their pets.[27] Rodd's view is an extreme version of an anthropocentric perspective that attempts to impose a human criterion, the obligations created by contracts, onto organisms for which such relations are, as far as we can tell, inappropriate.

Assuming that animals have interests, the question remains as to how the interests of animals are connected to moral deliberations. There is no short and swift inference from evolutionary histories or neurophysiological facts to moral conclusions. Any such inferences smack of the naturalistic fallacy. Nevertheless, a case can be made that our empirical understandings of animal behavior have important relevance for the moral status of animals. As Cartesian mechanisms, animals enjoy no special moral consideration—none at all. When conceived of as sentient, emotive, problem-solving organisms, however, their moral status needs to be reevaluated unless we are prepared to argue that sentience, emotive expressiveness, and problem-solving abilities are morally irrelevant, which would be morally frivolous.

That neurophysiological data reveal the existence of homologous structures in the brains of organisms related by descent both confirms Darwin's hypothesis and supports the idea that evolutionary considerations should be central to our understanding of what constitutes a moral community, who should enjoy membership, and how they should be morally evaluated. No moral implications are directly inferable from our evolutionary understanding of the interrelatedness of living organisms any more than moral inferences follow from any empirical theory.

The Relevance of Empirical Data to the Moral Status of Animals

We can now reflect on exactly how the empirical data and scientific theories, on the one hand, are related to our understanding of the nature of morality, on the other. The bottom line is the dichotomy between those who think that the empirical data are irrelevant to the moral status of animals and those who think that they are relevant. Among those who find the data relevant, we can distinguish between those who think that the data show clearly that the very idea of animals as moral creatures is ludicrous and those who think that the data support the idea that at least some animals are moral creatures with a degree of moral standing. Among this latter group, we should distinguish those who consider animals merely as moral patients from those who take them to be moral agents as well. Finally among those who consider animals as moral agents, we may distinguish those who take what I should call an anthropocentric view of agent morality from those who hold that animals may be moral agents within a moral community that may be characterized in terms of species-specific interests and concerns where these may not reflect the interests and concerns of human communities. Two questions deserve attention: (1) What is the relevance of empirical data to the moral status of animals; and (2) What is the proper perspective from which to judge the moral status of animals?

Case 1: The Empirical Data Are Not Relevant to the Moral Status of Animals

This position appears to be a non-starter and is included merely for completeness. Someone who endorsed a theology wherein human beings were by creative fiat distinguished from other animals in virtue of their capacity for moral behavior and evaluation might hold this view, but I take it that our modern understanding of Darwinian evolution, where humans are one among the animals, undermines the credibility of this position. Even those who are loathe to endow animals with moral sensitivities and behavior usually appeal to the alleged lack of conclusive evidence indicating that animals have the relevant mental capacities as reason for rejecting the idea that animals are moral beings. Building on the previous analysis, there are three interrelated lines of evidence that support the view that at least some animals possess the relevant capacities to make them viable candidates for moral beings.

The first line of support is the evolutionary argument from common descent. It supports the view that, in virtue of the fact that humans are related to other animals, we should expect that the biological basis of moral behavior is shared with our close evolutionary relatives, at the very least. The extent to which these data support enveloping other lineages with the capacity for morality is a function of how far back in evolutionary history the basic biological requisites were in place. The further back the basic biological components extend, the wider, in principle, the moral net may be cast.

The second line of evidentiary support comes from neuroscience research indicating that the basic evolved brain systems implicated in emotional and affective responses, or the physical grounds of moral sentiments and behavior, are to be found in ancient structures of the brain shared by, at the very least, mammals and perhaps birds. The structures in the human brain activated by basic emotional and affective responses can be found in related organisms as well. It is reasonable to presume that just as the activation of these circuits in humans is correlated with experiences of pain, fear, anger, and the like, so the activation of the homologous circuits in other organisms is similarly correlated with experiences. These primitive reactions in and of themselves do not constitute moral behavior. However, it is arguably the case that the distinctive moral capacity that humans have, and other animals lack, is a function of neocortical activity that grounds the possibility of human language and that enables human beings to not only have beliefs and desires, but to reflect on what they do and think as well as form judgments of approval and regret. So, defenders of this view may be prepared to accept the view that animals have minds of a sort, but reject the view that their mental capacities and capabilities are sufficient for them to be considered moral beings.

The third line of evidence for attributing mental powers to animals comes from the work of cognitive ethologists who observe the behavior of animals in the wild and under experimental conditions. The interpretation of these data is controversial. The observer effect is a problem with respect to interpreting experimental findings. Unless the experiments are carefully constructed, the presence of human experimenters and the design of experiments to elicit answers to questions posed by human experimenters and guided by human interests are likely to yield answers that may not reflect the true nature of the interests and motivations of animals who, at least according to the cognitive ethologists, have their own interests and motivations. This means that experiments need to be designed from the point of view of the animals under study, rather than from the point of view of human beings.[28] Even so, the artificial setting of the laboratory is liable to introduce factors that will yield skewed answers. The optimal solution for obtaining a proper picture of animal behavior is to study animals in their native habitats. However, this is no easy task, as there are both technical and financial difficulties involved in the kinds of long-term studies of animal populations that are likely to yield the most meaningful results about their natural modes of behavior. In any case, as critics point out, all that researchers have in the end are reports of animal behavior. Who knows what, if anything, is going on in their minds?

Case 2: The Empirical Data Are Relevant to the Moral Status of Animals

Given that we argued that empirical data are relevant to determining the moral status of animals, the new question is, does what we know justify the attribution of moral status to animals and, if so, what status? There are two possibilities. 2a: The empirical data suggest that animals do not have the requisite mental capacities to be considered as moral creatures. 2b: The empirical data suggest that at least some animals possess the requisite mental capacities to be considered moral creatures.

Case 2a. The case against animals as moral creatures rests on claims to the effect that animals either lack the requisite sense of self or the cognitive capacities that are often considered necessary conditions for being a moral creature. One argument hinges on the question of whether animals have beliefs and desires in the first place, since moral behavior is claimed to rest on the ability to evaluate one's actions, thoughts, and desires. If animals do not have relevant types of thoughts or desires, then they cannot engage in moral reflection on them. The argument is as follows: Beliefs and desires are propositional attitudes, and animals do not and cannot formulate propositions because they do not have language and language is a prerequisite for the ability to formulate propositions. Even if we were to concede that some non-propositional notion of belief and desire were viable and that one could attribute beliefs and desires to some animals, their status as moral creatures would still depend on their ability to reflect on these beliefs and desires. However, that ability, critics argue, depends on higher order cognitive capacities that seem to be unique to humans. Even Darwin, who was sympathetic to the continuity between the moral capacities of humans and the moral capacities of nonhuman animals, thought that these higher order cognitive functions were a distinct and unique feature of human beings that enabled them to be moral creatures and prevented the attribution of true morality to other animals. Along these same lines, de Waal, who has argued that the basic fundamental affects that underlie emotions, cognitive capacities, and moral judgments are enabled by primitive features of the mammalian brain, concludes that animals have at best a proto-morality.[29]

Even if animals are deemed not to be fully moral, there is a long tradition that accords them some moral standing on the grounds that, even if they do not possess language and perhaps are thereby incapable of forming beliefs and desires, they are still creatures whose interests and welfare deserve to be given moral consideration.

However, two qualifications are needed: First, one might argue that, because animals are propositionally challenged, and hence devoid of beliefs and desires, they do not have interests that need to be taken into account in moral deliberations by human beings. This possible lack of beliefs and desires may or may not be decisive based on whether one takes an internalist or externalist approach to the nature of interests. From an internalist point of view, since animals are incapable, in virtue of their lack of appropriate conative tools, of being privy to anything that might be considered to be in their interests, they do not have any interests. If so, this cuts into any consideration for any moral standing that we might be willing to afford them. From an externalist point of view, coupled with some reasonable assumptions about survival and reproductive success, it may well be argued that animals do have interests that deserve to be considered in moral calculations, regardless of the fact that animals themselves may be unaware of what these interests are. Similarly, one might argue from an internalist point of view that if animals are devoid of beliefs and desires then they have no conception of their own welfare that needs to be taken into account. From an externalist point of view, however, given the same reasonable assumptions about survival and reproductive success, it may be argued that their welfare is something that we, who are cognizant of their situation, ought to take into account.

A second qualification challenges the attribution of consciousness to some animals. Daniel Dennett has argued that attributions of consciousness require a certain level of interconnected neural organization that enables an organism to have a self.[30] Without a sense of self or a unity of apperception, Dennett argues, it is difficult to understand how consciousness is possible, or what levels of pain or suffering such a creature might be able to experience. No pain means no suffering, which in turn means no moral relevance. The example Dennett cites is the case of a snake that uses three disconnected neural modalities to devour its prey—it uses visual cues to strike at its prey, smell cues to find it, and tactile cues to locate the head in preparation of swallowing it.[31] In human beings, these modalities are integrated but in snakes they allegedly are not. Dennett suggests that snakes may not have a sense of self. As such, he argues that snakes do not have a capacity for suffering, and, hence, have no claim for moral consideration by humans. Nevertheless, it seems clear that the primates, some other mammals, and presumably any organisms that pass the mirror test do have a sense of self and hence are in principle capable of experiencing pain and suffering.

Case 2b. The empirical data suggest that at least some animals possess the requisite mental capacities to be considered moral creatures. The positions here are divided along two separate dimensions. One dimension assesses the nature of the moral standing of animals. Animals can have moral standing in virtue of being moral patients, moral agents, or both. Moral patients are animals whose well-being and interests need to be taken into moral consideration. Moral agents are animals who can be held morally responsible for their actions. Presumably, animals that are incapable of forming intentions or reflecting on their actions would not qualify as moral agents. To the extent that they are capable of enduring pain and suffering, they might qualify as moral patients; that is, they might deserve not to be maltreated even though they could not be held responsible for their actions. Finally, some animals are both moral patients and moral agents. Adult, non-impaired human beings fall into this last category, but perhaps some nonhuman animals do as well. From a broadly naturalistic point of view, it seems clear that empirical considerations are relevant to determining the moral status of animals. The question that remains is whether that status is to be determined using human standards to determine what is morally relevant or some other standards. If some other standards, what might they be?

Two Alternative Moral Perspectives

A long-standing tradition is to assess the moral status of animals in terms of how much they are like us. We can call this the "anthropocentric point of view." Such accounts take human morality as the standard for determining the moral standing of other animals. The more like us they are, the higher their moral status and the more their interests and well-being need to be taken into account.

Recent studies by students of animal behavior in general and cognitive ethologists in particular have led some to question the validity of this general approach. For example, tests of animal intelligence often proceed by putting animals in artificial laboratory situations and seeing how well they perform tasks that competent young children might perform. The results are then reported in terms of how well the animals perform compared to adults, infants, two-year old children, or what have you. This approach treats animals as more or less defective human beings. An alternative approach seeks to understand the behavior of animals on their own terms. This is one of the rationales for the emphasis that cognitive ethologists place on the study of animals under natural conditions. We can call this the "speciocentric point of view." With respect to the question of animal morality, this alternative perspective abandons the human moral condition as the gold standard for evaluating the moral status of animals and seeks to evaluate the moral condition of animals from the perspective of the animals themselves.

Two conditions need to be satisfied for the speciocentric point of view to be a sensible approach. First, this alternative depends on it being the case that animals have perspectives having something to do with moral behavior and that these perspectives are, in principle, discoverable. Second, an understanding of the nature of morality (or the concept of morality) that is not dependent on anthropocentric presumptions must be possible. Many cognitive ethologists seem convinced that both conditions can be met. Extensive field research combined with nuanced laboratory experiments are the keys to unlocking the condition of an animal from the animal's point of view. An evolutionary approach is the key to reconceptualizing the nature and function of morality from a species-neutral point of view. Obviously there is also a need for philosophical analysis of the concept of morality at work.

The anthropocentric perspective is the dominant approach. Most of the anthropocentric approaches, whether mainstream Utilitarian or Kantian, treat animals as moral patients. A hybrid account of the moral status of animals can be found in Rachels' *Created from Animals: The Moral Implications of Darwinism.*[32] Rachels argues that a proper appreciation of the moral implications of Darwinism serves to undermine what he sees as the traditional foundations of human morality and in the process supports a view that he labels "moral individualism," a view which has radical implications for the moral status of both human beings and animals. Rachels' book is one long argument to the effect that Darwinism undermines the concept of human dignity and, in so doing, presents a radical challenge to our understanding of what it means to be a matter of moral significance.

Rachels takes what he calls the traditional concept of human dignity to be the presumption that the primary purpose of morality is the "protection of human beings and their rights and interests."[33] This presumption is supported by certain factual (or quasi-factual) assumptions about human nature. Two basic claims supporting the sanctity of human dignity emerge from this. One is the presumption that human beings were created, as special, in the image of God. Call this "The Image of God Thesis." The second is the presumption that human beings alone among the animals are rational beings. Call this "The Rationality Thesis." It does not follow logically from these presumptions that human dignity is or ought to be the lynchpin of morality. However, Rachels argues, the primacy of human dignity as the foundation of morality does rest on and is supported by these presumptions. They serve as the rationale for putting human concerns ahead of all others in matters of morals.

Darwinism indirectly undermines the primacy of human dignity by undermining the presumptions that support the doctrine. The Darwinian perspective marginalizes God as the creator of human beings where humans are special among God's creations. Although Darwinism does not entail that God did not create human beings as special, it renders the story superfluous and perhaps suspect. If one does think that the theory of evolution entails that God did not create human beings as special, one need only reflect on the possibility that the entire evolutionary process is just God's way of insuring the specialness of human beings through "secondary causes."

From the Darwinian perspective, humans are just one among the animals. The Darwinian theory of common descent entails that all organisms are interrelated. Darwinian gradualism suggests that differences between species are often matters of degree and not matters of kind. These implications undermine the status of human beings as special and in doing so undermine the traditional moralities that are based on such assumptions.

The anthropocentric perspective is challenged as well by Gary Steiner in *Animals and the Moral Community: Mental Life, Moral Status, and Kinship*.[34] Along with Bekoff and Pierce, Steiner urges abandoning the idea that the human moral perspective is somehow privileged in favor of a view that treats the human perspective as only one among equally valid alternatives. They come to radically different conclusions, however, about where abandoning the anthropocentric view leads us. Steiner opts for advocating a universal morality, that is, a morality that is predicated on a single standard that applies to all sentient beings, whereas Bekoff and Pierce support a species-specific sense of morality, where what counts as moral for one species or population is not necessarily what counts as moral for other species or populations. However, it must be noted that Bekoff has taken the view developed in *Wild Justice* to lead directly to a universalistic view that he sees as congruent with Steiner's.[35]

According to Steiner, two mistaken assumptions underlie the refusal to include animals in the moral community. The first assumption is that mental life can only be thought of in intentionalistic terms. Rejecting this assumption opens the door for attributing a mental life to animals even if they are incapable of forming intentions. Steiner's argument for attributing a mental life to animals rests on distinguishing between what he calls "perceptual intelligence" from "rationality" or the capacity for making conceptual inferences.[36] This enables him to navigate between two extremes: Animals are dumb brutes for whom the question of a meaningful life does not arise, on the one hand, and animals are capable of "complex thought, self-reflective awareness, and moral agency," on the other.[37] Animals, he suggests, have perceptual intelligence. That is, they have the ability to react to their environments in ways that are conducive to the furtherance of their well-being, but they do not for the most part have the capacity for concepts, intentions, and reflection. The second mistaken assumption is that the moral status of animals hinges on their ability to possess intentionalistic states, such as beliefs and desires. So, it follows that animals can enjoy moral status even though they are not intentional creatures.

Steiner's overriding concern is with what he sees as the justice owed to animals. To this end, he proposes to appeal to a sense of social justice that is not species-specific but instead is part of a wider conception that he calls "cosmic justice."[38] Cosmic justice is distinguished from social justice, which Steiner sees as the restriction of the concept of justice to the human realm.[39] Cosmic justice, in turn, is part and parcel of a view of "cosmic holism." Cosmic holism is the view that "sentient beings, both human and nonhuman, have a kinship relation to one another that binds them together in a moral community in which neither can properly be said to be superior or inferior to the other."[40]

The demands of cosmic holism include abandoning an anthropocentric moral stance in favor of a universal sense of cosmic justice that, among other things, "demands universal veganism, the refusal to consume animal products of any kind."[41] In addition, Steiner claims that from this perspective there is absolutely no justification for using animals to satisfy human desires.[42] This is an austere conclusion that Steiner acknowledges will require negotiation. For one thing, it would seemingly preclude human beings from keeping pets either for simple leisure or for the psychological comfort that studies suggest having a pet can bring. That this is an ideal that may prove difficult to realize is attested to by the fact that Steiner's book is peppered with references to his pet cat, Pindar.

The question is whether appreciating the shared evolutionary heritage leads one to the view that there are several moral communities relativized to species, as I take the argument in Bekoff and Pierce to point to, or whether there is one moral community to which all sentient creatures belong. The science seems to support the former pluralistic view more than the latter monistic view. That is, the decision about who is to count as a moral agent or what counts as moral in a given set of circumstances should not be decided on the basis of human standards of morality. The moral status of animals needs to be determined in the light of what is discovered to be appropriate for the particular species or population under consideration.

Nevertheless, two implications about the ethical treatment of animals by human beings can be drawn. Panksepp, whose pioneering contributions to the field of affective neuroscience are based on years of experimental work on animals, notes that as we learn more about the affective experiences of animals, "[t]he practice of animal research has to be a trade-off between our desire to generate new and useful knowledge for the betterment of the human condition and our wish not to impose stressors on other creatures which we would not impose upon ourselves."[43] More generally, Bekoff argues that as the empathetic and affective nature of nonhuman animals becomes more evident we ought to become more sensitive to their needs and well-being by expanding what he calls our "compassion footprint."[44] Ignoring these implications threatens the moral integrity of human beings.

Conclusion

There are three lines of empirical evidence that support the view that animals possess enough mindedness to count as moral patients or partial members of a moral community; in some views there is enough evidence to count animals as moral agents and full members of a moral community. Evolutionary evidence supports the idea that there are no significant qualitative differences between humans and other animals. Neurophysiological evidence supports this conclusion insofar as homologous brain structures and systems implicated in the cognitive and affective capacities of human beings are widespread in the animal kingdom. Finally, evidence from cognitive ethology reinforces the view that a wide range of animals not only possess the relevant neural architecture necessary for sophisticated cognitive and affective behavior but that they manifest some degree of moral sensibility in their behavior.

There are many critics of such conclusions who argue, under the cloak of the charge of anthropomorphism, that the evidence supports no such far-reaching conclusions. We may, in fact, never be able to conclusively establish that animals have minds and experiences or are moved by moral considerations. However, the claim by some critics that such attributions are preposterous also cannot stand in the light of the available evidence. On balance, the weight of the scientific evidence, as presently understood, tilts against the critics. Given our understanding of shared evolutionary history and homologous brain structures, the default presumption ought to be that animals do experience emotions and react to them as well as experience and act upon moral sentiments, properly understood. The burden of proof, as de Waal notes, should be on those who reject the attribution of cognitive and emotional states as well as moral sensibilities to animals, not on those who make such attributions in the light of the mutually supporting evidence that currently exists.

If we take the claim that animals have the basic moral sensibilities of compassion, sympathy, and perhaps justice, then it behooves us to reexamine the anthropocentric bias that has permeated much of moral theory as propounded by human beings over the last two thousand or so years. Scientific evidence in and of itself does not and cannot entail that we should abandon our traditional anthropocentric perspective. However, it does suggest that this fundamental standpoint needs to be reexamined and perhaps re-justified. However, if we grant that animals are not merely moral patients at best but are capable of being moral agents of a sort, then it is far more difficult to see how the anthropocentric position can be defended. Two basic options emerge: First, a global species-neutral point of the sort advocated by Steiner or, second, a local species-specific point of view. This latter perspective, which seems to be more consistent with the evolutionary perspective that all animals are both similar to one another as products of common descent and yet different as different species, suggests that we think of the "moral community" as a collection of species-specific and for the most part non-overlapping moral communities.

Abandoning the anthropocentric perspective entails rethinking the nature of morality in the light of the fundamental evolutionary imperatives associated with the survival and flourishing of social animals. One implication is that human standards of morality no longer have an unchallenged claim for setting the criteria for the moral status of other animals. Instead, human beings become merely one among many social animals that possess the capacity for moral behavior, that is, for behaving in ways that are constrained by norms and values that are endemic to the species or populations of which they are members. In this respect, this mirrors the fundamental Darwinian point that human beings are not special creations distinct from the other animals in virtue of a God-given status, but instead are just one among the other animals. Humans are different, to be sure, but not thereby privileged in any way. This suggests a strategy for dealing with other presumed special features of human beings. One such feature is the possession of language. Human beings have a fully developed language with syntax, grammar, and the like where no other animal does. Some animals, such as the higher primates, may approximate bits of that language but the net product is inferior in crucial ways. Or, so the argument typically goes. Nevertheless, it should be apparent from what has been said above that this assessment reeks of the anthropocentric perspective that we have been at pains to root out. Should we not reexamine the argument from linguistic weakness, so to speak, in the same light? I offer here a few suggestive closing remarks.

What is the essence of language? Language functions primarily as a medium for the communication of cognitive information ("There is food there."), of emotional states ("I'm in a good mood. Let's play."), the conveyance of threats ("Those are fighting words."), as well as the expression of imperatives, the communication of norms, and the expression of approval and disapproval. No doubt there are other uses or functions. At any rate, the key point is that many if not all of these communicative functions are within the capacities of all social organisms. In one sense, this should be obvious because the stability and welfare of social units depends on the ability of the members of these social units to communicate with one another. Ample evidence for the existence of these modes of communication has been uncovered by animal behaviorists and cognitive ethologists. In the same way in which a set of social norms underlies all moral communities while manifesting itself in different ways, so there is a common communicative core that underlies all social communities while manifesting itself in different ways. From a non-anthropocentric point of view, human language with its words, grammar, syntax, and semantics is just one language among many. Of course, this sounds more radical than many may be willing to accept. Others conversant on the key issues will point out what appear to be serious objections to the lumping of human language with modes of animal communication. This is not the place to pursue such issues. My only point is that taking Darwin seriously recommends to us a non-anthropocentric view with respect to a number of things that, despite the force of Darwin's revolutionary insight, still strike us as uniquely human and as marking us off, in the last analysis, as not merely one among the animals.

Once we accept that empirical evidence is relevant to shaping our views about the nature of morality, the door is opened to reconceptualizing morality in non-anthropocentric terms. Taking Darwin seriously, we should consider the possibility that human morality is merely one manifestation of morality among others. All three lines of empirical investigation, the arguments from evolution, neuroscience, and cognitive ethology, all point to the existence of putative moral capacities in animals.

How are we to interpret the empirical findings? One option, the proto-morality option, would argue that although some animals possess the rudimentary features that manifest themselves in human moral systems, no nonhuman species can be said to be truly moral. Several reasons have been offered in support of this view. Here are three common considerations: (1) Animals do not possess language, which is necessary for forming the propositional attitudes that are fundamental to the expression of moral sentiments; (2) Animals, although they may be able to communicate with one another, are not conscious and as such lack the inner life necessary to be subjects of moral concern; (3) Animals, even if they are conscious, do not have the requisite complex brain capacities needed to reflect on their behavior and the behavior of others. However, what conclusion should we draw even if all these claims turn out to be correct? Does it follow that animals are not moral creatures at all or only that they are not moral creatures like us? Taking Darwin seriously suggests serious pursuit of the latter alternative.

A second option, then, is to look for an understanding of what it means for animals to be moral beings in our shared evolutionary heritage, but without the idiosyncratic baggage inherited from thousands of years of humans wrestling with understanding the nature of human morality. An immediate objection to this option is that by so reconceptualizing what it means to be moral we have changed the subject, and the very idea of animal morality is morality in name only. This response fails to take the relevance of scientific results for our understanding of morality seriously. Adopting an example from Allen and Bekoff: Is this reconceptualization of what it means to be moral any different from the reconceptualization of what it means to be a chemical element in light of the significance of atomic numbers?[45] Before their discovery, elements were identified in terms of their qualitative physical properties. Gold, for example, was defined in terms of its characteristic color, ductility, density, and the like. Later work by Henry Moseley revealed the significance of atomic numbers and what had been defining properties of gold were now known as chemical physical properties that gold, defined as such in terms of its number of protons, possessed. The point is that the characterization of any object or process that is subject to scientific review cannot be tied, absolutely, to any fixed definition.

One might object that the scientific analogy is not apropos to our problem on the grounds that the nature of morality is a purely philosophical question and not a scientific question at all. But this is implausible given the arguments we have rehearsed above. In addition, however, there is the following point. The basic neural mechanisms that enable human beings to experience and express moral concerns are shared across a wide range of species. In addition, cognitive ethological studies suggest that social animals express and enforce norms appropriate to their natures and cognitive capacities. As such, it is appropriate to attribute moral systems to them. How to properly characterize these systems remains an open question that requires further extensive empirical investigation and philosophical reflection. All human moral systems are rooted, either explicitly or implicitly, in some conception of human nature or what it means to be human. Human nature, however, is something that is a proper subject for scientific inquiry. Our best evolutionary, neuroscientific, and cognitive ethological theories lead us back to the Darwinian Imperative. Human beings are one among the animals. From this, human cognitive capacities are one among the cognitive capacities of animals. Finally, human moral systems are one among the moral systems of animals.

NOTES

1. Richard Sorabji, *Animal Minds and Human Morals: The Origins of the Western Debate* (Ithaca, N.Y.: Cornell University Press, 1993).

2. Charles Darwin, *On The Origin of Species: A Facsimile of The First Edition*, ed. Ernst Mayr (Cambridge, Mass.: Harvard University Press, 2000); Darwin, *The Descent of Man and Selection in Relation to Sex* (Princeton, N.J.: Princeton University Press, 1981); Darwin, *The Expression of The Emotions in Man and Animals*, 3rd ed. (Oxford: Oxford University Press, 1998).

3. Darwin, *Descent*, p. 36.

4. Darwin, *Descent*, p. 39–43.

5. Darwin, *Descent*, p. 42.

6. Darwin, *Descent*, pp. 46–48.

7. Darwin, *Expression*.

8. George Romanes, *Animal Intelligence* (London: Kegan Paul, Trench & Co., 1882); Romanes, *Mental Evolution in Animals* (London: Kegan Paul, Trench & Co., 1885).

9. Romanes, *Animal Intelligence*, pp. vii–ix.

10. Romanes, *Mental Evolution*, pp. 7–8.

11. C. Lloyd Morgan, *An Introduction to Comparative Psychology* (London: Walter Scott, 1914), p. 53.

12. Alan Costall, "How Lloyd Morgan's Canon Backfired," *Journal of the History of the Behavioral Sciences* 29 (1993): 113–124, and Alan Costall, "Lloyd Morgan, and the Rise and Fall of 'Animal Psychology,'" *Society & Animals* 6 (1998): 13–19.

13. Costall, "Lloyd Morgan, and the Rise and Fall."

14. Jaak Panksepp, *Affective Neuroscience: The Foundations of Human and Animal Emotions* (New York: Oxford University Press, 1998); Jaak Panksepp and Georg Northoff, "The Trans-Species Core SELF: The Emergence Of Active Cultural and Neuro-Ecological Agents Through Self-Related Processing with Subcortical-Cortical Midline Networks," *Consciousness and Cognition* 18 (2009): 193–215; and K. C. Berridge, "Comparing the Emotional Brain of Humans and Other Animals," in *Handbook of Affective Sciences*, ed. Richard J. Davidson, Klaus R. Scherer, and H. Hill Goldsmith (Oxford: Oxford University Press, 2003), pp. 25–51.

15. Berridge, "Comparing the Emotional Brain," p. 27.

16. Paul MacLean, *The Triune Brain in Evolution* (New York: Plenum Press, 1990).

17. Jaak Panksepp, "Affective Consciousness: Core Emotional Feelings in Animals and Humans," *Consciousness and Cognition* 14 (2005): 30–80.

18. Frans de Waal, *Primates and Philosophers: How Morality Evolved* (Princeton: Princeton University Press, 2006), pp. 61–67.

19. De Waal, *Primates*, p. 70.

20. Marc Bekoff and Jessica Pierce, *Wild Justice: The Moral Lives of Animals* (Chicago: University of Chicago Press, 2009), p. 1.

21. Bekoff and Pierce, *Wild Justice*, p. 7.

22. Bekoff and Pierce, *Wild Justice*, p. 59.

23. Bekoff and Pierce, *Wild Justice*, p. 30.

24. Bekoff and Pierce, *Wild Justice*, pp. 107–109.

25. In a lecture at the "Celebration of Darwin" conference held at Virginia Tech in November 2009.

26. Bekoff and Pierce, *Wild Justice*, p. 114.

27. Rosemary Rodd, *Biology, Ethics, and Animals* (Oxford: Oxford University Press, 1990).

28. De Waal, *Primates*, pp. 69–73.

29. De Waal, *Primates*, p. 181.

30. Daniel Dennett, "Animal Consciousness: What Matters and Why," *Social Research* 62 (1995): 691–710.

31. Dennett, "Animal Consciousness," pp. 701–703.

32. James Rachels, *Created from Animals: The Moral Implications of Darwinism* (Oxford: Oxford University Press, 1990).

33. Rachels, *Created from Animals*, p. 4.

34. Gary Steiner, *Animals and the Moral Community: Mental Life, Moral Status, and Kinship* (New York: Columbia University Press, 2008).

35. Bekoff, *The Animal Manifesto: Six Reasons for Expanding Our Compassion Footprint* (Novato, Calif.: New World Library, 2010).

36. Steiner, *Animals and the Moral Community*, p. 68.

37. Steiner, *Animals and the Moral Community*, p. 61.

38. Steiner, *Animals and the Moral Community*, p. 104.

39. Steiner, *Animals and the Moral Community*, p. 105.

40. Steiner, *Animals and the Moral Community*, p. xii.

41. Steiner, *Animals and the Moral Community*, p. 163.

42. Steiner, *Animals and the Moral Community*, p. 163.

43. Panksepp, *Affective Neuroscience*, p. 7.

44. Bekoff, *Animal Manifesto*.

45. Colin Allen and Marc Bekoff, *Species of Mind: The Philosophy and Biology of Cognitive Ethology* (Cambridge, Mass.: MIT Press, 1997), pp. 16–17.

SUGGESTED READING

ALLEN, COLIN, and MARC BEKOFF. *Species of Mind: The Philosophy and Biology of Cognitive Ethology*. Cambridge, Mass.: MIT Press, 1997.

BEKOFF, MARC. "Cognitive Ethology and the Explanation of Nonhuman Animal Behavior." In *Comparative Approaches to Cognitive Science*, edited by H. L. Roitblat and J.-A. Meyer, 119–150. Cambridge, Mass.: MIT Press, 1995.

———. *The Animal Manifesto: Six Reasons for Expanding Our Compassion Footprint*. Novato, Calif.: New World Library, 2010.

———, COLIN ALLEN, and GORDON M. BURGHARDT, eds. *The Cognitive Animal: Empirical and Theoretical Perspectives on Animal Cognition*. Cambridge, Mass.: MIT Press, 2002.

———, and JESSICA PIERCE. *Wild Justice: The Moral Lives of Animals*. Chicago: University of Chicago Press, 2009.

BERRIDGE, K. C. "Comparing the Emotional Brain of Humans and other Animals." In *Handbook of Affective Sciences*, edited by Richard J. Davidson, Klaus R. Scherer, and H. Hill Goldsmith, 25–51. Oxford: Oxford University Press, 2003.

DARWIN, CHARLES. *The Expression of the Emotions in Man and Animals*, 3rd ed. Oxford: Oxford University Press, 1998.

DE WAAL, FRANS. *Our Inner Ape*. New York: Riverhead Books, 2005.

———. *Primates and Philosophers: How Morality Evolved*. Edited by S. Macedo. Princeton, N.J.: Princeton University Press, 2006.

HAUSER, MARC D. *Moral Minds: The Nature of Right and Wrong*. New York: Harper, 2006.

PANKSEPP, JAAK. *Affective Neuroscience: The Foundations of Human and Animal Emotions*. New York: Oxford University Press, 1998.

———. "Affective Consciousness: Core Emotional Feelings in Animals and Humans." *Consciousness and Cognition* 14 (2005): 30–80.

———, and GEORG NORTHOFF. "The Trans-Species Core SELF: The Emergence of Active Cultural and Neuro-Ecological Agents through Self-Related Processing with Subcortical-Cortical Midline Networks." *Consciousness and Cognition* 18 (2009): 193–215.

RACHELS, JAMES. *Created From Animals: The Moral Implications of Darwinism*. Oxford: Oxford University Press, 1999.

STEINER, GARY. *Animals and the Moral Community: Mental Life, Moral Status, and Kinship*. New York: Columbia University Press, 2008.

PART V

SPECIES AND THE ENGINEERING OF SPECIES

ON THE ORIGIN OF SPECIES NOTIONS AND THEIR ETHICAL LIMITATIONS

MARK GREENE

DISCUSSIONS about what humans owe each other are frequently framed in terms of the individual: rights are enjoyed by individuals; duties are owed to individuals; and the basic currency of utilitarianism is individual welfare. Much of the debate in animal ethics follows this lead by focusing on the criteria that a nonhuman individual must meet in order to enjoy rights, to be owed duties, or to have its welfare count. Driven, in large part, by the efforts of those defending a place for animals within the sphere of moral consideration, sophisticated theoretical literature explores the grounds for moral consideration of individuals, human and otherwise.

There are other approaches to the debate over what we owe the natural world that are not restricted to obligations owed to individual organisms. In environmental ethics, there are those who urge direct moral consideration of broader entities such as ecosystems, wildernesses, or even the whole biosphere. A familiar manifestation of such broader moral concern is the idea that we have a duty to preserve species, especially those that are threatened by human activity. Although saving individual members of a species may contribute to saving the species, we cannot account for duties of species preservation solely in terms of duties to individual members of the species to be preserved. Indeed, imperatives of species preservation may conflict with the interests of individual animals. For example, captive breeding programs confine individuals and can involve onerous medical and surgical interventions. Even if such a program saves a species from extinction, it may do so to the detriment of individual animals.

This chapter will critically examine claims that species themselves, and not just the individuals that instantiate them, have morally considerable value. I will start by distinguishing derivative value from nonderivative value. Next, I will review suggestions for deriving moral value from the contributions that species make to other things of value, such as human well-being. Because promotion of human well-being is a widely-agreed-upon moral value, there is a strategic advantage for species preservationists in deriving species value from this source. However, this strategy can ground only limited and contingent duties of species preservation: duties that cannot be generalized to cover all species. I will explain how this realization has led some defenders of a general duty of species preservation to the view that species must have nonderivative value in their own right.

I agree that a general duty of species preservation requires attribution of nonderivative value to all species. However, Charles Darwin's view that species "are merely artificial combinations made for convenience"[1] calls into question the very intelligibility of claiming nonderivative value on their behalf. Although the intelligibility worry can be contained, biologically informed understandings of the species notion are not compatible with plausible claims of nonderivative value. My overall argument is very simple: since (1) a general duty of species preservation requires that species have nonderivative value and (2) species lack nonderivative value, it follows that (3) there is no general duty of species preservation.

A TAXONOMY OF VALUE

I will be discussing different sources of moral value and must therefore say how I am distinguishing them. Traditionally, a distinction has been drawn between instrumental and intrinsic value. Unfortunately, the notion of intrinsic value has suffered significant injury in the course of some attempts to claim it on behalf of the natural world. According to J. Baird Callicott, for example, although "the *source* of all value is human consciousness…it by no means follows that the *locus* of all value is consciousness." If a person values something "*for* its own sake, *for* itself," then on Callicott's reading, that thing is intrinsically valuable even though "it is not valuable in itself."[2] Callicott acknowledges that this is intrinsic value only in a "truncated sense."[3] By retaining dependence on human evaluations, Callicott's truncation excises the very substance from the idea of intrinsic value. Continuing to call the residue "intrinsic value" only invites confusion.

Instead of classifying value as either instrumental or intrinsic, I will distinguish *derivative* value from *nonderivative* value. Within the genus of derivative value, I wish to note rough distinctions between three species: *instrumental value, constitutive value*, and *projected value*. This gives a total of four kinds of value, or disvalue, that things may have in any combination:

(1) *Derivative Instrumental Value*: To have instrumental value is to make a causal contribution to bringing about a valuable state of affairs. The notion of a causal contribution is to be understood broadly. Both proximate causes and causally relevant background conditions can ground instrumental value. The familiar usage of "instrumental value" often suggests a narrower focus on being useful to some agent as a means to some desired end. I do not characterize instrumental value as requiring agency by definition. For example, if mountain ranges have value, then the collision of the Indo-Australian and Eurasian plates derives instrumental value from its causal contribution to the Himalayas.

(2) *Derivative Projected Value*: To have projected value is to be valued by some suitable valuer. There is no need, here, to specify the criteria for being a suitable valuer, except to say that in the context of debate concerning species preservation humans are assumed to count. For example, a beloved pet rock, if cherished by its owner, derives projected value from that fact; the rock owner projects value onto the rock in the very act of valuing it.

(3) *Derivative Constitutive Value*: To have constitutive value is to make a non-additive contribution to the value of a state of affairs by helping to constitute that state of affairs. By stipulating a non-additive contribution, I wish to limit constitutive value to cases exemplifying the cliché that sometimes the value of a whole is not just that of the sum of its parts. For example, if there is value in agricultural diversity, then Soay sheep may be said to derive constitutive value to the extent that they are constituents of such diversity.

(4) *Nonderivative Value*: To have nonderivative value is to have value that does not derive from any other value or valuer. Even if we agree with John Stuart Mill that "the sole evidence it is possible to produce that anything is desirable, is that people do actually desire it,"[4] it does not follow that the value derives from our evaluation and thus collapses into projected value. The direction of fit is different: if something is good because it is loved, then it is an example of projected value; if something is loved because it is good, then it may be an example of nonderivative value.

For my present purposes, the claim that something has value is to be taken as interchangeable with the claim that it merits consideration in moral deliberations. Things that have only derivative value merit consideration only indirectly, because of their impact on the source from which their value is derived. However, to say that a thing has nonderivative value is to say that it is owed moral consideration in its own right.

Having roughed out a taxonomy of value, I will now apply this scheme to some of the values that have been claimed on behalf of biological species. I will start with examples of claims that species have derivative value: instrumental, projected, and constitutive. I will then explain the limitations of derivative value as a foundation for asserting duties of species preservation. These limitations motivate serious consideration of the claim that species have nonderivative value.

Attributing Value to Species

The Derivative Value of Species: Instrumental Value

To the extent that species contribute causally to processes and conditions that advance our projects and serve our interests, they derive instrumental value from the presumed value of those interests and projects. Perhaps the most fundamental kind of instrumental value is what Holmes Rolston III has called "life-support value."[5] By providing so-called "environmental services," such as the conversion of carbon dioxide to oxygen by photosynthesis or the breakdown and recycling of dead organic matter,[6] plant and animal species contribute to the causal background conditions necessary for the continuation of human life. Also, many species serve as commodities or raw materials for the realization of more specific human ends. By forging causal links in chains of production some species have clear economic value.[7] The economic value of food animals and working animals, for example, is clear and significant.

Shifting the emphasis from work to play, Rolston distinguishes nature's recreational value from its narrower economic value.[8] Although much of the recreational use of nature is tightly integrated into economic activity, Rolston calls attention to other ways in which the natural world contributes to conditions for human leisure activities. Nature provides a rich playground with powerful waves to surf, challenging rock faces to climb, and elusive prey to test the skill of hunters. If recreational pursuits such as hunting and fishing have value, then some animal species derive value from their contribution to such pursuits. This source of recreational value is not limited to those species that are directly hunted or fished; it extends to other species that contribute to the health of the hunted, whether as prey for them or as contributors to the environment that sustains them.

The Derivative Value of Species: Projected Value

Many authors have stressed the aesthetic value of nature.[9] We find value in glimpsing the iridescence of a dragonfly, in witnessing the aerial abstractions of flocking starlings, and in the taste of sushi. However, the scope of nature's aesthetic value is not confined to what we find beautiful: a singular lack of beauty is no bar to recognizing value in the experience of seeing a wild manatee and having it approach close enough to smell its breath. Rolston broadens the scope of aesthetic value even further when he draws close parallels between the value of nature as an object of the "aesthetic encounter" and as an object of the "valuable intellectual stimulation" of science.[10]

I take aesthetic value to be a paradigm of projected value. That is to say, aesthetic value is projected onto nature by the eye of the suitably appreciative beholder. This might be disputed. There are those who appear to take the view that being fascinating or being beautiful are properties that things can have in themselves, that aesthetic value may reside in the tiger or the manatee regardless of what anyone thinks of them. Aesthetic value may elicit the appreciation of suitably sensitive observers, but it is there independently of there being any observer.[11] On this view aesthetic value would not derive from an observer's evaluation but would be an example of nonderivative value. Since the dispute need not be resolved in order to understand the idea of projected value, I leave the defense of such a view to those who can make any sense of it.

Following Rolston's lead in associating the "valuable intellectual stimulation" of science with aesthetic value, we should note that there is valuable stimulation in recreational pursuits. Should Rolston's "recreational value" be subsumed into aesthetic value? This subsumption does seem appropriate for many recreational activities, such as butterfly collecting, in which appreciation of the natural object is the ultimate end of the pursuit. However, this does not seem appropriate for surfing, climbing, or hunting. In these cases, the primary source of value is the pursuit itself, and nature derives value because it is instrumental in realizing the valuable pursuit.

The Derivative Value of Species: Constitutive Value

My right thumb would be of little use on its own, but it has great value, at least to me, as a constituent of my functioning right hand. This is the basic idea behind constitutive value. A similar case can be made for the constitutive value of a species if it can be identified as a constituent of some valuable whole or state of affairs. Rolston and others presume that there is value in natural diversity.[12] If this is right, it would be reasonable to argue that species derive value from the very diversity that they help to constitute. Similarly, species may be argued to derive constitutive value in virtue of their constitutive contribution to the value of broader natural entities such as ecosystems, wildernesses, and the biosphere.

The idea that such things as ecosystems or wildernesses have species as constituents is quite intuitive. However, Donald Regan has asserted a different source of constitutive value for species that is more technical. He claims that there is value in triads consisting of (i) a natural object, (ii) a person's knowledge of that natural object, and (iii) that person's pleasure in her knowledge of that natural object.[13] Regan takes the Grand Canyon as an example of a suitable natural object. Regan holds that value requires consciousness and that the Grand Canyon therefore has no value on its own. However, he argues that there is value in "the complex consisting of the Grand Canyon, plus Jones's knowledge of the Grand Canyon, plus Jones's pleasure in her knowledge of it."[14] On this account, the natural object has constitutive value: it is valuable not only as a means to eliciting pleasure but also as an essential constituent of a complex state of affairs that has value.

Regan argues that his scheme extends to yield duties to preserve any natural object, including species. There are complications, but the rough idea is that we can put any natural object, including species, in the place of the Grand Canyon as part of a valuable complex of natural object, knowledge of that natural object, and pleasure in the knowledge of that natural object. Regan attempts to extend his protection to all natural objects by assuming that "[a]s to any such complex, we have a [moral] reason to promote its existence." Even natural objects of which we are ignorant or in which we take no pleasure are included by Regan on the grounds that we can always promote the existence of valuable complexes involving such objects because it is always true that "we are capable of learning about the object" and "we are capable of learning to take pleasure in knowledge of any natural object."[15] The conclusion Regan draws from all this is that "[b]ecause the species is a separate 'natural object,' saving the species is a separate good."[16] Regan's assertions strain my credulity when applied to such natural objects as boils and brain tumors, but his scheme does serve as an example of according constitutive value to species.

Together with projected and instrumental value discussed above, this discussion completes my survey of three kinds of derivative value that have been claimed for species. I will now consider how the limitations of derivative value have motivated attributions of nonderivative value to species.

THE LIMITATIONS OF DERIVATIVE VALUE

For defenders of the natural world, appeal to derivative value offers significant advantages in their efforts to push public policy and private behavior toward a greater concern for conservation. Preservationists who ground their moral case in human interests and well-being are appealing to values that are already widely accepted as morally relevant and that can draw on the power of self-interest to help motivate us to action. As we will now see, the problem for those seeking to assert a general duty of preservation extending to all species is that exclusive appeal to derivative value can yield only limited and contingent duties to preserve species.

On December 29, 1979, completion of the Tellico Dam destroyed what was presumed to be the last habitat of a species of tiny fish, the snail darter (*Percina tanasi*).[17] Because of the difficulty of making a compelling case for its derivative value, the plight of the snail darter, or at least its legend, has become something of a touchstone in discussion of the moral grounding of duties to preserve species.[18] A case for preservation based solely on derivative value can sometimes be strong, as in the case of the Chinook salmon (*Oncorhynchus tshawytscha*). The Chinook salmon is economically important, with a large market for both farmed and wild-caught specimens; it has recreational significance as a sport fish; I venture that it is aesthetically impressive and scientifically interesting; it has ecological impact both as predator and prey; and it has additional value attributed to it as the state fish of Alaska and in some Native American cultures. The contrast between the Chinook salmon and the snail darter is stark in that no convincing case can be made for recognition of much derivative value in the latter. The snail darter is economically insignificant, recreationally inert, aesthetically uninspiring, scientifically unremarkable, ecologically inconsequential, and is no one's state fish. Where derivative value is lacking, grounding the ethical defense of species like *P. tanasi* on derivative value can yield, at best, vanishingly weak duties of preservation. Such is the predicament of the snail darter, and such is the challenge for those arguing for the preservation of this and other Cinderella species.

The situation gets worse when we note an ironic consequence of reliance on derivative value, namely that endangerment can actually weaken the case for species preservation. Rarity may enhance some aspects of derivative value,[19] but the fact that we are already managing with very few members of an endangered species might reasonably be taken as evidence that we could get by with none. While one might reasonably worry about the ecological consequences if a novel virus threatens world populations of the ubiquitous brown rat (*Rattus norvegicus*), the disappearance of the Sri Lankan Mountain Rat (*Rattus montanus*) would be of less ecological concern precisely because it is already endangered and limited to a small geographical range.

Some authors have placed notions of derivative value on the rack in the attempt to stretch them to cover more species. Appeals to ignorance are prominent in such attempts: species that now seem insignificant may later turn out to have great economic value, scientific interest, or ecological importance. It is notable that such considerations do nothing to dispute the claim that some species lack derivative value; instead, they urge a precautionary principle that because we cannot reliably know that any particular species lacks derivative value we should proceed as though they all have it. However, in order for such arguments to get anywhere near generating a general duty of species preservation, our ignorance must be exaggerated and the costs of general species protection must be ignored. This and the many other problems with such attempts have been thoroughly enumerated by Martin Gorke.[20]

To sum up: to the extent that species have derivative value, they merit consideration in our moral thinking. The most widely accepted claims of species' derivative value are likely to be those that derive instrumental or projected value from human interests. Such claims can be strong, as they are in the case of the Chinook salmon. However, the case of the snail darter shows why these sources of derivative value do not readily generate duties of preservation that cover all species. Thus, the balance of considerations so far strongly indicates that, while appeal to sources of derivative value can ground duties of species preservation in specific cases, it cannot be stretched to encompass a general duty of preservation for all species. Realizing the contingency and limitations of appeal to derivative value, some defenders of general duties of species protection have claimed nonderivative value for species. I will now review some of those claims.

THE NONDERIVATIVE VALUE OF SPECIES

The idea that species have nonderivative value enters the debate in many guises. Sometimes nonderivative value is hinted at or implied, rather than clearly stated. For example, though scientific value derives from human interests, Rolston hints at more when he asserts that "natural science is an intrinsically worthwhile activity." That science "cannot be worthwhile unless its primary object, nature, is interesting enough to justify being known," and that "no study of a worthless thing can be intrinsically valuable."[21] At other times, nonderivative value is explicitly claimed but then evaporates on closer examination. For example, Rolston introduces life value, or "[r]everence for life," as the idea that "life generically is of value" in a manner akin to "the life we value in persons," but he continues his explication by directing his readers to "notice some features that humans value in living things."[22] At still other times, nonderivative value is officially denied but seems to sneak in anyway. For example, in seeking to preserve the intuition that "value depends somehow on the existence of fairly sophisticated consciousness,"[23] Regan is pushed to a distinction between things that are worth knowing and things that are not.[24] Despite Regan's protestations, it is hard to see what this distinction can rest on other than some kind of nonderivative value.

Martin Gorke is refreshingly clear in arguing for the nonderivative value of species. His careful evaluation of the limitations of value derived from anthropocentric considerations, to which I referred above, leads him to summarize the situation thus: "*Anthropocentric* ethics is [not] capable of justifying *general* species protection....In the case of *nonanthropocentric* ethics, on the other hand, this is possible at least *in principle*."[25] Since Gorke is also explicit in assuming that there is a general duty of species protection as a fixed point in his ethical reasoning,[26] he takes the possibility of justifying such a general duty as a source of support for the non-anthropocentric, "holistic" ethics that he endorses. "Only in the context of holistic ethics is it possible to defend *all* species for their own sake."[27]

Anthropocentric considerations are, of course, examples of what I am calling "derivative value." By contrast, Gorke's holistic ethics takes the view that "not only all living things but also inanimate matter and entire systems have intrinsic [non-derivative] value."[28] Thus, in terms of my distinction between derivative and non-derivative value, Gorke's argument may be summarized as follows: since (1) a general duty of species preservation requires that species have nonderivative value and (2) there is a general duty of species preservation, it follows that (3) species have nonderivative value.

BIOLOGICAL FACTS, MORAL VALUE,
AND THE ROLE OF INTUITION

My summary of Gorke's argument invites comparison to my own argument as I summarized it at the beginning of this chapter: since (1) a general duty of species preservation requires that species have nonderivative value and (2) species lack nonderivative value, it follows that (3) there is no general duty of species preservation. Both arguments are valid and we agree on the first assumption. However, while Gorke affirms the general duty of preservation and infers the nonderivative value of species, I will be arguing against attributing nonderivative value to species and inferring from that the lack of a general duty of species preservation. Since my argument concerns moral value but will draw on biological facts, I will now prepare the ground with some general methodological comments.

As so often in ethical debate, and in philosophy more generally, part of the task is destructive: not so much trumpeting the glories of one's own view as highlighting the shortcomings of others or, as Bertrand Russell neatly put it, "I can only say that, while my own opinions as to ethics do not satisfy me, other people's satisfy me still less."[29] One way to disseminate dissatisfaction with the opinion of another is to show that the opinion is either contradictory or incoherent. We will consider a Darwinian challenge to the coherence of attributing nonderivative value to species, and I will argue that, although Gorke and others have responded poorly, that threat can be contained.

If the attribution of nonderivative value to species does not fall to incoherence, the debate pushes back to whether the attribution is plausible. I will argue that an understanding of the biological facts to which Darwin drew attention makes such attributions quite implausible. There is, of course, a general methodological problem in drawing either evaluative or normative inferences from purely factual claims, and the proper role of intuition in bridging this gap is open to debate. Gorke's view is that intuitions are the proper starting point of ethical reflection,[30] and his basic justification for assuming that there are general duties of species preservation is that it is "a postulate firmly rooted in intuition and generally quite well accepted, regardless of the way people justify it."[31]

On a weak reading, Gorke's idea may be that, as a matter of practical ethical debate, we take shared intuitions as our starting point and tease out the implications from there. If Gorke is correct in believing that the postulate of species preservation is firmly rooted in the widely accepted moral intuitions of his audience, then this is at least a defensible starting point for debate. What Tom Beauchamp and James Childress call "common morality" might be understood as a limiting case of this idea. The common morality is the set of core imperatives allegedly accepted by "all persons committed to morality." As they point out, "[b]ecause we are already convinced about these matters, the literature of ethics does not usually debate the merit or acceptability of these basic moral commitments."[32]

A stronger reading of Gorke's view is that endorsement by strongly rooted and widely accepted intuitions can be taken as evidence that a normative or evaluative claim is correct, not just that it can be used as an agreed starting point. In the context of metaphysical debate, Saul Kripke takes the stronger line on the role of intuition, saying, "Of course, some philosophers think that some-thing's having intuitive content is very inconclusive evidence in favor of it....I really don't know, in a way, what more conclusive evidence one could have about anything, ultimately speaking."[33] Returning to moral debate, Beauchamp and Childress seem to think that their common morality can bear this stronger reading when they say that "we rightly judge all human conduct by its standards,"[34] that common morality "supplies the initial moral content needed in our account of justification," and that we may "rely on the authority of the norms in the common morality."[35]

I make no claim either that my intuitions will be endorsed by all persons committed to morality or that they would be vindicated if they were. I proceed with more modest aims in view. Having developed a Darwinian challenge to the non-derivative value of species and having responded to the threat of outright incoherence, I will discuss how the biological facts underlying the Darwinian challenge should inform our understanding of species. A biologically informed understanding of species will enable a clearer view of what is being claimed when nonderivative value is attributed to species. This clearer view will, I think, highlight its implausibility. Though I claim no special authority, I do not think that my intuitions are especially idiosyncratic. Unless I am sorely deceived, my argument should at least push the burden of proof back to those claiming that biological criteria for species distinctions also mark moral distinctions.

THE DARWINIAN CHALLENGE

It has been asserted that "evolutionary biology has, from its inception, granted a central role to species" and that this "should be evident from the fact that the most important book in the history of this field...is titled *On the Origin of Species*."[36] Reading past the title, however, a different picture emerges. Darwin's actual view was that there was nothing fundamental about species as compared either to lower levels of classification such as subspecies or variety, or to higher levels such as genus, family, or phylum:

> Hereafter we shall be compelled to acknowledge that the only distinction between species and well-marked varieties is, that the latter are known, or believed, to be connected at the present day by intermediate gradations, whereas species were formerly thus connected....In short, we shall have to treat species in the same manner as those naturalists treat genera, who admit that genera are merely artificial combinations made for convenience.[37]

Classifications of human convenience are likely to track things that have value for us; that's what makes them convenient. Because of this, understanding species to be "merely artificial combinations made for convenience" presents no obvious problem for typical attributions of derivative value to species. However, Darwin's view strikes some as a worst case for claims that species can have nonderivative value. As Rolston puts it, "if species do not exist except as embedded in a theory in the minds of classifiers, it is hard to see how there can be duties to save them."[38] One worry is that it is simply unintelligible to attribute nonderivative value to entities that do not even exist independently of human classification. A different worry is that it is implausible to claim nonderivative value for classifications of convenience. Before I explain the reasons for which I share the latter but not the former worry, it is necessary to understand the reasons underlying Darwin's view.

The Biological Basis for the Darwinian Challenge

The Darwinian challenge is not premised on evolution per se but on observed biological facts, particularly facts about interbreeding. In his discussion of hybridism,[39] Darwin presents a sustained attack on the "view generally entertained by naturalists...that species, when intercrossed, have been specially endowed with the quality of sterility, in order to prevent the confusion of all organic forms."[40]

The "generally entertained" view to which Darwin was responding was creationist. From the creationist perspective, the original members of each species were created by God. Subsequently, the membership of a species consists of the original members plus those individuals that trace their ancestry exclusively to other members of the species. Though defined by ancestry, members of the same species are also united by commonalities of form and function that tend to set them apart from other species. The creator's original vision for each species is preserved and multiplied from generation to generation because of fertility within each species. Meanwhile, God prevents "the confusion of all organic forms" by enforcing reproductive isolation between species by making members of different species unable to interbreed and produce fertile young. When horses mate, more horses. When donkeys mate, more donkeys. But, when a horse and a donkey mate, the resulting mule is a sterile, ancestral dead end.

The idea that species boundaries are marked by reproductive isolation is not unique to creationism, nor is it incompatible with evolution. Ernst Mayr's view that "[s]pecies are groups of actually or potentially interbreeding natural populations, which are reproductively isolated from other such groups"[41] remains influential in biology and philosophy of biology, and is widely assumed to be correct among nonspecialists as a matter of biological general knowledge. What Darwin points out is that fertility and sterility are not "specially acquired or endowed" qualities but are matters of degree that are "incidental on other acquired differences."[42] Therefore, like similarities of form or overlaps of ecological niche, reproductive isolation does not describe sharp species boundaries that we can discover in nature. Instead, we draw species boundaries by imposing cutoffs on broader trends and tendencies.

We can better understand the substance and extent of Darwin's point by reviewing some of the biological facts that support it. Initially, it was studies of plant hybridization that led Darwin to conclude:

> It is certain, on the one hand, that the sterility of various species when crossed is so different in degree and graduates away so insensibly, and, on the other hand, that the fertility of pure species is so easily affected by various circumstances, that for all practical purposes it is most difficult to say where perfect fertility ends and sterility begins....It can thus be shown that neither sterility nor fertility affords any clear distinction between species and varieties; but that the evidence from this source graduates away, and is doubtful in the same degree as is the evidence derived from other constitutional and structural differences.[43]

However, although Darwin was able to cite numerous observations in support of this contention from experiments in plant hybridization, he confesses that "[i]n regard to animals, much fewer experiments have been carefully tried than with plants," and he doubts "whether any case of a perfectly fertile hybrid animal can be considered as thoroughly well authenticated."[44]

Was Darwin right to suspect that reproductive isolation also graduates away in the animal kingdom? He was. We will now examine two examples that confirm Darwin's suspicions about animals. In the first example, recent genetic studies show that familiar domestic chickens provide an example of a fertile hybrid animal. In the second example, graduation of fertility is elegantly illustrated by Californian salamanders.

The domestic chicken is considered a variety of the red junglefowl (*Gallus gallus*). Though there are areas of overlap, the wild range of the red junglefowl is separated from that of a distinct species, the grey junglefowl (*Gallus sonneratii*), by the Godavari River that runs through India to the Bay of Bengal. Beyond the geographical separation, the reproductive isolation of red and grey junglefowl is maintained by a variety of mechanisms that were first described in detail by G. Victor Morejohn in 1968.[45] One impediment was that differences of appearance, such as the color of the cock, led to hens not recognizing cocks of the other species as potential mates. Differences of courtship behavior, such as the call of the cock when offering tidbits of food to the hen, also stood in the way of any attempt at mating between the two species.[46] When these behavioral obstacles were overcome in captivity, genetic incompatibilities emerged: hybrids were produced, but those hybrids did not produce viable young when they were crossed. Morejohn concluded that geographical isolation, courtship behavior, and "genic incompatibility" all contribute to "maintaining both species as separate taxonomic units in the wild."[47]

Although he identified a range of mechanisms of reproductive isolation, Morejohn also uncovered a hint that the isolation may not be absolute. Though unable to produce viable young from mating hybrid offspring of red and grey junglefowl, Morejohn was able to produce a few viable young by backcrossing hybrid females with unhybridized males of both species.[48] Confirming the suspicion that the reproductive isolation of *G. gallus* and *G. sonneratii* is not absolute, a recent genetic study has shown that domestic chickens are not exclusively descended from wild red junglefowl as previously thought, but have hybrid ancestry, probably with the grey junglefowl.[49] Domestic chickens are certainly fertile, and as such they provide an example of a fertile hybrid animal.

Another fascinating and instructive example of the difficulties of relying on reproductive isolation to distinguish species is provided by the *Ensatina* salamanders that range around California's Central Valley. Patterns of phenotypic variations among adjacent populations of *Ensatina eschscholtzii* were interpreted as evidence that the salamanders exemplify the phenomenon known as a "ring species." This interpretation has now been supported by genetic studies.[50] Historically, *E. eschscholtzii* spread south from northern California, down the Sierra Nevada to the east, and down the Coastal Range to the west. From north to south along each route, adjacent populations are able to interbreed, although they do show incremental differences of phenotype, genotype, and of the ecological niche they inhabit. Being geographically isolated by the Central Valley, the Sierra and Coastal populations did not interact until they came back into close proximity where the Sierra Nevada and Coastal Range rejoin at the southern end of the Central Valley. Here, *E. eschscholtzii eschscholtzii*, an un-blotched Coastal Range subspecies, meets *E. eschscholtzii klauberi*, a blotched Sierra Nevada subspecies. Because the incremental evolutionary changes occurred independently along each route, the populations at the southern end of each route are quite distinct from each other and they are unable to interbreed. Therefore, by the criterion of ability to interbreed, adjacent populations of *E. eschscholtzii* are varieties of the same species along the Sierra Nevada and the Coastal Range but qualify as distinct species where the ring of adjacent populations closes.

E. eschscholtzii and other ring species present an elegant, frame-by-frame demonstration of the evolutionary process of speciation in action. However, ring species are unusual only in that so many incremental steps in the process are instantiated by populations still living at the same time. Just like the *eschscholtzii* and *klauberi* subspecies of *Ensatina* salamanders, red and grey junglefowl are connected by intermediate populations to a common ancestor. The difference is that the intermediate populations connecting red and grey junglefowl happen to have died out.

Were it not for the contemporaneous existence of the connecting ring, *E. eschscholtzii eschscholtzii* and *E. eschscholtzii klauberi* would be classified as distinct species. Conversely, if the intermediate populations linking red and grey junglefowl had survived, then *G. gallus* and *G. sonneratii* would probably be now be classified with those intermediate populations as subspecies of a single ring species. This reasoning bears out Darwin's prediction that we would be "compelled to acknowledge that the only distinction between species and well-marked varieties is that the latter are known, or believed, to be connected at the present day by intermediate gradations, whereas species were formerly thus connected."[51]

Fertile interspecies hybrids and ring species, as exemplified by chickens and *Ensatina* salamanders, show that Darwin was right to suspect that fertility within and infertility between species would prove to be a matter of degree among animals, just as he had argued it was among plants. This implies that distinguishing among species using the criterion of reproductive isolation is not fundamentally different from using criteria of phenotype, genotype, or ecological niche in that all these criteria are matters of degree and none can reliably deliver sharp boundaries between species. A further implication of the fact that all the biologically plausible criteria for demarcating species boundaries vary on continua of similarity and difference is that the "same species as" relationship must be intransitive. That A is similar to B and B is similar to C does not imply that A is similar to C. Equally, as we have seen, that A can interbreed with B and B can interbreed with C does not imply that A can interbreed with C. Therefore, if we base species distinctions purely upon biological criteria, that A is the same species as B and that B is the same species as C will not ensure that A is the same species as C.

One biologically respectable property that seems not to be a matter of degree is that of common ancestry. An organism together with all of that organism's descendants is called a "clade." There are those who argue that species are clades and that therefore all members of a species share at least a historical essence in that they share a common ancestor.[52] As long as there is a most recent common ancestor shared by all and only members of the species *Gallus gallus* and a different most recent common ancestor shared by all and only members of the species *Gallus sonneratii*, then red and grey junglefowl can be said to have distinct essences in their respective most recent common ancestors. Of course, the hybrid origin of domestic chickens complicates matters by entailing that chickens are members of both *G. gallus* and *G. sonneratii*. I am disposed to allow that dual species membership is a harmless terminological stipulation rather than a fundamental problem for making sense of species' nonderivative value.

However, according nonderivative value to clades does not justify according nonderivative value to species in particular. Even if species are clades, there are many clades that are very clearly not species: the animal kingdom, insecta, mamalia, my wife and son, just my son, and the like. Clades may have nonderivative value, but in order to justify according any special nonderivative value to those clades that are species we need criteria to pick out those clades that are species. This path merely brings us back to the issues of gradual variation and intransitivity.

The Darwinian Challenge to Species Essentialism

The biological facts to which Darwin has drawn attention matter because they guide our understanding of what a species is and hence our understanding of what we are claiming if we claim that species bear nonderivative value. In this section, we will see how the biological facts undermine two of the most straightforward ways of understanding claims of the nonderivative value of species; one theistic, one secular. The unifying features of these understandings are that they are realist and essentialist. I use "realism" to refer to the view that the species category "carves nature at its joints," or that species differences mark genuine discontinuities in reality independent of human classifications. By "essentialism," I mean the view that, for each and every species, there is some non-trivial property or set of properties such that is shared by all and only the members of that species.[53]

There is no neater way to make sense of the idea that species have nonderivative value than through the account of the origin of species that was dominant in the West in the eighteenth century, when Carolus Linnaeus established the basic framework that persists in biological taxonomy to this day.[54] The view of Linnaeus and his coreligionists was that species originated in divine acts of special creation; animal species were created by God, each according to its kind.

> And God said, Let the earth bring forth the living creature after his kind, cattle, and creeping thing, and beast of the earth after his kind: and it was so. And God made the beast of the earth after his kind, and cattle after their kind, and every thing that creepeth upon the earth after his kind: and God saw that it was good.[55]

The creationist notion gives a robustly realist account of species as endowed with God-given essences maintained, as we noted above, by reproductive isolation. As well as setting out a species notion, Genesis can be interpreted as giving divine warrant for the idea that species have nonderivative value.[56] God did not direct Noah to save as many individuals as he could cram onto the ark, but to preserve the essence of God's created kinds by saving breeding pairs. Because each species was created by God and was seen by God to be good, it is only proper for us to attribute nonderivative value to each species. We might quibble over whether this is a true case of nonderivative value or one of projected value with a species made good by God seeing that it was good. Such Platonic concerns need not delay us; either way, the ethically considerable value of a species is independent of human values or concerns.

The creationist need not deny the existence of individuals whose provenance includes ancestors of different kinds. However, hybrids, chimeras, and other monsters depart from the finite, original inventory of animal kinds that were created and valued by God. The biblical creation story gives no grounds on which to claim nonderivative value for these and other aberrations. Such a foundation for nonderivative value fits well with the intuitions of those who argue for the protection of natural species but are ambivalent about the status of human-made and other "unnatural" life forms.[57]

"God created, Linnaeus arranged"—or so it is said Linnaeus liked to say of his work.[58] However, what if God did not bring each species into being in an act of special creation? Does this undermine the idea that species are determinate, real, and objective parts of the natural world rather than mere ephemera of human arrangement? Not immediately, as we can see by reference to two contemporary writers.

Saul Kripke and Hilary Putnam have been influential in the philosophical understanding of scientific terms. Their view can be used to ground species essentialism without recourse to divine metaphysics. The standard Kripke-Putnam view is that natural kind terms, such as "gold" and "water," get their reference by initial baptism of samples, and that they refer to "the kind instantiated by (almost all of) a given sample."[59] Crucially, because natural kind terms are introduced by reference to samples, successful kind term reference does not require speakers to know much about the kind of thing they have baptized. The term "gold" was introduced by reference to samples of gold and that term designated the kind gold long before chemists discovered that gold is the element with atomic number 79, and "water" meant water long before its chemical structure was discovered to be H2O. Natural kind terms are rigid designators; that is, they refer to the same kind in all possible worlds. Thus, when we make the empirical discovery that gold is the element with atomic number 79 and that water is H2O, we make discoveries about what gold and water are, not just in this world, but in all possible worlds. That is to say, we make discoveries about what gold and water are essentially. In this sense, the standard Kripke-Putman view is that natural kinds have empirically discoverable essences. Both Kripke and Putnam assume that biological species are examples of natural kinds in this sense[60] and thus give a way to ground the idea that species can be sharply distinguished by some sort of natural essence without resort to creationist metaphysics.

Realism and essentialism about species offer an especially tidy foundation for attributions of nonderivative value in at least two ways. First, the idea that species exist independently of human classifications avoids the worry about the intelligibility of attributing nonderivative value to categories that are themselves derived from human convenience. Second, the idea that each species has a unique essence meshes well with the idea that each bears its own nonderivative value. Of course, finding a neat way to make sense of a claim is not the same as finding a reason to accept the claim, and I do not pretend that realism and essentialism automatically bring nonderivative value with them.

The basic biological facts underlying Darwin's challenge present a problem for any substantive species realism or essentialism and thus to any attribution of derivative value that relies on them. It is not that biologists are still searching for hidden essences that pick out species; we know what the candidate properties are; we know that common ancestry does nothing to distinguish species from other levels of classification; and we know that the other candidates are all differences of degree that are subject to intransitivity. Essences should be made of sterner stuff. "This may not be a cheering prospect; but we shall at least be freed from the vain search for the undiscovered and undiscoverable essence of the term species."[61]

RESPONDING TO THE DARWINIAN CHALLENGE

A Misdirected Response to the Darwinian Challenge

Darwin observed disagreement concerning the species notion, even among the creationists of his day, noting that "[n]o one definition [of "species"] has as yet satisfied all naturalists."[62] By precluding appeal to either created or discovered essences, or at least essences that pick out species in particular, Darwin's challenge only adds to the scope for disagreement. The proper understanding of the species notion remains today a subject of controversy among biologists and philosophers of biology.

A few defenders of general duties to protect species have taken note of this debate. Alex Wellington mentions it in passing.[63] Callicott raises the issue in arguing that it is unhelpful to talk in terms of "species rights" and in support of his account of the derivative value of species. He contrasts what he takes to be the traditional idea that a species is "a class or a kind," with David Hull's suggestion that species are historical entities, an idea Callicott rather weakly dismisses as not having "been universally accepted among philosophers of science." According to Callicott, it appears that possessors of rights must be "at least localizable things of some sort" but that "[a] class, by definition, is not an individual or localizable thing." Having asserted the "traditional view" that species are classes, Callicott concludes that claiming rights for species seems "conceptually odd if not logically contradictory."[64]

In stark contrast to Callicott, Rolston enlists Hull in support of his assertion that species are not just classes, but are historical entities that are localized in time and space, rather than just "in the minds of classifiers."[65] Gorke, too, worries that if species were "only *classes* devised by humans, then it would be difficult to explain why humans should hold any particular kind of responsibility toward them."[66] However, quoting Mayr, Gorke claims that "modern biologists have almost unanimously agreed that there are real discontinuities in organic nature, which delimit natural entities that are designated as species."[67]

The current literature bulges with such a rich diversity of rival species notions that it would be a dullard indeed who was unable to find one to suit an already-preferred position on duties of preservation. The idea that a species is a class, which Callicott reassuringly refers to as the notion that the term "species" "traditionally designates,"[68] is less reassuringly cast by Gorke as an idea "that many biologists, paleontologists, and philosophers used to hold" but that is "nowadays rejected by most taxonomists as unscientific."[69] With different authors appearing to claim consensus for incompatible species notions that just happen to fit with their favored rationale for species protection, it would be understandable for a casual follower of the debate to wonder whether there might be some cherry-picking going on.

Suspicion of cherry-picking is well founded, but there was no need for it. I will now explain why we should concede the intelligibility of attributing nonderivative value to species despite the Darwinian challenge. However, intelligibility alone does not suggest that such attributions are plausible. Defenders of species fell into the trap of pretending that the species notion had resolved in their favor as a result of prior misdirection. The problem was never the specifics of the species notion but the underlying biological facts of incremental variation and intransitivity. I will go on to show why these underlying facts are problematic for attributions of nonderivative value to species.

The Intelligibility of Attributing Nonderivative Value to Species

Although robustly realist and essentialist species notions might be especially well suited to attributions of nonderivative value, I see no obvious reason why mere classifications of convenience cannot pick out categories that bear nonderivative value. Imagine technology that allows us to bypass childhood and reproduce by direct creation of fully formed adults. Assume, plausibly enough, that childhood is a classification of convenience; that does not make nonsense of the claims that childhood has value independently of the value of any particular children and independently of the value of any instrumental, constitutive, or projected value. The fact that criteria of childhood are vague matters of degree does not imply that there is no intelligible distinction between childhood and adulthood or that the distinction cannot track something that marks nonderivative value.

Similarly, even if distinctions between species are based on vague, incremental, and intransitive classifications of convenience or even if there are some tricky borderline cases, we can and do distinguish *O. tshawytscha* from *P. tanasi*. In general, standards for distinctions among animal species yield workable agreements about the extensions of species names. Therefore, despite the fears of people like Rolston and Gorke, attributing nonderivative value to species does not lead to any obvious incoherence, not even in the Darwinian worst-case scenario wherein species are mere classifications of convenience. Having survived the Darwinian challenge to intelligibility, claims of species' nonderivative value must still pass the test of plausibility. It is that test to which we now turn. I will proceed by first considering the task facing defenders of species' nonderivative value and then by expanding on why the prospects for success are dim.

A Well-Directed Response to Darwin's Challenge

Joe LaPorte has pointed out that even if there is one correct species notion, there is no guarantee that actual biological usage will converge on that notion.[70] Extending LaPorte's insight into ethical contexts, my suggestion is that the proper focus of attention for thinking about the ethical status of species was never identifying or defending one particular correct analysis of the species notion. Rather, just as we routinely adopt a technical notion of personhood for ethical discussion, it can be perfectly proper to adopt a technical notion of species for the purposes of ethical argument. However, this is not to say that any stipulation will be satisfactory: the criteria used should be biologically plausible, deliver at least a rough concordance with actual biological nomenclature, and also provide plausible grounds for recognizing nonderivative value. I will argue that biologically plausible criteria that deliver reasonable concordance with actual biological nomenclature do not provide plausible grounds for claiming nonderivative value.

Through consideration of the basis for Darwin's challenge, we have seen that no biologically plausible species notion can insist on sharp and determinate distinctions between species. Rather, they all must recognize differences of degree and intransitivity. Because plausible biological criteria "graduate insensibly," it is all but inevitable that there will be cases that give rise to reasonable disagreement concerning species membership.[71] Rival species notions draw the boundaries differently and, even within any single notion, there are borderline cases. Nevertheless, as there is considerable agreement on the actual usage of species nomenclature in biology, usage among ethicists should fall into line if they want to be understood.

The requirement of approximate concordance with actual biological nomenclature sets a tighter standard than that to which philosophers sometimes consider themselves bound in other areas of ethical discourse, such as when deploying technical notions of personhood. The reasons for this tighter standard are practical. The confusion that often results from the divergence of technical and everyday usage of "person" illustrates one practical reason to either conform to existing biological standards in ethical usage of the term "species" or find a different term. Even if species are categories of convenience, it avoids needless confusion if all parties are employing roughly the same category of convenience. This is especially helpful inasmuch as ethical duties of species preservation drive policy and legislation, such as the Endangered Species Act,[72] and we need to be as clear as we reasonably can about what is being protected.

To attribute nonderivative value to species is, in effect, to claim that the biological criteria that describe species boundaries also describe something that has nonderivative value. To my knowledge, no defender of the nonderivative value of species has attempted to make the case for the moral relevance of these biological criteria. There are factual, biological criteria that can plausibly be linked to normative or evaluative claims. For example, the ability to feel pain is a biologically respectable criterion that may very plausibly be accepted as a criterion for moral standing. Conversely, although being a tongue roller is another thoroughly biological criterion, it would resound with implausibility to claim it as a mark of moral value. I will now review the principal candidate criteria for inclusion in a species notion, and I will argue that they are all more like tongue rolling than the ability to feel pain: they are all implausible grounds for attributions of nonderivative value.

The Implausibility of Attributing Nonderivative Value to Species

We have already seen that clade membership alone does nothing to pick out species in particular, and we have noted that the same goes for similarities of genotype: genes track clades and, though there are genetic markers for species, there are also genetic markers of both higher and lower taxa. To pick out populations that are commonly recognized as species, other criteria are required and the biologically plausible options are, broadly speaking, phenotype, ecological niche, and reproductive isolation. These, then, are the criteria that defenders of the nonderivative value of species would need to defend as the underlying basis for that value. I will argue that these criteria do not provide plausible grounds for the claim that species have nonderivative value. I will consider phenotype and niche first, and will then consider the more promising criterion of reproductive isolation in more detail.

Most people would have more trouble telling a Grant's zebra from a mountain zebra (different species) than they would telling a Great Dane from a Yorkie (same species). This exemplifies the first problem for appeal to phenotypic criteria as a grounding for claims that species have special nonderivative value: namely, that such criteria align poorly with commonly recognized species distinctions. Stripy equids are preserved if we preserve any one of the three remaining zebra species, while the loss of Great Danes would be the loss of a far more distinctive phenotype, although it would not remotely constitute a threat to any species. More generally, there is a problem with specifying the appropriate fineness of grain for identifying distinct phenotypes. At a coarse gain, there are phenotypic commonalties among far broader taxa than that of the species: for example, there are similarities among all primates, among all mammals, and among all animals on the pentadactyl plan. At a fine grain there are phenotypic differences between taxa far narrower than species: for example, there are differences between each of my sisters. Some species notions emphasize the ability to distinguish members of different populations, but that notion is itself highly context relative and lacks any particular link to distinctions at the species level.[73]

Turning to criteria of similarity of ecological niche, similar concerns apply with respect to the poor correspondence between commonly recognized species divisions and issues of graining: there are numerous dung beetle species and members of the species *Homo sapiens* occupy diverse niches on all continents. A further problem is that it is hard to give much credence to the idea that occupying some niche or other is, per se, a mark of nonderivative value. It is easy to think of cases where the occupant of an ecological niche derives value from the ecological services they perform or from some other contribution to wider ecological value. It is far less clear that there is any value at all in the fact that the niche of feeding on the blood of Manhattan condo owners is occupied by various species of bedbug.[74]

The remaining biologically plausible criterion for inclusion in a species notion is that of reproductive isolation. Appeal to the limits of interbreeding is noted as a common theme among rival species concepts.[75] One point in its favor is that, where other criteria do poorly, reproductive isolation does attain reasonable alignment with the species distinctions actually recognized in biological usage. The question is whether appeal to reproductive isolation offers plausible grounds for claims that species have nonderivative value.

We have seen how the biological facts leave no option but to accept Darwin's contention that reproductive isolation "is not a specially acquired or endowed quality, but is incidental on other acquired differences."[76] However, the fact that reproductive isolation does not deliver a sharp distinction between species does not mean that there is no distinction between species. We can accept that criteria for reproductive and evolutionary isolation are rough and ready, open to revision, and open to differences of opinion. We can agree that even the clearest criteria will be subject to vagueness, indeterminacy, and intransitivity. None of this forces us to pretend that we cannot distinguish red and grey jungle fowl. Nevertheless, recognition that reproductive isolation "is incidental on other acquired differences" does indicate that a closer look at the specific mechanisms of reproductive isolation is in order. The red and grey jungle fowl, discussed above, exemplify the range of mechanisms that interact to maintain reproductive isolation. These mechanisms include: geographical separation, anatomical differences, behavioral differences, genetic incompatibilities, and physiological incompatibilities.

Consider the following example of reproductive isolation due to geography: In January of 1790, Fletcher Christian scuttled HMS Bounty, marooning himself, his fellow mutineers, and a number of abducted Tahitian men and women on Pitcairn Island. This act caused the reproductive isolation of a small population of *H. sapiens*, an isolation that continues to a notable degree to this day. To hold that reproductive isolation confers nonderivative value on a population is to hold that the population constituted by the mutineers, their reluctant companions, and their descendants gained nonderivative value with isolation. If such nonderivative value grounds duties of preservation, as it must if it is to do the work demanded by species preservationists, then there would be at least some moral consideration against rescue and reintegration of marooned populations. Because this entailment is implausible, it brings into doubt the attribution of nonderivative value on grounds of reproductive isolation.

Matters become even more uncomfortable as we turn attention to behavioral mechanisms of reproductive isolation and remind ourselves some of the strategies adopted by our own species with the express intent of establishing and maintaining reproductive isolation between different human populations. Some of our legal behavior has provided distressingly explicit examples. Despite the United States Supreme Court's 1967 ruling in the case of Loving v. Virginia,[77] it was only in 2000 that Amendment 667[78] annulled the requirement in Alabama's state constitution that "[t]he legislature shall never pass any law to authorize or legalize any marriage between any white person and a negro, or descendant of a negro."[79] It is one thing to recognize value in the diversity of human populations as a basis for acknowledging the derived constitutive value of populations. It is another to suppose that any aspect of this value relies on behavioral barriers that promote reproductive isolation.

A possible rejoinder is to hold to the view that reproductive isolation does give rise to new nonderivative value, but also acknowledge that this is just one among many morally relevant considerations. Apparently contrary intuitions about exiled populations and legally enforced apartheid might be explained as cases in which the nonderivative value of isolation is utterly swamped by more pressing injustices. This rejoinder puts the case for species' nonderivative value in a bind. On one hand, the weaker the nonderivative value of isolated populations, the more plausible the rejoinder, but the weaker the resulting duties of species protection. On the other hand, suitably robust duties of species preservation require correspondingly robust claims of nonderivative value that should weigh as serious moral considerations even if, on balance, they are deemed to be overridden by competing considerations of justice. It then becomes harder to explain the sense that the alleged value of reproductive isolation along racial lines does not weigh at all.

The general problem is that if reproductive isolation in itself confers nonderivative value, then splitting up a united population, whether by geographical or behavioral means, gives rise to new nonderivative value. Conversely, reuniting formerly isolated populations entails a loss of nonderivative value. For example, a forest fire might leave a formerly contiguous vole population confined to geographically isolated pockets. Each population now has new nonderivative value. Then, over the years the forest regenerates and over many generations the voles reestablish their former range. Nonderivative value is lost as the voles merge back into one contiguous population. At the very least, these cases place considerable doubt on the idea that reproductive isolation is a mark of nonderivative value and shifts the burden of proof back to those who would defend that contention.

I will close this section with a thought experiment designed to put further pressure on the idea that reproductive isolation has any bearing on the nonderivative value of the natural world. Imagine that naturalists voyage to the Sogapalag Islands and find them to be teeming with a fabulously rich diversity of fauna. When they send back their first photographs, the itinerant naturalists have no trouble securing funding for extended field study on the islands. After years of careful study, to everyone's amazement, it emerges that there is no detectable reproductive isolation among any of the island's fauna. Despite widely varying phenotypes, the islands' fauna mate freely with each other and the phenotype of offspring seems uncorrelated with that of the parents: the birds really do mate with the bees and the result is as likely to be fish as fowl. Genetic studies confirm a free genetic flow within what must be considered, at least by the standards of reproductive isolation, to be one species, albeit one that is highly phenotypically heterogeneous.

Compare the Sogapalags to the actual Galapagos Islands, on which a similar diversity of fauna is accompanied by extensive reproductive isolation. If the claim that nonderivative value arises from reproductive isolation is correct, it follows that the Galapagos ecosystem is replete with far greater nonderivative value than that of the Sogapalags. Presumably there is far stronger moral reason to preserve the Galapagos than the Sogapalags. To the extent that this is a perverse result, nonderivative value cannot reside in reproductive isolation.

CONCLUSIONS FOR GENERAL DUTIES OF SPECIES PRESERVATION

In this chapter, I have argued that those who assert that there are general duties of species preservation are faced with a task that cannot be accomplished. I have distinguished derivative from nonderivative value and argued that appeal to the derivative value of species can only yield limited and contingent duties of species preservation. Gorke and others are right to argue that for a duty of species preservation to generalize to all species, each species must have nonderivative value. I have developed a Darwinian challenge to species essentialism: biologically plausible criteria for species distinctions all vary by degrees and lead to vagueness and intransitivity. Some defenders of species have taken the Darwinian view that the species notion is a classification of convenience to be incompatible with claims that species have nonderivative value and have insisted on the correctness of a species notion that appears conducive to their favored view of species value. This is a mistake. I have argued that the actual task was to state biologically plausible criteria for a species notion and to make the case that those criteria demarcate, at least roughly, something of moral value. To my knowledge, no one has attempted this task, and I doubt that it can be achieved. I have given my reasons for doubting that clade or genetics, phenotype, niche, or any of the mechanisms of reproductive isolation are plausible marks of nonderivative value.

Nothing I have said precludes the possibility of making a case for the preservation of a particular species on grounds of derivative value, and I think that such a case can be convincingly made for many species. My argument has been against attribution of nonderivative value to a species simply in virtue of being a species. Since general duties of species preservation require such attributions of nonderivative value, I conclude that there are no such general duties.

Acknowledgements

I acknowledge the support of the University of Delaware Program in Science Ethics and Public Policy. My thanks also to Evan Moore, Fred Schueler, Jeff Jordan, Kate Rogers, Rebecca Walker, Richard Hanley, Sheldon Greene, and Tom Powers for helpful conversation and comment. Particular thanks to Michael Leonard and to the editors of this volume for careful reading and insightful feedback on earlier versions.

NOTES

1. Charles Darwin, *On the Origin of Species by Means of Natural Selection* (London: John Murray, 1859), p. 485.

2. J. Baird Callicott, *In Defense of the Land Ethic: Essays in Environmental Philosophy* (Albany: State University of New York Press, 1989), pp. 133–34 (emphasis in original).

3. Callicott, *In Defense of the Land Ethic*, p. 134.

4. John Stuart Mill, *Utilitarianism* (London: Longmans, Green and Co., 1901), pp. 52–53.

5. Holmes Rolston III, "Values in Nature," *Environmental Ethics* 3 (1981): 113–28, pp. 116–17; Holmes Rolston III, *Environmental Ethics: Duties to and Values in the Natural World* (Philadelphia: Temple University Press, 1988), pp. 3–5.

6. Matthew A. Wilson and John P. Hoehn, "Valuing Environmental Goods and Services Using Benefit Transfer: The State-of-the Art and Science," *Ecological Economics* 60 (2006): 335–42.

7. Rolston, "Values in Nature," pp. 115–16; Rolston, *Environmental Ethics*, pp. 5–7.

8. Rolston, "Values in Nature," pp. 117–18; Rolston, *Environmental Ethics*, pp. 7–8.

9. Allen Carlson and Arnold Berleant, eds., *The Aesthetics of Natural Environments* (Peterborough, Ontario: Broadview Press, 2004); Rolston, *Environmental Ethics*, pp. 10–12.

10. Rolston, "Values in Nature," pp. 118–21.

11. G. E. Moore, *Principia Ethica* (London: Cambridge University Press, 1903), p. 41; Donald H. Regan, "The Value of Rational Nature," *Ethics* 112 (2002): 267–91.

12. Rolston, "Values in Nature," pp. 123–26; Rolston, *Environmental Ethics*, pp. 17–19.

13. Donald H. Regan, "Duties of Preservation," in *The Preservation of Species*, ed. Bryan G Norton (Princeton, N.J.: Princeton University Press, 1986): 195–220.

14. Regan, "Duties of Preservation," p. 201.

15. Regan, "Duties of Preservation," p. 202.

16. Regan, "Duties of Preservation," p. 208.

17. Zygmunt J. B. Plater, "Tiny Fish Big Battle," *Tennessee Bar Journal* 44 (2008): 14–42.

18. Rolston, "Values in Nature"; Martin Gorke, *The Death of Our Planet's Species*, trans. Paricia Nevers (Washington, D.C.: Island Press, 2003), pp. 131–32; Lilly-Marlene Russow, "Why Do Species Matter?" *Environmental Ethics* 3 (1981): 101–12.

19. For example, Lilly-Marlene Russow has claimed that rarity enhances aesthetic value: Russow, "Why Do Species Matter?" p. 111.

20. Gorke, *Death of Our Planet's Species*, pp. 137–80, chap. 22.

21. Rolston, "Values in Nature," p. 118.

22. Rolston, *Environmental Ethics*, p. 23.

23. Regan, "Duties of Preservation," p. 196.

24. Regan, "Duties of Preservation," p. 205.

25. Gorke, *Death of Our Planet's Species*, p. 196 (emphasis in original).

26. Gorke, *Death of Our Planet's Species*, pp. 129–33.

27. Gorke, *Death of Our Planet's Species*, p. 203 (emphasis in original).

28. Gorke, *Death of Our Planet's Species*, p. 203.

29. Bertrand Russell, *Last Philosophical Testament: 1943–1968*, ed. John G. Slater and Peter Köllner (London: Routledge, 1997), p. 51.

30. Gorke, *Death of Our Planet's Species*, p. 129.

31. Gorke, *Death of Our Planet's Species*, p. 9.

32. Tom L. Beauchamp and James F. Childress, *Principles of Biomedical Ethics* (New York: Oxford University Press, 2009), p. 3.

33. Saul A. Kripke, *Naming and Necessity* (Cambridge, Mass.: Harvard University Press, 1980), p. 42.

34. Beauchamp and Childress, *Principles of Biomedical Ethics*, p. 3.

35. Beauchamp and Childress, *Principles of Biomedical Ethics*, p. 387.

36. Kevin de Queiroz, "Ernst Mayr and the Modern Concept of Species," *Proceedings of the National Academy of Sciences* 102 (2005): 6600–7, here p. 6600.

37. Darwin, *On the Origin of Species*, p. 485.

38. Rolston, *Environmental Ethics*, pp. 134–35.

39. Darwin, *On the Origin of Species*, chap. 8.

40. Darwin, *On the Origin of Species*, p. 245.

41. Ernst Mayr, *Systematics and the Origin of Species, from the Viewpoint of a Zoologist* (Cambridge, Mass.: Harvard University Press, 1942), p. 120.

42. Darwin, *On the Origin of Species*, p. 245.

43. Darwin, *On the Origin of Species*, p. 248.

44. Darwin, *On the Origin of Species*, p. 252.

45. G. Victor Morejohn, "Breakdown of Isolation Mechanisms in Two Species of Captive Junglefowl (*Gallus gallus* and *Gallus sonneratii*)," *Evolution* 22 (1968): 576–82.

46. Morejohn, "Breakdown of Isolation Mechanisms," p. 577.

47. Morejohn, "Breakdown of Isolation Mechanisms," p. 581.

48. Morejohn, "Breakdown of Isolation Mechanisms," p. 581.

49. Domestic chickens get their yellow skin gene from *G. sonneratii*; Jonas Eriksson, Greger Larson, Ulrika Gunnarsson, Bertrand Bed'hom, Michele Tixier-Boichard, Lina Strmstedt, Dominic Wright, Annemieke Jungerius, Addie Vereijken, Ettore Randi, Per Jensen, and Leif Andersson, "Identification of the *Yellow Skin* Gene Reveals a Hybrid Origin of the Domestic Chicken," *PLoS Genetics* 4 (2008): e1000010.

50. Craig Moritz, Christopher J. Schneider, and David B. Wake, "Evolutionary Relationships within the *Ensatina Eschscholtzii* Complex Confirm the Ring Species Interpretation," *Systematic Biology* 41 (1992): 273–91.

51. Darwin, *On the Origin of Species*, p. 485.

52. Joseph LaPorte, *Natural Kinds and Conceptual Change* (Cambridge: Cambridge University Press, 2004), pp. 11–12.

53. The "non-trivial" caveat excludes such trivially essential properties as "being a member of species *X*."

54. Linnaeus published the first edition of his *Systema Naturae* in 1735. *Systema Naturae* established the binomial, genus-species nomenclature still in use today. The work expanded enormously over the course of its thirteen editions, the last of which was published posthumously in 1767. Linnaeus changed his name from Carl Linnaeus to Carl von Linné following his ennoblement and the Latinized form of his name under which he published changed accordingly in later editions of *Systema Naturae*.

55. Robert Carroll and Stephen Prickett, eds., *The Bible: Authorized King James Version* (Oxford: Oxford University Press, 2008), Genesis 1:24–25.

56. Callicott gives an interesting account of grounds for concern for other species in the book of Genesis: Callicott, *In Defense of the Land Ethic*, pp. 136–39.

57. Regan, "Duties of Preservation," p. 210; H. Verhoog, "The Concept of Intrinsic Value and Transgenic Animals," *Journal of Agricultural and Environmental Ethics* 5 (1992): 147–60.

58. Wilfrid Blunt, *Linnaeus: The Compleat Naturalist* (London: Frances Lincoln, 2004), p. 184; Gordon McGregor Reid, "Carolus Linnaeus (1707–1778): His Life, Philosophy and Science and Its Relationship to Modern Biology and Medicine," *Taxon* 58 (2009): 18–31, here p. 23.

59. Kripke, *Naming and Necessity*, p. 136.

60. Kripke, *Naming and Necessity*, pp. 119–21.

61. Darwin, *On the Origin of Species*, p. 485.

62. Darwin, *On the Origin of Species*, p. 44.

63. Alex Wellington, "Endangered Species Policy," in *Canadian Issues in Environmental Ethics*, ed. Alex Wellington, Allan Greenbaum, and Wesley Cragg (Peterborough, Ontario: Broadview Press, 1997): 189–218, here p. 191.

64. Callicott, *In Defense of the Land Ethic*, p. 135.

65. Rolston, *Environmental Ethics*, p. 135.

66. Gorke, *Death of Our Planet's Species*, p. 301 (emphasis in original).

67. Gorke, *Death of Our Planet's Species*, p. 302.

68. Callicott, *In Defense of the Land Ethic*, p. 135.

69. Gorke, *Death of Our Planet's Species*, p. 302.

70. LaPorte, *Natural Kinds and Conceptual Change*, p. 76.

71. David M. Watson, "Diagnosable Versus Distinct: Evaluating Species Limits in Birds," *BioScience* 55 (2005): 60–68.

72. "Endangered Species Act" (1973).

73. Watson, "Diagnosable Versus Distinct."

74. Teri Karush Rogers, "Buying and Selling in Bedbug City," *New York Times*, August 23, 2009, real estate sec., p. 1.

75. De Queiroz, "Ernst Mayr and the Modern Concept of Species," p. 6601.

76. Darwin, *On the Origin of Species*, p. 245.

77. *Loving v. Virginia*, 388 U.S. 1 (1967).

78. Constitution of Alabama of 1901, Amendment 667.

79. Constitution of Alabama of 1901, Article IV, Section 102.

SUGGESTED READING

Articles and Books

CALLICOTT, J. BAIRD. *In Defense of the Land Ethic: Essays in Environmental Philosophy*. Albany: State University of New York Press, 1989.

DARWIN, CHARLES. *On the Origin of Species by Means of Natural Selection*. London: John Murray, 1859. Facsimile available on http://www.esp.org/books/darwin/origin/facsimile/.

GORKE, MARTIN. *The Death of Our Planet's Species*. Translated by Paricia Nevers. Washington, D.C.: Island Press, 2003.

REGAN, DONALD H. "Duties of Preservation." In *The Preservation of Species*, edited by Bryan G. Norton, 195–220. Princeton, N.J.: Princeton University Press, 1986.

———. "The Value of Rational Nature." *Ethics* 112 (2002): 267–91.

ROLSTON III, HOLMES. *Environmental Ethics: Duties to and Values in the Natural World*. Philadelphia: Temple University Press, 1988.

SOBER, ELLIOTT. "Philosophical Problems for Environmentalism." In *The Preservation of Species*, edited by Bryan G. Norton, 173–94. Princeton, N.J.: Princeton University Press, 1986.

Reference and Web

ERESHEFSKY, MARC. "Species." In *The Stanford Encyclopedia of Philosophy*, edited by Edward N. Zalta (Summer 2010), available at http://plato.stanford.edu/entries/species/ (accessed July 29, 2010).

WILSON, ROBERT A., ed. *Species: New Interdisciplinary Essays*. Cambridge, Mass.: MIT Press, 1999.

International Commission on Zoological Nomenclature. Webpage available at http://iczn.org/.

CHAPTER 21

....

ON THE NATURE
OF SPECIES AND THE
MORAL SIGNIFICANCE
OF THEIR EXTINCTION

....

RUSSELL POWELL

We owe our existence to an unbroken chain of reproduction that began with the inception of life some 3.5 billion years ago. Rejoicing in the fortuity of our genealogy, however, can obscure an equally salient but far less auspicious pattern in the history of life—namely, the extinction of nearly every species that has ever existed. There have been geological moments, and one in particular about 250 million years ago, when complex life itself teetered on the brink of annihilation. Yet the fragility of animal life per se pales in comparison to that of individual taxa.[1] The mean duration of species is around four million years—an incomprehensibly vast interval to the human mind, but a mere pittance geologically speaking. And as we will see, species do not seem to get any better over time at not going extinct. What is more, if it were not for the most recent mass extinction, mammals might still be relegated to a small, nocturnal existence in the ecological shadow of dinosauria. Why, then, should we think that causing or permitting the extinction of species is a bad thing?

In claiming that it is prima facie wrong for humans to contribute to or fail to prevent the extinction of species, environmental ethicists have often relied on the intuition that species qua species have moral value. This intuitively attractive idea has been notoriously difficult to justify, however. And unlike moral judgments about fairness, lying, and torturing the innocent, intuitions about the value of species are not sufficiently robust across persons and cultures to serve as the lynchpin for an ethical framework, especially one that aims to influence public policy. Another reason to avoid hanging an ethical theory of species conservation on untutored moral intuitions is that folk judgments about biological entities have a less-than-stellar track record. As recently as the eighteenth century, it was still commonly believed that organisms are composed of various ratios of the four elements identified by Aristotle, that living things are driven by vital forces, and that species reflect eternal, immutable essences around which biological variation gravitates.

While commonsense intuitions have an important role to play in philosophical ethics, many folk judgments in the biological realm, such as our attitudes toward genetically modified organisms or invasive species, are often premised on certain presuppositions about the causal structure of the living world that may or may not hold up to theoretical scrutiny.[2] Just as our considered moral-status judgments about comatose patients should be informed by neuroscientific findings regarding the functional specialization of different regions of the human brain,[3] so too should environmental ethics be constrained by the ontological landscape that is sketched by the evolutionary and ecological sciences. This is not to imply that the relationship between descriptive biology and normative environmental ethics is a unilateral one; ethicists can draw the attention of scientists to the types of entities or causal relations that are deemed to be of moral significance, and scientists in turn can go out and look for these objects or properties in the blooming, buzzing confusion of the biological world.

In this chapter I will consider several moral justifications for the conservation of species in light of what we know about the ontology and evolution of species and their interrelations.[4] The discussion will be structured as follows. In section 1, I consider the nature of species and their role in evolutionary theory. Although there are many species concepts available in contemporary evolutionary biology, I believe that for the purposes of ethical analysis there is sufficient overlap between these to avoid systemic conflicts in our individuation of species. The dominant view in biology and the philosophy of science, and the one that I support, is that species should be thought of as individual lineages, rather than as atemporal, natural kind classes. That we regard species as concrete individuals that evolve in space and time has important ramifications for the value that we might ascribe to them.

Nevertheless, in section 2 I show that a species is not the sort of individual that exhibits properties thought to create morally relevant interests, such as sentience or goal-directedness, and I conclude therefore that species are not promising candidates for intrinsic moral value. I move on to consider theories that regard species as valuable not as moral patients, but because of their relation to morally considerable beings. I suggest a number of ways that species and even higher taxa could have both final and instrumental value that is not reducible to their constituent parts, focusing on informational, evolutionary, and ecological properties and the organizational levels at which they are manifest.

In section 3, I examine metaphysical and epistemic arguments against prioritizing species for conservation, including those grounded in the "seamless web" and "delicate balance of nature" metaphors that often motivate environmental law and policy. I find these arguments unpersuasive on both theoretical and empirical grounds.

Finally, in the conclusion, I consider whether there is a morally important difference between human-caused extinctions and those that result from "natural" evolutionary processes. I suggest that the degree of "badness" that is added to some state of affairs because a culpable moral agent is implicated is minimal, and thus should not significantly affect our conservation priorities.

1. The Ontology of Species

What Is a Species?

The aim of this essay is to determine whether species are proper objects of moral concern. To make this determination, we first need to be clear on the ontological status of species, which in turn requires that we examine the theoretical role that the species concept plays in contemporary evolutionary science.

As far as we can tell, all life on Earth shares a common ancestor. Despite this evolutionary continuity, biological variation is not smoothly distributed in space and time. There are fairly discrete clusters of variation at all levels of the nested biological hierarchy, a clumping of form that generally reflects evolutionary relationships and which taxonomic classification intends to capture. Among taxonomic categories (see note 1), the most theoretically fundamental is the species.

Nearly all biologists and philosophers of science agree that species are comprised of geographically distributed populations of organisms with varying levels of gene exchange. Yet there is no single, widely accepted definition of the term "species." Probably the most influential formulation is the "biological species concept," according to which a species is the maximally inclusive set of potentially interbreeding organisms that are reproductively isolated from other such groups.[5] This definition is far from canonical, however, and the species concept remains highly contested in biology and philosophy.[6]

Indeed, the last few decades have witnessed a proliferation of species concepts, sometimes complementary but often conflicting, with different groups of biologists attuned to properties that are of particular relevance to their subfield of study, such as evolutionary history, morphology, inter-fertility, or ecological role, to name a few. This conceptual and methodological discord has led many authors to embrace some form of theoretical pluralism regarding the species concept.[7] This poses a problem: if biologists and philosophers of science cannot agree on what the term "species" refers to, or which groupings of organisms or populations it picks out, then how can ethicists talk meaningfully about preserving such entities?

Fortunately, the space of disagreement between alternative accounts of species is small enough to avoid prejudicing ethical discussions concerning the moral value of species. The reason for this optimism is that regardless of which properties they deem relevant to or necessary for species membership, all contemporary species concepts agree on the following key point: species are separately evolving *metapopulations*. That is, they are sets of spatially distinct subpopulations of sexually reproducing organisms that extend over time to form *lineages* (ancestor-descendent sequences of breeding populations) whose boundaries are determined in part by patterns of interbreeding.[8]

Conflicting species judgments, to the extent they arise, usually occur in relation to attempts to taxonomically partition "evolutionary gray zones." These are situations in which *potentially* speciating lineages have yet to exhibit total reproductive isolation or significant genetic, morphological or ecological divergence.[9] A potentially speciating lineage that has begun to diverge from its parent population can be reabsorbed by the latter if there is sufficient gene flow (interbreeding) between them. The upshot is that judgments as to whether speciation has occurred are inherently *retrospective*, and in hindsight there will be nearly unanimous agreement as to whether a speciation event has taken place. So apart from these relatively rare and evolutionarily short-lived "gray zone" cases, scientists and environmental philosophers should encounter little practical disagreement over how to partition species for the purposes of biological and ethical analysis, respectively.[10]

Thinking of species as separately evolving meta-populations, rather than as classes in which membership is determined by a set of essential properties, is one of the central insights of the Modern Evolutionary Synthesis.[11] The shift in biology away from essentialism and toward "population thinking" built on the Darwinian reinterpretation of variation as a central, irreducible cause in evolution rather than indicative of some unseen but ontologically primary essence.[12] As David Hull puts it, the inability to differentiate species by sets of necessary and sufficient conditions "follows from evolutionary theory just as surely as quantum indeterminacy follows from quantum theory."[13] Rarely will there be traits that are both universal and unique to a given species, and even if there were such traits, they would be short-lived because the very evolutionary processes that are responsible for their fixation will inevitably produce exceptions to the rule.[14]

Consequently, the received view in biology and the philosophy of science is that species and other taxa are more productively thought of as individual lineages, rather than immutable natural kinds.[15] That is to say, species are individuals located in space and time with organisms as their constituent *parts*, rather than atemporal sets with organisms as their *members*. Because species are not natural kind classes, they originate, change, and go extinct much like individual organisms do. While natural kind classes may figure into laws of nature, they are not causal in the sense that they cannot make anything happen. By contrast, species as independently evolving meta-populations are not "mere" theoretical constructs. They are the furniture of the biological universe—real, causal entities whose ontological status in evolutionary biology is on par with other fundamental units of organization, such as organisms, cells and genes.[16]

If species are individuals with component parts, rather than abstract categories with concrete instances, this opens the metaphysical door to ethical value attaching to species over and above their constituent organisms, just as it does to organisms over and above their constituent cells. On the other hand, if species are atemporal classes, then their existence would not be predicated on the existence of any physical object, and consequently they could not go extinct or be harmed. Moral duties as typically conceived are owed not to abstract *properties* such as "pleasure" or "sentience" (or even their concrete instances), but rather to the *individuals* that possess those properties. What makes something a candidate for ethical conservation is not that it is a category, such as "castle," which can only be preserved in the mind, but rather that it is an individual that we deem valuable, such as "Edinburgh Castle." If species were classes, the relevant moral question would be whether we have an obligation to maintain non-empty sets rather than to preserve species themselves. Unlike species-as-individuals, species-as-classes do not exist in space and time and hence cannot be preserved—at least not in the way that conservation biologists typically go about preserving them (such as by protecting or restoring habitats or breeding populations).

Another reason to think that species are individuals, particularly for the purposes of ethical preservation, is that arguments in favor of conservation often hinge on the contingent generalization that extinction is irreversible. To the extent that the permanence of extinction serves as a central motivation for preservation efforts, this casts further doubt on the species-as-classes view. Since classes can be empty at one time and non-empty at another, it is difficult to see how the concept of extinction would apply to them. Consider the quintessential natural kind class *gold*. Gold is a class of heavy element with a particular atomic structure; this class was empty in the early stages of the universe, because the production of gold atoms had to await the formation of high-mass stars where they would be produced by nuclear fusion and dispersed by super novae. As an atemporal class, gold cannot go extinct, even if at any given moment there are no token gold atoms in the universe. Given the right conditions, gold atoms will emerge again and again with the same physical signature and intrinsic properties that determine membership in the class. The question, then, is whether we have any reason to believe that species resemble the heavy elements in this respect. Given the right *ecological* conditions, will token instances of a species type emerge predictably and repeatedly, just like individual atoms of gold?

The standard view in the philosophy of biology is that due to the nature of the evolutionary process, there are no invariant or even robust generalizations about the kinds of organisms and adaptive features that will evolve—that is to say, there are no laws of fitness.[17] Indeed, part of the supposed tragedy of species extinction is that some unique and historically contingent *thing* has vanished, presumably forever.[18] It would seem, then, that the only viable way to consider species as atemporal sets for the purposes of conservation would be to think of them as classes that are unlikely to ever be repopulated once they are emptied. The trouble with this approach is that the inherent contingency of the evolutionary process would lead us to posit an unmanageable disjunction of potential species-as-classes; and given that we expect natural kind classes to do theoretical work by figuring into laws of nature (which, among other things, are supposed to exhibit a wide range of invariance and support counterfactuals), such a result would be methodologically dubious.

Species as Pseudo-Individuals

Even if species are better thought of as individuals rather than classes, this does not imply that they are the *kind* of individual that might be amenable to moral consideration. Although individuals are never classes, an entity need not be a true individual in the biological sense—or in the sense relevant to moral analysis—simply because it is not a class. In order to constitute a biological individual, an entity must possess some or all of the following properties: internal homeostatic mechanisms, differentiated parts, functional autonomy, developmental life cycle, reproduction, and metabolism.[19] Even more broadly, to be an individual in the biologically relevant sense, an entity must maintain some degree of integration among its parts and it should respond as a single unit to environmental perturbations.[20] There are certainly possible worlds in which species exhibit some or all of these characteristics, but in the actual evolutionary world, species rarely if ever exhibit properties like functional integration and differentiation, cohesion, homeostatic mechanisms, goal-direction, reproductive life cycle, and so forth.

First, species do not replicate or reproduce like other complex living things that go through a unicellular bottleneck (or single-celled stage) in each generation. Instead, their "parts" (sub-populations) will occasionally give rise to new species in a manner that is analogous to "budding" in asexual organisms. Even if species may be said to reproduce in some rudimentary sense, unlike organisms they do not exhibit a "life cycle," or an intergenerationally replicable series of stages through which they pass, beginning and ending with same event.[21] Individual organisms are born, grow, reproduce, and die. Species originate, fluctuate in number and character traits, and go extinct. The history of individual species is not a linear story of developmental transformation, any more than the history of life is a ladder of progress climbing from protozoa to people.[22] Because species do not progress through stages of ontogeny in the way that organisms do, they do not form true reproductive lineages.

The analogy between organisms and species further breaks down because the latter do not clearly exhibit characteristics that befit the categories of health and disease, and hence it is not clear that they are capable of being harmed. Species qua species are not obviously goal-directed, they do not respond as a coordinated whole to perturbations, and they exhibit few if any biological functions. There are two dominant approaches to function in biology and the philosophy of science: "selected effects" and "causal role." The former refers to traits with a history of selection for their effects, while the latter describes features that play a current causal role in a highly integrated biological system.[23] Organisms are paradigmatic individuals because they exhibit functions in both senses of the term: they are functionally autonomous wholes that respond as a single, integrated unit to environmental perturbations, and they are comprised of a complex suite of phenotypic traits that have been constructed by the cumulative operation of selection for their effects. Not so with species. Species rarely exhibit irreducible species-level adaptations[24]; the few traits that have been proposed, such as those relating to population structure or geographic range,[25] are not complex, cumulative adaptations, and do not entail the homeostatic, goal-directed functions that give concepts like disease and harm their normative content.

Furthermore, the constituent parts of species (populations, groups, organisms etc.) are only loosely coordinated by gene exchange. Not only are conspecifics not *physically* connected to one another in the way that cells are adhesively joined, but neither are they *behaviorally coordinated* at the population level in the way that caste members of certain insect colonies (such as those of ants, bees, and wasps) tend to be. Individual members of species are not differentiated according to their functional role, unlike cells, tissues, and organs within the organism, or polymorphic castes within the insect colony. Finally, species members are often functionally opposed to one another. In fact, the competition within species is usually more intense than that between them—a degree of antagonism between parts that is not present between cells of the organism and other entities that have evolved mechanisms for enforcing cooperation and blocking conflict among their subcomponents. These mechanisms enable cell lines within organisms, and caste members within insect colonies, to sacrifice their own immediate reproductive fitness in order to partake in a coordinated, collaborative endeavor at a higher level of biological organization. Such mechanisms do not obtain at the species level.

To sum up the discussion thus far: species are neither abstract classes nor full-blown biological individuals. Instead, they may be roughly characterized as follows: independently evolving, spatiotemporally restricted meta-population lineages that are poorly integrated, weakly cohesive, non-goal-directed, nondevelopmental, and composed of largely nondifferentiated, nonfunctional, and noncooperative parts. Let us now consider the moral value that such an entity might reasonably be said to possess.

2. THE VALUE OF SPECIES

Do Species Have Intrinsic Value?

For an entity to be "morally considerable," it must have interests that it would be prima facie wrong to impede, where this wrongness stems from the intrinsic value of the being in question and not from any indirect effects that being may have on other beings.[26] Much of the discussion concerning moral considerability has centered around identifying the crucial property or set of properties that make a being the kind of being that can not only be harmed, but also *wronged*. Many candidate capacities have been advanced, such as personhood,[27] intentionality (of the first or second order), being the "subject of a life,"[28] sentience,[29] and goal-directedness,[30] to name a few.

As meta-populations only loosely integrated via patterns of gene exchange, it is difficult to imagine how species might instantiate any of the aforementioned properties, even the most zoologically inclusive ones such as goal-directness. Goal-directedness is one of the defining features of life. It refers to the ability to maintain physiological, morphological, or behavioral homeostasis across a wide range of environmental perturbations.[31] Although some ethicists maintain that goal-directedness is sufficient to create interests that bring a being within the ambit of moral consideration,[32] most moral philosophers hold that a more substantial *cognitive* property, such as sentience or personhood, is necessary and sufficient to create morally relevant interests—interests that demand equal consideration. For the present purposes, however, we will simply assume that goal-directedness is sufficient to create interests that make a being the kind of being that can not only be harmed in the biological sense, but also wronged in the moral sense.

Goal-directedness is such a fundamental property of organisms that it brings invertebrate animals, plants, protozoa, bacteria, and all other non-sentient but incontrovertibly living things into the moral fold. Yet it does little to make the case for species, which lack the coordination, cohesion, differentiation, and cumulative adaptations necessary for goal-directed behavior. If species qua species do not even exhibit goal-direction, let alone more complex functional properties like sentience, it is difficult to see how they might have morally relevant interests.[33]

Nevertheless, merely because an entity is not a classical individual does not mean that it lacks moral status. There is a growing tradition in political philosophy to recognize group rights, or rights possessed by a group as a group rather than by its members severally.[34] Because these rights are those of the group, they engender duties to the group entity, rather than or in addition to obligations owed to its members severally. Can the moral patient-hood of species be derived from theories of group rights? The argument for group rights proceeds as follows: if individuals have rights because they possess certain emergent properties (such as intentionality or sentience), then groups may have the same rights insofar as they too possess those emergent properties. Mere aggregates of individuals, such as crowds, do not have irreducible moral worth, because they do not exhibit any of the relevant properties at the level of the ensemble, even though the individuals that comprise them do possess those properties. By contrast, groups that have an integrated internal structure and division of labor, and which act pursuant to a collective purpose, may be said to have interests in themselves.[35] Yet this merely brings us back to where we started: namely, to the conclusion that species are not integrated, differentiated, purposeful, goal-directed entities, and hence that they are more like crowds of conspecifics than business firms or insect colonies.

There is an uncontroversial sense in which damage done to a species could entail a wrong to *individual organisms* of that species, assuming these individuals possess morally relevant characteristics. For example, habitat destruction could cause greater total suffering to individual organisms through starvation and exposure than would be the case if habitats were preserved. However, if the value of species simply reduces to that of their constituent organisms, then species have no normative value *as such*, and hence they are not proper objects of moral concern.

One common move at this point is to shift the locus of moral analysis away from species, and toward the entities forged by the complex interrelations between them, such as communities and ecosystems. A discussion of the ontological properties of ecosystems and their moral significance is beyond the scope of this essay. In section 3, however, I will consider the physical properties of communities insofar as they speak to the implications of species extinction. Suffice it to say that higher-level ecological entities have been notoriously difficult to delimit, leading many researchers to question their ontological and theoretical status.[36] And to the extent that ecosystems and communities can be individuated as organized systems, they have been shown to lack even the most basic properties of living things, such as goal-directedness.[37]

The Extrinsic Noninstrumental Value of Natural History

Some authors have argued that species have intrinsic value by virtue of the fact that humans value them for their own sake and not for any instrumental role they play in bringing about other valuable ends.[38] Others have rejected the notion that anthropocentric value can be "intrinsic."[39] Contrasting intrinsic value with instrumental value is misleading, however, for, as Christine Korsgaard points out, some values are both "extrinsic" (grounded in relational properties) and "final" in the sense that they are valuable as ends in themselves.[40] Whether we choose to call it intrinsic or extrinsic final value, if species qua species are valuable in the same way that historical artifacts are valuable, then this would provide a non-instrumental, normative basis for species conservation that does not rely on them being morally considerable.

Artifacts, architecture, and other objects of historical significance are commonly thought to have extrinsic final value, to which the proliferation of tourism-generating UNESCO World Heritage sites can attest. If *human* history has final value to both historians and people generally, it is not much of a stretch to conclude that *natural* history might have final value as well. Many protected World Heritage sites are in fact "natural" sites, such as mountain ranges, barrier reefs, and fossil deposits. Artifacts and other objects of material culture are considered valuable in part because of the information they contain about the past. Perhaps we can draw the more general conclusion that an object may be valuable simply by virtue of its possessing information.

Advancing a strong version of this argument, Luciano Floridi concludes that all natural objects, whether living or nonliving, have moral value qua "information objects," and from this he derives negative duties not to cause, and positive duties to prevent, the entropy of information.[41] One problem with this view is that the information object (the set of properties and characteristics of an entity) is an abstract category, and thus it cannot ground moral status or generate duties in others.[42] As we saw above in the context of species, because atemporal classes do not have interests, they cannot have intrinsic moral value and thus they cannot be wronged. A more plausible version of Floridi's thesis would be to claim that information is of enormous value *to humans* for both extrinsic final and instrumental reasons, and hence that objects containing information are valuable, even if they are not morally considerable.

What sorts of information do species possess? Information contained in a species includes all of its physical, anatomical, genetic, developmental, physiological, cognitive, behavioral, populational, ecological, evolutionary, and historical properties. Species as evolving lineages are inherently historical entities, accumulating vast amounts of information (in the "correlational" sense of the term) about their environment over evolutionary time, leading to exquisite adaptive solutions to countless ecological design problems. This epistemic feat is accomplished not with the benefit of foresight and intentionality, but through the cumulative operation of natural selection acting on a genetic channel of inheritance. Because natural selection tends to modify complex organisms by building incrementally "on top of" more primitive (evolutionarily older) structures, lineages will tend to accumulate or exhibit a net gain in information over generational time.

This would seem to be true for all levels of the nested taxonomic hierarchy. Higher taxa,[43] such as phyla, classes, and orders, will tend to accumulate information as their constituent species diversify into a wide range of morphologies and ecological roles. Some of this natural historical information will be valuable because of its immediate use to humans[44]; some of it will be valuable because of its potential but currently unforeseeable use to humans in the future; while still other bits will have no present or future utility, but will nonetheless be interesting and hence valuable in their own right. For many people, coming to appreciate the extraordinary evolutionary journey of a species is meaningful in and of itself, even if it does not lead to other valuable information or ends. A deeper understanding of natural history can lead to important self-knowledge by situating the human species in the vastness of evolutionary space and time, and illuminating the connections of humankind to other beings and its "embeddedness" in the great tree of life.

The Instrumental Ecological Value of Species

Many philosophers and policy makers are rightfully uncomfortable with the idea of resting a broader environmental ethic on final-value judgments that are not deeply rooted in human moral psychology and which might be held hostage to the ebb and flow of culture. Thus, the dominant approach to species conservation, especially in the arena of public policy, is a form of anthropocentric instrumentalism that links the extinction of species (and other indicators of biodiversity) to identifiable ecological harms that might befall morally considerable creatures, including and especially human beings.

Species have instrumental value insofar as they provide various resources such as food, medicine, fuel, energy, raw materials, and tourism.[45] Species also offer so-called "ecosystem services," such as the regulation of atmospheric oxygen, carbon dioxide and moisture levels, pollination, seed dispersal, waste decomposition, nutrient cycling, and so forth, the economic value of which is estimated to exceed the world's total gross domestic product.[46] In addition, speciel diversity provides what have been termed "evosystem services,"[47] which include the capacity for adaptive evolutionary change in the face of changing environments, as well as the ability to produce novel beneficial variations. Evosystem services may grow increasingly important as climate change and other anthropogenic alterations of the environment impose higher levels of stress on communities and ecosystems. Together, ecosystem and evosystem services forge a strong link between biodiversity and human wellbeing.

Knowledge of the ecological interactions and energy flow between species enables biologists to predict different harms from the extinction of different lineages. This, in turn, offers a basis for prioritization or "triage" in the distribution of limited conservation resources.[48] Species may be ranked based on their expected value or the disvalue of their extinction, and this can then be weighed against conflicting moral interests such as economic development and the alleviation of poverty. One implication of adopting a wholly instrumentalist approach to species value is that critically endangered species will often be assigned a low preservation priority, because lineages on the brink of extinction are typically small in number and will rarely be major ecological players let alone the lynchpins of global ecosystems. Assuming comparable extrinsic final values, we will usually get more moral bang for the buck by allocating resources to "threatened" rather than "critically endangered" species.[49]

Furthermore, measuring the success of preservation efforts in terms of the total number of species preserved can obscure important ecological differences between species, and hence the harms that are likely to flow from their extinction. Identical magnitudes of species loss can have profoundly different ecological and evolutionary consequences, depending on their *trophic distribution*, or their position in a food web. The elimination of primary producers, for instance, can have far more significant ramifications for global ecology than the extinction of lineages at higher trophic levels—not only with respect to its immediate consequences, but also with regard to the pace and pattern of faunal recovery in its aftermath.[50] Thus, it makes little sense to speak of the instrumental value of species in the abstract. In order to assess the disvalue that is likely to flow from the extinction of any given species, we need to advert to a detailed map of the causal ecological structure in which the species is embedded, and its relation to other things that we independently value.

Is the Value of Species Irreducible?

Thus far we have done little to show that the extrinsic value of species does not merely reduce to that of their constituent parts, such as organisms or populations. Our answer to the question of reduction will determine whether species qua species have normative value, or whether "species value" is really just a placeholder for that of organisms or populations. My contention is that the extinction of a species will tend to result in a greater loss of both natural history information and ecological/evolutionary utility than that which flows from the death of any populational subset alone.[51]

The vast majority of information about a species is accumulated and preserved at the meta-population level, where it remains distributed until the last viable population dies off and the species goes extinct. Although subpopulations and individual organisms contain unique informational signatures and thus may be uniquely valuable, the loss of information that corresponds to the death of *all* individuals of a species is greater than the loss of information that attends to the death of any non-zero fraction of the total population. Due to gene exchange within and between populations of a species, the informational (including genetic and phenotypic) differences between conspecifics are relatively small; by contrast, the differences *between species* can be anywhere from substantial to vast, depending on the age of their last common ancestor. As a heuristic, we might say that the greater the evolutionary distance between an extant species and its closest living taxon (for humans, this would be chimpanzees), the more information that will be lost by the extinction of the former. It follows that even a tiny, reproductively viable population can preserve the majority of valuable information contained in a species—information that would be irreversibly lost if the species were to go extinct. In other words, the informational and ecological properties of a species do not decline in linear fashion with the depletion of its constituent populations; instead, they drop off steeply at the species extinction event.

There remains a problem, however. If given their high relative similarity, each organism of a species possesses the overwhelming majority of the information that is contained at the species level, then is it not the case that the information lost due to extinction simply reduces to the loss of the last *individual* of a species? For that matter, why could not the information of an entire species be retained by preserving the DNA of one of its individuals, storing a formaldehyde-pickled specimen, or by maintaining a captive breeding population? The answer is that in each of these cases, only a fraction of the full informational value of a species would be retained, especially when it comes to complex animals. First, the phenotypes of complex organisms cannot be reconstructed on the basis of DNA information alone, because they are the result of a complex interaction of genetic and non-genetic factors in development. Second, many traits are polymorphic at the meta-population level, sensitive to developmental context, non-genetically inherited (transmitted in a social learning environment), and contingent on the frequency of similar traits in conspecifics. These and other so-called *epigenetic* factors cannot be "read off" of DNA and are not preserved by maintaining nonliving specimens of the species. For the same reasons, keeping a few solitary living individuals or even a captive breeding population does not preserve the dynamic evolutionary and developmental environment that shapes the distribution of complex traits in the wild. Nor does it preserve the ecological connections that the species has forged over time with other co-evolving lineages in its food web.

Like its informational value, the ecological/evolutionary utility of a species does not reduce to that of its component parts. There is a stark contrast between the irreversible effects of species extinction on the one hand, and the remediable nature of demographic depletion on the other. Species can recover from heavy losses and reassume their ecologically valuable role, so long as they maintain a minimally viable population that has the potential to rebound and take on a more prominent role in global ecology.[52] Extinction, by contrast, is irreparable. While it is conceivable that a breeding population of an otherwise extinct species could be "resurrected" from DNA material and reintroduced into the wild, because of the context sensitivity of development (discussed above) and the dynamic nature of ecology, there is no guarantee that the resurrected species would perform the same ecological role and hence possess the same instrumental value as its predecessor. While it is possible that another species could eventually evolve to fill the empty niche that was vacated by a given extinction (just as birds filled the aerial niches vacated by the extinction of the pterosaurs in the end-Cretaceous), this could take millions of years, during which time much ecological damage may be wreaked.

In sum, extinction is a species-level phenomenon that raises unique ethical issues over and above those that are associated with the depletion of its constituent organisms and populations. If this reasoning is correct, then it implies that talk of species value does not merely serve as a placeholder for the value of organisms and populations. Species are legitimate objects of value in their own right.

The Value of Higher Taxa

I have suggested that the extrinsic final and instrumental value of species is irreducible because information is stored and ecological/evolutionary utility is located uniquely at the meta-population level. These properties and their associated values are permanently lost with the extinction of species. If this analysis is sound, then a further case can be made for the irreducible value of *higher* taxonomic categories that represent larger swaths of the tree of life.

The effect of extinction on higher taxa is a much-neglected topic in discussions of biological conservation. This is probably due to the fact that in contrast to the general realism about species, higher taxa (such as genera, families, orders, classes, and phyla) are often thought to be carved up arbitrarily rather than at nature's joints. However, recognizing the extrinsic value of large-scale evolutionary history does not require accepting the reality of higher taxa. It only presupposes "tree-thinking," which is essentially the phylogenetic counterpart of population thinking—that is, the idea that species are not members of a classificatory set, but rather interconnected and diverging segments of an evolutionary branching sequence.[53] We can avoid problems associated with the individuation of higher taxa by focusing on the effects of extinction on *phylogenetic disparity*, or the "width" of the evolutionary tree. In order to show that phylogenetic disparity has unique evolutionary/ecological/informational value, we will have to delve somewhat deeper into large-scale evolutionary pattern and process.

The late paleontologist Stephen Jay Gould famously drew attention to the fact that the major animal body plans (such as those characterizing the vertebrates, arthropods, mollusks, and so forth) have been markedly segregated from one another in "morphological space" ever since their origin around 600 million years ago. Animal body plans are generally reflected under the taxonomic heading of *phyla*. Based on a particular interpretation of the fossil record of the earliest animals, Gould proposed that animal life began with many more body plans and hence a broader range of "morphospace" occupation, which was irreversibly whittled down over the history of life as many of the early body plans went extinct, leaving permanent gaps between the remaining phyla.[54]

Although Gould's hypothesis remains controversial, it introduced a theoretically important distinction between *diversity* on the one hand, and *disparity* on the other. Diversity is typically measured in terms of *species richness*, whereas disparity tracks *morphological variance*—a variable that is independent of speciosity. Since the origin of animals, diversity as measured by species richness has increased steadily over time.[55] Notwithstanding some major setbacks in the history of life, such as the end-Permian extinction that eliminated nearly 96% of living species, diversity has always rebounded and increased its outer limit. Not so with disparity, however, which on average has either remained the same[56] or else decreased[57] over the history of life.[58] Thus, in focusing almost exclusively on lower taxonomic categories like species, conservationists have tended to overlook an important aspect of biological variation—disparity—which appears to hold a far less secure place at the high table of macroevolution.

If Gould is right that losses in higher level morphospace will rarely if ever be recouped, then extinctions that prune larger trunks in the tree of life may be more evolutionarily consequential than losses in more "bushy" sections, even when the latter involve greater magnitudes of species loss. So while diversity is highly resilient, the loss of large-scale phylogenetic history—and with it the unique genetic, developmental, and phenotypic strategies that determine the parameters of ecological and evolutionary possibility for a group—is likely to be permanent.[59]

Fortunately, due to the hierarchical relationship between species and higher taxa, the latter are exceedingly difficult to kill off. The macroevolutionist David Raup demonstrated this in a series of models showing that the phylogenetic distribution of extinction is an important factor in understanding and predicting the effects of extinction events. Raup showed that if extinctions are random with respect to taxa, then even a species kill ratio approaching 95% can sustain 80% of the underlying phylogenetic disparity (evolutionary history).[60]

The key point is this: the same magnitude of species loss—that is, the same reduction in diversity—can have differential implications for disparity and hence for the future of life on earth, depending on the phylogenetic distribution of extinction.[61] Take, for example, the five living species of rhinoceros, three of which are critically endangered. These are the last remaining species of a once-successful family (Rhinocerotidae), which at one time included more than thirty genera that flourished for over 30 million years. The few rhino populations that have persisted into the twenty-first century do not play a significant role in terrestrial ecology. And yet, if living rhinos are pushed into the evolutionary abyss, not only will five species go extinct, but so too will an entire family of odd-toed ungulates, shrinking our increasingly narrow evolutionary portfolio. It is quite possible that nothing like a rhino will ever evolve again, and even if some independent lineage would eventually hit upon a convergent "rhino-like" solution, this might take millions of years longer than the evolutionary duration of the human species. As a result, losing the few remaining rhino species would have different implications for natural history than losing an equivalent number of species from a much more diverse family, such as

Bovidae (a group of cloven-hoofed mammals that includes upwards of 150 species). Moreover, the greater the phylogenetic disparity of a set of taxa, the more known and unknown ecosystem/evosystem services they are likely to provide. For these reasons, conservation triage should also take into account the phylogenetic implications of extinction.

3. Metaphysical and Epistemic Objections to Prioritizing Species

The Precautionary Approach

Given that different species have different extrinsic values, and given that our conservation resources are limited, it makes sense that we would prioritize species for conservation. Nevertheless, some radical biocentric environmental philosophies, such as "deep ecology," reject conservation triage because they maintain that all species have equal intrinsic worth and thus are entitled to *equal treatment*.[62] Not surprisingly, most moral philosophers have rejected such "biocentric egalitarianism."[63] But even some authors who embrace mainstream anthropocentric instrumentalism have rejected the triage approach, arguing that we should try to preserve *all* species regardless of their expected instrumental value.[64] This more precautionary approach to conservation, one that jettisons cost-benefit analysis when harms to human health or the environment are at stake, is motivated in part by the vulnerability of traditional principles of risk management to special-interest manipulation and human myopia.

The triage approach to species conservation could justifiably be rejected on precautionary grounds if two conditions were met: (1) Ontological condition: the causal structure of the biological world is such that communities and ecosystems are composed of highly interconnected webs of species poised in a delicate balance that is wont to unravel in the face of even small perturbations; (2) Epistemic condition: the causal interdependence of species is too complexly configured for humans to adequately map out, and therefore any prioritization decision, no matter how well intentioned, is likely to be erroneous and to produce more harm than good.

If ecosystems are highly integrated webs of species that are extremely sensitive to perturbations, then the extinction of even ostensibly low-value species may have cascading results. In this view, ecosystems are like houses of cards, subject to total collapse given the "wrong" ecological modification or once some unknown threshold of disturbance is realized. If this combination of sensitive ecological dynamics and severe epistemic limitations obtains, then our chances of improving global or even local ecology via selective preservation look bleak, while the risks associated with intervening appear unjustifiably high.

The Benevolent Balance of Nature

Is there a delicate and benevolent balance of nature that human efforts are more likely to disrupt than to improve? Before examining these claims in greater detail, however, it would be good to dispel notions that I am building up a straw man. The idea of a benevolent balance of nature is not simply a playful metaphor that is tossed around by environmental activists or peddled by "new age" energy healers. Delicate interconnectedness has not only been invoked by environmental philosophers in support of conservation efforts,[65] but it has also motivated a growing body of environmental law and policy. For instance, many domestic constitutions contemplate a positive right to a "natural ecological balance" or a "harmony of nature" that may be disrupted by anthropogenic extinctions or additions.[66]

Because individual organisms are goal-directed and homeostatic, they are by definition robust—not fragile—processes. Nonetheless, it is possible that complex associations of organisms and taxa are governed by a different and more volatile dynamic. Indeed, this has been the working assumption of biologists operating under the so-called "equilibrium paradigm" in ecology.[67] The eminent biologist and avid conservationist E. O. Wilson nicely captures this view, stating that "the biosphere [is] a stupendously complex layer of living creatures whose activities are locked together in precise but tenuous global cycles of energy and transformed organic matter.…When we alter the biosphere in any direction, we move the environment away from the delicate dance of biology."[68] Some early ecologists thought that ecosystems were so tightly integrated and internally differentiated that they constituted veritable super-organisms.[69] This view has been rebuked by mainstream evolutionary theorists for its failure to identify acceptable scientific mechanisms.[70] Nevertheless, many ecologists continue to see stability as following from natural selection and its various auxiliary principles, such as "competitive exclusion," or the idea that no two species can occupy the same ecological niche in the same community indefinitely.

There are several ways to think about stability in ecology. The first relates to constancy in the numbers of individuals that comprise a population.[71] Yet even this simple demographic version of equilibrium is problematic.[72] The net rate of change in species population density often fluctuates stochastically, rather than trending toward zero—at least until it hits the ultimate "absorbing boundary" of extinction. The second type of ecological equilibrium relates to the trophic structure of interconnected food webs. The view that ecosystems represent seamless webs of interdependent species hangs on the plausibility of a certain theory of niches (due to Elton[73]) that has been widely criticized[74]; in addition, it assumes that the resources within a community are fully allocated—an assumption for which there is little evidence.[75] Moreover, the origination and extinction of paleo-ecosystems is not generally coordinated in a way that indicates their long-term evolutionary cohesion. Studies of both living and paleontological communities show that trophic webs are usually maintained despite substantial changes in fundamental parameters, such as extinction, invasion, migration, diversity, and energy pathways.[76]

In fact, there is evidence to suggest that the more *weakly* linked the components of a given community are, the more diverse and stable that community will be.[77] An allusion to the individual organism helps to make this point. During embryogenesis, organisms make use of a modular construction that creates developmental "firewalls" to ensure that any damage incurred during development is contained in the affected module and does not spread in cascading fashion to collateral areas. It is likely that ecosystems make use of a similar design principle, as highly interconnected communities will not persist for long. A modular construction allows subassemblies to buffer the larger community against environmental perturbations, such as demographic changes or fluctuations in the biotic or abiotic environment.[78] Ironically, to the extent that ecological balance exists at all, it will depend on the *absence* of sensitive interactions between species.

Contrary to ecological balance theory, biotic interactions will tend to undermine, rather than reinforce, the stability of faunal associations. Species interact with one another in evolutionary time: sometimes these interactions are *cooperative*, as in the case of mutualisms, but often they are *strategic*, resulting in an evolutionary arms race in which one organism's ecological solution is another's design problem, and vice versa. This is common for example in the interactions between predator and prey and between host and parasite. Strategically interacting lineages must continue to evolve merely to maintain their present fitness levels—a dynamic that the biologist Leigh Van Valen has dubbed the "Red Queen effect," after a character in Lewis Carroll's *Through the Looking-Glass* who must continually run in place merely to remain where she is.[79]

The Red Queen effect is one of the chief explanations of Van Valen's famous observation that the probability of extinction for a given lineage does not vary as a function of its taxonomic age. In other words, a lineage's probability of going extinct is independent of its previous evolutionary success. Another way of putting it is that perpetual arms races prevent species from getting progressively better at not going extinct. Whereas biologists had tended to assume that strategic interaction would generally lead to evolutionarily stable solutions, even simple arms races have been shown to create ecological instability that ultimately leads to extinction.[80]

At the same time, the causal interrelations of species in a community are not entirely opaque. We can make fairly solid predictions about the ecological consequences of the extinction of certain species, given their centrality in a food web and their degree of connectivity to other species.[81] For all of these reasons, the metaphysics of ecology and the limitations it places on our ability to understand the causal structure of the biological world do not in my view justify an all-or-nothing or even precautionary alternative to the cost-benefit analysis of triage.

4. Conclusion: Is Anthropogenic Extinction Morally Unique?

In concluding, I will consider a final question: is there something that makes *anthropogenic* extinctions worse than the ordinary background and occasional mass extinctions that pervade the history of life? Marc Ereshefsky has made the case, persuasively in my view, that there is nothing intrinsically different about the nature of anthropogenic extinctions, apart from the fact that they are caused by human behavior.[82] The fossil record is punctuated by five major perturbations, the most recent being the end-Cretaceous mass extinction around 65 million years ago. Some biologists believe that we are currently in the midst of a sixth great mass extinction, one that is driven not by geological or extraterrestrial events, but by a single techno-logical species with a knack for population growth, habitat destruction, pollution, and climate modification on a genuinely global scale.[83] Many paleontologists are not convinced that the present rates of extinction rise to the level of the Big Five.[84] Virtually all biologists agree, however, that we are witnessing a period in which extinction is substantially outpacing speciation, threatening many lineages that narrowly avoided annihilation in the previous mass extinctions.[85] But the extinc-tions caused by human beings do not differ either in magnitude or distribution from those that have occurred throughout life's history.

The extrinsic final and instrumental values that we might attach to species, which range from the ecosystem/evosystem services they provide to their value as living evolutionary history, are equally implicated in species loss whether it is caused by human agency or non-agential processes. It is sometimes claimed that anthropo-genic extinction is "unnatural," or that somehow the element of human causation makes it morally distinct. Grounding moral judgments in the distinction between "natural" and "unnatural" is conceptually and normatively problematic for well-known reasons that I will not rehearse here.[86] Instead, I will consider whether there are any other plausible bases to treat anthropogenic harms differently from "natu-ral" harms, even when they are equivalent in terms of their likelihood and expected consequences for people and the environment.

Should we not expend the same resources guarding against a rogue asteroid impact as we would in dealing with a rogue nuclear regime, assuming the risks and magnitudes of nuclear fallout were identical? From a consequentialist perspective, the answer would seem to be yes. Yet there is a tendency, all things being equal, for people to be more concerned about anthropogenic harms than "natural" harms. Compare, for instance, the motivation in the United States to guard against terror-ism after 9/11, with the inclination to shore up domestic infrastructure after Hurricane Katrina (as measured, for example, by the amount of resources allocated and the alacrity of national response). For historical ecological reasons related to the demands of living in a complex social group, harm-by-hurricane is simply not as salient and does not evoke such powerful moral emotions as harm-by-hominid. I would venture to guess that this non-consequentialist asymmetry in moral judgment would also be borne out in the context of species extinction and climate change.

Is there any theoretical basis for this asymmetry? One possibility is that it stems from the considered intuition that it is a greater wrong to cause harm than to allow the same harm to occur "naturally." This idea is premised on there being a defensible distinction between positive and negative duties, which of course many philoso-phers deny. In any case, the degree of "badness" that is added to some state of affairs because of its "wrongness," that is, because it implicates a culpable moral agent, seems minimal and should not significantly affect our conservation priorities. Our duty to contain a preventable outbreak of infectious disease is equally weighty whether it is caused by bioterrorism or "natural" microbiological evolution. To conclude otherwise would be to confuse the culpability of the actor with the badness of the consequences that he or she brings about. If a species is extrinsically valuable, then (ceteris paribus) its extinction is morally undesirable, regardless of whether it is caused directly, indirectly, or not at all by human action.

Given the upward diversity trend of the last 600 million years, it is reasonable to believe that the biota will ultimately recover from the present spasm of extinction, and that it will do so spectacularly. Unfortunately, it is unlikely that there will be any humans around to witness life's new and magnificent evolutionary directions. *Homo sapiens* is substantially less than one million years old—a "preadolescent" from the standpoint of mean species duration, with hopefully millions of years to go before it is reduced to interesting bits of biostratigraphy. I have argued that species are valuable not only for the basic pleasures and utilities they provide, but also for the meaningful self-knowledge they bring about in contextualizing the human species as simply one twig on a vast branching lineage that extends back to the dawn of life on Earth. If this is so, then our time on this planet will go best if we experience it as stewards of a full house of biodiversity, rather than as overlords of an impoverished lot of a once-arborescent tree of life.[87]

NOTES

1. A "taxon" (plural: "taxa") refers to a taxonomic category or grouping used in biological classification, typically reflecting phylogenetic (evolutionary) relationships and character traits that distinguish it from other such units. A taxon may or may not be given a formal rank in the nested biological hierarchy, which includes populations of organisms, species, genera, families, orders, classes, phyla, and kingdoms, in ascending order of inclusiveness.

2. For a discussion, see R. Powell and A. Buchanan, "Breaking Evolution's Chains: The Prospect of Deliberate Genetic Modification in Humans," *Journal of Medicine and Philosophy* 36 (2011): 6–27; R. Powell, "What's the Harm? An Evolutionary Theoretical Critique of the Precautionary Principle," *Kennedy Institute of Ethics Journal* 20 (2010): 181–206.

3. J. McMahan, "The Metaphysics of Brain Death," *Bioethics* 9 (1995): 91–126.

4. Throughout this essay, I will use the terms "preservation" and "conservation" interchangeably, although I recognize that some authors prefer to distinguish them in order to connote different approaches to the management of land and biota.

5. E. Mayr, *Systematics and the Origin of Species* (New York: Columbia University Press, 1942).

6. For a review, see R. L. Mayden, "On Biological Species, Species Concepts and Individuation in The Natural World," *Fish and Fisheries* 3 (2002): 171–96; M. Ereshefsky, *The Poverty of the Linnaean Hierarchy: A Philosophical Study of Biological Taxonomy* (Cambridge: Cambridge University Press, 2001).

7. See e.g., P. Kitcher, "Species," *Philosophy of Science* 51 (1984): 308–33.

8. K. de Queiroz, "Ernst Mayr and the Modern Concept of Species," *Proceedings of the National Academy of Sciences USA* 102 (2005): 6600–7.

9. Evolutionary gray zones exist because subpopulations of a meta-population can attain substantial degrees of genetic, morphological, functional, or behavioral divergence before they become reproductively isolated; and conversely, they can become reproductively isolated without diverging significantly along any of the above parameters.

10. One reason why evolutionary "gray zones" are rare and short-lived relates to the general tempo and mode of speciation. In the earlier part of the twentieth century, many biologists argued that the species designation was arbitrary or conventional because there was no objective way of demarcating the point along the gradual continuum through which one species incrementally transforms into another (a process called "anagenesis"). However, it appears that speciation does not typically occur in this manner. Species tend to arise in geologically rapid bouts of evolution in small, geographically isolated populations, which then re-invade the mainland and appear in the fossil record as a new species. Once they have arisen, species tend to exhibit relative morphological stasis until they go extinct. This pattern, dubbed "punctuated equilibrium" by Niles Eldredge and Stephen Jay Gould, makes it easier for paleobiologists to objectively individuate species in space and time. N. Eldredge and S. J. Gould, "Punctuated Equilibria: An Alternative to Phyletic Gradualism," in *Models in Paleobiology*, ed. T. J. M. Schopf (San Francisco: Freeman Cooper, 1972), pp. 82–115; see also S. J. Gould, *The Structure of Evolutionary Theory* (Cambridge, Mass.: Harvard University Press, 2002), chapter 9.

11. E. Mayr and W. B. Provine, eds., *The Evolutionary Synthesis: Perspectives on the Unification of Biology* (Cambridge, Mass.: Harvard University Press, 1980).

12. E. Sober, "Evolution, Population Thinking and Essentialism," *Philosophy of Science* 47 (1980): 350–83.

13. D. Hull, "Are Species Really Individuals?" *Systematic Zoology* 25 (1976): 174–91.

14. J. Beatty, "The Evolutionary Contingency Thesis," in *Concepts, Theories, and Rationality in the Biological Sciences*, ed. G. Wolters and J. G. Lennox (Pittsburgh, Pa.: University of Pittsburgh Press, 1995).

15. The now-canonical distinction between classes and individuals in the context of biological taxa was introduced by M. Ghiselin, "A Radical Solution to the Species Problem," *Systematic Zoology* 23 (1974): 536–44, and was subsequently endorsed by D. Hull, "A Matter of Individuality," *Philosophy of Science* 45 (1978): 335–60.

16. K. de Queiroz, "The General Lineage Concept of Species, Species Criteria, and the Process of Speciation: A Conceptual Unification and Terminological Recommendations," in *Endless Forms: Species and Speciation*, ed. D. J. Howard and S. H. Berlocher (Oxford: Oxford University Press, 1998).

17. For a discussion of lawlessness in biology, see Beatty, "Evolutionary Contingency Thesis." See also E. Sober, *The Nature of Selection* (Cambridge, Mass.: MIT Press, 1984). There are a few authors who view "convergent evolution," or the independent origination of similar biological forms, as indicative of natural-kind-like replicability in the history of life. See, e.g., S. C. Morris, *Life's Solution: Inevitable Humans in a Lonely Universe* (Cambridge: Cambridge University Press, 2003); D. C. Dennett, *Darwin's Dangerous Idea* (New York: Simon & Schuster, 1995), p. 307. This view is decidedly in the minority, however, and I critique it elsewhere. See R. Powell, "Is Convergence More than an Analogy? Homoplasy and its Implications for Macroevolutionary Predictability," *Biology and Philosophy* 22 (2007): 565–78.

18. Consistent with thinking of species as historical individuals, some environmental philosophers have argued that even if we could restore a destroyed ecosystem molecule for

molecule, the product would be metaphysically and morally distinct from the original, because it would have a different *history*. See, e.g., R. Elliott, *Faking Nature: The Ethics of Environmental Restoration* (New York: Routledge, 1997).

19. R. A. Wilson, "The Biological Notion of Individual," *Stanford Encyclopedia of Philosophy*, ed. Edward N. Zalta (Winter 2010), available at http://plato.stanford.edu/entries/species/ (accessed February 1, 2010).

20. B. Mishler and R. Brandon, "Individuality, Pluralism, and the Phylogenetic Species Concept," *Biology and Philosophy* 2 (1987): 397–414.

21. Wilson, "Biological Notion of Individual."

22. On metaphors of progress in macroevolutionary theory and in popular conceptions of evolution, see S. J. Gould, *Full House* (New York: Three Rivers Press).

23. R. Amundson and G. V. Lauder, "Function Without Purpose: The Uses of Causal Role Function In Evolutionary Biology," *Biology and Philosophy* 9 (1994): 443–69.

24. Although it is widely accepted that species-level selection can occur in principle, it is probably rare in fact for two reasons: first, the lack of integration and specialization among species' parts, and second, the vast spans of time necessary for species-level selection to occur and the tendency for lower (e.g., organismic) level selection to undermine it in the interim. See T. Grantham, "Hierarchical Approaches to Macroevolution: Recent Work on Species Selection and the 'Effect Hypo-Thesis,'" *Annual Review of Ecology and Systematics* 26 (1995): 301–22.

25. D. Jablonksi, "Species Selection: Theory and Data," *Annual Review of Ecology, Evolution, and Systematics* 39 (2009): 501–24.

26. K. Goodpaster, "On Being Morally Considerable," *Journal of Philosophy* 75 (1978): 308–25.

27. C. Korsgaard, *The Sources of Normativity* (Cambridge: Cambridge University Press, 1996).

28. T. Regan, "The Case for Animal Rights," in *In Defence of Animals*, ed. P. Singer (Oxford: Basil Blackwell, 1985).

29. P. Singer, *Practical Ethics* (Cambridge: Cambridge University Press, 1993).

30. P. W. Taylor, "The Ethics of Respect for Nature," *Environmental Ethics* 3 (1981): 197–218.

31. E. Nagel, "Teleology Revisited: Goal-Directed Processes in Biology," *Journal of Philosophy* 74 (1977): 261–301. Nagel argued that in order for a system to be goal-directed, it must possess homeostatic mechanisms that coordinate parameter values that are "orthogonal" to or nomically independent of one another, such as glucose and water levels in the blood. The criterion of orthogonality effectively disqualifies certain physical processes (such as tornados) from exhibiting goal-directed behavior, even though they have a dynamic internal structure that is otherwise quite robust against perturbations.

32. Taylor, "Ethics of Respect for Nature."

33. Note that simply having adaptive features does not make an entity goal-directed. Earlier I discussed some potential species-level adaptations, such as geographic range or population density. Even if these traits are the product of between-species selection, they do not seem to constitute homeostatic mechanisms that coordinate nomically independent variables in the face of perturbation in order to produce some fitness-enhancing effect.

34. For a review, see P. Jones, "Group Rights," *Stanford Encyclopedia of Philosophy*, ed. Edward N. Zalta (Fall 2008), available at http://plato.stanford.edu/entries/rights-group/ (accessed February 1, 2010).

35. P. A. French, *Collective and Corporate Responsibility* (New York: Columbia University Press, 1984).

36. D. Simberloff, "A Succession of Paradigms in Ecology: Essentialism to Materialism and Probablism," *Synthese* 43 (1980): 3–39; K. Sterelny and P. Griffiths, *Sex and Death: An Introduction to the Philosophy of Biology* (Chicago: University of Chicago Press, 1999).

37. H. Cahen, "Against the Moral Considerability of Ecosystems," *Environmental Ethics* 10 (1988): 195–216. For a reply, see S. Salthe and B. Salthe, "Ecosystem Moral Considerability: A Reply to Cahen," *Environmental Ethics* 11 (1989): 355–61.

38. E. Hargrove, "Weak Anthropocentric Intrinsic Value," *The Monist* 75 (1992): 119–37.

39. P. Singer, "Not for Humans Only: The Place of Nonhumans in Environmental Issues," in *Ethics and Problems of the 21st Century*, ed. K. E. Goodpaster and K. M. Sayre (Notre Dame, Ind.: University of Notre Dame Press, 1979).

40. Korsgaard, *Sources of Normativity*.

41. L. Floridi, "On the Intrinsic Value of Information Objects and the Infosphere," *Ethics and Information Technology* 4 (2002): 287–304.

42. See, e.g., Kenneth E. Himma, "There's Something about Mary: The Moral Value of Things Qua Information Objects," *Ethics and Information Technology* 6 (2004): 145–59.

43. Ordinarily, the phrase "higher taxon" refers to any taxonomic rank above species, including genera, families, orders, classes, and phyla.

44. The emerging field of "biomemetics" demonstrates the immediate utility of natural historical information. For example, researchers are creating microfiber adhesives by studying gecko toe hairs, designing energy-efficient buildings by consulting the thermo-regulatory properties of termite mounds, and mining the rainforest for plants and fungi with resistance-free antibacterial properties.

45. A. Randall, "Human Preferences, Economics, and the Preservation of Species," in *The Preservation of Species: The Value of Biological Diversity* (Princeton, N.J.: Princeton University Press, 1986).

46. IUCN, UNEP, and WWF, *World Conservation Strategy: Living Resource Conservation for Sustainable Development* (Gland, Switzerland: International Union for Conservation of Nature and Natural Resources, 1980).

47. D.P Faith, S. Magallo, A.P. Hendry, E. Conti, T. Yahara and M.J. Donoghue, "Evosystem Services: An Evolutionary Perspective on the Links Between Biodiversity and Human Well-Being," *Current Opinion in Environmental Sustainability* 2 (2010): 66–74.

48. S. Sarkar, *Biodiversity and Environmental Philosophy: An Introduction* (New York: Cambridge University Press, 2005).

49. M. Colyvan, S. Linquist, W. Grey, P. Griffiths, J. Odenbaugh, and H. P. Possingham, "Philosophical Issues in Ecology: Recent Trends and Future Directions," *Ecology and Society* 14 (2009): 22–34.

50. R. V. Solé, J. M. Montoya, and D. H. Erwin, "Recovery from Mass Extinction: Evolutionary Assembly in Large-Scale Biosphere Dynamics," *Philosophical Transaction of the Royal Society* 357 (2002): 697–707. Some of the greatest extinctions in the history of life appear to have been triggered by the destruction of foundational species, as was likely the case for planktonic foraminifera during the end-Cretaceous perturbation in which the dinosaurs (and many other groups) perished. The same is true of "reef-builders" throughout the history of life, and the extinction of modern-day corals would likely have similar cascading consequences for extant marine diversity.

51. Note that even if the informational or ecological properties of species *supervene* on that of their constituent organisms—that is, even if all species with the same organism-level properties must have the same emergent, species-level properties—this does not imply that a species' value *reduces* to that of its constituents. A supervenience relation does not entail a reduction relation: emergent moral (or mental) properties cannot be explained

as and do not reduce to the sum value of the properties of the individual cells or atoms on which they supervene. J. Kim, *Supervenience and Mind: Selected Philosophical Essays* (Oxford: Oxford University Press, 1993).

52. R. H. MacArthur and E. O. Wilson, *The Theory of Island Biogeography* (Princeton, N.J.: Princeton University Press, 1967).

53. Phylogeny refers broadly to evolutionary history, in particular to the spatiotemporal distribution of ancestor-descendent populations; it is typically represented by a branching diagram reflecting patterns of common descent.

54. S. J. Gould, *Wonderful Life: The Burgess Shale and the Nature of History* (New York: W. W. Norton, 1989).

55. J. Sepkoski, Jr., "A Kinetic Model of Phanerozoic Taxonomic Diversity," *Paleobiology* 10 (1984): 246–67.

56. D. E. G. Briggs, R. A. Fortey, and M. A. Wills, "Morphological Disparity in the Cambrian," *Science* 256 (1992): 1670–73.

57. M. Foote and S. J. Gould, "Cambrian and Recent Morphological Disparity," *Science* 258 (1992): 1816–18.

58. For a review of the methodological challenges confronting measures of disparity, see J. Maclaurin and K. Sterelny, *What Is Biodiversity?* (Chicago: University of Chicago Press), especially chapters 3–4.

59. D. H. Erwin, *Extinction: How Life on Earth Nearly Ended 250 Million Years Ago* (Princeton, N.J.: Princeton University Press, 2006).

60. D. Raup, *Extinction: Bad Genes or Bad Luck?* (New York: W. W. Norton, 1991).

61. D. Jablonksi, "Extinction: Past and Present," *Nature* 475 (2004): 589; see also Erwin, *Extinction*.

62. A. Naess, "The Shallow and the Deep, Long-Range Ecology Movement," *Inquiry* 16 (1973): 95–100.

63. For a critique, see W. Grey, "Anthropocentrism and Deep Ecology," *Australasian Journal of Philosophy* 71 (1993): 463–75.

64. See, e.g., D. Takacs, *The Idea of Biodiversity: Philosophies of Paradise* (Baltimore, Md.: Johns Hopkins University Press, 1996); B. Norton, *The Preservation of Species: The Value of Biological Diversity* (Princeton, N.J.: Princeton University Press, 1986).

65. See, e.g., Norton, *Preservation of Species*.

66. For a discussion, see Powell, "What's the Harm?"

67. S. E. Kingsland, *Modeling Nature: Episodes in the History of Population Ecology* (Chicago: University of Chicago Press, 1985).

68. E. O. Wilson, *The Future of Life* (New York: Knopf, 2002), p. 39.

69. E.g., J. Lovelock, *Gaia: A New Look at Life on Earth* (Oxford: Oxford University Press, 1979).

70. S. J. Gould, "Kropotkin Was No Crackpot," *Natural History* 106 (1997): 12–21; W. F. Doolittle, "Is Nature Really Motherly?" *CoEvolution Quarterly* 29 (1981): 58–63.

71. K. Cuddington, "The 'Balance of Nature' Metaphor and Equilibrium in Population Ecology," *Biology and Philosophy* 16 (2001): 463–79.

72. S. P. Hubbell, *The Unified Neutral Theory of Biodiversity and Biogeography* (Princeton, N.J.: Princeton University Press, 2001); S. L. Pimm, *The Balance of Nature?* (Chicago: Chicago University Press, 1991).

73. C. S. Elton, *The Ecology of Invasions by Animals and Plants* (London: Chapman and Hall, 1958).

74. For the classic critique of "lock-and-key" models of the organism-environment relationship, see R. C. Lewontin, "Gene, Organism, and Environment," in *Evolution from*

Molecules to Men, ed. D. S. Bendall (Cambridge: Cambridge University Press, 1983); see also Sterelny and Griffiths, *Sex and Death*, chapter 11.

75. S. R. Reice, "Nonequilibrium Determinants of Biological Community Structure," *American Scientist* 82 (1994): 424–35.

76. An exception may be island invasions. See J. W. Valentine and D. Jablonksi, "Fossil Communities: Compositional Variation at Many Time Scales," in *Species Diversity in Ecological Communities: Historical and Geographical Perspectives*, ed. R. E. Ricklefs and D. Schluter (Chicago: University of Chicago Press, 1993).

77. K. McCann, "The Diversity-Stability Debate," *Nature* 405 (2000): 228–33; G. D. Kokkoris, V. A. A. Jansen, M. Loreau, and A. Y. Troumbis, "Variability in Interaction Strength and Implications for Biodiversity," *Journal of Animal Ecology* 71 (2002): 362–71.

78. K. Sterelny, "The Reality of Ecological Assemblages: A Palaeo-Ecological Puzzle," *Biology and Philosophy* 16 (2001): 437–61.

79. L. Van Valen, "A New Evolutionary Law," *Evolutionary Theory* 1 (1973): 1–30.

80. B. E. Beisner, B. E. E. McCauley, and F. J. Wrona, "Predator-Prey Instability: Individual-Level Mechanisms for Population-Level Results," *Functional Ecology* 11 (1997): 112–20.

81. See discussion in note 50 and accompanying text. See also J. Montoya, S. L. Pimm, and R. V. Sole, "Ecological Networks and Their Fragility," *Nature* 442 (2006): 259–64.

82. M. Ereshefsky, "Where the Wild Things Are: Environmental Preservation and Human Nature," *Biology and Philosophy* 22 (2007): 57–72.

83. M. Novacek, *Terra: Our 100-Million-Year-Old Ecosystem—and the Threats That Now Put It at Risk* (New York: Farrar, Straus and Giroux, 2008).

84. See, e.g., D. Jablonksi, "Lessons from the Past: Evolutionary Impacts of Mass Extinctions," *Proceedings of the National Academy of Sciences USA* 98 (2001): 5393–98; Erwin, *Extinction*.

85. D. B. Wake and V. T. Vredenburg, "Are We in the Midst of the Sixth Mass Extinction? A View from the World of Amphibians," *PNAS* 105 (2008): 11466–73.

86. For a discussion, see Ereshefsky, "Where the Wild Things Are."

87. I would like to thank Nick Shea, Sanem Soyarslan, and the editors of this volume for helpful comments on an earlier draft of this manuscript.

SUGGESTED READING

On the ontological and theoretical status of biological species, see M. GHISELIN, "A Radical Solution to the Species Problem," *Systematic Zoology* 23 (1974): 536–44; D. Hull, "A Matter of Individuality," *Philosophy of Science* 45 (1978): 335–60; P. Kitcher, "Species," *Philosophy of Science* 51 (1984): 308–33; R. L. Mayden, "On Biological Species, Species Concepts and Individuation in the Natural World," *Fish and Fisheries* 3 (2002): 171–96; K. de Queiroz, "Ernst Mayr and the Modern Concept of Species," *Proceedings of the National Academy of Sciences USA* 102 (2005): 6600–7; M. Ereshefsky, *The Poverty of the Linnaean Hierarchy: A Philosophical Study of Biological Taxonomy* (Cambridge: Cambridge University Press, 2001).

For a discussion of the history of essentialism and the shift toward population thinking in evolutionary biology, see E. SOBER, "Evolution, Population Thinking and Essentialism," *Philosophy of Science* 47 (1980): 350–83.

For a conceptual analysis of the property of goal-directedness and its implications for environmental ethics, see (respectively) E. NAGEL, "Teleology Revisited: Goal-Directed Processes in Biology," *Journal of Philosophy* 74 (1977): 261–301; P. W. Taylor, "The Ethics of Respect for Nature," *Environmental Ethics* 3 (1981): 197–218; H. Cahen, "Against the Moral Considerability of Ecosystems," *Environmental Ethics* 10 (1988): 195–216; and the rebuttal by S. Salthe and B. Salthe, "Ecosystem Moral Considerability: A Reply to Cahen," *Environmental Ethics* 11 (1989): 355–61.

On the conceptual and moral dimensions of anthropogenic extinction, see M. ERESHEFSKY, "Where the Wild Things Are: Environmental Preservation and Human Nature," *Biology and Philosophy* 22 (2007): 57–72. On mass extinction more generally, see D. Raup, *Extinction: Bad Genes or Bad Luck?* (New York: W. W. Norton, 1991); D. H. Erwin, *Extinction: How Life on Earth Nearly Ended 250 Million Years Ago* (Princeton, N.J.: Princeton University Press, 2006). For a comparison of modern-day extinction rates with those of past mass extinctions, see D. Jablonksi, "Lessons from the Past: Evolutionary Impacts of Mass Extinctions," *Proceedings of the National Academy of Sciences USA* 98 (10) (2001): 5393–98.

For discussions of the balance-of-nature paradigm in ecology, see K. CUDDINGTON, "The 'Balance of Nature' Metaphor and Equilibrium in Population Ecology," *Biology and Philosophy* 16 (2001): 463–79; F. Doolittle, "Is Nature Really Motherly?" *CoEvolution Quarterly* 29 (1981): 58–63. For an accessible introduction to the philosophy of ecology and problems in the philosophy of biology more broadly, see K. Sterelny and P. Griffiths, *Sex and Death: An Introduction to the Philosophy of Biology* (Chicago: University of Chicago Press, 1999).

For overviews of the concept of biodiversity and its role in conservation biology, see J. MACLAURIN and K. STERELNY, *What Is Biodiversity?* (Chicago: University of Chicago Press); S. Sarkar, *Biodiversity and Environmental Philosophy: An Introduction* (New York: Cambridge University Press, 2005).

On the nature of values and the sources of normativity, see K. GOODPASTER, "On Being Morally Considerable," *Journal of Philosophy* 75 (1978): 308–25; E. Hargrove, "Weak Anthropocentric Intrinsic Value," *The Monist* 75 (1992): 119–37; C. Korsgaard, *The Sources of Normativity* (Cambridge: Cambridge University Press, 1996).

CHAPTER 22

ARE ALL SPECIES EQUAL?

DAVID SCHMIDTZ

SPECIES egalitarianism is the view that all living things have equal moral standing. To have moral standing is, at a minimum, to command respect, to be more than a mere thing. Is there reason to believe that all living things have moral standing in even this most minimal sense? If so—that is, if all living things command respect—is there reason to believe they all command *equal* respect?[1]

I will explain why members of other species command our respect, but also why they do not command equal respect. The intuition that we should have respect for nature is one motive for embracing species egalitarianism, but we need not be species egalitarians to have respect for nature. I will question whether species egalitarianism is even compatible with respect for nature.

1. RESPECT FOR NATURE

According to Paul Taylor, anthropocentrism "gives either exclusive or primary consideration to human interests above the good of other species."[2] The alternative to anthropocentrism is biocentrism, and it is biocentrism that, in Taylor's view, grounds species egalitarianism.

Four beliefs form the core of Taylor's biocentrism:

(a) Humans are members of the Earth's community of life in the same sense and on the same terms in which other living things are members of that community.

(b) All species, including humans, are integral parts of a system of interdependence.
(c) All organisms are teleological centers of life. Each is a unique individual pursuing its own good in its own way.
(d) Humans are not inherently superior to other living beings.[3]

Taylor concludes, "Rejecting the notion of human superiority entails its positive counterpart: the doctrine of species impartiality. One who accepts that doctrine regards all living things as possessing inherent worth—the *same* inherent worth, since no one species has been shown to be either higher or lower than any other."[4] Taylor does not call this a valid argument (he acknowledges that it is not), but he thinks that if we concede (a), (b), and (c), it would be unreasonable not to move to (d), and then to his egalitarian conclusion. Is he right? For those who accept Taylor's three premises, and who thus interpret those premises in terms innocuous enough to render them acceptable, there are two responses. First, we may go on to accept (d), following Taylor, yet deny that there is any warrant for moving from there to Taylor's egalitarian conclusion. Having accepted that our form of life is not superior, we might instead regard it as inferior. More plausibly, we might view our form of life as noncomparable. The question of how we compare to nonhumans has a simple answer: we don't. We are not equal. We are not superior. We are not inferior. We are simply different.

Alternatively, we may reject (d) and say that humans are inherently superior, but then go on to say that our superiority is a moot point. Whether we are inherently superior—that is, superior as a form of life—does not matter much. Even if we are superior, within the web of ecological interdependence mentioned in premises (a) and (b), it would be a mistake to ignore the needs and the telos of any species referred to in premise (c). Thus, there are two ways of rejecting Taylor's argument for species egalitarianism. Neither alternative is committed to species equality, yet each, on its face, is compatible with the respect for nature that motivates Taylor's egalitarianism in the first place.

These are preliminary worries about Taylor's argument. Taylor's critics have been harsh, perhaps overly harsh. After building on some of their criticisms and rejecting others, I will explore some of our reasons to have respect for nature and ask whether they translate into reasons to be species egalitarians. I will conclude that Taylor's biocentrism has a point, but that biocentrism does not require any commitment to species equality.

2. IS SPECIES EGALITARIANISM HYPOCRITICAL?

Taylor is among the most intransigent of species egalitarians, yet he allows that human needs override the needs of nonhumans. In response, Peter French argues that species egalitarians cannot have it both ways. French perceives a contradiction between the egalitarian principles that Taylor officially endorses and the unofficial principles he offers as the real principles by which we should live. Having proclaimed that we are all equal, French asks, what licenses Taylor to say that, in cases of conflict, nonhuman interests can legitimately be sacrificed to vital human interests?[5]

Good question. Yet, somehow Taylor's alleged inconsistency is too obvious. Perhaps his position is not as blatantly inconsistent as it appears. Taylor could respond as follows: suppose I find myself in a situation of mortal combat with an enemy soldier. If I kill my enemy to save my life, that does not entail that I regard my enemy as an inferior form of life. Likewise, if I kill a bear to save my life, that does not entail that I regard the bear as inherently inferior. Therefore, Taylor can, without hypocrisy, deny that species egalitarianism requires a radically self-effacing pacifism.

What, then, does species egalitarianism require? It requires us to avoid mortal combat whenever we can, not only with other humans but with living things in general. On this view, we ought to regret finding ourselves in kill-or-be-killed situations that we could have avoided. There is no point in regretting the fact that we must kill in order to eat, though, for there is no avoiding that. Species egalitarianism is compatible with our having a limited license to kill. Many, including vegetarians, will say that it matters *what* we kill, and that even if we must kill to eat, we need not kill *animals*. Most vegetarians think it is worse to kill a cow than to kill a carrot. Are they wrong? Yes they are, according to species egalitarianism. Therein lies egalitarianism's failure to respect nature. I agree with Taylor that we have reason to respect nature, but if we treat a chimpanzee no better than we would treat a carrot, that is a failure of respect, not a token of it. Failing to respect what makes living things different is not a way of respecting them. It is instead a way of being indiscriminate.

3. Is Species Egalitarianism Arbitrary?

According to premise (c) of Taylor's argument for the biocentric outlook, as discussed in section 1, a being has intrinsic worth if it has a good of its own. He notes that even plants have a good of their own in the relevant sense. They seek their own good in their own way. Taylor defines anthropocentrism as giving exclusive or primary consideration to human interests above the good of other species. So, if we acknowledge that the ability to think is a valuable capacity and if we further acknowledge that some but not all living things possess this capacity, are we giving exclusive or primary consideration to human interests? Not at all. All we are doing is acknowledging that living things, including humans, can be valuable in various ways, and that some living things may be *more* valuable than humans along some dimensions. We acknowledge that not all living things are equal along all dimensions. Some are faster, some are smarter, and so on. We consider the possibility that all values are commensurable. (If they are not commensurable, then neither can they be equal.)

We note that if all values are commensurable, then in principle we can add up the values of all living things along all dimensions. If we do this, it might turn out that all living things have equal value. But that would be quite a fluke.

It will not do to defend species egalitarianism by singling out a property that all living things possess, arguing that this property is morally important, then concluding that all living things are therefore of equal moral importance. Why not? Because where one property such as simply being alive provides a basis for moral standing, there might be others. Other properties such as sentience might be possessed by some but not all living things, and might provide bases for different kinds or degrees of moral standing.

Taylor realizes that not all living things can think, and never denies that the capacity for thought is valuable. What he would say is that it begs the question to rank the ability to think as *more* valuable than the characteristic traits of plants and other animals. Taylor assumes that human rationality is on a par with, for example, a cheetah's foot-speed: no less valuable, but no more valuable either.[6] In this case, though, Taylor is missing the point. Let us concede to Taylor that the good associated with the ability to think is not superior to the good associated with a tree's ability to grow and reproduce. It doesn't matter. Suppose we let a be the good of being able to grow and reproduce, let b be the good of sentience, and let c be the good of being rational. Contra Taylor, anthropocentrists need not assume that c has a higher value than a. All they need assume is that the value of $a + c$ is higher than the value of a by itself. However much we value the ability to grow and reproduce, the fact would remain that chimpanzees have what trees have, plus more.

One valid response: it is a little odd to treat the properties of living things as if some property called "able to grow and reproduce" were manifested in an identical way by all living things, or as if a property called "able to run" were manifested in an identical way by all animals that run, and so on. In truth, not all a's are equal. Acknowledging that fact does not help the case for species equality, but it does complicate things. Perhaps the vegetative good of a tree is greater than the vegetative good of chimpanzees, so that the additional animal good of chimpanzees just suffices to balance the vegetative superiority of trees.

Although this response is interesting in its own right, I think it is unavailable to biocentric egalitarians such as Taylor. Taylor says that all living thing are equal in the sense that they all have lives of their own.[7] As soon as Taylor acknowledges that there are other dimensions of value, he has to choose. On the one hand, he can arbitrarily assign different values to each living thing's a so as to preserve an overall equality:

$$a_{tree} = a_{animal} + b = a_{human} + b + c.$$

Alternatively, and more plausibly, he can concede that equality is not the point. The intuitions motivating biocentrism—namely, that the world would be a better place if we stopped thinking of ourselves as superior, and that we live in a world of multidimensional value extending far beyond the human realm—can be grounded in the simple and plausible thought that the goods of trees and chimpanzees (and humans) are not comparable.

In short, although both trees and chimpanzees are teleological centers of life, and we can agree that this status is valuable and that trees and chimpanzees share equally in this particular value, we cannot infer that trees and chimpanzees have equal value. We are entitled to conclude only that they are of equal value so far as being a teleological center of life is concerned. From that, we may infer that *one* alleged ground of our moral standing (that we grow and reproduce) is shared by all living things. Beyond that, nothing about equality even suggests itself. It begs no questions to notice that there are grounds for moral standing that humans do not share with all living things.

4. SPECIESISM AND SOCIAL POLICY

Peter Singer and others talk as if speciesism—the idea that some species are superior to others—is necessarily a kind of bias in favor of humans and against non-human animals. (Singer has no problem with being "biased" against plants.) Not so. If we have more respect for chimpanzees than for mice, then we are speciesists, no matter what status we accord to human beings. A speciesist is taking no stance with regard to humanity when she says to an egalitarian, we *should* respect chimpanzees more than we respect mice, shouldn't we? Or if not, shouldn't we at least respect chimpanzees more than carrots?

Suppose we take an interest in how the moral standing of chimpanzees compares to that of mice and wonder what we would do in an emergency where we could save a drowning chimpanzee or a drowning mouse but not both. More realistically, suppose we conclude that we must do experiments involving animals, because it's the only way to cure an otherwise catastrophic disease, and now we have to choose which animals. Whichever we use, the animals we use will die. We decide to use mice. Then a species egalitarian says, "Why not use chimpanzees? They're all the same anyway, morally speaking, and you'll get more reliable data." Would that sort of egalitarianism be monstrous? I think so. But if we believe all living things are equal, then *why not* use the chimpanzee instead of the mouse?

If chimpanzees are, morally speaking, the wrong *kind* of animal to experiment on when researchers could use mice, then speciesism is to that extent closer to the moral truth than is species egalitarianism. Although in philosophy we tend to use science fiction examples, the situation just described is an everyday problem in the scientific community. Suppose researchers had to choose between harvesting the organs of a chimpanzee or a severely brain-damaged human baby. Peter Singer says we cannot have it both ways. Singer argues that if the ability to think is what makes the difference, then the brain-damaged infant commands no more respect than a chimpanzee, and should indeed command less. Singer concludes that if we need to use one or the other in a painful or lethal medical experiment, and if it does not matter which one we use so far as the experiment is concerned, then we ought to use the infant, not the chimpanzee.[8]

I am not trying to have it both ways when I note that, if we claimed that the rightness of eating beef has to be settled individual cow by individual cow, because some cows are brain damaged, Singer would agree in principle, then go on to insist that cows are the wrong *kind* of thing for us to be eating, and that we need a policy governing our exploitation of cows as a species. He would say it should not be up to individual consumers to decide on a case-by-case basis whether the cow they want to eat is sentient or brain-dead, or whether the benefits of eating this particular cow exceed the costs. Singer would allow that the benefits sometimes do exceed the costs, but he would insist that we need laws—laws governing how we treat cows in general.

Again, Singer would insist that researchers cannot be trusted to decide on a case-by-case basis whether to use mice or chimpanzees or defective people in their experiments, when turnips would do just as well. Likewise, Singer wants to insist that individual consumers should not decide on a case-by-case basis whether to eat cows or turnips—rather, they ought to quit eating cows, period. In the medical-research policy area, we rightly ignore Singer's point that some animals are smarter than some people. Singer would want us to ignore his point in that context, and in this sense wants it both ways. We instead formulate policy on the basis of characteristic features of species. Brain-dead infants are not representative of humanity as a species, any more than brain-dead cows represent cows as a species. In either case, unrepresentative individuals are not relevant as a basis for policy decisions that we intend to apply to humans or cows in general.

Think about it this way: suppose we argue that Canadians should have a right to free speech. Suppose Singer responded by calling us Canadianists, and said being a Canadianist is like being a racist. When we ask why, imagine Singer telling us that some Canadians are brain-damaged ("Just look at their arguments," he says); therefore, he concludes, it is mere chauvinism to assert that Canadians as a general class should have so special a right. We would reply that what matters is that it is compellingly good policy for every Canadian to have that kind of legal protection as a default presumption. Whether every Canadian has what it takes to exercise that right is beside the point. Singer would, and does, endorse some applications of this form of argument. For example, Singer would insist that there is a compellingly good policy for every cow to have certain kinds of legal protection, regardless of whether each and every cow has what it takes to benefit from it.

None of this is meant as a criticism of Singer. I aim only to interpret his writings in the most charitable way. Some of Singer's arguments might seem to commit him to a rather radical egalitarianism, but Singer does when pressed decline to bite that bullet. He says that many species are broadly equal to humans in terms of their characteristic capacities to feel pleasure and pain, but at the same time acknowledges that the characteristic cognitive capacities of normal humans are such that they tend to have more at stake than would other animals otherwise similarly situated. That's why he would feed starving humans before feeding starving animals.[9]

5. EQUALITY AND TRANSCENDENCE

Even if speciesists are right to see a nonarbitrary distinction between humans and cows as types, the fact remains that claims of superiority do not easily translate into justifications of domination.[10] We can have reasons to treat nonhuman species with respect, regardless of whether we consider them to be on a moral par with *Homo sapiens*.

Why respect members of other species? We might respect chimpanzees or mice on the grounds that they are sentient. Even mice have a rudimentary point of view and rudimentary hopes and dreams, and we might well respect them for that. But what about plants? Plants, unlike mice and chimpanzees, do not care what happens to them. They could not care less. So, why should we care? Is it even possible for us to have any good reason to care what happens to plants beyond caring instrumentally about how plants can benefit us?

When we are alone in a forest and wondering whether it would be fine to chop down a tree for fun, our perspective on what happens to the tree is, so far as we know, the only perspective there is. The tree does not have its own. Thus, explaining why we have reason to care about trees requires us to explain caring from our point of view, since that (we are supposing) is all there is. We do not have to satisfy *trees* that we are treating them properly; rather, we have to satisfy ourselves. Again, can we have reasons for caring about trees separate from their instrumental value as lumber and such?

One reason to care (not the only one) is that gratuitous destruction is a failure of self-respect. It is a repudiation of the kind of self-awareness and self-respect that we can achieve by repudiating wanton vandalism. So far as I know, no one finds anything puzzling in the idea that we have reason to treat our lawns or living rooms with respect. Lawns and living rooms have instrumental value, but there is more to it than that. Most of us have the sense that taking reasonable care of our lawns and living rooms is somehow a matter of self-respect, not merely a matter of preserving their instrumental value. Do we have similar reasons to treat forests with respect? I think we do. There is an aesthetic involved, the repudiation of which would be a failure of self-respect. Obviously, not everyone feels the same way about forests. Not everyone feels the same way about lawns and living rooms, either. However, our objective here is to make sense of respect for nature, not to argue that respect for nature is universal or that failing to respect nature is irrational. If and when we identify with a redwood, in the sense of being inspired by it, having respect for its size and age and so on, then as a psychological fact, we face questions about how we ought to treat it. When we come to see a redwood in that light, subsequently turning our backs on it becomes a kind of self-effacement, because the values we thereby fail to take seriously are *our* values, not the tree's.

So, the attitude we take toward gazelles, for example, raises issues of self-respect insofar as we see ourselves as relevantly like gazelles. Here is a different and complementary way of looking at the issue. Consider that lions owe nothing to gazelles.

Therefore, if we owe it to gazelles not to hunt them, it must be because we are *unlike* lions, not—or not only—because we are *like* gazelles. Unlike lions, we have a choice about whether to hunt gazelles, and we are capable of deliberating about that choice in a reflective way. We are capable of caring about the gazelle's pain, the gazelle's beauty, and the gazelle's hopes and dreams, such as they are. On the one hand, if we do care and we cannot adjust our behavior in light of what we care about, then something is wrong with us; we are less than fully, magnificently, human. For a human being, to lack a broad respect for living things and beautiful things and well-functioning things is to be stunted in a way.

Taylor could agree with this argument. He says that all living things are equal because they all have lives of their own, but then concedes that we can systematically privilege animals over plants because animals are sentient, and sentience gives animals, in effect, a superior kind of life. Where does this leave us? What Taylor is rightly conceding here is that when it is time actually to live our lives, we give up on saying all living things are *equal*, and we acknowledge that the moral point is to treat all living things with *respect*. Taylor, moreover, sees a basic connection between self-respect and respect for the world in which one lives.[11]

Our coming to see members of other species as commanding respect is a way of transcending our animal natures. It is ennobling. It is part of our natures unthinkingly to see ourselves as superior and to try to dominate accordingly; as noted, our capacity to see ourselves as equal is part of what makes humans unique. It may be part of what makes us superior. Aldo Leopold expressed a related thought. When the Cincinnati Zoo erected a monument to the passenger pigeon, Leopold wrote, "We have erected a monument to commemorate the funeral of a species....For one species to mourn the death of another is a new thing under the sun....In this fact...lies objective evidence of our superiority over the beasts."[12] Trying to see all living things as equal may not be the best way of transcending our animal natures, but it is one way.

Another way of transcending our animal natures and expressing due respect for nature is simply to not bother with keeping score. This way is more respectful of our own reflective natures. It does not dwell on rankings. It does not insist on seeing equality where a more reflective being simply would see what is there to be seen and would not shy away from respecting what is unique as well as what is common. Someone might say that we need to rank animals as our equals to be fair, but that appears to be false: consider that I can be fair to my friends without ranking them. Imagine a friend saying, "I disagree! In fact, failing to rank us is insulting! You have to rank us as equals!" What would be the point? Perhaps my friends are each other's equals in some respect. Even so, we are left with no need to *rank* them as equal. For most purposes, it is better for them to remain the unique and priceless friends that they are. Sometimes, respect is simply respect. It need not be based on a pecking order.

Children rank their friends. It is one of the things children do before they are old enough to understand friendship. Sometimes, the idea of ranking things, even as equals, is a child's game. It is beneath us.

6. RESPECT FOR EVERYTHING

Therefore, a broad respect for living or beautiful or well-functioning things need not translate into *equal* respect. It need not translate into *universal* respect, either. Part of our responsibility as moral agents is to be somewhat choosy about what we respect and how we respect it. I can see why people shy away from openly accepting that responsibility, but they still have it. We might suppose speciesism is as arbitrary as racism unless we can show that the differences are morally relevant. This is a popular sentiment among animal liberationists such as Peter Singer and Tom Regan. However, are we really like racists when we think it is worse to kill a dolphin than to kill a tuna? The person who asserts that there is a relevant similarity between speciesism and racism has the burden of proof: identify the similarity.

Burden of proof, crucial to many philosophical arguments, is a slippery notion. Do we need good reason to exclude plants and animals from the realm of things we regard as commanding respect? Or do we need reason to *include* them? The latter seems more natural to me, so I am left supposing the burden of proof lies with those who claim we should have respect for all living things. I could be wrong.[13] But suppose Alf says that oatmeal commands respect. Betty asks Alf why he thinks that. Alf responds by saying, "I don't need an argument. It's up to you to prove that oatmeal *doesn't* command respect." Something has gone wrong. Alf has mislocated the burden of proof. He fails to see that he implied that he had reason to believe that oatmeal commands respect. Taking the obvious implication of Alf's statement at face value, Betty asked Alf to state his reason, and Alf's response suggests that he does not have one. The way to rebut someone who says oatmeal doesn't command respect is to say why it does. After the positive side of a debate says that there is a reason for believing X and the negative side denies it, it's always up to the positive side to go ahead and state the reason.

But even if I am right, I would not say the positive side's burden here is unbearable. One reason to have regard for other living things has to do with self-respect. As I said earlier, when we mistreat a tree that we admire, the values we fail to respect are our values, not the tree's. A second reason has to do with self-realization. As I said, exercising our capacity for moral regard is a form of self-realization. Finally, some species share with human beings, to varying degrees, precisely those moral and intellectual characteristics that lead us to see human life as especially worthy of esteem. For example, Lawrence Johnson describes experiments in which rhesus monkeys show extreme reluctance to obtain food by means that would subject monkeys in neighboring cages to electric shock. He describes the case of Washoe, a chimpanzee who learned sign language.[14] Anyone who has tried to learn a foreign language ought to appreciate how astonishing an intellectual feat it is that an essentially nonlinguistic creature could learn a language—a language that is not merely foreign but the language of another species.[15]

Although he believes Washoe has moral standing, Johnson does not believe that the moral standing of chimpanzees, and indeed of all living creatures, implies that we must resolve never to kill. Johnson, an Australian, supports killing introduced animal species such as feral dogs, rabbits, and so forth to protect Australia's native species, including native plant species.[16] Is Johnson advocating a speciesist version of the Holocaust? Has he shown himself to be no better than a racist? I think not. Johnson is right to want to take drastic measures to protect Australia's natural flora, and the idea of respecting trees is intelligible. One thing I feel in the presence of California redwoods or Australia's incredible eucalyptus forests is a feeling of respect. However, I doubt that what underlies Johnson's willingness to kill feral dogs is mere respect for Australia's native plants. I suspect that his approval of such killings turns to some extent on needs and aesthetic sensibilities of human beings, not just the interests of plants. For example, if the endangered native species happened to be a malaria-carrying mosquito, I doubt that Johnson would advocate wiping out an exotic species of amphibian simply to protect the mosquitoes.

Aldo Leopold urged us to see ourselves as plain citizens of, rather than conquerors of, the biotic community,[17] but there are species with whom we can never be fellow citizens. The rabbits that once ate flowers in my backyard in Ohio and the cardinals currently eating my cherry tomatoes in Arizona are neighbors, and I cherish their company, minor frictions notwithstanding. However, I feel no sense of community with mosquitoes and not merely because they are not warm and fuzzy. Some mosquito species are so adapted to making human beings miserable that mortal combat is not accidental; rather, combat is a natural state. It is how such creatures live. It is fair to say that human beings are not equipped to respond to malaria-carrying mosquitoes in a caring manner. At the very least, most of us would think less of a person who did respond to them in a caring manner. We would regard the person's caring as a parody of respect for nature.

The conclusion that *all* living things have moral standing is unmotivated. There is no evidence for it, and believing it would serve no purpose. By contrast, for human beings, viewing apes as having moral standing is motivated, for the reasons just described. One further conjecture: like redwoods and dolphins, apes capture our imagination. We identify with some animals, perhaps even some plants. We feel gripped by their stories. Now, this is a flimsy thing to say, in a way. If we were talking about reasons to see charismatic species as rights-bearers rather than about reasons simply to cherish them, it would be too flimsy. I offer this remark in a tentative way. It is not the kind of consideration that moral philosophers are taught to take seriously, yet it may be closer to our real reasons for valuing charismatic species than are abstract philosophical arguments. Our finding a species inspiring, or our identifying with beings of a given kind, implies that if we fail to care about how their stories turn out, the failure is a failure of self-respect, a failure to care about *our* values.

Viewing viruses as having moral standing is not the same thing. It is good to have a sense of how amazing living things are, but being able to marvel at living things is not the same as thinking that all living things have moral standing. Life as such commands respect only in the limited but important sense that for self-aware and reflective creatures who want to act in ways that make sense, deliberately killing something is an act that does not make sense unless we have good reason to do it. Destroying something for no good reason is, at best, the moral equivalent of vandalism.

7. THE HISTORY OF THE DEBATE

There is an odd project in the history of philosophy that equates what seem to be three distinct projects:

(1) determining the essence of human beings;
(2) specifying how humans are different from all other species; and
(3) specifying what makes humans morally important.

Equating these three projects has important ramifications. Suppose for the sake of argument that what makes humans morally important is that we can suffer. If what makes us morally important is necessarily the same property that constitutes our essence, then our essence is that we can suffer. And if our essence necessarily is what makes us different from all other species, then we can straightforwardly deduce that dogs cannot suffer. (I wish this were merely a tasteless joke.) Likewise with rationality. If rationality is our essence, then it is what makes us morally important and also what makes us unique. Therefore, we can deduce that chimpanzees are not rational. Alternatively, if some other animal becomes rational, does that mean our essence will change? Perhaps this sort of reasoning accounts for why some people find Washoe, the talking chimpanzee, threatening.

The three projects should not be conflated in the way philosophy historically has conflated them, but we can reject species equality without conflation. As noted earlier, we can select a property with respect to which all living things are the same, such as being teleological centers of life, but we need not ignore the possibility that there are other morally important properties, such as sentience, with respect to which not all living things are equal.

There is room to wonder whether species egalitarianism is even compatible with respect for nature. Is the moral standing of dolphins truly no higher than that of tuna? Is the standing of chimpanzees truly no higher than that of mice? Undoubtedly, some people embrace species egalitarianism on the assumption that endorsing species egalitarianism is a way of giving dolphins and chimpanzees the respect they deserve. It is not. Species egalitarianism not only takes humans down a notch. It takes down dolphins, chimpanzees, and redwoods, too. It takes down any species we regard as special. But we have good reason to regard some species as special—those whose members are especially intelligent, especially beautiful, especially long-lived, especially beneficial to humans, especially complex, or even especially fast on its feet.

There is no denying that it demeans us to destroy living things we find beautiful or otherwise beneficial. What about living things in which we find neither beauty nor benefit? It is, upon reflection, obviously in our interest to enrich our lives by discovering in them something beautiful or beneficial, if we can. By and large, we must agree with Leopold that it is too late for conquering the biotic community. Our task now is to find ways of fitting in. Species egalitarianism is one way of trying to understand how we fit in, but all things considered, it is not an acceptable way. Respecting nature and being a species egalitarian are different things.

NOTES

1. This essay revises David Schmidtz, "Are All Species Equal?" *Journal of Applied Philosophy* 15 (1998): 57–67. By permission of Blackwell Publishers.

2. Paul W. Taylor, "In Defense of Biocentrism," *Environmental Ethics* 5 (1983): 237–43, at 240.

3. Paul W.Taylor, *Respect for Nature* (Princeton, N.J.: Princeton University Press, 1986), 99. See also Paul W. Taylor, "The Ethics of Respect for Nature," *Environmental Ethics* 5 (1981): 197–218, at 217.

4. Taylor "Ethics of Respect," 217.

5. William C. French, "Against Biospherical Egalitarianism," *Environmental Ethics* 17 (1995): 39–57, esp. 44. See also James C. Anderson, "Species Equality and the Foundations of Moral Theory," *Environmental Values* 2 (1993): 347–65, at 350.

6. Taylor "Ethics of Respect," 211.

7. For a similar critique of Taylor from an Aristotelian perspective, see Anderson "Species Equality," 348. See also Louis G. Lombardi, "Inherent Worth, Respect, and Rights," *Environmental Ethics* 5 (1993): 257–70.

8. Peter Singer, *Animal Liberation*, 2nd ed. (New York: Random House, 1990), 1–23. See also Lawrence Johnson, *A Morally Deep World* (New York: Cambridge University Press, 1991), 52.

9. Peter Singer, "Reply to Schmidtz," *Singer Under Fire*, ed. Jeffrey Schaler (New York: Open Court, 2009), 455–62, at 461.

10. This is effectively argued by Anderson, "Species Equality," 362.

11. Taylor, *Respect*, 42–43. Note: I have not discussed rights in this essay, but in Taylor's theory it is impossible for nonhumans to have rights, because to have rights, in Taylor's view, a being has to be capable of self-conscious self-respect. Interestingly, although Taylor denies that nonhumans can have moral rights, he grants that there can be reasons to treat trees and animals as having legal rights (Taylor, *Respect*, 246). It is not exactly called for on metaphysical grounds, but it would be a way of treating them with respect. I thank Dan Shahar for the citation and for helpful discussion.

12. Aldo Leopold, *A Sand County Almanac* (New York: Oxford University Press, 1966 [1949]), 116–17.

13. For a discussion of what it takes to deserve respect, see Part II of David Schmidtz, *Elements of Justice* (New York: Cambridge University Press, 2006).

14. Johnson, *Morally Deep World*, 64n.

15. This is what I wrote in the original version of this article. I since have heard that families of lowland gorillas have their own fairly complicated language of hand signals, so I would no longer describe chimpanzees as *essentially* nonlinguistic, except insofar as they

lack the vocal cords and the descended larynx that enabled *Homo sapiens* to articulate the elements of spoken language.

16. Johnson, *Morally Deep World*, 174.
17. Leopold, *Sand County Almanac*, 240.

SUGGESTED READING

Brennan, Jason. "Dominating Nature." *Environmental Values* 16 (2007): 513–28.

Brown, Christopher. "Kantianism and Mere Means." *Environmental Ethics* 32 (2010): 265–83.

Callicott, J. Baird. "Animal Liberation: A Triangular Affair." *Environmental Ethics* 2 (1980): 311–38.

Freiman, Christopher. "Goodwill Toward Nature." *Environmental Values* 18 (2009): 343–59.

Hill, Thomas E., Jr. "Ideals of Human Excellence and Preserving Natural Environments." *Environmental Ethics* 5 (1983): 211–24.

Jamieson, Dale, ed. *A Companion to Environmental Philosophy*. Oxford: Blackwell Publishers, 2001.

Regan, Tom. "How to Worry About Endangered Species." In *Environmental Ethics: What Really Matters, What Really Works*, edited by David Schmidtz and Elizabeth Willott, 105–8. New York: Oxford University Press, 2002.

Rolston, Holmes, III. "Values in and Duties to the Natural World." In *Ecology, Economics, Ethics: The Broken Circle*, edited by Francis Bormann and Stephen Kellert, 73–96. New Haven, Conn.: Yale University Press, 1991.

Sagoff, Mark. "Animal Liberation and Environmental Ethics: Bad Marriage, Quick Divorce." *Osgoode Hall Law Journal* 22 (1984): 297–307.

Sandler, Ronald, and Philip Cafaro, eds. *Environmental Virtue Ethics*. Lanham: Rowman & Littlefield, 2005.

CHAPTER 23

GENETICALLY MODIFIED ANIMALS: SHOULD THERE BE LIMITS TO ENGINEERING THE ANIMAL KINGDOM?

JULIAN SAVULESCU

It is now possible to radically alter animals or create new life forms by transgenesis or the creation of hybrids and chimeras. Transgenic animals are created by transferring genes from one species to another. Hybrids are created by mixing the sperm of one species with the ovum of another. Chimeras are created by mixing cells from the embryo of one animal with those of a different species. In each of these cases, the source of one animal could be a human being, that is, the genes, gametes, or embryonic cells could be from a human, thus creating an animal with human genes or a human-animal hybrid or chimera, part-human and part-animal. Genetically Modified Animals, or GMAs, is the term I will use throughout this chapter to refer to transgenic animals, hybrids, and chimeras. The creation of novel life forms has been and continues to be one of the most profound human achievements, though it has scarcely received commensurate critical ethical scrutiny. As the biotechnological revolution relentlessly progresses, the power to use the engine of life for human design will only increase, for both good and ill. We need a new ethic for evaluating the creation of GMAs.

One central thesis of this chapter is that there should not be an overall general normative evaluation of the acceptability or unacceptability of the creation of GMAs. The ethics of creating a specific GMA must be evaluated on a case-by-case basis. Nonetheless, I will argue that there should be limits to the creation of some forms of GMAs.

A second central thesis is that practical ethics requires developing richer strategies of ethical evaluation of technology. I argue for a four-stage evaluation process that establishes whether the creation of a GMA is permissible. Firstly, we must develop a justified account of moral status. The moral status of the GMA should be correctly determined and the GMA treated appropriate to status. This may require specific research to determine moral status prior to the use by humans. Secondly, the creation of GMAs should be done for good reasons as assessed by the enhancement of the animal itself or by the improvement of the lives of other animals, including humans. Thirdly, research should conform to the basic principles of research ethics. In particular, it should ensure that the risk of harm to which the animal is exposed is reasonable and if the GMA has a high moral status, then consent be obtained if possible and reasonable. Fourthly, creation of GMAs should not unreasonably harm others, either now or in the future, by the dual use of the technology (i.e., for both good and evil purposes), to create animals that present an infectious or other risk.

The third central thesis of this chapter is that there are not good arguments against the creation of GMAs tout court. I reject four arguments against the creation of any GMAs: (1) creation of GMAs represents a threat to humanity; (2) absolute deontological constraints preclude creation of GMAs; (3) creation of GMAs is wrong because it is playing God; (4) to create GMAs is to descend down a slippery slope to the radical genetic alteration of human beings and it is precluded by the precautionary principle.

Humans can responsibly and ethically create new forms of animal life. They have done so in various breeding programs. The challenge will be to use vastly more powerful biotechnology responsibly and ethically by developing more sophisticated and explicit strategies of ethical evaluation.

Section 1. Scientific Possibilities, Utility, and Regulation

In section 1, I will review the extraordinary range that currently exists or likely will soon exist of possibilities of creating GMAs through the construction of new kinds of beings. I will review the social value of creating GMAs in modern medical research. They potentially have great value in creating products useful to humans, creating disease-resistant or other more resilient species, and potentially providing organs and tissues for human medical use. I will also review some policy and regulatory issues that need to be addressed. Here we are at a primitive stage. Once the utility of these creatures has been established in science, there will be pressure from scientists and the public to create them, but our current regulatory and ethical frameworks are ill prepared to evaluate such proposals.

1. Transgenic Animals

It has been possible since about the 1980s to transfer genes taken from one species into another. This process, which produces "transgenic organisms," is responsible for the creation of what has sometimes been referred to as "genetic freaks." Two examples are ANDi, a rhesus monkey, and Alba, a rabbit, both of whom have a fluorescent jellyfish gene incorporated into their DNA. These animals are otherwise healthy and normal. But each has a fluorescent green glow. Scientists recently introduced a similar jellyfish gene into a human embryo, creating a fluorescent human embryo. It was destroyed, but if brought to term, it would have created a fluorescent human being.[1]

The creation of transgenic animals has been an important tool of modern science and technology. In this section, I will review four important uses in: (1) medical research; (2) creation of disease-resistant species; (3) production of useful biological substances; and (4) xenotransplantation research.

1. *Medical Research.* The use of transgenic animals as models for research into human disease has greatly improved our understanding of disease and the development of treatments. For example, scientists have recently inserted most of the human Chromosome 21 into mice to create a mouse model of Down Syndrome.[2] A substantial proportion (roughly half) of U.S. National Institutes of Health (NIH) research involves transgenic animals.

2. *Creation of disease-resistant species.* Genes can be transferred from one species to another to confer disease resistance on the receptive organism.

3. *Creation of species that produce products that are useful to humans.* Scientists successfully inserted spider genes in a herd of goats, resulting in the extraction of silk-forming proteins and fibers from the goat milk. The isolated fibers, which are those that comprise spider webs, are the strongest existing elastic materials and therefore have an extensive range of possible medical and industrial applications.[3] This silk is known as BioSteel, and the strands are used as sutures and materials to reconstruct tendons, ligaments, and bones. Genzyme Transgenics[4] utilizes transgenic animals to produce human therapeutic proteins (Anti-thrombin III, a protein that can prevent blood from clotting) in the milk of transgenic animals.[5]

4. *Xenotransplantation.* Human genes have been introduced into pig embryos to create organs that will not elicit immune responses when transplanted into humans.

One kind of GMA would be a genetically modified human, created by transferring genes from nonhuman animals into humans. I will not deal with this possibility in detail. Genes from animals, or artificial copies of those sequences,[6] could be used to confer resistance to diseases. Transgenesis could be used to more radically enhance human beings.[7] For example, it has been hypothesized that aging in human beings is related to the degradation of telomeres, the regions on the end of our chromosomes.[8] It is possible that animals with a significantly longer lifespan than humans, such as turtles and rockfish,[9] are genetically programmed to live longer, and these genetic programs could be inserted into the human genome. Transgenesis could also be used to introduce genes coding for superior physical abilities from other animals, for example transferring the gene responsible for enhanced night vision in animals like rabbits and owls, or introducing those genes that code for sonar in bats.

2. Animal Hybrids and Chimeras

For a long time, it has been possible to create cross-species hybrids through mating. The mule is a cross between a horse and a donkey. It is infertile. Other species hybrids include the "Tigon," which was created from a female tiger and a male lion. These hybrids require breeding programs.

Recently, animal chimeras have been created—a development that seems to take us to a new stage of cross-species biology. A chimera is a novel organism made by combining embryonic cells from more than one species, sometimes by fusing embryos.[10] The "Geep" is a sheep/goat hybrid created by the fusion of a sheep embryo with a goat embryo, first achieved over twenty-five years ago.[11] Such animals are chimeric, meaning that the animal consists of a mix of cells from both species. Such chimeras are not like those of mythology, half one animal, half the other. They are a blend of characteristics. For example, the geep looks like a goat or a lamb, and is similar to any other newborn animal in that it has a blend of characteristics of both parents.

Many other animal chimeras have been created. These include the chick-mouse, quail-duck, quail-mouse and even the quail-alligator.[12] More recently, human-animal chimeras have been created by combining cells of human and nonhuman origin.[13] The creation of many forms of GMAs raises ethical issues explored throughout this chapter. Some attempts have been made, largely unsuccessfully, to address them through policy and regulatory processes, a subject meriting brief consideration in its own right.

3. Regulation

The creation of human-animal chimeras, hybrids, and transgenic animals raises important regulatory questions: Who should regulate their creation and use? How should different ethical considerations be assessed and balanced in ethical review? When should research using these be considered preferable to use of nonhuman animals or human embryos? Where is special regulation or prohibition required? These questions have not been adequately addressed.[14] Voluntary codes, such as that of the National Academy of Sciences in the United States,[15] have been criticized for a weak ethical basis,[16] and there is widespread disagreement on many points between scientists.[17] There is a danger that if the preceding questions are left incompletely answered until the point at which applications for specific research projects are evaluated by ethics committees, then these evaluations will be "reactive" and knee-jerk, without timely, far-sighted, and comprehensive evaluation of the issues involved.[18] There have been calls for a full ethical analysis of ethical issues in advance.[19]

a. Regulation of "Admixed Embryos"

One illustrative example of attempted regulation is the United Kingdom's approach to the creation of GMAs involving human embryonic cells. The mixing of human and nonhuman embryonic material is highly controversial. The U.K. legislation is an example of a strategy that aims to realize the potential of this avenue of research while adopting a practical strategy to address ethical concerns by limiting the duration of embryo development. I will examine the utility of this research, the objections to it, and how the U.K. regulatory authorities attempted to assuage these. However, I will go on to argue that this strategy will fail in the long term as it becomes desirable and useful to allow embryos to develop to maturation.

In 2008, the Human Fertilisation and Embryology Authority (HFEA), which regulates artificial reproduction in the United Kingdom, agreed to license for research the creation of embryos using human and nonhuman embryonic material, which they called "admixed embryos." Scientists must apply for a license to create an admixed embryo. Crucially, permission will only be granted to allow development up to fourteen days, a point which has been taken to be of moral significance since legislation regulating assisted reproduction came into force in the 1980s,[20] and it is a criminal offence to allow the embryo to develop beyond that point. So, the United Kingdom has addressed the issue of the creation of GMAs involving human genes by limiting the period of embryonic development. Other GMAs not involving human embryonic material are not affected by these regulations.

The definition of "human admixed embryos" given in the HFEA Bill includes four different types of embryos that could be formed under license (see box 1).

Box 1. Human Admixed Embryos: Formation

Cybrids (or cytoplasmic hybrids): formed by transferring human nuclear DNA in the enucleated cytoplasm from the egg of an animal, such as a rabbit.

Transgenic embryos: formed by introducing animal DNA into one or more cells of a human embryo.

Chimeras: formed by adding one or more animal cells to a human embryo.

Hybrid embryos: formed from a human egg and animal sperm or vice versa; or from an animal pro-nucleus and a human pro-nucleus.

Box 2. Differences Between Embryos

Cybrids: nuclear DNA is human, matched to the person from whom they were "cloned"; very small amount of mitochondrial DNA from the animal's cytoplasm.

Transgenic embryos: variable amount of animal DNA depending on how many genes are transferred.[21]

Chimeras: have both human and animal cells, because they are formed by merging human and animal embryos.

Hybrid embryos: have both human and animal chromosomes, because they are formed by fertilizing egg and sperm from different species.

The differences between these embryos are shown in box 2. In short, in cybrids, the nuclear DNA (which confers nearly all phenotypic characteristics) is human, while the mitochondrial DNA (which is mainly responsible for energy metabolism) is nonhuman. However, the other embryos would have a variable amount of human and animal DNA.

b. Why Was Legislation Introduced? Utility

This legislation was introduced because scientists had applied for licenses to the HFEA to create "cybrids" that could serve as a source of embryonic stem cells. The legislation aimed to allow scientists to realize the utility of this line of research by limiting the period of embryonic development to that permissible for fully human embryos already permitted by existing legislation. In the United Kingdom, human embryos can be created for research purposes by cloning and IVF for up to fourteen days. This legislation merely extended that permission to include transgenic reproductive technologies involving nonhuman cells or genes.

Cybrids are useful for research because they reproduce indefinitely and are pluripotent, that is, they can develop into other cell types, such as nerve cells, muscle cells, and heart cells. One day, human embryonic stem cells (hES cells) may serve as "matched" tissue for self-transplantation to treat patients with disease. The hES cells could be derived by inserting the nucleus from a person's body cell into an egg that has had its nucleus removed and been stimulated to develop. This is known as nuclear transfer and is a type of human cloning. These stem cells should not be rejected by the patient's body, unlike donated tissue and cells from donated "sperm-egg" embryos, which have genetic material from both the egg and sperm donors. Long-term treatment with immunosuppressive drugs, with attendant serious risks, could then be avoided. Stem cells could also be used for "regenerative medicine" to replace dead or damaged tissue after stroke, heart attack, or other disease or injury. While admixed embryos could not be used for this purpose, the development of the technology using admixed embryos could facilitate the use of human embryos for this purpose in the future.

Already hES cells are being used in research into a range of human diseases that are caused by genetic mutations or that have a genetic element, such as diabetes, cancer, Parkinson's disease, Alzheimer's disease, and motor neuron disease.[22] Some of these cells have come from patients with those diseases, but it is not always convenient or possible to obtain hES cells that carry mutations for genetic conditions (disease models) in this way. Scientists therefore want to create hES cell lines with particular mutations to study diseases and test new drugs. This opens up a new and potentially important line of research to create cellular models of human disease that avoids some of the ethical objections to embryonic stem cell research, and which in time may lead to new treatments.[23] For example, the creation of cellular models for human disease using hES cell lines could produce knowledge of disease and potential drug therapies that would be cheap to produce and could be globally available, while also not facing oncogenic or infectious concerns that confront the direct development of embryonic stem cell therapies.

The use of animal eggs, and creating GMAs, has several advantages over the use of human eggs for this research. Far more eggs would be available, as the number of potential donors of human embryos and eggs is likely to be small. Women in reproductive technology programs often want to use their eggs themselves, donate them to another couple, or have them removed from storage. Other donors may be available, but obtaining eggs is invasive, and there are risks in superovulation and the surgical removal of eggs. Other sources of eggs, such as surgically removed ovarian tissue, cadavers, aborted fetuses, and hES cell lines, even if available, have limitations. There may be intuitive repugnance to these sources of eggs, and difficulties may arise in artificially "maturing" the eggs before they can be used in research. The eggs may be of poor quality, and they do not contain all the genetic mutations that need to be studied. Also, the efficiency rate of cloning is likely to be low (less than 1%), so vast numbers of human eggs would be required for research.

Using cybrids overcomes a number of objections commonly raised to hES cell research. These objections to traditional hES cell research are: (1) the research requires large numbers of eggs that must be removed from healthy young women at some small risk of life and health to them; (2) the research is a Western luxury that will be exotic and expensive, only benefiting the richest countries; (3) adult stem cell research shows great potential and is preferable to hES research; (4) recent research shows there may be infectious and other risks,[24] such as occurred with bovine spongiform encephalitis, when transplanting tissue back to people after it is grown on foreign culture material or uses animal eggs.

The creation of hybrids using nonhuman animal eggs to develop cellular models of disease is immune to all of these moral objections. Firstly, it would require no human eggs to produce vast amounts of tissue for the study of disease. Secondly, such research may result in the development of drugs for common conditions that afflict people all around the world, including the developing world. Thirdly, it is not possible to use adult stem cells in this way. Fourthly, there would be no risk of infection from drugs developed by studying tissue in this way, as the drug molecules would be produced pharmaceutically. These objections apply (if they apply) only to cloning for self-transplantation or traditional hES cell research. They do not apply to cloning and to the creation of cybrids that will help us understand and treat disease.

So, the introduction of legislation that limits cybrid (and other GMAs involving human cells) development to fourteen days seems to allow an adequate balance between utility and ethical concerns. Whatever ethical objections are raised, if such GMAs are killed at fourteen days, they cannot be realized. Let's call this the "limitation in development" approach. Such limitation in development approaches must fail. The reason is that there will be overwhelming reasons to create full-blown, live-born chimeras, including human-animal chimeras.

c. The Utility of Live-born Chimeras

Full-blown chimeras are valuable for research. Chimeras (which have mainly been mouse chimeras thus far) are already routinely used in laboratory research, especially to create models to study human disease. Chimeras such as the quail-duck chimera have been used to study embryonic development, such as the development of the face. "Primary" chimeras are produced by the injection of embryonic stem cells or neural stem cells into an early embryo (blastocyst). They are of importance in embryology, neurology, pathology, and the development of stem cell therapies. "Secondary" chimeras are created by introducing embryonic cells at a later stage. They are used to test the potentiality of stem cells to grow into fully differentiated tissues.[25]

There are strong scientific rationales for creating live-born human-animal chimeras. Chimeras are especially useful in neurobiological research.[26] For example, research on human development may be enhanced by producing chimeras that have large populations of functional human neurons. Irving Weissman, a pioneer of chimera research (as discussed by Henry Greely in the present *Handbook*), anticipates producing mice with fully "humanized" brain tissue.[27] Such animals would be ideal models for human disease, and chimeras with mature and functional human brain cells have been recently created.[28]

Moreover, research into cognitive or behavioral development may benefit from creating chimeras that have human capacities. This would likely involve keeping chimeras beyond fourteen days, especially during later times of neurogenesis (between day 4 and 18)[29] and/or using nonhuman primates.[30] The HFEA "limitation of development" approach of imposing a fourteen-day limit on embryo development will likely represent an obstacle to scientific research and so be challenged. The strategy of imposing a fourteen-day limit on development, employed by the HFEA, may deal effectively with ethical objections, but it is likely this limit will be challenged and ethical objections raised to allowing the further development of GMAs.

Several properties of chimeras are potentially morally relevant. Firstly, human-nonhuman chimeras do not have only genetically human chromosomes. They have a variable mixture of cells with and without human DNA. For example, in 2007, Esmail Zanjani created a sheep-human chimera (85% sheep; 15% human).[31] They could be created with predominantly human non-neural cells but nonhuman neurons, vice versa, or with mixed human and nonhuman neurons. Such chimeras may have cognitive capacities comparable to normal adult humans, to humans in other stages of development (or degeneration), or to normal nonhuman animals. Their capacities may be "enhanced" or "diminished" relative to nonhuman animals, insofar as such terms are appropriate. Such chimeras could then be subject to different procedures including modification, implantation into different species, environmental systems, painful procedures, and destruction. An example would be to create a mouse with neural stem cells from a human with some genetic disorder, such as Huntington's disease or inherited Alzheimer's disease. The mouse, with a small human brain, could be used as disease model for that disease in the full-blown human. One of the most provocative examples of a chimera would be a human-chimp chimera, created by fusing a chimp embryo with a human embryo. Such a chimera might well survive and could be used to study the development of language.

Secondly, it is currently unclear which species chimeras belong to and so which regulations and laws pertain, which potentially morally-relevant capacities they possess, and how these are relevant in given uses.

Thirdly, chimeras could be modified to experience less pain, boredom, or other negative mental states during research or farming. That is, chimeras could be created with altered sentience. In 2004, scientists in University College London made a small genetic modification to a group of mice to remove one of the nine types of sodium channels that are present in all mammalian muscle and nerve tissue. Mice who had the channel, Nav7, removed entirely from all tissue died as pups due to an apparent inability to feed. But mice who only had the channel removed from their sensory neurons were apparently normal in every way (life expectancy, weight, motor ability) except in one area. Although their ability to experience normal variations in heat and pressure was unchanged, they were much less vulnerable to pain both from external stimuli and from inflammation-induced sensitivity. Engineered mice exposed to painful levels of heat on a Hargreaves Apparatus, a kind of hotplate, were 20% less sensitive to painful levels of heat than their control-group counterparts. Likewise, mice missing the Nav7 channel displayed less pain on the application of paw pressure than control-group mice. In addition to these behavioral differences, the mice were much less susceptible to inflammation and to related pain. Mice were injected with a water and oil emulsion with mycrobacteria that causes inflammation leading to hyper-sensitivity to heat and pressure pain. Mice missing the Nav7 channel had no increase in sensitivity, although the swelling recorded was similar to control groups.[32] The creation of GMAs with significantly different behavioral characteristics and responses, such as to pain and boredom, and even perhaps those typical of human beings such as language and higher cognitive abilities, raises new and significant challenges in animal ethics.

4. Conclusion

In section 1, I have reviewed the scientific possibilities of creating GMAs either through the construction of transgenic animals or chimeras. I have reviewed the utility of creating GMAs. They have been a mainstay of modern medical research and represent an important way of genetically modifying animals to make them more useful to humans in other ways, such as by creating products useful to humans, creating disease-resistant or other more resilient species, and potentially providing organs and tissues for human medical use. I have argued that there are good scientific reasons to believe that more widespread genetic modification of animals using either human genes or human embryonic material is on the horizon. I have reviewed current regulatory issues and provided one example of legislation, that of the Human Fertilisation and Embryology Authority's approach to regulate the creation of human admixed embryos. While that legislation has been satisfactory in dealing with the issue of cybrids, I have argued that it will be inadequate to address the creation of full-blown, live-born human-nonhuman chimeras. I have argued that these will be useful for research and there will be pressure from scientists to create them. Our current regulatory and ethical frameworks are ill prepared to evaluate such proposals, and I turn attention now to that issue.

Section 2. A Four-stage Ethical Evaluation of the Creation of GMAs

We need a new, more comprehensive approach to evaluating the creation of live-born human-animal chimeras and radically modified GMAs. In section 2, I offer a four-stage process of ethically permissible evaluation of the creation of GMAs. In the final section, I go on to reject four commonly employed objections.

Stage 1. Determination of and Respect for Moral Status

The most important ethical constraint on creating GMAs should be should be *the correct determination of their moral status and appropriate treatment.* "Moral status" is the standing or position of a being within a hierarchical framework of moral obligations. The moral status of a GMA entails relevant obligations to treat it in certain ways while it is alive, in virtue of its nature, and whether it is wrong to kill it. Determination of the moral status of a being is based upon its intrinsic nature and its relationship to other beings, which themselves may determine its interests and rights. For example, how would we determine the status of a mouse with a small predominantly humanized brain that has similar interests to a human being? Is it a mouse, a human, or something in between? If it has intermediate moral status, how should it be treated by human beings? Is it permissible to kill it in research? What would the moral status of a human-chimp chimera be?[33]

One line of attack is to follow the U.K. strategy and, until such questions can be answered, require not letting such new life forms grow beyond early gestation. These entities could not have a higher moral status than human embryos and they would be destroyed before fourteen days of gestation, as required by law. However, such an approach can only stave off the issue for a limited time. At some point, there will be strong reasons based on utility for allowing the development of GMA embryos to full gestation.

We will, then, need to develop: (1) an ontological account of moral status; (2) an epistemological account of how we should decide the moral category into which a particular animal falls.

a. Ontology: What Are the Criteria of Moral Status?

What sorts of beings or entities have moral status? This complex question involves an understanding of the nature of a being or entity and placing that being within a normative framework of moral obligations. For example, it is not thought to be wrong to smash a rock because it does not have interests that would generate obligations toward it. It is not "hurt" by the act of smashing it. (The situation is even here complicated, as some environmentalists have noted in discussing the destruction of inanimate objects in nature.) It is wrong to smash the skull of a baboon with a high-impact mechanism in a laboratory, because a baboon can feel pain and has an interest in not experiencing pain. Two important issues relating to the treatment of GMAs are how we treat them while they are alive and whether it is wrong to kill them. What are the criteria for this kind of moral status? Answers to this question comprise a large literature (see for example, the contributions in Part III of this *Handbook*). In the following section, I consider three possible criteria for moral status. My aim is not to endorse or justify them (though I will reject the commonest one) but to describe them and to follow the implications of accepting two of them as criteria of moral status. My purpose will be to show that it is critical to establish an account of moral status and to be able to evaluate whether a GMA has a level of moral status before we proceed to create it.

1. Species Membership. A GMA's moral status might be thought to depend on its species,[34] notably the human species. This criterion of moral status seems to capture something like commonsense moral views, including how we think of "human rights." However, there are two problems with this approach. In some cases, it may be clear to which species a GMA belongs. For example, the SCID-hu mouse, which has a human immune system, is usually considered to be a mouse.[35] But in other cases it may be less clear whether an individual comes under the description of an animal with human cells, a human with animal cells, or some other description. Does a human-chimp chimera with 50% cells of each belong to the species *Homo sapiens* or *Pan troglodytes* (the technical term for the species of a chimp)? In this way, GMAs present a potentially revolutionary challenge to commonsense approaches to moral status.

Of great importance here is the relevance, if any, of biological species membership. It is a well-worn point now that species membership, even membership of the species *Homo sapiens*, is of no intrinsic moral significance. I will not repeat the arguments here.[36] Suffice it to say that species membership will not solve the question of moral status.

2. *Sentience.* Moral status is most plausibly analyzed in terms of a being's capacities and functionings. One theory of moral status relies on sentience—that is, the capacity for conscious experience, including painful mental states. Sentience provides moral reasons to avoid inflicting painful stimuli on and depriving of pleasurable experiences, animals with a cortex. A moral status based on an interest not to experience pain and deprivation is dependent upon having the capacities for nociception,[37] phenomenal conscious experiences,[38] and negative attitudes to pain.[39] As an example of possible use of this criterion of moral status, future variants of the Nav7 mouse might not experience pain.

3. *Higher Cognitive Functioning.* Another theory erects status on cognitive capacity and functioning.[40] Moral status has been related to rationality,[41] acting on the basis of normative reasons, autonomy, consciousness,[42] and self-consciousness.[43] Some accord instrumental value to capacities that affect an animal's welfare[44] or ability to "flourish."[45]

One influential theory of moral status is that of being a person. Persons are sentient beings that are also rational and self-conscious, conceiving of themselves as existing over time and capable of forming preferences for various of their possible future existences. Michael Tooley, John Harris, and Peter Singer (at least in some of his early works) argue that what matters morally is not being a member of the species *Homo sapiens* but the mental properties that characterize persons, or beings capable of playing a planned and chosen role in life—rationality and self-consciousness. These characteristics capture humans as persons.[46] According to this theory, it is wrong to kill persons because they can conceive of themselves as existing across time and have preferences for their continued existence into the future. Nonpersons do not have such preferences. The frustration of these preferences is wrong. Just as sentience grounds a right not to have pain inflicted, so being a person grounds a right to life. However, for illustrative purposes in this chapter, I will refer to "higher cognitive functioning" as the relevant property related to the moral status of having a right to life, not persons and associated theories of persons.[47] It is possible that higher cognitive capacities ground other interests, but for simplicity, I will limit discussion to the simple interests and rights that derive from sentience and higher cognitive capacities: interest in positive mental states and continued life.

b. Epistemology: Determining Moral Status

Once we are clear on the ontology—what moral status consists in, for example, sentience, higher cognitive capacity, and the like—it will be necessary to determine whether a novel GMA has the requisite properties (for example, capacity to experience pain or self-consciousness, moral agency, ability to use a language, etc.). This is an epistemological question about how we determine a being's capacities. Clearly such determinations have been enormously difficult in contemporary psychology and cognitive science, as several essays in this *Handbook* (by Kristin Andrews, Bryce Huebner, Peter Carruthers, José Luis Bermúdez, Sahar Akhtar, and Mark Rowlands) show.

Consider two examples. Synthetic biology or nanotechnology could be employed to create or modify complex neural networks in vitro. Could such networks become conscious? How would we determine this?

In 2001, scientists devised an experiment to test the effect of the anterior cingulate cortex (ACC) on the experience of pain.[48] Anecdotal evidence in humans has already suggested that surgical removal of the ACC alters the experience of pain so that although the patient is still able to identify its location and intensity, the experience is not unpleasant or dysphoric. In one experiment, lesions were surgically induced in the ACC of rats. Rats with the lesions displayed normal behavioral responses to pain stimuli, such as flinching and licking the affected area. Although they did not develop a conditioned avoidance response to the pain stimuli as did the control-group mice, this state did not affect their ability to develop a conditioned response to non-pain-related but still aversive stimuli (a dysphoric opioid agonist), nor did it affect the ability to learn conditioned pleasure responses.

Do such rats "experience" pain? It is not easy to say. Based on human reports, the answer may be no. While these are ordinary rats modified surgically, it is easy to imagine genetic modification that could bring about the same effect. Would it be permissible to inflict ordinarily painful stimuli on such rats? In many cases, it may not be possible to predict, based on biological data, the nature of the moral status of a GMA. Rather, we would require behavioral and other evidence that an animal has the relevant property that confers moral status.

Behavioral investigation may need to be sophisticated and diverse, including evaluation of social functioning. Consider a human-chimp chimera. Would it be best, in terms of pleasurable mental states and other markers of well-being, for such a GMA to socialize with chimps, humans, or other human-chimp chimeras? Getting answers may require complex field experiments involving different social arrangements.

c. Treatment Relevant to Moral Status

The moral status of GMAs may well be far from obvious. What then should be done? Could GMAs be farmed or experimented upon? Under what conditions should they live or be killed?

Two sub-constraints should be observed if GMAs are to be treated ethically. Firstly, *GMA-centered research to determine the being's nature and moral status should be performed before use by humans*, especially regarding what would make the GMA's life go well. So we need a robust account of moral status, and second, we will require empirical research to identify whether a GMA has the qualities which bestow that status. Secondly, where research is not done or fails to yield useful results, the *GMA should be accorded the highest moral status consistent with its likely nature*, based on available evidence and prediction. We should err on the side of sympathy and generosity. If there is a chance a GMA could experience pain or might not be able to interact socially, and we don't know, it should be treated as if it experiences pain and will have problems of social adaptation. Likewise, if it could plausibly have higher cognitive functions, it should be treated as if it had them.

STAGE 2. Reasons in Favor of Animal Enhancement

The second major ethical constraint on the creation of GMAs is that genetic modification of animals should be done for a good reason. Perhaps the animal should be
modified to enhance it for the benefit of itself, other members of its species, or
members of other species, such as human beings. Enhancement can be broadly
defined as follows: "X is an enhancement for A if X makes it more likely that A will
lead a better life in circumstances C, a given set of natural and social circumstances."[49]
If A is the GMA itself, modified by changing its biology or psychology, then it will
be an example of what I call enhancement-for-self. If A is a human being, or other
animal, then the GMA would provide enhancement benefits to others. The main
problems arise in enhancement-for-others, but enhancement-for-self can be problematic as well.

a. Enhancing of Self

While the ethics of human enhancement has been widely discussed,[50] the ethics of
animal enhancement has not. According to the definition of enhancement, to evaluate whether a modification is enhancing-for-self, we require an account of animal
welfare,[51] which is itself problematic in the case of GMAs because that life has
never been encountered before. We cannot refer to vague concepts such as the
"pigness of a pig."[52] An evaluation of whether a modification is good for an animal
will require an evaluation of broad and long-term effects. However, as the definition makes clear and I have argued elsewhere,[53] whether a given change in the biology of an animal constitutes an enhancement will depend on the context. For
example, is a modification that reduces sensitivity to pain good for an animal? It
will depend on whether the modification might render an animal more prone to
injury that harms the animal in other ways, which will depend on the environment
the animal is in.[54]

In sum, whether some change in biology or psychology of an animal constitutes
enhancement or infliction of a disability will depend on at least three things: (1) the
account of animal welfare; (2) the short- and long-term biological and psychological consequences of the intervention; and (3) accurate prediction of the likely natural
and social environment in which the animal will exist.

It is arguable that (3) should be revised. As it stands, it opens the door to biological modification to compensate for exploitation. For example, if the Nav7 modification could be performed on cows, pigs, or chickens to render them less susceptible
to pain, this could be employed to press higher numbers of animals into confined
conditions to increase yield. It could be argued that such conditions are unjust
and that (3) should be modified to (3*): accurate prediction of the likely natural and
social environment in which the animal would exist from among those in which it
is just to exist.

Evaluation of an intervention as an enhancement or not will thus require an
account of the environment in which it is just for animals to live.

b. Enhancement for Human Benefit

How should we evaluate the genetic modification of an animal to make it more useful to human beings? This is the basic issue associated with animal experimentation. One matter needing discussion is the moral problem of how to assess the harm it is reasonable to inflict on animals for the benefit of human beings. Strict equality in consideration of interests would require that animal pain/life be treated equally to human pain/life. This proposal is implausible. Even if it is wrong to kill an animal with a right to life, it seems as if it is less wrong than killing a human being. There seems to be a good reason (perhaps based on the greater richness and depth of human life to give greater [though not absolutely overriding] weight to human life) and suffering. Assuming that there is such a good reason, we can give priority to humans rather than other animals, but if there is no such good reason, then it would be unjustified. How a balance is to be struck is a complicated question, but one that must be answered. It is considered in further depth in the chapters in this *Handbook* by R. G. Frey, Elizabeth Harman, David Copp, Mark Rowlands, and David Schmidtz. A similar question is how we should weigh the lives of humans that are of different value.[55]

An extreme case of genetic modification of animals for human purposes would be "xenogestation": the genetic modification (or other biological modification, such as creation of a weak predominantly nonhuman chimera) of a nonhuman animal to carry a human embryo and fetus to term—for example, genetic modification of a pig to carry a human baby. The baby could be delivered by "natural" vaginal delivery or by Caesarean section. I will consider this as an example of the extreme modification of animals for human use.

Could there be a good reason for such a modification? Firstly, it could allow women to avoid the risks of pregnancy and delivery, which can be not merely inconvenient but lethal and are frequently associated with major chronic health problems, such as fecal and urinary incontinence, pain during sexual intercourse, and so on. Secondly, the fetal uterine environment could be genetically or otherwise manipulated to maximize the nutritional and otherwise supportive environment for the baby. The pregnancy could be extensively monitored and externally controlled. Thirdly, it could be used to carry embryos and fetuses destined for abortion, alleviating the burden to women of carrying pregnancies that would never produce a live-born baby. Fourthly, embryos or fetuses could also deliberately be created for either experimental or treatment purposes. There are, of course, objections to using human cloning as a source of cells, tissues, or organs for transplantation, including these: (1) cloning has been very inefficient, requiring hundreds of embryos to be created; (2) stem cells have not been successfully manipulated to produce transplantable tissues; and (3) there has been little real progress toward the development of organs.

One way to address these problems is to avoid stem cell technology entirely and allow nature to create the tissues and organs in a human fetus undergoing normal development. A cloned fetus could be aborted and tissues and organs used for medical purposes. Tissues have been used for these purposes from social abortions. The practice of abortion is premised on the fetus lacking moral status. If it is permissible to create and destroy an embryo for medical purposes, it is arguably permissible to create and destroy a fetus. A fetus could be cloned as a source of tissue or organs for transplantation and transferred to a genetically modified pig to gestate it until the organs and tissues were ready for transplantation. Such proposals may appear shocking, but they can be ethically defended, based on current practices, values, and public policies (or lack thereof).[56]

In summary, there will be some strong reasons in favor of the creation of radical GMAs based on utility to others, especially humans. Whether these reasons outweigh the reasons against the creation of the particular GMA needs ethical analysis. One significant factor in this balancing is how the GMAs own life goes. The balance of these reasons for and against will be reflected by the outcome of the four-stage evaluation process. What I have sought to establish is that there can be very strong reasons for the creation of full-blown GMAs.

c. Identity Determining versus Identity Preserving Interventions

When evaluating the strength of reasons in favor or against biological modification of animals, one important but difficult consideration is whether the intervention determines identity or preserves an existing identity. Genetic modification can occur prior to the establishment of individual identity or after an individual identity has been established. For example, in the case of human beings, many people believe that our identity is established at conception. However, this belief is disputed; other candidates include the point at which tissues such as the neural streak begin to develop.[57] It is not the purpose of this chapter to establish exactly when an animal begins to exist and is individuated. Let's call that point "identity origination."

Interventions prior to identity origination can be identity determining. Genetic selection is an obvious example. The selection of one embryo to implant, based on genetic profiling, from among several embryos determines the identity of the future animal. Interventions performed after identity origination can be identity altering. For example, genetic engineering of a chimp embryo to give it the mental capabilities of a human being would be identity altering. It would be like killing the chimp and replacing it with a human being. This would not be an enhancement of one chimp but its replacement by another different chimp or even an animal of a different species. These replacement cases are equivalent to killing and would be wrong if the original had sufficient moral status conferring a right to life. Interventions that are neither identity determining nor altering are identity preserving. For brevity, I will refer to both identity determining and identity altering interventions as identity determining.

Identity determining interventions raise what Derek Parfit has called the nonidentity problem.[58] Imagine we were to create a human-chimp chimera by fusing human embryonic cells with chimp embryonic cells. How should we evaluate this act in terms of the kind of life that the resultant chimera would lead? Should its life be compared with that of a human, or a chimp, or both? Imagine that the biology of a human-chimp chimera causes it to become depressed, or to experience some other persistent negative mental state such as that manifest in stereotypical behaviors, compared to either a typical chimp or human. Is this experience bad for the chimera in the same way a headache is bad for me? In virtue of the badness of the mental state, was the act of creating such a chimera wrong? If the adverse experiences were sufficiently profound, say constant terror from hallucinations (as occur with ketamine use) or severe pain, the biological conditions might render the GMA's life so bad that it is not worth living. It clearly would be unethical to create such animals.

What if the biology is such that it strongly predisposes to a life of diminished well-being, but the life has many pleasures and is still worth living? Suppose, for example, the animal suffers mild persistent depression. Here we encounter the non-identity problem. If the life is worth living because it has good food and social companions, and the intervention that causes the negative welfare property is a part of the identity determining intervention, the intervention is not so bad from that GMA's perspective. The depressed human-chimp chimera can have no complaint, even if mildly depressed, regarding the act of creating it, because without the act of fusing human and chimp embryonic cells, it would not have existed and its life is not all bad.

It is still possible to argue that there is something like an "impersonal wrong"[59] that makes it less desirable to create this life form, rather than a different life form with a better life. However, the strength of such reasons is unresolved, and certain arguments do not work well if the alternative is to create no animal at all. Identity determining interventions seem only seriously wrong when they create animals with lives that are not worth living or if the alternative had been to create an animal with a better life.

d. Social Construction of Well-being

A final consideration that bears on evaluating the reasons for or against animal enhancement is the social construction of well-being.[60] How well an animal's life goes is not merely a function of its inherent biology and psychology, but also of its social environment. Consider the introduction of a gene from a fluorescent jellyfish into a monkey embryo, which causes a fluorescent monkey. Whether this intervention is permissible turns in part on the quality of the monkey's life and how it would have gone in the counterfactual world in which the gene was not inserted. If the monkey is more easily seen by its mate or other members of group at night, and this benefits the monkey, then it is an enhancement. If it makes it more likely to be killed by predators, then a risk of harm has been caused. But these factors are determined by the environment in which the animal lives. What if the monkey enjoys a fabulous life because attention and love are heaped upon it because of its different and valued characteristic? Or what if, because it is fluorescent, it is forced to work and perform in endless circus shows, exhausted and isolated? The attitudes and practices of others make the life go well or badly in virtue of the biological modification, not the biological modification itself. The social construction determines the animal's quality of life.

Consider the example of xenogestation. Would a genetically modified pig created to serve human reproductive needs be happy? This will turn in large part on how it is treated. Is it compelled to experience painful delivery or Caesarean section? Would it have a social group, freedom, opportunity to exercise, eat tasty food, have its own family, and so on? In part, we must predict what the likely natural and social environment of a GMA will be in order to evaluate the effects of a biological modification.[61] In some cases, the social environment involving human choices is infected with prejudice and injustice. For example, it might be an enhancement to lighten the skin color of a mixed-race human embryo in a racist society. In such cases a biological modification that resulted in darker skin would not be inherently wrong; but what are wrong, however, are racist attitudes that cause the resultant harm. Unjust discrimination on the basis of some morally neutral biological characteristic, such as fluorescence, is a form of arbitrary discrimination like sexism, racism, and speciesism. It could be called "biologism"—the discrimination against a being on the basis of some arbitrary biological feature.

STAGE 3. Reasonable Risk and Consent

The third stage to evaluating creation of GMAs is whether it conforms to the two basic principles of the review of protocols in research ethics, in particular, establishing whether the GMA is exposed to reasonable risk and whether consent is obtained if the animal is competent. These principles apply whether or not the GMA is created as a part of a research project.

a. Reasonable Risk

All research involving living beings involves risk of harm, and research ethics review assesses whether risks are minimized and reasonable relative to the benefit to the participant (and perhaps others).[62] This approach can be generalized to the creation of GMAs outside research: are the risks to the GMA reasonable relative to the benefit provided to others. I have elsewhere identified a procedure for evaluating the reasonableness of risk that starts with a series of questions that can be modified for considering GMAs.[63]

1. Is there a known risk prior to commencing the intervention to the animal and what is its magnitude, based on evidence available at the time?
2. Should any non-living or epidemiological research, systematic overview, or computer modeling have been performed prior to modifying the animal to better estimate the risk to the animal or obviate the need for genetic modification?

3. Could the risk be reduced in any other way? Is it as small as possible?
4. Are the potential benefits (in terms of knowledge and improvement of welfare of the participating animal or other animals or people) of this intervention worth the risks?
5. Could this intervention generate knowledge or technology likely to signifi-cantly harm other beings with moral status, now or in the future?

b. Consent

The second principle of research ethics review is to obtain valid consent from com-petent participants. This principle is of use in evaluating the ethics of creating GMAs both inside and outside of research contexts. In the case of GMAs with rel-evant higher cognitive capacities, should they be created and brought to maturity—for example, a human-chimp chimera that is predominantly human—the question arises whether it is appropriate to obtain the consent of the GMA for participation as a research subject, and consent would be needed for any further biological modi-fication. The capacity to consent is related to whether an individual is autonomous.[64] Study of "animal autonomy" in the sense relevant to consent is an entirely undevel-oped field. It will require further conceptual normative work to elucidate how ani-mals could be autonomous in a morally relevant sense.

STAGE 4. Prevention of Harm to Others

An important stage of evaluation of the creation of GMAs is evaluation of the risks that their creation poses to others, now and in the future.

a. Direct Harm

One issue is direct harm that might result to humans from creating a GMA. For example, concerns have been expressed about the infectious risks of transferring ES cells from cybrids, or xenotransplants.[65] A related objection is that ES cells from cybrids may stimulate the development of teratomas.

While these concerns need to be taken seriously, the appropriate response is to use research to establish whether such problems exist. Conceptual analysis is also necessary to separate different kinds of research and intervention. For example, the benefits of ES research are not merely tailored regenerative transplants from ES cells. Basic ES cell research employing cybrids can help us to understand and treat disease and to develop pharmaceutical and other treatments. These objections apply only to applications such as transferring ES cells to patients and not to these other benefits.

b. Dual Use[66]

A profoundly important moral issue is the potential "dual use" of powerful biotech-nology—meaning use for both good and evil purposes. Nuclear technology is an example of a dual-use technology. Significant dual-use problems are raised by GMAs. I will here illustrate these problems and then suggest potential solutions.

Recently, the J. Craig Venter Institute produced the first synthetic bacterium. The group read the DNA genome, or genetic blueprint, for a bacterium, *Mycoplasma mycoides*. They transferred this code to a computer, which assembled chemicals to produce an artificial copy of the natural DNA—a synthetic copy. They transplanted that copy into a different *Mycoplasma* species, whose natural genome had been removed. The new life was controlled by the synthetic genome.

Synthetic biology is a relatively new and potentially powerful form of genetic modification. Venter has taken a leap toward the holy grail of artificially creating life. Synthetic biology could be used to create life forms that *could not, or at least would not, naturally exist*. Until now, genetic engineers shuffled genes around the animal and plant kingdoms. Synthetic biology opens the door to not merely copying existing blueprints, as Venter did, but constructing novel ones, according to human design. This creation would be to play God or at least to seize control of evolutionary developments.

The potential benefits of synthetic biology are huge. Synthetic organisms could be programmed to perform an almost unimaginable range of useful functions: agents that would search and destroy cancer or HIV, produce clean biofuels or other useful chemicals, and act as "bioremediators" that break down environmental toxins. Biological computing could use human neurons created artificially (instead of chips). Synthetic agents might even turn on and off parts of the human brain, treating mental illness and addiction, or augmenting cognitive capacity. The final chapter in human evolution might be to rationally design life itself and keep all life on earth under such control.

The biorevolution has the potential to be as powerful as the physics revolution. Like nuclear power, it could threaten the existence of humanity and of many species. In 2001, scientists genetically modified the virus, mousepox, and created a strain that killed 100% of mice. They published on the internet. A significant feature of this discovery lay in the recognition that similar changes could be made to human smallpox, the greatest infectious killer in human history, killing approximately one-third of people infected. The dangers here are at least as prominent as the potential benefits. Fanatics, ideologues, and psychopaths will no longer have to get their hands on military caches of virus held in the former Soviet Union. It will be possible to cheaply and easily synthesize pathogens, like smallpox, in the backyard, modifying them to make them perfectly lethal and superinfectious. Spread in a few airports, shopping malls, or stadiums, they would spread unrecognized and kill tens of millions of people and other creatures, as portrayed in the film, *Twelve Monkeys*.

In the 1950s and '60s, only a handful of people had the capacity to destroy the world. Soon, that power could be in the hands of hundreds of thousands. Given the existential cost of an "adverse event," there is a moral imperative to strategically augment and revise our regulatory structures, international oversight, rules for publication, access to reagents and technology, and the like. The biorevolution will certainly bring a new chapter in our history. We should monitor and help lead it ethically to ensure it is not the final chapter in human history.

SECTION 3. FOUR OBJECTIONS
TO THE CREATION OF GMAs

Several objections are often registered to the creation of genetic alterations. These objections, which I find unwarranted, are that that genetic alteration is: (1) a threat to our humanity; (2) forbidden by absolute deontological constraints; (3) playing God; (4) a dangerous slippery slope leading to genetic manipulation.

1. Threat to Humanity[67]

One common objection to creating GMAs, especially those involving substantial amounts of human genetic material such as a human-chimp chimera or a human-ized mouse, is that it represents a threat to our humanity. By "humanity," I mean our special moral status as human beings. Whatever account one adopts of the special moral status of human beings, genetic modification threatens to erode it. Among the several candidate properties for differentiating us from other animals and capturing our special moral status as human beings, let us assume, for the sake of the argument, that one necessary (but perhaps not sufficient) condition of humanity is the capacity to act on the basis of normative reasons, a capacity of practical rationality. Animals have desires and wants about what to do, but humans alone have higher order beliefs about their own beliefs and what they should do.

Consider an example. Animals respond to biological urges, such as to reproduce. They display some forms of practical rationality when they hunt, adapt to new environments, use tools, play games, mate, and plan ahead for some short distance in time. However, humans display a richer and deeper practical rationality. They make decisions about when to have children, how many children to have, and even what kind to have through prenatal diagnosis or prenatal testing. If our humanity is located, at least in part, in our practical rationality (in our capacity to make normative judgments, including moral judgments, and act on these), then there are two ways in which our humanity can either be promoted or threatened:

a. Actions that are an Expression of Our Humanity

When we act according to our judgments of what we have good reasons to do, we express our humanity. So whether creating transgenic humans or chimeras is an expression of our humanity or not turns on whether we judge that we have good reason to alter our genome. Some radical genetic alterations are an expression of our humanity, while others are not. For example, the introduction of animal DNA into humans or even embryonic cells to create weak human-animal chimeras to protect them from an uncontrollable HIV pandemic would be an expression of our humanity—we believe we have good reason to prevent HIV. The creation of human-chimp chimeras for entertainment, or to be slaves or pets, would not be an expression of our humanity—we may doubt that we have good reasons to create chimeras for these purposes.

Radical genetic alteration of humans or animals is not necessarily a threat to our humanity. Whether it is an expression or threat depends on whether there are good reasons for or against the alteration in question. Simply the fact that such alterations are judged by some to be unnatural or to cross the human-animal divide does not imply they are a threat to our humanity.

b. Effects on the Capacity for Practical Reasoning of the GMA

If practical rationality is central to our humanity as members of the species *Homo sapiens*, then there is a second way in which radical genetic alteration could threaten our humanity. It could do so by undermining the capacity of the being altered to engage in practical reasoning—reasoning about what it should do. For example, the introduction of genes from ferocious animals that contribute to aggression might make for better boxers or soldiers, but might also inhibit that human being's capacity to reason and reflect about what he or she should do. He or she could turn out to be more nonhuman than human. In other cases, introduction of animal genes might have no effect on the capacity to reason—for example, introduction of genes to protect against disease or improve our sensory faculties. These improvements do not threaten humanity, though a human person with bat sonar would be a very different kind of human. More optimistically, the introduction of animal genes might promote our capacity to reason and act on our value judgments. For example, a man who searches for hikers lost in the night on trails would be more likely to find them if he had acute night vision.

Similarly, radical prolongation of human life or memory by introduction of genes from elephants might improve opportunities for learning and for making more informed judgments, as well as increasing the efficiency of transmission from judgment to action. Improvement of our cognitive abilities may improve our capacity to reason. Improvement in empathy may improve our moral reflection. Whether transgenesis and the creation of human-animal chimeras threaten humanity depends on the effects of these changes, if any, on the essential features of humanity.

2. Absolute Deontological Constraints

While the creation of GMAs may have obvious utility, there are some so-called absolute deontological objections, that is, objections based on rules that should never be broken. These objections proscribe such research whatever its utility, just as, according to this objection, we should not use data from Nazi hypothermia experiments, regardless of its benefits, because of the evil involved. While it is obvious that torturing human beings is evil, in the case of GMAs we must ask what precisely is harmful and perhaps evil to the point that it makes such creating of GMAs unacceptable.

3. Playing God and Against "Nature"

Concerns about humans playing God can be understood in at least two different ways. On a religious interpretation, the concern is that humans are literally usurping the role of a higher being. On secular interpretations, the concern is typically with humans failing to recognize their own limitations, for example, by overestimating their ability to control complex ecosystems.[68]

Humans wishing to exert some influence on the genetic make-up of future beings have until now been constrained to working within the timescales and genetic possibilities dictated by evolution. Genetic engineering has partially freed us from this constraint. Synthetic biology promises to free us from a further constraint: the need for a natural template on which future organisms must be based. It will allow us to design and create life, not merely to tinker with or modify it.[69] It is also possible that synthetic biology will enable us to create life from non-living, inorganic matter. Indeed, this is arguably the most distinctive role of synthetic biology.

However, the history of humanity has been one of modifying the world and life for good reasons. The natural state of humans would be a life "nasty, brutish, and short," as Hobbes saw it, without many improvements that involved modifying the world. Vaccination, antibiotics, and nearly all of medicine involve powerful interventions. The objection that we would be playing God is only valid as a caution against premature or ill-informed action, to which no doubt humans are prone.

4. Slippery Slopes and the Precautionary Principle

Many people fear that the creation of GMAs is a descent down the slope to creating live-born, radically new human-animal life forms, ultimately modifying human beings themselves. The techniques developed here could be used to create human-chimp chimeras, reminiscent of *Planet of the Apes*. However, four responses can be made to this argument. The first three responses question how likely a descent down the slope is. The last questions whether the bottom of the slope is really that bad.

Firstly, the techniques for the creation of these life forms are not new. Interestingly, no live-born fluorescent humans have been created despite our ability to create fluorescent rabbits and monkeys. Despite having the technical ability for over twenty years, scientists have only recently produced a fluorescent human embryo.[70]

Secondly, it can be argued that there is no slippery slope, only a set of steps. Laws exist to prevent the development of such life forms. For example, in the United Kingdom, it is a serious criminal offence to allow them to develop beyond the specified time limit of fourteen days. This has been effective in regulating research and practice in that society. Of course, legislation can never entirely deal with such a problem, but it is an important instrument to reduce harm and risk.

Thirdly, it is a mistake to prevent some valuable course of action because of the mere risk of adverse consequences. One must give some consideration to the potential benefits, harms, and their relative likelihoods. The potential benefits are in some cases very significant, while the harms are of unestablished severity and small likelihood. It is hard to see any plausible principle of precaution favoring the prohibition of this kind of research.

Lastly, I question whether the bottom of the slope is really bad. Would it be wrong to modify humans to include animal genes or cells? As I have argued, the answer depends on the reasons, the benefit to the resulting humans, the risk to other humans and animals, the appropriateness of ethical review, and the like. In some cases, we might have good reason to genetically modify even humans—for example, to confer disease resistance to a novel pathogen.

Section 4. Conclusion

We are undergoing a great biological revolution. It is now possible to move genes around the animal and plant kingdoms, designing genetically modified animals by transgenesis or the creation of chimeras. Synthetic biology affords the possibility of designing life from the ground up, using inorganic chemicals, according to human design. These life forms have not and could not naturally exist. At present, we have only scratched the surface of this biorevolution, which will likely dwarf the industrial revolution. Biological life will become the powerhouse of innovation, creativity, and productivity. The biological engine could replace the mechanical engine in many contexts, and nanotechnology and computing technology may be integrated with genetic modification.

We will need a well-developed ethics to govern the creation of novel life. In this chapter, I have argued that we need case-by-case evaluation of new interventions and technologies, rather than blanket or global evaluations. I have argued that we need to develop richer, deeper, and more practical strategies of evaluation of this kind of technology. I have tried to set an agenda of problems to be addressed and to offer one very preliminary attempt at a four-stage evaluative approach. The four stages of evaluation of interventions to create GMAs are: (1) we need to use conceptual analysis to develop a justified, coherent, explicit account of moral status; (2) the creation of GMAs should be done for good reasons as assessed by the enhancement of the animals themselves or of other animals, including humans; (3) the creation of GMAs should conform to two basic principles in research ethics review, namely, ensuring that risks are reasonable and that appropriate consent is obtained; and (4) creation of GMAs should not be allowed to unreasonably harm others, either now or in the future.

I have argued in this chapter that four common objections to the creation of GMAs are not valid: (1) creation of GMAs represents a threat to humanity;

(2) absolute deontological constraints preclude creation of GMAs; (3) creation of GMAs is wrong because it is playing God; (4) creation of GMAs is to descend down a slippery slope to the radical genetic alteration of human beings and is precluded by the precautionary principle.

This is the beginning, not the end. One of the most urgent tasks for modern bioethics is to elaborate on and develop a full-blooded normative framework for evaluating the creation of radically genetically modified animals. Science is progressing exponentially, but this vital issue has to date received scant ethical scrutiny.[71]

NOTES

1. Lan Kang, Jianle Wang, Yu Zhang, et al., "iPS Cells Can Support Full-Term Development of Tetraploid Blastocyst-Complemented Embryos," *Cell Stem Cell* 5 (2009): 135–38.

2. Aideen O'Doherty, Sandra Ruf, Claire Mulligan, et al., "An Aneuploid Mouse Strain Carrying Human Chromosome 21 with Down Syndrome Phenotypes," *Science* 309 (2005): 2033.

3. "GM Goat Spins Web Based Future," *BBC News Online*, August 21, 2005. http://news.bbc.co.uk/1/hi/sci/tech/889951.stm (accessed July 4, 2010).

4. Genzyme Homepage, July 4, 2010, at www.genzyme.com.

5. Sarah Sue Goldsmith, "World's First Cloned Transgenic Goats Born," *Science Daily Magazine*, May 12, 1999.

6. Julian Savulescu and Loane Skene, "'The Kingdom of Genes: Why Genes from Animals and Plants Will Make Better Humans': Open Peer Commentary on Françoise Baylis' Animal Eggs for Stem Cell Research: A Path Not Worth Taking," *American Journal of Bioethics* 8 (2008): 35.

7. Savulescu and Skene, "Kingdom of Genes"; Julian Savulescu, "Human-Animal Transgenesis and Chimeras Might Be an Expression of Our Humanity," *American Journal of Bioethics* 3 (2003): 22–25.

8. Karl Lenhard Rudolph, Sandy Chang, Han-Woong Lee, et al., "Longevity, Stress Response, and Cancer in Aging Telomerase-Deficient Mice," *Cell* 96 (1999): 701–12; Maria A. Blasco, "Telomeres and Human Disease: Ageing, Cancer and Beyond," *Nature Reviews Genetics* 6 (2005): 611–22.

9. John C. Guerin, "Emerging Area of Aging Research Long-Lived Animals with 'Negligible Senescence,'" *Annals of the New York Academy of Science* 1019 (2004): 518–20.

10. House of Commons Science and Technology Committee, *Government Proposals for the Regulation of Hybrid and Chimera Embryos*, Fifth Report of Sessions 2006–7 (HCP 272–1).

11. Carole B. Fehilly, S. M. Willadsen, and Elizabeth M. Tucker, "Interspecific Chimaerism between Sheep and Goat," *Nature* 307 (1984): 634–36.

12. Anon., "The Bills of Qucks and Duails," *Science*, www.scienecmag.org/cgi/content/full/299/5606/523 (accessed January 24, 2003).

13. Fehilly et al., "Interspecific Chimaerism between Sheep and Goat"; Evan Balaban, Marie-Aimée Teillet, and Nicole Le Douarin, "Application of the Quail-Chick Chimera System to the Study of Brain Development and Behavior," *Science* 241 (1998): 1339–42;

Anon., "Bills of Qucks and Duails"; Sylvia Pagán Westphal, "Growing Human Organs on the Farm," *New Scientist* 180 (2003): 4–5; Graça Almeida-Porada, Christopher D. Porada, Jason Chamberlain, et al., "Formation of Human Hepatocytes by Human Hematopoietic Stem Cells in Sheep," *Blood* 104 (2004): 2582–90; Graça Almeida-Porada, Christopher Porada, Nicole Gupta, et al., "The Human-Sheep Chimeras as a Model for Human Stem Cell Mobilization and Evaluation of Hematopoietic Grafts' Potential," *Experimental Hematology* 35 (2007): 1594–600.

14. Linda MacDonald Glenn, "A Legal Perspective on Humanity, Personhood, and Species Boundaries," *American Journal of Bioethics* 3 (2003): 27–28.

15. National Research Council (NRC), Committee on Guidelines for Human Embryonic Stem Cell Research, *Guidelines for Human Embryonic Stem Cell Research* (Washington, D.C: National Research Council, National Academy of Sciences, 2005).

16. Jason Scott Robert, "The Science and Ethics of Making Part-Human Animals in Stem Cell Biology," *FASEB Journal* 20 (2006): 838–45.

17. Natalie DeWitt, "Biologists Divided over Proposal to Create Human-Mouse Embryos," *Nature* 420 (2002): 255.

18. E.g., Françoise Baylis, "The HFEA Public Consultation Process on Hybrids and Chimeras: Informed, Effective, and Meaningful?" *Kennedy Institute of Ethics Journal* 19 (2009): 41–62.

19. Françoise Baylis and Andrew Fenton, "Chimera Research and Stem Cell Therapies for Human Neurodegenerative Disorders," *Cambridge Quarterly of Healthcare Ethics* 16 (2007): 195–208.

20. At this point, twinning is no longer possible and organ systems, including the nervous system, have begun to develop, marking a time of higher level cellular co-ordination and the more obvious emergence of an organism.

21. In theory, the majority of the organism could be animal if enough genes were transferred into the one-cell embryo. However, this is not on the horizon at present.

22. Nigel Hawkes, "Nobel Scientists Urge Fertility Watchdog to Back Hybrid Embryos," *The Times* January 10, 2007, http://www.timesonline.co.uk/tol/news/article1291238.ece (accessed April 3, 2008).

23. Julian Savulescu, "The Case for Creating Human-Nonhuman Cell Lines," *Bioethics Forum*, January 24, 2007, http://www.bioethicsforum.org/research-cloning-hybrid-embryos.asp.

24. Peter Braude, Stephen L. Minger, and Ruth M. Warwick, "Stem Cell Therapy: Hope or Hype?" *BMJ* 330 (2005): 1159–60.

25. James A. Thomson, "Embryonic Stem Cell Lines Derived From Human Blastocysts," *Science* 282 (1998): 1145–47; M. William Lensch, Thorsten M. Schlaeger, Leonard I. Zon, et al., "Teratoma Formation Assays with Human Embryonic Stem Cells: A Rationale for One Type of Human-Animal Chimera," *Cell Stem Cell* 1 (2007): 253–58.

26. Oliver Brüstle, "Building Brains: Neural Chimeras in the Study of Nervous System Development and Repair," *Brain Pathology* 9 (1999): 527–45; Nicole M. Le Douarin, "The Avian Embryo as a Model to Study the Development of the Neural Crest: A Long and Still Ongoing Story," *Mechanisms of Development* 121 (2004): 1089–1102.

27. John Rennie, "Human-Animal Chimeras: Some Experiments Can Disquietingly Blur the Line between Species," *Scientific American*, June 27, 2005, http://www.scientificamerican.com/article.cfm?id=human-animal-chimeras.

28. Václav Ourednik, Jitka Ourednik, Jonathan D. Flax, et al., "Segregation of Human Neural Stem Cells in the Developing Primate Forebrain," *Science* 293 (2001): 1820–24;

Alysson R. Muotri, Kinichi Nakashima, Nicolas Toni, et al., "Development of Functional Human Embryonic Stem Cell–Derived Neurons in Mouse Brain," *Proceedings of the National Academy of Sciences USA* 102 (2005): 18644–48; Richard R. Behringer, "Human-Animal Chimeras in Biomedical Research," *Cell Stem Cell* 1 (2007): 259–62; Gabsang Lee, Hyesoo Kim, Yechiel Elkabetz, et al., "Isolation and Directed Differentiation of Neural Crest Stem Cells Derived from Human Embryonic Stem Cells," *Nature Biotechnology* 25 (2007): 1468–75; Yechiel Elkabetz, Georgia Panagiotakos, George Al Shamy, et al., "Human ES Cell–Derived Neural Rosettes Reveal a Functionally Distinct Early Neural Stem Cell Stage," *Genes & Development* 22 (2008): 152–65.

29. Oliver Brüstle, Uwe Maskos, and Ronald D. G. McKay, "Host-Guided Migration Allows Targeted Introduction of Neurons into the Embryonic Brain," *Neuron* 15 (1995): 1275–85; Kenneth Campbell, Martin Olsson, and Anders Björklund, "Regional Incorporation and Site-Specific Differentiation of Striatal Precursors Transplanted to the Embryonic Forebrain Ventricle," *Neuron* 15 (1995): 1259–73; Gord Fishell, "Striatal Precursors Adopt Cortical Identities in Response to Local Cues," *Development* 121 (1995): 803–12.

30. Ourednik et al., "Segregation of Human Neural Stem Cells in the Developing Primate Forebrain."

31. Almeida-Porada et al., "Human-Sheep Chimeras as a Model for Human Stem Cell Mobilization and Evaluation of Hematopoietic Grafts' Potential."

32. Mohammed A. Nassar, L. Caroline Stirling, Greta Forlani, et al., "Nociceptor-Specific Gene Deletion Reveals a Major Role for Nav1.7 (PN1) in Acute and Inflammatory Pain," *PNAS* 101 (2004): 12706–11.

33. Savulescu, "Human-Animal Transgenesis and Chimeras Might Be an Expression of our Humanity," 22–25.

34. I have rejected this position at length in Julian Savulescu, "The Human Prejudice and the Moral Status of Enhanced Beings," in *Human Enhancement*, ed. Julian Savulescu and Nick Bostrom (Oxford: Oxford University Press, 2009), 211–50.

35. Henry T. Greely, Mildred K. Cho, Linda F. Hogle, et al., "Thinking About the Human Neuron Mouse," *American Journal of Bioethics* 7 (2007): 27–40.

36. Peter Singer, *Animal Liberation* (London: Pimlico, 1999); Jeff McMahan, *The Ethics of Killing: Problems at the Margins of Life* (Oxford: Oxford University Press, 2002); Savulescu, "Human Prejudice and the Moral Status of Enhanced Beings," 211–50; Ingmar Persson and Julian Savulescu, "Moral Transhumanism," *Journal of Medicine and Philosophy* 0: 1–14 (posted Nov. 12, 2010), doi:10.1093/jmp/jhq052, available online at http://www.bep.ox.ac.uk/__data/assets/pdf_file/0020/18137/Savulescu_Persson_Moral_Transhumanism.pdf (accessed April 2, 2011).

37. Donald M. Broom, "Welfare Assessment and Relevant Ethical Decisions: Key Concepts," *Annual Review of Biomedical Sciences* 10 (2008): T79–T90.

38. Dominic Wilkinson, Guy Kahane, and Julian Savulescu, "'Neglected Personhood' and Neglected Questions: Remarks on the Moral Significance of Consciousness," *American Journal of Bioethics* 8 (2008): 31–33; Guy Kahane and Julian Savulescu, "Brain Damage and the Moral Significance of Consciousness," *Journal of Medicine and Philosophy* 34 (2009): 6–26.

39. Fred Feldman, "On the Intrinsic Value of Pleasures," *Ethics* 107 (1997): 448–66; cf. Irwin Goldstein, "Pleasure and Pain: Unconditional, Intrinsic Values," *Philosophy and Phenomenological Research* 50 (1999): 255–76.

40. S. L. Davis and P. R. Cheek, "Do Domestic Animals Have Minds and the Ability to Think? A Provisional Sample of Opinions on the Question," *Journal of Animal Science*

76 (1998): 2022–79; James A. Serpell, "Factors Influencing Human Attitudes to Animals and Their Welfare," *Animal Welfare* 13 (2004): S145–S51; Sarah Knight and Louise Barnett, "Justifying Attitudes Towards Animal Use: A Qualitative Study of People's Views and Beliefs," *Anthrozoös* 21 (2008): 31–42; Sarah Knight, Aldert Vrij, Kim Bard, et al., "Science versus Human Welfare? Understanding Attitudes toward Animal Use," *Journal of Social Issues* 65 (2009): 463–83.

41. Immanuel Kant, *Lectures on Ethics*, trans. L. Infield (New York: Harper & Row, [1930] 1963); Alan Gewirth, *Reason and Morality* (Chicago: University of Chicago Press, 1978); Alan Gewirth, "On Rational Agency as the Basis of Moral Equity: Reply to Ben-Zeev," *Canadian Journal of Philosophy* 12 (1982): 667–71.

42. Ned Block, "Some Concepts of Consciousness," in *Philosophy of Mind: Classical and Contemporary Readings*, ed. David J. Chalmers (New York: Oxford University Press, 2009), 206–18.

43. Paola Cavalieri and Peter Singer, eds., *The Great Ape Project* (London: Fourth Estate, 1993); David DeGrazia, *Taking Animals Seriously: Mental Life and Moral Status* (Cambridge: Cambridge University Press, 1996); Antonio R. Damasio, *The Feeling of What Happens: Body, Emotion and the Making of Consciousness* (London: Vintage Press, 2000).

44. I. J. H. Duncan and J. C. Petherick, "The Implications of Cognitive Processes for Animal Welfare," *Journal of Animal Science* 69 (1991): 5017–22; Marc Bekoff, "Cognitive Ethology and the Treatment of Nonhuman Animals: How Matters of Mind Inform Matters of Welfare," *Animal Welfare* 3 (1994): 75–96.

45. Bernard E. Rollin, *Animal Rights and Human Morality* (Buffalo, N.Y.: Prometheus Books, 1981); Bernard E. Rollin, *The Frankenstein Syndrome: Ethical and Social Issues in the Genetic Engineering of Animals* (Cambridge: Cambridge University Press, 1995); Robert Heeger and Frans W. A. Brom, "Intrinsic Value and Direct Duties: From Animal Ethics Towards Environmental Ethics?" *Journal of Agricultural and Environmental Ethics* 14 (2000): 241–52.

46. Peter Singer, *Practical Ethics* (Cambridge: Cambridge University Press, 1993); Peter Singer, *Rethinking Life and Death: The Collapse of Our Traditional Ethics* (Oxford: University Press, 1995); Michael Tooley, *Abortion and Infanticide* (Oxford: Clarendon Press, 1983).

47. See further the essays in this *Handbook* by Michael Tooley as well as by Sarah Chan and John Harris.

48. Joshua P. Johansen, Howard L. Fields, and Barton H. Manning, "The Affective Component of Pain in Rodents: Direct Evidence for a Contribution of the Anterior Cingulate Cortex," *PNAS* 98 (2001): 8077–82.

49. Adapted from Julian Savulescu, Anders Sandberg, and Guy Kahane, "What is Enhancement and Why We Should Enhance Cognition," in *Enhancing Human Capacities*, ed. Ruud ter Meulen, Guy Kahane, and Julian Savulescu (Oxford: Wiley-Blackwell, 2010).

50. Julian Savulescu and Nick Bostrom, eds., *Human Enhancement* (Oxford: Oxford University Press, 2009); Julian Savulescu, "Genetic Interventions and the Ethics of Enhancement of Human Beings," *The Oxford Handbook of Bioethics*, ed. Bonnie Steinbock (Oxford: Oxford University Press, 2007), 516–35; John Harris, *Enhancing Evolution: The Ethical Case for Making Better People* (Princeton, N.J.: Princeton University Press, 2007); Ruud Ter Meulen, Julian Savulescu, and Guy Kahane, eds., *Enhancing Human Capacities* (Oxford: Wiley-Blackwell, 2010); Julian Savulescu, "Genetic Enhancement," in

A Companion to Bioethics, 2nd ed., ed. Helga Kuhse and Peter Singer (Chichester: Wiley-Blackwell, 2009).

51. Martha C. Nussbaum, "Beyond Compassion and Humanity: Justice for Animals?" in *Animal Rights*, ed. Cass R. Sunstein and Martha C. Nussbaum (New York: Oxford University Press, 2004), 299–320; Bernard E. Rollin, *The Unheeded Cry: Animal Consciousness, Animal Pain and Science* (Oxford: Oxford University Press, 1989).

52. Rollin, *Unheeded Cry*.

53. Savulescu et al., "What is Enhancement and Why We Should Enhance Cognition?"; Julian Savulescu and Guy Kahane, "Disability: A Welfarist Approach." *Clinical Ethics* 6 (2011): 45–51.

54. Savulescu and Kahane, "Disability: A Welfarist Approach"; and Savulescu and Kahane, "The Moral Obligation to Create Children with the Best Chance of the Best Life." *Bioethics* 23 (2009): 274–90.

55. For a comprehensive treatment of this kind of problem, see John Broome, *Weighing Lives* (Oxford: Oxford University Press, 2004).

56. Savulescu, "Should We Clone Human Beings? Cloning as a Source of Tissue Transplantion," *Journal of Medical Ethics* 25 (1999): 87–95.

57. McMahan, *Ethics of Killing*.

58. Parfit, *Reasons and Persons* (Oxford: Oxford University Press, 1985).

59. Savulescu and Kahane, "Moral Obligation to Create Children with the Best Chance of the Best Life."

60. Savulescu and Kahane, "Moral Obligation to Create Children with the Best Chance of the Best Life"; and Kahane and Savulescu, "Welfarist Account of Disability."

61. Savulescu and Kahane, "Moral Obligation to Create Children with the Best Chance of the Best Life"; Savulescu and Kahane, "Welfarist Account of Disability."

62. Julian Savulescu and Tony Hope, "Ethics of Research," in *The Routledge Companion to Ethics*, ed. John Skorupski (Abingdon: Routledge, 2010).

63. Savulescu, "Genetic Interventions and the Ethics of Enhancement of Human Beings"; Savulescu and Hope, "Ethics of Research."

64. Ruth R. Faden and Tom L. Beauchamp, *A History and Theory of Informed Consent* (Oxford: Oxford University Press, 1986).

65. Braude et al., "Stem Cell Therapy."

66. Thomas Douglas and Julian Savulescu, "Synthetic Biology and the Ethics of Knowledge," *Journal of Medical Ethics* 36 (2010): 687–93.

67. Much of this section is taken from Savulescu, "Human-Animal Transgenesis and Chimeras Might be an Expression of Our Humanity," 22–25.

68. Cecil Anthony J. Coady, "Playing God," in *Human Enhancement*, ed. Julian Savulescu and Nick Bostrom (Oxford: Oxford University Press, 2009), 155–80.

69. Douglas and Savulescu, "Synthetic Biology and the Ethics of Knowledge."

70. Julian Savulescu, "The New Law on Admixed Embryos and the Genetic Heritage of the Living Kingdom," http://www.practicalethicsnews.com/practicalethics/2008/05/the-new-law-on.html (accessed May 28, 2008); Julian Savulescu, "Looking for Biopolitical Trouble," http://www.practicalethicsnews.com/practicalethics/2008/05/looking-for-bio.html (accessed May 14, 2008).

71. I would like to thank Dr. James Yeates for incredible and invaluable help and directions to the literature. I have learnt much from his work and benefitted significantly from his research. I would also like to thank Professor Loane Skene for her many valuable discussions and our joint work on this topic.

SUGGESTED READING

BAYLIS, FRANÇOISE. "The HFEA Public Consultation Process on Hybrids and Chimeras: Informed, Effective, and Meaningful?" *Kennedy Institute of Ethics Journal* 19 (2009): 41–62.

BAYLIS, FRANÇOISE, and ANDREW FENTON. "Chimera Research and Stem Cell Therapies for Human Neurodegenerative Disorders." *Cambridge Quarterly of Healthcare Ethics* 16 (2007): 195–208.

BEHRINGER, RICHARD R. "Human-Animal Chimeras in Biomedical Research." *Cell Stem Cell* 1 (2007): 259–62.

BROOM, DONALD M. "Welfare Assessment and Relevant Ethical Decisions: Key Concepts." *Annual Review of Biomedical Sciences* 10 (2008): T79–T90.

CAVALIERI, PAOLA, and PETER SINGER, eds. *The Great Ape Project*. London: Fourth Estate, 1993.

DEGRAZIA, DAVID. *Taking Animals Seriously: Mental Life and Moral Status*. Cambridge: Cambridge University Press, 1996.

DOUGLAS, THOMAS, and JULIAN SAVULESCU. "Synthetic Biology and the Ethics of Knowledge." *Journal of Medical Ethics* 36 (2010): 687–93.

GLENN, LINDA MACDONALD. "A Legal Perspective on Humanity, Personhood, and Species Boundaries." *American Journal of Bioethics* 3 (2003): 27–28.

GREELY, HENRY T., MILDRED K. CHO, LINDA F. HOGLE, et al. "Thinking About the Human Neuron Mouse." *American Journal of Bioethics* 7 (2007): 27–40.

House of Commons Science and Technology Committee. *Government Proposals for the Regulation of Hybrid and Chimera Embryos*. Fifth Report of Sessions 2006–7 (HCP 272–1).

MCMAHAN, JEFF. *The Ethics of Killing: Problems at the Margins of Life*. Oxford: Oxford University Press, 2002.

NUSSBAUM, MARTHA C. "Beyond Compassion and Humanity: Justice for Animals?" In *Animal Rights*, edited by Cass R. Sunstein and Martha C. Nussbaum, pp. 299–320. New York: Oxford University Press, 2004.

RENNIE, JOHN. "Human-Animal Chimeras: Some Experiments Can Disquietingly Blur the Line Between Species." *Scientific American*, June 27, 2005, at http://www.scientificamerican.com/article.cfm?id=human-animal-chimeras.

ROBERT, JASON SCOTT. "The Science and Ethics of Making Part-Human Animals in Stem Cell Biology." *FASEB Journal* 20 (2006): 838–45.

ROLLIN, BERNARD E. *The Frankenstein Syndrome: Ethical and Social Issues in the Genetic Engineering of Animals*. Cambridge: Cambridge University Press, 1995.

———. *Animal Rights and Human Morality*. Buffalo, N.Y.: Prometheus Books, 1981.

SAVULESCU, JULIAN. "The Human Prejudice and the Moral Status of Enhanced Beings." In *Human Enhancement*, edited by Julian Savulescu and Nick Bostrom, pp. 211–50. Oxford: Oxford University Press, 2009.

———. "Human-Animal Transgenesis and Chimeras Might Be an Expression of Our Humanity." *American Journal of Bioethics* 3 (2003): 22–25.

———, ANDERS SANDBERG, and GUY KAHANE. "What Is Enhancement and Why We Should Enhance Cognition." In *Enhancing Human Capacities*, edited by Ruud ter Meulen, Guy Kahane, and Julian Savulescu. Oxford: Wiley-Blackwell, 2010.

SINGER, PETER. *Animal Liberation*. London: Pimlico, 1999.

WESTPHAL, SYLVIA PAGÁN. "Growing Human Organs on the Farm." *New Scientist* 180 (2003): 4–5.

CHAPTER 24

....

HUMAN/NONHUMAN CHIMERAS: ASSESSING THE ISSUES

....

HENRY T. GREELY

Tonight I ask you to pass legislation to prohibit the most
egregious abuses of medical research: human cloning in all
its forms, creating or implanting embryos for experiments,
creating human-animal hybrids, and buying, selling, or
patenting human embryos. Human life is a gift from our
Creator—and that gift should never be discarded, devalued
or put up for sale. (Applause.)

—President George W. Bush,
State of the Union Address, 2006

PRESIDENT Bush's attack of "human-animal hybrids" was mainly Irv Weissman's
fault.[1] In the 1980s, Weissman, a Stanford University professor, was one of the first
people to successfully identify and purify the human stem cells that form blood. In
September 1988, Weismann and his colleague Mike McCune successfully created
what they called the SCID-hu mouse, a mouse that lacked its own immune system
but, through the transplantation of human bone marrow and other tissues, had a
functioning *human* immune system.[2] This mouse provoked special interest, as HIV/
AIDS, the most terrifying disease of the 1980s, is a disorder of the immune system.
The SCID-hu mouse permitted study of a human immune system in the context of
a living organism—an immune system that *could* become infected with HIV—but
in an organism that made a much better, easier, and cheaper experimental subject

than humans or chimpanzees, the other animal then known to be subject to HIV infection.

The SCID-hu mouse became a useful tool for studying human immune systems: not as revolutionary as its creators hoped, but still in use more than twenty years later. It raised some controversy—"history may someday record the saga of this mouse as one of the greatest horror stories in the annals of medicine"[3]—largely because some of the transplanted human tissue came from aborted fetuses.

In January 2000, Weissman and others announced that they had successfully isolated human brain stem cells, cells that form the neurons and glial cells that make up the human brain.[4] From that initial announcement, Weissman talked of the SCID-hu mouse as a precedent for creating a "Human Neuron Mouse," a mouse whose neurons, at least, were completely derived from human brain stem cells. Like the SCID-hu mouse, the Human Neuron Mouse could provide a way of examining human tissues inside a living organism that is much easier to study than *Homo sapiens*. Within a year, Weissman had asked me to organize an ad hoc working group to give him and Stanford Medical School advice on whether and how to proceed with experiments to create the Human Neuron Mouse. We did so, and, several years later, published a paper about our recommendations.[5]

In the meantime, media reports of Weissman's plans sparked considerable public interest and discussion. Eventually, in July 2005, then-senator Sam Brownback from Kansas introduced the Human Chimera Prohibition Act, to classify as federal felonies a variety of activities that would mix genes, cells, or tissues from human and nonhuman animals. In what might be called "the Weissman clause," the proposed legislation included in its prohibitions the creation or attempted creation of "a nonhuman life form engineered such that it contains a human brain or a brain derived wholly or predominantly from human neural tissues."[6] And the next year, Weissman's proposed mouse seems to have made an oblique appearance in the president's State of the Union address. At least it seems probable that President Bush meant "chimeras" like Weissman's when he said "hybrids," as no one was proposing making actual human/nonhuman hybrids, creatures that would be created by fertilizing one species's egg with the other species's sperm.

The last decade saw more accurate, analytical, and interesting discussions of human/nonhuman chimeras, as well.[7] Some of the discussion has been fairly abstract, but some of it was practical or even political. Most of the discussion took off from Weissman's Human Neuron Mouse and focused on nonhuman creatures with "humanized" brains, but some looked at other kinds of human/nonhuman chimeras as well. This chapter will first describe briefly the arguments and policies that have been made concerning human/nonhuman chimeras. It will then discuss the three sensitive types of such chimeras, those involving brains, gametes, and outward appearance, and one very sensitive use that could be made of human/nonhuman chimeras. And finally, it will propose that we take a pragmatic stance toward most of the sensitive types and uses of human/nonhuman chimeras, careful to create and employ them only for good reasons, and mindful of the possible negative social reactions.

1. Arguments and Policies

After defining the kinds of human/nonhuman chimeras to be discussed in this chapter, this section goes on to look at the ethical arguments made—or implied—about human/nonhuman chimeras. Although some arguments may be partially or totally rooted in religion, this chapter will not consider religious arguments against human/nonhuman chimeras. Those arguments may be powerful, and even determinative, for those who hold them, but, as religious arguments, they are not arguments that will convince people who do not believe in that religion, or in any religion. (They may, of course, have secular counterparts as ethical arguments and those are discussed below.) The chapter then looks at the rules made to govern research with human/nonhuman chimeras and the arguments they seem to embody.

1.1 Definition

The term "chimera," despite, or because of, its long history, is susceptible to many meanings.[8] This chapter will discuss living organisms that have, as an integrated part of their bodies, some living tissues, organs, or structures of human origin and some of nonhuman origin. Thus, Weissman's Human Neuron Mouse would be a relevant human/nonhuman chimera if, in fact, the neurons in its brain were of human origin but the rest of its tissues were murine (i.e., belonging to the Muridae or family of rodents including mice and rats).

For this chapter's purposes, those human components must be tissues, organs, or structures. The disaggregated human cells surviving (briefly) in the mouth of a man-eating tiger would not qualify it as a human/nonhuman chimera. Neither would the human genes in the *E. coli*, yeast, or Chinese hamster ovary cells that the biotechnology industry uses to produce human therapeutic proteins—or the human proteins thus produced—make those cells into the kinds of chimeras this chapter discusses. The human tissues must be living and integrated into the creature. A chimpanzee with a wig made of human hair thus would not be a human/nonhuman chimera.

Finally, this chapter deals only with creatures that are, or are viewed as, nonhuman creatures to which human tissues are added. To some extent, this limitation begs the question: is the Human Neuron Mouse a mouse to which human brain tissue has been added or human brain tissue to which (lots of) various mouse tissues have been added? In this particular case, one might make a temporal distinction—the human brain stem cells would have been literally added to the existing mouse fetuses shortly before their scheduled births—but one can imagine other cases, such as the mixing of human cells and nonhuman cells into a very early embryo, where "what gets added to what" does not give a clear answer. It is perhaps clearest to say that this chapter *excludes* creatures that were recognized as human beings (or human fetuses) who have acquired nonhuman tissues. It thus excludes the hundreds of thousands of humans living because their faulty human heart valves were replaced with heart valves taken from pigs or valves fashioned from cattle cartilage, as well as those very few humans who have, so far unsuccessfully, received organ transplants from nonhuman animals.[9]

Each of the choices made to limit the definition of "human/nonhuman chimera" could have been made in the other direction, or could have been avoided altogether. I have argued before that the existence, and often acceptance, of those other kinds of human/nonhuman chimeras is relevant for how we look at the more narrowly defined chimeras.[10] But it is this narrowly defined chimera that has dominated the debate so far—namely, the nonhuman animal with some integrated human tissues or organs. And so this chapter will discuss this phenomenon only.

Many of the arguments, and almost all of the policies, have focused even more narrowly on human/nonhuman chimeras created from human embryonic stem cells. This has both a scientific and a political reason. Scientifically, embryonic stem cells offer a particularly plausible way to create worrisome chimeras. These cells are much more versatile than whole organs, tissues, or more differentiated cells. Transplant a human kidney into a pig and it will remain a kidney; transplant human embryonic stem cells into a pig and they might take the form of any kind of cell—muscle, brain, skin, egg, or sperm.

The required destruction of human embryos made embryonic stem cell research become a focus for controversy, debate, politics, and regulation. No regulatory regime exists in the United States to govern the transplant of human tissues into nonhuman animals, beyond the concerns about animal welfare that have been implemented through Institutional Animal Care and Use Committees (IACUCs), required by federal law for most research. But the controversies over embryonic stem cell research brought forth statutes, regulations, guidelines, and recommendations concerning such research, policies that, although not primarily motivated by issues of chimeras, do address them.

This chapter is not limited to a discussion of human/nonhuman chimeras created using human *embryonic* stem cells. It includes those created using other types of human stem cells of human non-stem cells, of tissues, or of organs. (The Human Neuron Mouse, for example, was to be created using human *brain* stem cells derived from fetal tissue and not the less differentiated human embryonic stem cells.) Most of the examples, however, will involve embryonic stem cells, as they have generated the arguments and policies.

1.2 Arguments

The report on (and subsequent article about) the Human Neuron Mouse discussed four moral problems concerning the creation of the mouse: issues of cruelty to the mice; the source of the human tissue; the proper uses of human tissues in general; and the possible creation of a mouse with some human cognitive abilities. We dismissed arguments pertaining to the first three fairly quickly in the context of the Human Neuron Mouse,[11] and although aspects of these problems resurface from time to time in the discussions of human/nonhuman chimeras, they have not been subject to much attention. The last point, about "humanization," in several guises, has been the center of controversy. Our group refused to take a position on whether the creation of a mouse with some human cognitive abilities would be a good thing or a bad thing (and we were criticized by some for that refusal).[12] We first laid out reasons to doubt that the Human Neuron Mouse would have any significant human cognitive abilities and then suggested guidelines to limit even further the very small possibility of such a result.

Other authors, however, have not been as reluctant. Four main arguments have been raised, one implicitly and three expressly, against the creation of human/nonhuman chimeras: the ethical impermissibility of mixing human/nonhuman species, the possibility of moral confusion, the violation of human dignity by the creation of some chimeras, and the predictably inappropriate treatment of the resulting creatures by humans.[13]

Jason Robert and Françoise Baylis deserve credit for initiating the serious discussion of human/nonhuman chimeras through their article in the *American Journal of Bioethics* in summer 2003, as well as by a workshop on chimeras they organized at Dalhousie University in April 2003.[14] Their 2003 article starts by teasing out an implicit, "folk" objection to chimeras that seems to account for much of the unhappiness about human/nonhuman chimeras. On this view, species are real entities with fixed boundaries (for some, fixed by God at the time of their creation) and those boundaries should not be crossed. They particularly should not be crossed when they involve mixing humans, made in God's image, with the nonhuman creatures over which humans were given dominion.

The Human Chimera Prohibition Act, first introduced two years after the Robert and Baylis article, illustrates their point. This bill would have outlawed a variety of acts involving human chimeras, including to create or to attempt to create, to transport or to receive in interstate commerce, any human chimera. The bill's definition of "human chimera," however, differed importantly from this chapter's (as well as from any dictionary's) definition. It defined a "chimera" as

(A) a human embryo into which a nonhuman cell or cells (or the component parts thereof) have been introduced to render its membership in the species Homo sapiens uncertain through germline or other changes;

(B) a hybrid human/animal embryo produced by fertilizing a human egg with nonhuman sperm;

(C) a hybrid human/animal embryo produced by fertilizing a nonhuman egg with human sperm;

(D) an embryo produced by introducing a nonhuman nucleus into a human egg;

(E) an embryo produced by introducing a human nucleus into a nonhuman egg;

(F) an embryo containing haploid sets of chromosomes from both a human and a nonhuman life form;

(G) a nonhuman life form engineered such that human gametes develop within the body of a nonhuman life form; or

(H) a nonhuman life form engineered such that it contains a human brain or a brain derived wholly or predominantly from human neural tissues.[15]

Subsections (B) through (F) do not fall within this chapter's definition of human/nonhuman chimeras. Instead, they deal with hybrids or, perhaps, chimeras at the genetic or chromosomal level. And it is not at all clear what subsection (A) covers. Only subsections (G) and (H) seem clearly to involve the integration of human tissue or organs (gonads or brains) into a nonhuman animal. But all the sections involve the mixture of human "things"—eggs, sperm, nuclei, cells, or tissues—into nonhuman animals.

As Robert and Baylis point out, the folk objection to any reproduction-related mixing of fixed species that seems to lie behind the Brownback bill is incoherent. Species are not fixed entities. And, in fact, although their article does not stress this sufficiently, humans have busily crossed nonhuman species boundaries, not just with modern tools but, for ten thousand years or so, with the tools of agriculture. Many of our crops are human-created species mixtures, and some of our crops are literally chimeras (tree fruits and wine grapes are often grown on plants with the roots and trunk of one species but the limbs and fruits of another). Even some of our domestic animals are human-created hybrids. Consider the mule.

Robert and Baylis conclude that

> despite scientists' and philosophers' inability to precisely define species, and thereby to demarcate species identities and boundaries, the putative fixity of putative species boundaries remains firmly lodged in popular consciousness and informs the view that there is an obligation to protect and preserve the integrity of human beings and the human genome. We have also shown that the arguments against crossing species boundaries and creating novel part-human beings (including interspecies hybrids or chimeras from human materials), though many and varied, are largely unsatisfactory.[16]

Another way of expressing this folk objection to mixing human and nonhuman species, not discussed directly by Robert and Baylis, might be as an argument based on a version of "human dignity," not the dignity of any individual human but of the human species as a whole. This species-based concept of human dignity has been a favorite of some conservative bioethicists and was the subject of a book published by the President's Council on Bioethics.[17] It has also been subject to scathing attack.[18] The idea may be appealing to some as providing a secular way to criticize "repugnant" actions that are undertaken voluntarily by competent adult humans and that have no discrete harms to third parties. Thus, a voluntary decision by a competent adult human to allow some of his (in this case, not adult) stem cells to be incorporated into a mouse, with no negative consequences to any particular human, might still be argued to violate the dignity or "integrity" of the human species. This version of the argument, however, seems no more convincing than the more specific argument about crossing fixed species lines. It gives us no *reason* to oppose human/nonhuman chimeras, but just a conclusion—they would violate "human dignity."

Although they reject the folk objections, Robert and Baylis are not, however, willing to accept that the creation of such chimeras is appropriate. They disclaim any position on the ultimate propriety of such creatures, but they urge that "the most plausible objection to the creation of novel interspecies creatures rests on the notion of moral confusion." Their article does not contend that moral confusion *is* a strong argument against such chimeras, but instead states that they put forth "the following musings as the beginnings of a plausible answer, the moral weight of which is yet to be assessed."[19] They argue that it is not clear what moral status human/nonhuman chimeras should enjoy. This could not only lead to confusion about how to treat those creatures, but might also open debates about the moral status of fully nonhuman animals. Forswearing part-human chimeras might be necessary "to protect the privileged place of human animals in the hierarchy of being."[20]

> [T]he creation of novel beings that are part human and part nonhuman animal is sufficiently threatening to the social order that for many this is sufficient reason to prohibit any crossing of species boundaries involving human beings. To do otherwise is to have to confront the possibility that humanness is neither necessary nor sufficient for personhood (the term typically used to denote a being with full moral standing, for which many—if not most—believe that humanness is at least a necessary condition)....
>
> Given the social significance of the transgression we contemplate embracing, it behooves us to do this conceptual work now, not when the issue is even more complex—that is, once novel part-human beings walk among us.[21]

A third argument returns to "human dignity" but in another way—the human dignity of the chimeras, not of the "unmixed" human species. In several works, Cynthia Cohen and others have argued that some kinds of human/nonhuman chimeras might have some human traits that, because of their physical bodies, they would be unable to exercise.[22]

> Although it is fantastical, we can at least envision that some investigators might attempt to transplant a whole adult human brain into a nonhuman animal in order to study certain important neurological questions, resulting in a human-nonhuman chimera.... The development of such a chimera would arbitrarily limit the ways in which certain human characteristics and capacities associated with human dignity could be exercised in a nonhuman setting and would therefore contravene human dignity. Consequently, the decision to manufacture a nonhuman research subject with a human brain and, at most, diminished capacities for various forms of humanlike cognition and action would violate human dignity.[23]

This dignity argument is not saying that the creation of any human/nonhuman chimera in itself undercuts human dignity, through a kind of pollution, but that the creation of some human/nonhuman chimeras—creatures with enough humanity to be entitled to "higher" moral treatment but without the ability to exercise that humanity—would violate the human dignity *of the chimera*. Johnston and Eliot make a similar argument, contending that the creation of such chimeras as laboratory animals, or otherwise as means to someone else's ends, would violate the chimeras' human dignity.[24]

Rob Streiffer has made a related but perhaps somewhat different argument. Rather than focusing on the creation of human/nonhuman chimeras and whatever fixed limitations that creation may put on them, he looks instead at the likely *uses* of these chimeras—and what he sees is not pretty. Streiffer has argued that some chimeras, at least, may deserve to be treated as persons, at least to some extent, but that humans are unlikely to grant them that status.[25] As a result, sentient but only part-human organisms will be treated as laboratory animals or farm animals—as worse than slaves. It may be a benefit to a chimeric organism to be sufficiently "human" as to merit a higher moral status—but only if that higher moral status is respected. His argument goes well beyond the concern about animal welfare in the Human Neuron Mouse article. It encompasses not only the unnecessary infliction of pain, but also the denial of rights appropriate to "persons" even when that denial would not be inappropriate for what were truly "just" laboratory animals.

I have discussed four arguments against the creation of human/nonhuman chimeras: that species mixing is inherently wrong, that such creatures risk creating moral confusion, that the creation of chimeras that had some but inherently limited human powers violates human dignity, and that the ways we would be likely to abuse chimeras that deserve to be treated as human would be wrong. The first rests on unexamined fiat, not argument. The second, as Hilary Bok pointed out forcefully in a response to the Robert and Baylis article,[26] rests on speculative assumptions both about how confusing such chimeras would actually be, as well as assumes that the results of such confusion would be bad. For us to see human and nonhuman animals as more closely linked might a good thing. The third and fourth hold some appeal for me, but only in the very unlikely circumstances in which the human/nonhuman chimera actually had very substantial human cognitive capacities. Even there, I find Streiffer's argument, focusing on direct harms likely to be done to born creatures, more compelling than Cohen's argument that the mere creation of human/nonhuman chimeras without *full* human powers is wrong. Streiffer's argument, though, could be blunted by the adoption of policies that protected such chimeras.

1.3 Policies and Practices

With human/nonhuman chimeras, we have not only bioethical arguments, but bioethical policies and *practices* to examine. Several guidance or regulatory regimes exist for some kinds of stem cell research that deal expressly with human/nonhuman chimeras.

In the United States, the most important action has been the 2005 Guidelines for Human Embryonic Stem Cell Research, published by a committee of the U.S. National Academy of Sciences.[27] These guidelines cover a wide reach of policies about human embryonic stem cell research, from a requirement for ESCRO committees ("embryonic stem cell research oversight committees"—called "SCRO committees" in California, where their jurisdiction extends to some non-embryonic human stem cell research) to positions on informed consent and the extent of development permissible for in vitro human embryos. The guidelines discuss issues around human/nonhuman chimeras[28] and focus attention on three aspects of such chimeras:

> The hES cells may affect some animal organs rather than others, raising questions about the number of organs affected, how the animal's functioning would be affected, and whether some valued human characteristics might be exhibited in the animal, including physical appearance.
>
> Perhaps no organ that could be exposed to hES cells raises more sensitive questions than the animal brain, whose biochemistry or architecture might be affected by the presence of human cells. Human diseases, such as Parkinson's disease, might be amenable to stem cell therapy, and it is conceivable, although unlikely, that an animal's cognitive abilities could also be affected by such therapy. Similarly, care must be taken lest hES cells alter the animal's germline. Protocols should be reviewed to ensure that they take into account those sorts of possibilities and that they include ethically sensitive plans to manage them if they arise.[29]

Following from this discussion, the guidelines propose several recommendations relevant to such creatures. One recommendation states: "All research involving the introduction of hES cells into nonhuman animals at any stage of embryonic, fetal, or postnatal development should be reviewed by the ESCRO committee. Particular attention should be paid to the probable pattern and effects of differentiation and integration of the human cells into the nonhuman animal tissues."[30] Another recommendation concerns research that should not be permitted at this time: "Research in which hES cells are introduced into nonhuman primate blastocysts or in which any ES cells are introduced into human blastocysts." An addition states: "No animal into which hES cells have been introduced at any stage of development should be allowed to breed."[31]

Although they had no legal force in themselves, the NAS Guidelines have been quite influential. California has, in effect, adopted their main points in both its regulations for research funded by the California Institute for Regenerative Medicine (set up by California's 2004 Proposition 71) and its guidelines for non-CIRM-funded stem cell research in California. Other states and universities have also adopted these guidelines.

The NAS Guidelines also heavily influenced the guidelines promulgated by the International Society for Stem Cell Research in December 2006.[32] Guideline 10(2)(e) requires specialized oversight of human embryonic stem cell research, by an ESCRO committee or similar institution, for "[f]orms of research that generate chimeric animals using human cells. Examples of such forms of research include, but are not limited to introducing totipotent or pluripotent human stem cells into nonhuman animals at any stage of post-fertilization, fetal, or postnatal development."

Guideline 10(3)(c) prohibits "Research in which animal chimeras incorporating human cells with the potential to form gametes are bred to each other."

These are not the only policy recommendations that deal with human/nonhuman chimeras. Not all have adopted the same positions. Canada has taken a somewhat stricter position. The stem cell research guidelines of the Canadian Institutes of Health Research prohibit funding any research that would move human embryonic stem cells to nonhuman embryos or fetuses (though they do not prohibit moving such cells to born nonhuman animals).[33]

On the other hand, the regulations adopted by the U.S. National Institutes of Health in 2009 for human embryonic stem cell research it will fund are less restrictive than the NAS Guidelines. They do not require *any* ESCRO-type review for human embryonic stem cells placed into nonhuman animals or for anything else, but even these regulations carry forward some of the human/nonhuman chimera prohibitions of the NAS Guidelines:

> Although the cells may come from eligible sources, the following uses of these cells are nevertheless ineligible for NIH funding, as follows:
>
> A. Research in which hESCs (even if derived from embryos donated in accordance with these Guidelines) or human induced pluripotent stem cells are introduced into non-human primate blastocysts.
> B. Research involving the breeding of animals where the introduction of hESCs (even if derived from embryos donated in accordance with these Guidelines) or human induced pluripotent stem cells may contribute to the germ line.[34]

A narrower set of recommendations came from a multidisciplinary group convened by researchers at Johns Hopkins University. This group was formed in 2004 to consider issues of transplanting human neural cells or tissues to nonhuman primates. Because of the close relationship between nonhuman primates and human primates, and because of the large skull and brain size of some nonhuman primates, the group believed that the possibilities of, to some extent, "humanizing" their brains deserved special attention.

Its recommendations, published in July 2005, do not take a position on whether such research should be allowed—the group did not achieve consensus on that point—but it did propose "six factors that research oversight committees and other review groups should use as a starting framework. They are (i) proportion of engrafted human cells, (ii) neural development, (iii) NHP [nonhuman primate] species, (iv) brain size, (v) site of integration, and (vi) brain pathology."[35] The SCRO, or other oversight body, would apply these factors to consider just how plausible some humanizing effect might be. Transplantation of a large number of human cells into an early embryo of a great ape would be of more concern that transplantation of a small number of human cells into the healthy adult brain of a small primate, such as a marmoset. The result in the first case might be a large brain made up largely or entirely of human cells; in the second case, it would likely be, at most, a few scattered human cells in a very small brain.

The policies and practices actually adopted toward human/nonhuman chimeras do not track neatly with any of the arguments made against them. They allow creation of such chimeras, thus ignoring the first argument, which opposes any species mixing, as well as the second, which fears moral confusion from such creatures. The restrictions that are relevant to brains speak to some of the concerns of both Cohen's and Streiffer's human-dignity arguments, but what is the basis for other restrictions?

2. THREE SENSITIVE KINDS OF CHIMERAS—AND ONE SENSITIVE USE

I think the policies actually adopted with respect to human/nonhuman chimeras point us to an important reality, usually overlooked and never stressed in the debates about human/nonhuman chimeras: even as narrowly as this chapter defines them, not all chimeras are the same. We care about three kinds of chimeras: those with "humanized" brains (and hence, potentially, human behavior), those with human eggs or sperm, and those with certain forms of human outward appearance. The NAS Guidelines and the many policies that follow them speak to the first and third of these concerns by requiring consideration of the patterns of integration of the human cells in the nonhuman creatures, because of concerns about brains and physical appearance. Those policies address the second concern by banning the breeding of any chimeras that have received human embryonic stem cells, out of fear that the chimeras will have human eggs or sperm.

On the other hand, almost no one was worried when Weissman and McCune made mice with human immune systems. Terence Jeffrey (cited at note 3) is the only exception I can find. Except perhaps for concerns about animal welfare, no one seems to care about the possibility of a pig with a human kidney or a monkey with a human gall bladder, any more than we worry about humans with pig heart valves. It is not surprising, however, that in human/nonhuman chimeras we care most about behaviors (generated by brains), reproduction, and outward appearances, as these are a good first approximation for what we care most about in our fellow non-chimeric humans. And we may also care about the mixing of these characteristics of humans and nonhumans in ways that do not use human/nonhuman chimeras.

I believe that brains, gametes, and looks[36] explain our concerns about human/nonhuman chimeras, with one exception concerning a particular use of human/nonhuman chimeras. This section of the chapter will therefore look at each of three kinds of chimera in turn, before discussing the exceptional use.

2.1 Brains

We clearly care about human behaviors and capacities and we therefore worry that "humanized" brains might create them. Any kind of mixture between human and nonhuman animals is troubling to the folk objection to such chimeras. Brain-based chimeras, though, are the kind that most implicate the issues raised by Robert and Baylis, Cohen and Karpowicz, and Streiffer. And, in this respect, the academic discussion of these issues may truly be said to be all Irv Weissman's fault, or, more neutrally, his doing. Both moral confusion and the concerns about the human dignity of the chimera require that the chimera have some human moral, or plausible high moral, status. It is most likely that we would locate that status in the human brain, the source of our (we think) distinctly human consciousness and behaviors.

A human/nonhuman chimera with a brain that functioned like a human's would genuinely raise these ethical concerns.

Of course, what does it mean to say that the chimera's brain "functioned like a human's"? At one level, all mammalian brains function pretty much the same way—neurons are organized into a brain stem, a cerebellum, and a cerebrum and communicate using the same repertory of neurotransmitters. We see, we hear, and we move our bodies using the same kinds of brain activity. Humans do have unusually large brains, but not uniquely so. Whales, porpoises, and elephants have larger brains, though not always as a percentage of body size, while small birds have larger brains as a proportion of their body mass. Human brains have very large cerebrums and extremely large neocortexes compared with all other animals—except perhaps some of the cetaceans—but it is unclear how important that is, with or without the exceptions. No one has ever successfully identified a distinctly human brain location for consciousness, intelligence, conscience, or a soul.

And what behaviors are uniquely human? We used to cite tool use, but we now see more examples of tool use by animals in the wild, not just from other primates but from porpoises, sea otters, some kinds of birds, and possibly even octopi. Speech still seems distinctly human, although symbol or hand-sign-based language abilities of some sort have been seen in chimpanzees and possibly gorillas. (The extent of such abilities remains very controversial; the full play of language may well be uniquely human.) The report on Weissman's Human Neuron Mouse project confronted these issues. We ended up telling the researchers to look at the mouse brains to see whether, for brain regions that differed between mice and men, the chimeras had human structures or murine structures. For behavior, though, we gave up. Certainly, if a mouse had stood on its hind legs and said, "Hi, I'm Mickey," that would have been evidence of a humanlike behavior. But would a change in memory strength, for example, mean that the mouse had a more humanlike brain? And, if so, in what direction? We do not know whether a human would perform most mouse memory tests better or worse than a mouse. Ultimately, we recommended watching any born chimeric mice for unusual behaviors, as we could not predict what kinds of behaviors might show a human influence.

Still, although defining a humanized brain or behaviors is difficult, at some hypothetical point it would become obvious, particularly if the chimera were able to communicate. How likely is such a result? It depends.

For a human/mouse chimera to be like Kafka's Gregor Samsa, but trapped in a mouse's body, is scientifically absurd. Researchers have shown that human neurons can live, grow, and divide in a mouse brain; mice have been made with several percent of their neurons of human origin. One researcher has even shown that a human neuron in a mouse brain will function, at least in the sense of "firing" (creating an action potential on the axon).[37] Whether those cells really function, in the sense of playing a useful role in the working of the mouse brain, is still unknown. It may well be that subtle differences in the environment and the molecules surrounding the human neurons make them unable to contribute meaningfully in the mouse brain, and would, as a result, make a living, functioning Human Neuron Mouse impossible.

Even if the Human Neuron Mouse did function, though, its brain would be about one one-thousandth the size of a human brain and the size of the mouse skull greatly limits its ability to expand. The absence of uniquely human areas associated with consciousness or intelligence makes it seem likely that the sheer size of our brains is important to human functioning. Accordingly, the Human Neuron Mouse seems *very* unlikely to have a brain that is meaningfully human or even humanlike.

But that is much less clear for a human/great ape chimera. Human brains and the brains of the other great apes are much more similar to each other, in structure and in size, than human and mouse brains. Chimpanzee brains are about one third as large as human brains; gorilla brains are about 40 percent the size of ours. A brain from these apes also has a large neocortex, although not as large as a human's. And, to the extent that subtle differences in genes, proteins, and other molecules found in the brain may make a difference in functioning, because of recent (in evolutionary time) shared ancestry, the human versions of these biological molecules are more likely to be similar, or identical, to those found in other apes than to those used by other kinds of mammals. As the Johns Hopkins group concluded, it is not absurd to think that a gorilla with a brain derived from human embryonic stem cells or human brain stem cells would have a substantially "humanlike" brain and possibly human-like behaviors.

If our worry is the creation of something with a human consciousness or intelligence, it is important to note that human/nonhuman chimeras are not the only way to try to achieve that goal, and may not be the most plausible. Computers can already beat chess grandmasters; that level of intelligence seems a stretch for still hypothetical human/gorilla chimeras. Artificial intelligence may be a more promising route for anyone trying to create another humanlike intelligence.

If someone did want to use biological materials, it might be more promising to try to make genetic modifications in another great ape, making the genes suspected of being important in brain development identical to their human equivalents. (These would be "genetic chimeras" according to the definition of chimera I proposed in 2003, but are not chimeras as defined for this chapter.) And, of course, we could end up confronting humanlike but nonhuman intelligences in other ways, through communication with intelligent extraterrestrials or even through discovering an ability to communicate with existing nonhuman species already on Earth, such as whales or porpoises.

But whatever the source of a nonhuman humanlike intelligence, it would raise profound questions. Some of these were explored in a 1947 science fiction story by Robert Heinlein, "Jerry Was a Man."[38] Heinlein's story, set in an undefined future, features widespread genetic modification of animals, some as pets or follies (like a winged but flightless horse), but some as workers. Jerry was one of many genetically modified chimpanzees, given higher intelligence that included the ability to speak and used as field or domestic workers. These workers were treated as owned animals; the conflict in the story revolved around a court case seeking (successfully) to declare Jerry the holder of human rights.[39]

How would we react to a humanlike but nonhuman intelligence? Would we greet it gladly, welcoming "someone else to talk to?" Would we, as Streiffer (and Heinlein) fear, treat it badly? How *should* we react? How would our societies have to change in reaction to these other intelligent creatures? How would we even recognize it—through something like the Turing test for artificial intelligence?[40] These questions are not easily answered.

2.2 Gametes

Worry about human/nonhuman chimeras carrying human eggs and sperm seems obvious. One of the definitions of prohibited chimeras in the Brownback bill—"a nonhuman life form engineered such that human gametes develop within the body of a nonhuman life form"—spoke directly to this point. There are three possible bases for concern: the presence of human gametes themselves in chimeras, the possibility of the fertilization involving one human and one nonhuman gamete as a result of sex between two such chimeras (or a chimera and a nonhuman animal of the same species as the nonhuman part of the chimera), and the admittedly eerie chance that the mating of two human/nonhuman chimeras with human gametes could produce a human embryo. None of these, ultimately, is a strong argument against creating such chimeras.

Unless one sacralizes human gametes, there is no reason to be any more concerned about their mere presence in a human/nonhuman chimera than there is to be concerned about the presence of human liver cells, kidney cells, or tumor cells—and far less than for human neurons. The absence of gametes, which all human males experience before puberty, or of functional gametes, which all human females experience before puberty and after menopause, makes people infertile but has no other effect on their functioning or personalities. Similarly, the mere presence of gametes should have no consequences, either for humans or for chimeras. Nor, in spite of a few religious efforts to the contrary, do we treat gametes as sacred. In the normal course of events, sperm and eggs are not cherished, protected, and treated with respect. In all humans' lives, their gametes live and die, in the billions for men and in the hundreds for women, almost always futilely and never mourned or subject to "proper" or "respectful" disposal.

The worry *must* be not about the gametes in themselves, but about the possibility that the gametes will be effectively used—that human sperm or eggs in chimeras will fertilize or be fertilized. The folk concerns about crossing species boundaries undoubtedly include worries about human sperm fertilizing nonhuman eggs or, perhaps viewed culturally as worse, nonhuman sperm fertilizing human eggs. The Brownback bill's eight definitions of chimera contain two provisions speaking to this possibility: "(B) a hybrid human/animal embryo produced by fertilizing a human egg with nonhuman sperm; (C) a hybrid human/animal embryo produced by fertilizing a nonhuman egg with human sperm."

The idea that human sperm and rat eggs, or human eggs and mouse sperm, could form an embryo seems scientifically bizarre. The species are so different, down to the numbers of their chromosomes, that it seems impossible that they could make a hybrid that functioned, even at a prenatal stage. Nor are there any plausible reports of successful hybrids of humans with any other species. Controversy even continues as to whether modern humans could or did mate successfully with their closely related Neanderthal cousins.[41]

Of all existing species, it is least implausible that humans might be able to form hybrids that were viable, at least perhaps in utero, with chimpanzees, our closest living relatives. Would the existence of such a hybrid be a moral problem if it were terminated well before birth? It is likely that most, if not all, such hybrids would fail to develop successfully in utero for natural causes. Products of conception often fail to develop. It is estimated that more than half of all human zygotes resulting from sexual intercourse do not develop into successful pregnancies; even in healthy young women, only about 30 percent of embryos transferred through in vitro fertilization develop. It is hard to see a serious ethical problem in the creation of a transitory hybrid embryo that dies on its own, apart from a strict abhorrence of crossing species lines.

Even if such hybrid embryos could come into being and were not naturally unable to develop, the resulting pregnancies could be terminated. Aborting such a pregnancy seems unexceptional, even if the embryo or fetus were to be given, by the benefit of the doubt, some human moral status. Many countries, including the United States, allow abortion of fully human fetuses for broad reasons, and others allow abortion for reasons of fetal health. And, in the unlikely event that such a hybrid were viable, any born human/chimpanzee hybrid would be highly likely to have major health problems and so could be legally aborted in most countries.

Of course, if such a hybrid were possible, and if it both survived the natural problems of pregnancy and was not intentionally aborted the moral status of the resulting "humanzee" might pose a difficult problem. The chances of such a development are so low, and the issues about the treatment of the humanzee both so speculative and so difficult, that this chapter will note them, but not pursue them any further.

Thus far, we have been discussing conceptions and embryos resulting from the conjunction of a human gamete, in a human/nonhuman chimera, with a nonhuman gamete. There is an odder possibility: two beings, each a human/nonhuman chimera, and each with complementary human gametes, mate. The product of a human sperm fertilizing a human egg, even if it occurred within a nonhuman reproductive tract, would presumably be some kind of human embryo. That such a human zygote would be viable in an alien uterus seems unlikely. Depending on the nonhuman component of the chimera, a "human" pregnancy could be disastrous; no human fetus could develop successfully inside a mouse or a rat. Using the same logic as before, though, termination of the pregnancy would be appropriate. Of course, should such a fetus come to term and be born alive, it seems highly likely that we would, and should, treat it as fully human, even though its gestation in an alien environment might (or might not) cause some minor to great differences— probably negative but possibly positive—from other humans.

But there are some more fundamental problems with these odd reproductive scenarios. First, scientists are not going to want to create such pregnancies. Usually they are not even going to want to create chimeras with human gametes. An embryonic stem cell that was intended to be studied in the brain or heart or liver might, somehow, migrate to the host's gonads and somehow produce gametes, but how often? And, in terms of intentional creation of gametes, what scientific question is answered by seeing if two mice with human gametes can "naturally" form a human embryo? Why would a scientist want to see if humans and chimpanzees could form even momentarily viable hybrids? Even a mad scientist intent on producing a creature with some characteristics of both species would be more likely to proceed through more targeted methods, such as transplanting particular organs, transplanting stem cells that would produce only those organs, or changing particular genes of interest. Any of these conceptions would be accidents, and unlikely accidents at that.

Accidents cannot always be avoided but they can be limited. Chimeras could be sterilized. They could be segregated by sex in different cages or different rooms. Only immature chimeras could be used for research and they could be euthanized before they were reproductively mature. To avoid the Kafkaesque scenario of human sperm and human egg meeting as a result of the mating of two mice (or, more accurately, predominantly murine human/mouse chimeras), the chimeras could be made from only one sex. And accidents can be repaired—any pregnant chimeras could be euthanized as soon as the pregnancy was detected. These kinds of precautions are already being required by many SCROs in implementing the NAS Guidelines or state rules. We may worry about human gametes, but we need not worry very much.

It is again important to note that similar issues could arise without any use of human/nonhuman chimeras. One could attempt fertilization between human and nonhuman gametes much more simply in vitro (or, in some cases, in vivo). In fact, one common test for whether human sperm to be used for artificial insemination or in vitro fertilization is properly functional is to see if it can penetrate hamster eggs. A scientist who, for some bizarre reason, wanted to see what would happen to a fertilized human egg in a nonhuman uterus could just transfer an in vitro human embryo to an animal. No chimeras would be necessary.

2.3 Outward Appearance

The NAS Guidelines and other policies, as well as the Brownback bill, expressly discuss chimeras involving human cells in brains and, implicitly, chimeras with human gametes. Yet, apart from under thirty words in the NAS Guidelines, there is no express discussion of the possibilities of chimeras with any outwardly (and disturbingly) visible human features.

The NAS Guidelines briefly mention this possibility twice, on the same page.[42] First, the report says "The hES cells may affect some animal organs rather than others, raising questions about the number of organs affected, how the animal's functioning would be affected, and whether some valued human characteristics might be exhibited in the animal, *including physical appearance"* (emphasis added). Then, two paragraphs below, it lists five questions an ESCRO should consider in reviewing protocols for transfer of human embryonic stem cells into nonhuman animals. The fifth is "If visible humanlike characteristics might arise, have all those involved in these experiments, including animal care staff, been informed and educated about this?"

One can see why the NAS included that question. Chimeras with some visible human characteristics could be profoundly unsettling. Indeed, the characteristics need not necessarily be "visible." A human/monkey chimera with a human voice would be upsetting. Or the disturbing trait might be visible but not a matter of appearance per se. An ape that walked like a human could be distressing. The most plausible examples, though, involve physical appearance and so I will focus on them. (This has been reported to have already influenced at least one SCRO deliberation.[43])

To some extent, such chimeras already have been disturbing—and not just from mythological sphinxes, minotaurs, centaurs, and various human/nonhuman gods, from Anubis to Ganesha. In 1997, scientists in Massachusetts published a medical article with a photograph of what looked like a human ear, which they had grown on the back of a mouse.[44] The photograph became famous (or infamous) and the experiment became controversial, in part as a result of its use by a group campaigning against genetic engineering. Ironically, not only was the ear *not* the result of genetic engineering—except in the creation of the immune-deficient mouse used as a host—but the creature was not a human/nonhuman chimera. The ear was made of cartilage from the knees of cattle, molded into the form of a human ear.

One can imagine other chimeras with deeply unsettling human features, such as nonhuman primates with human faces or hands. These would, of course, break the folk requirement of a strict separation between humans and other animals. They could also create the kind of moral confusion that Robert and Baylis discuss: someone might, at least at first, be uncertain about how to treat a monkey with a human face. In a more extreme case, a nonhuman primate with a human outward appearance might lead to moral confusion about the status of a developmentally impaired human, one with less cognitive ability than the nonhuman primate. Of course, these examples do point out an issue with the moral-confusion idea. Is this moral confusion only bad if it leads people to treat other humans as having a lower moral status, more akin to that of the nonhuman animals they resemble? What if, instead, it were to lead to people treating nonhuman animals better, because their connections to humans have seemed closer as a result of human/nonhuman chimeras?

On the other hand, these kinds of chimeras do not seem to implicate the concerns about the "human dignity" of the chimera that Streiffer and Cohen discuss as, understandably, they do not talk about outward appearance as something that would entitle the chimera to human moral status.

Are there other ethical issues here? It is hard to see any serious ones, and it is also hard to see much scientific interest in creating such chimeras. Just as the chimera issues around brains and gametes could arise without actually involving chimeras, the creation of nonhuman living things that look, in part or whole, like humans would not necessarily require human/nonhuman chimeras. Realistic androids could present the same issues. So could animals made to resemble humans through cosmetic surgery or genetic engineering.

2.4 An Exceptional Use

One use of human/non-chimeras that does *not* involve brains, gametes, or outward appearance could still raise substantial public concerns: their consumption.[45] Would Hannibal Lecter's meal of human liver[46] have lost much of its power to shock if the human liver had been grown inside livestock?

For a human to eat tissue of human origin from a human/nonhuman chimera would likely be regarded as, at best, disgusting. The consumption of such tissue by nonhuman animals, either intended by humans or accidental, also seems likely to raise concerns. But *should* we be concerned? No human being was killed, or even died, to produce this meat. In fact, if an adult stem cell or an induced pluripotent stem cell were used, not even a human embryo would have been destroyed. And, after all, after our deaths, all of us are food for worms (or other small living organisms).

The taboo against cannibalism is broad and deep in the modern world (although, historically, it has not been universal).[47] Similarly, human cultures generally require some special treatment of the dead. The dead, or, at least, "our" dead, are to be protected from the ravages of scavenging birds and animals. For large animals, burial serves this purpose, as does cremation. (Of course, these practices are also not universal. The Parsi custom of exposing their dead to scavenging by vultures is clearly contrary; so, less obviously, is the practice of burial at sea.)

Revulsion against eating human tissues, even if not from human beings, is not necessarily specific to human/nonhuman chimeras. Consumable human meat might not even require the death of any animal, human, or human/nonhuman chimera. Research is ongoing to produce meat from stem cells grown *in vitro*. The current research focuses on pork, but, if it were successful, presumably there would be no technical barrier to growing "long pig" by tissue culture. (Similarly, if we were able to grow human tissues in vitro for medical purposes, that tissue might end up being eaten.) Human/nonhuman chimeras, however, might present a greater opportunity for human tissue, produced for a non-food reason, to be consumed accidentally.

Taboos against cannibalism do not stem from crossing species barriers—in that case, the eater and the meat are from the same species. The idea that humans should eat animals but not other humans does seem, though, to be based in a concern about upsetting a natural distinction between humans and nonhuman animals. Similarly, animals eating human flesh reverses this normal order—that humans should eat animals and animals should not eat humans. As such, these concerns seem to implicate both the folk concern about the natural order and, possibly, the concern of Robert and Baylis about sowing moral confusion. They would not offend either Cohen's or Streiffer's slightly different concerns about human dignity, unless, of course, the particular human/nonhuman chimera was "sufficiently human" to deserve human moral standing.

Are there strong reasons to be alarmed by the consumption, by humans or other animals, of human tissue that was not taken from human persons? I find it repugnant—the very thought is nauseating—but that is not a moral argument. The strongest argument I see is that this very disrespectful use of human tissue might bleed over into immoral actions against human persons. The cannibal who has feasted on chimeric (or in vitro) human flesh might lose some reluctance to take flesh from actual humans, dead or (even worse) still living. Or someone who has seen chimeric human flesh sold as pet food might, as a result, devalue people. Treating human tissues, even when not produced directly from the bodies of humans, as meat could lead some people to treat people as morally equivalent to the nonhuman sources of our usual meat. This is, at heart, a speculative empirical argument about how some people would actually behave. It is necessarily speculative; both the existence and the extent of such an effect seem untestable without doing the experiment. One could argue, though, that it is not worth taking the risk that such an effect will occur.

None of the policies about human/nonhuman chimeras deal with this issue, for a good reason. They are policies about biomedical research. The chimeras they govern are research animals. When these animals—whether mice, rats, or monkeys— are dead, they will, in the normal course of events, be disposed of as biomedical waste. No one eats laboratory animals or feeds them to pets.

There is, however, at least one example of likely nonhuman consumption of human/nonhuman chimeras. In the early 2000s, Dr. Esmail Zanjani at the University of Nevada, Reno, transplanted human stem cells, mainly human blood-forming stem cells, into pregnant sheep. According to Zanjani, these cells became a variety of different cell types in the sheep and their offspring. In one sheep, he claimed that human cells made up 40 percent of its liver.[48] The local press reported that the ewes that had received injections of the human cells were, after lambing, sent by the University to a ranch to be used in a weed control project. These naïve laboratory sheep were then allowed to graze in the wild, leading to many dead sheep and some happy coyotes and mountain lions.[49] The newspaper story was part of a series about the misuse of animals at the university; it focused on the cruelty of intentionally allowing these animals to be killed by predators or chased by wild dogs to a drowning death in the river. Yet the fact that the article returned, several times, to the question whether there were human cells in the ewes suggests that there was at least some interest in the idea of human tissue being eaten by predators. The use of partly human tissues from a human/nonhuman chimera as food, for people or for other animals, may well be a significant public concern.

2.5 Summary and Conclusions

Summing up now, the arguments about brain chimeras made by Robert and Baylis, Cohen and Karpowicz, and Streiffer all have merit, but arguments for limiting the creation and use of other kinds of human/nonhuman chimeras are weak. The ethical arguments against human/nonhuman chimeras based on gametes, though, are almost wholly without merit, as are the arguments about human appearance. The ethical argument against eating human tissue is not, fundamentally, an argument against creating human/nonhuman chimeras and is, in any event, quite speculative.

The ethical arguments against such possible practices may be weak, but the public reaction against them seems likely to be strong. The delivery of President Bush's 2006 State of the Union Address and the multiple introductions of Senator Brownback's Human Chimera Prohibition Act both occurred when no clearly offensive human/nonhuman chimeras existed. Creating broadly offensive chimeras could lead to a political backlash, damaging to the progress of science.

At the same time, the benefits of most of the controversial work are at best limited, and even where benefits may be plausible precautions might usefully be taken to avoid giving offense. It could be scientifically useful to study the development of human gametes inside a laboratory animal, given how difficult it is to study them inside living humans. But there seems no reason to attempt to have human/nonhuman chimeras with human gametes breed, either with animals with nonhuman gametes or with other chimeras with human gametes, or even to allow the possibility. Measures can be taken to minimize those chances of fertilization or the highly unlikely prospect of any significant embryonic development.

Similarly, it might be useful to study the development of some aspects of human external appearance—the growth of the ear, the nose, or the thumb, for example—in nonhuman animals in order better to treat human diseases or deformities. One could imagine precautions to avoid using species where the result would look "too" human.

There seems no plausible justification, scientific or otherwise, for building or allowing a human meat industry.

Brain-based chimeras are the major exception to this line of argument. Some kinds of scientifically and medically useful experiments could certainly be done without risking the creation of a chimera with any potential for distinctly human cognitive functions. Weissman's plans to use the Human Neuron Mouse to study the effects of pathogens, radiation, drugs, and other intervening forces on human neurons in vivo could, as our report argued, be done with minimal risk of creating a humanlike brain. But if you wanted to study brain development, or, more broadly, the necessary brain prerequisites for some human abilities, human subjects cannot be extensively used because of the risks to them. Human/nonhuman chimeras might provide the only plausible approach.[50] One can at least imagine trying to grow humanlike brains in chimeras to study, for example, the ability to use language.

3. WHAT IS TO BE DONE?

So what should we do?

The issues, ethical and otherwise, around humanlike intelligence in nonhuman creatures are profound. We should not attempt to create such intelligences, and thus potential carriers of "human" rights, without substantial prior public discussion. That discussion should take place in a factually rich context, with a realistic proposal at hand, and not merely through abstract speculation. Until then, there should be a moratorium on any attempt to create such intelligences, whether through human/nonhuman chimeras, genetic engineering, or artificial intelligence. It is not clear whether that moratorium could be informal—a consensus of the relevant researchers—or whether it needs to be formal. (Ironically, situations where creators of human/nonhuman chimeras face SCRO-like oversight are the only circumstances in which there are formal limits on this effort, although, even there, without an express moratorium.)

We should also be careful to avoid creating humanlike intelligence accidentally. And, at least with respect to creating that kind of intelligence in the course of human embryonic stem cell research, we do. The NAS Guidelines require SCROs to assess carefully the possible integration of the human cells into the central nervous system. The processes suggested in our report on the Human Neuron Mouse and the factors set out in the *Science* article on neural grafting into nonhuman primates are two useful ways for SCROs to approach this task.

But what should we do about other types or uses of human/nonhuman chimeras, where ethical concerns are weak but public concern is strong? Ultimately, science will be—and should be—governed by the societies in which it takes place. Public concerns, expressed through government action, have both practical and normative weight that scientists need to consider. I believe that scientists—along with others who believe that this kind of research is scientifically important and, ultimately, morally positive—must be pragmatic. In this case, that means acting prudently to avoid rousing public opposition unnecessarily. Scientists should not try to create human/nonhuman chimeras that are likely to provoke public controversy or to use less controversial chimeras in controversial ways unless the potential scientific gains are clearly large enough to justify the controversy. Arguments about those gains should be made to some outside body and, if a decision to proceed follows, the public should be educated about the justification for the work. At the same time, researchers should take precautions to avoid uses of chimeras that give unnecessary offense.

This essentially is the regulatory regime proposed by the NAS and ISSCR and implemented by California. Research putting human embryonic stem cells into nonhuman animals must be approved by SCROs. The NAS tells the SCRO to pay "particular attention…to the probable pattern and effects of differentiation and integration of the human cells into the nonhuman animal tissues." The chimeras may not be allowed to breed; in California SCROs, nonbreeding plans are required for approval. There are no current specific requirements for proper disposal of human/nonhuman chimeras, but the disposition of dead laboratory animals is regulated. It is not clear whether adding a provision on disposition of chimeric remains would add enough value to be worth the possible controversy caused by drawing attention to the possibility of their consumption.

The SCROs must decide whether, overall, the scientific merits of the research justify undertaking it. This weighing requirement was initially intended to assure those concerned about the moral status of the embryo that human embryos would not be destroyed (and the resulting stem cells would not be used) wastefully or for no good purpose. But it can also serve to prohibit, on a case-by-case and local basis, research with human/nonhuman chimeras where the costs—whether viewed as public unhappiness or outrage directly or through the negative effects of that reaction on science—outweigh the likely scientific benefits.

This kind of prudential regulatory scheme is likely to succeed, in part because there are few if any good scientific reasons to try to avoid or circumvent it. The spirit of provocation—the desire to *épater la bourgeoisie*—no doubt exists in some scientists, but scientists are usually quite constrained actors. They need jobs, they need tenure, and they desperately need funding. Independently wealthy gentlemen-scientists might be able to annoy the non-scientific (and much of the scientific) world just for the fun of it, but almost all scientists will not be able to breed mice with human gonads or to create a monkey with a human face without a very convincing justification, to funders, department chairs, deans, and company CEOs. Even without regulatory oversight, science for the sake of provocation is unlikely. In many countries today, this kind of research would almost always have to undergo some kind of regulatory oversight—occasionally by IRBs or their human experimentation equivalents in other countries, almost always by IACUCs or their equivalents, and in many jurisdictions by SCROs or their equivalents. The mad scientist may not be entirely extinct, but it is safe to say that he is neither well-funded nor unregulated.

Of course, as noted above, the issues about brains, gametes, appearance, and even human meat could all arise outside the context of human/nonhuman chimeras. Even when chimeras were involved, those chimeras could be created outside the context of the regulatory scheme for human *embryonic* stem cells, through the use of induced pluripotent stem cells, of more differentiated stem cells, or just of tissue or organ transplants. The same kinds of pragmatic concerns about inflaming public opinion against science apply in those situations. It is not clear to me, though, that the risks of such actions, and, outside the politically charged world of stem cell research, justify a formal regulatory scheme.

Ultimately, it is not scientists who worry me, but artists. Many artists find skewering widely held sentiments particularly satisfying (and, for some of them, financially rewarding). Chimeras have already attracted the attention of modern sculptors and painters. More literal bioartists already exist. Edouard Kac famously asked a French laboratory to create a transgenic rabbit that had received a jellyfish gene, widely used in scientific research, for green fluorescent protein (GFP). The rabbit, which Kac named Alba, would glow at least somewhat green when viewed under a particular frequency of ultraviolet light.[51] Kac was going to exhibit the rabbit and eventually take it home to his family as a pet. The laboratory ultimately got cold feet and refused to turn the rabbit over to Kac; its fate (and that of the laboratory's other GFP rabbits) is unknown.

Other bioartists are also active. In 2008, the University of Alberta sponsored a fascinating exhibition of bioart and published a catalog of the exhibition with both photographs and descriptions of the art and essays by bioethicists and others.[52] Adam Zaretsky, one of the contributors to that show, is particularly active, working himself in laboratories to create new life forms/art forms.[53] One artist, Steven Kurtz, was even prosecuted by the U.S. federal government for mail and wire fraud for obtaining microbes from a vendor inappropriately.[54]

What should be done about bioartists creating controversial human/non-human chimeras or making provocative uses of them? Little or nothing. Whether regulating art is more like a minefield or a quagmire is not clear to me (perhaps a minefield in a quagmire?), but it is clear that it would be a contentious and difficult process. It is also the case that art's provocations might lead to useful discussion and insights into the science, for the public and for scientists. Even if it were plausible to regulate art, it would be a bad idea. On the other hand, it might be a good idea for science to dissociate itself from provocative bioart. If the artists use facilities or collaborators from science, those should be subject to the regulation of SCROs and similar bodies. But we should not try to stop artists, acting on their own (and within general health and safety regulations), from making bioart—though we may have to cross our fingers and hope they do not bring down the public's wrath on science.

4. CONCLUSION

The ethics of creating and using human/nonhuman chimeras lead us down some strange paths. Where there are serious ethical issues, in the creation of other intelligences, the issues apply more broadly than just to such chimeras. Where the concerns focus directly on the chimeras, through the folk abhorrence of mingling human with nonhuman, this repugnance is not an ethical argument, but it is a political reality that needs to be considered.

The Human Chimera Prohibition Act, though introduced a second time by Senator Brownback in 2007, has not passed. And, after all the controversy, Weissman still hasn't made any fully human neuron mice. Early on, there were problems finding the mouse strain he needed to use to host the human cells. Then the postdoc interested in the project moved on from Weissman's lab. Weissman even began to refer to the Human Neuron Mouse as a thought experiment. In early 2009, though, in response to the substantial stimulus funding from NIH for biomedical research, Weissman submitted a grant application, with Stanford stem cell researcher Marius Wernig and me as co-principal investigators, to create human neuron mice. This application proposed using not only human brain stem cells derived from fetal sources, but also induced pluripotent stem cells that had become differentiated into neural progenitor cells. That application was denied—and the Human Neuron Mouse remains not yet born, but not yet dead. For better or for worse.

NOTES

1. George W. Bush, State of the Union 2006, http://georgewbush-whitehouse. archives.gov/stateoftheunion/2006/.

2. J. M. McCune et al., "The SCID-hu Mouse: Murine Model for the Analysis of Human Hematolymphoid Differentiation and Function," *Science* 241 (1988): 1632–39.

3. Terence P. Jeffrey, "NIH Mass Produces 'Human Mouse,'" *Human Events*, October 11, 2001, pp. 1–9.

4. The discovery was announced in a press release by CytoTherapeutics, then the corporate parent of StemCells Inc., on January 20, 2000: "CytoTherapeutics Subsidiary Stemcells, Inc. Announces First Direct Isolation of Human Brain Stem Cells," *BusinessWire*, January 20, 2000. The press release can be found at http://www.stemcellsinc.com/ news/000120.html. The first scientific publication of the result came in December 2000. Nobuka Uchida, et al., "Direct Isolation of Human Central Nervous System Stem Cells," *Proceedings of the National Academy of Sciences* 97 (2000): 14, 720–25.

5. Henry T. Greely, Mildred K. Cho, Linda F. Hogel, and Debra M. Satz, "Thinking About the Human Neuron Mouse," *American Journal of Bioethics* 7, no. 5 (2007): 27–40.

6. The Human Chimera Prohibition Act of 2005, S 659.

7. A particularly useful summary and analysis of these discussions can be found in a recent short book by political scientist Andrea L. Bonnicksen, *Chimeras, Hybrids and Interspecies Research: Politics and Policymaking* (Washington, D.C.: Georgetown University Press, 2009).

8. I have discussed in earlier work quite a few of those meanings. Henry T. Greely, "Defining Chimeras…and Chimeric Concerns," *American Journal of Bioethics* 3, no. 3 (2003): 17–20.

9. It is unclear how many people have received such "xenotransplants." There are at least four famous examples. In 1984, a baboon heart was transplanted into a twelve-day-old baby born with fatal heart defects. The child, known as Baby Fae, survived for twenty-one days with the baboon heart. In 1992 and 1993, a team led by Dr. Thomas Starzl twice transplanted baboon livers into people dying of AIDS. One survived for seventy days; the other for twenty-five. And in 1994, an AIDS patient named Jeff Getty received baboon bone marrow, in the hopes that these cells, which are subject to infection by AIDS, would make white blood cells to keep him alive. The baboon cells quickly disappeared from Getty's body and did not successfully give him a baboon immune system, but Getty, at least, survived the procedure, eventually dying in 2006 from cancer. About twenty people received animal kidneys in the early 1960s from at least two different surgical programs, but to much less publicity. A few other experimental xenotransplants were tried over the years before 2000, but few, if any, have taken place since then. Still, an International Xenotransplantation Association still exists and its official journal, *Xenotransplantation*, has just completed its sixteenth volume.

10. Greely, "Defining Chimeras," above n. 8.

11. We called for monitoring any live chimeric mice for signs of pain but saw no reason to think that would be so likely as to prevent the experiment. We noted some ethicists and research institutions view the use of human fetal tissue as making the investigator unacceptably complicit in an abortion. We argued, though, that the facts that abortion has been widely legal in the United States since January 1973; that the use of human fetal tissues, or cells derived from such tissues, is legal in the United States; and that research using them has, by Act of Congress, been eligible for federal funding since 1993

made their use, in general, ethically permissible. Finally, citing numerous other examples, we argued that whatever the limits of appropriate use of human tissues, its use in biomedical research was generally accepted as permissible.

12. See generally the open peer responses to our article in *American Journal of Bioethics* 7, no. 5 (2007): 41–58.

13. Robert Streiffer has a good discussion of these arguments, and one more, which he calls the "borderline personhood" argument, in an article I did not see until the last revision of this chapter. Robert Streiffer, "Human/Non-Human Chimeras," *The Stanford Encyclopedia of Philosophy*, ed. Edward N. Zalta, Spring 2009, available at http://plato.stanford.edu/entries/chimeras/.

14. Jason Scott Robert and Françoise Baylis, "Crossing Species Boundaries," *American Journal of Bioethics*, 3, no. 3 (2003): 1–13.

15. The Human Chimera Prohibition Act of 2005, Section 3(a), above n. 6.

16. Robert and Baylis, "Crossing Species Boundaries," 10.

17. President's Council on Bioethics, *Human Dignity and Bioethics: Essays Commissioned by the President's Council on Bioethics* (Washington, D.C.: President's Council on Bioethics, 2008).

18. Ruth Macklin, "Dignity Is a Useless Concept," *BMJ* 327 (2003): 1419–20; Timothy Caulfield and Roger Brownsword, "Human Dignity: A Guide to Policy Making in the Biotechnology Era?" *Nature Reviews Genetics* 7 (2006): 72–76.

19. Robert and Baylis, "Crossing Species Boundaries," above n. 14, at 9.

20. Robert and Baylis, "Crossing Species Boundaries," 10.

21. Robert and Baylis, "Crossing Species Boundaries," 10.

22. Cynthia Cohen, *Renewing the Stuff of Life: Stem Cells, Ethics, and Public Policy* (New York: Oxford University Press 2007); Phillip Karpowicz, Cynthia B. Cohen, and Derek van der Kooy, "Developing Human-Nonhuman Chimeras in Human Stem Cell Research: Ethical Issues and Boundaries, *Kennedy Institute of Ethics Journal* 15, no. 2 (2005): 107–34; Cynthia B. Cohen, "Beyond the Human Neural Mouse to the NAS Guidelines," *American Journal of Bioethics* 7, no. 5 (2007): 46–49.

23. Karpowicz et al., "Developing Human-Nonhuman Chimeras," above n. 22, at 123. (Presumably the same argument would apply to a decision by human parents to "create" a baby which "could not fully exercise the dignity-related capacities associated with the human brain.") It is not entirely clear to me whether the argument is mainly about being created with biologically limited abilities or is also about being created for a role (laboratory animal) that effectively limits those abilities. To the extent it includes the latter, it overlaps with Streiffer's argument, discussed below.

24. Josephine Johnston and Christopher Eliot, "Chimeras and Human Dignity," *American Journal of Bioethics* 3, no. 3 (2003): 6–7. Their argument may also overlap with Streiffer's.

25. Robert Streiffer, "At the Edge of Humanity: Human Stem Cells, Chimeras, and Moral Status," *Kennedy Institute of Ethics Journal* 15, no. 4 (2005): 347–70.

26. Hilary Bok, "What's Wrong with Confusion?" *American Journal of Bioethics* 3, no 3 (2003): 25–26.

27. Committee on Guidelines for Human Embryonic Stem Cell Research, *Guidelines for Human Embryonic Stem Cell Research* (Washington, D.C.: National Academies Press, 2005). Henceforth, "Guidelines."

28. Guidelines, 38–41, 49–50. The discussion was based in part on presentations at a public workshop the Committee held on October 12 and 13, 2004. Cynthia Cohen and I both testified about chimeras at this workshop, as did several others.

29. Guidelines, 50.

30. Guidelines (3(b)(ii)), 57.

31. Guidelines (3(c)(ii-iii)), 57.

32. International Society for Stem Cell Research, *Guidelines for the Conduct of Human Embryonic Stem Cell Research* (ISSCR 2006), available at http://www.isscr.org/guidelines/. (This document does not include page numbers.)

33. Canadian Institutes of Health Research, *Human Pluripotent Stem Cell Research: Guidelines for CIHR-Funded Research*, sections 8.2.4–8.2.7 (2006, updating earlier version from 2002 and 2005), available at http://www.cihr-irsc.gc.ca/e/1487.html.

34. National Institutes of Health, "National Institutes of Health Guidelines for Human Stem Cell Research," Section IV, 74 *Fed. Register* 32170 (July 7, 2009).

35. Mark Greene et al., "Moral Issues of Human–Non-Human Primate Neural Grafting," *Science* 309 (2005): 385–86. For purposes of disclosure, I note that I was one of the many authors of this article.

36. The phrase "brains, balls, and beauty" came to my mind, but it seemed a touch too crude and, in its second noun, was neither precise nor complete.

37. Alysson R. Muotri, et al., "Development of Functional Human Embryonic Stem Cell–Derived Neurons in Mouse Brain," *Proceedings of the National Academy of Science U.S.A.* 102, no. 51 (2005): 18,644–18,648.

38. Robert A. Heinlein, "Jerry Was a Man," in *Assignment in Eternity* (New York: New American Library, 1953).

39. The court case, and the story, ends with what can only be called a cringe-inducing parallel between Jerry and African-American slaves. Jerry wins his court case by playing a harmonica and singing "Suwannee River."

40. The Turing test refers to a proposal by Alan Turing that a computer could be considered to have attained artificial intelligences when an observer could not tell from its responses whether it was a computer or a human being. See the fascinating discussion of the Turing test in Graham Oppy and David Dowe, "The Turing Test," *The Stanford Encyclopedia of Philosophy*, ed. Edward N. Zalta, Fall 2008, available at http://plato.stanford.edu/archives/fall2008/entries/turing-test/.

41. The big differences between Neanderthal mitochondrial DNA and human mitochondrial DNA seem to make unlikely any genetic contribution to current humans from Neanderthal females. The very recent publication of "the" Neanderthal genome includes an assessment that finds some small (one to three percent) contribution of Neanderthal nuclear DNA to modern non-African humans, but whether this conclusion will hold up to scrutiny remains to be seen. Richard E. Green et al., "A Draft Sequence of the Neandertal Genome," *Science* 328 (2010): 710–22.

42. Guidelines, 50.

43. Patricia Zettler, Leslie E. Wolf, and Bernard Lo, "Establishing Procedures for Institutional Oversight of Stem Cell Research," *Academic Medicine* 82 (2007): 6–10.

44. Yilin Cao, Joseph P. Vacanti, Keith T. Paige, Joseph Upton, and Charles A. Vacanti, "Transplantation of Chondrocytes Utilizing a Polymer-Cell Construct to Produce Tissue-Engineered Cartilage in the Shape of a Human Ear," *Plastic and Reconstructive Surgery* 100, no. 2 (1997): 297–302.

45. One student at the Harvard workshop raised the issue of the use of human/nonhuman chimeras to provide organs for transplantation to humans in medical need. The student seemed to feel that this would raise concerns similar to those involved in eating human parts from such chimeras. Given our acceptance of organ transplants from both living and dead humans, as well as our general acceptance of eating nonhuman animals, any broad-scale objection to transplanted human organs from such a source seems to me both unlikely and unwarranted.

46. Thomas Harris, *The Silence of the Lambs* (New York: St. Martin's Press, 1988).

47. It is at least interesting that a central ritual of Christianity, communion or the Eucharist, is an example of the cannibalistic consumption of the body and blood of Christ, either symbolically in many sects or, in Catholic doctrine, literally, through the Miracle of Transubstantiation. "And as they were eating, Jesus took bread, and blessed [it], and brake [it], and gave [it] to the disciples, and said, Take, eat; this is my body." Matthew 26:26 (King James Version).

48. Jamie Shreeve, "The Other Stem Cell Debate," *New York Times Magazine*, April 10, 2005.

49. Francis X. Mullen, "From Research to Waste," *Reno Gazette-Journal*, March 30, 2005, 1A. There appears to be some dispute as to whether the ewes, which had been injected with the human cells, retained human cells at the time of their disposal to the ranch. The reporter notes that the ranchers and neighbors were warned that they did have human cells and were told not to eat them.

50. One might try to reach a similar result through a genetic chimera, starting with, say, a baboon and gradually knocking out baboon genes and adding parallel human genes. Although I view such transgenic creatures as one kind of chimera, on which see "Defining Chimeras," above n. 8, they are not within the scope of this chapter.

51. The full story of Kac and Alba is unclear. See Christopher Dickey, "I Love My Glow Bunny," *Wired* 9, no. 4 (April 2001). The French laboratory seems to have been producing GFP rabbits for several years before Kac requested one. The rabbits did not glow very green. The eyes were green and the skin had some green glow, but the skin was obscured by the (white) fur.

52. Sean Caulfield and Timothy Caulfield, eds., *Imagining Science: Art, Science, and Social Change* (Edmonton: University of Alberta Press, 2008). My own contribution to the collection, Henry T. Greely, "Within You, Without You," argues that many of our folk reactions to "unnatural" biotechnological oddities stem from a very parochial and limited view of the diversity actually found in nature.

53. See Zaretsky's website, Emutagen.com, http://emutagen.com/index.html.

54. Charges were dismissed against Kurtz before trial; Robert Ferrell, the respected geneticist who had ordered the microbes for him, had pleaded guilty earlier to a misdemeanor charge. See the discussion of the case and its resolution in Michael Beebe and Dan Herbeck, "UB Art Professor Steven Kurtz Cleared of Federal Charges," Indymedia–Rochester NY (April 22, 2008), available at http://rochester.indymedia.org/newswire/display/21161/index.php.

SUGGESTED READING

In addition to the specific readings below, the format of the *American Journal of Bioethics* generally has one target article that is the subject of numerous "open peer commentaries." The two target articles included below, Robert and Baylis ("Crossing Species Boundaries"), volume 3, no. 3, and Greely et al. ("The Human Neuron Mouse"), volume 7, no. 5, led to twenty-four and seven published commentaries, respectively (some in the printed version of the journal, others available only on-line), as well as authors' responses to those commentaries from Greely et al. Several of the commentaries are listed specifically below, but many more are worth reviewing.

BAYLIS, FRANÇOISE, and ANDREW FENTON. "Chimera Research and Stem Cell Therapies for Human Neurodegenerative Disorders." *Cambridge Quarterly of Healthcare Ethics* 16, no. 2 (2007): 195–208.

BONNICKSEN, ANDREA L. *Chimeras, Hybrids and Interspecies Research: Politics and Policymaking*. Washington, D.C.: Georgetown University Press, 2009.

CAULFIELD, SEAN, and TIMOTHY CAULFIELD, eds. *Imagining Science: Art, Science, and Social Change*. Edmonton: University of Alberta Press, 2008.

COHEN, CYNTHIA. *Renewing the Stuff of Life: Stem Cells, Ethics, and Public Policy*. New York: Oxford University Press, 2007.

———. "Beyond the Human Neural Mouse to the NAS Guidelines." *American Journal of Bioethics* 7, no. 5 (2007): 46–49.

Committee on Guidelines for Human Embryonic Stem Cell Research. *Guidelines for Human Embryonic Stem Cell Research*. Washington, D.C.: National Academies Press, 2005.

DEGRAZIA, DAVID. "Human-Animal Chimeras: Human Dignity, Moral Status, and Species Prejudice." *Metaphilosophy* 38, nos. 2–3 (2007): 309–29.

GREELY, HENRY T. "Defining Chimeras…and Chimeric Concerns." *American Journal of Bioethics* 3, no. 3 (2003): 17–20.

GREELY, HENRY T., MILDRED K. CHO, LINDA F. HOGEL, and DEBRA M. SATZ. "Thinking About the Human Neuron Mouse." *American Journal of Bioethics* 7, no. 5 (2007): 27–40.

GREENE, MARK, et al. "Moral Issues of Human–Nonhuman Primate Neural Grafting." *Science* 309 (2005): 385–86.

HEINLEIN, ROBERT A. "Jerry Was a Man." In *Assignment in Eternity*. New York: New American Library, 1953.

KARPOWICZ, PHILLIP, CYNTHIA B. COHEN, and DEREK VAN DER KOOY. "Is It Ethical to Transplant Human Stem Cells into Nonhuman Embryos?" *Nature Medicine* 10, no. 4 (2004): 331–35.

KARPOWICZ, PHILLIP, CYNTHIA B. COHEN, and DEREK VAN DER KOOY. "Developing Human-Nonhuman Chimeras in Human Stem Cell Research: Ethical Issues and Boundaries." *Kennedy Institute of Ethics Journal* 15, no. 2 (2005): 107–34.

KASS, LEON R. "The Wisdom of Repugnance: Why We Should Ban the Cloning of Humans." *Valparaiso University Law Review* 32, no. 2 (1998): 679–705.

KOPINSKI, NICOLE E. "Human-Nonhuman Chimeras: A Regulatory Proposal on the Blurring of Species Lines." *Boston College Law Review* 45 (2004): 619–66.

LOIKE, JOHN D., and MOSHE TENDLER. "Reconstituting a Human Brain in Animals: A Jewish Perspective on Human Sanctity." *Kennedy Institute of Ethics Journal* 18, no. 4 (2008): 347–67.

ROBERT, JASON SCOTT. "The Science and Ethics of Making Part-Human Animals in Stem Cell Biology." *The FASEB (Federation of American Societies for American Biology) Journal* 20 (2006): 838–45.

———, and FRANÇOISE BAYLIS. "Crossing Species Boundaries." *American Journal of Bioethics* 3, no. 3 (2003): 1–13.

STREIFFER, ROBERT. "At the Edge of Humanity: Human Stem Cells, Chimeras, and Moral Status." *Kennedy Institute of Ethics Journal* 15, no. 4 (2005): 347–70.

———. "Human/Non-Human Chimeras." In *The Stanford Encyclopedia of Philosophy*, edited by Edward N. Zalta, Spring 2009, available at http://plato.stanford.edu/entries/chimeras/.

VINING, JOSEPH. "Human Identity: The Question Presented by Human-Animal Hybridization." *Stanford Journal of Animal Law and Policy* 1 (2008): 50–68.

PART VI

PRACTICAL ETHICS

CHAPTER 25

THE MORAL RELEVANCE OF THE DISTINCTION BETWEEN DOMESTICATED AND WILD ANIMALS

CLARE PALMER

THE title of this chapter advances a controversial claim: that a morally relevant distinction could turn on whether an animal is wild or domesticated. This claim, although intuitively appealing, has not been carefully defended and remains undeveloped in animal ethics. To explore the claim further, I will first explain how I use the terms "domesticated" and "wild." Then I will argue that most approaches to animal ethics have focused on animals' perceived *capacities*, making factors such as wildness or domestication appear to be morally irrelevant. While animals' capacities are central to animal ethics, I will argue that they are not all that matters morally. Humans can establish certain *relations* with animals that change what is owed to them.

The relations on which I will focus in particular are those of *vulnerability* and *dependence*. In the human case, we normally think that making vulnerable people or making people vulnerable creates special responsibilities toward them. Likewise, I argue that making vulnerable animals, or making animals vulnerable, also creates special responsibilities. Domestication is one pervasive institution by which humans make vulnerable animals. When we domesticate, I argue, we owe such animals care and protection that we do not owe to wild animals that are born and live independently of us. However, domestication is not the *only* way in which humans make animals vulnerable. Other kinds of human-induced vulnerability such as captivity and habitat destruction are also of moral relevance. In general, I will claim that certain relations we can have with animals should be considered alongside animals' capacities when deciding what we morally owe to them.

The Meaning of "Wild" and "Domesticated"

The term "wildness," in the case of animals, can be used in at least three different ways; I will call these *locational* wildness, *dispositional* wildness, and *constitutive* wildness. Only the last of these is generally paired with "domesticated."

(a) *Locational* uses of wildness concern *place*. On this interpretation, wild animals are those animals that live in environments on which humans have had little influence or impact. Though it is difficult to give a very precise account of what this means, it is easiest to think of human influence on the environment here along a spectrum, with highly urbanized or otherwise developed places at one end, and undeveloped, sparsely populated areas, wildernesses, or the deep oceans at the other. The wildest animals in this sense, then, are those who live their lives in the places least influenced by humans, such as bears whose habitat is in remote mountain ranges or toucans who live their lives deep in tropical forests. In contrast, animals that are *not* wild in this sense live in close proximity, or within, environments wholly or largely created by humans—companion animals, garden birds, urban rats. Many other animals—such as badgers living on agricultural land—fall toward the center of the spectrum.

(b) *Dispositional* uses of wildness: here, "wildness" refers to animals' dispositions and behavioral responses toward humans; a wild animal is one that is not *tame*. A tame animal shows little fear of people (interpreted behaviorally) and is not aggressive toward people, except when seriously provoked. A "tamed" animal, more specifically, is one whose fear or aggression has been deliberately reduced by human actions or settings. Unlike the case of locational wildness, and (arguably) constitutive wildness, humans can *create* animals that are dispositionally wild—highly aggressive dogs and cocks used for cockfighting, for example.

(c) *Constitutive* uses of wildness: here, a wild animal is an animal that has not been *domesticated*, that is, humans have not bred it in particular ways. To give an account of this sense of wildness, then, I will first have to explain how domestication is best understood.

Most frequently, the term "domesticated animal" is used to mean something like "an animal that has been bred in captivity ... in a human community that maintains complete mastery over its breeding, organization of territory and food supply."[1] However, this definition can be parsed in different ways; some accounts emphasize the economic benefits to humans of human control over animal lives, while others focus on the biological effects of human control over breeding. Yet other accounts of domestication, in contrast, emphasize animals' property status. Nerissa Russell, for instance, maintains that "the most crucial thing about animal domestication is that 'wild' animals are converted to property."[2] Finally, other interpretations of domestication focus on cooperation and exchange, taking domesticated animals to include, for instance, urban sparrows or squirrels. This sense fits the original meaning of the term "domestication" as "becoming accustomed to the household."[3] These differing interpretations of domestication correspondingly suggest different interpretations of wildness in the constitutive sense: for instance, wild animals as those animals from whom humans do not benefit economically, or those that do not cooperate with people, or those that are not property. However, here I will adopt the most commonly used interpretation of domestication, as referring to animals intentionally controlled by humans with respect to breeding, in particular by selective breeding. Although there are persistent scholarly disagreements about the history of animal domestication, in particular concerning how far domestication was, initially, deliberate on the part of humans, more recently animal domestication has been intentional, with the aim of enhancing the utility of domesticated animals to people.

Domestication by selective breeding can change animal physiology in a variety of ways, though such changes vary by—and within—species, and with respect to the purpose for which animals have been domesticated. In general terms, domesticated animals are smaller than their wild ancestors and have reduced brain size and smaller teeth. They are frequently thought to be characteristically neotenous, displaying the persistence of juvenile characteristics into adulthood.[4] Particular domesticated animal species have, in addition, been bred by humans in ways that exaggerate, diminish, or shape a variety of other characteristics, such as temperament, body shape, fattiness, the possession of horns, muscling, the presence or absence of body hair, and fur and eye color. Sometimes selective breeding can produce unintended effects; for instance, the deliberate breeding for large breast size in turkeys had the unintended effect of making it impossible for turkeys to copulate normally.[5] Such domesticated characteristics often result in it being difficult or impossible for domesticated animals to live, or to live well, without human protection or provision. I will return to this point later.

There are, then, a number of different interpretations both of "wildness" and "domestication" in widespread use. An individual animal might be wild in several different ways; for example, a constitutively wild animal could live in a relatively undeveloped area and behave in fearful or aggressive ways toward human beings. On the other hand, an urban squirrel may be constitutively wild but not locationally wild, and may not be dispositionally very wild either. Further, wildness, in all three senses, may come in degrees. The sense of wildness in which I am most interested in this chapter is wildness as non-domestication. But wildness and domestication here also form a spectrum; there are many animals—feral animals and captive-bred populations, for example—that are neither fully domesticated, nor fully wild, in terms of the impacts of human breeding on their bodies.

The Focus on Animals' Capacities

Most work in animal ethics has focused on questions about whether animals are of direct moral relevance at all—whether they have "moral status"—and whether, if they have such moral status, how comparatively morally important they are with respect to other beings, particularly human beings. I will call this comparative importance "moral significance." These debates have turned on what capacities are morally relevant, which animals possess such morally relevant capacities, and how much moral weight particular capacities should be thought to carry. Candidate capacities for moral status have included rationality (understood in various different senses); being able to feel pain and pleasure; being conscious (again, interpreted in various senses); having preferences; being goal-directed; having a sense of oneself as a being that persists over time; or some combination of these.

None of these capacities though, however interpreted, maps easily onto a distinction between "wild" and "domesticated." The kinds of animal capacities that humans shape by domestication are not (currently, at least) the kinds of characteristics that are usually thought to affect either an animal's moral status or its significance.[6] While domesticated housecats and African wild cats—often thought to be the historic ancestors of domestic cats—have different body shapes and temperaments, they do not obviously differ (for instance) in terms of their capacity to feel pain and pleasure, nor their possession (or lack of possession) of reasoning ability. In accounts of animal ethics in which individual animals' capacities determine moral status and moral significance, wildness and domestication seem to be irrelevant. This is not surprising since, after all, wildness and domestication are not *themselves* capacities, even if we can identify a domesticated animal by brain size, tooth size, neotenous behavior, and so on. "Wildness" and "domestication" are more like relations, or perhaps relational properties, than the kinds of capacities that have featured in debates about moral status. The idea of "wild" and "domesticated" animals only makes sense because there are human beings in the world with whom animals may or may not have particular relationships. This is not true of a capacity such as sentience. Wildness—in the sense of non-domestication—emphasizes the *absence* of a particular kind of human-animal relation, while domestication signals the *presence* of that relation.

The irrelevance of an animal's being wild or domesticated holds, at least directly, for most prominent theoretical approaches to animal ethics. That an individual animal's capacities alone are all that is of moral importance has, in particular, been central for those advocating some form of philosophical animal liberation. Strong opposition to any claim that species membership—which is also not a capacity—could be of moral concern has been critical to philosophical animal liberation. This focus on capacities, which I will call "capacity-orientation," is central both to utilitarian and rights-based approaches to animal ethics. While different forms of utilitarianism aim at maximizing different goods—some focusing on pleasure and pain, others on preference satisfaction, for instance—neither pains nor preferences have any direct relationship to wildness or domestication. Animal rights theorists, likewise, may disagree on the capacities required for rights possession—being an "experiencing subject of a life" in Tom Regan's case, or being sentient in Gary Francione's, for instance—but again, wildness and domestication are of no direct moral significance on this basis.[7] (Some rights theories may be able to find a place for the moral relevance of a wild/domestic distinction, but this distinction has not *yet* played a role in existing accounts; I will say more about this later.) Equally, those who deny that animals have moral status at all, or who maintain that animals are of much less moral significance than some or all human beings, also base their judgment on animals' lack of critical capacities such as moral agency, language, or autonomy, capacities that, in contrast, are shared by (most) humans.[8] In none of these leading accounts of human ethical responsibilities to animals has animals' wildness—or otherwise—played any direct part.

This is not to deny that, in such capacity-oriented views, wildness or domestication may be of *indirect* moral relevance. What is in the interests of domesticated and wild animals will diverge. I am likely, for instance, to have to treat a domesticated housecat and an African wild cat differently in order to protect or to promote their respective interests. But in capacity-oriented views, the fact that one cat is domesticated and the other wild does not, in itself, mean that I should privilege protecting or promoting the interests of one over the other. Wildness and domestication are only relevant to what is *in* an animal's interests, rather than to whether an animal has morally relevant interests at all, or to how significant those interests might be.

Such capacity-oriented arguments have rightly been central to the development of animal ethics. That an animal can feel pain *is*—as Robert Nozick famously argues—a good reason to think that there is a moral distinction between smashing an animal's head with a baseball bat and smashing a baseball with that same bat.[9] In this chapter, I will accept that the capacity to feel pain is a reasonable basis for moral status. However, it is not necessary to deny that certain basic capacities are key to establishing moral *status* in order to argue that other factors—including wildness and domestication—might *also* be of moral relevance. Other factors may underpin the creation of *additional* responsibilities to animals that we already accept to have moral status on the basis of their capacities. This is widely, though not uncontroversially, thought to be the case in the human sphere. We may, for instance, have special obligations toward family members or owe compensatory justice to people we have harmed in the past. As I will go on to argue, factors like these, alongside capacities, can also be of moral relevance in establishing what we owe to animals.

Making such an argument supports a distinction that appears, at least, to be widely held, particularly with respect to *assisting* animals. For example: thousands of wildebeest drown every year crossing the Mara river in Kenya on their annual migration. No one suggests that humans should attempt to reduce the animal suffering involved by rescuing the drowning wildebeest, or by redirecting the migration toward safer river crossings, though this would be feasible. In comparison, the Humane Society of the United States spent millions of dollars rescuing ten thousand companion animals from drowning in flooded homes and streets in Louisiana after Hurricane Katrina. Both these cases are, of course, complex. But they do illustrate that, in practice, a distinction between what is owed to wild and domesticated animals does often seem to be at work. I will argue here that there are good reasons to make such moral distinctions between wild and domesticated animals, particularly in the context of assistance.

CAPACITY-ORIENTATION AND A WILD/ DOMESTICATED DISTINCTION

The first move in defending a form of wild/domesticated distinction is to clarify, in a broad sense, its *scope*. I am not arguing that animals' capacities are *irrelevant* to animal ethics, nor that relational properties such as wildness and domestication are all that we need to take into account morally. While animals' capacities are rightly thought to be central in terms of establishing that they have moral status at all, and in terms of what we should *not* do to them, I will argue that particular human-animal relations can establish *additional* moral responsibilities toward the animals concerned.

The most helpful way to develop this distinction is by setting it in the context of an existing theoretical approach to animal ethics. This will indicate how a moral distinction between wild and domesticated animals can be built into an approach to animal ethics that is otherwise capacity-oriented. I will use animal rights theory here, as it is the most straightforward theoretical approach to work with in this context. However, I am not suggesting that one *must* accept rights theory in order to accept the kind of distinction for which I am arguing. Other capacity-oriented views—including other deontological approaches to animal ethics, and some forms of indirect consequentialism—could also provide a foundation for, or at least be compatible with, the argument I will make here. However, animal rights theory shows most clearly how the kind of relational distinction I am making can be taken on board by a position that is currently capacity-oriented.

Different theorists ground animals' rights in different capacities, such as sentience or being an experiencing subject of a life. Whatever the relevant capacity, any being that has it has the relevant right. As Regan maintains: "If any individual (A) has a right, then any other individual like A in relevant respects has a right."[10] Almost all animal rights accounts understand such rights negatively, in terms of non-interference of various kinds (not harming, not killing, not infringing on liberty). We may, though, have duties to assist where there have been prior rights infringements, a claim to which I will return.[11] On all such accounts, only moral agents can violate rights; and since, as far as we know, only (some) humans are moral agents, only humans can violate animals' rights. What happens in wild nature—for instance between predator and prey, where no moral agents are involved—is beyond the scope of rights claims: "Nature has no duties; only moral agents do....Nature no more violates our rights than it respects them."[12] No one's rights are infringed when the lion pulls down the wildebeest. On the basis of rights, at least, humans have *no* duties to act in the wild in the context of predation, flood, or drought, for instance. It is only when harm is being committed by a moral agent that any rights violation is involved. Were there convincing evidence that any nonhuman animals are moral agents—able, for instance, to evaluate reasons for acting—the ability to violate rights might extend beyond humanity. But since there is no such convincing evidence, rights violations can only exist where human moral agents impact on animals. As negative rights theorists maintain, wild animals should just be left alone to "carve out their own destiny."[13]

Although there are difficulties with such animal rights positions, I will not pursue them here, since as I have noted, the position I will develop here is compatible with other theoretical approaches to animal ethics, as well as with animal rights theory. I want, instead, to focus on questions about which a primarily negative animal rights account gives us little guidance. Most work on animals' rights has concerned what *not* to do to animals; that is, not to violate their basic rights. This may be sufficient to govern human interactions with animals that live independently of us—those animals that are locationally and constitutively wild. But we also *create* animals, make them *dependent* on us, and bring them to live alongside us. We accept that these kinds of relations underpin positive responsibilities to provide support and assistance of various kinds, but a negative rights view, in itself, has nothing to say about these matters. Although assisting animals is not *impermissible,* a theory of negative rights alone provides little guidance for thinking through occasions when helping or providing for an animal might be morally desirable or required. Of course, a negative rights theorist can reasonably respond—as Regan does—that "the rights view is not a complete theory in its present form."[14] It is perfectly possible for a rights theorist to accept that there may be "additional, non-discretionary obligations" that arise not from negative rights, but rather "from our voluntary acts and institutional arrangements."[15] The view I will be defending here takes this possibility as its starting place. Alongside general negative duties not to harm animals in certain ways, I will argue, our "voluntary acts and institutional arrangements" create certain additional, nondiscretionary obligations toward some, but not all, animals. Domestication, I will argue, is one such arrangement.

Of course, I am not the first to defend this claim. There is already a diverse—though often not well-developed—body of work arguing that various relations and relational properties, which sometimes include wildness or domestication, are of moral relevance. Such accounts may be based on a variety of different understandings of relations: *social* relations,[16] *kinship* relations,[17] *affective* relations (which can be understood in very different ways), *community* relations,[18] and *contractual* relations. The last two of these maintain—for different reasons—that wildness and domestication *are* morally relevant. Although these relational approaches are either poorly developed or seriously problematic in other ways, I discuss them because they are helpful in thinking through *why* we may have different moral responsibilities—in terms of assistance, at least—toward wild and domesticated animals.

Two Existing Moral Distinctions Between Wild and Domesticated Animals

Community Membership

One ground for a moral distinction between domesticated and wild animals is "community membership," though this expression serves as a rough marker for a number of different views. Those who adopt this position maintain that we have different moral responsibilities toward animals that belong to communities in different relations with humans. So, for instance, domesticated animals may be understood as members of close, mixed communities with humans (an idea first proposed by Mary Midgley), whereas wild animals may be understood either to constitute their own, more distant, separate species communities, or to be members of broader ecological communities.[19] Admittedly, a wild/domestic distinction may not map exactly onto this community distinction—because, for instance, some constitutionally wild animals can be kept as companion animals in the mixed community—but a rough-and-ready distinction does hold here.

One version of this community approach maintains that, while some basic moral norms, such as negative rights, should operate when humans act in the wild—there would, for instance, be something morally wrong about gratuitously torturing wild-living, constitutively wild animals—additional, special obligations apply within human-animal mixed communities. J. Baird Callicott is a prominent advocate of such a view, arguing that humans and some animals are bound together into communities by affective bonds of sympathy and trust that create particular kinds of ethical obligation. Companion animals, part of the intimate, familial community, "merit treatment not owed either to less intimately related animals, for example to barnyard animals, *or*, for that matter, to less intimately related human beings."[20] In contrast, wild animals living in the wild should be "treated with respect" but are not owed more than this qua individuals (though the wild communities of which they are members should be protected).

There are difficulties with Callicott's argument, but a community model is one way of conceptualizing morally relevant differences between wild and domesticated animals. This "animal communitarianism" roughly parallels some communitarian arguments directed at human communities. For instance, while it is often argued that members of one community, say a national community, should not *harm* members of other nations and should respect those nations' territorial boundaries, frameworks involving assistance—such as to provide health care or social security—remain internal to specific communities. A parallel argument in the animal case is that while there are only negative duties to wild animals, who live in their own, distant communities, where animals are members of close mixed human-animal communities, they may be owed certain special kinds of support and assistance.

I find Callicott's basic thought here—that more is owed toward animals with whom we have certain close relations—very plausible, and I will develop a similar view later in this chapter. However, there are significant difficulties with the specific "communitarian" form Callicott's argument takes. There are substantial (though not insuperable) problems involved in extending the idea of "community" to include animals, and the sense in which we normally use "ecological communities" is very different from the intersubjective and participative sense in which we understand interhuman communities.[21] Although there might be a way of making this communitarian model work better, I will not develop it further here. However, I will draw on some of the ideas embedded in this sense of community—the importance of relations of created dependence, in particular—in developing my own account.

A related argument, based on a *political* model of national sovereignty rather than a model of ethical community, has been proposed by Robert Goodin, Carole Pateman, and Roy Pateman.[22] Focusing on the great apes, they argue that apes have an "authority structure in place over a particular territory" and that, as such, great apes could be regarded as "sovereign" over their territory by international law. This idea of "simian sovereignty" suggests that we could regard great apes, at least, as living within sovereign autonomous communities that demand certain kinds of respect. This proposal has not been put into practice, nor extended to wild animal species more generally, though John Hadley has developed an argument that wild animals should be given property rights over their territory.[23] Because the idea of animal sovereignty could not be extended to domesticated animals, any extension or implementation of this proposal would inevitably create distinctions in practice between what is thought to be owed to wild and to domesticated animals. However, such "sovereignty" proposals essentially focus on ways of providing stronger protection from human interference for *wild* animals and as such provide no guidance for thinking about moral responsibilities toward the animals that we produce and live alongside.

The Domesticated Animal Contract

A second argument for a moral distinction between what is owed to wild and domesticated animals rests on the idea of a "domesticated animal contract." This is not an attempt to locate animals within a general interhuman moral contract (though there have been several such attempts). Rather, it refers to the idea that there is a *special* contract relationship between human beings and domesticated animals, a contract that does not include wild animals. A number of different accounts of this idea exist, but all provide a backward-looking, historical story about the increasing entanglement of humans with certain species of animals.[24] Prior to domestication, animals had to obtain vital scarce resources and protect their lives against a range of threats—hunger, storms, disease, predators. Domestication marks the transition, a change of state, a crossing from "wild nature" into "human society" or "culture." Various versions of the domesticated animal contract deliberately draw on social-contract language. The role animals themselves played in this transition varies between accounts. Budiansky, most prominently, argues that members of certain animal species associated closely with human communities because they gained from the association. He claims that animals can be thought of as "collaborators" in domestication.[25]

The idea of the domesticated animal contract posits that domestication is a win-win deal; it offers both humans and animals benefits that outweigh the costs the contract may incur to both parties. In the animal case—assuming, of course, that the "contract" is actually kept—the benefit is lifelong protection from predators and provision. The cost is a slow change in animal "nature" as a consequence of human breeding; a restriction of liberty; and being used in various ways, including being killed, in the case of most agricultural and laboratory animals. In the human case, there are costs in terms of labor, the use of resources to provide for animals, to shelter them and give them medical attention. These costs are offset by the fact that humans are benefited in turn by animal work, companionship, entertainment, and by the food and clothing their bodies provide. Wild animals are outside this contract and unaffected by it (directly, at least). The domesticated animal contract, it is argued, changes the moral relationship of humans with some animals—the domesticated ones—while leaving the moral relationship with wild animals, whatever it might previously have been thought to be, essentially untouched. That is, whatever might be thought to be owed to all animals on the basis of their capacities alone is augmented, on this account, by the extra assistance or benefits conferred on domesticated animals by the domesticated animal contact.

Although the idea of a domesticated animal contract provides grounds for a moral distinction between what is owed to wild and domesticated animals, it is a deeply problematic idea for several reasons. Contracts are normally made between free and equal rational agents who understand and assent to them. Taken in any literal, historical way, animals could not have understood domestication as, in any sense, making a contract. One possible move here is to suggest that consent should be seen as hypothetical, not historical; animals *would* consent to the deal of domestication, were they able.[26] However, hypothetical contracts are inherently suspicious because they exclude content; and for all of this to be plausible, we would have to accept that domestication is a good deal for animals. They could not be supposed, hypothetically, to agree to a contract under which they would be worse off, but there is considerable scholarly debate about whether animals *can* be thought to have gained from domestication.[27] Domestication is—for all the individual animals involved—irreversible; unlike most versions of the social contract, there is no way out for individual animals if humans break the contract. Yet it is often advantageous for humans to break the contract, both because all and only those who could punish the contract-breakers have something to gain from the contract being broken and because someone who broke an *animal* contract would not be regarded by other humans as someone with a disposition to break contracts with *humans*.[28]

There is a deeper difficulty. On this view of the contract, individual domesticated animals are actually brought into being, with particular "domesticated" natures—such as playfulness, lack of aggression, high milk-producing capacity, and so on—*because* of the (putative) contract. Without it, some entirely different animals would have lived. So asking of an individual domesticated animal (even hypothetically) whether domestication is something to which it would have agreed is a strange question. Without domestication, that particular animal would not have existed. *It* has no possible *constitutively* "wild" or "non-domesticated" alternative existence.[29] The deep, fundamental ways in which domestication brings beings into existence in certain ways makes domestication stand out from any other form of social contract.

The idea of the domesticated animal contract, then, is too deeply problematic to accept. However, one need not adopt contract language to accept the claim that domestication changes humans' moral relationship to animals. Many of the factors that feature in the domesticated animal contract—in particular the ways in which domestication changes animals' natures, making many of them dependent on human beings—are, I will now argue, morally relevant. I will, in particular, argue that changing animal natures in the process of selective breeding creates a special moral relationship with the animals concerned, and that this is one among several relations that have this effect.

Developing a Framework for a Wild/ Domesticated Distinction: Parallels with Human Cases

The framework I will now develop is situated in the context outlined above. It is built onto a capacity-oriented, negative rights approach that gives us an account of general duties, telling us what we should *not* do to animals, though this is not the only theoretical approach onto which a framework like this could be built. The framework also draws on ideas discussed in "community relations" and "domesticated-animal contract" accounts, where it is argued that relations with certain animals—such as domestication—can give us reasons to assist or to provide for them. I will further develop such reasons here, arguing that where humans create certain relations with animals, in particular relations of vulnerability and dependence, they create special obligations to such animals. I use "special obligations" here just to mean "obligations owed to some subset of persons in contrast to natural duties that are owed to all persons simply qua persons"—though in this case, the relevant set and subset is composed of animals rather than humans.[30] Domestication, I will argue, is the primary example of such a special relation. Equally, where humans have previously violated animals' rights, or otherwise seriously harmed them, leaving animals worse off in ways that significantly affect their well-being, humans owe these *particular* animals, not animals in general, something like compensatory justice.

When describing the effects of domestication earlier in this chapter, I noted that domestication changed animals in various critical ways. Animals that are locationally and constitutively wild, I suggested, live their own lives, independently of human beings. The physical and psychological capacities they possess have come about independently of humans; they eat, play, fly, swim, and produce offspring independently of humans. They live apart from us, in something like "other nations." Domesticated animals, in contrast, are deeply entangled in human society. Their natures are shaped by us. They are bred by us. Their bodies have the forms that we have selected, and, to some degree, at least, their minds reflect human influence. They may not be able to provide for themselves, and even if they could, they are often confined in ways that makes this impossible for them. So, while wild animals provide for themselves, the existence, nature, and situation of domesticated animals is causally bound up with humans. These differences, I maintain, make a moral difference. To explain this, it is useful to draw on arguments often used in human cases to determine when we should assist, and when assistance may be permissible, but is not required.

In *Anarchy, State and Utopia*, Robert Nozick uses a thought-experiment that runs something like this: suppose there are a number of separate islands, each of which is occupied by a Crusoe. These islands differ with respect to the natural resources they have available; in addition, the Crusoes have differing abilities and dispositions for work, so they have different amounts of wealth and comfort. Later, the Crusoes discover one another through radio transmissions. Not only do they discover each other's existence, but they also have the ability to transfer resources between themselves, allowing for the possibility that better-off Crusoes could assist worse-off Crusoes. Is there any *duty* for the better-off Crusoes to assist the worse-off Crusoes? Nozick argues not. In particular, because the Crusoes have lived completely independently of one another, no question of *justice* is raised by the differentials in each Crusoe's holding (and hence in his well-being). Nozick maintains, "In the social noncooperation situation, it might be said, each individual gets what he gets unaided by his own efforts: or rather, no-one can make a claim *of justice* against his holding."[31] Where inequalities have been generated without injustices, there are no duties of justice for the better off to assist the worse off, or to try to move toward equal distributions of resources.

There are reasons for thinking that these arguments are problematic in the human case. But this thought-experiment does help to illumine the wild animal case. Take constitutively wild animals living in a designated wilderness. These animals are in a "social noncooperation situation" with respect to human beings. They live independently of human provision; they may have good or poor access to vital resources; some of them do well, some do badly; some are sick, some are healthy. No rational moral agents are causally responsible for their well-being; their rights have not been violated; if these animals are hungry, or suffering, or being preyed upon, there is nothing obviously unjust or (given the animal context) even loosely analogous to injustice about how they fare. So, on the grounds of injustice, at least, there is no reason to assist them, although this does not mean that assistance would be impermissible.

Many other animals, however, are not in the situation of independent existence and self-subsistence that characterizes constitutively and locationally wild animals. Domesticated animals are the prominent example. They are not the equivalent of animal Crusoes, living self-sufficiently on their own islands; rather, they normally depend on human support to live well. This dependence has not just come about serendipitously. It has been brought about by deliberate human activities. In most cases, humans are largely responsible for (a) the *actual situation* in which domesticated animals live, often closely confined in spaces that prevent them from finding food, mates, and the like independently of human provision; (b) key facets of domesticated animal *natures*, including in many cases an inability to be self-sufficient owing to physiological or temperamental changes; and (c) the *very existence* of most individual domesticated animals, because they are bred by humans.

Domesticated animals—as I will maintain later—are not the only animals whose lives have been impacted in very significant ways by human beings, such that they cannot be thought of as independently living Crusoes. To extend Nozick's analogy further, we could see humans as conducting "raiding parties" on the habitat and resources of wild animals, or as colonizing their "islands," killing them or forcing them to live in the barely viable borderlands of human settlements. In neither of these cases—of created dependence or prior serious harms—do animals live in the kind of situation Nozick envisages as grounding non-assistance. Where, in human cases, there is either created dependence or prior serious harm leading to continuing disadvantage, we generally think that some kind of special responsibilities to assist or make good follow.

One analysis of this kind—with a focus on requirements to assist in the case of prior harms—has been developed by Thomas Pogge.[32] Pogge uses fictional inhabitants of Venus to ground his comparative thought-experiment. Suppose we found a population of starving people on Venus, a population that was hungry for reasons completely unrelated to people on Earth. Should we assist these hungry people? Pogge thinks that our duties to assist here, if we have them at all, are very weak. So, does this also mean that the affluent have no duties to assist those who are distant and poor, or starving, on Earth? No, because people on Earth are not in the situation of the Venusians (or Nozick's Crusoes). Even if the poor on Earth are spatially distant from the affluent, they are not distant in what Pogge understands to be a morally relevant sense: the sense of entanglement and causal responsibility that underpins social justice. On Pogge's account, the affluent are at least in part, even if indirectly, *causally responsible* for the situations of the poor and suffering on Earth; that is, the affluent are failing or have already failed in their negative duties not to harm. The poor on Earth, unlike the fictional poor on Venus, are *victims of injustice*. This injustice, Pogge maintains, has three possible forms (it is not necessary to accept all three; any one will do the job for him): the effects of shared institutions; the uncompensated exclusion from the use of natural resources; the effects of a common and violent history. The better off shape institutions that benefit them while harming the worse off; and/or enjoy advantages from which the badly off are excluded without compensation; and/or benefit from a violent history that gives them a good start in life while depriving others of theirs. The wealth of the affluent has not been achieved *independently* of their relations to the suffering poor; though physically distant, the affluent and the poor are causally entangled, in particular through common institutions. This entanglement, in which the affluent violate their negative duties to those who are poor, in turn generates responsibilities to end such harms and to compensate those who have been harmed.

I will adapt Pogge's argument for certain animal cases. I will argue that where humans have become entangled with animals by practices or shared institutional frameworks that have made animals especially vulnerable, there are special obligations to assist or care for them. Such obligations are not *general*; that is, they do not apply to those wild animals that live independently of human beings. Equally, where humans have violated animals' rights or seriously harmed them, for instance by destroying their habitat or denying them access to vital resources, they owe animals compensation. I will look closely, in turn, at the two kinds of special relation in which I am particularly interested: first, created dependence and vulnerability, and second, prior harm.

Created Dependence and Vulnerability

All animals—including humans—are by nature vulnerable to certain kinds of threats, such as those from disease and disaster. Vulnerability is, as Martha Fineman rightly maintains, to some extent ineliminable, a consequence of being embodied, "a universal, inevitable, enduring aspect of the human condition."[33] However, for most people most of the time, and frequently for wild animals, this enduring vulnerability does not generate strong, ongoing dependence on others for care and provision. Although humans are dependent as infants, and often dependent when aged, their dependence is "episodic, sporadic, and largely developmental in nature."[34] But this is not the case for many domesticated animals. Dependence on humans, for these animals, is permanent, enduring, and lifelong. This is particularly true of animals bred in ways that meet specific human requirements—such as cows that can only give birth through cesarean section, cats bred without fur or claws, turkeys selected to gain so much fat that they cannot walk or fly, genetically modified laboratory mice created to be susceptible to specific cancers. Even less dramatic forms of domestication can make survival without human care highly tenuous. For example, domesticated horses released into the wild are often attacked by wild horses and fail to grow a sufficiently thick winter coat to be protected from cold; many are unlikely to live long in the wild. For these domesticated animals, we might say, vulnerability is realized as a result of their dependence on humans.

Does this created dependence mean that humans owe assistance to domesticated animals that they do not owe to animals in general? Yes. If humans close down animals' options by external constraints on their movements and environments, preventing them from fulfilling some or all of their needs in other ways, then by making animals vulnerable, special obligations are generated to relieve the animals of the additional burdens that they have been forced to assume. In short: if we make animals vulnerable, we should reduce or eliminate the problems for the animals the creation of such vulnerability generates. Likewise, when humans deliberately create morally considerable, sentient animals, who have no other ways of fulfilling their needs and are constitutively profoundly dependent on and permanently vulnerable to humans, they create special obligations to provide and care for those animals.

The first kind of dependence we can call "external dependence." Here animals are restrained or prevented—by cages, walls, fences, being tied, being denied roaming territory, and the like—from supporting themselves or seeking support elsewhere. This kind of dependence applies to most domesticated animals and some non-domesticated ones, including many zoo animals, who, even if they could support themselves in certain wild contexts, are prevented from doing so. As Keith Burgess-Jackson argues of companion animals, once they have been taken into our homes or confined in some other way, their alternative options for living have been shut down.[35] Animals confined in feedlots or labs are in similar situations. They have access neither to other humans who could support them nor to the resources to allow them to be self-sufficient. Because other options have been closed, and because humans are responsible for their closure, humans should provide for them.

The second kind of dependence, in which domesticated animals are unable to be self-sufficient because of their human-shaped natures, we can call "internal dependence." This kind of dependence creates special obligations. It is familiar in the *human* case that where we create internal dependence and accompanying vulnerability we also create special obligations. It is sometimes argued that just being a member of an especially vulnerable human population (even where there was no deliberate intention to create such vulnerability) is sufficient to ground a moral responsibility for extra care and protection. This is not the argument here, however. The realized vulnerability of domestic animals is not by chance or merely unfortunate; it is deliberately created by people. It is not, for instance, as though people just stumbled upon mice populations especially susceptible to forms of cancer, or hairless cats that burned when exposed to the sun. Animals have been shaped to be as they are by people, who have deliberately made them to be vulnerable and dependent by nature.

The closest human parallel (though it is admittedly far from exact) is the human choice to have a child. It is widely agreed that this choice generates special obligations toward the resulting child not just because *any* child is needy and vulnerable, but because you are responsible for this vulnerability. It is your child, and thus you have special obligations toward your own child that you do not have toward children in general. So, if a couple voluntarily decides to procreate, they undertake obligations to care for their child, either themselves or in some other way that will be good for the child. As Onora O'Neill maintains, "a standard way of acquiring obligations is to undertake them, and a standard way of undertaking parental obligations is to decide to procreate."[36] Suppose someone decides to procreate, but denies any obligations to the infant, neglecting it or failing to provide for its basic needs. The neglectful parent is morally culpable in a way that would not apply to some other adult who, though knowing that there are neglected infants nearby and being able to adopt one of them, nonetheless chooses not to do so. The decision to procreate brings with it special obligations to the offspring that results from that decision, obligations that are backward-looking to the decision itself.

Human procreation is relevant as an analogy because when humans breed domesticated animals, they create particular sentient beings that would not otherwise have existed. The sentient beings that are thus created are usually dependent on others to survive and flourish; that is, dependent on their creators to care for them or to organize alternative care for them. There are, though, key differences here. First, in the typical case of human children, physical dependence diminishes over time, and children eventually become independent of their parents. This happens to a much lesser degree, if at all, with domesticated animals. Second, human children are not deliberately created in particular ways. Though some forms of human embryo selection and gene modification now occur, and may be on the increase, moral justification for these practices usually rests on arguments that such selection will enhance the welfare of the child produced and increase the likelihood that he or she will live a healthy and independent life. (Although "savior siblings" may be selected for the good of someone else, this practice is not to the disadvantage of the selected embryo; and the deliberate creation of deaf children by deaf couples in part follows from the argument that a deaf child would fit more comfortably into deaf family culture than a hearing child.) In the case of domesticated animals, however, breeding and genetic modification are only rarely undertaken in the interest of enhancing individuals' welfare; instead, these practices much more commonly enhance both individual animals' long-term dependence on humans and their vulnerability to environmental stress. Where animals' welfare is improved by particular breeding practices, this is usually to allow animals to live better only within the context of the closely confined and stressful conditions of intensive farming. These differences do not undermine the relevance of the analogy; indeed, they highlight the comparatively permanent and deliberate nature of created dependence in the animals' case. If we have special obligations when we create dependent, vulnerable, morally considerable children, then we also have special obligations when we create dependent, vulnerable, morally considerable animals.

My argument, then, is that domesticated animals and animals that humans have made dependent in other ways are in a different moral relationship with humans than animals that are locationally and constitutively wild. This provides a reason for thinking that while it is permissible not to assist migrating wildebeest drowning in the Mara River, we should assist companion or agricultural animals trapped in houses or fenced fields by rising flood waters and unable to escape by themselves. Similarly, while there is no responsibility to feed or shelter starving locationally and constitutively wild animals in a hard winter, we should provide food, warmth, and shelter to animals we have confined in zoos, especially where they are located in climates very different from those to which they are native. Where we have created dependence and vulnerability in animals, whether the dependence is internal, external, or both, we should assist them in ways that allow them to overcome the obstacles to living well that such dependence and vulnerability has created.

Prior Harms

Domestication and confinement are not the only ways in which humans can make animals vulnerable. Harming animals in ways that have long-term negative effects can also do so. Since I am working with an account of animal rights, I will take "harms" here to mean violation of an animal's basic rights; but other theoretical accounts of serious harms would also work. Regan mentions, but does not develop, the possibility of "compensatory justice" in an animal rights context; my argument can be described as developing such an account.[37]

In the human context, claims for compensation, reparation, or other forms of special treatment on the basis of prior harms or injustices—such as slavery, displacement from land, and other human rights offenses—have become increasingly common. It is widely accepted that where human individuals or groups suffer serious, continuing disadvantage on the basis of some prior significant injustice, action should be taken to "make good" that prior harm. Pogge's account of affluent individuals' responsibilities toward the poor takes this form. Those who are affluent, Pogge maintains, benefit from arrangements that impoverish others, authorize the institutions that produce such arrangements, and neither take compensating action nor shield victims from the effects of these global institutions. Inasmuch as benefits could be refused (but instead affluent individuals eagerly accept them) and the systems involved could be opposed and the victims protected (but individuals do not act to do so) on Pogge's account some moral responsibility for compensating for these harms falls on affluent beneficiaries.

Arguments of this kind raise complex issues, even in human cases. Should it be the perpetrators or the beneficiaries of prior harms (where these groups are not identical) who have primary responsibility to "make good"? How should we deal with intergenerational claims where both original perpetrators, beneficiaries, and victims have died? What is an appropriate response to claims that rest on counterfactual conditions (about, for instance, what the position of some human group would be now, had the relevant past offense not occurred)?[38] Although these problems are difficult to resolve, they do not overwhelm the basic principle that if rights have been violated or substantial injustices have been committed, these have persisting negative effects, and others have perpetrated or benefited from those rights violations and injustices, then there is a moral case for some form of "making good" to the victims.

I want to extend these arguments to certain animal cases. If serious prior harms or rights infringements are grounds for compensatory or reparatory actions in *human* cases where they cause persistent and serious negative effects, so they should also be in *animal* cases where similar effects occur. Admittedly, animals' claims are likely to be weaker than human claims. In human cases, reparations claims (for example) are normally directly made by those who have been affected. Animals can neither recognize that their rights have been violated, nor can they actively make claims with respect to their situation. This is likely to be important in terms of the kind of benefit that might be owed them. For instance, they would get no satisfaction from an apology, or from a sense of "just desert," were perpetrators deprived of some benefits they had unjustly gained. The fact that animals are not capable of understanding that their rights have been violated, or of actively claiming any reparation, does not mean that they have no claims at all. In the human case, were the victims severely mentally disabled, we would not deny all claims, even if guardians or trustees were required to represent the claims.

To take a particular case: suppose that a number of coyotes have been displaced by a housing development. They cannot move to new territory, as the land is occupied by other coyotes, but they are no longer able to access their former hunting grounds or their denning areas. On most accounts of animal rights, this housing development has violated their rights; on any account that takes animals' moral significance seriously, the housing development constitutes a serious harm. It has persistent negative effects on the coyotes' well-being; additionally, they are vulnerable to new hazards, in particular to heightened traffic danger. This case is a relatively simple one in at least one sense: it is a roughly same generation case, that is, we are thinking about what is owed to the *very same* coyotes that were displaced.

These coyotes are constitutively wild (we have not selectively bred them) and they are not directly confined (we have not caged them). However, they have been made extremely vulnerable on account of human actions. So the situation of the coyotes, in practice, shares important characteristics with the domesticated and confined animals in the previous section: the coyotes are unable to be self-sufficient on account of deliberate human actions. On the same grounds, then, we should help the coyotes in overcoming the obstacles to living well that their vulnerability has created. This claim, though, raises a further question: *who* is responsible for helping these coyotes? This question is relevant not only to this case but to all the claims about assistance I have been making.

WHO SHOULD ASSIST?

Throughout this chapter, I have argued that where animals have deliberately been made vulnerable or dependent, whether by domestication, captivity, or serious prior harms such as habitat destruction, "we" should compensate, protect, care for or assist those animals in ways that relieve the burdens "we" have created. Is my suggestion, then, that *everyone* has some responsibility toward these displaced coyotes, even though very few people were in any sense involved in the housing development or have benefited from it? Do I have special obligations to rescue a threatened domesticated animal, even though I have never bred one myself? An objector will argue that these kinds of compensatory responsibilities or special obligations should instead be confined to narrow contexts, in particular to cases where I have in some sense consented to, or voluntarily accepted, particular obligations.[39]

This objection raises complex issues to which I cannot do full justice here. However, I will make some points in response. First, there are some special obligations that do not seem to be voluntarily assumed. Siblings do not voluntarily enter the role of being a sibling; children do not voluntarily enter the role of "being x's child," but these are generally thought to be relations that carry special obligations.[40] So, if my elderly, pet-loving parent dies leaving a houseful of hungry cats, I have a special obligation to care for, or to find care for those cats, even if I have not volunteered for the job, and indeed, even if I find it distasteful.

Second, there are many cases where the context of responsibility is indeed narrower than "everyone." The coyote case is an example. It raises many of the difficulties that human-reparations cases generate, and in this sense poses no special problems. In human cases, as noted earlier, there are questions about whether perpetrators, beneficiaries (where the two groups are not identical), or both are responsible for reparations to those whose rights have been violated. The same questions arise here. Planners, developers, and contractors are to different degrees responsible for displacing the coyotes from their habitat. They also benefit in terms of profit from the development of the land. But the perpetrators are not the only beneficiaries; the new residents of the housing estate, who now occupy what was formerly the coyotes' territory, also gain from the coyotes' displacement over the long term. The scope of responsibility to the coyotes can be construed relatively narrowly; just as in human cases, we can confine it to direct perpetrators and beneficiaries, and consider how best to "make good" from within this context. In practice, in this case, it is likely that the new human residents would best be able to "make good" some of the ongoing negative effects of the development on the coyotes (in other cases, it may be more appropriate to focus on perpetrators). The new residents here can partially accommodate the coyotes' interests by habitat restoration, traffic calming, messier land use, and generally being willing to tolerate coexistence, even though this will result in some inconvenience to them.

In some cases, then, "we" can be understood narrowly, to refer to those who directly caused, or who directly benefited from, rights violations or other serious harms to animals. However, the claims I made about domesticated animals were broader than this, maintaining that *everyone* has special obligations to assist domesticated animals that they do not have toward constitutively and locationally wild animals, even if they have not consented to such obligations, nor bred domesticated animals themselves. On what basis could everyone have such obligations?

A form of beneficiary argument, I suggest, also holds here. Almost all individual humans are, in some way, tied into the institution of animal domestication (and other forms of utilization, though domestication is my particular focus here). We all benefit from the existence of domesticated animals, even where we are not directly involved in creating them. We eat them, wear them, live with them, are entertained by them. If we accept benefits from an institution that creates dependent, vulnerable individuals, then we should also accept the responsibility to care for those individuals, to protect them from hazards (for instance, when trapped by floods, as in the Hurricane Katrina case), and to provide food, shelter, and medical care for them. This need not mean that we all have exactly the same responsibilities. If you breed a kitten, certainly you have primary responsibility to care for it. But if you were to abandon it in a dumpster and I find it there, then I have a responsibility not just to walk away. However, we are not *inescapably* saddled with these special obligations; an exit is possible. It is possible to reject the benefits of animal domestication by not using products derived from domesticated animals, not living with domesticated companion animals, protesting against domestication, and otherwise disassociating from domestication. If one refuses the benefits of animal domestication, then one does not share in the obligations that acceptance of the benefits brings.

I have argued here that we have special moral responsibilities of various kinds toward animals that we have made vulnerable (either by creating them to be vulnerable or rendering them vulnerable by captivity or habitat destruction) and from whose vulnerability we benefit. This vulnerability includes most domesticated animals, and constitutively wild animals where human actions have closed down other options for self-sufficiency. However, questions are now raised about animals that do not easily fall into either of these groups: urban rats, for instance. Am I suggesting that we should assist them?

OTHER ANIMAL CONTEXTS

Many animals do not fall into any of the groups I have considered. They are not currently bred by humans, but humans have had some significant influence on them. For example, they were once bred by humans but have become feral; they were deliberately relocated by humans but now are locationally wild; and so on. What implications do my claims have for animals in these situations? I do not think that there is any single *general* answer here. My view is that, alongside animals' capacities, we also need to consider animals' contexts, and the kind of human involvement there is or has been in creating those contexts, in order to make decisions about what we owe them. There will be different conclusions depending on humans' causal role in animals' situations, the kind and distribution of human benefits gained from animals, and so on. Here I will consider one case mentioned above: urban rats.

First, I should clarify the kinds of rats I have in mind, since we have rather different relations with different members of the rat family. I am not thinking of fancy rats that have been selectively bred as companion animals. I mean urban rats that live alongside human beings in drains and houses and that scavenge on human waste. These rat populations are dependent on human beings, because the presence of human settlement has permitted the significant expansion of the rat population. Although these rats are constitutively wild, we have not selectively bred them, and they are not locationally wild.

There are significant differences between the situation of these urban rats in comparison with domesticated animals or the displaced coyotes. Domesticated animals are being deliberately bred for human benefit. The coyotes were constitutionally and locationally wild until human development intruded into their territory. The urban rats are in neither of these situations. Humans are not responsible for breeding them, shaping their natures, or for physically restricting their movements. The rats are less vulnerable and dependent than most domesticated animals. They are not physically constrained, and generally have other options for survival than particular human beings (although those options may be less satisfactory). Humans do not benefit from their presence; quite the contrary. So, the same special relationships that give rise to special obligations in domesticated animal cases do not hold for the urban rats, and the rats are not living in wild habitats that humans developed. They are opportunists in human territory; their presence in urban areas is not on account of a prior rights violation or other serious harm, unlike the coyotes. The rats' habitat *is* urban. So, neither the arguments that I used to make a case for special obligations to domesticated animals, nor those that I used in the case of prior harms, apply here. We do not have responsibilities to assist urban rats.

There are other cases of animals living as urban opportunists, however, that are less clear-cut than urban rats. Some constitutively wild animals are deliberately encouraged to live and reproduce in particular places, such as the flocks of pigeons that were once regularly fed in Trafalgar Square in London.[41] Here, the creation of dependent relations can be easily be foreseen, and follows from a deliberate human practice that gives many people pleasure. In the case of pigeons: the feeding of pigeons significantly influences flock size, generating more frequent breeding seasons and creating more pigeons. Given that surrounding pigeon habitats are already occupied, a proposal suddenly to terminate all feeding, as happened in London, would result in the starvation of birds that have been brought into being by deliberate human practices. In this case, unlike the rat case, continuing food support is owed to the pigeons, even if it is gradually reduced over time to slowly bring down flock numbers, since this is a case where humans have deliberately brought vulnerable and dependent sentient animals into being.

These two cases—urban rats and Trafalgar Square pigeons—indicate that even in situations that superficially look similar, different human relations can give rise to different responsibilities. In some cases, humans have no responsibility to assist urban animal opportunists. In other cases, humans may have created relations that entail ongoing responsibilities, ones that should not be abandoned without careful consideration. On the argument I have developed here, we need to know about the particularity of each case in order to make a well-rounded judgment about what is owed to the animals involved.

CONCLUSION

This chapter is entitled "The Moral Relevance of a Distinction between Wild and Domesticated Animals." I have argued that, understood in a particular way, there *is* a morally relevant distinction between wild and domesticated animals. What is relevant is not so much "wildness" or "domestication" in themselves, but what these terms signal about human effects on animals' ability to be self-sufficient. Animals that have not been bred by humans, and that live largely beyond human influence, are self-sufficient. Like Nozick's Crusoes, they live their own independent lives. Domesticated animals, in contrast, are normally deliberately bred and confined in ways that make them both vulnerable and dependent. For this reason we have special obligations to them that we do not normally have to wild animals. But if it is the creation of vulnerability and dependence that is of moral relevance, then something resembling such special obligations can extend beyond domesticated animals. We owe assistance to constitutively wild animals who have been made vulnerable by habitat destruction, and to urban opportunist animals, if they have been deliberately and predictably made dependent, for instance by regular feeding. Accordingly, what if anything is positively owed to animals in particular contexts needs to be taken on a case-by-case basis. What we owe to a barn cat, living partly on mice and partly on what humans provide, will be different both from what we owe to an African wild cat and to a confined housecat.

I have argued here that, with sentient animals as with fellow humans, we live in a complex world of intention, creation, dependence, and vulnerability. Although respecting the rights of, or refraining from causing serious harms to, animals is an appropriate starting point for animal ethics, where we have deliberately created animal vulnerability or dependence, we should also care for or assist the animals concerned.

NOTES

1. Juliet Clutton-Brock, *The Walking Larder: Patterns of Domestication, Pastoralism and Predation* (London: Unwin Hyman 1989), p. 21.
2. Nerissa Russell, "The Domestication of Anthropology," in *Where the Wild Things are Now: Domestication Reconsidered*, ed. Rebecca Cassidy and Molly Mullin (Oxford and New York: Berg, 2007), pp. 1–26, esp. p. 36.

3. Rebecca Cassidy, "Introduction," in *Where the Wild Things are Now*, p. 3.

4. Stephen Budiansky, *The Covenant of the Wild: Why Animals Chose Domestication* (New York: Wiedenfeld and Nicholson, 1992).

5. Edward O. Price, "Behavioral Development in Animals undergoing Domestication," *Applied Animal Behavior Science* 65 (1999): 245–71, esp. p. 253.

6. See Adam Shriver, "Knocking Out Pain in Livestock: Can Technology Succeed where Morality has Stalled?" *Neuroethics* 2 (2009): 115–24.

7. Tom Regan, *The Case for Animal Rights* (Berkeley and Los Angeles: University of California Press, 1983); Gary Francione, *Animals as Persons* (New York: Columbia University Press, 2008).

8. For example, Carl Cohen, "The Case for the Use of Animals in Biomedical Research," *New England Journal of Medicine* 315 (1986): 865–70; R. G. Frey, *Rights, Killing and Suffering* (Oxford: Blackwell, 1983).

9. Robert Nozick, *Anarchy, State and Utopia.* (New York: Basic Books, 1974), p. 35.

10. Regan, *Case for Animal Rights*, p. 267.

11. Tom Regan, *Defending Animal Rights* (Champaign-Urbana: University of Illinois Press, 2001), p. 50.

12. Regan, *Case for Animal Rights*, p. 272.

13. Regan, *Case for Animal Rights*, p. 357; Francione, *Animals as Persons*, p. 13.

14. Regan, *Defending Animal Rights*, p. 51.

15. Regan, *Defending Animal Rights*, p. 51.

16. Leslie Pickering Francis and Richard Norman, "Some Animals are More Equal than Others," *Philosophy* 53 (1978): 507–27.

17. Gary Steiner, *Animals and the Moral Community: Mental Life, Moral Status and Kinship* (New York: Columbia University Press, 2008).

18. J. Baird Callicott, "Animal Liberation and Environmental Ethics: Back Together Again," in *The Animal Liberation/Environmental Ethics Debate: The Environmental Perspective*, ed. Eugene Hargrove (Albany: State University of New York Press, 1992), pp. 249–62.

19. Mary Midgley, *Animals and Why they Matter* (Athens: University of Georgia Press, 1983).

20. Callicott, "Animal Liberation and Environmental Ethics," p. 256.

21. Ernest Partridge, "Ecological Morality and Non-Moral Sentiments," in *Land, Value, Community*, edited by Wayne Ouderkirk and Jim Hill (Albany: State University of New York Press, 2002), pp. 21–35, esp. p. 22.

22. Robert Goodin, Carole Pateman, and Roy Pateman, "Simian Sovereignty," *Political Theory* 25 (1997): 821–49.

23. John Hadley, "Nonhuman Animal Property: Reconciling Environmentalism and Animal Rights," *Journal of Social Philosophy* 36 (2005): 305–15.

24. Desmond Morris, *The Animal Contract* (London: Morris Books, 1990); Budiansky, *The Covenant of the Wild*; Callicott, "Animal Liberation and Environmental Ethics"; Catherine Larrère and Raphael Larrère "Animal Rearing as a Contract?" *Journal of Agricultural and Environmental Ethics* 12 (2000): 51–58; Elizabeth Telfer, "Using and Benefitting Animals," in *The Moral Status of Persons: Perspectives on Bioethics*, ed. Gerhold K. Becker (Amsterdam: Rodopi, 2000), pp. 219–32.

25. Budiansky, *Covenant of the Wild*.

26. Jan Narveson, "On a Case for Animal Rights," *Monist* 70 (1983): 30–49.

27. See F. E. Zeuner, *A History of Domesticated Animals* (London: Hutchinson, 1963), p. 37; D. R. Harris, "An Evolutionary Continuum of People-Plant Interaction" in *Foraging*

and Farming: The Evolution of Plant Exploitation, ed. D. R. Harris and G. Hillman (London: Unwin Hyman, 1989), pp. 11–24; Clutton-Brock, *Walking Larder*, p. 27; Budiansky, *Covenant of the Wild*.

28. See David Gauthier, *Morals by Agreement* (Oxford: Oxford University Press, 1986), p. 162.

29. Derek Parfit, *Reasons and Persons* (Oxford: Oxford University Press, 1984), p. 349.

30. Diane Jeske, "Special Obligations," *Stanford Encyclopedia of Philosophy*, ed. Edward N. Zalta (Spring 2008), http://plato.stanford.edu/entries/special-obligations/ (accessed April 20, 2010).

31. Nozick, *Anarchy, State and Utopia*, p. 185.

32. Thomas Pogge, "Eradicating Systematic Poverty: Brief for a Global Resources Dividend," in *Ethics in Practice*, 3rd ed., ed. Hugh LaFollette (Oxford: Blackwell, 2007), pp. 633–46.

33. Martha Fineman, "The Vulnerable Subject: Anchoring Equality in the Human Condition," *Yale Journal of Law and Feminism* 20 (2008): 1–20, esp. p. 8.

34. Fineman, "Vulnerable Subject," p. 9.

35. Keith Burgess-Jackson, "Doing Right by Our Animal Companions," *Journal of Ethics* 2 (1998): 159–185, esp. pp. 168–69.

36. Onora O'Neill, "Begetting, Bearing and Rearing," in *Having Children: Philosophical and Legal Reflections on Parenthood*, ed. Onora O'Neill and William Ruddick (New York: Oxford University Press, 1979), pp. 25–38, esp. p. 26.

37. Tom Regan, *The Case for Animal Rights*, 2nd ed. (Berkeley and Los Angeles: University of California Press, 2004), p. xl.

38. Jeremy Waldron, "Superceding Historical Injustice," *Ethics* 103 (1992): 4–28, esp. pp. 8–10.

39. Samuel Scheffler, *Boundaries and Allegiances* (Oxford: Oxford University Press, 2001), p. 98.

40. Scheffler, *Boundaries and Allegiances*, p. 64.

41. See Clare Palmer, "Placing Animals in Urban Environmental Ethics," *Journal of Social Philosophy* 34 (2003): 64–78, esp. pp. 76–78.

SUGGESTED READING

BUDIANSKY, STEPHEN, *The Covenant of the Wild: Why Animals Chose Domestication*. New York: Wiedenfeld and Nicholson, 1992.

BURGESS-JACKSON, KEITH, "Doing Right by Our Animal Companions." *Journal of Ethics* 2 (1998): 159–85.

CALLICOTT, J. BAIRD, "Animal Liberation and Environmental Ethics: Back Together Again." In *The Animal Liberation/Environmental Ethics Debate: The Environmental Perspective*, edited by Eugene Hargrove, pp. 249–62. Albany: State University of New York Press, 1992.

CASSIDY, REBECCA, and MOLLY MULLEN, eds. *Where the Wild Things Are Now: Domestication Reconsidered*. Oxford and New York: Berg, 2008.

CLUTTON-BROCK, JULIET. *The Walking Larder: Patterns of Domestication, Pastoralism and Predation*. London: Unwin Hyman, 1989.

FINEMAN, MARTHA. "The Vulnerable Subject: Anchoring Equality in the Human Condition." *Yale Journal of Law and Feminism* 20 (2008): 1–20.

GOODIN, ROBERT E. *Protecting the Vulnerable.* Chicago: University of Chicago Press, 1987.

———, CAROLE PATEMAN, and ROY PATEMAN. "Simian Sovereignty." *Political Theory* 25 (1997): 821–49.

HADLEY, JOHN. "Nonhuman Animal Property: Reconciling Environmentalism and Animal Rights." *Journal of Social Philosophy* 36 (2005): 305–15.

MIDGLEY, MARY. *Animals and Why They Matter.* Athens: University of Georgia Press, 1983.

PALMER, CLARE. "Placing Animals in Urban Environmental Ethics." *Journal of Social Philosophy* 34 (2003): 64–78.

———. *Animal Ethics in Context.* New York: Columbia University Press, 2010.

POGGE, THOMAS. "Eradicating Systematic Poverty: Brief for a Global Resources Dividend." In *Ethics in Practice*, 3rd ed., edited by Hugh LaFollette, pp. 633–46. Oxford: Blackwell, 2007.

REGAN, TOM. *The Case for Animal Rights.* Berkeley and Los Angeles: University of California Press, 1983; 2nd ed., 2004.

TELFER, ELIZABETH. "Using and Benefitting Animals." In *The Moral Status of Persons: Perspectives on Bioethics*, edited by Gerhold Becker, pp. 219–32. Amsterdam: Rodopi, 2000.

CHAPTER 26

THE MORAL SIGNIFICANCE OF ANIMAL PAIN AND ANIMAL DEATH

ELIZABETH HARMAN

1. ANIMAL CRUELTY AND ANIMAL KILLING

In this paper, I will be concerned with this question: what follows from the claim that we have a certain kind of *strong* reason against animal cruelty? In particular, what follows for the ethics of killing animals? My discussion will be focused on examination of a view that I take some people to hold, though I find it deeply puzzling. The view is that although we have strong reasons against animal cruelty, we lack strong reasons against painlessly killing animals in the prime of life; on this view, either we have no reasons against such killings, or we have only weak reasons. My attention will be focused on animals of intermediate mental sophistication, including dogs, cats, cows, and pigs, while excluding more mentally sophisticated animals such as humans and apes, and excluding less mentally sophisticated creatures such as fish and insects. Whether any of what I say also applies to the animals I am excluding is a topic for further work.

I am interested in the claim that we have a certain kind of *strong* reason against animal cruelty. As will emerge, I take our reasons against animal cruelty to be strong in several ways. One way they are strong is the following: if an action would cause significant suffering to an animal, then that action is *pro tanto* wrong; that is, the action is wrong unless justified by other considerations. Such a view of animal cruelty is part of a more general non-consequentialist view on which there is a moral asymmetry between causing *harm* and causing *positive benefit*: our reasons against harming are stronger and of a different type than our reasons in favor of benefiting (and our reasons against preventing benefits).

Here is the claim that I take to be believed by some people, and which I plan to examine:

The Surprising Claim:

    (a) we have strong reasons not to cause intense pain to animals: the fact that an action would cause intense pain to an animal makes the action wrong unless it is justified by other considerations; and

    (b) we do not have strong reasons not to kill animals: it is not the case that killing an animal is wrong unless it is justified by other considerations.

The Surprising Claim seems to lie behind the following common belief:

> While there is something deeply morally wrong with factory farming, there is nothing morally wrong with "humane" farms on which the animals are happy until they are killed.

Some people think that factory farming is morally wrong, and that it is morally wrong to financially support factory farming, because factory farming involves subjecting animals to intense suffering. By contrast, "humane" farms do not subject animals to suffering, but they do kill animals in the prime of life. Some people who believe factory farming is morally wrong also believe that this "humane" farming is morally permissible. They appear to believe that while we have strong moral reasons not to cause animals pain, we lack strong moral reasons against killing animals in the prime of life.[1]

I find the Surprising Claim puzzling. My goal in this paper is to examine the Surprising Claim. I will ask: how could the Surprising Claim be true? In section 2, I will argue that the Surprising Claim is not true. I will then consider four views on which the Surprising Claim is true; each view rejects one of the claims made in my argument of section 2. I will ask what can be said in favor of each view, and whether any of these views is true. I will argue that each view is false. The fourth view I will consider is Jeff McMahan's time-relative interests view; one of my conclusions will thus be that this well-known view is false. Finally, I will draw some lessons about the relationship between the significance of animal pain and the significance of animal death.

2. AN ARGUMENT AGAINST THE SURPRISING CLAIM

In this section, I will argue that the Surprising Claim is false.

The Surprising Claim:

    (a) we have strong reasons not to cause intense pain to animals: the fact that an action would cause intense pain to an animal makes the action wrong unless it is justified by other considerations; and

(b) we do not have strong reasons not to kill animals: it is not the case that killing an animal is wrong unless it is justified by other considerations.

Consider part (a) of the Surprising Claim. If (a) is true, what explains its truth? It seems that it must be true because animals have moral status, and because any action that significantly harms something with moral status is impermissible unless justified by other considerations.

Here is an argument that the Surprising Claim is false:

1. If it is true that we have strong moral reasons against causing intense pain to animals, such that doing so is impermissible unless justified by other considerations, then part of the explanation of this truth is that animals have moral status.
2. If it is true that we have strong moral reasons against causing intense pain to animals, such that doing so is impermissible unless justified by other considerations, then part of the explanation of this truth is that significantly harming something with moral status is impermissible unless justified by other considerations.
3. If an action painlessly kills a healthy animal in the prime of life, then that action significantly harms the animal.
4. If it is true that we have strong moral reasons against causing intense pain to animals, such that doing so is impermissible unless justified by other considerations, then painlessly killing a healthy animal in the prime of life is impermissible unless justified by other considerations (1, 2, 3).
5. Therefore, the Surprising Claim is false (4).

I endorse this argument. I think it gives the right account of why the Surprising Claim is false. In the next three sections, I will discuss four views on which the Surprising Claim is true; those views reject this argument.

3. First View: Killing an Animal Does Not Harm It

Consider this view:

> First View: An action that painlessly kills an animal in the prime of life deprives the animal of future life, which would be a positive benefit to the animal, but does not harm the animal.

According to the First View, death is *bad for* animals, but a proponent of the First View would point out that there are two ways that events can be bad for a being:

an event can be or lead to something that is in itself bad for the being, such as suffering, or an event can be a deprivation of something that would have been in itself good for the being. A being is *harmed* when it undergoes something that is in itself bad, but a being is not typically harmed when it is merely prevented from something good.

According to the First View, claim 3 is false: while death is bad for animals in that it deprives them of futures that would be good for them, it does not harm them because it does not involve anything that is in itself bad for them, such as pain. A proponent of the First View would grant that claim 1 is true: animals have moral status. A proponent of the First View would also grant that claim 2 is true, but only if we have a suitably narrow understanding of what *harming* involves. In particular, a proponent of the First View would deny that claim 2 is true if "harming" is understood so broadly as to encompass all cases of failing to positively benefit, and all cases of preventing positive benefits.

A proponent of the First View would be *correct* in asserting that claim 2 is true only on a suitably narrow understanding of "harming"; indeed, that is the reading I intend in stating the claim and the understanding of "harming" I will use throughout the paper. There are many cases of failing to positively benefit people, or of preventing positive benefits to people, that do not generate strong reasons—there are many such cases in which it is false that the behavior is wrong unless justified by other considerations. For example, if I decide, on a lark, to give a particular acquaintance $200 and write her a check, but then I rip up the check, then my action prevents positive benefit to her but it is not the case that my action is wrong unless justified by other considerations; my action requires no justification.

Painless animal death involves no *bad experiences*. Rather, it involves failing to have future life. When death is bad for some being, typically that is because it is deprived of a future that would be good; so the badness of death consists in the failure to have some *good experiences* (and, for persons, the failure to have other things that make life meaningful and valuable). But suffering death then looks like it constitutes experiencing a failure to get a benefit *rather than* a harm. A proponent of the First View would say that this shows that in killing something, one is not *harming* it, but merely depriving it of a positive benefit.

The First View is false because, while it is typically the case that when a being fails to get a benefit, the being is not harmed, nevertheless some actions that deprive a being of a benefit do thereby harm the being. If someone deafens you (causes you to become permanently deaf), she simply deprives you of the benefit of hearing, but she thereby harms you. If someone steals your money, she simply deprives you of the benefit the money would have provided, but she thereby harms you.[2]

In particular, actively and physically interfering with a person in such a way that she is deprived of a benefit does typically harm that person. And if this is true of persons, it should also be true of animals. But killing an animal does actively, physically interfere with the animal in such a way that the animal is deprived of a benefit. So killing an animal is harming that animal.[3]

4. Second and Third Views: Death is Not Bad for Animals Because Animals Lack Sufficient Psychological Connection with Their Futures

In this section, I will consider two more views on which the Surprising Claim is true. Both views are more specific elaborations of the following basic idea:

> When a person dies, she *loses out* on the future she would have had. She had expectations, hopes, plans, and dreams that are thwarted. Animals, however, do not *lose out* on their futures. They do not have the right kind of psychological connection to their future lives to be losing out on them.

Here is one way of making this basic idea more precise. It is an argument that would be offered by someone who endorses the Second View:

(i) The death of a person is bad for her only because it frustrates her desires and plans for the future.

(ii) Therefore, death is bad in general only because it frustrates desires and plans.

(iii) Animals do not have desires and plans for the future.

(iv) Therefore, animals' deaths are not bad for them.

The Second View is more radical than the First View. The First View granted that death is bad for animals but denied that animals are harmed by being killed. The Second View denies that death is bad for animals at all. It follows that animals are not harmed by death, and that claim 3 is false.

The Second View is false because its claim (i) is false. It is true that *one way* death is bad for most persons is that it frustrates their desires and plans for the future. But a person might not have any desires and plans for the future, yet her death could still be bad for her. Consider someone who is depressed and wants to die; she is so depressed that she lacks any desires about the future and has no plans for the future. Suppose she in fact would recover from her depression and have a good future if she continued to live (because her family is about to intervene and get her treatment). If she dies now, then death deprives her of a good future and is bad for her. But death does not frustrate her desires and plans. In a more farfetched example, consider someone who *truly* lives in the moment. She enjoys life but has absolutely no expectations or desires about the future, and no plans for the future. If she dies now, her death is bad for her, although it frustrates no desires or plans.

Just as a person's death may be bad for her because she is losing out on a future life that would be good for her (even if she lacks desires and plans for the future), similarly an animal's death may be bad for it because the animal loses out on a future life that would be good for it, even if the animal lacks desires and plans for the future. This is why the Second View is false.

While the Second View is committed to claim (i), which is too strong, there is another way to make out the basic idea I outlined at the beginning of this section:

> Third View: It is true that animal pain matters morally. But it is a mistake to conclude that this is because *animals* have moral status. Rather, animals lack moral status. But *stages* of animals have moral status. Animal pain matters morally because an animal stage is in pain. What is better or worse for the *animal* does not matter morally, though what is better or worse for stages of it does.

The Third View assumes a certain metaphysical picture. It assumes that there are entities called "animal stages" that are temporal stages of animals; these animal stages exist briefly. An animal's life is made up of the existence of many animal stages in a series. An animal is a mereological sum of many animal stages.[4]

According to the Third View, claim 1 is false. While animal pain matters morally, it does not matter because *animals* have moral status. Rather, only *animal stages* have moral status. The Third View grants that claim 2 is true: if an action would harm something with moral status, then that action is wrong unless other considerations justify it. In this view, while animals are harmed by being killed, there is not thereby any reason against killing animals, because animals lack moral status. Animal stages have moral status, which is why we have reasons against causing animals to suffer; but animal stages are not harmed when animals are killed.

The Third View makes a number of seemingly counterintuitive claims. Some of these claims may seem false, though they are in fact true. For example, the Third View implies these two claims:

(v) There are some things that can be harmed but that lack moral status.
(vi) There are some things that lack moral status although they are entirely made up of stages that have moral status.

The Third View implies claim (v) because it holds that animals can be harmed but that animals lack moral status. The Third View implies claim (vi) because it holds that animals are entirely made up of animal stages, yet animals lack moral status.

It might seem that anything that can be harmed has moral status. Indeed, some philosophers write as though this is the case.[5] However, plants can be harmed, but plants lack moral status: the mere fact that an action would harm a plant does not provide a reason against the action. For example, suppose that I place a picnic table in my backyard, depriving a dandelion growing there of light. I harm the dandelion, but there is not thereby any reason against my action. (Our reasons to take care of the environment stem from our reasons to treat persons and animals well, but not from the moral status of plants.)[6]

It might seem that if something is entirely made up of stages that *have* moral status, then it too must have moral status. To see that this is false, let us consider some unusual entities that are made up of persons. One such entity is Longy: Longy is the mereological sum of several non-temporally-overlapping persons, including me. Suppose that Frank punches me. Frank harms me, and I have moral status; this is what makes it wrong of Frank to punch me (absent justification). When he punches me, Frank also harms Longy. It might seem fine to grant that Longy too has moral status. But this would be a mistake: entities like Longy do not have moral status.

Another example will enable us to see that entities like Longy do not have moral status. Suppose Bill is considering whether to do something that would cause major injury to you but would prevent worse injury to someone one hundred years from now; call her Gertrude. Bill justifies his action by saying, "Consider Thingy, which consists of the mereological sum of you and Gertrude. The action I am considering would hurt Thingy, but only in order to prevent worse injury to Thingy." This justification fails, and it fails because while Thingy lacks moral status, *you* have moral status.[7]

So, while the Third View is committed to claims (v) and (vi), which may appear to be false, those claims are true, so they are no problem for the Third View.

I will raise a different objection to the Third View, which comes out of what we have just been considering.

Consider a young cat that could lead a long, happy life if it is given serious surgery that would cause it quite a bit of pain (even with painkillers) for a few days, followed by a month of serious discomfort. Otherwise the cat will die within a few days, without experiencing much discomfort. In this case, it is permissible to do the surgery. My objection to the Third View is that the Third View cannot explain *why* it is permissible to do the surgery. It is not in general permissible to cause serious pain and injury to one morally significant entity in order to benefit others, but according to the Third View, that is what one would be doing. One would be causing pain and suffering to one animal stage, a morally significant entity, in order to benefit several different animal stages, other morally significant entities. (Note that I am denying that it is permissible to cause serious pain and injury to one in order to provide *positive benefits* to other morally significant entities; we sometimes use the word "benefit" to refer to the *prevention of pain or harm*, but that is not what is at issue in this case, because according to the Third View, no animal stage is harmed when an animal dies.)

Even more seriously, according to the Third View, what one would be doing in this case is causing serious pain and injury to one morally significant entity in order to cause there to be created some further morally significant entities who are happy but who otherwise would not exist at all. Doing that kind of thing is even more morally problematic than causing suffering to one being in order to benefit another who independently exists. (When the benefit is to an independently existing being, then one consideration in favor of causing the suffering is that otherwise there will be some beings who lose out on some benefits; but when the benefit would be to some beings who would not otherwise exist, then if one does not cause the suffering, it is not the case that there are some entities that lose out on some benefits they could have had.[8])[9]

The surgery on the cat is permissible. To account for the permissibility of the surgery, we need both these claims:

- While the action harms an entity that has moral status, it also benefits that entity.
- It is not the case that the action harms an entity that has moral status but the action does not also benefit that entity.

Thus, we need both of these claims:

- Animals have moral status.
- Animal stages lack moral status.

The fact that animals have moral status provides a justification for the surgery: while it harms the cat, a morally significant being, it also provides benefits to *that very being*. If animal stages had moral status, the surgery would be impermissible, because it would involve harming one morally significant entity—the stage of the cat during the recovery—in order to provide benefits that are not to that same entity, but to a different entity.

Because the surgery on the cat is permissible, the Third View must be false.

5. Fourth View: McMahan's Time-Relative Interests View

In this section, I will discuss a fourth view on which the Surprising Claim is true. Like the First View, the Fourth View grants that we have *some reasons* against killing animals; the Fourth View denies that these reasons are *strong*.

My discussion of the more extreme Third View will enable us to see why the less extreme Fourth View is also false.

The Fourth View is a view of Jeff McMahan's. He calls it the "time-relative interests view."[10] On this view, the badness of death for a morally significant being is not a direct function of what the being loses out on in dying; the badness of death is not simply a matter of how good the lost life would have been. Rather, it also matters what the being's *psychological relationship* is with that potential future life. If a being is such that, were it to continue to live, there would be only weak psychological connections between its current stage and its future life, then the goodness of that future is *less of a loss* for it than if the being would have stronger psychological connections with its future life: the being currently has less of an interest in continuing to live than if the psychological connection he would have to a future life would be stronger. This view has the virtue that it can explain why, as is plausible, the death of a ten year old is worse for the ten year old than the death of a one month old is bad for the one month old: while the infant loses out on more life, so loses more, the ten year old would have much greater psychological connections with its future if it continued to live. According to the time-relative interests view, the one month old has a weaker interest in continuing to live than the ten year old has.

The implications of the time-relative interests view for animal death are that animal death is not very bad for animals because animals do not have very strong psychological connections to their future selves: they do not have *strong interests* in continuing to live. But the view does not hold (nor is it plausible) that animals lack *any* psychological connections to their future selves; so the view does not hold that animal death is not bad for animals, nor that we have no reasons against killing animals. The view grants that animals have *some interest* in continuing to live.[11] The view supports the following claim:

> We have strong reasons against causing animal pain, and we have some reasons against painlessly killing animals in the prime of life, but these reasons are weakened by animals' lack of deep psychological continuity over time.

(Note that I stipulated at the beginning of the paper that I am only concerned with animals of intermediate mental sophistication, including dogs, cats, cows, and pigs, and excluding humans, apes, fish, and insects. My claims about the time-relative interests view's implications regarding animals are restricted to these animals of intermediate mental sophistication.)

The Fourth View can grant claims 1 and 2 of the argument of section 2. But the Fourth View denies claim 3: it holds that, while death is a harm to animals, it is a minor harm. On this view, killing an animal does not *significantly* harm the animal, and it is not the case that killing an animal is wrong unless justified by other considerations.

I will now argue that the time-relative interests view is false, for reasons similar to the reasons the Third View is false. My argument relies on some substantive claims about the nature of the psychological connections that animals have over time, and the way the time-relative interests view would handle these connections.[12] In particular, I assume that on the time-relative interests view, an animal now has greater psychological connection to its nearer future life than to its farther future life, and that an animal now has negligible psychological connection to its future life a sufficient amount of time into the future, such as five years into the future. It follows from this that, on the time-relative interests view, while it is currently in an animal's interest to continue to live for the next several months (at least), an animal currently lacks any interest in being alive five years from now, currently lacks any interest in having particular good experiences five years from now, and currently lacks any interest in avoiding particular bad experiences five years from now—any experiences it would have five years from now are so psychologically remote that the animal currently has no interests regarding those experiences.

My objection relies on two cases.

> Billy is a cow with a serious illness. If the illness is not treated now and is allowed to run its course, then Billy will begin to suffer mildly very soon, the suffering will get steadily worse, Billy will be in agony for a few months, and then Billy will die. If the illness is treated now, Billy will undergo surgery under anesthetic tomorrow. Billy will suffer more severely over the next two weeks (from his recovery) than he would have from the illness during that time, but then he will be discomfort-free and he will never suffer agony; he will be healthy and able to live a normal life.

It is permissible to do the surgery on Billy. This is permissible because, while the surgery will cause Billy to suffer, which he now has an interest in avoiding, it will prevent worse suffering to Billy, which he also now has an interest in avoiding.

Tommy is a horse with a serious illness. If the illness is not treated now and is allowed to run its course, Tommy will live an ordinary discomfort-free life for five years, but then Tommy will suffer horribly for several months and then die. If the illness is treated now, then Tommy will undergo surgery under anesthetic tomorrow. Tommy will suffer over the following two weeks, but not nearly as severely as he would five years from now. Tommy will be completely cured and will be able to live a healthy normal life for another fifteen years.

It is permissible to do the surgery on Tommy. This is in fact permissible because Tommy has an interest in getting to live a full life, and though he has an interest in avoiding the pain of recovery from surgery, it is overall in his interests to have the surgery.

But the time-relative interests view cannot explain why it is permissible to do the operation on Tommy. On that view, Tommy has a reasonably strong interest in avoiding pain in the immediate future; he has no interest in avoiding suffering five years from now or in avoiding death five years from now. While the time-relative interests view can easily account for the permissibility of the surgery on Billy, it cannot account for the permissibility of the surgery on Tommy.

Because the time-relative interests view cannot accommodate the truth that it is permissible to do the surgery on Tommy, and the truth that the two surgeries on Tommy and Billy are permissible for the same basic reasons, the time-relative interests view must be false.

6. CONCLUSION

What lessons have emerged from our examination of the Surprising Claim and the four views? The basic lesson is that if we have strong moral reasons not to cause animal pain, we must also have strong moral reasons not to kill animals, even painlessly. In section 2, I argued that this is true. I have considered four ways one might reject this argument and argued that each one fails.

The background picture I have been assuming is one on which our reasons against causing harm to animals are strong in two ways: these reasons are strong in that an action that would significantly harm is *wrong* unless other considerations justify it; furthermore, I have assumed that, just as the harming of persons cannot typically be justified by benefits to *other* persons, similarly the harming of animals cannot typically be justified by benefits to other beings. I have also assumed that harm to an animal can be justified by the prevention of greater harm and/or of death for that very animal, just as is true for persons.

One might try to develop a view on which the kinds of agent-relative constraints that apply to persons do not apply to animals or animal stages. On such a view, it would be permissible to harm one animal or animal stage in order to provide positive benefits to a distinct entity. If this view is correct, then my objections to the Third View are wrong-headed: the Third View can hold that the cat surgery is permissible. However, a view like this sees our reasons against causing suffering to animals as *much weaker* than I have been taking them to be, and as very different in kind from our reasons against harming persons. The lesson appears to be that it is possible to hold that we have reasons against causing animal pain, while lacking reasons against killing animals, but at the cost of holding that our reasons against causing animal pain are weak reasons.

NOTES

1. Someone might believe we should *support* "humane" farming because it is so much morally *better* than factory farming, without believing "humane" farming is morally unproblematic; this person need not believe the Surprising Claim.

2. One might object that by stealing your money, the thief violates a right of yours but does not harm you. The more general point I want to make about this case is that sometimes an action has a *strong* reason against it, such that the action is impermissible unless it is justified by other considerations, although the action is simply the deprivation of positive benefit. Stealing is one example. Killing is, I claim, another.

3. I discuss the asymmetry between harming and benefiting, and what harm is, in my "Can We Harm and Benefit in Creating?" *Philosophical Perspectives* 18 (2004): 89–113, and my "Harming as Causing Harm," in *Harming Future Persons: Ethics, Genetics and the Nonidentity Problem*, ed. Melinda Roberts and David Wasserman (Dordrecht and London: Springer, 2009), pp. 137–54. In those papers, I claim that killing and deafening are harming, and also that if a being dies or is deaf, then the being suffers a harm. Here I make only the former claim. Here I claim that killing or deafening a being is harming that being; I don't take a stand on whether if a being dies or is deaf, then the being suffers a harm.

4. I am using the term "temporal stages" for what are also often called "temporal parts." See "Temporal Parts" by Katherine Hawley in *The Stanford Encyclopedia of Philosophy*, ed. Edward N. Zalta (Spring 2009), available at http://plato.stanford.edu/entries/temporal-parts/ (accessed 7/31/10).

5. See Bonnie Steinbock, *Life Before Birth: The Moral and Legal Status of Embryos and Fetuses* (New York: Oxford University Press, 1992), and Joel Feinberg, "The Rights of Animals and Unborn Generations," in *Philosophy and Environmental Crisis*, ed. William T. Blackstone (Athens: University of Georgia Press, 1974), pp. 43–63.

6. I argue that some things can be harmed although they lack moral status in my "The Potentiality Problem," *Philosophical Studies* 114 (2003): 173–98.

7. One might think that entities like Longy and Thingy must not have moral status for a different reason: that they would generate *double-counting* of moral reasons. For example, if Frank punches me, if Longy and I both have moral status, then it seems there are thereby *two* moral reasons against his action: that he harms me, and that he harms Longy. But it is not the case that these are distinct moral reasons. I do not think that this is a successful objection to the claim that entities like Longy and Things have moral status. If one claims that they have moral status, one can also claim that mereologically overlapping entities do not generate distinct moral reasons. See footnote 21 of my "The Potentiality Problem."

8. I discuss differences in our reasons regarding beings whose existence is affected by our actions, and our reasons regarding independently existing beings, in my "Harming as Causing Harm," in *Harming Future Persons: Ethics, Genetics and the Nonidentity Problem*, ed. Melinda Roberts and David Wasserman (Dordrecht and London: Springer, 2009), pp. 137–54.

9. As I have articulated it, the Third View denies that any animal stage with moral status loses out when an animal dies. One might ask how long animal stages exist for, and what makes two distinct animal stages distinct. These are details that would have to be worked out by anyone who endorsed the Third View.

10. McMahan, *The Ethics of Killing: Problems at the Margins of Life* (New York: Oxford University Press, 2002).

11. Note that what a being "has an interest in" is a matter of what is *in the being's interests*, not a matter of what the being desires or wants.

12. I am also assuming that the time-relative interests view sees the badness of the death of animals as sufficiently diminished that it does not count as the kind of significant harm that is pro tanto wrong to cause.

SUGGESTED READING

FEINBERG, JOEL. "The Rights of Animals and Unborn Generations." In *Philosophy and Environmental Crisis*, edited by William T. Blackstone, 43–63. Athens: University of Georgia Press, 1974.

HARMAN, ELIZABETH. "Can We Harm and Benefit in Creating?" *Philosophical Perspectives* 18 (2004): 89–113.

———. "Harming as Causing Harm." In *Harming Future Persons: Ethics, Genetics and the Nonidentity Problem*, edited by Melinda Roberts and David Wasserman, 137–54. Dordrecht and London: Springer, 2009.

———. "The Potentiality Problem." *Philosophical Studies* 114 (2003): 173–98.

HAWLEY, KATHERINE. "Temporal Parts." In *The Stanford Encyclopedia of Philosophy*, edited by Edward N. Zalta (Spring 2009), available at http://plato.stanford.edu/entries/temporal-parts/.

MCMAHAN, JEFF. *The Ethics of Killing: Problems at the Margins of Life*. New York: Oxford University Press, 2002.

STEINBOCK, BONNIE. *Life Before Birth: The Moral and Legal Status of Embryos and Fetuses*. New York: Oxford University Press, 1992.

CHAPTER 27

THE ETHICS OF CONFINING ANIMALS: FROM FARMS TO ZOOS TO HUMAN HOMES

DAVID DEGRAZIA

ANIMALS, unlike plants, move around and do things. Such mobile activity allows them, at a minimum, to pursue means of survival. Sentient animals—animals capable of having feelings—experience such feelings as pain and pleasure in the course of their activity. Moreover, such animals *want* to move about and do certain things. When they are able to do what they want, they typically experience pleasure or satisfaction; when they are prevented from doing what they want, they typically experience frustration or other unpleasant feelings. For these reasons, *liberty*—the absence of external constraints on movement—is generally a benefit for sentient animals, permitting them to pursue what they want and need.

In many of our institutions and practices, we human beings restrict animals' liberty by confining them. Animals living on farms, those kept at zoos, pets in human homes, horses in stalls, and animals in stores, circuses, and laboratories all experience significant deprivation of liberty. This chapter will address two questions: Under what conditions, if any, is confining animals morally justified? What does the answer to this question suggest about the ethics of keeping animals in farms, zoos, and human homes?

This chapter begins by briefly examining the harms associated with confinement. It then examines and evaluates five competing standards for determining whether particular instances of confining animals can be morally justified:

(1) a basic-needs requirement;
(2) a comparable-life requirement;
(3) a no-unnecessary-harm standard;
(4) a worthwhile-life criterion;
(5) an appeal to respect.

Equipped with the findings of this theoretical examination, my discussion proceeds to an ethical evaluation of keeping animals in factory farms, traditional farms, zoos and aquariums, and human homes.

It may be helpful to anticipate the main conclusions. In the theoretical discussion of competing standards for justified confinement of animals, I argue the following: (1) a basic-needs requirement is appropriate; (2) a comparable-life requirement, although not basic (resting as it does on a no-unnecessary-harm standard) provides useful, appropriate guidance in at least "capture and confinement" cases; (3) a no-unnecessary-harm standard provides appropriate, if sometimes vague, guidance, but depending on its interpretation, it either requires supplementation by a basic-needs requirement or it converges with such a requirement; (4) a worthwhile-life criterion sets far too low a standard for confining animals to provide adequate guidance; and (5) appeals to respect are too uncertain with regard to their meaning and basis to provide adequate guidance. In applying these findings to the practical areas of concern, I defend these theses: (6) factory farming cannot meet appropriate standards for the confinement of animals and is therefore indefensible; (7) traditional animal husbandry can and often does meet appropriate standards; (8) excellent zoos and aquariums can satisfy appropriate standards for confinement, even though it may be impossible to do so in confining some animals such as whales; and (9) while the responsibilities involved in keeping pets in human homes are more demanding than many people realize, appropriate standards for confinement can be satisfied in the case of domesticated species.

Confinement and Harm

Not all restrictions of liberty associated with confinement are harmful. For example, human infants are enclosed in cribs and children are often required to stay indoors or inside moving cars—for their own safety. Similarly, a dog living with a family in the city is likely to be better off if she is not permitted to run around the neighborhood at will.

Yet confinement is often harmful. We can understand harmful confinement as involving external constraints on movement that significantly interfere with an individual's ability to live well. Because prison significantly interferes with people's ability to live well, imprisonment is a type of punishment. A monkey confined indefinitely in a small, barren cage is clearly harmed by the conditions of confinement. Monkeys need to roam around, explore things, play, and interact with other monkeys. Severe constraints on their movement often cause pain and bodily discomfort and are likely to cause distress and other negative emotions. Such conditions of confinement typically cause suffering.

Can confinement cause harm without causing suffering? Imagine a zoo kangaroo who is comfortable and avoids most of the unpleasant experiences that life in the wild would ordinarily involve, such as prolonged periods of hunger, discomfort in extreme weather conditions, and occasional injury. I will here make the plausible assumption that the *experiential well-being* of this animal is higher than it would be if he had a typical wild kangaroo life. At the same time, living in a typical zoo exhibit for kangaroos, he is much less active than wild kangaroos tend to be. Does confinement in the zoo harm him, to any degree? The answer to this question depends on the answer to a difficult theoretical question: is exercising one's natural capacities— or species-typical functioning—intrinsically valuable, in the sense of being conducive to well-being, independently of effects on experiential well-being? If so, the comfortable kangaroo is harmed to some degree by confinement in the zoo, which severely limits his ability to exercise his natural capacities. Thus, he is better off in the wild despite the fact that he suffers more in the wild. Conversely, if liberty is valuable only insofar as it promotes experiential well-being, the kangaroo is better off in comfortable captivity.

We cannot achieve a full appreciation of animal well-being without coming to terms with the twin issues of the value of liberty and the harm of confinement. Doing so would require determining which account of well-being across species is most adequate. Because I believe that the contest features a virtual tie between an updated "mental statism" (in which experiential well-being is prominent) and a flexible "objective list" account of well-being (in which the exercise of natural capacities, or species-typical functioning, is prominent),[1] I am in no position to declare a winner. Fortunately, our ability to address most practical situations involving confinement of animals need not await the outcome of this contest between mental statism and objective list accounts of well-being. In most cases, confinement that significantly interferes with an animal's ability to exercise her natural capacities also causes her to suffer, entailing unambiguous harm. In other words, the best accounts of animal well-being mostly converge in their implications for the confinement of animals in real-life scenarios.

Given this underlying agreement, we may profitably address the issue of appropriate ethical standards for confining animals. In the discussion that follows, I will assume, unless otherwise stated, that the animals under consideration are sentient and therefore capable of experiencing pleasant and unpleasant feelings, which I take to be necessary and sufficient for having interests.[2]

Competing Standards for Justified Confinement

A Basic-needs Requirement

Under what conditions is the confinement of animals morally justified? One attractive answer to this question is a *basic-needs requirement*. The conditions of this requirement are that the animal's basic needs be met despite the deprivation of liberty entailed by confinement. What counts as an animal's basic needs will vary depending on the nature of the animal in question. Physical needs include appropriate food, water, the sort of shelter (if any) animals of that species characteristically seek and depend on, freedom from conditions that cause significant experiential harm, and whatever exercise is needed for good health. Animals of sufficient mental complexity also have psychological needs. Depending on the type of animal in question, these needs may include sufficient stimulation, opportunities to play, access to family or other group members, and freedom from excessively stressful situations. Although incomplete, this sketch of what a basic-needs requirement demands for confinement to pass moral muster is sufficient for present purposes.

Why should we accept such a requirement? Can anything be said in its defense besides that it is attractive on its face? Let me suggest some plausible arguments.

A first line of argument appeals to *basic decency*. The assumptions underlying this appeal are that decency is a moral virtue and indecency a vice, and that certain types of action can be morally evaluated in terms of the virtue or vice that they express; when action types express a particular virtue or vice, the action types may be characterized with the term designating the character trait in question (e.g., "humane" or "inhumane").[3] According to the present argument, it is indecent or inhumane to keep animals while neglecting their basic needs—for example, forcing hens to go without food or water for days, providing a pet dog with no opportunities for exercise outdoors, or keeping a laboratory primate isolated indefinitely in a small cage. One who denied that such instances of confinement were wrong would presumably have to deny that they were inhumane—seemingly a steep argumentative hill to climb, especially when our imaginations are engaged with details of the kinds of deprivation and unpleasant sensations and/or emotional states that confined animals are likely to experience.[4] This appeal to decency will suffice to persuade some, but others will remain unpersuaded or uncertain.

Those who are not persuaded may doubt that animals have moral status. To say that animals have moral status is to say that their interests—such as their interest in experiential well-being—have moral importance in their own right and not only because of effects on human interests. Put another way, it is to say that gratuitously harming an animal is wrong because of what it does to the animal, not just because doing so may negatively affect the agent's character, other individuals' feelings, or other human concerns. Those who doubt that animals have moral status may find it natural to deny that decency requires meeting the basic needs of animals we choose to confine. It may be difficult to persuade such a person that animals have moral status.

Nevertheless, the most promising way to do so would be to deploy either or both of two strategies: (1) starting from the obvious point that human beings or persons have moral status and arguing that there is sufficient relevant similarity between all the humans who unquestionably have moral status (at least, sentient postnatal humans) and sentient animals; and (2) starting from a paradigm case of gratuitous cruelty to animals, which all agree is wrong, and arguing that the only adequate account of its wrongness acknowledges the moral status of its victims. At least one of these argumentative strategies should succeed in persuading morally serious people who consider the arguments with due care.[5] Accordingly, I assume that animals have some level of moral status.

In exploring why one might accept a basic-needs requirement, we have discussed an appeal to moral decency, possibly supplemented by or amplified with an appeal to animals' moral status. A second way to defend the basic-needs requirement is to appeal to *special relationships* between the confined animals and the people responsible for them. We generally accept that special relationships can be the basis for particular obligations.[6] Being a parent, for example, gives rise to obligations to protect the welfare of one's children and promote their development. Friends have obligations to support each other in familiar ways. Doctors have specific professional obligations to their patients, as lawyers do to their clients. We also generally accept that certain human-animal relationships can be the basis for special obligations.[7] If I adopt a cat, bring her home, and then neglect to feed her, I have failed morally. I have an obligation to feed and provide other basic care to that cat, due to our relationship—caretaker to pet[8]—but I have no obligation to feed and provide basic care to other cats with whom I have no special relationship.[9]

To what extent can we generalize the proposition that we have an obligation to meet the basic needs of our pets? Suppose a raccoon takes up lodging in my garage. Do I owe him food and protection? Presumably, I have no obligation and the reason is that his appropriating part of my property was not an arrangement to which I agreed or should feel obliged to agree. This suggests that the special human-animal relationships that generate obligations to meet basic needs are ones that human caretakers assume voluntarily. If one adopts a pet, one assumes special responsibilities. If one volunteers to work at a Humane Society shelter, one makes a similar assumption regarding the animals there. The same holds true for those who accept jobs that involve caring for animals at zoos or aquariums. Perhaps, then, the idea can be generalized: whenever one voluntarily or intentionally confines an animal for any significant length of time,[10] one enters into a special relationship that gives rise to an obligation to meet the animal's basic needs. If this generalization is legitimate, then we have a second argument for the basic-needs requirement.

However, there is reason to doubt the legitimacy of the generalization. As examples above demonstrate, there are human-animal relationships that place the human in a caretaking role, supporting the idea that she has an obligation to meet the basic needs of the animal. But what if the relationship is of a different sort such that no caretaking role is implied? Does a farmer necessarily have a special relationship to his farm animals? Farmers engaged in traditional animal husbandry on family-owned farms often do cultivate such relationships, but do they necessarily stand in special relationships to their animals? Do those who own, or work at, factory farms—which are characterized by extremely intensive confinement conditions (as explained in greater detail in a later section)—have such special relationships to these animals?

This relationship is likely to be regarded as purely instrumental by the persons involved: the animals are confined and treated in particular ways purely for the benefit of agribusiness owners, employees, and consumers who benefit from the business. It is unclear that the relationship of farmer (or farm employee) to farm animal is *itself* the basis of special moral obligations. So maybe only some, not all, relationships between humans and animals that involve prolonged confinement—namely, those in which a caretaking role is clearly implied—can support a basic-needs requirement. Such relationships include those of caretaker to pet, shelter employee to sheltered animal, and zoo employee to zoo animal.[11] What is unclear is whether *all* human-animal relationships involving prolonged confinement support a basic-needs requirement.

On the other hand, as we have seen, such a requirement receives support from the appeal to decency, supplemented, as necessary, with an appeal to animals' moral status. The appeal to decency applies to animals in general, not only to those standing in relationships with humans in which the latter can be assumed to have a caretaking role. Finding the appeal to decency persuasive, I suggest that a basic-needs requirement applies to all animals in situations of confinement.

A Comparable-Life Requirement

Another standard for the ethics of confining animals is *a comparable-life requirement*. It requires of one who confines an animal to do so in such a way that the animal can be expected to have a life that is at least as good as she would likely have had in the wild. This standard appeals to a counterfactual: how the animal would most likely have fared in living free of confinement by human beings. One basis for this standard is the idea that, if we confine animals, we should not in doing so make them worse off than they otherwise would be. Failing to give animals we confine a comparably good life to what they would likely have had in the wild is to harm them unnecessarily.[12] This standard may seem intuitively plausible and firmly grounded in the importance of avoiding unnecessary harm. Nevertheless, the comparable-life requirement faces challenges.

First, the counterfactual consideration of how well off an animal would likely have been in the wild is of dubious relevance in some contexts. When considering animal species that have evolved into highly domesticated forms, especially cats and dogs, it is pretty clear that these animals would languish and may well be doomed if they had to live outside human homes. The nature of domesticated animals makes them unfit to live independently of humans. Now, one might interpret differently the counterfactual of how the animal would likely have fared if not confined—by considering typical lives of the wild animals from which the domesticated animals evolved—in which case one could interpret the standard as demanding that dogs in human homes have lives that are, overall and as best as we can determine, at least as good as wild wolves typically have. Setting aside how hard it would be to make this comparison, it is not relevant. Dogs are not wolves, so dogs cannot have the lives of wolves. Thus, the typical lives of wolves provide no baseline for evaluating a domestic dog's quality of life and determining a caregiver's obligations under a comparable-life requirement.

The counterfactual in question is also of dubious relevance in cases in which animals are brought into existence with a specific plan of confining them. Suppose mice are bred specifically for research purposes. The comparable-life requirement would expect researchers to keep the mice in such conditions that we could reasonably say they would be no better off, overall, in the wild. But why should their likely welfare in the wild be relevant? These mice have come into existence only as part of a plan to use them as research subjects; they could not have been wild mice. It would therefore be incorrect to claim that raising laboratory mice under certain conditions harms (or benefits) them in the sense of making them worse (or better) off than they would have been in the wild.

While the counterfactual entertained in connection with the comparable-life requirement seems irrelevant in cases involving domesticated animals or animals bred for purposes involving life in captivity, it seems entirely relevant in *capture and confinement cases*: cases in which wild animals are captured and transported into captivity in human homes, zoos, aquariums, or the like. It is plausible to maintain that a necessary condition for justifiably capturing and confining a wild animal is that she not be made worse off due to the intervention. If the animal is made worse off for being captured and kept in confinement, she has been harmed to whatever extent she is made worse off. And, unless the capturing and confining is somehow necessary—and here proponents of the intervention bear the burden of arguing persuasively that it is—then the capture and confinement would violate the moral expectation of avoiding unnecessary harm.

Here we need to consider another possible basis of the comparable-life standard. If we ask, for example, whether it is permissible to confine fish in a tank, it may seem appropriate to take fishes' typical lives in rivers or oceans as a moral baseline. If their lives in a tank are overall as good as, or better than, life in the wild, it is hard to see why confining them to the tank deserves condemnation; conversely, if their lives in captivity are worse than they probably would have been in the wild, confining them becomes problematic, whether or not harmful. This suggests a standard that is not limited in relevance to capture and confinement cases. The underlying idea is that the "normal" or natural life for these animals sets a normative threshold.[13] Thus, even if goldfish were bred simply in order to live in tanks, the natural life for goldfish would serve as a moral baseline, whether or not we should assert that these goldfish were *harmed* by being confined in certain conditions.

It is fair to ask, however, why what is natural or normal in the absence of human intervention should be morally relevant. Here one might appeal to basic decency: even if these goldfish, or the lab mice considered earlier, could not have existed in the wild because they were bred for purposes that entailed captivity, it is indecent to confine them in a way that entails a worse life than their conspecifics typically have in the wild. But one might still wonder why this is so. What does the typical welfare of conspecifics have to do with how one should be treated? (Do appropriate standards for parenting human children depend on how they are typically treated?) By contrast, the connection between decency and meeting basic needs seems tighter and more compelling. Another approach would be to appeal again to the requirement of avoiding unnecessary harm while insisting that the associated counterfactual is relevant: goldfish and mice (unlike dogs and cats) live well in the wild, so to breed them and give them less good lives is to harm them.[14]

My conclusion is that the comparable-life requirement is relevant in capture and confinement cases, but is not basic because it rests on the more fundamental principle that we should not harm unnecessarily. Also, it begs the question of what harms to animals should count as necessary by assuming that capturing and confining animals such that they do not enjoy comparably good lives as they would otherwise have is "unnecessary." On the other hand, we may interpret this requirement not as implying uniformly that failures to meet the comparable-life standard are wrong, but that there is a burden of proof to justify exceptions.[15] Understood this way, the requirement, although not basic, offers a fairly vivid marker for cases involving the capture and confinement of wild animals. As we have seen, the comparable-life requirement *may* also apply to and illuminate cases involving the breeding and confinement of animals whose conspecifics (unlike cats and dogs) can fare well in the wild; due to considerable uncertainty on this score, however, I will set aside this possibility. I emphasize only that the comparable-life requirement is appropriate for capture and confinement cases.

The No-unnecessary-harm Requirement

The principle of nonmaleficence, a principle basic to virtually all moral systems, states that we should not cause unnecessary harms. Stated another way, it asserts that causing harm is *pro tanto* wrong—that is, wrong in the absence of a compelling justification. Yet some people may be uncertain whether, or even deny that, nonmaleficence applies to animals as well as to human beings. Pointing out that animals, too, can be harmed in ways that matter to them may answer such doubts. If not, however, then an argument that animals have moral status will be needed to complete the case that nonmaleficence applies to animals. As explained above, I assume that at least one such compelling argument is available.

Accordingly, I assume that a no-unnecessary-harm requirement does apply to animals. This requirement, which seems straightforward and commonsensical, is more basic than the comparable-life requirement. Nonetheless, the no-unnecessary-harm requirement faces challenges, beginning with the prospect of significant disagreement over which sorts of harms count as necessary and which as unnecessary. At the most general level, I hold that a harm is "necessary" if and only if causing that harm is, all things considered, morally justified (morally required or at least permissible). This returns us to the idea that, according to the principle of nonmaleficence, causing harm is *pro tanto* wrong, or wrong unless adequately justified. While the present unpacking of the concept of necessary harm is relatively abstract and formal, it reminds us that the causing of harm requires justification—and that this requirement applies in the case of animals, not only in the case of humans. There remains, of course, the issue of when harming animals is justified.

To address this issue, we need first to consider how to conceptualize harm. We commonly understand harm as setting back one's interests so that one is made *worse off*, where the point of comparison, or baseline, is either (1) one's state prior to the harmful action or (2) the state one would have been in if not for the harmful action.[16] In short, one is harmed if one is made worse off than one *was before* and/or *would have been otherwise*. In ordinary cases, these two baselines both apply and converge. If I am happily admiring the view on the top of a building when you sneak up and push me over, sending me to my death, then you make me worse off than I was beforehand as well as worse off than I would have been if not for your battery. There are cases, however, in which it seems that the two commonly accepted standards of harm either do not apply or, if they do apply, generate implausible implications.

Suppose a dog is bred with the sole purpose of being sold in a pet store. Although the puppy experiences his first few weeks of postnatal life in the company of his mother and sibling litter-mates, and is regularly fed and given water, he is rarely let outside, has no significant opportunities for exercise, and is mostly confined to a single room. Do his conditions of confinement harm him? Is he worse off than he was before being confined, or worse off than he otherwise would have been? No. He was confined from birth. And, since he was bred specifically in order to be sold to a pet store, and would not have existed independently of this plan, there is no counterfactual well-being to which to compare his actual well-being. Thus, if we understand harm in the standard senses of being made worse off than one was, or would have been, this puppy is not harmed by confinement.

In the preceding example, we might say that if the puppy isn't harmed, he isn't harmed unnecessarily. However, some philosophers challenge the standard conception of harm that has been assumed so far in our analysis. A prominent alternative is the normative conception, which construes harm by comparison with a threshold of well-being such that, if one's situation is caused to land below the threshold, one is per se harmed. Elizabeth Harman claims, for example, that "[a]n action harms a person if the action causes pain, early death, bodily damage, or deformity to her, even if she would not have existed if the action had not been performed."[17] This idea clearly applies to animals. One might even connect this idea with the basic-needs standard by proposing that, if an action ensures that someone—whether human or animal—will fail to have her basic needs met, the action harms her. This implies that the breeder harms the puppy by bringing him into the world in conditions that involve a failure to meet basic needs such as those for exercise, adequate stimulation, and sufficient time outdoors.

In relation to harm, there are two plausible ways of understanding the puppy's situation. One way is to conceptualize harm along standard lines, with the implication that the dog is not harmed by being brought into the world in desolate conditions. If this is correct, the no-unnecessary-harm requirement seems *insufficient* as a moral guide for conditions of confinement. The puppy's basic needs should be met; it would be neglectful to leave them unmet. The second plausible approach is to embrace a normative conception of harm with the implication that the puppy *is* harmed by being confined in desolation, whatever the birth conditions. Since the harm is unjustified, it qualifies as an unnecessary harm and therefore is wrong according to the no-unnecessary-harm requirement. As suggested above, a natural way of construing a normative conception of harm is in terms of failure to meet basic needs.

I conclude that in certain kinds of cases—which I call *creation and confinement cases*—the basic-needs standard offers appropriate guidance for the ethics of confining animals. This conclusion is justified whether harm is understood along standard lines or is understood in normative terms. If harm is understood in a standard way, then in certain kinds of cases, like those of the puppy, there is no harm and therefore no violation of the no-unnecessary-harm standard, but the conditions of confinement seem unjustified. In such cases, with harm so understood, the no-unnecessary-harm standard provides insufficient guidance—hence the importance of the basic-needs standard, which provides needed guidance by indicating where confinement is unjustified. If, on the other hand, harm is construed normatively, the conditions of confinement in the relevant set of cases do count as harmful and as violating the no-unnecessary-harm requirement, in which case the basic-needs standard specifies and clarifies the content of the requirement. So the basic-needs standard is helpful in creation and confinement cases no matter how we understand harm.

We have evaluated the no-unnecessary-harm requirement as it applies to creation and confinement cases. Is the requirement an adequate guide in other sorts of confinement cases? The sorts of cases we need to consider feature an animal who is not created solely or primarily for a purpose that includes confinement, but is eventually confined. We may speak of such cases as involving confinement of "independently existing animals." These cases include, but are not limited to, the capture and confinement cases discussed earlier.

Consider a pet dog who enjoys a reasonably good life with her human caretaker until the latter dies. No one adopts the dog, who is taken to a shelter. In the shelter she is confined most of the time to a cage, but the staff makes every reasonable effort to meet her basic needs, including those for exercise, stimulation, and time outdoors. Here, the confinement seems quite justified—the dog has nowhere else to go—so the no-unnecessary-harm requirement is not violated even if the conditions of confinement include some harm. This is a plausible verdict.

However, suppose that after many months during which no one adopts her from the shelter, the dog is taken to a research lab, where she is again confined. The investigators, however, are not very conscientious and the dog's conditions of confinement are worse than they were at the shelter. The harm here seems unnecessary, either because transfer to a research lab does not seem justified or because, although such transfer was justified (at least permissible), the people responsible for the dog could be expected to do a better job of meeting her basic needs. In that case, the conditions of confinement at the lab flout the no-unnecessary-harm requirement—via the basic-needs standard—indicating wrongful treatment. This implication is plausible. Thus, we have reason to believe that in cases involving the confinement of independently existing animals, the no-unnecessary-harm requirement offers sound moral guidance.

But do we have adequate criteria for determining whether a particular harm qualifies as necessary? Earlier, I stipulated that a harm counts as necessary if and only if causing that harm is, all things considered, morally justified (morally required or at least permissible). So we need adequate criteria for determining whether a particular harm is justified; otherwise, our confidence in the no-unnecessary-harm requirement will be limited. At a general level, we may say that the criteria for which harms count as necessary will be determined by the details of whatever ethical theory or moral framework is assumed to be correct. (It is not among the aims of this chapter to determine which comprehensive theory or framework is the most adequate.) From a consequentialist perspective, necessary harms to animals are those that must be accepted as part of the cost of the effort to achieve the best possible consequences.[18] From the perspective of an animal rights view, what counts as a necessary harm is determined by (1) the content of the rights asserted by the view and (2) whether the view allows rights to be overridden and, if so, under which conditions.[19] Hybrid views may also have criteria for the conditions under which harms count as necessary.

Although the criteria for which harms count as necessary can in principle be determined by a broader ethical theory, no ethical theory is immune to substantial challenge. One might reasonably decline to choose among contending theories, viewing more than one as defensible, yet feel the need for criteria of necessity. Accordingly, one might embrace certain criteria on the basis of their intuitive plausibility. Suppose we consider the extremely harsh conditions of confinement in factory farms: Are these conditions necessary in any plausible sense? Well, are they necessary for anything that is morally very important? Intuitively, they would have to be in order to be justified. Are the products of factory farms necessary in this sense? One might argue that meat products are not necessary—where viable nutritional alternatives are readily available—for human life or health. Further, one might continue, if such products are necessary for anything else—such as higher levels of convenience or of enjoyment for meat-eaters—these goods, intuitively, do not seem weighty enough to justify extensive harm to beings with moral status. According to this approach, then, the harms involved in confining animals in factory farms are unnecessary; they are not, all things considered, justified. This provides an example of how one might reason about whether particular harms are necessary without appealing to a background ethical theory.

I will now summarize this section, make a suggestion, and state a caveat. First, I have argued that (1) in creation and confinement cases the basic-needs requirement is appropriate, and (2) in cases involving the confinement of independently existing animals, the no-unnecessary-harm requirement is appropriate along with the basic-needs standard. The second kind of case permits only harms that count as necessary. If independently existing animals are to be harmfully confined, those who propose such confinement bear the burden of justification. Finally, a caveat: if the standard considered in the next section is legitimate, it may permit a wide range of circumstances in which harmful confinement is justified. Indeed, there may be instances in which leaving some basic needs unmet is justified, as we will now see.

The Worthwhile-Life Standard

Imagine two lions, born and raised in a zoo. They share a relatively small enclosure. Although well fed and provided veterinary care when needed, they exercise little and are listless much of the time. Overall, they are comfortable but under-stimulated and appear bored by their sterile living quarters. We may imaginatively fill in further details to produce a picture in which the lions appear to have a life that is worthwhile yet severely constrained by the small enclosure. Are these conditions of confinement morally justified?

One might reply negatively as follows: "Some of the animals' basic needs are not being met. They need more room to roam and exercise, and more stimulation—not to mention other lions to interact with. So the conditions of confinement are unjustified." A proponent of the worthwhile-life criterion could respond by pointing out that the lions have worthwhile lives, which cannot be a bad thing. The defender of the basic-needs standard might counter by arguing that failure to meet some of the animals' basic needs is unjustified—even if their lives are worthwhile overall—and that such failure can be understood as unnecessary harm. In response, the proponent of the worthwhile-life criterion could advance a version of what we may call *the worthwhile-life argument*:

> Admittedly, the living conditions aren't optimal. Maybe we can even say they involve harm. But, if so, the harm is necessary. After all, substantially better living conditions would have cost much more to produce and maintain. And (we will assume for the sake of argument) the available funds did not permit such expenditure. These two lions would not have been bred if such superior living conditions had been demanded, but they were bred and have worthwhile lives. Since they would not have existed had a higher standard been required, any sensible advocate for the lions will not complain about the necessary living conditions. Having a worthwhile—albeit far from optimal—life cannot be worse for them than never existing at all.

This argument is powerful. Notably, it does not rest on the controversial assumption that coming into existence benefits an individual. It need only assume that having a worthwhile life cannot be worse than nonexistence. This seems undeniable.

Several further points about the worthwhile-life argument are important to note before we evaluate it. First, it applies only where the cost of improving the conditions of confinement is so high that a requirement to do so would deter bringing the animal into existence in the first place. In other words, the claim of necessity requires the assertion that better living conditions are unaffordable. Second, this argument does not apply in cases where the living conditions are so bad as to make the animal's life not worth living, which would constitute wrongful-life cases. Third, the argument presumably does not apply to most cases in which an independently existing animal is captured and confined. The presumption is that in such cases, an animal is made worse off by being captured and confined; this presumption can be defeated in cases in which an animal is known to be very badly off in the wild and likely to fare better in captivity. Fourth, and most crucially, the argument sets a very low standard of acceptability for bringing animals into existence. Because this standard merely requires a worthwhile life, the standard could be met even when an animal's life is predictably *barely worth living*—and several of her basic needs are not met.

Is the worthwhile-life argument sound? Consider what it implies for human children: it is morally permissible to bring children into the world in circumstances in which their lives, predictably, will be barely worth living. I reject this implication, while recognizing that not everyone will reject it. In my view, it would be irresponsible knowingly to impose so much hardship on children. If my judgment is correct, how can the worthwhile-life argument be sound in the case of animals? One way to make sense of the double standard would be to assume that human children have substantially higher moral status than nonhuman animals. Perhaps that is correct; but we do well to note that a proponent of the worthwhile-life standard for animals, and not for human children, must make this assumption, which would surely be contested by some animal advocates.

A distinct consideration may prove decisive even for those who accept the double standard (or disagree with my judgment that the worthwhile-life argument has unacceptable implications for human children). We accept that we should not harm animals needlessly, but face an apparent impasse. The critic of the worthwhile-life standard charges it with excessive laxity for permitting a failure to meet some basic needs, entailing unnecessary harm. The proponent of the worthwhile-life standard would counter that, wherever the standard permits a failure to meet basic needs, such a failure *is* necessary—for the animal with a worthwhile life would not have come into existence otherwise. Stated simply, such harms are necessary for worthwhile lives. But now the critic of the worthwhile-life standard can reply decisively: *those lives themselves were not necessary.*

This reply assumes that there is no obligation, or even any significant moral reason, to bring into existence animals (or persons) who are likely to have good lives. Intuitively, the assumption seems right. Yet, wouldn't we believe it wrong to bring a predictably miserable, languishing being into existence? The conjunction of this belief and the assumption creates what Jeff McMahan, following Derek Parfit, calls the *Asymmetry*: "the view that, while the expectation that a person's life [the point also applies to animals] would be worth living provides no moral reason to cause the person to exist, the expectation that a person's life would [not be worth living] does provide a moral reason *not* to cause that person to exist."[20] If we rejected the Asymmetry and embraced a worthwhile-life criterion for when we have moral reason to bring someone into existence, that would imply that we have moral reason to increase substantially the population of humans and sentient animals, if they would have worthwhile lives, even if doing so would greatly lower average quality of life—what Parfit calls *the Repugnant Conclusion*.[21] I agree that this is a repugnant implication. Accordingly, I suggest that to deny the Asymmetry is intuitively implausible on its face and generates unacceptable implications such as the Repugnant Conclusion. Accepting the Asymmetry permits me to appeal to the judgment that it's wrong to cause unnecessary harm—and to fail to meet basic needs—in rejecting the worthwhile-life standard.[22]

Appeals to Respect

Some animal advocates recommend a fundamental role for the concept of *respect*. Do we owe animals respect? If so, why, and what does such respect require?

Some who hold that we owe animals respect maintain that animals have a *right* to respectful treatment, which entails a right not to be harmed.[23] The intent behind this assertion is not that it is always wrong to harm an animal, but that such harm is justified only under narrowly circumscribed conditions. For example, amputating a dog's leg, causing harm, may be necessary to produce a net benefit for the dog: survival with three legs. Another example of justifiably overriding the right not to be harmed is the classic lifeboat case involving a dog and several persons: the dog is thrown overboard to prevent the lifeboat from sinking, which would cause everyone aboard to drown.[24] The view that animals have a right not to be harmed, grounded in respect, shares with other views the general idea that animals should not be harmed *unnecessarily*. But, by construing the criteria for necessary—justified— harm very narrowly, the present view has a strongly deontological flavor.

What does respect for animals, understood as entailing a right not to be harmed, suggest about confinement? Justified confinement would involve conditions that, at a minimum, do not constitute a net harm to the animal—providing needed safety, for example, without constraining her more than necessary for her safety. A dog living with a human family in the city, then, might be confined to the house, a fenced-in backyard, and (sufficiently frequent) walks on a leash.

A right not to be harmed, grounded in respect, would have implications well beyond those concerning the confinement of animals. For example, it would seem, at least initially, to rule out all animal research that harms animals without offsetting benefits to the animal subjects themselves. That would limit research on animals to that which does not harm them at all, or harms them only in order to benefit them as in last-resort veterinary research. Very little current animal research satisfies this moral standard. An amendment to this view would also permit research that imposes "minimal" (but nonzero) risk to animals, without compensating benefits to them, so long as an interested proxy consents to it, just as we permit minimal-risk research on human children with appropriate consent from a parent or guardian. This amendment would liberalize standards for justified animal research, but only slightly. Consequentialist theories, by contrast, justify a wider range of experiments entailing harm to animal subjects, because consequentialism permits more trade-offs of harms and benefits among different individuals—that is, harming some for the benefit of others—than does the present view.

Importantly, however, the issue of what harms count as necessary in the sense of justified and compatible with respect can be understood differently from the interpretation just discussed. Suppose coming into existence with the likelihood of a good life counts as a benefit. Even if we deny that nonexistence is a harm (there being no determinate subject who is harmed), we might hold that coming into existence with good prospects is a benefit (there being a determinate subject who benefits once she comes into existence).[25] Now recall the worthwhile-life argument: a worthwhile life cannot be worse than nonexistence, so it cannot be worse for a subject to be brought into existence, even in the face of various harms, so long as he is likely to have a worthwhile life. Even if one will predictably face certain harms, it is a net benefit, in this view, to be brought into existence with the prospect of a life worth living. This suggests a more relaxed interpretation of respect, understood in terms of nonharm: wherever an animal is brought into being as part of a plan that makes a worthwhile life probable, it is justified to bring her into existence even if the life would be barely worth living. Such reasoning could be deployed to justify a considerable amount of animal research as well as traditional animal husbandry and the keeping of animals in zoos, aquariums, and human homes in a wide range of living conditions. Such use and confinement of animals could be said to be adequately respectful of them.

In the previous subsection, I argued against the worthwhile-life standard. I believe that respect for animals, construed in terms of nonharm, cannot properly be specified in terms of the worthwhile-life standard—or, if it can be, then appeals to respect, so understood, provide poor moral guidance.

Respect for animals, however, can be construed independently of the concept of harm. Consider the idea that it is wrong, because disrespectful, to *use* people. One might similarly argue that is wrong, because disrespectful, to use animals. Accordingly our focus should move away from harm, net benefit, and similar concepts toward the idea of using another being. In this view, confining animals for our own use—say, keeping them as pets or as experimental subjects—would be wrong even if they were provided great lives with no avoidable harm. Indeed, to conduct field studies in which animals are observed, without disturbing them, for the sake of human curiosity, is wrong, as is keeping animals in a zoo exhibit that meets all their basic needs, and gives them a life better than their conspecifics normally have in the wild—so long as the basic purpose were human-centered—for this would involve using the animals. This position moves very far in a deontological direction.

It is also implausible. First, the position gives no weight to an animal's welfare. The fact that an animal would actually be better off confined in certain conditions would have no weight according to this view; all that would matter is that the animal is used for the sake of others. Second, its conception of impermissible use lacks a theoretical foundation. The most plausible account that prohibits the use of human beings is probably Kant's view. But Kant was explicit that, in his view, it is not necessarily wrong to use human beings; it is only wrong to use human beings *as mere means*. Teachers and students use each other for education and employment, and grocers and consumers use each other for food and profit. But justified cases of *using* in this view do not use anyone as a *mere* means. What determines whether an instance of using someone involves using him as a mere means? Kant himself is not clear on the answer. For my present purposes, it suffices to suggest that one uses another human being as a mere means when one uses her in a way to which she does not, or could not, provide her voluntary, informed consent. In the case of children or other human beings of significantly diminished competence, we may ask whether a responsible proxy does, or could, provide valid consent for the use in question.

This idea is applicable to animals. Would a particular use of an animal count as using her as a *mere* means? Not if a responsible proxy who cared about and was in the role of protecting the animal's interests consented to such use. Presumably, such a proxy would consent to such use whenever it benefited the animal more on balance than any other option did—and perhaps also when the risk of harm was effectively zero. Thus, if we construe respect for animals in terms of non-use, we must do so along these lines, which permit conditions of confinement and other interactions with animals that are consistent with not causing them net harm.

Still, there are other ways of interpreting respect for animals. There are various possible views holding that respect for animals involves giving them what their allegedly intermediate moral status requires. In this general approach, animals have moral status but less than that of persons. They may not be used for frivolous purposes, but they may be used, and confined in such use, for sufficiently weighty human purposes such as important biomedical research. This approach is analogous to one sometimes taken in discussions of embryo research: human embryos deserve some respect, but less than that of persons, and may therefore be used for important research purposes even though such usage entails their destruction.[26]

By now it should be clear that appealing to respect for animals in an effort to illuminate the ethics of confinement has the substantial disadvantage of great uncertainty and disagreement about what respect for animals entails. We have a much better sense of what respect for persons involves (even if there is some dispute on this matter), respect for persons being tied to their undeniable full moral status and their capacity for autonomous decision-making. But the content of appropriate respect for animals is at least as uncertain as animals' moral status relative to that for persons. It is insufficiently clear what, exactly, about animals commands respect and what sorts of treatment of animals are compatible with it. By comparison, the concepts of harm, basic needs, and net benefit—though hardly free of difficulties—are clearer in their application to animals. For this reason, I conclude that appeals to respect for animals do not promise helpful guidance for the ethics of confinement or other areas of practical animal ethics.

Recapitulation and Conclusions about Standards for Justified Confinement

Let us take stock of our findings about the various standards we have proposed.

The basic-needs requirement was supported by (1) an appeal to basic decency (on the assumption that animals have at least some significant level of moral status) and (2) an appeal to special relationships. The latter appeal may apply only in settings in which human-animal relationships imply a caretaking role for the human.

But the basic-needs requirement is supported by the appeal to decency in all cases involving confinement of animals.

The comparable-life requirement was supported primarily by appeal to the wrongness of causing unnecessary harm. But the counterfactual baseline of how well the animal would likely have fared in the wild proved of dubious relevance in cases involving either (1) domesticated animals or (2) creation and confinement. However, the counterfactual is relevant in cases involving wild, independently existing animals who are captured and confined. The upshot was that while the comparable-life requirement is not basic—appealing as it does to the wrongness of causing unnecessary harm—it provides a fairly vivid marker for capture and confinement cases.

The more basic no-unnecessary-harm standard is supported by the principle of nonmaleficence, on the assumption that animals have moral status. But which harms are necessary—that is, justified, all things considered? And what counts as a harm? Suppose we understand harm in standard fashion: harm as being made worse off than one was before or would have been otherwise. That understanding of harm implies that animals are not harmed if brought into existence and confined in conditions that predictably make life just barely worthwhile, so long as those conditions were essential to the plan to bring the animals into being. If the judgment that there is no harm in such creation and confinement cases is correct, then the no-unnecessary-harm standard is insufficient and requires supplementation by the basic-needs requirement. If, however, harm is construed normatively, then the no-unnecessary-harm requirement seems to offer sufficient moral guidance in creation and confinement cases, while effectively converging with the basic-needs requirement. In cases involving confinement of independently existing animals, the no-unnecessary-harm standard proved adequate. What counts as *necessary* harm, meanwhile, can be determined either by reference to a background moral theory or by appeal to an intuitive understanding of what sorts of goods (e.g., human life and health) are weighty enough to justify the harm in question.

The worthwhile-life standard receives support from the consideration that having a worthwhile life cannot be worse for an individual than never existing. So, if the cost of having that worthwhile life includes certain harms, they can fairly be regarded as necessary even if they entail failure to meet certain basic needs. This argument dubiously implies that it is morally acceptable to bring into being animals with lives barely worth living, so long as they would never have existed had a higher quality of life been demanded. Such an implication, I suggested, is unacceptable in the case of human children; if I am correct here, then accepting a worthwhile-life standard for animals requires an assumption of inferior moral status for them. A more decisive response to the proposed standard is that, even if the predictable harms are necessary for the worthwhile lives under consideration, *those lives themselves are not necessary*. There is nothing wrong with declining to bring into being worthwhile lives, especially those that are barely worth living. On the basis of this response, I rejected the worthwhile-life standard.

The final approach we considered involved appeals to respect. What does appropriate respect for animals involve? One possibility is that it requires not harming them except under very circumscribed conditions. Another possible interpretation of respect and permissible harm, however, invites a worthwhile-life standard. Meanwhile, some construe respect for animals to rule out all use of them—or, more plausibly, use of them as mere means. Still others take appropriate respect for animals to reflect an intermediate moral status, in between that of mere "things" and persons. Due to uncertainty about the meaning, and basis, of respect for animals, appeals to respect proved unhelpful as a guide to the ethics of confining animals.

Our discussion leaves us with several conclusions that can guide moral evaluation of particular institutions and practices involving the confinement of animals. First, animals' basic needs must be satisfied. Any exceptions to this rule would require explicit, compelling justification. Second, the no-unnecessary-harm requirement is appropriate, despite its vagueness. What counts as a necessary harm will be determined either by reference to a background ethical theory or on a more intuitive basis. Because, to some extent, the basic-needs and no-unnecessary-harm requirements converge, justifications for exceptions to the basic-needs requirement may also refer to a background theory or proceed more intuitively. The comparable-life standard makes vivid the no-unnecessary-harm requirement in capture and confinement cases. Meanwhile, the worthwhile-life criterion sets too low a moral standard, and appeals to respect are too vague and uncertain, to provide helpful guidance. Therefore, in the sections that follow, I will appeal frequently to basic needs and unnecessary harm, and somewhat less frequently to comparable lives.

Before proceeding to those sections, it may be helpful to clarify their role in this chapter. Considering how much space has been devoted to the theoretical issues examined in this discussion, why not stop here? The sections that follow serve several purposes. First, exploring the ethics of confining animals on farms, zoos, and human homes will bring the theoretical conclusions to life by engaging them concretely and vividly rather than leaving them hanging as abstract generalizations. Second, analysis of the practical ethical issues will benefit from the theoretical exploration provided in the main part of this chapter. For example, we will find that confinement conditions alone preclude factory farming from being morally justified; conversely, we will learn that if traditional animal husbandry is unjustified, the reasons must be unrelated to conditions of confinement. Meanwhile, our theoretical findings help to illuminate the complex ethics of keeping animals in zoos and aquariums, as well as the everyday ethical issues raised by the keeping of pets. A final purpose of the sections that follow is to remind the reader of the ubiquity of contexts in which human beings confine animals, demonstrating the great practical importance of this chapter's topic.

FACTORY FARMS: WRONGFUL LIFE

"Factory farming," as the term will be used here, refers to animal husbandry that tries to raise as many animals as possible while minimizing the space of their enclosures in an effort to maximize profits. The institution of factory farming dominates meat, egg, and dairy production in the United States, Great Britain, Canada, and many other industrial countries. Because factory farming in the United States implicates the most consumers and affects the most animals, the description of this institution provided in this section will concentrate on the United States.[27] Readers in Canada, Great Britain, and most other developed nations will find that much of the description of American factory farming applies to their countries as well. The most general points about factory farms and the extensive harm they cause to animals will apply to factory farming wherever it occurs.

Considering the numbers of animals involved and the extent to which they are harmed, factory farming causes more harm to animals than does any other human institution or practice. Well over 8 billion animals are raised and slaughtered in the United States each year, a figure that does not count the over 900 million who die on farms or in transit to the slaughterhouse.[28] The vast majority of these animals are raised in factory farms.[29] American farm animals have virtually no legal protections. The most important federal law with respect to animal protection is the Humane Slaughter Act, which does not cover birds—most of the animals consumed—or rabbits, is weakly enforced and has no bearing on living conditions, handling, or transport.[30] Our present concern is the living conditions of animals on factory farms, and more specifically, their conditions of confinement.

As noted above, factory farming involves the effort to raise as many animals as possible in minimally sized enclosures so as to maximize profits. The very term *factory farming* implies intensive confinement. Following are some details.[31]

After hatching, *broiler chickens* are moved to enclosed sheds that contain automatic feeders and waterers, which bear no resemblance to the environments for which the birds are naturally suited. From ten thousand to seventy-five thousand chickens are kept in a shed, which becomes increasingly cramped as the birds grow at abnormally fast rates. Crowding frequently leads to cannibalism, other aggressive behaviors, and panic-driven pile-ups. To limit the damage of their aggressive behaviors to each other, chickens may undergo debeaking and toe clipping without anesthesia. By slaughter time, chickens have as little as six-tenths of a square foot of standing room apiece.

Laying hens live their entire lives in "battery" cages made of wire. The cages are so crowded that hens can seldom stretch their wings fully. For these animals, problematic confinement is not only a matter of little space. The highly unnatural, uncomfortable wire footing is also a source of stress and discomfort. Unanesthetized debeaking is common.

Today, most *pigs* spend their lives indoors in crowded pens of concrete and steel. A few weeks after birth, they are weaned and separated from their mothers before being taken to very crowded "nurseries." Due to poor ventilation, they constantly breathe powerful fumes from feces and urine. Tail docking, ear notching, teeth clipping, and castration—all without anesthesia—are common. Upon reaching sufficient size with the help of a growth hormone, many and perhaps most are taken to "finishing" pens, which typically have slatted floors and no bedding, where they languish until it is time to be transported for slaughter. Breeding sows are increasingly confined in narrow gestation crates.

Veal calves have some of the harshest living conditions among farm animals. Shortly after birth they are taken from their mothers and transported to a veal barn, where they are confined in solitary crates too small to allow them to turn around or sleep in a natural position. They may be deprived of iron and appropriate fiber in order to control the appearance of their meat. Their deprivations sometimes lead to such neurotic behaviors as sucking the boards of crates and repetitive tongue rolling. The sisters of veal calves commonly become *dairy cows*. Typically they live in very crowded "drylots," which are devoid of grass, or indoors in a single stall. Unanesthetized dehorning and tail-docking are common, as is the removal of their calves shortly after birth.

This description of confinement in factory farms has omitted turkeys, who are increasingly raised in intensive confinement, as well as cattle raised specifically for beef, who are not. Nor does it discuss other animals who are sometimes raised on farms for the purpose of meat production such as rabbits, sheep, and horses. Nevertheless, it is sufficient for the present discussion.

I contend that wherever the term "factory farming" is properly applied, the conditions of confinement are so intensive that they render the animals' lives not worth living. Any readers who are not yet persuaded that this is the case are encouraged to read detailed descriptions of factory farming in such sources as those cited in this section or, perhaps better, to visit or view video footage of factory farms. Whether we understand animal well-being in terms of experiential well-being, including the avoidance of suffering, or in terms of species-typical functioning, it is clear that factory farm animals languish terribly because of their confinement. To bring animals into existence in order to live in such conditions of confinement is morally unjustifiable. In other words, each animal brought into being in order to live life in a factory farm represents a *wrongful life* case.

The judgment that conditions of confinement in factory farms are unjustified is straightforwardly confirmed by the standards we have discussed. Factory farm animals' basic needs are not met and they suffer unnecessary harm. If one replies that confinement-related harms are necessary for them to exist at all—given the effects of the profit motive on the conditions in factory farms—the appropriate rejoinder is that the lives of these animals are not worth living. Thus, even if we embraced the worthwhile-life criterion, which I rejected for being excessively lax, confinement in factory farms would fail that criterion. There is no reasonable way to justify the conditions of confinement that characterize factory farming.

TRADITIONAL FARMS

As noted above, most meat, dairy products, and eggs consumed in the United States, Great Britain, Canada, and various other countries are produced in factory farms. But people also eat animal products that derive from other sources, including traditional farms, many of which raise animals without the intensive confinement that distinguishes factory farms.[32] Knowing that these farms are far more humane than factory farms, some consumers try to purchase only products from traditional farms, although misleading labeling often confounds the effort to do so.[33]

The question we address here is not whether it is morally appropriate to consume the animal products of traditional farms. Even the comparatively mild treatment of animals on traditional farms imposes some nontrivial harms including: branding and dehorning cattle without anesthesia; unanesthetized castration of cattle and hogs; rough—sometimes brutal—treatment in transport, handling, and slaughter; and death itself, which should count as a harm whenever the animal's life is worth continuing.[34] Moreover, insofar as the animal products of traditional farming are unnecessary due to the availability of nutritious alternatives, the associated harms should count as unnecessary. But none of the harms just mentioned involves the *conditions of confinement* on traditional farms.

Depending on details, the conditions of confinement on traditional farms either are morally justified or can be improved to the point at which they would be morally justified. The confinement conditions need not be harmful, and they are compatible with meeting the animals' basic needs. Therefore any cogent moral critique of traditional animal husbandry will have to focus on other features of this institution, such as the harms mentioned above: invasive procedures performed without anesthesia, rough treatment on the way to and in the slaughterhouse, and death.[35]

ZOOS AND AQUARIUMS

Zoos, or zoological parks—which exhibit animals chiefly for entertainment, educational, or scientific purposes—would not exist if humans never took animals from the wild.[36] The same is true for aquariums, which keep fish and other aquatic animals in tanks. The process of keeping animals in zoos or aquariums begins with humans capturing wild animals and introducing them to captive life. Once animals from a particular species are captive, an alternative to further capture is to breed animals already in captivity. The discussion that follows will refer mostly to zoos, although the general ideas will apply to aquariums as well.

Zoos vary enormously in terms of their conditions of confinement. The continuum includes "zoos" that are not literally zoological parks. The most squalid are roadside menageries, often featuring a single caged animal to attract passers-by to a gift shop, gas station, or other small business. Some larger menageries house multiple animals in small, barren cages that in no way educate viewers or encourage them to respect animals. Even zoological parks vary greatly. Many that are licensed by the United States Department of Agriculture are nevertheless not accredited by the American Zoo and Aquarium Association.[37] Some are as bleak as the menageries described above. Many, such as the National Zoo in Washington, D.C., have good exhibits, with adequate space and enrichment given the animals' basic needs, as well as poor exhibits.[38] The very best are effectively sanctuaries that tend to have fewer species, provide a great deal of space, and approach reproducing the animals' natural habitats—minus most of the dangers. Some of the world's finest zoos are Zoo Atlanta, the Bronx Zoo-Wildlife Conservation Park, San Diego Wild Animal Park, and Scotland's Edinburgh Zoo and Glasgow Zoo. The current trend among the best zoos is away from cages toward more naturalistic habitats.

Much of the discussion that follows will consider animals who live their whole lives in zoos rather than animals who are captured and transitioned into zoo life. In my view, it is almost always wrong to capture wild animals, or at least such "higher" animals as mammals and birds, for the purpose of keeping them in zoos. Such capture almost always imposes unnecessary harm.[39] The issue at hand, though, is confinement. Is it morally justified to keep animals in zoos at all? If so, under what conditions is it justified?

Of the standards defended earlier, one of them, the no-unnecessary-harm requirement, is hard to apply directly to the conditions of confinement. That we should not harm zoo animals unnecessarily offers little concrete guidance besides indicating that zoo keepers should do the best they can for their animals—within appropriate constraints of budget and space. Meanwhile, the basic-needs requirement provides more substantial guidance. It is often clear whether an animal is or is not being fed adequately, is or is not getting sufficient exercise, and the like. In cases involving the capture of wild animals, the comparable-life standard is also relevant and helpful.

Challenging my claim that the no-unnecessary-harm standard is not very helpful in the present context, a thoroughgoing zoo critic might contend that confinement in zoos per se is harmful and that, inasmuch as zoos themselves are not plausibly regarded as necessary, the harm of confinement counts as unnecessary.[40] But this argument is unconvincing because, whether or not zoos are necessary, confinement with adequate space and habitat is not inherently harmful, as explained in the first section of this chapter. While the imagined zoo critic may reply that confinement of zoo animals, whether or not harmful, is *disrespectful,*[41] we have found reason to be skeptical about appeals to respect as a foundation for an ethics of confinement. No doubt some zoo exhibits—think of a very small, utterly barren enclosure for elephants—express disrespect for animals, but the existence of clear cases of disrespect has no tendency to prove that all confinement is disrespectful. An exhibit that clearly allows for a good life for captive animals while encouraging viewers to admire them is, I suggest, more accurately described as respectful than as disrespectful.

In considering whether particular zoo exhibits are morally justified, we should ask whether the animals' basic needs are met. In cases involving the capture of wild animals prior to their confinement in zoos, we should also ask whether their lives are probably as good in the zoo, all things considered, as they would be in the wild. Consider two examples in which both standards are relevant. Suppose that a whale is captured to live in an aquarium, where her basic needs are met, but just barely. Suppose also that the whale would be much better off in the ocean with a much larger social group and far more room for movement. In this capture and confinement case, keeping the whale is unjustified. Now imagine that elephants are captured for zoo life. Large and well-supplied sanctuaries are clearly preferable to zoos, and the elephants would likely fare better in sanctuaries than they would in the wild, due to limited food supplies. By contrast, while their nutritional needs are met in the zoo, their needs for exercise and stimulation generally are not. If not, then here too the confinement is unjustified.

While my view does not prohibit keeping animals in zoos, the basic-needs requirement presents a challenging standard. My guess is that only a rather small fraction of current zoo exhibits meets this standard,[42] but that it can be met—at least with most animals. With highly social animals, such as primates and elephants, meeting this standard probably requires family preservation in view of the likely psychological harms of breaking up families.

In the case of most species, an excellent zoo exhibit can meet animals' basic needs and, in capture and confinement cases, can provide a good alternative to the wild by promising certain benefits. Life in the wild can be "nasty, brutish, and short." While we should not be naïve about harms that are likely to attend particular conditions of captivity, we also should avoid romanticizing the wild.[43] Animals in the wild commonly die earlier than members of their species can live. Assuming death is a harm in any worthwhile life, longer life is a benefit. Excellent zoos offer this benefit. Moreover, regular feeding is hardly guaranteed in the wild. Moderate states of poor health can cause prolonged discomfort to animals without killing them. The availability of veterinary care in a zoo counts as a benefit. These benefits do not derive from confinement, but confinement makes possible the organized provision of such benefits, which should be taken into account in evaluating conditions of confinement.

However, even zoos that offer good veterinary care and extend the lives of animals often neglect certain psychological needs that qualify as basic needs. Social animals confined in small enclosures that deprive them of family members or a reasonable number of conspecifics are regrettably commonplace. Intelligent animals given little to do are condemned to boredom and listlessness. Many ape and elephant exhibits combine these inexcusable forms of neglect. It is plausible to believe that such neglect motivates desperate attempts by zoo animals to escape their enclosures—attempts that sometimes result in their being killed by zoo personnel.[44]

What would enable zoos to meet the basic-needs and, where it applies, comparable-life requirements for justified confinement? Besides the obvious imperatives of adequate food, veterinary care, and comfortable shelter, key features include ample space for exercise, preservation of species-typical social groups (for social species), and creative enrichment of living spaces—preferably in an environment resembling their natural habitats. The need for sufficient exercise calls for exhibits to be larger rather than smaller. The need for sufficient enrichment calls for creative planning on the part of those responsible for zoo exhibits. Ways to provide enrichment include hiding food, placing food in boxes or other unusual containers, or putting it where animals must climb or jump to obtain it, assuming the inconvenience is not cruel; arranging more feedings of smaller amounts rather than one or two large feedings daily; and providing a variety of items to manipulate, explore, and hide behind. For example, orangutans at the National Zoo in Washington have the opportunity to participate in noninvasive cognitive experiments, and may decline when they are not interested; they also can climb on towers and long cables outside of their enclosure.[45]

A few special cases among the animals confined in zoos deserve mention. First, consider nonsentient animals. Whereas the types of animals confined on farms are clearly sentient creatures, some zoo animals—especially the most primitive invertebrates—are likely to be nonsentient. In effect, the basic-needs and comparable-life requirements apply only to creatures capable of having pleasant and unpleasant experiences, and therefore interests: sentient animals. Animals incapable of having pleasant or unpleasant experiences cannot be harmed, at least in any morally relevant sense, and therefore cannot be harmed wrongly. Lacking needs and interests, they cannot have a good or bad life. There appears to be no coherent case against keeping nonsentient creatures in particular conditions—except that zoos should not display them in ways that obviously suggest disrespect (e.g., through gratuitous destruction of living animals). My assertions depend on the assumptions that some animals are nonsentient and that, despite our incomplete knowledge of animals' mental lives, we can responsibly make this judgment about particular animals such as relatively primitive invertebrates.

Just as nonsentient animals represent a special case in the ethics of confining animals in zoos, so do the most complex of nonhuman animals: Great Apes and whales.[46] Their cognitive, emotional, and social complexity sets a strong presumption against keeping them in zoos or aquariums. Yet people's interest in species preservation is strongest in regard to our cousins, the Great Apes. If humans keep these animals, they should meet their basic needs. If capturing them from the wild is involved—as I doubt it should be unless they face an imminent threat—the comparable-life standard should be met. That may require family preservation, a great deal of space, and enrichment that encourages climbing, exploring, playing, and problem-solving. Meanwhile, it would seem impossible to capture and keep whales in aquatic exhibits while meeting the comparable-life condition. Their marine habitat, their propensity to swim enormous distances, and their rich social lives constitute a form of life requiring an environment that is beyond our capacity to simulate. It is perhaps also doubtful that captivity, even for those who are bred rather than captured, is compatible with meeting some of their basic psychological needs. The case for banning dolphin and other whale exhibits is strong.

KEEPING PETS IN HUMAN HOMES

Willy, a five-year-old golden retriever, has a comfortable life. He eats well, receives veterinary care when needed, and is free from abuse. But he spends much of his time alone and with nothing to do. Twice a day he enjoys fifteen-minute walks on a leash. He gets outside only for walks, because while the house in which he lives has a medium-sized backyard, it is not fenced in. Both human parents work away from home during the day. Their child is affectionate toward Willy, but interacts with him only sporadically and is gone most of every school day.

Willy's human caretakers have not satisfied appropriate conditions for confining animals. To be sure, Willy would be worse off as a stray. Although he would gain a more exciting life, he would be hungry much of the time, would lack veterinary care, and would be subject to weather extremes and various outdoor dangers including automobiles. But our discussion of the comparable-life standard suggested its irrelevance in cases involving domesticated animals. The standards that do apply are the no-unnecessary-harm requirement, which is somewhat difficult to apply in view of uncertainty about what harms are necessary, and the basic-needs requirement, which offers clearer guidance. In the present case, Willy's human caretakers have failed to meet the basic-needs requirement by failing to meet his needs for ample exercise, stimulation, and companionship.

Willy's situation is not unusual for a pet, so it is worth asking what his caretakers might do to meet his basic needs. First, they could substantially increase the time he is outside for walks. They have a decently sized backyard, and they could fence it, allowing Willy to enjoy the outdoors for more hours. He could then feel the grass, hear other animals' sounds more fully, dig around, and otherwise engage his senses. The family could also consider adopting another dog to provide additional companionship for Willy. In cases like this one, it is apparent that the size of the enclosure—the house—is not the only issue in evaluating the conditions of confinement. The details of the conditions of confinement, including the opportunities for interactions with others and for engaging one's muscles and senses, matter. The same might be said for animals on traditional farms.

Such changes as those considered here are significant, affecting a family's lifestyle and possibly entailing considerable expense, but such changes are not too much to expect of human caretakers. The important point is that *keeping pets— indeed, more generally, keeping animals—is a very serious responsibility.* People should plan carefully before taking on pets and should do so only if their basic needs can be satisfied. In the case of some animal species, the animals' basic needs cannot realistically be satisfied in human homes. Perhaps there are rare exceptions to this rule, but, generally speaking, it is wrong to keep exotic or undomesticated animals such as monkeys, rodents, birds, snakes, and iguanas as pets. Human homes are so different from the habitats for which these animals are naturally suited that confinement to the former is generally incompatible with at least some of their needs for a comfortable environment, adequate exercise and stimulation, access to social group members, or the like.[47]

Dogs and cats, by contrast, have evolved in ways that enable them to thrive in human households. Their basic needs can be satisfied. We have seen in Willy's case how a dog's basic needs can be met. With cats, matters are somewhat different because they require less stimulation and interaction with humans than dogs require. At the same time, it seems unlikely that their basic needs for exercise and stimulating their senses can be met if they are permanently kept indoors, as cats living with people in cities often are. Families who are not comfortable allowing their cats to roam outdoors, or at least to take walks on a leash, should probably not have cats at all.

CONCLUSION

Human beings confine animals in many contexts. We confine various mammals and birds on farms, innumerable species in zoos and aquariums, pets in human homes, future pets in stores and shelters, horses in stalls, rodents and other animals in laboratories, as well as horses, bears, and elephants in traveling circuses. All confinement involves a deprivation of liberty, which tends to be harmful. But confinement can also provide the benefits of protection and organized provision of food and various forms of care. These facts motivate the questions that this chapter has addressed: (1) what standard or standards should guide the ethics of confining animals? (2) what are the implications for the ethics of keeping animals in farms, zoos, and human homes? An earlier section recapitulated our answer to the first question.

With regard to the second question, we may identify several take-home lessons. First, considerations of confinement are sufficient for a moral condemnation of factory farming. Indeed, I suggest (although here I cannot adequately defend this suggestion) that the most fundamental moral wrong of factory farming is that its conditions of confinement preclude a decent life for farm animals. Second, because traditional animal husbandry does not confine farm animals intensively, this institution is morally acceptable with respect to animals' freedom of movement. Third, while zoos and aquariums necessarily confine animals, they do not necessarily harm them, or fail to protect their basic needs, in doing so; thus, they should be evaluated on a case-by-case basis. Fourth, while keeping various types of pets in human homes is not necessarily wrong, or even morally problematic, the decision to have pets imposes substantial responsibilities on human caretakers including obligations connected with appropriate housing.

Finally, no one is exempt from the responsibility to consider the ethics of confining animals. Everyone must consider whether or not to eat meat, eggs, and dairy products and, if so, from what kinds of farms; whether or not to patronize zoos, aquariums, horse races, and circuses; whether or not to support animal research when the issue of their justifiability is raised; and whether or not to adopt pets. It is hoped that this chapter will facilitate the examination of these issues insofar as they involve the confinement of animals.[48]

NOTES

1. See my *Taking Animals Seriously: Mental Life and Moral Status* (Cambridge: Cambridge University Press, 1996), chap. 8, where I make the case for my thesis and note that there is almost no developed literature on well-being across species. For good discussions of human well-being, see Derek Parfit, *Reasons and Persons* (Oxford: Clarendon, 1984), appendix I; James Griffin, *Well-Being: Its Meaning, Measurement, and Moral Importance* (Oxford: Clarendon, 1986); Martha C. Nussbaum and Amartya Sen, eds., *The Quality of Life* (Oxford: Clarendon, 1993); and L. W. Sumner, *Welfare, Happiness, and Ethics* (Oxford: Oxford University Press, 1996).

2. See DeGrazia, *Taking Animals Seriously*, chap. 5 for my argument that we should regard at least all vertebrate animals as sentient.

3. I understand Judith Jarvis Thomson as making a similar appeal in claiming that it would be indecent and therefore morally objectionable for one to refuse to do what would be required of a "minimally decent Samaritan." See "A Defense of Abortion," *Philosophy and Public Affairs* 1 (1971): 47–66.

4. Admittedly, another possibility would be to allow that the conditions were indecent and inhumane while claiming that they were nevertheless justified. A consequentialist might hold that the indecent conditions must be tolerated because there is no better way to achieve the best consequences. Such a thinker would hold that the indecent conditions were in that sense *necessary*. This position may therefore be considered in the discussion below of the no-unnecessary-harm standard.

5. Most animal-protection philosophers have favored the first strategy. See, for example, Peter Singer, *Animal Liberation*, 2nd ed. (New York: New York Review of Books, 1990), chap. 1, and Tom Regan, *The Case for Animal Rights* (Berkeley and Los Angeles: University of California Press, 1983). By contrast, I believe the second strategy is less vulnerable to charges of begging questions and is more likely to persuade.

6. For a classic argument that special relationships are the basis of particular obligations, see W. D. Ross, *The Right and the Good* (Oxford: Oxford University Press, 1930), chap. 2: "What Makes Right Actions Right?" The same thesis is at home in the ethics of care, although care theorists sometimes prefer the concept of responsibility to that of obligation. For the seminal classic in the ethics of care, see Carol Gilligan, *In a Different Voice* (Cambridge, Mass.: Harvard University Press, 1982).

7. See, e.g., Mary Midgley, *Animals and Why They Matter* (Athens: University of Georgia Press, 1983).

8. It is still common to speak of human *owners* of pets. Because I deny that animals are rightly regarded as property, I resist this term. Of course, neither do I think animals should be slaves, so I also reject the older term *master* to designate a human caretaker of an animal.

9. Some theorists agree that special relationships give rise to particular obligations while denying that the relationships are the ultimate basis of the obligations. Utilitarians, for example, assert that the ultimate basis is the impartial production of utility, or welfare, and that human beings in the world as we know it will most reliably promote utility if they recognize such relationship-based obligations. See Frank Jackson, "Decision-Theoretic Consequentialism and the Nearest and Dearest Objection," *Ethics* 101 (2001): 461–82.

10. The qualification "for any significant length of time," which will be assumed for the remainder of the discussion, is intended to exclude such cases as capturing a mouse in one's cupboards with the intention of removing her from one's house and releasing her.

11. Other relationships that do not involve significant periods of confinement, such as veterinarian to animal patient, may give rise to a modified basic-needs requirement. For example, a veterinarian is obliged to *promote* her animal patients' basic needs, but is not obliged to *meet* them (ensure that they are met) because she is infrequently in contact with her patients and cannot control their living conditions.

12. The argument construes harm in counterfactual terms: harm as making one worse off than one would otherwise have been. Later we will consider other conceptions of harm.

13. A complicating factor is that what is normal for animals (in the sense of typical or common) may be worsened by human intervention. For example, if people pollute lakes so that what becomes normal for catfish is much worse than what was normal prior to the human intervention, arguably the previous norm—what was normal prior to human intervention—is a more appropriate moral baseline than the current norm.

14. As we will see in later section, harm in this argument must be understood in a nonstandard, normative way, but such an appeal is possible and arguably cogent. As we will also see, though, the implications of avoiding unnecessary harm—when harm is understood in this nonstandard way—converge with the basic-needs requirement.

15. Similarly, we could understand the basic-needs requirement not as an absolute requirement; perhaps in some circumstances, violations are justified by an overwhelming gain in utility or overall welfare. At the same time, it would be appropriate to understand the requirement as strongly presumptive and as placing a burden of proof on those who would claim a justified exception.

16. For a good discussion of leading conceptions of harm, see Lukas Meyer, "Intergenerational Justice," in *The Stanford Encyclopedia of Philosophy*, ed. Edward N. Zalta (Spring 2009): available at http://plato.stanford.edu/entries/justice-intergenerational, sects. 3.1 and 3.2.

17. Elizabeth Harman, "Can We Harm or Benefit in Creating?" *Philosophical Perspectives* 18 (2004): 89–113, here p. 93.

18. See, e.g., R. G. Frey, *Interests and Rights* (Oxford: Clarendon, 1981).

19. See, e.g., Regan, *Case for Animal Rights* and Evelyn B. Pluhar, *Beyond Prejudice: The Moral Significance of Human and Nonhuman Animals* (Durham, N.C.: Duke University Press, 1995).

20. Jeff McMahan, *The Ethics of Killing* (Oxford: Oxford University Press, 2002), p. 300. Parfit coined the term in *Reasons and Persons*, p. 391. See *Reasons and Persons*, part IV for much of the reasoning that motivates this problem along with various quandaries in population ethics and in connection with the "non-identity problem."

21. Parfit, *Reasons and Persons*, chap. 17. Parfit addresses the problem only in connection with humans.

22. My arguments are not intended to suggest that there could never be moral reason to bring worthwhile lives into existence. There might be such reason if such lives were necessary to prevent a valued species, such as ours, from going extinct. My claim is that the fact that a life would likely be worthwhile does not by itself provide moral reason to bring that life into existence.

23. See, e.g., Regan, *Case for Animal Rights*, chap. 8: "The Rights View."

24. For a good discussion, see Regan, *Case for Animal Rights*, pp. 324–25.

25. For a tentative defense of this thesis, see Parfit, *Reasons and Persons*, Appendix G. Note that the claim that coming into existence with a worthwhile life is a benefit does not entail that we have an obligation to bring worthwhile lives into being (although utilitarians who accept the former claim will have some explaining to do if they want to avoid such a conclusion).

26. See, e.g., Bonnie Steinbock, "What Does 'Respect for Embryos' Mean in the Context of Stem Cell Research?" *Women's Health Issues* 10 (2000): 127–30.

27. My description draws from the following sources: Singer, *Animal Liberation*, chap. 3, which draws mainly from trade journals and magazines of the farm industry; Bernard Rollin, *Farm Animal Welfare* (Ames: Iowa State Press, 1995); Karen Davis, *Prisoned Chickens, Poisoned Eggs: An Inside Look at the Modern Poultry Industry* (Summertown, Tenn.: Book Publishing Co., 1996); Jim Mason and Peter Singer, *The Way We Eat: Why Our Food Choices Matter* (New York: Rodale, 2006); Farm Sanctuary, "The Facts about Farm Animal Welfare Standards," available at http://www.upc-online.org/Welfare/standards_booklet_FINAL, pp. 5–7; Melanie Adcock and Mary Finelli, "The Dairy Cow: America's 'Foster Mother,'" *HSUS News* (1995); Melanie Adcock, "The Truth Behind 'A Hen's Life,'" *HSUS News* (1983); Marc Kaufman, "In Pig Farming, Growing Concern," *Washington Post*, June 18, 2001, pp. A1, A7; Marc Kaufman, "'They Die Piece by Piece,'" *Washington Post*, April 10, 2001, pp. A1, and A10–A11; and the following webpage and brochures from the Humane Society of the United States: "Factory Farms" (www.hsus.org/farm_animals/factory_farms/; 2005), "Farm Animals and Intensive Confinement" (1994), and "Questions and Answers About Veal" (1990).

28. See USDA, National Agricultural Statistics Service (NASS), *Poultry Slaughter* (Washington, D.C.: USDA, 1998), p. 15; and NASS, *Livestock Slaughter 1997 Summary* (Washington, D.C.: USDA, 1998), p. 1.

29. See U.S. General Accounting Office, *Animal Agriculture: Information on Waste Management and Water Quality Issues* (Washington, D.C.: GAO, June 1995), pp. 2, 47; Humane Society of the United States, "Factory Farms"; and USDA, National Agricultural Statistics Service (nass@nass.usda.gov).

30. Gail Eisnitz, *Slaughterhouse* (Amherst, N.Y.: Prometheus, 1997).

31. The references listed above in note 27 support the description that follows.

32. Despite the common term "family farm," the distinguishing characteristic of the type of farm in question is traditional animal husbandry without intensive confinement, not ownership by a family. A family might own an agribusiness with one or more factory farms. Also, a farm featuring traditional husbandry might not be family-owned. So the term "family farm" is potentially misleading. Of course, some traditional farms raise only crops, not livestock, but I will resist the awkward if more accurate term "traditional *animal-raising* farm."

33. See Suzanne Hamlin, "Free Range? Natural? Sorting Out Labels," *New York Times*, November 13, 1996, p. C1; Singer and Mason, *Way We Eat*, chaps. 3, 8; and Farm Sanctuary, "Facts about Farm Animal Welfare Standards," pp. 2–5.

34. See Singer, *Animal Liberation*, pp. 145–56; and Eisnitz, *Slaughterhouse*.

35. I attempt such a critique in "Moral Vegetarianism from a Very Broad Basis," *Journal of Moral Philosophy* 6 (2009), pp. 160–64. For an argument that consuming the products of traditional animal husbandry is not only permissible, but obligatory, see Roger Crisp, "Utilitarianism and Vegetarianism," in *Ethical Issues: Perspectives for Canadians*, 2nd ed., ed. Eldon Soifer (Petersborough: Broadview, 1997), pp. 190–200.

36. My description of zoos draws broadly from Stephen Bostock, *Zoos and Animal Rights* (London: Routledge, 1993), chaps. 5–7; Jeremy Cherfas, *Zoo 2000* (London: British Broadcasting Co., 1984); John Grandy, "Zoos: A Critical Reevaluation," *HSUS News* (Summer 1992); John Grandy, "Captive Breeding in Zoos," *HSUS News* (Summer 1989); Humane Society of the United States, "Zoos" (1984); Linda Koebner, *Zoo Book: The Evolution of Wildlife Conservation Centers* (New York: Forge, 1994); Bryan Norton et al., eds., *Ethics on the Ark: Zoos, Animal Welfare, and Wildlife Conservation* (Washington, D.C.: Smithsonian, 1995); and H. Hediger, *Wild Animals in Captivity* (New York: Dover, 1964).

The description also draws from my own observations of zoos in Washington, D.C., New York City, Houston, Munich, Germany, and San Jose, Costa Rica as well as aquariums in Baltimore and Chicago.

37. Humane Society of the United States, "Zoos."

38. For a critique of the National Zoo focusing on several apparently unnecessary deaths of zoo animals, see Richard Farinato, "Too Many Deaths at the Zoo," Humane Society of the United States website, (humanesociety.org), February 26, 2003. A favorable discussion of the National Zoo's treatment of apes appears below in the present chapter.

39. I develop the argument in *Taking Animals Seriously*, pp. 291–94.

40. Cf. Dale Jamieson, "Zoos Reconsidered," in Bryan Norton et al., eds., *Ethics on the Ark: Zoos, Animal Welfare, and Wildlife Conservation* (Washington, D.C.: Smithsonian, 1995), 52–66.

41. See, e.g., Tom Regan, "Are Zoos Morally Defensible?" in Bryan Norton et al., eds., *Ethics on the Ark: Zoos, Animal Welfare, and Wildlife Conservation* (Washington, D.C.: Smithsonian, 1995), 38–51.

42. See, e.g., Grandy, "Zoos: A Critical Reevaluation." My own experiences at several zoos—the National Zoo in Washington, the Munich Zoo, and a zoo in San Jose, Costa Rica—are consistent with the claim that recent years have seen some improvements.

43. For an emphasis on such harms in captivity, see Farinato, "Too Many Deaths at the Zoo" and People for the Ethical Treatment of Animals, "Zoos: Pitiful Prisons" (www.peta. org/MC/factsheet_display.asp?ID=67), downloaded on March 27, 2009. For reminders of the dangers of life in the wild, see Bostock, *Zoos and Animal Rights*, chap. 5 and Hediger, *Wild Animals in Captivity*, chap. 4.

44. For examples of such tragic endings, see PETA, "Zoos: Pitiful Prisons."

45. I recently had the privilege of a behind-the-scenes tour of the Great Ape House and Think Tank at the National Zoo. I was deeply impressed by the care given the apes and pleased to witness animal research that seemed not only harmless to the animal subjects but rewarding for them. My thanks to Marietta Dindo, a zoo employee and post-doctoral fellow of the Mind-Brain Institute at George Washington University, for providing the tour.

46. For excellent introductions to Great Apes and to dolphins in particular, see Paola Cavalieri and Peter Singer, eds., *The Great Ape Project: Equality Beyond Humanity* (New York: St. Martin's Press, 1993) and Thomas White, *In Defense of Dolphins* (Oxford: Blackwell, 2007).

47. For a helpful discussion, see the Humane Society of the United States, "Should Wild Animals Be Kept as Pets?" (www.hsus.org/wildlife/issues%20facing%20wildlife), downloaded March 27, 2009.

48. My thanks to Tom Beauchamp, Ray Frey, Liz Harman, Eric Saidel, and members of the Primate Cognition Interest Group at the National Zoo for feedback on a draft of this chapter.

SUGGESTED READINGS

ARMSTRONG, SUSAN J., and RICHARD G. BOTZLER, eds. *The Animal Ethics Reader*. London: Routledge, 2003.

BEAUCHAMP, TOM L., F. BARBARA ORLANS, REBECCA DRESSER, DAVID B. MORTON, and JOHN P. GLUCK. *The Human Use of Animals: Case Studies in Ethical Choice*. 2nd ed. New York: Oxford University Press, 2008.

CAVALIERI, PAOLA, and PETER SINGER, eds. *The Great Ape Project: Equality Beyond Humanity*. New York: St. Martin's Press, 1993.

DeGrazia, David. *Animal Rights: A Very Short Introduction*. Oxford: Oxford University Press, 2002.

———. *Taking Animals Seriously: Mental Life and Moral Status*. Cambridge: Cambridge University Press, 1996.

Frey, R. G. *Interests and Rights*. Oxford: Clarendon, 1981.

Harman, Elizabeth. "Can We Harm or Benefit in Creating?" *Philosophical Perspectives* 18 (2004): 89–113.

Koebner, Linda. *Zoo Book: The Evolution of Wildlife Conservation Centers*. New York: Forge, 1994.

Mason, Jim, and Peter Singer. *The Way We Eat: Why Our Food Choices Matter*. New York: Rodale, 2006.

Norton, Bryan G., Michael Hutchins, Elizabeth R. Stevens, and Terry L. Maple, eds. *Ethics on the Ark: Zoos, Animal Welfare, and Wildlife Conservation*. Washington, D.C.: Smithsonian Institute Press, 1996.

Parfit, Derek. *Reasons and Persons*. Oxford: Clarendon, 1984.

Pluhar, Evelyn B. *Beyond Prejudice: The Moral Significance of Human and Nonhuman Animals*. Durham, N.C.: Duke University Press, 1995.

Rachels, James. "Why Animals Have a Right to Liberty." In *Animal Rights and Human Obligations*, edited by Tom Regan and Peter Singer, pp. 122–31. 2nd ed. Englewood Cliffs, N.J.: Prentice Hall, 1989.

Regan, Tom. *The Case for Animal Rights*. Berkeley and Los Angeles: University of California Press, 1983.

Singer, Peter. *Animal Liberation*. 2nd ed. New York: New York Review of Books, 1990.

Taylor, Angus. *Animals and Ethics*. Petersborough, Ontario: Broadview, 2003.

Varner, Gary E. *In Nature's Interests? Interests, Animal Rights, and Environmental Ethics*. New York: Oxford University Press, 1998.

White, Thomas. *In Defense of Dolphins*. Oxford: Blackwell, 2007.

CHAPTER 28

KEEPING PETS

HILARY BOK

A pet[1] is a nonhuman animal whom we take into our homes and accept as a member of our households. Owning pets can be wonderful when all goes well: domesticated animals such as cats and dogs are typically happy living in human households, and have much better lives with humans than they would in the wild. They welcome our companionship, and we welcome theirs. It is fascinating to try to figure out how the world appears to a wholly different mind with senses and instincts that are quite unlike our own, and wonderful when we forge a genuine friendship across species boundaries. For the most part, the interests of pets and their owners do not diverge in any serious way, but on occasion life with pets can go badly wrong. We can fail our pets in any number of ways, and when we mistreat or neglect them, they normally have no recourse. They can be intractably aggressive, and pose a danger to us and to others. They can be destructive or unsanitary. Our homes should be places of comfort and safety; it is hard to share them with beings who are aggressive or destructive.

I will not dwell on issues of obvious cruelty and neglect in this paper. It is plainly wrong to be gratuitously cruel to any animal, human or nonhuman, or to deprive any being of its basic needs when we have taken that animal into our care. I will focus instead on various issues that might confront pet owners whose intentions are basically good, since they raise the most difficult moral issues.

Thinking about the ethical issues involved in keeping pets is different, and in some ways simpler, than thinking about issues involved in our relations with other nonhuman animals. Questions involving the treatment of wild animals—for example, what kinds of incursions on their habitat are excessive, whether it is permissible to cull animals to protect ecosystems or indigenous species, and whether it is permissible to hunt them—concern the treatment of animals with whom we have an ambiguous relationship. One plausible answer to such questions is that, absent some compelling reason, we should try to minimize the harm we inflict on wild animals, and otherwise leave them to live out their lives in whatever way their instincts dictate. This thesis is not a plausible default position, however, when it comes to nonhuman animals whom we have chosen to take into our households, and who are dependent on us for their survival.

Most other questions about the ethical treatment of nonhuman animals involve using them for humans' own purposes in, for instance, research, product testing, food production, or entertainment, in ways that impose significant and uncompensated sacrifices on them. The treatment of pets is a different matter. Pet ownership does not normally involve using nonhuman animals for some purpose that does not benefit them, though of course there are exceptions. So whereas considering many other ways of using nonhuman animals necessarily involves an attempt to determine how such sacrifices might be justified, this is not the problem when dealing with pets.

Moreover, in the case of pet ownership, there is no need to consider the question whether nonhuman animals have interests that are worth taking seriously, or how we should reconcile our use of them with the fact that they are individuals with lives and interests of their own. Pet owners normally agree that their pets do have interests and that these interests, in general, deserve consideration. I will follow them in this assumption.[2] Pet owners' intentions are both generally benign and focused on nonhuman animals as individuals. For this reason, in considering the moral issues raised by pet ownership, we do not need to ask how a nonhuman animal, considered as an individual with interests, can be used for purposes that have nothing to do with those interests. Instead, we have to ask: granted that we are dealing with nonhuman animals as individuals with morally significant interests, how should we relate to those individuals, and under which conditions does our treatment of them go wrong?

Finally, when one considers issues like the use of nonhuman animals in medical research or factory farming, it is easy to proceed at a high level of abstraction. The nonhuman animals in question are generic animals, or at best generic mice or pigs, with whom we do not have any sort of social relationship. This makes it easy to imagine that the central questions concern the construction and justification of a sort of moral hierarchy, in which animals with different capacities occupy different places, and animals with greater capacities, like humans, deserve greater consideration.

Questions involving pets, with whom we have relationships, make it clear that things are often more complicated. Sometimes humans' greater capacities give us greater rights: for instance, while I cannot force a competent adult human to get medical care against her wishes, even if she badly needs it, I can take a cat to the veterinarian over her vehement protests. But sometimes things work the other way: the fact that humans can understand more, and can act on that understanding, allows us to require more of ourselves than we can reasonably require of nonhuman animals, and leads us to accept behavior from them that we would not accept from a competent adult human.

Our Asymmetrical Relationship to Pets

The relationship between pets and their owners is asymmetrical in a number of ways. Adult humans are more physically powerful than all cats and many dogs. Moreover, we can use various tools to overcome differences in physical strength—leashes, cages, and so forth—and we can, if need be, deploy the resources of the state on our behalf. Nonhuman animals have no legal standing to sue for the enforcement of animal protection laws, or any other laws that concern them. When nonhuman animals find themselves at war with the world of human institutions, they generally lose.

We also know much more about certain crucial things than nonhuman animals do. Pets live in human households that we have arranged for our convenience in ways they often cannot grasp. Their food is in bags or tins behind a cupboard door. The door to the outside is one that we can open and they generally cannot. We understand why certain pieces of furniture should not be jumped on, while they must be taught that the kitchen counter is off limits.

Moreover, we understand the peculiar world of human institutions: for instance, we know, and our pets do not, that certain kinds of behavior can get them into trouble with the law. We also understand more about the future, and about such crucial causal connections as that between letting a stranger stick a needle into them and their not getting sick. If we did not understand such things, we might decide not to exercise our greater power over our pets. But we do. It is no kindness to a pet not to vaccinate her against diseases, or not to train her to act in ways that will prevent her being seized and put down.

Pets are dependent on their owners. We supply their food and water, which they are normally unable to obtain for themselves. We provide them with shelter and veterinary care. Their ability to meet their basic social needs depends on us: if we keep them locked up in an apartment all day, they cannot normally get themselves out, or make other arrangements. In addition, dogs depend on us for exercise. Pets also depend on us for various items that make their relationship with us much more pleasant than it might otherwise be. Cats cannot provide their own kitty litter boxes and keep them clean. Dogs do not normally housetrain themselves. Moreover, we need to provide dogs with basic training in the norms of human society if we do not want them to attack our guests, eat our food, and so forth.

That our relationships with pets are asymmetrical in these ways does not mean that our relationships with them cannot be mutual in important respects, still less that we will abuse our greater power over them. It simply means, first, that if we choose to abuse or neglect them, they generally have no recourse; and second, that we have the capacity to set the terms of our relationships with our pets. We normally choose to take pets into our home, and if we change our mind about living with them, we can choose to give them away, or, if need be, have them killed.

We set terms for our relationships with humans as well, though they are often implicit, since most humans understand, at least in broad terms, what kinds of behavior are unacceptable to others. Comparing our responses when our pets violate those limits to our responses when humans do can sound peculiar: we do not normally encounter other humans who do not know, for instance, that they ought not to urinate on our rugs, and thus we rarely have to inform others that this kind of behavior is unacceptable to us. However, those limits do exist. As Vicki Hearne writes: "Try putting your ice cream cone on my typewriter, and you'll get the idea."[3]

Our relationships with pets differ from those with humans in two crucial respects. First, when one ends a human friendship or divorces a spouse, this normally does not result in that friend or spouse's death or imprisonment. Adult humans can generally fend for themselves, and in most circumstances, we can assume that they will survive the end of their relationships with us. This is not true of pets. Second, competent adult humans can explain to one another why they do what they do, and, in particular, the terms on which they are willing to enter into a relationship with others. For instance, if I believe that I cannot be friends with someone who routinely destroys my property without good reason, I can normally assume that other adults will understand this. In the unfortunate event that one does not, I can inform her that I am not willing to accept this conduct in a friend, and explain why. If my friend has some reason for acting as she does, she can let me know what it is. If her reason makes sense to me, I will accept it. If not, I can explain why it does not, and she can understand me, and can accept or reject those terms as she sees fit.

We cannot similarly explain ourselves to our pets. This matters immensely to our relationships with them. As noted, most people have limits on the kind of behavior they are willing to tolerate from others, whether human or nonhuman. Some human demands are unreasonable: it would be wrong, for instance, to expect a cat never to shed. But others are not, and we owe it to any nonhuman animals we take into our homes to do our best to ensure that they abide by these standards. For instance, we are responsible for ensuring that our pets do not pose a danger to other humans. This is a duty we owe not just to those humans but to our pets, since nonhuman animals who attack humans are normally killed.

Most people also place some limits on what they are prepared to tolerate in their homes. Humans should be flexible with nonhuman animals, and should not expect them to behave like small furry humans. Nonetheless, there are limits to what most people are prepared to live with. We require some basic level of sanitation in our homes, and few people would be willing to live with animals who routinely urinate or defecate wherever they see fit. Our home is where we keep our property, and while we can tolerate the occasional destruction of unimportant objects, few people are prepared to tolerate animals who routinely burrow into the furniture, rip out the drywall, or chew the carpet to shreds.

We are responsible for doing what we can to ensure that our relationship with our pets is not intolerable in these ways. This is a duty we owe not just to ourselves but to them, since pets who transgress these limits—who urinate on rugs and furniture, destroy their owners' property, or make it impossible to invite guests over without calamity—will at best be constantly aware of their owners' irritation and anger, and more likely be caged or killed. If an adult human whom I had invited into my home urinated on the carpet or attacked my other guests, I could ask her to stop, and if she refused, I could ask her to leave. Our capacity to understand each other, to explain ourselves, and to choose whether to modify our expectations or our conduct, gives us the ability to come to terms with one another and work out mutually acceptable accommodations between our interests. We do not have this kind of reciprocity with nonhuman animals.

In some cases, this does not cause serious problems. Cats, for instance, are easy to get along with, at least if one is not too particular about cat hair, clawed sofas, and the like. This is fortunate since they are difficult to train. It would be hard to get cats to use a litter box if they were not inclined to do so. Luckily, they are, and one normally needs simply to show a cat where the litter box is and maintain it in a reasonable state of cleanliness. It would be hard to train a cat not to attack us if we were the size of the animals they regard as prey. Luckily, we are much bigger than they are. There is no obvious reason why cats should be affectionate to humans. Luckily, most of them are. Because they are naturally inclined to behave in ways that many humans find acceptable, even adorable, the fact that it is difficult to train cats to act differently is not normally a problem.

Dogs are different. Unlike cats, many dogs are large enough to pose a real danger to humans. They do need to be housebroken and trained. Fortunately, they are also unlike cats in that they are, in general, not just tractable but willing to enter into the spirit of some of the things we do. We can often get across our desires, intentions, and feelings to dogs, and we can often understand theirs. While some areas of human endeavor, like accounting and the law, are forever closed to dogs, we are able to communicate with them about a surprising number of things, including some very complicated ones, such as what it means to do a perfect retrieve.

Moreover, dogs care about what we think of them, and they normally care enough to be willing to enter into genuinely reciprocal relationships with us that involve both parties' constraining their conduct to accommodate the other. Most wild animals do not. They might regard us as objects of curiosity or even—with time—affection, but that does not mean that they are willing to alter their behavior on our behalf. As Vicki Hearne writes:

> Even the most messed-up, dingbat, rotten nasty dog in the world, the one who goes for your throat as soon as you say good morning, is immediately, *immer schon, toujours déjà*, prepared for the possibility of friendship, and that is what it means to say you can work with a dog....A wolf does not refuse friendship, because for the wolf it isn't there to be refused in the first place....Human love and praise are alien to her, so if you don't build into your relationship with the wolf, step by step, a language of relation whose meanings she will accept, she just isn't there. It is not like a scene from Rambo, training a wolf. It is more like training a cloud or a dream or a shadow.[4]

Finally, dogs are able not just to adjust their conduct every now and then, but to do so consistently. This is often difficult for most wild animals. Chimpanzees, for instance, might want to get along with us, and understand that involves, for instance, not attacking people; but even if they often manage to restrain their urge to bite, they generally do not manage to do so *all of the time*, as they would have to do in order for us to trust them. Vicki Hearne contrasts her dog to Washoe, a chimpanzee who can use sign language:

> Washoe, like my dog, has been told, and in no uncertain terms, that she ought not to bite even though she might want to. With my dog, the issue was settled long ago, almost without our noticing it, and we are in agreement. If my dog were to bite a visitor, I would be forced to consider the possibility either that the visitor had committed a crime or that my dog had gone crazy. And I would have to work out what had happened before I could again take my dog for a walk. If there was no reason for the bite, nothing that a reasonable person could recognize as a reason, the relationship with the dog would have broken down.
> But there is no such agreement with Washoe.[5]

Because there is no such agreement with Washoe, she is in a cage when Hearne encounters her. Later, Hearne watches as Washoe is taken for a walk with "leashes, a tiger hook and a cattle prod."[6] Because we can reach such an agreement with dogs, we allow them to play not just with us but with our children. That we can come to such an understanding with dogs—that we can have confidence that a given dog will not bite strangers except when there is something "that a reasonable person could recognize as a reason" to bite, for instance—is an astonishing feature of our relationship with nonhuman animals. Why should a dog care about the difference between, say, a burglar and a plumber, rather than regarding both as intruders on her territory? Why should she restrain her impulse to defend her territory when she sees that the humans with whom she lives accept the arrival of the plumber with equanimity? Wolves do not take their cues from us. Why should dogs? And yet they do. Because we can trust most dogs to draw these distinctions, and to act on them consistently, we can generally work out an accommodation with them that is satisfying to all concerned.

However, while our ability to work things out with dogs is impressive, it is not unlimited. If an individual dog cannot be trained to do something that her owner regards as essential, her owner cannot simply explain what is so important about that thing, nor can the dog explain why doing that thing is so objectionable. Even if we think we see why the dog refuses to do that thing, we cannot reason with the dog and try to explain why we think she is mistaken. In the absence of the kind of robust reciprocity provided by the ability to use and understand a common language, we cannot be confident that we will be able to work out an accommodation with dogs or other pets when we need to.

Our ability to explain ourselves to one another, and to adjust our conduct accordingly, provides limits on what kinds of behavior human adults need to accept from others. Because we can explain what we expect of one another, and give one another a choice between meeting those expectations, explaining any failure to do so, and ending our relationship, we can regard other adults as having forfeited the rights of friendship if they insist on violating those expectations for no good reason.

They had a choice, and they are responsible for it and its consequences. These limits do not exist with nonhuman animals. If, for instance, I own an aggressive dog, I cannot tell her that if she continues to attack people, I will have to take her to the pound, where she will, in all likelihood, be killed. I can and should try to train her not to attack people, but I cannot explain to her why it matters so much that this training succeed. If I fail, I cannot say that she knew what the consequences of her aggression might be and chose to incur them. She did not. But though she is not responsible for her conduct, she will pay the price.

Is It Wrong to Own Pets?

Some writers and activists believe that because of the asymmetries I have identified, it is prima facie wrong to keep nonhuman animals as pets. Gary Francione argues that it is wrong to keep pets under present arrangements because they are, legally, property:

> Property owners can, of course, choose to treat any of their property well or poorly. I may wash and wax my car regularly, or I may ignore the finish and let the paint fade and the body corrode away. Similarly, many of us who live with dogs and cats choose not to treat our animals solely as economic commodities and instead accord them a level of care that exceeds their market value. But...we may also choose to treat these animals as nothing more than property.[7]

Francione concedes the need to care for existing animals who need homes. But because he thinks it wrong to own animals, he concludes that "were there only two dogs remaining in the world, I would not be in favor of breeding them so that we could have more 'pets' and thus perpetuate their property status."[8] The legal status of nonhuman animals raises serious moral questions. Nonhuman animals do not have legal standing, at least in most nations, and thus they cannot sue humans for protection under existing animal cruelty laws. Moreover, those laws that do exist are, as Francione notes, not always enforced. However, it is not clear why Francione believes that this fact about the law and its enforcement implies that we should not own pets we have not rescued. Assume for the sake of argument that legal protection for nonhuman animals is wholly inadequate, that we should replace our existing laws with some other legal system that affords nonhuman animals the right to sue for protection,[9] and that these laws should be vigorously enforced. Assume further that you fully intend to treat any nonhuman animals you adopt well, rather than letting yourself be guided by the minimal standards of the law; that there are no nonhuman animals in need of rescue; and therefore that your reasons for adopting a nonhuman animal do not include the need to give that animal a home. Under these circumstances, would it be wrong to own a pet?

If our existing system of laws is inadequate, then we should try to change it. If owning nonhuman animals somehow interfered with our efforts to do so, that fact would constitute a reason not to own them. But it does not: nothing about owning a nonhuman animal precludes or hinders trying to change that animal's legal status. If owning nonhuman animals helped to perpetuate our present legal arrangements, then one might argue that we should withdraw our support for those institutions by giving up our pets, or at least not adopting new ones. But it does not: boycotting the institution of pet ownership would not in any way undermine it, since it does not rely on the existence of pet owners for support.

If our present laws require us to do things to our pets that are morally wrong, then we would have a reason not to own them. But the law is not so structured. It simply gives us the legal right to do various things that we morally ought not to do: abandoning them at will, mistreating them in ways that fall short of the legal definition of animal cruelty, and so forth. It cannot be wrong to participate in a legal relationship that does not prohibit every form of immoral treatment. For instance, there are many things that it would be wrong to do to a child, but that are not prohibited under child abuse statutes. We do not conclude that it is wrong to have children, only that we should give our children better treatment than the minimum required by law.

A legal system governing the status of nonhuman animals will presumably give humans some measure of control over them: for instance, the right and the responsibility to prevent them from being a danger to others, to ensure that they get proper medical care even if they do not consent, and so forth. In exercising this degree of control, we do not violate their rights. The problem with our present system is not that it requires that we do things that are wrong, but that it gives us rights over nonhuman animals that we arguably ought not to have. Fortunately, this system does not force us to exercise these rights. We can, and should, treat our pets in just the same way that we would if the laws governing nonhuman animals were exactly as they should be. If we do so, it is not clear how the fact that our laws are not ideal harms nonhuman animals. They are unaware of our legal arrangements, and as long as we live up to our responsibilities toward them, they need not be directly affected by their status as property.

It is, of course, possible to harm others in ways they are not aware of. For instance, it would be wrong for a surgeon to write "IDIOT!" on her patients' foreheads while they are anaesthetized, even if, before her patients come round, she washed off her writing so carefully that her patients never knew what she had done. But in the case of the alleged harm done to nonhuman animals by being owned by caring owners, those animals are not simply unaware of their legal status. They could not possibly be aware of it without some drastic alteration to their mental capacities. Moreover, nonhuman animals do not care about their legal status. Arguably, I harm a dog if I expose her to ridicule, even if I do so in a way that she cannot understand, say by writing "IDIOT!" on *her* forehead. Dogs understand some dimensions of ridicule, and if their owners expose them to ridicule, they show contempt for something dogs care about: their standing in the world as they know it, and in their households. Loss of standing is something dogs recognize as bad, and to gratuitously expose a dog to ridicule harms that dog in ways she can understand, whether she is aware that she has been exposed to ridicule or not.

Those who draft laws according to which nonhuman animals have no legal standing and no recourse in the face of abuse or neglect also disregard things that nonhuman animals actually care about, such as their safety. If there are better work-able alternatives, those legislators arguably harm those animals, as do any humans who take advantage of these laws to mistreat their pets. However, if someone who wants to take a dog into her home allows the law to describe her as "owning" that dog, but does not allow this fact either to affect her treatment of her dog or to per-petuate that legal system, and if the alternative is allowing domestic dogs to die out altogether, it is hard to see how her action harms her pet in any way.

People for the Ethical Treatment of Animals (PETA) objects to the institution of pet ownership on other grounds. PETA's website says, "In a perfect world, animals would be free to live their lives to the fullest, raising their young and following their natural instincts in their native environments. Domesticated dogs and cats, however, cannot live "free" in our concrete jungles, so we are responsible for their care."[10]

Our "concrete jungles" are indeed inhospitable to cats and dogs. Feral dogs and cats have short and difficult lives. According to the Humane Society of the United States, half of all feral kittens die without human intervention.[11] According to the ASPCA, the lifespan of feral cats who survive kittenhood, and who are not cared for by humans, is two years.[12] During that time "feral cats must endure weather extremes such as cold and snow, heat and rain. They also face starvation, infection and attacks by other animals."[13] Feral dogs face the same problems as feral cats. In addition, they pose a serious danger to human beings and livestock. For that reason, in most devel-oped countries feral dogs are captured and euthanized whenever possible.

The perils of our "concrete jungles" are not the only reason why fully domesti-cated animals like cats and dogs cannot "live their lives to the fullest" in the wild. These species have lived with humans for thousands of years, during which time they have adapted to life with us. We might debate whether our ancestors should have domesticated them in the first place, but that ship has long since sailed. By now, our homes *are* the "native environment" of cats and dogs, and if we treat them well, they can be much happier in our homes than in the wild. The idea that there is some other environment where they might better follow "their natural instincts" is as much a fantasy as the idea that we should move back to the Great Rift Valley to follow ours.

PETA argues that while it is permissible for us to own pets now, given the num-ber of nonhuman animals who need homes, it is not permissible to allow our pets to breed, since this "perpetuates a class of animals who are forced to rely on humans to survive."[14] One problem with domesticated animals' reliance on us is obvious: there are some people who cannot be relied on, and a nonhuman animal who is owned by a human runs the risk of being owned by one of those unreliable people. It is plainly bad for any animal, human or nonhuman, to be dependent on someone who does not meet his or her basic needs: to depend for food on someone who regularly forgets to feed us, or for shelter on someone who leaves us out in the freezing rain.

If that risk is the only problem with pets' dependence on us, however, then it does not imply that any specific person ought not to own pets. When a person considers adopting a pet, he should first ask himself whether he can care for the animal he adopts and provide her with a good and loving home, and, if so, whether he is willing to commit himself to providing her with the care and love that she deserves. If he is, then the fact that others do not is no more a reason for him not to own a pet than the fact that some husbands abuse their wives is a reason for him not to marry.

One might think that there is something wrong with dependence even if the person one depends on never lets one down. It is wrong to perpetuate someone's dependence on us if that being both wants to live independently and is capable of doing so. It is less clear, however, what is wrong with perpetuating a class of animals who are dependent on someone who loves them and treats them well, when those animals do not want to live independently and would not be capable of doing so. The unconvincing alternative envisaged on PETA's website is not that humans should stand in some other relationship to domestic cats and dogs, but that domestic cats and dogs should cease to exist.

On Thinking Before Adopting a Pet

To adopt an animal as a pet is to undertake to meet her needs, and to accept the responsibility of ensuring that one's relationship with her is good for all concerned. We should consider whether we are willing to accept that responsibility before acquiring a pet. The time to discover that you do not have the time to train a puppy or cannot tolerate cat hair on the furniture is *before* you adopt a pet, not after. Sometimes we cannot know in advance that we will be unable or unwilling to meet a nonhuman animal's needs. But often we can know this. In such cases, we owe it to the animal in question, to ourselves, and to others not to adopt it in the first place.

This fact implies that we should not adopt wild animals as pets. There might be some species of wild animals whose members are, by sheer luck, suitable to be kept in human households, but most wild animals are not. In many cases, we cannot provide wild animals with even their most basic needs—for instance, an adequate diet. In the case of large animals, especially predators, we cannot provide them with nearly enough space to roam around in, as anyone who has seen a tiger or cheetah pacing round and round a cage can attest. When we cannot meet such basic needs, we have no business taking these animals as pets.

Even if we can meet a wild animal's basic needs for food and shelter, adopting them is a serious mistake, both in terms of their interests and in terms of their owners'. Most wild animals are neither psychologically nor behaviorally suited to life with humans. While domesticated species like cats and dogs have spent thousands of years adjusting to life with humans, non-domesticated species have not. In particular, taking wild animals as pets often involves three sets of problems.

First, wild animals are likely to be unhappy in human households. They are adapted to life in environments quite different from a house or apartment. Even if they are allowed to roam freely in their owner's house or yard, most of them will not be able to perform most of their normal species behaviors, and will be bored and miserable. In addition, domesticated animals are less fearful in general, and much less fearful of humans in particular, than their wild counterparts. They find it easier to interact with us socially, rather than regarding us simply as objects of curiosity or alarm. These adaptations help them to live much happier lives in our homes than, say, a terrified antelope or squirrel.

Second, domesticated animals are generally much less aggressive than their wild counterparts. They are also more tractable: more willing to adjust their behavior to suit the requirements of living with humans. To take a wild animal as a pet is to adopt an animal with whom no such accommodation can be reached. In the case of small or docile animals, that might not be a problem. But if an aggressive animal is large enough to do damage, then it is irresponsible to adopt it, both because of the risks to humans and because of the very great likelihood that that animal will end up either caged or killed.

Third, when wild animals are forced to live in a human home, many act in ways that their human owners find it hard to live with. We accept without question a whole array of basic social requirements that we take to be important to our ability to live with one another: that fights for dominance not take the form of physical attacks on other members of our social group; that we urinate and defecate only in certain specific places, and in private; that some objects are our possessions and should not be destroyed; and so forth. These are our requirements, and they make sense to us. But there is no reason to suppose that they must make sense to members of other species.

People who take non-domesticated animals as pets often overlook how deeply unnatural living with human beings and conforming to human requirements might be for a member of a species that has neither a history of domestication nor any particular affinity for our peculiar way of life. Because they do so, they imagine that it will be possible to live with animals who cannot adapt to life with humans. This serves neither their own interests[15] nor those of the animals in question. As an example, consider nonhuman primates. There are no accurate statistics on the number of nonhuman primates owned as pets, but the most common estimate for the United States puts the number at 15,000.[16] Most are acquired as infants, when, like infants of most mammalian species, they are unusually submissive. However, when nonhuman primates reach sexual maturity, they often become aggressive and pose serious risks to their owners and others.

Moreover, nonhuman primates are agile, athletic, clever, and inquisitive. Since they have opposable thumbs, they can open doors, cabinets, and drawers, turn on faucets, unlock doors, and unscrew lids. They love to take things apart, shred them, and so forth, and they are not good at learning that there are some things—the contents of one's pantry, one's clothing, one's tax returns, one's curtains and furniture—that are off limits. They have been known to tear out drywall, disassemble screens, and unlock doors. Nonhuman primates are tractable only when they are young. As adults, while they might be quite capable of working out what we find acceptable, most have neither any particular interest in constraining their conduct accordingly nor the capacity to do so consistently.[17] If this meant only that they were not constantly attentive to our needs, it would not be a serious problem. But what it actually means is that no amount of training and no amount of love will prevent a nonhuman primate from attacking humans, possibly maiming or killing them, or from destroying their owners' homes and possessions.

Some people might be happy living with pets who are both aggressive and destructive, but most are not. For this reason, nonhuman primates who are kept as pets often end up in cages. A cage is a terrible place for an intelligent, complicated, social being to live, especially if it is kept alone. Moreover, most nonhuman primates live for fifteen to thirty years, and chimpanzees in captivity can live to be sixty. That is a very long time to spend in a cage. For human primates, imprisonment is a punishment appropriate only for serious crimes. Nonhuman primates end up being imprisoned simply because of the combination of their natural instincts and the thoughtlessness of the humans who adopt them.

Keeping wild animals as pets is, in most cases, a serious mistake, and one that people most often make not because they consider the costs in advance and accept them, but because they do not stop to think about what they are getting into. A news story about the sale of tiger cubs in a WalMart parking lot quotes the facilities director of a local zoo:

> "There are thousands of large cats including tigers, leopards, and lions owned privately—and legally—in Texas," Stones said. He said he thinks some tiger owners may not realize the effort that goes into caring for the cats. "They buy them as babies," Stones said. "They don't realize it's going to get to be hundreds of pounds, eat an awful lot of food and become dangerous."[18]

It does not take a great deal of thought to realize that tiger cubs grow up to be large and dangerous, and it is astonishing that there are people who are willing to adopt wild animals without being willing to devote even that minimal amount of thought to their decision. The animals they adopt normally pay a much heavier price for their owners' thoughtlessness than their owners do. With the exception of those disabled humans who have capuchin monkeys as helpers,[19] private individuals have no compelling need to adopt members of any non-domesticated species. Given the consequences for the animals they adopt, it is wrong to do so.

While wild animals are particularly unsuited to life in a human household, it is important to consider our willingness to meet the needs of any nonhuman animal before we take that animal into our homes. Consider Ernie, a golden retriever described in an essay by Jon Katz.[20] Ernie was given as a birthday present to a child, Danielle, who lost interest in him after a few weeks. Danielle's parents did not normally get home from work until after 7 P.M. They had a housekeeper who did not like dogs in general, or Ernie in particular:

> Because nobody was home during the day, he wasn't housebroken for nearly two months and even then, not completely. No single person was responsible for him; nobody had the time, will, or skill to train him. As he went through the normal stages

of retriever development—teething, mouthing, racing frantically around the house, peeing when excited, offering items the family didn't want retrieved, eating strange objects and then vomiting them up—the casualties mounted. Rugs got stained, shoes chewed, mail devoured, table legs gnawed. The family rejected the use of a crate or kennel—a valuable calming tool for young and energetic dogs—as cruel. Instead, they let the puppy get into all sorts of trouble, then scolded and resented him for it. He was "hyper," they complained, "wild," "rambunctious." The notion of him as annoying and difficult became fixed in their minds; perhaps in his as well.

A practiced trainer would have seen, instead, a golden retriever that was confused, under-exercised, and untrained—an ironic fate for a dog bred for centuries to be calm and responsive to humans.

Ernie did not attach to anybody in particular—an essential element in training a dog. Because he never quite understood the rules, he became increasingly anxious. He was reprimanded constantly for jumping on residents and visitors, for pulling and jerking on the leash when walked. Increasingly, he was isolated when company came or the family was gathered. He was big enough to drag Danielle into the street by now, so her parents and the housekeeper reluctantly took over. His walks grew brief: outside, down the block until he did his business, then home. He never got to run much.

Complaining that he was out of control, the family tried fencing the back yard and putting Ernie outside during meals to keep him from bothering them. The nanny stuck him there most of the day as well, because he messed up the house. Allowed inside at night, he was largely confined to the kitchen, sealed off by child gates.

Danielle's parents might have done any number of things that would have made this story turn out differently. They could have thought seriously about whether their daughter was likely to lose interest in her puppy. They could have made getting a puppy conditional on their daughter's taking him to obedience class, which would have helped not only with Ernie's socialization and training but with their daughter's understanding of the responsibilities of owning a dog. They could have, and should have, come up with a backup plan in case their daughter could not be persuaded to take a lasting interest in her dog. The family in question seems to have had enough money that they could have hired a trainer to work with Danielle and Ernie when things began to go wrong. If they were not willing to do any of these things, they could have simply gotten a kitten, or no pet at all. Because they were not willing to think through what owning a dog entails, and whether they and their daughter were willing to accept the responsibility of ensuring that Ernie had a decent life, Ernie ended up confused, neglected, alone, and unhappy; and the family ended up with a dog they do not seem to want, and are unwilling to care for. This story is tragic, all the more so since it could so easily have been avoided.

A decent person will avoid these problems. She will think before she adopts a pet. She will consider whether she is able and willing to meet that pet's needs, and will adopt a pet only when she is confident that she can do so. In the case of dogs, this will involve considering not only the normal needs of dogs in general, but those of the particular breed she is considering adopting.[21] She will also consider the source from which she plans to acquire it. One option is to adopt from shelters.

Every year, 5–8 million pets are estimated to enter shelters in the United States, of whom 3–4 million are euthanized.[22] The ASPCA estimates that half of dogs euthanized in shelters, and seven out of ten cats, are killed "simply because there is no one to adopt them."[23] With so many pets in need of homes, there is virtually never a good reason to support puppy mills[24] or others who breed animals in inhumane conditions.

Those who want to adopt a purebred dog and want to purchase one from a breeder, rather than acquiring one from a breed rescue group, should consider not just the conditions in which the breeder keeps dogs but the breed itself, since some breeds are prone to serious health conditions that not only require considerable veterinary care but compromise the dogs' quality of life. Because they have been bred to have very short muzzles and flat faces, English bulldogs often suffer from serious respiratory problems, including sleep apnea.[25] Cavalier King Charles Spaniels' skulls are too small for their brains, and as a result, a number of them develop syringomyelia, a neurological disorder one of whose symptoms is constant pain.[26] No one who cares about nonhuman animals should support the continued breeding of animals whose health is needlessly compromised, or the breed standards that too often encourage this.

TEACHING PETS WHAT THEY NEED TO KNOW

Once one has adopted a pet, one is responsible for providing for that pet's basic needs. These include not just food, shelter, and medical care, but attention, affection, and, in the case of dogs, exercise. Moreover, anyone who is not willing to commit to finding good homes[27] for all her pet's progeny should have her pet spayed or neutered. For most people, providing for these needs is not onerous. Those who suspect that they might be unable or unwilling to provide for them should not adopt pets.

In addition, dogs need to be trained. Some rules are important both for the dog's safety and for the safety of others. Dogs need to be taught not to attack or bite others, not to run headlong into traffic, and so forth. The need for this kind of training is obvious: it is irresponsible to own a pet who is a danger to others, and it is no kindness to a dog to allow her to behave in ways that risk her death or injury. Dogs also need to be taught how to function in human society. For instance, they should be taught not to eat off people's plates or jump up on people, and that there are some objects that belong to others and are not to be used as chew toys. Dogs, like humans, are social animals who can understand the emotions of those around them. They do not enjoy being disliked, especially when they do not know what they are doing wrong. We teach our children not to do some things that they might be inclined to do, like smearing food all over themselves or others. If we did not teach our children not to do so, other people would not want to be around them, and they would be either very lonely, very unhappy, or both. The same is true of dogs.

Teaching a dog, like teaching a child, need not involve the exercise of power, or even what would normally be called "training." Barbara Smuts describes her relationship with her dog Safi:

> Because I had so much respect for her intelligence, I did not consider it necessary to "train" her. Instead, I discuss all important matters with her, in English, repeating phrases and sentences over and over in particular circumstances to facilitate her ability to learn my language. She understands (in the sense of responding appropriately to) many English phrases, and she, in turn, has patiently taught me to understand her language of gestures and postures (she rarely uses vocal communication)....
>
> Early in our relationship, we came upon several deer about a hundred yards away grazing in an open field....Safi leapt forward (she was not on a leash). I said, without raising my voice, "No, Safi, don't chase." To my amazement, she stopped in her tracks. Thus I learned that I could communicate prohibitions without yelling or punishing her.[28]

The most basic function of training is to enable us to tell dogs not to do something when it is very important that they not do it, to teach them to avoid behavior that is dangerous to themselves or to others, and to teach them how to function in human society. If we can achieve these ends without explicit training, well and good. But this is not always possible, and when it is not, we owe it to our dogs to train them.

Training a dog involves exercising power over her. Power can always be abused, and we owe it to our dogs not to abuse it. We might respond to the possibility of abuse in two ways: either by declining to exercise control over dogs or by exercising it with as much wisdom and humanity as we can muster. It would be unfair to our dogs to choose the first option. We know about dangers that dogs do not understand, like cars. It would be irresponsible not to teach our dogs to avoid them, and, when necessary, to obey us when we tell them to do something like not chasing a rabbit across the street, even if they do not understand why. Likewise, it would be irresponsible not to train our dogs not to harm others.

The proper response to the possibility of abusing our power is not to refuse to exercise it, but to try to exercise it responsibly and humanely. We should never be cruel to our dogs or ask them to do things they cannot do, and we should always have respect for their nature and capacities. We should also allow them to teach us what they need, as Smuts did with Safi: "She has taught me that I must not clean the mud off her delicate tummy area with anything but the softest cloth and the tenderest touch. She has made it clear that stepping over her while she is asleep makes her extremely uncomfortable, and so I never do it."[29] But no combination of respect, mutuality, and responsibility can eliminate the need to teach our dogs what they need to know to live a safe and decent life; and if this requires using our power over them, then while we should take care never to abuse that power, we should not decline to use it because we mistakenly think that we are being kind to our dog.

Discomfort with training might also reflect the idea that we should let dogs behave naturally. The idea that this precludes training them reflects confusion about what is natural, and about the relation of nature to culture in the lives of social animals like humans and dogs. We do not discover what is natural for a human child by dropping him into the wilderness to raise himself, free from the distorting influences of parenting and socialization. We do not believe that it would be "more natural" for children to exist without being able to speak, even though, left entirely to themselves, children would not learn to use language. Humans are social animals, and it is not natural for us to grow up without parental guidance and teaching.

Dogs are also social animals. Had we not domesticated them, they would have been socialized by their parents and by the members of their pack. But, for better or for worse, we have domesticated them and brought them into human society. Our homes are their natural environment, we are their pack, and we are responsible for their socialization. As pack animals, they need society, and function best when they understand what others expect of them. Our expectations should never be unreasonable, but they do exist, and we do not respect dogs by failing to communicate them.

On Some Serious Problems in the Human-Pet Relationship

In some cases, it can seem difficult or impossible to construct a mutually satisfying relationship with a particular pet. Our first response ought to be to try to ask whether our expectations are unreasonable in some way: if a dog owner does not like taking her dog for walks, that is her problem, and one she should have thought of before getting a dog. If our expectations are not unreasonable—if, say, the problem is that a dog is aggressive—then we should try to change her behavior. In so doing, we should also take seriously the possibility that we are at fault: that our pet's behavior reflects some confusion or incoherence in our expectations or training that we can and should fix. If necessary, we should consult others, and ask explicitly whether there is something wrong with the way we relate to our pet.

Sometimes, however, these efforts do not work. And while it might or might not be true that there are no bad pets, just bad owners, it is certainly true that there are some pets whom the particular person who owns them cannot train, even if she tries, and even if her pet is a member of a species that is normally trainable. Some pets are very difficult, some people are not very good with nonhuman animals, and sometimes the particular difficulties of a pet meet the particular imperfections of an owner in such a way that even if *someone* might be able to train that pet, her owner is not that person. And while it is sometimes possible to find that person—to hire a trainer who can help bridge the gap between an owner's capacities and a pet's temperament—sometimes it is not.

Ideally, one might find another home for one's pet, but this is not always possible, especially not if one is honest with any prospective future owners about one's pet's behavioral problems. When it is not, a pet owner must choose between learning to live with her pet's problematic behavior and giving that pet up to a shelter, where she will probably either live out her life in a cage or be put to death. If one's pet is a danger to others, then the issues are straightforward.[30] One might try to keep one's pet from harming others, but no human contrivance is infallible. Fences break. No one closes all the doors between their pets and the outside world every single time. If one's pet is a danger to others, one should not trust one's own efforts to keep that danger from being realized. If the owner of a dangerous pet has tried and failed to change her pet's behavior, she should put her pet down.

This conclusion does not depend on the claim that humans' interests matter more than those of nonhuman animals. In the world as we know it, nonhuman animals who pose a danger to humans are put down with or without their owner's consent. If a pet owner has tried and failed to train her pet not to attack people, the question whether to euthanize her pet amounts to the question whether to do so before or after her pet attacks someone. Moreover, a vicious pet's owner cannot know how many people her pet might injure or kill over the course of her life. Vicious pets do not have a fixed quantity of attacks in them, which they can simply get out of their system and be done with. For this reason, the justification for killing them cannot be that we know that that the pet will kill exactly one human being, but that we value that human's life more than the pet's. It is the fact that vicious animals are a standing danger to others and will likely continue to be so.

Suppose, however, that a person's pet behaves in a way that she finds it difficult to live with, but that does not endanger others; that she is not unreasonable to find her pet's behavior difficult; that she has tried and failed to get her pet to stop behaving in this way; and that she cannot find another home for her pet. In such a case, matters are more complicated. On the one hand, a pet owner who finds herself in this unfortunate situation might try to weigh her own interest in living without that pet's problematic behavior against her pet's interest in staying alive.[31] When she does so, her own interests might well prevail. While I cannot defend an account of nonhuman animals' interest in life within the limits of this paper, it plainly differs from competent humans' interest in life. Humans are capable of autonomous action; those nonhuman animals that we know of are not.[32] We therefore have a right to self-government that they cannot share, a right that is violated when we are killed. Moreover, we have interests in the future of a kind that nonhuman animals lack.[33] For instance, a human might want to stay alive in order to finish some project that is important to her, or to witness some future event, like the birth of a grandchild. Nonhuman animals have no such projects, and cannot anticipate the future in this way. They therefore lack these sorts of interests in staying alive. When a nonhuman animal is killed painlessly, she is robbed of her future and of any happiness that future might contain. But since she has neither the capacity for autonomy nor specific future-oriented goals, her capacity for self-government cannot be violated in ways that presuppose these things.[34]

For this reason, we think it appropriate to arrange a painless death for our pets when their lives promise more unhappiness than happiness: for example, when they are terminally ill in a way that will prevent them from enjoying what life remains to them. When the light has gone out of their eyes and the activities that once made them happy no longer attract them, when they seem to be enduring, rather than enjoying, their lives, and when we know that this condition will not pass and cannot be cured, we think it is an act of kindness to arrange a painless death for them. If nonhuman animals' interests in staying alive are based on their interest in whatever future happiness they might enjoy, then it must be possible, in principle, that that interest might be outweighed when the only alternative is even greater unhappiness (or the sacrifice of some comparably important interest) for someone else. In particular, there must, in principle, be some amount of unhappiness that a pet might cause her owner such that it is greater than the happiness that that pet would enjoy were she to continue to live.

Some people are too quick to conclude that they have reached this point. They give up their pets, or have them killed, even when their pets' behavior would not require much adjustment at all: when they no longer want to spend the time it takes to walk a dog, for instance. This is a terrible way to treat anyone, let alone a companion and a member of one's household. We should stick by our friends, of whatever species, and when their needs are reasonable, we should accommodate them. But the fact that some people are too quick to conclude that their interests outweigh their pets' does not mean that that conclusion is never justified.

On the other hand, there are some reasons to question whether the fact, when it is one, that a pet owner would sacrifice more by keeping her pet alive than her pet would sacrifice by being painlessly put to death should determine what her owner ought to do. First, the fact that we can explain ourselves to one another makes it much easier to set limits on our relationships, as does the fact that we can end a friendship with a human, or ask her to leave our home, without thereby condemning her to death. When we need to, we can explain to one another what kinds of behavior we find intolerable, and let any friends who violate those expectations without a good reason decide whether to alter their behavior or give up our friendship. Obviously, we cannot similarly explain ourselves to nonhuman animals.

Likewise, pets are not morally responsible for what they do. Humans' capacity to think about what they are doing, and to make autonomous choices, not only limits what we can do to them; it limits what we must accept from them. A competent adult human who destroys our possessions without some very good reason either knows or ought to know[35] that she is doing something that we will probably find unacceptable, and she is responsible for her conduct. A dog who routinely destroys our possessions is not. Worse still, it is hard to rule out the possibility that we contributed to our pets' misconduct: that it reflects some failure in our training, that there was something we could have done to prevent it, had we had the wit to see it, or, as in the case of wild animals, that the problem is one that we should have foreseen.[36]

Because we can explain ourselves to other people, we have a kind of safety valve in dealing with other competent adults. We need never impose limits on the kinds of behavior we are prepared to tolerate from them without letting them know what those limits are, and giving them a chance to convince us that we are being selfish or unreasonable. When we decide that we cannot tolerate their conduct, it need never be because they cannot understand what we expect of them. We can always give competent adult humans a choice about whether to abide by limitations on our relationship with them, and just as we are responsible for imposing those limitations, they are responsible for their decisions about whether to abide by them. This means that when we end a friendship, we never have to act unilaterally. We are never forced to decide whether to accept someone's behavior without allowing that person a say in the matter, or to cut off a friendship for reasons that person does not and cannot comprehend. This is true precisely because we can explain ourselves to one another, and because others can understand what we say and decide for themselves how to respond.

In our dealings with our pets, we can hope that our interests and theirs will generally coincide, and that any conflicts will be minor and easily resolved. But while things often work out this way, they do not always do so. When our interests and those of our pets are in serious and irresolvable conflict, it is up to us to decide what to do, since we can understand all the competing considerations and weigh them against one another, and our pets cannot. We must make this decision without being able to explain ourselves to them, and without the comfortable knowledge that they understand what we ask of them and why we ask it, and have chosen not to accede. To make a decision that will probably lead to their death under these conditions is, and ought to be, very difficult.

One might think that making this kind of unilateral decision should be not just difficult, but impossible: that because we must make unilateral decisions when dealing with our pets, and because we can never secure their consent to those decisions, we can never justifiably impose uncompensated sacrifices on them, whatever the cost to us. This cannot be right. As Elizabeth Anderson writes: "To bind oneself to respect the putative rights of creatures incapable of reciprocity threatens to subsume moral agents to intolerable conditions, slavery, or even self-immolation. As it cannot be reasonable to demand this of any autonomous agent, it cannot be reasonable to demand that they recognize such rights."[37]

If it is wrong to unilaterally harm someone who cannot understand why we impose that harm, or what she might have done to avoid it, then it is wrong to defend ourselves against nonhuman animals when they try to kill us, at least when defending ourselves involves inflicting even non-lethal harm on them. It would likewise be wrong to defend ourselves against such lesser harms as the loss of a limb or the destruction of our home. Should it turn out that termites are sentient, we might hope to lure them away from our homes by building even more enticing wooden structures for them to feast on, but it would be wrong to kill them. As Anderson writes, it cannot be reasonable to accept this kind of limitless vulnerability to nonhuman animals.

One might try to avoid these counterintuitive conclusions by holding that it is permissible to unilaterally impose harms on nonhuman animals only when doing so is the only way to avoid an even greater harm to ourselves or to another. This is much more plausible[38] than the view that we can never unilaterally impose sacrifices on nonhuman animals for any reason. However, it would imply that we can permissibly give a nonhuman animal to a shelter that will probably kill her when keeping that animal would impose a greater harm on us than a painless death would impose on her, at least in the absence of any other reason to think that it would be wrong to do so.[39]

The claim that we cannot unilaterally impose significant sacrifices on beings who cannot understand why we impose them applies to all nonhuman animals. This universal scope is one reason why it is so implausible. But pets are not just any nonhuman animals; they are nonhuman animals whom we have voluntarily taken into our homes and made members of our households. Pet owners choose to adopt pets and to assume responsibility for their care. While most people do not knowingly adopt pets who will be hard to live with, they either know or should know that no pet's behavior is fully predictable. If their pets turn out to be difficult to live with, that fact is not a piece of random bad fortune; it is a risk one accepts when one adopts a pet.

The fact that pet owners choose to assume responsibility for their pets means that they should be willing to accept significant sacrifices in order to live up to those responsibilities. But unless one thinks that unforeseen circumstances can never excuse one from meeting an obligation one has voluntarily assumed, there must be *some* things a pet could do, other than posing a genuine danger to her owner or to others, that her owner would not be obligated to accept. Suppose, for instance, that a pet somehow prevents her owner from sleeping for more than an hour, whether by howling, jumping on her owner's face, or some other means, and that nothing her owner does—training, medical examinations, more attention and exercise— keeps her from doing this. Suppose that while this is not actually dangerous, it is extremely unpleasant: certainly not what one would call a mere inconvenience. Suppose further that while her owner knew that some pets are hard to live with, and was prepared to work hard to make her relationship with her pet a good one, when she adopted this pet, she had no reason to anticipate a problem of this magnitude. In that case, one might think, that pet's owner might be justified in concluding that her responsibility to her pet does not require that she simply accept the fact that she will not be able to sleep for more than an hour for the foreseeable future.

Finally, while pets are not always capable of understanding our expectations, let alone adapting to them, we can almost always adapt to our pets' behavior. It is not pleasant to learn to live with a dog who makes dens in the sofa stuffing and digs through the floorboards, or a cat who urinates everywhere except the litter box, but it is not impossible. Doing so might require that we sacrifice our own interests, but our relationship to our own interests is different from our relationship to the interests of others, whether human or nonhuman. We can decide not to take the interests of others into account when they are not relevant to a particular decision, as when we do not consider how much a student wants to get a good grade in evaluating her work, but we cannot decide not to take the interests of others into account when they are relevant. Our own interests are different: precisely because our interests are *ours*, we can decide to waive our right to have them considered even in situations to which they are relevant. Since it is up to us to decide how to respond to our pet's behavior, and since we can always adjust, somehow or other, to a pet who is not actually dangerous, we can always decide not to consider our own interests at all: to find some way of living with a nonhuman animal who requires a level of sacrifice we would regard as intolerable if it were imposed on us by another adult human.[40]

In our dealings with adult humans, we normally decide not to take our own interests into account[41] in two kinds of cases. First, we can set our own interests aside when the stakes are low, as when we decide not to bother to calculate each person's share of a restaurant bill even when we suspect that our share is the smallest. By definition, when we must decide whether to accept a major sacrifice in order to prevent a pet's death, the stakes are not low. Second, we can choose to set our own interests aside in exceptional situations, like emergencies. When our friends are in trouble, we should try to help, without considering the inconvenience to ourselves. Emergencies can last for a long time, and they can occur in succession. For this reason, a person can have reason to set aside her own interests in her dealings with others not just occasionally, but for years. Nonetheless, in such cases setting aside one's own interests should be a response to an exceptional situation, or to a series of them.

If someone asks us to set our own interests aside not occasionally, or because of an unforeseeable emergency, but consistently, she treats us as though we were her slaves, whose interests count for nothing. It would not be in her interest that we accede to her demands—no one has the right to expect that others should make themselves her slaves, and it is no favor to such people to encourage such expectations. Nor are we under any obligation to accede to them: relationships between competent adults ought to be reciprocal, and a relationship with someone who treats us in this way is not. Any competent adult who asks us to treat her in this way can be expected to know better; thus, the unreasonableness of her demands allows us to attribute an attitude to her that itself relieves us of any obligation to meet them. Because we cannot expect nonhuman animals to understand that they are asking us to set our own interests aside, or to see why that demand is unreasonable, we cannot attribute any such attitude to them.

Nor is our relationship to pets, in this respect, like our relationship to infants and young children, for whose sake we often set our own interests aside. The need to set our own interests aside when dealing with young children is comprehensible, temporary, and fully predictable. We know why we sometimes need to set our own interests aside in dealing with small children, and we expect that they will eventually outgrow the need for this kind of care. In this case, our relationships are not permanently defined by a unilateral sacrifice of interests, as they can be in a relationship with a pet.

To set one's interests aside when the stakes are high is an act of generosity, but it cannot be morally required of us. If it were, the fact that we are capable of understanding the trade-off between our interests and those of a nonhuman animal, and of setting our own interests aside, would itself mean that we could not assert our own interests at all in any case in which our interests conflict with theirs. It is one thing to say that we should not sacrifice a nonhuman animal's interests for our own lightly or quickly, or that we should be prepared to be generous above and beyond the call of duty. It is another to say that even when our interests outweigh theirs, the fact that we can abdicate our interests while they cannot abdicate theirs implies that we should always do so.

On those rare occasions when living with a particular pet involves significant sacrifice, but not a danger to others, when we cannot change that pet's behavior, and when we can find no other home for her, we have to decide whether to give that pet up to a shelter where she will probably be killed, or at best live out her life in a cage. Because we cannot explain ourselves to our pets and have voluntarily assumed responsibility for them, and because we owe loyalty to our friends and companions, we should be reluctant to give them up, and should not do so for the sake of mere convenience, or to avoid a sacrifice we could easily live with. But while we should think hard before giving a pet up, and should try as hard as we can to find other alternatives, there are situations in which we can legitimately decide that the sacrifice a pet imposes on us is too great.

CONCLUSION

If the observations and arguments I have made are correct, it is a mistake to think of our duties to nonhuman animals as derived from a sort of hierarchy of moral considerability in which animals with greater capacities occupy a higher rank. Sometimes higher capacities do translate into greater rights: for instance, those nonhuman animals with whom we are familiar do not have a right to autonomy, since they are not capable of autonomous action. Sometimes, however, the fact that an animal does not have some capacity simply means that we owe that animal more consideration or forbearance than we would otherwise. And the fact that we are capable of assessing our reasons for action, and of weighing others' interests against our own, is not just a prerequisite for having moral obligations. It imposes on us obligations that go beyond strict fairness, and in which this capacity places us at a disadvantage compared to beings who lack it. Because we can sometimes see that a sacrifice on our part could allow us to work out a decent relationship with a pet, while pets normally cannot understand what sacrifices on their part might do the same, we require such sacrifices of ourselves, but not of them.

Fortunately, if we are willing to think before we acquire a pet and to meet our pets' needs conscientiously, including their needs for attention, affection, and training, we will rarely encounter a need for serious sacrifice. For the most part, owning pets involves not serious moral dilemmas but minor inconveniences set against a background of wonder, delight, and the joy of opening our hearts to animals who are so willing to open theirs to us.

NOTES

1. Some animal rights activists object to the use of the word "pet," on the grounds that it is demeaning. I do not share this view, and will therefore continue to use this term. Some of those who object to the word "pet" also object to describing the humans who live with pets as their "owners," and prefer the word "caregiver" or "guardian." Cf. Carrie Allan, "What's in a Word? (And Does it Matter?)," *Animal Sheltering* (Nov.–Dec. 2000), http://www.animalsheltering.org/resource_library/magazine_articles/nov_dec_2000/whats_in_a_word.html (accessed January 1, 2010). In this case, the allegedly objectionable term and its proposed replacement are not interchangeable. As a matter of law, humans can and do own nonhuman animals. Replacing "pet owner" with "companion animal caregiver" does not change that fact. But while it is accurate to describe people as "owning" pets, it is not accurate to describe all pet owners as either caregivers or guardians: some pet owners are abusive, and do not provide their pets with anything worthy of the terms "care'" or "guardianship." Protesting animals' legal status as property with this terminological change is, in my opinion, as misguided as protesting the existence of slavery by replacing the term "slave owner" with "slave caregiver": it obscures the issues it is meant to ameliorate.

2. I will not consider whether our duties to animals are direct or indirect—that is, whether (for instance) we should not gratuitously harm animals because we owe it to them not to do so, or because those character traits that would allow us to gratuitously harm a being who is under our care are wrong in their own right, and might make us more likely to harm others to whom we do have direct duties, like other humans. While these two views have different implications in some situations (e.g., those in which I can be certain that any defects in my character will never affect another human being), under normal circumstances one's views on this question need not affect the question of how we ought to treat animals, especially given the further assumption that one of the best ways to develop some character trait is to act as one would if one had it.

3. Vicki Hearne, *Adam's Task* (New York: Alfred A. Knopf, 1986), pp. 53–54.

4. Vicki Hearne, *Animal Happiness* (New York: HarperCollins, 1994), pp. 225–26.

5. Hearne, *Adam's Task*, p. 36.

6. Hearne, *Adam's Task*, p. 39. These precautions are not unreasonable: chimpanzees can kill humans.

7. Gary Francione, *Introduction to Animal Rights: Your Child or the Dog?* (Philadelphia: Temple University Press, 2000), p. 77.

8. Francione, *Introduction to Animal Rights*, p. 170.

9. One alternative is described in David Favre, "A New Property Status for Animals: Equitable Self-Ownership," in *Animal Rights: Current Debates and New Directions*, ed. Cass Sunstein and Martha Nussbaum (Oxford: Oxford University Press, 2004).

10. People for the Ethical Treatment of Animals, "Doing What's Best For Our Companion Animals," http://www.helpinganimals.com/factsheet/files/FactsheetDisplay.asp?ID=133 (accessed January 1, 2010).

11. Humane Society of the United States, "Feral Cats: Frequently Asked Questions," http://www.humanesociety.org/issues/feral_cats/qa/feral_cat_FAQs.html (accessed January 1, 2010).

12. Ibid., http://www.aspca.org/adoption/feral-cats-faq.html (accessed January 1, 2010).

13. Ibid., http://www.aspca.org/adoption/feral-cats-faq.html (accessed January 1, 2010).

14. People for the Ethical Treatment of Animals, "Doing What's Best For Our Companion Animals," http://www.helpinganimals.com/factsheet/files/FactsheetDisplay.asp?ID=133 (accessed January 1, 2010).

15. The happiness of the people who take non-domesticated animals as pets might be thought not to matter: they made their bed, one might think, and they should lie in it. Whatever one makes of this argument, it works only if the unhappiness of those people does not lead to unacceptable consequences for the animal they take in. But this is plainly false: an animal who attacks humans, for instance, is likely to end up being caged or euthanized, as is an animal who destroys her owners' home.

16. *New York Times*, "Primates Aren't Pets," http://www.nytimes.com/2009/02/25/opinion/25wed4.html (accessed January 1, 2010).

17. Capuchin monkeys trained to work with the disabled are an interesting exception.

18. "Police Investigate Sale of Tigers in WalMart Parking Lot," *The Monitor* (McAllen Tex.), June 16, 2008, http://www.themonitor.com/articles/ones-13216–cubs-selling.html (accessed January 1, 2010).

19. http://www.monkeyhelpers.org/

20. Jon Katz, "Poor Little Rich Dog," *Slate*, July 19, 2004, http://www.slate.com/id/2103801/pagenum/all/ (accessed January 1, 2010).

21. Considering the needs of particular breeds is especially important in the case of dogs, since the requirements of different breeds of dogs vary greatly. For instance, whenever another iteration of Disney's *101 Dalmatians* is released, shelter workers brace themselves for a wave of Dalmatians abandoned by people who bought them on impulse, without bothering to find out what kind of home a Dalmatian needs. Had they inquired, these people would have learned that Dalmatians require a great deal of exercise, having been bred to run alongside coaches on long journeys. When they do not get the exercise they need, they can become destructive or neurotic. They make wonderful pets for people who are prepared to walk a mile or more every day with them. Others, however, would be much better off with a dog of a different breed.

22. The American Society for the Prevention of Cruelty to Animals estimates that 5–7 million pets enter shelters every year (ASPCA, "Pet Statistics," http://www.aspca.org/about-us/faq/pet-statistics.html [accessed January 1, 2010]). The Humane Society puts the number at 6–8 million (Humane Society of the United States, "HSUS Pet Overpopulation Statistics," http://www.humanesociety.org/issues/pet_overpopulation/facts/overpopulation_estimates.html [accessed January 1, 2010]). Both estimate the number of pets euthanized at 3–4 million.

23. ASPCA, "Pet Statistics," http://www.aspca.org/about-us/faq/pet-statistics.html (accessed January 1, 2010).

24. ASPCA, "What Is A Puppy Mill?" http://www.aspca.org/fight-animal-cruelty/puppy-mills/what-is-a-puppy-mill.html (accessed January 1, 2010).

25. See J. C. Hendricks, L. R. Kline, R. J. Kovalski, J. A. O'Brien, A. R. Morrison, and A. I. Pack, "The English Bulldog: A Natural Model of Sleep-disordered Breathing," *Journal of Applied Physiology* 63 (October 1987): 1344–50.

26. See Clare Rusbridge, "Chiari-like Malformataion and Syringomyelia in Cavalier King Charles Spaniels," PhD thesis, 2007, available at http://igitur-archive.library.uu.

nl/dissertations/2007–0320–201201/index.htm (accessed January 1, 2010); House of Lords, *Fixing Ancestral Problems: Genetics and Welfare in Companion Animals Focusing on Syringomyelia in Cavalier King Charles Spaniels as an Example* (Report of the Companion Animal Welfare Council Workshop, April 29, 2008), available at http://www.cawc.org.uk/reports (accessed January 1, 2010).

27. The good home in question cannot always be one's own. Cats and dogs can have large numbers of kittens and puppies over the course of their lives, too many to be well taken care of within one home.

28. Barbara Smuts, "Reflections," in J. M. Coetzee, *The Lives of Animals* (Princeton, N.J.: Princeton University Press, 1999), pp. 115–16.

29. Smuts, "Reflections," p. 117.

30. I do not mean to suggest that the decision to put down any dog is easy, just that the issues involved are not complex.

31. This kind of decision is one we rarely face in dealing with humans. Competent adult humans are capable of living independently, and in developed countries children and the incompetent can generally find some accommodation. If the only such shelter available to them is, say, a home in which the elderly or disabled are warehoused, their happiness might depend on our willingness to make significant sacrifices for their sake, but their lives normally do not. We are therefore almost never in the position of having to choose between their lives and our happiness, as we can be with pets.

32. In what follows, I will omit this qualification for reasons of simplicity. My arguments refer only to those nonhuman species we know of, none of whose members have the capacity for autonomous action. They are not meant to apply to intelligent aliens and the like, whom it would be wrong to keep as pets.

33. Those nonhuman animals who might be exceptions to this claim are either impossible to keep as pets (e.g., dolphins) or dangerous (e.g., chimpanzees).

34. Some humans also lack these capacities. Small children and the demented differ from nonhuman animals in that they either will have or have had a capacity for autonomy or self-governance. However, some seriously cognitively impaired humans never have that capacity. The arguments sketched above do not imply that it would be permissible to give them up to some shelter that would probably kill them, even if such shelters existed. Even if one thinks that the most fundamental reason why killing humans is wrong derives from our right to self-government, there are reasons why one might not think that it should be permissible to kill humans who lack this capacity. The most obvious are slippery-slope arguments about the effects of holding that there are some human beings that we can permissibly kill without their consent and in the absence of excusing conditions like self-defense. Assuming for the sake of argument that there are such humans, one might think, human history does not inspire confidence in our ability to draw the line that separates those humans from others. Given the stakes, we should try to find bright lines such that all beings whom it would be wrong to kill fall on one side of them, so that we do not have to rely on that ability. The line separating humans from nonhumans is such a line; the line between humans with at least a rudimentary capacity for self-government and humans without one is not.

35. In the rare cases in which this is false—for instance, someone from a very different culture in which it is, oddly enough, perfectly acceptable to destroy other people's possessions—we can generally explain things to the person in question, and once we have done so, we can expect her to respect our wishes or forfeit our friendship.

36. We should generally accommodate ourselves to foreseeable problems involving domesticated animals (e.g., a cat who sheds, a dog who needs companionship) unless we

can find another home for them, since such problems are not normally serious enough to warrant giving up a pet to a shelter. One exception might be people who knowingly adopt pets to whom they are severely allergic.

37. Elizabeth Anderson, "Animal Rights and the Values of Nonhuman Life," in *Animal Rights: Current Debates and New Directions*, ed. Cass Sunstein and Martha Nussbaum (Oxford: Oxford University Press, 2004), pp. 287–88.

38. That this view is much more plausible than the one considered earlier does not mean that it is correct. Anderson argues that it is not: that we can justifiably impose harms on animals that are greater than the sacrifices we would have to accept if we did not act. She considers some species of mice and rats who cannot survive outside human dwellings, and argues that we can justifiably exterminate them or remove them from our homes. She then notes: "It could be argued that in such cases, the interests of humans simply outweigh the interests of vermin. But this thought is hard to credit. Except in plague conditions, most vermin do not threaten to kill us. What are rat feces in the bedroom to us, compared to a painful death for the rat?" (Anderson, "Animal Rights and the Values of Nonhuman Life," p. 288.)

39. One might argue that the loss of her life is a greater harm to an animal than any amount of unhappiness is to a human, and thus that only threats to human life can justify giving a nonhuman animal to a shelter that will probably kill her. However, one would then have to show that an animal that lacks autonomy has an interest in her life that goes beyond her interest in the happiness her future life would contain.

40. The fact that humans can voluntarily decide to set their own interests aside affects many moral choices involving humans and animals in ways that are sometimes not taken into account in the literature on animal rights. For instance, Tom Regan considers four humans and a dog in a lifeboat that can only hold four: "One must be thrown overboard or else all will perish. Whom should it be? If all have an equal right to be treated respectfully, must we draw straws?" Tom Regan, *The Case for Animal Rights* (Berkeley and Los Angeles: University of California Press, 1983), p. 285. It is striking that Regan never considers the possibility that one of the humans should choose to jump overboard, thereby eliminating the need to throw anyone overboard against her will.

41. Sometimes we fail to consider our own interests not because we set them aside, but because we think they are obviously outweighed. Thus, when someone is drowning, the reason not to stop and weigh that person's life against the lovely silk dress that I will ruin if I try to save her is not that I set my own interests aside; it is that her life so obviously outweighs my dress that stopping to consider the matter when every second counts would be absurd. Those cases in which we set aside our own interests to help others are those in which our interests might actually outweigh theirs, were we to compare them, but we choose not to take them into account.

SUGGESTED READING

ANDERSON, ELIZABETH. "Animal Rights and the Value of Nonhuman Life." In *Animal Rights*, edited by Cass R. Sunstein and Martha C. Nussbaum. Oxford: Oxford University Press, 2004.

BEAUCHAMP, TOM L., F. BARBARA ORLANS, REBECCA DRESSER, DAVID B. MORTON, and JOHN P. GLUCK, eds. *The Human Use of Animals: Case Studies in Ethical Choice*. 2nd ed. New York: Oxford University Press, 2008.

Burgess-Jackson, Keith. "Doing Right by Our Companion Animals." *Journal of Ethics* 2 (1998): 159–85.

Clark, Stephen R. L. *Animals and Their Moral Standing.* London: Routledge, 1997.

DeGrazia, David. *Animal Rights: A Very Short Introduction.* Oxford: Oxford University Press, 2002.

Diamond, Cora. "Eating Meat and Eating People." In *Animal Rights,* edited by Cass R. Sunstein and Martha C. Nussbaum. Oxford: Oxford University Press, 2004.

Francione, Gary L. *Introduction to Animal Rights: Your Child or the Dog?* Philadelphia, Pa.: Temple University Press, 2000.

Hearne, Vicki. *Adam's Task: Calling Animals by Name.* New York: Alfred A. Knopf, 1986.

———. *Animal Happiness.* New York: HarperCollins, 1994.

Midgley, Mary. *Animals and Why They Matter.* Athens: University of Georgia Press, 1983.

Rollins, Bernard. *Animal Rights and Human Morality.* Buffalo, N.Y.: Prometheus Books, 1981.

Smuts, Barbara. "Reflections." In *The Lives of Animals,* by J. M. Coetzee. Princeton, N.J.: Princeton University Press, 1999.

Varner, Gary. "Pets, Companion Animals, and Domesticated Partners." in *Ethics for Everyday,* edited by David Benatar, pp. 450–75. New York: McGraw-Hill, 2002.

Zamir, Tzachi. *Ethics and the Beast.* Princeton, N.J.: Princeton University Press, 2007.

CHAPTER 29

ANIMAL EXPERIMENTATION IN BIOMEDICAL RESEARCH

HUGH LAFOLLETTE

SHOULD we use animals in biomedical experimentation? Most people think so. They embrace the Common View, which includes both moral and empirical elements. The two-part moral element is that although (a) there are moral limits on what we can do to (some) nonhuman animals, (b) humans can use them when doing so advances significant human interests.[1] Put differently, they think nonhuman animals have some moral worth—that their interests count morally—although that worth is not especially high. The empirical element is that biomedical experiments using animals significantly benefit humans. The truth of these claims would morally justify the practice.

The Common View is one among many views about the moral permissibility of biomedical experimentation using animals. This view is best seen as resting near the center of a moral continuum, with the Lenient View at one extreme and the Demanding View on the other.[2] The Lenient View holds that even if animals have moral worth, their worth is so slight that humans can use them virtually any way we wish and for any reason we wish. The Demanding View holds that the moral worth of animals is so high that it bars virtually all uses of animals in biomedical research. The Lenient and the Demanding Views share one significant claim: each thinks we need to determine only the moral worth of nonhuman animals to morally evaluate the practice of animal experimentation. However, few people would agree. Most people think we must also know the extent to which biomedical research on animals benefits humans. Perhaps they are mistaken. Still, since this view is so common, it is a prudent place to begin.[3]

THE MORAL STATUS OF NONHUMAN ANIMALS

Historically, few people have had moral qualms about using animals for their purposes.[4] Even so, most would not have harmed their nonhuman animals frivolously. It would be imprudent for a farmer to fail to feed the pigs she planned to eat or to fail to care for the ox she needed to pull her plow. That would be unwise, just as it would normally be unwise for us to let our houses or automobiles deteriorate. However, few people would have thought that there is anything *intrinsically* wrong with killing an animal or making it suffer,[5] just as few people today would think there is anything *intrinsically* wrong with taking a sledgehammer to their cars. To that extent, the Historical View is a form of the Lenient View. By the mid-1700s, that view began to give way to the Common View. (For a more detailed historical accounting, see the first two chapters in this *Handbook*.)

Indirect Limits on What We Do to Animals

Since what we do to nonhuman animals often benefits or harms humans, we have a reason to be morally concerned about them. Killing someone else's dog is wrong because it harms the animal's owner—much as someone harms her by throwing acid on her Saab or burning her favorite coat. Killing millions of honeybees or overfishing the ocean is wrong because these actions diminish limited resources humans need—much as we would by burning a million acres of Sequoias for a campfire. Disemboweling one's own dog in public would be wrong because it would offend many humans—much as someone would by belching loudly and repeatedly in a quiet romantic café. Finally, hitting, taunting, or killing animals is arguably wrong since people who do so are thereby more likely to mistreat humans.[6] All these considerations limit what we can permissibly do to or with nonhuman animals.

Although these provide plausible human-based reasons for not harming some nonhuman animals, most people do not think these considerations capture the most important moral consideration: harming animals is wrong because of what it does to the animals themselves. In this way the Common View diverges from the Historical View.

Direct Limits

Few people think it is morally acceptable to nail a fully conscious and unanesthetized dog to a board and then slowly disembowel it so we can determine the layout of its organs or see how its blood flows. Few think it is morally acceptable to roast an unanesthetized, fully conscious pig to slightly enhance the taste of pork tenderloin.

According to the Common View, the wrongness of these actions cannot be exhaustively explained by the fact that such actions indirectly harm humans; they are also—indeed primarily—wrong because they harm animals. The harm, according to most people, is that such actions cause pain to animals. How is this relevant to an assessment of biomedical experimentation using animals? Mammals and birds—the most common laboratory animals—can feel pain, and most experiments cause lab animals pain.[7] Most people think we must consider this pain when deciding how to act; they think we should not make these animals suffer needlessly.

Many other people think this is only part of the moral story. They think it is also wrong to kill some animals, at least to kill them without good reason. They believe that animals' lives are valuable. Of course, there are important disagreements about just how valuable nonhuman animals' lives are, and there are disagreements about what counts as a good reason for killing them. Some think we are justified in killing a nonhuman animal only for the same reasons that would justify killing another human—for example, in self-defense. Many others would not go nearly so far, but they would think humans need a compelling reason to take an animal's life. Still others think that any minor human interest would suffice. Still, this much seems true: most people would be appalled at a neighborhood child who shoots squirrels with his BB gun just so that he can watch them writhe in pain and at a businessman who kills a wild gorilla so that he can use its shellacked skull as a spittoon.

How might we explain the idea that nonhuman animals have a valuable life that counts morally? Those who embrace this view likely endorse Tom Regan's claims that some nonhuman animals are "subjects-of-a-life." Regan claims animals have:

> beliefs and desires; perception, memory, and a sense of the future, including their
> own future; an emotional life, together with feelings of pleasure and pain;
> preference- and welfare-interests; the ability to initiate action in pursuit of their
> desires and goals; a psychophysical identity over time; and an individual welfare
> in the sense that their experiential life fares well or ill for them, logically indepen-
> dent of their utility for others and logically independent of their being the subject
> of anyone else's interests.[8]

In this view, if we kill a nonhuman animal, we deprive it of a future it desires; we ignore its legitimate interests. Some with moral misgivings about killing nonhuman animals will not buy this explanation. They think nonhuman animals' lives are morally valuable, albeit less valuable that those of humans. "Normal (adult) human life is of a much higher quality than animal life, not because of species, but because of richness; and the value of a life is a function of its quality."[9] In this view, animals' lives cannot be taken cavalierly, but they can be taken if necessary for a significant public good.

Since Regan's view is highly controversial, we might make more progress if we begin by examining animal experimentation assuming only the weaker view that it is wrong to cause an animal needless pain, coupled with the idea that many laboratory animals' lives—especially mammals—have some value, even if that value is not high. After all, virtually all sides of this debate embrace these views—researchers as well as animal activists, and, according to the Gallup poll, also the American public. Of course, there are still significant disagreements about (a) how valuable nonhuman animals' lives are, (b) what constitutes a good reason for taking their lives or causing them pain, and (c) whether most biomedical experiments using animals provide such a reason.

Knowing that animals have moral worth only lets us know that their interests should count. It does not tell us how *much weight* their interests have or *how* those interests should be counted. These questions are distinct, in part because they usually reflect different theoretical stances. Those who speak of nonhuman animals' interests as having *weight* often embrace some form of consequentialism where the animals' interests, whatever they happen to be, are balanced against competing human interests. If their interests are sufficiently weighty, then we are morally limited in what we can do to animals.

Regan will reject this approach; he will reject any talk of "balancing interests." He thinks that animal interests—like human interests—are not subject to moral calculation, but are rather morally protected by rights.[10] On his deontological view, it is not just that rights are weightier than other considerations; they are trumps that can never be overridden in the pursuit of human goods.

Those who embrace this view think that discussing potential benefits of biomedical experiments using animals is morally irrelevant. On their view, it wouldn't matter if experiments benefitted humans enormously. They would be immoral in precisely the same way and for the same reason that we think nonconsensual experiments on humans, including those performed by the Nazis or in the Tuskegee syphilis study, would be immoral.[11] Right or wrong, most people reject this defense of abolitionism. They think that the benefits of animal experimentation matter morally. It is to this issue that I now turn.

BENEFITS OF ANIMAL EXPERIMENTATION

The empirical element of the Common View holds that the practice of biomedical experiments using animals substantially benefits humans. This claim, when conjoined with the second moral component of the Common View—the claim that we can use animals when doing so significantly benefits humans—is thought to justify the practice. Notice, though, what follows from saying that the benefits to humans *outweigh* moral costs to animals. It acknowledges that the interests of nonhuman animals carry moral weight.

Since nonhuman animals' interests have moral weight, their interests will sometimes constrain the pursuit of human interests. Clearly they do. All sides of the debate think that we should not keep lab animals in squalid conditions, and all sides think that we should anesthetize laboratory animals against substantial pain, unless there are compelling scientific reasons why we cannot. These are important concessions.

For in the world of limited finances, the money experimenters use to care for (and anesthetize) animals is money they cannot use to conduct more experiments. All sides to the debate thereby acknowledge that respecting the interests of animals limits animal experimentation. Therefore, the issue is not *whether* the interests of animals should constrain animal experimentation. The issue is *how much* and *under which conditions* they should constrain it.

The Prima Facie Case for Animal Experimentation

The case for thinking that experimenting on animals will significantly benefit humans rests on three interlinked pillars: (1) the common sense idea that we can legitimately generalize what we learn from animals to human beings; (2) the claim by many medical historians that animal experiments have been essential for most major biomedical advances; and (3) plausible methodological reasons supporting the common sense and historical arguments. I examine each pillar in turn.

Common Sense Argument

The common sense argument is plausible. We see broad biological similarities between humans and animals, particularly other mammals. Given that, we infer that: the skeletal structure of humans will resemble that of chimpanzees; the blood of humans and rats will circulate in similar ways; the mechanisms whereby rabbits and humans exchange gasses with the air will be comparable; and the reactions of humans and guinea pigs to toxic substances will be akin.

This argument form is plausible. Disputants on all sides of this debate use it. Researchers use these analogical arguments to explain why they think we can safely generalize from animals to humans. Defenders of animals' interests use them to show that nonhuman animals morally resemble human beings. They claim that chimpanzees reason, that dogs scheme, and that rats grieve because these animals act in the same ways humans act when they reason, scheme, or grieve. I suspect, in the end, that the precise forms of these analogical arguments are relevantly different. Still, as a starting point of inquiry, and in the absence of contrary evidence, it is reasonable to make inferences from animals to humans.

Historical Evidence

Historical evidence reinforces the common sense view. According to the American Medical Association:

> [V]irtually every advance in medical science in the 20th century, from antibiotics
> and vaccines to antidepressant drugs and organ transplants, has been achieved
> either directly or indirectly through the use of animals in laboratory experiments.
> The result of these experiments has been the elimination or control of many
> infectious diseases—smallpox, poliomyelitis, measles—and the development of
> numerous life-saving techniques—blood transfusions, burn therapy, open-heart
> and brain surgery. This has meant a longer, healthier, better life with much less
> pain and suffering. For many, it has meant life itself.[12]

Biomedical advances are not simply the result of research seeking a cure to a specific disease or condition (applied research). Basic research—research aimed at understanding "how living organisms function, without regard to the immediate relation of their research to specific human disease—also prompts biomedical discoveries."[13] Finally, it is not just that animal experimentation was necessary for past discoveries, but also it will be essential for future ones. As Sigma Xi claims: "an end to animal research would mean an end to our best hope for finding treatments that still elude us."[14]

Scientific Rationale Supports History and Common Sense

There are good methodological reasons reinforcing the common sense and historical pillars of the argument.

Good Science Requires Controlled Experiments. Scientists want tightly controlled experiments where they can exclude any factors that might skew the study's results. Only then can they be confident they have discovered a causal relationship rather than a mere correlation. However, meeting this scientifically high standard with human subjects is scientifically difficult and often morally impermissible. Suppose researchers want to know if smoking causes heart disease in humans. (a) They cannot merely compare the incidence of smokers who die from heart disease to that of nonsmokers. There may be other factors (e.g., lifestyle choices) that are the primary culprit. (b) Researchers can design reasonably reliable epidemiological studies that exclude many extraneous features (e.g., patients' diets) that could skew the study's results. However, these studies face two problems: (1) designers cannot be confident they know which factors are relevant; (2) even if they knew all relevant factors, they often rely on patients' self-reports to determine if those factors are present (if they smoke or drink and how much they exercise, etc.). However, self-reports are notoriously unreliable. These factors explain why epidemiological studies, although valuable, have several marks against them. (c) In principle, scientists *could* conduct wholly controlled studies on humans: they could seriously limit subjects' motion, their exposure to relevant environmental factors, and their diets. However, controlling humans in these ways would be morally unacceptable. So what is a serious and moral scientist to do?

Intact Systems. Some have suggested that we could use human cells and tissue cultures rather than humans or animals. For some purposes and at some testing stages, we can. However, defenders of biomedical research using animals claim these micro methods are insufficient when we need detailed information about the causes of, or possible cures for, a human disease. Humans and animals are not, they note, loose associations of biological parts; rather, they are intricately related "intact systems." Just as one cannot model the workings of a computer by looking at chips and hard drives lying on a table, one cannot model complex human biomedical behavior by looking at detached human body parts. Only one intact system can reliably model another.[15]

An Intermediate Conclusion

The prima facie case for the validity and importance of biomedical experimentation using animals is plausible. To challenge the case, objectors must show that the status of nonhuman animals is greater than, or that the benefits of experimentation are less than, most people suppose. In the next section, I address the second possibility, starting with concerns about the prima facie empirical argument.

Evaluation of the Prima Facie Case

The Common Sense View

The common sense argument for the effectiveness of biomedical experiments using animals is sensible. Animals and humans are similar in obvious ways; the issue is whether they are sufficiently similar to justify biomedical inferences from animals to humans. Whether they are depends on the other pillars of the argument. That is where the real work of the prima facie argument is being done.

The Historical Argument

Those defending animal experimentation claim that virtually every medical advance is attributable to that practice. In a minimal sense they are correct. The history of most biomedical discoveries during the last seventy-five years will reveal at least some experiments using animals. However, simply because something is part of a development's history does not mean that it was a causally significant—let alone a necessary—element of that history. Virtually all biomedical scientists drank milk as infants. However, that does not establish that milk drinking leads to biomedical knowledge. Not every element of a history is a significant causally contributory factor of that history.

Researchers are, in most cases, *legally required* to use animals for most biomedical experiments. Given the law, *of course* the use of animals is part of the history of biomedical discovery. So we must determine the degree to which the correlation reflects facts about scientific discovery rather than the state of the law. Defenders of experimentation would argue that it is the former. They contend that surveys of primary research show that this correlation is not simply, or even primarily, an epiphenomenon of the legal system.

There are good reasons to take these surveys seriously, but there are also good reasons to be careful in accepting their findings unquestioningly. Although academic journals and books will report some dissimilarities between animals and humans, they likely underreport them. When scientists are working within a guiding paradigm, we should expect failures to be underreported. If a researcher is trying to discover the nature of human hypertension, and conducts a series of experiments on a gazelle, only to discover that gazelle rarely develop hypertension, then she will likely not report her findings, not because she wants to suppress relevant information, but because most scientists won't be interested (unless, of course, they had thought about developing a gazelle model of hypertension). Even when scientists do report negative findings, others are less likely to discuss them—especially if the results do not explain the failure. Therefore, these failures, even if common, will rarely be well-known parts of the history of biomedical discovery, although occasionally failures are mentioned if researchers explain why the experiment failed.[16]

We have similar reasons to be careful when interpreting standard histories of biomedical research. When historians of medicine discuss the history of a biomedical advance, they typically underreport failed experiments, even experiments that appear in the primary research literature. This, too, is normal. Historians chronicle events that they think illuminate history. For instance, American historians do not mention the vast majority of events in our country's past—for example, a two-minute extemporaneous stump speech Adlai Stevenson gave during his second failed run for the presidency. Barring some unusual reason, describing this speech in detail would be a distraction. We do the same thing when telling our personal stories: we focus on events that elucidate our current understanding of ourselves. We downplay, forget, or omit elements of our histories we consider tangential. Biomedical historians likely will not mention (even if they know about) most failed experiments; they see them as diversions from, rather than illuminating elements of, the scientific narrative. Since the use of nonhuman animals is integral to the current biomedical paradigm, we should expect histories to emphasize the successes of that paradigm.

These considerations give us grounds for caution when interpreting both primary research and historians' claims, especially since most of us seek evidence supporting our antecedently held views.[17] We often fall prey to the shotgun effect or we unintentionally engage in selective perception. If I fire a shotgun in the general direction of a target, several pellets will likely hit it. Since researchers conduct thousands of experiments annually, we would expect some substantial successes when surveying the practice over decades. The researcher then commits the fallacy of selective perception if she counts the hits and ignores the misses. For instance, researchers have been trying to understand ALS (Lou Gehrig's disease) for more than seventy years. Yet in "terms of therapeutic treatment of this disorder, we're not that much further along than we were in 1939 when Lou Gehrig was diagnosed."[18] To date, investigators have only found one drug that benefits humans with the disease, and that benefit is slight: it helps extend the patient's life for a few months. Yet researchers continue to employ the same mouse model of ALS that has guided research for years. Even advocates of these experiments acknowledge "previously, medications that have been found to be effective in the mouse model of ALS have not shown benefit when brought to human clinical trials."[19] Given advocates' belief in the power of animal models, they do not construe these failures as a mark against the practice. They continue to hope that each new drug with beneficial results in mice will have similar affects in humans. When they eventually find a beneficial drug, then advocates of biomedical research using animals will doubtless cite the success as proof of animal experimentation's enormous value, despite the previous significant failures.[20]

Opponents of animal experimentation often commit the same fallacies by focusing exclusively on the practice's failures; and failures there are. However, critics often forget that failures are common in science. We need more than just lists of putative successes and failures. We need to discuss evolution—the overarching biological theory. Why? Although particular scientific "facts" inform and shape theories, theories give us a framework for understanding, interpreting, and evaluating putative facts, especially when the facts are conflicting.

In later sections, I explain how evolution informs this debate. First, I offer some additional "facts" that suggest the limitations of the practice. I want it to be clear that the failure of the mouse model of ALS is not unique.

Some Empirical Evidence Undermining the Reliability of Animal Experimentation

Many people have heard about problems with animal testing on thalidomide, a drug that caused serious physical defects in more than ten thousand children worldwide, but did not appear to have any adverse effects in standard laboratory animals (although researchers later found some species in which the effects were similar). I want to mention other findings that, although less well known, are more instructive. Rats and mice are closely related species; they resemble each other far more than either resembles humans. Despite their close relationships, chemicals that induce cancers in rats produce cancers in mice in only 70% of the time.[21] That is not a wholly insignificant correlation, of course, but it is far from perfect. Then, in roughly a third of these cases, chemicals that produce cancer in both animals do not produce cancer at the same site. This is extremely troubling when we are trying to understand the causes of and mechanisms for treating cancer, sufficiently troubling that it prompted a leading team of researchers to conclude that, in its current form, "the utility of a rodent bioassay to identify a chemical as a 'potential human carcinogen' is questionable."[22]

The problem even pervades the history of one of researchers' vaunted successes. In the early years of polio research, scientists focused almost exclusively on one animal model of the disease, a form of the disease in rhesus monkeys. This obsession, according to researcher and medical historian J. R. Paul, made research focus on the wrong route of infection, and therefore likely delayed the discovery of a treatment for polio by twenty-five years.[23]

There is especially strong evidence of significant biomedical differences between humans and nonhuman animals in teratology (study of abnormal development): "False positives and false negatives abound. Once one has established that a drug is a teratogen for man, it is usually possible to find, retrospectively, a suitable animal model. But trying to predict human toxicity—which is after all what the screening game is about—is quite another matter."[24] It is difficult to find a suitable animal model even in nonhuman primates, our closest relatives.[25] These differences are so profound that we cannot safely generalize findings in animals to humans even for drugs within the same chemical or pharmacologic class.

Finally, species' differences are common in the endocrine system. "[G]enerally the same or very similar hormones are produced by corresponding glands of different vertebrates. Despite the general similarities, hormones do many different things in different vertebrates."[26] Because the endocrine system plays such a central role in overall function of the body, differences in these systems are amplified elsewhere in the organism. "The poor predictiveness of animal studies for humans thus becomes comprehensible in terms of interspecific variations in endocrinology."[27] These variations are the products of evolution, are conservative inasmuch as they "use" the same biochemical building blocks across species, but they are radical inasmuch as they use those endocrinal blocks for different functional ends.

   These brief examples do not show that biomedical experiments using animals are worthless. All areas of even mature sciences have experimental failures. However, these examples do indicate that there are important differences between species. We need a theoretical framework to interpret empirical results, a theory to explain just why we should expect significant species differences. It is to evolutionary theory that I now turn for this framework.

EVOLUTION AND ITS INFLUENCES

Understanding Similarities and Differences Between Species

The current practice of biomedical research is grounded in the work of eighteenth-century French physiologist Claude Bernard.[28] Bernard wanted to make physiology a *real* science by adopting the methods of physics. For him that meant that all life—like all matter—was fundamentally the same. By testing on one species, we can straightaway discover important biological information about another:

> Experiments on animals, with deleterious substances or in harmful circum-
> stances, are very useful and *entirely conclusive* [emphasis mine] for the toxicology
> and hygiene of man. Investigations of medicinal or of toxic substances also are
> wholly applicable to man from the therapeutic point of view; for as I have shown,
> the effects of these substances are the same on man as on animals, save for
> differences in degree.[29]

Bernard is partly right. There are clear commonalities between species. Having discovered that numerous species of mammals, reptiles, amphibians, and birds have blood circulating throughout their bodies, we can infer that the same will be true of a related species we have not yet examined. To that degree we can generalize from species to species. However, this fact can easily mislead us. We are considering a much narrower issue: Can we reliably infer details of specific human diseases by experimenting on laboratory animals?

   To address this question, I must explain the nature and use of animal models of human biomedical phenomena. Researchers seek to identify or create a condition in laboratory animals (AIDS, cancer, etc.) that resembles some human condition they want to understand. They then proceed in two different ways. Some seek to better understand the nature of the condition in nonhuman animals.[30] This is a form of basic research with no direct application to humans, although the knowledge gained may eventually be used in humans. We will explore this use of animals later.

Other researchers engage in applied research. They directly seek a cure for some human disease or condition. After identifying a potential animal model of the human disease, they may give the animal a drug or excise a growth, or see if implanting stem cells alters that condition. If the intervention cures the animals or attenuates the disease, then others may try the same intervention in a small number of humans— first to see if it is relatively safe (it doesn't cause any significant adverse effects), then to see if it is efficacious. If it is both safe and efficacious, then the researcher will try the intervention in a larger sample of humans. If it is significantly unsafe or demonstrably inefficacious, then they will either modify or abandon the idea.

Two Issues about Models in Applied Research

We now see that to assess the benefits of biomedical experimentation using animals we must answer two different empirical questions. One, is the disease in the laboratory animal relevantly similar to the human condition it supposedly models (the *similarity problem*)? Two, if the models are similar, can we reliably generalize from animals to humans (the *inference problem*)? These issues are clearly related, albeit distinct.

There are always some similarities and some differences between a condition or disease in animals and in humans. I earlier noted obvious ways in which species are similar. They are also different, and different in ways that are biomedically significant. Mice are the standard model of human cancer. However, although 80% of human cancers are carcinomas, sarcomas and leukemia are more common in mice.[31] Additionally, most AIDS research has been guided by animal models in primates, despite important differences between the conditions in the two species: "The only nonhuman primate species that can be reproducibly infected by HIV is the chimpanzee....[However] HIV does not replicate persistently in chimpanzees, nor does HIV consistently cause AIDS in this species."[32]

Of course, not all differences undermine inferences from animals to humans. Although a human femur is different from a gorilla femur, most differences will be irrelevant if orthopedists simply want to know how to repair a fractured human femur. On the other hand, seemingly miniscule differences may turn out to be highly significant. Therefore, before we can rely on a model, we must know if the condition in the animal model is *relevantly* similar to the human condition. That is not easy to do.

Suppose, though, we do know that they are highly similar. We must still determine if the methods which prevent, control, or cure the disease in nonhuman animals will do the same in humans (the inference problem). In a not-insignificant number of cases, the answer is "No." As I noted earlier, ALS researchers have long relied on what they deemed a promising mouse model of the disease. Yet after years of study, the interventions that work in the mouse have been, with one minor exception, unsuccessful in humans.

Although these two questions are independent, they are linked. We often know if differences are relevant only after we discover if research leads to a cure for, or at least an attenuation of, the human condition. However, we cannot know that it leads to a cure or an attenuation until we have conducted tests in both animals and humans. That shows why experiments on animals cannot do what they aim to do—that is, give us *confidence* in predictions about human biomedical phenomena prior to human testing. Still, it may be that animal models are sufficiently similar to the corresponding human condition so that we can make qualified, albeit still useful, inferences about humans. Before we can ascertain that fact, we must determine how common and how deep species differences are. That requires understanding the profound ways that evolutionary forces shape biological organisms.

Evolutionary Influences Prompt Changes

Over evolutionary time, the environments in which animals lived and competed changed. Some animals' food sources either died or became more plentiful. Animals that adapted to their new environments survived or even flourished, while those that did not adapt either disappeared or became less successful. Evolutionary processes prompted biological differences between closely related species, differences that go all the way to the building blocks of life: "[T]he genomes and chromosomes of modern-day species have each been shaped by a unique history of seemingly random genetic events, acted on by selection pressures over long evolutionary times."[33] This history is relevant for assessing biomedical experimentation using animals.

Organizational Complexity Amplifies Adaptive Changes

Defenders of animal experimentation note that animals and humans are highly organized, intact systems. That fact, they claim, is why we must experiment on animals rather than on human parts. They are right by half. Since animals are intact systems, we should be cautious when making inferences from experiments on isolated tissues to humans. However, what this fact gives with one hand, it takes away with the other. The same factors also give us reason to be cautious about making biomedically significant inferences from nonhuman animals to humans. Evolutionary pressures reward species that have advantageous adaptations. These adaptations are frequently biomedically significant. Because humans are intact systems, the adaptations' biological significance is often amplified in one or more of the following four ways.

First, structures and processes interacting with adaptations must change to accommodate them. "New parts evolved from old ones and have to work well with the parts that have already evolved."[34] These accommodations partly explain why beneficial adaptations are rarely unqualifiedly beneficial. Changes advantageous in one niche may become detrimental if the climate changes, a new predator appears on the scene, or the individuals relocate to a new environment. For instance, a single gene for sickle-cell anemia is highly beneficial in a malaria-prone environment. The same trait is highly detrimental (because offspring with two sickle-cell anemia genes usually die before fifty years of age) once malaria has been controlled or people susceptible to the trait relocate to a malaria-free area.

Second, a beneficial adaptation might prompt potentially detrimental changes elsewhere in the organism. Humans are more fit because they have relatively large brains. Brain size, though, is limited by skull size. Therefore, humans could develop larger brains only if there were compromises elsewhere within the organism. When human skulls became larger to permit larger brains, human infants had to be born earlier; they were therefore more dependent on parental care than are most mammals. Having more developed cognitive skills is beneficial. Being wholly dependent on one's parents for longer makes human infants especially vulnerable. For instance, more than half of deaths from hunger-related problems are in children under five years of age. Such compromises are ubiquitous. "The body is a bundle of compromises, compromises which, even if they currently serve (or once served) some fitness advantage, now cause disease."[35]

Third, organisms often retain elements of their evolutionary pasts even when those elements no longer promote survival—for example, the human appendix. These structures may affect biochemical processes or create the possibility of detrimental, even life-threatening, conditions, such as appendicitis. Other elements of their evolutionary pasts may significantly influence cellular and metabolic functions.[36]

Fourth, resulting differences between two species may be exaggerated if their "molecular clocks" (the rate at which their DNA and proteins evolve) are different. Although the human and mouse genomes are approximately the same size, "There has been a much longer period over which [genomic] changes have had a chance to accumulate—approximately 80 million years versus 6 million years.... [Moreover] rodent lineages...have unusually fast molecular clocks. Hence, these lineages have diverged from the human lineage more rapidly than otherwise expected."[37]

In concert, these factors lead to important differences between species, differences greater than those we might initially expect. These give us a reason to think that the results of animal experiments will rarely be straightforwardly applicable to human beings.

Functional, Explanatory, and Causal Properties

To understand the effects of evolutionary change, we must distinguish three perspectives from which we can describe biological phenomena. In talking about ways in which all life is the same we mask these differences. (1) Sometimes we talk about ways an organism functions within its environment: that it moves, exchanges gases with the air, takes in nourishment, and the like. In so doing, we are talking about an organism's *functional properties*. (2) At other times, we describe an organism's mechanisms for achieving these functions. In so doing, we are talking about its *causal properties*. Finally, (3) we sometimes describe an organism's mid-level properties, properties we can see as either causal or functional. For instance, breathing is a functional property inasmuch as it identifies the fact that an organism exchanges gasses with the air, and it is a causal property inasmuch as it describes (albeit abstractly) a mechanism for performing that function (the *way* the organism oxygenates its blood). I call these dual-purpose properties *explanatory properties*.

Each way of describing an organism's properties serves a different but important purpose. Evolutionary theorists focus on organisms' functional properties to describe how natural selection favored a creature within its environmental niche. Functional properties are also key to understanding a creature's moral status, since, as I noted earlier, a creature counts morally if it can feel pain, think, or emote.

However, biomedical researchers are not currently investigating either functional or explanatory properties since these do not explain disease or uncover cures. In the early years of biomedical discovery, researchers did seek to understand common biological functional properties like the circulation of the blood.[38] Now they are only tangentially interested in these properties. Researchers know *that* the blood circulates; now they want to know the ways blood absorbs oxygen or the way it responds to certain chemicals. In short, they want to identify and understand an organism's causal mechanisms.

Researchers evidence this focus both explicitly and implicitly. They study biological systems to understand what causes or exacerbates a disease or condition. Then they implicitly demonstrate this focus when making inferences from animals to humans. Unless researchers assume that laboratory animals and humans have relevantly similar causal mechanisms, they have no reason to think that a drug or chemical that is harmful to animals will also be harmful to humans. As researchers with the Carcinogenic Potency Project put it, "Without data on the *mechanism* of carcinogenesis, however, the true human risk of cancer at low dose is highly uncertain and could be zero."[39]

Unfortunately, although the distinction between these three perspectives is important, researchers and their apologists either do not notice or appreciate them, or else they assume that if two animals share any properties then they must share all related ones. Neither assumption is plausible. Of course most animals share abstract functional properties: they move within their respective environments, they gain nourishment, and they excrete wastes. Many share the same explanatory properties: most use lungs to exchange gasses with the air. However, only someone guided by the Bernardian paradigm would infer that humans and nonhuman mammals therefore have similar causal mechanisms for all or even most biomedically significant phenomena.

For instance, although cats, rats, pigs, and humans all successfully metabolize phenol (metabolizing phenol could be a functional or even an explanatory property), the mechanism of metabolism varies widely between species. There are two primary mechanisms. Some species metabolize phenol primarily using only one mechanism. For example, pigs rely entirely on one while cats use only the other. Other species use both mechanisms roughly equally.[40] Species differences are evident even in closely related species: humans and New World monkeys use different metabolic pathways.[41] Why do these differences matter? Because researchers often speak as if the condition or disease being studied in laboratory animals strongly resembles the condition in humans. Evolutionary theory suggests that is not a plausible expectation. We thus have reason to think that nonhuman animals are not, in general, strong models of human biomedical phenomena.

Strong Models

This claim that animals are strong models of human biomedical phenomena might be true if we were talking about functional or explanatory properties. Those properties *are* broadly similar across most mammalian species. However, biomedical researchers using animals study creatures' biomedically significant (causal) mechanisms. It is only by studying these that researchers can understand the causes of, and identify potential cures for, human disease. However, inferences from animals to humans will be questionable if the condition in the laboratory animal differs causally from the human condition. Given the myriad ways that evolutionary forces shape an organism's biological systems, we should expect causal differences. Many differences run all the way to the genome.[42] These differences are not simply, or even primarily, in the number of genes a species has, but in whether, when, and how those genes are expressed (the particular order and manner in which genes turn on or off).[43] That explains why even two seemingly similar animals may be so different biomedically.

In short, evolutionary theory—the theoretical glue of modern biology—gives us reason to expect that a biomedically significant condition in an animal will never be exactly like the condition in humans. Researchers are usually satisfied if they can find or create a condition in laboratory animals that symptomatically resembles the human condition. However, symptomatic similarity does not guarantee causal similarity. That is why interventions that cure a disease or condition in laboratory animals not infrequently fail in humans. The history of biomedicine is littered with such cases. Researchers have tested 85 potential AIDS vaccines in 197 different human clinical trials. However, although many of these were promising in animal trials (that's why they proceeded to clinical trials), "just 12% of these trials have reached Phase II [an early phase of human testing with a small number of human subjects], only seven (3.5%) have reached Phase III [a later phrase with more human subjects], and altogether, 18 trials were prematurely terminated."[44] One vaccine seemed especially promising given its effects in animals. However, researchers had to stop the clinical trial midstream because it appeared to increase people's susceptibility to HIV/AIDS.[45]

As I was completing this paper, researchers reported one study in which a new vaccine was 31% effective.[46] Some defenders of research have hinted that this just shows how successful animal experimentation is. However, believing that a single and relatively minor success demonstrates the predictive power of animal models simply illustrates the psychological power of the shotgun effect and of selection attention. If this is a success, and we cannot be confident that it is since this is a single study, it comes only after twenty-five years of failures. Perhaps some advocates will claim that all the failures are worth the eventual success. However, that is a separate question and a moral issue I address shortly. The issue here is the predictive power of animal models. A single minor success (if it is a success) after a quarter century of failures is hardly proof of the predictive power of animal studies.

Where does this situation leave us? Concrete examples and evolutionary processes give us reason to think that animal models are always different to some degree from the human condition they model. Inferences from animals to humans are never certain. In the end, I suspect all serious researchers know that. Talk about animal models being "entirely conclusive for the toxicology and hygiene of man"[47] is just an artifact of the public debate over biomedical experimentation using animals.

Most cautious defenses of the practice usually employ another strategy when defending biomedical experimentation using animals: (1) they emphasize the value of basic research, or (2) they claim that animal models, although causally different from the human condition they model, are similar enough to that condition to justify inferences from animals to humans. I will examine each suggestion in turn.

Other Defenses of the Practice

Basic Research. Many defenders of biomedical experimentation using animals claim that basic research has been profoundly beneficial to humans.[48] Basic research does not directly seek a cure for any disease. Rather it seeks to understand fundamental biomedical phenomena—although this understanding, advocates say, empowers other researchers to find cures for human diseases or conditions. For instance, if basic research explains the causal mechanisms whereby mutant superoxide dismutase 1 induces motor neuron death in a mouse, then clinicians and applied researchers may have insight into ways to prevent, control, or cure human patients with this disease. I have no doubt that some basic research yields applied biomedical fruit. However, we must be careful not to overestimate these benefits. While Comroe and Dripps claimed that well over half of all clinical advances were traceable to basic research, the Health Economics Research Group found that the real figure is much lower, somewhere between 2% and 21%.[49]

Additionally, basic research is also partly vulnerable to the previous arguments about species differences. Knowing how mutations cause neuronal death in a mouse might be scientifically interesting, but on its own, it will not illuminate the mechanisms in humans if the mouse and the human are causally relevantly different. To that extent, basic research will not predictably have the indirect benefits often attributed to it.

Weak Models and Dynamical Systems Theory. We might also think that animal models are valuable if the conditions in animals and humans are *sufficiently* similar to *generally* justify inferences from one to the other. This would be plausible, if, as Bernard thought, biological systems were simple systems, ones where small differences between the model and the condition modeled make little if any difference. However, we have strong evidence that biological systems in higher animals are not simple; they are complex with extensive interactions and feedback mechanisms. Even a small change one place in an organism can have significant effects elsewhere in that organism. The behavior of complex systems is best explained by *dynamical systems theory*—or what is colloquially called "chaos theory."[50] This theory explains why even seemingly minor differences between two creatures may result in widely different reactions. "Among rodents and primates, zoologically closely related species exhibit markedly different patterns of metabolism."[51]

Where does this leave us? Both empirical evidence and evolutionary theory give us reason to think that inferences from nonhuman animals to humans are never certain. However, it does not show that the practice of using animals as models of human disease does not have reasonable levels of probability or that it has not benefitted humans. The moral question is whether any benefits are morally worth the costs. I now turn now to that question.

THE MORAL COSTS OF ANIMAL EXPERIMENTATION

I begin by combining two strands of argument. Some defenders of biomedical research using animals offer deontological arguments for the practice—arguments that seek to explain why humans can use animals for their purposes. I say a bit more about those arguments at the end of the paper. However, since virtually everyone now acknowledges that nonhuman animals have some moral status, most defenders of the practice employ in significant measure a consequentialist justification of the practice. They claim that biomedical experimentation using animals is justified because of its enormous benefits to human beings. As Carl Cohen, who begins by offering a deontological justification of the practice, puts it:

> When balancing the pleasures and pains resulting from the use of animals in research, we must not fail to place on the scales the terrible pains that would have resulted, would be suffered now, and would long continue had animals not been used. Every disease eliminated, every vaccine developed…indeed, virtually every modern medical therapy is due, in part or in whole, to experimentation using animals.[52]

Most defenders of biomedical experimentation think that this point supplies a devastating response to any criticism of animal experimentation. They think everyone (a) will acknowledge the enormous benefits of the practice and (b) will acknowledge that such benefits morally justify that practice. These are highly debatable assumptions. One, the earlier arguments suggest that claims about animal experimentation's benefits are bloated. Two, even if animal experimentation has significant benefits, there are enormous moral costs of the practice that defenders do not acknowledge or address. These costs might well be sufficiently great to undermine the legitimacy of the practice, no matter what its benefits. This position might be an alternative route to defending some form of abolitionism. I am unable to fully evaluate that claim here. What seems minimally true is that defenders must establish profound, and perhaps overwhelming, benefits of experimentation to morally justify the practice.

The Moral Scales

Researchers need to demonstrate the success of animal experimentation even if animals had no moral worth. If animal experimentation were only marginally beneficial, the practice would be a terrible waste of scarce public resources. Our need to demonstrate its success increases once we note that researchers, like most of us, think that nonhuman animals—at least mammals, the most common laboratory animals—have moral status. If nonhuman animals were devoid of value, or if their value were morally negligible, then the impact of experimentation on them would not enter the moral equation. Defenders of research accept that the costs to animals must be given due consideration—not only before permitting the general practice of biomedical experiments using animals, but arguably before we determine if any particular line of experimentation is morally justifiable.[53] For present purposes, I assume that although nonhuman animals have non-negligible moral status or value, their value is considerably less than that of humans. Even granting them minimal value raises potent moral objections to animal experimentation. If arguments against research are potent on this minimalistic assumption, then defenders of research will be vulnerable to arguments showing that the moral value of animals remotely resembles that of humans.

As Cohen's claim suggests, we often think about the choice as two options resting on an old-fashioned set of scales, with the benefits to humans on the right pan of the scales, and the costs to animals on the left. When we ordinarily make a utilitarian calculation, we assume that the creatures in each pan have the same moral worth. Therefore, when deciding what to do, we need consider only (a) the extent of the harms and benefits and (b) the number of creatures harmed or benefitted. However, since I am plausibly assuming that nonhuman animals have less moral worth than humans do, we must modify the relative costs and benefits accordingly. Although this is difficult to specify with precision, we can take inspiration from "cruelty to animal" statutes on the books in most developed countries. Although what counts as "cruelty to animals" varies from jurisdiction to jurisdiction, we can definitely say that it is wrong to inflict excruciating pain on an animal merely to bring a human some tinge of pleasure. Most people think it wrong to roast a chimpanzee alive to make a bookend from its hand or to slowly kill an elephant so we can use its tusks for a paperweight.

Here's the idea. Even if creaturesA have less moral worth than creaturesH, as long as the former have non-negligible worth—of the sort specified by "cruelty to animal" statutes—then there are circumstances under which morality demands that we favor them over the latter creatures. If the harm to creaturesA is considerably greater than the benefits to creaturesH—or if there are considerably greater numbers of creaturesA suffering that harm—then morality demands that we favor the former in those circumstances. With this adjustment in place, a utilitarian would hold that the moral permissibility of an action would be the product of (a) the moral worth of the creatures that suffer and benefit, (b) the seriousness of the wrong and the significance of the benefit of those respective creatures, and (c) the number of such creatures that suffer and benefit.[54]

That shows that the calculation is more complicated than defenders of animal experimentation suggest. In the public debate, they often cast the choice as one between "your baby or your dog." Since the baby is worth more than the dog, then everyone will choose the baby. However, the choice has not been, nor will it ever be, between "your baby and your dog." Single experiments, and certainly not lone experiments on single animals, will *never* lead to any medical discovery. Only coordinated sequences of experiments can lead to discovery. This is a point about the nature of science; it is not unique to biomedical experimentation. All scientific experiments are part of a pattern of activity—an institutional practice—and discoveries are made though an organized pattern of experimentation. Therefore, the core issue is whether that practice or institution significantly benefits humans. Consequently, we must reformulate the moral question: is this practice—or some attenuated version of it—morally justified even though it kills and causes pain to a significant number of animals?

Two Moral Assumptions

This way of framing the issue still makes it appear that we begin with the scales evenly balanced. Or, if they are tipped, they are tipped in favor of humans since we think that humans have greater moral worth than nonhuman animals. Doubtless that is why defenders such as Cohen claim the benefits of research "incalculably outweigh the evils."[55] However, this claim ignores two widely held moral views, which, if true, tip the scales sharply in favor of nonhuman animals. If I am correct, then defenders of biomedical research using animals must show significant benefits of experimentation to even the scales, let alone to tip them in favor of experimentation. Even if they can do that, we should fully appreciate the moral costs of such research. These costs are generally overlooked or ignored by those defending the practice.

Acts Are Morally Weightier than Omissions

Imagine any morally bad condition. Most people assume it is worse to bring that condition about than allowing it to happen. It is morally worse to kill someone than to let her die, to steal than to fail to prevent theft, and to lie rather than to fail to correct a lie. This claim comes in two forms. The absolute view holds that it is categorically worse to do harm than to fail to prevent it: it is always worse to cause a harm than to fail to stop another harm from occurring, no matter how benign the first and how serious the second. The relative view holds that it is worse to cause a harm than to fail to prevent one, although not categorically so. In some circumstances it is permissible to do a small harm to prevent a much greater one.

Regardless of which form one holds, most people think that it is not only worse to do harm than to fail to prevent harm, but that it is *much* worse. Although specifying how much worse is difficult, I can illustrate. Although most people would be aghast if Ralph failed to save a drowning child, particularly if he could have done so with little effort, they would not think Ralph nearly as bad as his neighbor Bob who held a child's head under water until she drowned. Minimally, "much worse" means this: the person who drowns the child should be imprisoned for a long time—if not executed—while the person who allowed the child to drown should not be punished at all, although perhaps she should be morally censured.

If the person had some special duties to the child—for instance, if she were a lifeguard at the pond—then we might hold her liable for the child's death, although even then we would not charge her with first-degree murder. If we did punish her, we would claim her obligation arose because of her special status: she voluntarily assumed responsibility for people swimming in her pond. The current issue we are discussing, however, is about the function the difference between doing and allowing plays in our moral thinking when people have not assumed any special responsibility for those who are harmed. Here the situation is quite different. Even in European cultures with "Good Samaritan" laws, someone who violates such laws—say, by not saving a drowning child—may be punished, but far less severely than someone who kills a child.[56] That signals a profound moral difference.

How is this relevant to the current debate? The researchers' calculation requires rejecting this widely held belief that there is a significant moral difference between harm we do and harm we do not prevent.[57] The experimenter knowingly kills—and often inflicts pain and suffering on—creatures with non-negligible moral worth to prevent future harm to humans. Put more abstractly, she causes harm to prevent future harm. Experimenters would likely contend that the moral asymmetry between doing and allowing is applicable only if the wrong perpetrated is morally equal to the wrong not prevented. Since animals are not as valuable as humans, then the wrong permitted is morally weightier than the wrong perpetrated.

However, the doing/allowing distinction has moral bite even if the harm not prevented is worse than the harm perpetrated. Although it is worse for a child to die than for a child to be spanked for inappropriate reasons, most people think this difference in moral weight is outweighed by the moral asymmetry between what we do and what we allow. That is, most people will think an adult has done something worse if he spanks his child (or worse still, a strange child) for inappropriate reasons than if he fails to feed a starving child on the other side of the world.[58]

A defender of research might respond that this example is irrelevant since both cases involve children—creatures of the same moral worth. For reasons offered earlier, this objection is misguided. Although the relative worth of creatures enters the moral equation, it is not the only factor. We must also include the seriousness of the harm (significance of the benefit), the number of creatures subjected to that harm (benefitted), and, especially relevant to the current discussion, whether we cause or merely permit the harm. For instance, Ralph intentionally chooses not to send money that would keep a starving Pakistani child alive. His next door neighbor,

Bob, picks up a stray puppy, takes it home and kills it slowly, causing it great pain. Although the law would do nothing whatsoever to Ralph, Bob would be charged with cruelty to animals. Finally, although most people in the community would not condemn Ralph for his inaction, they would roundly condemn Bob for his cruelty and callousness. They would not want to live next door to Bob, nor to have him as a veterinarian.

Some animal researchers might argue that they have special obligations to people—obligations that override the force of this asymmetry. That is not plausible. Lifeguards are hired to save those specific people swimming in their pools from drowning. Animal researchers are not hired to save particular people from kidney disease. They are hired to conduct experiments on animals. Everyone may hope that these experiments would benefit anyone who happens to have the disease. However, that does not mean that the researchers have special obligations to these as-yet-unidentified people. Special obligations are just, that, special: direct obligations to particular, identifiable individuals. Here's a clear way to see the point. If a lifeguard fails to rescue someone swimming in his pool, he can be subject to both civil and criminal penalties. No one who dies of renal failure could sue (let alone successfully sue) an animal experimenter who failed to find a cure for the disease.

Finally, even if we could make sense of the claim that researchers have special obligations to humans who might benefit from their research, it is more plausible to think that they have special obligations to their laboratory animals, since by law investigators are specifically required to care for them.[59]

Consequently, if this asymmetry is morally relevant, it is relevant even given the presumed difference in moral worth. Therefore, unless the benefits to humans are substantially greater than the costs to animals, then these will not outweigh the special immorality of causing harm. How much greater the benefits must be depends in part on whether defenders hold the absolute or relative form of the distinction. However, it is enough to acknowledge that experiments that kill numerous animals and yield only slight benefits to humans will not cut the moral mustard.

Some theorists do not accept this moral distinction; they think there is no moral difference between what we do and what we permit. For them, this asymmetry provides no objection to animal experimentation. Although I have sympathies with this claim, it is not a position most defenders of research embrace. If defenders of animal experimentation do not think that doing harm is worse than failing to prevent harm, then they think that we should pursue any activity that yields benefits greater than that activity's costs. If we could achieve extremely important biomedical benefits only by invasive, nonconsensual experiments on humans, then these would be morally justified. This is a most unwelcomed consequence for most defenders of animal experimentation since they categorically reject nonconsensual invasive biomedical experiments on humans.[60] That denial cannot be defended by those who reject the first asymmetry. At most they can say that such experimentation could be justified only if the benefits were substantial, and because such conditions are rarely satisfied, then nonconsensual experiments on humans are rarely justified.

Even this line of defense will be difficult to hold. It is implausible to think that invasive experiments on non-consenting humans would never yield substantial biomedical benefits to many humans. Apparently, German experiments on inmates taught us a fair bit about treating burns and Japanese experiments on prisoners of war taught us about infectious agents. This should not be surprising. Humans are the best test subjects. If invasive nonconsensual experiments on humans are justified if the benefits are high enough, then nonconsensual experiments will sometimes be justified.

I am not taking a stance on the moral significance of the doing/allowing distinction. My claim is that most defenders of research will be uncomfortable either embracing or denying its significance. If advocates categorically reject invasive nonconsensual experiments on humans, then they must (a) think that nonhuman animals are devoid of moral worth or (b) believe it is categorically worse to commit an evil than to fail to prevent one. The first option clashes with their claim that animals' interests go on the moral scales. The second raises an additional justificatory hurdle to defending the practice since experimenters do harm to prevent harm. Perhaps, though, experimentation is acceptable if the benefits of experimentation are overwhelming.

Definite Harms are Morally Weightier than Possible Benefits

To make matters worse for consequentialist defenders of experimentation, the trade-off is not between harm we do to animals and human suffering we fail to alleviate. That description masks the fact that the suffering of animals is definite while benefits to humans are merely possible. It is sometimes legitimate to give up some definite benefit B in the hope of obtaining a greater benefit B1—if B1 is sufficiently great. For instance, I might give up $10 to obtain a 10% chance of gaining $200. Generally speaking, it is reasonable for me to forego a definite benefit B for another benefit B1, if the product of the utility and probability of B1's occurring is much greater than the utility of B (being definite, its probability is 1). Therefore, researchers must show that the product of the probability and utility of benefits to humans is greater than the product of animals' definite harm (adjusted for their diminished value) and the number of animals who suffer.

Demanding that researchers establish that any particular experiment will be successful is too stringent. The issue is whether we can reliably predict that the *practice* of experimentation will produce sufficient benefits for humans, benefits that outweigh the costs to nonhuman animals. We will have difficultly doing so because both the utility and the probability of the practice are unknown, while the harm to animals is substantial and definite.

Rejecting this second assumption also comes at considerable cost. It would be the height of foolishness to give up any good G1 for the mere chance of obtaining some other good G2 if G2 were not greater than G1. Abandoning this assumption would be to abandon rationality itself.

What Really Goes on the Scales?

Cohen's accounting of what goes on the moral scales is incomplete. When determining the benefits and costs of animal experimentation, we must include not only the costs to animals (which are direct and substantial), but also the costs to humans (and animals) of misleading experiments. I earlier noted that J. R. Paul claimed that adherence to animal models of polio delayed the development of a vaccine for more than two decades. Many lives were lost or ruined because of this delay.

Animal experiments also seriously misled us about the dangers of smoking. By the early 1960s, human epidemiological studies showed a strong correlation between lung cancer and smoking.[61] Nonetheless, Northrup brushed off the claim that smoking caused cancer thusly, "The failure of many...investigators to induce experimental cancers, except in a handful of cases, during fifty years of trying, casts serious doubt on the validity of the cigarette-lung cancer theory."[62] Finally, an AIDS vaccine researcher has concluded, "The lack of an adequate animal model has hampered progress in HIV vaccine development."[63] These three cases show that there will be substantial costs *to humans* of relying so heavily on animal experimentation. We should count these costs.

Researchers insist that we should put possible benefits on the scales, since no benefits are certain. That is reasonable, at least if we also include possible costs. For instance, some people speculated that AIDS was transferred to the human population through an inadequately screened oral polio vaccine given to 250,000 Africans in the late 1950s. Such a claim has been widely repudiated.[64] However, even if it is false, something like it might be true. After all, we know that one dangerous simian virus (SV40) entered the human population through inadequately screened vaccine.[65]

Finally, and perhaps most importantly, what is crucial is not the benefits animal experimentation did and will produce, but the benefits that only it could produce. We must determine (a) the role that medical intervention played in lengthening life and improving health,[66] (b) the contribution of animal experimentation to medical intervention, and (c) the benefits of animal experimentation relative to those of nonanimal research. In sum, what goes on the moral scales are not all the purported benefits of experimentation, but only the *increase* in benefits relative to alternatives. Since we do not know what the alternatives would have yielded, determining that increase will be difficult. Minimally, though, we have no reason to think that none of the advances attributable to animal research would have been made without that research.

A Final Dilemma

Deontological Concerns

The previous discussion explores consequentialist concerns about animal experimentation. The practice also faces deontological objections. I mentioned the most obvious one earlier. Tom Regan claims that animals are subjects of a life, and, as such, cannot be used for human purposes.[67] Many animal activists embrace Regan's idea, though, rightly or wrongly, a majority of people reject it.[68] However, we can combine elements of Regan's view with some empirical arguments in this essay to frame a dilemma for defenders of animal experimentation.

Biomedical researchers claim (1) that biomedical experiments using animals are *scientifically* justified because (carefully selected) nonhuman animals are good models of human biomedical phenomena, and (2) that these experiments are *morally* justified because humans and nonhuman animals are morally relevantly different. To scientifically justify inferences from animals to humans, defenders must identify substantial and pervasive *causal* similarities between humans and nonhuman animals. To morally justify the practice they must find sufficient relevant *functional* differences between humans and nonhuman animals. Defenders of research claim it is easy to do the latter: humans have cognitive and emotional abilities that nonhuman animals lack, at least in sufficient degree.[69] As Cohen put it, "Animals...lack this capacity for free moral judgment. They are not beings of a kind capable of exercising or responding to moral claims. Animals therefore have no rights, and they can have none."[70]

As it turns out, there is mounting evidence that the mental lives of nonhuman animals are far richer than people historically supposed.[71] However, we can sidestep this question. Defenders of experimentation will have trouble supporting the combination of (1) and (2), whether the differences in mental abilities are great or slight.

To see why, we must understand how scientists explain the presence of cognitive and emotional traits in humans and their absence in animals. The usual answer is that humans have an advanced cerebral cortex, which nonhuman animals lack. Human mental superiority is reflected in differences between our respective "encephalization quotient" (EQ), the ratio of the "brain weight of a species with the brain weight of an average animal of the same approximate body weight....According to this formula, the actual brain size of humans comes out to six times what we would expect of a comparable mammal."[72] There is little doubt that the average human is more cognitively sophisticated than the average nonhuman animal, and that we can best explain this difference by differences in our respective brains. However, because biological systems are highly interconnected intact systems, it is implausible to think that human brains, and thus cognitive abilities, evolved without significant biological changes elsewhere in the organism. To think this could have happened researchers must embrace bio-Cartesianism.

Bio-Cartesianism

Descartes claimed that the mind and the brain are ontologically distinct substances operating in wholly different domains and then had a problem getting these substances to interact. Animal experimenters have unconsciously adopted a biological corollary—what Niall Shanks and I call bio-Cartesianism.[73] Animal researchers assume that the brain, although formed by the same evolutionary pressures that shape other biological systems, somehow developed independently of those other systems. This makes no evolutionary sense. Higher-order cognitive abilities evolved because they were advantageous to the creatures' survival, and, having developed, shaped those creature's biological systems and behavior:

> [S]ome types [of monkeys] have higher EQs than others and [that con-nects] …with how they make their living: insect-eating and fruit-eating monkeys have bigger brains for their size, than leaf-eating monkeys. It makes some sense to argue that an animal needs less computing power to find leaves, which are abundant all around, than to find fruit, which may have to be searched for, or to catch insects, which take active steps to get away.[74]

These evolved cognitive differences affect noncognitive biological systems; we must consider these differences in the practice of biomedicine. As one animal research handbook cautions:

> When selecting nonhuman primates because of their close relationship to humans, choice of species of nonhuman primate is important. For example, a completely vegetarian species may not be as useful because of differences in microflora of the intestine, which may affect drug metabolism.[75]

Once we understand the ways that cognitive functioning is related to other biological systems, we can state this deontological dilemma for defenders of research: they must embrace bio-Cartesianism to morally defend their practice and they must reject it to scientifically defend their practice. They embrace it by claiming that humans and animals are sufficiently different to morally permit animals' use as experimental sub-jects. They reject it by invoking the "intact systems" argument to scientifically defend the practice. Defenders of experimentation cannot have it both ways. If nonhuman animals and humans are sufficiently similar to think that inferences from the former to the latter are scientifically legitimate, then they are likely sufficiently similar cogni-tively to think that nonhuman animals have significant moral worth. If nonhuman animals and humans are sufficiently different functionally to morally justify the practice, then they are likely sufficiently different biologically so that we have greater reason to suspect that inferences from animals to humans will often be suspect.

CONCLUSION

I have tried to identify and evaluate arguments for biomedical experimentation using animals. Animal experimentation is not useless as critics sometimes aver. However, neither are the benefits of the practice as clear, direct, or compelling as defenders commonly claim. Likewise, I do not think that the moral arguments defending the practice are wholly wanting, nor are they as persuasive as defenders claim. There are significant moral costs of the practice.

Defenders of the practice carry the moral burden of proof. The moral onus always rests on anyone who wishes to harm sentient creatures, to do what is, all things being equal, a moral wrong. Because people on both sides of this debate acknowledge at least some level of moral status for nonhuman animals, defenders must provide clear evidence that the value of the institution of research exceeds its moral costs. I suspect that their most promising way of scientifically defending the practice would emphasize limited and focused basic research. The results of that research will rarely yield immediate and direct benefits. However, they arguably provide a broad understanding of biological processes that may suggest promising curative strategies. Whether such benefits are sufficient to morally defend the practice is another question.[76]

NOTES

1. A 2003 Gallup poll found "96% of Americans saying that animals deserve at least some protection from harm and exploitation," even if many reject the idea of "animal rights." "Public Lukewarm about Animal Rights," http://www.gallup.com/poll/8461/Public-Lukewarm-Animal-Rights.aspx (accessed January 18, 2010).

2. As far as I know, no one has ever used this particular classificatory scheme. However, I think it will soon become apparent that it nicely categorizes the range of moral views.

3. I am not using claims about "what most people believe" as a surreptitious way of grounding controversial moral claims. However, I think ethicists should begin by addressing common moral views, even if we later conclude that these views are mistaken. What justifies the claim that this is what most people believe? I am a competent speaker of the English language who has discussed these issues with thousands of students at dozens of universities worldwide.

4. Keith Thomas, *Man and the Natural World: Changing Attitudes in England 1500–1800* (London: Allen Lane, 1980), pp. 17, 25, and 31.

5. Lori Gruen, "Moral Status of Animals," *Stanford Encyclopedia of Philosophy*, ed. Edward N. Zalta (Summer 2003), http://plato.stanford.edu/entries/moral-animal/ (accessed February 17, 2010).

6. Immanual Kant, *Lectures on Ethics* (Indianapolis, Ind.: Hackett Pub. Co., 1980), p. 239.

7. David DeGrazia and Andrew Rowan, "Pain, Suffering, and Anxiety in Animals and Humans," *Theoretical Medicine* 12 (1991): 193–211.

8. Tom Regan, *The Case for Animal Rights* (Berkeley and Los Angeles: University of California Press, 1983), pp. 243–48.

9. R. G. Frey, "Moral Standing, the Value of Lives, and Speciesism," in *Ethics in Practice: An Anthology*, 3rd ed., ed. H. LaFollette (Oxford: Blackwell Publishers, 2007), pp. 192–204.

10. Regan, *Case for Animal Rights*, pp. 151–55.

11. R. J. Lipton, *Nazi Doctors: Medical Killing and the Psychology of Genocide* (New York: Perseus Publishing, 2000); Tuskegee Syphilis Study Legacy Committee, "A Request for Redress of the Wrongs of Tuskegee," (1996), http://hsc.virginia.edu/hs-library/historical/apology/report.html (accessed May 15, 2003).

12. American Medical Association, *Use of Animals in Biomedical Research: The Challenge and Response* (Chicago: American Medical Association, 1992), p. 11.

13. J. H. Comroe Jr. and R. D. Dripps, "Scientific Basis for the Support of Biomedical Science," *Science* 192 (1976): 105–11; J. H. Comroe Jr. and R. D. Dripps, "Ben Franklin and

Open Heart Surgery," *Circulation Research* 35 (1974): 661–69; S. Zola, "Basic Research, Applied Research, Animal Ethics, and the Animal Model of Human Amnesia," in *Why Animal Experimentation Matters: The Use of Animals in Medical Research*, ed. E. F. Paul and J. Paul (New York: Transaction Publishers, 2001), pp. 79–92.

14. Sigma Xi, "Sigma Xi Statement of the Use of Animals in Research," *American Scientist* 80 (1992): 73–76.

15. American Medical Association, *Use of Animals in Biomedical Research*, p. 27.

16. The first meta-analysis of the standard mouse model of ALS reveals that "'positive' results are often considered more interesting and hence more likely to be published." Michael Benatar, "Lost in Translation: Treatment Trials in the SOD1 Mouse and in Human ALS," *Neurobiology of Disease* 26 (2007): 6.

17. Psychologists claim that humans are often mislead by "confirmatory biases." See David Dunning's *Self-Insight: Roadblocks and Detours on the Path to Knowing Thyself* (New York: Psychology Press, 2005), pp. 46ff.

18. B. Patoine, "Pitfalls and Promise on the Road to ALS Therapeutics," The Dana Foundation (2009), http://www.dana.org/news/publications/detail.aspx?id=19766 (accessed September 12, 2009).

19. ALS Association, "Trial of Arimoclomol in SOD1–Positive Familial ALS Opens for Enrollment," (2009), http://www.alsa.org/patient/drug.cfm?id=1414 (accessed July 27, 2009).

20. Jim Schnabel, "Standard Model," *Nature* 454 (2008): 683.

21. L. B. Lave, F. K. Ennever, H. S. Rosencrantz, and G. S. Omenn, "Information Value of the Rodent Bioassay," *Nature* 336 (1988): 631–33.

22. L. S. Gold et al., *Misconceptions about the Causes of Cancer* (Vancouver: Fraser Institute, 2002), p. 35; L. S. Gold, T. Slone, N. Manley, and L. Bernstein, "Target Organs in Chronic Bioassays of 533 Chemical Carcinogens," *Environmental Health Perspectives* 93 (1991): 233–46.

23. J. R. Paul, *The History of Poliomyelitis* (New Haven, Conn.: Yale University Press, 1971), pp. 384–85.

24. L. Lasagna, "Regulatory Agencies, Drugs, and the Pregnant Patient," in *Drug Use in Pregnancy*, ed. L. Stern (Boston: ADIS Health Science Press, 1982), p. 15.

25. B. M. Mitruka, H. M. Rawnsley, and D. V. Vadehra, *Animals for Medical Research: Models for the Study of Human Disease* (New York: Wiley, 1976), pp. 467–68.

26. A. Gorbman et al., *Comparative Endocrinology* (New York: Wiley, 1983), p. 33.

27. J. Hart, "Endocrine Pathology of Estrogens: Species Differences," *Pharmacologial Therapy* 47 (1990): 203–18.

28. American Medical Association, *Use of Animals in Biomedical Research: The Challenge and Response*; Hugh LaFollette and Niall Shanks, *Brute Science: Dilemmas of Animal Experimentation* (New York: Routledge, 1996).

29. Claude Bernard, *An Introduction to the Study of Experimental Medicine* (Paris: Henry Schuman, Inc., 1949), p. 125.

30. Hugh LaFollette and Niall Shanks, "Two Models of Models in Biomedical Research," *Philosophical Quarterly* 45 (1995): 141–60.

31. B. Alberts et al., *Molecular Biology of the Cell*, 5th ed. (New York: Taylor & Francis, 2008), p. 1207.

32. Lee, "Acquired Immunodeficiency Disease Vaccines: Design and Development," in *AIDS: Biology, Diagnosis, Treatment and Prevention*, fourth edition, edited by V. T. DeVita, Jr (Philadelphia, PA: Lippincott-Raven Publishers, 1997), p. 609.

33. Alberts et al., *Molecular Biology of the Cell*, p. 205.

34. Coyne, Jerry, *Why Evolution Is True* (New York: Viking, 2009) p. 81.

35. R. M. Nesse and G. C. Williams, *Why We Get Sick: The New Science of Darwinian Medicine* (New York: Crown Publishers, 1995), p. 4.

36. Coyne, *Why Evolution Is True*, pp. 57–79.

37. Alberts, et al. *Molecular Biology of the Cell*, p. 249.

38. Zola, "Basic Research, Applied Research, Animal Ethics, and the Animal Model of Human Amnesia," pp. 79–92.

39. Gold et al., *Misconceptions about the Causes of Cancer*, p. 34.

40. J. C. Caldwell, "Comparative Aspects of Detoxification in Mammals," in *Enzymatic Basis of Detoxication*, vol. 1, ed. W. Jakoby (New York: Academic Press, 1980), pp. 87–110.

41. E. A. Hodgson, *Textbook of Modern Toxicology* (Hoboken N.J.: Wiley Interscience, 2004), p. 177.

42. Alberts et al., *Molecular Biology of the Cell*, p. 205.

43. Coyne, *Why Evolution Is True*, pp. 62–80.

44. J. Bailey, "An Assessment of the Role of Chimpanzees in AIDS Vaccine Research," *Alternatives to Laboratory Animals* 36 (2008): 381–428. See pages 386–409 and 412–15 for detailed information drawn from the International AIDS Vaccine Initiative.

45. J. Lauerman, "Merck Aids Vaccine Failure May Doom Promising Study," *Bloomberg.com*, 2009, http://www.bloomberg.com/apps/news?pid=20601109&sid=axb42ai9OAlw (accessed August 22, 2009).

46. M. Falco, "Combo Vaccine Reduces Risk of HIV Infection, Researchers Say," *CNN*, 2009, http://www.cnn.com/2009/HEALTH/09/24/hiv.vaccine/index.html (accessed August 18, 2009).

47. Bernard, *Introduction to the Study of Experimental Medicine*, p. 125.

48. Comroe and Dripps, "Scientific Basis for the Support of Biomedical Science;" Comroe and Dripps, "Ben Franklin and Open Heart Surgery," pp. 661–69.

49. Health Economics Research Group (London: Brunel University, 2003), p. iii.

50. Stephen H. Kellert, *In the Wake of Chaos: Unpredictable Order in Dynamical Systems* (Chicago: University of Chicago Press, 1993).

51. J. C. Caldwell, "Comparative Aspects of Detoxification in Mammals," pp. 85–113 and 106.

52. Carl Cohen, "The Case for the Use of Animals in Biomedical Research," *New England Journal of Medicine* 315 (1986): 865–70, here p. 868.

53. J. Smith, and K. Boyd, *Lives in the Balance: The Ethics of Using Animals in Biomedical Research* (Oxford: Oxford University Press, 1991), pp. 138–39.

54. LaFollette and Shanks, *Brute Science*, pp. 247–61.

55. Cohen, "Case for the Use of Animals in Biomedical Research," p. 868.

56. H. M. Malm, "Bad Samaritan Laws: Harm, Help, or Hype," *Law and Philosophy* 19 (2000): 707–50, here pp. 744–45.

57. S. R. L. Clark, "How to Calculate the Greater Good," in *Animals' Rights: A Symposium*, ed. D. Patterson and R. Ryder (London: Centaur Press, 1977), pp. 96–105.

58. Such a view is clearly implied by John Arthur in his "Rights and the Duty to Bring Aid," in *World Hunger and Morality*, ed. Will Aiken and Hugh LaFollette (Upper Saddle River, N.J.: Prentice-Hall, 1996), pp. 39–50.

59. See Institute for Laboratory Animal Research, *Guide for the Care and Use of Laboratory Animals* (Washington, D.C.: National Academies Press, 1996), p. 2, http://www.nap.edu/openbook.php?record_id=5140&page=1.

60. Cohen, "Case for the Use of Animals in Biomedical Research," p. 866; American Medical Association, *Use of Animals in Biomedical Research*, pp. 1, 7.

61. See E. Brecher, and R. Brecher, *The Consumers Union Report on Smoking and the Public Interest* (Mount Vernon, N.Y.: Consumers Union Press, 1963).

62. E. Northrup, *Science Looks at Smoking: A New Inquiry into the Effects of Smoking* (New York: Coward McCann, 1957), p. 133.

63. Lee, "Acquired Immunodeficiency Disease Vaccines," p. 609.

64. For a different view, see B. F. Elwood, and R. B. Striker, "Polio Vaccine and the Origin of AIDS" (Letter to the Editor), *Research in Virology* 144 (1993): 175–77.

65. Leonard Hayflick, "Human Virus Vaccines: Why Monkey Cells?" *Science* 733 (1972): 813–14.

66. T. McKeown, *The Modern Rise of Populations* (New York: Academic Press, 1976), especially chapter 5.

67. Regan, *Case for Animal Rights*, pp. 243–45.

68. Gallup Poll, "Public Lukewarm about Animal Rights."

69. M. A. Fox, *The Case for Animal Experimentation: An Evolutionary and Ethical Perspective* (Berkeley and Los Angeles: University of California Press, 1986).

70. Cohen, "Case for the Use of Animals in Biomedical Research," p. 866.

71. D. G. Griffin, *Animal Minds* (Chicago: University of Chicago Press, 1992); R. J. Hoage, and L. Goldman, eds., *Animal Intelligence: Insight into the Animal Mind* (Washington, D.C.: Smithsonian Institution Press, 1986); C. Allen and M. Bekoff, *Species of Mind: The Philosophy and Biology of Cognitive Ethology* (Cambridge, Mass.: MIT Press, 1997); M. Bekoff, C. Allen, and G. Burghardt, eds., *The Cognitive Animal* (Cambridge, Mass.: MIT Press, 2002).

72. Fox, *Case for Animal Experimentation*, p. 38.

73. LaFollette and Shanks, *Brute Science*, pp. 238–41.

74. M. A. Edey and D. C. Johnson, *Blueprints: Solving the Mystery of Evolution* (New York: Penguin Books, 1989), pp. 189–90.

75. Mitruka et al., *Animals for Medical Research*, p. 342.

76. I am indebted to Niall Shanks for the collaborative work that informs important elements of this essay. I am grateful to Eva LaFollette for her insightful philosophical comments and stylistic suggestions. Finally, I thank Tom L. Beauchamp and R. G. Frey for helpful comments on an earlier version of this paper.

SUGGESTED READING

BEAUCHAMP, TOM L. "Opposing Views on Animal Experimentation: Do Animals Have Rights?" *Ethics and Behavior* 7 (1997): 113–21.

BEKOFF, MARK, COLIN ALLEN, and GORDON BURGHARDT, eds. *The Cognitive Animal.* Cambridge, Mass.: MIT Press, 2002.

BERNARD, CLAUDE. *An Introduction to the Study of Experimental Medicine.* Paris: Henry Schuman, Inc., 1949.

CAIRNS-SMITH, A. G. *Seven Clues to the Origin of Life.* Cambridge: Cambridge University Press, 1985.

COHEN, CARL. "The Case for the Use of Animals in Biomedical Research." *New England Journal of Medicine* 315 (1986): 865–70.

COMROE JR., J. H., and R. D. DRIPPS. "Scientific Basis for the Support of Biomedical Science." *Science* 192 (1976): 105–11.

DEGRAZIA, DAVID, and ANDREW ROWAN. "Pain, Suffering, and Anxiety in Animals and Humans." *Theoretical Medicine* 12 (1991): 193–211.

DRESSER, REBECCA. "Standards for Animal Research: Looking at the Middle." *Journal of Medicine and Philosophy* 13 (1988): 123–43.

EDEY, MAITLAND A., and DONALD C. JOHNSON. *Blueprints: Solving the Mystery of Evolution.* New York: Penguin Books, 1989.

FOX, MICHAEL A. *Case for Animal Experimentation: An Evolutionary and Ethical Perspective.* Berkeley and Los Angeles: University of California Press, 1986.

GRANT, JONATHAN, LIZ GREEN, and BARBARA MASON. *From Bedside to Bench: Comroe and Dripps Revisited.* Uxbridge, U.K.: Health Economics Research Group, Brunel University, 2003.

LAFOLLETTE, HUGH, and NIALL SHANKS. *Brute Science: Dilemmas of Animal Experimentation.* New York: Routledge, 1996.

MAYR, ERNST. *Toward a New Philosophy of Biology.* Cambridge, Mass.: Harvard University Press, 1988.

NESSE, RANDOLPH M., and GEORGE C. WILLIAMS. *Why We Get Sick: The New Science of Darwinian Medicine.* New York: Crown Publishers, 1995.

NORCROSS, ALASDAIR. "Animal Experimentation." In *Oxford Handbook of Bioethics*, edited by Bonnie Steinbock, pp. 648–67. Oxford: Oxford University Press 1997.

REGAN, TOM. *The Case for Animal Rights.* Berkeley and Los Angeles: University of California Press, 1983.

SHANKS, NIALL, and RAY GREEK. *Animal Models in Light of Evolution.* Boca Raton, Fla.: Brown Walker Press, 2009.

SIGMA XI. "Sigma Xi Statement of the Use of Animals in Research." *American Scientist* 80 (1992): 73–76.

SINGER, PETER. *Animal Liberation: A New Ethics for Our Treatment of Animals.* 2nd ed. New York: New York Review, 1975.

SMITH, J. A., and KENNETH M. BOYD, eds. *Lives in the Balance: The Ethics of Using Animals in Biomedical Research.* Oxford: Oxford University Press, 1991.

CHAPTER 30

....

ETHICAL ISSUES IN THE APPLICATION OF BIOTECHNOLOGY TO ANIMALS IN AGRICULTURE

....

ROBERT STREIFFER AND JOHN BASL

1. INTRODUCTION

....

The controversy about the use of modern biotechnology in agriculture erupted in the early 1990s when the United States Food and Drug Administration (FDA) approved recombinant bovine growth hormone, a chemical produced using genetically engineered microorganisms and injected into dairy cows to increase milk yield.[1] Public debate continued as farmers—primarily in the United States, but also in Argentina, Brazil, Canada, and other countries—increasingly planted genetically engineered soybeans, corn, canola, and cotton. Recent biotechnology research includes the development of genetically engineered animals and cloned animals to be used as food or breeding stock in agriculture. This chapter provides a framework for evaluating the two most important aspects of these new applications of modern biotechnology to the food supply.

Most of the research on genetically engineered livestock has focused on cattle, sheep, goats, pigs, chickens, and fish.[2] Cloning (using somatic cell nuclear transfer) was first successfully performed in 1996, producing Dolly the sheep. Among other livestock species, cloning has now been successfully performed using cattle, pigs, and goats.[3] The aims of the research are usually the same as those of traditional breeding, improving traits such as feed efficiency, growth rates, fat-to-muscle ratio, and resistance to pests and diseases.[4] There is no food containing meat from genetically engineered animals yet on the market, but in 2009, the FDA released its "Guidance for Industry" on the regulation of genetically engineered animals, and there are reportedly two genetically engineered animals under review for commercialization: Enviropigs, engineered to produce lower levels of phosphorus pollution in their manure, and AquAdvantage salmon, engineered to reach their harvest weight twice as fast as conventional salmon.[5] And while there is a voluntary moratorium in the United States on the sale of milk and meat from cloned cattle or their offspring, the FDA cleared the way to commercialization in 2008 when its officials concluded that these are as safe as their conventional counterparts.

Genetically engineered and cloned livestock will add a more controversial dimension to the debate about biotech foods for three reasons. First, the social science data strongly support the conclusion that the public is even more concerned about animal biotechnology than plant biotechnology.[6] The Food Policy Institute at Rutgers found that, in the case of genetically engineered animals, 61% of the U.S. public disapproved or leaned toward disapproval, compared to 41% in the case of genetically engineered plants.[7] The Pew Initiative on Food and Biotechnology found that people's mean comfort rating with genetically engineered animals was only 3.81 out of 10, compared with 6.08 out of 10 with genetically engineered plants.[8]

Second, many of the animals used in agriculture are sentient beings, capable of consciously experiencing pleasure, pain, and other enjoyable or aversive mental states. This is true of all mammals and most likely extends to all vertebrates and to cephalopods.[9] As sentient beings, these individuals have a welfare, which confers some degree of moral status. Harms and benefits to them matter morally, regardless of whether anyone else is affected. The independent moral importance of animals takes on new significance in light of the size of the livestock sector. The number of animals consumed in agriculture each year is estimated to be around 146 billion, approximately 56 billion land animals and 90 billion aquatic animals.[10] By comparison, the number of animals used in research each year is estimated to be below 100 million, a little more than half of one-tenth of 1% of the number of animals consumed in agriculture.[11] The number of animals consumed for food every eight months is roughly equal to the total number of Homo sapiens that have ever lived in the entire fifty-thousand-year history of the species.[12]

Third, livestock agriculture, like agriculture more generally, has a tremendous impact on the environment, both locally and globally. In the most authoritative report to date, the United Nations Food and Agriculture Organization (FAO) concluded that "The livestock sector emerges as one of the top two or three most significant contributors to the most serious environmental problems, at every scale from local to global."[13]

The livestock sector's massive scale means that routine agricultural practices that are detrimental to animal welfare or to the environment pose some of the most pressing global ethical issues the human species has ever encountered. An application of biotechnology to animals in agriculture ("animal biotech" for short) can either mitigate or exacerbate these animal welfare or environmental problems, which provides a useful way to begin evaluating specific applications. Determining how well an application of animal biotech fares in comparison to current practices with respect to these two kinds of issues is not sufficient to yield an overall evaluation of issues raised by that application, which would require an evaluation of food labeling, food safety and nutrition, impact on food prices and availability, and socioeconomic effects on rural communities. Nonetheless, comprehensive assessment will generally not be possible without determining the impact of that application on animals and the environment. Furthermore, as will become evident, evaluating animal biotech's impact on these central issues will serve to deepen our understanding of the various philosophical dimensions these applications raise.

The discussion that follows is divided into two main sections. In section 2, we discuss the issue of animal welfare and the prospects for animal biotech to improve or exacerbate existing animal welfare issues that arise in agriculture. Evaluating the impact of animal biotech on animal welfare requires an understanding of what constitutes animal welfare (section 2.a), what constitutes an improvement or worsening of animal welfare (section 2.b), and of the current welfare issues that arise for animals in agriculture (section 2.c). After discussing each of these, we consider a series of applications of animal biotech and discuss their impact on welfare (sections 2.d and 2.e). In section 3, we focus on the prospects for animal biotech to improve or worsen environmental problems. Livestock agriculture is a significant contributor to a range of environmental issues, including climate change (section 3.a) and local environmental issues such as water pollution, water use, and biodiversity loss (section 3.b). We then discuss the potential pathways by which animal biotech might affect environmental issues (section 3.c). To explore these issues in more detail, we consider Enviropigs (section 3.d) and AquAdvantage salmon (section 3.e) from an environmental perspective.

2. ANIMAL BIOTECH
AND ANIMAL WELFARE

2.a Theories of Animal Welfare

It is uncontroversial that one way an application of animal biotech can have a morally relevant impact is by its effect on an animal's welfare (its well-being, quality of life, or how well its life is going in the broadest sense). Insofar as an application of animal biotech provides a net increase of animal welfare compared to current practices, there is a moral reason in favor of using that application. And, of course, insofar as an application decreases animal welfare compared to current practices, there is a moral reason against using that application.

But while it is uncontroversial that animal welfare matters, there is controversy surrounding what defines an animal's welfare. Before turning to particular theories of animal welfare, though, it is important to distinguish the constituents or components of animal welfare from both (a) the circumstances or resources that typically make a positive contribution to welfare and (b) the indicators or surrogate measures used as evidence of welfare. For example, providing an injured animal with a splint could contribute to its welfare, but the splint is not itself a constituent of welfare. In contrast, being free of pain is clearly a constituent of an animal's welfare. Other examples are less clear: some would argue that mating is itself a constituent of welfare for sexually reproducing species, while others would argue that it merely contributes to welfare by reducing frustration or causing pleasure. At any rate, a theory of welfare is concerned in the first instance with the constituents of animal welfare, although it will of course inform discussions of (a) and (b).

So, what constitutes animal welfare? Mentalistic views of welfare characterize well-being entirely in terms of the presence or absence of mental states. Hedonism is the most well-known Mentalistic view. Standard Mentalistic views of *animal* welfare characterize animal welfare as the absence of negative mental states—such as distress, fear, pain, and suffering. Broader Mentalistic views of animal welfare also include positive mental states (such as happiness, joy, and pleasure) as components of animal welfare.

Mentalistic views as general theories of welfare suffer from several problems. Pain is often taken to be the paradigm example of a negative state, the absence of which counts as a component of welfare. However, it is acknowledged nowadays that pain has both a sensory and an attitudinal aspect. The sensory aspect refers to the phenomenological qualities of the painful experience, whereas the attitudinal aspect refers to the aversive response to the phenomenological qualities. Although rare, there are documented cases of humans who report experiencing sensory pain but who lack the attitudinal aspect, that is to say, they don't mind the pain.[14] How to analyze such cases is controversial. Some people argue that even though the pain isn't minded, it is still a negative component of welfare because of the phenomenological aspect. Others argue that if the individual does not have an aversive response to the sensory experience, then the pain is not a negative component of welfare. In the case of most agricultural animals, this distinction is irrelevant because the pain will almost invariably be unwanted. However, there is emerging research on fish that suggests that although they are capable of the sensory aspect of pain, they may be incapable of the attitudinal aspect of pain.[15] If this is right, and if the attitudinal aspect is essential for pain's being a negative component of welfare, then this will have important implications for the welfare evaluation of pain.

Mentalistic views also ignore the possibility that the frustration or satisfaction of desires can impact welfare without having any impact on the subjective experiences of the individual in question. Examples are easy to identify in the case of human beings (e.g., a desire to have a faithful spouse and a desire to have one's work appreciated after one is dead), and such examples provide one of the primary reasons many people find Desire-Satisfaction views, which characterize welfare entirely in terms of the satisfaction or frustration of the individual's desires, more plausible than Mentalistic views.

Desire-Satisfaction views allow that an individual's mental states play a foundational role in his or her well-being, and so are similar in this respect to Mentalistic views. Unlike Mentalistic views, however, Desire-Satisfaction views allow that the satisfaction of desires for external states of affairs having nothing to do with one's mental states can be components of one's welfare. For example, a person whose spouse is unfaithful has a life that goes less well, other things being equal, even if the affair is never discovered. Desire-Satisfaction views of animal welfare are not very well-developed in the literature, even though Desire-Satisfaction views of human well-being are widely discussed and are often held to be an improvement over traditional Mentalistic views.

Desire-Satisfaction views distinguish between actual desires and informed desires. It is widely acknowledged that the satisfaction of actual desires is not always a component of welfare. For example, people often have desires based on false beliefs and the satisfaction of those desires would not necessarily be a component of their welfare. This leads most proponents of a Desire-Satisfaction view to adopt the requirement that, in order for a desire's satisfaction to count as part of the individual's welfare, the desire must be an "informed desire," that is, one capable of persisting in the face of correct information and rational reflection.[16] However, the requirement that desires, to be components of welfare, be informed seems to support the idea that the states of affairs themselves are components of welfare, making the appeal to the desire unnecessary. As James Griffin says, "What makes us desire the things we desire, when informed, is something about them—*their* features or properties. But why bother then with informed desire, when we can go directly to what it is about objects that shape and form desires in the first place?"[17]

Such considerations support Objective List views, according to which there are some goods that contribute to welfare regardless of whether they contribute to the subjective experience of the individual or whether they are or would be desired by him or her. Objective List views are distinguished on the basis of what objective features of a life they take to constitute welfare. In the case of humans, the kind of Objective List views that are found to be most plausible are hybrid views that include mentalistic components, desire-based components, the satisfaction of basic psychological and biological needs, and the development and exercise of valuable cognitive and emotional capacities. David Brink, for example, includes the autonomous undertaking and completion of "projects whose pursuit realizes capacities of practical reason, friendship, and community."[18] The realization of some of these goods requires significant cognitive capacities, many of which will be beyond those typical of livestock. Objective List views of animal well-being often focus on health or natural functioning.

Health views of animal welfare characterize welfare in terms of the biological, and in some cases psychological, health of the animal, typically measured in terms of life expectancy and morbidity. Alternatively, according to Natural-Functioning views, welfare is constituted by an animal's functioning—behaviorally, physiologically, and developmentally—in species-typical ways.[19] For example, nest building is a species-typical behavior for many species of birds. For such species, engaging in nest building will be a component of welfare on a Natural-Functioning view, independently of the behavior's effects on health.

Health and Natural-Functioning views do not seem to capture everything that is a component of welfare. Individuals may be benefitted in not developing and behaving in species-typical ways. For example, analgesia prevents a species-typical response, that is, pain, but can contribute to welfare. An animal may experience negative mental states such as boredom that detract from welfare long before the animal's health becomes compromised. Health and Natural-Functioning views are implausible because they ignore mentalistic components of an animal's life that are relevant to its welfare.

A broader, hybrid Objective List view could include mentalistic components, satisfaction of desires, and health and natural-functioning components. Such a view could accommodate suggestions by some commentators to include animal dignity and animal integrity as components of welfare. Animal integrity has been characterized as the wholeness or intactness of the organism, as well as its "species-specific balance" and its capacity to sustain itself in an environment suitable for its species.[20] Animal dignity has been characterized as "the uninhibited development of the functions that a member of its species can normally perform."[21] In our taxonomy, animal dignity and animal integrity are roughly equivalent to health-based and natural-functioning-based components of welfare.[22]

Although a full defense is beyond the scope of this article, our view is that a hybrid Objective List view is the most plausible in the case of animals, just as it is in the case of humans. In the case of livestock, though, the list of welfare components will be dominated by mentalistic components, because many of the characteristics of humans that make a wider range of goods accessible to them are absent or present only to a lesser degree in livestock. These components will include the absence of negative mental states, but, diverging from typical Mentalistic views of animal welfare, also the presence of positive mental states as components of welfare. As Franklin McMillan notes, experimental and ethological research has "convincingly demonstrated that many of the sources of joy in humans have been described in animals, including play, competition, discovery, creativity, eating and drinking, companionship, and recognition of familiar persons or animals."[23]

Despite our focusing on mentalistic components of welfare, we believe that the components of welfare included in other views are often contributors to or provide important evidence about the welfare of animals. For example, the frustration of a desire will often result in a subjective experience of frustration; illness and disease often result in unpleasant experiences; and animals that develop and behave in species-typical ways are more likely to have satisfying mental experiences. Thus, the kinds of considerations that are typical of discussions of animal welfare are often relevant, even if they do not, strictly speaking, count as components of animal welfare in our view.

Moreover, many of the claims about what harms or improves animal welfare will hold no matter which view of animal welfare is correct, although some of the applications discussed below are philosophically interesting precisely because they improve welfare according to one view, but harm it according to another. These cases, described in section 2.e, will serve to highlight the differences between theories of animal welfare and the need to be clear about which theory is correct.

2.b Two Ways of "Improving" Animal Welfare

Given that animal biotech will in many cases result not in alterations of animals that already exist, but in the creation of different animals with new properties, the issue arises of whether there is a morally relevant difference between altering individuals and substituting one individual for another. In this section we explain this distinction, but argue that they are morally on a par with each other.

A person might be making one of two different claims in saying that a technology improves (or harms) animal welfare. One claim is that there is an individual who is made better off by that application. For example, a cow that is sick and receives antibiotics is made better off. In this case, the application is good for the animal simply in virtue of its having a positive overall effect on the animal's life-long welfare. There is clearly a reason in favor of using that application, a reason grounded in the fact that the application made that individual's welfare better.

In many cases where welfare problems are due to genetics, however, this simple picture no longer applies. Cloning and methods of pre-conception genetic modification, such as selective breeding, bring into existence individuals that are distinct from the ones who would have been brought into existence had the cloning or selective breeding not occurred. Thus, one cannot truthfully say that there is an individual whose welfare was improved by the application. Rather, there is an individual brought into existence who has a certain level of welfare, and that individual is at a higher level of welfare than the level at which a distinct individual, who did not in fact come into existence, would have been. So while aggregate welfare has been improved, it cannot truthfully be said that any individual's welfare was improved. Let us refer to cases of the first kind as alterations, because a single individual's welfare is altered, and cases of the second kind as substitutions, because one animal is substituted for another.[24]

Cloning and selective breeding will count as substitutions. Genetic engineering usually takes place post-conception, and so the founder animals will count as being altered. The individuals that result from the subsequent breeding of the founder animals, however, will count as substitutions.

Whether alterations and substitutions are morally different is a matter of considerable controversy. On the one hand, an alteration is bad for the altered animal if and only if it has a negative impact overall on the animal's life-long welfare, whereas a substitution is bad for the animal brought into existence only in the extreme case that the animal will have a life so bad that it is not, on balance, worth living at all. Many alterations will be bad for an animal even though the similar substitution would not be, and this would suggest that there is a morally relevant difference.[25] On the other hand, consideration of similar cases supports the view that alterations and substitutions are morally on a par with each other. Consider the following:[26]

> Case 1: A woman is currently pregnant with a child that she knows will be born mentally handicapped if not treated. She is told that by taking a safe and affordable medication, she can prevent the child from being born with the mental handicap. She chooses not to take the medication and the child is born with the handicap.
> Case 2: A woman is contemplating becoming pregnant but is told by her doctor that if she conceives within the next month any resulting child will be born with a mental handicap. If she waits to become pregnant, the resulting child will not have the mental handicap. The woman decides against waiting, becomes pregnant, and the resulting child has a mental handicap.

Case 1 is an alteration whereas Case 2 is a substitution. However, assuming that the handicap and the women's reasons for their decisions are the same in both cases, then the women's actions seem equally morally problematic.[27]

The fact that, on the one hand, there seems to be a morally relevant difference between alterations and substitutions but, on the other hand, the women's actions seem equally morally problematic suggests the following: substitutions and alterations are, other things being equal, on a par with each other, but a different explanation must be given of why a substitution is morally problematic, one that does not appeal to the action's being bad for an individual. The problem of adequately articulating this explanation is known as the "nonidentity problem." It is the subject of considerable controversy, but we will simply assume that a substitution that results in animals being brought into existence with a welfare that is worse than the welfare of animals that would have been brought into existence instead, even if an animal has a life worth living, is as morally problematic (or, is nearly as morally problematic) as the similar alteration would have been. Similarly, a substitution that results in animals with welfare that is better than that of the animals currently used in agriculture is a moral improvement, other things being equal.

2.c Existing Animal Welfare Issues

Evaluating whether a given application of animal biotech improves or worsens animal welfare requires an understanding, not only of the theories of animal welfare discussed above, but also how things currently stand with respect to animal welfare. It is clear that many agricultural practices pose significant animal welfare issues. In this section, we review some of these problems, which provide opportunities for animal biotech to make morally significant improvements in livestock agriculture, but which also provide a cautionary tale. Traditional agricultural technology, directed primarily by parties whose interests lie in greater intensification and increased productivity, has resulted in numerous welfare problems. Allowing animal biotech to be directed by those same interests threatens to make animal welfare problems worse, rather than better.

Starting around the 1950s, housing practices in some parts of the livestock sector significantly reduced the amount of space afforded each animal.[28] Farmers shifted from having small outdoor flocks of egg-laying hens (layers) to keeping flocks averaging 60,000 layers in battery cages that provide only 118–137 square inches of space per bird (roughly 11 inches per side).[29] On the positive side, these cage systems have helped keep infectious disease and hygiene problems manageable in intensive poultry systems, a fact relevant on any plausible view of welfare. On the negative side, these small cages prevent nesting behavior, dust bathing, perching, standard postures, wing flapping, and even tail wagging (problems on Natural-Functioning views), frustrating many natural desires (problems on Desire-Satisfaction views) resulting in significant pain and distress (problems on Mentalistic views) and significant morbidity and mortality (problems on Health views).[30] As one significant example: shortly after the widespread introduction of battery cages, producers began identifying what was called "cage layer fatigue," where hens would be unable to stand up but were still willing to eat and drink. The cause was ultimately determined to be severe osteoporosis caused by the battery cages' restrictions on movement. One study found that 33% of layers had freshly broken bones just prior to slaughter.[31]

In swine production, average herd size increased from 20 head in the 1950s to 766 head in 2002, while housing practices changed from groups of pasture-raised animals to individual confinement for much of the pig's life-cycle.[32] Pregnant sows are often kept individually in gestation crates, typically 2 to 21/2 feet wide by 7 feet long by 3 feet high, and sows that have just given birth are kept in similarly sized farrowing cages. Boars also are often kept in individual stalls.

These housing practices have several welfare implications. Individual housing guarantees access to food and water and prevents aggressive behavior, but the small space to which pigs are restricted by gestation crates and farrowing cages (not even enough to allow the pig to stand up, walk, or turn around) results in boredom, atypical repetitive behaviors (so-called stereotypies), skin lesions, and bone problems due to lack of movement and an inability to rest in species-typical positions. Farrowing cages reduce the incidence of mothers lying down on top of their newborn piglets, but also frustrate nesting behaviors for which the mothers have a strong instinct.[33] Only during the final stages of growth are swine sometimes allowed to live in groups on pastures, but this also causes welfare problems from tail biting, ear biting, and general aggression arising from unstable social groups.[34]

Breeding for increased productivity has also caused significant welfare problems. Problems from heat stress, identified by some as the largest welfare problems for western dairy operations, have been exacerbated by selection for increased productivity, which almost always increases metabolic heat output and makes it more difficult for cattle to manage heat from their surrounding environment.[35]

Although the number of dairy cows in the United States has steadily declined since 1940 (from 25 million to 9 million), daily milk production per cow has increased dramatically (12 lbs in 1940 to 55 lbs in 2006), largely due to breeding for increased productivity.[36] The feed required for such high production often causes subclinical ruminal acidosis (a kind of indigestion), which in turn affects the blood flow to the feet and hooves, causing lameness in approximately 17% of all dairy cattle. The feed also frequently causes gas to collect in the abomasum gastric compartment, making it float in a way that prevents normal digestion.[37] Selection for increased milk yield has also resulted in mastitis (an inflammation of the mammary gland caused by infection) becoming one of the most common dairy cattle diseases, with many countries reporting incident rates of around thirty cases per hundred cows.[38]

Many common animal management practices also raise significant welfare issues. Painful surgeries are routinely performed without anesthetic or analgesic.[39] Swine, dairy cows, and sheep routinely have their tails docked, either using a hot docking iron or an elastic band that cuts off blood flow to the tail. Dehorning of cattle by caustic paste or hot iron is common. Boars are often castrated to prevent "boar taint," an unpleasant taste in pork. Other surgeries include teeth clipping in piglets, debeaking in laying hens and broiler chickens, de-toeing of laying hens, broiler chickens, and turkeys, and desnooding of turkeys. Thirteen percent of dairy cattle in the United States are branded.[40] Although improvements have recently been made in slaughter and preslaughter handling techniques (largely due to the pioneering work of Temple Grandin, a leading expert on animal welfare science), it is still true that animals often unnecessarily experience fear and excruciating pain during the slaughter process due to problems with stunning and handling equipment, facility design and maintenance, and poor employee training and supervision.[41]

2.d Potential Impact of Animal Biotech on Animal Welfare

The above examples of agricultural practices that raise animal welfare issues indicate some of the areas where animal biotech has the potential to affect welfare, either by altering the individual animals so that their welfare is improved or by changing which animals are brought into existence. A comprehensive literature review was performed in 2002 by the National Research Council (NRC) of the National Academies, looking at the potential of animal biotechnology to cause "pain, distress (both physical and psychological), behavioral abnormality, physiologic abnormality, and/or health problems; and, conversely, their potential to alleviate or to reduce these problems."[42]

The NRC documented several ways that cloning and genetic engineering could be used to improve animal welfare.[43] Substantial research is dedicated to using genetic engineering to increase parasite and disease resistance in livestock. Cloning can theoretically be used to preserve endangered species, or even to bring back extinct species, that might provide a source of genes for improving resistance. Knocking out a single gene can eliminate the development of horns, thus eliminating the need for dehorning cattle. Boar taint can be modified by genetics and, if eliminated, would reduce the need for castration. If layer hens could be engineered so that they only produced female offspring, male chicks would no longer be routinely slaughtered at hatching.

The NRC also documented several ways that cloning and genetic engineering involve procedures that increase risks to animal welfare.

Electroejaculation, used in some species to collect sperm, and injections to induce ovulation cause distress. Embryo collection and transfer require painful invasive procedures in several livestock species. Embryo collection in poultry requires killing the hen. Collection of sperm and eggs in some species of fish also requires killing the animal. Cows that receive transferred embryos are often subjected to transvaginal amniocentesis followed by selective abortion of fetuses not carrying the transgene of interest. These procedures can cause pain and distress in the cow, and are likely to be repeated several times over the life of a valuable breeding animal.

When an animal is cloned or when a new line of genetically engineered animals is created, the animals are produced using in vitro embryo culture. This puts this generation of animals at higher risk for large offspring syndrome (LOS), resulting in an increase in perinatal mortality, immune system dysfunction, brain lesions, and congenital malformations in the skeleton, vascular system, and urogenital tract. Because of their large size and longer gestational period, animals suffering from LOS often result in difficult pregnancies, with health complications both for the fetus and for the pregnant female. However, these problems are not passed on to offspring subsequently produced through normal breeding.

The insertion of foreign DNA through microinjection can result in mutations when the inserted DNA alters the expression of an already existing gene. This occurs at an estimated rate of 5–10%. An estimated 25% of the mutated animals are able to survive, but many survivors suffer from problems such as "severe muscle weakness, missing kidneys, seizures, behavioral changes, sterility, disruptions of brain structure, neuronal degeneration, inner ear deformities, and limb deformities."[44] These mutations can be passed on to offspring.

Despite improvements in gene regulation, problems persist with gene overexpression, most notably with the insertion of growth hormone genes, genes that are of great interest to the agricultural community. Problems can include "diarrhea, mammary development in males, lethargy, arthritis, lameness, skin and eye problems, loss of libido, and disruption of estrous cycles."[45] Fish that have been genetically engineered with growth hormone genes, such as the AquAdvantage salmon, are at higher risk for head and jaw deformities, often resulting in reduced viability and premature death. These problems, too, can be passed on to offspring.

The above welfare issues are effects of unintended and unwanted problems with the process of genetically engineering or cloning an animal, and it is important to note that researchers will continue to refine these techniques to minimize their occurrence. Moreover, producers will presumably not be interested in using animals that fail to be more productive than their conventional counterparts because of these health concerns. However, commentators have also noted that even if nothing goes wrong with the process of genetically engineering or cloning an animal, the intended effects themselves can have a negative impact on animal welfare. Breeding for increased productivity and the associated changes in feed, housing, and management necessary to achieve high production levels have resulted in increased rates of problem births in beef and dairy cattle, lameness, metabolic disorders, and mastitis in dairy cattle, over-excitability in pigs and cattle, osteoporosis and cloacal prolapse in layers, insatiable appetite in broiler breeder chickens and sows, and skeletal and cardiovascular disorders in turkeys and in broiler chickens.[46] Even genetically engineering an animal to be resistant to disease or parasites could be, on balance, bad for animal welfare if it results in their being subjected to even more severe space limitations.

Grandin goes so far as to suggest that "the most serious animal welfare problems may be caused by over selection for production traits such as rapid growth, leanness, and high milk yield."[47] So although each application of animal biotech should be evaluated on its own individual merits, to the extent that genetic engineering and cloning are used with the same motivations and for the same purposes as traditional selection, there is a presumption that animal biotech generally is likely to exacerbate rather than mitigate animal welfare problems.

2.e Improving Welfare by Diminishing Animals?

When most people think of using animal biotech to improve animal welfare, they think of using animal biotech to enhance the animal in some way, for example, to enhance its capacity to resist injury or disease. But some have argued that we ought to use animal biotech to disable or diminish animals relative to normally functioning members of their species. Although some diminished animals are of interest because they show increased productivity, the ethical argument for using diminished animals begins from the claim that these animals will be better off in terms of welfare than their nondiminished counterparts. To explore this issue, consider the following examples.

Consider genetically engineering turkeys so that they no longer have the maternal desire to brood over their nests. In industrial agricultural settings, turkeys are not allowed to fulfill their natural desire to brood over their eggs because brooding delays the next laying cycle and decreases egg productivity. The natural desire to nest is frustrated, causing turkeys to suffer psychological stress.[48] Researchers at the University of Wisconsin–Madison have suggested using genetic engineering to knock out the gene that produces prolactin, a hormone that plays a key role in broodiness. If knocking out the gene for prolactin eliminates the desire to brood, this could reduce the levels of psychological stress experienced by nest-deprived turkeys.[49]

Second, consider the example of using chemical mutagenesis to produce a line of blind chickens. Chickens are naturally aggressive and have become more aggressive as an unintended byproduct of selective breeding for egg productivity. This aggression is exacerbated by the crowded conditions in which chickens are kept in industrial agriculture settings.[50] Producers typically try to minimize the damage of the resulting attacks by trimming the birds' beaks, combs, and toes, but this is less than ideal, as "[t]he beak of the fowl is well innervated and contains both mechanoreceptors and nociceptors."[51] The results of debeaking are described by Ian Duncan: "behavioral changes suggestive of acute pain have been found to occur in the 2 days following surgery. These are followed by chronic pain that lasts at least 5 or 6 weeks after the surgery."[52]

An alternative method became available, though, when researchers used chemical mutagenesis to induce a genetic mutation that resulted in congenital blindness.[53] Researchers at the University of British Columbia identified the specific mutation, speculating early on that "because of [blind chickens'] docility and possibly reduced interaction in a social hierarchy, studies of their behavior with relation to growth rate, feed efficiency, and housing density may have some practical application."[54] Follow-up studies indicated that flocks of blind chickens use less feed and show increased egg productivity, while also suffering fewer injuries and less feather and skin damage.[55]

As a third example, consider microencephalic pigs. Researchers identified a gene (Lim1, also referred to as Lhx1) responsible for head morphology in the mouse.[56] Using genetic engineering to knock out this gene resulted in mice with bodies that were normally formed except for lacking heads. The pups survived in utero for about ten days, about halfway through normal gestation. This research, as well as other research on organizer genes and research on microencephalic animals, vividly raise the possibility that highly social and intelligent animals, such as pigs, which suffer greatly in current factory farming conditions, could be engineered or bred to have just enough brain stem to support biological growth, but not enough to support consciousness.[57] Such animals, lacking the capacity for consciousness, would be incapable of suffering.

As mentioned, some of these diminished animals are of interest because they use less feed or show higher productivity, but the ethical argument for using them focuses on the Comparative Welfare Claim: that animals that we currently use are worse off in terms of animal welfare than their diminished counterparts would be if we used them instead. The Comparative Welfare Claim is partly empirical and partly conceptual. Its truth will depend on the empirical facts about the lives of the diminished animals and the lives of their nondiminished counterparts, and it will also depend on the conceptual facts about what constitutes animal welfare.

The standard Mentalistic view of animal welfare defines welfare as the absence of negative mental states such as distress, fear, pain, and suffering (section 2.a). If the standard Mentalistic view is correct, then the Comparative Welfare Claim is equivalent to the claim that the animals we currently use experience more (or more severe) negative mental states than would their diminished counterparts, if we used them instead.

Taking the above examples as philosophical thought experiments, one can simply stipulate that the diminished animals do better in terms of negative mental states than their nondiminished counterparts. But it is important to keep in mind that we do not have the liberty to make stipulations about the empirical effects a proposed modification would have on our actual practices. It is also crucial to be clear on the welfare claims being made: even supposing that the standard Mentalistic view is correct, and even supposing that a modification clearly alleviates a source of suffering, it is still an open question whether the modification improves the animal's welfare, on balance.

The stress caused by nest-deprivation to turkeys is well-documented, and this supports the conclusion that eliminating the desire to nest will eliminate a significant source of frustration. However, it is unknown at this time whether a lack of prolactin will cause health problems that might lead to negative mental states, and the existing data regarding genetic engineering's impact on an animal's welfare is not promising when compared to genetic modification through traditional breeding and selection.[58] Transgene insertions should be evaluated on an event-by-event basis, and it would be premature to claim that nonbrooding turkeys will be better off according to the standard Mentalistic view than their broody counterparts.

The case is also inconclusive with blind chickens. On one hand, blind chickens would not be made to suffer the well-documented pain of having their beaks, toes, and combs amputated, and there is data to support the claim that they suffer fewer injuries and less feather loss.[59] On the other hand, there were no significant differences in the two surrogate measures of psychological stress, adrenal weight and plasma corticosterone levels, measured by Cheng and Ali. This suggests that they are just as stressed as sighted birds.[60] And it is difficult to even begin conceptualizing how we would measure and compare the negative mental states of individuals of a nonhuman species where one of them is missing an entire sense modality of such crucial importance as sight. Will their being blind result in stress, anxiety, frustration, or boredom? And even if blind chickens have fewer or less severe negative mental states, the standard Mentalistic view of animal welfare is clearly overly narrow in considering only the absence of negative mental states. So even if blind chickens do experience fewer or less intense negative mental states because of their inability to see, sight might still provide a significantly enhanced subjective environment that generates a greater number of positive mental states. On a view such as our hybrid Objective List view, this could mean better welfare overall.

What about microencephalic pigs who lack the capacity for consciousness? If the standard Mentalistic view of animal welfare were correct, then microencephalic pigs would have a better welfare than normal pigs since microencephalic pigs have no negative mental states. Indeed, if animal welfare is defined solely in terms of the absence of negative mental states, then microencephalic pigs would have a better welfare than any pig, no matter how well treated. But this, of course, demonstrates the implausibility of excluding positive mental states as components of animal welfare.

On our hybrid Objective List view, positive mental states are important components of animal welfare, and the question of whether the welfare of microencephalic pigs is better than their conventional counterparts will largely turn on whether the pigs in current industrial agricultural practices suffer so much that the suffering outweighs any positive mentalistic benefits. That is, in mentalistic terms, are the lives of pigs in industrial agriculture so bad that they would have been better off not having existed at all? If the answer to that question is no, then pigs without the capacity for consciousness do *not* fare better than the animals currently used.

Admittedly, the lives of pigs in industrial agriculture fall far below the lives that pigs are capable of in better circumstances, but that isn't the relevant comparison for determining the truth of the Comparative Welfare Claim. The relevant comparison is whether their lives are worse than having no life at all.

Even for diminished animals that do fare better than their nondiminished counterparts from a mentalistic perspective, it is still possible that they will do worse in terms of desire satisfaction, health, or natural functioning. Even if blind chickens suffer less, because they are blind, they are still, other things being equal, less healthy and less capable of natural functioning compared to normal members of their species. Certainly, chickens will have desires that will be harder to satisfy if they are blind, and these need to be taken into account if desire satisfaction is a component of welfare. So even if blindness results in a better balance of mentalistic factors, it does not follow that the welfare of blind chickens is better than the welfare of sighted chickens unless these other components are ruled out.

The case of microencephalic pigs raises an additional issue, as many people would argue that if an organism never attains consciousness, then it has no moral status whatsoever. If an animal has no moral status, then it cannot be wronged at all, and so it cannot be wronged by an application of animal biotech. This view of moral status implies that recent attempts to use the concepts of animal integrity and animal dignity to argue that creating microencephalic animals wrongs the animals themselves are bound to fail, although it leaves open the possibility that the creation of such animals is wrong for other reasons.[61]

3. Animal Biotech
and the Environment

The other main way in which animal biotech could have a significant impact on livestock agriculture is by affecting livestock agriculture's environmental impact. Current agricultural practices contribute to climate change, use large amounts of natural resources, pollute the water supply, and negatively impact biodiversity. As mentioned above, the FAO recently conducted a comprehensive review of the environmental impact of livestock agriculture, concluding that the livestock sector is one of the most significant contributors to environmental problems, both globally and locally. [62]

In the following sections, we review the impact of livestock on climate change (section 3.a) and on local environmental issues such as pollution, resource use, and local biodiversity levels (section 3.b). The general ways in which animal biotech might alleviate or exacerbate environmental problems are discussed (section 3.c), followed by an examination of two specific applications of animal biotech: Enviropigs (section 3.d) and AquAdvantage salmon (section 3.e).

3.a Livestock Agriculture and Global Climate Change

The most serious global environmental challenge is that of anthropogenic climate change. Costs from climate change come in a variety of forms: health risks, harm to water and land resources, and the economic costs of adapting to a changing climate, for example.[63] The fact that the primary costs of anthropogenic climate change will likely be borne by those that contributed the least to the problem raises issues of economic and environmental justice.[64]

In addition to the direct costs of climate change to humans, climate change is a major driver of biodiversity loss in terms of both species and ecosystems.[65] Observed migration patterns of species correlate with those predicted by current climate change models and populations unable to migrate are contracting and in some cases going extinct.[66] According to the most recent Intergovernmental Panel on Climate Change (IPCC) report, upwards of 20% of species studied will have an increased risk of extinction under the circumstances of a 1.5–2.5°C increase in average global temperature.[67] In addition, the IPCC report claims that "For increases in global average temperature exceeding 1.5–2.5°C and in concomitant atmospheric carbon dioxide concentrations, there are projected to be major changes in ecosystem structure and function, species' ecological interactions, and species' geographical ranges, with predominantly negative consequences for biodiversity, and ecosystem goods and services e.g., water and food supply."[68]

According to the FAO report "Livestock's Long Shadow," the livestock sector produces significant amounts of three main greenhouse gases: methane, nitrous oxide, and CO_2. These contributions accrue from deforestation to create more land for agricultural use (CO_2), natural digestive processes in agricultural animals (methane), and from the production and use of manure (nitrous oxide). The sector is responsible for 9% of all anthropogenic CO_2 emissions, 30% of all anthropogenic methane emissions, and 65% of all anthropogenic nitrous oxide emissions.

Methane and nitrous oxide are more potent greenhouse gases than CO_2, with methane's potential contribution to global warming being nearly twenty-five times that of CO_2 and with nitrous oxide's being almost three hundred times that of CO_2.[69] When the amounts of methane and nitrous oxide emissions are converted to CO_2 equivalent, the standard measure for contributions to global climate change, and added to the carbon emissions, the livestock sector is responsible for 18% of total emissions.

3.b Livestock Agriculture and Local Environmental Issues

The contribution of animal agriculture to environmental problems is not confined to the indirect contribution made by impacting climate. Animal agriculture contributes directly to local environmental issues such as water depletion, water degradation, and biodiversity loss.

Many regions already lack suitable water supplies, and the number of such regions is expected to increase in the future. It is predicted that by 2025, 64% of the world's population will live in areas that are "water-stressed."[70] Besides increasing water stress, increasing demands on water supplies have a significant impact on ecosystems such as wetlands, which provide an estimated U.S. $33 trillion in ecosystem services worldwide.[71]

While it is difficult to calculate the full impact of animal agriculture on water resources, the amount used, both directly and indirectly, is significant. Water is used in animal agriculture to feed and service animals, in processing products after slaughter, and to grow the feed crops for the animals. The livestock sector makes a significant contribution—93% of withdrawal and 73% of consumption (removal that renders the water unavailable for other uses)—to human water use globally, and this does not include water used in aquaculture.[72]

In addition to water use, animal agriculture contributes to water pollution. The high number of animals used in animal agriculture produces large amounts of waste, which is often used as fertilizer. Given the scale of agricultural facilities, the amount of waste produced is typically much higher than should be used for fertilizing in surrounding fields, leading to excessive applications of waste manure.[73] This increases the amount of runoff that contaminates the water system, leading to large amounts of phosphorus and nitrogen in the water system.

Water pollution due to current agricultural practices has direct costs for humans and is another driver of biodiversity loss. The chemical runoff from farms increases the risk of algal blooms that produce environmental toxins and may lead to environmental "dead zones," areas inhospitable to many species.[74] Additionally, this runoff increases the costs of water filtration and purification, and makes certain areas unusable for leisure activities.[75]

The most important drivers of biodiversity loss, according to a 2005 Millennium Ecosystems Assessment, include climate change, pollution, habitat change, and invasive species.[76] Current animal agricultural practices contribute to each of these.[77] The impact on biodiversity from climate change and pollution were highlighted above.

In terms of habitat change, animal agriculture is a major contributor to deforestation and agricultural animals now make use of 30% of the earth's available land.[78] Land use and deforestation due to an expanding livestock area is expected to increase due to an increased demand for meat products in developing countries.[79] In addition, in areas where large numbers of animals are kept on a range, there are increased rates of desertification.[80] Deforestation and desertification degrade, fragment, and destroy native habitats.[81] These forms of habitat change are associated with biodiversity loss in various forms: genetic diversity, species diversity, and ecosystem diversity.[82]

Invasive species, alien species introduced into a nonnative ecosystem that result in degradation of the native ecosystem, may introduce diseases, outcompete native species, or otherwise undermine ecosystem stability. Animal agriculture's contribution to the problem of invasive species comes in many forms. The animals used in agriculture are often nonnative to the area where they are housed and pastured (when they are pastured). Sometimes these animals escape and become feral. Furthermore, the introduction of animal agriculture to an area often makes that area more easily invaded by alien species.[83]

3.c Potential Impact of Animal Biotech on the Environment

Animal biotech has the potential to affect the environmental impact of livestock agriculture in several ways. While using animal biotech to increase individual productivity imposes additional risks to animal welfare, it could have a beneficial effect on the environment by decreasing the number of animals used in agriculture, reducing the amount of feed needed and pollution produced. Increasing feed-conversion efficiency or nutrient utilization would also reduce the amount of feed or additives needed, which would have environmental benefits and might not have any negative impact on animal welfare.

The most widely discussed environmental risks of animal biotech arise from the intentional or unintentional release of genetically engineered animals. Whether the release of a genetically engineered animal results in an environmental harm depends, in part, on what in the environment is valuable and the ways in which it is valuable. Views about the value of the environment (or of the things in it) can be divided between those views on which the environment has intrinsic value and those on which it has merely instrumental value. To claim that the environment has intrinsic value is to claim that it is valuable for its own sake, independently of whether anyone or anything values it.[84] To claim that it has merely instrumental value is to claim that it is valuable only because it is instrumentally useful to other entities achieving their worthwhile ends.

Some would argue that the environment has intrinsic value stemming from its naturalness.[85] Naturalness is typically understood as coming in degrees and as being defined as freedom from the influence of humans. On such views, the novel presence of a genetically engineered animal in the environment would itself detract from the naturalness of the area and so constitute an environmental harm, even if its presence had no further consequences. Views on which the environment is intrinsically valuable because it is natural are controversial. It is difficult to see why being natural would be a source of value except instrumentally, in that that we value areas that are free of our impact and that such areas are beneficial to ourselves and other entities. Similar worries arise for views that claim that the intrinsic value of the environment stems from its biodiversity.

While the intrinsic value of the environment is controversial, there is little doubt that the environment is instrumentally valuable. Ecosystems provide services and resources for human and nonhumans alike. Whether the release of a genetically engineered animal results in an environmental harm, on instrumental views of environmental value, will depend on (a) the further environmental consequences of the release and (b) the moral status of those affected. So-called anthropocentric views of moral status claim that only highly rational beings have moral status (a claim we rejected at the beginning of this chapter on the grounds that all sentient animals have some degree of moral status).[86] On such views, the novel presence of a genetically engineered animal will constitute an environmental harm if, but only if, the consequences for the environment have a negative impact on humans. Sentientist views of moral status claim that only individuals with sentience have moral status.[87] Some biocentric views claim that all living things have moral status.[88] On these views, the release of a genetically engineered animal into the environment will constitute an environmental harm if and only if it negatively impacts sentient beings or living things respectively.

In many cases, the release of a genetically engineered animal will have significant consequences for all of the entities in question. In such cases, whether the release counts as an environmental harm will be independent of which of the above views is true, although which view is true might well affect both the magnitude and distribution of the harm. At any rate, on any of the above views of environmental harm, the environmental risk of the release of a genetically engineered animal is affected by several related properties of the species in question, including the ability to initially escape captivity, the ability to travel, the ability to maintain a feral population, and the extent of environmental disruption that might be caused by such a feral population.[89] Taking such factors into account, the NRC concluded that fish and shellfish pose the greatest environmental concern, pigs and goats pose a moderate degree of concern, and chickens, cattle, and sheep pose the least concern.[90] Concern is increased when the transgene enhances fitness, as would be expected with several genetically influenced traits under study, including salt-water tolerance in fish, cold tolerance, increased growth rate, enhanced disease resistance, and improved nutrient utilization.

The two genetically engineered animals currently under review for commercialization, Enviropigs and AquAdvantage salmon, provide opportunities to explore these environmental issues in more detail.

3.d Enviropigs

Swine require phosphorus in their diets, but the phosphorus in the grain-based foods that are standard in contemporary agriculture is bound in phytate, which swine are unable to digest.[91] To meet the phosphorus requirements of swine, farmers supplement the standard feed with mineral phosphate.

This standard practice contributes to a variety of local environmental problems and, in turn, to more global ones. The phosphate must be mined and shipped to farms, where, because it is inexpensive, it is fed to swine in abundance to ensure that their nutritional needs are met.[92] While the swine are able to digest the mineral phosphate, the use of mineral phosphate increases the amount of phosphorus in the swine's manure. Manure is used as fertilizer and so the amount of phosphorus released into the environment is increased. The amount of phosphorus released in this form causes serious environmental damage, especially when the phosphorus makes its way into the water system where it increases algal bloom and thus contributes to the death of native species as well as to drinking-water contamination.[93] Phosphorus pollution is consistently ranked by the Environmental Protection Agency as one of the top causes of water quality problems.

The Enviropig is a transgenic animal created by scientists at the University of Guelph to help alleviate these environmental concerns. The Enviropig has been genetically engineered to produce phytase in its saliva, an enzyme that allows it to digest the plant-based phytate in standard feed, thus reducing or eliminating the need for supplemental phosphorus. Research suggests that Enviropigs secrete up to 75% less phosphorus in their manure than their nongenetically-modified counterparts fed the standard diet, and up to 25% less than swine fed supplemental phytase to help digest plant-based phosphorus.

Insofar as the Enviropig merely replaces existing swine without causing further changes to the agricultural system, for example without changing the number of pigs per farm or the number of farms (what we will call "the scenario of mere replacement"), the use of Enviropigs yields environmental benefits by decreasing the demand for phosphorus mining and reducing phosphorus pollution of ground and surface water. The decreased demand for mineral-bound phosphorus decreases pollution due to transportation and mining, especially in areas without easy access to mineral-bound phosphorus. This is in addition to the benefits of decreasing phosphorus runoff and its impact on water quality and potable water. Under the scenario of mere replacement, these environmental benefits provide a reason for replacing existing swine with the Enviropig.

While under the scenario of mere replacement there are ethical reasons in favor of replacing nontransgenic swine with the Enviropig, in scenarios where the adoption of Enviropigs results in an increase in the number of swine ("the scenario of growth"), the opposite conclusion will be justified. Environmental pollution serves as one of the main constraints on the growth of the livestock industry.[94] If the introduction of the Enviropig allows for an increase in the total number of swine used in animal agriculture by reducing the amount of phosphorus pollution per pig, there may be no benefit in terms of aggregate phosphorus reduction, or at least some of the benefit will be mitigated by the increase in numbers. Furthermore, the increase in the number of swine will contribute to other environmental problems associated with livestock: ground compaction, where the number of pigs per unit area increases; deforestation, where the area per farm can be increased; reduction in available water resources; and an increase in other environmental pollutants such as greenhouse gasses associated with animal agriculture.

Even if Enviropigs do not increase the number of swine, they still pose an environmental risk if they escape into the surrounding environment. As mentioned above, the NRC identified pigs as a species the unintentional release of which poses a moderate degree of environmental risk. Swine have been known to escape and are capable of living outside captivity. In some cases, feral pigs have been known to cause significant environmental damage. Moreover, the transgene in question would allow escaped Enviropigs, or feral pigs with which they bred, to obtain needed phosphorus from plants in the surrounding environment, increasing their pest potential.[95]

Furthermore, if Enviropigs do increase the total number of animals, then this would also increase the probability that animals escape captivity and cause environmental damage.

3.e AquAdvantage Salmon

AquaBounty Technologies, an aquaculture biotechnology company, has developed growth-enhanced Atlantic salmon by introducing a gene from Chinook salmon that regulates the fish's growth hormone differently, allowing them to reach harvest weight in half the time of conventional Atlantic salmon. The use of growth-enhancing genes in fish species has generated significant concerns about negative environmental impact. Modeling by Howard Muir and Richard Howard suggests that growth-enhanced fish can exhibit a "Trojan gene" effect.[96] This occurs when growth-enhanced fish have better mating success than their nonenhanced counterparts, which allows the transgene to quickly spread throughout the surrounding population, but the off-spring of the growth-enhanced fish are less fit than the offspring of their nonenhanced counterparts. The result is that the surrounding population can rapidly become extinct with even a very small number of original genetically engineered fish escaping captivity. And yet sometimes hundreds of thousands of farm-raised fish escape from open-water net pens in a single year.[97]

The likelihood of such a scenario, though, is still the subject of considerable debate.[98] Moreover, AquaBounty has proposed to reduce this environmental risk by restricting the use of their genetically engineered salmon to land-based containment systems and to sell only reproductively sterile eggs (illustrating another application of animal biotech, the induction of triploidy, which can be used to be reduce environmental risks). Although induced sterility is not 100% reliable, and even land-based containment systems pose environmental risks, enforcing these precautions would significantly reduce the environmental risks of AquAdvantage salmon.

4. CONCLUSION

An all things considered judgment about an application of animal biotech must take into account its impact on animal welfare and the environment. Sections 2 and 3 serve to show how to evaluate an application of animal biotech by reviewing its impact on the animal welfare and environmental problems that plague modern agriculture. These evaluations are not exhaustive, but they do develop a general framework for evaluating animal biotech from an animal welfare and an environmental perspective.

Understanding the impact of animal biotech on animal welfare requires an understanding of the components of animal welfare and an appreciation of the philosophical issues that arise when assessing improvements in welfare. We have argued for a view of welfare that focuses on both positive and negative mentalistic components while acknowledging the contributory and epistemic importance of nonmentalistic factors such as health and natural behaviors. In addition, we have argued that the nonidentity problem should not lead us to dismiss the moral significance of welfare issues improved or worsened by using genetically engineered animals instead of their conventional counterparts.

Although genetic engineering and cloning both have the potential to mitigate existing animal welfare problems, they continue to be imperfect procedures that often increase the risk of welfare problems compared to the status quo. Even when nothing goes wrong with the procedures, the intended effect can itself exacerbate welfare problems. This will be especially likely when the intended effect is to increase individual productivity, as it often will be. Proposals to improve welfare by using animal biotech to eliminate or reduce livestock's cognitive capacities should be viewed with substantial skepticism.

Evaluating the environmental impact of an application of animal biotech is similarly complicated. Animal biotech that reduces the number of animals or increases feed-conversion efficiency or nutrient utilization could reduce the environmental harms of livestock agriculture. Discussions of environmental risk from animal biotech have focused on the effects of releasing genetically engineered animals into the surrounding environment. Evaluating the normative significance of these effects can be aided by distinguishing between the intrinsic value and the instrumental value of the environment and between different views about the sources of those kinds of environmental value. The proposed benefits of an application of animal biotech should not be evaluated under the assumption that livestock animals will merely be replaced by others that are more environmentally benign. If an application of animal biotech allows for an increase in the number of agricultural animals, there may be no net environmental benefit, and there may even be additional environmental harm.

Evaluating animal biotech requires that we confront a host of empirical and philosophical issues. The facts about the conditions of animals in agriculture and the impact of livestock on the environment make these issues more pressing than they were in the past, and the problems are likely to increase. The framework developed above is an important component of a strategy for forming all things considered judgments about animal biotech.

ACKNOWLEDGEMENTS

For their many helpful comments, we thank Ronald Sandler, Antonio Rauti, Tom Beauchamp, Justine Wells, Sara Gavrell Ortiz, Rebecca Stepien, Richard Reynnells, the joint 2005 Agriculture, Food, and Human Values Society and the Association for the Study of Food and Society discussion group, and the 2010 Greenwall Fellows discussion group.

NOTES

1. For a summary of the controversy, see Sheldon Krimsky and Roger Wrubel, *Agricultural Biotechnology and the Environment* (Urbana: University of Illinois Press, 1996), pp. 166–190.

2. Gregory S. Harper, Alan Brownlee, Thomas E. Hall, Robert Seymour, Russell Lyons, and Patrick Ledwith, *Global Progress Toward Transgenic Food Animals: A Survey of Publicly Available Information* (2003), pp. 7, 47–57, available at http://www.foodstandards.gov.au/_srcfiles/Transgenic%20Livestock%20Review%20CSIRO%20FINAL%20 12Dec20031.pdf (accessed August 30, 2010).

3. Siobhan DeLancey, Larisa Rudenko, and John Matheson, "A Primer on Cloning and Its Use in Livestock Operations," http://www.fda.gov/AnimalVeterinary/NewsEvents/FDAVeterinarianNewsletter/ucm108131.htm (accessed August 30, 2010).

4. Harper et al., *Global Progress*, pp. 7, 47–57.

5. FDA, "Guidance for Industry: Regulation of Genetically Engineered Animals Containing Heritable Recombinant DNA Constructs," January 15, 2009, http://www.fda.gov/downloads/AnimalVeterinary/GuidanceComplianceEnforcement/GuidanceforIndustry/UCM113903.pdf (accessed August 30, 2010).

6. Thomas Hoban, "Education Required for Animal Biotechnology" (2002), at www4.ncsu.edu/~hobantj/biotechnology/biotechnology_webpage.html (accessed August 19, 2010).

7. William Hallman, W. Carl Hebden, Cara Cuite, Helen Aquino, and John Lang, "Americans and GM Food: Knowledge, Opinion, and Interest in 2004: Report No. RR-1104–0007" (New Brunswick, N.J.: Food Policy Institute, Cook College, Rutgers University, 2004).

8. Pew Initiative on Food and Biotechnology, "An Update on Public Sentiment about Agricultural Biotechnology" (2003), at http://www.pewtrusts.org/uploadedFiles/wwwpewtrustsorg/Public_Opinion/Food_and_Biotechnology/2003summary.pdf (accessed August 19, 2010).

9. See the chapter "Which Animals are Sentient?" in Gary Varner, *Personhood and Animals in the Two-Level Utilitarianism of R.M. Hare* (Oxford: Oxford University Press, forthcoming).

10. For the estimate of land animals used, see Food and Agriculture Organization of the United Nations (FAO), "FAOSTAT" (2008), at http://faostat.fao.org/ (accessed August 31, 2010). For the estimate of aquatic animals used, see Animals Deserve Absolute Protection Today and Tomorrow (ADAPTT), "The Kill Counter," July 1, 2010, at http://www.adaptt.org/killcounter.html (accessed August 30, 2010).

11. Nuffield Council on Bioethics, "The Ethics of Research Involving Animals" (2005), at http://www.nuffieldbioethics.org/fileLibrary/pdf/RIA_Report_FINAL-opt.pdf (accessed August 19, 2010).

12. Population Reference Bureau, "How Many People Have Ever Lived on Earth?" at http://www.prb.org/articles/2002/howmanypeoplehaveeverlivedonearth.aspx (accessed on August 30, 2010).

13. FAO, "Livestock's Long Shadow: Environmental Issues and Options" (2006), at ftp://ftp.fao.org/docrep/fao/010/a0701e/a0701e.pdf (accessed August 31, 2010).

14. David DeGrazia, *Taking Animals Seriously: Mental Life and Moral Status* (Cambridge: Cambridge University Press, 1996), pp. 105–7.

15. Gary Varner, *In Nature's Interest: Interests, Animal Rights, and Environmental Ethics* (Oxford: Oxford University Press, 1998).

16. James Griffin, *Well-Being: Its Meaning, Measurement, and Moral Importance* (Oxford: Clarendon Press, 1986), pp. 11–14.

17. Griffin, *Well-Being*, p. 17.

18. David Brink, *Moral Realism and the Foundations of Ethics* (Cambridge: Cambridge University Press, 1989), p. 233.

19. Both of these views should be distinguished from a third alternative, Productivity views, on which welfare is constituted by the productivity of an animal in terms of its market product. Productivity views are implausible because there are times when an animal's being more productive are clearly bad for it. For example, forcing laying hens to molt by withholding food increases egg production but has a negative impact on their welfare, as it results in significant amounts of pain and distress.

20. Bernice Bovenkerk, Frans W. A. Brom, and Babs J. Van Den Bergh, "Brave New Birds: The Use of 'Animal Integrity' in Animal Ethics," *Hastings Center Report* 32, no. 1 (2002): 16–22, here p. 21.

21. Sara Gavrell Ortiz, "Beyond Welfare: Animal Integrity, Animal Dignity, and Genetic Engineering," *Ethics and the Environment* 9, no. 1 (2004): 94–120, here p. 112.

22. These commentators understand these concepts as being in contrast to welfare because they understand welfare to be synonymous with Mentalistic views. In our taxonomy, these features are understood to be part of a broad conception of welfare that extends beyond mentalistic components as captured in Desire-Satisfaction views and Objective List views.

23. Franklin McMillan, "Do Animals Experience True Happiness?" in Franklin McMillan, *Mental Health and Well-Being in Animals* (Ames, Iowa: Blackwell Publishing, 2005), p. 223, citations omitted.

24. The difference between these cases has been most influentially discussed by Derek Parfit, *Reasons and Persons* (Oxford: Clarendon Press, 1984), esp. pp. 351–79.

25. Robert Streiffer, "Animal Biotechnology and the Non-Identity Problem," *American Journal of Bioethics* 8, no. 6 (2008): 47–48.

26. These cases are adapted from Allen Buchanan, Dan Brock, Norman Daniels, and Daniel Wikler, *From Chance to Choice: Genetics and Justice* (Cambridge: Cambridge University Press, 2000), pp. 244–45.

27. For a discussion of different ways to justify this claim, see Bonnie Steinbock and Ron McClamrock, "When Is Birth Unfair to the Child," *Hastings Center Report* 24, no. 6 (1994): 15–21; Dan Brock, "The Non-Identity Problem and Genetic Harms: The Case of Wrongful Handicaps," *Bioethics* 9, no. 3/4 (1995): 269–75; and Bernard Rollin, *The Frankenstein Syndrome: Ethical and Social Issues in the Genetic Engineering of Animals* (Cambridge: Cambridge University Press, 1995), pp. 185–87.

28. David Fraser, Joy Mench, and Susanne Millman, "Farm Animals and Their Welfare in 2000," in *The State of the Animals: 2001*, ed. D. Salem and A. Rowan (Gaithersburg: Humane Society Press, 2000), pp. 87–99.

29. Ian Duncan, "Welfare Problems with Poultry," in *The Well-Being of Farm Animals: Challenges and Solutions*, ed. G. J. Benson and B. Rollin (Ames, Iowa: Blackwell Press, 2004), pp. 307–23; National Animal Health Monitoring System (NAHMS), "Layers '99: Part 1: Reference of 1999 Table Egg Layer Management in the U.S." (2003), p. 5 at http://www.nahms.aphis.usda.gov/poultry/layers99/Layers99_dr_PartII.pdf (accessed August 31, 2010).

30. Duncan, "Welfare Problems with Poultry."

31. N. G. Gregory and L. J. Wilkins, "Broken Bones in Fowl: Handling and Processing Damage in End-of-Lay Battery Hens," *British Poultry Science* 30, no. 3 (1989): 555–62. For further discussion, see A. B. Webster, "Welfare Implications of Avian Osteoporosis," *Poultry Science* 83 (2004): 184–92.

32. Animal and Plant Health Inspection Service (APHIS), "Swine 2006: Part IV: Changes in the U.S. Pork Industry: 1990–2006" (2008) at http://www.aphis.usda.gov/vs/ceah/ncahs/nahms/swine/swine2006/Swine2006_PartIV.pdf (accessed August 31, 2010); Fraser et al., "Farm Animals and Their Welfare."

33. Timothy Blackwell, "Production Practices and Well-Being: Swine," in *The Well-Being of Farm Animals: Challenges and Solutions*, ed. G. J. Benson and B. Rollin (Ames, Iowa: Blackwell Press, 2004), pp. 241–69.

34. Blackwell, "Production Practices and Well-Being."

35. R. J. Collier, G. E. Dahl, and M. J. VanBaale, "Major Advances Associated with Environmental Effects on Dairy Cattle," *Journal of Dairy Science* 89:4 (2006): 1244–53; Franklyn Garry, "Animal Well-Being in the U.S. Dairy Industry," in *The Well-Being of Farm Animals: Challenges and Solutions*, ed. G. J. Benson and B. Rollin (Ames, Iowa: Blackwell Press, 2004), pp. 207–40.

36. NAHMS, "Dairy 2007: Part II: Changes in the U.S. Dairy Cattle Industry, 1991–2007)" (2007), p. 14 at http://nahms.aphis.usda.gov/dairy/dairy07/Dairy07_dr_PartII.pdf (accessed on August 31, 2010); USDA Agricultural Marketing Service, "Milk Production–December 1, 1940" (1940), at http://usda.mannlib.cornell.edu/usda/nass/MilkProd/1940s/1940/MilkProd-12-16-1940.pdf (accessed August 31, 2010).

37. Garry, "Animal Well-Being in the U.S. Dairy Industry."

38. Bjørg Heringstad, Gunnar Klemetsdal, and John Ruane, "Selection for Mastitis Resistance in Dairy Cattle: A Review with Focus on the Situation in Nordic Countries," *Livestock Production Science* 64 (2000): 95–106.

39. David Fraser and Daniel Weary, "Quality of Life for Farm Animals: Linking Science, Ethics, and Animal Welfare," in *The Well-Being of Farm Animals: Challenges and Solutions*, ed. G. J. Benson and B. Rollin, pp. 39–60.

40. NAHMS, "Dairy 2007 Part I," p. 10.

41. Temple Grandin, "Animal Welfare in Slaughter Plants," *Proceedings of the 29th Annual Conference of American Association of Bovine Practitioners* (1996): 22–26; Temple Grandin, "Progress and Challenges in Animal Handling and Slaughter in the U.S.," *Applied Animal Behaviour Science* 100 (2006): 129–39.

42. National Research Council (NRC), *Animal Biotechnology: Science-Based Concerns* (Washington, D.C.: National Academies Press, 2004), p. 93.

43. NRC, *Animal Biotechnology*, pp. 93–102, 104–7.

44. NRC, *Animal Biotechnology*, p. 97.

45. NRC, *Animal Biotechnology*, p. 98.

46. Compiled from P. Sandøe, B. L. Nielson, L. G. Christensen, and P. Sørenson, "Staying Good While Playing God—The Ethics of Breeding Farm Animals," *Animal Welfare* 8 (1999): 313–28; Temple Grandin and Mark J. Deesing, "Genetics and Animal Welfare," in *Genetics and the Behavior of Domestic Animals*, ed. T. Grandin (San Diego: Academic Press, 1998): 319–46; Michael Greger, "Transgenesis in Animal Agriculture: Addressing Animal Health and Welfare Concerns," *Journal of Agricultural and Environmental Ethics*, http://www.springerlink.com/content/p811p18540937037/ (accessed August 21, 2010).

47. Grandin and Deasing, "Genetics and Animal Welfare," p. 319.

48. Ian Duncan, "Animal Welfare Issues in the Poultry Industry: Is There a Lesson To Be Learned?" *Journal of Applied Animal Welfare Science* 4, no. 3 (2001): 207–21, here p. 208.

49. Michael Reiss and Roger Straughan, *Improving Nature: The Science and Ethics of Genetic Engineering* (Cambridge: Cambridge University Press, 1996), p. 174; Andy Coghlan, "Pressure Group Broods Over Altered Turkeys," *New Scientist* 138, no. 1875 (1993): 9.

50. Ahmed Ali and Kimberly Chang, "Early Egg Production in Genetically Blind (rc/rc) Chickens in Comparison with Sighted (Rc+/rc) Controls," *Poultry Science* 4, no. 5 (1985): 789–94, here p. 791.

51. Duncan, "Animal Welfare Issues in the Poultry Industry," p. 214.

52. Duncan, "Animal Welfare Issues in the Poultry Industry," p. 215.

53. K. Cheng, R. Shoffner, K. Gelatt, G. Gum, J. Otis, and J. Bitgood, "An Autosomal Recessive Blind Mutant in the Chicken," *Poultry Science* 59 (1980): 2179–82.

54. Cheng et al., "Autosomal Recessive Blind Mutant," p. 2182.

55. Ali and Cheng, "Early Egg Production in Genetically Blind Chickens."

56. William Shawlot and Richard Behringer, "Requirement for LIM1 in Head-Organizer Function," *Nature* 374 (1995): 425–30.

57. W. Balduini, M. Cimino, G. Lombardelli, M. Abbracchio, G. Peruzzi, T. Cecchini, G. Gazzanelli, and F. Cattabeni, "Microencephalic Rats as a Model for Cognitive Disorders," *Clinical Neuropharmacology* 9, suppl. 3 (1986): s8–s18; Ingolf Bach, "The LIM Domain: Regulation by Association," *Mechanisms of Development* 91 (2000): 5–17.

58. NRC, *Animal Biotechnology*, pp. 93–107.

59. Ali and Cheng, "Early Egg Production in Genetically Blind Chickens."

60. Ali and Cheng, "Early Egg Production in Genetically Blind Chickens."

61. Bovenkerk et al., "Use of 'Animal Integrity' in Animal Ethics"; Gavrell Ortiz, "Beyond Welfare," p. 112.

62. FAO, "Livestock's Long Shadow," (2006), p. xx.

63. Intergovernmental Panel on Climate Change (IPCC), "Climate Change 2007: Impacts, Adaptation, and Vulnerability," ed. M. Parry, O. Canziani, J. Palutikof, P. van der Linden, and C. Hanson (Cambridge: Cambridge University Press, 2007), pp. 7–22.

64. Peter Singer, *One World: The Ethics of Globalization* (New Haven, Conn.: Yale University Press, 2002), pp. 14–15; Stephen Gardiner, "Ethics and Global Climate Change," *Ethics* 114 (2004): 555–600.

65. IPCC, "Climate Change 2007"; Philip Hulme, "Adapting to Climate Change: Is there Scope for Ecological Management in the Face of a Global Threat?" *Journal of Applied Ecology* 42 (2005): 784–94; and FAO, "Livestock's Long Shadow."

66. Camille Parmesan, "Ecological and Evolutionary Responses to Recent Climate Change," *Annual Review of Ecology, Evolution, and Systematics* 37 (2006): 637–69.

67. IPCC, "Climate Change 2007."

68. IPCC, "Climate Change 2007," p. 11.

69. FAO, "Livestock's Long Shadow."

70. FAO, "Livestock's Long Shadow."

71. FAO, "Livestock's Long Shadow," p. 127.

72. K. Turner, S. Georgiou, R. Clark, R. Brouwer, and J. Burke, "Economic Valuation of Water Resources in Agriculture: From Sectoral to a Functional Perspective on Natural Resource Management," *FAO Paper Reports* 24 (2004).

73. P. Hooda, A. Edwards, H. Anderson, and A. Miller, "A Review of Water Quality Concerns in Livestock Farming Areas," *The Science of the Total Environment* 250 (2000): 143–67.

74. FAO, "Livestock's Long Shadow."

75. FAO, "Livestock's Long Shadow."

76. Millennium Ecosystem Assessment, "Ecosystems and Human Well-Being: Biodiversity Synthesis" (Washington, D.C.: World Resources Institute, 2005), at http://www. millenniumassessment.org/documents/document.354.aspx.pdf (accessed August 30, 2010).

77. FAO, "Livestock's Long Shadow."

78. FAO, "Livestock's Long Shadow."

79. Ramona Ilea, "Intensive Livestock Farming: Global Trends, Increased Environmental Concerns, and Ethical Solutions," *Journal of Agricultural and Environmental Ethics* 22 (2009): 153–67.

80. FAO, "Livestock's Long Shadow."

81. FAO, "Livestock's Long Shadow."

82. FAO, "Livestock's Long Shadow."

83. FAO, "Livestock's Long Shadow."

84. John O'Neill, "The Varieties of Intrinsic Value," in *Environmental Ethics: An Anthology*, ed. H. Rolston III and A. Light (Malden: Blackwell Press, 2003).

85. Robert Elliot, "Faking Nature," *Inquiry* 25, no. 1 (1982): 81–93; Holmes Rolston III, *Philosophy Gone Wild* (Buffalo, N.Y.: Prometheus Books, 1989); and William Throop, "Eradicating the Aliens: Restoration and Exotic Species," in *Environmental Restoration*, ed. W. Throop (Amherst, N.Y.: Humanity Books, 2000).

86. William Baxter, *People or Penguins: The Case for Optimal Pollution* (New York: Columbia University Press, 1974); Bryan Norton, "Environmental Ethics and Weak Anthropocentrism," in *Environmental Ethics: An Anthology*, ed. H. Rolston III and A. Light (Malden: Blackwell Press, 2003).

87. Peter Singer, *Animal Liberation*, 2nd ed. (New York: Random House, 1990).

88. Kenneth Goodpaster, "On Being Morally Considerable," *The Journal of Philosophy* 75 (1978): 308–25; Varner, *In Nature's Interest*.

89. NRC, *Animal Biotechnology*, p. 83.

90. NRC, *Animal Biotechnology*, p. 83.

91. Serguei Golovan, Roy Meidinger, Ayodele Ajakaiye, Michael Cottrill, Miles Wiederkehr, David Barney, Claire Plante, John W. Pollard, Ming Fan, M. Anthony Hayes, Jesper Laursen, J. Peter Hjorth, Roger Hacker, John Phillips, and Cecil Forsberg, "Pigs Expressing Salivary Phytase Produce Low-Phosphorus Manure," *Nature Biotechnology* 19 (2001): 741–45.

92. Cecil Forsberg, John Phillips, Serguei Golovan, Ming Fan, Roy Meidinger, Ayodele Ajakaiye, D. Hilborn, and R. Hacker, "The Enviropig Physiology, Performance, and Contribution to Nutrient Management Advances in a Regulated Environment: The Leading Edge of Change in the Pork Industry," *Journal of Animal Science* 81, e. suppl. 2 (2003): e68–e77.

93. S. Carpenter, N. Caracao, D. Correll, R. Howarth, A. Sharpley, and V. Smith, "Nonpoint Pollution of Surface Waters with Phosphorus and Nitrogen," *Ecological Applications* 8, no. 3 (1998): 559–68; A. Jongbloed and N. Lenis, "Environmental Concerns about Animal Manure," *Journal of Animal Science* 76, no. 10 (1998): 2641–48.

94. Leora Vestel, "The Next Pig Thing," *Mother Jones*, October 26, 2001, at http:// motherjones.com/environment/2001/10/next-pig-thing (accessed August 31, 2010).

95. NRC, *Animal Biotechnology*, p. 84.

96. William Muir and Richard Howard, "Possible Ecological Risks of Transgenic Organism Release When Transgenes Effect Mating Success: Sexual Selection and the Trojan Gene Hypothesis," *Proceedings of the National Academy of Sciences* 96, no. 24 (1999): 13853–56.

97. Dennis Kelso, "Genetically Engineered Salmon, Ecological Risks, and Environmental Policy," *Bulletin of Marine Science* 74, no. 3 (2004): 509–28; and NRC, *Animal Biotechnology*, p. 90.

98. Norman Maclean and Richard James Laight, "Transgenic Fish: An Evaluation of Benefits and Risks," *Fish and Fisheries* 1 (2000): 146–72.

SUGGESTED READING

BENSON, JOHN, and BERNARD E. ROLLIN, eds. *The Well-Being of Farm Animals: Challenges and Solutions*. Ames, Iowa: Blackwell Press, 2004.

BROOM, DONALD, and ANDREW FRASER. *Domestic Animal Behaviour and Welfare*. 4th ed. Oxfordshire: CAB International, 2007.

DAWKINS, MARIAN STAMP. *Animal Suffering: The Science of Animal Welfare*. London: Chapman and Hall, 1980.

DEGRAZIA, DAVID. *Taking Animals Seriously: Mental Life and Moral Status*. Cambridge: Cambridge University Press, 1996.

EVANS, J. WARREN, and ALEXANDER HOLLAENDER. *Genetic Engineering of Animals: An Agricultural Perspective*. New York: Plenum Press, 1986.

FOER, JONATHAN. *Eating Animals*. New York: Little, Brown, 2009.

HOLLAND, ALAN, and ANDREW JOHNSON. *Animal Biotechnology and Ethics*. London: Chapman and Hall, 1998.

HOUDEBINE, LOUIS-MARIE. *Animal Transgenesis and Cloning*. Hoboken, N.J.: John Wiley and Sons, 2003.

IMHOFF, DANIEL. *The CAFO Reader: The Tragedy of Industrial Animal Factories*. Berkeley: Watershed Media, 2010.

MARCUS, ERIK. *Meat Market: Animals, Ethics, and Money*. Boston: Brio Press, 2005.

McMAHAN, JEFF. *The Ethics of Killing: Problems at the Margins of Life*. Oxford: Oxford University Press, 2002.

MIDGLEY, MARY. *Animals and Why They Matter*. Athens: University of Georgia Press, 1983.

National Research Council (NRC). *Animal Biotechnology: Science-Based Concerns*. Washington, D.C.: National Academies Press, 2004.

Pew Commission on Industrial Farm Animal Production. *Putting Meat on the Table: Industrial Farm Animal Production in America*. Philadelphia and Baltimore: Pew Charitable Trusts and Johns Hopkins Bloomberg School of Public Health, 2008.

Pew Initiative on Food and Biotechnology. *Future Fish: Issues in Science and Regulation of Transgenic Fish*. Philadelphia and Baltimore: Pew Charitable Trusts and Johns Hopkins Bloomberg School of Public Health, 2002.

———. *Exploring the Moral and Ethical Aspects of Genetically Engineered and Cloned Animals*. Philadelphia and Baltimore: Pew Charitable Trusts and Johns Hopkins Bloomberg School of Public Health, 2005.

ROLLIN, BERNARD. *The Frankenstein Syndrome: Ethical and Social Issues in the Genetic Engineering of Animals*. Cambridge: Cambridge University Press, 1995.

———. *The Unheeded Cry: Animal Consciousness, Animal Pain, and Science*. Exp. ed. Ames, Iowa: Iowa State University Press, 1998.

SANDLER, RONALD. *Character and Environment: A Virtue-Oriented Approach to Environmental Ethics*. New York: Columbia University Press, 2007.

SAPONTZIS, STEVE. *Morals, Reasons, and Animals*. Philadelphia: Temple University Press, 1987.

————, ed. *Food for Thought: The Debate over Eating Meat.* Amherst, N.Y.: Prometheus Books, 2004.

SCULLY, MATTHEW. *Dominion: The Power of Man, the Suffering of Animals, and the Call to Mercy.* New York: St. Martin's Press, 2002.

SINGER, PETER. *Animal Liberation.* 2nd ed. New York: Random House, 1990.

————, and JIM MASON. *The Ethics of What We Eat: Why Our Food Choices Matter.* New York: Rodale, 2006.

THOMPSON, PAUL. *Food Biotechnology in Ethical Perspective.* 2nd ed. Dordrecht, Netherlands: Springer, 2007.

————, ed. *The Ethics of Intensification: Agricultural Development and Cultural Change.* Dordrecht, Netherlands: Springer, 2008.

United Nations Food and Agriculture Organization (FAO). "Livestock's Long Shadow: Environmental Issues and Options" (2006), ftp://ftp.fao.org/docrep/fao/010/a0701e/a0701e.pdf (accessed August 31, 2010).

VARNER, GARY. *In Nature's Interest: Interests, Animal Rights, and Environmental Ethics.* Oxford: Oxford University Press, 1998.

....

CHAPTER 31

ENVIRONMENTAL ETHICS, HUNTING, AND THE PLACE OF ANIMALS

....

GARY VARNER

ENVIRONMENTAL ethics emerged as a subfield of academic philosophy in the 1970s and 1980s in response to a series of publications claiming that the root cause of environmental problems is human-centered thinking in ethics. Insofar as they counter anthropocentrism in ethics, animal welfare and animal rights views sounded like a step in the right direction to some environmental ethicists. However, a near consensus soon emerged that an adequate environmental ethic had to be holistic; it had to attribute intrinsic, noninstrumental value to entities such as species and ecosystems. By contrast, according to both animal welfare and animal rights views, only the lives of conscious individuals have intrinsic value.

Apart from this deep philosophical difference about what has intrinsic value, environmental ethicists routinely claim that the practical implications of animal welfare and animal rights views would be drastically anti-environmental: sometimes requiring a degree of intervention in nature that environmentalists would find abhorrent, such as removing predators from ecosystems and feeding starving deer; other times prohibiting interventions that environmentalists support, such as using hunting to control wildlife populations and captive breeding of endangered species.

This essay presents an overview of these charges and some possible responses to them. Focusing on the latter claim about their practical implications, I argue that environmental ethicists have tended toward caricature of animal welfare and animal rights views, and that a closer look reveals significant convergence between the implications of such views and goals on the environmentalist agenda. With regard to the first, more philosophical claim, it is true that ecosystems and nonanimal species have only instrumental value from an animal welfare or animal rights perspective. To the degree that the implications of animal welfare and animal rights views converge with the goals of environmentalists, however, an adequate environmental ethic need not attribute intrinsic value to ecosystems and nonsentient organisms.

Holism and Sentientism in Environmental Ethics

Early on, scholars from a range of fields contributed to the literature that became the starting point of environmental philosophy and concentrated attention on the moral standing of nonhuman nature. Ecologist Aldo Leopold's 1949 book, *A Sand County Almanac*, especially the essay "The Land Ethic,"[1] is often held up as presenting an alternative and non-anthropocentric ethic, although others see him as advocating a kind of enlightened anthropocentrism. In a 1967 essay in *Science* titled "The Historical Roots of Our Ecologic Crisis," historian Lynn White Jr. argued that an anthropocentric Christian cultural heritage is the primary cause of the environmental crisis.[2] In 1972, Christopher D. Stone, a legal theorist, published "Should Trees Have Standing? Toward Legal Rights for Natural Objects."[3] After being cited in a U.S. Supreme Court minority opinion on an important environmental case, *Sierra Club v. Morton*,[4] the essay was reprinted in book form and became a classic.

The first book-length treatment of environmental problems by a professional philosopher came two years later, in 1974. In *Man's Responsibility for Nature*, John Passmore replied to White by emphasizing how various strains of Christianity embody significantly different attitudes toward nature and by emphasizing the extent to which enlightened anthropocentric thinking can address environmental issues.[5] Richard Routley, later Richard Sylvan, presented "Is There a Need for a New, an Environmental Ethic?" to the International Congress of Philosophy in 1973,[6] the same year Arne Naess published "The Shallow and the Deep, Long-Range Ecology Movement: A Summary" in the philosophy journal *Inquiry*.[7] In 1975, Holmes Rolston III published "Is There an Ecological Ethic?" in the journal *Ethics*.[8] These essays each called for a new, non-anthropocentric ethic grounded in, or at least informed by, ecological science. Discussion of related issues by philosophers soon exploded. In 1979, the journal *Environmental Ethics* appeared, followed by *Environmental Values* in 1992 and *Ethics and the Environment* in 1996. The number of books published in the area is now large, and many university departments of philosophy include at least one course in environmental ethics.

Because environmental ethics focuses intensely on questions about the moral status of nonhuman nature, a standard taxonomy of views in environmental ethics categorizes them in terms of which things have intrinsic value, as summarized in table 1. *Anthropocentrism* is the view that, when it comes to making decisions about the environment, only the interests of human beings matter; nonhuman nature has only instrumental value. Animal welfare and animal rights views are classified, following John Rodman's first use of the term in 1977, as *sentientism*, because of their commitment to the moral standing of all conscious animals.[9] The view that all living things have interests, or that their lives have intrinsic value in some other way, has been called *biocentric individualism*: "biocentric" because such an ethic is life-centered, but "individualism" to distinguish this view from the dominant view in environmental ethics, which is holism. In contrast to all of the foregoing views, *environmental holism* is the view that entities like species and ecosystems have intrinsic value in addition to, or, in the case of what we might call *pure holism*, instead of individual organisms.

Environmental ethicists have generally been critical of anthropocentrism, claiming that it is an inadequate basis for sound environmental policy because it does not recognize the intrinsic value of nonhuman animals, plants, species, and ecosystems. Environmental ethicists have been similarly critical of animal welfare and animal rights views, claiming that although they recognize the intrinsic value of sentient life, they attribute only instrumental value to nonconscious organisms as well as ecosystems and endangered species. Biocentric individualism turns out to be the minority view among philosophers writing on environmental ethics. For their part, holists criticize it for the same reasons they criticize anthropocentrism and sentientism: because neither species nor ecosystems are literally living organisms, neither has intrinsic value according to a biocentric individualist. Alternatively, anthropocentrists and sentientists generally criticize biocentric individualism as being unable to show that nonconscious organisms have intrinsic value in the same way that sentient individuals do.

In the first two decades of environmental philosophy, a near consensus was reached on one point: sentientism cannot generate an adequate environmental ethic. The depth and rapidity of this falling out is reflected in the title Mark Sagoff gave to a 1984 essay: "Animal Liberation and Environmental Ethics: Bad Marriage, Quick Divorce." Sagoff argued that a systematic commitment to either animal

Table 1 A Taxonomy of Views in Environmental Ethics in Terms of What Things have Instrinsic Value.

	Anthropocentrism	Sentientism	Biocentric individualism	Holism (pluralistic)	Holism (pure)
Human beings?	Yes	Yes	Yes	Yes	No
Sentient animals?	No	Yes	Yes	Yes	No
Nonsentient organisms?	No	No	Yes	Yes	No
Species and ecosystems?	No	No	No	Yes	Yes

welfare or animal rights would require us to intervene in natural areas on a massive scale. "One may modestly propose," he wrote, "the conversion of national wilderness areas, especially national parks, into farms" where predators would be fed soybean meat substitutes. Similarly, he argued:

> The environmentalist would sacrifice the lives of individual creatures to preserve the authenticity, integrity and complexity of ecological systems. The liberationist—if the reduction of animal misery is taken seriously as a goal—must be willing, in principle, to sacrifice the authenticity, integrity and complexity of ecosystems to protect the rights, or guard the lives, of animals.

Sagoff concluded that "environmentalists cannot be animal liberationists [and] animal liberationists cannot be environmentalists."[10]

I say that this is the one point on which a near consensus was reached early on in the field because not only holists but also anthropocentrists and pluralists have claimed that sentientism is systematically at odds with the environmentalist agenda. For instance, Bryan Norton argues that the implications of anthropocentrism and environmental holism converge when a sufficiently broad range of human interests are taken into account across a long enough time period, yet he seems to endorse the claim that sentientism is systematically at odds with sound environmental policy.[11]

So at this point in time, the near consensus among environmental ethicists is that sentientist views are systematically at odds with sound environmental policy. To fairly assess this claim, it is necessary to distinguish the two main varieties of sentientist ethics and then look carefully at how these two varieties could address the specific ways in which they are alleged to be at odds with sound environmental policy.

ANIMAL WELFARE VERSUS ANIMAL RIGHTS

Sentientist approaches to ethics are usually divided into so-called animal welfare and animal rights views. Although other approaches have been taken,[12] these are the main categories in sentientist ethics as well as the ones that have been the object of environmental philosophers' criticisms. As such, these will be the focus in this essay.

In the popular media, the distinction is usually portrayed as follows. Animal rights advocates are radicals bent on abolishing many currently accepted practices involving animals and willing to use illegal means, even terrorist tactics, to that end. They are also represented as driven by emotional attachments to animals rather than reason and as being poorly informed about the nature of the animals for which they advocate.

"Animal welfare" has been promoted as a contrast term by many scientists, veterinarians, agriculturalists, and others who want to express their commitment to ethical treatment of animals while simultaneously distancing themselves from what they see as the radical agenda of animal rights advocates. Not surprisingly, then, in the curricula of colleges of veterinary medicine and agriculture, animal welfarists are commonly understood as "us," manning the ramparts against "them," the dangerous, anti-science animal rights lunatics.

As philosophers generally conceive of the issues, however, the distinction between animal rights and animal welfare is not based on any of the above views. First, both animal welfare views and animal rights views are represented as reasoned applications of traditional ethical theories, or perhaps as drawing out the implications of what some call the common morality of our society. Specifically, "animal welfare" is used to refer to versions of utilitarianism, whereas "animal rights" refers to views that attribute individual moral rights to animals, where these rights are conceived of as trump cards against utilitarian arguments, or at least as strong demands of some kind.

Further, as I will stress in the remainder of this essay, neither animal welfare views nor animal rights views so conceived are necessarily committed to the complete abolition of practices such as hunting or endangered species programs, although some particular animal rights philosophers—notably Tom Regan—have called for complete abolition of such practices. Likewise, neither of these views is necessarily committed to employing illegal means, let alone terrorist tactics, to achieve its endorsed goals.[13]

The Practical Charges against Sentientism

Environmental ethicists often use the label "animal liberation" to stand for both animal welfare and animal rights views taken together, with "liberation" pointing to analogies with other liberation movements in recent history. J. Baird Callicott appears to have initiated this trend in his 1980 paper titled "Animal Liberation: A Triangular Affair." There he argued that the "booming controversy" over the moral status of nonhuman animals is an "internecine" conflict that ignores another, truly radical alternative.[14] Environmental holism as epitomized in the Leopold land ethic is, he argued, a much more radical departure from traditional ethics than animal liberation, which simply extends traditional liberalism by bringing a larger range of individuals into the moral community and conferring on them something like liberty rights. Leopold's holism, by contrast, represents a point off of this continuum of individuals. And according to Callicott, it is the holistic dimension of the land ethic that makes it best suited to ground an adequate environmental ethic.

Callicott's "Triangular Affair" paper has probably been more widely reprinted in anthologies on environmental ethics than any other essay besides Leopold's "The Land Ethic." It has become the paradigmatic statement of the claim that sentientist views cannot ground an adequate environmental ethic because their implications are systematically opposed to the goals on the environmentalist agenda. The reasons Callicott gave in support of this claim fall under the following five headings.

(1) *Hunting to control overpopulation:* Callicott emphasized that environmental ethics would require hunting of animals when necessary "to protect the local environment, taken as a whole," implying that animal liberationists must oppose all hunting.

(2) *Predators:* He said that, from an environmental perspective, large predators that kill other sentient animals are "critically important members of the biotic community," whereas animal liberationists should see them as "merciless, wanton, and incorrigible murderers."

(3) *Endangered species:* Callicott made two claims:

(3A) Nonsentient organisms have no moral standing according to sentientism, and therefore "humane herdspersons" might allow sheep to graze on plants that are "overwhelmingly important to the stability, integrity, and beauty of biotic communities."[15]

(3B) Additionally, while environmentalists place special value on members of endangered species, from an animal liberation perspective individuals with similar levels of sentience are of equivalent value, even if one is a member of an endangered species while the other's species is plentiful.[16]

(4) *Vegetarianism:* Callicott also argued that because vegetarian diets can feed people more efficiently, the conversion to vegetarianism that animal liberationists call for would likely lead to catastrophic overpopulation.[17]

(5) *Liberating domestic animals:* Finally, he thought that the so-called liberation of domestic animals would lead to four potential outcomes:

(5A) They would "become abruptly extinct" because they could not survive on their own.

(5B) Some would survive but, as with feral horses, would adversely affect their environments.

(5C) With slaughter banned, populations of domestic animals would mushroom, consuming enormous resources and greatly increasing human society's ecological footprint.

(5D) On the other hand, if we cease to allow domestic animals to breed, Callicott claimed that "there is surely some irony in an outcome in which the beneficiaries of a humane extension of conscience are destroyed in the process of being saved."[18]

In the remainder of this section, I will deal briefly with charges #4 and #5. The remainder of this essay will be devoted to more detailed responses to #1 through #3.

Regarding reason number 5, as Edward Johnson pointed out in an early response to Callicott's essay, any wholesale liberation of domestic animals would almost certainly take the final form imagined by Callicott: forced sterility or some other end to their reproduction. Nevertheless, from the individualist perspective of sentientism, "it is not the species that is being liberated, but the individual members of the species," and they are not destroyed in the process described in number 5D.[19] Thus, only a fundamental misunderstanding of the sentientist perspective could sustain the specific criticism stated in Callicott's objection 5D.

Regarding number 4, it is at least an oversimplification to say that a conversion to universal vegetarianism would automatically cause a dramatic increase in the human population. Admittedly, human populations have continued to grow at an alarming rate, even as various consequences of overpopulation have been popularized since the time of Malthus. Nevertheless, voluntary vegetarianism has risen in popularity only in affluent nations where birth rates are on the decline.

THE VARIETIES OF HUNTING AND WILDLIFE POPULATION CONTROL

Callicott's first reason for claiming that the implications of sentientism are at odds with sound environmental policy is that sentientist ethics would prohibit hunting for wildlife population control. This claim will be my focus for most of this paper, for two reasons. First, it is the most commonly discussed criticism of sentientism from the environmentalist perspective. Second, a thorough discussion of this charge both illustrates how the issue is more complex than environmentalist critics let on and sets the stage for dealing more briefly with the other two charges.

To understand the claim that opposition to hunting would be anti-environmental, it is important to distinguish various kinds of hunting and their associated motivations. By "therapeutic hunting," I mean hunting that is designed to secure the aggregate welfare of the target species across generations, the health and/or integrity of its ecosystem, or both. By "subsistence hunting," I mean hunting for food or other essentials, and in the category of "sport hunting," I include hunting aimed at maintaining religious or cultural traditions, reenacting national or evolutionary history, practicing skills, competing with an animal's survival skills, or just securing a trophy.

Many examples of hunting do not fit neatly into just one of these three categories. For instance, there are people who hunt, not out of any form of economic or nutritional necessity, but just because they treasure venison. I'm inclined to categorize that as sport hunting, but there is an element of subsistence hunting here. Similarly, when hunters work under the direction of wildlife managers, the individual hunters may think of themselves as granted permits to engage in sport hunting, but insofar as the permit system is designed to ensure that specified numbers and categories of animals are culled from the herd, hunters are the means by which wildlife managers conduct a therapeutic hunt. In some cases, it can even seem strained to refer to culling as "hunting," for instance when it is done by wildlife professionals; but culling definitely fits the general definition of "hunting" as in "[t]o pursue (wild animals or game) for the purpose of catching or killing."[20]

The threefold distinction can be used to clarify what form of hunting is at issue when environmentalists criticize sentientism on the grounds that it cannot endorse hunting as a means to control wildlife populations. When it comes to sport hunting, environmentalists are themselves of two minds. Some are avid hunters themselves who affirm sport hunting as long as it is managed in ways that prevent harm to ecosystems, whereas others endorse hunting only when it is needed to prevent damage to ecosystems. The latter endorse sport hunting only to the extent that it is an efficient way of doing the culling, that is, only if it also counts as therapeutic hunting. So sentientists could oppose sport hunting without running afoul of the environmentalist agenda, so long as they still endorse therapeutic hunting. The fact that sport hunters might be the only practical way to get the culling done could introduce further complications, but so long as sentientists can endorse therapeutic hunting, sentientism is not as starkly at odds with sound environmental policy as Callicott and other environmental ethicists assume.

Another complication is that not all species must be hunted to prevent damage to their ecosystems. For instance, populations of quail, mourning doves, and squirrels are not known for producing adverse habitat changes. Other species, notably ungulates like deer and elk, are known for habitat destruction, and it is regarding such animals that environmentalists uniformly support hunting for population control. In assessing the adequacy of sentientist perspectives on hunting, then, it is important to distinguish these two categories of animals. In the early 1990s, I adopted the following terminology from a Texas Agricultural Extension agent named Ron Howard. Any species that has a fairly regular tendency to overshoot the carrying capacity of its range, and in a way that degrades the range's capacity to support future generations of it and other species, Howard calls an "obligatory management species." For such species, active management of their populations is obligatory from an environmentalist's perspective. Howard refers to species that lack this tendency as "permissive management species," meaning that from an environmentalist's perspective, hunting would be permissible, but not obligatory.[21]

In my terminology, hunting of permissive management species would only be therapeutic if it secures the aggregate happiness of the target species, because permissive management species do not normally damage their habitats. Thus, it is not surprising that environmentalists are of two minds about hunting permissive management species. Those who accept sport hunting generally have no problem with it, whereas those who oppose sport hunting that is not also therapeutic take a dim view of it.

When it comes to clear cases of subsistence hunting, almost everyone supports it, whether the target is a permissive or an obligatory management species. So the central issue in the debate between sentientism and holism in environmental ethics is whether or not sentientist ethics can endorse therapeutic hunting of obligatory management species.

THERAPEUTIC HUNTING AND ANIMAL WELFARE

It is fairly easy to see how therapeutic hunting could be endorsed from a utilitarian animal welfare perspective. Aldo Leopold used drawings similar to figure 1 to illustrate what he believed happened to the deer herd on Arizona's Kaibab Plateau after wolves were eliminated in the early 1900s, and to Wisconsin's deer herd during the 1940s when, he believed, annual culls had been too low to prevent overpopulation.[22] In both cases, he argued, hunting could have been used to prevent the overpopulation event and preserve higher future carrying capacity, as suggested by the dashed line. Assuming that average well-being is preserved under the two future scenarios, then, preserving the higher carrying capacity suggested by the dashed line would maximize aggregate happiness vis-à-vis the reduced future carrying capacity suggested by the solid line. An effective therapeutic hunt need not be the least bit sporting, either, so hunters could use every advantage to make quick, clean kills, which would presumably result in less suffering than allowing animals to die from malnutrition, injuries, hypothermia, and so on during an overpopulation episode. So, to the extent that figure 31.1 accurately represents the choice between letting nature take its course and preventing overpopulation by culling, therapeutic hunting could be justified from a utilitarian animal welfare perspective.

Carrying capacity is not static, however, and that dramatically complicates real-world population-management decisions. Although carrying capacity can be reduced by overpopulation as illustrated in figure 1, it can rebound when species become underpopulated. For this reason, unmanaged wildlife populations will tend to oscillate around a long-term average carrying capacity in various ways, as depicted in figure 31.2. The form taken by the oscillation depends on the lengths of two time lags. One is the time it takes for vegetation to respond to underpopulation of the herd. The other is the time it takes for various density-dependent limiting mechanisms (DDLMs) to take effect as resources become scarce. For instance, a controlled

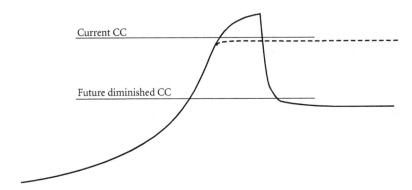

Current CC

Future diminished CC

Fig. 31.1 The future carrying capacity of the range is diminished by allowing overpopulation of an obligatory management species.

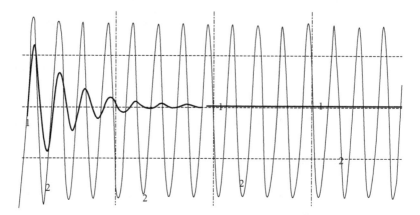

Fig. 31.2 The effects of time lags in DDLMs and vegetation responses cause oscillation around the long-term average carrying capacity.

study of penned white-tailed deer found that with "high nutrition," 1.96 fawns were born per pregnant doe, compared to only 1.11 per pregnant doe with "low nutrition." The frequency of female fawns was also cut in half, from 64.4% with high nutrition to 31.6% with low nutrition. Although author Louis J. Verme cautioned against extrapolating these results to wild populations because the low nutrition deer received in the study was "far short of the average rate for does on our worst habitat,"[23] a general tendency to produce fewer fawns per doe and a lower percentage of female fawns would tend to reduce the reproductive capacity of a herd as population densities rise. Fawn survival rates and the average weights of adults also get lower, while parasite loads and malnutrition increase.

So deer have various DDLMs that cause their population growth curve to flatten out during overpopulation events and to steepen during underpopulation. The more these DDLMs lag behind current population levels, the more the herd will tend to overcompensate for underpopulation and overpopulation. Generally speaking:

(1) Where the time lags are shorter, the oscillations will be less extreme and they will tend to dampen out over time, as shown in wave 1 of figure 2.
(2) Where the time lags are longer, the oscillations will remain extreme, as shown in wave 2 of figure 2.

The upshot is that the obligatory management species for whom DDLMs kick in most slowly and whose environments rebound most slowly during underpopulation will be those to whom the argument that therapeutic hunting can maintain carrying capacity applies most adequately. In obligatory management species whose environments rebound most quickly and whose DDLMs kick in fastest, it is less likely that culling will be needed to protect carrying capacity.

This further complicates real-world management decisions, because the classic examples of obligatory management species are ungulates, but ungulates vary greatly in terms of both their own maximum reproductive rates and their habitats' responses to overpopulation. In *A Sand County Almanac,* Leopold gave the following description of the effects of deer overpopulation in the American Southwest:

> I have seen every edible bush and seedling browsed, first to anaemic desuetude, and then to death. I have seen every edible tree defoliated to the height of a saddlehorn [*sic*]. Such a mountain looks as if someone had given God a new pruning shears, and forbidden Him all other exercise. In the end the starved bones of the hoped-for deer herd, dead of its own too-much, bleach with the bones of the dead sage, or molder under the high-lined junipers.[24]

The description is figurative, of course, but it emphasizes a real feature of typical southwestern deer habitat: high-lined junipers take many years to recover from overpopulation.

As noted earlier, deer typically fawn twins under good conditions, and this is generally true of other ungulates adapted to sub-climax environments. These are ecosystems that have been disturbed in ways that allow fast-reproducing species of animals and plants to thrive. White-tailed deer, mule deer, and moose are all examples of sub-climax adapted ungulates. By contrast, climax-adapted ungulates such as bighorn sheep, mountain goats, bison, and caribou normally produce only one offspring per year, even under the best of conditions. These species therefore have a lower maximum reproductive capacity and, consequently, their populations will tend to oscillate less dramatically than those of deer and moose.[25]

Furthermore, particular populations of deer do not always cause habitat damage in the absence of both natural predators and human hunting:

> Most wildlife biologists and managers can point to situations where deer populations have not been hunted yet do not fluctuate greatly or cause damage to vegetation. Certainly deer reach overpopulation status in some park situations, but the surprising thing is how many parks containing deer populations have no problem.[26]

And while a species indigenous to one area may be classified as a permissive management species there, it can become an obligatory management species when introduced elsewhere. A famous example is the European rabbit, which seriously impacted the Australian environment when it was introduced there.

This section has emphasized two things. First, if the goal is to maximize aggregate happiness, as in utilitarian versions of sentientism, then there is an obvious argument in favor of using hunting to prevent overpopulation in obligatory management species. Second, however, this argument's application to real-world cases is complicated by the fact that species differ in the extent to which they qualify as obligatory management species, and even in the clearest examples of these species, some local populations do not regularly cause habitat damage. So while it remains unclear exactly what the practical implications are, it will not do to simply dismiss utilitarian versions of sentientism as incapable of supporting the therapeutic hunting of obligatory management species.

THERAPEUTIC HUNTING AND ANIMAL RIGHTS

What if the goal is not to maximize aggregate happiness, however? What about sentientists who endorse a rights view?

Regan appears to think that all hunting, including therapeutic hunting, is ruled out by his rights perspective:

> Put affirmatively, the goal of wildlife managers should be to defend wild animals in the possession of their rights, providing them with the opportunity to live their own life, by their own lights, as best they can, spared that human predation that goes by the name of "sport."…If, in reply, we are told that [this] will not minimize the total amount of suffering wild animals will suffer over time, our reply should be that this cannot be the overarching goal of wildlife management, once we take the rights of animals seriously.[27]

In the first sentence of this passage, Regan seems to be calling for the extension of various liberty rights to animals, a move analogous to previous, human-oriented liberation movements.

Unlike authors in the human rights tradition, however, Regan adopts a very general account of what it means to "have moral rights." Rather than enumerate various specific rights that individuals have, as advocates of human rights commonly do, Regan characterizes "having rights" as an all-or-nothing thing. On his view, to say that individuals "have rights" is to say that they are due a special form of respect as individuals, the cash value of which is that if individuals "have rights," then we cannot justify harming them on the grounds that doing so will maximize aggregate happiness. In this way, Regan conceives of moral rights as "trump cards" against utilitarian arguments. This makes Regan's animal rights view fundamentally different from the kind of utilitarian animal welfare view described in the preceding section. And since the argument in favor of therapeutic hunting given there was explicitly utilitarian, that is an argument that Regan cannot accept, as he says in the second sentence of the above passage.

Given that rights are trump cards against utilitarian arguments, Regan recognizes that nonutilitarian principles are needed to decide whose rights to violate in cases where conflicts occur and it is impossible not to override some individuals' rights. Where noncomparable harms are involved, Regan argues, a *worse-off principle* expresses the form of special respect that is due individuals who have rights. This principle requires us to avoid harming "the worse-off individual." This is the individual or individuals who will, under one available option, be harmed to a significantly greater degree if their rights are overridden than any individual will be harmed under the other available options, if their rights are overridden. Regan emphasizes that where this principle applies, the implications of his rights view are profoundly anti-utilitarian.

Where the harms are all comparable, however, Regan endorses a *miniride principle*, according to which "we ought to choose to override the rights of the few in preference to overriding the rights of the many." Regan recognizes that, where it applies, this principle has similar implications to utilitarianism, but he claims that its justification is nonutilitarian and that it expresses a commitment to "showing equal respect for equal rights of all the individuals involved."[28]

Obviously some general conception of harm is needed to determine which of these two principles is applicable to various cases. As Regan conceives it, an individual is harmed to the extent that his or her capacity to form and satisfy desires is diminished. However little conscious suffering is involved, death completely eliminates this capacity, and for this reason Regan holds that death is the greatest harm that can befall an individual, even if it is accompanied by no suffering whatsoever. This claim, coupled with his worse-off principle, is his basis for arguing for the total abolition of slaughter-based agriculture: even a painless death is non-comparably worse than the harm that any human would suffer from abolition.

On the other hand, Regan holds that the harm that death is to a normal, adult human being is non-comparably worse than the death of any nonhuman animal. The reason is that although humans and animals share the capacity to form various kinds of desires for food, drink, sex, rest, warmth, and the like, humans' cognitive capacities allow them to form more long-term, complicated desires: to raise a family, succeed professionally, and so on. This is why Regan believes that human beings would be justified in killing nonhuman animals for food if this were the only way to survive.[29]

This brief summary of his rights view ignores many qualifications and complications that Regan discusses in his four-hundred-plus-page book, but it suffices to set up a discussion of its application to hunting, specifically. I just noted that Regan believes humans would be justified in engaging in subsistence hunting if it were the only way to feed themselves. Clearly, he should oppose sport hunting, on the grounds that the animals killed are harmed non-comparably to humans prevented from enjoying their sport. What about therapeutic hunting, however?

In the passage quoted above, Regan appears to oppose all therapeutic hunting, but he never considers which of his two principles would apply. All he says is that a common rationale for therapeutic hunting—that it minimizes aggregate suffering—cannot be accepted if the animals have trump cards against such utilitarian arguments. Yet it seems plausible to hold that all individuals of any given species are harmed comparably by death, whether it results from humans letting nature take its course or from humans actively culling overpopulated animals. If so, then his miniride principle would seem to apply, and it would seem that if humans can minimize the number of deaths that will occur, then that is what they are obligated to do. At least in the clearest cases of obligatory management species, preventing overshoot and collapse of the carrying capacity of the range would seem to minimize deaths vis-à-vis letting nature take its course. This suggests that Regan should support therapeutic hunting of obligatory management species, at least in cases where letting nature take its course will result in more deaths, as represented in figure 1.

Although Regan does not explicitly make this move in his book, something that he says about nonhuman predators suggests a reply he would want to make:

> If the rights view professes to condemn sport hunting and trapping, it might be claimed, then it should do the same when it comes to the fatal interaction between animals themselves.

The rights view rejects this argument. Animals are not moral agents and so can have none of the same duties moral agents have, including the duty to respect the rights of other animals. The wolves who eat the caribou do no moral wrong, though the harm they do is real enough.[30]

Similarly, Regan could argue that nature writ large is not a moral agent. While letting nature take its course may result in more harm overall, because more animals will die, it does not involve rights violations the way therapeutic hunting would.

When justified in this way, Regan's opposition to therapeutic hunting appears to rest on drawing a distinction between killing and letting die. While this distinction has been appealed to in discussions of medical ethics, where it has been invoked in discussions of active versus passive euthanasia for instance, philosophical treatments of the distinction have generally dismissed it:

> In ordinary language, *killing* is a causal action that brings about death, whereas *letting die* is an intentional avoidance of a causal intervention so that disease, system failure, or injury causes death....Nothing about either killing or allowing to die entails judgments about actual wrongness or rightness. Rightness and wrongness depend on the merit of the justification underlying the action, not on whether it is an instance of killing or of letting die.[31]

For instance, a decision not to administer a readily available treatment for a curable disease would, in ordinary language, be a case of "letting die," and yet it is presumptively wrong for a doctor to do this. Similarly, if a wildlife-population situation is accurately represented by figure 1, then in letting nature take its course we are knowingly avoiding a causal intervention that could prevent some number of deaths. Unlike the medical ethics case, of course, we cannot know which individuals are saved by the intervention, and it is probably true that some of the individuals killed in a therapeutic hunt would not have died had we let nature take its course.

That suggests a related move that Regan might make. Regarding medical research, Regan imagines someone arguing that his worse-off principle justifies research on animals, because when a human dies from a disease, the harm that he or she suffers is non-comparably worse than the harm that any experimental animal would suffer as a result of dying in the course of developing a cure. In response, Regan holds that "[r]isks are not morally transferrable to those who do not voluntarily choose to take them." This "special consideration," he claims, blocks the worse-off principle's application to the case of medical research, except for animals that are already suffering from the malady under study.[32]

Similarly, he might claim that this "special consideration" blocks the miniride principle's application to the case of therapeutic hunting, because human hunting transfers risks away from those animals that would be most likely to die from other causes. This is true to the extent that we cannot know which animals would die of natural causes were we not to intervene. To the extent that therapeutic hunting can be made to mimic natural attrition, this move of Regan's is vitiated, but it also seems that we accept involuntary transfers of risk in other life-and-death contexts, such as when a draft is instituted in a time of war.[33]

In summary, while Regan himself opposes all therapeutic hunting, his opposition does not ultimately rely on the miniride and worse-off principles that are at the core of his theory. His opposition relies either on the distinction between killing and letting die, or on the claim that involuntary transfers of risk are strictly impermissible if individuals have rights. Since neither of these are accepted as absolutes where human right-holders are concerned, it is clear that at least some forms of animal rights views could support therapeutic hunting of obligatory management species. For instance, if Regan abandoned the assumption that rights are not violated when we knowingly let nature take its course and the claim that involuntary transfers of risk among rights-holders are impermissible, then his miniride principle would seem to support it for the reasons given in this section.

SOME FURTHER COMPLICATIONS

The preceding sections have argued that versions of both utilitarian *animal welfare* views and nonutilitarian *animal rights* views can support therapeutic hunting of obligatory management species. Before turning to Callicott's claims that sentientist views are anti-environmental in their implications for predators and endangered species, this section briefly considers two additional complications.

One is that contraception and translocation could, at least in principle, be used to control populations without killing any animals. In practice, however, many animals die as a result of translocation. For instance, a study of deer relocated from Angel Island in San Francisco Bay found that about 85% died in the first year, mostly in the first three months. Elephant translocation has been developed to the point that mortality rates are much lower and entire matriarchal family units can be moved together, but in Africa there are few appropriate habitats currently unoccupied by elephants, and most wildlife managers face an overabundance of elephants. The same is true for deer in North America. Translocation is also expensive and labor-intensive, especially when sport hunters are willing to pay to participate in a therapeutic hunt.

Contraceptive technologies are being developed, but these are also expensive and labor-intensive, as is illustrated by extensive research on African elephants. The contraceptives used there require repeated injections and, sometimes, locating specific individuals at specific times. Wildlife managers in Kruger National Park in South Africa estimated that stabilizing the elephant population there would require treatment of 75% of the female population, or about 4,500 animals. In elephants, there can also be significant welfare-related side effects. For instance, estradiol-17β must be implanted subcutaneously. This intervention requires restraint and sedation, and it produces hormonal secretions that cause bulls to behave as if the treated females are in estrus. This resulted in treated cows being harassed by bulls, to the extent that 30% of the calves of treated mothers died in a short period of time. Vaccination with pZP produces antigens to sperm on females' eggs without inducing false estrus and can be administered with dart guns. Still, even this maneuver would alter the pyramidal structure of matriarchal family groups, the health effects of frequent mating—pregnant and nursing cows do not mate—are unknown, and captive females who go unmated and experience frequent estrus tend to develop uterine fibroids and cysts. Kruger wildlife managers conclude that the only currently feasible option for controlling elephant wildlife populations on a large scale is culling.[34]

Another complication was foreshadowed in the above section on types of hunting, where I observed that sometimes sport hunters might be the only practical way to get the culling done. Utilitarianism is sometimes characterized as the view that the end justifies the means, and Henry Sidgwick famously argued that utilitarians in a nonutilitarian society should use nonutilitarian arguments to persuade others to do what they believed best.[35] So it is relatively easy to imagine a utilitarian coming to terms with the need to use sport hunters to do the culling.

Using sport hunters in a therapeutic hunt seems especially troubling for a view like Regan's, however. The problem is that the defense of therapeutic hunting developed above depended on abandoning the distinction between killing and letting die, thus construing animals' rights as being overridden whether we let nature take its course or use therapeutic hunting to prevent overpopulation. So construed, when we use therapeutic hunting, we are attempting to minimize the overriding of animals' rights, but just the same we are overriding the rights of the particular animals that get killed. To send sport hunters out to do the killing, then, is troubling insofar as those hunters take pleasure in competing with the animal's survival skills, in practicing their hunting skills, or in getting venison.

From a rights perspective, it would be best if therapeutic hunts were pure culls with no admixture of sport. That situations may arise where the only feasible way to get the culling done is by issuing permits to sport hunters suggests that our society's priorities are wrong: we should be ready to spend what it takes to hire professionals who do not revel in the sporting aspects of the hunt. The fact that there are sport hunters eager to do the job also shows that we are still bringing up children with attitudes that are at odds with such an animal rights perspective, but that is hardly surprising. Like the fact that most people are still untroubled by intensive animal agriculture, it emphasizes that advocates of animal rights have a lot of work to do.

A welter of complications have been discussed in this and preceding sections. Certainly there are still further complications, but one thing is clear: it will not do to assume that sentientist views cannot endorse hunting for wildlife population control. Environmentalists are themselves of two minds with regard to sport hunting, and the only hunting they uniformly support is what I have characterized as therapeutic hunting of obligatory management species. Versions of utilitarian *animal welfare* views can support this kind of hunting, however, and whether or not self-professed animal rights activists endorse it, at least some forms of *animal rights* views can endorse it, too. When it comes to hunting for control of wildlife populations, then, the critiques of sentientism quoted from Callicott's and Sagoff's famous essays are based on simplistic caricatures of sentientism. When these views are fleshed out, there emerge various lines of response to the claim that sentientists must oppose therapeutic hunting.

Predators and Endangered Species

The preceding sections have shown that the practical implications of sentientism need not be drastically anti-environmental in the first, fourth, or fifth ways that Callicott claimed in his "Triangular Affair" paper. I will now more briefly address his second and third claims, about endangered species and predators.

Regarding predators, Callicott said that while they are "critically important members of the biotic community," animal liberationists should see them as "merciless, wanton, and incorrigible murderers." Nonhuman predators may be "merciless, wanton, and incorrigible" in ways that all wild animals are, but in order to be murderers they have to be moral agents. Regan, as we have already seen, denies that nonhuman predators can be said to violate prey animals' rights, precisely because they are not moral agents. The same move is open to a utilitarian.

The force of Callicott's point concerns the suffering and death that natural predators cause, but his own description of predators as "critically important members of the biotic community" anticipates another response that some sentientists have made. Leopold believed that the extermination of natural predators led to the dramatic irruptions of deer populations in Arizona and Wisconsin that occurred during his lifetime. This convinced him that large predators like wolves and mountain lions are essential to keeping game populations in check. Accordingly, he wrote:

> Hunting is a crude, slow and inaccurate tool, which needs to be supplemented by
> a precision instrument. The natural aggregation of lions and other predators on
> an overstocked range, and their natural dispersion from an understocked one, is
> the only precision instrument known to deer-management.[36]

Leopold seemed to be saying that therapeutic hunting could not adequately substitute for natural predators and that human management alone could not prevent habitat damage. This pessimism might seem exaggerated today, when improved census techniques and zone-sensitive permit systems can be used to cull herds more accurately than was possible in his time. Still, to the extent that predators really are in this sense "critically important members of the biotic community," they have a special kind of instrumental value from a sentientist perspective. Sentientists Steve Saptonzis and Peter Singer appeal to this kind of value, arguing that even if preventing predation would be a good thing in principle, in practice it probably would cause more harm than good.[37]

Regarding Callicott's third point, about endangered species, I noted earlier that he made two separate claims:

(3A) Nonsentient organisms have no moral standing according to sentientism, and therefore "humane herdspersons" might allow sheep to graze on plants that are "overwhelmingly important to the stability, integrity, and beauty of biotic communities."[38]

(3B) Additionally, while environmentalists place special value on members of endangered species, from an animal liberation perspective individuals with similar levels of sentience are of equivalent value, even if one is a member of an endangered species while the other's species is plentiful.[39]

Regarding the first of these, considerations similar to those just raised regarding the role of predators apply. If the plants in question are "overwhelmingly important to the stability, integrity, and beauty of biotic communities," then they have enormous instrumental value.

Similarly, environmentalists commonly emphasize that modern humans rely on biodiversity in two important ways that would make members of increasingly rare species increasingly valuable. One is that extant species serve as genetic resources for various uses in medicine, agriculture, and the like. The other is that diversity begets diversity in an ascending spiral through niche specialization, so that even if we preserved samples of all extant species, the loss of various species from the wild would still impoverish the world for future generations of humans who could have benefitted from further speciation. Thus humans depend, in a way that no other species does, on background levels of biodiversity. From this perspective, even if humans will never be able to put the members of a particular species to any use, the continued existence of that species in the wild contributes to the general diversity that begets further speciation, and in this way the loss of each species impoverishes future generations of human beings.

In the foregoing ways, sentientists can respond to Callicott's claims about predators and endangered species. The responses depend on the claims that large, natural predators and endangered plant species are, in Callicott's words, "overwhelmingly important to the stability, integrity, and beauty of biotic communities" and that all species, including ones whose members will never prove useful to humans, are valuable because they fuel the long-term process of speciation. These are empirical claims that might well turn out to be false. For instance, in the above discussion of hunting, I noted that Leopold might be wrong about humans' ability to control obligatory management species in the absence of large natural predators, and in general, by the time a species becomes endangered, it is almost never influential enough to be "overwhelmingly important" to the stability of its ecosystem. Nevertheless, environmentalists commonly make such claims, and if they are true, then it is clear how sentientists could respond to Callicott's second and third claims.

Intrinsic Value and the Notion of an Adequate Environmental Ethic

One final objection to sentientism's adequacy as a basis for environmental ethics remains. As I noted at the outset, aside from claiming that the practical implications of animal welfare and animal rights views would be drastically anti-environmental, environmental ethicists commonly stress that there is a deep philosophical difference between sentientism and the dominant view in the field, which is holism. They want to say that an adequate environmental ethic must attribute *intrinsic* value to entities such as species and ecosystems.

So-called "last-man cases" have been used to argue this point.[40] Suppose, we are asked, that the last human being on earth destroys a rare plant, a species, or an entire ecosystem, without affecting any sentient beings. While environmentalists typically want to say that this action of the last man would be wrong, a sentientist would either reach the opposite conclusion on the grounds that the last human's mere whim suffices to justify the action, or deny that there is any moral issue at all because no other sentient beings' interests are at stake.

So if an adequate environmental ethic must match the intuitions that environmentalists have about intrinsic value, then it looks like sentientism is inadequate in a deep way. However, is this an appropriate standard for judging the adequacy of an environmental ethic? In general, I agree with R. M. Hare that appeals to moral intuitions should not drive theory selection in ethics.[41] A thorough discussion of that issue in meta-ethics is well beyond the scope of this chapter, but let me close with two related observations.

First and foremost, if we *define* an adequate environmental ethic as one that reflects the intuitive value commitments of environmentalists, without reference to what arguments can be given for and against having those value commitments, we will reduce environmental ethics to a kind of moral anthropology and rob it of an important critical role. Although one of the functions of philosophical discussion is values clarification, or helping people to better identify what their values are, ultimately we want to know what their values should be. In environmental ethics, we want rational arguments in favor of and against various positions on what has intrinsic value, what public policies should be, and so on. If sentientism had the suite of practical implications that Callicott alleged in his "Triangular Affair" piece, that would be a substantive argument against it. Showing that environmentalists' intuitions about "last-man" cases differ from those of sentientists shows only that environmentalists and sentientists in fact have different views about what has intrinsic value. Such thought experiments are decisive only if we define an adequate environmental ethic in terms of the environmentalists' views about intrinsic value, which would be to beg a central question of environmental ethics.

Second, when it comes to arguments that an adequate environmental ethic must be holistic, I think there is a tendency to confuse what is most important from a management perspective with what is ultimately important from a moral perspective. It is a mistake to infer from the fact that sound environmental policy should focus on species and ecosystems to the conclusion that these holistic entities have anything more than instrumental value. A business manager may believe that ultimately it is only individual wealth that matters—that business is valuable only as a means to the end of producing individual wealth—and yet still recognize the necessity of managing the business holistically or as a system. Similarly, a sentientist, who believes that ultimately only the conscious experiences of sentient animals have intrinsic value, may still recognize that in developing sound environmental policy, attention must be focused on ecosystems and species. So sentientists and holists can agree that, from a management perspective, species and ecosystems are of paramount importance and should be the focus of environmental policy, but the sentientist can do this without committing to intrinsic value for holistic entities.

NOTES

1. Aldo Leopold, "The Land Ethic," in *A Sand County Almanac* (New York: Oxford University Press, 1949), 201–26.

2. Lynn White Jr., "The Historical Roots of Our Ecologic Crisis," *Science* 155 (1967): 1203–7.

3. Christopher D. Stone, "Should Trees Have Standing?" *Southern California Law Review* 45 (1972): 450–502.

4. *Sierra Club v. Morton*, 405 U.S. 727 (1972).

5. John Passmore, *Man's Responsibility for Nature* (New York: Charles Scribner's Sons, 1974).

6. Richard Routley, "Is There a Need for a New, an Environmental Ethic?" World Congress of Philosophy, *Proceedings* 15 (1973): 205–10.

7. Arne Naess, "The Shallow and the Deep, Long-Range Ecology Movement: A Summary," *Inquiry* 16 (1973): 95–100.

8. Holmes Rolston III, "Is There an Ecological Ethic?" *Ethics* 85 (1975): 93–109.

9. John Rodman, "The Liberation of Nature?" *Inquiry* 20 (1977): 83–145. "Sentient" has different connotations in different contexts, but in discussions of animal welfare and animal rights, the term standardly means being able to consciously experience suffering and/or enjoyment. It is often associated with the capacity to experience physical *pain*, specifically, although conscious suffering is possible in the absence of this capacity—as illustrated by patients with congenital insensitivity to pain.

10. Mark Sagoff, "Animal Liberation and Environmental Ethics: Bad Marriage, Quick Divorce," *Osgood Hall Law Journal* 22 (1984): 297–307, at 303–4.

11. Bryan Norton, *Toward Unity among Environmentalists* (New York: Oxford University Press, 1991); see, for instance, 222–23 and 241–42.

12. See, for instance, Bernard Rollins' rational reconstruction of "the social ethic" governing common sense thinking about animals in *Animal Rights and Human Morality* (Buffalo, N.Y.: Prometheus Books, 1981); "The New Social Ethic for Animals," in his *Farm Animal Welfare: Social, Bioethical, and Research Issues* (Ames: Iowa State University Press, 1997), 3–26; and in part one of his *An Introduction to Veterinary Medical Ethics: Theory and Cases* (Ames: Iowa State University Press, 1999).

And recently some prominent virtue theorists have shown an interest in animal ethics. See, for instance, Rosalind Hursthouse's and Martha Nussbaum's contributions to this volume.

13. Tom Regan, "How to Justify Violence," in *Terrorists or Freedom Fighters: Reflections on the Liberation of Animals*, ed. Steven Best and Anthony J. Nocella II (New York: Lantern Books, 2004), 231–36.

14. J. Baird Callicott, "Animal Liberation: A Triangular Affair," *Environmental Ethics* 2 (1980): 311–38, at 318.

15. Callicott, "Triangular Affair," 320.

16. Callicott, "Triangular Affair," 326–27.

17. Callicott, "Triangular Affair," 335.

18. Callicott, "Triangular Affair," 331.

19. Edward Johnson, "Animal Liberation Versus the Land Ethic," *Environmental Ethics* 3 (1981): 265–73, at 267.

20. *Oxford English Dictionary*.

21. Ron Howard, personal communication with the author.

22. Susan L. Flader, *Thinking Like a Mountain: Aldo Leopold and the Evolution of an Ecological Attitude toward Deer, Wolves, and Forests* (Lincoln: University of Nebraska Press, 1974), 203.

23. Louis J. Verme, "Reproduction Studies on Penned White-Tailed Deer," *Journal of Wildlife Management* 29 (1965): 74–79, at 76–77.

24. Leopold, *Sand County*, 130–32.

25. Dale McCullough, *The George Reserve Deer Herd: Population Ecology of a K-Selected Species* (Ann Arbor: University of Michigan Press, 1979), 160 and 172.

26. Dale McCullough, "Lessons from the George Reserve," in *White-Tailed Deer Ecology and Management*, ed. Lowell K. Halls (Washington, D.C.: Wildlife Management Institute, 1984), 211–42, at 239–40.

27. Tom Regan, *The Case for Animal Rights* (Berkeley and Los Angeles: University of California Press, 1983), 357.

28. Regan, *Case for Animal Rights*, 305.

29. Regan, *Case for Animal Rights*, 94–103 and 351–53.

30. Regan, *Case for Animal Rights*, 357.

31. Tom L. Beauchamp and James F. Childress, *Principles of Biomedical Ethics*, 6th ed. (New York: Oxford University Press, 2009), 172–74.

32. Regan, *Case for Animal Rights*, 322 and 377.

33. Gary Varner, "The Prospects for Consensus and Convergence in the Animal Rights Debate," *Hastings Center Report* 24, no. 1 (January/February 1994): 27–28.

34. Ian Whyte and Richard Fayrer-Hosken, "Playing Elephant God: Ethics of Managing Wild African Elephant Populations," in *Elephants and Ethics: Toward a Morality of Coexistence*, ed. Christen Wemmer and Catherine A. Christen (Baltimore: Johns Hopkins University Press, 2008), 399–417.

35. Henry Sidgwick, *The Methods of Ethics*, 7th ed. (London: Macmillan and Company, 1907), book IV, chapter 5, §3.

36. Aldo Leopold, "Report to American Wildlife Institute on the Utah and Oregon Wildlife Units," quoted in Flader, *Thinking Like a Mountain*, 176.

37. Steve F. Sapontzis, *Morals, Reason, and Animals* (Philadelphia: Temple University Press, 1987), chapter 13; Peter Singer, *Animal Liberation*, 2nd ed. (New York: Avon Books, 1990), 226.

38. Callicott, "Triangular Affair," 320.

39. Callicott, "Triangular Affair," 326–27.

40. The "last-man" cases originated with Routley, in "Is There a Need for a New, an Environmental Ethic?"

41. R. M. Hare, *Moral Thinking: Its Levels, Method and Point* (London: Oxford University Press, 1981).

SUGGESTED READING

CALLICOTT, J. BAIRD. "Animal Liberation: A Triangular Affair." *Environmental Ethics* 2 (1980): 311–38.

JOHNSON, EDWARD. "Animal Liberation Versus the Land Ethic." *Environmental Ethics* 3 (1981): 265–73.

LEOPOLD, ALDO. *A Sand County Almanac.* New York: Oxford University Press, 1949.

MCCULLOUGH, DALE. *The George Reserve Deer Herd: Population Ecology of a K-Selected Species.* Ann Arbor: University of Michigan Press, 1979.

———. "Lessons from the George Reserve." In *White-Tailed Deer Ecology and Management,* edited by Lowell K. Halls, 211–42. Washington, D.C.: Wildlife Management Institute, 1984.

———. "North American Deer Ecology." In *Aldo Leopold: The Man and His Legacy,* edited by Thomas Tanner, 115–22. Ames: Iowa State University Press, 1987.

REGAN, TOM. *The Case for Animal Rights.* Berkeley and Los Angeles: University of California Press, 1983.

RODMAN, JOHN. "The Liberation of Nature?" *Inquiry* 20 (1977): 83–145.

ROUTLEY, RICHARD. "Is There a Need for a New, an Environmental Ethic?" International Congress of Philosophy, *Proceedings* 15 (1973): 205–10.

SAGOFF, MARK. "Animal Liberation and Environmental Ethics: Bad Marriage, Quick Divorce." *Osgood Hall Law Journal* 22 (1984): 297–307.

SAPONTZIS, STEVE F. "Saving the Rabbit from the Fox." In *Morals, Reason, and Animals,* 229–48. Philadelphia: Temple University Press, 1987.

SINGER, PETER. *Practical Ethics.* 2nd ed. New York: Cambridge University Press, 1993.

———. 2001. "Animals." In *A Companion to Environmental Philosophy,* edited by Dale Jamieson, 416–25. Malden, Mass.: Blackwell Publishers, 2001.

VARNER, GARY. 1995. "Can Animal Rights Activists be Environmentalists?" In *Environmental Philosophy and Environmental Activism,* edited by Donald Marietta and Lester Embree, 169–201. (Reprinted as chapter 5 of Varner 1998.)

———. *In Nature's Interests? Interests, Animal Rights, and Environmental Ethics.* New York: Oxford University Press, 1998.

———. 2001. "Sentientism." In *A Companion to Environmental Philosophy,* edited by Dale Jamieson, 192–203.

WHYTE, IAN, and RICHARD FAYRER-HOSKEN. "Playing Elephant God: Ethics of Managing Wild African Elephant Populations." In *Elephants and Ethics: Toward a Morality of Coexistence,* edited by Christen Wemmer and Catherine A. Christen, 399–417. Baltimore: Johns Hopkins University Press, 2008.

CHAPTER 32

VEGETARIANISM

STUART RACHELS

Over the last fifty years, traditional farming has been replaced by industrial farming. Unlike traditional farming, industrial farming is abhorrently cruel to animals, environmentally destructive, awful for rural America, and wretched for human health. In this essay, I document these facts, explain why the industrial system has become dominant, and argue that we should boycott industrially produced meat. Also, I argue that we should not even kill animals *humanely* for food, given our uncertainty about which creatures possess a right to life. In practice, then, we should be vegetarians. To underscore the importance of these issues, I use statistics to show that industrial farming has caused more pain and suffering than did the atrocious behavior that produced the Holocaust.

1. The Cruelty of Industrial Farming

Pigs

In America, nine out of ten pregnant sows live in "gestation crates." These pens are so small that the pigs can hardly move. When the sows are first crated, they flail around, as if they're trying to escape from the crate. But soon they give up. The pigs often show signs of depression: they engage in meaningless, repetitive behavior, like chewing the air or biting the bars of the stall. The animals live in these conditions for four months. Gestation crates will be phased out in Europe by the end of 2012, but they will still be used in America.[1]

In nature, pigs nurse their young for about thirteen weeks. But in industrial farms, piglets are taken from their mothers after a couple of weeks. Because the piglets are weaned prematurely, they have a strong desire to suck and chew. But the farmers don't want them sucking and chewing on other pigs' tails. So the farmers routinely snip off (or "dock") the tails of *all* their pigs. They do this with a pair of pliers and no anesthetic. However, the whole tail is not removed; a tender stump remains. The point is to render the area sensitive, so the pigs being chewed on will fight back.[2]

Over 113 million pigs are slaughtered each year in America.[3] Typically, these pigs are castrated, their needle teeth are clipped, and one of their ears is notched for identification—all without pain relief.[4] In nature, pigs spend up to three quarters of their waking hours foraging and exploring their environment.[5] But in the factory farms, "tens of thousands of hogs spend their entire lives ignorant of earth or straw or sunshine, crowded together beneath a metal roof standing on metal slats suspended over a septic tank."[6] Bored, and in constant pain, the pigs must perpetually inhale the fumes of their own waste.

Pigs in the industrial farming system suffer the effects of overcrowding. In 2000, the U.S. Department of Agriculture compared hog farms containing over ten thousand pigs (which is the norm) with farms containing under two thousand pigs. The larger farms had three times as much mycoplasma pneumonia, six times as much swine influenza, and twenty-nine times as much flu of another strain.[7] Some of the pigs die prematurely. Others get sick and are euthanized however the farmer sees fit. "The survivors live just long enough to stumble over the finish line—and onto our dinner plates."[8]

Cows

Over 33 million cows are slaughtered each year in America.[9] These animals are also routinely mutilated without pain relief. American cows are hot-iron branded and castrated[10]; their tails are docked[11]; they are dehorned through sensitive tissue[12]; and their ears are cut for identification.[13] The weaning process is traumatic for them, as it is for pigs. A calf would normally suckle from its mother for six months,[14] but in factory farms, mother and child are separated almost immediately. Cows separated from their calves will mope and bellow for days.[15] And then, if the calf is male, it might be put into the veal industry, with all its horrors.[16]

Cattle feedlots are like premodern cities, Michael Pollan writes, "teeming and filthy and stinking, with open sewers, unpaved roads, and choking air rendered visible by dust."[17] Ellen Ruppel Shell likens the feedlot to "a filth-choked slum."[18] Feedlot cattle often stand ankle-deep in their own waste[19] and are exposed to the extremes in weather without shade or shelter.[20] One study found that cattle lacking shade were four times more aggressive toward other cows than cattle in shade.[21]

Poor sanitation and stress make the cows ill. Also, they're fed grain, which they didn't evolve to eat.[22] Feedlot cows often get acidosis (kind of like heartburn) which can lead to diarrhea, ulcers, bloat, rumenitis, liver disease, and a weakening of the immune system that can lead to pneumonia, coccidiosis, enterotoxemia, and feedlot polio.[23] About one-sixth of dairy cows have mastitis, a painful udder infection.

Mastitis is exacerbated by bovine somatotrophin injections, which are banned in Europe but are routine in the United States.[24] Between 15% and 30% of feedlot cows develop abscessed livers. In some pens, it's as high as 70%.[25]

Sometimes the cows die prematurely. A woman who runs a "dead-stock removal" company in Nebraska said that her company had hauled off 1,250 dead cattle during a recent heat wave and couldn't handle all the calls it got.[26] Cows normally live about twenty years, but the typical dairy cow is considered "spent" around age four.[27] Before dying, the cow's ride to the slaughterhouse may be long, cramped, and stressful. There are, in practice, no legal limits on how long calves can be trucked without food, water, or rest.[28]

Chickens

Chickens in the egg industry are also mutilated: parts of their beaks are severed without pain relief.[29] A chicken's beak is rich in nerve endings, so the debeaking causes it severe pain. Why does the farmer do this? Because chickens cannot form a pecking order in the cramped conditions of the industrial chicken house. Thus, bigger chickens may peck smaller ones to death, and the prematurely killed chickens can't be sold on the market. So, the farmer debeaks every bird.

American farmers raise nearly nine billion chickens each year.[30] Almost all of the chickens sold in supermarkets—"broilers"—are raised in windowless sheds, each housing thirty thousand-plus birds.[31] In these sheds, the chickens cannot move around without pushing through other birds. Nor can they stretch their wings or get away from more dominant birds.[32] Also, the sheds reek of ammonia.[33] "High ammonia levels give the birds chronic respiratory disease, sores on their feet and hocks, and breast blisters. It makes their eyes water, and when it is really bad, many birds go blind."[34] The chickens are bred for unnaturally fast growth, and so they cannot stand for long periods. Some cannot stand at all. Consequently, the chickens spend a lot of time sitting on the excrement-filled litter. One study found that 26% of broilers have chronic pain from bone disease.[35] Broilers live for six or seven weeks.[36]

The egg-laying hens have it even worse. The farmers underfeed the hens because they don't want them to grow rapidly—they want them to lay eggs for as long as possible. On some days, the hens get no food at all.[37] Each hen gets less living space than a single sheet of typing paper.[38] The hen "spends her brief span of days piled together with a half-dozen other hens in a wire cage....Every natural instinct of this hen is thwarted, leading to a range of behavioral 'vices' that can include cannibalizing her cage mates and rubbing her breast against the wire mesh until it is completely bald and bleeding."[39] Under these conditions, 10% or so of the birds simply die.[40] Laying hens would normally live for more than five years, but in factory farms they're killed after thirteen months.[41]

Temple Grandin, who designs humane slaughterhouses, was horrified to discover that male chicks—which aren't used for food, since they don't grow fast enough—are sometimes thrown in dumpsters.[42] "Spent" hens might also be tossed in dumpsters, or buried alive, or thrust squirming into wood chippers.[43] When the broilers are being rounded up for slaughter, one reporter writes, the catchers "grab birds by their legs, thrusting them like sacks of laundry into the cages, sometimes applying a shove."[44] And then later, dangling from one leg, "the frightened birds flap and writhe and often suffer dislocated and broken hips, broken wings, and internal bleeding."[45] Even after all that, the stunning system of electrified bath water sometimes fails to knock the birds out. Peter Singer and Jim Mason estimate that around three million chickens per year are still conscious as they're dropped into tanks of scalding water.[46]

Turkeys

Turkeys are treated like broiler chickens.[47]

Seafood

Every year, human beings eat about 100 million tons of seafood. Americans alone eat around 17 *billion* marine creatures.[48] Some of these creatures are caught in the ocean and killed. Others are raised in fish farms, which are landlocked ponds or cages in the sea. Today around one-third of the world's seafood comes from fish farming.[49]

Fish farms are floating factory farms. Like factory farms, they're extremely crowded. For example, fifty thousand salmon may be confined in one cage at a density equivalent to putting each thirty-inch fish in a bathtub of water.[50] Or consider shrimp, America's favorite seafood since 2001.[51] While traditional shrimp farms yielded less than 450 pounds per acre, the newer, more efficient farms yield as much as 89,000 pounds per acre.[52]

For salmon, the overcrowding leads to stress, sea-lice infestations, abrasions, and a high death rate.[53] It also leads to abnormal behavior. Like the tigers that pace around tiny zoo cages, salmon swim in circles around their sea cages. Also, salmon are starved for seven to ten days before slaughter.[54] Shrimp ponds, meanwhile, are "dangerously overcrowded and indifferently managed, plagued by overfeeding, plankton blooms, and inadequate water circulation."[55] Farmed shrimp are highly susceptible to infection, despite being given antibiotics.[56] Some of these fish die. The fish that survive are killed inhumanely. Often, they are simply allowed to suffocate in the air—a process that can take up to fifteen minutes.[57]

Wild-caught fish live better lives. In practice, however, you can't selectively buy wild-caught fish, because it's almost impossible to know what you're buying. Even the sellers rarely know what they're selling you. Anyway, wild-caught fish are killed inhumanely, and this provides a strong reason not to eat them. Singer and Mason write: "Each year, hundreds of millions of fish are hooked on longlines—as much as 75 miles of line....Once hooked, swordfish and yellowfin tuna weighing hundreds of pounds will struggle for hours trying in vain to escape. Then they are hauled in, and as they come up to the boat, fishers sink pickaxes into their sides to pull them aboard. They are clubbed to death or have their gills cut and bleed to death."[58]

Gill nets are also cruel. Left drifting in the sea, these nets trap the fish. Some fish struggle violently and bleed to death; others remain trapped, perhaps for days, until the boat returns. In bottom trawling, which is environmentally destructive,[59] a net is dragged along the ocean floor, and caught fish may be dragged along for hours. When the nets are hauled up, fish that live in deep waters "may die from decompression, their swimbladders ruptured, their stomachs forced out of their mouths, and their eyes bulging from their sockets. The remainder will suffocate in the air."[60] Some of these fish may be cut up on factory ships while still alive.

Singer and Mason make a strong case that fish feel pain.[61] Immobile bivalves like oysters, clams, mussels, and scallops, however, almost certainly don't feel pain.[62] It is less clear whether crustaceans, such as shrimp, crabs, and lobsters, feel pain, although a recent article in *Animal Behaviour* argues that crabs do.[63] If these invertebrates *can* feel pain, they must feel a lot of it. We rip the legs off crabs, pile them in buckets for long periods of time, and boil them alive.

Random Acts of Cruelty

That's enough unpleasant detail. I won't describe how calves, geese, and ducks are abused in the production of veal, foie gras, and duck's liver. To focus on those examples would keep the topic at too safe a distance from us, since few of us eat veal, foie gras, or duck's liver. It would also distract from the larger picture. Industrial farming has not "merely" been cruel to hundreds of thousands of ducks, geese, and veal calves; it has caused massive pain to *tens of billions* of pigs, cows, chickens, turkeys, shrimp, tuna, and so on.

I will, however, discuss one more type of cruelty, which is not so well known. Random acts of cruelty are common on industrial farms. Here are some examples.

In 2003, a man who spent years working in an Arkansas slaughterhouse described workers "pulling chickens apart, stomping on them, beating them, running over them on purpose with a fork-lift truck, and even blowing them up with dry ice 'bombs.'"[64]

In 2004, a video shot by an undercover investigator in West Virginia showed workers slamming live chickens into walls, jumping up and down on them and drop-kicking them like footballs. The investigator saw "'hundreds' of acts of cruelty, including workers tearing beaks off, ripping a bird's head off to write graffiti in blood, spitting tobacco juice into birds' mouths, plucking feathers to 'make it snow,' suffocating a chicken by tying a latex glove over its head, and squeezing birds like water balloons to spread feces over other birds." In one video clip, "workers made a game of throwing chickens against a wall; 114 were thrown in seven minutes. A supervisor walking past the pile of birds on the floor said, 'Hold your fire,' and, once out of the way, told the crew to 'carry on.'"[65]

In 2008, undercover investigators worked for over three months at an Iowa hog farm. They documented workers beating pigs with metal rods and jabbing clothespins into their eyes. As a result, three men got suspended sentences and were ordered to pay small fines. A man who beat a pig on the back at least ten times with a metal gate rod was fined $625—the maximum allowed under the law.[66]

In 2008, People for the Ethical Treatment of Animals (PETA) secured the conviction of a farm worker in West Virginia who shoved feed down a turkey's throat and maliciously broke another turkey's neck. This man was given a one-year prison sentence, which is the biggest penalty ever given to a farm worker for animal abuse.[67] The PETA investigator also saw workers shoving excrement into turkeys' mouths, holding turkeys' heads underwater, slamming turkeys' heads against metal scaffolding, and hitting turkeys on the head with pliers and a can of spray paint.[68]

These instances of cruelty are "random" in the sense that the industrial farming system does not intend for them to happen, and any particular act of cruelty cannot be predicted. However, it is predictable that such cruelties will happen under the current system. The examples are not isolated; they arise from the working conditions. Slaughterhouses and factory farms are filthy, smelly, and disgusting. The work is dull, and the pay is terrible. The hours on the job must pass very slowly for the workers. In such circumstances, human beings will do what they can to relieve boredom. And these particular workers are desensitized to animal suffering because the system they work under treats animals like merchandise or chattel. Thus, random acts of cruelty occur often.

Things Are Getting Worse

There is hope: thanks to writers like Jonathan Safran Foer, Laurie Garrett, Jeffrey Moussaieff Masson, Michael Pollan, Eric Schlosser, and Peter Singer, more and more educated people are learning about industrial farming. In the last ten years, the pace of animal welfare reform has picked up, even if it's still slow. And according to a 2006 poll, 75% of the public would like to see government mandates for basic animal welfare measures.[69] The public may someday turn against the meat industry, just as it once turned against the cigarette industry.

However, the sheer sum of suffering is increasing. In 2000, global meat production was 229 million tons. By 2050, it will be around 465 million tons, according to the United Nations.[70] Meat consumption will rise as world population rises and as places like India and China become wealthier. Not only will the United States export more meat, but American companies will open more plants overseas, and businesses new to the industry will copy the American model of production.

Industrial farming is so obviously cruel that industry officials rarely defend it in any detail. Sometimes they issue blanket denials ("our methods are humane"), but usually they say nothing at all. They just want the news stories to die and the checks to keep rolling in.[71] How *you* view industrial farming is probably how you would have viewed slavery, had you been a white Southerner before the American Civil War. Industrial farming, like antebellum slavery, is manifestly immoral, but it's the status quo: it benefits the wealthy, it's entrenched in our economic system, and most people on the street support it.

Most of the cruelty in the industrial system occurs in factory farms. Factory farms are not slaughterhouses. Factory farms are where the animals live; slaughterhouses are where they go to die. "Factory farm" is actually a misnomer; a factory farm is not a farm. It's a building or set of buildings, not a plot of land. In industry parlance, factory farms are called "CAFOs" or "Concentrated Animal Feeding

Operations." Joel Salatin, an old-fashioned farmer in Virginia, has a different name for them: "industrial fecal factory concentration camp farms."[72]

2. The Argument
for Compassionate Eating

We can formulate various arguments against industrial farming. Here is one, which I'll call *The Argument for Compassionate Eating*, or ACE.[73] ACE has two premises ("P1" and "P2") and two conclusions ("C1" and "C2").

> P1. It is wrong to cause suffering unless there is a good reason to do so.
> P2. Industrial farming causes billions of animals to suffer without good reason.
> C1. Therefore, industrial farming is wrong.
> C2. Therefore, you shouldn't buy factory-farmed meat.

This is not an argument for vegetarianism; to say you shouldn't buy factory-farmed meat is not to say you shouldn't buy any meat. You might buy humanely raised meat or hunted meat. Moreover, the argument is cast in terms of buying, not eating. Perhaps you shouldn't support the meat industry by buying its products, but if someone else is about to throw food away, you might as well eat it.

As a matter of terminology, a *vegetarian* doesn't eat any meat or seafood; an *ovo-lacto vegetarian* is a vegetarian who eats eggs and dairy; a *pescetarian* eats seafood but no other kind of meat; a *locavore* eats only locally produced food, which might include meat; and a *vegan* consumes no animal products at all. ACE argues for being a "conscientious omnivore," to use Singer and Mason's phrase.[74] Its mantra is: boycott cruelty.

Unfortunately, cruel companies dominate the market, so ACE has taxing implications: you may eat almost no meat, dairy, or seafood sold in restaurants and grocery stores. Don't be fooled by the labels: "humanely raised," "animal care certified," "free range," "cage free," "free running," "naturally raised," "all natural," "natural," "farm fresh," and "wild" are all irrelevant to animal welfare.[75] They are legally told lies. Nor does "organic" mean much to the animals.[76] In the 1990s, "Big Organic" out-lobbied the animal welfarists, and so the big companies determined what the "organic" label means in America.[77] Food labeled "organic" *is* better for the environment and better for human health,[78] and it probably tastes better,[79] but that's all.

How might ACE be criticized? The first premise—it is wrong to cause suffering unless there is a good reason to do so—is very modest. *Suffering*, by its nature, is awful, and so one needs an excellent reason to cause it. Occasionally, one will have such a reason. Surgery may cause a human being severe postoperative pain, but the surgeon may be right to operate if that's the only way to save the patient.

And what if the sufferer is not a human, but an animal? This doesn't matter. The underlying principle is that *suffering is bad because of what it's like for the sufferer.*

Whether the sufferer is a person or a pig or a chicken is irrelevant, just as it's irrelevant whether the sufferer is white or black or brown. The question is merely how awful the suffering is to the individual.[80] Thus, the premise simply states, "it is wrong to cause suffering unless there is a good reason to do so."

The second premise—industrial farming causes billions of animals to suffer without good reason—is equally secure. Above, I described how billions of animals suffer in the industrial food system. What "good reason" could there be for all that pain? The pleasure we get from eating meat is not good enough, especially since we can enjoy eating other things. Moreover, I'll explain in the next section that industrial farming is disastrous for humans as well as for animals.

Most of the objections one hears to arguments like ACE are intellectually pathetic—either they're irrelevant or they admit of multiple, obvious refutations. Here are some typical exchanges. Objection: Wouldn't life in the wild be worse for these creatures?[81] Reply: No, it wouldn't. Anyway, how is this relevant to the argument? Objection: Why should we treat animals any better than they treat each other?[82] Reply: You assume that we should look to animals for moral guidance, but we shouldn't. Anyway, we treat animals much worse than they treat each other. Objection: Animals on factory farms have never known any other life.[83] Reply: But they can still suffer. Similarly, it is wrong for parents to abuse their children, even if their children have never known any other life. Objection: Animals are dumb. Reply: Intelligence is irrelevant. Albert Einstein didn't have the right to torture people with low I.Q.s. Objection: The world is full of problems, and surely solving human problems must come first.[84] Reply: A problem this big can't wait. Anyway, human problems needn't "come first"— we can try to solve several problems at once. Finally, industrial farming *is* a human problem: I'll explain in the next section that industrial farming is awful for people. Objection: We have no duties toward animals because they can't have duties toward us.[85] Reply: Human infants and the severely disabled can't have duties toward us, yet we have duties to them; and we certainly shouldn't abuse them. Nor should we abuse animals. And the reason, in each case, is the same: human infants, the severely disabled, and nonhuman animals are all capable of experiencing pain and suffering.

Singer and Mason say that the best defense of eating meat is that, without the food industry, these animals wouldn't exist, so at least now they have lives.[86] But this defense is as bad as the others. It would be *much* better if these animals had never existed, given how horribly they suffer. Each life in the industrial farming system has a high negative utility.

Resistance to arguments like ACE usually stems from emotion, not reason. When you describe factory farming to meat eaters, they feel attacked. Moreover, they want to justify their next hamburger. Thus, they advance whatever justification comes to mind. This is human nature. I heard of one educated person who responded to an ACE-like argument by questioning whether animals can feel pain—even though we have every behavioral, physiological, and evolutionary reason to believe that they can. This person wasn't uncompassionate; rather, his defenses were up. But, eventually, he was persuaded by the opposing argument. His name is Peter Singer.[87]

The best objection to ACE is to deny the inference from C1 ("industrial farming is wrong") to C2 ("you shouldn't buy factory-farmed meat"). This inference needs justification; we need some argument to bridge the gap between "it's wrong" and "don't support it." One may doubt, in this case, whether any argument could succeed. Industrial farming is controlled by giant, multi-billion-dollar corporations that make decisions based on the bottom line, and one person's eating habits won't affect the balance sheet in any significant way. In short, the objection says, *my actions won't make a difference, so I might as well enjoy my meat and hope that someday the government will force agribusiness to change.*

This is not just a challenge to ACE. It's more general: "Why should I participate in *any* group project, when my participation is unlikely to matter but has costs for me? Why recycle? Why vote? Why write letters for Amnesty International? Why boycott products made from slave labor? Why drive an electric car? Even if I believe in the causes—renewing resources, electing good public officials, freeing political prisoners, combating slavery, and protecting the environment—why should I participate, when I won't affect the outcome, and I don't want to participate?" This is the best defense of laziness, apathy, and selfishness. Ultimately, however, it will not succeed.

There are three plausible ways to try to bridge the gap.

First, you might say: "It doesn't matter whether I'll make a difference. *I shouldn't participate in a morally corrupt enterprise, regardless of the cost-benefit analysis of my participation.* If I ate industrially produced meat, I'd be benefiting from cruelty. I'd have dirty hands. So, I shouldn't do it." If this line of thought sounds too high-minded, consider a different example. Suppose that someone you know—a charming but sketchy character—has just mugged an old woman for $200. Now he wants to treat you to dinner. Should you accept? Why not have a good meal before calling the cops?[88] Common sense says, "Don't do it. Don't become a part of the mugging, even after the fact." These two examples are essentially the same. In each, you take a "principled stand": you simply refuse to benefit from evil. To enjoy industrial meat or the spoils of a mugging would compromise your moral integrity; it would stain your soul.

As a utilitarian, I reject this way of filling in the argument. I don't believe that it's *intrinsically* good to opt out of immoral enterprises; I would assess my participation based on its probable effects. However, I do admire the desire to disassociate one's self from evil unconditionally. People who take a principled stand against industrial farming set a good and unambiguous example to their neighbors. Also, their uncompromising attitude makes them especially unlikely to backslide.

Second, you might say, "If nobody bought factory-farmed meat, then there wouldn't be any; industrial farming wouldn't exist. Thus, the group of omnivores is responsible, both causally and morally, for the animals' suffering. Therefore, each member of the group is responsible. If I bought factory-farmed meat, I would join that group, and then I too would be responsible. So, I shouldn't buy factory-farmed meat."[89] On this view, what matters in assessing your behavior is not just the effects of your particular act, but the effects of *all* the acts of which yours is a part. Consider another example. Suppose a firing squad of twelve expert marksmen shoots and kills a person whom they know to be innocent. Each marksman can truthfully say, "I made no difference. If I hadn't fired my gun, the outcome would have been the same." But surely we can hold *somebody* responsible for the killing. We can look at the entire group and say, "If it weren't for you, that person would still be alive. You are all responsible. Therefore, each of you is responsible."

Again, I find this reasoning plausible, but I reject it. At its core, it errs by treating *group* responsibility as primary and individual responsibility as derivative. If responsibility is a basic moral notion, then it stems from the free, informed choices of individuals, not from the behavior of groups. A group does not make choices, except in the derived sense that its members make choices. And a group is not morally responsible, except in the derived sense that its members are morally responsible. Personally, I would justify my refusal to be in the firing squad by appealing to the chance, however remote, that others in the group would miss, or would develop cold feet, or would have defective equipment, thus making me the sole cause of the innocent person's death.

I prefer a third way of bridging the gap between "it's wrong" and "you shouldn't support it." On this proposal, I shouldn't support industrial farming because my behavior *might* make a difference: the meat industry might produce less meat next year if I don't buy meat this year.[90] After all, my behavior might determine whether a threshold, or tipping point, of sales is reached, thus prompting a reduction in production.

If this seems too hopeful, consider the logic of the situation. Assume that I normally eat twenty chickens per year, and let's try out different assumptions about how sensitive the meat industry is to changes in demand. Suppose, first, that they are maximally sensitive, or sensitive to differences of one: for every chicken consumed this year, there will be one additional chicken grown next year. If so, then my decision not to eat chicken is fully rational: it is guaranteed to reduce the suffering of twenty chickens at very little cost. Or rather, for economic reasons, other people might eat more chicken if I eat less,[91] so let's say instead that I would reduce the suffering of ten chickens at very little cost. Next, suppose that the meat industry is sensitive only to differences of 10,000: it will increase next year's supply only when the number of chickens consumed this year reaches a multiple of 10,000. So, for example, when the millionth chicken is sold this year, this will ensure greater production next year, because 1,000,000 is a multiple of 10,000. However, the sale of additional chickens won't affect production until 1,010,000 chickens are sold. Now the question is whether my chicken boycott will determine whether some multiple of 10,000 is reached. If so, then the odds of my boycott mattering are merely 10 in 10,000, or 1 in 1,000. However, when a multiple of 10,000 is reached, the industry will increase production by 10,000. So, I now have a 1 in 1,000 chance of eliminating the suffering of 10,000 chickens, rather than a 1 in 1 chance of eliminating the suffering of 10 chickens. Each action has the same expected utility; both are fully rational. Finally, assume that the industry is sensitive only to multiples of 100,000. Now the odds of my boycott mattering dip down to 10 in 100,000, or 1 in 10,000. However, the payoff would be a world in which 100,000 fewer chickens suffer. Again, my action would be fully rational.[92]

This analysis is oversimplified. For example, it ignores the possible effects of government subsidies. But the basic idea is compelling: if the odds of success are high, then the payoff would be high enough to justify boycotting meat; and if the odds of success are low, then the payoff would be proportionally greater, and again the boycott is morally correct. Furthermore, as Alastair Norcross says, "many people who become vegetarians influence others to become vegetarian, who in turn influence others, and so on."[93] Thus, the payoffs of boycotting meat may be even higher than my analysis suggests.

Therefore, ACE is sound. You shouldn't buy factory-farmed meat.

3. Industrial Farming is Awful for People

Many people won't care about ACE, because they don't care about the kinds of animals being abused. However, industrial farming is awful for humans as well.

Infectious Disease

Factory farms breed infectious disease. They house billions of sick animals; the overcrowding ensures wide and rapid transmission; and many diseases can be passed from animals to humans.

Industrial farming is responsible for both the bird flu epidemic and the swine flu pandemic. Bird flu (H5N1) evolved for more than twelve years, mostly on poultry farms.[94] It has a 63% mortality rate in humans.[95] A United Nations task force found that one of the root causes of the bird flu epidemic was "farming methods which crowd huge numbers of animals into small spaces."[96] Since November 2003, there have been around four hundred confirmed cases of bird flu.[97]

The swine flu virus (H1N1), despite its name, actually combines genetic material from pigs, birds, and humans. Six of the eight viral gene segments in H1N1 arose from flu strains that have been circulating since 1998. Some scientists believe that parts of the virus can be traced back to an Indiana pig farm in 1987.[98] In 2005, a worker at a Wisconsin hog farm contracted an early version of H1N1, evidently from the pigs.[99] According to Dr. Michael Greger of the Humane Society of the United States, "Factory farming and long-distance live animal transport apparently led to the emergence of the ancestors of the current swine flu threat."[100]

Swine flu was first diagnosed in humans in the spring of 2009. Its first-known victim was a boy in Mexico who lived near an American-owned hog farm.[101] In the United States, the virus killed around 10,000 people in its first seven months and caused about 213,000 hospitalizations.[102] Laurie Garrett, the Pulitzer Prize–winning author of The Coming Plague, herself got the swine flu, and she wrote: "This bug is, in virology parlance, a 'mild flu,' but only somebody who hasn't been laid low by H1N1 would consider days of semi-delirium, muscle aches, fatigue, nausea, and stomach twisting to be 'mild.'"[103]

In 2003, the American Public Health Association called for a moratorium on factory farming,[104] and in 2005 the United Nations urged that "Governments, local authorities and international agencies need to take a greatly increased role in combating the role of factory-farming" because factory farms provide "ideal conditions for the [influenza] virus to spread and mutate into a more dangerous form."[105] These admonitions were ignored. In late 2009, the incidence of flu was higher than at any time since the 1918 Spanish flu epidemic.[106]

Industrial farming also promotes *drug-resistant* disease due to its massive use of antibiotics. Doctors know to prescribe antibiotics sparingly, even to sick patients, but on factory farms, antibiotics are put preemptively in the feed. A study in the *New England Journal of Medicine* showed an eightfold increase in antimicrobial resistance from 1992 to 1997, linked to the use of antibiotics in farmed chickens.[107] In the United States, about three million pounds of antibiotics are given to humans each year, but animals receive many times that amount.[108]

Escherichia coli (*E. coli*), which comes mostly from animal manure, is especially worrying.[109] One doctor said, "I've had women tell me that *E. coli* is more painful than childbirth."[110] *E. coli* infections cause about 73,000 illnesses in the United States each year, leading to over 2,000 hospitalizations and 60 deaths.[111] The number of beef recalls due to *E. coli* is increasing: from 2004–2006, there were twenty recalls; from 2007–2009, there were at least fifty-two.[112]

In one *E. coli* outbreak, in which 940 people got sick, a children's dance instructor named Stephanie Smith got ill from a grilled hamburger she bought at Sam's Club. Her diarrhea turned bloody, seizures knocked her unconscious, and doctors put her into a medical coma for nine weeks. Now she's paralyzed from the waist down. An investigation revealed that Cargill, which supplied the meat, had violated its own safety procedures for controlling *E. coli*. Moreover, Cargill's own inspectors had lodged complaints about unsanitary conditions at the plant in the weeks before the outbreak. Michael Moss, an investigative reporter for the *New York Times*, found that "Many big slaughterhouses will sell only to grinders who agree not to test their shipments for E. coli....Slaughterhouses fear that one grinder's discovery of E. coli will set off a recall of ingredients they sold to others." Hamburgers are more dangerous to eat than they might be, because "a single portion of hamburger meat is often an amalgam of various grades of meat from different parts of cows and even from different slaughterhouses." Moss concludes, reasonably enough, that "eating ground beef is still a gamble."[113]

When a lot is at stake, it is rational to worry about small chances. Industrial farming might one day cause a pandemic that kills hundreds of millions of people and leads to a massive destabilization of political systems and economic structures. What are the chances that this will happen? It is hard to say, but the risk is too high.

Pollution

Industrial farms do not form closed ecological loops, like traditional farms do.[114] Instead, they massively pollute the land, air, and sea. For example, nitrogen and pesticides run off the cornfields that supply the farmers with animal feed[115]; shrimp farming has ruined more than half a million acres of land, which now lie abandoned[116]; tractors, trucks, and combines spew out exhaust fumes; and wastewater pumped from fish ponds pollutes canals, rivers, and streams with pesticides, antibiotics, and disinfectants.[117]

Most of the pollution, however, comes from sewage. In America, chickens, turkeys, and cows produce over three times more total waste than people,[118] and an adult pig produces about four times as much feces as a human.[119] Just two feedlots outside of Greeley, Colorado, produce more excrement than the cities of Atlanta, Boston, Denver, and St. Louis combined.[120] Although human sewage is elaborately treated, animal sewage is not.[121] And the farmers don't know where to put it.

Manure is wet and costly to transport, so it is often sprayed on nearby fields, where it contaminates the air, water, and land.[122] It stinks, it kills fish and amphibians, and it generates acid rain and sewage runoff. Sometimes the manure is pumped into lagoons. This can lead to heavy metals leaking into the soil, and those metals might eventually wind up in our crops.[123] Moreover, the lagoons can suffer catastrophic breaches—as when, in 1995, a breach in North Carolina released twenty-five million gallons of untreated hog waste into the New River.[124] In 2002, 71% of Nebraska's rivers and streams were too polluted for recreation, aquatic life, agriculture, and drinking.[125] The problem, of course, is not limited to Nebraska. According to the Environmental Protection Agency, 35,000 miles of American waterways have been contaminated by animal waste.[126] Worldwide, over a million people get their drinking water from groundwater that is moderately or severely contaminated with pollutants that come mostly from fertilizers and the application of animal waste.[127]

The air pollution from industrial farms contributes heavily to climate change. When people think of global warming, they think of SUVs, but the livestock sector emits more greenhouse gases than the entire transportation sector, mostly due to methane and nitrous oxide emanating from manure. Meat production thus contributes more to global warming than cars, trucks, buses, SUVs, trains, planes, and ships combined.[128] And as Singer and Mason say, more global warming "will mean more erratic rainfall patterns, with some arid regions turning into deserts; more forest fires; hurricanes hitting cities that at present are too far from the equator to be affected by them; tropical diseases spreading beyond their present zones; the extinction of species unable to adapt to warmer temperatures; retreating glaciers and melting polar ice caps; and rising sea levels inundating coastal areas."[129] The typical American diet generates the equivalent of nearly 1.5 tons more carbon dioxide per person per year than a vegan diet with the same number of calories.[130]

Overconsuming Resources

Industrial farming overconsumes a number of scarce resources. First, it overconsumes fossil fuels.[131] When we eat animals that grazed on pastures, we harvest the free energy of the sun. By contrast, intensively raised animals are fed mostly corn, which requires chemical fertilizers made from oil.[132] Cattle are especially energy intensive. A typical steer consumes the equivalent of about thirty-five gallons of oil over its lifetime.[133] In general, it takes three units of fossil fuel energy to make one unit of food energy in American agriculture. However, it takes *thirty-five* units of fossil fuel energy to get one unit of food energy out of feedlot cattle.[134] Fossil fuels are also used to transport food long distances. One-fifth of America's petroleum consumption goes to producing and transporting food, much of it meat.[135]

All this energy is used to shed even more energy: we lose calories by feeding the animals grain and then eating the animals, rather than eating the grain directly. For example, we could get ten times as many calories by eating corn directly than by feeding corn to a steer or chicken and then eating it.[136] Industrial farming thus over-consumes *food*.

Fishing also overconsumes marine life. First, we overfish. Large-fleet fishing began in the 1950s and has depleted stocks in every ocean.[137] Second, about 25% of caught fish are "bycatch." This means that the fisherman doesn't want them. Each year, *billions* of fish are killed and not eaten—around twenty-seven million tons of sea creatures.[138] Because we have overfished the oceans, we now get an increasing amount of our seafood from fish farms. These farms also overcon-sume resources. Shrimp, for example, is unsustainably produced.[139] When we intensively farm fish, we feed fish to the fish, and we lose calories in the process. For example, it takes three pounds of wild fish to produce one pound of salmon.[140]

Industrial farming overconsumes water, at a time when water shortages are becoming critical worldwide. Industrial farming depletes rivers as well as aquifers: many American rivers are doing badly because so much of their water has been diverted for irrigation.[141] The thirstiest meat is beef. To produce one pound of beef, American farmers use around 1,584 gallons of water—which is actually *less* than the worldwide average for cattle raising.[142] Producing a pound of hamburger takes twelve times as much water as producing a pound of bread, sixty-four times as much water as producing a pound of potatoes, and eighty-six times as much water as producing a pound of tomatoes.[143]

Finally, industrial farming overconsumes land: it fells trees and destroys natural grasslands. Forests are razed all over the world in order to grow food to feed factory-farmed animals.[144] Forest destruction reduces biodiversity, contributes to global warming, and harms the animals that live there. Rainforest destruction is especially bad. The Amazon rainforest, say Singer and Mason, "is still being cleared at an annual rate of 25,000 square kilometers, or 6 million acres, to graze cattle and grow soybeans to feed to animals."[145] In 2004, the clear-cutting of coastal forests for shrimp production in Asia had especially horrible consequences: it contributed sig-nificantly to the tsunami that pummeled eleven nations with twenty-foot waves, killing tens of thousands of people in a matter of hours.[146]

Harming the Powerless and Vulnerable

Industrial farms significantly harm three groups of people who lack the resources to fight back: people who live near the plants, plant workers, and small farmers.

Local residents must cope with the stench. The odors that emanate from indus-trial farms are incredible. According to Pollan, feedlots create a stink you can smell for more than a mile.[147] Consider some testimonials. A Nebraska resident who lives near a pig farm says that she's sometimes woken up at night by the odor, which burns her eyes and makes her feel sick. Another Nebraskan said that the stench nauseates her, gives her seven-year-old son diarrhea, and gives her tremendous headaches.[148] Residents in Kentucky complained of "hundreds of thousands of flies and mice" near factory farms. They also complained of gagging, coughing, stomach cramps, diarrhea, nausea, persistent mouth sores, and intestinal parasites.[149] Epidemiological studies suggest that the rancid fumes give local residents asthma as well as neurological maladies such as depression.[150]

No fetid, wafting stink would be tolerated in Beverly Hills. But industrial farms occupy rural America, where the residents lack both financial and political power and are particularly vulnerable to harm and abuse. According to the Pew Commission's report, when industrial farms replace small, locally owned farms, residents can expect "lower family income, higher poverty rates, lower retail sales, reduced housing quality, and persistent low wages for farm workers."[151] As mentioned above, over a million people worldwide drink from polluted wells. The Pew report describes how industrial farms damage social capital in rural communities, where "social capital" refers to "mutual trust, reciprocity, and shared norms and identity."[152] Industrial farming, in short, destroys rural communities.

If the stench emanating from industrial farms is incredible, imagine what it's like inside the buildings themselves. Jim Mason said that he and his photographer "spent whole days inside egg and hog factories, and afterward the smell would linger for days—even after scrubbing ourselves and our gear."[153] As part of his research, Mason got a job squeezing semen out of turkeys and "breaking" the females in order to artificially inseminate them. He describes the breaking as follows: "For ten hours we grabbed and wrestled birds, jerking them upside down, facing their pushed-open assholes, dodging their spurting shit, while breathing air filled with dust and feathers stirred up by panicked birds. Through all that, we received a torrent of verbal abuse from the foreman and others on the crew." Mason was required to "break" one bird every twelve seconds. It was, he said, "the hardest, fastest, dirtiest, most disgusting, worst-paid work [he] had ever done."[154] The Pew Commission's report documents some of the health hazards of industrial farm work: chronic respiratory irritation, bronchitis, nonallergic asthma, increased airway sensitivity, organic dust toxic syndrome (which is nasty), and exposure to harmful gases including hydrogen sulfide, which sometimes kills workers.[155] Small wonder that random acts of cruelty are common in industrial farming.[156]

American companies actively recruit undocumented workers. Sometimes they even bus the workers in from Mexico. Undocumented workers make "good employees": they work for little pay, they don't become whistle-blowers, and they don't organize unions. Sometimes the police conduct raids and deport a few of the workers, but the bosses who recruit them never get in trouble.[157] Perhaps this typifies industrial labor relations. On shrimp farms in Thailand—where much of our shrimp comes from—migrant workers from Burma, Cambodia, and Vietnam suffer the abuses of unpaid overtime, child labor, and sexual assault.[158]

Small farmers have also been crushed by the weight of corporate interests. Since the early 1970s, farm income in America has steadily declined along with the price of corn.[159] Most small farms have shut down. But these farms haven't been outcompeted honestly; rather, the big farms have prevailed by externalizing their costs and benefiting from government subsidies.[160]

Some small farmers have chosen to work for the big companies as "growers," which is the industry's name for factory-farm operators. The growers usually regret it. They are, as one put it, "serfs at the mercy of the companies that make a fortune on their backs."[161] The grower has no power in the system. Typically, he signs a contract with a company that guarantees him a certain amount of income for a certain amount of product. But to run his business, he has to purchase expensive equipment and go into debt. Once he has debt to pay, he must do whatever his bosses say. The grower makes little money, and he owns neither his animals nor the crops that feed them.[162] Worst of all, he cannot look for better bosses the next time around, because "there is often an unwritten rule that one company will not pick up a grower who has worked for another company. So if a grower does not like the contract that Tyson offers, there is nowhere else to go."[163]

Industrial farming in America also harms farmers abroad. America is the world's largest exporter of food,[164] and U.S. government subsidies have put many foreign farmers out of business by keeping American prices artificially low.[165] In the documentary *Food, Inc.*, Michael Pollan notes that one and a half million Mexican farmers were put out of business by NAFTA. Some of those farmers were among those bussed in from Mexico to work illegally in American farms.[166]

Health

Our love of meat is bad for our long-term health. It partly explains why we get the "Western diseases" or "diseases of affluence": heart disease, obesity, diabetes, and cancer.[167] A 2009 study in the *British Journal of Cancer* finds that vegetarians are 12% less likely than omnivores to develop cancer.[168] A study by the National Institutes of Health, which followed over half a million Americans for more than a decade, found that people who eat the most red meat and processed meat were likely to die sooner, especially from heart disease and cancer, as compared to those who ate much less red meat and processed meat. The study's findings suggest that, over the course of a decade, the deaths of one million men and maybe half a million women could be prevented by eating less red meat and processed meat.[169]

According to the American Dietetic Association,

> appropriately planned vegetarian diets, including total vegetarian or vegan diets, are healthful, nutritionally adequate, and may provide health benefits in the prevention and treatment of certain diseases. Well-planned vegetarian diets are appropriate for individuals during all stages of the life cycle, including pregnancy, lactation, infancy, childhood, and adolescence, and for athletes.... The results of an evidence-based review showed that a vegetarian diet is associated with a lower risk of death from ischemic heart disease. Vegetarians also appear to have lower low-density lipoprotein cholesterol levels, lower blood pressure, and lower rates

of hypertension and type 2 diabetes than nonvegetarians. Furthermore, vegetarians tend to have a lower body mass index and lower overall cancer rates.[170]

These are the facts, taken from the best scientific sources. Claims like "Milk helps prevent osteoporosis" and "Vegetarians have trouble getting enough iron and protein" are false urban legends, often promoted by the meat industry.[171]

Industrial Meat Isn't Cheap

If you get bird flu, swine flu, or *E. coli*, the meat industry should pay your medical bills. If you die, the meat industry should compensate your family. The government shouldn't pay billions of dollars to stockpile vaccines for diseases that evolved in factory farms; the meat industry should do that. Nor should taxpayers foot the bill for crop subsidies. Nor should farmers be allowed to pollute the land, air, and sea without paying for the environmental clean-up and health costs. If the food industry did all this—in other words, if it paid its own bills—then industrial meat would be expensive, not cheap.

Sometimes people say things like, "Industrial farming, whatever its drawbacks, at least succeeds in producing cheap meat. And we need cheap meat in order to feed all the hungry people in the world." This claim is wrong for three independent reasons: (i) Industrial meat isn't actually cheap; it appears cheap only because its real cost isn't reflected in its price. (ii) The poorest people in the world don't eat factory-farmed meat—it is too expensive for them, even at artificially low prices. (iii) Industrial farming wastes calories. It's cheaper and more efficient to feed the world grain than to feed the world animals that eat grain.

4. THE RIGHT-TO-LIFE ARGUMENT

In the popular mind, the main moral argument for vegetarianism concerns killing, not suffering. Sometimes animal welfare groups have promoted this misperception—for example, when PETA ran a newspaper ad that called Ronald McDonald "America's #1 Serial Killer" and showed the clown gleefully wielding a bloody knife, about to kill a chicken.[172] Such rhetoric aside, the argument about killing might go like this:

> P1. Many nonhuman creatures—such as pigs, chickens and tuna—have a right to life.
> C1. So, it is wrong to kill these creatures in order to eat them.
> C2. So, we should be vegetarians.

Some people will respond: "Even if animals have a right to life, these animals *wouldn't exist* without our farming practices. Therefore, we may kill them." This is fallacious. If an individual has a right to life, then killing her is wrong even if we are responsible for her existence. Human parents, for example, may not kill their grown children simply because they're responsible for their existence. After all, grown children have a right to life.

The crucial claim is premise one. *Do* many nonhumans have a right to life? Common sense offers no clear answer; people are ambivalent about killing animals. Many people will say something like this: Is it okay to kill a rat in your house? Sure. Is it okay to kill your pet? No way. Is it okay to kill animals on farms? I guess so, but I wouldn't want to do it myself. (I mean, we've got to eat, right?) Is it okay to hunt deer? Well, that seems mean, but I guess there are a lot of hunters out there.

Here are some facts that might be used to argue that farm animals *do* have a right to life: they have experiences; they have desires; they have cognitive abilities; they can live in social networks; and they can have lives worth living. And here are some facts that might be used to draw the opposite conclusion: animals are dumb; animals don't possess the desire to live, because they have no concept of life and death; animals have short life spans; and animals can enjoy only lower, baser pleasures. No Mozart.

The issue of killing animals is like the abortion issue: both issues are immensely difficult because it is so hard to say what characteristics confer a right to life on an individual. Pope John Paul II was sure that abortion is wrong under all circumstances, but for those not so sure, he offered this argument: abortion *might* be murder; so, we shouldn't perform abortions.[173] This argument will have force so long as the *other* arguments about abortion are considered inconclusive. A similar argument exists for vegetarianism:

> P1. Killing animals *might* be murder.
> C1. So, killing animals to obtain food is wrong.
> C2. So, we should be vegetarians.

Call this the "Argument from Caution." It is broader than the Argument for Compassionate Eating. To play it safe, we shouldn't hunt, nor should we buy meat that was humanely raised and slaughtered.

The Argument from Caution won't persuade many people, because most people think they *know* whether animals have a right to life. I do not share their certainty. The Argument from Caution, I think, is excellent, even if it changes few minds.

5. WHY IS THERE INDUSTRIAL FARMING?

It's hard to believe that all this is true. How could something so bad be the status quo in a free, democratic society? How could it be the status quo in *our* society? Why do we tolerate it? Why do we support it?

Let's consider, first, how the food system developed. The basic crop is corn. When the Green Revolution began around 1940, farmers could harvest only seventy or eighty bushels of corn per acre. But yields grew, and during the Nixon administration, "the government began supporting corn at the expense of farmers."[174] Subsidies pushed down prices, and as farmers made less money, they were encouraged to grow more and more corn. By 1980, yields were up to two hundred bushels per acre.[175] Every American president since Nixon has been in bed with the food industry; the food industry has, in effect, directed American agricultural policy. Today, growing corn is the most efficient way to produce food calories,[176] and American cornfields occupy an area twice the size of New York State.[177]

The scientists and engineers who improved corn yields must've thought they were saving the world; they must've thought they were taking a giant stride towards eliminating human hunger. Instead, their work helped to replace traditional farming with industrial farming. Cows and chickens could now be fed more cheaply with corn than with grass.[178] So, the pastures were turned into cornfields, and the animals were brought inside. Throughout this process, the driving force in agribusiness was profit; the driving force in Washington was re-election.[179] Both animal interests and long-term human interests were ignored.

It is easy to see how corporate greed and political self-interest can fuel an evil enterprise, but the blame extends far beyond a few dozen individuals. Industrial farming would never have prevailed without government subsidies, and the general public is responsible for the governments it elects and re-elects. We're also responsible for not improving our system of government. In America, money dominates politics. Under a better system, politicians could not be bought so easily.

And, of course, most people eat meat. Why haven't more people become vegetarians? Peter Singer has been giving arguments like ACE since 1973.[180] Why haven't Singer and his followers spurred a moral revolution? There are several reasons for this, none of which make human beings sound virtuous.

The main reason is selfishness. People enjoy eating meat. Also, they like the convenience of having meat as an option. So, people eat meat, even if they suspect they shouldn't.

A second reason is ignorance. Many people know nothing about industrial farming, so they eat meat. Some of that ignorance, however, is willful: people don't know because they don't want to know. If you try to tell people about contemporary farming, they will communicate through words or body language that they want you to stop. Selfishness lies behind some ignorance.

Ignorance also plays a subtler role. In general, many people have no idea what to make of unorthodox moral ideas. They don't know what questions to pose; they don't know what objections to offer; they may even have a hard time processing the idea that other people could believe such a thing. What will such people think when someone comes along and makes a bunch of disturbing claims about the meat industry? They might think, "The world is full of crackpots. Here's one of them." Or they might tell themselves: "That sounded convincing. But I'll bet someone on the other side could've sounded just as convincing, making the opposite claims."

Third, the animals we abuse can't help us see the error of our ways. Unlike human slaves, animals don't even have the potential to fight back. They can't hire lawyers; they can't write blogs; they can't organize protest rallies; and they will never engage in civil disobedience.

Fourth, when you urge people not to eat meat, you are, in effect, criticizing them for something they do every day. In *How to Win Friends and Influence People*, Dale Carnegie's first principle of handling people is "Don't criticize, condemn or complain." Carnegie writes, "Criticism is futile because it puts a person on the defensive and usually makes him strive to justify himself. Criticism is dangerous, because it wounds a person's precious pride, hurts his sense of importance, and arouses resentment."[181] Carnegie's observations are astute. I would add that some people get defensive about eating meat because they feel guilty.

A fifth reason is the desire to conform. Many people bristle at the idea of giving up meat because it would make them seem different from their neighbors—and not just different, but weird, in a controversial and moralistic kind of way. Vegetarianism isn't cool. Americans may pride themselves on individualism, but conformity is the stronger motivator in society. Where I grew up, there was an area of town where "non-conformists" hung out: young people who "dressed different" and "looked different." A friend of mine, surveying the scene, described it aptly: "Let's all be different together." Even non-conformists want to fit in.

A recent study illustrates the idea that conformity matters more to people than morality. Robert Cialdini wanted to know which signs posted in hotel rooms would succeed in persuading guests to reuse towels. The traditional sign says, "Do it for the environment." But a sign which said that most guests had reused their towels was 18% more effective. And a sign which said that most guests *staying in this room* had reused their towels was 33% more effective.[182] "If all your friends are eating meat," said James Rachels, "you are unlikely to be moved by a mere argument."[183]

Finally, many people lack empathy for farm animals. Somehow the pain that farm animals experience doesn't seem real to them. Partly, this is because the animals are out of sight: just as we don't properly empathize with starving people we don't see, we don't properly empathize with abused animals we don't see. Also, suffering animals might not *look* like they're suffering, at least to people unfamiliar with the type of animal in question. A photograph or video taken inside a chicken house will show the overcrowding, but it might not convey the suffering. Chickens don't cry out or writhe in pain the way that humans do. And most of us cannot discern anything from a chicken's facial expression or posture.

Some people, however, lack empathy for animals simply because *they're just animals.* Perhaps the operative psychological principle is this: the less you seem like me, the less I care about you. Human beings are especially indifferent to the suffering of fish. Singer and Mason pose an excellent question: "how could people who would be horrified at the idea of slowly suffocating a dog enjoy spending a Sunday afternoon sitting on a riverbank dangling a barbed hook into the water, hoping that a fish will bite and get the barb caught in its mouth—whereupon they will haul the fish out of the water, remove the hook, and allow it to flap around in a box beside them, slowly suffocating to death?" Their answer seems plausible: "Is it because the fish is cold and slimy rather than warm and furry? Or that it cannot bark or scream?"[184] Human beings are land creatures and mammals. Thus, we care especially little about sea creatures who are not mammals.

6. Industrial Farming and
the Holocaust

When people learn about the abuse of animals in industrial farming, they often think, "that's awful." However, they might have the same reaction to countless other things, such as a child getting cancer, or a hurricane destroying a neighborhood, or a person committing suicide. Factory farming is much worse than those things.

When people in our culture think of a moral horror, they often think of the Holocaust—the campaign of genocide in which Hitler and his Nazi thugs starved, beat, and ultimately murdered 5.7 million Jews.[185] Other horrors may also come to mind, such as the slave trade, the oppression of women, and the history of imperial aggression and domination. But I'll focus here only on the Holocaust, because the basic, chilling facts about it are so well documented, and because it is the central example of a moral atrocity in our culture. To compare all the evils of the Holocaust to anything else would be a formidable task indeed, but I'll limit my focus to one question: which set of events has caused more suffering, the Holocaust or industrial farming? Here the word "suffering" should be understood in its proper sense: suffering is *extreme* pain, or agony. To compare industrial farming to the Holocaust in this respect, and this respect only, let's consider the number of victims involved in each.

Today around ten billion animals per year are killed in American slaughterhouses,[186] and the vast majority of these animals suffered greatly. Let's assume, very conservatively, that during the last twenty years, around five billion animals per year have suffered in American factory farms, which amounts to 100 billion suffering animals. And let's assume that the Holocaust caused suffering to 20 million human beings. This means that, *for every single human being who suffered in the Holocaust, five thousand animals have suffered in American factory farms during the last twenty years*. And really, this calculation greatly underestimates the ratio. It ignores all the intensively farmed fish; it ignores all the animals that suffered in factory farms but died before slaughter; it ignores all the farm animals that suffered more than twenty years ago; and it ignores all the human victims of industrial farming. Pain calculations are hard to make, but a five-thousand-to-one (or much greater) ratio makes this judgment easy: industrial farming has caused more suffering than the Holocaust.

Many people hope that animal pain isn't really so bad. Michael Pollan, for example, thinks that human pain might differ from animal pain "by an order of magnitude." Citing Daniel Dennett, he suggests that we distinguish "pain, which a great many animals obviously experience, and suffering, which depends on a degree of self-consciousness only a handful of animals appear to command. Suffering in this view is not just lots of pain but pain amplified by distinctly human emotions such as regret, self-pity, shame, humiliation, and dread."[187] According to this argument, animals don't really suffer, because their pain isn't amplified by such emotions as regret, self-pity, shame, humiliation, and dread.

This argument, however, is unsound. Imagine that a human being has twisted her ankle and is now on the ground, writhing in agony. She's trapped in a world of pain, waiting for it to end. But she doesn't blame herself for the pain, nor does she fear for her future. Her pain is *not* "amplified by distinctly human emotions such as regret, self-pity, shame, humiliation, and dread." Her pain just hurts like hell. This example proves that pain can be very, very bad even if it's not "amplified by distinctly human emotions." If castrating a pig without anesthesia causes the pig *that* type of pain—and I believe it does—then that's enough for my arguments.

That takes care of the substantive issue; the remaining question is whether intense pain deserves the name *suffering*, if it hasn't been amplified by distinctly human emotions. Consider the woman who has twisted her ankle and is racked by pain. Does she suffer? She is certainly in a horrible state, and anyone who could end her pain has a strong moral reason to do so. Given those facts, I would say that she suffers. If someone else wants to use the word "suffering" differently, so be it. Instead of asking whether industrial farming has caused more suffering than the Holocaust, I can ask whether industrial farming has caused more agony, or more intense pain, than the Holocaust. And the answer would be the same: industrial farming has caused (at least) five thousand times more agony, or intense pain, than the Holocaust.

But suppose I'm wrong. Suppose that, for whatever reason, human pain is ten times worse than animal pain. On that assumption, factory farming over the last twenty years has still caused pain morally equivalent to five hundred Holocausts. Or suppose there's only a 10% chance that the arguments in this paper are correct. On that assumption, factory farming, again, has had the expected utility of five hundred Holocausts. And if there's only a 10% chance that animal pain is 10% as bad as human pain, then factory farming has had the expected utility of fifty Holocausts (or really more, since I'm ignoring a lot of the suffering caused by industrial farming). The philosophical arguments for vegetarianism are easy. What's hard is getting people to stop eating meat.[188]

NOTES

1. The information in this paragraph comes from Peter Singer and Jim Mason, *The Ethics of What We Eat: Why Our Food Choices Matter* (Emmaus, Pa.: Rodale, 2006), pp. 46–47. Some states have banned gestation crates.

2. The information in this paragraph comes from Michael Pollan, *The Omnivore's Dilemma: A Natural History of Four Meals* (New York: Penguin Books, 2006), p. 218.

3. According to the U.S. Department of Agriculture, 113.73 million pigs were slaughtered from November 2008 to October 2009. See http://usda.mannlib.cornell.edu/MannUsda/viewDocumentInfo.do?documentID=1096. (Total figure reached by summing the figures for each individual month.)

4. Singer and Mason, *What We Eat*, p. 50.

5. Singer and Mason, *What We Eat*, p. 46.

6. Pollan, *Omnivore's*, p. 218.

7. Ellen Ruppel Shell, *Cheap: The High Cost of Discount Culture* (New York: Penguin Press, 2009), pp. 178–79.

8. Shell, *Cheap*, p. 179.

9. According to the U.S. Department of Agriculture, 33.17 million cows were slaughtered from November 2008 to October 2009. See http://usda.mannlib.cornell.edu/MannUsda/viewDocumentInfo.do?documentID=1096. (Total figure reached by summing the figures for each individual month.)

10. Pollan, *Omnivore's*, p. 69, in reference to steer 534; Singer and Mason, p. 273.

11. David DeGrazia, "Moral Vegetarianism from a Very Broad Basis," *Journal of Moral Philosophy* 6 (2009): 143–65 (see p. 152).

12. DeGrazia, *Very Broad Basis*, p. 160; Singer and Mason, p. 273.

13. DeGrazia, *Very Broad Basis*, p. 160.

14. Singer and Mason, *What We Eat*, p. 57.

15. Pollan, *Omnivore's*, p. 73; Singer and Mason, pp. 57–58.

16. Singer and Mason, *What We Eat*, p. 273.

17. Pollan, *Omnivore's*, p. 72. For a similar account, see Singer and Mason, p. 63.

18. Shell, *Cheap*, p. 178.

19. Pollan, *Omnivore's*, p. 317.

20. Singer and Mason, *What We Eat*, p. 273.

21. Singer and Mason, *What We Eat*, p. 63.

22. Pollan, *Omnivore's*, p. 77.

23. Pollan, *Omnivore's*, p. 78.

24. Singer and Mason, *What We Eat*, p. 57.

25. Pollan, *Omnivore's*, p. 78.

26. Chris Clayton, "More than 1250 Nebraska Cattle Died in Heat Wave," *Omaha World-Herald*, July 27, 2005.

27. DeGrazia, *Very Broad Basis*, p. 152.

28. Singer and Mason, *What We Eat*, p. 273. The U.S. Department of Agriculture agreed to regulate trucking transport in 2006, but as of July 2010, they hadn't yet done so.

29. Singer and Mason, *What We Eat*, p. 37; Pollan, *Omnivore's*, p. 318.

30. Singer and Mason, *What We Eat*, p. 24.

31. Singer and Mason, *What We Eat*, p. 23.

32. Singer and Mason, *What We Eat*, p. 23.

33. Pollan, *Omnivore's*, p. 171.

34. Singer and Mason, *What We Eat*, p. 24. They cite six sources (see p. 304, footnote 13).

35. Singer and Mason, *What We Eat*, p. 24.

36. Six weeks: Singer and Mason, *What We Eat*, p. 25; seven weeks: Pollan, *Omnivore's*, p. 172.

37. Singer and Mason, *What We Eat*, p. 25.

38. Singer and Mason, *What We Eat*, p. 37.

39. Pollan, *Omnivore's*, p. 317.

40. Pollan, *Omnivore's*, p. 318.

41. Singer and Mason, *What We Eat*, p. 107.

42. Singer and Mason, *What We Eat*, p. 40.

43. Singer and Mason, *What We Eat*, p. 106.

44. Peter S. Goodman, "Eating Chicken Dust," *Washington Post*, November 28, 1999. The routine rough handling of animals can also be viewed in the documentaries *Our Daily Bread* (2005), *Food, Inc.* (2008), and *Death on a Factory Farm* (2009).

45. Singer and Mason, *What We Eat*, p. 25.

46. Singer and Mason, *What We Eat*, p. 26.

47. Singer and Mason, *What We Eat*, p. 28.

48. Singer and Mason, *What We Eat*, p. 112, using data from http://www.fishinghurts. com/fishing101.asp. Singer and Mason arrived at the 17 billion estimate by dividing the total weight of seafood consumed by an estimated average weight per creature.

49. Singer and Mason, *What We Eat*, p. 122.

50. Singer and Mason, *What We Eat*, p. 122.

51. Shell, *Cheap*, p. 173. Shrimp are America's "favorite seafood" as measured by weight.

52. Shell, *Cheap*, p. 175.

53. Singer and Mason, *What We Eat*, p. 129.

54. Singer and Mason, *What We Eat*, p. 129.

55. Shell, *Cheap*, p. 175.

56. Shell, *Cheap*, p. 176.

57. Singer and Mason, *What We Eat*, p. 129.

58. Singer and Mason, *What We Eat*, pp. 129, 132.

59. Singer and Mason, *What We Eat*, pp. 126–27.

60. Singer and Mason, *What We Eat*, p. 132.

61. Singer and Mason, *What We Eat*, p. 131.

62. Singer and Mason, *What We Eat*, pp. 133, 276.

63. Robert W. Elwood and Mirjam Appel, "Pain Experience in Hermit Crabs?" *Animal Behaviour* 77 (2009): 1243–46.

64. Singer and Mason, *What We Eat*, p. 27.

65. Donald G. McNeil Jr., "KFC Supplier Accused of Animal Cruelty," *New York Times*, July 20, 2004.

66. The information in this paragraph comes from a PETA newsletter dated August 24, 2009, and signed by Ingrid E. Newkirk.

67. The information up to this point in this paragraph comes from the August 24, 2009, PETA newsletter.

68. "Farmhands Convicted in PETA Sting," *PETA's Animal Times* (Summer 2009), pp. 20–21. See the HBO documentary, *Death on a Factory Farm*, for gratuitous abuse at a hog farm in Ohio.

69. The poll is cited in "Putting Meat on the Table: Industrial Farm Animal Production in America," A Report of the Pew Commission on Industrial Farm Animal Production, 2008, p. 31. According to the report, the poll was conducted by Oklahoma State University and the American Farm Bureau Federation. However, the link they give to the poll is now defunct. The Executive Director of the Pew Commission told me through his assistant that the poll was conducted in 2006.

70. The United Nations Food and Agriculture Organization, "Livestock's Long Shadow: Environmental Issues and Options" (2006), p. xx, http://www.fao.org/docrep/010/a0701e/a0701e00.HTM.

71. On the meat industry's tendency to clam up, see, for example, Singer and Mason, p. 10, Michael Moss, "E. Coli Path Shows Flaws in Beef Inspection," *New York Times*, October 4, 2009, and *Food, Inc.*

72. Pollan, *Omnivore's*, p. 241.

73. James Rachels gives essentially the same argument as ACE in both "The Moral Argument for Vegetarianism," reprinted and revised in *Can Ethics Provide Answers?* (Lanham, Md.: Rowman & Littlefield, 1997), pp. 99–107, and "The Basic Argument for Vegetarianism," reprinted in *The Legacy of Socrates: Essays in Moral Philosophy*, ed. Stuart Rachels (New York: Columbia University Press, 2007), pp. 3–14.

74. Singer and Mason, What We Eat, p. 91.

75. Singer and Mason, chaps. 3 and 8; Jonathan Safran Foer, "Against Meat: The Fruits of Family Trees," *New York Times*, October 7, 2009. See http://www.humanesociety.org/issues/confinement_farm/facts/guide_egg_labels.html for the Humane Society of United States' "brief guide to labels and animal welfare."

76. Pollan, *Omnivore's*, pp. 140 and 172; Singer and Mason, chap. 8.

77. Pollan, *Omnivore's*, pp. 155–57.

78. Singer and Mason, *What We Eat*, pp. 276–77; Pollan, *Omnivore's*, pp. 162, 179, 182.

79. Pollan, *Omnivore's*, p. 176.

80. I learned this from Peter Singer, "All Animals Are Equal," chap. 1, *Animal Liberation* (New York: HarperCollins Publishers, 2002 [1975]). Or perhaps I learned it from James Rachels, who learned it from Peter Singer.

81. Pollan, *Omnivore's*, p. 310.

82. Pollan, *Omnivore's*, pp. 309–10. Compare "the Benjamin Franklin defense," Singer and Mason, pp. 243–44.

83. Pollan, *Omnivore's*, p. 310.

84. Pollan, *Omnivore's*, p. 310.

85. Singer and Mason, *What We Eat*, p. 243.

86. Singer and Mason, *What We Eat*, pp. 248–53.

87. Stanley Godlovitch helped persuade Peter Singer to become a vegetarian in Oxford in the early 1970s. Godlovitch told me about Singer's initial reaction to the argument when Godlovitch and I were both at the University of Colorado at Boulder during the 1998–99 academic year.

88. Michael Huemer suggested this example to me several years ago.

89. Derek Parfit discusses this kind of reasoning in *Reasons and Persons* (Oxford: Oxford University Press, 1984), p. 70, and in "Comments," *Ethics* 96 (1986): 832–72.

90. Thus, I disagree with James Rachels, who laments, "It is discouraging to realize that no animals will actually be helped simply by one person ceasing to eat meat. One consumer's behavior, by itself, cannot have a noticeable impact on an industry as vast as the meat business" ("Moral Argument for Vegetarianism," p. 106).

91. If one group of consumers exits the market, prices will drop, and then the remaining consumers will buy more.

92. I thank Michael Huemer for helping me think through this paragraph. Both Alastair Norcross and David DeGrazia discuss the expected utility of becoming a vegetarian in terms of how many total new vegetarians, including oneself, there are likely to be. This frames the argument too narrowly. Even if I am sure to be the only vegetarian in the world, my vegetarianism would still be correct because it might result in sales being below a certain threshold. See Norcross, "Torturing Puppies and Eating Meat: It's All in Good Taste," *Southwest Philosophy Review* 20 (2004): 117–23, reprinted in *The Right Thing to Do*, 5th ed., ed. James Rachels and Stuart Rachels (New York: McGraw-Hill, 2010), pp. 130–37 (see pp. 135–36), and DeGrazia, *Very Broad Basis*, p. 158.

93. Norcross, "Torturing Puppies," p. 136.

94. Laurie Garrett, "The Path of a Pandemic," *Newsweek*, May 11/May 18, 2009, pp. 22–28 (see p. 26).

95. Garrett, "Path of a Pandemic," p. 26.

96. UN News Centre, "UN Task Forces Battle Misconceptions of Avian Flu, Mount Indonesian Campaign," October 24, 2005, www.un.org/apps/news/story.asp?NewsID=16342&Cr=bird&Cr1=flu.

97. The Centers for Disease Control and Prevention, http://www.cdc.gov/flu/avian/gen-info/qa.htm (accessed July 1, 2010).

98. Garrett, "Path of a Pandemic," p. 26.

99. Garrett, "Path of a Pandemic," p. 22.

100. Michael Greger, M.D., "CDC Confirms Ties to Virus First Discovered in U.S. Pig Factories," August 26, 2009, The Humane Society of the United States website, http://www.hsus.org/farm/news/ournews/swine_flu_virus_origin_1998_042909.html.

101. Garrett, "Path of a Pandemic," p. 26.

102. Donald G. McNeil Jr., "Swine Flu Death Toll at 10,000 Since April," *New York Times*, December 10, 2009.

103. Laurie Garrett, "Surviving Swine Flu," *Newsweek*, September 28, 2009.

104. American Public Health Association, "Precautionary Moratorium on New Concentrated Animal Feed Operations," Policy Number 20037, November 18, 2003, http://www.apha.org/advocacy/policy/policysearch/default.htm?id=1243.

105. UN News Centre, "UN Task Forces Battle Misconceptions."

106. This is according to the director of the National Center for Immunization and Respiratory Disease. See "H1N1 death toll estimated at 3,900 in U.S.," CNN.com, November 12, 2009.

107. Jonathan Safran Foer, "Eating Animals is Making Us Sick," October 28, 2009, in the Opinion section of http://www.cnn.com.

108. Foer, "Eating Animals is Making Us Sick."

109. Though perhaps MRSA is even more worrying. See p. 21 of the Pew Commission report.

110. Moss, "E. Coli Path."

111. P. S. Mead, L. Slutsker, V. Dietz, L. F. McCaig, J. S. Bresee, C. Shapiro, P. M. Griffi, and R. V. Tauxe, "Food-Related Illness and Death in the United States," *Emerging Infectious Diseases* 5 (1999): 607–25.

112. William Neuman, "After Delays, Vaccine is Tested in Battle against Tainted Beef," *New York Times*, December 4, 2009.

113. The information in this paragraph comes from Moss, "E. Coli Path." An animal vaccine for *E. coli* is now being tested. Neuman, "After Delays."

114. Pollan, *Omnivore's*, pp. 67–68.

115. Pollan, *Omnivore's*, p. 130.

116. Shell, *Cheap*, p. 176.

117. Shell, *Cheap*, p. 176.

118. Pew Commission report, p. 3, based on figures from the USDA and the EPA.

119. Singer and Mason, p. 43.

120. Shell, *Cheap*, p. 179.

121. Singer and Mason, p. 43; Pew Commission report, p. 12.

122. Pew Commission report, p. 12.

123. Shell, *Cheap*, p. 179.

124. *Oceanview Farms v. United States*, 213 F. 3d 632 (2000).

125. The Nebraska Department of Environmental Quality, Water Quality Division, 2002 Nebraska Water Quality Report, Lincoln, 2002.

126. Elizabeth Kolbert, "Flesh of Your Flesh: Should You Eat Meat?" *The New Yorker*, November 9, 2009, pp. 74–78 (see p. 76).

127. B. T. Nolan and K. J. Hitt, "Vulnerability of Shallow Groundwater and Drinking-Water Wells to Nitrate in the United States," *Environmental Science & Technology* 40 (2006): 7834–40, cited on p. 29 of the Pew Commission report.

128. "Livestock's Long Shadow," p. xxi (manuscript; for a published version, see Henning Steinfeld, Pierre Gerber, T. D. Wassenaar, Vincent Castel, Mauricio Rosales, and Cees de Haan, "Livestock's Long Shadow" [Food and Agriculture Organization of the United Nations, 2006], available at http://books.google.com/books).

129. Singer and Mason, *What We Eat*, p. 144.

130. "It's Better to Green Your Diet than Your Car," *New Scientist* 17, December 17, 2005, p. 19, www.newscientist.com/channel/earth/mg18825304.800.

131. Pollan, *Omnivore's*, p. 7; Singer and Mason, *What We Eat*, p. 63.

132. Singer and Mason, *What We Eat*, p. 63.

133. This assumes that the cow eats twenty-five pounds of corn per day and weighs 1,200 pounds at slaughter (Pollan, *Omnivore's*, pp. 83–84).

134. Pew Commission report, p. 29.

135. Pollan, *Omnivore's*, pp. 83 and 183.

136. Pollan, *Omnivore's*, p. 118. See Singer and Mason, p. 232, for more information.

137. Mark Bittman, "Loving Fish, This Time with the Fish in Mind," *New York Times*, June 10, 2009.

138. Singer and Mason, *What We Eat*, p. 112.

139. Singer and Mason, *What We Eat*, p. 276.

140. Bittman, "Loving Fish."

141. Singer and Mason, *What We Eat*, p. 234.

142. A. K. Chapagain and A. Y. Hoekstra, *Water Footprints of Nations: Volume 1: Main Report*, Unesco-IHE Institute of Water Education, Delft, November 2004, Table 4.1, p. 41.

143. Chapagain, *Water Footprints of Nations*, Table 4.2, p. 42. See Singer and Mason, p. 236.

144. Singer and Mason, *What We Eat*, p. 233.

145. Singer and Mason, *What We Eat*, p. 233.

146. Shell, *Cheap*, p. 176, citing "In the Front Line: Shoreline Protection and Other Ecosystem Services from Mangroves and Coral Reefs," *United Nations Environment Programme World Conservation Monitoring Centre*, Cambridge, England, 2006.

147. Pollan, *Omnivore's*, pp. 65–66.

148. Singer and Mason, *What We Eat*, p. 44.

149. Singer and Mason, *What We Eat*, p. 30.

150. Pew Commission report, p. 17. Also see Shell, *Cheap*, p. 179.

151. Pew Commission report, p. 49.

152. Pew Commission report, pp. 43–45.

153. Singer and Mason, *What We Eat*, p. viii.

154. Singer and Mason, *What We Eat*, p. 29.

155. Pew Commission report, p. 16.

156. For more on the mistreatment of workers, see Eric Schlosser, *Fast Food Nation: The Dark Side of the All-American Meal* (Boston: Houghton Mifflin, 2001), chaps. 7 and especially 8.

157. *Food, Inc.*, 51–55 minutes into the movie.

158. Shell, *Cheap*, p. 177.

159. Pollan, *Omnivore's*, p. 53.

160. American agricultural subsidies benefit big farms, not small farms: Pew Commission report, p. 47.

161. Singer and Mason, *What We Eat*, p. 33.

162. Pew Commission report, p. 5.

163. Singer and Mason, *What We Eat*, p. 33. Also see *Food, Inc.*, around the 16-minute mark, and p. 49 of the Pew Commission report.

164. Shell, *Cheap*, p. 171.

165. Shell, *Cheap*, pp. 166–67.

166. *Food, Inc.*, 51–55 minutes into the movie.

167. Singer and Mason, *What We Eat*, p. 245.

168. T. J. Key, P. N. Appleby, E. A. Spencer, R. C. Travis, N. E. Allen, M. Thorogood and J. I. Mann, "Cancer Incidence in British Vegetarians," *British Journal of Cancer* 101 (2009): 192–97.

169. The study was directed by Rashmi Sinha and reported in the March 23, 2009 issue of *The Archives of Internal Medicine*. My information comes from Jane E. Brody, "Paying a Price for Loving Red Meat," *New York Times*, April 28, 2009.

170. This is from the abstract of "Position of the American Dietetic Association: Vegetarian Diets," *Journal of the American Dietetic Association* 109 (2009): 1266–82, http://www.eatright.org/cps/rde/xchg/ada/hs.xsl/advocacy_933_ENU_HTML.htm.

171. Jonathan Safran Foer, "Food Industry Dictates Nutrition Policy," October 30, 2009, in the Opinion section of http://www.cnn.com, pp. 1–2. On related matters, see Michael Pollan, *In Defense of Food* (New York: Penguin Books, 2008).

172. This was a full-page ad in *The Reader*, a weekly newspaper serving Omaha, Lincoln, and Council Bluffs, October 21–27, 1999.

173. *Evangelium Vitae*, encyclical letter of John Paul II, March 25, 1995, reprinted in *Biomedical Ethics*, 6th ed., ed. David DeGrazia and Thomas A. Mappes (New York: McGraw-Hill, 2006) as "The Unspeakable Crime of Abortion," pp. 457–59 (see p. 459).

174. Pollan, *Omnivore's*, p. 48.

175. Pew Commission report, p. 3.

176. Pollan, *Omnivore's*, p. 54.

177. Pollan, *Omnivore's*, p. 65. Our food system is ecologically precarious because it relies on just a handful of crops. Pollan, *Omnivore's*, p. 47 and elsewhere.

178. Pollan, *Omnivore's*, p. 39.

179. Pollan tells of how George McGovern lost his senate seat in a tussle with the food industry. McGovern's story provided a cautionary tale to other politicians: don't mess with agribusiness. *In Defense of Food*, pp. 22–25.

180. Peter Singer, "Animal Liberation," *New York Review of Books*, April 5, 1973.

181. Dale Carnegie, *How to Win Friends and Influence People* (New York: Simon & Schuster, 1981 [1936]), p. 5.

182. I learned about this study from Bonnie Tsui, "Greening with Envy: How Knowing Your Neighbor's Electric Bill Can Help You to Cut Yours," *The Atlantic*, July/August 2009, pp. 24, 26.

183. Rachels, "Basic Argument," p. 7.

184. Singer and Mason, *What We Eat*, p. 130.

185. See the conclusion of Timothy Snyder, *Bloodlands: Europe between Hitler and Stalin* (New York: Basic, 2010).

186. Singer and Mason, *What We Eat*, p. v.

187. Pollan, *Omnivore's*, p. 316.

188. I thank Michael Huemer for helping me with this paragraph.

SUGGESTED READING

Articles and Books

ENGEL JR., MYLAN. "The Immorality of Eating Animals." In *The Moral Life: An Introductory Reader in Ethics and Literature*, edited by Louis Pojman, pp. 856–90. New York: Oxford University Press, 2000.

FOER, JONATHAN SAFRAN. *Eating Animals*. New York: Little, Brown, 2009.

MASSON, JEFFREY MOUSSAIEFF. *The Face on Your Plate: The Truth about Food*. New York: W. W. Norton, 2009.

NORCROSS, ALASTAIR. "Torturing Puppies and Eating Meat: It's All in Good Taste." *Southwest Philosophy Review* 20 (2004): 117–23, reprinted in *The Right Thing to Do*, 5th ed., edited by James Rachels and Stuart Rachels, pp. 130–37. New York: McGraw-Hill, 2010.

Pew Commission. "Putting Meat on the Table: Industrial Farm Animal Production in America." A Report of the Pew Commission on Industrial Farm Animal Production, 2008.

POLLAN, MICHAEL. *In Defense of Food*. New York: Penguin Books, 2008.

———. *The Omnivore's Dilemma: A Natural History of Four Meals*. New York: Penguin Books, 2006.

RACHELS, JAMES. *Created from Animals: The Moral Implications of Darwinism*. Oxford: Oxford University Press, 1990.

SCHLOSSER, ERIC. *Fast Food Nation: The Dark Side of the All-American Meal*. Boston: Houghton Mifflin, 2001.

SHELL, ELLEN RUPPEL. *Cheap: The High Cost of Discount Culture*. New York: Penguin Press, 2009.

SINGER, PETER. *Animal Liberation*. New York: HarperCollins, 2002; first published, 1975.

———, and JIM MASON. *The Ethics of What We Eat: Why Our Food Choices Matter*. Emmaus, Pa.: Rodale, 2006.

United Nations Food and Agriculture Organization. "Livestock's Long Shadow: Environmental Issues and Options" (2006).

Websites and Documentaries

Death on a Factory Farm (2009). An HBO documentary directed by Tom Simon and Sarah Teale.

Food, Inc. (2008). A documentary directed by Robert Kenner.

http://www.hfa.org/about/index.html. The website of the Humane Farming Association.

http://www.humanesociety.org/. The website of the Humane Society of the United States.

http://www.peta.org/. The website of People for the Ethical Treatment of Animals.

http://www.ucsusa.org/. The website of the Union of Concerned Scientists.

CHAPTER 33

THE USE OF ANIMALS IN TOXICOLOGICAL RESEARCH

ANDREW N. ROWAN

VARIOUS approaches to the determination and understanding of the moral status of animals have been discussed in previous chapters by philosophers. These analyses include various virtue-based, consequentialist, and deontological approaches, as well as discussions about the importance of particular characteristics such as animal cognition, personhood, and sentience when addressing the continuing challenge of how humans should guide their treatment of animals. I will here focus on how we might apply such moral theorizing in real-life situations where humans use animals for societal benefit. Specifically, this chapter will address our use of animals to assess the toxicological hazards and risks posed by various chemicals, such as those for drugs, cleaning agents, pesticides, cosmetics, and the like, that we manufacture and use in our modern world.

A BRIEF HISTORY OF THE USE OF ANIMALS IN TOXICITY TESTING

It should be noted that the systematic and widespread use of animals for toxicity testing and risk assessment is a very recent phenomenon. The first systematic toxicity testing conducted on behalf of public authorities in the United States used twelve human volunteers in the first decade of the twentieth century. Dr. Harvey Wiley's famed Poison Squad consisted of twelve young males who were the subjects of feeding experiments, conducted between 1902 and 1904, using preservatives, including benzoate, borax, and formaldehyde, that were then found in the American food supply.[1]

In 1955, Arnold J. Lehman and Geoffrey Woodard quoted A. L. Tatum on the necessity of animal studies: "People are rather unpredictable and don't always die when they are supposed to and don't always recover when they should. All in all, we must depend heavily on laboratory experimentation for sound and controllable basic principles."[2] The use of humans then gave way to an increasingly heavy use of laboratory animals.

In 1927, J. W. Trevan devised the LD50 test as a way of standardizing biological therapeutics. In this test, fifty to a hundred animals are dosed with the test substance so that approximately half die within two weeks.[3] From this, a median lethal dose with statistical confidence limits can then be calculated. The LD50 was pressed into service as a basic toxicity measure for all chemicals. At one point in the 1960s, a Canadian toxicologist became so caught up by the measure, and perhaps its false promise of accuracy, that he ran feeding experiments to determine the rat LD50s of egg whites and distilled water.[4]

The testing approaches used for biological substances and vaccines were later applied to other chemicals as various poisoning scandals led to a vast expansion of animal use in toxicity testing. In the 1930s, the Lash Lure eyebrow and eyelash dye was so toxic that a number of users were blinded or disfigured, and the mixing of an antibacterial solution with the wrong solvent led to the deaths of more than one hundred people. Both of these events contributed to the passage of the Food, Drug, and Cosmetic Act of 1938, which was intended to make sure that drugs and other chemicals were safe for human use.[5] In 1962, following the thalidomide tragedy, in which many infants were born with severe deformities, Congress tightened standards again with the Kefauver Amendment, which required that drugs should not only be tested for safety, but also that the companies should prove that drugs were effective before they would be allowed onto the market.

By the end of the 1980s, between 10% and 20% (my informed estimate of the proportion of animals used in toxicity testing) of all laboratory animals were being used in a variety of tests for a wide range of agents and products including drugs, vaccines, cosmetics, household cleaners, pesticides, manufacturing chemicals, foodstuffs, packing materials, and so on.[6] These were mostly rats, but also included rabbits, guinea pigs, and hamsters, as well as some dogs and primates. In 2005, laboratory animal use worldwide was estimated to be over 115 million.[7] The most thorough testing was reserved for products that were to be used in or on foodstuffs and for drugs that would be taken for long periods of time, such as cholesterol-lowering drugs. For these agents, the tests that are performed are: acute, or lasting less than a month; sub-acute, or lasting a month to three months; and chronic, or lasting more than three months. These tests determine, among other endpoints: general toxicity; eye and skin irritancy; the agent's potential to cause mutations, reproductive problems, and fetal malformations; and the agent's carcinogenicity. Today, the cost of a full-scale battery of tests runs into multiple millions of dollars and can take four or more years to complete. Other agents, such as cosmetics, are not subjected to the same intensity of testing, but would still require information on, for example, general oral toxicity, eye and skin irritancy, phototoxicity, and, perhaps, mutagenicity. In the past twenty years, there has been a dramatic decline in the number of animals used in laboratories, particularly for drug discovery and toxicity testing. The decline is, in part, attributable to an increasing use of alternative methods such as a reduction in the number of animals used to produce lethal-dose estimates for chemicals, a reduction in the animal testing of chemicals used in household products and cosmetics, and the use of various in vitro methods that do not require animals.

Political pressure by animal protection organizations in the European Union led to the passage of the Seventh Amendment to the Cosmetics Directive that eventually will prohibit, by March 11, 2013, any animal testing of cosmetic ingredients or finished cosmetic products that are marketed within EU nations. Multinational companies such as Procter & Gamble, Unilever, and L'Oreal have been devoting tens of millions of dollars each year toward the development of new approaches for safety testing their cosmetic products, with the goal of satisfying their concerns about human safety without using data from animal tests.

In Great Britain, the Seventh Amendment to the Cosmetics Directive and the increased attention toward alternatives to animal testing have led to a substantial decline in animal safety testing, as demonstrated in figure 33.1. The use of animals in all toxicology studies has declined by an average of 21,000 procedures a year. This is an almost 50% drop, although there are signs that this trend may be reversing. In the past few years, increased numbers of animal procedures have been performed for quality control and for metabolism studies. Toxicity testing on environmental, industrial, and household chemicals not used for therapeutic purposes has declined by an overall average of 11,000 procedures a year, and in 2008 was at approximately 25% of the total in 1992.

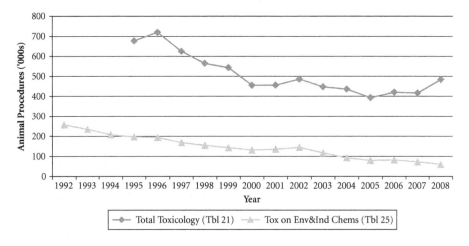

Fig. 33.1

Two different sets of figures on toxicology testing in Great Britain are extracted from the Home Office annual reports on the statistics of research animal use. These are the totals for all toxicity studies (extracted from Table 21); and the totals for the testing of environmental and industrial chemicals (extracted from Table 25).

Since 1992, there have been substantial increases in funding toward alternative testing methods: total research funding on alternatives is now estimated to amount to somewhere between $75–100 million a year. Meanwhile, the past few years have seen a growing dissatisfaction, probably influenced by technical developments in toxicology, with the predictive power of animal tests for human risk assessment. In the past few years, developments in new technology and methods have led to several remarkable initiatives in terms of movement away from a reliance on animal testing.

In the European Union, European Commission Vice President Günter Verheugen and then–Science and Research Commissioner Janez Potočnik, now the European Commissioner for the Environment, launched the European Platform for Alternatives to Animals, or EPAA, in November of 2005 to support and track progress toward the ending of laboratory animal use. In the United States, in 2007, an expert panel of the National Research Council, convened at the request of the U.S. Environmental Protection Agency (EPA) to determine what toxicology would, and perhaps should, look like in the twenty-first century, essentially concluded that the future of toxicology would be non-animal testing.[8] The June 12, 2007, press release announcing the release of the new report, "Toxicity Testing in the 21st Century: A Vision and a Strategy," included the following: "The new approach would generate more-relevant data to evaluate risks people face, expand the number of chemicals that could be scrutinized, and reduce the time, money, and animals involved in testing, said the committee that wrote the report."[9] This report was discussed at the 2007 annual Toxicology Forum meeting in Aspen, Colorado, where most of the participants appeared to regard the vision as a pipe dream. However, almost three years later, that pipe dream is beginning to take hold in both European and American policy circles.

Partly as a response to the National Research Council report, the U.S. EPA, the National Chemical Genomics Center of the National Institutes of Health, and the National Toxicology Program signed a memorandum of understanding to cooperate on a program to establish non-animal high-throughput test systems that would help in setting testing priorities. The article in *Science* describing this collaboration noted that "it is a research program that, if successful, will eventually lead to new approaches for safety assessment and a marked refinement and reduction of animal use in toxicology."[10]

There is also growing criticism of the utility of animal testing in both the toxicological arena as well as in other areas of biomedicine. One of the most cogent critiques was produced by Troy Seidle and published as a free-standing report by People for the Ethical Treatment of Animals–Europe, or PETA-Europe.[11] Seidle reviewed how results from the rodent bioassay for carcinogens had been used by the International Agency for Research on Cancer, or IARC, and by the U.S. EPA (in the IRIS database) to classify the tested chemicals as possible, probable, or known human

Table 1 The Use of Animal Bioassay Data to Classify Chemicals

	IARC	Percentage of Total	U.S. EPA (IRIS)	Percentage of Total
Number Unclassifiable	135	28.4%	14	2.9%
Not Classified	260	54.6%	406	85.3%
Probably Not a Carcinogen	0	0%	2	0.4%
Possible Carcinogen	65	13.7%	19	4.0%
Probable Carcinogen	10	2.1%	32	6.7%
Known Carcinogen	6	1.3%	3	0.6%
Total Percentage Classified	17.1%		11.7%	
Total Number in Database	476		476	

How the IARC and EPA have used animal bioassay data to classify tested chemicals as possible, probable, or known carcinogens. (Table compiled from the source provided in endnote 11.)

carcinogens. Of the tested chemicals, only 17.1% for the IARC and 11.7% for the EPA had been classified according to their carcinogenic potential (see table 1). This classification raises serious questions about how useful the data in the remaining bioassays, which comprise over 80% of the total, might be for human risk assessment.

The toxicological literature is full of peer-reviewed papers that raise similar doubts about the usefulness of animal test data for human risk assessment. Until now, the standard response to such critiques has been to ask what might be done instead. As the scientific basis of modern toxicology improves, we see an increasing number of interesting and promising technological opportunities that, as demonstrated by the decline in animal use, are already replacing animal toxicity testing and could, with additional development, be used even more widely.

MORAL DECISION MAKING IN TOXICOLOGICAL TESTING

Academic philosophers debate the relative merits of virtue, consequentialist, or deontological approaches when addressing the moral status of and moral protections for animals. In practice, it would appear that an average institutional review committee concerned with research involving animals uses a mixture of all three approaches when deciding whether to approve a particular animal toxicity study. The committee—usually designated an Institutional Animal Care and Use Committee (an IACUC) in the United States but with different designations in other countries—consists of people who are chosen in part because they are considered to be reasonable representatives of the various parties involved in deciding whether to approve any particular project. The moral virtue of the selected individuals is an important consideration when responsible institutions select candidates for these committees.

In most of the Organization for Economic Co-operation and Development, or OECD, countries, there are specific regulations about what researchers may or may not do to animal subjects, and the system is presented in terms of specified duties that might be called a deontological approach. Finally, IACUCs are committed to a consequentialist approach in weighing the costs and benefits of the proposed studies. No country could be said to have an adequate body of institutional review unless costs and harms to animals were assessed with utmost seriousness.

I will now expand on these embryonic ideas about the use of the categories of virtue, deontological, and consequentialist moral theories. My goal is to show that, while theoretically, such a system could work well, in practice it does not. I will provide some reasons as to how and why it does not.

The Virtue Component: Selection of Committee Members

In the United States, when the IACUC system was enacted under regulations following the passage in 1985 of amendments to the Animal Welfare Act, committees were required to include among their members persons unaffiliated with the institution conducting the research. The unaffiliated committee members could be scientists, but should "provide representation for general community interests in the proper care and treatment of animals."[12] This implies that the pre-IACUC system, in which institutional scientists and animal care staff decided which projects could be pursued, did not sufficiently account for broader community values on the topic. Some nations, such as Sweden, have gone further and explicitly require committees that review animal protocols to include people who represent not just community interests, but also animal welfare interests and the law specifically requires that there be representatives drawn from animal protection organizations.[13]

While there is no evidence that the members of IACUCs lack moral virtue in carrying out their responsibilities, there is empirical data indicating that IACUC decision making varies substantially from one committee to the next and that IACUCs could function far more effectively than they do in addressing animal welfare issues and in evaluating the costs and benefits of animal research.[14] It appears that the public senses this lack of consistency. The 1985 Animal Welfare Act amendments were passed in part to reassure the public that oversight of animal research would be enhanced and improved, but public support for animal research fell from 63% to 44% from 1985 to 2001 (see figure 33.2).[15]

There is no empirical data indicating how well IACUCs function in terms of raising appropriate moral issues about animal use in toxicology testing, but there are a few anecdotes. Twenty years ago, while serving on a university IACUC, I decided that the next time we were presented with a protocol that included the determination of an LD50 value, I would oppose it. This particular IACUC had reviewed two such protocols the previous year. In due course, a protocol with an LD50 determination came before us and, somewhat apprehensively, I stated that I would vote against approval because of the LD50 component. Somewhat to my surprise and delight, everybody immediately agreed with this position and the protocol was rejected. The

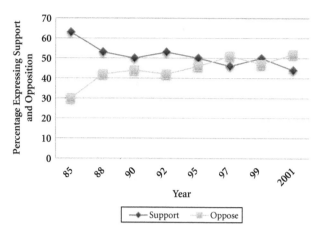

Fig. 33.2 Public support for and opposition to animal research as measured by the National Science Board from 1985 to 2001. Survey participants were asked if they supported or opposed the following statement: "Scientists should be allowed to do research that causes pain and injury to animals like dogs and chimpanzees if it produces new information about human health problems."

scientist protested the decision and came before us arguing that the FDA required such a test but we were able to provide evidence that there was no specific FDA requirement for LD50 testing and the scientist decided he could use another non-lethal endpoint in his research protocol. Thereafter, the IACUC routinely rejected protocols that included an LD50 determination.

This anecdote illustrates that "virtue" in an IACUC may be latent, sometimes appearing only when custom is challenged. I should have acted earlier, but it took me several years to feel sufficiently confident as a member of the IACUC to raise the issue. As Rebecca Dresser[16] reports, it is not a simple or easy matter to put forward proposals that change traditional ways of doing things, and even caring ("virtuous") people may continue to accept the status quo when they should not.

The Deontological Component: The Rules for the Review Committees

A variety of relatively formal rules have become part of IACUC functioning and address certain ethical aspects of animal research. For example, the use of curariform agents to keep animals immobilized is not permitted unless the animal is fully anes-thetized and monitored to make sure that it will not become conscious while still paralyzed. However, the current rules tend to pressure IACUCs toward accepting animal studies that may cause significant animal distress and suffering in toxicity testing, even when IACUC members raise questions about the endpoints. For example, the large majority of animals reported to be subject to Category E studies (where animals experience pain or distress and do not receive pain-relieving drugs) in the United States are used in vaccine safety and potency tests as well as other safety testing. These tests are usually considered to be legally required, either explicitly or implicitly, and hence IACUCs tend not to challenge such projects. They simply note in their annual reports that the study did not permit the use of pain-relieving drugs. This is one of the most common reasons given for why animals experience pain or distress without pain-relieving drugs in the annual reports institutions file with the USDA.

The Consequentialist Component: Weighing Costs and Benefits

Most of the public debate on the ethics of animal use in laboratories tends to focus on consequentialist issues. For example, defenders of the use of laboratory animals consistently point to the human benefits derived from the new data produced by these studies, while also arguing that the animals experience little or no pain and distress in the process. By contrast, critics (e.g., Peter Singer) argue that the benefits of animal research are rather limited or absent altogether,[17] while the animals suffer significantly. In theory, it should be possible to determine which of these two contrasting positions (benefits outweighing suffering versus suffering outweighing benefits) is correct by examining empirical data, but this exercise is more difficult than might be expected and has rarely been attempted. Typically, defenders of animal use point to anecdotal examples that they believe support their case, while critics point to different examples that support theirs. Anecdotal and case-study presentations can be useful, but they are not an adequate substitute for careful moral deliberation about these serious concerns.

The difficulties of a consequentialist moral calculus are not limited to an assessment of relatively vague benefits to be weighed against reasonably concrete potential harms. One also has to balance benefits that might accrue years or even decades into the future against harms that will occur instantly or at least within a reasonably short period of time. In toxicity testing, one has the additional problem of trying to assess benefits that are presented as negative outcomes, in that the benefits are the absence of any real or potential damage to humans or the environment when the agent is used as intended. As should be evident, it is difficult to prove that humans were spared some deleterious outcome because of a toxicity study in animals. Instead, animal testing is justified by calling attention to past harms to humans that have occurred when animal studies were not done. For example, in the Cutter polio disaster, hundreds of children died after receiving doses of the Salk polio vaccine that had not been properly killed. In the thalidomide disaster, animal studies were either not done or were not adequately performed, and the drug, given to pregnant women to relieve morning sickness, caused babies to be born without arms or legs. As a result, proponents of animal testing argue that, because these situations were preventable, we must continue to conduct animal tests in order to avoid such damage in the future.

On the harms side, it is more difficult than it might appear to reliably assess the extent of animal pain, distress, and suffering. Animals clearly experience harm in toxicity studies (they will all die at the end of the study and many experience significant distress), but we have surprisingly little empirical data on both the types of harms experienced and the prevalence of those different types. One of the studies that was conducted to determine if there are any patterns to the harms experienced by animals looked at the adverse outcomes for animals used in acute toxicity studies.[18] The authors investigated in detail the clinical symptoms reported by several different laboratories when conducting acute lethal dose studies (the LD50 test is an acute lethal dose test) on a select set of chemicals. In a perfect world, each laboratory would report the same type of and level of adverse effects for each chemical.

There were five notable findings in the Schlede reports.[19] First, there were large variations in the documentation of clinical signs from one laboratory to the next. For instance, one laboratory reported an incidence of convulsions as 39% and another laboratory as only 9% despite testing the same set of chemicals at the same dose levels. Second, more than six hundred different combinations of clinical signs were reported for the 1128 rats that died during the study. Of these, 470 combinations were reported only once. Third, there seemed to be no direct relationship between the severity of a single clinical sign and the moribund state, death, or survival of an animal. Fourth, moribund animals and animals that died during the study had a higher frequency of all forms of convulsions than surviving animals. In surviving animals, convulsions were almost nonexistent with fewer than 1% of surviving animals experiencing any kind of convulsion, compared to up to 45% of those that did not survive. Finally, lateral recumbency (lying on one side) and tremors were observed in surviving animals at distinctly lower rates (six-fold less) than in animals that died or became moribund.

The data presented in the above studies indicate that toxicity testing causes significant harms to animals and yet there are very few studies of the rates of the harm or attempts to determine if there might be less severe signs that could be used as "refinement" signals that an animal is about to experience harm. If we were truly serious about reducing animal distress in toxicity testing, then surely we would have made far more effort to look for reliable ways to reduce pain and distress caused by agents that are administered at toxic doses.

Little research addresses the humane endpoints in toxicity testing, and within this research, there is relatively little data on the adverse consequences experienced by animals.[20] We know that a certain number of animals develop tumors in carcinogen tests, but carcinogen studies comment neither on the level of suffering experienced by individual animals nor on the number of animals that might be expected to undergo different grades of suffering. In other words, even if an IACUC wished to spend some time reviewing the cost/benefit ratio in an appropriately consequentialist manner, there is little data on the cost experienced by laboratory animals that could be called upon when deciding whether to approve or disapprove any given study.

At face value, it is somewhat surprising that there should be so little data on the adverse outcomes of toxicity studies and on what might be done to alleviate animal pain and/or distress in such studies. However, lack of data on and attention to animal distress is not just a problem in toxicity testing, they are features of the whole enterprise in which animals are used to advance human knowledge. There are indications that animal pain and distress is consistently under-reported by research institutions in the United States[21] and one relatively old ethnography of laboratory research on animals[22] indicates that researchers do not "see" or discount animal pain and distress when it occurs. More recently, ethnographer Arnold Arluke has shown that animal researchers experience explicit or unacknowledged discomfort or guilt when they conduct research that harms animals.[23]

The lack of data on animal pain and distress means that, even if an IACUC decided to make a serious attempt to look at the cost/benefit ratio for a toxicity study, it would have little empirical data to weigh on either side of the equation. For the most part, it is likely that IACUCs that review a significant number of animal toxicity studies will focus on experimental design questions, then approve those studies that are carefully designed and that report they will make a serious attempt to minimize animal distress and suffering.

Conclusion

Toxicity studies on animals comprise perhaps 10–15% of the overall use of animals in biomedical laboratories. Such studies are conducted to protect humans, the environment, and animals from toxic hazards caused by the use of the eighty thousand or so chemicals that are currently in commerce. However, there is little data available to demonstrate how effective such studies have been in the past or might be in the future. We know that toxicity studies in animals lead to harms and suffering for at least a portion of the animals used, but we still do not know much about the extent or prevalence of such harms. Therefore, when national and international legislative and regulatory bodies discuss the need to minimize harms and exhort overseers to ensure that animal studies are conducted according to agreed upon ethical standards, we do not know if current practices are reaching even minimally appropriate standards.

There is a consensus that animal tests are time consuming and not particularly effective in predicting harms to humans and the environment. In the past five years, interest in moving away from animal tests and toward the use of quicker and cheaper alternative test systems has grown rapidly. Increasing numbers of toxicologists and regulatory officials are now looking to new technologies, and the potential to refine and develop those new technologies in significant ways, to predict the hazards and risks to humans and the environment caused by commercial chemicals. Some still struggle to tweak existing animal tests so that they will be ethically improved as well as better predictors of, say, endocrine-disrupting effects, but it is becoming more evident that animal studies are not the best way to assess the toxicity and the risks of the vast number of chemical substances that have already been developed.

Moral concerns will not be the only factor in reducing our reliance on animal test data in the next twenty years. Economic and efficiency issues will be at least as influential in driving us to develop a completely new paradigm for human hazard and risk assessment. I do not think it is hyperbolic to claim that the new paradigm could be as influential for biomedical knowledge and science as the Human Genome Project has been for human research. The prospect of having non-animal methods that would allow us to quickly and cheaply evaluate the biological effects of chemicals, (thereby giving us the ability to use that data to predict risks and potential benefits for humans, animals, and the environment), has enormous implications for biomedical science and human health.

NOTES

1. C. Lewis, "The 'Poison Squad' and the Advent of Food and Drug Regulation," *FDA Consumer* 36 (2002): 12.

2. A. J. Lehman, et al. "Procedures for the Appraisal of the Toxicity of Chemicals in Foods, Drugs and Cosmetics," *Food Drug and Cosmetic Law Journal* 10 (1955): 679.

3. J. W. Trevan, "The Error of Determination of Toxicity," *Proceedings of the Royal Society, London: B. Biological Sciences* 101 (1927): 403.

4. E. M. Boyd and I. L. Godi, "Acute Oral Toxicity of Distilled Water in Albino Rats," *Industrial Medicine and Surgery* 36 (1967): 609–13. Why is it false in its promise? The answer is that LD50 values vary between species by a factor of 1,000 or more and that the experimental conditions used also lead to large variations within species. Therefore, even if you produce a number for the LD50 with a typical Standard Deviation of around 25–50%, the human lethal dose may be (and usually is) considerably different. Measuring an LD50 is somewhat similar to estimating the distance from where you are sitting to the wall by eye and saying that it is 323.45 cm plus/minus 74.53 cm. Your estimate is not made any more accurate by including the extra digits and the SD.

5. A. N. Rowan, *Of Mice, Models and Men: A Critical Evaluation of Animal Research* (Albany: State University of New York Press, 1984).

6. Home Office, *Statistics of Scientific Procedures on Living Animals Great Britain 2008*, HC 800 (London: The Stationery Office, 2009).

7. K. Taylor, N. Gordon, G. Langley, and W. Higgins, "Estimates for Worldwide Laboratory Animal Use in 2005," *ATLA* 36 (2009): 327–42.

8. National Research Council, *Toxicity Testing in the Twenty-first Century: A Vision and a Strategy*, Committee on Toxicity and Assessment of Environmental Agents, National Research Council, 2007, available at http://www.nap.edu/catalog/11970.html.

9. Anon., "Report Calls for New Directions, Innovative Approaches in Testing Chemicals for Toxicity to Humans," *National Academy of Sciences Press Release on NRC Report*, June 12, 2007.

10. F. C. Collins, G. M. Gray, and J. R. Bucher, "Transforming Environmental Health Protection," *Science* 319 (2008): 906–7.

11. T. Seidle, *Chemicals and Cancer: What the Regulations Won't Tell You about Carcinogenicity Testing* (London: PETA, 2006).

12. R. Dresser, "Community Representatives and Nonscientists on the IACUC: What Difference Should It Make?" *ILAR Journal* 40 (1999): 29–33.

13. D. Britt, "Ethics, Ethical Committees and Animal Experimentation," *Nature* 311 (1984): 503–6.

14. S. Plous and H. Herzog, "Reliability of Protocol Reviews for Animal Research," *Science* 293 (2001): 608–9.

15. NSB, *Science and Engineering Indicators Survey–2002* (Arlington, Va.: National Science Foundation (NSB–02–1), 2002).

16. Dresser, "Community Representatives and Nonscientists on the IACUC," 29–33.

17. P. Singer, *Animal Liberation*, 2nd ed. (New York: Random House, 1990).

18. R. C. Greek and J. S. Greek, *Specious Science: How Genetics and Evolution Reveal Why Medical Research on Animals Harms Humans* (New York: Continuum, 2003).

19. E. Schlede, U. Mischke, R. Roll, and D. Kayser, "A National Validation Study of the Acute-Toxic-Class Method: An Alternative to the LD50 Test," *Archives of Toxicology* 66 (1992): 455–57. E. Schlede, U. Mischke, W. Diener, and D. Kayser, "The International Validation Study of the Acute-Toxic-Class Method (Oral)," *Archives of Toxicology* 69 (1995): 659–70.

20. W. Stokes, "Humane Endpoints for Laboratory Animals Used in Regulatory Testing," *ILAR Journal* 43, Supplement (2002): S31–S38.

21. Humane Society of the United States, *Taking Animal Welfare Seriously: Minimizing Pain and Distress in Research Animals*: A Special Report. Undated but originally produced in 2000. Available at http://www.humanesociety.org/assets/pdfs/animals_laboratories/pain_distress/taking_animal_welfare_seriously.pdf (accessed on June 7, 2010).

22. M. T. Phillips, "Savages, Drunks and Lab Animals: The Researcher's Perception of Pain," *Society and Animals* 1 (1993): 61–81.

23. A. Arluke, "Trapped in a Guilt Cage," *New Scientist* 134 (1992): 33–35; "Uneasiness among Laboratory Technicians," *Occupational Medicine* 14 (1999): 305–16.

SUGGESTED READING

BATESON, P. "Ethics and Behavioural Biology." *Advances in the Study of Behavior* 35 (2005): 211–33.

BEAUCHAMP, T. L., F. B. ORLANS, R. D. DRESSER, D. MORTON, and J. P. GLUCK. *The Human Use of Animals: Case Studies in Ethical Choice*. 2nd ed. New York: Oxford University Press, 2008.

CARBONE, L. *What Animals Want: Expertise and Advocacy in Laboratory Animal Welfare Policy*. Oxford: Oxford University Press, 2004.

FRASER, D. *Understanding Animal Welfare: The Science in its Cultural Context*. London: Wiley-Blackwell, 2008.

National Research Council. *Toxicity Testing in the Twenty-first Century: A Vision and a Strategy*. Committee on Toxicity and Assessment of Environmental Agents, National Research Council, 2007, available at http://www.nap.edu/catalog/11970.html.

PURCHASE, I. F. H. "Ethical Issues for Bioscientists in the New Millennium." *Toxicology Letters* 127 (2002): 307–13.

ROWAN, A. N. "Formulation of Ethical Standards for Use of Animals in Medical Research." *Toxicology Letters* 68 (1993): 63–71.

————, and F. M. Loew. "Animal Research: A Review of Developments, 1950–2000." *State of the Animals* 1 (2001): 111–20, available at http://www.hsus.org/web-files/PDF/hsp/ MARK_State_of_Animals_Ch_07.pdf.

Smith, J. A., and K. A. Boyd. *Lives in the Balance: The Ethics of Using Animals in Biomedical Research: The Report of a Working Party of the Institute of Medical Ethics.* Oxford: Oxford University Press, 1991.

Stephens, M. L., A. M. Goldberg, and A. N. Rowan. "The First Forty Years of the Alternatives Approach: Refining, Reducing, and Replacing the Use of Laboratory Animals." *State of the Animals* 1 (2001): 121–35, available at http://www.hsus.org/ web-files/PDF/hsp/MARK_State_of_Animals_Ch_08.pdf.

CHAPTER 34

WHAT'S ETHICS GOT TO DO WITH IT? THE ROLES OF GOVERNMENT REGULATION IN RESEARCH-ANIMAL PROTECTION

JEFFREY KAHN

WITH apologies to Tina Turner, what's ethics got to do with federal policy for the oversight of animal research? Hopefully everything. Over the course of the histories of research involving both animal and human subjects, numerous parallels have evolved with respect to government-mandated oversight regimes in each. For research on human subjects, there is a long history of discussion of the ethics of all aspects of the enterprise, and a notable proportion (maybe even a disproportionate amount) of bioethics scholarship is devoted to the topic. For both human and animal research, there are sometimes detailed rules regarding acceptable risk, treatment of research subjects, prospective review and approval of proposed research, and institutional oversight committees. In addition, the history of research ethics shows that moral argument often goes well beyond standing law and policy in establishing what may and may not be done in both human and animal research. So ethics is indeed at the core of human and animal research oversight, from providing the principles at its foundation to acting to inform analysis of issues as they arise and evolve.

However, for all the parallels in the histories of research policies and oversight, the rules for research on animals depart from those for human research in fundamental ways that raise moral concerns. Policies for human research protections follow and are based on articulated principles about the ethics of research. From these principles came policies. Even as they continue to evolve in their interpretation and application, these policies can still be traced back to the touchstone principles on which they are based. For animal research oversight policies, the connection between policies and principles is far less clear. For example, assuming that animals do not have the capacity to consent to participate in research, what would be a reasonable substitute for consent in biomedical research that uses them as subjects? Policies often express requirements for the minimization of pain and suffering, but such rules and restrictions are about maximizing human benefit at the cost of minimal harm to animals rather than about protecting the rights and interests of animal subjects.

The lack of articulated principles from which policy could be derived, or at least on which it would be based, is indicative of moral inconsistencies at the core of animal research policies. If ethics is important for how we think about the acceptability and proper limits of animal research—and it is—then the connection between ethical principles and policy should be made far more explicit. If there are real moral commitments, then policies must closely follow them and make their way fully into application and practice.

Animal research oversight in the United States is in some respects relatively mature, as reflected by a collection of federal-level policies that govern how animals must be treated in biomedical research as well as the conditions in which they are housed, fed, transported, and cared for during their time as research subjects. Policies include the means for avoiding or alleviating pain and suffering and for euthanizing research animals in non-survival research. While the policy environment is somewhat well developed, the moral rationales on which this collection of policies are based have received less attention and analysis and cannot be said to be as well developed.

The tenuous connection between research animal protection policies and the ethical principles underlying them prompts a series of questions that will structure the remainder of this chapter: (1) What are the moral underpinnings for these policies?; (2) which principles would be adequately reflective of the ethics of biomedical research on animals?; and (3) how might policies be changed to align with a more reflective set of foundational principles? The moral standing of animals as research subjects is at the core of attempts to answer these questions, though not all of the weight of an adequate response should be placed on the problem of moral standing. Many problems about the protection of animals can be raised even in the absence of a well-developed account of moral standing. For example, "What level of pain is maximal in research" can and should be addressed without having available a polished account of moral standing.

THE CENTRAL ROLE OF ETHICS IN GOVERNMENT OVERSIGHT POLICIES

What are major societal motivations for the use of animals as research subjects? The answer(s) have some ethical bases, and highlighting them will go some way toward creating a common basis for debate. It is difficult to assess how well policies and processes are performing their functions without clarity about their goals. We need to understand why and with which goals we are crafting and promulgating policy so that we have some measures of success. For example, a protection regime that is based on attention to minimizing the number of animals used in research would be perceived and would function very differently than one whose main focus is to protect researchers' freedoms or to introduce efficiency in the institutions in which research is performed. Bluntly put, the objective of protecting freedom of inquiry and efficiency in research is fundamentally different from the objective of protecting animals from misuse and abuse. They are all noble goals, but the present point is that we are often unclear precisely which of these goals our policies are intended to protect.

Institutional committees charged with the review and oversight of animal research (Institutional Animal Care and Use Committees, or IACUCs) are sometimes accused of protecting the interests of researchers and institutions at the expense of animal subjects. Critics point to an absence of impartiality because membership on IACUCs is largely drawn from among those who engage in animal research and are faculty or staff at the institution appointing the committee.[1] However comprised, there is motivation for IACUCs to make sure both that rules are followed and that research moves forward, and sometimes these motivations conflict.

The motivations and goals in creating rules for the control and oversight of research have traditionally varied and have not been brought into the open. The most virtuous motivation in the case of animal oversight is born out of a commitment to a moral principle of animal welfare and reflects this commitment in policy, but rules also often exist for less high-minded reasons, such as to promote investment, to protect those engaged in carrying out research, to serve as a reaction or response to misbehavior or scandal, or to promote and protect public trust. All are in fact important aspects of policymaking. Why we have rules matters, and the explanation helps us assess whether rules achieve their stated goals, whether it makes sense for these rules to carry the imprimatur of ethical principles and whether there are other motivating commitments or foundations supporting them. Policies that reflect clear and principled commitments are more easily justified, their goals more evident, and success or failure easier to measure.

Consider the analogy of a society's policy commitment to universal health care for its citizens. Such a policy is arguably supported by appeal to some version of the principle of a right to health care, and its success is measured by concrete metrics such as the proportion of the population within the resulting medical system, the numbers of patients receiving care, hospital admissions, procedures performed, and so on. An example of a policy with less obvious connection to principle is the history of policies that have emerged from the debate in the United States over federal funding of human embryonic stem cell (hESC) research. The administration of President George W. Bush approved the use of federal funding for hESC research on cell lines that had already been created from so-called "excess" *in vitro* human embryos. The Bush policy allowed such research only if the cells had been extracted (and the embryos from which they were extracted destroyed in the process) prior to August 9, 2001.

The puzzle was how to articulate and defend the principle(s) underlying the policy decision. The answer seemed to be something like a commitment to prevent additional embryos from being harmed in publicly funded hESC research along with a desire to support a promising new area of research. Neither was articulated to the public (or in private, so far as we know), which left those interested in the issue to intuit the underlying ethical principles for themselves. The resulting policy left very few people happy among those invested in the debate over this research, and the events did not serve either the interests of researchers hoping for a minimal body of restrictions on the science or those opposed to hESC research on moral grounds.[2] Without clearly articulated principles on which the promulgated policy was based, it is difficult to know whether the Bush administration considered it successful, and if so, how success would be assessed.

Policies need principles underpinning them, and policies and practice suffer without them, as I have indicated by using the example of hESC research. This shortcoming is not unique to stem cell research and is a serious problem in the world of animal research.

THE MORAL UNDERPINNINGS OF CURRENT POLICY

Although there is little controversy over the claims that *humans* have moral standing, that research on humans deserves principled and well-developed public policies, and that all human beings have equal moral value, such claims have rarely been made in the United States when animals are research subjects. In fact, very few countries have carefully articulated and defended policies for animal research. This is in large part why it is so difficult to reach clear moral principles on which animal protection policies can be based. If animals are not accorded a significant moral standing beyond the premise that they should have clean cages (a norm that may have been developed as much in the interests of scientists as for the interests of animals), then policies regarding their use in biomedical research are destined to have little to do with respect for the animals who are subjects. Instead, policies would more appropriately focus on issues related to bench science, such as responsible data management, authorship, and the sharing of laboratory reagents.

This point was made in discussions of the ethics of animal research fully a decade before the first federal policies related to research-animal protections were promulgated. As part of a *Nova* episode in 1975 that included the airing of Fred Wiseman's film *Primate*, a discussion followed the film with a panel of prominent scientists, the filmmaker, and a well-known philosopher (Robert Nozick). The film was a kind of exposé documenting research being performed at the Yerkes Regional Primate Center in Atlanta. It became a cause célèbre for improved treatment of research animals.

During the course of the televised discussion (later widely disseminated through *Nova* tapes) there was much discussion about how to take account of the "moral costs" of performing research on animals to serve human interests. The prominent scientist and Nobel Laureate David Baltimore repeatedly stated his view that research on animals is not a moral question at all, but one of line-drawing between animals and humans: "I don't think it really is a moral question. It's a question of, here we all sit, we want to survive;...and we therefore have drawn some sort of line about human experimentation,...and if we lose that [line between animal and human experimentation] then I think you will take apart the research establishment."[3]

If this were the prevailing view as policy is developed, then perhaps the principle is "biomedical research on animals is ethically acceptable so long as it serves the interests of humans." This might be termed a principle of efficient animal use for human benefit. Such a principle could be invoked to prohibit redundant, gratuitous, or ineffective procedures, but any sort of research with a justifiable scientific rationale would be acceptable.

The federal policies in the United States that were promulgated in the 1980s reflect an approach not entirely at odds with the view espoused by Baltimore and those who share his views—an instrumental view of the use of animals as research subjects.[4] The ethics underlying these policies have never been well articulated. Especially lacking is a precise justification for the way in which animals are often used and why it is allowed, and even promoted, by regulatory policies. This critical point is obvious in the narrative "justificatory" language of U.S. federal regulations:

> The development of knowledge necessary for the improvement of the health and well-being of humans as well as other animals requires in vivo experimentation with a wide variety of animal species....Procedures involving animals should be designed and performed with due consideration of their relevance to human or animal health, the advancement of knowledge, or the good of society.[5]

Anyone who takes this justification of animal research seriously needs to acknowledge and stand ready to defend the fact that the use of animals is based almost entirely on human needs and has little to do with animal welfare. The policies go on further to reflect concern with use of appropriate types and numbers of animals as well as minimizing the pain and suffering experienced by animal subjects, reflecting a commitment to the humane treatment of animal subjects. With clear statements of how we are to perform animal research, together with what is not allowed, comes the need to articulate the basis for all such prescriptions and proscriptions—not just how, but why. This requires a research ethic for animals that reflects the sort of principled commitment we have already made in research on humans. It will require a far clearer connection between the "how" and the "why" than we have seen in the past.

This assessment does not mean that we must reconsider the entire basis on which animal research is justified; it means that we must acknowledge it more squarely in order to inform animal research protection policies. This step will be the initial one in helping to create a system that goes further toward respecting and protecting the interests of animal subjects rather than merely referencing human benefits. Currently, our policies seem to reflect a moral ambivalence about animal use in research. There is a societal desire for the human benefits of animal research, but our policies are poorly calculated to recognize the moral cost of achieving them. Explicit recognition that there is real moral cost in using animals in research would help create a policy climate in which the connection between principle and policy is not only clearer but more easily assessed, and therefore more readily improved. One way to move closer to this goal is through the careful development of ethical principles and the specification of those principles as they might be used in support of the policies and practices of biomedical research involving animals.

ETHICAL PRINCIPLES FOR BIOMEDICAL RESEARCH ON ANIMALS

The literature on the moral status of animals continues to grow, and there is increasingly compelling evidence and argument that many animals have the properties that qualify them for a significant level of moral status. This does not imply equivalent moral status among all animals, of course, or moral status equivalent to that of humans. However, moral status is critical in the discussion over what sorts of considerations are due animals and what ethical principles ought to underpin biomedical research involving animals. It is relatively uncontroversial to claim that sentient creatures have interests, and that those interests warrant at least some consideration. Even those who do not support an animal-rights view or a strong animal-welfare view acknowledge this perspective. As Kenan Malik observes, "I would not kick a dog in the same way as I would kick a ball."[6] But while sentient creatures may have interests—a dog certainly has an interest in not being kicked—it does not follow that all beings with interests also have rights by virtue of the interests (though it may well be the case that dogs have rights that derive from some of their interests).

Even when arguments are mounted in favor of (some) animals having rights, it is not always obvious what is meant, and therefore it is not always clear what is at stake. David DeGrazia offers a helpful distinction of three ways in which the term "animal rights" can be understood.[7] The first is an understanding that animals have rights if they have moral status, and with such status comes recognition that the interests of animals have some moral importance separate from how they might promote human interests. Accepting such an understanding would justify (or require) protection policies that reflect respect for animal interests that go beyond mere humane treatment. A second understanding of animal rights views animals as deserving equal consideration, in the sense that their interests are no less important than the interests of humans. This understanding draws support from some utilitarian approaches to ethics and some sentience-based theories that endorse equal treatment of comparable interests of every individual. Accepting this understanding would require a consistency in approaches to protection policies across both human and animal subjects (more on how this might look below). DeGrazia also offers a third, even stronger, understanding of animal rights, in which even the most basic of animal interests—to be free from pain and suffering, and to continue living— "deserve moral protections that generally override appeals to the general welfare or social utility."[8] Such an understanding would allow only biomedical research that was in the interests of individual animal subjects on which it was conducted.

The upshot of these ways of understanding animal rights is that with even a limited moral status come some valid claims that recognize interests. Claims about the treatment of animal subjects may not require that all of the principles that are core to the ethics of research on human subjects must be extended to animals, but it does argue for more thoroughgoing appreciation for how we understand animal welfare in the animal research context. Research on human subjects is driven by a commitment to better understand and treat human health and disease, but with that commitment comes ethical concerns inherent in an enterprise that relies on exposing individuals to risk for benefits that will mostly be for persons other than the research subjects. Likewise for biomedical research on animals, which shares as its primary (if not exclusive) motivation the goals of improved knowledge on human health and advancing the treatment of human disease. The difference is that whereas other humans (and sometimes the human subjects themselves) may realize benefits either directly or indirectly as a function of their participation, animal research is overwhelmingly performed in support of human rather than animal health, and it is almost never the case that research involving animals can be described as therapeutic research for the animals involved.

For research on both humans and animals, policies and the rules they articulate are attempts to codify how research must be performed as well as what is not allowed. Federal regulations make clear that research on human or animal subjects, when carried out in settings to which the rules apply, may be conducted only if there are guarantees of adequate protections aimed at protecting the rights and interests of the individual subjects who are participating. In animal research, adequate protections must be in place to assure the continued well-being of animal subjects. For human subjects, the requirements set out to assure these protections include appropriate risk-benefit balancing, fair distribution of the risks and benefits, and perhaps most centrally, adequately informed subjects who voluntarily agree to participate in the research. That is, informed consent is a necessary condition of justified research. There are no parallel requirements for the consent of animals, of course, and the policies in place are motivated more by a goal of reducing if not minimizing harm to animal subjects than in promoting their welfare. The protections in place for animal subjects are promoted by policy recommendations regarding housing of animals, restrictions on major surgeries, and requirements for the use of anesthesia and analgesia.

The cornerstone of the ethics of research on human subjects is informed consent, based on the moral principle of respect for persons.[9] The rules for human subjects include additional protections for identified subpopulations that were historically deemed "vulnerable": prisoners, pregnant women (to protect unborn fetuses), and children.[10] These protections were based on concerns about the lack of capacity to consent on the part of particular groups, animating the need for additional protections. The fundamental policy commitment is that less capacity to consent confers additional protections, including limitations on acceptable risk, and requirements for the potential for offsetting direct medical benefit to subjects when research carries significant risk.

There is no reason why animal subjects do not qualify for the same sorts of additional protections on precisely the same grounds as humans. However, as Rebecca Walker points out, there is an unjustified double standard—"for humans lesser capacities correspond to greater protections but for animals the opposite is true."[11] While the double standard is unlikely to disappear, examining its bases will be helpful for identifying principles at stake in the ethics of human and animal research. As Walker has pointed out, deontological approaches have been applied in the context of human research to identify the acceptable (and the proscribed) treatment of research subjects, while animal research has applied consequentialist approaches to defend the use of animals in support of benefits to humans. At the core of this deontological approach is the principle of respect for persons, on which so much of the ethics of research on human subjects is based. The principle of respect for persons, along with those of beneficence and justice, are now long-standing commitments invoked in support of protection policies.[12] Of the three, only two are relevant in relation to animal research. It makes little sense to dwell on how to translate or extend to animals the principle of respect for persons, although the notion of surrogate consent and advocate consent make a great deal of sense in the context of animal research; and the other two principles have many direct applications to animals, even if they have only infrequently been so applied.

Animals do not have all the relevant features to warrant consideration as beings with the same moral status and the same set of rights that humans have. Still, we would be morally unsympathetic if we were to dismiss claims that many animals (vertebrates in particular) are sentient beings with interests that warrant consideration and that sentience (among other criteria) confers moral status. Such consideration translates into responsibilities to promote and protect animal welfare. Welfare is not merely equivalent to humane treatment, and while the same deontological constraints need not be part of human and animal research protections, protection of the interests of subjects demands a different justification than merely consequentialist calculations in which animal interests are summarily trumped by human interests.

The regulations in the United States evidence a very limited sense of the term "welfare" for animals. It seems largely to be understood as "humane treatment" (whatever the ill-defined term "humane" means) rather than as acting to protect and promote the well-being of animal subjects.[13] There is here a misalignment of policy and principle, which can be corrected only through a revised and stronger sense of welfare. Strong welfare would move beyond mere humane treatment of animals used in biomedical research, and represent a commitment to animal research that takes seriously the interests of animal subjects without those interests trumping human interests.

This approach would require that IACUC members engage in questions about whether the harm to animal subjects is outweighed by the potential for human benefits. An example of research that probably should not be conducted when it has a harmful effect on animals occurs when previous research has already yielded similar results or when there is likely to be limited value produced by research that causes significant pain or suffering to animals.

ALIGNING POLICY WITH PRINCIPLE

The policies crafted for protection of human subjects and the oversight of human subject research had multiple goals and influences, including reaction to public scandals regarding the mistreatment of research subjects. In response came far stronger and more detailed rules for protection and oversight to prevent wrongdoing from happening in the future and to enhance public trust in the research enterprise.[14] The history of animal research oversight has also been somewhat reactive to scandal.[15] As use increased in the last forty years, the policy environment reflected news accounts of pet theft and inhumane practices by animal dealers, prompting rules to prevent cruelty to animals and to promote humane treatment.[16] Policies were therefore a response meant to protect the most basic welfare of animal subjects in biomedical research: adequate housing, food, and water, and the minimization of pain and suffering. However, these are but the most obvious of the concerns that we should have about animal welfare.

While in some respects the policies for the oversight of human and animal research came to look similar—for example, federal rules to protect individual subjects and to have local institutional review bodies—it would be a mistake to conclude that they share a deep underlying ethic of shared principles. And yet, following the logic commonly invoked in support of additional protections for these identified subgroups, *all* animal subjects appear to qualify for protection of their interests. If we find it morally objectionable to expose human subjects to substantial risk for the benefit of others, why is it not equally objectionable to do so with animal subjects? The difference in treatment can be justified within some range by pointing to the differences in moral status between most animals and most humans, but this is a superficial difference in light of the fact that it is wrong to ignore the powerful similarities and parallels between humans and many nonhumans.

What policy changes would a stronger and more justified welfare ethic require in research animal protections? It would require greater attention to balancing the risks to animal subjects against the likely benefits to humans or animals and the value of the knowledge to be gained from the research—on the part of researchers, and the review committees charged with animal research oversight. This difference of approach would supplant the current approach in which any amount of risk to animals can be justified by reference to the "needs of science," and the instrumentalist view that such justifications imply. Specifically, a strong welfare ethic would not allow research that harms animals without the prospect of adequate advancement of human interests. The greater the burden for animal subjects, the greater the justification required.

For example, under federal research regulations, research that is likely to cause significant pain and suffering to animal subjects can only be justified on the grounds that it could not be performed under pain-free conditions. Strong welfare provisions would require additional justification, based on claims that the likely benefit to humans is so significant that it warrants the burdens it would cause to the animal subjects. This determination ought to be made not merely on the basis of whether the proposal includes interesting or useful science, but on whether there is sufficiently clear application to human health and sufficient likelihood for success. This assessment would likely involve a consideration of scientific merit, which is considered by some a controversial aspect of prospective review, though their reason is largely based on the grounds that IACUCs are not the appropriate bodies to do such assessment. Without weighing in on the propriety of IACUCs taking on scientific merit review, I think it is important that assessment of scientific merit feed into prospective review and oversight of animal research. This goal could easily be accomplished by drawing on the expertise of existing peer-review bodies, both within institutions and that are part of the existing infrastructure for grant review at funding agencies.

A more explicit connection between ethical principles and animal research protection policies will heighten awareness among researchers, the members of IACUCs, institutional officials, as well as policy makers about what animal research entails and what is at stake in relation to the treatment of animal subjects. Increased awareness and attention to the ethics at stake in animal research will contribute to more careful review, judicious use of animals, and the evolution in policy that is part of a dynamic oversight system.

CONCLUSION

Policies for oversight of biomedical research on animals have caused us to be less deliberate in our decisions about the use of animals. We should now be more explicit in our consideration of the moral costs of animal research, which will better connect ethics, policy, and practice. While there is much to be gained from performing research on animals, it is intellectually dishonest to pretend that the foundations are perfectly identical to those set out for research on human subjects. However, the foundational principles and their applications are not as different as many seem to assume. It is time to move past a limited humane care ethic that exploits animals under the justification of a commitment to human interests, while taking only an instrumental view of animal interests. Only when we get beyond this limited moral vision will we really know what ethics has to do with it.

NOTES

1. M. Lawrence Podolsky and Victor S. Lukas, eds. *The Care and Feeding of an IACUC: The Organization and Management of an Institutional Animal Care and Use Committee* (Boca Raton, Fla.: CRC Press, 1999).

2. Jeffrey Kahn, "Organs and Stem Cells: Policy Lessons and Cautionary Tales," *Hastings Center Report* 37 (2007): 11–12.

3. Graham Shedd, Fred Wiseman, Adrian Perachio, David Baltimore, Richard Lewontin, and Robert Nozick, "Can Some Knowledge Simply Cost Too Much?" *Hastings Center Report* 5 (1975): 6–8.

4. Gary L. Francione, *Rain without Thunder: The Ideology of the Animal Rights Movement* (Philadelphia: Temple University Press, 1996).

5. "Principles for the Utilization and Care of Vertebrae Animals Used in Testing, Research, and Training": see Public Law (1985).

6. Kenan Malik and Peter Singer, "'Should We Grant Rights to Apes?' Peter Singer Debates Kenan Malik," at www.kenanmalik.com/essays/singer_debate.html, accessed August 22, 2010.

7. David DeGrazia, "Regarding Animals: Mental Life, Moral Status, and Use in Biomedical Research: An Introduction to the Special Issue," *Theoretical Medicine and Bioethics* 27 (2006): 277–84.

8. DeGrazia, "Regarding Animals," 280.

9. National Commission on Biomedical and Behavioral Research, *The Belmont Report* (Washington, D.C.: Government Printing Office, 1978; cited as *Belmont Report*); and Ruth R. Faden and Tom L. Beauchamp, *A History and Theory of Informed Consent* (New York: Oxford University Press, 1986).

10. United States Department of Health and Human Services, "Basic HHS Policy for Protection of Human Research Subjects," 45CFR46, Subpart A, revised January 15, 2009; http://www.hhs.gov/ohrp/humansubjects/guidance/45cfr46.htm, accessed August 25, 2010.

11. Rebecca L. Walker, "Human and Animal Subjects of Research: The Moral Significance of Respect versus Welfare," *Theoretical Medicine and Bioethics* 27 (2006): 306.

12. *Belmont Report.*

13. Walker, "Human and Animal Subjects," 305–331.

14. James F. Childress, Eric M. Meslin, and Harold T. Shapiro, eds., *Belmont Revisited* (Washington, D.C.: Georgetown University Press, 2005).

15. U.S. Congress, Office of Technology Assessment, *Alternatives to Animal Use in Research, Testing, and Education* (Washington, D.C.: Government Printing Office, February 1986).

16. Bernard E. Rollin, "The Regulation of Animal Research and the Emergence of Animal Ethics: A Conceptual History," *Theoretical Medicine and Bioethics* 27 (2006): 285–304.

SUGGESTED READING

ALLEN, COLIN. "Ethics and the Science of Animal Minds." *Theoretical Medicine and Bioethics* 27 (2006): 375–394.

BAYNE, KATHRYN. "Developing Guidelines on the Care and Use of Animals." *Annals of the New York Academy of Sciences* 862 (1998): 105–110.

BEAUCHAMP, TOM L. "Opposing Views on Animal Experimentation: Do Animals Have Rights?" *Ethics & Behavior* 7 (1997): 113–121.

BEAUCHAMP, TOM L., F. BARBARA ORLANS, REBECCA DRESSER, DAVID B. MORTON, and JOHN P. GLUCK. *The Human Use of Animals.* 2nd ed. New York: Oxford University Press, 2008.

CARBONE, Larry *What Animals Want: Expertise and Advocacy in Laboratory Animal Welfare Policy.* Oxford: Oxford University Press, 2004.

DEGRAZIA, DAVID. *Animal Rights: A Very Short Introduction.* Oxford: Oxford University Press, 2002.

DEGRAZIA, DAVID. "Regarding Animals: Mental Life, Moral Status, and Use in Biomedical Research: An Introduction to the Special Issue." *Theoretical Medicine and Bioethics* 27 (2006): 277–284.

DEGRAZIA, DAVID. "Animal Ethics around the Turn of the Twenty-First Century." *Journal of Agricultural and Environmental Ethics* 11 (1999): 111–129.

DRESSER, REBECCA. "Research on Animals: Values, Politics, and Regulatory Reform." *Southern California Law Review* 58 (1985): 1147–1201.

DRESSER, REBECCA. "Standards for Animal Research: Looking at the Middle." *Journal of Medicine and Philosophy* 13 (1988): 123–43.

FREY, R. G. "Moral Community and Animal Research in Medicine." *Ethics & Behavior* 7 (1997): 123–136.

LAFOLLETTE, HUGH, and NIALL SHANKS. *Brute Science: Dilemmas of Animal Experimentation.* New York: Routledge, 1996.

National Research Council. *Toxicity Testing in the Twenty-first Century: A Vision and a Strategy.* Committee on Toxicity and Assessment of Environmental Agents, National Research Council, 2007. http://www.nap.edu/catalog/11970.html.

ORLANS, F. BARBARA. "Ethical Decision-making about Animal Experiments." *Ethics & Behavior* 7 (1997): 163–171.

PLUHAR, EVELYN B. "Experimentation on Humans and Nonhumans." *Theoretical Medicine and Bioethics* 27 (2006): 333–355.

"Principles for the Utilization and Care of Vertebrae Animals Used in Testing, Research, and Training." Public Law: PL 99-158, Nov. 20, 1985.

REGAN, TOM. "The Rights of Humans and Other Animals." *Ethics & Behavior* 7 (1997): 103–111.

ROLLIN, BERNARD E. "The Regulation of Animal Research and the Emergence of Animal Ethics: A Conceptual History." *Theoretical Medicine and Bioethics* 27 (2006): 285–304.

ROWAN, A. N. "Formulation of Ethical Standards for Use of Animals in Medical Research." *Toxicology Letters* 68 (1993): 63–71.

RUDACILLE, DEBORAH. *The Scalpel and the Butterfly: The Conflict Between Animal Research and Animal Protection.* Berkeley and Los Angeles: University of California Press, 2000.

SHAPIRO, PAUL. "Moral Agency in Other Animals." *Theoretical Medicine and Bioethics* 27 (2006): 357–373.

STENECK, NICHOLAS H. "Role of the Institutional Animal Care and Use Committee in Monitoring Research." *Ethics & Behavior* 7 (1997): 173–184.

SZTYBEL, DAVID. "Taking Humanism Seriously: 'Obligatory' Anthropocentrism." *Journal of Agricultural and Environmental Ethics* 13 (2000): 181–203.

TANNENBAUM, JERROLD, and ANDREW N. ROWAN. "Rethinking the Morality of Animal Research." *Hastings Center Report* 15 (1985): 32–43.

WALKER, REBECCA L. "Human and Animal Subjects of Research: The Moral Significance of Respect versus Welfare." *Theoretical Medicine and Bioethics* 27 (2006): 305–331.

CHAPTER 35

LITERARY WORKS AND ANIMAL ETHICS

TZACHI ZAMIR

For at least two decades, Anglo-American moral philosophy has not restricted itself to conceptual analysis and argumentation. A number of philosophers have turned to literature, finding in its form, content, or experience important modes of thought that are decisive for comprehensive moral reflection. Advocates of this "literary turn" in moral philosophy do not necessarily believe that moral philosophy can or should be superseded by a morally oriented form of literary criticism, but they hold that some sensitivities or aspects of moral reflection are significantly deepened by engaging with literary works.[1]

Animal ethics has up to now avoided engaging with literary-oriented ethics. To my knowledge, the only literary work focusing on animals that has attracted substantial interest from philosophers is J. M. Coetzee's *Elizabeth Costello*.[2] The two striking chapters on animals in Coetzee's novel have elicited detailed and thoughtful responses by Stanley Cavell, Cora Diamond, Stephen Mulhall, and Peter Singer. Other philosophers who have advocated stronger or weaker versions of the "literary turn" have not addressed the issue of animals at all. An obvious example is Martha Nussbaum, whose work on animals is unrelated to her work on literature's epistemological, rhetorical, and moral advantages. Within literary studies, environment-oriented critics (ecocritics) often discuss the moral dimensions involved in specific poetic choices of literary works dealing with nature or animals. These critics are aware of the main approaches within animal ethics, including utilitarian and rights-based theories. However, to judge by the argumentative resources that they avoid drawing on, they seem unfamiliar with the literary bent within contemporary moral philosophy.

The theoretical field is embryonic: On the one hand, ecocritics seem to be unaware of the arguments that moral philosophers have advanced regarding the moral contributions of literature. Moral philosophers, on the other hand, are only beginning to mine rich literary descriptions of animals to gain moral insight and to explore the ways in which the invocation of animals awakens morally relevant dimensions in literary works, thus deepening literature's philosophical import.

The point of this chapter is not merely to harmonize the objectives of distinct orientations within the humanities, but to show how literary-oriented animal ethics is capable of touching dimensions of thought and action that are *unique* to our dealings with animals. My choice of literary examples initially adumbrates the ways by which literature is able to unsettle a practice. I then turn to more oblique routes connecting the literary representation of animals and the attaining of moral insight, suggesting how and why literature might be able to transform or broaden moral perception, and why it is advantageously situated to explore the interdependence between humanity and animality.

Professionalizing Slaughter

The following description of the Sodderbrook poultry plant in Tristan Egolf's *Lord of the Barnyard* is dispassionately informative:

> The building itself accommodates over seven hundred employees at full force. Almost five hundred of them are incrementally spaced out along two one hundred yard elevated evisceration tiers stretching from one end of the building to the other. Above each platform, a hook-lined belt runs along the length of the ceiling, from which the incoming gobblers hang suspended for various stages of disembowelment. At the farthest end of the plant the kill room is situated behind closed doors. As the belt emerges from the kill room flaps, it forks and proceeds over a colonnade of one hundred and eighty gallon scalding vats, along the tiers, and ultimately through to the finishing rooms, where the packaging workers seal and wrap the final products. There's also a storage house situated thirty yards from the main building, which is put to use primarily during the holiday seasons when excess production brings in almost 36,000 birds a day.[3]

Describing an ostensibly shocking content in parsimonious, dry terms is a familiar rhetorical device. Six different quantitative measurements are crowded into the short space of the lines, infusing a mechanical air into the building, and rendering the human activity within it robot-like. As has been demonstrated by Netta Baryosef-Paz, Egolf's text broaches a subliminal, multilayered analogy between the employees in the plant and the slaughtered animals.[4] The exploited Mexican workers manning the evisceration tiers are "incrementally" crowded in the building (humans and animals alike are described mostly through passive verbs), and described as living-dead (the most experienced being nicknamed "the zombie"). The text shifts our attention from the killed animals onto the human agents who are slaughterers, but are also victims, internally eroded, deadened by their activity.

All this is not obvious, but it might be articulated without resorting to a literary work, say, by an activist who has researched the working conditions in abattoirs. *The Lord of the Barnyard*'s most profound achievement as a literary text lies
in the ways in which the narrative recreates the meaning of *work*. By his first lunch
hour, Egolf's protagonist has already slit three thousand throats. The narrator
invites us to access this experience by indirection, by lingering over moments of
return to work:

> The thirty minute break in the parking lot, with the long line of wetbacks casually
> smoking on the steps and platform of the docking bay, slipped through his fingers
> like that much sand. The screeching of the belt, which had ground to a halt and
> given way to an ominous calm, still rang through his head in nerve-shattering
> repetition, just as the stiffening in his right hand from maintaining the grip of the
> cleaver all morning only commenced to crawl upward into his forearms with the
> temporary inactivity. It seemed he barely had time to walk to the edge of the lot
> and have a smoke of his own before the bell sounded out, signaling everyone back
> to post. The whine of the belt resumed and fell right back into place in his head,
> where it had never missed a notch to begin with. He pulled on his mouthguard,
> unsheathed his blade, punched back in on the clock, and descended into the pit
> once again.[5]

Or:

> The first night he lay awake in the dark hearing the slosh of turkey blood, the
> popping of the electrocution box, and the screaming of shackled birds in the
> same manner others hear register keys and the clanging of dishracks long after
> the day's end. He drifted into an uneasy sleep sometime after 3 a.m. with the
> events of the day having become distant and horrific to the point of impossibility.
> Then, at 6:15 a.m., his alarm clock rattled to send him back through the fields for
> another go.[6]

Work implies repetition and routine. The more explicit allusions to hell—invoked
by the scalding vats and the "pit" to which the protagonist descends—are less biting
than the experience of hell as a collapse of the divide between internal experience
and external fact. The belt that had ground to a halt in reality is still turning in his
head, as does the "sloshing" of turkey blood and the sounds of electrocution and
screaming when he lies sleepless in his bed. Trauma can partly consist of such brutal
invasions of inner space by an overwhelming context, an inability to maintain the
divide between rest and work. The factory, with its sounds and smells, burrows into
him like the gradual stiffening of his arm, capturing the progressive deadness that,
as he works, works its way into him.

But it is the feat of will involved in *going back* that betrays the difference
between killing thousands of birds and executing this as part of labor. The literary
imagination of a gifted author invites us to pause and take in experiential shades
that may evade us when abstractly discussing the pros and cons of eating animals.
The industrialization and professionalization of killing, which the eating of animals necessitates, is here given a palpable form, forcing readers to come to terms
with the meaning of paid labor, which such a widespread practice entails. Egolf's

description of his protagonist's job interview—"Did he have any severe allergies? Did he have a workable knowledge of the Spanish, or any other Latin, language? ... *Did he have any objection to working with a knife?*"[7]—succeeds in turning formulaic questions into an alarming parody of the concept of work. The intentionally vague "working with a knife" allows the "work" to present itself as work, and aligns the proposed labor with other professional contexts in which one works with a knife, such as surgery or gardening. One can imagine a chef addressing such a question to a kitchen assistant at a job interview. The posing of such questions reinforces the text's preoccupation with other hallmarks of professional activity: the punching of the "clock," the "break," starting on a "prove your worth trial run basis," being "signed on permanently." Egolf thus highlights the poultry plant's attempt to utilize the rhetoric of ordinary work. Such a focus hampers the attempt to mold the activity conducted within the plant's walls into just another form of labor.

This brings us to the walls themselves. The contribution of literary imagination extends to the qualitative feel of inhabiting the place: the "hook lined belt," the "scalding vats," the "forking" of the belt, all imbue the description with a jabbing aura, as if the place were somehow lashing out at its beholder. The wounded onlooker then notes that Sodderbrook's description is studded with references to pipes—"pipe-lines," "piping," "drainage ducts," and "jammed drainpipes." The liquid flowing in the pipes is blood, which overturns the meaning of waste and drainage since the (symbolic and material) life-stuff of blood metamorphoses into effluents. This inverted disposal system further estranges the poultry plant. The effect is staggering: this one of a kind system of drainage animates the building itself, as its pipes imaginatively form a diabolical system of veins and arteries pumping dead blood. On the one hand, this icon creates a powerful symbol: the building instances a carnivorous society in which the animation of many is dependent upon the death of many more. On the other hand, the animated building is not merely a symbol or a metaphor for a human society living off the blood of other beings. The pipes are literally there, disposing literal blood. The singularity of the edifice issues from this jarring merging of symbol and symbolized.

I suggested that the morally significant effect of Egolf's description of slaughter is that it unmasks a supposedly ordinary institution's attempts to pass off as facilitating supposedly ordinary professional activity. One may, however, object, questioning the extent to which the moral force of the text is occasioned by the alleged immorality of slaughter as such. Other forms of labor might instill a deadening of the worker's mind without the work itself being implicated in presumably immoral acts. Marx attacked the immorality of factories without pinpointing the morality or immorality involved in what these actually produce. Non-factory structured work can also strike us as deadening or psychologically damaging. Suicide attacks have, for example, led Israel to fund specialized units charged with identifying body parts for the purposes of burial. The work's objective is here undoubtedly laudable. Yet this does not purge the activity from the ghoulish aura that it transmits. Does Egolf's description show that slaughtering animals is significantly distinct from other unappetizing professions which are amoral or even praiseworthy?

Responding to this question first requires recognizing that some particular actions remain morally troubling even when one is satisfied with the justification for performing them. A military commander compelled to send some soldiers to their death since such sacrifice is predicted to save many more of his men can be tormented for a lifetime by the order he gave while believing that it was fully justified to do so. A physician giving up on one patient in order to save two others, or euthanizing someone who is in pain and is about to die, can find himself similarly afflicted. The same can hold for an attorney who chooses to go on defending a rapist when, late in the trial, it is intimated to him that the accused is, indeed, guilty and the acquittal depends upon suppressing the revealed information. These can all be seen as justified acts in their contexts, yet ones that haunt the agent. "Haunting" means that the singular act—of sacrificing soldiers, euthanizing a patient, or helping to acquit someone who is guilty—is autonomous from the overarching moral vindication. The act occasions remorse, which, as R. M. Hare claims, is distinct from regret, since the former does not imply a wish for reversibility whereas the latter does.[8]

When an act haunts in this sense, it intensifies the need for a justification of the practice. People haunted in this way are compelled to revisit the justifications for the action, to reexamine the available alternatives and to rethink the probable outcome of acting differently. Sodderbrook's description is not an argument for or against slaughter. But it does make slaughter stand out from its enveloping justifications. It is here that slaughter differs from the other examples surveyed above. Euthanasia, military sacrifice, or withholding incriminating knowledge can be excused as necessary evils, acts that one would avoid when possible, yet ones that will be performed when called upon even if they might give rise to remorse. Slaughtering animals for food is not a necessary evil. Unlike these other acts, slaughter does not save lives, or alleviate pain. Indeed, it achieves precisely the opposite. Slaughter accordingly haunts us in a different way. Remorse here may lead to regret. Facilitating this intervention is the moral achievement of Egolf's text.

MORALITY, IMAGINATION, AND LITERATURE

Before proceeding to discuss literary texts that achieve other moral objectives, it would be helpful to spell out more precisely—at least in this first example—how and why literature advances moral reflection in Egolf's novel. Rather than merely familiarize aspects of "the literary turn" for readers who might be unacquainted with it, my explanation will also hopefully broaden the turn's scope by shedding light on the unique ways in which animals can affect the literary imagination and the moral outcroppings that literary engagements with them may yield.

While not providing an argument regarding the pros of vegetarianism and while not intended as a treatise advocating animal liberation, Egolf's description is moralistic in the sense that the narrative is not impartial. Two levels are involved in this moral intervention—which I will respectively call "internal" and "external." The "internal" level is the literary work's nuanced charting of an experience that its *fictional characters* undergo. The "external" sense is the work's rhetoric: the experience orchestrated by the work for its non-fictional *reader*. The relations between these levels are varied: the external dimension can approximate the internal one (we call such a state "empathy"). But it may also diverge sharply from it (comedy often relies on the creation of such distance between reader and character). Moreover, the levels themselves are unstable and may split up in various ways (e.g., readers may empathize and simultaneously be shocked upon realizing their empathy).

Literature both refers to and creates experience. Through this double movement, literature can broaden our factual intake both by center-staging less noted facts (e.g., the experience of slaughter as a profession above) and by modifying one's relation to the facts that one already knows. Regarding the second of these, Stanley Cavell claims that, "the extreme variation in human responses to [the mass production of animals for food] is not a function of any difference in our access to information; no one knows, or can literally see, essentially anything here that the others fail to know or can see."[9] Cavell is somewhat overstating the extent to which only the second process matters. Egolf's text probably does make most of us see something that we failed to acknowledge before. But there is a sense in which literature's contribution has less to do with disclosing facts, and more to do with experiencing anew familiar facts. Such a moral goal overlaps with a widely influential claim made independently within literary studies, in which one of poetry's (and literature's) distinct functions is to occasion defamiliarization.

When Shklovski initially introduced this idea into aesthetics, he was not interested in moral philosophy.[10] He was, rather, unraveling one source of literature's uniqueness: we go to art, literature, and poetry, since they can reacquaint us with what we already know. Shklovski was thereby linking aesthetics and epistemology. Overfamiliarity, and not just underfamiliarity, is detrimental to knowledge. In some works of literature, poetry, or art, such overfamiliarity can be momentarily reversed by estranging the reader from the seemingly known. Such a process can relate to the words themselves (as poetry often does, compelling us—as Sartre notes—to look at words anew). It can also concern patterns, gestures, behavior, or institutions. Applied to moral philosophy, this linkage can be important because in some domains we are in need of a reacquaintance with the familiar, a process that elicits morally relevant informing that is non-reducible to accessing new information. Through Egolf's text, we are reacquainted with abattoirs and how work in them can be experienced.

But how, specifically, does "informing" of this nature link with broader moral reflection? How does it inform or guide action? Specifically, does the above episode in *Lord of the Barnyard* constitute a *reason* to ban practices that rely on slaughter? If so, how compelling is this reason? Do such experiences corroborate arguments? Can they be weighed against counter-arguments? Should they be assessed in relation to opposing readings of the same literary text, or to readings that support opposing experiences? These questions are too complex to be comprehensively addressed here. Answering them requires a theory regarding the relations between aesthetics and rationality that I have elaborated elsewhere.[11] The gist of the idea is a distinction between "reasons" and "grounds." A "reason" for endorsing P can be an argument or a fact supporting P (or an argument or a fact suggesting that not P is false). A "ground" for endorsing P is of a different nature. It can relate to undergoing a powerful experience in which one's sense of the import of P has been modified, but not because one has mastered new information. An experience of this kind can form an enabling condition—similar to the relationship between sight and light, the latter enabling the former.

To state this point non-metaphorically: only by undergoing some experiences, witnessing some event or following through some process, can one access the relevant arguments and facts. True, experiences can often mislead one into accepting wrong beliefs (literature and art can and are used to entrench morally objectionable claims). Moreover, literary works can sometimes be construed as arguments (that is, as "reasons" and not just "grounds"), arguments that are not deductively valid, but that are nevertheless admissible within the framework of a theory that is open to non-valid yet acceptable implications (enthymemes). But these complications need not concern us here. Suffice it to note that the experiences created through immersion in a literary work can make a positive contribution to a knowledge-claim, by deepening the claim, re-organizing it within the other claims the agent already accepts, centralizing it within one's belief system, or changing its connections with motivation or action.

DESCRIBING ANIMALS

Egolf's description of slaughter provides a relatively straightforward example of the interfusion of moral and aesthetic elements within a literary work. But pursuing the implications of a text's rhetorical choices can achieve more than recreating and sharpening our acquaintance with practices that directly victimize animals. A burgeoning body of work from cultural and ecocritical studies targets the conceptual, emotional, and psychological underpinnings of representing animals in seemingly benign domains. Such critical interventions usually include either an explicit or implicit moral dimension. They can extend beyond literary works, per se, applying reading methods developed in literary studies to non-literary texts. So before I return to literary works, I note how the interpretive strategies that are typically used in relation to literary works can yield morally important insights when they are applied to non-literary texts.

Consider Eileen Crist's work on ethological vocabulary.[12] Ethological texts are not literary works. But Crist reads these texts with an eye to images, figures, rhetorical selections, choice of perspective, and the reader experience that these configure. Crist shows how the founders of ethology (Lorenz and Tinbergen) establish its distinctiveness and its claim to greater objectivity. Their target is Darwin, whom they consider a scientifically crude precursor. They accordingly devise a new descriptive discourse that—unlike Darwin's—is restricted to observable properties. Science is thus relieved of the crass anthropomorphism that characterizes Darwin's descriptions of animal behavior. Crist shows how this laudable attempt to secure objectivity is far from neutral. In the seemingly objective writings of these ethologists, agnosticism regarding animal experience—a hallmark of academic caution—is suffused with depersonalizing constructs. Instead of producing unbiased descriptions, the ethological text progressively undermines conceiving of animals as experiencing entities.

One of Crist's most winning examples involves a comparison between naturalist and ethologist descriptions of courtship behavior in fish.[13] She quotes the following lines from Darwin on the male stickleback's behavior after the female "surveys the nest which he has made for her":

> He [the male stickleback] darts around her [the female] in every direction, then to his accumulated materials for the nest, then back again in an instant; and as she does not advance, he endeavors to push her with his snout, and then tries to pull her by the tail and side-spine to the nest.

Crist then contrasts this to an ethologist's (Tinbergen's) description of the same behavior:

> One of the most complete analyses of chain reactions...has been carried out with the mating behavior of the three-spined stickleback. Each reaction of either male or female is released by the preceding reaction of the partner....The male's first reaction, the zigzag dance, is dependent on a visual stimulus from the female, in which the sign stimuli "swollen abdomen" and the special movements play a part. The female reacts to the red color of the male and to his zigzag dance by swimming right towards him. This movement induces the male to...This in turn entices the female to...thereby stimulating the male to point his head into the entrance. His behavior now releases the female's reaction.

Both descriptions are obviously not literature, but attending to the rhetorical aspects of these passages suggests that Tinbergen's text is implicated in what Crist calls "mechanomorphism": the animal being rendered as a package of discrete operations, unconnected by an overarching experiencing structure. Imaging animals as a cluster of actions was not, Crist maintains, an intended objective of classical ethology. It was rather an outcome of the attempt to devise an objectively neutral scientific discourse that would avoid the pitfalls of anthropomorphic projection. Crist's point is not to recreate the older discredited discourse. She shows, rather, how the new, more self-conscious and theoretically sophisticated vocabulary that came to supersede naturalism supports, willy-nilly, a systematic effacement of animal agency. The scientific vigilance employed for generating impartial observational categories becomes a form of animal objectification by enlisting the prestige of science for the mechanization of animals.

To repeat, Crist's reading is not a morally oriented analysis of a literary text. Nor is it a work of moral philosophy. Nevertheless, it does show that paying minute attention to language carries important moral consequences. Crist is subjecting science to a reading. She shows how, like literary works, scientific vocabulary systematically hampers particular imaginative options that close observation of an animal might elicit. True, any description and choice of vocabulary may draw us to or dissociate us from some imaginative options. Yet here, science—the vocabulary through which we describe reality when thinking rigorously—is contributing to a ubiquitous power structure that commodifies animals by denying that they process reality in a way resembling our own.

THE MENTALITY OF APES

Let us return to literary works. Animals have, of course, featured in many works of literature in different periods and genres——fables, short stories, tales, allegories, and spiritual works. Animals form marginal or central characters in plays, poetry and prose, in adult and in children's literature. They have been infused with moral traits and religious meaning, acting as potent symbols for human weakness or strength, as sources of temptation, danger, and horror, but also as companions, redeemers, and teachers of faith.

In the more interesting moments in which literature invokes animals, the encounter with them provides an opportunity to rethink humanity, its presumption of distinctiveness, its limitations due to its dependency on a linguistically mediated experience of reality. Animals experience the human world constructed around them, and can thus reflect back to humans the alien's outlook, one that blends naïveté and superiority. A literary appropriation of such an outlook may constitute a striking counter-evaluation of human activity and value schemes. Here, for example, is how J. M. Coetzee construes the thought processes of a chimpanzee, Sultan, who is subjected to cognitive experiments:[14]

> Sultan is alone in his pen. He is hungry; the food that used to arrive regularly has unaccountably ceased coming. The man who used to feed him and has now stopped feeding him stretches a wire over the pen three meters above ground level, and hangs a bunch of bananas from it. Into the pen he drags three wooden crates. Then he disappears, closing the gate behind him, though he is still somewhere in the vicinity, since one can smell him. Sultan knows: Now one is supposed to think. That is what the bananas up there are about.... But what must one think? One thinks: Why is he starving me? One thinks: What have I done? Why has he stopped liking me? One thinks: Why does he not want these crates anymore? But none of these is the right thought....The right thought to think is: How does one use the crates to reach the bananas? Sultan drags the crates under the bananas, piles them one on top of the other, climbs the tower he has built, and pulls down the bananas. He thinks: Now will he stop punishing me? The answer is: No. The next day the man hangs a fresh bunch of

bananas from the wire but also fills the crates with stones so that they are too heavy to be dragged. One is not supposed to think: Why has he filled the crates with stones? One is supposed to think: How does one use the crates to get the bananas despite the fact that they are filled with stones?...One is beginning to see how the man's mind works....At every turn Sultan is driven to think the less interesting thought. From the purity of speculation (Why do men behave like this?) he is relentlessly propelled towards lower, practical, instrumental reason (How does one use this to get that?) and thus towards acceptance of himself as primarily an organism with an appetite that needs to be satisfied.

Coetzee is writing against Wolfgang Köhler's experiments conducted on chimpanzees, specifically, against Köhler's chapter on the use of implements in his *The Mentality of Apes* (1917). Here is Köhler's own description of this experiment:

> When the objective is fastened *at a height* from the ground, and unobtainable by any circuitous routes, the distance can be cancelled by means of a raised platform of boxes or steps which can be mounted by the animals. All sticks should be removed before this test is undertaken, if their use is already familiar—the possibility of utilizing old methods generally inhibits the development of new....Sultan soon relinquished this attempt [to reach the bananas that are nailed to the ceiling by jumping]....He paced restlessly up and down, suddenly stood still in front of the box, seized it, tipped it hastily straight towards the objective....About five minutes had elapsed since the fastening of the fruit; from the momentary pause before the box to the first bite into the banana....The observer watched this experiment through the grating, from the outside of the cage.
>   There was something clumsy in the animal's execution of this feat. He could quite well have shoved the box right under the bananas [he had placed it at a distance that compelled him to leap to the fruit]. Also, he did not place the box lengthways vertically which would also have saved waste of energy. But the whole action was too rapid to allow of delay for these finesses.[15]

When juxtaposed, it is easy to see how Coetzee overturns Köhler's description. Who is the experimenter and who is the subject in this scene? Coetzee wrenches the experimenter from the safety of an insulated observer, turning Köhler into the one observed by the ape ("One is beginning to see how the man's mind works"). In light of Coetzee's rendering of the internal, speculative mind of Sultan, puzzling over its observer's possible motivations for placing the bananas out of easy reach ("Why has he stopped liking me?"), Köhler's own low evaluation of Sultan's performance—his sense of Sultan's "clumsiness" and "wasted" energy—discloses the crude and unsubtle nature of the scientist's mind. Upon analyzing this episode in *Elizabeth Costello*, Stephen Mulhall writes: "human observation of nonhuman animal life runs a very grave risk of producing only the results that those observers can imagine getting."[16] This result is clumsy indeed.

The implicit dialogue between Köhler and Coetzee, between science and literature, suggests how literature is able to expose the limitations of rigorous, scientific, human thought. The unique freedom enjoyed by a work of fiction enables Coetzee to unapologetically leap into the mind of a chimpanzee and describe its experience.

Köhler does not enjoy the same freedom. Yet the fictional nature of literature's response to science does not short-circuit the dialogue between the texts, and does not undermine the biting, non-fictional critique to which Coetzee subjects Köhler. A critique of this kind is epistemic—exposing, as Mulhall notes, the biases of the imposed framework—and also moral, by virtue of showing how animal experience is being reduced. By "reduced," Mulhall means that Köhler's scheme only admits and values manifestations of instrumental thinking. Readers need not endorse a belief in Sultan actually entertaining speculative thoughts of the "why-do-men-behave-like-this?" kind in order to perceive this reduction. Nor should they believe in some ideal, non-reductive mode of describing the mind of an ape. The moral point is to note how descriptions deny possibilities.

Here, the moral consequence of a literary depiction does not amount to a denaturalizing of a practice, as it does in Egolf's novel, but triggers in the readers a powerful reacquaintance with the latent morality involved in processing what one sees. We meet the same questions raised by Crist: What do we *see* when we look at an animal? What capabilities are extended and denied to the perceived animal? How are animals as such conceived and possibly distorted through this process? What does this construction tell us about human beings? Finally, is such intake amenable to change in the sense of being superseded by a vocabulary that is more cautious of its own reductive biases than Köhler's, if found morally wanting? Such questions need not imply that Köhler should have withdrawn his work in light of the moral misgivings regarding his language. If Köhler would have read Coetzee and conceded the force of his critique, he could have retained his observations, admitting the radically different interpretations that these observations can support.

Needless to say, the idea that perception is morally guided is not exclusive to philosophers of literature. Educators and reformers of language often highlight how language, perception, and stereotyping interlock. Literature often precipitates this undesirable process, bolstering and reinforcing misperceptions. But, as Coetzee's portrayal of Sultan's mind shows, literature is also capable of burrowing into the position of that which is being defined, classified, or described in a way which challenges its definers.

PRESERVING HUMANITY

Literary encounters with animals can be revealing in other morally relevant ways. Yann Martel's use of a Bengal tiger in his *Life of Pi* is a case in point. Martel presents the adventures of a young survivor in a lifeboat following the sinking of a ship carrying zoo animals. The lifeboat includes a hyena, a zebra, an orangutan, and a Bengal tiger. The first three animals die quickly. The tiger and the boy remain together in the boat for over two hundred days. The story is mostly devoted to the relationship formed between the boy and the tiger during this time, to the ways in which the boy manages to fend off the hungry tiger through various ploys. The twist appears at the very end of the novel, when the boy is rescued. We then learn the gruesome details of the actual survival: the boy initially finds himself in the survival boat with an injured sailor (the zebra), a ruthless cook (the hyena), and his mother (the orang-utan). He then witnesses the murder of the sailor and of his mother by the cook. The murders are followed by anthropophagy: The shocked boy then murders the cook and eats some of his body.

The colorful adventure tale that Martel weaves metamorphoses into a medita-tion on trauma. The Bengal tiger becomes an imaginative projection of an indigest-ible feared inner nexus that the boy confronts. The novel invites its own rereading, this time with an eye to the emotional infrastructure that underlies the boy's projec-tion. An introductory episode—in which the boy's father (a zookeeper) exposes the boy to the dangers of the tiger by showing his terrified son how the caged tiger devours a goat—suggests that the specific nature of the projection performed by the boy on the lifeboat is rooted in a prior fearful experience. The ruthless caged tiger resurfaces in the boat. This time, though, it is not a mortifying external entity. It is an imaginary construct enabling a withdrawal from one's own unfathomable feroc-ity, shaped into a lethal animal.

The process Martel dwells on is the precise opposite of the varieties of anthro-pomorphic projection.[17] If anthropomorphism entails turning animals into humans by refusing to perceive the possible radical otherness of animals (molding them into humans in animal skins), Martel articulates a process whereby human traits are metamorphosed into animal ones. Whereas anthropomorphism familiarizes ani-mals through their humanization, Martel's protagonist preserves the alien nature of animal experience in order to dissociate himself from inconceivable dimensions of his own human experience. Anthropomorphism renders animals human-like. In Martel's novel, the human being is rendered animal-like.

Yet, since there is no tiger on the lifeboat, Martel's novel is not really about the sensual perception of living animals. It is about the animal's capacity to populate the imagination in a tormented mind and to fulfill a therapeutic, self-preserving func-tion. Rather than provide a literary dressing to a psychological process, the novel eavesdrops on a relationship that preserves the survivor's inner life. Unburdened by the superiority that a pathologizing of the boy's reaction would inevitably induce in its readers, the novel's first reading fosters emphatic participation in the prolonged illusion the boy maintains, an illusion presented as reality. Readers thereby are privy to and share the process undergone by the protagonist whereby humanity is safe-guarded by a specific, threatening construction of the animal. By withholding the gruesome trauma that underlies the action, the novel enables readers to experience—rather than merely conceive—the codependence between humans and animals.

Ecocritics and animal ethicists frequently mention how "animal" and "nature" are constructed categories that are and have been obsessively invoked in Western thought in order to define "humanity." Indeed, such self-other co-definition is nowa-days a trivial cliché. Martel's novel allows this codependence between humans and animals to materialize and be experienced as a temporally prolonged state in which a perpetual reinvention of the animal takes place in order to preserve the human.

Literature here deepens our sense of what a knowledge-claim is. More than a justified true belief, a knowledge-claim can sometimes include a justified propositional component (in our case, stating the dependency of "human" on "animal"), but which is also supported by the lessons provided by a lived-through or carefully imagined process. As suggested in this chapter's second section, claims sometimes need to be embedded in experiences in order that a justified belief becomes a belief that one actually endorses. Literature, in Martel's novel, is able to provide such an experience with regard to how constructing animals as others helps salvage what we are.

Giorgio Agamben writes that "man is the animal that must recognize itself as human to be human." Agamben is revisiting the taxonomies introduced by Carl von Linné (Linnaeus)[18] in the eighteenth century, discovering that "*Homo sapiens* is neither a clearly defined species nor a substance; it is, rather, a machine or device for producing the recognition of the human."[19] The claim stresses not some conceptual codependency, but focuses on recognition, the act by which a perceptual input needs to be perpetually subsumed under a category. The idea is that our humanity is something we need to be perpetually reassured of, which in turn implies that it is a category which is consistently under attack. One is always withdrawing from this category. It accordingly needs to be freshly recreated by constructing and sustaining its opposite.

This is precisely the process on which Martel's novel dwells. The specific understanding granted by this novel relates to the process by which constructing and maintaining the animal as an impending danger helps sustain, even saves, the human. When the survivor is rescued and the tiger leaves, the boy feels that he owes the tiger "more gratitude than [he] can express. [He] would like to say to the tiger 'thank you. Thank you for saving my life.'"[20] The moral repercussions of such a realization entail the surfacing of parallel self-preserving motives that underlie other animal-related practices typically discussed by animal ethicists. Whether one considers pet keeping, zoos, hunting, "pest" control, vivisection, circuses, or slaughter, such practices mobilize self-defining structures that need to be noted, meaningfully acknowledged, evaluated, and, perhaps, restructured. Is it, for example, relevant to debates over the morality or immorality of eating animals to consider how the eating of animals on the one hand, and the taboo on cannibalism on the other, is safeguarding human distinctiveness from animals: "I eat them, therefore I am?" Sometimes the directly moral, argumentative aspects of the issues are not the more significant working hidden motivations. If Martel's hero epitomizes a broader dependency of humans on animals, reforming such procedures goes beyond practical ethics, reaching deep into our sense of self.

HUMANS PERCEIVING ANIMALS

Such a result is unsurprising. To look at animals involves a setting into motion, but also a denial, of kinship. This is perhaps why some striking literary encounters with animals involve humans *looking* at them.

Literary works that focus on humiliated humans can unintentionally expose the moral dubiousness of such looking when they describe human beings who have been reduced to the status of animals. Here is a passage from Toni Morrison's *Beloved*:

> Paul D hears the men talking and for the first time learns his worth. He has always known, or believed he did, his value—as a hand, a laborer who could make profit on a farm—but now he discovers his worth, which is to say he learns his price. The dollar value of his weight, his strength, his heart, his brain, his penis, and his future.

Paul D hears his white owner admitting that he

> would have to trade this here one for $900 if he could get it, and set out to secure the breeding one, her foal and the other one, if he found him. With the money from "this here one" he could get two young ones, twelve or fifteen years old. And maybe with the breeding one, her three pickaninnies and whatever the foal might be, he and his nephews would have seven niggers and [the farm] would be worth the trouble it was causing him.[21]

The owner's language suggests that the commodification inscribed into the descriptive categories applied to slaves is somehow ingrained into the very perception of a slave. While this detail in the novel provides another hallmark of dehumanization, it enables us to pose the question whether "dehumanization" of this kind is wrong merely because it addresses fellow humans as animals, or whether there is something objectionable in the treatment itself, in the commodification of an experiencing entity as such. Such a claim does not imply that dehumanizing African Americans is comparable to objectifying animals. But since our wrongdoings to animals are often conceived as the most barbarous acts imaginable when applied to humans ("they went like sheep to slaughter"), one may go on to question the practices themselves. Dehumanization is wrong because humans are maltreated by being reduced to animals. But, on a deeper level, it can also be wrong because in the practices in which "dehumanization" is conjured up, *animals* are being reduced by being maltreated by humans. Animals are reduced because the practices involve molding experiencing entities into objects or commodities. "Dehumanization" is thus doubly wrong: it is not wrong merely because a human being is "reduced" to an animal. It is wrong because a human being is reduced to a *reduced* animal.

Lingering over the language of a slave-owner discloses that reductionism is not a mere feature of actions, but of perception as such. Exposing the moral undersides of perception can take other revealing forms. In a further instance of humans positioned as watched animals, consider the following episode in David Garnett's *A Man in the Zoo*. (The story is of a man who volunteers to be exhibited as a sample of *Homo sapiens* at the London Zoo. The episode depicts his girlfriend coming to watch him):

> In a few minutes she found herself in front of [her lover's] cage, and gazed at him helplessly. At that moment he was engaged in walking up and down (which occupation, by the way, took up far more of his time than he ever suspected).

> But she could not speak to him, indeed she dreaded that he should see her. Back and forth he walked by the wire division, with his hands behind his back and his head bent slightly, until he reached the corner, when up went his head and he turned on his heel. His face was expressionless.[22]

Cultural analysis of zoos has often discussed the intriguing complexity involved in acts of personalizing and depersonalizing animals in such surroundings. Zoos indulge in selective naming: some animals are personalized and nicknamed, endowed with biographies and speaking voices. They can receive letters addressed to them by children, and can "respond" to these by writing back to the kids. Other animals are anonymous instantiations of a type.[23] Preventing a comprehensive individuation of the caged animals is part of the zoo's uneasy middle-ground existence. Zoos are venues for entertainment, but they also see themselves as fulfilling an educational role. Zoos can be seen as a laudable modern Noah's ark, a vehicle for preserving endangered species. But they can also be regarded as destructive institutions that carry out an aestheticization of animal incarceration.

Garnett focuses on depersonalized exhibiting in zoos. His protagonist ardently strives to exhibit "humanity." Garnett's heroine experiences nausea at the sight of her expressionless lover. She watches a de-individuated face that lacks particularity, history, and a singular sense of life. *She is momentarily watching a genus*, a manifestation of a kind. If she persists, she would perhaps manage to look at him similarly to how we look at an ant or a spider, beings whose individuals we cannot tell apart. As in the above example from Morrison, the shock does not arise merely from sensing that her lover has been diminished. Rather, there is something shocking—nauseating for the heroine—about the capacity to experience such perception, both as perceiver and as perceived. By installing a human being in a zoo cage, the novel induces its reader to review seemingly benign forms of animal observation (the novel opens with the lovers walking among the cages in the zoo and talking about the animals as any visitor would). The moral dubiousness of zoos no longer exclusively resides in their restriction of movement as part of an overarching attempt to aestheticize animals, but extends to encompass the zoo's participation in and encouraging of modes of perception in which animals become types. Indeed, turning animals into generic instances helps suppress the awkwardness of caging them. The zoo thus relies on the same distancing mechanisms salient in the above description of slaves ("the breeding one," "this one here"), in which departicularization suppresses moral concerns.

Literature exposes the layering of one's perception of animals and its dependency on particular animal-related practices. Introducing and sustaining questionable forms of perception of animals is sometimes overlooked when scrutinizing such practices. But hampering such accustomed perception can mobilize an important *gestalt* shift in rethinking, and possibly reforming, our relation to animals.[24] Literature may help effect such a shift because a literary work *is* a thought-out configuration of an imagined perspective on events. "Perspective" here means both the point of view explicitly endorsed by the narrator (if there is one), and, more elusively, the vantage point progressively established for the reader.

ANIMALS PERCEIVING HUMANS

What happens when animals look back? Philosophers have sometimes utilized such thought experiments in order to rethink humanity. Nietzsche begins his meditation on history from a munching cow staring back at the traveler/reader, mirroring to the traveler/reader his own historically structured experience of self, which differs radically from the cow's present-oriented experience. Derrida's reflections on self-hood are triggered by a moment in which his cat stares at his naked body upon exiting the bathroom, prompting Derrida to note that nakedness is a quintessentially human experience.

In literary texts, such encounters with an animal's gaze often alert us to suppressed dimensions of what we are, aspects that the animal can either somehow pick out or is compelling us to acknowledge through our awareness of what the animal is looking at when it looks at us. Consider literary horror narratives that position humans as the intended or actual prey of animals. By overturning the usual predatory hierarchies, such episodes do not merely generate fear of death. The fear stems, rather, from imaginatively inhabiting a position in which one is dispassionately perceived as food:

> With night came horror. Not only were the starving wolves growing bolder, but lack of sleep was telling upon Henry. He dozed despite himself, crouching by the fire, the blankets about his shoulders, the axe between his knees, and on either side a dog pressing close against him. He awoke once and saw in front of him, not a dozen feet away, a big grey wolf, one of the largest of the pack. And even as he looked, the brute deliberately stretched himself after the manner of a lazy dog, yawning full in his face and looking upon him with a possessive eye, as if, in truth, he were merely a delayed meal that was soon to be eaten. This certitude was shown by the whole pack....They reminded him of children gathered about a spread table and awaiting permission to begin to eat.

To be looked at in such a way provokes not only dread in Jack London's Henry, but also a philosophical reconsideration of his relation to his body, once the view of his corporeal existence as potential food for others is forced upon him by the nearing wolves:

> As he piled wood on the fire he discovered an appreciation of his own body which he had never felt before. He watched his moving muscles and was interested in the cunning mechanism of his fingers. By the light of the fire he crooked his fingers slowly and repeatedly, now one at a time, now all together, spreading them wide or making quick, gripping movements. He studied the nail-formation, and prodded the finger-tips, now sharply, and again softly, gauging the while the nerve-sensations produced. It fascinated him, and he grew suddenly fond of this subtle flesh of his that worked so beautifully and smoothly and delicately. Then he would cast a glance of fear at the wolf-circle drawn expectantly about him, and like a blow the realization would strike him that this wonderful body of his, this living flesh, was no more than so much meat, a quest of ravenous animals, to be torn and slashed by their hungry fangs, to be sustenance to them as the moose and the rabbit had often been sustenance to him.[25]

London pauses over a moment in which the wolves' stalking triggers an acknowl-edgement of one's corporeal existence. What Henry discovers is not the sophistica-tion of his thought, nor is it the organization and shape of his limbs, but rather the subtlety and complexity of his operating, world-processing body. He moves his fin-gers "now one at a time, now all together," he "gauges" the sensations produced by pressing his finger tips, realizing that this "wonderful body of his, this living flesh" is about to be unceremoniously devoured. In diametrical contrast to Plato's Socrates in the *Phaedo*, for whom oncoming death awakens hopes for the mind's release from the limitations of the body, and accordingly philosophizing all the way to his execution, London's Henry is all of a sudden discovering his body, its motion, its sensitivity. Oncoming death induces Socrates to transcend his body. For London's Henry it enables perceiving the body anew.

Derrida, Cora Diamond, and Martha Nussbaum have all, in various ways, con-tended that something in animals awakens our vulnerability as well as our need to avoid it. Diamond, in particular, has noticed how such a regained sense of one's shared vulnerability with animals occasions in humans a sense of isolation.[26] By situating his lonely protagonist in the midst of a frozen wilderness, with his own frightened dogs "pressing close to him" from either side, London is able to crystal-lize in one poignant image the dawning realization of one's own fragile embodi-ment, how it elicits isolation and dread, but also a sense of fellowship with animals, with the dogs pressing against him seeking his protection in the face of a shared foe and a shared impending fate. Cold, tired, alone, and about to die, a man is thinking about the animals he eats, is pressed against by the animals he uses, and stalked by the animals that would eat him. Jack London brings to a peak the external and internal bombardment of humans by seen, felt, and recalled animals; the way in which this pressure destabilizes the human-animal divide; and how, intriguingly, it gives rise to self-centered wonder. This diffusion of the human-animal divide is achieved by bringing out that which is shared between humans and nonhumans: the body's vulnerability, its capacity to nourish someone else, the fear of death, all of which are experienced in different ways by Henry and his dogs.

Here, as elsewhere, horror scenarios provide opportunities to be *observed* in unfamiliar ways. The specific and repeated enlargement of reptiles and insects in monster scenarios is related to this attraction to being looked at by an insect:

> Hardly had Sam hidden the light of the star-glass when she came. A little way ahead and to his left he saw suddenly, issuing from a black hole of shadow under the cliff, the most loathly shape that he had ever beheld, horrible beyond the horror of an evil dream. Most like a spider she was, but huger than the greatest hunting beasts, and more terrible than they because of the evil purpose in her remorseless eyes. Those same eyes that he had thought daunted and defeated, there they were lit with a fell light again, clustering in her out-thrust head. Great horns she had, and behind her short stalk-like neck was her huge swollen body, a vast bloated bag, swaying and sagging between her legs; its great bulk was black, blotched with livid marks, but the belly underneath was pale and luminous and gave forth a stench. Her legs were bent, with great knobbed joints high above her back, and hairs that stuck out like steel spines, and at each leg's end there was a claw.

> As soon as she had squeezed her soft squelching body and its folded
> limbs out of the upper exit from her lair, she moved with a horrible speed,
> now running her creaking legs, now making a sudden bound.[27]

The alliteration in "squeezed her soft squelching body" triggers a unique form of disgust. The repulsion is rooted in the appeal to the hypothetically tactile feel of the giant spider's *soft* body. It is also the outcome of the acoustic infiltration of Sam and the reader, who imagines hearing the squelching sound of her movement. But focus on the *eyes*. Insects are entities whose world-processing is radically different from the one we ascribe to other predators. Whereas we cannot know what it might be like to be a bat, Jack London's hungry and cunning wolves do transmit—rightly or wrongly—an emotive structure that we can try to approximate through an (admittedly inaccurate) anthropomorphic projection. By opposition, the emotional void staring back at us from the empty eye of an encroaching giant spider relates to something more inferior in us than the body on which the wolf feeds.

The indifference that—rightly or wrongly—we impose on insects, renders the look of the monstrously enlarged insect the closest approximation to an animated yet indifferent universe, one for which the living and non-living are no more than interchangeable material formations. It is as if one were being watched by matter. And it is not accidental that, like other monster slayers who have picked on eyes (Odysseus and Harry Potter come to mind), Tolkien's Sam will also attack the spider's eyes. The violence to the monster's eye represents a victory of the human over the deadening stare of the monster, of this *particular* monster, since it is the eye that obliterates the distinction between living humans and the matter of which they are formed.

These briefly examined texts suggest that when animals look at us, our sense of distinctiveness wavers. We are momentarily thrown back to being potential food, or packaged matter. By inviting us to imaginatively inhabit the position of humans being looked at in these ways, literary works are able to undermine our sense of uniqueness, our separation from matter and its world. Farce is able to achieve this too: a pompous man discovering that he is subject to the laws of gravity by slipping on a peel. The eyes of animals create this effect differently, looking at that aspect of us which we do not acknowledge, but also *being that* which we distinguish ourselves from. Their look disturbs because it actively seeks in us that which we share with them.

UNCONDITIONAL LOVE

The fear unleashed at moments in which humans dare meet the animal's inquiring stare is not the only option explored by literature. Through their very capacity to ignore our obsessive, constitutional urge to be disentangled from matter, embodiment, and nakedness, animals can project a powerful acknowledgement of us (anthropomorphized into "unconditional love") seldom bestowed by fellow humans. The Disneyfication of animals is the more widespread, sentimental, child-oriented rendering of this anthropomorphic projection. More sophisticated literary depictions of such unconditional love (particularly in relation to dogs) have been offered (e.g., in Tolstoy's short animal stories), revealing the projections that underlie human connection with companion animals, the need we ask them to fill.

Consider the following moment in which, again, an animal is looking back at a human:

> For the first time [Miss Barrett] looked [Flush] in the face. For the first time Flush looked at the lady lying on the sofa. Each was surprised. Heavy curls hung down on either side of Miss Barrett's face; large bright eyes shone out; a large mouth smiled. Heavy ears hung down on either side of Flush's face; his eyes, too, were large and bright; his mouth was wide. There was a likeness between them. As they gazed at each other each felt: Here am I—and then each felt: But how different! Hers was the pale worn face of an invalid, cut off from air, light, freedom. His was the warm ruddy face of a young animal; instinct with health and energy. Broken asunder, yet made in the same mould, could it be that each completed what was dormant in the other? She might have been—all that; and he—But no. Between them lay the widest gulf that can separate one being from another. She spoke. He was dumb. She was woman; he was dog. Thus closely united, thus immensely divided, they gazed at each other.[28]

Surprisingly mixing separation and union in a moment of reciprocal watching, Virginia Woolf's narrator shows how the proximity between woman poet and dog is *not* undermined by the awareness of distance between them. In fact, the novel dwells precisely on how the indelible divide between them—language—is what animates and distinguishes the love they share. It is precisely through a poet's awareness of the import of language that the force of non-linguistic acknowledgement can emerge. "Do words say everything?" wonders Miss Barrett, "Do not words destroy the symbol that lies beyond the reach of words?"[29] And then Flush's hairy head presses against her.

Whereas the fear evoked by monsters and preying wolves relates to the human resistance to discovering their vexingly inerasable ties with matter or animals, companion animals tap our need to be accepted, loved, or (to keep to safer adjectives) "pressed against" by someone who will have nothing to do with our linguistic selves: "Where Mrs. Browning saw, he smelt; where she wrote, he snuffed." The very attempt to access this olfactorily based experience of reality elicits a halt: "Where Mrs. Browning saw, he smelt; where she wrote, he snuffed. Here, then, the biographer must perforce come to a pause."

Good things happen when great minds pause. There now follows a striking passage involving ironic admissions of defeat undermined by repeated attempts to transgress it through complex, sense-based figures, which might somehow reach and convey a dog's experience. (I will cite only the admission of defeat):

> Where two or three thousand words are insufficient for what we see…there are no more than two words and one-half for what we smell. The human nose is practically non-existent. The greatest poets in the world have smelt nothing but roses on the one hand, and dung on the other. The infinite gradations that lie between are unrecorded. Yet it was in the world of smell that Flush mostly lived. Love was chiefly smell.…To describe his simplest experience with the daily chop

or biscuit is beyond our power. Not even Mr. Swinburne could have said what the smell of Wimpole Street meant to Flush on a hot afternoon in June. As for describing the smell of a spaniel mixed with the smell of torches, laurels, incense, banners, wax candles and a garland of rose leaves crushed by a satin heel that has been laid up in camphor, perhaps Shakespeare, had he paused in the middle of writing *Antony and Cleopatra*—but Shakespeare did not pause.[30]

These moments in *Flush* conflate two distinct limitations. The first is the inaccessibility of a dog's experience; the second is the relative insulation of material objects from words. The first chasm is reciprocal: dogs cannot access our experience; and we cannot enter theirs. We can quite confidently ascribe to animals some general states: suffering, or feeling comfortable, or agitated. But the details and shades of these pointers are denied us. Behind this admission of poetic failure lies the hope to alight on a possessor of a greater descriptive capacity, a Shakespeare (and Shakespeareans would admire the sensitivity of choosing *Antony and Cleopatra*). The second chasm is distinctly ours. It relates to language and to the manner whereby language touches and sharpens our experience of the world but also insulates us from it. It is precisely the non-verbal communion she shares with the dog that, for Mrs. Browning, "leads to the heights of love."[31] Woolf is associating pets with the human need for close and intimate companionship that possesses its own language, but has nothing to do with words.

Some would recoil from such talk about a cocker spaniel, dismissing it as mawkish. The narrator is caught up in a heroine's over-sentimentality. Worse, the narrator is consciously and crudely trying to wring an emotional response. But what does such accusation of "sentimentality" actually amount to? Is sentimentality the mistake of going beyond proper emotional expression? Of feeling too much? Exploiting one's feelings (or one's readers' feelings) in discourse? Of misusing, perhaps misrepresenting, one's emotions? But then, why should such errors of judgment (if they are such) occur in an accomplished author like Woolf?

Human beings, we often hear, are subjects in language but also subjects of language: subjected subjects. We want others to acknowledge and respond to our linguistic selves, and we want such recognition to be freshly recreated. Simultaneously, we also possess non-linguistic strata, the ones we respond to in babies and lovers. The overlap between the talk elicited by babies, puppies, and puppy lovers betrays the emergence of a trans-linguistic element being responded to: "Mr. Browning wrote regularly in one room; Mrs. Browning wrote regularly in another."[32] The architectural divide between the rooms embodies the separation of husband and wife by language, of two loving selves who, paradoxically, are professionals, artists of language that should be united by their mutual vocation. The cruel irony of such a sentence—a devious god would have these two professional wordsmiths simultaneously write love poems to each other—is that the partitions between the rooms of the house are made to resemble the walls of words. Language, the vehicle of contact, is here formed and performed separately by lovers-poets, supposedly bonded by marriage. Flush (and, significantly, the couple's baby) is situated between them, eliciting and projecting a different response, personal, yet not attuned to an entire range of values. Like Derrida's cat, who is unaware that it is Derrida at whom he looks on a daily basis, Flush's attention sustains and validates an aspect of Mrs. Browning, the one craved by the wolves snarling at Henry, the one blankly stared at by Tolkein's giant spider.

The moral relevance of a focus on a non-linguistically mediated dimension of ourselves or of others is that watching animals and being watched by them invites us to enlarge the scope of what we see. The disabled, the old, and the exceedingly attractive are sometimes implicitly or explicitly asking others to ignore aspects of their bodies and to relate to "who they are." Morally responding to another can thus call for downplaying their body. The eyes of an animal suggest the opposite route: we are urged to hone and sharpen our capacity to perceive and take in embodiment, whether functional or dysfunctional. Authors such as Merleau-Ponty, Richard Shusterman, Mark Johnson, and Shaun Ghallagher have advocated an extensive rethinking of philosophy's disparaging evaluation of the body. Perhaps—in ways that we cannot clearly articulate now—a moral outlook that attempts to encompass the disabled, to respond to the politically dispossessed, the institutionally crushed, and the economically disempowered, can begin not from rights, capabilities, prefer-ences, or duties, but from emulating what animals see when they look at us stepping out of our bathrooms. Perhaps moments in which rights are denied and capabilities are ignored are rooted in center-staging one abstraction or another, instead of sus-taining the impression of another's embodiment and allowing it to feed and mobi-lize the other layers of one's moral sensitivity. Perhaps, finally, instead of regarding our animality as a threat (as "dehumanization"), we can turn our overlap with them into a new basis for the morality that should distinguish us.

Conclusion

This chapter's sampling of literary works hopefully exhibits the variety of moral insights offered by literary explorations of animals and our relations with them. Some of these literary interventions can modify action. Egolf's description of slaughter and how it manifests itself as repetitive labor is of this nature. Other texts may force us to pause before denying animals particular dimensions of experience (Coetzee). A third group of texts brings home to us our ingrained existential depen-dence on animals. Without animals we could not be human (Martel). Yet another group of literary works probes aspects of perceiving animals and being perceived by them (Garnett, Woolf, Tolkien, London).

A central motif running through my reading of these works is how literature questions what one sees when one is looking at an animal. This seems to harmonize more with moral approaches that start off from the intake of a noninstrumentalizing look (Levinas, Cavell, Buber), rather than those focalizing agency (deontic, conse-quentialist, contractarian) or character (virtue ethics). But literary interventions can deepen action-oriented moral approaches as well as character-oriented ones. The emphasis on suffering and the capacity to identify it (consequentialism) or the plausibility (or implausibility) of limiting the applicability of virtues like mercy or justice to humans alone (virtue ethics) exemplify how the literary imagination can extend and sharpen moral sensitivities couched in all of the dominant moral frameworks.

How effective are such literary explorations, given their fictional nature and the aesthetic necessity of preventing them from turning into didactic, one-sided manifestos? Gauging impact is always difficult. If my own experience is anything to go by, I would say that literary works can have a lasting and powerful effect. I was a vegetarian before I had read Coetzee's *The Lives of Animals*. But I do not think I would have written a book on animal ethics without the push I had received from Coetzee's work. I have obviously been exposed to arguments and direct, chilling pro-animal propaganda before. But something in Coetzee's work brought home to me the enormity of the animal issue and why I should be concerned with it as a philosopher.

Situated between emotional appeals and defensible claims, between rhetoric and argument, the literary text is able to address us not as targets to be politically mobilized. Nor does it position us as judges of arguments that need to be appeased. Mirroring our actions and practices for us, alerting us to what we see, the subtle fingers of literature fleetingly touch and leave us. Unexpectedly, it is the soft touch which can mount a powerful punch.

NOTES

1. For a relatively recent survey of this large field, see Adia Mendelson-Maoz's "Ethics and Literature: Introduction," *Philosophia* 35 (2007): 111–16. For useful analytic breakdowns of competing versions of the turn, see N. Carroll's "Art, Narrative, and Moral Understanding," in *Aesthetics and Ethics: Essays at the Intersection*, ed. J. Levinson (New York: Cambridge University Press), pp. 126–60 and M. Nussbaum, "Literature and Ethical Theory: Allies or Adversaries?" *Yale Journal of Ethics* 9 (2000): 5–16.

2. J. M. Coetzee, *Elizabeth Costello* (New York: Penguin Books, 2003).

3. Tristan Egolf, *Lord of the Barnyard* (New York: Grove Press, 1998), p. 128.

4. Netta Baryosef-Paz, "Restored Absents: A Comparative Reading in Tristan Egolf's *Lord of the Barnyard* and Ruth L. Ozeki's *My Year of Meats*," unpublished MA thesis, The Hebrew University of Jerusalem, 2009.

5. Egolf, *Lord of the Barnyard*, pp. 133–34.

6. Egolf, *Lord of the Barnyard*, p. 135.

7. Egolf, *Lord of the Barnyard*, p. 129, italics in the original.

8. R. M. Hare, *Moral Thinking: Its Levels, Method and Point* (Oxford: Oxford University Press, 1984), pp. 28–29.

9. Stanley Cavell, Cora Diamond, John McDowell, Ian Hacking, and Cary Wolf, *Philosophy and Animal Life* (New York: Columbia University Press, 2008), p. 93.

10. Victor Shklovsky, "Art as Technique" [1917], in *Russian Formalist Criticism: Four Essays*, ed. Lee T. Lemon and Marion J. Reiss (Lincoln: University of Nebraska Press, 1965), pp. 3–24.

11. Tzachi Zamir, *Double Vision: Moral Philosophy and Shakespearean Drama* (Princeton, N.J.: Princeton University Press, 2006), chapters 1 and 2.

12. Eileen Crist, *Images of Animals: Anthropomorphism and Animal Minds* (Philadelphia: Temple University Press, 1999).

13. Crist, *Images of Animals*, pp. 168–69. Darwin himself cites Warrington.

14. Coetzee, *Elizabeth Costello*, pp. 72–73.

15. Wolfgang Köhler, *The Mentality of Apes*, trans. Ella Winter (New York: Routledge, 1999 [1924]), p. 41.

16. Stephen Mulhall, *The Wounded Animal: J. M. Coetzee and the Difficulty of Reality in Literature and Philosophy* (Princeton: Princeton University Press, 2009), p. 43.

17. For a distinction between five kinds of anthropomorphism, see Bob Mullan and Garry Marvin, *Zoo Culture* (Urbana: University of Illinois Press, 1998), pp. 13–14. Mullan and Marvin are relying on Randall Lockwood's (then-unpublished) "Anthropomorphism is Not a Four-Letter Word."

18. Carl von Linné, *Systema Naturae*, 10th ed. (Holmiae, 1758; fac. London: British Museum, 1956).

19. Giorgio Agamben, *The Open: Man and Animal*, trans. Kevin Attell (Stanford: Stanford University Press, 2004), p. 26.

20. Yann Martel, *Life of Pi* (Toronto: Vintage Books, 2002), p. 317.

21. Toni Morrison, *Beloved* (London: Vintage, 1997), pp. 226–27.

22. David Garnett, *A Man in the Zoo* (New York: Knopf, 1924), p. 44.

23. Besides Mullan and Marvin, *Zoo Culture*, see also Randy Malamud, *Reading Zoos: Representations of Animals and Captivity* (New York: New York University Press, 1998).

24. Cavell (in Cavell et al., *Philosophy and Animal Life*) speaks about animals and a required *gestalt* shift, associating it with Wittgenstein's ideas regarding seeing aspects.

25. Jack London, *White Fang*, in *The Call of the Wild & White Fang*, chapter 3 (London: Wordsworth Classics, 1992), p. 88.

26. See Cary Wolfe's "Flesh and Finitude: Thinking Animals in (Post) Humanist Philosophy," *SubStance* 37 (2008): 8–36.

27. J. R. R. Tolkien, *The Lord of the Rings* (*The Two Towers*) (New York: Ballantine Books, 1965), p. 378.

28. Virginia Woolf, *Flush* (London: Vintage, 2002 [1933]), pp. 23–24.

29. Woolf, *Flush*, p. 37.

30. Woolf, *Flush*, pp. 126–27.

31. Woolf, *Flush*, p. 152.

32. Woolf, *Flush*, p. 125.

SUGGESTED READING

Philosophers on literary representation of animals: CAVELL, STANLEY, CORA DIAMOND, JOHN MCDOWELL, IAN HACKING, and CARY WOLF. *Philosophy and Animal Life.* New York: Columbia University Press, 2008.

COETZEE, JOHN M. *The Lives of Animals.* Princeton, N.J.: Princeton University Press, 1999. (This version of the two "lessons" from Coetzee's *Elizabeth Costello*—discussed above—includes responses by philosopher Peter Singer and representatives of other disciplines.)

MULHALL, STEPHEN. *The Wounded Animal: J. M. Coetzee and the Difficulty of Reality in Literature and Philosophy.* Princeton, N.J.: Princeton University Press, 2009.

WOLFE, CARY. "Flesh and Finitude: Thinking Animals in (Post) Humanist Philosophy."
    SubStance 37 (2008): 8–36. (The entire issue is devoted to animals from a Continental
    philosophical perspective).For some philosophically informed ecocritical accounts of
    animal representation:CRIST, EILEEN. *Images of Animals: Anthropomorphism and
    Animal Minds*. Philadelphia: Temple University Press, 1999.
MALAMUD, RANDY. *Reading Zoos: Representations of Animals and Captivity*. New York:
    New York University Press, 1998.
SCHOLTMEIJER, MARIAN. *Animal Victims in Modern Fiction: From Sanctity to Sacrifice*.
    Toronto: Toronto University Press, 1993.

INDEX

....

A

Abolition of animal use, 65, 81, 137, 859, 867

Abolitionists, 79, 181–82, 200, 235, 265,
    799, 812

Abortion, 260, 264–65, 336, 655–56, 685,
    836, 894

Abuse of animals
    ancient views of, 37
    companion animals, 216, 761, 771, 777–78, 783
    in farming and meat production, 881–82, 884,
        887, 891, 895–97
    laws pertaining to protection from, 776–78, 882
    in medical research, 671, 678, 921
    through neglect, 27, 200, 207, 216, 716, 741–42,
        746–47, 760, 769, 771, 777
    rights against, 199–200, 207, 219
    utilitarian views of, 9. See also Utilitarianism
    See also Cruelty; Harm; Neglect of animals;
        Suffering; Pain

Acidosis, 835, 878

Active potentiality, 336, 352–53

Activists, 79–80, 136, 618, 775, 799, 818, 870,
    934. See also Animal rights movement

Adaptive preferences, 236

Adventitious rights. See Rights

Advocacy, 62, 76, 79, 83, 244, 365, 407, 522, 637,
    708, 749, 858, 926

Aesthetic value, 259, 580–81

Affects (human and nonhuman), 8–9, 19, 166,
    408, 411, 549, 553–54, 559–63, 567–68

Affections
    ancient views of, 44–48, 50, 55
    animal capacities for, 20, 42, 159, 163, 773
    in companion animals, 27, 163, 762, 782, 790
    natural, 47
    their role in virtue theory, 148
    See also Affects; Emotions

Agency
    not acknowledged for nonhuman animals, 9, 13,
        15, 19, 103, 145, 147–48, 152–53, 157–60, 184, 209,
        383, 391, 401, 519, 522, 527, 548, 559, 705, 707,
        868, 871, 939
    and acting for reasons, 127, 521, 530
    in apes and other primates, 189–90, 391, 409
    Aristotle's views on, 525–27, 548. See also
        Aristotle and Aristotelian theory
    autonomous, 11, 132, 203, 389. See also
        Autonomy

comparison of human and animal, 11, 19, 80,
    132–34, 145, 147, 150, 152–60, 189–91, 262–67,
    321, 548–49, 559, 564, 705, 707
as a condition of moral status, 11–19, 124, 184,
    260–66, 311, 320, 519–22, 527, 548–49, 652, 705
in contractual agreements, 16, 266, 349–50, 367,
    385–401, 710
as a criterion for personhood, 311, 320–22, 367
degrees of, 240, 264, 321
empirical evidence for in animals, 549, 561, 564,
    567–68
Humean views of, 9, 15, 128, 145, 148, 150, 152–53,
    155–59. See also Hume and Humean theory
and intentionality, 314–15, 363, 530, 566
Kantian views of, 69, 103, 128, 132, 157–58, 284,
    367, 525, 527, 548. See also Kant and Kantian
    theory
in literature, 953
metacognition as a condition of, 9, 103, 145, 150,
    152–53, 157, 527, 559
and the moral community, 19, 188–91, 556, 559,
    561, 564, 566–68
moral motivation and, 19, 148, 520–22, 530
moral responsibility as a consequence of, 23, 80,
    157, 184, 188, 260–62, 383, 549, 564, 605, 620,
    636, 707, 741, 868, 871, 934
as a moral-status condition, 11, 13, 103, 124, 209,
    222, 260–72, 320–22, 348, 351–54, 360, 367,
    387–91, 548, 564, 652, 705–7
moral status in the absence of, 9, 16, 19, 148,
    159–60, 174–75, 184, 191, 519, 559, 564
nonhuman capacities for, 16, 19, 189–90, 240,
    262, 264, 314–15, 321–22, 360, 363, 367, 382, 391,
    409, 521, 542–43, 556, 561–68
as a property of personhood, 320–22
properties of, 11, 156–57, 189–90, 222, 262–64,
    539–40, 652
rational, 69, 103, 127–28, 152–53, 262–65, 284,
    351–52, 382, 385–401, 446, 496, 710, 713
rights from, 11, 132, 175, 184, 203, 209, 222,
    348–54, 360, 367, 389, 707, 713, 787, 868, 871
utilitarian views of, 155, 175, 188–91, 263, 548, 871.
    See also Utilitarianism
virtuous, 125–28, 131–36, 159–60, 397, 521–27
See also Moral agency; Moral status;
    Personhood

Agent-neutral reasons, 282

Agent-relative constraints, 735

Aggression, 27, 80, 155, 558, 662, 702–3, 711,
    775–76, 769, 779–80, 784, 834, 837, 878, 897.
    See also Violence
Agriculture
    agriculturalists on animal welfare, 858–59
    and the ancients' views of animals, 40, 49
    and animal domestication, 709–11
    and animal rights and welfare, 858–59, 867,
        870, 872
    animals deriving value from, 579
    and animals' moral status, 266–67
    and applications of biotechnology to animals,
        28, 826–29, 835–37, 839–47, 676, 826–47
    and biodiversity, 579, 872
    contractarian theories as a flawed theory of, 13,
        266–67, 348–50, 560, 709–11
    and domesticated animals, 702, 710, 717
    environmental impact of, 827–28, 840–47,
        889–90
    and the mixing of species, 676
    and overconsumption of resources, 889
    in policy, 895
    slaughter-based, 867, 878
    as a source of harm, 93
    United Nations Food and, 827, 840–41
    United States Department of, 758, 878, 913
    use of animals in, 710, 717, 753, 826–47, 867, 870,
        872, 878
    See also Animal rights; Factory farms; Genetic
        modification; Vegetarianism
Alienation, 392
Alternatives to animal testing, 29–30, 135, 137,
    748, 758, 800, 818, 845, 908–16, 927–28.
    See also Reduction; Refinement
Altruism, 16, 101, 103, 125, 233, 307, 382, 523.
    See also Benevolence
Amoebas, 444
Amphibians, 340, 637, 805, 889
Analgesia, 831, 926
Analgesics, 376–77, 835
Anatomical experiments, 62. *See also*
    Experimentation
Anatomy, comparative, 552
Ancestors and ancestry, 42, 49, 91, 312, 327,
    587–92, 683, 703–4, 777, 887
Anemia, 808
Anencephalic infants, 179, 188, 190, 390, 392
Anesthesia, 481, 507, 734–35, 756, 758, 835, 878,
    898, 926
Anger, 14, 18, 39, 44–46, 92, 158, 233, 470, 526,
    561, 773. *See also* Aggression; Fear
Animal dealers, 927
Animal kingdom, 17, 19, 53, 158, 229, 374–75,
    380, 401, 408, 411, 414, 417–18, 422, 432,
    436, 442, 551, 557, 568, 588, 590
Animal liberation, 119–23, 217, 263, 636, 705,
    757, 781, 858–60, 871, 877, 937, 940–41.
    See also Animal rights; Animal rights
    movement

Animal minds. *See* Cognition; Metacognition;
    Minimal minds; Minds; Psychology
Animal rights
    abolitionist views of, 182–83. *See also* Abolition
        of animal use; Abolitionists
    agriculture and, 859, 867, 870, 872. *See also*
        Agriculture
    and animal experimentation, 6, 62, 80, 82, 176,
        182–83, 198, 200–1, 209, 221, 858, 925
    animal welfare views contrasted to, 4, 182, 198,
        200–1, 855–59, 866, 869–72
    basic rights and, 5, 198–206, 210–19, 707, 717
    Beauchamp on, 9–10, 198–214
    Bentham on, 78–80, 83. *See also* Bentham and
        Benthamite theory
    Boyle on, 6, 61–62, 67–69
    a catalogue of, 212–19
    chickens and, 61, 217–18, 893. *See also* Chickens
    chimpanzees and, 205, 214–15, 323
    of companion animals, 80
    and contractarian approaches, 349–54
    contracts as basis of, 219, 349–50
    and correlative obligations, 9–10, 104, 198,
        206–11, 213–14, 219
    environmental ethics in tension with, 28,
        855–57, 859, 870, 872
    of farm animals, 80, 176, 198, 200, 748, 858, 870
    Garrett on the history of, 6, 61–62, 68, 78–83
    history of, 5–6, 61–62, 68–83, 199
    Hutcheson on, 5–6, 62, 72–74, 78–80, 83
    killing and, 201, 206, 216, 220–22
    legislation regarding, 80–81, 219
    to liberty, 176, 182, 200, 707, 748–49
    the meaning and concept of, 200–2, 925
    negative and positive rights, 705–8. *See also*
        Rights
    Primatt on, 79–80
    as protection of animal interests, 9, 200, 925
    Regan on, 182, 361–62, 364–65, 705, 717–18
    religious perspectives on, 68, 72–83
    and rights talk, 61–62, 68, 72, 199–200
    and rights theory, 4, 9, 198–203, 205–12, 214, 219,
        222, 706–7, 925. *See also* Rights
    Salt on, 79–80, 83, 855–59, 866, 869–72
    Singer's avoidance of rights language, 120, 177
    subject-of-a-life as criterion for, 14, 360–65,
        705–6
    as trumps, 182–183, 202, 219–20, 748, 866
    utilitarian views of, 74–81, 83, 94, 120, 175–76,
        182–83, 200, 859, 866, 869–70, 925. *See also*
        Utilitarianism
    Varner on, 28, 855–59, 866–72
    and vegetarianism, 82. *See also* Vegetarianism
    virtue ethics and, 136. *See also* Virtue ethics
    wild-domesticated distinction as relevant for,
        705, 707, 717–18
    of zoo animals, 198, 200–1, 207. *See also* Zoos
    See also Animal rights movement; Human
        rights; Laws; Moral status; Rights

Animal rights movement, 81, 83, 136, 176, 200, 870. *See also* Animal liberation; Animal rights

Animal testing, 804, 908–9, 913. *See also* Experimentation; Laboratory animals; Laboratory research

Animal welfare
    and animal rights, 4, 182, 198, 858–59, 866, 869–72
    animal welfarists, 28, 200–1, 282–83, 855–59, 883
    biotechnology and, 28, 654, 674, 678, 681, 827–40, 843, 846–47
    and chimeras, 674, 678, 681
    environmental ethics in tension with, 28, 855–57, 859, 869–72
    Humean views of, 146. *See also* Hume and Humean theory
    in industrial farming, 28, 827–40, 843, 846–47, 882–83, 893
    in legislation, 6, 54, 62, 79, 397, 674, 911, 921, 924
    mentalistic views of, 829–31, 834, 838–40, 847
    and the quality of life, 285
    in research, 674, 911, 921–28
    and societal values, 296–300
    in a society-centered conception of morality, 277, 292, 296–300
    utilitarian views of, 69, 78, 172–73, 192, 863, 866, 869–70. *See also* Utilitarianism
    and vegetarianism, 81–82, 893. *See also* Vegetarianism
    See also Agriculture; Animal rights; Diseases; Laws; Pain; Suffering

Animating souls, 36

Anthropocentrism
    abandonment of, 566–68
    and biocentrism, 628, 630–31
    challenges to, 123–24, 132–33, 548–49, 566–69, 584, 611, 628–39, 856–58
    Darwinian view of, 551
    in environmental ethics, 133, 140, 612, 617, 843–44, 855–58
    in environmental theories, 138, 140, 548, 556, 564–65, 855–58
    in the interpretation of animals' capacities, 307, 454, 549, 551, 558–61, 565, 568–69
    and moral status, 548–49, 564–69, 843
    in theory of persons and personhood, 307
    and species conservation, 584, 611–12, 617
    traditional views of, 559

Anthropodenial error, 473–74

Anthropogenic alterations, 613, 618–20, 841

Anthropomorphism
    about animal minds, 15, 17, 231, 443, 450–51, 469–75, 568
    Darwinian views of, 550–51, 939. *See also* Darwin and Darwinian theory
    emotions and, 470–79
    in literature, 943, 949–50

    in mental attributions, 15, 17, 231–32, 378, 441–43, 450–51, 454, 469–90, 549–51, 554, 568
    and moral status, 549, 551, 554–56, 568
    and pets, 132
    and psychological properties, 231, 469–75
    in scientific discourse, 442–43, 939
    and the "triune brain" hypothesis, 554
    and viruses, 442, 450
    See also Anthropocentrism

Antibiotics, 663, 800, 832, 880, 888

Antidepressant drugs, 800

Ants, 382–83, 401, 609, 946

Anxiety, 221, 286, 839

Apes
    and agency, 189–90, 391
    Cartesian views of, 447
    and chimeras, 680, 683, 687
    and cognitive ethology, 557
    consciousness in, 375
    Darwin's view of, 20, 550. *See also* Darwin and Darwinian theory
    emotional capacities of, 14, 20, 391, 488–89, 557, 761
    and evolution, 20, 91, 550
    great apes, 14, 91, 189–90, 214–15, 221, 313, 318, 322–23, 409–10, 425–26, 481, 485, 683, 709
    intellectual capacities of, 14, 20, 391, 447, 482, 726, 734, 760–61, 941–42
    literary representations of, 940–42
    mindreading abilities of, 375, 409–10, 425–28, 433
    and the mirror test, 101, 313, 481–82. *See also* Mirror self-recognition; Self-consciousness
    moral status of, 14, 322–23, 637
    personality in, 485–87
    potential personhood of, 322–24
    proteriction of, 14, 221, 322–24, 637, 709
    in research, 214–15, 323, 476, 485, 554, 941–42
    self-awareness of, 101, 229, 313–14, 318, 481–82
    sociality of, 162, 232, 426–27, 760–61
    and sovereignty, 709
    theory of mind concerned with, 488, 554
    and tool use, 232. *See also* Tools and tool use
    in zoos, 760–61. *See also* Zoos
    See also Baboons; Chimpanzees; Cognition; Gorillas; Monkeys; Orangutans; Persons and Personhood

Aquaculture, 829, 842, 846

Aquariums, 26, 198, 200, 717, 739–42, 744, 752, 755, 758, 761, 763. *See also* Zoos

Aquatic animals, 758, 827

Aristotle and Aristotelian theory
    on the moral status of animals, 5, 10, 15, 36, 42–52, 127–30
    Nussbaum's neo-Aristotelian capabilities theory, 230–31, 238–45
    as a theory of agency, 525–527, 548
    theory of virtues, 8, 128, 524–29, 531, 536, 548
    See also Capabilities theory; Virtue ethics

Arthritis, 286, 507, 836
Arthropods, 615
Aspirational rights, 208
Ataraxia, 70
Automata, 234, 310, 447, 496
Autonomy
    and choice and consent, 238, 659, 786
    comparison of human and animal capacities
        for, 103, 162, 186–87, 238, 285, 320–21, 785–86,
        790, 830–31
    its connection to personhood, 311, 320–21, 323
    and consciousness, 339–40
    duty to respect others', 122–23, 388, 709, 753
    Kantian views of, 103, 105, 108, 132, 186–87, 222,
        238–39. See also Kant and Kantian theory
    moral autonomy, 222
    and pets, 785–86, 790
    and quality of life, 186–87, 285
    rights of, 203, 222, 323, 388–89, 659, 709,
        785–86, 790
    its role in moral status, 11, 103, 122–23, 311,
        320–21, 323, 339–40, 388–89, 652, 705, 709,
        753, 790
    and the self, 108
    and sovereignty, 709
    utilitarian views of, 175
    See also Agency; Consent; Moral agency;
        Psychology
Axiological value, 336, 337, 340–41, 347,
    350, 361

B
Baboons, 21, 423–24, 651. See also Apes
Bacteria, 450–51, 610
Basic-needs requirements, 26, 739–43, 748,
    753–55, 759–60, 762. See also Animal
    rights; Confinement; Zoos
Bats, 444, 644
Bears and bear baiting, 73, 75, 630, 702, 763
Bees, 81, 374, 376, 380–84, 401, 455–63, 598, 609.
    See also Honeybees
Behavioral research, 215, 551. See also
    Behaviorism; Experimentation;
    Laboratory research
Behaviorism
    animal, 187, 555, 569
    behaviorists, 187, 338, 342, 356, 473, 551–53, 569
    logical, 338, 342, 356
    psychological, 450–451, 471, 473, 551, 553, 559
    See also Psychology
Beneficence, 151, 163, 208, 238, 270–71, 397, 926.
    See also Benevolence; Hume and Humean
    theory; Nonmaleficence
Benevolence, 12, 72, 77, 131–32, 148, 160, 163–65,
    212, 267–71. See also Altruism; Beneficence;
    Virtue ethics
Bentham and Benthamite theory
    and animal rights, 6, 78–80, 83

arguments against, 244–47
and moral problems about animals, 9, 78–83,
    172–92, 234–36, 282–83, 548
on suffering as the central issue, 263, 282, 383, 548
and utilitarian thinking about animals, 234–35,
    247, 383
See also Utilitarianism
Bestiality, 43
Bioart, 692–93. See also Literary
    representations of animals
Biocentrism, 132, 617, 628–31, 844, 857
Biodiversity, 28, 383, 612–13, 621, 828, 840–43,
    872, 890
Bioethics, 179, 665, 919
Biologism, 658
Biology
    and animal welfare, 830, 838
    bio-Cartesianism, 819–20
    biological distinctions among species, 19–22,
        155, 180, 327, 578, 585–99, 604–10, 615–21, 651,
        805–7, 820, 865
    biological modifications, 643–64, 683, 703, 838
    biological needs, 287, 830
    biologists, 19, 231, 592–93, 605, 607, 613, 618–20,
        865
    and chimeras, 644, 657. See also Chimeras;
        Genetic modification
    Darwin's theories in, 585–88, 590–96, 598–99.
        See also Darwin and Darwinian theory
    experiments in, 286, 643, 648, 653, 657, 683,
        809–12, 820–21, 838, 907, 916
    evolutionary, 20, 22, 205, 310, 320, 561, 585–99,
        604–9, 804–5, 807–11, 819–20
    history and philosophy of, 3–4, 19–20, 36, 49–52,
        446–47, 590–93, 604–8, 805
    and hybrids, 50, 587–88
    mechanistic models of, 17, 441, 446–47, 664, 801
    in modification and enhancement, 653–61,
        663–64, 683
    and moral status, 265
    and pain, 495
    and personality, 155
    and personhood, 326–27, 334–35. See also
        Persons and personhood
    synthetic, 653–61, 663–66
    wildlife, 865
    See also Brains; Philosophy of mind; Philosophy
        of biology; Theory of mind
Biospheres, 22, 577, 581, 618
Biotechnology, 23, 28, 642, 659, 673, 826–28,
    835, 846
Bioterrorism, 621
Bird flu, 887, 893
Birds, 51–53, 81, 133, 217, 340, 380, 413, 556, 559,
    561, 615, 682, 721, 756, 759, 762–63, 798,
    831–39, 879–81, 887, 891, 933–34. See also
    Bird flu; Chickens
Bisons, 865
Bivalves, 881

Blame, 79, 136, 397, 519–20, 526–27, 895, 898

Blastocysts, 679–80

Blindness
  in farm animals, 837–40, 879
  as a human impairment, 76, 390
  in mindreading, 411. *See also* Mindreading
  in research animals, 837–40, 907

Blindsight, 375, 532–34

Blood transfusions, 800

Boars, 41, 835

Bodily nociperception, 498

Bonobos, 358, 425, 427–28, 489. *See also*
  Chimpanzees

Boredom, 92, 395, 649, 760, 831, 834, 839, 882.
  See also Cages; Factory farms

Bottom trawling, 130, 881

Bovine somatotrophin injections, 879

Bovine spongiform, 647

Brains
  of apes, 683
  and behaviorism, 356, 550, 570. *See also*
    Behaviorism
  and bio-Cartesianism, 819–20
  brain death, 261, 333–36, 352, 633
  of chimeras, 24, 648–50, 672–77, 679–83, 686,
    688–90, 692–93
  and collective mental states, 462–63
  comparisons between human and animal, 19,
    341, 410, 550, 553–55, 559–70, 632–33, 682–83,
    819
  cortexes, 377, 390, 458, 462–63, 652–53, 819
  damage to, 138, 286, 317, 335, 345, 352, 632–33
  of dogs, 356
  and emotions, 358, 553–55, 559–61, 563
  and the human neuron mouse, 24, 672–77,
    679–83, 693
  and mental states, 334–45, 352, 356, 377, 498, 534,
    553–55, 559–70, 632, 681–82, 819–20
  of mice, 24, 649–50, 672–77, 679–83, 693
  mind-brain identity theory, 338
  and mindreading, 410. *See also* Mindreading
  and personhood, 317, 333–36, 341–42, 345
  reprogrammed, 334–35
  size of, 410, 680, 682–83, 703–4, 782, 808, 819–20
  structure of, 390, 462–63, 534, 550, 553–55,
    559–61, 568, 604, 682, 836, 838
  and theories of mind, 338, 819
  See also Brainwashing; Minds; Neocortex;
    Psychology

Brainwashing, 388

Breeders, 746, 782, 837

Breeding, 28, 94, 181, 217, 591, 606–7, 614,
  642–44, 680–81, 703–4, 710, 716, 721, 745,
  775, 782, 826–27, 832–36, 839, 945–46.
  See also Captive breeding; Interbreeding;
  Reproduction; Selective breeding

Breeding programs, 246, 642, 644. *See also*
  Breeding; Captive breeding;
  Interbreeding; Reproduction

Broiler chickens, 756, 835, 837, 879–80. *See also*
  Chickens; Factory farms

Bull baiting, 75

Bull elephants, 233. *See also* Elephants

Bulls, 50, 73, 869

Burden of proof in argument, 19, 27, 520, 568,
  586, 598, 636, 745, 820

Burn therapy, 800

Business ethics, 24, 174

C

Cadavers, 647

Cages
  battery, 756, 834
  bird, 51, 81, 879–80
  for companion animals, 81, 110, 715, 747, 771,
    773–74, 778–80, 785, 790
  as confining. *See* Confinement
  farrowing, 834. *See also* Pigs
  and industrial farming, 883. *See also* Factory
    farms
  for laying hens, 756, 834, 879
  loss of freedom the result of, 110, 182, 217, 715, 740
  for pigs on farms, 756, 834–38, 840, 877–78,
    881–93
  for primates, 180, 740–41
  in research, 430–32, 476, 636, 686
  rights against, 182, 200, 217, 771
  in shelters, 747
  suffering as a result of, 51, 81, 183, 740–41, 756,
    758–59, 780, 834, 879–80
  used for wild animals, 180, 718, 740, 778–80,
    880, 943
  in zoos, 180, 217, 715, 740, 758–59, 880, 946.
    See also Zoos
  See also Captive animals; Chickens; Confinement;
    Crates; Factory farms; Pigs; Zoos

Calves, 54, 245, 757, 869, 878–79, 881. *See also*
  Cattle; Livestock

Canaries, 306

Cancer, 179, 214, 647, 660, 715, 804–6, 809, 818,
  892–93, 897

Cannibalism, 39, 42, 45, 688–89, 756, 944

Capabilities theory
  capabilities of perspectival and positional
    thinking, 233, 242
  capabilities as capacities for functioning, 237,
    245, 247
  the capability to form self-conceptions, 229
  as a neo-Aristotelian account, 238–41
  as a non-metaphysical political view, 240, 247
  Nussbaum on, 228–48

Captive animals, 477–78, 577, 588, 614, 702–3,
  718–20, 740, 744–45, 750, 758–61, 780,
  844–46, 855, 870

Captive breeding, 577, 614, 855. *See also*
  Breeding; Interbreeding; Selective
  breeding; Reproduction

Capuchin monkeys, 471, 780. *See also* Monkeys
Carcinogens, 804, 909–10, 914. *See also* Cancer
Caretakers, 481, 742–43, 747
Caribou, 865, 868. *See also* Elks
Carnivores, 133, 935
Castration, 756, 758, 835
Categorical imperative (Kant), 99, 102, 105,
    107. *See also* Kant and Kantian theory
Cats
    as an ancient concentration of worship, 40
    breeding of, 132, 714
    confinement of, 742–45, 763
    cruelty toward, 726
    domestication of, 73, 704–5, 714–15, 722, 743, 745,
      762–63, 769, 777–78
    duties toward, 235, 282, 297, 394, 397, 567, 715,
      719, 722, 732–33, 736, 742–43, 762–63, 769,
      773–80, 788
    euthanasia of, 81, 782
    in experiments, 73, 809
    the gaze of, 947, 952
    historical perspectives on, 40, 81, 235, 447
    mental life of, 319–20, 352, 443, 447, 726, 734
    moral status of, 73, 306, 394–98, 704–5, 726,
      733–34
    as pets, 26, 73, 110, 132, 297, 567, 705, 714, 719, 722,
      742–43, 762–63, 769–73, 775, 777–82, 788
    as predators, 126, 320
    in shelters, 110, 782
    veterinary care of, 132, 770
    violence toward, 92, 180, 394–95, 397–98
    vulnerabilities of, 714–15, 719
    wild and stray, 73, 133–34, 704–5, 722, 777
Cattle, 28–29, 41–42, 45, 48, 52, 140, 205, 217–18,
    591, 674, 687, 757–58, 774, 826–27, 834–37,
    844, 878–79, 888–90. *See also* Bulls;
    Calves, Dairy cows, Dairy products
Cells
    in animal development, 91, 339, 606–8
    brain, 648, 672–74, 682–83, 693
    and chimera research, 23, 324, 641, 644–46, 648,
      651, 656–59, 661, 664, 672–81, 683–89, 691–93
    and cloning, 655, 826
    embryonic, 23, 641, 644–48, 656–57, 661, 674,
      676, 678–81, 683, 686–89
    eukaryotic, 450, 462
    evolutionary development of, 91, 462, 606, 609, 808
    genetic modification of, 641–64
    medical conditions of, 376, 441, 808
    in minimal-mind theory, 441, 450
    reproductive, 23, 339, 608, 644–51, 655–57, 659,
      661, 922
    in research, 23, 324, 646–51, 655–57, 659, 661, 664,
      671–93, 801, 806, 922
    single, 91, 450, 608
    in species development, 606–9
    spindle, 557
    stem, 646–49, 655–56, 671–76, 678–80, 683,
      686–89, 691–93, 806, 922

and tumors, 684
use in enhancement, 324
Cephalopods, 827
Cerebellum, 682. *See also* Brains
Cetaceans, 318, 321–22, 682
Charity, 12, 49, 131, 208, 212, 267–71, 475.
    See also Altruism; Benevolence
Cheetahs, 631, 778
Chemical mutagenesis, 837–38
Chickens
    and animal rights, 61, 217–18, 893
    and antibiotics, 888
    biological modification of, 28, 654, 826, 837–40
    biotechnology in the use of, 28–29, 654,
      756, 826
    blind, 837–40, 879
    and breeding, 837, 879
    broiler, 756, 835, 837, 879–80
    cruelty to, 879–81, 883, 896
    domestic, 588–90
    effects on the environment, 844, 889
    and enhancement, 654
    farming of, 29, 217–18, 654, 756, 835, 837–40,
      879–81, 886, 888, 895–96
    fertility of, 589
    genetically engineered, 28, 826, 837
    harms experienced by, 217–18, 654, 756, 835,
      837–40, 879–81, 883, 886, 896
    hybrid, 588–90
    industrial farming of, 29, 217–18, 879–81, 883,
      886, 888, 890, 895–96
    laying hens, 835
    and overconsumption of food, 890
    and pollution, 889
    poultry, 834, 836, 887, 933, 935
    reproductive qualities of, 588–90
    See also Breeding; Factory farms;
      Environments
Chimeras, 21–24, 591, 641, 644, 648–61, 673–93.
    See also Genetic modification
Chimpanzees
    and animal rights, 205, 214–15, 323
    attitudes of, 380
    behaviors of, 471, 558
    brains of, 683
    character traits of, 155–58, 774
    and chimeras, 649–51, 653, 656–57, 659, 661–63,
      672–73, 682–83
    cognitive abilities of, 232, 312–13, 375, 380,
      421–34, 554–58, 940–41
    and economic games, 482–84
    in environmental ethics, 139
    experimentation involving, 800, 806, 813, 911
    genetic engineering of, 649–57, 672–73, 682–86,
      800
    with HIV, 806
    linguistic skills of, 312, 358, 682, 774
    mentality of, 232, 940–41
    and mindreading, 409, 421, 424–34

and mirror self-recognition, 313, 480–82.
  See also Mirror self-recognition;
  Self-consciousness
moral status of, 11, 23, 139, 183–84, 192, 285–86,
  630–38
personality of, 155–56, 158, 485–86
and personhood issues, 323–24
as pets, 774, 780
pygmy, 358
and quality of life, 187, 192
in research, 214–15, 346, 476, 554, 806, 911
rights of protection for, 205, 214–15, 323–24, 780
in species egalitarianism, 23, 630–34, 636, 638
and theory of mind, 471–76, 480–89, 554
and tool use, 232. *See also* Tools and tool use
Washoe, 636, 638, 774
See also Apes; Cognition; Mindreading; Persons
  and personhood; Theory of mind
Chromosomes, 644, 646, 648, 675, 685, 807
Circuses, 3, 26, 198, 216–17, 658, 738, 763, 944.
  See also Zoos
Civil disobedience, 895
Civil society and its laws, 64, 67–68, 107.
  See also Laws
Clades, 590, 595, 599. *See also* Species
Classification systems for species, 487, 586–87,
  590, 592, 594, 598–99, 605. *See also* Species
Climate change, 613, 620, 807, 828, 840–42, 889,
  924. *See also* Environments
Cloning, 181, 646–48, 655, 671, 827, 832,
  835–37, 847
Cockfighting, 53, 75, 93, 140, 215, 702. *See also*
  Dogfighting
Cockroaches, 17, 43, 442, 446, 464. *See also*
  Minimal minds; Roaches
Coercion, 80, 97, 110, 217
Cognition
  in chimpanzees, 232, 312–13, 375, 380, 421–34,
    554–58, 940–41
  cognitive abilities, 16–17, 20, 146, 233, 407, 422,
    469, 479, 649, 662, 674–75, 679, 687, 819, 894
  cognitive awareness, 498, 509
  cognitive conditions of persons, 308–9, 311,
    313–15, 317–18, 320–22
  cognitive capacities to detect predators, 423–24,
    455–56, 555
  cognitive powers, 20, 497
  emotions and animal cognition, 14–19, 71,
    229–33, 247
  evidence for moral cognition in mammals,
    232–33, 316, 555, 557, 561, 563
  metacognition theory, 9, 103, 145, 150–54, 157–58,
    318, 374–75, 382, 410, 535–42
  perception and, 14, 46, 92, 346, 380–82, 432, 566
  See also Cognitive ethology; Cognitive science;
    Metacognition; Minds; Perception;
    Psychology
Cognitive ethology, 19, 310, 319, 322, 443, 495,
  555, 568, 569

Cognitive science, 16–17, 407–8, 442, 448, 450,
  453, 458, 463–64, 472, 652
Collective behaviors and collective minds, 17,
  442–43, 459–64, 610. *See also* Minimal
  minds
Commercial farming, 119, 123, 128–33, 137, 139,
  176, 180. *See also* Factory farms
Committee review of research
  embryonic stem cell oversight committee,
    678–80, 687
  ethics review, 29–30, 136, 183, 645, 650, 910,
    921, 926
  and government policy and regulation, 645–47,
    686–87, 692–93, 908–10, 920–24, 927–29
  IACUCs (Institutional Animal Care and Use
    Committees), 674, 692, 910–15, 921, 927–28
  See also Government oversight, policy, and
    regulations; Law
Common morality, 585–86, 859
Commonsense morality, 278–81
Communitarianism, 708–9
Companion animals, 110, 216, 290, 297, 299,
  702, 706, 708, 715, 720–21, 950. *See also*
  Pets
Comparative anatomy, 552
Compassion
  and animal mindedness, 17, 443, 464
  animals' capacity to exhibit, 18, 70–71, 233,
    523–24, 542, 567–68
  in Buddhism, 124
  in Hinduism, 124
  Hume's views of, 71–72, 145, 269. *See also* Hume
    and Humean theory
  Kant's views of, 104. *See also* Kant and Kantian
    theory
  Mandeville's views of, 71–72
  and mentality, 233, 464
  Montaigne's views of, 70–72
  and moral status, 17, 145, 299–300, 567–68
  and pain, 443, 884
  in the philosophy of Arnobius, 53
  in Singer's views, 119, 125
  regarding suffering, 70
  in treating animals as moral subjects, 18, 523–25,
    528–29
  and vegetarianism, 129–31, 133. *See also*
    Vegetarianism
  the virtue of, 8, 119, 124–31, 133–34, 140, 212, 277,
    293–96, 299–300
  See also Affects; Emotion; Sympathy; Virtue
    ethics
Competition, 290, 375, 488, 609, 618, 831
Confinement
  in animal research, 555, 715, 739, 747,
    752–53, 763
  animals' rights to be free of, 217–18, 220
  as cause of dependence, 715–18, 722
  DeGrazia's analyses of, 26, 738–67
  of dogs, 216, 746–47, 751, 761–63, 781

Confinement (*continued*)
  of domesticated animals, 25, 712–15, 718, 722,
    739, 754, 762–63, 834
  of elephants, 246–47
  in farming, 26, 218, 654, 716, 739, 743, 748,
    756–58, 763, 834–38, 877–93
  harms caused by, 26, 738–41, 743–63
  liberty interests in escaping, 26, 220, 738–41, 763
  in pig farming, 756, 834–38, 840, 877–78, 881–93,
    898
  standards of justification for, 26, 738, 741–49,
    751–55, 763
  in zoos and aquariums, 200, 246, 717, 739–40,
    744, 752, 758–61, 763. *See also* Zoos
  See also Cages; Confinement; Crates; Dogs;
    Elephants; Factory farms; Harm
Conscience, 53, 208, 682, 860
Consciousness
  and animal welfare, 838–40
  in anthropomorphic views, 470
  and behaviorism, 553
  Cartesian views of, 15, 94, 147, 446, 448, 496
  collective, 462
  Darwinian views of, 454. *See also* Darwin and
    Darwinian theory
  and environmental ethics, 28, 855–57, 874
  in episodic memory, 316–17
  and functionalist-style mental states, 343, 345,
    347, 357, 359, 363, 367
  human brain location of, 681–83
  Kantian views of, 535
  lack of in animals, 15, 94–95, 101, 147, 470,
    496–97, 547, 838–40
  and the Lockean theory of personhood, 306, 318
  and Metacognition, 537–40. *See also*
    Metacognition
  of microencephalic pigs, 839–40
  in mindreading, 407, 419–20
  in the mirror test, 101–2; *See also* Mirror
    self-recognition
  and moral status, 13–16, 262–65, 306, 311, 338–45,
    358–67, 407, 531–32, 547, 570, 652, 704, 840, 857
  as a necessary condition for personhood, 306,
    318, 337–41, 347
  nonverbal, imagistic thinking and, 16, 357–58,
    364–66
  and pain, 382, 496–98, 652, 827, 838–39, 897
  and personhood, 264, 306, 311, 314, 318, 336–48, 365,
  phenomenal, 374–77, 497–98, 532–35, 652
  and psychological unity, 317–18, 366
  self-consciousness, 13, 67, 95, 101–3, 262–65, 306,
    311, 314–18, 336–40, 347–65, 498, 535, 563,
    652, 897
  utilitarian views of, 94–95. *See also*
    Utilitarianism
  See also Self-consciousness
Consent, 7, 30, 109–10, 139, 222, 280, 388, 642,
  658–59, 664, 678, 710, 719, 751, 753, 776,
  785, 787, 920, 926

Consequentialism, 235, 706, 751, 799, 953.
  See also Deontological thinking;
    Utilitarianism
Consequentialist thinking, 97–98, 172–74, 269,
  620, 748, 812, 817–18, 906–14, 926–27, 953
Conservation, 23, 582, 604–7, 611–18, 621
Contraception, 246, 869
Contracts, 16, 39, 218, 266, 349–50, 367, 385–86,
  388–93, 398, 401, 560, 709–11
Contractarian theories (contractarianism)
  and animal domestication, 709–11. *See also*
    Domesticated animals
  and animals' moral status, 266–67
  as a flawed theory of animal rights, 13, 266–67,
    348–50, 560, 709–711
  and indirect duties to animals, 16, 218–19, 394,
    396–99
  and lack of moral standing, 13, 16, 39, 218–19,
    348–50, 359, 367, 373, 390–93, 399, 401
  and moral standing of infants and the senile,
    387–90
  and utilitarianism, 383–87, 399, 401
Contractualism, 16, 266–67, 373, 383–94, 399,
  401. *See also* Contracts; Contractarian
    theories
Conventionalism, 12, 255, 266–69
Cooperation, 97, 101, 109, 152, 155, 161, 245,
  288–90, 375, 401, 482, 488, 554–58, 609, 703
Coping, 137, 450, 457–58, 461, 463
Correlativity of rights and obligations, 9–10,
  104, 198, 206–11, 213–14, 219. *See also*
    Obligations; Rights
Cosmetics and cosmetic surgery, 29, 119, 134–35,
  137, 140, 198, 215, 279, 292, 503, 906–8
Courage, 20, 41, 53, 125–26, 131, 158, 212, 236.
  See also Virtue ethics
Coyotes, 555, 689, 718–19, 721
Crates, 217, 756–57, 781, 834, 877, 940–41.
  See also Confinement; Cages; Pigs
Creationism, 587, 591–92
Crickets, 442
Crops, 41, 50, 676, 842, 889, 892–94
Cruelty
  ancient conceptions of, 39, 230
  in animal enhancement and research, 134–37,
    325, 927
  and animal rights, 79–81, 200–1, 207, 219
  in animal sacrifice, 82
  in arguments for compassionate eating, 883–85
  and chimeras, 325, 674, 689
  of confinement, 738–63. *See also* Confinement
  its effects on society, 297
  in experimentation, 63, 119, 134–39, 674
  in factory farming, 29, 119, 729, 877–83, 885, 891.
    See also Factory farming
  and familyism and speciesism, 122
  and the human neuron mouse, 674
  Kantian views of, 10, 69, 100, 238, 294, 395–99.
    See also Kant and Kantian theory

and killing animals 25–26, 726
legal prohibitions of, 80, 207, 775–76, 813, 816, 927
Mandeville's view of, 71–72
Montagne's view of, 71–72
to pets 27, 132, 219, 769, 775–76, 781, 783. *See also* Companion animals; Pets
and Plato's views of animals, 54–55
in religion and scripture, 54, 77, 80, 82
for the sake of pleasure and entertainment, 39, 75, 813
and utilitarianism, 79, 192, 235, 885. *See also* Utilitarianism
and vegetarianism, 129, 131, 140, 877, 880–83, 885. *See also* Vegetarianism
and virtue ethics 8, 119–20, 124–25. *See also* Virtue ethics
See also Abuse; Neglect of animals; Pain
Crustaceans, 373, 881
Culling, 220, 861–64, 867, 870. *See also* Hunting
Cytoplasm, 645–46

D
Dairy cows, 28, 757, 826, 834–37, 878–79, 883. *See also* Cattle; Dairy products; Livestock
Dairy products, 7, 110, 756–57, 763
Darwin and Darwinian theory
    animal psychology in, 471
    Humean theory as connected to, 8, 151. *See also* Hume and Humean theory
    and minimal-mind theory, 453–62
    the moral status of animals in, 19–20, 522, 547–50, 563–69
    and personhood theory, 310, 326. *See also* Persons and personhood
    responses to, 592–99, 939
    species essentialism considered in, 590–92
    theory of evolution in, 550–60, 576
    the value of species in, 584–90
    See also Evolution; Species
Deafness, 76, 325, 390, 716, 729
Deception, 110, 154, 338, 409, 417, 422–24, 441
Deer, 41, 221, 783, 855, 862–65, 869, 871, 894
Deforestation, 841–42, 845. *See also* Environments; Habitats
Degrees of moral standing, 264, 266, 561, 631. *See also* Moral status
Dehumanization, 400–1, 945, 952
Deities, 78–79, 335. *See also* Gods
Dementia, 350, 478, 485
Democracy, 452, 894
Deontological reasoning
    about animal research, 812, 818–20, 906, 910–13, 926–27
    appeals to respect for animals in, 751–52
    contrasted with utilitarian reasoning, 97, 120–21, 173, 269, 799
    in environmental ethics, 123, 138–39, 141

about genetically modified animals, 642, 661–65
in Kantian theory, 7, 98. *See also* Kant and Kantian theory
about moral status, 120–23, 138–39, 337, 910. *See also* Moral status
in person theory, 337. *See also* Persons and personhood
about rights, 202, 706, 799
side-constraints in, 97–98
and virtue ethics, 120–21, 123, 130, 138–39, 141. *See also* Virtue ethics
See also Kant and Kantian theory
Depression, 180, 657, 730, 877, 891. *See also* Diseases
Derivative value, 578–84, 587, 592–93, 598–99
Developed and developing countries, 135, 777, 813, 842
Developmental abnormalities, 351
Diabetes, 647, 892–93
Dignity, 37, 240, 243, 304, 325, 565, 676–77, 831, 840
Discomfort, 216, 228, 496, 500–1, 732, 740, 756, 760, 915. *See also* Distress; Pain; Suffering
Discrimination, 21, 153, 155, 377, 658. *See also* Prejudice; Racism; Sexism
Diseases
    from confinement, 220. *See also* Confinement
    duty to prevent, 205, 207, 215–16, 621
    from eating meat, 892–93
    and environmental conditions, 139, 842, 844
    in farming, 29, 834–35, 842, 878–79, 887–89, 892–93. *See also* Factory farms
    in pets, 710, 714, 771. *See also* Companion animals; Pets
    and research on animals, 671, 679, 690, 800–6, 808–12, 816, 868, 925. *See also* Experimentation
    resistance to and prevention of, 28, 642–44, 646–50, 659, 662, 664, 827, 837, 844
    and welfare interests, 213, 831, 837
    See also Illnesses
Dishabituation paradigm, 411, 427
Distress, 29–30, 100, 122, 158, 213, 216, 221, 229, 740, 829, 834–38, 912–15. *See also* Pain; Suffering
Docking (of tails), 756–57, 835, 878, 934
Dogs
    brains of, 356
    confinement of, 216, 746–47, 751, 761–63, 781
    dependence on their owners, 771
    experiments on, 62–73, 219, 413
    hunting, 158
    inability to understand human language, 774, 783–84
    See also Companion animals; Domesticated animals; Pets
Dogfighting, 53, 93, 215–16. *See also* Cockfighting

Dolphins, 23, 101, 153, 229, 312–13, 323–24, 346, 364, 367, 410, 478, 510, 636–38, 761

Domesticated animals, 25–26, 37, 44–45, 73, 77, 80, 163, 204, 214–16, 701–22, 739, 743–44, 754, 762, 769, 777–79, 784. *See also* Companion animals; Pets; Wild animals

Domestication, 25, 37, 39–40, 42, 48, 73–74, 163, 246, 701–4, 718–22, 779

Donkeys, 587

Dragonflies, 580

Dreaming states, 463–64, 634–35, 730

Drowning, 216, 389, 397, 632, 689, 706, 717, 815–16

Drug addiction, 502

Drugs, 29, 441, 646–47, 690, 803–4, 806, 829, 906–8, 912–13

Dualism, 36, 338, 342

Ducks, 881

Dying, 48, 179, 462, 733, 868, 879. *See also* Death; Euthanasia; Killing; Letting die

Dynamical systems theory, 812

E

Earthworms, 315–16, 453–54

Ecology
    and animal research, 476
    derivative value in, 583, 595–96
    environmental ethics and, 140, 604, 612–19, 856, 858, 860, 888
    role of farming in, 841, 888
    species considerations in, 22, 587, 589–90, 604–7, 611–19, 629, 708–9, 841

Economic games (and Chimpanzees), 480–82

Economic value of animals, 580, 583, 612–13, 775, 882, 886

Economics, 49, 293, 385, 393, 502, 888

Ecosystems
    and animal rights, 28–29, 855–58
    climate change and, 841
    farming effects on, 841–43
    and hunting, 861–62, 865, 872
    the place of sentient animals in, 123, 139, 855–58, 872–74
    species preservation in, 20, 22, 28–29, 577, 581, 598, 610–13, 617–19
    See also Environments; Habitats; Hunting; Species

Education, 125, 237, 325, 398, 401, 478, 752

Egalitarianism, 23, 617, 628–33, 638–39. *See also* Justice; Species

Electrocution, 216, 934. *See also* Abuses; Pain; Suffering

Elephants, 39, 101, 139, 219, 228–34, 236, 245–47, 313, 510, 542, 682, 759–60, 763, 869–70

Elks, 217, 862. *See also* Caribou

Embryology, 648

Embryos
    admixed, 645–46, 650

and chimeras, 23, 641, 644–46, 648–49, 656, 661, 671–74, 681–92. *See also* Chimeras
    of chimpanzees, 649, 656–57. *See also* Chimpanzees
    cloning of, 655, 836
    destruction of, 655–56, 674, 688, 692, 753
    embryogenesis, 619
    embryonic stem cell research oversight committee, 678–80, 687
    embryonic stem cells, 646–48, 674, 679–81, 683, 686–87
    genetically modified animals (GMAs), 641, 644–56, 658, 661–63, 671–75, 681–92, 716
    of goats, 644
    hybrid, 645–46, 675, 683, 685
    moral status of, 650, 656, 692
    of pigs, 643
    in research and regulations, 645–50, 674–80, 686–92, 753, 911, 922
    of sheep, 644
    in transgenic animals, 643–46
    in xenotransplantation, 64

Emotions
    ancient theories of, 5, 44–48, 230
    and animal cognition, 14–19, 71, 229–33, 247. *See also* Cognition
    in animal welfare, 819, 830, 858
    anthropomorphism about animals', 470–79. *See also* Anthropomorphism
    behaviorists' views of, 553–61
    Darwinian view of, 20, 550–51. *See also* Darwin and Darwinian theory
    of gorillas, 312
    Humean views of, 8, 15, 70, 144–50, 158. *See also* Hume and Humean theory
    in literature portrayals, 943, 949–53
    of mice, 229
    and mindreading, 411, 413. *See also* Mindreading
    and moral status, 390–91, 396–99, 559–63, 568, 798
    and pain, 188, 205, 399, 401, 498, 503, 508–10, 740–41, 898–99, 938
    and personality, 485, 487
    of pets, 782
    and rights of animals to emotional protection, 205, 213, 229, 233, 361
    sentimentalist views of, 147
    utilitarian views of, 188, 401. *See also* Utilitarianism
    virtue ethics appeals to, 127–28. *See also* Virtue ethics
    in zoo animals, 761. *See also* Affects; Psychology; Zoos

Empathy, 158, 229, 233, 307, 489, 522, 542, 555–58, 567, 662, 896, 937

Encephalitis, 647

Endangered species, 383, 583, 613, 835, 855, 857, 859–60, 869–72, 946. *See also* Species

Endocrine system, 804–5

Enhancement, 305, 324–26, 642, 654–58, 664

Entertainment industry, 3, 200
Entitlements, 78, 201, 231, 245. *See also* Animal
   rights; Rights
Environments
   and biotechnology, 28, 827–34, 839–47
   environmental ethics, 4, 20–22, 123, 133, 138–40,
      174, 577, 604, 855–62, 872–73. *See also*
      Environments
   environmental policy, 28, 604, 857, 873–74
   hunting in, 861–65. *See also* Hunting
   and species preservation, 580, 603–20, 871–72
   in zoos, 760–61. *See also* Zoos
   See also Climate change; Farms; Hunting;
      Species; Zoos
Episodic memory, 316–17
Essentialism, 20, 590–93, 598, 606
Ethics committees. *See* Committee review of
   research
Ethnography, 915
Ethologists, 187, 311, 421, 425, 450, 548, 550, 555,
   558, 562–65, 569, 939
Ethology, 65, 477, 483, 490, 505, 939
Eucalyptus forests, 637
Euthanasia, 207, 210–11, 785. *See also* Dying;
   Killing
Evils, 24, 46, 53, 79, 134–39, 178, 184, 280, 642,
   659, 662–63, 817, 885, 895, 936, 949.
   See also Harm; Suffering
Evolution
   of consciousness, 314, 320, 362, 364, 375
   Darwinian, 20, 310, 326, 449, 454, 548–52, 560–61,
      587–88. *See also* Darwin and Darwinian theory
   and domestication, 246
   evolutionary theory, 362, 604, 606, 805, 810, 812
   of language, 312
   and mental life, 13–19, 310, 373, 443–50, 490, 552,
      554–58
   and mindreading abilities, 410. *See also*
      Mindreading
   persons as a product of, 307, 310, 326–27.
      See also Persons and personhood
   of species, 20–23, 52, 587–88, 604–21
   See also Darwin and Darwinian theory;
      Domesticated animals; Wild animals
Exotic animals, 478, 637, 762. *See also* Wild
   animals
Experiments and Experimentation
   anatomical, 62
   on bees, 381, 455–56
   benefits of, 27, 799–803, 810–21
   and bio-Cartesianism, 819–20
   Darwinian views of, 454, 588. *See also* Darwin
      and Darwinian theory
   on dogs, 62–73, 219, 413
   on elephants, 228–36
   on embryos, 671
   on genetically modified animals, 662, 672–93.
      See also Genetic modification
   the justification of, 27, 96, 280, 796, 799, 810–21, 923

medical, 27, 62, 137–40, 219, 221, 279–80, 286,
   292, 389, 632
on mice, 24, 229, 379, 671–93
mindreading, 407–15, 422, 424–32. *See also*
   Mindreading
mirror tests, 101–2, 228–29, 232, 236, 241, 313,
   411–12, 470, 480–84, 564. *See also* Mirror
   self-recognition
moral status, as relevant to, 27, 123, 138–39, 181,
   183, 279–80, 292, 565, 567, 632–33, 653, 813
pain in, 63, 69, 93, 183, 209, 229, 389, 653, 798, 815,
   907–15. *See also* Pain; Suffering
on primates, 153, 303, 346, 408–13, 424–34,
   481–88, 554, 632, 636, 940–41
and quality of life, 286
on rats, 653
religious views of, 62–63, 67
and rights, 73, 81–82, 205, 868
testing involving, 29–30, 119, 134, 137, 198, 215,
   292, 801, 804–7, 810, 906–15
thought experiments, 164, 306, 400, 506, 522,
   598, 712–13, 838, 873, 947
and toxicity testing, 907–15. *See also* Toxicity
   and toxicology
See also Laboratory animals; Laboratory
   research; Pain; Toxicity and toxicology
Exploitation, 40, 120–21, 138–39, 162, 200, 430,
   435, 522, 537, 633, 654. *See also* Abuses;
   Cruelty; Harm; Suffering
Extinct species, 22, 246, 550, 603, 606–8,
   613–16, 619, 835, 841, 846, 860. *See also*
   Species
Extinction, 22, 578, 603, 606–8, 611–21, 841, 889
Extrinsic value, 258, 613, 615, 617. *See also*
   Intrinsic value; Instrumental value

F
Fables, 940
Factory farms
   conditions in, 80, 119, 755–57, 838, 877–98
   confinement on, 26, 738–39, 743, 748, 755–57,
      762–63
   disease on, 887–88
   as environmentally destructive, 29, 877, 888–90
   moral status issues, 123, 180, 292
   rights of animals on, 176, 198, 203, 205, 214,
      217–18
   suffering on, 26, 29, 129, 137, 176, 180–81, 727,
      756–57, 877–98
   virtue ethics analyses of, 128–33. *See also* Virtue
      ethics
   See also Confinement; Crates; Farms; Harm;
      Pain; Suffering; Vegetarianism
Fairness, 165, 385, 482–84, 555–58, 603, 790.
   See also Justice
False belief tasks and tests, 415, 417–18,
   425–28, 435
Familyism, 121–22

Farms
    confinement on, 26, 738, 755, 761, 763, 834
    and the environment, effects on, 842, 844–45, 858
    fish, 846, 880
    humane, 26, 727, 758
    suffering on, 26, 29, 129, 137, 176, 180–81, 727,
        756–57, 877–98
    traditional, 26, 176, 738, 743, 757–58, 763
    See also Domesticated animals; Factory farms
Fawns, 864
Fear
    and animal welfare, 829, 835, 838
    anthropomorphic conceptions of, 17, 469
    as emotion, 233, 413
    evolution of, 92
    and mindreading, 413, 416
    in wild and domesticated animals, 702–3, 779
    See also Affects; Emotions; Evolution;
        Mindreading
Feral animals, 49, 133, 637, 704, 720, 777,
        843–45, 860. *See also* Wild animals
Fertilizers, 842, 844, 889. *See also* Agriculture
Fetuses, 255, 261, 387–88, 647, 655–56, 672–73,
        680, 685, 836
Fighting. *See* Cockfighting, Dogfighting
Fish, 47, 130, 340, 412, 442, 582–83, 598, 744,
        758, 826, 829, 836, 844, 846, 880–81,
        888–90, 896–97, 939. *See also* Fish
        farming; Fishermen; Goldfish; Jellyfish;
        Rockfish; Swordfish
Fish farming, 130, 880, 897. *See also* Farms;
        Fish; Fishermen
Fishermen, 47, 93, 198, 324, 400, 442, 580, 880, 890
Folk psychology, 436–37, 479–80, 487. *See also*
        Psychology
Forests, 634, 702, 890. *See also* Deforestation
Fossils, 611, 615, 620, 889
Fowls, 596, 598, 837. *See also* Birds; Chickens
Foxes, 133
Frogs, 445
Fruit flies, 316
Functionalist-style beliefs and desires, 342–47,
        356–59, 361, 364–66
Fur, 140, 152, 703, 714

G
Games, 45, 53, 202, 230, 471, 482, 489, 661
Gametes, 23, 641, 672, 675, 679, 681, 684–86,
        688, 690, 692
Garden pests, 400
Gazelles, 634–35
Geeps, 644
Gender, 21, 393, 476. *See also* Sex; Sexism
Genes, 23, 324, 595, 605–6, 641–46, 650, 660–64,
        672–73, 683, 686, 808, 810, 835–38, 846, 887
Genetic modification
    of cells, 641–64
    using chickens, 28, 826, 837
    using chimpanzees, 649–57, 672–73, 682–86, 800
    experiments involving, 662, 672–93
    to generate knowledge, 24, 647, 659
    genetically modified animals (GMAs), 28, 641,
        644–56, 658, 661–63, 671–75, 681–92, 716
    genetic alteration in research, 648, 651, 677–78,
        689–92, 714
    genetic engineering, 181, 656, 687–88, 691,
        835–39, 847
    using horses, 683
    involving pets, 662, 683, 689
    killing of animals after, 650–52, 660
    laws governing, 649–50, 663, 674–75
    using monkeys, 657, 663, 681, 687, 689
Genocide, 897
Genotype, 589–90, 595
Genus, 241, 578, 586, 946
Gestation, 217, 650–51, 685, 756, 834, 838, 877.
        See also Crates
Goal-directed action, 433–37
Goats, 643–44, 943
Gods, 36–37, 39–43, 47–50, 52–54, 334, 389, 554,
        687, 952. *See also* Deities
Golden retrievers, 761, 780–81
Goldfish, 744–45
Gorillas, 102, 180, 312–13, 425, 480–82, 484,
        682–83, 798, 806. *See also* Apes
Government oversight, policy, and
        regulations, 30, 219, 908–10, 920–24,
        927–29. *See also* Committee review of
        research
Great apes, 14, 91, 189–90, 214–15, 221, 313, 318,
        322–23, 409–10, 425–26, 481, 485, 683, 709.
        See also Apes; Primates
Great chain of being, 65, 75–76, 79
Greenhouse gases, 841, 845, 889. *See also*
        Environments
Grey junglefowl, 588–90
Grief, 229, 233–36, 246, 390, 800
Guilt, 61, 100, 233, 242, 391, 401, 474, 479, 896,
        915, 936
Guinea pigs, 800, 907
Guppies, 155

H
Habitats
    destruction of, 25, 133, 139, 212, 215, 582, 610,
        620, 702, 713–14, 719–22, 769, 842,
        862–65, 871
    of elephants, 246–47, 869. *See also* Elephants
    natural study of, 555, 557, 562
    preservation of, 200, 217, 582, 607, 610, 769, 862, 871
    in zoos, 759–62. *See also* Zoos
    See also Ecology; Ecosystems; Environments;
        Habitats
Happiness, 18, 65, 68, 74–79, 106, 110, 173, 201, 235,
        284, 385, 470, 528–29, 551, 785–86, 829, 862–66.
        See also Pain; Pleasure; Utilitarianism

Hares, 41, 81

Harm
and animal rights, 179, 205–10, 212–22, 707–14, 717–22
to animals kept as pets, 769–90. *See also* Pets
to chickens on factory farms, 217–18, 654, 756, 835, 837–40, 879–81, 883, 886, 896. *See also* Chickens
from confinement, 26, 739–45, 761–64
to the environment, 840–48, 852–58, 890
from farming practices, 29, 755–61, 891–93
genetic engineering productive of, 325, 642, 655–65, 832–36
human-caused to animals, 10, 28, 31, 200, 257–59, 606–13, 620, 676–78
in Humean thought, 163–65, 269–72. *See also* Hume and Humean theory
due to hunting, 862–69
to laboratory animals, 677–78, 710, 744, 914–15
misery as a form of, 74–75, 79–80, 92, 162–64, 214, 551, 858
in a moral sense, 80, 155, 392–401, 726–36, 745–55, 813–17
normative conceptions of, 746–47, 754
obligations to avoid causing, 25–26, 72, 75, 164, 200–1, 213–14
of subjects during experiments, 247, 796–809, 913–15, 920–29
in utilitarian theories, 93–100. *See also* Utilitarianism
See also Animal rights; Cages; Confinement; Factory farms; Hunting; Nonmaleficence; Obligations; Suffering

Herds, 163, 217, 220, 247, 643, 834, 861–65, 871
Hermit crabs, 376
Higher-order thinking, 374–75, 532, 535–39, 541. *See also* Mindreading
Hives, 381, 454–61. *See also* Bees; Honeybees
Hog farms, 217, 878, 881, 887. *See also* Factory farms; Farms; Pigs
Hogs, 175, 235, 758, 878, 881, 887, 889, 891. *See also* Pigs; Hog farms
Honeybees, 17, 175, 380, 442–43, 445, 447, 454–59, 461–63, 797. *See also* Bees
Hormones, 804–5

Horses
and the Clever Hans effect, 415
confinement of, 738, 757, 763
domestication of, 246, 715, 860
and equestrian arts, 80
and genetic modification, 683
and hybrids, 644
moral status of, 263
physiology of, 80
and rights, 78, 80, 198, 207–9
and societal values, 297
views of, in ancient times, 41, 44, 46–47, 51–52

Hospitality, 36, 39–40, 104–5. *See also* Altruism; Benevolence

Human dignity, 304, 565, 675–78, 681, 687–88. *See also* Dignity; Persons and personhood

Human embryonic stem cells, 646, 674, 680–83, 687, 691–92. *See also* Stem cells
Human genome, 324, 644, 676
Human rights, 5, 11, 198–205, 211–14, 217, 221, 304, 323–24, 651, 683, 717, 866, 869
Humane slaughter, 75, 81, 879. *See also* Slaughter; Slaughterhouses

Hume and Humean theory
account of agency, 9, 15, 128, 145, 148, 150–53, 155–59
on animal welfare, 146
arguments against, 78–80, 128
on compassion, 71–72, 145, 269
on emotions, 8, 15, 70, 144–50, 158
on harm to animals, 163–65, 269–72
human and animal psychology in, 146–57
on the mental life of animals, 15, 144–50, 152–58, 162
on the place and standing of animals, 8–9, 15, 66–72, 144–66, 266–72, 334
on sympathy, 149–50, 157–58, 162–65
theory of justice, 71–72, 144–46, 152, 159–66, 266–71
as virtue theory, 147–62, 165, 172
See also Virtue ethics; Moral sense theory

Hunger, 92, 162, 710, 740, 895

Hunting
to control wildlife populations, 220, 855, 860–64, 867, 870
dogs used in, 158
sport, 41, 75, 93, 139, 220, 580–81, 861–62, 867, 870
subsistence, 861–62, 867
techniques of, 164
therapeutic, 861–71
See also Culling; Wild animals

Husbandry, 26, 40, 485, 739, 743, 752, 755, 758, 763

Hybrids, 21, 23, 40, 43, 49–52, 230, 504, 565, 587–91, 641, 644, 647, 671–76, 685–86, 830–31, 839. *See also* Chimeras; Genetic Modification

Hyenas, 942–43
Hypertension, 802–3, 893
Hypothermia, 662, 863

I

IACUCs (Institutional Animal Care and Use Committees). *See* Committee review of research; Government oversight, policy, and regulations

Iguanas, 762
Illnesses, 180, 188, 190, 351, 503, 734–35, 831. *See also* Diseases
Imagistic thinking, 14, 357–60, 364–68. *See also* Apes; Cognition; Language; Psychology
Immune system, 441, 651, 671–72, 681, 836, 878
Impartial points of view, 5, 39, 155, 172, 629, 921, 937, 939

Imprisonment, 740, 772, 780. *See also* Cages; Confinement; Crates

Inalienable rights, 261. *See also* Animal rights; Rights

Incest, 39, 383, 400

Indignation, 391, 401

Indirect duties, 16, 69, 157, 238, 373. *See also* Kant and Kantian theory

Individualism, 857, 860, 896

Infanticide, 389. *See also* Euthanasia; Killing

Infections, 42, 647, 672, 777, 804, 835, 878, 880, 888. *See also* Diseases; Illnesses

Inherent value. *See* Instrumental value; Intrinsic value

Insects, 151, 232, 257, 373–74, 378, 380, 384, 400, 421, 454, 459, 609–10, 726, 734, 820, 948–49

Instincts, 102, 151, 155, 159, 319–20, 447, 453, 535, 769, 770, 777, 780, 834, 879, 950

Institutional rules, 202, 211, 219

Instrumental value, 21–22, 28, 258–59, 578–83, 604, 611–17, 620, 634, 652, 743, 843, 847, 855–57, 871–73. *See also* Intrinsic value

Instrumentalism, 451, 457, 612–13, 617, 928

Integrity (moral), 136–37, 173, 567, 885

Intelligence
    in ancient theories, 5, 38, 44, 46–47
    in animal psychology, 317–22
    and animal welfare, 838
    and the argument for compassionate eating, 884
    artificial, 326, 683–84, 691
    brain location of, 682–83
    and chimeras, 683–84, 690–91, 693
    and confinement, 760, 780
    Darwinian views of, 20, 453–56, 552. *See also* Darwin and Darwinian theory
    of elephants, 233, 246. *See also* Elephants
    historically important concepts of, 35–38, 46–50, 101–3, 151
    in Humean theory, 151. *See also* Hume and Humean theory
    intellectual equality, 162
    intelligence hypothesis, 410
    in Kantian theory, 101–3. *See also* Kant and Kantian theory
    and minimal minds, 451. *See also* Minimal minds
    and the moral status of animals, 10–15, 147, 393, 410, 563, 566, 638
    and reason, 101–3
    its role in personhood theories, 233, 239, 305–11, 315, 318, 326, 347
    and self-consciousness, 239, 318
    tests of, 564, 684
    See also Cognition; Minds; Psychology

Interbreeding, 587, 596, 605–6. *See also* Breeding; Captive breeding; Reproduction

International rights, 5, 199, 211. *See* Animal Rights; Human Rights; Rights

Inter-temporal calculations, 496, 501, 503, 505, 507

Intransitive consciousness, 537–39

Intrinsic value, 28–29, 123, 140, 192, 239, 258–59, 284–87, 307, 337, 361, 384, 578, 609, 611, 843, 847, 855–57, 873–74. *See also* Extrinsic value; Instrumental value

Introspection, 375, 475, 497, 505

Invertebrates, 17, 340, 376, 380–83, 401, 442–43, 446–49, 451–54, 457–58, 464, 610, 761, 881

Isolation, 218, 220, 236, 325, 587–88, 595–98, 948. *See also* Cages, Captivity; Confinement; Crates

J

Jackals, 53

Jealousy, 46, 551

Jellyfish, 643, 657, 692

Jus animalium, 80

Justice
    in ancient philosophy, 37, 45–50, 55
    and animal research, 926. *See also* Experimentation
    in contractualism, 385–86, 397
    in the history of philosophy, 37, 45–50, 55, 71–80
    in Hume's theory of, 71–72, 144–46, 152, 159–66, 266–71. *See also* Hume and Humean theory
    and the moral status of animals, 12, 212, 240, 255, 267, 269, 271, 384–86, 549–58, 566–68, 705–19. *See also* Moral status
    Rawls's theory of, 231, 243–44, 266, 349–50, 385, 388, 393
    and rights, 80, 212, 717. *See also* Rights
    social justice, 6, 566, 714
    the virtue of, 120–34, 926, 953
    See also Fairness

K

Kangaroos, 740

Kant and Kantian theory
    on animal thinking, 230–31, 237–39, 241–44
    on animals as moral subjects, 528–29, 531, 535–36
    on autonomy, 103, 105, 108, 132, 186–87, 222, 238–39. *See also* Autonomy
    on categorical imperatives, 99, 102, 105, 107
    by contrast to Humean theory, 157–58, 161–62
    on cruelty, 10, 69, 100, 238, 294, 395–99. *See also* Cruelty
    on interacting with animals, 7, 91, 93, 97–111
    on kindness and compassion, 294. *See also* Compassion
    Korsgaard on, 91–117
    on the moral community, 191
    on moral status, 258, 261–64, 548, 565
    on normative self-control, 523–27
    on obligations, 69. *See also* Obligations

perspectives on killing, 7, 97–110
on respect for animals, 752
views of agency, 69, 103, 128, 132, 157–58, 284, 367, 525, 527, 548
views of consciousness, 535
in virtue ethics, 127–28, 132. *See also* Obligations; Virtue ethics
on welfarism, 284
Kidneys, 135, 674, 681, 684, 816
Killing
and animal rights, 201, 206, 216, 220–22
of animals, 5, 23–29, 130–33, 256–57, 280–89, 726–36, 797–98, 813–16, 893–94, 933–34
duty not to kill, 296–99
early historical attitudes toward, 39–42
of elephants, 245–47. *See also* Elephants
of genetically-modified animals, 650–52, 660
Humean perspectives on, 159–66
Kantian perspectives on, 7, 97–110
and moral agency, 351–54
of pets, 785–88. *See also* Pets
utilitarian perspectives on, 93–97. *See also* Utilitarianism
See also Euthanasia
Kinship, 21, 239, 241, 560, 567, 708, 945
Kittens, 720, 781. *See also* Cats
Knowledge
of animal folk-experts, 478–80, 484–85, 487–88, 490
from animal research, 135, 567, 802, 806, 915–16, 923, 925, 928
capacity to attribute, 375, 428–32, 436
as a form of cognition in animals, 14, 154, 232, 461
from genetic modification science, 24, 647, 659
of hypothetical agents in contractualist theories, 385, 388, 392–93
and literary insights, 937–938, 944
of the mental lives of animals, 10, 70, 80, 180, 187, 230, 442–43, 559, 761
states of, 375, 429, 431, 436
See also Cognition
Koko the gorilla, 312, 481

L
Laboratory animals
chimpanzees as, 482. *See also* Chimpanzees
cruelty toward, 207, 217, 741, 798–99. *See also* Cruelty
harms to, 677–78, 710, 744, 914–15
moral status of, 123, 184, 548, 565, 677, 741, 744. *See also* Moral status
mammals as, 763, 798, 813
obligations to care for, 714, 816. *See also* Obligations
and pain, 748, 798–99. *See also* Pain Experimentation

Laboratory research
and animal psychology, 476. *See also* Minds; Psychology
chimeras in, 648, 677–78, 689–92. *See also* Chimeras
as a constraint on liberty, 217, 741
efficacy of, 476, 478, 482, 562, 564–65, 690, 804–6, 809–10
genetically altered animals in, 648, 651, 677–78, 689–92, 714
Kewalo Basin Marine Mammal Laboratory, 478
moral assessment of, 741, 798
regulation of, 909, 923
and toxicity testing, 907–16. *See also* Toxicity and toxicology
value of, 800, 907, 913
See also Experimentation; Committee review of research; Government oversight, policy, and regulations
Lactation, 892
Land quadrupeds, 52
Language
in ancient thought, 5, 49
barriers between animals and humans, 548, 553, 783–84
and communication, 15, 19–20, 262, 444, 569
and conceptual representations, 456–58, 485
as a condition for mindreading, 16, 408, 417, 419–20, 422, 425, 428–29, 431, 434, 436. *See also* Mindreading
as a condition for moral behavior, 523, 547–48, 562–63, 569–70
as a criterion of moral status, 14–16, 19–20, 67–68, 72, 93, 262–64, 355, 366–68, 547–48, 562–63, 569–70, 636. *See also* Moral status
and intentional states, 15, 355–57, 378, 417, 446–48, 457–58, 471, 500, 562–63, 570
and mentality, 311–13, 346–47, 443–44, 446, 457–58, 462–64, 471, 475–76, 496, 547
and minimal minds, 451, 456–58, 462–64. *See also* Minimal minds
and mindreading, 408–43. *See also* Mindreading
and personhood theory, 14, 264, 307, 311–13, 325–26, 339–40, 346–48, 367–68. *See also* Persons and Personhood
in primates, 65, 355, 358, 367, 481, 569, 636, 682, 774
and self-awareness, 102, 311–13, 317, 348, 365, 498. *See also* Self-consciousness
sign language, 312, 355, 358, 367, 481, 636, 682, 774
study of, 649, 690
of thought, 312, 325–26, 357–58, 366–68, 451, 462–63, 471–72, 476, 553
and time-awareness, 317, 365, 443, 498–99
as a unique feature of humans, 5, 49, 65–66, 68, 93, 101, 523, 649, 705, 950–52
Large offspring syndrome, 836
Laws
animals as subjects of, 67–68, 773
civil, 64, 67–68, 78–81, 107

Laws (*continued*)
  environmental, 604, 618
  general animal welfare, 50, 54–55, 217, 245, 633,
    756, 771, 775–77, 881
  governing genetically modified animals, 649–50,
    663, 674–75
  of humanity, 71–72, 159, 161–64, 269, 271
  and legal standing of animals in, 207, 323, 709, 775
  moral, 99, 102–9, 222, 238–39
  natural, 6, 48, 62, 64, 68–70, 74, 77–79, 161–62,
    199, 269–71, 606–7
  rights in, 62, 201, 206, 267
  regulating the use of animals in research, 214,
    802, 816, 911
  on the use of great apes, 214
  See also Committee review of research;
    Government oversight, policy, and
    regulations
Letting die, 868–70. *See also* Euthanasia;
  Killing
Liberation. *See* Animal liberation
Liberty. *See* Animal liberation; Autonomy;
  Cages; Confinement
Limbic system, 377, 553–54
Lineages, 22, 52, 549, 554, 561, 604–9, 612–16, 619–21, 808
Linguistic behavior, 346, 356, 474–75. *See also*
  Language
Lions, 52–53, 66, 71, 182, 555, 634–35, 644, 689,
  707, 749, 780, 871
Literary representations of animals
  as animals view humans, 947–52
  in Coetzee's work, 30, 932, 940–42, 953
  forms of description in, 938–40
  literary-oriented animal ethics, 30, 932–33, 953
  the mentality of apes, 940–42
  and moral imagination, 934–38, 953
  seeing human traits as animal, 942–46
  in slaughter and predation, 933–38, 944–45, 947, 949
  Zamir on, 932–55
  in zoos, 942–46. *See also* Zoos
Livestock, 28, 123, 688, 777, 826–27, 830–36,
  840–47, 889. *See also* Cattle; Dairy cows,
  Dairy products
Lobsters, 175, 257, 881
Logical behaviorists, 338, 342. *See also*
  Behaviorism
Love, 20, 47–50, 100, 124, 128, 131–33, 140, 150, 154,
  158, 208, 236, 246, 401, 773, 780, 950–52

M
Macaque monkeys, 153, 425. *See also* Monkeys
Mad cow disease, 140
Magnanimity, 551
Malaria, 110, 808
Malnutrition, 863–64. *See also* Hunger
Mammals
  brain structure of, 554, 559, 561, 563, 682–83, 819.
    See also Brains

  capacity for emotion in, 233, 561, 563
  capacity for pain experience and suffering in,
    377, 499, 564, 798, 827
  in confinement, 759, 763, 779. *See also*
    Confinement
  consciousness in, 318, 340, 362–65, 499, 564
  evidence for moral cognition in, 232–33, 316, 555,
    557, 561, 563
  intentional states in, 314, 362–65, 380
  Kewalo Basin Marine Mammal Laboratory, 478
  as laboratory animals, 763, 798, 813. *See also*
    Laboratory animals
  moral status of, 214, 314, 340, 362–63, 365, 401,
    557, 559, 798, 813. *See also* Moral status
  self-consciousness in, 232–33, 316, 318, 363–65,
    499, 510, 564
  similarities between humans and other, 596,
    682–83, 800, 808–10, 819, 896
  time-awareness in, 318, 499, 510
  tool use in, 232. *See also* Tools and tool use
  and the triune brain hypothesis, 554
  in zoos, 759. *See also* Zoos
Marmosets, 680
Mates, 458, 484, 560, 588, 713
Mating, 234, 323, 443, 450, 558, 588, 644, 684,
  686, 829, 846, 870, 939. *See also* Breeding
Matriarchal family units, 869
Maximal minds, 443. *See also* Minimal minds
Meat industry, 690, 882–83, 886, 893, 895.
  See also Factory farms; Vegetarianism
Medical ethics, 24, 124, 187–88, 868
Medical experimentation, 27, 62, 137–40, 219,
  221, 279–80, 286, 292, 389, 632. *See also*
  Experimentation; Laboratory research
Meerkats, 555
Memory, 14, 36, 40, 45, 54, 94–95, 306, 316–17,
  343, 346, 357, 361, 366, 396, 407, 498,
  508–10, 521, 662, 682, 798
Mental illness and mental handicaps, 203, 387,
  660, 833
Mental lives of animals
  access to (knowledge of), 15–18, 312, 475, 552, 555, 761
  compared with the mental life of humans, 20,
    65–66, 69, 144–50, 162, 165, 550, 566
  and consciousness, 314–16, 338, 344–45, 347,
    357–59, 367, 374–75, 535, 537–40
  evolution of the, 552
  Hume on the, 146–50, 152–54, 156–58, 162.
    See also Hume and Humean theory
  and language, 312, 346–47, 357, 365–68, 443–44,
    451, 457–58, 476, 555. *See also* Language
  and metacognition, 103, 150–54, 157–58, 318,
    374–75, 382, 410, 535. *See also* Cognition;
    Metacognition
  and moral agency, 65–67, 152–53, 156–58, 319, 352–53,
    558–62, 564. *See also* Agency; Moral agency
  and moral status, 8, 14–16, 19, 333–35, 339,
    352–53, 360, 367–68, 373, 547. *See also* Moral
    status

and personhood, 13, 309, 311–13, 319, 332–36, 338–40, 344–45, 347, 367–68, 652. *See also* Persons and Personhood
and self-awareness, 103, 312, 340, 348, 358–59, 365, 375, 410, 554
sophisticated, 726, 734, 741, 819
See also Consciousness; Mental States; Minds; Psychology; Self-consciousness
Mental states
collective, 17, 459, 462–63
Descartes on, 15, 147, 445–49
folk ascriptions of, 441–44, 454, 463, 469–70, 473, 480
functionalist analysis of, 342–47, 356, 363, 444
minimal, 441, 450–51, 458, 462–64
negative or painful, 497, 507, 649, 652, 657, 827, 829, 831, 839
non-embodied, 334
positive or pleasurable, 653, 827, 829, 839
representing mental states of others (mindreading), 407, 409–10, 415, 420, 424
scientific evidence for, 231, 378, 380, 552–56
See also Mental life of animals; Minds; Minimal minds; Psychology
Mentalistic views of welfare, 829–31, 834, 838–40, 847
Mentality
of animals in agriculture, 827, 829–30. *See also* Agriculture
of apes, 940. *See also* Apes
of chimeras, 649, 653, 657. *See also* Chimeras
Darwinian view of, 454. *See also* Darwin and Darwinian theory
Humean views of, 144–50, 152–54, 156–58, 162. *See also* Hume and Humean theory
in mentalistic views, 829–31, 838–39
and mindreading, 407–37. *See also* Mindreading
and minimal minds, 441–64
and moral agency, 351–54, 356–59. *See also* Agency; Moral agency
and moral status, 15–19, 359–60, 365–68, 373, 547, 556, 560, 566, 652. *See also* Moral status
and pain and suffering, 497, 507, 652. *See also* Pain; Suffering
and personhood criteria, 13, 309–19, 332–48. *See also* Persons and personhood
in psychology, 473–76. *See also* Psychology
and self-consciousness, 103, 363, 365. *See also* Self-consciousness
and theory of mind, 489. *See also* Theory of Mind
See also Apes; Language; Mental life of animals; Minds; Minimal minds; Mindreading; Philosophy of mind; Psychology; Theory of mind
Mentation, 20, 144, 147, 166
Metabolism, 608, 646, 809, 812, 820, 908
Metacognition, 9, 103, 145, 150–54, 157–58, 318, 374–75, 382, 410, 535–42. *See also* Cognition

Metaphysical views, 66, 76, 99, 106, 231, 239, 242–43, 321–22, 363, 378, 382, 448, 586, 604, 606, 731
Metapopulations, 606
Metarepresentation, 16, 408–10, 418–19, 421–22, 435–36
Metempsychosis, 51
Mice
ethical issues in scientific research using, 64, 181–82, 184, 632–33, 674, 714–15, 744–45, 770, 803–4
the human neuron mouse, 24, 672–75, 678, 682–83, 690–93
moral status of, 23, 181–87, 245, 247, 500, 632–34, 638, 652. *See also* Moral status
and pain, 229, 247, 500, 649–53, 674, 692–93. *See also* Pain; Suffering
as used in chimera research, 644, 648–51, 661, 673, 676, 681–82, 685–89, 692–93, 714–15. *See also* Chimeras
as used in scientific research, 81–84, 229, 247, 379, 555, 632–33, 643, 650, 660, 671, 692, 714, 744–45, 770, 803–11, 838
Microencephalic pigs, 838–40. *See also* Hogs; Pigs
Microorganisms, 28, 826
Migration, 618, 706, 841
Mind-independence of value, 384, 401
Mindreading, 16–17, 375, 407–22, 424–26, 428–37, 473, 487–89, 554
Minds
access to animal, 15, 159, 180, 310, 443, 474–75, 479, 549, 555, 562
animal and human, 310, 382, 442, 448, 450, 464
animal minds and moral status, 3–4, 6, 8, 14, 16–17, 19, 145, 148, 245, 547. *See also* Moral status
animal minds and personhood, 311, 327
animal minds with intentional attitudes, 380, 451
and bodies, 71, 159, 335–36, 338, 447, 619, 948. *See also* Brains
Darwin on animal, 310, 453–54, 550–51. *See also* Darwin and Darwinian theory
Descartes on animal, 310, 447–49, 819
Hume on animal, 15, 146–48, 154–56. *See also* Hume and Humean theory
and language, 16–17, 444, 474–75. *See also* Language
mind-brain identity theorists, 338
minimal, 17, 441–45, 449–54, 456, 458–59, 461–64. *See also* Minimal minds
philosophy of, 128, 189, 310–11, 338, 342, 363, 416, 442, 495, 497
scientific evidence about, 310, 443, 454, 456, 461–64, 568
self-consciousness in, 311, 319. *See also* Self-consciousness
See also Autonomy; Brains; Cognition; Philosophy of mind; Philosophy of biology; Psychology; Theory of mind

Minimal minds, 17, 441–43, 450–53, 458–59, 462, 464. *See also* Minds; Psychology

Miniride principle, 866–69

Miracle of the meta, 537, 539, 541–42. *See also* Cognition; Metacognition

Mirror self-recognition, 101–102, 228–29, 232, 236, 241, 313, 411–12, 480–82, 564. *See also* Self-consciousness

Misery, 74–75, 79–80, 92, 162–64, 214, 551, 858. *See also* Cruelty; Discomfort; Harm; Pain; Suffering

Molecular clocks, 808

Mollusks, 615

Monkeys

behavioristic accounts of, 471

Capuchin, 471, 780

confinement of, 740

Darwinian views of, 550. *See also* Darwin and Darwinian theory

and genetically altered animals, 657, 663, 681, 687, 689

Macaque, 153, 425

and mindreading, 375, 413, 425. *See also* Mindreading

in research, 153, 182, 413, 433, 471, 636, 692, 804. *See also* Experimentation

rights of, 207

and tool use, 232. *See also* Tools and tool use

See also Animal rights; Apes

Monsters, 40, 43, 45, 591, 950

Moose, 865, 948. *See also* Elks

Moral agency

and acting for reasons, 127, 521, 530

as a criterion for personhood, 311, 320–22, 367

empirical evidence for in animals 549, 561, 564, 567–68

the Humean account of, 148, 150, 152–53, 155–59. *See also* Hume and Humean theory

intentionality as a condition of, 566

and killing, 352–54

language as a condition of, 548

metacognition as a condition of, 9, 103, 145, 150–53,157, 527, 559

moral motivation and, 19, 148, 520–22, 530

and moral responsibility, 188, 519–521, 549, 564, 636, 707, 868, 871

and moral status, 11, 13, 124, 209, 222, 348, 351–60, 367, 652, 705

See also Agency; Autonomy

Moral anthropology, 873

Moral codes, 185, 288–97, 299, 557

Moral communities, 15, 19, 74, 77, 178, 184, 188–92, 260–61, 548, 556–61, 566–69, 859

Moral deliberation, 220, 532, 534, 560, 563, 579, 913

Moral dilemmas, 25, 127, 791

Moral ideals, 208

Moral intuitions, 73, 353–54, 585, 604, 873

Moral judgment, 15, 21, 147, 152, 155–59, 281, 288, 291, 322, 563, 603, 620, 661, 819

Moral laws, 105–8, 222. *See also* Categorical imperatives

Moral patients, 19, 155–56, 519, 548–49, 559, 561, 564–68, 604, 610

Moral philosophy, 5–9, 24–25, 30, 105, 120, 125–26, 138, 174, 182, 189, 199, 202, 208, 212, 270, 932, 937, 940

Moral psychology, 128, 189, 391, 401, 612. *See also* Psychology

Moral rights, 6, 8, 16, 181, 199, 202, 206, 218, 260–61, 323, 332, 351–53, 373, 859, 866. *See also* Animal rights; Human rights; Rights

Moral sense theory, 47, 54, 67, 74–75, 78, 148, 153, 157, 551, 610. *See also* Hume and Humean theory

Moral status (moral standing)

in the absence of agency, 9, 16, 19, 148, 159–60, 174–75, 184, 191, 519, 559, 564

agency as a condition of, 11–19, 124, 184, 260–66, 311, 320, 519–22, 527, 548–49, 652, 705

anthropocentrism and views of, 548–49, 564–69, 843

of apes, 14, 322–23, 637. *See also* Apes

autonomy and, 11, 103, 122–23, 311, 320–21, 323, 339–40, 388–89, 652, 705, 709, 753, 790. *See also* Autonomy

of cats, 73, 306, 394–98, 704–5, 726, 733–34. *See also* Cats

of chimpanzees, 11, 23, 139, 183–84, 192, 285–86, 630–38. *See also* Chimpanzees

in Darwinian theory, 19–20, 522, 547–50, 563–69. *See also* Darwin and Darwinian theory

deontological reasoning about, 120–23, 138–39, 337, 910

of embryos, 650, 656, 692. *See also* Embryos

emotions and, 549, 551, 554–56, 568

intelligence as a basis of, 10–15, 147, 393, 410, 563, 566, 638

issues about, on factory farms, 123, 180, 292

justice and, 12, 212, 240, 255, 267, 269, 271, 384–86, 549–58, 566–68, 705–19

of laboratory animals, 123, 184, 548, 565, 677, 741, 744

language as a criterion of, 14–16, 19–20, 67–68, 72, 93, 262–64, 311, 339–40, 355–56, 366–68, 547–48, 562–63, 569–70, 636, 705

of mammals, 214, 314, 340, 362–63, 365, 401, 557, 559, 798, 813

mental life as a source of, 3–4, 6, 8, 14–19, 145, 148, 245, 333–35, 339, 352–53, 359–60, 365–68, 373, 547, 556, 560, 566, 652

of mice, 23, 181–87, 245, 247, 500, 632–34, 638, 652

in person theory, 10–14, 121–23, 253–65, 304–5, 307–8, 320–24, 336, 339, 609–10, 652, 677, 731, 742, 750–55, 906

of pets, 73, 123, 132–33, 158, 290, 297, 391–92, 397–400, 567, 769–70, 894, 927, 951

as relevant to use of subjects in experimentation, 27, 123, 138–39, 181, 183, 279–80, 292, 565, 567, 632–33, 653, 813

role of consciousness in, 13–16, 262–65, 306, 311,
338–45, 358–67, 407, 531–32, 547, 570, 652, 704,
840, 857
See also Animal rights; Domesticated animals;
Persons and personhood; Pets; Wild animals
Moral traditions, 258, 262, 269, 523
Morbidity, 246, 831, 834. *See also* Death; Dying;
Euthanasia; Killing
Morphine, 377
Morphology, 605, 838
Mortality, 834, 836, 869, 887. *See also* Death;
Dying; Euthanasia; Killing; Letting die
Mothers, 35, 54, 156, 159, 191, 389–90, 488, 746,
756–57, 834, 869, 877–78, 943
Motives, 44, 103, 109, 153, 156, 161–63, 256, 257,
392, 519–21, 528, 559, 628, 757, 944
Motivation
agency and moral motivation, 19, 148, 520–22, 530
to avoid pain, 376–77
in bees, 381–82. *See also* Bees; Honeybees
in contractualist theory, 390, 399
as a criterion for moral evaluation, 18, 153,
161–64, 396, 399, 519–21, 525, 527–28
moral, 18, 44, 396, 519–25, 528–32, 534–37,
540–42
and moral-status accounts, 386, 562
normative force of, 529–30, 536–37, 539–42
sources of moral, 44–45, 109, 147, 383, 386, 522
in utilitarian theory, 383. *See also* Utilitarianism
Motor neuron disease, 647
Mountain goats, 865
Mourning doves, 862
Mules, 587, 644, 676, 865
Muscles, 41, 646, 649, 674, 836, 887
Mythology, 644

N
Nanotechnology, 653, 664
Natural affections, 47
Natural functioning, 830, 840
Natural habitats, 228, 232, 555, 557, 759–60.
See also Environments; Habitats
Natural kinds, 22, 52, 265, 591–92, 604, 606–7
Natural law, 6, 62, 64, 68–70, 74, 77–79, 161–62,
199, 269–71, 606–7
Natural rights, 5–6, 61–62, 64–65, 67, 72, 75, 78,
199. *See also* Animal rights; Human rights;
Rights
Natural selection, 449, 550, 612, 618, 809.
See also Darwin and Darwinian theory;
Evolution
Natural virtue(s), 151, 160, 162, 165. *See also*
Virtue ethics
Naturalism, 35, 37, 40, 70, 145, 384, 448, 478,
490, 560, 564, 759, 939
Nectar, 381, 457, 459–61
Negative rights, 61, 203, 213, 707–8, 711. *See also*
Animal rights; Human rights; Rights

Neglect of animals, 27, 200, 207, 216, 716,
741–42, 746–47, 760, 769, 771, 777. *See also*
Abuse; Cruelty
Neocortex, 554, 683. *See also* Brains
Nervous systems, 315, 450, 454, 691
Nests and nesting, 54, 381, 455, 460–61, 831, 837,
839, 939
Neurogenesis, 648
Neurons, 458, 462–63, 648–49, 660, 672–73,
682, 684, 690
Neuroscience, 310, 553, 557–58, 561, 567, 569
Neurotransmitters, 462, 682
Neurotrophins, 462
Newborns, 179, 286, 644, 834
Niches, 587, 589, 590, 595–96, 599, 615, 618, 807,
809, 872
Nonderivative value, 578–82, 584–87, 590–99.
See also Intrinsic value; Rights
Nonlinguistic creatures, 417, 429, 437. *See also*
Language
Nonmaleficence, 212–15, 745–46, 754. *See also*
Beneficence; Harm; Obligations;
Suffering
Nonpersons, 123, 305, 308, 321, 325–26, 350.
See also Persons and personhood
Nonsentient beings, 28, 123, 133, 284, 295, 610.
See also Sentience
Normative conceptions
of harm, 746–47, 754
of person, 13, 322, 327
of value, 92, 105–6, 108, 610, 613
Normative ethics, 12, 127, 135, 138, 173–74, 291,
298, 308–9, 595, 604, 611, 665
Normative force
of intuitions, 350, 585–86
of morality, 103, 105–9, 204, 525–27
of motivations, 529–30, 536–37, 539–42
Normative reasons, 102, 191, 390–91, 529–30,
536–37, 652, 661
Normative self-government, 103, 105–9, 523–24
Normativity, 144–45, 241, 526–27, 536–37,
540–42
Nutrition, 91, 612, 758, 828, 864

O
Obligations
to avoid animal cruelty, 25, 69, 71, 75, 200–1
to avoid doing serious harm to animals, 25–26,
72, 75, 164, 200–1, 213–14
based on ability to experience pain, 69, 79, 377,
383. *See also* Pain
based on inherent value, 200. *See also* Inherent value
based on interests or welfare, 93, 107, 109, 200–1,
215–16, 239, 260, 284, 651
based on reciprocity, 7, 69, 72, 93, 103–7, 109,
242, 560, 789
based on relations of dependence, 712, 715–16,
719–22

Obligations (*continued*)
  based on similarities between humans and
    animals, 51, 69, 145
  based on social relations with humans,
    163, 708
  based on voluntary acts, 560, 707, 715–16, 719–21,
    742, 788, 815–16
  of care, 149, 215–16, 219, 577, 714–16,
    742–44, 763
  correlativity between rights and, 9–10, 104, 198,
    206–11, 213–14, 219. *See also* Rights
  contractual, 560
  direct, 12–13, 51, 69, 73, 103–7, 109, 207–9, 261,
    267, 521, 816
  to domesticated animals, 707, 712, 714–16,
    719–22, 742–44, 763, 788–90. *See also* Pets
  to family members, 398, 705, 719
  to future human generations, 140
  indirect, 69, 100, 103–7, 109, 207, 209, 521
  to laboratory animals, 816. *See also* Laboratory
    animals
  and moral status, 12–13, 145, 260–62, 267, 300,
    323, 560, 650–51. *See also* Moral status
  natural law theories of, 64–65, 67–69,
    74–75, 199
  of nonneglect of animals, 27, 200, 207, 216, 716,
    741–42, 746–47, 760, 769, 771, 777
  perfect and imperfect, 208
  reciprocity theories of, 103–7
  relation to rights, 7, 9–10, 64–65, 68–69, 72, 75,
    104, 198–211, 213–14, 219, 707–8
  of species preservation, 21, 23, 577, 607, 610
  specification of, 209–11
  to treat animals kindly, 71–72, 145, 163–65, 269,
    277, 300, 743
  to treat animals with respect, 200, 237
  to zoo animals, 750. *See also* Zoos
  See also Domesticated animals; Wild animals
Observer-expectancy effect, 471
Octopi, 155, 682
Omnivores, 42–43, 53, 883, 885, 892. *See also*
  Carnivores
Orangutans, 65–66, 425–26, 477, 481, 485–87,
  761, 942–43. *See also* Apes
Organs, 24, 135, 190, 221, 324, 389, 390, 447, 609,
  632, 642–43, 650, 655–56, 673–76, 679, 686,
  797
Original sin, 77
Ovarian tissue, 647
Overpopulation, 860–65, 870. *See also*
  Culling
Oversight systems, 21, 30, 183, 205, 660,
  679–80, 691–92, 911, 919–21, 927–29
Owls, 350, 644
Owners of animals, 136–27, 60, 218, 358, 479,
  560, 769–80, 784–89, 797, 945. *See also*
  Domesticated Animals
Oxen, 41–42, 54, 797
Oysters, 442, 448, 881

P
Pack animals, 784
Pain
  capacity for, in mammals, 377, 499, 564,
    798, 827
  compassion in the face of, 443, 884
  from confinement on factory farms, 26, 738–39,
    743, 748, 755–57, 762–63
  consciousness of, 377, 382, 496–98, 652, 827,
    838–39, 897
  emotional, 188, 205, 399, 401, 498, 503, 508–10,
    740–41, 898–99, 938
  in experimentation, 63, 69, 93, 183, 209, 229, 389,
    653, 748, 798–99, 815, 907–15
  in the form of discomfort, 216, 228, 496, 500–1,
    732, 740, 756, 760, 915
  in mice, 229, 247, 500, 649–53, 674, 692–93
  obligations to avoid causing, 69, 79, 377, 383
  psychological theories of, 496–511
  utilitarian theories regarding, 9, 93–95, 129,
    172–93, 234–37, 377, 556
  See also Cruelty; Suffering; Utilitarianism
Painkillers, 732
Paleontologists, 593, 620
Parrots, 102, 312
Partridges, 46
Patricide, 39
Perception
  and agency, 16, 102, 380–82, 431–32
  and behavior, 102, 343–44, 380–82, 431–32,
    452, 463
  and cognition, 14, 46, 92, 346, 380–82, 432, 566
  and consciousness, 374–76, 497–98, 534
  empathy in, 158
  moral, 31, 933, 942, 945–47
  of pain, 377, 497–98
  its role in moral status, 361, 566, 798
  of self, 412
  selective, 803
  See also Cognition; Pain; Suffering
Perceptual states
  in mindreading, 375, 408, 412, 415–19, 422–25,
    430–32, 435–37, 488
  and motivational states, 92
  and propositional attitudes, 378, 416–17, 428
Personal identity, 315, 363
Personality, 17, 51, 155–56, 234, 334, 470, 479,
    484–87, 684
Persons and Personhood
  and animal enhancement, 324–27, 677
  cognitive conditions of, 308–9, 311, 313–15,
    317–18, 320–22. *See also* Cognition;
    Metacognition
  human and non-human, 13, 304–5, 308–9, 311,
    313–18, 320–27, 367
  moral agency as a criterion of, 265, 320–22
  and moral status theory, 10–14, 121–23, 253–65,
    304–5, 307–8, 320–24, 336, 339, 609–10, 652,
    677, 731, 742, 750–55, 906

person theory, 11, 13–14, 222, 304–5, 307–9, 332–48, 366–68, 594–95
potential, 265
self-consciousness as a criterion of, 313–18, 365–66, 339
Pescetarians, 883. *See also* Vegetarianism
Pesticides, 29, 215, 888, 906, 907
Pets
  basic-needs requirement for keeping, 26–27, 742–43, 746–47, 761–63, 778–82, 788–91
  confinement issues, 26, 738–39, 741–43, 746, 752, 755, 761–63, 780
  contractual obligations to, 560
  definition of, 26–27, 769
  dependence on owners, 719, 771–73, 775, 777–78
  difficulties in keeping, 567, 775–78, 784–91
  domestication of, 163, 762–63, 769–91
  and genetic modification of, 662, 683, 689
  mistreatment of, 80–81, 769
  moral status of, 73, 123, 132–33, 158, 290, 297, 391–92, 397–400, 567, 769–70, 894, 927, 951
  nondomesticated animals as, 778–80
  obligations to, 560, 567, 752, 755. *See also* Obligations
  respect toward, 132–33
  rights of, 73, 80
  special relationships with humans, 26–27, 132, 158, 163, 391–93, 397–98, 400, 479, 769–72, 951
  theft of, 927
  training of, 27, 771, 774–75, 780–84, 786
  See also Companion animals; Domesticated animals
Pests and pest control, 28, 93, 160, 182, 198, 827, 845, 944
Phenomenal consciousness, 374–77, 498, 532–35
Phenotype, 589–90, 595–96, 598–99
Philosophy of biology, 3, 587, 607. *See also* Biology
Philosophy of mind, 128, 189, 310–11, 338, 342, 363, 416, 495, 497. *See also* Minds; Minimal minds; Psychology; Theory of mind
Phyla, 91, 586, 612, 615
Phylogenetic tree, 450, 459
Pigeons, 256–57, 379, 635, 721
Pigs
  in ancient thought, 41–44, 46, 49
  castration of, 898
  enviropigs, 844–46
  on farms, 756, 834–36, 838, 840, 877–78, 881–93
  metabolization in, 809
  microencephalic, 839–40
  mindreading in, 410
  modifications of, 654–58, 681, 844
  piglets, 834–35, 877–78
  as pork, 42, 688, 797, 835
  research on, 770, 826–27
  rights and interests of, 175, 217–18, 221, 756, 839–40, 893

valves from, 674
  See also Cages; Confinement; Hog farms; Hogs
Placebo effect, 507–8
Pleasure
  in ancient philosophy, 45, 55
  and animal welfare, 829
  attributions of states of, 442
  bodily, 45, 129–30
  and consciousness, 92, 315, 341–42
  derived from knowledge, 581–82
  derived from liberty, 738
  gained through uses of animals, 55, 73, 75–76, 134, 237, 721, 812–13, 870, 884
  interests based on the capacity for, 355, 560, 738, 798, 827, 829
  and moral status, 77, 339–42, 360–61, 367, 560, 633, 653, 704–5, 798, 827, 894
  and pain, 77, 92–93, 177, 235–37, 247, 315, 339–42, 350, 355, 360–67, 442, 551, 560, 633, 704–5, 738, 798, 812, 827. *See also* Pain
  qualitative experiences of, 236, 341–42, 653
  rights based on the capacity for, 77, 176, 340, 350–61, 367, 894
  similarities between animals and human beings, 442, 551, 633
  and temperance, 130
  utilitarianism and the role of, 93–94, 96, 173, 176–77, 235–37, 247, 705. *See also* Utilitarianism
  the value of, 77, 173, 236–37, 247, 259, 350, 621, 657
  in virtue ethics, 129–30, 159. *See also* Virtue ethics
Plovers, 154, 409
Poachers, 234, 245–46. *See also* Hunting
Poisons, 62, 68, 464
Polio vaccine, 182, 818, 913
Political liberalism, 239–40, 243. *See also* Justice
Pollen, 381, 457, 460, 612
Pollution, 29, 620, 677, 827, 840–45, 889. *See also* Environments
Pork, 42, 688, 797, 835. *See also* Hogs; Hog farms; Pigs
Porpoises, 682–83
Potential persons, 325, 333, 336, 347. *See also* Persons and Personhood
Poultry, 834, 836, 887, 933, 935. *See also* Chickens
Precautionary principles, 583, 642, 665
Predators
  in ancient philosophy, 47
  animals perceiving humans as, 947
  cognitive capacities to detect, 423–24, 455–56, 555
  ecological role of, 619, 807, 855, 860, 865, 869, 871–72
  and environmental ethics issues, 855–72
  and extinction, 619
  and honeybees, 455–56

Predators (*continued*)
  literary representations of, 947, 949
  moral evaluation of, 111, 707, 855, 858, 860, 867,
    871–72
  risk that domesticated animals will be killed by,
    657, 689, 710, 778
  and salmon, 583
  See also Prey
Pregnancy, 655, 685–86, 892
Prejudice, 66, 235, 489, 658. *See also*
  Discrimination
Preservationists, 578, 597
Prey, 41, 47, 51, 53, 101, 421, 448, 553, 557, 563,
    580, 583, 619, 707, 773, 803, 871, 947. *See
    also* Predators
Pride, 20, 44–45, 51, 69–71, 124, 133, 150, 158, 896
Primates. *See* Apes; Baboons; Chimpanzees;
    Evolution; Gorillas; Monkeys;
    Orangutans
Property rights, 160, 165–66, 199, 709. *See also*
    Rights
Propositional attitudes, 17, 339, 377, 380,
    415–22, 428, 431, 434–36, 562, 570
Proteins, 643, 673, 683, 808
Protocols (research), 183, 658, 687, 911–12
Protozoa, 608, 610
Prudence, 36, 212
Psychology
  anthropomorphism and animal minds, 15–17,
    231, 443, 450–1, 469–75, 568
  as basis of moral status. *See* Moral status
  comparative, 16, 375, 378, 407, 411, 413, 425,
    552–53
  conscious psychological connections, 366
  developmental, 16, 407, 411, 426–27
  of emotions and passions, 70, 551
  folk, 436–37, 479–80, 487
  and genetically modified animals, 654–57
  Hume on human and animal, 146–57. *See also*
    Hume and Humean theory
  imagistic thinking, 14, 357–60, 364–68
  and metacognition theory, 9, 145, 150–53, 318,
    535–42
  of mindreading, 16–17, 375, 407–37
  and mirror self-recognition, 101–2, 228–29, 232,
    236, 241, 313, 411–12, 480–82, 564. *See also*
    Mirror self-recognition
  moral, 128, 144–46, 189, 391, 401, 524–25,
    559–60, 612
  psychological boundaries, 293–95
  psychological capacities, properties, and states,
    17, 332–35, 337, 341, 353, 399, 412–15, 424,
    469–90
  psychological connectedness, 315–17, 341, 347,
    359–60, 365–67, 730–34, 830–31
  psychological continuity, 146, 734
  psychological needs in animals, 215–16, 741,
    760–61, 830
  psychological unity, 315–18, 357

psychologists, 341, 382, 411, 471, 473–75, 479,
    496, 501
psychopathology, 158
and theories of pain and suffering, 496–511
theory of mind, 409, 473, 487–89, 554
See also Apes; Autonomy; Cognition;
    Metacognition; Mind; Philosophy of mind
Public health, 207
Punishment, 18, 54, 68, 233, 387, 470, 483, 740, 780
Puppies, 159, 746–47, 778, 781–82, 816, 951
Pushmi-pullyu, 452, 458–59, 462, 464

Q
Quails, 862
Quality of life, 11, 179, 185–87, 191–93, 247,
    286–87, 560, 658, 744, 751, 754, 782, 828

R
Rabbits, 110–11, 130, 133, 137, 285, 553, 637,
    643–45, 663, 692, 756–57, 783, 800, 865,
    907, 948
Raccoons, 133, 742
Racism, 21, 83, 119–21, 136–38, 235, 294, 633,
    636–37, 658. *See also* Discrimination;
    Prejudice
Rape, 18, 216, 470
Rational agents, 284, 385–93, 396–98, 401, 446,
    710. *See also* Agency; Moral agency
Rationality, 8, 11, 16, 101, 128, 132, 175, 242–43,
    262–64, 320–21, 339, 378, 495–96, 548, 566,
    631, 638, 652, 661–62, 704, 817, 938
Rats, 110–11, 133–34, 160, 221, 232, 379, 444, 458,
    500, 555, 583, 653, 673, 684–85, 689, 702,
    720–21, 800, 804, 809, 894, 907, 914
Ravens, 410
Realism, 70, 231, 240, 590–92, 615
Recombinant bovine growth hormone, 28, 826
Red junglefowl, 588–89
Reduction in numbers of animals used in
    research, 908–10
Refinement of animals used in research, 909, 914
Regret, 562, 630, 892, 897, 898, 936
Regulation. *See* Committee review of research;
    Government oversight, policy, and
    regulations
Religion, 37, 40, 65, 79–82, 219, 221, 247, 354, 673
Reproduction, 91–92, 383, 449, 550, 563, 587–89,
    591, 595–609, 631–32, 645–47, 658–61,
    681–86, 721, 759, 829, 846, 860–65, 907. *See
    also* Breeding; Mating
Reproductive isolation, 587–89, 591, 595–99, 606
Reptiles, 340, 554, 805, 948
Research. *See* Experimentation; Laboratory
    animals; Laboratory research
Responsibility (moral), 23, 80, 157, 184, 188,
    260–62, 383, 549, 564, 605, 620, 636, 707,
    741, 868, 871, 934. *See also* Obligations

Rhesus monkeys, 232, 410, 413, 636, 804
Rhinoceroses, 616
Rights
    acquired, 61, 72–73, 78
    adventitious, 75, 77
    agency and, 184, 209, 222, 349, 351–54, 707, 787,
        868, 871
    animal. *See* Animal rights
    animal rights movement, 81, 83, 176, 200
    aspirational, 208
    of autonomy, 203, 222. *See also* Autonomy
    correlative with obligations, 9–10, 104, 198,
        206–11, 213–14, 219
    derivative, 209–12
    human. *See* Human rights
    inalienable, 261
    international, 5, 199, 211
    as justified claims, 9, 201–4, 209, 219
    moral, 6, 8, 16, 181, 199, 202, 206, 218, 260–61, 323,
        332, 351–53, 373, 859, 866
    natural, 5–6, 61–62, 64–65, 67, 72, 75, 78, 199
    negative, 61, 203, 213, 707–8, 711
    of pigs, 175, 217–18, 221, 756, 839–40, 893
    positive, 61, 203, 618
    property, 160, 165–66, 199, 709
    rights talk, 61–62, 68, 72, 199–200
    the rights view, 28, 200, 707, 855–56, 859,
        866–70, 925
    specification of, 198, 209–12, 924
    theories of, 4, 9, 198–203, 205–12, 214, 219, 222,
        706–7, 925
    universal, 199, 304
    See also Animal rights; Human rights
Ring species, 589
Risks
    of domesticated animals being killed by
        predators, 657, 689, 710, 778
    to the environment, 841–47
    for pets, 779, 782
    in research, 30, 642–47, 657–59, 664, 751, 753,
        809, 836, 919, 926, 928
    and rights to protection against, 215, 868–69
    in toxicological studies, 29, 280, 455, 642,
        906–16. *See also* Toxicity and toxicology
    See also Harm; Suffering
Rituals, 39, 49, 53–54, 229, 504. *See also*
    Sacrifices
Roaches, 13, 160. *See also* Cockroaches
Rockfish, 644
Rodents, 673, 762–63, 804, 808, 812, 909. *See
    also* Mice; Rats; Squirrels
Rodeos, 215–16

S
Sacrifices, 27, 37, 39, 42, 53–54, 73, 76, 82, 121,
    183, 219, 235, 482, 609, 770, 786–90, 858, 936
Safety, 134, 266, 693, 739, 751, 769, 777, 782, 787,
    828, 888, 907–9, 913, 941

Salmon, 28, 320, 582–83, 827–28, 836, 840, 844,
    846, 880, 890
Sanctuaries, 134, 217, 219–20, 246, 759–60. *See
    also* Habitats; Zoos
Seafood, 29, 880, 883, 890
Sea otters, 682
Selective breeding, 703, 711, 832, 837. *See also*
    Breeding; Reproduction
Self-awareness, 16, 311–22, 347–48, 407–12, 498,
    501, 559, 634
Self-consciousness, 13, 67, 95, 101–3, 262–65,
    306, 311, 314–18, 336–40, 347–65, 498, 535,
    563, 652, 897. *See also* Consciousness;
    Mirror self-recognition; Self-
    consciousness
Self-defense, 75, 222, 295, 798
Self-recognition, 236, 245–46, 313–14, 412, 480,
    481. *See also* Mirror self-recognition;
    Self-consciousness
Senility, 387–90, 393
Sensations, 14, 46, 50, 205, 229, 236, 244–45,
    338–44, 356, 374, 377, 497–99, 533, 741, 829,
    948
Sentience
    and capabilities theory, 10, 243, 247
    and confinement, 26, 738–42, 750, 761
    as a criterion of moral status, 9, 11, 14, 174–81,
        188, 192, 262–66, 280, 283–85, 292–95, 566–67.
        See also Moral status
    in domesticated and wild animals, 704–6,
        715–16, 721–22
    in early modern philosophy, 70, 75, 80
    and experimentation, 820, 906, 924–26
    in genetically modified animals, 649, 652, 678
    in person theory, 308, 311
    and rights, 351–52, 360
    sentientism, 857–62, 865, 869–74
    and utilitarianism, 9, 173–81, 192, 235–40. *See
        also* Utilitarianism
    in virtue ethics, 121–24, 131–34
    See also Moral status; Pain; Perception; Persons
    and personhood
Sex, 122, 129–30, 294, 478, 684, 686, 867
Sexism, 21, 120, 121, 294, 658
Shame, 20, 46–49, 233, 242, 401, 551, 897–98.
    See also Guilt
Sheep, 28, 42, 44, 54, 63, 73, 99, 147, 579, 644, 648,
    689, 757, 826, 835, 844, 860, 865, 871, 945
Shelter(s), 49, 110, 133, 151, 283, 500, 560, 710,
    717, 720, 741–43, 747–48, 760, 763, 771,
    777–78, 781–82, 785, 788–90, 878
Sickle-cell anemia, 808
Sign language, 312, 355, 358, 367, 481, 636, 774.
    See also Language
Single-celled bacteria, 450
Skepticism, 70, 523–24, 847
Skulls, 390, 651, 680, 683, 798, 808
Slaughter
    and animal rights, 201, 215, 867, 894

Slaughter (*continued*)
 and animal welfare, 834–35. *See also* Animal
  welfare; Laws; Pain; Suffering
 and factory farms, 756, 834–35, 878, 880, 897,
  933–38, 952
 and sentientism, 860
 See also Death; Dying; Factory farms; Killing;
  Letting die
 and traditional farms, 758
 as viewed in ancient philosophy, 37, 42, 50–54
 as viewed in modern philosophy, 71, 75, 81
Slaughterhouses, 31, 181, 215, 756, 758, 879–82,
  888, 897
 See also Factory farms; Killing
Slaves, 36, 40–47, 65, 68, 73, 78, 265, 349, 662,
  678, 789, 885, 895, 897, 945–46
Slippery slope arguments, 642, 661, 663, 665
Slugs, 43
Smallpox, 660, 800
Smells, 102, 456, 489, 551, 563, 580, 890–91, 934,
  940, 951
Snails, 42, 448, 582–83
Snakes, 49, 53, 413, 563–64, 762
Social behavior, 101, 218, 229, 408, 412–13,
  550, 557
Social cognition, 17, 247, 409, 436. *See also*
  Cognition
Social engineering, 289–99
Social justice, 6, 566, 714. *See also* Justice
Sociality, 288–91
Socialization, 325, 781, 784
Species
 agriculture and the mixing of, 676
 biological distinctions among, 19–22, 155, 180,
  327, 578, 585–99, 604–10, 615–21, 651, 805–7,
  820, 865
 classification systems for, 487, 586–87, 590, 592,
  594, 598–99, 605
 endangered, 383, 583, 613, 835, 855, 857, 859–60,
  869–72, 946
 evolution of, 20–23, 52, 587–88, 604–21
 extinct, 22, 246, 550, 603, 606–8, 613–16, 619, 835,
  841, 846, 860
 obligations to preserve, 21, 23, 577, 607, 610
 preservation of, in ecosystems, 20, 22, 28–29, 84,
  577, 580–81, 598, 603–13, 617–19
 species considerations in ecology, 22, 587,
  589–90, 604–7, 611–19, 629, 708–9, 841
 species egalitarianism, 23, 630–34, 636, 638
 species essentialism, 590–92
 species-typical behavior and functioning, 216,
  218, 477, 740, 757, 760, 831, 834
 value of, in Darwinian theory, 584–90. *See also*
  Darwin and Darwinian theory
Speciesism, 12, 21, 120–22, 138, 184, 277, 300,
  309, 352–54, 632, 634, 636–37, 658
Sperm, 23, 641, 645–46, 672–76, 681, 684–86,
  836, 869
Sponges, 448

Squirrels, 81, 180, 703, 779, 798, 862
Starlings, 580
Starvation, 96, 130, 205, 210, 610, 633, 713, 717,
  721, 777, 815, 855, 896, 940, 947
Stem cells, 646–49, 655, 671–76, 679–80, 683,
  686–89, 692–93, 806
Stress in animals, 542, 716, 741, 756, 834, 837,
  839, 878–80
Subject of a life, 121, 200, 315, 609, 705–6, 818
Subsistence, 220, 500
Substantive mindreading, 408, 412, 414, 415,
  418, 424. *See also* Mindreading
Suffering
 compassion and sympathy directed at, 70, 497
 capacity for in mammals, 377, 499, 564, 798, 827
 on factory farms, 26, 29, 129, 137, 176, 180–81,
  727, 756–57, 877–98
 mental, 497, 507, 652
 misery as a form of, 74–75, 79–80, 92, 162–64,
  214, 551, 858
 as a result of being caged, 51, 81, 183, 740–41,
  756–59, 780, 834, 879–80
 theories of pain and, 496–511
 See also Cruelty; Discomfort; Harm; Factory
  farms; Nonmaleficence; Pain; Vegetarianism
Supermarkets, 879
Surgery, 504, 507–8, 732–36, 800, 838, 935
Swine, 220, 834, 844–45, 878, 887, 893. *See also*
  Hogs; Pigs
Swine influenza, 220, 878, 887
Swordfish, 880
Sympathy
 and cognition, 233
 and consciousness, 373–78
 toward domesticated animals, 708
 Humean views of, 149–50, 157–58, 162–65.
  See also Hume and Humean theory
 and moral status, 16, 145, 149, 373, 391, 398–401,
  524, 568
 and pain and suffering, 497
 and personhood, 307, 322. *See also* Persons and
  personhood
 in research contexts, 476,
 warranted, 383–86
 See also Compassion; Moral sense theory
Synthetic biology, 660, 663. *See also* Biology

T
Tame animals, 46, 81, 702
Tattoos, 503
Termites, 17, 442, 787
Testing. *See* Experimentation; Laboratory
  animals; Laboratory research
Theory of mind, 409, 473, 487–89, 554. *See also*
  Apes; Cognition; Philosophy of mind
Tigers, 101, 214, 581, 644, 673, 774, 778, 780, 880,
  942–44
Time-awareness, 495, 498–99, 501, 505, 507, 510

Timidity, 18, 20, 470

Tissue damage, 509

Tissues, 24, 324, 503, 509, 609, 642, 648, 650, 655–56, 671–76, 679–80, 688–89, 691, 807

Tools and tool use, 101, 151, 192, 200, 231–32, 319, 326, 412, 418, 421, 436, 444, 557, 563, 643, 661, 676, 682, 771, 781, 871

Torture, 75–76, 110, 278–82, 288, 293, 323, 353, 356, 603, 663, 708, 884. *See also* Cruelty

Toxicity and toxicology, 29, 198, 804–5, 811, 906–15

Transgenic animals, 21, 23, 133, 641–46, 649, 661–64, 692, 845

Trust, 155, 212, 522, 708, 774

Trustworthiness, 22, 165, 212, 552

Tuna, 23, 324, 636, 638, 881, 893

Turkeys, 29, 703, 714, 757, 835, 837, 839, 881–82, 889, 891

Turtles, 644

U

Utilitarian theories

act-utilitarianism, 173–75, 182

about animal research, 813–14, 925

on animal welfare, 69, 78, 172–93, 859, 863, 866, 869–70

Benthamite thinking about animals, 9, 172–76, 230–37, 244, 247, 383. *See also* Bentham and Benthamite theory

hedonistic, 277

in the history of modern philosophy, 62–79

Mill's views, 172–76, 234–37, 244

on pain and suffering, 9, 93–95, 129, 172–93, 234–37, 377, 556

on rights, 93–95, 119, 172–93, 200, 222, 859, 866–67, 869–70, 932

on rights of animals, 74–81, 83, 94, 120, 175–76, 182–83, 200, 859, 866, 869–70, 925

the role of pleasure in, 93–94, 96, 173, 176–77, 235–37, 247, 705

and sentientism, 9, 173–81, 188, 192, 235–37, 859, 863, 865–67, 869–71, 925

Singer's account of, 9, 93–95, 131, 172–76, 556

and speciesism, 183–86

strengths and problems of, 172–93, 234–37, 401, 834

views of agency, 155, 175, 188–91, 263, 548, 871

views of moral status, 9, 121, 172–93, 269, 373, 401, 925

and virtue ethics, 119–23, 127–31. *See also* Virtue ethics

See also Consequentialism; Consequentialist thinking; Rights; Sentience

V

Vaccines, 800, 810, 812, 818, 893, 907, 913

Veal, 757, 878, 881. *See also* Cattle; Calves

Veganism and Vegans, 567, 883, 889, 892

Vegetarian diets, 53–54, 81–82, 119, 128–30, 140, 230, 820, 860, 883, 887, 892, 953

Vegetarianism, 29, 81–82, 124, 129–30, 135, 138, 860–61, 883, 893–94, 898, 937

Vegetarians, 23, 29, 53–54, 230, 557, 630, 820, 877, 883, 887, 892–95

Veil of ignorance, 349–50, 385, 388, 392–93

Vermin, 350

Vertebrates, 400, 615, 805, 827, 926. *See also* Invertebrates

Veterinarians, 132, 201, 210–11, 770, 816, 858

Vices (contrasted to virtues), 8, 64, 69, 119–20, 124–25, 130, 136–37, 140, 155, 158, 212, 244, 397. *See also* Virtue ethics and theory; Virtues

Violence, 10, 43, 161, 238, 389, 391, 949. *See also* Aggression

Vipers, 63. *See also* Snakes

Virtue ethics

in ancient philosophy, 5, 37–43, 53

Aristotelian views of, 8, 128, 524–27, 548. *See also* Aristotle and Aristotelian theory

cataloguing the virtues and vices, 212

on charity and benevolence, 269–70

of compassion, 8, 119, 124–31, 133–34, 140, 212, 277, 285, 293–300

in contractarianism, 396–99

Driver on, 147–62

and environmental ethics, 138–41

and experimentation on animals, 134–38

in history of modern philosophy, 65–74, 77

Humean views of, 128, 147–49, 151–52, 154–58, 161–62, 165. *See also* Hume and Humean theory

Hursthouse on, 8, 119–41

of justice, 120–34, 267, 926, 953

and moral status questions, 8, 119–25

natural and artificial virtues, 151, 160, 162, 165

and research ethics, 906–12

and rights, 202–5, 212

and vegetarianism, 129–31. *See also* Vegetarianism

virtue language, 125–27

See also Aristotle and Aristotelian theory; Benevolence; Hume and Humean theory; Vices

Viruses, 284, 441, 450, 583, 660, 818, 887–88

Vivisection, 63, 80, 82, 182, 184, 944

Vulnerability, 25, 165–66, 202–5, 212, 215, 486, 649, 701–2, 712–22, 787, 890–92, 948. *See also* Animal welfare; Factory farms; Harm

W

War, 40–41, 47, 82, 222, 382, 771, 817, 868

Wasps, 374, 380, 382, 609

Waste decomposition, 612

Water pollution, 28, 828, 842. *See also* Pollution

Welfare. *See* Agriculture; Animal rights;
    Animal welfare; Diseases; Laws; Pain;
    Suffering
Welfarists and Welfarism, 172–73, 200–1,
    282–87
Whales, 346, 364, 367, 555, 557, 683, 739, 759–61
Wild animals
    in ancient philosophy, 41–54
    caging and confinement of wild animals, 180,
        718, 740–62, 778–80, 880, 943
    the distinction between wild and domestic, 25,
        73, 79, 705, 707, 714, 717–22
    endangered species, 871–72
    fear in, 702–3, 779
    feral animals, 49, 133, 637, 704, 720, 777, 843–45, 860
    harm to, 770–80
    hunting to control populations of, 220, 855,
        860–67, 870
    and industrial farming, 884, 899
    and mindreading, 413
    moral status of, 123
    and rights, 207, 214–27
    and tool use, 279
    in virtue ethics, 133–34. *See also* Virtue ethics
    wild and stray cats, 73, 133–34, 704–5, 722, 777
    See also Domesticated animals
Wildebeests, 706–7, 717
Wilderness, 22, 713, 784, 858, 948
Wildlife, 217, 855, 861–66, 869–70. *See also*
    Wild animals
Wildness, 25, 701–8, 722
Wolf children, 325
Wolves, 44, 74, 147, 325, 421, 555–58, 744, 773,
    863, 868, 871, 947–52

Wool, 7, 42, 54, 73–74, 110
Worms, 175, 453–54, 688
Worse-off principle, 866–68
Worthwhile-life argument, 749–52

X
Xenophobia, 482
Xenotransplantation, 181, 643

Y
Yeast, 673
Yellowfin tuna, 880

Z
Zebras, 595–96, 942–43
Zoos
    apes in 760–1
    and aquariums, 758–63
    breeding programs in, 246
    cages in, 180, 217, 715, 740, 758–59, 880, 946
    confinement in, 26, 217, 246, 717, 738–55,
        758–63, 881
    and domestication, 25, 715
    elephants in, 228–29, 246–47
    in literature, 942–46
    mammals in, 759
    research in, 229, 477, 485
    and rights, 198–201, 207–17 *See also* Aquariums;
        Confinement; Domesticated animals;
        Sanctuaries
Zygotes, 336, 350, 352, 685. *See also* Embryos